The Informatics Handbook

Telecommunications Technology and Applications Series

Series editor: S. Sharrock

Titles available

1. **Coherent Lightwave Communications Technology**
   Edited by Sadakuni Shimada

2. **Network Management**
   Concepts and tools
   Edited by ARPEGE Group

3. **The Informatics Handbook**
   A guide to multimedia communications and broadcasting
   Stewart Fist

The Informatics Handbook

A guide to multimedia communications and broadcasting

by

Stewart Fist
Freelance technical writer and journalist, Australia

CHAPMAN & HALL
London · Glasgow · Weinheim · New York · Tokyo · Melbourne · Madras

Published by Chapman & Hall, 2-6 Boundary Row, London SE1 8HN, UK

Chapman & Hall, 2-6 Boundary Row, London SE1 8HN, UK

Blackie Academic & Professional, Wester Cleddens Road, Bishopbriggs, Glasgow G64 2NZ, UK

Chapman & Hall GmbH, Pappelallee 3, 69469 Weinheim, Germany

Chapman & Hall USA., 115 Fifth Avenue, New York, NY 10003, USA

Chapman & Hall Japan, ITP-Japan, Kyowa Building, 3F, 2-2-1 Hirakawacho, Chiyoda-ku, Tokyo 102, Japan

Chapman & Hall Australia, 102 Dodds Street, South Melbourne, Victoria 3205, Australia

Chapman & Hall India, R. Seshadri, 32 Second Main Road, CIT East, Madras 600 035, India

First edition 1996

Printed in England by Clays Ltd, St Ives plc.

ISBN 0 412 72530 4

A Catalogue record for this book is available from the British Library

∞ Printed on permanent acid-free text paper, manufactured in accordance with ANSI/NISO Z39.48-1992and ANSI/NISO Z39.48-1984 (Permanence of Paper)

Acknowledgements

To Jillian Fist

Simply to thank my wife and partner Jillian for her help and patience during the production of this book, would be to understate her involvement. For the last year we have both been working most of each week upgrading our paper and electronic databases, checking information, subbing copy, keying in the corrections, and formatting the Camera Ready Copy. For Jillian, this has been full-time involvement; this book is as much the product of her work, as it is of mine.

I couldn't (and wouldn't) have done this alone: the project would have been abandoned half-way without her constant help and support. S.F.

I am a independent writer, so the editors I work with in Australia have also made valuable contributions. They have both supported me over the years this book has been a 'work-in-progress', and encouraged me to publish what was originally just an enormous jumble of personal reference files.

Mark Smeaton of ***Australian Communications*** magazine has been particularly encouraging over the many years we've worked together, and also Chris Moriatry of ***LAN Magazine*** and Jeremy Horey of ***The Australian*** newspaper. Stuart Sturrock of ***Mobile Communications International*** in the UK took an early draft to Chapman & Hall and convinced technical editor Daniel Brown the project was worth pursuing. They are obviously both men of insight and intelligence!

And then there's the production staff at Chapman & Hall; particularly Alexandra Stibbe and Penny Sucharov who have laboured over the style, the cross-referencing, and the problems I have with American spelling. To the uninitiated, trying to correct this stuff must have been about as alien a task as subbing Japanese Kanji. I thank them all for their efforts and tolerance.

Lastly, there's the electronic forums on CompuServe and the Internet. There's no better place to ask questions and get answers than here. And just by reading telecommunications and multimedia forums, digest and Usenet material regularly, you soon get a taste for the different ways terms are being used within these rapidly developing international electronic communities. I always received polite answers to what must have often been rather stupid questions. Patrick Townson (Telecom Digest) and Scott Loftesness (Telecommunications Forum) are the two sysops who most obviously come to mind here. They run useful and lively discussion groups.

Lastly, I must thank the thousands of marketing people whose job it is to confuse all discussion in this area by inventing new and more exotic terms and acronyms. Without them, this book would have only been half the size.

Technical History

This book began life as a series of notes on an Apple II. My background is in television current affairs and film production, and I came into the computing and communications industries late in life. I found then, as I find now, that the terminology used in these industries is often totally confusing and contradictory, and this was a way of tracking and understanding.

The serious development of the database only really became possible when Bill Atkinson wrote that marvellous program, HyperCard, and I mounted it on my Macintosh. I still don't know whether to call it a database or system software, but whatever it is, HyperCard is the best possible way for an information magpie like me to store information. And, as I later found, HyperCard is also the best way to create intelligent sorting algorithms (sorry, 'handlers') and complex file-extraction programs, when the time came to create camera-ready copy.

The output of the HyperCard files were passed through to FrameMaker 3.0 which is probably the best book publishing program ever written. It creates the many styles and formats, and automatically tracks page references, and creates the running heads at the top of each page. The typeface is Garamond Narrow, which was forced on us by the sheer volume of words (the original files have over a million, and this book has 650 000 words). Without a condensed typeface, the book would have had many more pages and been much more expensive. The CRC output was from a Hewlett Packard LaserJet IIP printer, directly onto A4 copy paper, using 12 point type. This was photo-reduced (linear 80%, which is a 1.6-times resolution improvement) in the plate-making process to the Crown Quarto book size.

Introduction

This is not a dictionary — and nor is it an encyclopedia. It is a reference and compendium of useful information about the converging worlds of computers, communications, telecommunications and broadcasting. You could refer to it as a guide for the Information SuperHighway, but this would be pretentious. It aims to cover most of the more important terms and concepts in the developing discipline of Informatics — which, in my definition, includes the major converging technologies, and the associated social and cultural issues.

Unlike a dictionary, this handbook makes no attempt to be 'prescriptive' in its definitions. Many of the words we use today in computing and communications only vaguely reflect their originations. And with such rapid change, older terms are often taken, twisted, inverted, and mangled, to the point where any attempt by me to lay down laws of meaning, would be meaningless. The information here is 'descriptive' — I am concerned with usage only.

The headwords:

This book therefore contains key-words and explanations which have been culled from the current literature — from technical magazines, newspapers, the Internet, forums, etc. This is the living language as it is being used today — not a historical artifact of 1950s computer science.

Purists can contest many of these definitions, but I'm not overly concerned with being 'correct', only with reader comprehension — with current usage being readily understood. And that often means navigating the inconsistencies and the mis-uses, and understanding the implications, because sub-text is often more important than the overt usage itself.

One problem all lexicographers face, is in choosing which 'headwords' to include and which to ignore. My computer database has many more than you'll find here, and if you check technical dictionaries in a bookstore there are thousands of acronyms and terms listed. Like bacteria, they divide and duplicate every second day. But the words and acronyms included here, I believe, are the most important. They are the key to understanding the literature on convergence and its consequences in the immediate future. And this is what Informatics is all about.

The technical level:

This is obviously not a reference for people just buying their first PC. It is for those who already having a reasonable background knowledge — they will already be readers of books and magazines on informatics technologies. And this area of convergence is, by definition, populated by people with a high level of knowledge in one technical area, but perhaps little or none in others. So I've tried to pitch the explanations at the level appropriate to the terms — rather than predict the knowledge of the reader. If you need to find out what 'bandwidth' means, then obviously your information requirements are relatively basic. If you want an explanation of 'TINA', however, you are probably already seriously involved in telecommunications.

As a technology writer, I've also collected a whole range of useful facts over the years — data-rates, channel numbers, frequency ranges, bandwidths, etc. This sort of information is amazingly difficult to find in much of the literature, so it has been included here, wherever possible. This information will not be 100% correct; it is changing all the time, and it is often varies in different countries. The information in this book has been checked and rechecked to the best of my ability, but this is both an ephemeral and moving target. Updating and correcting information in the primary database is a daily event, so this book is already out-of-date.

On-line and CD-ROM:

It is our intention to create a CD-ROM version, and to put the full database on-line in the near future. The details have still to be finalized. However, it is important that we get feedback from readers who find discrepancies, or who wish to suggest additions — or who perhaps have a better and more useful explanation for some of the items. My contact details are: S.F.

Stewart A. Fist

Internet: fist@ozemail.com.au

CompuServe: 100033,2145 — Facsimile: +612 416 4582

User's Guide

Author's abbreviations:

#	—	generic substitute for any number.
aka	—	also known as.
sync	—	synchronous.
async	—	asynchronous.
sic	—	as it is written.

Bit-byte numbers and rate:

b	—	bit.
B	—	Byte (8-bits, unless otherwise specified).
kB	—	kiloByte (thousands of Bytes).
MB	—	MegaByte (millions of Bytes).
GB	—	GigaBytes (thousand million Bytes).
kb/s	—	kilobit/sec. (thousands of bits per second).
Mb/s	—	Megabit/sec (millions of bits per second).
Gb/s	—	Gigabit/sec (thousand million bits per second).

Frequency:

Hz	—	Hertz (cycles per second).
kHz	—	kiloHertz (thousands of Hertz).
MHz	—	MegaHertz (millions of Hertz).
GHz	—	GigaHertz (thousands of millions of Hertz).
THz	—	TeraHertz (million million Hertz)

General:

n	—	is often use as any integer — as in 'n x 64'.
p	—	also often means any integer — as in 'p x 64'.
ms	—	millisecond (thousandths of a second).
μs	—	microseconds (millionths of a second).
mm	—	millimeter (1/25th of an inch).
cm	—	centimeter (about 4-inches).
m	—	meter (roughly 40-inches).
km	—	kilometer (5/8ths of a mile).
in	—	inch (25 mms)

Indexing

The terms are generally retained in the normal word order rather than strict alphabetic order of the letters — but there are some exceptions and some traps.

For instance, the sequence of head-words will be:

page
page description languages
page printer
paged virtual memory
pagination
paging

You will find that this is different in some dictionaries because they give the alphabetic order, more weight than the word divisions.

However the word-approach sometimes breaks down with acronyms and abbreviations. For instance, 'BISDN' (also written 'B-ISDN') could be in the front of the B section (as a distinct word), along with' B-channel', and 'B-spline', but it is back in the main body, with 'BiSync' and 'bistable'.There is no right and wrong way to handle these.

Numbers are sorted ahead of alpha characters. So the head-word 'B8ZS' will precede 'back-porch'. Pure numbers, such a '486' and '5/6 tone' are in their own Numbers section, preceding the A-listings. They are in true number order — not on a character-by-character basis.

Word splits are a particular problem, so look out for non-standard splits or for words that may be joined. If you don't find word in one place, check the other. 'Halftone' could also be written as 'Half tone' (but it generally isn't) and be further down in the listing so check both.

Acronyms/abbreviations:

In this industry, the acronym is widely used, while the full-out term is rarely heard. I have tended to list under the term most used — and this tends to be the acronym (but not always). You'll find ATM under the acronym, not under 'Asynchronous', however the industry uses 'BiSync' rather than BSC.

The obvious:

I've tried also to avoid filling up the book with the bleeding obvious. You wont find Network Adminstrator, for instance, because I assume you are all intelligent enough to realise that this person administers a network. You will, however find Network Administrat*ion*, since it spells out the common functions which come under this heading.

Cross references:

Cross references are not just to closely related items; think of these as hypertext links. I've tried to give you cross-references to other useful or interesting items. These might be an alternate technology, or other technologies that share a similar characteristic, or where there is a concept which (to me anyway) has a relationship of some kind. With these references you should be able to track your way through the listings even if you don't quite know what you are looking for.

0TLP

Zero Transmission Level Point. The reference point in a transmission system, from which relative losses are measured.

1/4 (one-quarter)

- Audio tape standards: See Quarter-inch.
- Computer tape standard: See QIC.

1/2 (half)

- **Code:** A one-half code is one where one in every two bits carries information and the other is redundant and used for error correction. It is a term used mainly in the paging industry and for deep-space transmissions where you can't ask data to be re-sent. FLEX and POCSAG are 2/3 codes, while GSC is a 1/2 code — so GSC is relatively inefficient — but it may also be more robust and able to correct errors when the owner is in a lift-well in a steel-and-concrete building.
- **Tape systems.** See half-inch tape.

2/3 (two-thirds)

Code: A two-thirds code is one where two out of every three bits carries information, and one in three is redundant. It is a term used mainly in the paging industry. FLEX and POCSAG are 2/3 codes, while GSC is a 1/2 code — so GSC is relatively inefficient.

3/4 (three-quarter)

Tape standard: See U-matic.

1:1 interleave

When a magnetic disk has an interleave factor of only one, it means that the head can read every sector on one track in a single disk rotation. The disk controller must be able to process the data at a fast enough rate so that consecutive sectors can be handled without needing a pause. This is either a very fast controller, or a slow disk. See RLL.

1B1Q

See HDSL and 2B1Q.

1Base-5

This is the 1Mb/s version of 802.3 Ethernet over unshielded twisted pair(UTP). It was originally promoted as StarLAN by AT&T. It is a CSMA/CD system using twisted pair telephone cables. The advantages of this approach are that a company's LANs wiring will be installed along with its telephone wiring in the same cables, all radiating from a central hub.

- 10Base-T should be considered as a highspeed version of this technology.

1GL

First generation languages. Machine code which is not really a language at all but rather a way of accessing memory locations using mnemonic codes. See language generations

1MB barrier

See 640kB barrier.

2:7 interleave

See RLL.

2B1Q

Two binary, one quaternary. A very simple baseband line-coding system used in the USA for ISDN. It is a multi-level line code which uses two positive, and two negative voltages — so the receiver must detect both amplitude and polarity of the signals, but not the phase.

It tends to be susceptible to impulse noise. For this reason QAM techniques are generally more suited to higher data-rates (see CAP), although 2B1Q works extremely well in the local loop at 160kb/s.

In America, both ISDN and HDSL use similar line-coding because 2B1Q cuts the overall bit stream line-rate in half, since 2-bits are carried in each symbol. This also has advantages when bundling HDSL pairs and other (T-1) pairs, in that the power spectrum is centered much lower, which minimizes crosstalk.

- See quaternary, ISDN and TCM.

2GL

Second generation languages. Assembly languages which use cryptic mnemonic commands. See assembly language and language generations.

2-wire

- **Telephony:** This means the typical local loop twisted-pair connection. It applies even to the 'quad' wire that probably carries the signals into your home. See wires.
- **Video:** This means component systems of the Y/C type which requires two different connections between machines in any post-production facility (one for the luminance, and one for the color). Of course they also need another for audio, but this isn't counted. 'Wire' here means a full cable. See 3-wire.

3+ NOS

A network operating system from 3Com.

3-axis stabilized

These are geostationary communications satellites which have large flat 'wing' panels of solar cells (as distinct from the 'barrel-shaped' spin-stabilized type). The large wings on these satellites can generate more power than the cylindrical type so they are widely used for direct broadcasting where output power is essential. Gyroscopic stabilizers are provided to handle small angle corrections (up to 5°), and larger changes are made by judicial use of the thrusters. See nadir face.

3-D

Despite common perceptions, the third dimension in video is not just achieved by presenting slightly different views of a scene to the two eyes. This is called stereopsis, and it is only one of the mechanisms used by the brain to judge 'depth' in a scene, and then normally only for objects within a few meters.

Other image-factors which help the 3rd Dimension illusion of 'depth' are generally present in all movies. They are:

- **Relative size.** Objects closer to the camera appear larger than objects at a distance. They also move more slowly across the screen.
- **Relative height.** Unless your eye or camera is at ground level, objects higher in the image plane (but only to the limit of the horizon line) will be further away than objects lower in the image.
- **Cut-off.** Near objects tend to block off (obscure from view) parts of more distant objects.
- **Shade and shadows.** Shading of light on a spherical object, for instance, gives unambiguous information about its shape. Textures can be seen on close surfaces, but not on distant.
- **Aerial haze**. Distant mountains take on a bluish coloration.
- **Parallax:** If the camera moves even slightly sideways, the relative movement and obscuration of close vs. distant objects gives the viewer excellent judgement of depth.
- **Stereopsis:** This is only present in the special 3D movies using red-blue or polaroid glasses to separate the two images. A slightly different view is then received by each eye, because the two images were recorded by cameras which were horizontally separated by a few inches.

3D graphics were only really possible on powerful workstations until recently. Now these techniques are moving onto PCs.

There are a number of different approaches in the labs, but currently the *de facto* standard must be OpenGL 3D libraries and APIs, which are supported by Microsoft and IBM. This is now standard in Windows NT and is an option in Windows 95. Warp has support also.

- Sharp has demonstrated a 'twin-LCD' form of 3D TV screen which has two superimposed flat-screens in a sandwich. The viewer sees one screen layer with the right eye, and the other with the left. You don't need special polaroid or coloured glasses, but your eyes must be at a standard distance from the screen.

- The Nintendo system uses two displays, each consisting of a one-dimensional Red LED array and a synchronized scanning mirror. This is the Private Eye technology of RTI. The image looks to be about 30 cm away and directly in front of the viewer: it has four levels of tone in red against a black background.

- There's a new Sanyo 3D system that doesn't require glasses. It is correctly called 3-D LCD.

3-D sound

This differs from surround sound (with four or five speakers), but it achieves much the same results from two speakers...so they say! 3D sound is a feature on many of the new PC sound cards. See Dolby, Prologic and wave-table.

3DO

A video games console and standard which will probably have applications in new interactive set-top boxes in the future cable television systems. It incorporates MPEG2 compression and CD-Audio and CD-ROM standard. Scientific Atlanta is now using it in its new digital set-top boxes.

3DO has been developed by a partnership of Electronic Arts (the world's largest producer of video games), Matsushita (the world's largest manufacturer of consumer electronics), Time Warner (the world's largest cable distribution company), and AT&T (the world's largest telephone company). It was also boosted as the world's greatest breakthrough in games and superhighway standards, but it seems over the last few years to have become the world's greatest non-event.

Their new M2 64-bit game unit has a graphics performance of one million polygons per second (peak) and a sustained rate of 700 kilopolygons per second. It runs a PowerPC 602 64-bit processor at 66MHz, and has a dedicated graphics IC, and special I/O control ICs. A floating point processor handles anti-aliasing, shading and other image processing. It also provides an MPEG-1 decoder and a 24-bit DSP for audio. 4MBytes of DRAM are provided as main memory.

3GL

Third Generation Language. These are high-level languages such as Basic, Fortran, Cobol, and C, etc. which are written using English-like commands, and can be made transportable between various platforms. They need to be interpreted or compiled to run. See language generations

3-wire

A video production term used with YUV and YIQ component editing and post-production equipment. The three-wire (actually three separate connections) is necessary to carry the different video components (luminance and two color-difference signals) which remain as separate signals through the post-production process. This is the best way to keep the video at the highest quality, but most post-production facilities can only handle composite video.

4:1:1

A version of the 4:2:2 component video standard which is used in news gathering where color quality is not considered so important. Two thirds of the bits are devoted to the definition of the luminance image, and only one third divided between two color-difference signals. See color difference and 601 CCIR.

4:2:0

A variation on the 4:2:2 component video standard where a composite color signal is only included with every second line.

Another source says that the Red (R-Y) signal will be transmitted in one field, and the Blue (B-Y) in the next field (similar to SECAM in analog). The MPEG2 main level/main profile encoders use this coding technique to reduce the bit rate to 8Mb/s. See color difference and 601 CCIR.

4:2:2

Four parts of luminance, and two parts each of the two color-difference signals. This is the ITU-R Recommendation 601 for digital component video production and distribution (but not transmission) without compression.

The standard is now widely used around the world and both 50Hz and 60Hz versions exist. Both have a luminance sampling frequency of 13.5MHz using 8-bit linear PCM (the '4' signal), and two chrominance difference signals (R-Y and B-Y) sampled at half that rate (6.75MHz).

The D-1 videotape recorder standard was based on this recommendation but with 10-bits per sample (8-bits is now thought to be insufficient for primary sampling), producing a serial bit-rate of 270Mb/s. The samples are then rounded down to 8-bits for in-house transmission.

• The ETSI/EBU-300.174 television transmission standard is designed to allow compression of these signals to 34Mb/s with virtually no loss of quality. From 1996 on, however, the most common output will be as MPEG2 compressed video. See 601 CCIR and widescreen.

4/6GHz satellites

See C-band.

4-ary

See quaternary.

4B3T

4-Binary with 3-Ternary code. This is a form of AMI where binary is mapped into a ternary transmission system (zero, plus both + and – voltages). A group of four binary bits is mapped into true three-state ternary code using look-up tables — achieving a baud-rate reduction of 25%. This line-code technique is used for Basic Rate ISDN in European countries.

4B5B

This is a code translation technique used in FDDI and other transmission systems where the code is converted from 4-bit nibbles to 5-bit symbols before NRZI line-coding. The idea is to maintain clock synchronization by ensuring that there can never be a long string of zeros or ones on the line.

In FDDI, the use of 4B5B means that to achieve an actual 'throughput' rate of 100Mb/s, the network must operate at 125Mbp/s. This is also used more recently in IsoEthernet.

4GL

See fourth generation languages.

4mm tape

See DataDAT and DAT.

4-phase modulation

Any modulation technique which yields dibits for each baud. So each transmitted 'event' or 'symbol' can produce dibits of 00, 01, 10, or 11. This doesn't necessarily refer to 'phase' as in sine-wave cycle, it can sometimes mean just 'four-level'.

4-wire

See four-wire and wire.

5

The number five, on its own, is sometimes being used to indicate a Pentium processor. You may see it during discussions about new computers, in the form: "... the Acme 5/100..." which means the Acme PC has a Pentium running at 100MHz.

5/6 tone

An early tone paging code developed by Motorola, which relies on unique addresses using any five of ten possible frequencies. The sixth tone, was a way of later extending the address to provide priority messages, coded instructions etc.

This standard has now been superseded by POCSAG and GSC coding systems with tone, numeric and later alpha standards included. See CAP, RDS, TAP, ETC, PET, IXO and FLEX.

5GL

Fifth generation language. See.

5-level code

The Baudot code or ASCII Alphabet No.2 (See)

7E1

The old asynchronous modem standard for communicating with mainframes. This is 7-bits, even parity, and 1 stop-bit (plus 1 start-bit which is assumed).

8-to-14 coding

See EFM and channel coding.

8-bit CPU

See 32-bit CPU.

8-bit toggle

This is a command for some modems which allows them to switch the 8th bit in each byte off (set it to 0). This is sometimes necessary to prevent gibberish on the screen, resulting from the transmission of a 7-bit ASCII byte, but with the 8th bit set to 1 for some internal communications reason. This causes the PC to reproduce the extended ASCII character rather than the normal alphanumeric character.

8B/10B

IBM's line-coding system used in Fiber Channel and some ATM implementations. Eight data bits are translated to a 10-bit group, with the two extra bits being 'disparate control' (error detection and timing recovery).

8mm tape

A very popular tape back-up standard for computers derived from Super-8 video, which uses video helical scan recording techniques. Only the physical mechanism and the deck now remain the same as video; for computer storage different cassettes, different tape formulations and different electronics are used. 8mm tapes have about ten times the capacity of the older 3480 half-inch cartridge systems, and they are cheaper and smaller in size.

These units use a three-head configuration on the rotating arm (read, write and servo) and have a 90° tape wrap. The heads rotate at 1800 rpm, and the tape moves at 11.1 mm per second.

Currently the standards extend to 7GBytes of native storage space on each tape (at a transfer rate of 30MByte/min), with up to 70GByte (uncompressed) in a 10-tape library/silo. Compression systems are often used. The main challenge to this standard comes from the 4mm DAT machines, which are about half the cost, and more recently from DLT; both outperform 8mm. See DataDAT also.

8N1

Shorthand for the most commonly used communications protocol with PCs. It means 8-bits, no parity, 1 stop-bit. (plus 1 start- bit assumed) See 7E1.

8PSK

An 8-level Phase Shift Keying technique for carrying digital signals over analog systems. It is used in digital radio and television broadcasting. 8PSK has eight phase angles, and it can therefore provide 3-bits per symbol, with one symbol per Hertz of bandwidth. In practice, by the time that FEC is added, it only provides about 2-bits per Hertz. It may be used together with a spectrum-spreading technique such as COFDM. See QPSK and 16-level QAM.

9-track

This is the old open-real half-inch tape mainframe recording standard for data. It stored the data in 9 parallel tracks at 800, 1600, 3200 and 6250 bits per inch.

10Base-#

These are all forms of IEEE 802.3 Ethernet running at the maximum data-rate of 10Mb/s using baseband transmission. The final number/character identifies the type of media or indicates the maximum length of a segment.

The options are:

- **10Base-2:** This uses RG58U thin Ethernet coax with a segment length up to 200 meters. It is also called Cheapernet or ThinNet (a slightly thinner cable still). The stations are daisy-chained.
- **10Base-5:** This uses RG-8 'yellow' thick Ethernet coax (max. 500 meters). This is the original standard Ethernet often called 'Thicknet'. It uses a bus topology.
- **10Base-F** (or FO): Ethernet over fiber optic cable. This is for point-to-point links. See also FOIRL.
- **10Base-T:** Ethernet on unshielded twisted pair. This is for segment distances up to 100 meters.

With 10Base-T, two twisted pairs are terminated in an RJ-45 type connector. The typical installation uses four-pair Cat. 3 UTP wiring to the desktop, and 25-pair Cat.3 in the wiring closet. Radiating (star) networks are essential with this standard, which is why recent emphasis has been on collapsed backbones and wiring hubs.

• See all of the above and Ethernet.

• Note: there are 1Base-5 and 10Broad-36 implementations of 802.3 also.

10Broad-36

This is the Broadband implementation of 802.3 Ethernet. It requires coaxial cable (75 ohm) and uses DPSK modulation at 10Mb/s max. It has a maximum segment length of 3.6 km (this is its main value).

12/14GHz satellites

See Ku-band.

16-bit CPU

See 32-bit CPU.

16QAM

A 16-level Quadrature Amplitude Modulation technique for carrying digital signals over analog systems. It is used in digital radio and television broadcasting. 16QAM can theoretically provide 8-bits per Hertz of bandwidth — but in practice it appears to be limited to 4 or 5 after FEC and other overheads are discounted. It may be used together with a spectrum spreading technique such as COFDM. See QPSK and 8PSK.

16-VSB

16-level Vestigial Sideband. A modulation and transmission technique developed by Zenith in the USA, and once thought likely to be used for future terrestrial HDTV systems. It was designed for applications requiring large amounts of non-critical digital data on coaxial cable systems. It transmits 43Mb/s in a single 6-MHz wide cable channel (7-bits per Hz), which is one-third more data than the competing 64-QAM proposal (which now appears likely to be the standard).

22AWG

The standard American Wire Gauge most widely used for telephone twisted-pair.

32-bit CPU

Microprocessors are often described as being 8-bit, 16-bit or 32-bit wide. This can actually refer to two different things:

— It can mean that the internal data register is 32-bits wide, or

— It can mean that the data bus is 32-bits wide.

These days these are generally the same, but they aren't necessarily, and they weren't for many of the early computers.

• The significance is in the fact that it gives an indication of how many bytes will be shifted in each memory access. 32-bit wide systems will bring 4-Bytes into memory in one action.

40-column

See 80-column.

80-80

A type of telex and telegraph circuit where the signals are transmitted by changes in voltage from +80V to –80V.

80-column

This refers to the character size on the screen. The first PCs used screens primarily for programming, and they provided 40 very wide capital letters to the line. This was the limit which could be resolved on most TV sets and monitors at that time.

Later, personal computers became word processors, and new character-generator ROMs and special 80-column video cards (which later became circuits on the motherboard) changed the text output presentation to the present character style and quality, allowing 80 characters to a line. This is the common standard for PC displays.

40-column displays are still used for videotex and teletex, and you need this size character for low quality monitors and TV set display. With higher quality monitors it is now possible to read 132 characters or more along the average text line

80 filter

In film production, this is the special color of a glass filter (or roll of celluloid filter material) which converts the normal color temperature of tungsten light to match that of daylight. It is a light-blue color and is more correctly known as a Wratten 80.

85 filter

In film and video production, this is a special light-orange colored glass filter (or roll of celluloid filter material) which converts the normal color temperature of daylight to match that of tungsten light. Professionals tend to use indoor (tungsten) film as a standard, and add the filter when working outdoors. These filters are correctly known as Kodak Wratten 85.

88open

A consortium which certifies Unix software compatible with Motorola's 88000 RISC processor.

100Base-4T+

100Base-4T+ is one of two signaling systems being used with 100Base-T (the other is 100Base-X) which use the standard Ethernet frame format. It amalgamates CSMA/CD with a new signaling scheme which requires four pair of Cat.3 (also 4 or 5) wire. This implementation of 100Base-T is half-duplex over the Cat.3 cable. The data is encoded as eight binary, six ternary (8B/6T) code with a bandwidth limit of 30MHz.

Of the four pair, three are used for data in each direction with the fourth pair for collision detection. Like 10Base-T, 100Base-4T+ uses the first and second pair for data and collision detection each in one direction only. But unlike 10Base-T, 100Base-4T+ uses the third and fourth pair in a bidirectional way to carry data.

• You can't use 4T+ with 25-pair/Cat.3 bundles in the wiring closet. See 100Base-X

100Base-T

This is a Fast Ethernet standard being developed by the IEEE 802.3 committee and the Fast Ethernet Alliance. 100Base-T uses CSMA/CD multiple access techniques, and depends on a Media Access Control which is the same as standard Ethernet. It has the same frame format and length. Adaptors are usually switchable between 10 and 100Mb/s.

The two signaling schemes under development are 100Base-X and 100Base-4T+. Both support distances of 100 meters from hub to workstation (UTP), and distances of 10 meters between hubs (however the maximum distance along the Ethernet backbone is then reduced to 210 meters); fortunately this doesn't restrict the topology in collapsed backbones. Each wiring closet is connected to a central data-pump that serves as the collapsed backbone.

There are a number of variations at the physical layer:

- 100Base-X requires 2-pair of Category 5 UTP cable.
- 100Base-4T+ will run over 4-pair of Cat 3, 4 or 5 UTP cable.
- 100Base-FX requires two optical fibers. This is the same as the 100Base-X but over fiber.
- 100Base-T4 is essentially the same as 4T+.

Routing and switching between 10Base-T and 100Base-T is relatively easy and fast, and the Fast Ethernet cabling also uses the same PIN-outs as the older system. See Nway.

• The disadvantages of 100Base-T are with the relatively inflexible architecture and the fact that Fast Ethernet only allows two repeater hops (each of 15 meters) per segment. It also requires workstation and server upgrades, and some of the older wiring schemes won't support it. See Fast Ethernet and 802.3.

100Base-VG

100Mb/s over Voice-Grade (aka AnyLAN). A new standard being promoted by the IEEE 802.12 committee for fast LANS at 100Mb/s over voice-grade cables. This is the protocol proposed by IBM, AT&T and HP for multimedia to the desktop.

It is 're-engineered Ethernet' which combines the simply network access characteristics of Ethernet with the control and deterministic elements of Token Ring. It uses a tree-topology from a hub over four pairs of Cat.3 cable. It is designed to also work over Cat.5 UTP wire, STP, or fiber; in fact it will work over virtually any cabling system for distances equal to 10Base-T.

AnyLAN now accommodates both Ethernet and Token Ring frame formats (hence the change of name). It assigns to the hub the job of contention resolution, and DPAM (Demand Priority Access Method) replaces Ethernet's CSMA/CD.

The network can span up to 1220 meters radially from the hub, with individual hub-to-hub and hub-to-workstation distances of 100 meters over 4-pair/Cat. 3 cable (or 200 meters over Cat. 5). There can be up to four repeater hops per segment.

A speed-matching translator bridge is needed to connect a 10Base-T network into AnyLAN, and an old network will also require workstation and server upgrades.

• IBM, Apple, Microsoft, Novel and Cisco support this protocol on and off, but now they also support 100Base-T and have placed greater emphasis on ATM. Recently, however, there's been another surge of enthusiasm for AnyLAN.

• Many commentators have said that 100Base-VG is technically superior to 100Base-X, but that it may be too expensive. It does have advantages with Token-Ring users where it's only high-speed competition comes from FDDI.

100Base-X

100Base-X is one of the two new signaling systems being used with 100Base-T Ethernet which have the standard Ethernet frame format (the other is 100Base-4T+). 100Base-X is an amalgamation of CSMA/CD and the PMD FDDI physical layer which will run over only two pairs of Cat. 5 wire, STP or fiber.

It uses four binary, five binary (4B5B) line coding in five-bit code groups that are mapped to symbols representing 16 data values, four control codes, and an idle code. The actual data-rate after the 4-bit/5-bit conversion is 125Mb/s.

100VG-AnyLAN

See 100Base-VG.

147

Eureka-147 radio broadcasting protocols. See DAB.

240M standard

An ANSI/SMPTE standard for high definition television studio production which seems to have been abandoned. It was derived from the NHK MUSE system. See 601 CCIR.

286

The short-name for the Intel 80286 family of micro processing chips which were used firstly in the IBM-ATs and compatibles, and later also in the PS/2 Models 50 and 60.

The 286 was formally announced in Feb 1982 and shipped in volume in 1993. The initial speed was rated at 1MIPS, but it later ran much faster.

Computers based on the 286 chip are generally used as faster (than XT) MS-DOS machines, however, the processor has two modes: the 'real mode' which imitates the 8086 PC with its 640kB/1MB limit, and the 'protected' mode with a 16MB memory limit. This second mode was originally used with different operating systems.

The Intel 286 chip has 130,000 transistors, and it took 3.5 years to design. It is best characterized as a single-tasking 16-bit processor which is only really suitable for simple DOS-based applications. The clock rate grew over time between 6 and 25MHz.

• With EMS (bank-switching) the working memory with MS-DOS on 286 machines was expandible up to 8MB.

370

A popular range of IBM mainframes. This is their current top-of-the line machine for general applications. The 370 runs one of two operating systems — either VM or MVS.

386

The correct number for this Intel processor family is 80386. Like the 286, it gives its name to a generic range of IBM-compatible personal computers.

Intel announced the 386 in Oct 1985 and it was shipped in volume in 1986. 44.5 million were eventually sold.

With this processor, Compaq stole the march on IBM for the first time. Members of the 386 family of chips have about 275,000 transistors and, with the right operating system, they feature true multi-tasking. Initially the chip was rated at 5MIPS.

This is a 32-bit, 1-micron, partial-CMOS processor which runs at 16—33MHz. It took Intel three years to design the 386, but AMD later cloned it in 0.8 micron, full-CMOS form, which ran cooler and faster. AMD's version had some special innovative features — like a sleep-mode for laptops, also.

The 386 has three operating modes:

- Virtual 8086 mode. See virtual 8086 mode.
- Protected mode (also in 286) which could now deal with 16MB of RAM. This allowed the EMS functions to be emulated in normal memory. See protected mode. Unix and OS/2 operated in this mode.
- Native mode could handle up to 4GB of memory with the right operating system.

• **386DX:** The original version of the chip is now known as the 386DX and has a 32-bit external data path. It is rated at above 5 MIPS and has a clock speed of 33MHz.

• **386SX:** This reduced 16-bit (data path) version of the Intel 80386 chip with a 16Hz clock speed was widely used as a replacement for the older 286 (which essentially has the same layout). It was for entry-level Windows-based machines only. Later versions of the 386SX ran at 33MHz.

• There is a famous story about how a user found a maths error in the early 386DX — the result of a badly designed transistor — and how it took Intel two weeks of frantic work to revise the design before the story got out. So the Pentium wasn't the first!

386 extended mode

See protected mode.

386 Smart Cache

An Intel chip which integrated a cache subsystem into one chip. It had the control logic, a write buffer, line buffer, and 16kB of SRAM. It was designed to work also with the 486.

486

Intel's family of PC microprocessing chips; correctly named the 80486, but also known as the i486. Intel announced this chip in Apr 1989 and it shipped in volume in 1990 — eventually 75 million were sold. These are 1-micron processing chips with 1.2 million transistors. They were initially rated at 20MIPS.

There are two basic versions: the DX and the SX and various speed options. These are not RISC chips, but they have many RISC-like functions. All of the 486 chips are really hybrid versions which include the old 80386DX CPU, and so remain fully backwardly compatible with the 386. This change was supposedly to provide 'mainframe' power while retaining OS/2 and MS-DOS compatibility.

- **486DX:** This chip incorporates a 386DX processor, a 387DX maths coprocessor, a paging and memory management unit, 8kB of data and instructional cache, and it runs a full 32-bit wide external bus. The chip has a burst mode which allows 128-bits of data to be read with only a single data request.

The first version ran at 25MHz and the second at 33MHz, the third at 50MHz, and the fourth at 66MHz. Even the earliest versions operated at 15 to 20 Vax MIPS, which makes the 486DX about 3—4 times as fast as the 80386.

Intel's new design process made it possible to compile the chip from basic circuits held in a library, so the chip took only 2 years in the design stage. The chip was renamed i486 in an attempt to circumvent the generic use of the terms 486 or 80486 which Intel couldn't control, however the company claims that the 'i' means 'integrated'.

- **486SX:** Originally the SX suffix indicated a 16-bit external bus — but not in this chip. The 486SX combined the 386DX, 8kB of internal cache and the memory management module onto a single chip. The first version ran at 20MHz, the second at 25MHz and the third at 33MHz. The 486DX was almost identical except for an integrated 387 math co-processor, however the 387 co-processor can be purchased separately and used together with the SX to synthesize 486DX operations.

- **486DX2:** This is a later version where the 2 indicates a 'Clock Doubler'. This clocks the chip at twice the speed for internal functions, as for external. The 486DX2/66 will only run at 33MHz when accessing memory and other logic-board components, but internally it churns along at 66MHz.

- See Pentium and PowerPC, RISC and CISC.

- AMD now has a version running at 120MHz.

488

IEEE 488. See GPIB.

500

- **Cable:** This is Cable TV coaxial distribution cable. The 500 comes from the outer dimensions of 0.500-inches, as against the trunk cable which is usually 0.750-inches ('750'). Half-inch coax costs about US$1.50 a meter.

- **Telephony:** In the USA the 500 prefixed numbers (0-500 or 1-500) are reserved for Personal Number Services. This is the first implementation of UPT, and these numbers supposedly follow you wherever you live in the USA — they should even be portable between carriers.

This last point is significant because, until now, the 'ownership' of a number has remained with the carrier. Now, for the first time, it appears as if you actually own your own number. There will be some interesting legal battles over this.

528MB barrier

The PC architecture running the DOS operating system with standard IDE devices (the ATA interface) imposes a 528MB barrier on hard disks, so a 1GB disk needs to be treated as two separate physical drives.

Most standard disk drives run up against this barrier because of slightly incompatible techniques used for addressing sectors on the disk using the cylinder+head+sector approach.

When an application accesses a hard drive for reading or writing, a DOS function-call will be sent to BIOS, which then relays it to the hard disk. Unfortunately, the ATA interface which links the hard drive with BIOS has strict limitations which are different from those of the INT13 interface which links BIOS and DOS.

The INT13 standard uses a cylinder address of 10-bits, a head address of 8-bits, and a sector address of 6-bits. However the ATA standard can handle cylinder addresses to 16-bits even though it only gets 10, a head address of 4-bits even if INT13 has 8, and a sector address of 8-bits even though it only gets 6.

Since this is a two-stage process, the highest numbers which can get through the system are 10-bits of cylinder address, 4-bits of head address, and 6-bits of sector address. This combination of $2^{10} \times 2^4 \times 2^6$ is the 528.48 MB limit.

562

The EIA/TIA 562 specification is for a serial interface to replace the old RS-232C. It is intended that this will work with the new low voltage (3.3V) components. The Specification provides backward compatibility with RS-232 and it guarantees that new devices can operate up to 64kb/s.

568

The EIA/TIA-568 compliance standard for a range of cable types used in LANs in the 10—150Mb/s range. It provides only a physical definition of the cable and connectors. At the time of writing these are being drafted. They will be the performance characteristics which can be expected within networks. A related standard, 569, defines the installation methods.

569

The EIA and TIA specifications for cable installation. See 568.

586

- **Microprocessor:** The Intel 80586 chip became the Pentium — although some of the literature continues to use the term P5. See Pentium and P# chip.

- **Wiring:** The EIA/TIA 586 standard which specifies commercial building wiring for voice and data communications. See Category # wiring (for Category 1 to 5).

601 CCIR

A CCIR (now ITU-R) recommendation adopted around the world for uncompressed digital video used in studio television production. This standard is also known as 4:2:2 and is the foundation standard for all digital video.

The original recommendation specified a data delivery rate of 165Mb/s using three-wire circuits (one for luminance, and one each for two color-difference signals). It is based on an NTSC screen format of 512 x 480 pixels, and a PAL screen format of 704 x 576 pixels. In the original form it did not allow for wide-screen or HDTV.

The standard image formats are now extended horizontally to provide 720 x 480 pixels at 30Hz (ex-NTSC) and 720 x 576 pixels at 25Hz (ex-PAL and SECAM).

Uncompressed digitized NTSC image and sound video stream now require 210Mb/s, and PAL requires 216Mb/s. This is a pixel-depth of 20-bits — 10-bits for luminance, and 5-bits each for color-difference signals. It is now possible to compress these images to about 12Mbit/s without substantial subjective loss of picture quality using MPEG2.

— The 4:2:2 version is the studio production standard which reserves 4-bits for luminance, and 2-bits each for color-difference signals.
— There are also other versions. The 4:1:1 is an electronic news-gathering format which reduces the bandwidth by cutting the color quality. See 4:2:2, 4:1:1, and 4:2:0.

• The 601 resolution can be at different levels, and for different aspect ratios.

— 8-bit = 243Mb/s for 4:3 uncompressed
— 10-bit = 270Mb/s for both 4:3 and 16:9 (with 25% lower horizontal resolution).
— 10-bit = 360 Mb/s for 16:9 full resolution. (wide-screen, but not HD).

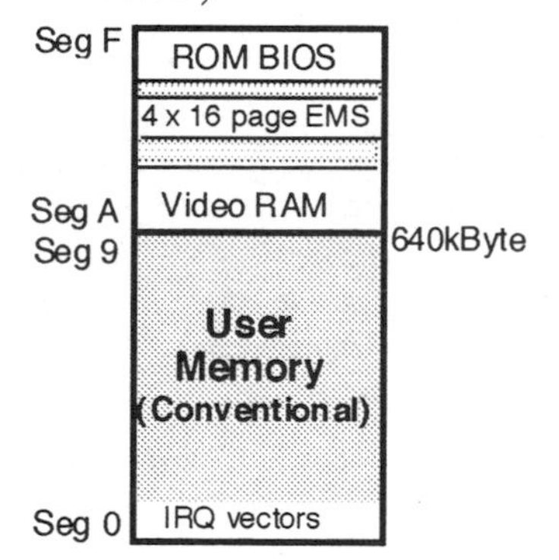

640kB barrier

The 640kB barrier was set by software, while the 1MB barrier was established by hardware. See above.

The early PCs only had a 20-bit wide data path, so the maximum that could be addressed as a single action was 2^{20} memory locations which is the 1MB limit. However, they decided to reserve the upper part of this memory for video RAM (for the bit-map driving the screen) and BIOS (the ROM which handles input and output).

They could equally have put this all at the bottom of memory, but in a fit of pure madness, they chose to put it at the top — since no one would ever want more than 640kB of memory anyway!

So a line was then drawn at 640kB, and locations below this barrier held working (conventional) memory for the operating system, the application (singular then), and data. Between 640k and 1MB the locations held the BIOS, video RAM and some device drivers. See EMS and UMB.

656

Recommendation 656 of the CCIR defines serial and parallel interfaces to Rec. 601 component video. It will handle both 8- and 10-bit resolution.

750

• **Microprocessor:** Intel's new twin video processor chip-set which implements the DVI (Indeo) data expansion techniques.

• **Cable:** This is Cable TV coaxial trunk cable. The 750 comes from the outer dimensions of 0.750-inches, as against the feeder cable which is usually 0.500-inches. Three-quarter inch coax costs about US$3.00 a meter.

800 service

Aka Watts or freephone lines: In the US and elsewhere the 800 telephone access code has been reserved for automatic reverse charge calling — also called freephone. The owner of the number pays the bills. In the US these numbers often need to be dialed with a prefixed 1- access code (1-800).

In most of the world, these freephone services are only cost-effective if you are a large company. Discount rates of 50% or more, then apply. However the popularity of this service in the USA lies with the fact that a single 800 number is recognized nationwide. So it allows the owner to be rung from anywhere, and the connection can be made to anywhere in continental USA. So when an American changes his/her home or business location, the 800 number follows. This depends on a sophisticated intelligent network which is part way to UPT. Recently the 500 numbers have gone even further.

The 800 service is only a free phone service when the owner so decides to make it free (but most do). However, be aware that this doesn't necessarily mean that the information obtained will be free — which many people discover to their cost. 800 numbers are traps used by sex-line operators; the call may be free, but the young-lady's time will be charged at $5 a minute. The FCC is expected to rule against 800 charging systems.

• AT&T has made a fortune out of the 800 service; 13 billion calls (40%) of their annual network traffic are now freephone calls. This is obviously not just used for telemarketing.

802.x

The IEEE's series of LAN protocols. The ISO versions of these standards have an extra '8' on the front, so Ethernet in ISO terms, is 8802.3, etc.

The main players in local and wide area networking are:

– **802.1**: A specification for spanning tree bridges and the spanning tree protocol. This is a hardware level network management standard for Ethernet MAC-layer bridges and the Heterogenous LAN Management specification for both Ethernet and Token Ring hubs.

– **802.2:** A Logical Link Control protocol based on HDLC for LAN and MAN link-level control. It specifies the transmission of data between two stations at the Data-link layer. It defines the handling of errors, framing, flow control and the Layer 3 service interface. It is used in both 802.3 and 802.5 LANs.

– **802.3**: A modified form of CSMA/CD for Ethernet in its various forms; 10Base-5 is virtually the original Ethernet. This protocol specifies the Physical layer and the MAC sublayer of the Data-link. See CSMA and 10Base-#.

– **802.4:** The Token-passing bus specification. This protocol specifies the Physical layer and the MAC sublayer of a broadband system. It is physically a linear bus or tree, but logically the stations on the ring are aware of their neighbors. It uses coaxial cable for robustness, and token-passing for maximum predictability. It is used in some manufacturing networks.

– **802.5:** Token Ring specification. This comes in both 4 and 16Mb/s rates. The physical wiring is shielded twisted pair in a star topology with a hub at the center (although it is a logical ring). An addendum to this Protocol specifies source-routing bridges also. It is almost identical to IBM's Token Ring.

– **802.6:** DQDB for MANs. See DQDB.

– **802.7:** Defined for broadband LANs which can carry video, data and voice. It uses radio frequencies over coaxial cable, and generally has a tree topology.

– **802.9:** Interface for LAN-to-ISDN links.

– **802.11:** Wireless LAN standards (line-of-sight infra-red, and spread spectrum) with up to 1000 nodes and 20Mbit/s data rates.Currently it operates at 2Mb/s in 2.4–2.5GHz range (ISM). It uses collision avoidance and has a max. range of about 100 meters. See Hiperlan also.

860

Intel's 'superchip' which was to have been the basis for some new and cheap supercomputers. The 860 has a RISC core and MMU on board. It use a 64-bit architecture and combines both memory and the central processor in a single chip. The design largely eliminated the need for supporting services, and this has allowed optimization of the communications channels between chips for parallel processing.

When grouped in a parallel array 860s can perform super-fast parallel operations. The design includes some routing capabilities in hardware for messages between the processors. Current transputers need to have software written for this function. See transputer.

900 services

In the USA, these are a form of enhanced 800 service. But here you should expect to be charged.

960

Intel's family of 32-bit superscalar 'megaprocessors', each with 600,000 transistors capable of performing up to 150 MIPS — which is 4 to 8 times that of the 486 (which has twice the number of transistors).

The 960 family has chips which are rated at 10, 24, 33 and 75MHz clock speeds. They all have a parallel processing architecture that can theoretically execute up to 3 instructions per cycle (but average about 2). These were the first Intel chips which could perform two or more operations at the same time. The market for them is probably as embedded controllers in electronic equipment. Both 5 and 3.3 volt operations are possible with later models.

3002 channel

An old US term for a standard four-wire connection. It means 2 pairs of wires, each with 3000Hz in bandwidth. See wires.

3010

An IBM intelligent display terminal.

3174

IBM's new version of the old 3274 cluster-controller for terminals.

3270

The IBM terminal designation number 3270 has become a generic name for a family of system components including terminals, printers, terminal cluster-controllers and front-end processors for mainframes. Each of these will have a unique number along the lines 327x. They are widely used with SNA protocols.

Originally, this was a popular IBM mainframe 'dumb' terminal type which is often emulated by PCs running special software. They use EBCDIC character sets. Millions of 3270-type terminals are still in use for airline bookings, insurance claim processing, credit-card transactions, etc.

These terminals typically hook up to the mainframe through a cluster controller which stores and maintains the contents of the terminal screen, and sends it only to the host when the user presses the Enter key. Both BSC and SDLC protocols were originally used, but now SNA is in fashion. See 7171.

• The key difference between 3270 and ASCII terminals (and terminal emulation) is that 3270s only transmit data when the Enter key is hit. It is a 'batch' process, while an ASCII terminal sends a character every time one is typed.

• Note: 3270 terminal emulation actually appears to the host as if it were a 3278 Model 2 terminal.

3274/6

A pair of commonly-used cluster-controllers for the 3270 family. As many as 32 terminals and printers can be linked through these controllers to a mainframe's front-end processor. The cluster controller stores the screen information for each, and only sends it to the host when the user presses Enter. A polling operation is used to access the individual units. See 7171.

3278

The most popular IBM terminal type in the 3270 family. It has a monochrome display and only a limited graphics set, and it communicates with the mainframe via a complex protocol (often BSC or SNA). A PC can emulate the 3278 terminal, but to do so it must simulate some extra keys and lights. See 7171.

3279

A color version of the 3278 terminal in the 3270 family.

3705

A front-end communications processor which can be used to link several 3274 cluster-controllers to a 370 mainframe. It can also provide a limited number of dial-up ports.

3725

An IBM front-end processor used to link groups of cluster-controllers to a mainframe.

3745

A communications controller that combines cluster-controller and front-end processing functions. It can link with up to eight Token Ring networks for a total of 512 terminals.

3820 series

This series of IBM terminals used 8-bit EBCDIC not ASCII. You needed a protocol convertor. See BiSync.

4004

The world's first microprocessor (from Intel). It was a 4-bit chip released in 1971 for program process control (mainly refrigerators) — but it was soon used to build the first 'home-brew' personal computers. It wasn't much more programmable than the single-chip TI calculators (the TMS1000) of the time. Some people would credit the 8008 with being the first real microprocessor. See MPU

7171

The IBM 7171 communications controller allowed ASCII terminals such as PCs emulating VT-100 or something similar, to replace the IBM 3270 terminals (although the old name 3270 often still survived). The 7171 translates between the ASCII and the EBCDIC character sets, and performs the other functions of the cluster controller.

8008

Possibly the first real microprocessor. It was made by Intel. Don't confuse this with the later 8080 or 8086 which were the first 'successful' Intel microprocessors used in the PC. See 4004

8086

The second Intel chip used by IBM in its XT line of personal computers (but the earliest of the two to be fabricated). This was the CPU also used in the IBM PS/2 Models 25 and 30.

The 8086 chip was first released in April 1979 and the 8088 appeared nine months later as a cut-down version. The main difference between the two is that the 8086 chip uses a 16-bit data bus while the 8088 has only an 8-bit data bus. Both chips have a 20-bit address bus (which can address 1MB) and they handle data in 16-bit slices. The 8086 had a clock rate between 6 and 12MHz.

The 8086 shares the same instruction set (108 microcode commands) with the 8088, but the 8-bit chip was chosen for the first IBM PC because of the higher costs associated with true 16-bit architecture (mainly the cost of peripherals). The 8088 chip was slower (which made it easier to design for) and had the necessary support chips and memory architecture at the time.

There is a CMOS version of the 8086 which has the characteristic of being able to run at variable clock rates from zero to 5MHz. The later 8086-2 ran at about 7MHz.

8088

This was the cut-down 8086 Intel chip which became the foundation for the IBM PC line of computers. The chip was a 16-bit processor, with a clock speed of between 4.77 and 9.5MHz. It used an 8-bit data bus but overcame this by handling 16-bit data in two sequential bytes. It had a 20-bit address bus, however, which allowed access to 1MB of memory.

In the early IBM PCs, the 8088 only ran at 4.77MHz but later there was a slightly faster version called the 8088-2. In later PCs (the XT) this chip gave way to its parent the 8086 and, later still, to the 80286 and 80386. See 8086.

8514/A

An IBM standard for graphics adaptors which was designed for high resolution graphics. It was introduced as an extension to VGA in 1989, and it supported 256 colors (from a palette of 262,000) at 1024 x 768 pixels of resolution.

It employs a custom graphics co-processor and special system software (8514 AI) and was introduced with IBM's Micro Channel Architecture release.

This was an interlace standard, and it operated much faster than the top-end Super-VGA boards of the time because all critical graphics processes were performed in hardware by special chip circuitry, rather than by ROM-resident software.

It failed as a standard because it was incompatible with CGA, EGA and VGA, and the standard has now almost been forgotten. See XGA, Super VGA and TIGA.

880#.#

The ISO versions of the IEEE LAN standards. The ISO numbering system for these standards simply adds an '8' in front of the IEEE numbering system. So the IEEE's 802.3 standard becomes the ISO's 8802.3. The IEEE standard numbering system is widely used, so I've stuck with this.

9660

The ISO standard for file formats and directories on a CD-ROM disk. This is an operating-system independent file system — so it takes a least-common-denominator approach. This means that the file system just dumps any special vendor-specific features such as the Macintosh's resource forks which give Mac files the icons and Mac 'look-and-feel'. Most disk manufacturers conform to the standard, but produced distinct Mac and IBM-PC versions at the same time. See High Sierra and CD-ROM.

68050

There was no Motorola 68050 — they abandoned it because it was only a small leap into the future, but it was planned to have a 64-bit bus, integrated FPU, built-in graphics accelerator, and a streamlined instruction set — but it wouldn't have been superscalar. See 68060.

68060

This Motorola superscalar chip is still under development at the time of writing, and it may even have been abandoned. Motorola is now fully occupied with RISC.

680x0

The Motorola 680x0 CISC chip family is used in the Macintosh and a number of other computers and workstations.

— **68000:** The original chip used in the first Macs. This was a 16/32-bit hybrid. The chip ran at 7.8MHz, and the computer had 128kB of RAM.

— **68020:** A true 32-bit chip which ran at 16MHz which was first used in the Macintosh II series (1986).

— **68030:** An even faster (20MHz) 32-bit chip which is still used. The computers using this chip have virtual memory management

— **68040:** This uses a semi-RISC instruction set, and so it is faster in instruction processing (even at the same clock speed) than the 030. It is a 32-bit chip with 1.2 million transistors, and with performance capabilities around 20 MIPS.

Motorola is now making RISC chips. See 88000.

80x86

Intel's range of processor chips which followed the 8088. The first member of this family was the 8086 chip used in the XT. This was followed by the unsuccessful 80186 which was never used in a PC. The later versions were the AT's 80286 (called the 286), the 80386 (386), the 80486 (486) and the 80586 (the P5 or Pentium). See all.

80860

Intel chip. See 860.

88000

Motorola's RISC chip. See 88open.

A# (paper sizes)

The most common European paper sizes are the A-series. You will see that there's a repetitive pattern in the dimensions caused by each of the larger sizes being folded in half across the longer sides to produce the next. In each case the aspect ratio (length vs. width) remains constant:

A0	1189 x 841 mm	46.81 x 33.11 in.
A1	841 x 594 mm	33.11 x 23.39 in.
A2	594 x 420 mm	23.39 x 16.54 in.
A3	420 x 297 mm	16.54 x 11.69 in.
A4	297 x 210 mm	11.69 x 8.27 in.
A5	210 x 148 mm	8.27 x 5.83 in.
A6	148 x 105 mm	5.83 x 4.13 in.

See B# (paper sizes), book sizes and paper sizes also.

A5#

The A51 algorithm was a highly secure system used in the early European GSM mobile phones. However, in the interests of the national security services being able to listen in on private conversations, it was judged far too secure for non-Europeans. Therefore, outside Europe, GSM is fitted with a version called A50 (which is really a non-functioning version). Later these can be upgraded to A52, which the European security services know how to break.

A-format video

An early attempt by Ampex to produce a professional one-inch analog videotape standard. The original design ideas were later incorporated into the C-format standard.

A-Law

The current ITU (European) standard for telephone pulse-code modulation, known correctly as PCM Encoding Law — Version A. Actually this is a companding standard which is used in the ADC conversion process.

The audio signal is filtered to restrict it to a 300—3400Hz bandwidth, and then sampled at 8kHz. The 'law' which is then applied is that of 'pseudo-logarithmic quantization', where an 8-bit word is assigned to each sample according to the ITU's G.711 recommendation (7-bits value and a 1-bit +/– sign).

The A-Law PCM standard is widely used outside the USA, while the American version is called μ-255 or μ-Law. (See mu-Law.) It is slightly incompatible with A-Law for best quality.

A-party

The caller in any phone conversation, as distinct from the B-party who is the receiver.

• 'A-party Clear' telephone systems are those in which a call is only cleared when the A-party puts the phone on-hook. See call clearing.

A to D conversion

Analog to Digital conversion.

AAL

ATM Adaptation Layer. A key component in the new protocol standards of ATM cells. AAL lies above the basic ATM layer, and serves to adapt the different telecommunications service requirements (application bit-rates and data unit structures) to the ATM network. It's prime task is to segment and re-assemble signaling messages and/or frames into ATM cells without introducing errors. It is divided into two sub-classes; the SAR (Segmentation and Reassembly) and the Convergence (which deals with error control).

Four classes have now been defined:

• **AAL-1** (Class A) for connection-oriented services requiring a constant bit rate, and requiring timing relationship between ends.

• **AAL-2** (Class B) for the above but variable bit rate.
- AAL-3 (old Class C) for the above but no timing relationship required.
- AAL-4 (old Class D) for connectionless variable bit rate services with no timing relationship required. Note 3 and 4 have now been merged into:

• **AAL-3/4** — which now becomes Class C. This is intended for LAN traffic.

• **AAL-5** (The new Class D) General purpose basic applications which don't take any of the payload for formatting. It is the simplest to implement.

A&B bit signaling

A bit-robbing technique used in T1 transmissions after the introduction of the D2 channel banks. There are numerous versions in the T1 specifications, but generally one bit is stolen from every 6th and 12th frame in each of the 24 subchannels. These bits carry supervisory information — off-hook, on-hook, ringing, busy, conditions.

Bit-robbing isn't used with the latest T1 systems — those using B8ZS line codes instead of AMI. It gets even more complicated with more modern T1 systems using D4/ESF. See ESF and ABCD.

A/B roll editing

A technique of film production, later used in video, to allow dissolves, fades and superimposition during the post-production processes by alternately copying images from two (the A and B) sources. Sometimes three or more sources will be used.

• In film production, the cut original film will be spliced into two reels during the editing/matching process, with each reel holding alternating shots (with black spacer in between). Each roll can then be printed separately onto a single 'composite master' negative which is used for the final printing. Double exposures, dissolves and superimpositions are printed as overlapping images at various intensities.

• In video production the same basic idea is used to introduce special effects into a production during the editing and dubbing stages. Two separate source tapes are constructed with the appropriate shots included (identified by 'timecodes'). Both tapes are then replayed simultaneously using sync-lock through a special effects generator (SEG — video switcher) under program control. This controller alternates between the sources (using time-code triggers) and creates dissolves, fades and superimpositions by mixing the output from both.

A/B switch

A generic term for any switch which can alternate either between two inputs or two outputs.

These are commonly used to join two or more computers to a single printer or modem. Some of these are complex (A,B,C,D...) multi-connector switches.

• By law, all American cable television set-top boxes must be equipped with an A/B 'input selector' switch which allows them to take their signal source from the cable or from an off-air antenna. The early rented boxes were designed in such a way that it was difficult for a householder to change from one signal source to another, so the cable operator was able to exclude local broadcasting stations and steal their advertising revenues. The compulsory A/B switch and the 'must-carry' rule were designed to counteract this anti-competitive practice. See must-carry.

abbreviated dialing

A technique used with on-line communications (mainly message- and packet-switching). The full network X.121 address of a commonly-used host will be stored at the system's primary node (the first network computer) and provided with an easily-remembered 'alias'. A user need only then enter this alias, and the node will look up and use the full address to establish the connection. See X.121.

ABCA

Joint American, British, Canadian and Australian standards. Mainly for military hardware and supplies.

ABCD

A form of bit-robbing used for signaling in the US T1 channels. This is also called ESF. Every 6th byte the channel is robbed of a single (least significant) bit and this is designated as a signaling/supervision indicator for a DC state on the local loop side of the channel unit (on-hook or off-hook indicators).

This results in a 1.3kb/s in-band signaling slot/sub-channel, which is divided into A&B, and sometimes into C&D sub-sub-channels. This provides many more bits than necessary, because MF tones are generally used for carrying the dialled number. See ESF and A&B bit signaling also.

Abekas

A special hard-disk recorder for video. It stores a few minutes of video at a time and is used for instant replays of sports action. It will also holds digital images while they are being processed in a digital-effects laboratory. See Henry also.

abend

An ABnormal END to a program. A crash or system bomb.

ABI

Application Binary Interface. This is the hardware-level interface which allows the construction of object-code compatible systems. ABIs provide application portability. For instance, the ABIs supposedly allow Unix software to be portable across any processor architecture — including the various SPARC designs and the Mac/Motorola. To be entirely portable, however, software must comply both with the ABI requirements and with Open Look. See Unix International.

PowerPC has also announced a ABI that would 'abstract low-level hardware' to allow software from Macs, DOS and Unix machines to run on the same PowerPC system.

• Note: another source maintains that 'ABIs only allow compatibility between products built on the same microprocessor architecture.'

ABIST

Automatic Built-in Self-test. This is an ideal, more than a reality. The idea is that the new VLSI devices and future VLSI can actually test themselves before they get out into the marketplace.

ablative pit

This is one of the three current WORM (optical disk) recording technologies (others are bubble formation and dye polymer). A high-powered laser burns pits in a reflective layer sandwiched between two clear plastic disk surfaces. Lasers on a low-power setting can read whether the light is reflected from the surface (the 'land') or not (from the 'pit').

Ablative techniques are not reversible; you can't erase and record these disks. However you can erase data by overwriting an area, or, alternately, you can prevent access to areas of data by modifying the directories. WORM technology is well suited to applications that require an audit trail — but only if full overwriting isn't used to destroy previous data.

ABM

Asynchronous Balanced Mode. An HDLC communications mode which supports peer-oriented communications between two stations; either one can initiate the transmission. This provides a high degree of independence in initiating actions, since any device can send Data-link commands at any time. This is used by IBM on Token Ring LANs (LLC) and SNA Data-links.

ABR

Available Bit Rate. In ATM systems, this is the flow control technique which better suits the original 'asynchronous' packet idea of making bandwidth available on a first-come, first-served basis. This is an alternative to the reserved fixed-bandwidths provided by CBR and FBR (see) and Variable Bit Rate (VBR)

There are a number of ABR proposals being considered for ATM. Generally, with ABR, the terminals don't request the allocation of any specific bandwidth. They assume that the network will make its best-effort to deliver whatever is sent, and they are willing to tolerate the occasional cell-dumping due to overload.

Obviously, in these circumstance, flow-control and priority play a very important part in network management. Successful operations depend on the network having a rate-based congestion control scheme, and the ability to distinguish between packets which are essential and those which are dispensable.

The carrier can have high average utilization on ATM trunks with ABR because full bandwidth doesn't need to be reserved. The contrast is with UBR (Unspecified Bit Rate), VBR (Variable Bit Rate), and CBR (Constant Bit Rate).

ABS

• **Absolute value.** A programming function that removes the sign from a number. For example ABS(–34) = 34.

• **Adaptive Block Size.** A HDTV video compression technique where the DCT algorithm is applied to blocks of pixels which vary in size from 16x16, down to 2x2. Larger blocks result in better compression ratios, while smaller blocks can provide better resolution and fewer artifacts. Two bits on average is needed as overhead to handle block-size.

absolute address

• Located at a fixed address in memory (ROM or RAM) — a machine address. The term is used to distinguish between two different techniques of accessing RAM — absolute and relative (or offset) addressing.

– **Absolute addressing**: Also called 'direct', or 'specific' addressing. When program instructions need to access data stored in a particular memory location they will provide the CPU with the actual fixed (absolute) address of that location.

This is a single-step operation which keeps CPU operations very fast, but it is only possible when the application is always loaded into the same section of memory (addresses locations) every time. With multi-user and multi-tasking computers this isn't always possible.

– **Relative addressing**: The program instructions provide the CPU with an 'offset' from another location, usually the 'base' address for that application. The CPU can calculate the required location by always adding an offset to the base address. See base address and offset.

• In spreadsheets, an absolute address is a fixed row-column location. In Lotus 1-2-3 the $ sign denotes an absolute row/column address, so a reference to C6 is a reference to the fixed cell location C6.

• The term is also applied to the addressing of peripherals. For instance, Disk drive C: Cylinder X: Sector Y is an absolute address for a block of data on a hard disk.

absolute RGB encoding

Absolute RGB techniques encode the information needed to control each screen pixel in two, three or four bytes (between 16 and 32 bits overall). The distinction being made is that between absolute encoding systems, which always provide a full range of spectrum colors (although with varying degrees of color range), and those systems that depend on a limited 'palette' of colors selected through a Color Look-up Table (CLUT).

With a two-byte absolute pixel 'depth', the computer will allocate 5 bits each as control for the Red, Green and Blue levels, and use the 16th bit as a transparency code (it is used in an on—off fashion when combining this image with other graphic images). This technique gives a total of 32,768 possible colors for each pixel, but at the cost of needing 615kBytes of video memory for each full-screen image at the high-resolution of a 640 x 480 VGA screen.

With 32-bit encoding, usually one byte (8 bits each) is reserved for R, G and B, and the fourth byte is used for transparency and other qualities. So this standard is usually (and more correctly) referred to as a 24-bit color standard. It provides 16.7 million colors at a cost of 1.2MBytes of video memory per VGA screen if all 16.7 million were to be available to each pixel. See CLUT and pixel depth.

absolute value

See ABS.

absolute vector

A screen or graphics vector with beginning and end points, designated by x/y coordinates — as distinct from a 'relative' vector which is defined in offset terms from some point.

abstract

In information retrieval, abstracts are summaries of the major points in articles or reports. These are usually formal database records which include details of title, author, publisher, date etc. Librarians distinguish between 'indicative' abstracts (roughly, what the book/article is about) and 'informative' abstracts (methods and conclusions).

• **Abstract symbols** in programming are those which haven't been strictly defined for general use — they can be used by each particular application for its own purposes.

• **Abstract numbers** are 'theoretical numbers'.

• **Abstract syntax** means the description of data structures which are quite independent of the hardware and encoding.

AC

Alternating current. Flows of electrical current which periodically reverse voltage (direction). Alternating current has the advantage of passing across condensers and being modified by transformers, transistors, etc. Direct current is much more limited in applications for conveying information.

• The term is usually applied to the mains power supply. The frequency with which the current reverses is now measured in Hertz (previously in cycles-per-second), and the two major standards are (generally):

Europe/S.E.Asia/Australia	230/240 volt	50Hz.
North & South America/Japan	115/120 volt	60Hz.

• As a guide: the Australian national grid is at 132,000 volts, which is then broken down to 11,000 volt at suburban zone stations (about 100 for a large city), which then feed sub-stations (1500 for a city) which produces 415 volt three-phase current, on wires running down two or three suburban streets. This is a three-wire feed, and 240 volt comes from any two of the wires.

• Baseband signal transmissions where the direction of electron-flow reverses periodically. Telephone voice signals, for instance, are carried through the network as AC frequencies which reverse in accordance with the analog audio input.

• Radio and broadband LAN transmissions are also AC in character, but of very high frequencies.

AC-bias

A high frequency AC current (about 100kHz) which is applied to analog magnetic recorder heads (typically in audio tape recorders) to energize the tape particles and 'linearize' the recording process. Without bias, the frequency response will be distorted.

AC15

An old standard interface for PBXs. See DPNSS.

Academy format

The name generally applied to the original 4:3 aspect ratio image (width-to-height) used for 35mm cinema projection and film production. This is the screen ratio now used by television.

• In common use in the film industry, 'Academy' simply means that the film has not been shot in widescreen or Cinemascope ratios.

• More strictly: with 35mm film 'Academy format' was applied only to silent movies where the image also occupied the area now reserved for sound. This format provided a 4:3 ratio also. See widescreen, Cinemascope format and HDTV also.

ACC

Accumulator. See.

accelerator

• The term is usually applied to an add-in card which contains a faster central processor to boost slow, older computers. These cards generate their own clock pulses, and often have their own RAM, so they virtually use the old computer only as a basic I/O device. You will need to remove the old processor chip, and plug in a connector which carries the processor links back to the card. Dedicated logic accelerators often use ASICs and DSPs to execute special, but more limited, sets of functions.

• The term has also been applied to special chip-upgrades on the motherboard and sometimes to a special form of piggy-back chip.

• The term is sometimes also applied to graphic and numeric co-processors which assist the CPU in computationally-intensive tasks. Video accelerators are also now popular with the move into multimedia.

• In Windows 3.0, an accelerator is a keystroke which has special meaning to an application. It generates a command and is a key-stroke alternative to using the menu for indicating choices — it 'accelerates' by reducing the number of mouse movements.

accent

The small marks over (or under) the letters in some languages — the term covers the grave, acute, cedilla, tilde and circumflex. Also called diacritic.

acceptance angle

In fiber optics, the photons from the driving laser (or LED) must strike the end of the fiber at an angle almost parallel to the axis. With very fine single-mode fibers, this angle is critical and therefore these fibers are said to have a narrow acceptance angle, and so they generally require a laser driver. The larger multimodal fibers, however, have a wider acceptance angle which allows them to gather more light and work with LEDs.

access

To be able to use:

• In computer technical terms, you can access a memory location in RAM, ROM or on a disk or tape. Generally you will do this to read or write from/to a memory location.

• In security terms, it means to be allowed to (or able to) use a service. This is usually a database made generally available to paying customers, or a piece of hardware available on a network. Unless you have both an identification number/name and a password you will be denied access.

• Older computer and video publications often distinguished between random-access storage systems and sequential-access systems. Tape is a sequential storage system; if you want some data at the end of the tape you've got to spool through everything before you can gain access. Disk and drum systems are called 'random' access, but this term should be reserved strictly for chip-memory storage.

Disks and drums still need to wait while a sequence of data passes under the head before they can find the sector they require, so these are strictly 'quick', but not true random access techniques (admittedly, they are often 'close-enough'!)

• In telecommunications, 'equal access' or 'open access' means that you can select (on a long-term basis) any long-distance carrier you wish, and use that service without needing to dial a special prefix each time. See equal access.

• An access line, is a channel in a major bearer, or a local loop connection which gives the customer access to the telephone network.

• All LANs or shared media, need an access technique. In Ethernet this is CSMA/CD, in Token Ring it is token passing, in GSM mobile phones it is TDMA. See channel access.

access arm

The arm holding the read/write head on a disk drive. See voice coil and stepper.

access charge/fee

The amount charged by a communications network or database utility for use. In addition to the access (entry) charge, often a per-character charge will be imposed, and there may be additional time charges also.

With on-line interactive systems, the use of the term is more general. The distinction is often between the charges levied by the host who supplies the content, and those 'carriage' (access) charges which will be levied by the network provider.

access code

• In PBXs, access codes are the preliminary numbers that may need to be dialled to get through the switchboard to use a public-switched telephone line. Usually this is a 0 or a 9.

• In a public-switched telephone system the term is often applied to any numbers preceding those needed for local calls e.g. the area- and country-codes. However, it more correctly means area-like codes which actually instruct the carrier to perform some special service.

In the USA both the 0 and the 1 prefixes have been reserved for access (they can't be used as the first digit in an area code), and in much of the world the 0 and the 00 have special significance for interstate or international calls, or as an indication that special services are being dialled. The US system is in the process of being changed to require a prefixed 1 access code to be dialled with all long-distance calls. The 500, 800 and 700 numbers are also correctly 'service access codes' even though they look and act like area codes.

• In countries without a monopoly carrier, sometimes a special access code needs to be dialled to select your preferred long-haul carrier.

• In some e-mail services, the access code is a prefixed address code that allows you to make connection to a specific secondary service (such as telex or facsimile) through the electronic mail facility.

access control

• Security measures, such as passwords, call-back systems, etc. used to prevent unauthorized entry into a computer system.

• There are also physical access controls such as magnetic keys and card systems which must be inserted to enter a room.

• Media Access Control (see MAC protocols). There are two main techniques: deterministic (polling or tokens) or contention. See Ethernet and Token Ring.

• An access controller is another name for a multiplexer. In video-conferencing it may also be an inverse multiplexer.In radio telephony and satellite services, this can mean the 'sharing' techniques used as well as the methods of policing them — FDMA, TDMA or CDMA techniques. See all.

• When LAN data is accessing a backbone or WAN, the two main special features of access management are:

— **Priority Handling:** A priority code may be allocated to each user-circuit so that a particular grade of service (high, medium or low priority) can be supplied to that user. Traffic with a higher priority is then given precedence in the queuing system.

— **Quota Control:** This is also needed in priority systems to prevent high priority data from completely locking out low-priority users. So with quota control, the network bandwidth is allocated on the basis of the priority gradings but only within the limits of quotas. Sixty percent of the bandwidth may be given to high-priority users, thirty percent to medium priority, and ten percent to low-priority users.

access method

• **Computer storage**. Any technique (in a program or the operating system) that defines how data is read, written and updated. Different access methods are used by different storage media. Tapes, for instance, store data sequentially, while disks provide random access, and usually need an index and a FAT.

• **Computer input**. See DMA and channel (processor I/O).

• **Communications:** See access protocols and access control.

• **SNA**: This is set by software in a special SNA processor which controls the flow through the network.

access network

In telephony, this is the local loop — the copper twisted-pair between your phone handset and the local telephone exchange — and about 40% of the capital investment of a telephone network is here.

While the 'backbone' network (exchanges and inter-exchange connection) are shared, the access network links are generally exclusive to each customer.

With the advent of telephony over Cable TV networks and cellular wireless, access techniques are now changing. And in the long-term with the introduction of fiber to the curb or fiber to the home they could change even more. Wireless local-loop developments could, in fact, replaced wire with radio links.

access point

• In a database record, this is the key field. This is essential in relational database management for linking files — it must exist in all of the databases if they are to be related. See DBMS.

• In networking of any kind, this can mean either the physical or the logical point of access.

• In telephony, this refers to the local-loop boundary which separates the responsibilities of the carrier, from those of the subscriber. This is also called the demarc. Usually, this is the point where the cable crosses the property boundary of the subscriber, but may be within the premises. The access point has particular implications with modern ISDN connections where equipment is needed at the line-termination point, and also with future fiber-to-the-home systems which can't draw current from the exchange. Power then needs to be provided at the access point.

access protocol

The agreed standard used in LANs and communications systems for moderating (arbitrating or controlling) data entry to the system. It may be:

— A routine under the control of the front-end processor in a mainframe system. (VTAM etc.)

— The basic control and management technique used for sharing the resources of a local area network (CSMA/CD, Token Ring etc.). These ensure equitable access, and prevent data being lost through collisions and overwriting. The access protocols are the network's 'traffic cop' and they must be built into the NIC. See Ethernet, Token Ring, contention.

— The techniques used in cellular radio and satellite communications for handling multiple transmissions in point-to-multipoint duplex situations. See TDMA, FDMA and CDMA.

access time

The waiting time: the time it takes a data storage device to retrieve (and sometimes store) data. It is measured from the initial call to the time of delivery. Alternately, it is the time it takes data buffered in a network transceiver to get access to the network. With disk drives it is the sum of the disk-drive's average rotational latency and the average seek time.

• When comparing hard disk drives, the term is generally applied to the speed of data retrieval (read), however it can also be used for data storage (write) times. There are two complications when measuring average disk-read access times:

— Sequential disk access: Where the blocks of data are to be found (in an interleaved sequence) on disk segments which are adjacent to, or on, the same track. With most hard disks this may take about 20ms.

— Random disk access: Where data is retrieved from anywhere on the disk. This naturally includes greater access arm movement, etc. and may well take 40—60ms.

The vast majority of disk access time is spent finding the data and only a small fraction will be spent on the transfer phase. Generally 85ms is seen as a very slow hard-disk access time, and 15ms, a fast one. See RLL and interleave ratios also.

• CD-ROM access typically takes from 300ms to over 1 sec.

• For semiconductor devices (typically RAM or ROM) this is the time taken for the chip to deliver the data onto the data bus. Semiconductor access is measured in nanoseconds.

— The old silicon memory devices had access times in the order of 450ns.

— DRAM typically measures in the 30 to 100ns range (generally about 80ns).

— SRAM is between 15 and 30ns. This is very fast.

— BiCMOS SRAM chips have been produced with a 6ns access time.

— EEPROM is typically around 150ns. See Wait states also.

• With LANs, WANs and MANs this is the time a node needs to wait (on average) before it can put data onto a network. The bigger the MAN/WAN usually the longer the access time, so access controls and methods become very important. LANs are segmented to improve access times.

accession number

A unique number given to each record in a database in strict order of entry.

accounting

• International telecommunications (and regional companies) must account to each other for services provided. Internationally, carriers will often share ownership on submarine cables or trans-border links so they will have established accounting rates. Including charges for units, or minutes, and quotas. Quotas are determined by the authorities beforehand.

• Accounting software for business. The modules required include:

— Data entry transactions.

— Data posting to accounts.

— Maintenance of accounts receivable and accounts payable records.

— Year-end reports — balance sheets and income statements.

— Security module, including an audit trail.

accounting management

This is a network management term for the collection, by the central network management console, of network data on the usage of resources. This is one of the five principal roles of network management. See CMIP.

accumulator

• This is the major register (working memory cell) in a computer processing chip's Arithmetic Logic Unit (ALU). The accumulator holds the result of any computation within the microprocessor. The number of bits simultaneously handled by the accumulator (generally 8-, 16- or 32-bits) establishes the primary dimensions and data-handling abilities of the CPU, and therefore the power of the computer as a whole.

• The term is used in programming for any active memory location where the results of arithmetical operations are stored.

• The correct name for any rechargeable 'battery'. Batteries are actually use-and-discard devices, while accumulators are rechargeable.

accuracy

Don't confuse accuracy with precision. Accuracy is a measure of how close you are to the target, while precision defines the fineness of divisions in the measuring device.

ACD

Automatic Call Distributor. ACD telephone equipment queues the in-coming calls and distributes them to operators (or other 'agent-staff') who answer questions from the public or take orders, etc. The switch will automatically route the calls to the next available agent in the 'line hunt' group.

In telephony, ACDs are usually stand-alone units which large companies use to support their main telephone switchboard. However, call distribution functions are included in many of the new digital PBXs.

ACE

A workstation standard being devised by the Advanced Computing Environment group which includes DEC, Compaq, Microsoft, MIPS, Sony and about 180 others. Intel aligned itself briefly with the group also.

Their overt aim was to produce a RISC (MIPS R4000s) high-performance workstation using an operating system based (initially) on the SCO flavor of Unix, and including Microsoft's OS/2 version 3.0 (32-bit, based on Windows NT).

It was actually an attempt to wrestle control away from IBM, Apple, Motorola and Sun and, as you'd expect, these companies (and HP also) were not members of ACE. Sun was seen as the dominant force against ACE, and the later Apple/IBM deal to develop the PowerPC and Kaleida operating system was interpreted as a direct threat to the ACE group.

Compaq pulled out in 1992, and the name ACE hasn't surfaced in the literature for a couple of years, however the ARC program has some connections. See ARC.

ACFG

AutoConFiGuration. The PnP extensions of BIOS are known as the ACFG BIOS extensions. See PnP.

achromatic

Free of color. Without unwanted color.

• Chromatism in lenses produces color fringes (chromatic aberrations) which reduce image quality. So 'achromatic' refers to optical systems that don't have a bias for one color over the others. This requires compound lenses using two types of glass.

• Good camera lens systems comprise a number of elements using different types of glass so that one counteracts the color-splitting actions of the other. These are called 'achromatic doublets'.

• By analogy, the achromatic idea of being 'without color bias' is carried over into acoustics to mean 'without specific frequency emphasis'. Here 'achromatic' means that the sound system is exhibiting its expected 'flat' (even-handed for all frequencies) response.

• The term has also extended into electronics and telephony, carrying with it the same concept of 'even-handed for all frequencies', or perhaps 'without specific frequency emphasis'.

ACIA

Asynchronous Communications Interface Adaptor. A device used to control and format data in asynchronous serial communications — usually between a PC and a modem. It performs the functions of a UART.

ACID test

Atomicity, Consistency, Isolation and Durability test — the four ideal characteristics that must be possessed by a good transaction processing system (where short messages are exchanged between, say, an ATM and the bank's host mainframe computer). All OLTP applications must pass the ACID test.

— **Atomicity** refers to the fact that the exchange is performed or not performed, in an all-or-none fashion. The transaction is not recorded unless all prescribed events occur.

— **Consistency** suggests that all operations must be performed accurately, correctly and with validity. There must be uniformity; and if a transaction is aborted in the middle, both ends must reflect the same information. This ensures that one account isn't credited without another being debited.

— **Isolation** must exist from other processes until the transaction is complete. This is essential. Each record in the database is only available for one transaction at a time, so it can't receive conflicting directions.

— **Durability** against failure. If it crashes, it can reset itself to the same place without losing instructions or information. It must be fault-tolerant.

ACIPR

Adjacent Channel Interference Protection Ratio.

ACIS

American Committee for Interoperable Systems. They have a solid-modeling kernel for 3-D CAD/CAM. See wire modeling.

ACK

• ASCII No.5 — Acknowledge (Decimal 6, Hex $06, Control-F). A transmission control character that is sent (in reply) by the receiver of the data as a positive response.

— In file communications it says that the receiver is ready to accept more.

— It is also an affirmative response to the sender's request or poll, and is often used in association with EOB. See NAK.

• Acknowledgment of flags are part of TCP.

ACK-Ahead

This is a variation on the XModem protocol which speeds up transmission of files across good quality links.

ACK-NAK

An slow block-by-block error-checking scheme used in XModem and Kermit file transfer protocols. The transmitter sends a block and then waits for an acknowledgment (an ACK character) before sending the next. If a NAK is received the last block is retransmitted. In later versions the block size was increased to overcome some of the inherent delays. See ARQ.

acknowledgment

See ACK and ARQ.

ACL

Advanced CMOS Logic. See CMOS.

ACLE

Analog Component Link Equipment. A television studio system for distributing component video signals using a time-division (TDM) technique. S-MAC is another similar technique.

ACMS

Automated Connection Management Service. In a switched virtual network, this is the part which establishes the connection. It is an intelligent directory and authentication service. Requests to establish a connection are sent directly from the workstation to the ACMS.

acoustic couplers

A form of modem that doesn't make direct electrical connection with the phone lines, but relies instead on cups being placed over the earpiece and microphone of the conventional telephone handset. The tones are transmitted acoustically across the air-gap.

Couplers were once used by travellers when direct access to the phone lines was not always possible (in a phone booth or hotel room). These days we bare the wires at the socket or make the connection in the microphone with alligator clips if they don't supply a standard RJ-11 plug. Couplers were only really suitable for lower rates (300b/s).

acoustic delay

If a signal is fed to a piezoelectric element attached to the end of a quartz crystal, the signal will take a finite time to propagate acoustically (physically) along the length of the crystal. The signal can then be recovered by another piezoelectric element and converted back to its original electronic form. These techniques are used to create very short delays (sometimes called 'transient memory') for both digital and analog signals.

In VCRs, acoustic delays are designed to retard the video signal by exactly one line-scan period. If the machine detects a 'drop-out' (lack of signal from the tape replay) in a line, a rapid switch can then insert the output of the delay into what would otherwise be a white flash on the screen. This is a form of analog error correction by 'interpolation' or 'substitution', which is not acceptable for data.

acquisition time

- Electronic circuits attached to a network take a finite time to react to signals. The acquisition time is a measure of the ability of a particular circuit (on a LANs or elsewhere) to react with a high degree of reliability. It is also, therefore, a measure of the system's upper capacity limits.
- In networks, the acquisition time will be the sum of the propagation delay and the time it takes the transceiver to react to the signal.
- The term is sometimes used as a synonym for 'access time'.

Acrobat

Adobe's Acrobat is a file-viewing application that lets anyone using different hardware and software look at a document in its original format and with graphics. It is a page format language which builds on the company's Multiple Master fonts and PostScript. It uses the Multiple Master fonts to emulate the metrics of virtually any font, and it calls on a reduced instruction set of PostScript to compress its files. The Acrobat file standard is PDF (Portable Document Format). There are inexpensive document-viewers to handle this format. See HTML, RTF and SGML also.

acronym

You may be surprised to find that we actually use very few acronyms in the communications and computer industries, but we use a lot of abbreviations. An acronym is an abbreviation which makes a word in the language — so SWIFT (Society for World Interbank Financial Transactions) is an acronym, while NASA (National Aeronautics and Space Administration) is just an abbreviation. In common use, however, we tend to call them all acronyms.

ACS

- **Asynchronous Communications Server**. This is most probably a dedicated PC with built in asynchronous modem and communications software which is mounted on the LAN to act as a dial-in/dial-out server.
- **Access Control System.** See access control.

ACSE

Association Control Service Elements of the OSI's Applications layer. The aim of the ACSE specification is to allow applications on a network to establish connections (called associations), maintain and terminate connections, and transfer messages between their Application layers. In an open standard; the commands ('elements') must be common to many types of applications.

We can divided them broadly into three categories:

— Association control (used by MAP and TOP),

— Context control,

— Commitment concurrency/recovery control.

See CASE, SASE, ROSE and CCR.

ACSSB

Amplitude Companded Single SideBand. A narrowband technique used for Public Mobile Radio (PMR). It's an improved version of SSB. ACSSB only occupies 5 or 6.25kHz of bandwidth compared with the 12.5kHz (minimum practicable) spacing of common FM analog channels.

For dispatch and fleet management, ACSSB compares favorably with TDMA digital on a voice-channel-per-bandwidth basis, but it has other problems that make it less suitable for cellular telephony. It is currently a favorite technique for conserving spectrum in analog PMR, and it provides a low signal-to-noise ratio for use with satellite communications to mobiles.

In the USA it can be used in the 150—174MHz band and also between 220 and 222MHz. The UK allocation is 208—225MHz, and Australia has used the band 148—174MHz (with 6.25kHz channel spacings).

ACTS

• ACTS is a $1.5 billion, 4-year European program of research into implementation of the global information infrastructure (GII). This is an EC research program which is open to third country participation. It replaces RACE.

• Advanced Communications Technology Satellite. This will use the Ka-band.

action frame

In computer-aided learning, this is a frame (a screen) of information that requires a decision (a test answer or choice) by the student.

action line

See eyeline and reverse.

active documents

A new and rather vague concept of electronically exchangeable documents which include many of the attributes of hypertext and of hot-linked data.

It is claimed that active documents will eventually have the intelligence to change themselves (based on rules, standing instructions, conditions met, events, or on the arrival of new information) when conditions change. Information in these documents will be dynamic, and information may be added or subtracted (customized) according to the user's needs.

active file

The file currently open and in use. Computer files generally need to be opened to be read or changed, and they should be closed when inactive. See record locking and file locking.

active hub

A multiport device that repeats and amplifies LAN signals. See hubs and concentrators. The distinction is with passive hubs, which are wiring points. An active hub is not necessarily an intelligent hub – it might just be a repeater.

active lines

Television and monitors. The lines making up a television frame are not all used to create the image. Each scan line represents a sequence of signal variations within a certain time slot, but the image is not being created constantly by scans across the surface of the screen. These take place in strict order from top to bottom.

A certain number of scan lines are then wasted during the vertical-blanking interval (the vertical fly-back — each 1/50th or 1/60th sec.) and it is this time interval which creates the black bar at the top/bottom of a television image. The active lines are those used for image only (but including those parts of the image not seen behind the bezel).

• NTSC has 510 active lines out of a total of 525, with about 483 being clearly visible.

• PAL and SECAM have 588 active of the total of 625, but usually only 575 lines are seen by the viewer.

See pitch, VBI and Teletext.

active-matrix screens

A type of flat panel LCD screen (monochrome and color) which is essentially a very large silicon chip coated with a thin film of transistor-type material. These are also called 'thin-film transistor' screens or TFTs.

Behind each pixel on the screen are two or four transistors which are responsible for switching the power to the liquid crystal cells. This technique allows for higher polarizing currents, giving the screens excellent (up to 100:1) contrast ratios.

The transistor-switches make very fast-acting displays which are capable of good graphic operations, and which react fast enough to be used with a WIMP interface. They have switching speeds of 20 to 40ms, as compared to normal passive LCDs with 300ms switch rates.

This screen technology is important to overcome a problem of cursor 'submerging' (the cursor temporarily disappears if you move the mouse too quickly). Unfortunately the screen cost is still relatively high (because of a low manufacturing yield) when compared to Supertwist LCD screens and some other techniques.

The current state of the art with TFT color screens produces a screen of 20 x 25 cm with a yield of about 50%, at an ex-factory cost of about US$800. See LCD, Supertwist LCD and TSTN.

active nodes

See passive nodes.

active pillar

This is a telephone pillar (or pedestal — street connection point) which has electric power and active electronics, as distinct from one which is completely passive and is used only for solder connections.

active video inversion

This, and sync suppression, are two popular ways of scrambling analog video images. They can be run together. Information about whether the video is normal or inverted in voltage is carried in the vertical blanking interval. So the scrambling mode can be changed as often as once per field. See line shuffling.

activity

• A measure of the number of records used (read, changed or added) in a database. It is an average (an activity ratio or percentage) that indicates when a database should be upgraded and/or purged of obsolete material.

• In PERT and CPM project-management techniques, an activity is any task which consumes time and resources.

actual

Actual is sometimes used in computer jargon as a synonym for 'absolute'.

actuator

In a hard-disk unit, this is the rotary-positioning unit which consists of an arm with the read/write head, and the rotary voice-coil or stepper motor. The precision with which the actuator moves, defines the number of cylinders or tracks that can be accessed on a disk (or disk stack) and is, therefore, a major factor in limiting the overall storage capacity.

ACTV

Advanced Compatible Television. These were three related experimental 16:9 ratio analog TV systems from the NBC/Sarnoff Laboratories. They included both extended-definition (ED) and high-definition (HD) proposals. They were receiver- and channel-compatible with the existing NTSC television standard, relying on augmenting (adding extra information to) the existing TV signals, and using sophisticated receivers to improve performance. This was in contrast to other HDTV proposals which require radical modification to present standards.

ACU

Acknowledgment Unit. See ACK.

A-D converter

Analog to Digital Conversion. See.

ADA

Asynchronous Dial Access.

adapter

See adaptor

adaptive

In computers and telecommunications, this term is often used to mean that a system, technique, or application, changes automatically in response to needs or requirements in some way. Sometimes there's the sense that it can learn from what has happened in the past; sometimes just the sense that it adjusts dynamically.

- In some forms of digital telephony, for instance, adaptive techniques choose to use the best possible coding/sampling scheme for voice quality on a varying basis, depending on the bandwidth available and the type of voice being sampled. These techniques reduce the bandwidth requirements.
- In other adaptive forms of digital telephony, just the quantization scale will change to suit the conditions.

adaptive channel allocation

A multiplexing technique where the allocation of bandwidth to the channel is not predetermined but is assigned on demand. It is also called dynamic channel allocation, bandwidth-on-demand, or rubber-bandwidth. See adaptive duplex and statistical multiplexing also.

adaptive compression

An automatic data compression technique used in many modern modems. These adjust compression to match the file type and standards available, and the choice of technique is conveyed between the modem pair in a way that is transparent to the users.

Compression systems can be totally mechanical (as in the 2:1 interlace of analog television, or Huffman's code) or 'adaptive' (variable) — depending on the type of material being transferred. See MNP and LZW.

Adaptive Delta Modulation

ADM is a more complex version of the Delta PCM approach to bandwidth reduction in digital telephony. It includes adaptive principles to avoid compliance errors.

The encoder keeps track of the rate at which the differences between the samples change. It therefore maintains an adaptive concept (recent average) of the 'step size' (see Delta PCM) which will change over time. Samples are taken, and the difference is calculated. The sample-difference is then encoded in terms of the number of steps of the current step-size.

If sound quality needs to be increased, the step-size can be reduced (the sampling frequency increased) and the bandwidth will change. Similarly, if the bandwidth needs to be reduced for any reason, a decision to increase the step-size can be transmitted over the channel, and both ends will then adapt to a more coarse sampling pattern. See delta, ADPCM, CD-ROM/XA, CD-i, HD-MAC and particularly CVSD.

Adaptive Differential PCM

See ADPCM.

adaptive duplex

In modem communications, information flows are rarely balanced; more data travels in one direction than the other. This imbalance in the transfer is said to make the link 'asymmetrical'. In full-duplex operations it therefore makes little sense to reserve equal bandwidths for both directions (but most modems do this).

Adaptive duplex techniques rely on the use of modem pairs which can constantly assess the amount of data in each buffer, communicate information about the buffer load to each other, and jointly allocate bandwidth accordingly. See DAMQAM and adaptive channel allocation.

adaptive equalizer

An analog public telephone network device that counteracts delay distortion. See equalization.

Adaptive PCM

APCM reduces the sampling rate (and therefore the bandwidth) when the communications system experiences load. It naturally also lowers the quality of the audio signal. This technique is used in some multiplexed telephone links to handle potential overload conditions. Similar techniques are VRAM and TASI.

Note: This is an adaptive form of the standard A-Law PCM, not the differential variety. See ADPCM.

adaptive-predictive PCM

An ADPCM technique that uses a feedback signal to ensure that quantization errors don't accumulate. See ADPCM.

adaptive routing

See dynamic routing.

adaptor

Aka adapter. In computing and communications, any thing or person that adapts is called an adapter/adaptor. Usually, the adapter (-er) is the person who adapts, while adaptor (-or) is a mechanical device of some kind.

• **Computers:** This is a general name for any plug-in PC card/board. The term covers just about anything from expanded memory, to a CPU accelerator, to a Network Interface Card (NIC) for LAN connection. The term is used casually, but originally there were three defined types of add-in cards:

— Accelerators: which provided extra processing power and/or reduced the load on the main CPU.

— Expansion cards: which originally carried extra memory. See expanded memory.

— Adaptors: which were fundamentally drivers to allow the use of peripherals and communications devices.

New adaptors are designed to support more sophisticated peripherals, and most also have processing on-board, so it is often difficult to categorize them. Video-card adaptors allow the attachment of a higher standard of video monitor, while video accelerators handle MPEG decompression and may have their own video RAM ('expansion').

• **Cables:** Any hardware unit which changes from one physical standard to another — or which allows the interconnection of different standards. e.g. gender-bender on a cable, which change a male connection to female (and vice-versa).

• **LAN:** In networking circles an adaptor will almost certainly be a Network Interface Card (NIC). The NIC is the add-in card which provides the electrical connections and transceiver functions on the network.

• **Power:** In the supply of mains power, adaptor probably means a multi-plug cable box, or power-point 'doubler'.

ADB

Apple Desktop Bus. A serial bus standard that Apple uses for connecting its mouse and keyboard. Other peripherals (like tablets and modems) can be daisy-chained from the keyboard using this bus. Don't confuse this with the SCSI connection, but they operate in a very similar way.

• The ADB can handle up to 16 devices with a transfer rate of 4.5kb/s from each.

ADC

Analog to Digital conversion.

ADCCP

Advanced Data Communications Control Procedures; aka ADCP. An ANSI version of SDLC for error control which is widely used at the Data-link layer of the OSI model. It is functionally equivalent to HDLC. This is a bit-oriented, data-link control protocol. See Data-link controls, and also HDLC and Binary Synchronous.

ADCP

Advanced Data Communications Protocol. See ADCCP.

add-drop multiplexers

A form of multiplexer which allows one channel (or a group of channels) to be added to, or dropped from, broadband digital TDM data streams.

The aim is to add one or more channels to, or extract one or more channels from, a multiplexed stream without disrupting the onward transmission of the other channels. Until recently, this was only possible by demultiplexing the bearer hierarchy to the lowest level, extracting/adding the new channel, then remultiplexing the lot — a tedious and expensive process. See plesiochronous transmission, Sonet, SDH and drop-and-insert multiplexers.

addend

A maths term. In an addition, this is the operand (the number added). It's the number which is added to the augend to form the sum. If a + b = c, a is the augend and b is the addend.

adder-subtracter

A logic-gate device which adds or subtracts binary numbers. It is an essential part of the ALU in the central processor of a computer. See accumulator.

additive primaries

In video, any color or hue can be obtained by mixing various amounts of three primary colors. The distinction being made here is between additive color techniques (Red, Green, Blue, which together as light sources make white) used in light-generation systems such as monitors, and the subtractive color techniques (Cyan, Yellow and Magenta, which when mixed together as inks make black) used in printing (subtractive-reflectance systems). See RGB and CYMK.

address

A number or code which represents the origin or destination location of data being moved or stored in some way.

• The identifying number of a user-node on a network — used on LANs, packet-switching or message-switching network. Typically a packet traveling on a network will contain both the sender and destination addresses. Addresses must be unique, and any device on the network to which data is sent, or from which data is received, must have an address — each computer, server, terminal, printer or controller.

So an address is a data structure which identifies a unique entity — a station on a network or a process. Don't confuse this with a VCI (virtual circuit identifier) which specified a route to reach an address. And some devices on a network have names, which then need to be translated into addresses through some form of look-up table. The name will stay with the device if it moves, and then when it is in a new place on a WAN it will need a new address. See naming services.

• Many communications systems will use both local addresses and extended network addresses (known as global addresses).

• In computer systems a memory address is an identifying number (usually expressed in hex) that is unique to one Byte of RAM or ROM held in a memory location. Addresses begin at hex $0000 0000. See memory map.

• A wider use of the term for networking is: those bits in a packet, a frame or a cell that will be used to direct the packet to its destination. With a datagram this will always be an identifying address (usually the E.164 standard), but with ATM and some connection-oriented packet systems, there is likely to be a virtual-circuit or virtual path/virtual channel identifier at the cell (or physical) level. The full 'packet' address information will then be carried as payload within the cells. See ATM, VCI and VPI.

• Address messages are often sent over the network to carry all the routing information necessary to establish a call or a data-link. A router needs to maintain look-up tables of addresses, and associate each of these with one of its ports. It routes by checking the packet address against the look-up table, then using this to select the output port.

• The identity number of a peripheral on a daisy-chained serial connection like SCSI is also called an address. With SCSI, the number 0 is usually given to the CPU and 7 to the internal hard-disk. Numbers between 1 and 6 are available for peripherals (these will need to be mechanically set on the device).

• In international networks, a unique address number must be assigned to every device on the network. The Internet, without any central authority, has a major problem here, but standards exist which ensure that the device address is unique, worldwide. See IP address.

• It is often important to distinguish physical addresses from network addresses. The term physical address usually means a Data-link layer (MAC) address. The network address is also called a protocol address and is at higher layers.

• In telephone networks, both address-complete and address-incomplete signals may be sent on the supervisory back channel on a switched public network.

• Symbolic addresses are the names of operands used in a programming language.

address bus

One of the two main parallel communications channel between the CPU and the RAM main memory in a computer, across which the address is transmitted in order to activate a location so as to transfer data over the data bus to/from the CPU. The width of the address bus (not the number of bits which the CPU can crunch simultaneously) sets the amount of primary memory a computer can use without bank switching. The Apple II, for instance, used 16-bit addresses (although it only had 8-bits stored at each location). This allowed direct addressing of 64kBytes, then by adding bank-switching, they reached 128kBytes.

• IBM's early 8088- and 8086-based PCs had a 20-bit address bus which allowed them direct access to 1MByte of memory, although MS-DOS only permits 640kBytes to be used for working memory (for the operating system, applications and data). See addressable memory, expanded memory and also bank-switching.

address resolution

Methods used for resolving differences between addressing schemes; usually for mapping Layer 3 (Network) addresses to Layer 2 (Data-link) addresses.

address space

The amount of memory available for use by application programs (and possibly the operating system combined). This has become quite confusing over the last few years because:

— Some equipment vendors and retailers included the video RAM space above MS-DOS's 640kB limit, which was strictly true, but also deliberately deceptive. A 286 PC with 1MB of address 'space', may only have 640kB of usable memory.

— Address space in terms of 'working memory' now includes extended and expanded memory.

— In multitasking environments, 'address space' can mean only that space devoted to each application and its data. OS/2 protects this address space, while Windows 3.x doesn't. See addressable memory.

addressability

The screen elements of monitors can't always be individually controlled by the video control system. Physical picture elements don't necessarily have a one-to-one relationship with video memory locations.

Often it was necessary to treat small groups of screen elements together as single 'pixels' (picture elements) because of video memory limitations. This was especially so with older PCs, and it is also a problem in on-line videotex and broadcast teletext where data rates (and therefore screen resolution) need to be kept low.

• The early IBM PC line limited the requirements for video RAM by storing only the character bytes of text in memory rather than the fully rasterized bit-mapped images of these characters. Therefore one byte of text controlled the switching of a 9 x 14, or 8 x 8 matrix of pixels on the screen. The addressability of the early PCs in text mode was therefore only 24 lines by 80 characters.

• With graphics, any group of screen elements that switches together will collectively be known as one pixel — so strictly, you can't specify screen resolution by the number of pixels. The term should only be applied to graphic systems as a whole, including software screen drivers.

• You will also find this confusion between pixels and addressability in claims about EDTV/HDTV where line-doubling techniques are used. If you automatically double-display each line, are you doubling the number of pixels? The promoters of the system will say that you are, but common sense says that you are not increasing the resolution at all — you are just making it more difficult to see scan lines (assuming screen dimensions are the same). The addressability hasn't changed.

addressable memory

The maximum working area of memory that the microprocessor can access without special 'bank-switching' or extended memory techniques. Usually it means the working area of memory (with operating system, applications and data) and the area reserved for video RAM, I/O and device drivers.

• With early 8-bit computers using 16-bit addressing, the physically addressable memory space was from decimal 0 to 65,535 (or hex $0000 0000 to $FFFF FFFF) which is 64kBytes of working memory. Later bank switching effectively doubled this by tricking the computer into reusing addresses with different blocks of memory. This is how the early Apple IIs provided 128kBytes of memory even though the CPU is only capable of addressing 64k locations.

• With the early IBM PCs and compatibles, the 20-bit address bus allowed address 'space' up to 1MByte from the CPU's point-of-view. However from the user's, or programmer's, viewpoint the operating system imposes a limit on working memory of only 640kB (without the LIM extensions, etc.).

• 24-bit/bus computers can address 16MB, and 32-bit computers, up to 4GB in 'native' mode. See address space and memory map also.

• A 32-bit computer will not use all the 32-bits available for memory location purposes. Very often the top two bits will be set aside as 'flags'. They will indicate whether the address mode is (00) User (01) Supervisor (10) Executive or (11) Kernel. The next bit may also be a flag indicating whether the location is main memory or I/O.

addressing flexibility

On a LAN, it is often desirable to send messages to users in groups, rather than individually. Sometimes we want to broadcast to all users on the network simultaneously. Some LAN addressing schemes provide these broadcasting functions while others allow only individual terminals to be addressed.

addressing level

In machine code and assembly programming there are three levels of addressing:

— Level 0: the address part of the instruction is the operand.

— Level 1: the address part is the location at which the operand will be found (direct addressing).

— Level 2: the address part of the instruction is the location at which an address for the operand will be found (indirect addressing).

adjacency

Routers and end nodes on a common network media segment can sometimes establish a relationship for exchange routing information. Adjacency is the relationship formed between them. See adjacent nodes also.

adjacent channel interference

Any two channels which have a common border in the radio frequency spectrum (no substantial guard band between) are adjacent channels. In UHF/VHF analog television transmission, you can not use adjacent channels in the same area unless both transmit at the same power levels from the same antenna, and the receivers have modern 'cable-ready' saw-tooth filters (which most modern TV sets now do).

If adjacent channels are used at different powers from different antenna, there will always be some homes in the area where the power of the sound frequencies will overwhelm the power of the image frequencies close alongside.

In the cellular phone industry you can't transmit using adjacent frequencies in adjacent cells because of similar interference problems (except in CDMA). The cluster must be designed to keep a one cell (or sector) separation between adjacent channels.

In fact, in an AMPS cellular system channels up to four numbers apart can cause interference in some circumstances — especially with the control channels. This type of interference often causes blocking of incoming calls, and requires frequencies to be redistributed within the cellular cluster.

adjacent nodes

SNA uses the term, as you would expect, to mean neighboring nodes without another node between. However other systems simply mean by adjacency, that the nodes share a common segment.

ADLC

Adaptive Lossless Data Compression. An IBM chip-based data compression system which can compress and decompress data at 5Mb/s. No additional external memory is required.

The chip is small enough for PCMCIA Type 1 cards, and it doubles or triples memory capacity (averaging 3:1 compression, supposedly) without degrading CPU performance. A second product called ADLC Macro is said to operate at 40Mb/s. The system is based on the Lempel Ziv 1 algorithm

ADM

• **Adaptive Delta Modulation.** See also CVSD.

• **Add-Drop Multiplexer.** See SDH.

Adobe Type #

This is in reference to Adobe's PostScript implementations:

— Type 1. These were the original PostScript fonts. Adobe Type Manager implements the basic translation process needed for the conversion of these outline fonts to bit-mapped images. Apple will build this level of type into future Mac operating systems along with TrueType.

— Type 2. Introduced with PostScript 2, but completely compatible with Level 1.

— Type 3. These fonts don't have the full set of hinting information and are not usable with Adobe Type Manager. See hinting and outline fonts.

• There's also Adobe Type M which are the Multiple Master fonts used to emulate the metrics of virtually any font used in Acrobat. See.

ADP

• **Automatic Data Processing.** The old term for data processing (DP) or electronic data processing (EDP) which just means any computer operation. These are large-corporation, mainframe-oriented terms like IT and MIS.

• **Application Development Platform** (of IBM).

ADPCM

Adaptive Differential Pulse Code Modulation. This is now a very popular audio compression system, widely used in telephony and in disk storage devices. ADPCM is also used in the new CD-ROM/XA and CD-i systems, modems, telephone transcoders, with ISDN networks, and for satellite HDTV. There are a number of ITU standards.

ADPCM uses a modified DPCM sampling technique where the encoding of each sample is derived by a complicated procedure that includes calculating a 'difference' (DPCM) value, and then applying to this a complex formula which includes the previous quantization value. The result is a 4-bit code which can recreate the almost same audio quality as the older 8-bit PCM at the same average sampling rate. The ITU has recommendations G.726 and G.727 which cover ADPCM at rates of 40, 32, 24 and 16kb/s.

See Embedded ADPCM, ADM and A-Law also.

• ADPCM has a couple of different standards. The ITU-T's G.721 (widely used in the USA and Canada on 'pair-gain' subscriber circuits) seriously affects the phase of modems, and so V.29, V.32, V.32 bis, and V.34 modems don't work at all well. Other algorithms such as G.726 will slow down modem connections, but create less problems.

ADSI

Analog Display Services Interface. This is an enhanced CLID protocol for analog phones which Bellcore has developed mainly to carry names, phone numbers and a few standard messages over the link to the home/office phone.

The ADSI protocol also has fields for the data-type, and this allows messages to be transferred over the POTS or ISDN lines in a standardized way to a LCD display or PC. This has applications for telephone-linked services like electronic home banking, Pay TV selection, home-shopping, etc. There are three levels of device:

— Level 1: Calling name and number delivered after the first phone ring tone.

— Level 2: Calling name and number are both delivered with call waiting.

— Level 3: A telephone with a display screen.

Level 3 devices have screen-sizes of 20-40 characters wide by 6-or-more rows deep, and four to six softkeys. Software menu trees can be downloaded and stored as macros. The user sees a plain English display and can press the appropriate softkey to activate the features — and these features can be dynamically changed over the phone line. High-end ADSI phones will have things like pull-out QWERTY keyboards for data entry, PCMCIA Type I slots, Smartcard slots, and magnetic strip card readers.

ADSI is designed for general use by the public, so anyone can have an ADSI server. Screen phones like this may soon be used for all sorts of information and shopping applications. See CLID and CLASS.

• Currently ADSI only appears to be in use in trials which are essentially just enhanced versions of CLID. They deliver name and number.

• US ADSI trials (called Delux CLID) extract the name from the billing information held in the local exchange. The British trials allow you to select your own name for transmission (once only, not on a call basis).

ADSL

Asymmetrical Digital Subscriber Line. This is a general name applied to a couple of different major modulation technologies aimed at delivering television quality video over the existing copper telephone wires.

ADSL has now become a generic term: the original modulation technology (CAP) didn't reach the early claims and there's now a totally new multi-frequency idea under development (DMT) although AT&T have continued to support CAP.

• The original claim by AT&T Paradyne in early 1983 was that their 16-CAP technology (a form of 16-level QAM) could carry video over copper in the local loop at speeds up to 2Mb/s from the exchange to the customer, and also provide a slow-speed return path. The test rates achieved were about 1.5Mbit/s for distances of 2 to 5km and higher rates for shorter distances. It has since improved.

• In mid-1993 it was decided to change tack and introduce an entirely new approach using DMT multi-frequency (218 different tones) which was very similar to COFDM. Data-rate standards have been defined from 1.5Mb/s to 9.6Mb/s over distances up to 6km. See HDSL, DMT and CAP also.

• More recently, in 1994, a number of ADSL classes have been defined to support either video or video and data.

— Video Class 1 is simplex-video at 6.144Mb/s and duplex-data at 160—576kb/s. Class 4 is simplex-video at 1.536Mb/s and duplex-data at 160kb/s..

— Data Services are for Class 5 at rates between 6.2Mb/s (down) to 576kb/s (up), and Class 10 for 160kb/s (down) and 16kb/s (up).

In terms of distance, the specification calls for total ADSL transmission rates of 6.784Mb/s (including uplinks), to be carried over 3000m of 26 gauge wire with a noise margin of 6dB.

• Politically, ADSL is an attempt by the telephone carriers to move into the cable television industry for video dial-tone. However it appears to be a much too expensive solution to compete with coaxial and hybrid fiber/coax on a large scale. It may have application at the fringes of the cable areas and for special purposes (especially HDSL).

ADSR

The characteristics of a sound in a MIDI system. It means Attack, Decay, Sustain and Release.

advertising

Routers sometimes maintain their lists of usable routes available, by regularly sending and receiving routing updates along these routes: this is advertising. See flapping.

aerial

The British term for antenna. Aerials both send and receive. See antenna.

AES

Applications Environment Specifications. See OSF.

affine transformations

A library of mathematical functions which are used to code and decode the underlying patterns of shapes within shapes (fractals) and so achieve high compression ratios for image files. These transforms are combinations of rotations, scaling and translation of the coordinates of points of an image.

AFM

Audio Frequency Modulation. A Super-8 video term, meaning that the audio has been recorded along with the video in a complex modulated signal by the helical scan system. This is generally superior to the older technique of recording audio as a slow-speed linear track on the side of the tape. Some video recorder systems do both.

AFP

AppleTalk Filing Protocol. This protocol allows applications to use native file-system commands to manipulate files on a Mac node.

AFT AFT

An error detection and correction system which is built into some high-speed modems.

agate line

A printing measure used by some old-fashioned newspapers and printers. There are 14 agate lines to an inch.

AGC

Automatic Gain Control (also called AVC for Automatic Volume Control) — electronic feedback circuits widely used in consumer electronics to moderate variations in the signal reception strength.

agent (general)

• In data communications, an agent is a process which provides a uniform interface to all hosts on the network; it translates protocols. In this definition, a device driver is an 'agent'.

• A quite different technical description now being used is software which processes queries on behalf of an application, and returns replies.

—There's a distinction here between advisory agents (which offer instructions and advice) and assistant agents (which learn about you and help actively).

• In telecommunications, it can also be a small application which sits on the PC desktop and manages one aspect of modemized communications. For instance, you might have a Fax-agent, which knows how to handle incoming fax messages, and a Data-agent which looks after straight data calls.

• See following entries for agent (personal) and agent (network management).

• The key operating environment for agents is General Magic's Telescript.

agent (network management)

In network management, software agents reside in all routers, bridges, hubs, etc. to collect and report errors and statistics (values of specific variables) to the centralized management console. Obviously, the device must be intelligent enough to run the software.

Agents send back alerts and reports, and they often deal with other components through MIBs. They will usually be specific to the hardware and to the network operating system. Agents are often introduced into bridges and routers as special adaptor cards. See SNMP.

• HP says about NewWave, that an agent is a type of scripting ability that records both keystrokes and underlying logic — a form of object linking. If you understand this, you're a better man than I, Gunga Din.

• Replication agents keep files on file-servers aligned. See replication.

agent (personal)

In some modern computer applications and operating systems, an agent is a customized part of the interface between the user and the machine. It is used to automate repetitive tasks (such as logging on to remote computers, etc.) and it provides us with a basic form of customization.

It is a self-contained software program which can act on its own, and 'learn' without being taught. Marvin Minsky of MIT first used the term in this way — and many of the computer gurus and designers believe that this represents the next stage in user-interfaces past the CUI and GUIs to Agent-User Interfaces (AUI).

It is anticipated that agents will increase in functionality very substantially over the next few years, and that eventually the main task of your desktop PC will be to run the agent software and maintain the agent AI databases. Processing will be done on-line.

Your agents will eventually probe the outside world actively looking for information in which you will be interested; it will act as your surrogate/assistant. Information is collected for you without you needing to issue a query (push vs. pull).

There are two main components to future software agents:

— The generic part which interacts with other agents, and undertakes the tasks of communications, cooperation and negotiation.

— The role-specific part, with specialized tasks.

• In the more futuristic ideas of some computer gurus, an agent becomes an almost-human personal assistant. It will be the part of the operating system which holds information about your likes and dislikes, your interests, your contacts, etc. and applies these in an intelligent way to tasks. It may even have a name, and you may talk to it like a human 'butler' on the screen. This is Apple's Knowledge Navigator concept of an agent.

aggregate bandwidth

In routers and switches, this is the total capacity of all the ports communicating through the device. With 10 ports each of 10Mbit/s, the aggregate capacity would be 100Mb/s.

aggregation

In systems like ISDN where a number of B-channels can be supplied over a standard link (30 with PRA over E1) it is sometimes necessary to group channels together to gain enough bandwidth in a single stream for, say, videoconferencing.

The problem is that each B-channel might be taking a different path across the network. The timing may therefore be different, and each may have different line characteristics. So channel aggregation requires special equipment. Some of the bandwidth will be treated as overhead to supply timing control.

Aggregation is the technique of providing timing corrections so that n x 64kb/s channels behave as if they were one. There are now a number of standards. See BONDing and Q931W.

aggregator

In the American FCC terminology, an aggregator is a hospital, hotel, university or any entity which provides telephone services for the public or transient users. They are regulated in a way that ensures equal access to long distance carriers.

agile receiver

A receiver capable of adapting to various conditions, frequencies, loads and/or standards.

• In personal pocket-phones, the new DECT and PCN systems will have agile electronics which will enable them to select the best communications channel, and to channel-hop if a better one becomes available in the middle of a conversation.

• In High Definition TV, this means a digital receiver capable of making changes on the fly to accommodate both 50Hz and 60Hz standards, different line numbers, and different picture 'frame' rates (most emphasis is placed on the different picture-frame rates). The proposed HDTV receiver design (which is just an idea at present) would need to be based on digital transmissions and DSP technology.

• In paging, this is a device which can find its information on a number of different channels, automatically selecting the correct one for the location/time, etc. This is often necessary with paging since different countries have allocated different frequencies — and there are a number of different, and incompatible, standards of transmission. See Ermes.

aging

In communications systems, some information 'ages' very rapidly (in microseconds). In the buffer of a communications device, a line of information which has been held up for any reason can quickly reach the point that it makes sense to throw it away rather than allow it to load down the system, and deliver it late. So certain timers will be watching each block of information and, depending on the data-type, they may make a decision to discard data.

• Two-thirds of the entries in this book aged between the time of original entry, and the final copy sent to the printer. Some will have aged even more by the time you read the information, and a year or two from now, the aging of this information will become obvious once again. That's why we need your help in updating these entries. If you have comments, criticism, corrections, additions, or just suggestions, please contact me at:

fist@ozemail.com.au

AI

Artificial Intelligence. A general term that covers a range of advanced ideas, all to do with computers approximating or simulating human thinking. Central to the concept is probably 'adaptability' — the idea that the computer can modify its behavior in the light of answers to questions. This suggests that it can take a creative approach to problem solving, rather than just following steps designed by the programmer.

Another avenue of AI research is in the development of natural language (written English) programming. Other areas are computer vision and robotics.

There are dozens of definitions as to what AI 'really is' including the original one of Turing (see Turing test). However, most would agree that the term suggests an AI computer can learn (modify its strategy) from experience (heuristics) and perform other human-like tasks such as pattern recognition. See natural language, expert system.

• Some people define AI simply as the construction of programs that utilize a knowledge base.

AIA

Applications Integration Architecture. This is another attempt to define a global blueprint for programming interfaces. It is a related standard to NAS. AIA sets the software framework standards that will allow interoperability and portability for applications, while NAS provides the toolkit and libraries.

AIM

Advanced Informatics in Medicine — a program of the European Union examining the role of computers and communications technology in medical applications.

AIN

Advanced Intelligent Network. A term applied recently to those highly intelligent network controls being devised to handle future broadband services. IN (Intelligent Networks) is the term being applied to current analog and ISDN services (like Centrex, Caller Line Identification) while AIN is applied to future UPT and UMTS-type services which will allow you to use any phone (and bill you personally), find you anywhere, and contact you on a personal phone number. It would probably be fair to say that (mobile-phone) GSM's wireline network is an early version of an AIN.

• Bellcore published an AIN specification in May 1991 which described how US carriers should develop links between exchanges to process the new features and services.

air interfaces

The term is applied to the range of techniques adopted for the radio link standards in cellular, PCN and cordless telephone systems. There are currently three major analog air interfaces (AMPS, NMT and TACS) and at least three digital systems are emerging (GSM, CDMA and NADC).

The problem with the CT-2/Telepoint's introduction in the UK was that the networks were up and running before an agreed air interface standard was defined — and this made all the equipment incompatible outside a single network. The Telepoint operators in the UK were later forced to provide a Common Air Interface (CAI) radio link standard in parallel with their proprietary systems, but by then it was too late and the technology died. See also DECT, PCN, CT2 and CAI.

AIX

• Advanced Interactive eXecutive. IBM's proprietary Unix for PCs and 370 mainframes. AIX is based on System V Unix with Berkeley extensions. A large part of the Unix System V kernel was rewritten to produce AIX, but the most noticeable change is in the security provisions.

AIX has now been ported down to the PS/2 level, and the new OSF/1 operating system is largely based on it. AIX Release 3.0 complies with the XPG3 specification and Apple has recently done a deal with IBM for AUX to be used as a shared interface in the Pink operating System with AIX as the core. See Unix, X-Open and Xenix also.

• AIX/ESA is an IBM promise to produce an operating system based on OSF/1.

AKM

Apogee Kick Motor. The rocket motor fired at the highest point of transfer orbit to circularize the orbit and inject the satellite into the Clarke orbit.

ALC

• Assembly Language Coding (an IBM generic term).

• Automatic Level Control. See AGC.

alert box

A small window that appears in the computer screen when an error has been detected. Usually they tell you nothing you don't already know. The box remains on the screen, usually providing an error code or explanation, until you click on a button. These are called 'dialog boxes' in some systems.

algebraic syntax

In high-school algebra we always execute the parts of an equation in strict order:

— those within brackets first,

— then multiplications and divisions,

— then additions and subtractions.

This is not always the case in computer programming. See Reverse Polish Notation.

algoristic

Strict systematic (step-by-step) methodology which must yield an exact answer. By extension, the word implies that there are clear-cut rules or processes for solving the problem. The emphasis here is on the logical basis of the operation and the systematic progression through every stage, so a distinction is being made with heuristic approaches.

algorithm

A sequence of logical operations — set rules or processes. An algorithm is always a reasonably self-contained section of a program (a set of instructions) devoted to the performance of one activity. Good programmers keep libraries of useful algorithms which can be reused when the need arises.

An algorithm provides:

— a prescribed series of rules, routines, instructions or processes that,

— perform a single operation, and

— provide a solution to a single problem,

— in a finite number of steps.

Application programs really only consist of two elements — algorithms and data.

alias

There are two distinctly different concepts here:

• The spurious result of overlapping dissimilar structures which interact with each other. When two different cycles,

standards or signals conflict, resulting in a spurious third — then this unwanted signal is called the alias.
— The 'beat frequency' caused by slightly mismatched (out of tune) audio notes, is an alias.
— In electronics, the term is applied to 'artifacts' (unwanted ghost images or sound frequencies) which appear in an electronic system due to incompatibilities between two standards. Hence, the term can be applied to:
— The distortion that occurs in A-to-D and D-to-A conversions when the original sampling rate was less than twice the maximum frequency required (the Nyquist limit). In digital sound sampling, this overlapping and the consequent creation of beat (difference) frequencies, causes distortions.
— On a text or graphic computer screen, aliasing of graphics appears as jagged edges (the step effect) along diagonals or curves which are enlarged or reduced by non-integer amounts (related to the pixel size). The term, strictly, shouldn't be applied to the original bit-mapped image (which will also have the step-effect), but only to the result of non-integer scaling. If you double the dimensions of an image, there is actually no aliasing, simply an enlarging of the existing pixel steps — but the term is often mis-applied to this enlarging process. The term is also used incorrectly to refer to the process of getting rid of these effects.
— The term is also incorrect applied to artifacts in TV. Vertical aliasing appears as movement of fine detail on a screen consisting of raster-scanned lines. It is exaggerated by the limited persistence of the image on the phosphor of the video tube. It produces flickering effects on the horizontal lines of tennis courts, and moire patterns on herringbone suits and paisley.
— Temporal aliasing is the classic wagon-wheel 'backward rotating' effect in films and video. The wagon wheels haven't rotated enough between each frame of image, so there is a mismatch between the image frame rate and spoke rotation rate.
• A 'nick-name' which refers to a real named object.
— In electronic mail, an alias is an alternative name for some object (which can mean any person, machine, or computer storage mailbox) to which mail may be addressed. For instance, 'Data Widgets' may be the trading name (and an alias) for a division of Acme Widgets Inc. (which is the correct name of the company). In any international electronic telecommunications directory we will need both name and alias references, but only the name reference should contain the full directory data.
— A pointer to the key record. (See X.500)
— System 7.x in the Mac provides aliases. These are special icons which substitute for other icons and perform the same functions when double-clicked.
• Alias is also a 24-bit image format from Alias Research, used for rendering images from AliasAnimator, a high-end 3-D application.

aliasing filters

In analog phone systems, these filters are provided to prevent aliasing in the form of an objectionable beat frequency. These appear in the spectrum whenever the Nyquist limit is exceeded by any component of the signal.

all-points addressable

The maximum resolution possible in the graphics mode of a PC, as distinct from text or alphanumeric mode. See addressability.

ALM

Application Loadable Modules. These are in libraries of reusable software modules used in componentware. You select component 'objects' and plug them together to build an application. See application builders and componentware.

Aloha

The original collision-sense multiple access (CSMA) system developed by the University of Hawaii, and later taken up and used as the core protocol in Ethernet. Stations transmit whenever they have data to send, and any unacknowledged transmissions will be repeated.

• Pure Aloha stations just send packets whenever they want to, and then wait a fixed time for an acknowledgment, or it sends again.

• Random Aloha is widely used in satellite systems. The intending data sender listens for a gap in transmission. When such a gap arrives, it seizes the opportunity and sends a packet to announce its presence and establish priority; however it must also listen to the system to detect possible collisions with others who may have transmitted at the same time.

If a collision has occurred, both senders cut transmission, wait a random amount of time (milliseconds), then try again. Random Aloha is an asynchronous (try at any time) system. Unfortunately, Aloha has an average peak utilization of only about 20% of the channel capacity.

• Slotted Aloha establishes frames or time-slots in which an attempt to transmit can be made. A frame must be transmitted exactly at the beginning of a time period (at a frame boundary). This increases the efficiency of the system to about 40% channel utilization.

Aloha Multiple Access

Aloha Multiple Access is a VSAT variation of the above. With VSATs the problem is with the delay in time for the data to travel to and from the satellite before a collision can be detected. Some trunked radio systems also use this technique. See Aloha, TDMA and DFSA.

alpha

• Anything to do with the language characters, as distinct from the numeric characters. It is never clear whether 'alpha' refers to punctuation as well. See alphanumeric string.

• The letter A — the first in a sequence.

• In the sense of testing software and hardware; this is the inhouse testing phase, with beta testing being a phase where the trial product has limited release to outsiders (industry and media specialists) for further testing before formal release.

• The Alpha is DEC's 64-bit wide RISC chip. It is superscalar and super-pipelining. See PALcode.

alpha channel

A concept that arose out of Lucusfilms (Star Wars), and which has been taken up in some high-end graphic workstations (like NeXt and Apple's Mac). While color images are created in a computer by bits for Red, Green, Blue, a fourth 'channel' is also included to hold values of transparency/opacity. This is used either for masking or for additional color information.

So in a 32-bit system of image mapping, 8-bits will each be reserved for the RGB (a total of 24-bits), and another 8-bits for the alpha channel. See compositing operators.

alpha-geometric videotex

Alpha-geometric videotex was devised by the Canadians and taken up by the US as part of the NAPLPS standard. The computerized screen controller is instructed to draw the screens using primitives (such as polygon, circle or line) by a series of Picture Description Instructions (PDIs). The technique creates much better graphic images than the alpha-mosaic approach, but at the cost of greater bit requirements for each screen. See videotex, alpha-mosaic and alpha-photographic.

alpha keypads

The new international telephone keypad standard is the same as the US, but with the Q and Z added. All future US phones will carry the 7 on the Q key and 9 on the Z key.

There are at least six different versions of alpha keyboards in use around the world. But the international standard is:

International Alpha-keys

1 = nil	2 = ABC	3 = DEF
4 = GHI	5 = JKL	6 = NMO
7 = PQRS	8 = TUV	9 = WXYZ
* = nil	0 = nil	# = nil

• Most US financial systems and many voice mail systems use the 1 for Q and Z. However AT&T long ago standardized on Q on 7 and Z on 8, as with the UK and (some) European standards. However in Europe and in the UK until 1994, the addition of alpha characters to phones had almost disappeared except, for some strange reason, on cellular phones.

alpha-mosaic

The simplest system of transmitting videotex which treats the screen primarily as tiled blocks like a chess board, and inserts either text or one of a standardized graphic character set into these blocks. Prestel in the UK provides a 40 x 24 array of 'tiled' text character blocks. It then subdivides these text areas to create an 80 x 72 matrix for finer graphic resolution.

There were also 63 pre-defined graphics characters, each using a character matrix of 2 x 3 elements and with these Prestel could draw primitive diagrams and maps. Prestel's Level I system only supported crude graphics but it is very economical on bit-rate. See videotex, alpha-geometric and alpha-photographic.

alpha-photographic

The best example of this approach is the Japanese 'Captain' videotex system, but the technique is also used by the British Picture-Prestel. It operates virtually as slow-scan video, and provides the fine graphic resolution necessary for Japanese characters or high-quality images, but at the cost of many bits per screen. This requires a very wide-bandwidth channel. See videotex, alpha-geometric and alpha-mosaic.

Alphabet No.#

See ASCII.

alphabetic order

This is a common indexing problem with computer applications, and the solution is not as obvious as it may seem.

Computers sort on an ASCII number sequence and so place decimal numbers, spaces, etc. before the alpha characters. In a true 'alphabetic' ASCII sort, the capitals will also precede the lower-case alphabet — whereas in real-life we tend to group them together in an index. Numbers also create problems: in any ASCII sort, the number 110 will be placed ahead of the number 2.

Many sorting systems have been specially designed to ignore the upper/lower case problem (usually by dropping one bit from the binary code), but the problem of sorting strings which contain numbers, symbols and spaces remains difficult to solve. This was a major problem I faced in maintaining this dictionary's database. And we face similar problems in card or paper-file organization. For instance:

— Does S/N ratio precede SAA? or should it follow? And where does S100 fit? Is S100 to be treated as S One Hundred and slotted in just before the word sort, or would people look for it as S(number) just before SAA?

— What do we do about spaces?

— If I treat numbers as numerals and place them (as the computer does) ahead of all alphabetic characters, will people expect to find the '8086' chip reference before or after the '80286'?

There is no standard solution to these problems. See collating sequence and alphanumeric string.

alphameric

This just means alphanumeric.

alphanumeric string

A series of characters, all of which are either letters of the alphabet, numbers or special symbols. Spaces, punctuation marks, $, %, &, etc. may (or may not) be included in this definition — usually not.

The term is often used to indicate that a string like 65843 (say a zipcode) is not a number but a numeral (effectively a word) and therefore is not usable for arithmetical manipulation. Number sorts will create quite a different order to numerals in alphanumeric string sorts — and true alphanumeric sorts can be quite different to ASCII code sorts (the capitals and lower case will be treated differently, for a start). See alphabetic order.

We should probably identify these three categories as core:

— Alphabetic characters A—Z and a—z. (I would add the space character here also.)
— Numeric characters 0—9. (They are not numbers in this context.)
— Special characters: printable symbols and punctuation marks.

But more disputable, would be the inclusion of:

— Essential control characters: which initiate phrasing or logical actions, such as carriage return, line feed, and tab.
— National characters which exist in different forms in different countries. For instance, the @, # and $.

Alt

• **Key:** The Alternate key on an IBM-PC keyboard is a 'shift' key — a modifier. The Alt key gives the other key a new meaning, and this is usually software dependent (often the combination calls up a macro).

Alt-key combinations output double bytes (not a single ASCII code) where the first byte is a Null (zero) and the second byte identifies the command. Its use is therefore reasonably non-standard and specific to each system or application. The Macintosh equivalent to Alt is the Option key. See Extended ASCII also.

• **Newsgroups:** These are the most varied newsgroups on Usenet and they are the least controlled.

Altair

• Currently a wireless LAN device.

• Historically, the great breakthrough in personal computers. It was promoted as a project in ***Electronics*** magazine, and available from a two-man company (MITS) in Albuquerque, New Mexico. When constructed it was a blue box with nine toggle switches (the machine code input), an 8008 processor (the 8080 was a later upgrade!) and 256 bytes of memory. Bill Gates later sold them his Basic on paper tape.

alternate inversion

This is a digital technique used in long-haul communications where the synchronized pulse trains are interpreted alternately as pulse = 0 (space), then pulse = 1 (mark). Normally without inversion, a long chain of 0s would be represented as an absence of pulses (a chain of zeros) over a period of time but, if the chain is too long, the receiver and sender could then lose synchronization.

This clever idea of alternating interpretation of pulses and non-pulses as zeros means that the receiver will not lose synchronization during long delays. Don't confuse this with AMI where only the binary 1 alternates between positive and negative voltages — and the binary 0 remains at zero volts.

alternate mark inversion

See AMI.

alternative path propagation

See multipath effect.

ALU

Arithmetic Logic Unit. The ALU is the central part of any processing unit. It controls all arithmetic and logical operations of the CPU. See accumulator.

Alvey program

A five-year (1983-88) $500 million UK research program into advanced information technology, which has now been incorporated into the wider Esprit program in Europe.

AM

Amplitude Modulation. The technique of modulating information onto a carrier by changing the carrier's amplitude in sympathy with the incoming data stream. The signal can either be in analog (infinitely variable) or digital (integer-state) form.

• In AM radio, the analog modulating signal from the microphone or other audio source, changes the amplitude of the carrier. With AM, therefore, you can talk about the composite signal occupying a single (more correctly, 'narrow') carrier frequency, while FM occupies a wider channel of frequencies. However AM always has side-bands.

AM commercial radio stations transmit at frequencies between 530 and 1700kHz. In Europe, AM channels are spaced 9kHz apart, and in the USA at 10kHz. Don't confuse AM with ASK. See single sideband also.

• Digital modulation of the signal's amplitude in the form of amplitude shift keying (ASK) is used in QAM digital modem transmission. This was also the technique used with the early data-cassette recorders before home PCs gained disk drives. When the carrier (radio or audio) is switched on or off (full or zero modulation) to send the binary code, it would be better called OOK (see).

• Alternatively, the carrier amplitude can be switched between two, three or more intermediate levels to send ternary, quaternary, etc. codes. See dibit, tribit and quadbit.

• Digital AM (ASK) with two or more amplitude levels can be coupled with PSK to increase data rates in faster modems. See QAM, Trellis Code Modulation and DAMQAM.

ambient

The surrounding environment. Usually the computer environment, but it may mean temperature and humidity.

• Ambient Temperature — this usually means the temperature of the air or water into which the heat of the device must be dissipated. All electronic devices suffer problems when the ambient temperature exceeds the design specifications.

Amdahl

• A computer company that makes IBM-like mainframes.

• A measure of performance 'balance' in computers. Gene Amdahl once postulated that a computer needed a megaByte of memory for every million instructions per second. So the MIPS/MB ('Amdahl') ratio should be 1:1.

• Amdahl's Law was a mathematical proof that you could never use more than 100 processors tied together in a supercomputer configuration. The maths were correct, but the basic assumptions weren't.

One expression of Amdahl's Law says that 1Mb/s of I/O capacity is required for every MIPS of processor performance.

AMI

Alternate Mark Inversion. A digital transmission technique used mainly to allow transformer (AC) coupling of transmission lines. It has three voltage states, although it is a binary, not a ternary system.

This is a form of alternate inversion (see), but here the zero amplitude represents 0, and both positive and negative voltages represent the logical 1. The AMI 'rule' is that each alternate 1-bit (the 'mark') is sent inverted with respect to the last, thus maintaining an average zero DC voltage on the line and allowing transformer coupling (segments to be linked through transformers) since signals appear on the line as AC-like changes.

However, a chain of zeros will still result in a period of no voltage variation and this will create clocking (synchronization) problems, so the maximum number of consecutive zeros allowed is usually limited to 15.

Note that any change of state will normally be between a positive, zero and a negative voltage (or vice versa) — so any change which is positive-zero-positive or negative-zero-negative is a code violation (see). This provides either a form of error detection, or a way of signaling.

There are alternative formats known as HDB3 (the most common form used) and B8ZS. See these, alternate inversion and pseudoternary also.

AML

Added Main Line. These can be either analog or digital. The term refers to the fact that the telephone company has run out of twisted-pairs in a certain area of town, and is therefore using 'pair-gain' techniques (simple frequency division multiplexing) to put two or more 4kHz phone links down a single pair to the local pillar, and then split them to a number of houses.

• DAML is a Digital AML unit. Most use standard PCM. However in some cases, voice calls are now being delivered over pair-gain systems by using 32-bit ADPCM. This seems to restrict the data rate on modem operations to about 4.8kb/s.

AMLCD

Active matrix LCD. See LCD and active matrix.

amorphous

Without structure or shape; unorganized.

• Amorphous materials do not have a regular crystalline structure and so the molecules group in a relatively random way.

• Normally a LCD element on a portable screen switches at the junction of a vertical and horizontal matrix line. Amorphous LCD screens are a new form which don't have the rigid cross-check pattern. It may be possible to make flat-panel screens much larger and cheaper using the amorphous approach.

• The term is often applied to a type of reversible optical disk technology where the reflectivity of the surface is modified during recording. The surface changes from being crystalline to amorphous when heated by a high-powered laser, and this slight change in reflectivity and polarization can be read by a low-powered laser. See magneto-optical disk.

amp

See ampere.

ampere

The amp is the unit for measuring electric current. A current of 1 amp, is defined as having 6.25×10^{18} electrons flowing every second, and such a current has a charge of 1 coulomb. So 1 amp = 1 coulomb/sec.

ampersand

The & sign.

amplification

The strengthening of a weak signal. The contrast here is with attenuation.

• This is a measure of gain, and gain is the ratio between the output power of a signal and the input power. It is usually expressed in dB.

• Amplification produces gain in both signal and noise, so the S/N ratio is not affected (although additional noise may be added by the amplifying electronics). In a chain of amplifiers, each stage of amplification can be expected to have more noise in proportion to the signal.

• Regeneration of digital signals is often classed as amplification, but it is strictly not. The S/N ratio is changed dramatically in regenerators. See repeaters and regenerator.

amplitude

The size or maximum value of something. The strength of the signal.

• The volume or 'loudness' of a sound.

• The magnitude of a direct current (DC) measured as the deviation from the zero point.

• The magnitude of alternating current (AC) measured as the deviation from the zero point — but note, there are both peak and RMS values. See.

• When viewing a signal graphically, amplitude is the distance the prescribed point on a waveform is away from the base line. Amplitude is measured from the zero-point, not from the highest positive to the lowest negative. This is the sample dimension at any time interval during quantization.

amplitude group delay distortion

This is telecommunications distortion which affects (or upsets) the time relationship between various frequency components in the transmitted signal. Different frequencies actually travel at slightly different speeds down wires and through optical fiber cables. This rarely affects voice, however it does cause problems with high-speed data. See delay distortion.

amplitude modulation
See AM.

amplitude shift keying
See ASK.

AMPS
Advanced Mobile Phone Service. The US standard for the current FM analog cellular mobile telephone system widely used in other parts of the world (TACS is a variation).

This was the original cellular system invented by Bell Laboratories back in the mid-1970s. It is designed for the 800MHz frequency band (but a 450MHz version was once available) with a channel spacing of 30kHz (there is an optional 25kHz).

The first commercial system began in Chicago in November 1982: prior to this, a 44 channel system had been trialed. There are now 832 channels in all. The frequency allocation is for duplex 25MHz bands (824—849MHz and 869—894MHz) to be subdivided into A and B bands for two operators.

The duplex separation is 45MHz and channel numbering is complex because there have been two allocations. The first provided A-Channels 1—333 and B-channels 334—666. Later channels 667—799 were added to the top, and 991—1023 were added at the bottom of the band. There's a deliberate gap in the numbering here.

The standard has three output-power recommendations: 10 watts for cell base station sites, 3 watts for vehicular mobiles, and 1 watt for transportables. Lower power (0.6 watt) handheld transceivers are also widely used.
See NAMPS, NADC (D-AMPS), TACS, NMT, CT2, GSM and DECT also.

AMT
Advanced Manufacturing Technologies. These are factory technologies which include the following concepts:
- Manufacturing Resource Planning (MRP/MRPII): planning and control.
- Computer Integrated Manuf. (CIM): an umbrella concept.
- CAD/CAM: design and manufacture (see).
- Computer Numerical Control (CNC): lathes; drill presses.
- Flexible Manufacturing Systems (FMS): (groups of CNC machines).
- Distributed Numerical Control systems (DNC): materials handling.
- Computer Aided Process Planning (CAPP).

Also various inventory control systems, labor, and machine optimization planning, etc.

A/N mode
Alphanumeric mode, as distinct from graphics mode. This is an older term which was widely used when only character-based computer terminals were available (before the Macintosh). At that time relatively primitive graphics were provided by a special graphics mode — so the distinction was made between this and the A/N mode. Sometimes the term is now used to refer to the use of a character-based user-interface (CUI) on a PC, as distinct from a graphic-based one (GUI).

analog
A copy, likeness, or imitation of something in strict proportion. The implication in electronics is that analog signals copy the original source of variation in a continuous fashion, not in discrete steps. Digital signals vary only in discrete (discontinuous) amounts, usually identified by an integer number.
• All our old electronic communications systems were analog with the exception of telegraphy and telex.
• The major problem with analog communications and computing techniques is that it is difficult for electrical circuits to distinguish between changes of the signal and those produced by noise in the system. So when you amplify or copy an analog signal, you also amplify and copy the noise. See generation loss.
• In communications, the concepts of analog vs. digital is often confusing because the two techniques may be used in harmony. For instance, a modem accepts digital signals in a binary DC form from the computer, and modulates them onto an analog/AC audio carrier for carriage across the phone lines. The result is discrete digital variation of an analog carrier. At the foundation, most systems are analog — television, sound systems, radio, light in fibers — but they may be digitally modulated.

analog channel
This is a channel designed to carry audio (voice) frequencies. A standard telephone connection (not ISDN) from your home to the exchange is an analog channel. At the exchange it may be digitized and multiplexed over interexchange and trunk lines.

The irony here is that, in most data communications through a telephone network, the computer's digital signals will be modulated by an modem onto an analog carrier for the local loop. This composite digital/analog signal/carrier will then be digitized as a PCM signal and multiplexed for onward transmission at the exchange. At the other end it will need to be demultiplexed, then converted back to an analog signal, which is then transmitted over the local loop and demodulated by the receiving modem, to get access to the original digital signals. This complexity explains why the idea of a fully-digitized telephone network appears so attractive to everyone. See ISDN.

analog computers
The first type of functional computers were analog. They were used in a variety of tasks, from the calculation of shell trajectories in military ballistics, to compensating for drop-outs in professional 2-inch videotape. Analog computing techniques are still used for a wide variety of scientific tasks.

analog component video
Analog component video treats or transmits the various components of an analog video image separately. The distinction is

with composite video and television which we generally know by the terms NTSC, PAL or SECAM.

There are a number of types of component video, depending on the way luminance and color elements are treated.

- Pure component signals from a 3 tube (or CCD) color camera are RGB.
- Derived component signals combine RGB to produce separate luminance (Y) and chrominance (C) color signals. The simplest of these is called a Y/C system. This common form of analog component approach is used in S-VHS and Hi-8 video camcorder standards. They output the Y (luminance) signal on one cable connection and the C (chrominance) on the other.
- There are also varieties of derived component video which have a single luminance (Y) signal and two color-difference signals (U and V). This approach is used in professional video equipment. Most commonly, the chrominance is treated in the form R-Y and B-Y (primary color, less the luminance value).

Whenever video is handled in component form through the post-production phase, the recorders will record and reproduce these Y and C signal/s separately. Ideally all editing equipment, switches, TBCs, etc. will be designed to handle the components separately to preserve the best possible quality throughout the production chain. Problems arise when parts of the post-production chain use component video, and other parts use composite.

At least one type of composite output will be provided on most professional video gear. See MAC, component, composite, YUV, YIQ, and RGB.

analog monitors

Analog monitors require that the computer's digital signal must first be converted to analog in the video adaptor card. The analog signals are then transmitted to the monitor (usually in RGB form for color) as base band continuously variable signals which drive the electron guns. The value of the analog RGB approach over digital monitors is in higher image quality and the relative independence of the monitor from the video adaptor card.

Strictly, all monitors are analog, since all require the same form of varying voltage to drive the electron guns of the CRTs. The real distinction here is in where the A-to-D conversion is taking place. In analog monitors, all conversion happens inside the computer (or on the adaptor card) and this can then be simply changed or upgraded; in digital monitors it happens inside the monitor and so is fixed.

analog repeater

An amplifier used in long-haul communications cables to boost analog signals for onward transmission. Note that analog repeaters do not regenerate the signals (restore them to original form), both the noise and the signal on the line are amplified, although noise may be reduced by filtering.

analog to digital conversion

A-to-D conversion or ADC. To be converted into a digital form, an analog (constantly variable) signal must be periodically sampled, and at these sample points the waveform deviation from zero must be measured for amplitude (quantized). The higher the sampling rate, the more accurately the digital number sequence can represent the original waveform. Normally, the sample rate will be twice the highest frequency required (the Nyquist rate). The process of assigning numbers to the waveform amplitude is known as 'quantization'.

The standard ADC telephone technique is called PCM, and the sampling rate here is 8000 times a second (twice the 4kHz of analog telephony) with 8-bit quantization.

- The quantization can be a simple linear scale, or it can be a companded scale (logarithmic). Companded signals give better results with the same bit-rates. See A-Law.
- 8000 samples with the 8-bit quantization results in the standard 64kb/s data rate of a 'standard digital voice channel'. This rate was thought to be essential for good quality telephony in the 1980s – however most voice telephony is now carried long distance using ADPCM at half this rate, and for undersea cables at one-quarter the rate. See ADPCM and CELP.
- For CD-Audio quality in music reproduction, the sampling rate is set much higher at 44,100 a second, and 16-bit quantization is used (for each of two channels, in stereo).
- For video, the waveform sampling rates obviously need to be much higher, since the higher the rate, the better the image definition. So rates of millions of samples per second are used. Usually the Red, Green and Blue signals are sampled independently, each to a 5- or 8-bit quantization level.
- Various sampling techniques are used in ADC:
  - A quantization number is generally assigned to each sample depending on its amplitude (height above the zero point). This is the PCM approach.
  - Instead of an absolute value, a 'changed' value can be calculated by comparing the new sample with the previous (difference techniques). This approach reduces the bit-rate needed for the same voice quality by about 50%, because shorter binary numbers will generally be involved (e.g. 4- as against 8-bits). This is the ADPCM approach.
  - Delta techniques also reduce the bit-rate needed by about 50%. These assign bit-values only (not sample bytes), and the logic (1 or 0) of the bit either says 'up' or 'down'.
- Both differential and delta techniques are very efficient in reproducing average voice waveforms in an almost lossless way, however neither handles gross changes in waveform (phase-changes) well, and so neither is successful with modern high-speed modems. The data-rate limit for modems appears to be about 9.6kb/s across the 'differential' systems.
- The sampling rate may be fixed in the PCM, Differential PCM or Delta PCM standards or it can be variable or 'adaptive'. This variability may depend on how critical the reproduction quality needs to be for that particular period, or it could be set

by overload conditions in broadcast or telephony systems. See ADPCM.

• The bit-length assigned to each sample can also be set to 8-bits for PCM, or to either 16- or 14-bits for CD-Audio. Conversion from higher to lower bit rates requires throwing away or rounding the 'least-significant bits'. Rounding produces better quality.

See DAC, A-Law, mu-Law, PCM, quantization errors, ADPCM, delta modulation.

anamorphic

Not of equal physical dimensions. Graphical distortion, in the sense of not having the height and width reproduced proportionally.

• In Cinemascope film projection and camera systems the horizontal angle of image compression is much more than the vertical angle: the Cinemascope aspect-ratio of 2.3:1 is horizontally compressed to fit onto a film which normally has a aspect-ratio of only 1.33:1.

This is only possible through the use of specially-made anamorphic lenses which incorporate cylindrical (rather than entirely spherical) lens elements. A circle photographed through an anamorphic lens, becomes a vertical oval.

You sometimes see the anamorphic compression effect during the opening titles of Cinemascope movies shown on TV, when the programmers allow the 'squeezed' image to be shown so that titles and credits aren't truncated.

• In video, anamorphic operations can also be performed electronically by changing the horizontal scan distance in proportion to the vertical scan. This can give the same wide-screen effect — but naturally the horizontal resolution suffers in proportion to the increased aspect ratio.

• In sound, the term is used to indicate variable treatment of audio qualities. Dolby B is said to be the audio equivalent of anamorphic. It emphasizes some aspects of the sound during recording, then equalizes the modified sound during reproduction.

• In computer graphics the term refers to the ability to stretch a drawing horizontally or vertically. Often, if the 'shift' key is held down while the drag operation is performed, no anamorphic change is possible — so circles remain circles.

AND (Boolean)

A specific logical connector used in text selection (information retrieval) systems and in other operations involving Boolean algebra. You may wish to select database records containing 'asbestos AND health', and the AND here specifies that both terms must be present in each record (or abstract) before it will be selected.

In Venn diagrams, the AND area is the region or coverage where the A and the B circles overlap. The distinction is with Boolean OR and NOT operators.

AND-gate

A logical transistor device (chip element) used in basic computer IC design. As with most logic gates, an AND-gate has two inputs and one output. When both or either input is at zero, the output is zero. Only when both inputs are 1 is the output 1. See Boolean algebra.

ANDF

Architecture Neutral Distribution Format. This is a European (OMI) idea for a software format which is capable of being directly compiled into machine-specific code at each target machine. It won't matter whether the chip is Intel or Motorola, or RISC or CISC.

angle modulation

A collective term for frequency and phase modulation. See FSK, phase modulation and PSK.

angstrom

A measure used for the wavelength of light. One angstrom is equal to a wavelength of 10^{-10} meters. The symbol is a Swedish capital Å. Nowadays, the tendency is to refer to light wavelengths in terms of nanometers (nm) rather than angstroms.

• Visible violet light is about 4000Å or 400nm, and red extends up to 7200 Å or 720nm.

• The reference yellow light of a sodium-vapor lamp has a wavelength of 5890Å or 589nm.

• A 'radio' wave of 1Å would be in the middle of the X-ray spectrum.

ANI

Automatic Number Identification. ANI is similar to, but not identical to, Caller-ID. ANI is a common method of tracking calls within the telephone exchange-to-exchange network, and is provided by the carriers for their own use. In the US, it is automatically sent between telephone companies for billing purposes, and is used to identify the A-party to the B-party in the case of free-phone or emergency services. It pre-dates both CCS#7, LASS and CLASS.

ANI can be carried in packets over CSS#7 signaling networks and as DTMF or MF tones over POTS lines. It is invariably coupled with the number being called in a ANI-DNIC string. Free-phone services in the USA can arrange to have ANI delivered even if the CLID is blocked. See DNIC, CLID and intelligent networks

ANIM

An Amiga animation format which is actually a series of Amiga IFF still images compressed into a large file. There are many variations. OPT 5 is the most common, but OPT 7 and 8 are able to take advantage of 32-bit memory in the later Amigas.

anisochronous

The translation is: an (without) iso (equal) chronous (time). It is a term that has become totally (and confusingly) misused over the years.

• Generally it is used as a synonym for 'asynchronous'. This refers to the fact that the bits forming each character are carefully clocked in time, but the time-gap between each character is totally variable (which is why start- and stop-bits are used).
• More correctly, it means 'erratic delivery' — which asynchronous terminals may, or may not, provide. With key-board entry, the output on an async link will also be anisochronous — but with file transfers, the output will be as regular in delivery times as a synchronous system. This makes no sense.

The implication of not having delivery regularity or predictability is therefore only partly true with asynchronous systems — which is why ATM is also an isochonous delivery system.

anisotropic

Showing different properties in different planes or along different axes. This is applied to light polarization, to electricity, and to the production and manipulation of images — all in slightly different ways.
• With images, the term has the sense of different dimensions on different (x vs. y) axes — or 'rubber-sheet' geometry. Imagine painting an image on a rubber sheet, then distorting it horizontally or vertically by stretching. This is the anisotropic or anamorphic effect.
• Anisotropic/anamorphic lenses are used in the CinemaScope process to horizontally squeeze an area of view with an aspect ratio of 2.5:1 down to fit a conventional 'Academy' frame with a ratio of 1.33:1. A circular object would look oval (longer axis vertically) on the film or if projected through a normal lens, and all actors look extremely tall and thin. Another anisotropic lens on the projector is used to unsqueeze the image in the cinema.
• In computers, anisotropic mapping is a graphic process where you can drag or stretch images in one or more directions. Most draw programs allow this.

ANL

Automatic Noise Limiter. A device used to cut ignition noise in a transceiver or receiver. The term is actually wider than this and covers a range of devices, but these are the most common.

ANN

Artificial Neural Networks 'governed by fixed-point attractor dynamics in terms of a Hebbian learning matrix among bifurcated neurons' may be the way to understand and model complex neurodynamic patterns. Each node 'generates a low-dimensional bifurcation cascade towards chaos, but together they form collective ambiguous outputs; e.g., a fuzzy set called the Fuzzy Membership Function (FMF).'

This feature becomes particularly powerful for real-world applications in signal processing, pattern recognition and/or prediction/control, so they say! I only report this stuff, I don't necessarily understand it.

anode

The positive electrode. Electrons (which carry a negative charge) actually leave a battery from the negatively charged cathode and flow to the anode. However, for historical reasons, we think of the current flowing from plus to minus (anode to cathode). The anode is mostly identified by a plus sign, but in some schematic diagrams, the anode is often represented by a triangle. See cathode.

anomaly

Anything unexpected — especially if it can't be explained or it happens erratically. It carries the sense of deviating from the expected course of events.

anonymous ftp

File library transfers (downloads) which are available to be used by anyone without prior arrangement. These are usually files stored in large university or government computers, but which they make freely available to the public.

You do not need to be registered on the computer service to use these files so you don't need a password — although the database may generally require one for other services. When asked for a username, give 'anonymous'. When asked for the password, give your e-mail address. See ftp, Gopher, Archie, Veronica and WWW.

ANSI

This organization was originally the ASA which gave its name to the standard for film sensitivity (now ISO). ANSI is one of the top global standards-setting organizations, and the principal standards development organization in the USA.. It doesn't directly develop standards, but accredits other groups to do so.

ANSI also acts as the clearing house and coordination agency for standards. It represents the US position on international bodies like ITU and ISO. Agreement within ANSI is generally easier to achieve than it is within the ISO, so many preliminary ANSI standards later evolve to international ISO or ITU standards.

ANSI has a number of standing committees which, in the digital data processing area, have the 'X + number' identifier.

This is a hierarchical numbering system. For instance X3 (established in 1960) identifies numerous sub-committees working on computer standards and information processing.

Within each X-area, are more specific Technical Groups which are also given a number, as are the 'Task Groups'. Sometimes only the Technical group is specified, as in the T1 transmission standards committees.
• The term ANSI is also used for a type of computer terminal which conforms to ANSI's terminal emulation standards. See VT-# terminals.

ANSI graphics

This is an old technique which used the upper 128 (8-bit) PC character set (of graphic primitives and symbols) together with ANSI Escape sequence controls to create what appears to be graphics, while the computer remains in text mode. This was similar to videotext.

ANSI.SYS

A console device-driver with MS DOS that performs many of the old BIOS functions. There are 15 separate functions under five categories: cursor control, screen erasure, video mode setting, video attribute setting, and key redefinition. The driver can give you some control over your system's display and let you create simple keyboard macros by redefining your function keys.

• If the CONFIG.SYS file contains a line:

```
DEVICE = ANSI.SYS
```

and all commands are sent through DOS rather than BIOS services, the PC will obey ANSI screen control codes.

ANSI terminals

See VT-# terminals.

ANSI X.12

A widespread standard (originally US) for EDI which is now slowly being superseded by EDIFACT. X.12 is actually the identification of the committee which oversaw the development of the EDI standards.

answer mode

Modems at rates up to 9.6kb/s transmit and receive using different audio frequencies. They must therefore be set to the answer mode (also called receive) for receiving and the originate mode for sending — one of each. It doesn't matter which modem uses which mode as long as both don't use the same, but the convention is that the calling party should choose the originate frequencies.

However some on-line database services will always transmit originate frequencies even when called, so modems must be able to switch (or be switched) from one mode to the other. Generally an 'R' added to the end of the AT dialing sequence code will reverse the expectations, so that the caller then uses the answer mode.

answer signals

Telephone systems exchange a lot of supervisory signals internally during call establishment and tear-down phases. American systems, with multiple independent carriers and RBOCs, need to exchange many of these signals between themselves whenever calls cross business boundaries. Monopoly carriers often deal with these matters internally.

• An answer signal is sent on the backchannel when a call is answered.

• An answer-charge signal is sent on the backchannel to indicate that a charge should be made.

• An answer-nocharge is sent with emergency numbers and those which don't incur a charge.

answerback

This was a standard terminal identification feature on Telex machines which gave them a high degree of legal standing since it could be reasonably established that a message had been sent to the correct machine if answerback was received.

Telex machines have an inbuilt circuit which sends back this pre-determined message when interrogated. When the calling terminal makes connection, it sends a 'Who are you?' request, and the destination replies with its answerback. Some modems and communications programs have this facility also. Don't confuse this with call-back.

• Answerback simulators (hardware and software) are sometimes used to spoof the sender into thinking it is dealing with a teleprinter.

antenna

The US term for aerial. Antenna is used for both the transmitter and the receiver. Antenna design is a highly skilled discipline. It must take into account the bandwidth, the type of wave being generated (sky wave, surface, etc.), the polarization of the signal, the height above ground and the angle of radiation, the impedance, field-strength, and antenna gain.

— The basic form of antenna is a half-wave dipole. This is a single wire whose length is approx. half the transmitted wavelength. The radiation from a dipole is not isotropic (equal in all directions). It is strongest in directions vertical to the wire. The same antenna can be used on several bands of frequencies by operating it on harmonics.

— Vertical quarter-wave antennas are often used when low-angle radiation or reception is required (along the ground). A cellular phone has this kind of antenna.

— The ground plane antenna is a vertical quarter-wave antenna using an artificial metal ground. This is the type you see vertically mounted on a car roof. It increases the gain in the horizontal plane.

— The Yagi is the classic TV antenna; it's a folded dipole type. It has different elements for different frequencies, and it utilizes harmonics.

— Dish systems act as concentrators to an antenna (actually to a detector inside the feed horn). See also phased array.

antenna gain

With satellite direct broadcasting, this is a measure (in dBi) of the effective concentration of the signals produced by the reflecting surface of the antenna dish. The bigger the dish, the higher the gain and the higher the pointing accuracy. Parabolic surfaces, however, should have a greater gain than spherical curves, and the material covering the dish surface will also make a difference.

anti-aliasing

A technique used to make hard-edged 'stepped' objects on computer screens or printed graphics appear to be of a higher resolution by filling in (or averaging out) pixels adjacent to those in the problem areas. This was mainly a problem before outline fonts were widely used.

'Aliasing' in this context is the dreaded 'jaggies' — the stepped effects on letters and lines which are slanted or curved, especially when larger sizes have been generated from small bit-maps. To iron out these 'steps', pixels on the screen in the 'corners' of the steps may be partly illuminated (or lightly printed). The edge is therefore softened slightly, and this makes its shape appear smoother and less step-like.

This is a form of dithering and these fonts are known as 'fuzzy fonts' — however PCs now use outline fonts and much smaller dot-size in their laser printers, so the problem has been largely solved in other ways.

Antiope

A French combined teletext and videotex standard which emphasizes the mode of presentation of the alphanumerics and graphic pages. The transport mode was either wire-carried viewdata (over the telephone network) or broadcast teletext (in the TV blanking interval), and the basic technology was alpha-mosaic and very similar to Prestel. Teletel is the French viewdata service based on Antiope. See Minitel also.

antistatic

Desktop mats/pads and wristbands which are electrically conductive and wire-connected so that any electro-static charge will flow away from the hands and create a voltage equilibrium with components and devices before they are touched. These are used to prevent electro-static damage through components during manufacture or repair.

It is important that all components reach equilibrium before they are touched, but it is not important that this equilibrium be the same as any outside earth potential — *exactly the opposite in fact.*

So do not leave a computer plugged into a power-point (despite the claims of most of the industry) since this destroys equilibrium, and increases the likelihood of static damage.

- ***I repeat: Do not leave computers plugged in to a power point while you insert or remove cards or components.*** Unplug the computer, then touch the top of the power supply — you will have then established electrostatic equilibrium with the computer, and have created a situation least likely to create high-voltage current flows.

AnyLAN

See 100Base-VG in the numbers section.

AO

Acoustic-Optical devices. These are one-dimensional SLMs used in optical computing. See SLM.

AOR

Atlantic Operating Region — for satellites. See IOR and POR.

APA

All Points Addressable. In computer graphics this means a system where you can control every pixel on the screen. This is common now, but it wasn't a few years ago. See EGA.

APCM

Adaptive PCM.

APCO-25

Association of Public Safety Communications Officials standard for a new worldwide emergency service digital trunked mobile radio system. Project 25 of APCO has been the development of a digital public safety ('critical') mobile phone system which is now being widely accepted around the world.

The core standard is FDMA access with QPSK-C modulation and embedded signaling at 2.4kb/s. They provide a channel bandwidth of 12.5kHz which can carry a data-rate of 9.6kb/s. It's a digital trunking standard which provides DES encryption.

APD

Avalanche Photo-Diode receiver.

aperture

An opening or hole:

- In cameras, the aperture is the central hole in the iris diaphragm which admits (and restricts) the light passing through the lens, and therefore controls the brightness of the image on the CCD, video tube, or photographic film surface. It is equivalent to the 'pupil' of the eye.

  Camera aperture is measured in 'f-stops' which are calculated as focal length-over-diameter. This is a theoretical calculation only, and doesn't take into account light losses in the lens caused by absorption and interference. The practical equivalent is called a 't-stop' and it is always less (requiring a wider aperture, or lower equivalent number). For accurate exposure control, the t-stop should always be used.

  Aperture also has a strong effect on the visual appearance of the image because of the way it manipulates the depth of field (the planes that are in focus). Larger apertures give less depth of field (more critical focus).
- In satellite communications, the aperture is the dish diameter.
- There is also a satellite use of the term related to time — as in 'The aperture available for the Shuttle launch is between 1600 and 2100 hours'.
- Aperture cards hold a frame of microfilm and allow normal library card index boxes to be used to store microfilmed data.

API

Application Programming Interface. These are the common rules which allow applications software to communicate with the operating system. They are software specifications which application developers need before they can write programs for a particular operating system or architecture.

In practice, an API is a collection of commands ('calls') around which the application programmer can design ways for software to communicate; it provides the way for applications to gain access to system resources from within a programming language, like C.

A small API, say for the control of a CD-ROM, may involve only a few lines of simple instruction, while the API for control of telephone systems (TAPI) may involve hundreds of different calls on functions or subroutines, all of which need to be explained to the applications programmers. They are usually written in a language like C.

The writers of operating systems now provide libraries of programming functions, calls and interfaces with access to a particular network layer. For instance, in the specific area of terminal emulation, an API provides keystroke simulation and I/O to the device buffer (presentation space).

An API may also send and receive structured fields: when you want to move data from one PC application to another you do it through the APIs which perform the translation functions. So they are used to transfer data between the operating system, application, database management systems and other control programs in computer operations.

In the past, APIs have been fairly narrow in their application, but now Microsoft and others are attempting to write wider APIs for messaging (MAPI) and telephony (TAPI) to bring standard office and telco equipment into the domain of the local network.

- Undocumented APIs have been the cause of some vigorous criticism directed at Microsoft over the years. It is said that they had 'undocumented features' in their operating systems (and Windows) which allowed the inhouse applications writers to create lean-and-mean applications, while competing software writers could only use those which were documented.
- In networking, APIs are best seen as proprietary interface languages that will run only on the network-server and file-server of one particular LANs vendor. They provide functions such as the file, print, communications and administrative services accessible on the network. Supplementary APIs may also provide functions for peer-to-peer communications or for terminal emulation.
- If you want to migrate a Window's application to OS/2, for instance, the main task would be to make numerous API-call translations.
- Microsoft's Win32 is being promoted as the standard desktop API in the future. Currently Win16 has a enormous user base.

See TAPI, MAPI, CPI communications, CMA, APPC, and LU6.2.

API/CS

Application Programming Interface/Communications Services. This is a generic term for communications APIs, however the term is usually applied to a specific set of open interfaces which provide a link between Netview/PC and SNA.

APNSS

Analog Private Network Signaling System. For PBXs. This is the analog version of DPNSS, where only leased analog lines are available for linking company premises.

APOC

Advanced Paging Operator Code. A 1993 upgrade to the POCSAG paging protocols, which was developed by Philips. It can be implemented in four separately-defined parts, and all four remain compatible with POCSAG 1200.

APOC offers a very robust form of synchronization, extensive bit interleaving, reduced message length and, therefore, repeated transmission opportunities. It uses a text compression system to reduce alpha message length.

On a shared POCSAG channel, the bit rate can be gradually increased from 1.2kb/s to 6.4kb/s because APOC and POCSAG share the same synchronous characteristics. APOC modulation is essentially similar to both ERMES and FLEX modulation systems, with either 2-PAM/FM or 4-PAM/FM (depending on the data-rate). See RAMP also.

apogee

In satellite orbits: the point in any non-circular orbit or ballistic curve, farthest from Earth. Perigee is the lowest point.

- In spacecraft and in satellite launching, the perigee stage boost is used to take the satellite from low-earth orbit to the higher (perhaps geosynchronous) orbit. The thrusters are fired while the satellite is in low-earth orbit in order to gain the additional velocity needed: in the process, this stage produces an elliptical-shaped orbit. The apogee stage involves the firing of the thrusters in reverse at the top of the orbit to slow the satellite, and convert the oval path into a circle.

The preliminary boost from the low 'parked' orbit to the higher one is called, confusingly, either the perigee-thrust stage or the apogee-boost stage.

APPC

Advanced Program-to-Program Communications. These are IBM's peer-to-peer communications protocols (related to LU6.2) which allow a PC or mid-range host software to communicate directly with mainframe or other software. APPC lets SNA applications communicate directly with peer SNA applications so it can be used over SNA, Token Ring, Ethernet and X.25. It is for multi-vendor, multi-platform environments.

The APPC protocol suite is a collection of SAA commands (an API) that developers can use within applications. The alternative was to deal only with a central mainframe in a host-terminal relationship. It replaces IBM 3270 emulation for PC-to-mainframe links, and was introduced by IBM in 1985 as part of SNA. It requires a higher level of software than the older NetBIOS, however, NetBIOS is still widely used (OS/2 LAN Server remains a NetBIOS application). Some applications written for NetBIOS are not portable to the APPC environment — and vice versa. See APPN, SAA, API, CPI communications and LU 6.2.

APPC/PC

The PC version of APPC. See APPC.

append

To add to the end of a file or list. The distinction is with insert, replace or update.

APPI

Advanced Peer-to-Peer Internetworking. See APPN.

Apple key
On an Apple GS or Macintosh computer keyboard, this is the key variously called 'Open Apple', 'Snowflake', or 'Command' (as distinct from Control). It's on the left-hand bottom of the keyboard. It is a 'shift key' and is used in a similar way as the IBM Alt key, in conjunction with the normal alphanumeric keys to perform extended control and character-creation functions.

Applesoft
A form of Basic built into the early Apple II machines in ROM.

applet
A sub-application. This is a mini-application (virtually a utility) which provides the minimum amount of code needed to function — it will have no bells or whistles at all.

AppleTalk
These are Apple's Macintosh networking protocols (released in 1984) which use the LocalTalk data-link hardware system built into every Mac. However AppleTalk work-group LANs are commonly connected to Ethernet as a major corporate backbone, but they can also interconnect with Token Ring and FDDI.

Physically, AppleTalk is a CSMA/CA (Collision Avoidance) network. It has separate protocols which correspond to all seven layers of the OSI model. Two of the most important of these protocols are Dynamic Node Addressing (DNA) and Distributed Name Service (DNS) which are now making AppleTalk popular for wireless LANs, since they don't require constant management and monitoring for people logging on and off the network.

Apple has since released AppleTalk Internet Router software to connect up to eight AppleTalk networks into a transparent internet. AppleTalk Phase II has also increased addressing bits and provided additional internetworking features.

• In the early Mac literature you will find that AppleTalk refers to both the physical components of the system and the higher protocols. Later the term LocalTalk was introduced for the physical components and lowest protocols, to make a clear distinction. EtherTalk and TokenTalk can also be used with AppleTalk (often needing plug-in cards).

• The main components of the AppleTalk protocol suite are:
— Datagram Delivery Protocol — network functions.
— AppleTalk Session Protocol — transport functions.
— Name Binding Protocol — addressing.
— AppleShare — file sharing.
— AppleTalk Remote Access — report access functions.

application
A word now so wide in meaning that it is almost meaningless. In its broadest sense it is anything useful you can do with a computer (other than programming). Any of these could fit:

• Programs which are used to create and modify information, or perform various other human-useful and substantial tasks. The distinction is with operating systems which do useful things but only for the computer itself, and utilities and TSRs which are considered too minor to qualify for the 'applications' title.

Supervisory programs and some of the larger accessory (TSR) programs are a real problem here since they are part-application yet not substantial enough or are too general to qualify for the term. Games are also a problem: normally you wouldn't refer to a computer game as an application.

• A programmer's definition is: those software programs (or packages) which make calls to the operating system and manipulate data files, thus allowing a user to perform a specific job. This covers just about everything!

• There is also non-specific use of the term, to refer to the job being performed — the job for which a computer program is needed. So here it refers to the task or problem to which the computer program is applied, rather than the program itself.

• Application 'packages' are coherent sets of programs and data files, some of which may reside on disk and only come into RAM when called for. For instance, a word processor with a spelling checker and thesaurus.

application builders
This sometimes seems to be a wider term than application generators. At one end of the scale these are application development systems with their own modules and programming language, such as Visual Basic; at the other, this is componentware. See application generators.

application development systems
These consist of a programming language together with the associated utilities needed for program debugging. They may also provide libraries of 'routines'. See componentware and ALM.

application generators
Application generators are special programming tools which supposedly require entry of only a description of the problem, rather than detailed coding. They use a high-level (AI-type) language to generate the required code.

Simple query languages have developed over a period of time into full 4GLs which were able to produce their own computer code, and about 1988, 4GLs evolved into CASE tools. These were marketed under the 'applications generator' banner. CASE has since become a dishonored term because of the ridiculously excessive claims that were made about its increase in programming productivity ('ten-times' was often conservative!). See CASE and 4GLs.

Application layer
The seventh and highest of the OSI's 7-layer model which provides a 'window' between applications in order to exchange information. The name is misleading because this layer doesn't contain the ultimate application processes — it just services those application processes. This level deals with the network operating system and interacts with the end-user's application software.

This is the most visible of all the seven layers because it is the layer that an application sees. Software at the Applications layer handles file-sharing, database management, electronic mail, print-spooling, etc.
The OSI standards define four essential elements in the layer:
— Association control,
— Reliable transfer,
— Remote operations,
— Commitment, Concurrency and Recovery.
This is where you'll also find a number of main interface standards. These will create the basic building blocks for future applications so that they can interwork easily: ACSE, FTAM, MHS, VTP, MMS, X.400 and X.500. See.

application services
In the VAN and VAS area, application services generally means the on-line provision of EDI and perhaps a range of international billing and accounting services for large international corporations. On-line conveyancing, international airline booking, financial transactions, and customs clearance are other examples.

applications builder
This is software which takes you step-by-step through the development of an application customized to your requirements. This will usually be the creation of a database system. These are closely related to form-builders and other 4GL systems which work in similar ways. They will often output a source code in C++ which will then require compilation.

applications processor
The mainframe or mini that is running the applications — as distinct from the front-end processor (FEP) that may be handling the communications.

applications-specific languages
Languages which are designed to perform a narrow range of tasks extremely well. Fortran was originally designed for scientific computing and Cobol for business data processing, but both of these languages are now much wider in their application. There are languages which are specifically designed only for programming telephone switches, for instance.

applicative language
The same as functional programming languages. See.

applique
Appliques translate and transpose the serial signals of a communications protocol (say of RS-232) to another (say V.35). They are mounting plates containing connector hardware.

APPN
Advanced Peer-to-Peer Networking software — IBM's distributed-processing LAN/WAN strategy based on LU6.2 and Type 2.1 network node protocols (in competition to TCP/IP). It was originally developed for networks of AS/400 minicomputers, but it has since been enhanced to support routing over internets and 3174 cluster controllers. There are both MS DOS- and OS/2-server versions. IBM says it has better flow-control, response times and data integrity than TCP/IP.

APPN is the key to distributed computing in an IBM environment; it promises transparent access to all SAA computing resources across a multinode network. It can be used over SNA, Token Ring, Ethernet and X.25. The early versions used a routing algorithm called label-swapping, but the new APPN+ uses source-routing.

At one time a rival group, led by Cisco, was promoting APPI ('Internetworking') standard for carrying SNA and APPN over TCP/IP. This has now been abandoned.

• APPN+ is the real name of the improved routing variation which abandoned label-swapping (see) for source-routing. IBM say that it has better flow control and error recovery also. The + has since been dropped from the name.

• Gigabit APPN is effectively ATM with an IBM name; it will be able to handle ATM fixed-length cells as well as longer data cells. This is the next stage of APPN. See IS-IS and OSPF.

• APPN/HPR (with High Performance Routing) is IBM's client-server solution for companies with mainframe-dominated network environments. See APPC.

approved device
In telecommunications, this is any terminal unit that can be legally connected to the telephone system. All such equipment needs technical approval to ensure it won't be dangerous or destructive to the system.

apps
Applications.

APS
Asynchronous Protocol Specification. This is a specification being promoted in the ITU to allow the linking of mail-enabled applications over any type of wireless networks. The members of the group include Microsoft, Apple, DEC and Intel.

AQA/EEMS
The AST/Quadram/Ashton Tate version of expanded memory (EMS 3.2) which came to be known just as EEMS. It didn't last long; AST later joined the LIM group to create EMS 4.0 which was a joint specification. APA could concurrently run multiple applications if they supported it. See EMS and EEMS.

arbitration
A set of rules for settling disputes and allocating resources — memory, peripherals, etc. This arbitration is important (but not essential) in multi-user or multi-tasking systems, and also on serial bus/links like SCSI. Some systems, such as CSMA don't use arbitration, while others do. See SCSI and bus arbitration.

ARC
• **Advanced RISC Computing** group, which evolved from the ACE consortium. They tried to establish a common standard based on the MIPS RISC platform which was to be scalable from laptops to mainframes. It was to complement the ACE CIRC 486 'industry standards'.

• **ARC ('Archive')** was a very popular early form of MS-DOS file compression (also available as LHARC). It has now been replaced by ZIP as the ***de facto*** standard.

Note: Archiving compressed files makes them much more vulnerable to tape/disk degradation than archiving uncompressed files. The more you compress them, the more vulnerable they generally are. Ideally you would archive using a FEC error control system like Reed-Solomon.

Archie

A directory service on the Internet which indexes file titles (Note: titles only) for a large number of anonymous ftp (see) sites, and stores them on an Archie host. There are a few dozen Archie hosts worldwide.

You can (1) run the Archie program (on an Archie server) as a client, or (2) you can Telnet to a nearby Archie host and use keywords, or (3) you can e-mail an Archie server and receive the response by e-mail. The program searches its lists and returns the results, providing you with likely sites offering ftp access. The reply-file is likely to be huge if you don't tightly define your search. If you want to search on key-words within the files, not just on titles, use WAIS.

Archimedes

A European plan to orbit four (two pairs) communications satellites in a highly elliptical orbit for DAB. Each satellite would be able to broadcast up to 100 stereo programs simultaneously. The launch of the first pair is planned for late 1995.

architecture

Generally, this word vaguely describes the components of a system in a way that stresses either their inter-relationships or functional purpose.

• **Computer hardware:** More specifically, it means the underlying design and structure of the computer's hardware and operating system that define storage methods, operations and compatibility with other systems.

The main components of a computer architecture would be:

— The type of central processor and its microcode.
— The ALU's accumulator capacity (in bits — 8, 16, or 32).
— The width of the address bus (how much memory it can directly access) and data bus.
— The clock rate, wait-states, and length of access lines.
— The provision of expansion slots and an expansion bus.
— Coprocessors, and ancillary support systems.
— Memory management.
— How I/O is handled; BIOS, interrupts, DMA

The key division is between architectures which are closed (proprietary) and open (available to anyone — but maybe only on licence).

• **Open architecture** computers use available components, BIOS and O/Ss which are openly available, and they generally provide standardized slots into which adaptor cards can be inserted to extend and modify the basic operations. This means that competition tends to keep the hardware cheap.

Unfortunately, unless the rules of the architecture are strictly policed, random changes often introduce incompatible variations (this was very common in the early days with PC clones). Many add-ins may also have unconventional functions which create conflict problems.

• **Closed architectures** such as the Macintosh have the advantage of a single authority (the manufacturer) which completely controls the design and operations of the computer, and can thus preserve continuity and consistency of the interface. They have no competition within their standard, so prices tend to be higher — but more profit perhaps means that more will be invested in R&D and they 'try harder' because they are isolated.

With the early Macs, the term 'closed architecture' also used to mean that the computers had no slots, so third-party developers couldn't takeover and drive developments.

• **Software:** In software, open architecture is applied to the willingness of the proprietary company or consortium to publish details about how their operating system or middleware works. They will provide a fully documented set of APIs.

Microsoft has long been known to have 'undocumented features' in their operating systems, but generally MS-DOS is an open operating system because APIs are jointly developed and published. Application software is slave to the operating system — so it is vital that all developers are fully conversant with all the operating system's features.

• **Networks:** With networks, the term architecture is applied both to the physical topology — whether it is a bus, tree-branch, ring or star — and to the networking standards.

• **Protocol stacks:** These are also network architectures. So SAA and the OSI model are both 'architectures' in the sense that they are sets of defined standards with layers designed to interwork and support each other.

archive

A long-term storage space for important information. Archival copies must be made on special tapes, film or disks and kept under climate-controlled conditions. Back-ups are the most common archive function — however the term 'archive' carries the implications of longer life and the preservation of all versions; back-up is just concerned with the immediate future.

The two modern archival media which have proved to be the most long-lasting are punched-tape and B&W archival film. A punched-tape doesn't have the storage capacity needed today. However, B&W film has a proven life of at least 100 years under the right conditions, while many magnetic video tapes have only lasted 20 or so — which doesn't auger well for floppy disks. The nickel stampers used in the manufacture of CD-ROM disks may have potential for long-term (1000 year) storage.

• **Archival copies** are usually specially made or processed for longevity. See father, grandfather and backup.

• **Archive programs** for small business or domestic use are very often just file compression utilities, to save on disk space

— but be aware that compressed files will be more vulnerable to disk defects than uncompressed files.

ARCnet

The first commercially available LAN, developed by Datapoint in 1977. It had a transmission rate of 2.5Mb/s over either ring or bus topology networks. It has since been upgraded to 20Mb/s. ARCnet is a token-passing system which can run over thin-coaxial, twisted pair and (more recently) optical fiber. It is still widely used in the US, but is on the decline and it does not have IEEE endorsement.

ARDIS

Advanced Radio Data Information Service. A national packet-radio network in the USA, promoted by IBM and Motorola. The service was designed by Motorola specifically for IBM field staff, and later became a public offering. Motorola build the Info-TACS radio modem for this service. Currently it has a data-rate of 19.2kb/s, and is migrating to use DataTAC.

Its main competitor is Mobitex (see) although CDPD is now becoming significant.

areal density

The packing density. This can mean the number of pixels on a standard screen area, the number of transistors on a chip of a certain size, the number of bits stored on a disk surface — all measured in elements per sq.mm or something similar.

argument

A programming term which has a number of slightly different applications:

- An independent variable. It is a value that is passed between programs, subroutines or functions. (In some cases this is just called a 'parameter'.)
- A variable which sets the value of a function. A function can only return a result when it has been given an argument (sometimes a 'dummy' one) which is a number. For instance, in the Basic function, TAB(9), the '9' is the argument. See function.
- A variable which is used to reference a table. It specifies the location of a particular item.
- In floating-point systems, the argument is the fixed-point part of the number.

Aries

See ECCO.

arithmetic coding

A technique which improves on Huffman's coding by ignoring the concept of fixed-length bytes in the original message. Instead, it treats the whole message stream as if it were a continuous floating point numeral with an unlimited (potentially) number of bits. The algorithm must be based on a good probability model of the data being coded.

arithmetic functions

These are stored routines (algorithms) which operate on a number, to produce a result. Functions are more complex than 'operators' (like plus, minus, divide by); they include ABS, INT, SQR, SGN, and usually need an argument. See functions.

arithmetic logic unit

See ALU.

arithmetical operators

Arithmetical operators are used to manipulate numbers: to add, subtract, equate with, compare size, etc.

$+ - * /$ are the primary arithmetic operators.

$< > = \neq$ are secondary arithmetic or relational operators.

Note that sometimes command-words like 'DIV' or 'GE' are used instead of the divide (/) or greater-than (>) signs.

ARM

- Originally 'Acorn RISC Machine', now **'Advanced RISC Machine'** — a type of low-power, high-speed RISC processor used by Apple in their Newton PDA. It combines a RISC processing module with an ASIC. Apple partly owns the UK-based company. There's also a new asynchronous (clockless) ARM chip which is raising much interest.

The current ARM is version 7. It has an advanced RISC, 32-bit architecture with a geometry of 0.4 microns. (The original Newton's ARM 6 chip was 0.6 micron.) The chip consumes less than 1.5 milliamps per megaHz, and will operate from a 3-volt power supply, compared with the 5-volts of its predecessor.

The main advantage of the ARM designs seem to be low power consumption for the processing power they provide. The chips are now made both in the UK and the USA.

- **Asynchronous Response Mode.** A mode of HDLC which involves one primary, and at least one secondary node. Either can initiate communications.

Armonk

A town in New York where IBM has its headquarters. Often used as a synonym (by over-smart reporters) for IBM's decision makers.

ARP

Address Resolution Protocol is used on Internet TCP/IP systems to translate between ('bind') the IP address and a real Ethernet 802.2 address. ARP determines whether the packet's source and destination addresses are in the format needed for dial-link or IP control. RARP (Reverse) translates the other way, from IP back to Ethernet. This is defined in RFC 826.

ARPA

Advanced Research Projects Agency (of the US Department of Defense). Later ARPA became DARPA ('Defense'). ARPA was the main funding source for a lot of research into AI and communications.

ARPANET

Advanced Research Projects Agency (of the US Department of Defense) Network. The original packet-switching network, which dates from 1973, ran on a NCP (Network Control Protocol) suite, but changed to TCP/IP in 1983.

ARQ

Automatic Repeat reQuest. An error-control system which is very widely used (with a few variations). In the original ARQ system, an ACK (acknowledgment) is returned whenever the checksum matches the addition of the block, and a NAK is sent when it doesn't.

With more reliable transmission methods these days (mainly because of optical fiber) it is often better to transmit a NAK only when an error has been noticed. Compromise systems will allow many blocks to be sent before an ACK is received, instead of the system waiting for an ACK after each one (these are sliding windows).

Full-duplex error-control uses ARQ standards like HDLC and SDLC. See ACK-NAK, FEC and sliding windows.

- There are three main types of ARQ:

— Stop-and-wait: The original ARQ systems were often called 'half-duplex', because the transmitter waited for acknowledgment before sending the next block.

— Go-back-N: The originator sends blocks determined by a window size, but it is not required that every block be acknowledged. If a NAK is sent, the sender steps back a set number of blocks and begins again.

— Selective-reject: The only blocks retransmitted are those identified with a NAK. The receiver must be able to insert the replacement block into the correct sequence, since other blocks will have been received in the meantime.

array

A set of matrixed variables under one name. These are related through the common 'identifier' (variable name) and a number of 'subscripts' (number sequences). For instance the list of variable names may be: Company$(1), Company$(2), Company$(3), etc.

Array variables usually only have one dimension — but the term is used loosely to encompass matrixes which can have a number of dimensions, such as Company$(1,1) and Company$(1,2).

Arrays are used when tables of data must be kept in memory for instant access. Many programming languages require you to dimension an array (declare the maximum number) before use, and the subscripts must sometimes be carried in square brackets.

array processor

Aka vector processor. A hardware add-on, or a stand-alone computers (some with mainframe power), which provides a quick method of performing repetitive arithmetical functions in signal (and other) processing.

They use a number of linked processors, each handling one part of the array. So they can sometimes process the whole of a large array in a single step — with a single machine instruction, using what amounts to a wildcard for referencing the array subscript.

This technique is used mainly in graphics and video image processing at the low level, and in scientific areas like fluid dynamics, CAT scans, aircraft engineering with supercomputer-level processing. See multi-processing and vector architecture.

ARS

Automatic Route Selection. Aka Least-Cost Routing. This is a feature on most modern PBXs, where long-distance and international calls are placed through carriers and resellers offering the lowest possible rates on a call-by-call basis. ARS PBXs can be programmed to take into account, time of day, day of week, distance — even quality — and select from a range of carrier options. This type of selection requires a reasonable degree of intelligence and complex rate tables. See least-cost routing.

artifacts

Aka artefact. Errors or unwanted features generated by the system — in video, graphics or audio. Artifacts are defects in an image/sound reproduction caused by the transmission, storage or reproduction process. In television, artifacts are any unwanted image disturbances (flickering areas, electronic noise, ghosts, etc.) on the screen produced by the electronics or outside interference. In hi-fi, artifacts may be evident as a disturbing buzz of the speaker when it is reacting in a non-linear way to certain frequencies.

The term has also been extended to incorporate any form of visual image disturbance introduced by aliasing in the recording or transmission systems, and many artifacts are the result of filtering and processing operations done to overcome other problems.

The main artifacts in analog television are:

— Vertical aliasing: where the letters become unreadable as they scroll.

— Cross color: when herringbone coats or striped shirts generate rainbow colored moire patterns.

— Cross luminance: where colors intrude into the brightness signal, forming sparking spots around objects.

— Poor-relative color resolution: which is noticeable mainly as reds tending to flood into nearby areas.

— Chrominance shift: where the colors blur to the right.

— Interline flickering and large-area flickering.

See aliasing and distortion also.

artificial intelligence

See AI.

artificial reality

Artificial reality devices process images captured by a video camera (or some other less intrusive measuring device) to define your position in space (and the orientation of your head) as a way to control the 'reality' environment.

This is a way of interacting with a computer without the cumbersome paraphernalia associated with virtual reality (data gloves/helmets). Information about your movements is coordinated with the computer graphics so that they respond and change according to your actions.

AS

Autonomous System. Networks with a common routing strategy which are under common administration. A 16-bit unique address is assigned to these systems by the DDN Network Information Center. See border gateway and IGP.

AS/400

IBM mid-range (mini) computer series which has some software compatibility with the System/36. It has proved to be very popular. The machine will support up to about 500 users depending on the precise system architecture, and it runs the OS/400 operating system.

ASA

- **American Standards Association.** This organization established the original film sensitivity rating, now known as ISO. It was an arithmetical rating (unlike DIN), and so a film with an ISO rating of 400 will be twice as sensitive to light as film rated ISO 200, and therefore require one stop less light.
- **American Statistical Association.**

ASC

A Basic statement that returns the ASCII No.5 code of a character. The command ORD is used in Pascal.

ascender

In desktop publishing and printing, this is the part of a printed character which rises above the body. The letters b, l, h and k all have ascenders, while g, q, and p all have descenders. The significance of this is for readability. Ascenders on one line and descenders on the line above should not touch.

The essential space between is called the 'leading' and it should be at least 1 point. For this reason, type is usually defined as, say, '10-on-11' — meaning 10 point type with 1 point of additional space (leading). The line space is therefore 11 points.

Fonts which have relatively short ascenders and descenders will be more readable for a set point size. Their x-height will be larger. See x-height and point.

ASCII files

ASCII No.5 'print' characters will be recognized by nearly all modern personal computers and software. Most word processors provide a 'text-only' ASCII file-creation function for creating files which can be transferred between machines and between applications. In making an ASCII-file, machine-specific or program-specific formatting codes will be stripped out and abandoned.

There's not total agreement on what should remain in, and what should be dropped from an ASCII file. Many MS-DOS programs still leave the carriage return + line feed (CR+LF) within the file at the end of every line (not just at the end of a paragraph). This creates problems when these MS-DOS files are used by a Macintosh — the extra screen-control characters need to be stripped.

- On the Internet or in any situation where text exchange is not controllable, you should always use the 'Unformatted' or 'Text-only' or 'ASCII-file' file-type selection. See RTF.

ASCII No.2

American Standard Code for Information Interchange No.2. This was the old 5-bit code used primarily for telex and telegraphy. To these 5-bits were added a start and stop-bit.

With ASCII No.2 there were 52 alphanumeric characters (including symbols) plus two format-effectors (carriage return and line feed) and two modifiers (figure- and letter-shift).

Since 5-bits can only generate 32 possibilities, the figure-shift and letter-shift were used to indicate 'everything following is alpha' or 'everything following is numbers and symbols'.

ASCII No.5

American Standard Code for Information Interchange No.5 — as defined in ANSI X.3.4 and ISO646. If only the words 'ASCII text' are specified, you can assume it is the No.5 version. It is pronounced 'a-ski'.

The ASCII Alphabet No. 5 is the 7-bit standard for characters, numeral, punctuation, space, carriage return, line-feed, and a few control and command keys (a total of 128), which is now the standard for computing and telecommunications. The alphabet was specified by ANSI in 1977.

The ISO adopted 7-bit ASCII as the standard ISO 646 (with only minor modifications). Later they added a (semi-) standard extensions for the 8th bit — and therefore another 128 characters called Latin1. This extension includes letters for most European special characters (umlauts, diacriticals, etc.). Other countries have developed other extensions for their own languages.

An 8-bit variation which was standardized for IBM-compatibles also exists, called 'Extended ASCII', but not all PCs conform to this standard. Many tend to use the extended 128 characters in their own way.

ASCII protocol

This implies that only the basic alphanumeric print and control characters are transmitted (the 7-bit 128 characters), and therefore that the file will not be formatted in any way (other than to include Carriage Returns, Line Feeds, and Tabs). Note, however, that RTF files are still ASCII files.

- The use of this term often also implies that no effective error checking (with the possible exception of parity) exists; that the file is being sent over a network without any special protocols to handle block transmission errors.

ASIC

Application-Specific Integrated Circuits. These are large-scale chips made specifically to perform particular tasks — as distinct from most ICs (particularly processing and memory chips), which are made for a wide range of applications.

ASICs are used in computers to handle jobs such as controlling the video screen or running the SCSI port. Being specially designed for the task, they reduce the number of components, increase operational speed, lower costs, and reduce the box size. Generally the design of the chip will be compiled from a library of elements.

ASIC is a general term which can incorporate gate-arrays, digital signal processors, etc.

ASIC density, which until recently was in the 25 to 50k range of 'gate equivalents', has now reached 500k to 1 million gate equivalents. At these element densities, and with high clock rates, the fabricators were forced to lower voltage to 3.3 volts because the chips will get too hot. CMOS is often used for this reason also. Low-voltage ASICs come in three flavors: gate-arrays; embedded gate-arrays and standard cells.

ASK

Amplitude Shift Keying. A modulation (modem) technique of varying the amplitude of the carrier in accordance with the modulating signal — but only in integer amounts for digital. Don't confuse this with AM. (although it is a sub-set).

With modems, the carrier is an audio tone sent end-to-end through the phone network, and ASK digital modulation changes this tone in amplitude. This technique is rarely used alone these days, because amplitude is often highly variable on a phone network.

The main types of ASK are:

— Binary: A single audio tone is simply switched on and off. This is more correctly called On-Off Keying (OOK). This technique is now used in fiber-optic systems to modulate the laser or LED light into pulses.

— Tri- or quad-bits can be transmitted by varying the carrier between three or four preset levels.

• Amplitude changes are often used in association with phase changes to transmit additional information. See QAM and DAMQAM.

ASM

Assembler. See.

ASME

American Society of Mechanical Engineers. See ANSI also.

ASMP

ASymmetrical MultiProcessing. This is one of two key architectures used mainly for multimedia servers where multiple CPUs are working together in the same computer. The other is SMP (Symmetrical).

The ASMP system assigns each CPU to a specific, individual task. One may handle network traffic, another runs the NOS, a third handles disk I/O, etc. Each processor works independently. The main CPU handles the operating system, interrupts and core application, while one or all of the other processors handle I/O tasks. This gets over the problem of constantly assigning tasks (since they are pre-assigned) and the architecture allows special attention to be paid to speeding up bottlenecks, such as disk access — an important consideration in multimedia. Dedicated I/O processors on the storage system leave the server CPU free for other tasks.

Note that SMP and ASMP aren't mutually exclusive; there are hybrids which combine both approaches.

ASN.1

Abstract Syntax Notation.1. The 'language' of OSI applications (IS 8824) and a Presentation layer data-unit specification. It is widely used in TCP/IP related standards and is used to define the format of PDUs.

ASN.1 is a notation for abstract data types and their values. It has now become an essential part of the X.500 directory standard and ROSE. ASN.1 provides X.500 with a way of identifying the use to which packets/blocks of data should be put.

There are four categories of data-types:

— Simple: atomic types with no components.

— Structured: has components.

— Tagged: types derived from other types.

— Other.

The basic language structure is the module, and each of these modules is then a formalized statement in the ASN.1 language. The language is far too complex to discuss here., however, ASN.1's application in X.500 directories offers a glimpse of the possibilities:

• **X.500:** The idea is that certain information characters will be packaged with a couple of bytes of 'header' — these identify the type and the length of the block. Universal tags have been defined in the standard and are used in X.500.

These Basic tag-types are: Boolean, Integer, Bit-string, Byte-string, Real, Enumerated. Character string types are: NumericString, PrintableString, TeletexString, VideotextString, IA5String, GraphicsString, VisibleString, and GeneralString.

So, for instance, the first byte in the data module says that everything in the following packet is information for a Teletex machine; the second byte specifies the length of the data payload, and the rest is data. Alternately, the first header byte may carry the instruction that the data should be forwarded on to a telex machine.

• ASN.1 has a syntax which is encoded by either BER (Basic Encoding Rules) or DER (Distinguished Encoding Rules).

ASP

Audio Signal Processors. These are nothing more than DSPs used in sound cards.

ASPEC

See Musicam.

aspect ratio

The ratio of the width of something in relationship to its height (or thickness).

— The aspect ratio of the average computer monitor screen and television set is 4:3 or, to put it another way, 1.33-to-1.

— Wide-screen cinema has an aspect ratio of 1.66:1 or 1.85:1.

— Cinemascope has an aspect ratio of 2.4:1

— Wide-screen (EDTV) and high-definition (HDTV) seem to have settled on an aspect ratio of 16:9 (1.77:1). Note that this is 4^2–by–3^2.

ASPI
Advanced SCSI Programming Interface.

ASR
Automatic Speech Recognition — a possible future I/O device for computers — at least for dictation or transcribing speeches. But will we feel stupid talking to computers?
There are three factors of importance:
— vocabulary size,
— training requirements,
— continuous speech/discrete speech.
Continuous speech seems to be the main problem: computers have great difficulty in detecting the end of words. Discontinuous speech in the form of a limited range of commands, is relatively easy and has proved successful for computer applications that require the operator's hands to be free.

Obviously there's an important application here also with the handicapped. Automatic translation telephones depend on the combination of ASR, language translation and speech synthesis. AT&T are a leader in this technology. See speech recognition.

assemble editing
With videotape, assemble editing is the process of adding shots strictly in sequence, with each new one appended to the last. The contrast is with insert editing where a shot is inserted into the middle of another, or into an already-completed sequence.

The distinction is not quite so important today, but it was very important with non-professional videotape editing a few years ago, because any change required the loss of a generation, which decreased quality.

assembler/assemble
A program that translates a program from its assembly language symbolic (source) code to the computer machine (object) code, which is very close to the binary code the computer can execute. This process may be long and involved, requiring a number of repeated operations to perform the translation, allocate storage, compute absolute addresses, etc. The key processes of the assembler are:
— It breaks down each instruction into its constituent parts.
— It calculates addresses based on the first location (offset).
— It translates the operation mnemonics into machine code.
— It translates the addressing information into machine code.
— It defines the memory space required for storage of each part of the program.
There is a wide variety here. The old CP/M assembler converted the source code only to a hex format. It then required a LOAD program to make the final translation to binary.
• The distinction must be made between an assembler and a compiler. Compilers translate high-level languages into an assembly form first, and then the assembler changes it into binary code. See object code and source code.
• The assembly language itself is sometimes called an assembler. This is how some mainframe people will use the term.

assembly language
Assembly languages are low-level symbolic languages, which produce source code — not object code. Don't confuse the output of an assembly language with machine code.

However assembly languages are also CPU-specific because each is tied to the microcode of the processing chip They are one step up from the machine code, and therefore from the binary that the computer uses for its internal operations.

All assembly languages have letter codes which are mnemonics for an operation, and each mnemonic has three (and only three) letters. Each assembly statement produces one machine instruction (an operation code). A single space is often used to separate field labels from operation codes, and commas are another common form of separator.

Assembly languages allow symbolic addressing where the computer itself decides the location of data in memory, and this is essential in multi-tasking computers.
• You need to understand how the CPU and other hardware elements of the computer operate before you can write programs in assembly language, but the benefit of using this approach is in having programs that will run very fast. (Although nowadays compiled 'C' programs are just as fast.)

assembly mnemonics
Assembly languages use easy-to-remember three letter abbreviation for computer instructions, directives, etc. Each of these abbreviated commands corresponds to a single machine language instruction on a one-to-one basis. These are not the same mnemonics which will be used in machine code

A program in machine code refers to specific locations in memory and specific registers. Once a program has been written in machine code (object code), any change of even one byte in one location with throw it into confusion.

Assembly language, is a source-language or symbolic language, which removes much of the rigidity from machine code, while retaining most of its better characteristics.

However different computers use different assembly languages (although these have many characteristics in common). They are all based on three letter abbreviations; for instructions, fields, directives. Addressing and operatives depend on the system.

assign
In computer terms this means:
— To add a value to a variable. To insert a string (text) or value (number) into a symbol which represents a variable. LET A$ = "Country" is an assignation of the word 'Country' to the string variable A$.
— To make one thing equal to another.
— To reserve blocks of memory space for specific purposes.

assignment
One of the major characteristics of high-level languages running on von Neumann architectures. Values are assigned to specific locations in computer memory and then sequenced.

The computation results depend on the order in which the statements are executed — on the 'procedure'. This is in contrast to non-procedural AI languages like LISP.

associations

In transaction processing (database operations), associations are the communications paths between two applications.

asterisk

• In telephony, this is also known as 'Star'. It is one of 12 DTMF tone-pairs, used in telephone signaling.

• In most text-programs and in information retrieval, the asterisk is often used as the wildcard (see). It means anything.

ASVD

Analog Simultaneous Voice and Data. AT&T.Paradyne's technology to carry a voice signal and a data signal simultaneously over a single telephone line. It provides better voice quality than the competitive DSVD standard, but it only transmits data at 14.4kb/s.

asymmetrical

Unbalanced — not evenly distributed.

• In communications —it means that the traffic in one direction will be much higher than in the other. Cable TV is totally asymmetrical (one-way), while a two-way telephone conversation is roughly symmetrical (although you could say it is asymmetrical in alternating directions — which is why DCME is possible). See ADSL also.

• Host-terminal and client-server systems are asymmetrical, while peer-to-peer computing is symmetrical.

• In processing, multiprocessing architectures can be either symmetrical or asymmetrical. See SMP and ASMP.

• In modems, the term means that the data-flow is predominantly one-way; this is true of most BBSs and of information retrieval where large-file downloading is the norm. The bulk of the channel's bandwidth will therefore be reserved for the host computer's transmissions, while the low data-rate back-channel is only to be used for command, control and ACK signals sent from the PC. The videotext 1200/75 bps standard is a typical asymmetrical modem technique with downloading at 1200 bps, and uploading from the keyboard at only 75 bps.

• In encryption, asymmetrical systems are those that are encoded with one key but can only be decoded with another. These are public-key encryption systems (see RSA and PGP). Symmetrical systems are those which are coded and encoded with the same key. These are private key systems (see DES).

asymmetrical multiprocessing

See ASMP.

asynchronous

Without external timing control. It generally means 'without clocked timing' or it refers to transmissions which have a variable time-interval between successive bytes or events.

An asynchronous communications channel is one that can transport data, but doesn't carry a timing signal. This is confusing in telecommunications because two quite different and well known technologies each use the term in a different way, and there's still the general sense of 'sequential, but without timing' which is used indiscriminately:

• ***One-user-per channel — a byte-framing term.***

This still needs clock timing, but the timing only needs to remain accurate for the duration of one byte. It takes its clock timing from the arrival of the first bit (called a start bit) which is then ignored. The leading edge of the bit is the timing point.

Asynchronous modem communications came about because the receiving computer could never predict when it would receive a byte (a keystroke-generated character) from a keyboard. It therefore could never be sure that the CPU wasn't occupied on other matters, and it needed time to direct attention at the arriving byte and get its clocks working so that the bit intervals could be sampled to check for 0 or 1 bits. Characters sent directly from a keyboard must always be asynchronous because it is impossible to depress the keys regularly.

In fact, today's modems are synchronous between modems, but asynchronous across the RS-232 cable between the modem and the PC only. See asynchronous communications.

• ***Many-users-per-channel — a multiplexing term.***

Asynchronous Transfer Mode (ATM) in fast-packet switching uses a strictly timed continuous sequence of 'slots', so it uses a true synchronous line transmission system — but each node has 'asynchronous' access to these slots. The node grabs and uses whatever empty slots are available, so access to the network is asynchronous, even if the network itself is synchronous. The contrast here is with STM. See ATM and isochronous also.

• ***Sequential — without timing requirements.***

— In IBM's SNA terminology the term refers to store-and-forward type operations, as distinct from real-time online operations.

— In computer programming the term is sometimes applied to a system that queues the stages of a program so that the completion of one, initiates the next stage.

— In data file-types, an asynchronous file is one where the importance isn't tied to a specific data-transfer rate. The distinction is between, say, a text file which has characters that can arrive at any time, and a real-time music file which needs segments delivered at a constant rate so that there isn't a sudden break in the sound delivery.

— In microprocessors, the new 'asynchronous' ARM chip (a RISC) avoids much of the problems associated with the need for clock logic and allows them to design more efficient circuits.

asynchronous communications

When computers are transmitting one character at a time (say, from a keyboard) the arrival of the data at the other end of the communications link will be erratic and 'unexpected'.

In such circumstances, the receiver doesn't automatically have a timing point at which to sample the wave-form, to decide whether the bit is a 0 or a 1.

So a start-bit needs to be added to each byte to alert the receiver to the arrival of the information. The start-bit therefore both warns the computer of the arrival of a byte and provides a timing mark for the sampling. Once the edge of the start-bit has been detected and the timer is running, the start-bit is then ignored. The 'stop bit' is nothing more than a minimum period of time for the system to reset, waiting for the next byte.

Since the line is generally held in the 'high' state when not in use, the start-bit will usually be a logical zero (low voltage), while the stop bit will be a logical one (high voltage) — and it is usually held for at least one normal bit-length to ensure isolation.

Synchronous transmission between computers (usually mainframes) is faster than asynchronous, since the overhead of start and stop bits can be dropped and the bytes transferred regularly according to synchronized clock pulses.

• Modern modems are often asynchronous devices between the PC and the modem, and synchronous devices between the modems across the phone line. So each 8-bit byte will pass out of the PC through the UART where start and stop bits are added, cross the RS-232 cable to the modem as a 10-bit byte, be reconverted back into 8-bits, passed through compressing and error-checking processes, then passed as a compressed file in block form to the modem-pump for transmission. The reverse sequence obviously happens at the other end.

• Asynchronous, as applied in ATM and ATDM is a different concept entirely. It deals with the lack of any rigid control to blocks or packets of information being allocated frames/slots on an otherwise synchronous system. It is the lack of rigid access control that makes these systems asynchronous.

asynchronous devices

The early asynchronous modems were asynchronous through all stages — from PC to modem, modem to modem, and modem to PC.

• Modern synchronous modems are often but necessarily async between the modem and the terminal, they will most likely be communicating with each other across the telephone lines in a synchronous way. Whenever phase changes are being used for transmission, the links are self-synchronizing.

• Asynchronous terminals (PCs) and modems usually communicate with each other via UARTs and RS-232 cables but there's no necessary relationship between RS-232 and async communications.

• Asynchronous line drivers allow the attachment of remote asynchronous data terminals to a multiplexer or computer using baseband digital techniques. They are used in telecommunications for distances below a couple of miles. See line-driver.

• Don't confuse ATDM (asynchronous TDM) with ATM.

asynchronous mode

Applied to the common PC communications systems where one character is being sent at a time, with an added start- and stop-bit. Don't confuse this with Asynchronous Transfer Mode.

Asynchronous Time Division

This is a form of cell-switching which later became ATM; it was proposed by the French as the new international telephone standard. ATD was to have used a short, fixed-length packet having a payload of only 32 bytes. See Asynchronous Transfer Mode.

Asynchronous Transfer Mode

ATM is the standard protocol for a form of fast-packet switching now accepted by the IEEE and ITU as the basis of future broadband telephone networks (with SDH support in the interim).

ATM is similar to TDM, but individual circuits are not allocated to specific time slots in the ATM frame — it's a case of first-come, first-served (except when priority is provided), and this is the 'asynchronous' (unpredictable) function.

ATM frames are synchronous, but they are dynamically filled with packets of whatever information is available at a node, so the bandwidth of each channel is flexible. This will allow variable bit-rate services into the phone network. Like all packet-switched systems, ATM is physically always connected to user terminals, but in normal use, it requires physical call establishment, so it is classed as a connection-oriented service. There is also a connectionless form. See AAL.

ATM is the most important new communications Protocol today, and it offers all the advantages of conventional packet-switching coupled with the high-rates and predictability of circuit-switching. It is extremely flexible, and can carry all types of traffic, voice, data and video — and multiple channels of each, on a single physical connection. In fiber or coaxial, this is a true broadband network, with true integrated services (hence B-ISDN) which is why it is the ideal technique for the public telephone/television/data networks of the future. It is also an excellent private network protocol.

The key to ATM is the use of a short packet known as a cell, which are 53 bytes in length with a 48 byte payload. The 5 byte header contains virtual circuit and virtual path identification (VCI and VPI — see) and HEC only. This is cell-relay system, where nodes don't check for packet payload errors. ATM is designed primarily to run on optical fiber, but it will also run on copper twisted pair (25Mb/s), coaxial, and possibly over radio links as well (with added error checking).

ATM cells move across the network from one switching node to the next in a very rapid store-forward way. The switches are classed as 'fast' because they are fabricated in silicon, but they process each cell before dispatching it to the next node. The processing consists of checking the header integrity using the CRC in the HEC byte, then translating the addressing information (the VPI and VCI) using switch-routing tables, before shunting it though the switch fabric.

ATM switches are fast-packet, high capacity 'virtual circuit' switches which are currently designed to handle data at rates from 25Mb/s to 2Gb/s or more. Terabit/sec rates are projected for interexchange links before the year 2000. True commercial public-network ATM switches at the 620Mb/s data rate are expected in 1995–6 time-frame and most advanced carriers already have plans to deploy these in the next few years.

Private ATM networks are already popular with about a half-dozen vendors selling switches, although full specification of the public protocols aren't yet completed. However, the key definition of the Adaptation Layers (AALs — see) is nearing completion.

The acceptance of ATM as the future foundation of all public and private networking is now almost universal. It is accepted as such by computer and communications equipment manufacturers, network providers, and user groups (defense, government and private).

- An ATM 25Mb/s data rate standard (promoted by IBM) has now been accepted, and at this rate the protocol can be implemented and provide a full-duplex service over existing twisted-pair cabling. Adaptor cards for this low data rate are already rivalling Ethernet NICs in cost.
- Currently ATM 155Mb/s data rates can be extended over an WAN for private network use.
- See ATM Service Classes.

AT (computer)

Advanced Technology. IBM's AT computers using the Intel 80286 chip were introduced in Aug. 1984. Initially these processors ran at 6MHz (compared with 4.77MHz for the XT).

The AT used a 16-bit data path (the AT-bus) and it had a 24-line address bus, so the chip was able to address 16MB of memory. The machine was given a new keyboard (the AT-keyboard) and it was introduced together with the 1.2MB floppy disk standard.

These machines introduced 'modes' to improve functionality, while remaining backwardly compatible with older PCs. When running MS-DOS, AT machines used the 'real' mode (which imitated the 8086 XT) but they also had a 'protected virtual mode' which could be used for the new OS/2 operating system and Xenix.

- **AT bus**: This is now called the ISA bus. It's the standard 16-bit interface (data-lines), connectors (for adaptor cards), and general expansion system used by the AT and most later computers running MS-DOS.

IBM attempted to replace it with MCA, and the clone makers attempted to upgrade it to EISA, but it still survives — often supplemented by a VL-bus and more recently by PCI when high throughputs are required.

- **AT bus interface.** See IDE and ATA.
- **AT keyboard:** The standard 84-key keyboard introduce by IBM with the AT personal computer in 1984. It resolved many of the design mistakes of the early PC keyboard, but introduced one of its own with a small backspace key. Later versions have 101 keys, including separate cursor keys and 12 (rather than 10) function keys.

AT command set

This has nothing to do with the IBM computer called the AT.

Most communications software for PCs will communicate with an 'intelligent' modem using the Hayes AT command set (the ***de facto*** standard for asynchronous devices) or the ITU's V.25bis standard (mainly for synchronous devices).

This standardized set of command (each beginning with the characters 'AT') is used by most modern modems. However there are many variations on the standard — but fortunately, most use the basic set in a reasonably consistent way.

Modems and PCs interact in two ways. When the modem is in Command-mode, it interprets bytes arriving over the 'Tx' link as commands intended for its own configuration and control. When the modem is in Data-mode, it interprets all bytes arriving over the Tx line as data needing to be modulated and passed on through the telephone network.

In Data-mode, only one command will be recognized on the Tx line. This is the +++ sequence which says 'escape to Command-mode'. In the Command-mode, each command string (there can be more than one command, concatenated into a string) must be preceded by the characters 'AT' (for 'Attention') and followed by a carriage return.

Commands are variable names followed immediately by a parameter. The command 'ATM2E1' would place the value 2 into the variable M, and the value 1 into the variable E. (Check your manual to see what this does!) If a variable name is not followed by a value, then 0 is assumed.

Later variable names for some of the more esoteric functions are differentiated by %, & and $ (as in %A or &A). Other more basic variables are in the form S1, S2, S37(so-called 'registers'). Wherever a number is found in a variable name, an equal sign must be used as a separator from the value (so 'ATS4=10').

- The +W extensions are packet-radio (wireless) extensions to the AT modem protocols.

At Work

Microsoft At Work is a Windows-like operating system and a general software architecture. It was designed for office equipment, to enable all office devices to interconnect — PCs, printers, copiers, telephones, fax machines and PDAs. The key concept was to devise a form of universal storage of information that every product could understand and use.

- This common use appears to be some time in the future.
- Novell has a rival system called NEST.

ATA

AT Attachment. This is form of processor I/O (PIO) which is used by IDE for device control. It is a specification of I/O Channel Ready which serves to transfer data, status and control information between a PC and a hard disk drive. There are three levels:

• ATA is used with IDE disk drives and was originally known as the IDE interface. It has a limited data throughput rate of 2 to 3 MB/s (peak of 4.1MB/s).

• Fast ATA is another implementation of the ATA interface. This is possible by way of direct connection to the CPU bus (a local bus connection), bypassing the expansion bus completely. Fast ATA is too fast for the ISA bus. It apparently 'pushes the data transfer rate up to 6.67MByte/s, or 10MByte/s for a cache-hit burst', and can thus be used with VL-Bus or PCI.

• Fast ATA-2 is a new interface standard being proposed (now before ANSI), with burst rates of over 11MB/s and 16.6MB/s (see Multiword mode 2 and PIO mode 4). This should not be confused with Enhanced IDE.

• Fast-ATA and Fast ATA2 need the system BIOS and the hard drive to support the PIO and DMA protocols.

ATAPI

ATA Packet Interface. This is an extension of the ATA protocol (a minor BIOS modification) designed to support CD-ROM with IDE. It is now the ***de facto*** IDE CD-ROM standard.

ATAPI brings a single command set and a single register set to CD-ROMs, and it can be adapted also for other storage devices. It was designed to use packet commands based on SCSI-2 so that peripheral manufacturers could easily create ATAPI drivers. Several of these commands are specific to CD-ROMs

ATD

• Asynchronous Time Division. See.

• Asynchronous Time Division is also a term used synonymously with statistical multiplexing. See ATDM.

ATDM

Asynchronous Time Division Multiplexer. The distinction is with (the normal) 'synchronous' TDMs where the time slots are fixed in a very rigid way. In normal TDM, slots will be wasted (unfilled) if all terminals aren't transmitting constantly.

ATDMs are much more efficient if the traffic is bursty. They will dynamically adapt the slot allocations, so that time slots are provided to the various channels only when needed. Channels with a higher requirement can therefore be given larger slots or more slots in each cycle.

This technique reduces bandwidth wastage with bursty traffic, but it comes at the expense of control overhead. See statistical multiplexing, ATD and ATM.

Athens agreement

An agreement between digital recording hardware manufacturers (CD disk and DAT tape) and the music recording companies to withhold digital recording technology until copy-protection schemes could be devised.

Atlas

The Unix International approach to distributed processing which is still in the process of being defined. It can be considered as an extension of DCE and it focuses on features that DCE doesn't supply — such as distributed transaction processing and fault-tolerant support services. See distributed computing environment.

ATM

• **Asynchronous Transfer Mode**. See.

• **Adobe Type Manager:** This is the module of the Adobe PostScript interpreter, responsible for the rasterization (bit-mapping) of the various outline fonts. It was extracted and packaged independently to allow raster devices (mainly monitors) and non-PostScript printers to work with Adobe's Type I outline typefaces.

The current version is ATM 3.0, which now supports multiple master typefaces which generate entire families of type from a single master face. See PDL, multiple master, TrueType and outline fonts.

• **Automatic Teller Machine** — in banks.

ATM Forum

A consortium of users, vendors and service providers who are working on industry-wide standards for ATM as private networking protocols. Some of the activities of this forum extend into the telephone area also, but the carrier activities tend to be left to the ITU-T. However, the ATM Forum have certainly led the way in the last few years.

ATM Service Classes

ATM is the keystone of what was once called Broadband -ISDN. The definition of service classes, was the first stage in establishing the universal application of these cell-switched protocols to all future requirements. There were originally four defined service classes, and a fifth, virgin class:

— Class 1 is for constant bit-rate services.

— Class 2 is for variable bit-rate services, mainly video.

— Class 3 is for connection-oriented data.

— Class 4 is for connectionless-oriented data.

— Class 5 is pure ATM without modification.

These correspond to the AALs.

A more recent approach is to base service classes entirely on the flow control mechanisms used:

— CBR: Constant Bit Rate. This is the most straight-forward. The application negotiates a peak cell rate — and it is given this, no more, and no less. Above this rate any data will be dropped. These systems are used when guaranteed bandwidth is needed, and where needs to be very little variation in cell delay.

— VBR: Variable Bit Rate. This approach imposed a tight limit on cell rates and delays. VBR is therefore ideal for video codecs which may have a less stringent delay requirement than CBR.

VBR will also tolerate data bursts which lie outside the negotiated CBR rate.

— USR: Unspecified Bit Rate. With this approach there are no guarantees, so it requests only that the network provide 'best-effort' to get the data through. It exploits unused network capacity, and is therefore cheaper.

— ABR: Available Bit Rate. This is the latest to be defined. It is essentially VBR with some attempt at controlling possible cell loss. It has a negotiated minimum rate.

ATM25 Alliance

See Desktop ATM25 Alliance.

atom

A single element— the non-divisible particle.

• In computing, it is an element or operation that can't be subdivided.

• In database up-dating and transaction processing, it means the whole record.

• There's also a use in Windows which means a short-integer which has been equated to a string. Atoms are established by kit routines.

atomic (transaction)

• **Atomic Lock:** A hardware lock which acts in combination with a software lock to preserve integrity in a database.

• **Atomic Update:** These are all-or-none updates to records. See ACID test and transaction processing.

atomic clusters

This refers to a very futuristic technology which may have an impact on computer architecture in a decade or so. These clusters are minute clumps of semiconductor materials which exhibit different properties, dependent on the cluster size. At this sub-micron size, electrons begin to obey the laws of quantum physics rather than conventional physics.

ATRAC

Adaptive TRansform Acoustic Coding. The compression algorithm used in Sony's MD-Audio (Mini Disk). It achieves a 5:1 compression without noticeable subjective drop in music quality. It is essentially similar to Philips PASC used for DCC, however ATRAC is based on Fourier Transform Coding. See MiniDisk.

ATSC

Advanced Television Systems Committee of America. The organization which is directing the development of HDTV in the USA. It is a voluntary standards-setting committee which is also involved in developing anti-ghosting measures for conventional TV.

attach

In networking, this means to access a network server — or to access additional servers after logging on to the first.

attack

In electronic music the attack is the profile of the 'envelope' at the start of a music note. When sampled and analyzed, musical notes can be divided by time into four phases; attack—sustain—release—decay, and each of these elements can be manipulated to create new audio effects.

ATTC

Advanced Television Test Center in Washington DC. Responsible for testing US HDTV standards for the FCC.

attention code

The AT code used in modem control. See AT command set.

attenuate

To reduce in magnitude. This term is applied to both current and voltage levels. It means the loss of signal strength through a system, measured as the difference between the transmitted signal level and the received power level. In communications, attenuation is usually expressed in dB per kilometer. The opposite is gain.

• In baseband systems, attenuation is mostly caused by the resistance of the circuit.

• In many coaxial cable systems the signal loss over distance is mainly caused by skin-effect and dielectric loss. See.

• The attenuation of telephone wires is related to the resistance (R), the inductance (L) and the capacitance (C) between the pair along the length. Repeaters are used to overcome attenuation problems.

atto-

In the SI units, the prefix atto- represent 10^{-18}. The symbol is the small a.

attribute

A property possessed by something:

• General — a characteristic imposed on a device, or on data. Screen characters have attributes of color, hue, intensity, etc. often set by an 'attribute' byte.

• Any programming variable will have attributes which specify what type of variable it is. It may be a string variable, or a fixed-point variable, or a floating-point variable.

• The attributes in device drivers are those algorithms that tell the driver how to handle data, control strings, etc.

• In printing and screen graphics, the term is often applied to the underlining or boldfacing of words.

• In some files, attributes can be used to keep a file hidden, or make it read-only. This is set by one or more attribute bytes. In MS-DOS, the ATTRIB command can set a file to: read-only; system; archive; or hidden. Four bits are involved as flags, so in each case they are set to Yes or No.

• In a database, attributes are specifications or restrictions assigned to a certain field to ensure the correct entry of data. For instance, an alphanumeric-type field might be chosen to hold phone numbers, and so the attribute would be set to permit numbers only — alphabetic characters would be rejected.

Another attribute could also specify that the numbers in this field can only be entered in particular groupings — say, the access code in brackets (NXX), followed by three digits, a space, then the last four digits. Other very useful field attributes for databases are No Duplicates and No Unique.

• In X.400/X.500 messaging, attributes are flexible fields which can hold user-defined information. These are messaging attributes (ATs) which perform a wide variety of referral functions and other tasks. The ITU X.500 recommendations specify attribute-types: System, Labeling, Geographic, Organizational, Explanatory, Postal Address, Telephone, Telecommunications, Preference, Relational, and Security.

• In networking, an attribute describes access to, and properties of, files and directories within a filing system. Attributes include read, write, create, delete, execute only and hidden.

attribute byte

In the early IBM PCs and clones, text characters were held in screen memory as paired 8-bit bytes stored side-by-side. The first of the pair contained the ASCII value of the character itself, and the second contained the attributes which will be applied to this character when it is displayed on the screen.

The attribute byte must be read on a bit-by-bit basis, numbering from the least-significant bit:

— The first four attribute bits control 'foreground' color (those of the characters themselves): bits-0, 1, 2 control monitor hardware color switches (blue, green, red) (1 = on), and bit-3 sets the intensity to high (1) or low.

— The next three bits control 'background' color: bits-4, 5, 6 control blue, green and red (1 = on), and

— Bit-7 controls flashing (1 = on).

This standard was modified in later color systems.

ATV

Advanced Television systems, and also the name of a committee which is the American equivalent of DVB in Europe. It is a general name used in the US for technologies which encompass IDTV, EDTV and HDTV developments, based both on analog and digital techniques. There are two approaches to improving domestic television services:

• To create entirely new systems using the best digital technologies now available without reference to current analog standards. This approach will create problems in persuading people to abandon old equipment and buy new.

• To augment existing systems with additional channels and extra information — and thereby enshrine the inefficiencies of the system, but provide an evolutionary path (see PAL Plus).

The original four HDTV systems under evaluation in the US have now been reduced to one composite proposal under the ATV task force, now renamed the Grand Alliance (see). Commercial operations are due to begin just before the Atlanta Olympics in 1996.

audio

Sound frequencies between about 50Hz (cycles per second) and 20,000Hz, which is at the top end of the range that young people can hear. Older people lose the ability to hear high frequencies progressively throughout life — although young people are losing their high frequencies very rapidly these days. The average 25 year old today has high-frequency discrimination no better than his/her 50 year old parent.

• Telephone audio quality is usually between about 100Hz and 3600Hz. However telephone bandwidth is set at 4000Hz to allow for some guard-band.

audio cassette tape

There are four basic formulations for audio cassette tapes, and the hardware needs to be adapted to get the best results from each.

— Type I uses ferric oxide particles and is the cheapest. These cassettes are used for voice recording.

— Type II uses chrome dioxide or some similar material with higher coercivity (see). These give better music reproduction, especially at the higher frequencies.

— Type III is no longer used.

— Type IV is metal particle tape which provides the best performance, but is very expensive. The hardware needs a higher bias setting to run these tapes.

• For high-band video recordings, metal evaporated tape is also used. This is probably classed as Type V.

audio conferencing

Usually multiple telephone link-ups used for business discussions. Most telephone companies provide this service, and many carriers now allow self-dialing of conference calls.

In the USA it is now common to have three-way calls which are established by using 'hold' features while dialing the next number. This is part of the CLASS services.

• Large-scale audio conferencing, between groups of people in a number of different rooms around the country, usually needs to be set up by an operator as a special service. To avoid echo and feedback problems, each group will usually be seated around a table with a special multi-microphone unit which automatically attenuates all microphones not directly in use. Conferences of this size normally need a moderator. See both teleconferencing and electronic whiteboards.

audio-visual

A/V. Any presentation system which combines both vision and sound — it is a term which predated 'multimedia'. Video, TV, films, overhead projectors, slide-shows with music, are all audio-visuals — and the term suggests that the A/V is used to support a live presenter, rather than replace him/her.

When computers merged with these technologies and began to take over the storage and display functions, the word 'multimedia' appeared.

audiotex

This is the voice and DTMF version of videotex, which gives access to database information. You are asked to reply to specific questions, or press the tone numbers to narrow down the query. It is most often used in booking and ordering systems.

One more recent application is for a voice-mail service that allows normal text-mail to be delivered over the telephone by voice synthesis.

audit trail

Tracking the progressive changes made to data (transactions), from the current state of the records, back to the first input — step-by-step.

In many businesses (especially in finance, accounting and banking) you must be able to verify all past transactional activities, and be sure (at a legal level) that records haven't later been modified. So a 'trail' must be left through any changes. Records must not be 'erased' in the sense that they are overwritten, and not recoverable.

This means that floppy disks are not suitable for historical records; magnetic disks destroy their audit trail by constantly over-writing and reusing disk sectors. However WORM (Write Once, Read Many-times) optical disks retain their trail, and for this reason alone WORM technology could become mandatory in some areas of computing. Note that new CD-R (erasable-rewritable) systems can become WORM just by using the appropriate blank disk.

augend

A maths term. In an addition, it is the number added to the addend (the operand) to form the sum.

augmentation

Assisting, or adding something to a basic standard.

• The term was widely used in US analog HDTV discussions where it was once felt desirable to transmit a basic 4:3 NTSC signal, and then add augmentation information by simultaneously transmitting another channel to increase the resolution for those users who had the special TV sets. These were the augmentation channels.

In television there are at least five ways to augment an existing TV signal to improve its image quality:

— Frequency: through a new subcarrier.

— Time: interleaved transmission of HD picture.

— Space: compressed image detail into those areas hidden by the bezel at either side of the picture.

— Modulation: phase modulation of the main carrier.

— SNR: reduce the signal-to-noise ratio and overlay additional information.

• Motion compensation is a different form of augmentation used in highly compressed digital television systems. See motion compensation and spectrum folding also.

AUI

• **Agent-User Interface.** See agent (personal)

• **Attachment Unit Interface**. The interface (cable) between the transceiver (the MAU) on the network and the networked device (the DTE). Most modern network devices (hubs, routers, bridges, etc.) now have AUI ports for connection to Ethernet cables. This port will make direct contact with the Media Access Unit (MAU) of the transceiver, and is therefore specific to the network cable type (unshielded twisted-pair, coaxial, etc.).

authentication

The process of checking that the user is allowed to use the system, or that the sender of a message is who they say they are.

• Passwords and PIN numbers are the principal means of authentication. Often these will be supplemented by random questioning of the user, and checking answers against a previously recorded list ('Mother's maiden name?')

• Call-back is used in some systems to check that the access is being made from a phone line which is listed as being acceptable. (Note this can be defeated in some telephone networks.)

• In cellular phones, the carriers are now experimenting with reading the 'electronic' fingerprints of the devices (the claim is that each unit has regular deviations from normal operations, and that this can be detected, recorded and checked against known lists). I'll believe this when I see it!

• At higher levels of security, private encryption codes can exchange authentication messages. See DES.

• For general public use, public key encryption techniques such as RSA and PGP are increasingly being used for digital signatures (see).

• Biometrics, including finger-print and hand-print, retinal pattern, etc. provide high degrees of security, but are proving difficult to implement.

• In EFTPOS and X.400 messaging, authentication also extends to checking the network sub-systems (domains) to make sure that an unauthorized computer system hasn't been nefariously added to the network. See smartcards also.

authoring

Forms of high-level programming used to control interactive audio-visual equipment when creating interactive games, and educational (CAI) programs. Authoring is the act of tying together all the different elements of multimedia with the interactive input from the user.

• In the older A/V world this meant correlating the actions of the computer program, video-feeds (from disk or tape), slide-changers, the audio-feeds (perhaps a number of tracks), generated screen text, defining decision points, creating branching instructions, etc. especially in interactive videodisk systems — usually those with touch-screen inputs. It was a particular form of programming, and it had its own particular problems and special languages (or procedures).

• With computer-based multimedia, the job has become more simple, although the tasks being undertaken are now much more complex.

authoring system

A collection of programs, utilities and equipment needed to write interactive multimedia/audio-visual programs.

authority zone

This is the subsection of the domain over which a name-server holds authority. It is associated with DNS in the Internet. See.

authorization code

See authentication.

auto-abstracting

A method of creating abstracts of database records or files automatically. The computer extracts full lines of text directly from the file on the basis of keywords — so it allows you to read these in context (to a limited degree). You will read something like this:

```
... the opinion of the government's
adviser on asbestos says it...
... for Australian asbestos workers
is in doubt. The Minister said...
```

Usually you will be able to judge from these fragmented lines of text whether the article is worth reading. See auto-indexing.

auto answer

A modem feature which detects the ring of an incoming call and answers without assistance from the software. Modern modems are inevitably auto-answer these days, and a parameter in the S-register can be set to select on which 'ring' they will answer.

auto-dimensioning

In CAD — where dimensions on a drawing can be calculated and added automatically. The program calculates the actual size by using the scale of the drawn objects. It will then usually add arrows, and inserts the dimension numbers.

Often the graphics side of auto-dimensioning is selectable; the program may simply display the dimensions of objects outside the drawing window.

auto-indexing

The automatic selection and indexing of keywords, usually by reference to a previous list of important words used in a particular field of science. This process involves full-text indexing and the creation of inverted files (see). When you search these indexes you are given the number of 'strikes', and you will then narrow the search down by combining the original search term with another, using a Boolean operator — such as:

```
FIND Asbestos.
F1 = 287 (strikes)
FIND F1 AND HEALTH
F2 = 38 F1 AND HEALTH
```

The distinction is with normal human abstracting and keyword selection. Auto-indexing produces many false strikes when compared with human indexing (but see Cyc), however it is cheap.

auto linefeed

With this feature switched on, each carriage return has a line-feed automatically added. Some machines output a carriage return alone — while others always use a carriage return and line-feed together. This is usually written as CR+LF

auto recalc

Spreadsheets often need to automatically recalculate long chains of formula whenever a value is changed. A simple change to one cell may send a ripple of change through hundreds of cells. So auto recalculation is essential before any major decision is made on the results.

However, if large spreadsheets are left running with auto recalc on, while minor adjustments are being made, the time wasted with recalculations constantly made, can be infuriating.

You need to strike a balance between constant recalculations, and the confidence that they have been made.

auto repeat

In most computers, if a key is held down for more than a fraction of a second, the key repeats. Usually you can set the delay time before a repeat occurs, and you can also set the repeat rate.

When you strike a key on a keyboard, a code (not necessarily an ASCII code) is sent to the keyboard controller (usually a CPU in its own right) — which then looks up a table, and outputs the correct ASCII character.

When you lift your finger off that key, another code (usually the same, but with the 8th bit set to 1) is sent. If the second 'release' code is not received within a preset time, the controller repeats its ASCII output.

auto start

See AUTOEXEC.BAT.

AutoCAD

A professional CAD software which has a majority of the PC market, and is the ***de facto*** standard in this area. Autodesk also produces a range of rendering and sketching products which work with AutoCAD.

AUTOEXEC.BAT

The batch file in MS DOS which is automatically executed on startup. It can be used to initialize settings, load RAM-resident programs, and boot a specific application. The OS/2 equivalent is STARTUP.CMD.

autoexposure

The early autoexposure systems were basic reflected light meters which read the whole area of the image, averaged it, and opened the camera's aperture until the average intensity equalled the predefined setting (established by the ISO or ASA of the film, together with shutter speed). These worked well for average scenes, but not for anything unusual: they always average the scene to mid-gray tones.

Later systems place more emphasis on high-lights and dark areas (paradoxically, by discounting them in the measurement) and this generally produces better results.

autoflow

The technique available in some page-layout programs where the program will continue inserting text blocks into columns, from one page to the next, without operator intervention. Since newsletters and newspapers define areas available for each story, the more sophisticated page-layout programs will allow you to specify which columns of text should flow to which other column spaces.

autofocus

Modern cameras have a range of different systems for auto-focus. Many earlier devices used ultrasonic sound echoes by detecting delays; now almost all have a through-the-lens system. Ultrasonic focus systems caused millions of tourists on trains and buses to shoot out-of-focus shots when the camera calculated the scene distance as three-inches to the window glass.

Some later systems also used infrared. But today's systems use complex optical principles where a microprocessor will examine contrast in the image by detecting edges of objects. Some also have eye-movement controls, so that when the camera operator looks at the subject, the camera reads the direction of gaze through the viewfinder, and checks focus only in that area of the image.

automata

Anything to do with robotics or automatic machines which mimic human behavior or appearance. Its an archaic sci-fi term.

automatic gain control

See AGC.

automation

The discipline involved in the design and development of machines, processes and systems which aim to reduce the level of human involvement in their day-to-day control. Automation is not just robotics, it includes a wide range of control applications, from self-defrosting refrigerators to basic numerical control lathes.

autonomous

An action which hasn't been directed by an external management system.

autonomous system

See AS.

auto-ranging

• In modems, this is the ability of the modem to scan across two or three possible communications standards (or groups of standards), and select the one that matches the incoming call. In the AT-command set, auto-ranging parameters are set by the B-mode command.

• In power supplies, auto-ranging is the ability to select the correct setting for the voltage of the local mains source without needing to depend on correct selection by the user. In 115 volt countries they draw 115 volts, and in 230 volt countries they draw 230 volts without exploding.

autotrace

See trace.

A/UX

Apple's Macintosh version of Unix. A/UX is based on System V with Berkeley extensions.

AUX

Auxiliary: With audio or video equipment this is a cable connector or mini-phone jack. AUX sometimes appears above both input and output connection points. These are known also as line-level connectors and they must be distinguished from MIC (microphone) inputs which supply much lower signals levels (millivolt), and have different impedances (usually 600 ohms). AUX will input/output a signal at a level and impedance suitable for connection to another amplifier, turntable, or radio tuner at its AUX point.

• Don't connect an Aux Out to a Microphone Input unless you want to see smoke rise.

auxiliary

• **Storage:** See secondary storage.

• **Data-field:** The last 288 bytes of a sector in a CD-ROM or CD-i disk. It is used for secondary Reed-Solomon error detection and correction in mode 1, and for user-data in mode 2. See CD forms and modes.

AV

Audio-Visual. At one level these are slide-shows with sound, or over-head projector transparencies. At another, the term is synonymous with multimedia and virtual reality. See audio visual.

• **Drives:** AV drives are high capacity, quick-access, disk drives with gigaByte capacity. They must have uninterrupted data flows or very large buffers. They will usually carry AVI or MPEG files.

• **Amplifiers:** These are audio processing amplifiers which are generally capable of providing surround sound using Dolby Stereo or ProLogic standards. See.

availability

• In terms of maintenance and reliability, availability is measured by mean-time-between-failures (MTBF).

• In multi-user systems, availability can mean the processing time each user has for his/her applications.

avalanche

The process where an electron or other charged particle collides with other molecules and releases new electrons, which, in turn, have more collisions. This is like a fission reaction — a cascading process which builds catastrophically.

avalanche photo-diodes

APDs are a type of optical fiber communications receiver used to convert optical signals to electrical. When a photon of light hits the APD, every primary electron which results from a photon interaction, produces several secondary electrons, and these, in turn, produce more — hence 'avalanche'. These receivers excel when detecting data at high bit rates in the 1.5μm part of the infra-red spectrum. The switching time is often in the order of a trillionth of a second. See PIN also.

AVI

Audio-Video Interleave. The video and sound file format used by Video for Windows for multimedia applications; it is also the animation format for Video for Windows.

It interleaves digitized or computer generated video frames and synchronized audio. These files need data rates of about 1Mb/s, but are much lower in quality than MPEG; and are generally independent of screen resolution and palette.

AVI is Microsoft's attempt to match the Macintosh Quicktime, but Apple has now released a Quicktime for Windows which seriously challenges AVI.

Avid

A company and trade name for very sophisticated electronic video editing equipment. The Avid equipment is now so widely used in television production that the name is beginning to be used generically. Silicon Graphics Inc. now owns Avid.

AVL

Automatic Vehicle Location system. This can mean a system which reports its position only to the driver, or a system that reports both to driver and to the home base. It is generally applied only to the latter.

- In the past, these systems have used transponders on board road trucks, and perhaps a dozen or so fixed transmitter sites around the city. This provided triangulation to locate local delivery vehicles in most cities. Nowadays they use a variety of distance measuring techniques, and operate by paging in the lower UHF bands.
- New GPS-GEO satellite systems have a high degree of positional accuracy (100 meters). However, they need to be coupled with a return channel via the communications satellite to report the vehicle's location to the base. Omnitracs is probably the best known of these.
- Position location in LEO systems can be either of two types:
  - The Doppler effect can be used to calculate the position of the vehicle, with low accuracy (say, half a kilometer). This type of location is suitable for ships or containers.
  - GPS can be used to calculate the position, and it will automatically report back over the messaging channel. Orbcomm will provide both of these services, as will later LEOs. See Geostar and L-band also.

AWG

American Wire Gauge. A copper wire-size specification. The higher the number, the smaller the diameter. AWG26 copper is a single strand of 0.4mm diameter and can carry 1 watt. AWG22 is 0.64mm in diameter and can carry 5 watts. See SWG.

AXE

A fully computerized (SPC) telephone exchange switch designed by Ericsson but made under license by a number of companies. Both analog and digital versions exist, although AXE has almost become a generic term for digital exchange switch in some parts of the world. See ARE and SPC.

axial ratio

This is a measure of circularity of polarization in satellite communications. It is related to cross polarization discrimination.

AZERTY

A type of keyboard layout used in Europe, particularly France. As with the conventional QWERTY layout, the name refers to the keys on the upper left row. These are designed to handle the diacritical characters.

azimuth

In general, this is any angle measured from a reference point in the horizontal plane.

- With satellites, this is the angular direction from North or South that the dish needs to point to find the satellite. It is the angle between the antenna beam and the meridian plane (but only measured along the horizontal plane). See declination and elevation also.
- In tape and disk recording, this is the alignment of the recording and playback head (or heads). Ideally both will be aligned at right-angles to the flow of the tape or the movement of the disk. But more importantly, (if there are different heads for record and replay) both must be aligned to match each other.
- In video recording, with helical scan systems having two rotating video heads, it is common to align the heads deliberately at offset angles. This tends to reduce the effect of one head on the track of another (if they overlap). Provided the same head angle is used for playback, it doesn't reduce the quality of the recording. In VHS, the audio track head azimuths are set alternately at $+30^{\circ}$ and -30°, and the video heads at $+6^{\circ}$ and -6°.
- Azimuth loss, is the loss of signal quality when the replay heads are out of alignment with the record heads. A slight angular difference will make a big difference to the signal.

B

B# (paper sizes)

Along with the A-series, the B-series of paper sizes is now almost universal. As with the A-series, each smaller sheet size is made by folding the larger page across the longer sides. The aspect ratio of all these sizes is the same.

B0	1414 x 1000 mm	55.67 x 39.37 in.
B1	1000 x 707 mm	39.37 x 27.83 in.
B2	707 x 500 mm	27.83 x 19.68 in.
B3	500 x 353 mm	19.68 x 13.90 in.
B4	353 x 250 mm	13.90 x 9.84 in.
B5	250 x 176 mm	9.84 x 6.93 in.
B6	176 x 125 mm	6.93 x 4.92 in.

See A# (paper sizes), book sizes and paper sizes.

B (or b)

While the computer industry and the telecommunications industry remained separate entities there wasn't much of a problem in the use of bit and byte. The communications people tended to refer to a byte as an 'octet' which made the distinction even easier.

However with the current convergence of these industry sectors, bits-per-second and Bytes-per-second confusion is becoming a major problem. So I'm establishing, what I think are some common-sense rules here:

- **B.** The capital B should be used to indicate 'byte' which is a single text character or 8-bits. Historically Bytes are used when computers are being discussed, and bits are usually assumed in communications — but let's now make this clear by capitalization.
- **b.** The lower-case 'b' is used to indicate 'bit'. However this is only an emerging publishing standard and not too many editors worry about making the distinction, which is why things are often totally confusing.
- In this reference, I've taken it upon myself to always capitalize the B in Bytes — just to constantly make the point. So kB should mean 1024 Bytes, and kb should mean 1024 bits — and I intend to even use the capital in the middle of such words as megaBytes, just to make this obvious. Now that we have convergence between computer and communications, this bits vs. Byte confusion is a constant source of irritation.
- In much of the literature, you'll also often find references to memory as being 1M or 256k, and this always assumes Bytes. However the chips used to create this memory are usually 'bit-wide' and so each chip may only hold 1 Megabit or 256 kilobits — but then you need eight of them in parallel to have any useful working memory. See by-#.
- Incidentally, in communications articles you'll often find 'per second' is assumed also. You might read that a modem handles data at 1.2kb, when they mean 1.2kilobit/second or 1.2kb/s.

B-channel

In ISDN telephony, the B-channels are the 64 kb/s switched circuits 'Bearer' full-duplex channels which are designed to carry voice and data. In the USA, some ISDN services are being marketed with only 56kb/s bearer channels, but this is not part of the international standard.

The distinction is with the 'D-channel' which is the packet-switched signaling and User Data channel.

Bearer channels are independently switched through the telephone network, so two circuits, both being switched to the same end customer, may travel over quite different paths. This makes it difficult to aggregate the bearer channels if, say, a wider bandwidth of 128kb/s is required for videoconferencing. See PRA, BRA, D-channel, H-channel, Q931W and aggregation .

B-format video

A 1-inch helical-scan videotape television production standard from the mid-1970's, which proved to be less successful than the C-format. The 'B' came from Bosch which devised the standard. It used a drum diameter of 50.3mm, with an 'alpha' tape wrap (full loop). There were two video and two erase heads on the drum, and the machine wrote six tracks for one field. There were three linear audio tracks and one control track. It required an analog time-base corrector to reproduce broadcast quality images.

B-ICI

See Broadband InterCarrier Interface.

B-ISDN

See BISDN page 79.

B-party

The person who is called when setting up a phone conversation, as distinct from the A-party who is the call originator.

- B-party Clear telephone systems are those in which a call is only cleared when the B-party puts the phone on-hook. See call clearing.

B-spline

An outline technique used with modern graphics and page layout programs, which defines curves by using multi-variable (cubed) equations. The technique is an alternative to using Bezier curves and outlines.

The B-spline approach needs more points to define a simple curve than Bezier, but the rasterization conversion phase is quicker. However when the shapes become exotic, so do the polynomials that are being used to define the shapes.

TrueType uses quadratic B-splines, while Adobe's Postscript uses Bezier.

B-tree indexes

The 'balanced tree' technique of organizing indexes and directories which allows only two pointers at any node (except the leaf). When index items are inserted or deleted, the data structure realigns its hierarchy to keep each of the two search paths uniform in length.

The B-tree directory continually maintains a balance between each directory node at each level, in order to keep the search process optimal — each node will maintain a similar number of keys. A version known as B+-tree also allows you to link to the data sequentially.

B3

One of the highest security level classifications for message security by the US Department of Defense.

B8ZS

Bipolar with Eight Zeros Substitution line-coding. This is a form of AMI (a bi-polar line-code) used in the American T1 transmission systems (and also with PRA ISDN) to improve the synchronization between sender and receiver.

The principle is the same as HDB# line-coding (see) but with an 8-bit string which is not confined to byte boundaries. After substitution, there can be no more than 7 zeros in a row. Code violations are also used (see). See A&B bits, bit robbing, violations, scrambling and HDB3 .

babble

Heavy cross-talk on a communications network.

BABT

British Approvals Board for Telecommunications.

back channel

See backward channel.

back door route

A route for border gateways (specified by IGP) to reach a non-local network. See border gateway and IGP.

back-end

• A slang term meaning a host computer (particularly one that handles a database).

• In client-server, it means the node or server-software which provides services to the front end (client).

• It can also mean, generally, any applications program on a host computer or a file server. Back-end computers/programs often communicate with the user's PC by the SQL query language. The distinction is then with the front-end of the system which includes the data-entry devices and the screens, keyboard and printers.

back focus

A camera lens will have the normal focus controls which allow it to vary from close to distant objects. However, it will often have a 'back-focus' adjustment, which is to position the active image element (the CCD or video tube) at the right distance behind the lens.

— When the lens is set to infinity, the camera should normally be focused on far distant objects; if it is not then the back-focus needs adjustment.

— If a camera is carefully focused on a nearby object with the camera zoomed in, but the focus changes by itself as the camera zooms out, then the back focus needs adjustment. Set the lens on infinity first.

back lobe

In a directional antenna such as a normal TV Yagi, there is always some reception sensitivity to signals arriving from the side (side lobes) and the back (back lobes).

back porch

In a television signal, this is the part between the trailing edge of the horizontal scan line's sync pulse, and that of the corresponding horizontal blanking pulse.

• The front porch is that part of the video waveform between the beginning of the line-blanking pulse, and the leading edge of the corresponding line sync pulse.

back pressure

• In large internetworks this is the propagation of information about network congestion, from further downstream.

• In smaller LANs, it is a technique of flow-control used on Ethernet with CSMA/CD, effected by simulating collisions on a network segment whenever the buffers are nearly full.

backbone

Ethernet and some of the higher-speed network protocols are suitable for use as a 'central spinal cord' through a factory, office or college campus. Backbones handle traffic which is sourced from, and destined for other networks.

In the past, broadband technology on coaxial cable was enough to handle the load of most backbones, but lately, optical fiber has become popular for the physical backbone transmission medium (often using FDDI or Fast Ethernet).

In a campus network, the backbone provides the long-distance, high-speed communications to which the work-group LANs are linked through bridges or routers. The links to departmental networks are sometimes known as ribs. See distributed backbones and collapsed backbones.

backface

That part of a graphic image which is cut off by foreground surfaces from view. This is also known as the backplane.

backfile

In databases, backfiles are older versions of the database usually kept in off-line storage. See grandfather.

background

• **Background Processing** (task): Any operating system that permits pseudo-multitasking, always has one foreground application and one or more background applications which have a low priority. Background tasks are stalled while in the background, although all their parameters are preserved. When the task is brought to the foreground (by clicking on the appropriate window) the application will continue from where it left off.

In true multitasking operating systems, applications in the background may continue to function while work continues in the foreground. Usually the foreground will have a higher priority, however. Background database sorts, for instance, may continue while a word processor is being used to write a letter. See pre-emptive multitasking.

• **Background Color:** The base color of a screen as distinct from the text in foreground. The screen's background color is sometimes called 'paper color'.

• **Background ROM:** The contents of Background ROM are copied into RAM, and used from RAM.

• **Background Image:** Part of the display which doesn't change. Usually a changeable form or card,(to which data can be added) will overlay the background. HyperCard has both background fields and buttons (common to all cards in the stack) and card fields and buttons (specific to that card). Both can coexist, so background here almost means 'global'.

backhaul

• In cellular phone systems, the backhaul is the link to a remote repeater site. A remote repeater is like a dumb terminal; it doesn't have the normal intelligence of a base station so the signal needs to be bought back (via lines or microwave) to interface with the normal cellular phone network.

• Satellite operators speak of the backhaul in two quite confusing ways:

— It is the terrestrial link connecting an earth station to its local switching center, or

— It is the earth-station to satellite links (both uplink and downlink) as distinct from the satellite to mobile links. The term then also include the terrestrial links between the satellite's earth-station and its local switching center or network operator.

backing store

That part of the hard-disk which contains temporary programs and data not needed immediately by RAM — it is virtual memory. This an old term which was applied to the use of mass storage facilities to temporarily 'back-up' (or more correctly 'supplement') the more volatile and scarce RAM memory in large time-sharing systems.

However now that the price of RAM is decreasing, the idea of using magnetic media to store data that will be required later in the processing cycle, has become far less important. In virtual memory operations, disk storage is treated as an extension of the normal machine RAM, and so this is a correctly termed a backing store operation.

backlit LCD

A type of flat-panel screen which has better contrast and low-light capabilities because the screen is illuminated from behind. The cost, however, is in increased battery consumption. These screens are usually used on portables. See Supertwist LCD.

backoff

In any network scheme relying on contention, the stations which are attempting to gain access will detect the collision, and then backoff for a random period. This randomizes the delay before they attempt retransmission.

backplane

The back of the computer box, where cables emerge and connectors plug in, etc. By extension, the term has come to mean the more mundane levels of the connection processes, including the slots and buses in a PC. In an LCD screen the backplane is just the common electrical connection between all elements.

Don't confuse backplane with 'back panel' which just means the plate at the back of the computer, to which the switches and plugs are attached. The term 'backplane' has slightly different meanings to different technical groups.

• Historically, this is the part of a mainframe involved with computer-to-computer or computer-to-peripheral functions, as distinct from the front-plane; this is the man-machine interface.

Typically, large computers consisted of many backplane slots into which communications boards were inserted. The backplane supplies the bus or communications channel through which the boards communicate with the CPU. There are 'active backplanes' which involve some processing within the bus network itself, and 'passive backplanes' which just supply power and connections.

• The term is also applied to a new generation of modular electronic devices (like TV sets) into which you can plug interface and translation cards to convert the set, say, from NTSC to PAL.

• In CAD systems which create wire-frame images, the backplanes are those which will be covered by the frontplane images if the image is converted to solid modeling. This is also known as the back face.

• In LAN hubbing, the experts talk of backplanes and sub-backplanes. The front plane is that part of the network which is being serviced by the hub, and backplanes are those systems which link the hub to external LANs or WANs.

backslash

Generally, in computing, the \ sign. This sign is sometimes used to denote an integer division operation in computer arithmetic while the normal forward slash is used for decimal division.

There's a lot of confusion about which symbol is correctly called a backslash. Many references refer to the 'normal' slash (used between words like Jan/Feb and in fractions like 1/2) as the backslash, while others call this the 'slash' and the reverse-angled version the 'backslash'. The second is obviously correct.

• The backslash is used to separate directory\subdirectory\-filenames in MS-DOS.

backspace

Don't confuse backspace with delete. Backspace just moves the cursor/insertion point back a character position; it may or may not, then overwrite the old character. See BS.

• Tape storage also can backspace.

• Daisywheel printers only provided underlined characters and bold characters by backspacing and then overprinting. The bold was created by overprinting the same character, and the underline by overprinting an underline character.

backtracking

• In programming, this is a technique that allows the program to recover after reaching a dead-end in a search path. Special algorithms will cause a retreat from dead-ends and then the search will recommence along an entirely new path.

• In hypertext, backtracking is a navigation technique. Since hypertext provides no defined pathways through the maze of information modules, it is necessary to provide the user with a way of retracing his/her path.

• Backtracking is built into languages like Prolog. See backward chaining.

backup

A copy of data (or of an application) made in case the system crashes or the disks are damaged.

There are three ways to backup:

— A full copy of all the files you select.
— An incremental backup copies only those which have been changed since the last full or incremental backup.
— A differential backup, which copies only those which have changed since the last full (not incremental) backup.

You should always use a back-up methodology which maintains three versions of the files. This is the grandfather, father, son technique. See all.

backup tape systems

There are numerous tape systems intended primarily to provide backup for computer systems. The old 9-track techniques are still widely used, and there are new DC2000 and DC600 standards. Also the video Super-8 standard has been modified for computer use, as has the audio DAT. See 8mm, DDS, Data-DAT, DC2000, DC600 and streaming tape drives.

Backus-Naur

Also called Backus Normal Form (BNF). A type of standardized shorthand which can be used to write computer statements. This was the first 'meta-language' for computing, and one which can be used to precisely describe the rules of grammar of another language. It was developed before Cobol days by J. Backus and P. Naur.

A sentence, which in English, would read as 'A sentence can be a noun followed by a verb', can be expressed in BNF as:

```
S —> NV.
```

BNF is used to define and summarize the syntax of written phrases. For instance, a vertical slash sign indicates that the following is optional; square brackets enclose something that is optional; the asterisk indicates repetition.

• Many modern languages now contain syntax rules that can't be expressed in BNF. They are sometimes used with parsing.

backward chaining

In expert systems it is possible to work the logical process both ways — forward chaining takes you towards a conclusion by way of the knowledge base. However to validate this conclusion, you may want to see how it was derived, so backward chaining is a way of checking.

Whenever a hypothesis is formulated, the rules appropriate to, and corresponding with this hypothesis should then be selected from the knowledge base by backward chaining. If enough of these rules can be matched to assertions in the database, then the original hypothesis is considered to be valid. See forward chaining.

• Don't confuse this with backward learning.

backward channel

A supervisory or 'return' channel not used as the main communications channel, but provided mainly for error checking and control (supervisory). The term is sometimes used in a vague way. For instance, the V.23 'hybrid' videotex modem standard is often said to have a 1.2kb/s forward channel and a 75b/s backward channel. But the 75b/s channel is primarily used to carry your keyboard requests to the host computer, so it's not supervisory in the accepted sense of the word — this is really an asymmetrical full-duplex channel.

• Some high-speed data systems use half-duplex transmission in a full-duplex way. While two computers are linked to exchange stored e-mail, for instance, they may each transmit to each other at the same time. In this case each forward channel will be the backward channel of the other.

backward compatibility

Ensuring that changes to computer hardware and software will not leave previous owners/users out on a limb. New programs for new devices should still run on old machines of the same make, if claims of backward compatibility are made.

However, don't accept most compatibility claims at face value — new versions very often won't work in complete harmony with other applications and/or other resources. New versions are very rarely 100% compatible with old versions — and they are rarely compatible with legacy resources that were, themselves, compatible with the previous version.

backward error correction

These are automatic retransmission schemes (such as ARQ) for error detection and subsequent correction. The most popular protocol here is LAP-B. The distinction is with forward error correction. See FEC.

backward learning

In network management, information about remote nodes can be surmised by assuming that packets will travel equally well both ways over the network. So if a node (A) receives a packet from (B) via the intermediate node (C), it will assume that this is the best path. Later, when it needs to, it will transmit to B, via C. It assumes that the network is symmetrical — which it may not always be.

backward supervision/learning

• Some packet-switching networks have intelligent nodes which are capable of deducing the congestion on distant parts of the network by noting the source and number of packets passing through. This information is then used by the network management to divert packets around the congested nodes.

• In general communications, both of these terms are also applied to supervisory messages that are passed from the slave, back to the master station. 'Supervision' in this context often means nothing more than checking that a circuit is functioning — but it can also mean 'control'.

bad sector

An unreadable patch on a disk — a defect — which makes data storage within that sector inadvisable. During the formatting process, many computers will mark bad sectors of a disk in the sector allocation tables, as 'used', so that they will be ignored for any future storage. Unfortunately, some sectors only become recognized as bad after the disk has been in use for some time, or become corrupted by physical damage (coffee is the main culprit). Often the data can still be recovered by special disk-aid programs.

BAK file

A backup file which is automatically created in some systems (CP/M and early MS DOS) when a file is updated using the older text editors. The old copy of the file remains on the disk and is renamed with a .BAK extension. This happens automatically whenever a new file is created.

In all disk recording systems, the new updated file version is always written to the disk before the older version is 'modified' (destroyed). However, with most computer systems only the pointers to the older version in the directory structure are destroyed. The file ceases to exist only when it is overwritten by the next file-save operation.

In early MS-DOS systems, these original files were temporarily maintained by the .BAK name-change — so there was always a father-son relationship between the .TXT file and the .BAK file of the same name.

BAL

• **Branch and Link;** a subroutine instruction.

• **Basic Assembly Language** for IBM mainframes.

balanced

• In audio distribution and high-rate data communications systems where there are two symmetrically-active wires, usually surrounded by an earthed shield. Both wires will be loaded equally (but with opposite polarity) and hence have equal impedance to the earth.

At any point of time, one wire is as positive as the other is negative. This reduces the effect of interference, and allows long circuits without greatly increasing noise on the system. Studio and professional sound cabling is always balanced.

Typically a balance system can be identified because it employs two wires (or four for stereo) within a shielding, while an unbalanced system will have only one wire (or two for stereo) and use the shielding for the return path. See differential signaling.

• Communications circuits are balanced when the terminating network's impedance matches the impedance of the incoming line. The line is terminated with a matching (balanced) load.

• A balanced amplifier has the output voltage level in the quiescent state of near zero. There is then less likelihood these circuits will suffer problems due to the voltage difference with the earth potential.

balloon

An on-screen 'dialog box' which is used to provide help associated with various icons on a Macintosh screen.

balun

A contraction of BALanced and UNbalanced, but the significance is with impedance-matching.

• A passive device that connects balance lines to imbalanced lines (two-wire + shielding to single-wire + shielding). For instance, baluns allow equipment intended for twisted pair media to run on coaxial — or vice versa. The balun is also acting as an impedance-matching device here.

• In television reception a balun is used to join the balanced output of a Yagi antenna to unbalanced coaxial cable. Some baluns are also used to join UHF and VHF outputs together.

band

• **Communications:** A range of frequencies used in communications for one purpose. A band can be referred to by the range (width) of the frequencies in terms of identifying both its upper and lower limits — or it may be identified only by its central frequency.

Often you will find that the bandwidth is assumed, according to a known standard — and sometimes it is assumed that an equal amount is left for both the go and the return channel. See duplex offset.

— Telephony used a voiceband between about 300Hz and 3600Hz.

— GSM mobile phones operate in the 900MHz band.

Baseband systems are those which have unmodulated DC (digital) or audio frequency AC (analog) signals.

Broadband systems are those which require signals to be modulated using higher frequencies (radio frequencies).

The width of the band (bandwidth) is taken from the lowest, to the highest frequency (including any guard band separation between bands). See narrowband, voiceband, wideband, and broadband.

• **Disk:** On magnetic disks, a band is a (usually contiguous) group of tracks that are logically treated as one unit. On some disk or drum systems, say, eight tracks can be written in parallel (simultaneously with eight heads) and this produces a band.

• **Printers:** Band printers were impact printers that used a horizontal belt with the type along it.They were a form of daisy-wheel without the daisy-wheel.

band-stepper

See stepper.

banding

• A defect in video output from a Quad videotape recorder which causes a different band of coloration across the screen, about one quarter of the screen height.

• During the printing process, the computer (or the printer) may not have enough memory to hold simultaneously the bit-mapped information needed to output a large graphics image and the applications and text. The banding process is a way of breaking the bit-map down into a series of horizontal bands which will then be printed one at a time. Device drivers may or may not have banding capabilities.

• A visible stair-step-like series of shades in a larger area of color printing that should appear as a smooth gradient. This is a quantization problem in desktop publishing,

bandpass

• Amplifiers and circuits which have a defined (but wide) frequency band with a flat response. All frequencies within that band are amplified or treated equally.

The new Cable TV amplifiers, for instance, have a bandwidth from 40MHz to 1GHz (150 analog channels), and provide an 18dB gain, with a worse-case frequency response across that bandwidth of only +/–0.3dB.

• In one sense, all networks have bandwidth limitation, so all networks act as bandpass filters — and the frequencies they allow through (reasonably) unattenuated are 'passband' frequencies.

• Television tuners are a form of bandpass filter in that they reject most of the channel frequencies, but allow a passband (NTSC 6MHz and PAL 7–8MHz) through.

• Graphic equalizers used in hi-fi equipment are considered to be a parallel arrangement of bandpass filters.

• In telecommunications, bandpass filters are often used to remove or reduce unwanted frequencies. Such a filter will allow only a relatively narrow range of frequencies (say only 4kHz for voice) to pass through by attenuating those higher and lower than the frequencies required. The non-attenuated frequencies are then known as bandpass or passband frequencies. See high-pass filters and low-pass filters.

bandwidth

Specifically: the difference between the lowest and highest frequencies being used (in one contiguous band, with acceptable attenuation) for a specific purpose. The range of frequencies occupied by a signal, or passed by a channel.

General: The information-carrying capacity of a band or service (including digital as well as analog services).

• Sometimes the term refers to the range of frequencies reserved by the network management system for use by a customer, and sometimes it refers to the amount actually consumed by the user. Some modern communications systems are variable in their bandwidth consumption — see VBR and statistical multiplexing.

• The term can apply to individual baseband signals or to the individual (or combined) channels in multiplexed and modulated systems.

• The term is more correctly applied to analog radio or audio signals, but it is equally applied these days to digital data traffic over these carriers — so often it can be expressed in two forms — both Hertz and bits/second (b/s or kb/s or Mb/s). The type of modulation techniques used, defines the ratio between these two; there is no fixed relationship. Ten bits to each Hertz appears to be possible, but currently we can send only about 8-bits for each Hz of spectrum, and 4- and 6-bits/Hz relationships are more common.

• Nominal bandwidth includes the guard bands needed to prevent one user from interfering with its neighbor.

• In many cellular systems they quote bandwidth in terms of the transmit channels only (an equal receive channel is assumed). For full bandwidth consumption, you therefore need to double these figures. But beware; TDD systems use the same frequencies for both transmit and receive. See.

• Shannon's Law sets the theoretical limit to the amount of binary data that can pass through a channel (given a certain noise value) — the channel's theoretical and finite capacity.

• Digital signals will usually be carried any distance on an analog carrier (whose bandwidth is expressed in Hertz), but the digital signal bandwidth is still quoted in terms of bits per second.

• In a TV monitor, it is the bandwidth of the video circuits which drive (energize) the electron beam, which ultimately determine the horizontal resolution of the image. The higher the rate-of-change possible in these circuits, the better the distinction between image elements across the horizontal scan lines, and the finer the final image resolution. This applies to both analog and digitally-originated systems, and to digital video which has been decompressed.

The vertical resolution is set by the number of scan lines and is fixed. But horizontal bandwidth is often expressed in terms of the number of (vertical) test lines or cycles which are resolvable by eye in the test pattern. See cycles.

bandwidth (guide)

Some common bandwidth requirements and limitations are:

- **Telephone systems.**

— Today's telephone circuits provide for a bandwidth of about 3.4kHz (from 300Hz to 3700Hz), but with guard-bands they are treated as being 4kHz wide. Normally just under 4kHz is provided through all modern switching and multiplexing systems in an analog exchange — but you can only depend on the 3.4kHz internationally.

— The general use of the 64kb/s PCM standard for interexchange digital telephony, gives a normal home modem a finite throughput limit in the order of 32kb/s across a network. This is because the digital signal is modulated onto an analog carrier, which is then converted again (by ADC sampling) to digital 64kb/s for on-carriage — then the process is reversed at the other end. When ADPCM is used for interexchange carriage, the theoretical limit probably comes down even more to about 28.8kb/s.

Inflated claims are often made for modem data rates. Usually they attempt to confuse the issue with digital compression and the throughput of the modem-to-modem connection.

— With analog telephony over twisted-pair and through an analog exchange network (assuming average line-noise) the theoretical limit for a modem is about 40kb/s, although this will probably never be achieved. To go faster than this requires expanding the 4kHz bandpass filtering necessary for interexchange multiplexing (see Shannon limit).

— Cheap telephone coaxial cable can handle signals economically at multiplexed frequencies between 60kHz and 60MHz for long distances, but they need repeaters every few kilometers (sometimes every 2km).

— Voice: Digital exchange and multiplexing equipment is designed around 64kb/s PCM voice channels. The analog 4kHz voice output from the phone handset is converted to 64kb/s PCM at either the exchange (POTS) or at the source (ISDN). However, ADPCM at 32kb/s and 16kb/s offers almost as good voice transmission quality (although it is not as good for modems) and this is being widely used where long-distance voice capacity is a problem.

— With the new ADSL multi-carrier techniques, telephone twisted pair seem to be able to carry a 1MHz (or about 6Mb/s) for a few kilometers in the local loop to the exchange. This bandwidth can not pass through the normal telephone switches or conventional interexchange network (limited to 4kHz) so it must make connection to a broadband distribution network in the exchange.

- **Cable networks:**

—The new hybrid fiber/coax systems which are rapidly being deployed now around the world, will provide future Cable TV installations with a bandwidth up to 1GHz. They should have no more than 4 amplifiers along any feeder, and the feeder coax should be 0.75-inch, for lengths of less than 1km.

— Submarine fiber using the new erbium optical amplifiers can now carry 5Gb/s per fiber comfortably.

— Wavelength Division Multiplex (WDM) systems using solitons may easily provide hundreds of Gigabit/s in each fiber within the decade.

— Standard terrestrial fiber cable is rated at 2.4Gb/s today using single 1550nm infrared lasers over distances of 150km or more between repeaters. Connections at these Gigabit/sec rates are now being offered to corporate customers by some carriers.

— The best optical fiber available today in the laboratories has a bandwidth of about 10 terabits per second with a single laser. One experiment in the UK has proved that up to 600 different laser wavelengths can be used down a single fiber (but probably four to six will be enough).

- **Radio and Television:**

— The common bandwidths needed for certain services are:

AM radio 10kHz each (usually in MF band between 530kHz and 1.6MHz);

FM radio 200kHz each (in the VHF TV spectrum between 88 and 108MHz);

Television, 6MHz for NTSC, 7MHz for VHF PAL, and 8MHz for many UHF PAL systems (7MHz for others).

— As a rough guide, in the bandwidth consumed by an analog TV channel we could carry 2000 telephone calls, 800 AM radio stations, and 40 FM radio stations. This would be possible in a cable, however co-channel and adjacent channel interference would reduce this greatly if these stations were broadcasting in the radio spectrum.

— The standard FM audio accompanying a PAL TV picture has a bandwidth of about 200kHz. It is separated from the video by nearly 1MHz, which is why, in a 6MHz-wide NTSC signal, the video may only occupy about 4.5MHz.

If the audio is too close to the video, it begins to effect the picture. For this reason, it is common to leave adjacent TV channels unoccupied in a location, so the sound carrier doesn't also intrude into the next video carrier. These are called taboo channels.

— Video signals for baseband PAL TV need about 5.5MHz, but the overall TV signals (with guard bands and sound) requires 7 or 8MHz (depends on country). For NICAM sound (see) PAL needs 8MHz.

— Compressed digital video of the same quality as PAL can usually be carried at 2Mb/s for pre-compressed material, and about 6Mb/s for real-time action sports, etc. See MPEG.

- **Consumer electronics:**

— For baseband (direct along wires or fibers) high-quality music needs about 20kHz of analog bandwidth.

— With some baseband digital systems such as audio CD. The standards call for the signal to be sampled at 44.1kHz (twice the highest audio frequency required). This produces stereo (twice x) 16-bit samples which need a data stream of 150kBytes/s (1.2Mb/s). But modern compression systems can reduce this greatly with no appreciable quality loss. See DCC and PASC.

- **Satellites:**

— Satellite transponders are traditionally 27, 36, 54, 64 or 72MHz wide and can generally carry two or more standard analog TV signals for distribution to re-transmitter mastheads. For DTH, they will only carry one signal, because of transponder power limitations. A transponder with 50—150 watts of power is considered high-powered — yet with only the radiating power of a light-bulb at 32,000km.

— Normally a 27MHz transponder will carry only one analog television signal, while a 54MHz transponder can carry two. However with digital transmissions (MPEG) you can multiply each of these by 6 for about the same reception quality.

— With digital MPEG compression, a single transponder will normally carry between 4 and 8 channels, but these will vary from 2 to 6Mb/s depending on the precompression time and the amount of screen action involved. Currently the channel bandwidths need to be pre-set (at the beginning of the program) but from next year statistical multiplexing will allow dynamic bandwidth-on-demand (on a per second basis) within the overall MCPC limitations.

bandwidth compression

Any band of frequencies carrying information can be compressed by removing anything 'unnecessary' or 'redundant'. There are four main types of compression in this context:

- Interlacing. In analog television we only transmit half the image in any one field, and the other half in the next field. This reduces the bandwidth while only decreasing the subjective quality marginally. See Interlace and 4:1:1.

- Information which is repeated. A frame of any moving picture, is very like the next frame — so a large part of the information in the second frame is redundant. This is inter-frame coding. Similarly, in an area of blue sky one pixel is almost identical to another, so intra-frame coding can reduce the data-stream. See MPEG and JPEG.

- Information which can't be perceived. If an object moves across the screen at a rapid rate, the eye can't detect fine detail. Similarly, the ear can't hear high-frequency components when they are masked by louder low frequency sounds. So some compression systems will remove high-frequency components at the top end of the bandwidth when they are not needed. See MPEG, DCT and DSS.

- Our everyday 'coding' systems have not been designed for economy. So English text can usually be reduced by at least 50% and completely restored again to the original. See Huffman codes, LZW, MNP, RLE and compression.

Note that some compression systems are lossless (the original can always be restored in full) and some are lossy (where the slight loss is judged acceptable).

bandwidth on demand

Communications channels which provide wider bandwidths when necessary. This is not variable bit-rate (VBR), but rather a much slower version in which channel bandwidths can be augmented or redefined on a minute-by-minute basis.

This include three different concepts here:

- Dial-up ISDN (or other) links which can augment the private tie-lines when required. Used for PBXs and WANs.

- Statistical multiplexing, is used with private line TDM channels which can be reconfigured on-the-fly to provide wider channels for some services. This usually involves some priority scheme.

- Packet or cell-based networks which work on an on-demand basis. ATM, frame-relay and X.25 packet switching networks all provide bandwidth on demand, up to set limits.

bandwidth reservation

In both circuit-switched and cell-based network architectures, bandwidth is reserved and guaranteed during the call establishment phase. These are connection-oriented systems. With data-gram systems and most packet/frame networks, the bandwidth is limited, but not necessarily guaranteed.

- Bandwidth reservation can also mean that, in a circuit-switched network, some bandwidth has been reserved for high-priority calls.

bang

Unix and MS-DOS slang for the exclamation point! Note that this is used to signify factorials.

bang path

The old UUCP addressing specification on the Internet.

bank switching

A system of expanding the 'virtual' workspace to handle much more RAM than a microprocessor can theoretically address. It is a way of getting around the limitations imposed by 16-bit address buses in the early PCs.

The width of the address bus defines how many address locations a microprocessor can directly access. A 16-bit bus can only handle 2^{16} = 64kBytes of RAM space. However it is possible to spoof the CPU in using other banks of memory — using a window, into which external data is paged (shifted in and out).

Access must always be via one or more blocks of memory locations that lie within the normal range of the CPU — within its normal addressable limits. Typically the banks are switched in and out in 16kB or 32kB amounts, and they must always leave the RAM occupied by the main operating system intact.

Typically you would keep one bank for the data and program, and use the other for the O/S, firmware (ROM), and screen RAM.

• The famous Apple IIe was widely known as a 128kB machine, yet its 6502 microprocessor was theoretically only able to address 64kB of memory; it just switched between two banks. Similarly, expanded memory in MS-DOS machines (using more than the 640kB limit) used bank switching, and it was also used on the IBM EGA and other color graphics cards.

• There is a more general use of the term also. It means that one section of a system is not available whenever the other is used, and vice versa. The term 'bank' implies that the alternate sections are of the same type and perform the same functions.

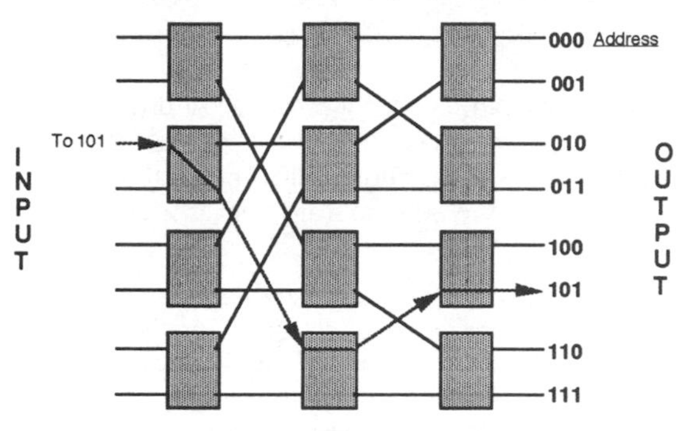

Banyan 3-element switch matrix

Rules: Any of 8 inputs can pass packets to any of 8 outputs.
Each switch element uses only one binary address digit at a time.
If this is a 0, it takes the highest output: if a1, it takes the lowest.
More than one packet can cross the matrix simultaneously.

banyan

A class of ATM-type delta switches which provide fast-packet switching. There are two main sub-types:

• Buffer-banyan: The incoming cells are buffered at the point of contention before being delivered to the banyan switching component.

• Batcher-banyan: These have an array to minimize contention. This array queues and sorts incoming cells before they reach the banyan switch's gate point and are released to cross the switch fabric.

The banyan switch is a mesh of parallel gates, each of which makes only one of two connections based on a single bit in the address — a logical 1 will take the cell along one path, and a logical 0 will take it along another. The first bit of the packet address will be read by the first switch and directed along one path to the next switch element. The preliminary digit will then be discarded, and the next digit read by the next switch which will direct the packet along another path, and so on.

At each succeeding gate the packet is directed along an upper or lower path according to the preliminary binary digit, until it crosses the whole switch fabric. This banyan technique produces a very fast non-blocking way of routing fast-packets of data. Other similar approaches are the Omega, CLOS and Delta switches. See cross-connects also.

BAPCo

Business Applications Performance Council. They have a benchmark for business applications. The companies involved include Microsoft, Intel, Compaq and IBM.

bar chart

The representation of statistical or numerical information in the form of bars on a graphic. Both horizontal and vertical bar charts are common.

barcode

A sequence of black and white lines of different thicknesses which carry a numeric and/or alphabetic code. There are dozens of variations.

Between two and four different bar-code (line) widths are often used, but if a code has too much variation in line width it becomes difficult to print (or photocopy) and retain accuracy. Header and trailer bars are provided (like start- and stop-bits) to allow the scanner to calculate the scanning speed. And since the head-code can be identified, most bar-codes can be read in either direction — provided the software allows for this.

There are many different barcode standards:

– 3 of 9 Code is the most popular for alphanumeric information. It provides upper case, numbers and a few special characters only.
– Code 39 is a variation on 3 of 9 Code.
– Extended 3 of 9 extends the system to lower case.
– Interleaved 2 of 5. The most popular for number-only applications. The first character is represented by bars and the next by interleaving spaces.
– Industrial 2 of 5. A variation on the above which is easier to print.
– Codabar. Used in industrial applications and libraries for numbers and these $-:/.+ symbols.
– EAN-13. A 13-digit numeric code, but very widely used for retail packaging. This is also called UPC and JAN.
– UUC/EAN Code 128. Aka UPS. A 128 character code used for product numbering. This is the most important barcode in the future:
  – It can be applied to all products by all suppliers.
  – It is recognized at all points in the distribution chain.
  – It can be used for in-house systems.
  – It generates, such info as sales and movement data.
  – It is the basis for EDI communications.
– Application Identifiers (AI) of 2—4 digits are supplementary to the above.

• The 'Big Four' codes are UPS, Code 39, Interleaved 2 of 5, and Codabar.

barcode reader

An optical device with a light-source and a photo-sensitive cell which reads the light reflected from a printed barcode and outputs a sequence of binary digits (according to the width of the lines) in the form of electrical pulses. Barcodes can be read with wands, scanners or in-counter readers. These days the barcode can be read in any direction, and at any orientation.

barker channels

Cable TV and satellite channels which carry program guides.

base

- The basic foundation of a numeric system — the radix. Our normal 'decimal' system (correctly called denary) is base-10, while computers use base-2 (binary) and many communications systems use base-3 (ternary). Octal is base-8 and hex is base-16. The term radix is out of favor.
- In an exponential function such as: $x = y^z$ the 'y' is the base.
- In film and magnetic media, the base is the underlying material onto which the binder and the emulsion of magnetic material is then layered. The quality of the base defines the possible distortions (mainly stretching) of the media with temperature variations or age conditions. This has a direct association with the archival life of the material.
- The base is the strata material of a semiconductor. In transistors, the base is one of the three contacts, with the emitter and the collector being the others. In the normal 3-contact transistor, the base is inevitably the central contact.

base address

The starting point in memory for a program or block of data. Since applications may be loaded into different parts of RAM at different times, this is the key point of reference as to where an application begins.

Within the program, it is likely that only relative addresses will be provided, and so offsets will need to be added to the base address to locate the data.

This makes addressing a constant two-step (plus calculation) approach, but without this, a computer would not have any flexibility in its use of memory, and multiple applications (including TSRs) in memory would probably be impossible.

base-displacement

The flexible technique of allowing an applications program to reside in any part of working memory. See base address.

base station

This is often central to the whole two-way radio network, but in cellular systems the base is only central to a small part. A cellular phone base-station is the transmitter site at the center of just one cellular area; although it can have many sectors and each one is logically a cell. The base station houses the radio equipment.

The distinction may be made with a repeater which has no intelligence and relies on a base station. See backhaul.

baseband

Without modulation or modification and, also, the raw signal remaining after demodulation. The opposite of baseband is probably something like 'broadcast' or 'R/F modulated', not broadband.

Baseband signals are the pure electrical variations generated at the backplane of a computer or at the output of an electrical amplifier, or from a microphone. Digital baseband signals are usually intermittent DC, while analog baseband signals are AC at the original frequencies.

Usually only one baseband signal can exist at a time on any electrical circuit. Unless something is done to modify them (modulate and/or multiplex them) more than one would create conflict in a single circuit.

Another implication of baseband, is that the signal will probably not travel very far without attenuation — although this is relative. Before valve amplification became possible, it was still common for telephone baseband signals to carry surprising distances.

However, for carriage over any substantial distance these days, the signals are always modulated into higher (usually radio) frequencies. By modulating different signals into different frequencies, it is possible to share a circuit or a bandwidth. This is FDM.

- In telecommunications and broadcasting, baseband always refers to the transmission of signals at their original frequencies. A telephone system in the local loop (but not necessarily between exchanges) is a baseband signal; the wires to an loudspeaker carry the original baseband signal of a radio station.
- Computer baseband signals are found at the UART of the PC in serial form, or at the Centronix port in parallel form. When a modem is being used, these baseband signals are carried over the RS-232 link to the modem. Here the signal is modulated onto an audio frequency.

  Paradoxically, the combined signal will still be considered baseband from the viewpoint of the telephone system, because both signal and carrier are themselves at audio frequencies.
- In today's LANs, the packets are transmitted at baseband and are not modulated to a higher frequency within the LAN. At one time, Broadband LANs used radio-frequency carriers to carry many different signals simultaneously.
- Baseband video is another name for video in the form that exists in the cable between the camera and the recorder, or between the VCR output and the direct input to a monitor (not the R/F output, where you need to use a tuner to select the VCR). It can exist in a couple of forms: composite, RGB, YUV, YIQ or Y/C.

Baseband LANs

Today almost all LANs are baseband, although that wasn't the case a few years ago.

- **Baseband LANs** only carry one message packet at a time using different forms of time-sharing on the network. Distance is often limited to only a couple of kilometers.

• **Broadband LANs** are 'frequency-division' systems, treating the cable as if it were a broadcast medium and transmitting a number of different channels simultaneously. Broadband LAN receivers need to tune to a selected frequency — as with radio and TV broadcasting, and they need to demodulate the selected channel to extract the original baseband signals.

Broadband LANs have been largely displaced by high data-rates and fiber optics. See MAP.

baseband modems

Relatively short distance 'modems' (or rather, line-coders) for private use; which are usually not intended to be connected to the public switched telephone network. These devices are also called line drivers, line adaptors and Data Service Units (DSU). See.

BASIC

Beginner's All-purpose Symbolic Instruction Code. A very popular high-level language developed by John Kemeny and Tom Kurtz at Dartmouth College in 1965. It was based on Fortran.

Basic was originally an interpreted language which made it an ideal language to learn, but kept execution slow. Lately there have been both compiled and extended versions, which make Basic a valuable programming language for business applications also.

Most Basic isn't structured, so it is easy to get into bad habits with this language (although Quick Basic is structured). Some PCs have the ROM Basic interpreter within the machine and can therefore run simple programs direct, other machines have the Basic language itself also in ROM.

• There are hundreds of versions of Basic. GW Basic became very popular because it introduced graphic controls for CGA adaptors. Quick Basic was the first to move into structure programming, 3rd party libraries and compilation.

• MS-DOS has recently introduced QBasic to replace GW Basic. IBM's Basica is virtually identical to QBasic.

• Microsoft has a very interesting, programming language called Visual Basic.

Basic Cable

The primary subscription service which was originally called CATV (Community Antenna TV). The subscription payment is for the delivery of a clean signal — the carriage not the content. Basic Cable subscription services usually consist of program channels which could be obtained free in other ways — such as over-the-air broadcasting. Don't confuse this with 'Premium' or Pay TV. See STV.

basic mode link controls

The standard set of control characters found in the ASCII No.5 set: SYN, ACK, BEL, etc.

Basic Rate Access

BRA. The single twisted-pair connection between the customer's premises and the nearest local digital exchange which provides three ISDN channels each way: 2B+D. This is the ISDN service for homes and small businesses.

It represents the extension over the local loop of the core digital network of the telephone service. This core network (in all advanced countries) is now entirely digital and based on 64kb/s PCM voice channels, using CCS#7 signaling (over a separate packet-switched network).

The 2B+D of BRA refers to the fact that a multiplexed pair of independent 64k/s circuit-switched B-channels is being provided over the twisted-pair, along with a single 16kb/s D-channel. The B-channels are switched into the core network direct, while the D-channel packets are primarily intended for switching and signaling, and link to the CCS#7 network.

The D-channel can also provide packet-links for User Data while can carry X.25 connection. See Primary Rate Access, ISDN, B-channel and D-channel.

basic service

A term used to distinguish the fundamental voice and data network services of the telephone carriers from the more computational-intensive value-added services and networks they may also provide. See BSC.

BASICA

The command BASICA in MS-DOS calls up an interpreter for a special version of Basic stored in ROM. Only IBM use this, the clone makers use GW-BASIC.

basis weight

See paper weight.

BAT files

MS DOS files having the extension BAT are 'batch' command files which are used to establish a sequence of automatic actions. These will usually consist of a series of DOS commands (written as an ASCII file with a word processor or line editor) and saved with the .BAT extension. When you 'call' a .BAT file (by typing its name) it executes the commands in order (although some can branch). See AUTOEXEC.BAT.

batch

The accumulation of similar tasks so that they can all be performed during one session.

• A type of file used in MS-DOS, Unix and other systems to automate a number of procedures during start-up. See AUTOEXEC.BAT.

• This concept of performing a sequence of operations in order, is also extended to 'batch' suites of computer applications programs. This use of the term carries the sense that each application's process must be completed before the next begins.

• Originally the term carried the sense of 'performing all associated operations at one time', rather than progressively. The distinction was then with interactive or 'real-time' processing. Most early mainframe computers only ran batch operations, and they used punched cards for input of commands and data. You left your bundle of cards with the EDP computing staff and they ran the program when time became available.

• In file transfer, some protocols such as XModem, can only handle one file at a time, while YModem, and particularly ZModem, handle groups of files in batches.

• In data communications, 'batch' can sometimes mean sending a large file (or more often, a group of files) over the network without bothering about error checking. You just dump them into the system and they are conveyed to the other computer. This is really 'Batch-ASCII'.

batch server

A process found on some LANs where tasks can be distributed to designated batch-processing workstations. For instance, a programmer may send source code for compilation to a batch server, and then continue working on his/her workstation while the compilation is done. Other time-consuming tasks, like OCR or database sorting, can similarly be redirected to specially equipped (but unattended) batch-server stations on the network.

batcher-banyan

See banyan.

batteries

Actually, rechargeable batteries are correctly called accumulators — but, since people rarely use the term, I'll let it pass.

— Alkaline batteries: Numerous tests have shown that all the brands are pretty much the same in storage capacity, so buy only on price.

— Ni-cads: These have been in popular use for rechargeable systems for years, but they have a problem with 'memory'. After a while they begin losing capacity if they aren't fully discharged. You must regularly discharge these cells fully by shorting across the anode/cathode with a length of wire.

— Nickel-hydride: These are the best of the common batteries for devices like mobile phones. They cost a bit more, but they are probably worth it. However when Lithium-ion batteries become available these will take second place.

— Lithium-ion: These new solid polymer batteries have extra capacity and a single-cell voltage of 3.6. See lithium ion.

• The IEC in Geneva sets battery size standards (IEC86).

• See individual entries of the names above.

battery-backed RAM

A form of memory architecture where the memory is maintained in the chips even if the mains power fails or is switched off. Some memory extension cards use this battery-support, and some of the first laptop portables use full battery-backed CMOS RAM as an alternative to providing a disk drive.

On startup, computers with battery-backed RAM may boot with amazing speed because neither the system nor the application needs to be read in from disk. This is treated as a form of non-volatile memory — although the term more correctly applies to EPROM and EEPROM. The development of FRAM is aimed at providing the intrinsic value of battery-backed RAM without needing the battery.

baud

Named after Baudot who invented a telegraphic code. Baud is often confused with bits per second, but, although the two often coincide at many lower modem speeds, they are not the same.

Baud actually refers to the number of changed states ('events') that are transmitted every second — the so-called 'symbol-rate'. Symbol is used here in the normal way. Each character I am typing is a symbol — and each has 26 possible alpha-states, while decimal numbers have only ten.

If a transmission system only has two possible states then each symbol can represent either a 0 or 1. However if a system has three possible states (say +1V, –1V and 0) then either a binary or a ternary code can be used, and with ternary each symbol then transmits more than one bit during each event.

Even with binary systems, it is possible to discriminate between a number of phase-angles and amplitude changes simultaneously, and decode these to give 5- or even 6-bits per event.

In the early FSK modem systems, the output alternated between two audio tones, so only one bit was carried by each change of state. A 300 baud modem therefore produced 300 bits per second. But in most higher speed modems PSK or QAM is used (see). Here a number of bits can be carried by signal phase changes by angle difference (which can be accurately measured now). If two, three or four bits are carried in each single 'event' (by measuring the angle and/or amplitude at each phase change), then bits per second and baud assume a 2:1, 3:1 or 4:1 relationship.

• The terms kilobaud (kbaud) and kilosymbols/sec (ks/s) are occasionally used to express modem speed, but you've got to know whether dibits, tribits, etc. are being transmitted before you can convert these to kilobits/sec (kb/s). However usually the person is just mistaken, and is using 'baud' incorrectly.

• Baud also traditionally includes any signaling information transmitted in the channel, so it doesn't necessarily relate directly to the data-carriage (throughput) rate. See PSK, QAM, TCM, dibit, tribit and quadbit.

Baudot code

A simple alphanumeric code for telex machines and teleprinters which uses 5-bit characters (plus a start and stop bit). It is also known as the Murray code (which was a slight variation).

This is the International Telegraphic Alphabet also known as ASCII No.2. Since five bits can only produce 32 unique codes, two codes (FIGS and LTRS) are use to 'shift' between a figures function mode (with 31 characters) and a letters mode (with another 31 characters).

BB

Broadband. See.

BBS

Bulletin Board System.

BCC

Block Check Character. The BCC is the result of an error checking algorithm — a type of checksum. It is accumulated during the transmission of a block of data, then appended to the end of the block — sometimes in truncated form. See LRC, CRC and FCS.

BCD

Binary Coded Decimal. A 4-bit code widely used in financial transactions in the early days. The four bits of the nibble can represent numbers from 0 to 15, but only those from 0 to 9 were used. So the number 23 is represented by the two nibbles 0010 followed by 0011,

Normally the binary of the decimal 143 would be represented by a single byte (10001111), but in BCD it needs one and a half. So the BCD approach takes more memory space but it can result in higher arithmetical precision.

- This later briefly became a 6-bit binary code, which was later still extended to create IBM's 8-bit EBCDIC code. See.

BCF

Backus-Nur Form. See.

BCH

- **Bids per Circuit per Hour**. See bid.
- **Coding:** A matrix coding technique used for forward error correction. The information is read into the matrix along the horizontal axis, and read out in the vertical. This results in a correction system capable of handling burst errors. See Reed-Solomon.

BCP

Byte Control Protocols. Another name for character-controlled protocols. A way of regulating how information is sent and received across a communications link — in this case, within frames which are flagged by synchronization characters. Bisync is a BCP.

BCPL

A high level language halfway between Algol and C.

BCS

Basic Carrier Services. A term used to distinguish the monopoly network services of a dominant telecommunications carrier from the higher level services (HLS) such as packet-switching, and value-added services (VAS). In many countries these non-BSCs are open to competition, however it is often very difficult to decide which are BSCs and which are not.

BDC

Block DownConverter. See LNB also.

beacon

- **LAN:** When a node on a Token Ring LAN fails to receive a token after a long delay it will transmit a 'beacon' every 20 milliseconds. This is designed to disable the MSAU of the next node higher upstream and create a bypass. The node makes the assumption that there has been a malfunction — usually a break in the ring.
- **Satellite:** A communications satellite transmits a beacon radio signal so that moveable earth stations can lock on to, and track, the movement of the satellite. Unmodulated beacons are used for propagation tests and tracking, modulated beacons are used with telemetry.

bead

- A single short routine in some languages. These beads can be threaded together. The term is used in Forth.
- The insulating layer in a coaxial cable.

beam

A directional transmission of light, electrons or radio.

- In a monitor or TV set the beam/s are generated by the electron gun/s at the back of the tube and deflected by horizontal and vertical magnetic coils (or electrostatic plates) to sweep across the phosphor surface of the tube. At the tube's phosphor surface, the beam becomes a raster scan. The electronics of the monitor's drive circuits modulate the brightness of the beam to create the images — and the highest bandwidth that can be modulated onto the beam establishes the horizontal resolution of the image.
- In light or radio frequency transmission, the beam is a directional signal (expressed by an angle), although engineers still talk about omni-directional (all directions) beam.
- In satellites, the same bandwidth of available frequencies can be used and reused if they are applied to spot beams which don't overlap on the Earth's surface. Some of the new LEO projects use an antenna system which 'projects' a cell-like structure on the ground, where each cell reuses the same CDMA frequencies.
- **Beam splitters:** In optics, beam splitters can be used to separate different colors. They are usually gratings.
- **Beam steering:** See phased array.

beam benders

Small, cheap, wideband amplifier-transmitters (transponders) on poles which are used by UHF and MDS systems to push some extra signal into shadow areas. The frequency remains unchanged.

beamwidth

When a signal is sent or received using any sort of focusing (parabolic) dish the narrowness of the beam is expressed as beamwidth. It is expressed as the angle between the points where the beam drops to half power (3dB).

bean

A single narrow band of carrier frequencies in a multiple-carrier line-coding or transmission system. This is also sometimes called a 'finger'.

In the DMT form of ADSL, these beans are spaced every 4kHz along the available spectrum from about 20kHz to 1.024MHz. Each of these becomes a carrier which is then modulated by a form of multi-level QAM. By then combining the signals carried over this combination of multiple beans (with up to 10-bits per Hertz), these systems are able to carry video bandwidths over copper twisted pair.

bean counter

Accountant or a financial executive, usually one who is tight with the money and is seen as lacking creativity or imagination.

bearer

• In telecommunications the bearer was originally the medium of transmission — wires on poles, underground cable, microwave links, optical fiber, etc. It is used today as a vague generic for bulk 'transmission lines'.

• Nowadays, the term is generally applied to the high-bandwidth multiplexed channels between exchanges. You can have voiceband bearers, carrier bearers, or broadband bearers.

— Voiceband bearers are the twisted pair in the local loop, or the two pair (4-wire) links which carry voice between exchanges.
— Carrier bearers are carried by cables capable of handling about 120 voice channels.
— Broadband bearers are coaxial, microwave links or fiber optics carrying thousands of voice calls.

Some bearers are still analog, but most are digital.

• **Bearer services** are techniques like the X.400 message handling system, which will be widely used to carry EDI and EFT services. These services carry the information without modifying the message structure.

In the USA, this term is used to distinguish data services from voice telephony. The bearer service will be tariffed differently, and offer a different grade of service.

• **Bearer channels** in ISDN are the 64kb/s B-channels.

beat/BEAT

• If two signals of almost the same frequency are 'beat' together, the result will be a 'difference' signal. For instance if audio tones of 2000Hz and 2050Hz are sounding simultaneously, the observer will also hear a regular 50Hz surge as the two frequencies regularly coincide and cancel each other. This is the beat frequency. Usually it is annoying and unwanted.

It results from a positive/negative summation of the signals; sometimes the waves will be added together, sometimes they will be in conflict. The principle is put to use in heterodyne receivers to tune (select individual channels from) both radio and optical fiber carriers.

• **Bistable Etalon with Absorbed Transmission:** One of the devices which is a photonic equivalent of a transistor. The other is the SSEED.

BECN

Backward Explicit Congestion Notification which is pronounced bek'n). In frame relay, BECN is carried as a bit which is set in the header of a control frame to indicate that congestion exists. The node that detects the problem sets the bit, and it is transported back over the network in the direction of the frame's source. It is a warning that any data flows,above the limits set by the source's CIR, may be discarded by the network.

bel/BEL

• See decibel.

• ASCII No.5 — Bell (Decimal 7, Hex $07, Control-G). If you send this control code to a terminal it generates a warning 'bell' which these days is probably an electronic beep to call for human attention.

— Sometimes it is used to command alarm or security devices.
— The same character is also used in some phototypesetting systems as a flag.

bell compatible

Those modems which meet the standards used by Bell modems. America now conforms to the rest of the world in using the ITU's V.standards. See modem standards.

Bell standard modems

The early US modem standards which were quite different from those of the CCITT (now ITU). The major standards are:

— Bell 103A: Asynchronous, full-duplex at speeds up to 300b/s.
— Bell 113: The same as 103 but either originate or answer, never both.
— Bell 201B: For full-duplex synchronous data at 2.4kb/s.
— Bell 201C: Half-duplex version of the above.
— Bell 202A: Async to 1.8kb/s, but half-duplex. 4-wire needed for FDX.
— Bell 208: Synchronous 4-wire at 4.8kb/s for leased lines.
— Bell 209: Synchronous 4-wire at 9.6kb/s.
— Bell 212: Synchronous at 1.2kb/s using 300 baud.
— Bell 212A: Async and full-duplex at 1.2kb/s using 600 baud.
— Bell 2400: Identical to the ITU's V.22bis 2.4kb/s standard which uses 600 baud and quadbits.

• Generally any US modem rate above 1.2kb/s will also provide the ITU's standards. However, there are some differences in the provision of error checking, compression and fall-back procedures in some of the lower rate machines. See MNP and V.42bis.

Bellcore

One of the successors to the famous Bell Laboratories after AT&T was divested. Bellcore is now owned and run by the seven RBOCs (some of which have their own research labs in addition to their Bellcore involvement). It has coordination and standards roles in addition to its R&D function, and some of the RBOCs now want to sell it. Don't confuse Bellcore with AT&T's current Bell Laboratories.

Belling-lee

A simple slip-on connector widely use on video coaxial cables. It is used on TV antenna fittings.

benchmark

Standardized programs used to test certain characteristics of hardware or software. For any sort of realistic assessment, a 'suite' of benchmark tests will be employed — you can't judge on just one.

Benchmarks are usually calculated by the time the device/ application takes to perform some operation. But these figures are very suspect in most cases. They almost always depend on the implementation of a particular programming language or a highly specific task so, at best, they can only be considered 'rule-of-thumb' guides.

The major benchmark results for computer processors are given in MIPS, MFLOPS and more recently SPECmarks. Other well-known benchmarks are Dhrystones, Whetstones, Linpacks and Khornerstones. I'm not joking!

If you want to test complete systems or software, quite different benchmarks are needed — and there are many available — some of which make sense, and some often don't.

For instance, a benchmark used for database managers which measures the time taken to sort, say, 10,000 records certainly does give an indication of performance, but only in a very limited way. However, it may be highly relevant to the company buying that machine.

A benchmark for a word processor which registers how many keystrokes the user needs to a) save a file and b) resume keying, is also significant. But a multiple-action problem of this kind may easily be overcome by the use of macros or it may disappear with a GUI interface.

• Synthetic benchmarks (for hardware). These attempts to make the tests more relevant to the tasks for which the computer will be used. They do this by mimicking a suite of average applications, but they run up against problems with the variability in compilers.

See all listed above and TPC, BAPCo, SPEC, GPC, NAS, Perfect, and iCOMP.

benchmarking

• Specific computer terminology, see benchmark.

• General business; the process of comparing your organization and its processes to world-class companies (both within and outside your industry). The idea is to identify and use the best-of-class business practices you can find for goal setting and process improvement. See TPC and TQM.

bend loss

When an optical fiber is bent beyond its design parameters, there is a loss of signal power. There are two types of bending, long curvature of the fiber beyond the restricted limits, and microbend caused by very local pressures on the fiber.

bent-pipe transponders

In satellites, this is a transponder which does nothing to the up-link signal except amplify it and retransmit it on the down-link. It doesn't change the frequencies, and there is no on-board processing.

Bento

A multimedia file format (a document storage technology) being co-developed by Apple, Lotus, Borland, WordPerfect and Kaleida. It is closely related to the Script X multimedia object-oriented language and run-time environment (which is platform independent). It is designed for single-user compound documents, rather than for networks.

Benutzerfreundlichkeit

The German version of the hackneyed phrase 'user-friendly'. It has a better sound to it, don't you think?

BER

• **Bit Error Rate**: The number of bad bits in the transmission or storage of data, usually expressed in terms of 10^{-x}. BER is the digital equivalent of signal-to-noise ratio (S/N) in analog systems.

A BER of 10^{-3} is one incorrect bit in every thousand. Some confusion arises from the fact that modern storage and transmission systems (especially modems) often have built-in error correction, not just detection. See error rate also.

— The 'BER threshold' is the point at which the BER exceeds the error correction capabilities, and therefore error-free communications fail.

— The BER for a cellular phone channel is about 2000-times that of a wireline service.

• **Block Error Rate** (more correctly BLER).

• **Basic Encoding Rules** (in relationship to OSI): This is one method of encoding the data-units which make up the ASN.1 syntax; the other is DER (Distinguished Encoding Rules). See ASN.1.

Berkeley Unix/System Distribution

See BSD.

Bernoulli box

A removable large-capacity disk drive (from Iomega) widely used for general computer storage and for archiving. This is not a removable hard-disk, but rather a very large special type of floppy. The disk spins at about 1500 rpm (5 x normal floppy rate) and yet is very reliable. These units have a fast access time and used Reed-Solomon error correction codes and SCSI connections. Disks are available for storage from 20 to 44MB, and there are dual drive back-up systems with 88MB.

BERT

Bit Error Rate Test. This is the process used in many communications systems (mainly modems, etc.) to set or reset the data-rate to be used across a link. A trial amount of data is transmitted and if the error rate is too high the data-rate will be lowered during the handshake period — and/or during normal transmission — if the controls exist.

best-effort delivery

Networks which don't use acknowledgment and retransmission to ensure delivery of the data.

beta

• The last phase of hardware and software testing before release. This is supposedly, testing conducted in a real world environment. But another way of looking at beta-software or beta-hardware is to say that it means 'not thoroughly tested yet', so expect it to bomb.

Alpha tests, are in-house testing procedures, although alpha hardware is sometimes secretly released to potential third-party developers for evaluation and experience.

• In VCRs, this standard was originally called Betamax (see).

• In professional video production, this means Betacam — and there are both analog and digital versions. They use the half-inch Betamax cassette, but run at a faster tape speed.

Betacam

The Sony 1/2-inch cassette professional ENG camera system which later evolved to a metal-tape standard called Betacam SP and, more recently, became the world's first digital camcorder (Betacam SP DSP).

This family of camera-recorders has proved to be highly successful in professional television production, both for ENG and general use; it has about 90% of the professional market around the world. Don't confuse Betacam with the old VHS-like Beta standard.

• Betacam was originally an analog component (YUV) system which used time-division and some compression. In the magnetic tape recorder, the luminance information was stored on one track and the two chroma signals were time-compressed and stored on a second track.

Betacam was designed to use the same cassette and the same basic transport mechanism as the original Beta(max) home VCR and it also used an identical head-drum diameter (74.5mm) — so Sony was able to benefit from these mass-produced components. However the recorders ran tape at six-times the speed of the consumer version and so tape capacity was usually limited to 20 minutes.

The first Betacam cameras had a single Trinicon color tube, but most of those sold later were three-tube cameras.

• Modern **Betacam SP** cameras cost in the region of $50,000 and the new digital equipment costs more. And for all this, you can't even see the recording; Betacam doesn't have playback facilities in the camcorders.

• Generally, today, a low-cost video program will be shot in the analog Betacam component format, then edited direct on to the 1-inch Type C composite machines. This is called 'inter-format editing'. Component Beta always has a time-code recording capability.

• The new **Digital Betacam** (SP DSP) system available from Sony uses mild bit-rate reduction (2.3:1 DCT-type compression) with component recording. This is an 8-bit system which many professionals feel is not good enough for full-scale television production, however it has sold well since its release in October 1993. There's a new half-inch metal particle tape for this camcorder.

It records a 4:2:2 component digital signal onto a 14 micron tape. With a two-hour load in a single cassette, this is low-cost by professional standards. The camera is also capable of 16:9 widescreen recording, and some models now have Betacam and Betacam SP playback also. See M-11 and DVC.

• Sony are now developing a 'Son of Digital Betacam' system, which is 'Level 3' with a totally new digital video architecture. They aim to integrate a high-speed video network and distributed server architecture, with hybrid data storage using both tape and disk. See SDI and SDDI.

Sony will offer three levels of bit-rate reduction:

— Level 1 is D-1 quality at 90Mb/s.

— Level 2 is studio quality at rates between 30 and 40Mb/s, and is suitable for most post-production equipment.

— Level 3 outputs 18Mb/s and is designed for ENG. This is 'Son of...' and is equal to the analog Betacam SP format.

Betamax

Sony's domestic Beta(max) 1/2-inch VCR standard was finally pushed out of the market by the success of VHS. The Betamax system evolved from the Sony U-matic 3/4-inch cassette system. Betacam is the professional version of Betamax (but only in terms of using the same cassette); otherwise the two are totally incompatible. See Component Beta.

betaware

The supposedly 'almost-finished' software that a company may release to other developers and technical journalists. The idea is to ensure that there aren't too may bugs and conflicts with other software, and also to begin the marketing process during the debugging phase. They can always disclaim responsibility for major computer crunches when software is still beta.

At the beta-release phase, the software is supposed to functions as the final product should — with only a few bugs still to be removed. They say that the beta version is 'beta than nothing', but don't be too sure! It usually crashes regularly, sometimes wipes out other programs, and runs at half the speed of the final product. It might cost you a lot.

bezel

The frame around a monitor or video screen which cuts off the edges of the picture. The area of video hidden by the bezel is known as the overscan.

Bezier curves

A mathematically described curve which joins two or more points. It is used in graphics, tracing and numerical machine control.

Bezier curve objects can be manipulated on the screen; the curves are controlled by a series of 'anchor' points. At both the take-off and landing points ('nodes') the curves are tangential lines which establish the direction and shape, and it is this information that is stored or used. These shapes constitute the foundation of outline text fonts, which is why these font files are highly compressed and infinitely scalable.

Very complex shapes can be drawn using Bezier curves in a chain. The Bezier points have both 'direction' and depth unless they are 'endpoints' or 'uncontrolled points'. But note that Bezier control points do not necessarily touch the curve.

In structured graphics, freehand drawings are often converted to a series of Beziers, and the outlines of text characters can also be mapped as Beziers (as is PostScript). See B-spline.

BGP

Border Gateway Protocol. This is an inter-domain routing protocol defined by RFC 1105 for the Internet. It is seen as a potential replacement for EGP.

BHCA

Busy Hour Call Attempts — a carrier's measure of the telephone network efficiency. The distinction is with the traditional Erlang.

bias

A tendency one way, rather than the other.

- A current applied to an electrical device (usually a magnetic head) to provide a reference level. See AC bias.
- A DC voltage applied to a transistor's control electrode to establish the desired operating point. (In vacuum tubes a similar system is called grid bias).
- The amount by which a signal or value departs from a reference value.
- In number theory, bias is anything that tends to distort a process. For instance bias can distort random number generation making numbers more likely to appear than others.
- In transistor manufacture the silicon is biased for either negative (N) or positive (P) mobility, thus producing the difference between, say, NMOS and PMOS transistors. CMOS has both N and P biased sections.

bibliographic database

A database which holds information (usually abstracts) of books, reports, magazine articles, etc. The distinction is with databases used for commercial purposes or for statistical information.

BICI

See Broadband InterCarrier Interface.

BiCMOS

An advanced form of semiconductor technology which combines bipolar devices and CMOS on the same chip. It is potentially capable of doubling the existing speed of processing chips and overtaking the common TTL-interface standard.

bid

An attempt by a possible sender to claim control of the network for transmission. When you pick up the phone to initiate a call, you are bidding for a circuit. If you don't get dial tone, you are blocked.

The term is used in LANs where contention schemes are in use. It is also used in telephone networks where it is measured and expressed as Bids per Circuit per Hour (BCH). It is a test for the pressure on a network.

bidirectional

- **Printing:** Most impact printers save time when printing lines of text by printing the first line from left to right, then the next line from right to left, and so on. The printer must have enough memory to hold at least one line of text.
- **Communications:** This term can be applied both to full duplex communications (both ways simultaneously) and also to fast-turn-around half-duplex communications where the two parties alternate automatically. Usually it would suggest full-duplex. See TDD also.
- **Bus architectures:** Within a computer the address bus is uni-directional. However a data bus and the bus employed for communicating with a peripheral, will usually be bidirectional.
- **Microphone:** Able to accept sounds from both sides. This type of microphone is used for radio interviews where the mike can be swung between the two speakers. These mikes have a figure-8 sensitivity graph.
- **Interpolation:** In video compression this means motion compensation interpolation schemes which result in improved compression rates because they are designed to compare past and future reference frames of image (from a storage source). MPEG-2 is of this type.

BIDS

Broadband Integrated Digital Star — a fiber-to-the-home network topology and technology which relies on active optoelectronic switching equipment in local street pillars. Each pillar supplies a separate fiber feed to 32 homes with four channels for each — telephone, two TV channels and a control. See BPON and TPON.

bifurcate

To split into two branches — as with a fork in a road.

Big 8

See RS-232 Big 8.

Big Blue

A now-very-irritating cliche for IBM. The covers on the early IBM mainframes were blue.

big-end(ian)

When storing or transmitting the data, the big-endian systems always transmit or present the most significant bit (or byte) first. The contrast is with little-endian systems, where the least significant bit or byte is transmitted first, since this is the least destructive to the data if it is lost or changed.

big iron

Mainframe computers.

Big LEO

Low-earth orbiting satellite systems working in the L-band, and providing both voice and data. These have special licenses from the FCC in the USA, quite separate from Little LEOS (data-only in UHF) and operate in frequencies allocated by ITU-R.

The current Big LEO proposals are Iridium from Motorola, GlobalStar from Loral-Qualcomm, Odyssey from TRW, Ellipso from Ellipsat/MCHI, and Constellation/Bell Atlantic with a proposal called ECCO. Inmarsat-P is an international proposal.

• Not all of these are true LEO proposals: Odyssey, Ellipso and Inmarsat-P are MEOs.

• The FCC has said that the Big LEO systems will operate in the 1610—1626.5 and 2483.5—2500MHz frequency bands. They plan to assign the CDMA systems 11.35MHz of shared bandwidth at 1610 —1621.35 MHz and TDMA system 5.15MHz of dedicated bandwidth at 1621.35 —1626.5MHz. See all names above and Little LEO also.

bilateral control

Control which is equal in either direction on a communications link. This is usually a telecommunications control system where Station A provides the sync signals for Station B, while B provides the sync for A. There is no master-slave clock relationship.

billboard

Billboard services are those offered through low-power radio or infrared transmissions to people only in the immediate vicinity of a transmitter. Talking billboards are generally very low-power radio transmitters which provide information to passing motorists (or pedestrian tourists), on a specific frequency, for distances up to a few hundred meters.

billi(on)

The prefix billi now means one thousand million. The US billion is 10^9, but the old British billion (10^{12}) is still in use. Most of the world now follows the American usage — especially in technical areas. In SI terms a billion is represented by the prefix giga (G), and one billionth is represented by nano (n).

BIM

Beginning of Information Mark. A code that marks the start of the data stream.

BiMOS

A hybrid of bipolar and MOS techniques. BiCMOS is a variation which has proved to be useful in the fabrication of high-speed memory chips.

binary

Having only two states — on or off, 0 or 1, Yes or No. In normal computer usage, the basic binary notation for a nibble (4-bit byte) extends from zero = 0000, 1 = 0001, 2 = 0010 to 15 = 1111. However there are a number of different forms of binary notation used in computers and communications for special purposes. See gray binary and symmetrical binary also.

• Binary Characters usually refers to two-state data-types. Y(es) or N(o), or perhaps T(rue) or F(alse).

binary chop

The quickest way to zero-in on some desired element in an ordered list. This search technique is used with index searches which cannot be directly related to an address or to an offset. You first halve the list, choose one of the halves, then halve again, etc. until the data is reached. This approach is also called binary searching.

binary digit

A bit — the 'atom' of digital techniques. See bit.

• A bit is written as a 1 or a 0, but really inside the computer it is represented as an electrical voltage which can be detected on not detected — which is ON or OFF.

• On a magnetic disk or tape it is a magnetic polarization which has either a North or South orientation. This can be detected during the 'read' process, and converted into electrical signals.

Unless otherwise stated, there are always 8-bits to a byte. Sometimes only the less-significant (right-hand) 7-bits are specified, as in the ASCII No.5 alphabet, but this is simply converted to 8-bits by leaving the most-significant bit (the left-hand bit) at zero.

• Note, that while there are eight bits to a byte, some later computer and communications systems also added a parity bit to equal nine.

• Also, in asynchronous communications, you should usually assume 10-bits to a byte to allow for 'book-end' start-bits and stop-bits.

binary files

All files are binary, but the distinction being made here is between files which need to use all 256 possibilities (all 8-bits) available in a byte, and simple text files (ASCII files) which only use the 128 ASCII characters (only 7-bits). The older paging systems, for instance, will only handle the 7-bits, so they can't transmit binary files.

'Binary file' most likely signifies a couple of things. Quite likely this is an application, or a graphics file, or one which has complex text/graphic formatting. Also sound and compressed image files will need all 8-bits, and therefore need to be handled as binary.

Binary therefore also suggests that the transfer must be error-free, since just a single bit error can crash an application; 99.99% accuracy isn't good enough.

For this reason, binary files are almost always transmitted using special file-transfer protocols which can perform error checking and correction functions, while 7-bit text is often transferred 'in-the-raw' without bothering about error checking protocols.

- In the Macintosh, the Macbinary selection will transfer both the Resource and Data forks so that the file will arrive complete with its icon and formatting.
- Many of the world's older communications networks will only handle 7-bit bytes because they were designed exclusively for text.
- In X.25 networks, before 8-bit files can be transmitted, some flow control and checking mechanisms must be turned off by the selection of a special profile or a change of parameters.
- 8-bit bytes can always be transferred over a 7-bit system by breaking each byte into two 4-bit nibbles, then adding parity checking to bring them back to 7-bit bytes. See Hamming code.

binary fraction

A binary fraction mimics the conventional decimal fraction, but uses the base of 2. Therefore a binary fraction of .011 would mean zero halves, one quarter, plus one eighth. This is the way 'real' numbers are stored in most computers.

binary phase shift keying

BPSK. A technique used in telecommunications where the information is carried by a 180° inversion of the carrier waveform. Bit/second and baud match with BPSK. It is often used where a very robust (not prone to error) system of line-coding is required; CDMA uses it in mobile telephony. See PSK.

binary search

See binary chop.

binary synchronous

BSC (or Bisync) communications was IBM's earliest data-link control standard for medium and high-speed data communications. It was the predecessor of HDLC, SDLC and SNA.

Bisync was a half-duplex system for character-orientated sync data. It sent messages as a header, information field, and CRC. It requires both sender and receiver to maintain various counts and to decode control characters. Both point-to-point and multi-point operations were supported. It can be expressed in 8-bit EBCDIC, 7-bit US ASCII or 6-bit transcode.

binary tree

A branching structure which only has two branches at each hierarchical level. Binary trees are used for searching data structures. See binary chop and B-tree indexes.

bind

- **Computers:** To link modules/systems together. To create a cause-effect relationship or a two-way association of elements.

— After a source program is compiled, the object code must still obtain all the addresses of the routines and variables it will use. This is called linking or binding. Early binding, means it does this when the program is compiled. Late binding, means it only does this at runtime.

— Icons on a GUI also need to be 'bound' to their applications or documents in some way.

— In programming, binding can be the assignment of a value to a variable or parameter (also assigning an address).

- **Printing:** The process of stitching, stapling, or gluing sheets of paper together to make a book. See saddle-stitch and perfect bound.
- **Communications:** A term used vaguely in electronic mail systems, SNA and general telecommunications. A 'bind' is the result of a request from one computer/device to establish a communications link with another. When the other commits to receiving the message it is 'bound' and a session has been established.

biometrics

The security processes associated with personal (not machine) identification performed by measuring, or analysing, parts of the body. Fingerprints, retinal veins, voice characteristics, the general dimensions of the hands, and the vein structure on the back of the hand have all been proposed as possible sources of identification information.

bionic

Patterned after human or animal parts. A term used in medicine, robotics and far-out television series.

BIOS

Basic Input/Output System. BIOS is a collection of service and support routines (programs) associated with the operating system, usually stored in ROM. These perform basic operations with the hardware.

Although BIOS is stored in ROM, it is usually transferred to RAM during the boot process, and it resides in MS-DOS machines above the 640kB barrier. Most software in the PC must call BIOS routines as well as those in DOS software.

BIOS assists the CPU with such functions as reading the keyboard information, reading or writing to disk, placing the graphics and characters on the screen, and communicating with other attached devices such as printers. So BIOS constitutes a distinct section of the operating system.

The routines handled within the operating system (the 'DOS calls') can't support graphics, so PC software will usually need to use call some routines which are supplied by BIOS as well as those available in DOS.

The idea of this 'modular' specification of BIOS first arose with CP/M. The original intention in breaking BIOS away from the disk-based operating system and treating it as a separate module, was to provide for a totally standardized exchange between the hardware and the software in similar (but not identical) systems.

IBM wanted to make it easier to port operating systems from the original 8086 to new processors and it didn't anticipate the dominance of MS-DOS. There were a number of other OS contenders at the time, and the idea was to have a common BIOS owned by IBM.

By each machine's hardware having its own BIOS (in a ROM chip), BIOS can still be incorporated into the operating system to standardize it, and IBM would control the system.

Where some of the problems arise with BIOS, is that software writers can access some functions (such as writing characters to the screen) either through BIOS, or through the operating system (which is slower), but BIOS must be perfectly standard if all software is to work and behave perfectly.

With the original IBM PCs, this was the only key component in the system which was owned and controlled by IBM (Microsoft owned the bulk of the OS) and, until the BIOS chip was 'reverse engineered', the clone-makers couldn't make machines which were 100% IBM-compatible.

• BIOS provides the connections between physical devices (keyboards, video screens, disk-drives, printers, etc.) and the CPU, which functions at the core of the operating system. So the sub-routines that drive these devices constitute BIOS, and the CPU must communicate with each of these devices through a BIOS 'driver', specific to the device. When a PC is booted, it first runs the autostart routine to test memory, then searches for any of these additional BIOS components that may be found on plug-in cards. If it finds any, it will establish pointers to these for later use.

• BIOS clones have now been 'standardized' (in late 1993) by a group of clone-makers led by Compaq and Phoenix Technologies, with Intel and Microsoft also involved. See Phoenix BIOS.

• When major changes involving peripherals are made (such as Plug-and-Play), BIOS will need to be modified; PnP needs an upgraded BIOS.

• OS/2 doesn't use BIOS: all I/O is provided by OS/2 itself.

biphase

Originally these were line-coding schemes developed for Ethernet in order to remove all DC energy. They are bipolar systems in one sense, since they alternate between positive and negative voltages — but in biphase, the mark/space conditions aren't signalled by the actual voltages during the bit-interval, but by the direction of transition from one state to the other. Timing is extracted from the synchronous data stream without needing separate clock links with these systems.

All biphase systems require twice the bandwidth for transmission as simple bipolar systems — so they are grossly inefficient. However the advantages are:

— easy automatic synchronization because there is always a predictable transition in each bit interval (these are known as self-clocking codes);
— no DC component, and therefore no coupling problems;
— error detection; and
— some signaling via code violations.

See Manchester coding, Differential Manchester and bipolar.

bipolar

Two polarities — having both positive and negative polarity, or having two (and only two) different states.

• **Bipolar ICs** have bipolar transistors which are used in all TTL devices. Bipolar transistors are either n-p-n or p-n-p devices, but there are now many variations on these. See ECL.

In transistors and ICs, this is both a generic term for a range of similar devices, or reference to a very fundamental process of chip fabrication.

There are two basic forms of transistor — unipolar and bipolar. Bipolar transistors have emitter, base and collector connections (Unipolar have source, gate and drain.). The problem with bipolar technology is that the transistors occupy a relatively large area of the chip surface and consume a significant amount of power.

Furthermore, bipolar transistors consume power both in the on- and the off-states — whereas MOS transistors only consume power during the transition. However, recent experience suggests that bipolar technologies scale-down better than MOS to submicron sizes. They are certainly faster.

• **In communications:** Bipolar signals (aka 'polar') have both positive and negative values. This is a wide term which covers a few distinctly different techniques. For instance:

— The 'mark' (1) bit will be represented by, say, a +3 voltage, and the 'space' (0) bit represented by –3 volts — rather than by zero volts. Another approach is taken in biphase.
— The 'mark' (1) bit will be represented by either a +3 volts or a –3 volts, while the 'space' (0) bit is represented by zero volts. This is the form used in AMI, and it is also known as 'pseudoternary' since zero is seen as a 'state'.

Both of these schemes have the advantage of acting like AC signals while providing no DC component (the +ve and –ve pulses average out). The distinction is with unipolar signals. See AMI also.

birds

• A communications satellite: GEOs and LEOs.

• Birdies are whistles and chirps in an electronic system.

bis

In reference to the ITU standards — 'bis' is French for two, second or 'encore'. It refers to the fact that this is the first variation on the basic standard; the next variation is 'ter' (third).

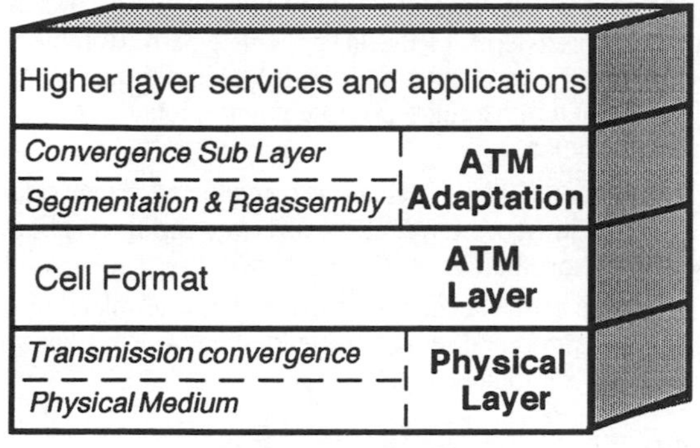

The Broadband ISDN Model

B-ISDN

Broadband ISDN. The term was popular only a few years ago to identify the proposed future telecommunications system based on ATM. It was a 'service definition', or 'total network specification which included the definitions of access, user interfaces, services, etc.

It is quite incorrect to assume any relationship between ISDN (sometimes called Narrowband ISDN) and B-ISDN. The only remote relationship is that B-ISDN extended the integrated concept of ISDN — everything down the one wire. It is more closely related to:

— optical fiber, not copper twisted pair,
— high megabit/sec rates, not low kilobit/sec rates,
— packet-switching, not circuit switching,
— complete replacement of whole network, not just an access technology,
— meshed networks and redundant paths, not star networks.

B-ISDN public networking was designed for two-way telephone and data, moving images (videophones and video conferencing, video security). It included one-way broadcast TV and radio program distribution, and high-bandwidth interactivity.

The technology is based on ATM cells for multiplexing and switching (Originally, it was intended to use STM, rather than ATM!) and it was once seriously proposed that each home would be serviced by fiber carrying data at two rates — two-way flows at 155Mb/s for voice telephony, videophones, etc, and the one-way supplementary video downlinks of 622Mb/s rates.

It is noticeable that the term B-ISDN has now largely dropped out of use, and the proposed rates of 155–622Mb/s have migrated from proposals for the rates over the local loop, to those for main bearers. Today people tend to talk about ATM systems rather than B-ISDN because the old service-rates appear to be a ridiculous overkill (see MPEG).

The B-ISDN model is defined in three layers:
— ATM Physical Layer
— ATM Layer
— ATM Adaptation Layer (AAL)

These correspond to the first three of the OSI's 7-layer model. B-ISDN was also strongly associated with radiating (star network) fiber to the home, while ATM is now seen potentially as a 2Mb/s to 45Mb/s protocol for shared coaxial as well as for meshed networks. See ATM, fast-packet switching and SDH.

bistable

The characteristic of being stable in either of two conditions. An electric light switch is bistable — it is always either on or off — there's no intermediate stage.

Similarly a flip-flop (the core technology of static memory) is also bistable, either holding a bit, or not. The term refers generally to any compound, component or device which can be used to store binary digits.

BiSync

Binary Synchronous. See.

bit/BIT

- **Binary digIT.** See.
- **BIT** — Built-In Test.

bitblt

Bit-block transfers. Either a hardware- or software-based graphics operation which is used in video adaptor cards to move blocks of bits, usually from the main memory into video memory very rapidly. However, source and destination can be bit maps anywhere in working or high-memory, or even display devices. Alternate transfer processes tend to slow down the video operations.

bit density

See packing density.

bit error rate

See BER.

bit field

Sometimes an 8-bit byte in a computer is not treated as a simple byte (character). The group of bits will be subdivided and the bits used individually or in sub-groups for special flag, marking or attribute purposes.

bit map

A bit-mapped image is formed by a rectangular grid of pixels on the screen which is raster-scanned. Each pixel of this image matrix is directly related to a memory location (one or more bits). The precise point, brightness and color of any pixel on the screen is accurately reflected (by the bits/bytes which control it) at one single memory location. Usually this memory 'bitmap' is in Video RAM. See VRAM.

For monochrome bitonal images, each memory location will consist of a single bit, set either to a zero or a one. For color images, each memory location will hold from 3- to 24-bits (or 32-bits with an alpha channel). The pixel-depth of these bits (how many to each location) controls the hue and color of each pixel.

- The contrast is with images stored as mathematical formulae, or those using some encryption/compression process which requires processing before it can be displayed. Vector graphics and outline fonts are stored as formulas, while compressed images require preprocessing before a bit-mapped image can be created.

• When the output of any device is to be via a 'raster-scanning', any sophisticated form of image representation must first be converted to a bit-map of some kind, so the screen can be continually refreshed in the easiest possible way. Rasterization (the scanning process) is used by conventional monitors and printers; without it, the CPU would need to do complex conversions every 60th of a second.

• While bit-mapped images are the most basic form of graphic representation in a computer, they are not the ideal way to treat images which may need to be scaled up or down — except when the scaling is by integer amounts. When non-integer scaling is performed the image resolution suffers and proportions are often distorted. See hint and alias.

• Bit-maps are sometimes incorrectly called memory maps, but this is confusing. The distinction is with vector-graphic and outline systems. See blitter and jaggies.

• Characters or fonts: see bitmapped fonts.

bit mode

See byte and block multiplexing.

bit-order

See little-endian and big-endian.

bit-oriented protocols

This is a class of communications protocols where frames are transmitted, without regard for the framing. Bit-oriented protocols are generally believed to be more efficient and more reliable than byte-oriented or block-oriented protocols.

In practice, the term appears to be used in two quite different ways, for slightly different processes:

• Communications protocols which use individual bits within the bytes for control during data transmission. Protocols such as SDLC insert individual control bits (zero-bit insertion) into the frames for control, while alternate techniques use full bytes for control. See byte-oriented protocols.

• These are protocols which do not impose character assignments on to the transmitted data bits; they treat the bits as a stream.These are time-division multiplexing protocols in which the individual slots are only one bit in length, rather than one byte.

bit plane

In some computer systems the color information in video RAM is segmented into bit-planes. The first bit of each video pixel will be stored together in one segment as a 1 pixel-depth bit-map; all the second bits will be stored in another segment as another bit-map; the third bits in another, etc. until the pixel-depth is exhausted.

The number of 'planes' (individual bit-maps) will therefore equal the pixel-depth of the image. This type of distributed storage is still classed as a form of bit-mapping, although all the bits associated with a single pixel are not now found at a single memory location.

The planar technique enables the computer to change color modes rapidly because it is either able to chain these planes (segments) together, or use them in parallel.

The more normal technique is that of storing all information associated with one pixel at a single location, which is known as packed-pixel architecture.

bit position

A normal byte has 8 bits which are numbered from 0 (the least significant bit) to 7 (the most significant bit). Bit positions are numbered from right to left, and they traditionally use zero to reference the first bit. So the fifth bit in a byte, is bit number four. Life wasn't meant to be easy!

bit-rates

Data-rates. There are a number of popular misconceptions which create constant confusion:

• Data-rates are usually measured in bits (not bytes) per second, if the subject is communications. However, when dealing with computer devices (such as data transfers from hard disks or CD-ROMs) the rate is often expressed in Byte/sec. With modems, you'll find both are thrown about with gay abandon and there will often be some confusion as to whether 8-bits should be allocated to a normal Byte, or 10-bits to an asynchronous Byte.

• Bit-rates should not be confused with baud. The measure, bit/sec, is a way of expressing data transmission speeds, while baud (not bauds/sec) is a measure of the rate of change in line condition — which may or may not coincide with the bit-rate measurement. Modems and high-speed communications equipment transmit information by creating changes in line conditions in a number of different ways — voltage levels, audio levels, phase changes, frequency changes, etc. and more than one technique can be used at the same time.

These changes or events are known as symbols, and the rate at which the changes occur on a link is its baud rate. However, each change (each symbol) may carry more than one bit of information. If a symbol carries two bits of information, it is known as a dibit, if it carries three it is a tribit, and so on. So the bit-rate can be two, three, four or more times the baud rate.

This is the same multi-bit relationship that we find between the number of characters there are in this line of text, and the information content of each line. In written English, each character can be any one of 30-odd alphanumerics, so the information-rate (the bit rate) is more than 30 times the symbol-rate. (baud).

• Bit-rates are not necessarily throughput-rates. Throughput can be lower or higher than the bit-rate over a channel.

— If a link is constantly subject to errors, and there's a need to regularly ask for data to be resent, then the actual throughput will be much less than the bit-rate suggests.

— If, in order to prevent errors, the protocol adds substantial overhead (redundancy) to the data, then the bit-rate quoted may be deceptive. Golay FEC systems, for instance, double the number of characters which need to be sent; while Reed-Solomon FEC adds between 11% and 16%.

— If data compression is being provided in the communications device (the modem) then the throughput might be double the bit-rate on the telephone link. These days the modem can be too fast for the UART of the PC to handle when line conditions are good and the file is highly compressible.

bit robbing

A technique widely used in T1 lines in the USA for providing supervision information. It is also widely used for handling overloads on submarine cables. (See bit stealing)

In the first T1 systems, the 8th bit was stolen from every frame for control signaling. This was overkill. Later they began to use only the 8th bit in frames 6 and 12. This is known as A&B bit control. Now there is ABCD robbing.

Bit robbing is only permissible when the channel is being used for voice. Text can only be carried in 7-bit form (56kb/s). The 8th bit is then usually set to 1 to indicate that data is being carried.

bit significance

The design of the ASCII No.5 code has meant that some bits carry special significance. For instance, the binary representation of lower-case and upper-case characters has been chosen so that conversion from one to the other can be made by just switching the 6th bit (bit position No.5). The capital A = 0100 0001 and thc lower-case a = 0110 0001.

Bit position No.5 therefore has a significance beyond its numerical value. Similarly in Apple II computers the 8th bit was used to indicate that the byte has been generated by the keyboard. IBM PCs generated a special 7-bit code when the keyboard key was pressed, and then repeated the code with the 8th bit ON to indicate that the key has been released.

bit-slice technology

A mini and mainframe way of handling data in (usually) 4-bit nibbles rather than 8-bit bytes or 16-bit words. By linking a number of 4-bit microprocessing chips in parallel, it was possible to construct computers which would handle data in 'words' which were 8, 12, 16, 24, 32, or even 64-bits wide. This approach was said to give them a much higher processing rate. This is a mini and mainframe technique only, and it results in a fast but very costly processor.

bit-stealing

On some digital voice cables the carriers can temporarily gain extra capacity by reducing the number of bits used to carry the voice. This is a technique used on submarine cables as a way of handling peak-load conditions.

Voice is generally being transferred across these systems as a series of 8-bit PCM bytes or 4-bit ADPCM nibbles. If the bandwidth demand suddenly exceeds the supply, the system will discard the low-order bits (least significant), reducing the PCM voice to a 7-bit sample, or the ADPCM code to a 3-bit difference value.

Provided this only happens for a few seconds at a time it doesn't degrade the quality too much, and the extra capacity can be used to support other channels. It allows them to operate their cables at rates nearer the overload limit — but it is murder on fax. Normally, when bit-stealing is being done on a cable, the fax user will dial a special 'International Fax' prefix to warn the system that stealing bits from his channel is prohibited. See DCMS.

bit string

Where a nominal byte is being used to provide eight independent bits, each of which will serve a different purpose. See bit field and flag.

bit stuffing

I have a few subtle differences in definition here for communications:

- A frame or packet often needs to be an exact length. If the available data doesn't fill the packet, the packet will be stuffed with extra zeros. The receiver will know to remove and discard these.

- When data streams are being merged in a TDM, it is necessary to match the clock timing of each (mainly for switching purposes). So redundant bits are often inserted into data streams to increase the rate. The presence of these redundant bits is then signalled to the other end where they are removed. This is necessary with PDH transmissions over SDH.

- A technique where frames of data are delimited by a special bit pattern (usually 0111 1110). If this is to be readily identified as a delimiter, the same bit pattern obviously can't be used within the file material — yet this is a standard ASCII byte (the tilde ~).

To avoid confusion therefore, a rule is established which says that after any five consecutive 1s a zero must always be added. Obviously, it will then be extracted and discarded at the other end — however, this technique means that the delimiter can always be recognized because it is the only byte with six 1s.

I suspect this is a misuse of the term 'bit stuffing'; it would be better called 'zero insertion'.

bit synchronization

Timing synchronization needs to be maintained between any two communicating devices (except async devices) by clocks and timing signals. Usually in phase-change systems the receiving station will adjust its clock to match transitions on the line created by the sender rather than use a special clock signal. This is bit-synchronization.

Special synchronization characters (SYNs) are inserted into the data stream at the beginning of transmission to provide a reference and SYNs are often used to maintain synchronization while the channel is idle.

bitmapped fonts

These are fonts of a single size and character. You need one set for Times 12 point Roman, another for Times 12 point Italic, another for Times 9 point Roman.

These font files are the sequential collection of bits which represent (in memory) the raster-scan of the character image which is to be displayed.

If you take Times 9 point Roman and enlarge it to Times 12 point Roman, the dimensions may look OK, but it will likely have jagged edges. See aliasing. If you take Times 9 point Roman and enlarge it only marginally, to say Times 10 point Roman, it will probably look quite out of balance.

The distinction is with 'vector' or outline fonts, which are individually calculated from a formula to fit each size.

bitonal

Images consisting of two tones only (usually white and black). The term refers mainly to line art. The distinction is generally between bitone and halftone images.

bitstream

• **Audio:** Bitstream is also called 'single bit'. It is a development in very fast digital-to-analog conversion, specifically designed to drive hi-fi amplifiers from standard CD-Audio disks. This approach to reproduction is said to give much better subjective sound quality than the original DAC. Bitstream apparently improves accuracy when reading and converting the digital signals, and it avoids the need to use some of the more variable analog components.

The bitstream distinction is with the original CD reproduction techniques which were based on 16-bit stereo samples which had quantization errors. These later moved to 18 and 20 bits (during reproduction only) in order to overcome DAC rounding problems (by a process of 'dithering' to cover the step-like changes in value).

If you simply transfer the CD's output of 44.1k of stereo 16-bit samples to a DAC every second, spurious frequencies are generated in the stepped reconstruction of the waveform (quantization artifacts). In early CD players, analog filters were sometimes used to remove these spurious frequencies, but these filters result in phase-shift distortions.

For best results, some form of 'smoothing' must be done at the digital stages, and bitstream uses a technique known as oversampling where the output of bits is at a higher rate than the input. This can be done, for instance, by automatically interposing an average between every two samples, to reduce the abrupt change. Another way is to use delta (difference) techniques — and bitstream is the extreme case of this, with some variations.

Bitstream replay machines use special one-bit DACs that provide a stream of single bits (not the normal 16 or 20-bit words) — but these bits are flowing at very high rates, usually at a 256-times oversampling rate. This is not as extreme as it sounds: it means 256 bits (in one-bit delta 'samples') are being generated from one stereo sample of 32-bits, which is a ratio of only 8:1.

There are two competing techniques:

- Matsushita's bitstream technique using PWM (Pulse-Width Modulation) is called MASH, and is very similar to the Philips PDM system (below).
- Philips PDM (Pulse-Density Modulation) technique multiplies the 44.1kHz samples to create a 11Mb/s data stream.

• **Fonts:** Bitstream is also the name of a company which produced an outline font definition similar to TrueType and the Adobe fonts. They have now licensed the TrueType fonts.

• **Cable:** Bitstream is also a telecommunications technique used in undersea cables. They use an analog supergroup (see) to carry a digital 1.5Mb/s stream of traffic.

• **Video:** In the new digital video recording systems (mainly D-VHS) bitstream is a way of recording compressed or processed (i.e. encrypted) signals, such as those of digital broadcasts, on a tape directly as digital data. Later they will be output in the same state as they were input.

The idea is to bypass the need to decompress and recompress received files. Bitstream just treats them as a stream of binary digits and doesn't attempt to find out what they are.

black box

• An aircraft flight parameter recorder.

• Almost any strange and unusual piece of equipment — especially one that can't be opened for investigation.

• A company specializing in telecommunications equipment.

• In phone phreaking — it's a tone generator used for making international phone calls. See blue box and red box.

• The conceptual treatment of computer (and other) systems as a group of individual modules, without needing to understand how the technologies within these modules work. The only requirement is to look at the way the modules act on input and produce output, and how they interface with each other.

With the black-box approach to technology (common among decision makers) we know what the boxes do, but we don't necessarily know how they do it. Many benchmark tests look at the computer in this way.

black holes

There are three slightly different uses of this term:

• On large LANs, 'black holes' occur when packets are accidentally sent to an address that doesn't exist.

• Any area of an internetwork where packets enter, but don't emerge.

• Sometimes packets can circulate endlessly through the network causing a string of routing errors. This can also be compounded because error messages can also circulate, ultimately causing a whirlpool of packets. The resources of the network will be swallowed up, until a time-out occurs and the packets die away. This is more related to packet storms.

B-LAN

Bridged LAN. See bridge.

blank character

The space (ASCII 32) character is considered a blank character (it is 'printable' — even though you don't use ink) while the null (ASCII 0) is considered a control character.

blank string

This is a "string" with nothing in it; it is not necessarily a string of nulls (zero bits).

blanking

In TV, video and computers (except in some drafting applications) we currently use "raster" monitors, where the image is created by the electron gun/s scanning horizontal lines, beginning at the top left of the screen and ending at the bottom right. After each horizontal scan the electron gun beam is switched off (blanked) while the beam is repositioned for the next horizontal scan. This is horizontal blanking.

When the beam reaches the bottom of the screen and needs to reposition to the top to begin the next image frame, there is a much longer vertical blanking interval.

Blanking is actually a voltage reversal process which leaves the screen black (unexposed). It can also be considered a time delay, or time wasted. Numerous attempts have been made to insert various forms of information content into these blanking periods, and so use the transmission facilities to full advantage. See teletext and closed captioning.

bleed

• In desktop publishing, you bleed a picture on a page layout when it runs right out to the edge of the page — occupying margin space outside the area of text.

• Bad printing with too much ink, or slow drying, also causes the characters to 'bleed' on the page. OCR has a lot of trouble reading newsprint where this has been allowed to happen.

BLER

Block Error Rate. This is the number of blocks received with at least one erroneous bit, as a ratio of the blocks received. This is sometimes confusingly written as BER which is bit-error rate.

blind copy

A facility in some electronic mail systems which allows you to 'carbon' copy (c.c.) an electronic message without notifying the recipient of the main message that you are doing so. It is considered to be unethical.

blitter

A manipulator of bit mapped images in a computer.

BLOBs

Binary Large OBjects. A semi-humorous name applied to very large contiguous files which may need to be transmitted across a network, and which can therefore be expected to hog the network resources for a considerable period of time.

A BLOB in a database record, however, is not restricted in length. This is something associated mainly with CAD, engineering and automation.

The ATM cell switching defeats BLOB problems by breaking them down into short cells, which gives every other user a chance to access the network. BLOBs are in the news recently because of the potential problems we may experience on the data superhighway with video-on-demand or similar very large files. These will probably travel across the network in MPEG-2 compressed form, but they will still consume resources equivalent to about 2Mb/s — and that's for each home.

block

The term is often used as synonymous with "frame", but frame also has other meanings which can be confusing. Block is used to mean a contiguous sequence of data bytes or characters, which are all part of a file, or from a single source. Blocks of data are handled as a single unit.

• This can be a file's basic physical unit, especially in magnetic storage: a sequence of bytes treated as a single entity; a frame in transmission, or a cluster of sectors on a disk. The term is often associated with disk and tape storage, and it implies that the block will be addressed as one unit, and written and read as one unit. Between blocks are blank areas known as interblock gaps.

• In communications, block refers to a group of characters which are treated as one for the purpose (usually) of error checking. The check is made at the end of each block, and the checksum becomes the end of the block. These blocks might be 128 bytes long, or much longer.

• A section of text in a document, selected and marked (highlighted) to be shifted or changed in some way.

• A physical unit in a computer architecture — i.e. memory block or I/O block. This is often 512 bytes in size.

• A substantial section of a program, designed to perform a special task. This is most likely a procedure which can be logically treated as one unit in a block diagram.

block codes

These are error control systems performed on blocks (as distinct from parity checks on bytes, etc.). The emphasis here is on detection, not correction — although correction is always possible by calling for retransmission. LRC is a block code, as is the normal CRC checking. See ARQ and X-modem.

block device

A computer device that deals in blocks of data rather than streams of bits or bytes. A disk drive is a block device. The distinction is often with character (byte) devices like keyboards.

block diagram

A 'black-box' diagrammatic approach, illustrating the conceptual architecture of computer/electronic hardware or software.

block downconverter

This is a device mounted usually behind the satellite dish, to convert the entire satellite band (usually 500MHz in bandwidth) to a lower intermediate frequency so that it can, a) travel the distance required down the antenna lead, b) pick up much less noise and interference, and c) be in a frequency range which makes it easier to tune and demodulate.

Low Noise Block (LNB) converters are fixed-frequency local oscillators, and they produce intermediate frequency (IF) output.

block error correction

• During CD-ROM premastering, an error correction code (ECC) is appended to each block (2048 bytes) of data. The distinction is with the other forms of error correction also used at various levels.

• **Block checksums** are the same as longitudinal parity checking (see LRC). The checksum is appended to the end of the block of data. CRC is a more complex and better technique.

block gap

The unrecorded space between blocks of data on a disk or tape. It is wasted storage space, but it allows insert additions to be made, and it is easier for parts of the tape to be found when spooling rapidly.

block move

See block operation.

block multiplexing

In time division multiplexing, the MUX may interleave the channels by transmitting only one bit at a time from each, or one byte at a time, or one block at a time. All approaches have advantages and disadvantages — however bit and byte modes are the most common.

block mux channel

See BMC.

block operation/transfer

• In word processing, this means selecting, cutting and later pasting a phrase or set of phrases as one unit.

• In computer memory management it means transferring a section of data to a new location. Blitter is a block operation.

block rules

These define the way a language's programs will run. In Fortran, a block is either the compulsory main program or independent sub-programs which are not nested. PL/1 programs can also be divided into procedure or begin blocks.

blocking

A condition where no communications paths are available to establish a circuit. They are in limited numbers, and all are currently in use — so even though the receiver is free to take a call, one can't get through.

Networks are blocking or non-blocking. Non-blocking networks will let any call through, provided the receiver is free to take the call.

• In telecommunications, blocking is the measured ratio of unsuccessful call attempts as a ratio to the total number of first-call attempts (the sum of the successful and the unsuccessful) expressed as a percentage. A figure of 2% blocking means that 1:50 of the time people will be unable to find a vacant channel.

• Some telephone networks and services are 'non-blocking' so a newcomer is stopped from making a call. Packet-switching networks are often of this type, as are CDMA wireless telephony systems. When these systems overload, the quality of all suffers (from delays or voice quality), but newcomers are not denied access.

• During telephone maintenance, blocking signals may be sent deliberately to stop calls being made from an exchange (or node) while allowing them still to accept incoming signals.

• Blocking signals may be sent to 'reserve' a circuit for special use.

• Blocking and non-blocking modes are provided with pipes in client-server applications. The blocking mode causes the pipe to hold its read-cycle until data is available — effectively stalling the system temporarily if the data is delayed.

• The blocking factor is the number of records in each block of data on a disk.

blowing

To add data to an PROM or EPROM. You blow a PROM once, and once only. However you can burn new information into an old EPROM after first 'washing' it clean, by exposing it to UV light for about five minutes. The process is usually done in a PC using a special card.

blown fiber

This refers to the idea of leaving a hollow tube threaded through a building or down a street for later optical fiber. When increased capacity is required, the fiber will simply be blown through the tube by air pressure.

BLT

Block transfers. Generally known as bitblts. See.

BLTZ

A data compression technique derived (by British Telecom) from Lempel-Ziv, and used in the ITU's V.42bis international standard for high-speed modems. See LZW.

Blue-book Ethernet

The early Ethernet specification established by DEC, Intel and Xerox. It was also known as DIS Ethernet.

blue box

In phone phreaking, this is the type of tone-box invented by Steve Wozniac and manufactured by him in association with Steve Jobs (both of Apple fame).

It generates MF tones which simulate a new origination for the call after it leaves the billing office, thereby by-passing the billing procedure. The newer system ignore these tones since they signal in a digital bit-stream (see CCS#7). However Blue boxes can still cause trouble if one section of the link uses the older techniques. Many networks now have a detector, which sets of alarms if a Blue box is used. See red box and black box.

blues

In film and video production these are a form of transparent film of a medium deep blue color, which is used to change the color temperature of film lights to that of daylight. Blues come in large rolls, and they are clipped onto the barndoors of lights (sometimes in frames), and sometimes stapled across windows. The opposite of the blue (changing daylight to match quartz light) is the orange 85 filter.

B-MAC

Scientific Atlanta's variation on the MAC satellite analog television standard. It is only one of the MAC (Multiple Analog Component) family; C-MAC, D-MAC, D2-MAC, are all variations on a theme. The video is identical in each system, but the handling of audio and digital data channels changes. B-MAC used a data/ audio channel of 1.8Mb/s for data or sound in any combination: ADPCM sound techniques were also used. See MAC.

BMC

Block Multiplexer Channel. An IBM channel which implements the US channel standard, FIPS-60. This also carries other names: OEMI channel, 370 block multiplexer channel, and block mux channel.

BMP

Bit Mapped. BMP was Microsoft's simple bitmapped format for Windows. It is a standard image format which supports monochrome, 16 colors, 256 colors and 24-bit color. This later become a subset of the Windows DIB formation, but it does not support image compression.

BMUX

B-channel Multiplexer. This separates the various signals in ISDN into voice and data streams which can be switched to different locations. It is also responsible for aggregating streams.

BNC connector

Bayonet Nut Coupling: A small bayonet-type connector for coaxial cable. It is extensively used in video and it is the standard connector for 10Base-2 Thinnet and Cheapernet.

board

Usually short-hand for an expansion board, adaptor, or add-in card, but it could mean the motherboard or the PCB board material. See all.

BOCs

Bell Operating Companies (also called 'Baby Bells' or telcos).

These are the 22 local telephone companies (LECs) that figured in the break-up of AT&T's American telephone monopoly. The 22 BOCs were merged into 7 RBOCs with regional responsibility. There were also thousands of independent small local telephone companies, and about 1000 of these remained as independents. The largest independent LEC (not a BOC) is the GTE system. See RBOC and LATA.

body

- The main part of a program.
- The bulk of the text in a word processor file. Hence, body-type is the major type font used when publishing an article.
- The body size of type is the full height measured from the top of the tallest ascender to the bottom of the lowest descender (as distinct from the x-height). It is the 'point size'.
- In newspapers, this is the main bulk of the text of an article, also called the copy. The distinction is with headlines, captions, illustrations, etc. See copy.

body stabilized

A non-rotating type of satellite. See 3-axis stabilized.

BOF

Beginning of File.

boilerplate

To patch together from basic information stored in a library.

- In word processing, boiler-plating it is a technique of writing letters by selecting standard 'canned' paragraphs.
- With files: Boilerplate files are prepared 'electronic' from segments assembled by the operator. This is a lego approach to writing reports, letters, programs, etc. by slotting together modules of text or algorithms already written.

bold(face)

In desktop publishing and word processing, any typeface (font) which prints blacker than usual. These fonts are very often simply known as 'bold'.There was often a boldface attribute in the old PCs 16-bit bytes. See attribute.

- Impact printers often make letters bold by double-striking.
- Some typefaces have a special bold version. This was always true of the bit-mapped fonts, but some outline fonts also have these. Others can calculate a bold version when required.

BOM

Beginning of Message. This term is sometimes used in ATM discussions along with COM (Continuation of Message) and EOM (End of Message).

bomb

A disaster.

• A system bomb: A disastrous problem has occurred, a crash or abend — you'll need to reboot the computer.

• A virus: A 'bomb' is a deliberately planted bug which may destroy your files or crash your computer. A Logic bomb is one that hides in the system, and is triggered by some event - usually a date.

BONDing

Bandwidth ON Demand INteroperability Group. An American group proposing an ISDN aggregation standard (mainly used for videoconferencing) which is byte-oriented and runs on special hardware.

• Multilink is a cheaper software-base method which is also widely used. See aggregation and Q921W.

book sizes

The worldwide book trade has accepted new paper sizes and has devised an abbreviated system for specifying the trimmed sizes of a book. Note that some of these do not square precisely with paper sizes; this is due to different forms of trimming. The major book sizes are:

A4	300 x 210 mm	11.75 x 8.25-in.
A5	210 x 150 mm	8.25 x 6-in.
A6	150 x 105 mm	5.875 x 4.125-in.
B5	250 x 175 mm	9.875 x 6.875-in.
B6	175 x 125 mm	7 x 5-in.
B7	125 x 90 mm	5 x 3.5-in.
C4	255 x 190 mm	10 x 7.5-in.
C8	190 x 125 mm	7.5 x 5-in.
D4	290 x 220 mm	11.25 x 8.75-in.
D8	225 x 145 mm	8.75 x 5.625-in.
Ffol	345 x 220 mm	13.5 x 8.5-in.
F4	215 x 170 mm	8.25 x 6.75-in.
F8	170 x 110 mm	6.75 x 4.25-in.
M8	30 x 150 mm	9 x 5.75-in.
M4	295 x 23 mm	11.5 x 9-in.
P8	210 x 135 mm	8.25 x 5.25-in.
R4	320 x 255 mm	12.5 x 10-in.
R8	255 x 160 mm	10 x 6.25-in.
SF4	200 x 170 mm	7.75 x 6.75-in.

• In Europe the most economic book size to print (using web offset) for runs of, say, 5000, is probably the 'Trade size' of 210 x 137 mm (trimmed size), followed by A5, B5 (250 x 176) and American Quarto (280 x 210).

• Many of the smaller-run printing houses (1000 copies or less) will use sheet-fed presses which may require a different broadsheet size, and therefore be most economical at a size different from the above. See paper sizes also.

Boolean

• **Algebra:** A branch of mathematics dealing with logical statements and propositions which forms the basis of the operations of computer hardware and some software. Boolean algebra deals with classes and propositions, and uses such operators as AND, OR, NOT, EXCEPT, IF..THEN...ELSE.

• **Fields:** In databases, this means a field that can carry a true or false (Y or N) indicator only.

• **Operations:** This refers to the basic logical function of the electronic gates used in computers. For instance, with the AND function, the output signal will only be 1 only if both input signals are 1. Boolean operations function according to truth tables. See.

• **Operators:** The application of Boolean arithmetic to the problem of selecting items from a database. AND, OR and NOT are the primary Boolean operators.

• **Search:** The use of Boolean operators to combine search items into a search strategy in information retrieval. Booleans allow you to narrow down the number of records retrieved, and hone in on those most likely to fit your requirements.

A typical Boolean search term is something like:

```
>FIND ASBESTOS AND MINING NOT HEALTH
```

booster

• **Satellites:** In satellite terminology, the booster is the last stage of the multi-stage launch vehicle, or it can be the first stage of a satellite thruster which has been launched from the shuttle.

With GEOs, the booster will shunt the satellite from the low transfer orbit, out to the 22,438 mile (Clarke's) orbit necessary for geostationary operations. Boosters are usually powered by cryogenic fuels (liquid hydrogen and oxygen). The terminology is confusing; these are called perigee thrusters, or apogee boosters.

• **Television:** In television reception, a booster (sometimes called a 'reflector') is a signal amplifier/retransmitter which transmits in the same wide frequency band as it receives (as distinct from a translator). These are also called beam-benders, and they are used to fill shadow-pockets where signals don't normally reach.

boot

• To load the operating system into computer memory from a disk or tape. This is a shortening of the original term 'bootstrap', as in 'lift yourself up by your own bootstraps'.

This was a semi-humorous reference, originally applied only to the small built-in program which initiated the primary action of bringing the operating system into memory — at which stage the operating system takes over and runs the computer itself.

Boot now has come to mean the action of turning the machine on, including the OS's self-testing phase.

— 'Cold boot' implies that the machine has been started from scratch, while 'warm boot' suggests that the machine is already in use, but that something has caused sufficient problems with the application or the operating system that a semi-restart is necessary.

• We also now talk loosely of 'booting' an application, which can suggest loading both the operating system and the application ready for use.

• The routine in ROM that performs the above.

boot blocks/sector

The sectors on a disk that are accessed first by the drive during startup. These sectors contain the basic operating system data. The initial boot program is in ROM or PROM, but this provides only enough information for the machine to access the boot blocks on the disk. The terms 'boot program' and 'boot sector' also apply to these tracks on the disk.

bootP

This is a protocol which can be used by a network node to discover the IP address of its Ethernet interfaces in order to effect network booting.

BOP

Bit-Oriented Protocols. These are data communications protocols which do not assign any character (eg. ASCII) to the flow of data bits. They will treat the traffic as a simple stream of bits. The comparison is with BCP.

border gateway

A gateway-router which communicates with routers in other autonomous systems. See AS and IGP.

boresight

Roughly, the satellite beam center — the center of the footprint — of the 'isobar' contours. Very often the power of a satellite beam will be quoted in terms of the boresight level, the assumption being that it will fall off at a reasonably regular rate the further it is away from the boresight. This does not always happen.

A typical C-band boresight level for distribution to rctransmission mastheads might be 37dBw, while with total continental coverage it may be as low as 25dBw (for a hemispheric beam). For analog direct broadcasting satellites intended for home dishes under 1 meters, the level must be in the 45dBw range at present. See LBC and EIRP.

• The boresight is along the axis of symmetry of a microwave's parabolic reflector.

BORSCHT

A popular mnemonic for the technical features and services the phone company must supply to a customer over the local loop:

— Battery (talk power),
— Over-voltage protection,
— Ringing (bell signaling power),
— Supervision (to detect when the phone is picked up, and what number was dialled),
— Codec (for linking analog to digital circuits),
— Hybrid (for two-wire to four-wire connection),
— Testing.

boson

A class of particles that includes photons. Bosons are uncharged particles, and therefore photons beams can pass through each other without one influencing the other. Electrons can't do this.

bot

A virtual person, agent, or thing on the network. The term is a shortening of knowbot. It's a on-line agent (software) which can perform simple remote tasks for you on networks (particularly The Internet). Some on-line forums use bots quite extensively. But never run a bot that has been sent to you — it could encapsulate a Trojan Horse.

bottleneck

The weak link in the chain. The narrow part of the artery.

• In computing, the bottleneck is often the time it takes for RAM memory to provide information to the CPU. This is the famous von Neumann bottleneck. See wait-states and parallel processing.

• In communications, it can be any node or process which obstructs the full flow of the communications channel.

• In LAN it is now often the bandwidth of the main-server connection, or the server itself.

bottom-up

Techniques that begin at the lowest level of abstraction and proceed towards the highest level. See top-down.

bounce

When e-mail is not delivered for any reason, it is usually bounced back to the sender. The bounce may include the original message plus details of the path it took through the Internet. If you haven't made an obvious addressing error, consult your Postmaster.

bound(ed)

• **Bound data:** Data which has been 'reserved' for possible change while a transaction is in progress. It is bound in the sense that it can't be accessed by another computer/terminal until it is either updated, or restored. Other users must be locked out until the transaction is completed. See ACID test.

• **Bound(ed) delays:** Delays which are kept within some definite limits. This is a term used in ATM to make the point that, while delays might happen to ATM cells passing across the network, these delays are kept within tight limits — they are not 'unbounded', as they are in many other packet-switched systems.

• **Bounded medium:** Bounded media are those which confine and retain signals — such as wires, optical fiber, and wave-guides. The distinction is with radio broadcasting which is not bounded.

Bounded media implies that the physical path of the communications is determined by the medium. Don't confuse this with 'bind'.

boundary function

Some SNA subarea nodes (typically in IBM 3745 devices) can provide protocol support for any attached peripheral nodes. This is the boundary function.

boundary representation

See graphics modeling.

boundary routing

This is a network system architecture which simplifies the job of handling remote locations from a central router. It has been developed by 3Com as an open standard, which consists of a backbone interface and remote client specifications.

With boundary routing, the remote router only needs to decide whether a packet is local or for the WAN — and if it is for the WAN it will pass it on using PPP. This means that the remote router can be cheap and simple, and the management can be concentrated at the central device.

bounding box

The imaginary rectangle that can surround an image. This is a term used in page-layout programs to define the space within text-copy which needs to be left free for a later illustration.

boustrophedon

As the oxen plows. The scanning of a screen or a line of print from left to right, then right to left, then left to right again.

Bourne shell

A standard command-interpreter for Unix System 7. This is a standard user-interface shell which is very fast and compact. C- and Korn-shell are two others.

bozo filter

See kill.

BPDU

Bridge PDU. This is a hello packet in spanning-tree.

BPF

Bandpass filter. See bandpass.

BPON

Broadband on a Passive Optical Network. This is the basically the same technology as B-TPON, and the evolutionary form of TPON. These were optical fiber networks which European carriers planned to use in the streets ('fiber to the home') in the UK and Europe.

They use a passive splitter in the street pillar, and 20 to 30 channels of PAL video could be carried in analog form over the fiber using Wavelength Division Multiplexing (multiple lasers of different wavelengths) along with a 64kb/s interactive (single ISDN) channel for telephony. The digital revolution caught up with these plans.

bps, Bps or BPS

- **bps** — bits per second. It is now much more correct to write this as 'bit/s' or 'b/s' — and therefore also 'kbit/s' or 'kb/s', 'Mbit/s' or 'Mb/s'. See bit-rates.

- **Bps** — Bytes per second (used often in computing when dealing with the content, rather than the technologies). This should be written (but often isn't) as Byte/s or with the capital-B as B/s (and kB/s or MB/s).

Note that some sources invert this, and use the small b for Bytes, and the capital B for bits. I find this choice weird, and beyond all comprehension.

- **BPS** — Broadband Packet Switching. A Broadband-ISDN term for ATM.

BPSK

Binary Phase Shift Keying. See.

BRA

Basic Rate Access. See.

brackets

A term loosely used both for the curved ( ) parentheses and the square [ ] true brackets. The term often also encompasses curly { } brackets. In UK English, the word brackets often refers to all types, while American English makes a distinction between parentheses and the others.

BRAM

Broadcast Recognition Access Method. A LANs transmission protocol. See BRAP.

branch

The point at which a program can take one of many (two or more) paths.

- Conditional branching is a process in computer programming where diverging directions are available to the computer. One path is chosen over the other on the basis of some pre-set condition — usually the size of a number in a variable, or the match or non-match of an alphanumeric string.

- Branching also occurs in interactive video programs where new pathways are chosen through the available A/V material, depending on the answer to a question, or by the user's deliberate choice.

- The term applied to the hierarchical tree structure of directories, etc. — root, trunk, branch, twig, leaf.

- In programming — a jump or GOTO instruction.

branched path applications

A CAI term for 'non-linear' training programs that take the student through an ordered hierarchy of branching modules. Branched training modules can be either:

— a) tree-like (constantly diverging from the trunk), or

— b) branch-and-return. In the second case they have many affinities with linear structures.

The distinction is mainly made between linear progression and branching programs, and also between these more traditional structures and the freedom of hypertext (mesh-networked) structures. Hypertext often has no discernible structure.

BRAP

Broadcast Recognition with Alternating Priorities. A queuing contention protocol for LANs. Each time a station fails to gain access to a network it must wait for a time set by a variable held in memory. But each time it fails, that number-value reduces by a set amount. So the longer the node has waited, the more chance it has to take control of the network. This creates a random variable time delay across a large network.

breadboard

A way of trying out an electronic circuit without having to solder all the components together. These days the breadboard is a plastic mesh which holds electronic components temporarily.

break

• A key on the keyboard used to stop the execution of a program. Or any interrupt signal used to stop a process.

• On older asynchronous terminals, this was a key which didn't output an ASCII character but changed the line voltage for a short period of time (usually dropped it to zero for more than one character length) and it therefore caused the system to disengage or be alerted in some way. The receiving end would then seize control of the circuit.

This is a hardware control system: there is no standard ASCII equivalent, but Control 1 is often used to simulate the break.

• A series of logical 0s transmitted by a remote station when it wants to break the link. This can be detected by some modems and adaptor cards and is interpreted as a break.

• All keys on the keyboard make and break electrical contact, and sometimes put out a scan code at the time of the break.

breakout

A connection box which is inserted into multiple line circuits (especially serial cables) to show activity on each line (through LEDs) and also to enable outputs on one line to be diverted to another. Breakouts are commonly used to fix problems with RS-232 cables.

BRI

Basic Rate Interface of ISDN. See Basic Rate Access.

bridge

• To deliberately or accidentally make an electrical connection from one circuit to another.

• A combined hardware and software device (often a PC with plug-in cards, but more often a dedicated box) which is used to interconnect LAN segments. Interconnect devices (bridges and routers) are used to link separate segments into a single network. Repeaters are used within segments to extend the length. Routers are used to connect LANs into WANs.

Bridges are essentially simpler devices than routers — but they perform some basic routing functions (see spanning tree algorithms). They base their filtering (selection) and forwarding (redirection) functions on the unique MAC address which is found on all network interface cards (routers use a network address), and they build a database of addresses against which to check each incoming packet, before forwarding that packet (or returning it to the same segment). They require a tree-type topology (without loops) and they need network management.

Bridges are used to:

— **Segment** an over large network, mainly to improve network performance by eliminating unnecessary traffic. Each side of the bridge then becomes a LAN in its own right. However they connect the two sides into a single logical network (the stations don't need to know that the division has been made), while keeping local traffic confined to its own local segment — filtering out those packets on the first LAN segment that aren't meant for the second.

— **Extend** a LAN and its resources beyond its topological limits. Bridges regenerate the binary information, so the distance limitations imposed in LAN standards only apply to segments and to segments connected by repeaters.

— **Extend** the LAN's resources to other networks (usually of the same signal type, if not using the same media). They are often used to connect networks which use different media; e.g. fiber-optics to coaxial.

• There are three basic types of bridge:

— Simple bridges which have a fixed look-up table and follow it blindly,

— Dumb (source-routing) bridges which need to be told which packets to relay and where,

— Spanning-tree bridges which self-learn by experience.

Smart bridges monitor the source address of packets and automatically build routing tables. If entries in the table aren't used within a certain time (usually about 5 minutes), most smart bridges will delete the entry to prevent the database table becoming flooded.

• The intelligence of bridges is very variable. Most now have learning capabilities, so they know, without being told, which addresses are local and which are remote. Some can do filtering of special protocols, but not in such a flexible way as routers. Some can even make connections, when required, over dial-up phones and bridge over long distances.

• With packets intended for a remote location, LAN bridges broadcast the message to all networks, unlike routers which send the message specifically to the one destination. So, when bridges are being used, each node on the internetwork must have a unique address.

• Bridges usually operate at the Data-link layer (within the MAC sub-layer), which means that they are protocol-transparent and can forward all high-level protocols. They handle data as 'packets', rather than as bits or messages.

• Remote bridges or 'half-bridges' are available for long-distance connection. These may use physical circuits (usually modems over analog telephone lines), virtual circuits (X.25 packet-switching) or datagrams (DQDB-type MANs).

• Bridges have traditionally been stand-alone devices but

increasingly they are being 'internalized' as plug-in cards for network servers.

Bridge disk standard

The CD-ROM/XA variations on the original Yellow Book Standard, which provides a bridge between the CD-ROM format and CD-i. It includes three new levels of audio and image definitions. This is also used by Photo CD. See CD-ROM/XA and CD-i.

bridge-router hybrid

Aka brouter. This is a device that combines the functions of a bridge with that of a multiprotocol router. Routers can send messages directly to their destination, while bridges broadcast the message to all networks in an internet. However while bridges are protocol-transparent, most routers can handle only one type of network protocol. A brouter combines these attributes. It will route some protocols (TCP/IP and OSI) and bridge the others (such as XNS).

Because they handle more than one protocol, brouters are often confused with multiprotocol routers. However a true brouter will decide whether a protocol should be routed or bridged. Brouters are therefore ideal for small-load LANs, but unfortunately they are complex, expensive, and difficult to install. However they provide an ideal solution for heterogenous networks having internetworking problems.

bridge tap

The remnant of a previous connection left dangling on a communications cable — unused and un-terminated. This often happens with old telephone and coaxial cable networks, and it is the cause of numerous problems. Bridge taps cause impedance mismatches in the cable, and they create reflections (echoes) on the line. They can also radiate RFI in cable TV networks. Usually there is an allowance made for cumulative bridge tap lengths on a cable system.

• The US line code standard for ISDN, 2B1Q, was specifically chosen because it is resistant to problems created by these bridge taps.

bridgeware

A translation program or hardware used to transfer data or applications from one system to another.

BRM

Basic Reference Model. The OSI 7-layer reference model. See.

broadband

This term is a little confusing because it is applied to two different concepts.

— The first is to the provision of a very wide channel which may have only one user, or may have multiple users with separate time-divided channels.

— The second is to 'multiband' systems where a number of users may share the wide spectrum, with each having its own frequency band allocation (FDM). This should be more correctly referred to as 'multiband' but most people call it 'broadband'.

• In analog telecommunications, traditionally the term baseband or narrowband was applied to the normal channel-width provided for voice-grade links (up to 4kHz). Anything which used a wider frequency range than narrowband was called broadband, although lately some other distinctions have been made:

In modern PSTN and ISDN terminology, any digital data-rate below 64kb/s is called narrowband; up to 2Mb/s is often called wideband, and above 2Mb/s is called broadband. In the USA the dividing point is set at 1.5Mb/s. This is a digital distinction.

• Telephone technicians talk about the narrowband and the broadband networks — meaning with narrowband, the local loop and the exchange switching, and using broadband to mean the multiplexed bearers used for long-distance.

• In the early days of LANs, broadband networks were used as backbones in large enterprises. Separate frequency-divided (FDM) channels (including both transmit and receive paths) were allocated to the different services using analog R/F carriers which would typically operate up to the 450MHz range over a single coaxial cable. Usually modems were needed for all computer devices needing to access the network.

• There are five modern digital public communications techniques which are now considered to be in the broadband category: frame-relay, Sonet/SDH, FDDI, DQDB and ATM/B-ISDN. See baseband also.

broadband bearer

A transmission system for broadband signals — usually provided by coaxial cables and repeaters, or microwave radio links. Note that twisted pair in the streets or in the office can carry broadband, but the term is usually not used in this way. See ADSL and UTP.

Broadband InterCarrier Interface

B-ICI. This follows the UNI, and precedes the DXI implementation agreements for ATM. It defines how two public ATM network will be connected for Private Virtual Circuit (PVC) operations. Eventually these interfaces will evolve to support SVC also.

broadcast

To send messages simultaneously to a number of people/stations. There is a technical sense here and also a logical sense.

— The technical sense of the term means to send the signal to everyone, which is in contrast to selective distribution or point-to-point links. This also sometime carries the implication that the transmission is provided without prior arrangement.

— In the second case, the term means transmitting signals that will appeal to a wide group, which is in contrast to narrowcasting which appeals only to a selected group. However, note how vaguely the terms are used! Usually the distinction between broadcasting and narrowcasting is that between mass-appeal content, and niche programming.

• In radio and TV the term broadcasting suggests that the material is available to anyone in the reception zone without additional equipment. Free-to-air TV using VHF-UHF is said to be broadcast, while terrestrially transmitted Pay TV using MDS is often claimed as narrowcasting.

• On some LANs the term suggests that the message will be delivered to all users currently on the system. A broadcast frame is a Data-link layer frame addressed to all stations attached to that data-link. A special 'broadcast address' is reserved for sending to all stations simultaneously.

• In data communications over large internetworks, broadcasting often just means to send a message to a number of different (but usually specified) users, locations, or mail boxes. The distinction is with multicasting and narrowcasting, two equally vague terms.

broadcast storms

See packet storms and black holes also.

broadcast videotex

See teletext.

broadsheet

A variety of paper-sizes, all about the dimensions of a single sheet (a page) from a large ('broadsheet') morning newspaper. The historic inch-based sizes are:

Crown	20 x 15 in.
Demy	22.5 x 17.5 in.
Medium	23 x 18 in.
Royal	25 x 20 in.
Large Post	21.0 x 16.5 in.
Foolscap	17.0 x 13.5 in.

My two local broadsheet newspapers measure 17 x 23.5 in. (425 x 600 mm) and 16 x 22.5 in. (405 x 575 mm). Obviously it depends on their printing press and source of paper.

broker

A component that links the client (the user application) with the server (the resource).

bromide

In printing, this is a halftone — a screened photograph or illustration at the correct size ready for insertion into the camera-ready copy.

brouter

A bridge-router hybrid. See.

brownout

Abnormally low voltage in the mains supply. It is usually applied to long-term (hour-long) changes. See sag.

browse

• Scanning through information with no particular target in mind. See also hunting and grazing.

• In a more formal database, the browse function allows you to scroll both vertically and horizontally through the database tables, and possibly edit the text as if it were a word processor.

• Hacking into a communications networks.

• In object-oriented programming you will browse to display and edit the class hierarchy of the objects.

• An Internet, hypertext or CD-ROM browser provides a way of wandering through modules of information (records, cards, etc.) in an entirely arbitrary way: not proceeding on a plan towards an objective. Browsers are either specific applications or sections of search and retrieval applications. The contrast is with systematic navigation.

BRR

Bit Rate Reduction. This is the same as compression, except that it is possibly a more correct term for what goes on in most compression systems. There are two functions of a bit-rate reducer:

— To translate to another coding system, so that the information is coded in the most efficient way. This is entropy coding, and it is lossless.

— To throw away information that won't be noticed, or which may only be slightly noticeable. This is lossy.

Compression can mean both of these functions, or it may only apply to the first. Obviously, with critical data, you don't just throw away information, however minor.

brute-force

The use of raw computer power, rather than applying an elegant intellectual solution to the solution of a problem.

BS

ASCII No.5 — Backspace (Decimal 8, Hex $08, Control-H). This character moves the cursor back one position on the screen — but doesn't necessarily delete it. It is one of a group of non-print characters known as format effectors.

• Don't confuse backspace with delete (DEL).

BSC

Binary Synchronous Communications. See binary sync.

BSD

Berkeley System Distribution. The version of Unix devised by the University of California at Berkeley. BSD 4.3 is essentially AT&T's System V Unix with a bunch of extensions which provide a rich set of networking and other utilities. They changed the kernel to improve file-naming conventions and networking also. Most Unix implementations are able to handle the discrepancies between Berkeley and System V.

Most of the implementation of TCP/IP was also done at Berkeley as part of BSD Unix. And since it was freely available, BSD Unix quickly became the reference implementation of TCP/IP. See Posix and Mach also.

BSI

British Standards Institution.

BSS

Broadcast Satellite Service.

BT

British Telecom.

BTA

Basic Trading Area. Most large cities are MTAs and most of the larger US towns are classed as BTAs. See MTA

BTAM

Basic Telecommunications (or Terminal) Access Method. A technique that permits read-write communications with a remote computer.

B-TPON

Broadband Telephony over a Passive Optical Network. The video distribution version of TPON fiber-to-the-home. It is also called BPON.

BTV

Business Television. This is a subset of tele-conferencing which is often confused with video-conferencing. Business TV is a one-way presentation of video material over cable or satellite, often with the return voice-only channels being supplied by the telephone network.

btw

Internet shorthand for 'by the way'.

bubble

• **Bubble Memory:** A form of electronic memory made by creating a chain of tiny magnetic bubbles. Bubble memories had a very high packing density, but the data needs to be read sequentially. For a while it appeared as if bubble memory would prove to be ideal for portable computer mass storage since the bubble system didn't need battery power to retain its memory. However, it has other problems — like cost!

• **Optical disk:** In some forms of optical disk recording, a bubble is formed at the boundary between two disk layers. This is a different technology to the above. There are a couple of different bubble-formation methods, but dye-polymer is the best known. See dye-polymer optical disks.

• **Bubble jet:** A type of ink-jet printer; they are talking about micro-bubbles. To print a small 10 point character size at high (almost laser) quality, a matrix of dots (36 x 48) is used with a bubble-jet head having 14 nozzles to the millimeter. These nozzles are about 30 microns in diameter, which is finer than a human hair.

The printing process involves bubble formation (heating), growth, contraction (the creation of discrete ink bubbles in the nozzle) and ejection. Ink bubbles are ejected 6000 times a second.

• **Bubble sort:** A very slow method of computer sorting of information where the computer ripples through all records in a database, comparing only two at a time and swapping position (if necessary) on a one-to-one basis. It takes many passes (called 'strokes') before the list will be in order. The comparison is with a shell sort or QuickSort.

bucket

• An area of RAM that is treated as a whole by some addressing systems. It can be large or small.

• Slang for a variable in programming.

• A store into which data can flow — a buffer.

• Leaking buckets are those which are permitted to overflow (data is lost). This is a technique used for handling overload in ATM networks.

buffer

• **Memory:** An area of memory (a 'holding' area) typically used for the temporary storage of information in transit between the computer and peripheral devices, or between, say, the keyboard and its CPU. Buffers are used to compensate for the different (or rather variable) rates at which devices process data. They are FIFO (first-in, first out) devices — and very often they are directly addressed.

Most buffers are merely partitioned areas of the main RAM memory or space reserved in extended or enhanced memory. But sometimes they are special add-on hardware units, or hardware memory built-in to the peripheral. See print spooler and look-ahead.

Buffer contents usually disappear when you switch off the power — which is where most problems occur; buffer contents must be saved to disk (flushed) before power-down.

— The PC keyboard has a buffer of 16 bytes, so it will hold the last 16 keystrokes if they are made while the CPU is occupied elsewhere.

• There's really no difference between a buffer and a cache in such areas as disk-drives. Generally, however, a buffer will be at a memory address known to the application. Cache is related to the device itself.

• A circuit placed between other circuit elements to prevent direct interaction or to match impedances. Buffers of this kind may be inverting or non-inverting.

• **Register:** A buffer register is a register that loses its contents when read. It is used in an ALU to latch data. In mainframes, a buffer is often the final 'holding register' on the output (the C registers).

• **Fiber:** The thin protective plastic sheath around an optical fiber within a cable sheath (not the plastic cable surround). Don't confuse this with cladding (see).

buffer banyan

See banyan.

buffered terminals

Terminals which don't transmit instantly. In polling systems, the information must be temporarily stored in the terminal, awaiting the poll from the host.

bug

• A malfunction of any kind — either hardware or software —

but mainly software. Popular myth has it that this was named by software pioneer Grace Hopper after she found a moth trapped between two relay contacts in an early computer, but the term was widely used well before then. Bugs are intrinsic errors in the design, while glitches are temporary errors that may be generated by random external events.

• Bugs are most probably software problems. In programming there are logical errors, and syntax errors.

• A semi-humorous term for an IC. See DIP.

• In security, to listen in without authorization. Usually to tap into a telephone wire, or to place a listening device in a room. The device is also called a bug.

built-in test

BITs. Many communications devices now have built-in tests. Modems have a number of loopback tests, for instance, and you should learn to use them.

bulk storage

Any form of mass storage, especially those suitable for archiving. The implication is that bulk storage devices don't need fast access times — so tape is used more than disk because of its lower cost and higher capacity. Bulk storage may be on-line, off-line or near-line.

• This usually means tape or cassette storage, or recently, optical disk.

bullet

• **Printing:** The large dot which begins this paragraph.

• **Cable TV:** A bullet is a signal sent down a cable TV network as part of the normal data stream to disable any set-top box which is not acknowledged by the cable company (stolen or pirate). Most pirate equipment depends on simulating the 'broadcast' mode, so the bullet tells the converter to shut down if it isn't specifically enabled (authorized). Obviously, it is preceded by an intense period of authorization of legitimate sets.

Despite the common propaganda, there is no harmful voltage spike, or anything which could destroy electronics.

bulletin board service

BBS. These are usually non-profit database and e-mail services run by computer clubs or individuals. BBSs are dial-up, text-based, information and computer conferencing facilities — usually based around electronic mail — and some now providing access to the Internet. Some bulletin board systems are free, some need payment of a subscription, and some are for business in-house use only.

There's a lot of shareware and freeware software around which can be used to establish a BBS, and most will run on an older PC — however, a reasonably-sized hard disk is needed. The main cost in establishing a BBS is in the telephone lines (although most calls are inward) and modems. BBSs are either single-user (one caller at a time) or multi-user systems, and the person who manages (and often owns) a BBS is called a 'sysop' (system operator).

Usually you can ring the number and try out a BBS without formally registering, although your access may be restricted. You can often register on-line, also, if you wish. See Fidonet.

BUNCH

In the 1980s, the main competition for IBM mainframes came from the BUNCH group — Burroughs, Univac, NCR, Control Data and Honeywell. How times have changed! Univac (Sperry) and Burroughs became Unisys; AT&T bought NCR, Control Data declined, Bull bought Honeywell and their machines are now made by NEC.

Of the other non-BUNCH contenders: Amdahl, is now half-owned by Fujitsu; HDS is now 80% owned by Hitachi, and the UK's ICL has been taken over by Fujitsu.

bundle

• A number of optical fibers or twisted-pairs grouped together in a separate sheath.

• To group a number of different related applications together into one package for marketing reasons.

• To sell hardware and software together, or hardware and services — often in such a way as to ensure that no normal consumer without a pocket calculator and half-an-hour to spare, can ever calculate which is the best deal available.

California has recently prohibited the bundling of cellular phones and long-term service contracts — and, with a bit of luck, other enlightened governments will follow their lead.

• Bundled software is a term sometimes misused to mean integrated software.

bureau

A commercial operation supplying computer and communications services on a time or usage basis.

burn in

To test equipment by running it for a specific time. Most hardware problems with modern digital electronic systems occur in the first few hours of operations, so it makes sense to insist that a new computer is burned-in by the retailer for a few days before purchase.

burner

A circuit (usually on a plug-in board) which allows data to be added to a PROM or EPROM. See blowing.

burst

Erratic (asynchronous) transmission of data in groups.

• A set of characters, bytes or bits grouped together for transmission.

• A type of modem that handles data in interleaved bursts — mainly used for communicating data (e-mail) over satellites.

• A description of the type of traffic on a communications system where the information is erratic and predominantly one way (but alternating), such as normal voice conversations. The data-rate on each circuit changes in a very short time period, and one side of the conversation is always listening while the other is talking.

LAN traffic, is of course, even more bursty than voice. It is entirely erratic in most cases. Bursty traffic is often an indication that packet-switching or statistical multiplexing is required.

• To separate fan-fold paper out into individual sheets.

burst error

The distinction is with impulse noise and spikes which can create an error in a communications system of about 1- or 2-bits in length. Burst errors are caused by sustained interference.

• On a CD disk a scratch or fingerprint can cause the loss of many bits during detection. Error correction codes can overcome most of these problems (thousands of bits). See Reed-Solomon code.

• In communications, burst errors are often taken as errors in more than one adjacent bit.

• In LANs, malfunctioning cards can produce packet-storms and black-holes because these create bursts of error messages which then overload the network.

burst mode

A special multiplexer may be able, under certain conditions, to briefly suspend normal time-division operations and utilize the full channel bandwidth for a single high-speed transmission of data in burst mode. This is not common.

burst modem

A modem designed for TDMA operation and also found in digital satellite ground stations.

Bus & Tag

An old channel standard widely used by IBM mainframes. Until recently it had a top speed of 4.5MB/s. This is a parallel system with rates now up to 36Mb/s for distances of 120 meters. See ESCON.

bus (computer)

A wider term than its most common computer use. It applies to shared connections, which are used for transferring electrical power, signals of any kind, and data specifically. It can apply to the wires (the hardware) or the signals.

In computers and communications, it usually means a common/shared data path for digital electronic device transmissions. It suggests that all stations on the network will see all the data transmitted (although each may select only those packets addressed specifically to it).

A bus can be:
- serial or parallel,
- unidirectional or bidirectional,
- high or low speed,
- local or general,
- arbitrated or free-for-all,
- internal or external,
- specific (for, say, address or data) or general-purpose.

• In computing, the term is applied mainly to the internal parallel paths on the mother board which link the CPU to the memory, adaptor card slots, the UART, parallel interface, and via these to any external peripheral buses.

• SCSI is recognized both as a bus and as a serial connection also. The distinction is that a bus usually has some sort of arbitration or conflict resolution. SCSI has this; RS-232 doesn't.

• The CPU in a microcomputer usually controls an address bus, a data bus, and a control bus, and it has internal buses itself within the chip. Some people think it is more correct to talk about a computer having a single address/data bus with separate address and data channels.

These buses or channels all use parallel paths.

— An 8-bit wide data bus can carry a single byte when transferring data. A 16-bit wide data bus will transfer data as 16-bit words (usually two related bytes — see attribute byte).

— The width of the address bus usually defines the amount of memory that can be directly accessed by the computer. Generally the bus-width will match the internal capabilities of the processor.

— To access 64kBytes of memory requires a memory address of 16-bits. A 20-bit address bus theoretically has a 1MByte limit for single-cycle operations. Similarly, modern computers addressing gigaBytes of locations require a 32-bit bus.

— Memory addresses greater than the single-state address-bus limitation, however, can be addressed by bank-switching or by multiple-state addressing.

See address bus, local bus, expansion bus, data bus, SCSI, PCI, ISA, NuBus, etc.

bus (network topology)

With LANs, bus topology means a shared serial link along which nodes are attached. A LAN bus topology can just be a single line or it can be a more complex structure in a tree-branch configuration. See tree.

The distinction here is with ring or star architectures. However you need to be careful here; Ethernet is said to be a bus technology, but it can also run on switched hubs in a star configuration. Each segment, however, is a bus.

bus arbitration

On some peripheral connections systems like the modern SCSI bus, the master control system must provide arbitration between the peripherals — it must make decisions as to which peripheral gets access.

bus enumerator

A fancy term for a device driver.

bus extender

• This is usually a stand-alone box which connects to one slot on the PC motherboard, and into which more adaptor cards can be added. This is a way of extending the number of slots available. Often bus extenders need their own power supply because they impose too much of a load on the existing computer power feed.

• The same term has been applied in LANS to repeaters, bridges and routers.

bus master

This is a fast approach to the transfer of data from a peripheral (or network) to system memory. It is a feature of both MCA and EISA, and it is now finding its way on to the ISA bus as a popular way to boost the performance of PCs acting as servers on small LANs.

The plug-in controller boards (for servers) by-pass the CPU by taking control of the bus for short bursts. The bus then transfers data directly into memory. Peripherals can pump data through the system 3 to 5-times faster than would normally be possible. A master algorithm supervises all data flowing through the bus, and generates the sync timing signals.

bus mouse

A mouse that needs a controller-adaptor which is plugged into one of the expansion slots. The distinction is with a mouse that has its controller built into the motherboard or keyboard , and with a serial mouse that plugs into a standard serial port.

business graphics

Pie-charts, bar-charts, graphs, etc. — as distinct from photographic images and free-hand drawings used for illustration purposes.

busy hours

The peak traffic periods (usually an hour in the morning and another in the afternoon) in a telecommunications network. The average volume of the busy hour (measured usually in Erlangs) is the key factor in deciding when a network upgrading is necessary. Telephone networks are engineered almost entirely to cope with these busy hours.

button

There are soft-buttons and hard-buttons.

• In software it is a Macintosh/HyperCard term which has now become used more generally. A button is an icon/object on the screen (which you can often create yourself) which has an in-built routine (a macro) which you may also write or create yourself. When you click on the button, it performs some function by following its underlying script. Buttons are widely used in hypertext and hypermedia. They can be transparent, and so be placed over text or graphics.

• In hardware, these are switches or keys such as the 'mouse button'.

BVU

Broadcast Video Unit — a name applied by Sony to early hi-band (semi-professional) U-matic video. It applied to both the normal hi-band form and to the SP U-matic. BVU implies that the equipment is broadcast quality when in fact it is substantially below the requirements of the major city networks or production houses. These recorders are used in business, universities, and small country news gathering operations.

BX.25

A recent Bellcore enhancement to X.25.

by-#

Memory chips are made in different configurations. A chip can be manufactured to provide only a single bit in each byte, which means that eight of these will be needed for normal 8-bit wide RAM, or 16 if the data-bus is 16-bits wide. Other chips are manufactured with a four-bit width, which means that only two are needed to create a byte (or four for a 16-bit data-bus); and some are eight-bits wide — or 'byte-wide'.

Look on memory as an array where the vertical dimension established the number of memory locations, and the horizontal dimension sets how many discrete chips will be needed to create a byte (or two bytes for a 16-bit wide data-bus).

So chips which can hold one million bits can be configured as: 1024k by-1, 512k by-2, 256k by-4, or 128k by-8 depending on whether they can receive, hold and transmit one, two, four or eight bits at a time. See byte-wide.

by-line

In page layout, this is the name of the author.

bypass

• In electronics, a bridge, shunt or connection made around a circuit or hardware module.

• In telephony, bypass is the technique of making a connection directly to a trunk exchange without passing through the local exchange (using leased lines or microwave). In the USA, these long-haul networks will be operated by different companies, so there is considerable savings in going direct. Large organizations can sometimes negotiate direct connections at reduced costs for their long distance traffic. SMDS may also by-pass the local carrier — especially if it carries voice.

• A direct radio link — usually to a satellite from the company's own premises through their own dish — rather than using a public earth station. VSATs are a way of bypassing both local and long-distance carriers, as are microwave links.

• On an FDDI or Token Ring network, a node can be ejected from the link, or it may de-insert (remove itself). It is then said to be in bypass mode.

byte

An 8-bit byte can represent one character, or represent part of a graphic, or constitute a control code. In binary mathematics, 8-bits creates 256 different code possibilities (but numbered from 0 to 255).

For this reason, there are 256 distinct codes in all in extended ASCII alphabet, of which the lower 128 (using only 7-

bits) are standardized in the ASCII No.5 alphabet. Originally the eighth bit (controlling the upper 128 characters) was reserved for parity; but later in communications systems these codes became used for network control. However the upper 128 characters are now used for graphic, symbols, etc. in most PCs and are now relatively standardized within the DOS world. (Apple has a different standard.)

Generally a byte is now considered to be 8-bits, but in strict use, the term applies to any number of bits which can be considered as a unit. At one time the term applied also to 16-bit and 32-bit groups in mainframes, but now 'word' is more often used. A 'word' is usually two bytes in length (but it can be more) and a nibble is half a byte (4-bits).

The term 'byte' and 'character' are generally interchangeable when dealing with text using the standard ASCII code-set, but recently we have seen the emergence of 16-bit Unicodes and the ISO's 10646 which require two- or four-bytes for each character. See Unicode.

byte addressable

If a computer stores only 8-bits at one address location it is byte addressable. If it stores more than 8-bits (often 16 or 32) then it is said to be word-addressable. In word addressable systems, a single access of a memory location will then bring more than one byte into memory. However this may still only contain one alphanumeric character. See Unicode and attribute.

byte mode

See byte multiplexing.

byte multiplexing

A term used in time-division multiplexing where the slots are allocated to the various channels (sources) a byte at a time. The technique is used with low-speed input or output devices. The distinction is with bit multiplexing and block multiplexing, or possibly with the use of longer packets or messages. This is also called byte mode.

byte-order

In a serial or parallel connection, where information is stored as multiple-related bytes, it is usual for the most important bytes to be transmitted first — but this is not necessarily so. It may be that the first byte is more vulnerable to interference; so the least-significant byte may be transmitted first. See bit-order also.

byte-oriented protocols

There seems to be two slightly different uses of this term:

- These are systems which use full bytes as frames and control codes. The distinction is with bit-oriented protocols. See binary synchronous.

- Data-link protocols which use a specific character (one from the ASCII character set) to delimit the frames. This has consequences in terms of zero-bit insertion, etc.

byte-serial transmission

Each byte is transmitted in sequential order. Note that this does not prevent parallel transmission of the bits in each byte — it only refers to the bytes themselves. I have no idea why they make this distinction!

byte-wide

Most memory chips are of the 'by-1' type which means they can send and receive data only one bit at a time. Each address location in a chip only holds the one bit of data. For an 8-bit byte, therefore, eight memory chips are needed in parallel. However, four-bit ('by-4') chips are now being made for computers, and 'by-8' or 'by-9' (with parity) special chips are made for calculators and portables. By-8 and by-9 chips are 'byte-wide' chips. See by-# also.

C

C
Coulomb: The SI unit of electrical charge.
- C and C++: The programming languages: See C language.
- Conditioning: See.

C#
The term 'C' followed by a number can mean many different things. For instance:
- **C0** is a standard code series in videotext. It provided the control functions such as cursor control and format effectors.
- **C0 — C8** are international paper sizes. They are often used for envelopes.
- **C2** is a channel designation for circuit-switched channels. See C-channel.

— **C-2** is a line-conditioning standard. It means that the line's impedance, etc. has been corrected at least enough for reliable terminal operations. C-5 is a much higher standard. See conditioning.
— In the US Department of Defense (DOD) **C-2** is a security classification. The DOD has levels A, B, C and D where A is the most stringent. Most corporate software (with password protection) is classified as C, sublevel 2. See DOD.
- **C4** is a group of CAD-related activities: CAD, CAM, CAE and CIM. See all.

— In the army **C-4** means Command, Control, Communications, and Computers.
— And **C4** is also an envelope size of 229 x 324mm. It can take A4 paper flat.
- **C-5** in cabling is Category 5 UTP. See Category # cable.
- **C-6** is a common envelope standard of 114 x 162mm. It will take A4 paper folded in fourths.

C4FM
Quaternary FM transmission using QPSK-C modulation. It needs a CFDD-compatible receiver.

C&D bits
Control and Display bits. This is terminology from the CD-Audio disk format. These bits are used by the audio disk players to cue up the next selection, but they can also carry text of lyrics or titles if the replay unit has some form of text display.

C-band
Satellites: C-band satellites transmit at around 4GHz and receive around 6GHz. The actual international band set aside for C-band satellites is 3.7—4.2GHz for downlinks, and 5.925—6.425GHz for uplinks.

Until the late 1970s, the C-band carried most of the satellite traffic in the world (this changed at WARC in 1979), but C-band is mainly used today for TV delivery, both at an industry level (for retransmission), and also occasionally for semi-DBS systems (Star TV across Asia).

With C-band you typically need a much larger dish (about 1.5m to 3.0m for home DBS) than Ku-band, but it proves to be more reliable in bad weather. New LNA developments have greatly reduced the dish size. However, the EIRP of C-band satellites is strictly limited by WARC because they share their frequencies with terrestrial microwave systems.

The C-band is also being regularly extended. From 1969, the frequencies for downlinking have been 3700—4200MHz with the Russian Gorizont series using 3675—3650MHz (channel 'One-minus'). Now recent satellites have extended the bandwidth to 3400—4200MHz. This now provides a frequency range for 30 transponders, each of 26.7MHz. See also the L-, X-, Ka-, Ku- and S-bands and radar bands.
- **PCS:** There is a C-band in the US PCS auctions. It was set aside for minorities and disadvantaged organizations, and has no relationship to the satellite C-band.

C-channel
The term for a circuit-switched digital channel within the 802.9 standard. Channel capacity is signified by Cx designation.

These are wideband isochronous channels for high-speed packet- and circuit-mode services, which were intended to be available for high-speed data transfers, video services and image transfers. The relationship between these and ISDN standards are:

C1	64kb/s	(a standard B-channel)
C6	384kb/s	(a standard H0-channel)
C24	1.536Mb/s	(an H11-channel)
C30	1.920Mb/s	(an H12-channel)

C:drive
In most MS-DOS and OS/2 machines, the C: drive is the primary hard disk.

C-format video
This is the most widely used 1-inch helical-scan videotape standard in the professional television production industry. It dates from 1977, and it superseded the old 2 inch quad machines after being unsuccessfully challenged as the new world standard by the B-format.

Ampex and Sony combined to create this standard, and C-format video has been formally accepted as a standard by the SMPTE and EBU. C-format video produces analog composite images and it has sufficient bandwidth for magnetic recordings that can withstand 4 or 5 levels of generation-loss without noticeable degradation of picture quality.

It uses a single-head which draws a 40cm track on the tape for each field. The tape is in an 'omega' wrap (nearly 360°) around the 13cm diameter drum; it doesn't close the circle, but leaves a 12 line gap which lies in the vertical blanking interval.

Ampex invented dynamic-tracking for C-format, which enabled this standard to produce high-quality still images and two-field replay even at still or slow tape speeds. When later combined with a 'read' head in addition to the single record head, it was possible to have 'confidence' replay (DRAW) and see the images 1/50th second after recording (rather than before, as with most systems).

C language

The C programming language was developed by Bell Laboratories in the 1970s. Kernigan and Richie created it to write Unix, and a C compiler is included in the Unix operating system.

It is a reasonably structured language that combines the versatility of assembly with the speed of writing applications usually only found in a high-level language. At the time it was unique in giving the programmer complete access to the internal (bit-by-bit) representations of all data-types in memory.

C is not machine-dependent, so programs can be transferred to other machines with little difficulty. There are a number of versions: ANSI C is a variation on the original, and Objective C and C++ are two object-oriented versions.

• C++ was written by Bjarne Stroustrup. It is object-oriented, and provides reusable software libraries and management utilities. Lately it has become a very important language for computers and communications systems.

C-MAC

A variation on the basic MAC satellite TV system which offers up to 8 audio channels. C-MAC was the original IBA-designed (British) analog component television delivery system. It used standard MAC video techniques, and had a separate digital data stream running at 20.25Mb/s. However the system was costly to implement and not very suitable for cable. B- and D-MAC systems were more widely used. See MAC.

C shell

A standard user (interface) shell for Unix. Bourne and Korn shells are two others.

C-stand

In film and video production, this is a light-weight tripod stand (like a light stand) which has a cross arm and clamp. It is used to hold flags, gobos and background materials. It is a general-purpose holding device.

cabinet

In telephony, cabinet will usually refer to a street pillar, where the thick cables from the exchange are solder-connected to quad (two-pair) individual cables which run to each home. These are passive boxes.

Cabinet can also mean the large metal box in city streets and small country towns which houses amplification and multiplexing equipment. These cabinets are active boxes; they come in standard sizes, and require mains (or sometimes solar) power — and generally battery-backup.

Cabinet design is important in hot climates; you can't always equip these boxes with air conditioning, yet they must be able to withstand the harshest of conditions. The general name for all telephone cabinets, boxes and pillars is 'street furniture'.

cable

• An assembly of one or more conductors of any kind of an insulating protective sheath.
— Co-axial cable is a special type used for high-frequency signal transmission where interference can be expected. It is shielded by a woven wire net or an aluminum foil sleeve.
—Twisted pair cable is the type usually used for local loop telephone connections; quad is a particular version which is used for the last dozen meters.
— Optical fiber cable (from 2 to 400 pairs) provides the interexchange telephone network and long distance connections these days.

• Historically, cables were overseas telegrams sent by Morse code.

cable matcher

A gender bender. See.

cable modem

A cable modem is a device that transmits and receives modulated data over the two-way version of the coaxial cables used for TV distribution. Ethernet is sometimes handled in this way. Intel has produced a cable modem chip with a 30Mb/s rate in the downstream direction, and 128kb/s rate upstream.

cable pressure

In many of the older high-speed and large-capacity terrestrial cable installations for telephony, the cable sheath was pressurized by compressed air. It's a technique which keeps water out, and also warns of early damage through loss of pressure.

cable terminator

With almost all daisy-chained cable networks, long-haul lines, or high-speed local connections (like SCSI), a terminator must be added to the far end of the chain/line to 'soak-up' excess signals and prevent electronic echoes being returned down the line.

Cable TV

The distribution of video and TV programs via land-lines using a common tree-bus network. Cable began in the USA as the Community Antenna TV services in about 1948. It migrated from CATV to Cable TV when it began to provide additional channels imported via satellites, microwave relays and land-lines.

Later, the addition of racks of VCRs and small studios at the 'head-end', enabled local programming to be initiated. The FCC has had varying requirements over the years for some local programming in the form of news, weather, and a 'public access channel'. This experiment in social engineering has not been a success.

Later in the USA, communications satellites enabled national Cable-only channels to evolve, and Premium (Pay) services began to be introduced, over and above those advertising-supported channels provided free with the Basic Cable subscription.

The TV network affiliates and local independent (over-the-air) broadcasters initially welcomed Cable systems as a way to extend their advertising reach, but later Cable became a substantial competitor to broadcast services because it diluted the number of viewers.

• Cablecast is the term used for programs originated over the cable network, as distinct from those taken from broadcast services.

• Wireless Cable is the term applied to MMDS and LMDS radio transmission systems.

• The term 'cable channel' can refer to the set of frequencies used on a local coaxial system to transport a program stream, but more often it refers to a nationwide program distribution service owned by Time-Warner, HBO, etc. These companies are also called MSOs. (see)

The most widely-viewed Premium cable channel in the USA, is ESPN which is the major national sports service. The Disney channel is another popular channel, as is CNN, the Turner news channel.

These are all Cable channels, in the sense that you won't see the same programming on the four national free-to-air broadcasting networks — but note that these same program streams can reach you via street-coaxial, DirecTV satellite, or MMDS microwave broadcasting.

You'll find these same programs in many of the packages offered by all service providers — no matter what the distribution medium. So here 'Cable channel' refers to content, rather than the carriage mechanism.

• Cable networks have traditionally been a franchised operation in the USA, where both programming plans and the price to be charged is renegotiated every five or so years with local authorities. The FCC also has a technical regulatory role.

Cable companies have, until recently, been prohibited from carrying telephony — while telephone companies have been prohibited from carrying video. This is all now changing.

• Cable networks have traditionally been tree-branch networks originating from the 'head-end' (which has the satellite dishes, antenna, encryption equipment, and distribution amplifiers). The FCC controls technical standards.

The main trunk feeds were 0.75-inch ('750')coaxial, and the distribution feeds were 0.5-inch ('500') coaxial. Passive splitters are then connected to (typically four) house drops, feeding the signals to the home.

Amplifiers are separated by distances of about half a mile (depending on the load) and up to 40 amplifiers can be in the signal path for houses at the far end of the network. This all adds noise, problems, and decreases service reliability.

• The new hybrid fiber-coaxial (HFC) networks use optical fiber trunks (with no in-path amplification) to reach up to 50km from the head-end. As a rough guide: these terminate in an active hub, which then splits into 12 coaxial distribution feeds to service a total of 500 homes.

In the worse case, a home at the end of the chain will now only have 4 coaxial amplifiers in the signal path. These systems can handle two-way (interactive) traffic. They have bandwidths of 1GHz, enough for 150 analog channels or perhaps 600 digital channels.

• In late 1994 the US still had about half the world's cable connected homes (134 million worldwide). The UK had only about 0.6 million. The US Cable TV figures in 1994 were:

- 11,385 Cable networks.
- 59.1 million homes connected (62.5% of all homes — up 3.5% from 1993).
- Average payment for cable US$30 per month.
- 40% of all set-top boxes were individually addressable (capable of supporting Pay-per-view).
- The two largest MSOs were TCI with 10.3 million homes and Time/Warner with 7.1 million homes.

• **Cable on Demand** is an older name for Video on Demand. See VOD.

• **Cable Ready** television sets are those equipped with a modern SAW filter which enables them to tune all UHF and VHF frequencies, and to separate adjacent channels.

cache

The word means 'hiding place'. It is a temporary fast-storage memory area (a buffer) for data that may be required shortly, but which is hidden from your application/CPU while it is engaged in some other process.

The idea of cache is always to improve the performance of the machine/software by holding large blocks of instruction or data temporarily (along with the location information) in a secondary memory area. The cache is then checked first by the CPU before it accesses main memory; so the success of cache depends on the skill of the cache controller in having the right data available to the CPU at the right time. This involves 'prediction' and 'locality' (the assumption that nearby data will be wanted next, or the assumption that recently used data will be wanted again). See look-aside cache.

All cache systems attempt to anticipate what data the processor will require in the next few cycles, and they use different forms of prediction. Disk and RAM cache, for instance, shift data into cache in large blocks rather than individual bytes. This avoids the speed problems of writing to, or reading from, disk or RAM memory, one element at a time — and it assumes that nearby bytes will be required shortly.

Cache memory controllers are responsible for intercepting calls to a disk or to a memory location, and for checking in the cache first. If the data is there, it is known as a 'cache hit'.

• The four main types are: internal cache (on-board

memory), external cache (fast memory or RAM cache), disk cache and, more recently, LAN cache.

— **On-board** memory cache is designed in to the processor chip, it can't be added later. A chip can have a cache for either/or data and instructions. Generally only a small amount of SRAM is provided, however in the 486 chip, a full 8kBytes of internal memory cache is included in the design. Transputers gain most of their speed from on-board cache. [Chip cache should always refer to Bytes (never bits), even if they leave the Byte off.]

— **Fast memory** or **RAM cache** is similar to the above, but a special area of the main addressable memory is set aside for cache use (often up to 64kB). This part of the memory map is fitted out with SRAM chips, and the cache controller (usually a special chip like the Intel 82385) will shift blocks of data from the data or application area into this cache, where it will then be accessed directly by the CPU.

— **Disk cache** can be added later. Both hardware disk caches and software caches are used. A hardware cache will normally have its own pool of high-speed RAM on a special adaptor board. Software cache will simply be an allocated area in the current computer DRAM.

— **LAN cache** is a recent development in client-server models where the server predicts the next request of a workstation and down-loads the data first. See memory cache, disk cache, coherence.

• Cache is now classified by levels. Level-1 is on-chip, Level-2 is high-speed memory within the memory map, and Level-3 cache is for I/O to peripherals.

— DEC's Alpha 21164 chip, for instance has on-board Level-1 instruction and data caches of 8kB each; a Level-2 three-way set-associative with writeback cache of 96kB; and an off-chip Level-3 CPU-support cache of 64MB. There's also a four-instruction-issue superscalar, instruction pipeline, and dual integer ALUs on the chip. The peak throughput is 1200MIPS.

— Intel has announced that it will provide 256kBytes of SRAM Level-2 cache with its P6 processor and have 16kB of Level-1 cache on the chip.

cache controller

A cache controller is an intelligent I/O device which attempts to predict what data the CPU will need next. These operate at various levels, but they all bring into cache blocks of data in a single move. They then spoof the CPU into believing that the location next in line, is in the cache rather than in regular memory. The data is therefore close at hand, and can be used instantly.

The efficiency of a cache controller depends on the accuracy of its guesses in keeping data in-hand. See look-aside cache, and also below.

cache hits

If the CPU is designed to work at top speed with its RAM, the cache controller subsystem will typically predict the next data required, and bring it into the cache, perhaps 80% to 90% of the time. Cache hits (expressed in percentages) are the ratio of memory-read operations that are successfully supplied from cache, as against the number of times the CPU needs to access slower system memory.

CAD

• **Computer Aided Design:** Most CAD systems use very high quality workstations, with superb monitors, but some of the lower-level applications will run on conventional PCs.

CAD with its graphics emphasis employs different peripherals to a normal PC, however. Input is often by graphics tablets and scanners, and output is more likely to be by vector printer/plotters.

CAD software has both horizontal and vertical markets. There are many generic CAD applications, but many more specialized programs for electrical circuit design, IC chip design, PCB design, mechanical process design, etc.

• **Computer Aided Drafting:** This term is not now used. See CADD.

CAD/CAM

Computer Aided Design/Computer Aided Manufacture. The philosophy of linking the computer design processes to the real-time control of the manufacturing machinery. This was more an integration ideal, than a product. The direct integration of numerical control (NC) machinery (lathes, milling machines, etc.) showed much promise about five years ago, but it is less in the news now. However, NC is common — it is just that the real-time integration didn't ever eventuate.

CAD-E

Computer Aided Design — Engineering. A variation on CAD-CAM. Also CAE.

CADD

Computer Aided Design and Drafting. This is CAD with extra drafting features, such as dimensioning. However since these features now come as standard in CAD software, the extra D has evaporated.

caddy

A flat plastic case which encloses a disk (usually a CD-ROM) to protect it while being handled. When you insert the CD disk together with the caddy into some CD-ROM drives, the drive will capture the disk but eject the container. Other drive systems hold both the caddy and the disk during use.

Libraries like the first type of drive, so that the caddy can be removed by the librarian. This then prevents the CD being ejected and stolen. Don't confuse caddy with the cheaper jewel case which protects CD-disks in storage.

CAE

• **Computer Aided Engineering:** This is mainly applied to equipment which aided in the development of new products, rather than in the manufacturing process.

• **Common Applications Environment**: The professed aim of the X/Open groups is to allow easy transference of applications across different architectures. The foundation of CAE is the XPG components. See.
— IBM also have a similar approach called 'CAE for SAA'.

CAFE
Conditional Access For Europe. A 3 year ESPRITE program in Europe to develop a smartcard which includes (among other features) such functions as electronic purse, driver's license, and payphone payment facilities. 13 companies are participating in trials. See SVC, STT and VME.

CAI
• **Computer Aided (or Assisted) Instruction:** The term CAI usually implies an interactive program of some sort — probably using a videodisk player under the control of a computer program. However there are many different CAI techniques, some of which involve still-images only; some which use audio only — and with varying degrees of interactivity.

CAI can be 'linear' in nature (taking the student step by step through a pre-determined path) or 'branching' (choice of multiple paths). Branching programs use either tests or choices to define a new path at each decision point (node).

• **Common Air Interface:** See.

Cairo
Microsoft's working title for a new and upgraded object-oriented version of the Windows NT operating system. It will incorporate a version of OLE.

CAL
• **Computer Assisted Learning** — see CAI.
• **Conversational Algebraic Language.**

calculator
The term goes back to the days of Pascal and the early developers of mechanical (gear-chain) counting devices. The first true electronic calculator was probably that sold under the Bowmar Instruments label in 1971, and it used the Texas Instruments TMS1000 4-bit chip. In the late 1970s, within about three years, the price of a hand-held electronic calculator dropped from a few hundred dollars, to twenty dollars.

calibration
The process of fine-tuning the equipment to a standard measure, so as to produce consistent and reliable results.

• Calibration bars in printing are strips of tones added to the negative or viewed on a proof, to check the printing quality.

call
To contact a specific node, function, routine, or person.

• A mechanism by which the execution of a program is temporarily suspended while a subroutine is executed. Execution of the main program will restart at the instruction after the call, but only when the subroutine finishes. This restart address is often kept on the stack. See GOSUB.

In programming jargon you 'call' a certain subroutine (or the memory location where that routine resides) in order to activate it (pass control to it). APIs are little more than the names of routines and information about how to call them. Some calls allow a machine language subroutine to be accessed from a higher-level program. There are:
— far calls, which jump from one memory segment to another, and
— near calls, where all jumps are within a memory segment.

• In some operating systems (like Unix) one program can call another, and that can call another, and so on — a process which is then called chaining.

• In telecommunications, the process of calling sends off-hook (request for connection) and dialing signals/messages through the system. The party which initiates the call is known as the A-party and the receiver is the B-party. See in-band signals, out-of-band signaling, channel associated signaling, pulse dialing, DTMF, and CCS#7.

• The term is also used for the total communications process in switched networks (not just the set-up phase). Here calls are divided into three phases: call establishment, message transfer and call clearing. This sequence holds for all connection-oriented systems (switched and packet networks) but not for connectionless services. See connectionless.

call-back
• A modem technique used to foil hackers. When you sign on to a host computer, the host accepts your identification and password, but then cuts the connection. It will then dial your number. Only when your modem answers can you make contact with the host or network.

Unfortunately, there are often simple ways (but only with some telephone networks) to fool the host computer into thinking it is making a call-back when it isn't.

• In modern telephone networks with intelligence (AIN), special CLASS features such as automatic call-back can easily be implemented. If you (the A-party) call a number and find it engaged, you will dial a special call-back code and terminate your call. As soon as the B-party's phone becomes free, the network immediately calls both A and B, makes the connection, and lists the charges against A's account. Both POTS and ISDN have this potential.

• Call-back is also a semi-legitimate way to defeat high international telephone call costs. Call-back companies operate in slightly different ways, but generally they originate in the USA where international prices are very low.

If you wish to call from outside the USA, you will dial a special US call-back number (and perhaps enter a PIN), then drop the call immediately. The US company will then automatically call you back and offer dial-tone for you to make contact anywhere in the world.

Your calls are therefore made at the USA-based prices. Note that two international calls are involved, but generally the call-back company gets such good bulk deals from the international carriers that overall costs are still lower.

call-blocking
The ability to automatically block incoming calls from being directed to a particular telephone extension or line. This is an exchange facility, available since the analog days in some circumstances. It originally used the ANI (the billing number) which was carried through the network, and it could be set to reject individual numbers, groups of numbers, or area codes.

Call blocking of a slightly different type is now an important part of Caller Line ID (CLID) where many people refuse to accept calls from certain numbers based on the CLID delivered to their home unit. Or they will refuse to accept calls which don't present an identifying CLID code. CLID blocking can be supplied by the carrier at the exchange, or by special CPE.

• Call-blocking is a term also applied to a restriction put on outgoing calls so that, for instance, only local calls can be made.

• In modern exchanges offering CLASS features, you can sometimes have a list of up to 20 numbers which will be blocked at the exchange and not passed on to the customer. The customer can add to this list any number at any time, just by post-dialing a star-code immediately after receiving a call from that number.

call clearing
Aka clear-down or tear-down. The orderly disengagement of a call in a switched network. The procedures used for call clearing in analog telephone networks (analog between exchanges) and those used in modern digital interexchange networks are quite different. See CCS#7 and tear down.

• Note that three different control approaches are used in telephone systems:
— the **calling party clear** approach, where the call is only cleared when the calling (A-party) puts down the handset,
— the **called party clear** approach, where the call is cleared when the B-party puts down the handset, and
— the **first-party clear**, where the call is cleared when either puts the phone down.

The A-party-clear approach means that the receiver can put down the handset and move to another phone extension without losing the call, however it leaves the B-party vulnerable to having a phone unable to receive calls if the A-party does not hang up correctly. There are also implications in hacker-scams involving entry into systems requiring call-back procedures.

call control signals
• Public switched telephone networks are a hot-bed of signaling activity which often remains transparent to the user. These are some of the signals exchanged in establishing a voice or data call.
— call request (includes the numbers and your ANI),
— call answered signal (when the handset is lifted),
— call connected signal,
— call blocking, if the trunks are all in use,
— call engaged signal,
— call not accepted signal,
— call failure signal,
— call clearing signal.

• In packet-switched networks these call packets are usually exchanged:
— call request packet,
— call accepted packet,
— call progress packet,
— call connected packet,
— call clearing packet.

call diverter
There are two variants, both of which are usually protected by keypad passwords.

• The first is a form of free-phone line supplied by a company, so that their executives can call in and access an outgoing line for long distance and international calls. The call would then be charged to the company rather than to the individual. This type gives you dial-tone.

• The second form is a simple connection to a company line which automatically outdials to a set location and passes your call through; the caller has little or no control.

Phone phreaks often have ways of getting access to call diverters, and there are ways of defeating the limitation of automatic outdiallers.

call forwarding
Both a PBX and a carrier-supplied service which allows you to divert an incoming call to another extension, or another telephone number. It is part of intelligent networking (see CLASS) and stored-program control switching.
— Direct call forwarding will divert your calls immediate, without giving you time to answer.
— Some systems offer call forwarding if not answered.
— Some offer call forwarding on busy. See hunting.
— Some can set up diversions which forward calls on the basis of time of day and day of week.

• With ISDN and X.400 messaging systems, call-forwarding may eventually be available globally so you can have phone calls and messages re-directed to your location anywhere in the world. But beware. Sometimes a traveller receiving a call from a building only one block away will involve two international links — both paid for by the recipient.

• Most mobile phone systems permit 'roaming', and calls to a mobile phone will be forwarded through the networks, to the last-known location of the mobile. GSM has a call-forwarding (international roaming) system built in.

See CLASS and tromboning.

call management

This refers to different levels of control exerted over the use of the phone system by large corporations and sometimes small phone-intensive companies.

• These are services and networks used by large distributed corporations and organizations to handle large numbers of phone enquiries. They will also allow monitoring of traffic, and probably compile historical data to identify peak periods and allow better staffing decisions. See ACD.

• At another level is the recording of billing information and general phone activity in professional offices. This is usually done by software at the desktop in a PC, but some PBXs also have this feature.

call priority

Priority can be provided to a call in switched LANs and in connection-oriented telephone systems of any kind (including ATM). In LANs having bandwidth reservation features, call-priority defines which calls have prior access, and in what order.

call-request packet

In a packet switching network the primary user node will first transmit a special call-request packet to establish a virtual circuit between itself and the required host. The response is a call-accepted packet which carries the desired routing plan and delivers virtual circuit information to intermediate nodes. From that point on, normal data packets are used.

call set-up

This term covers the full process, from the identification of the address required, to the route selection and the final acceptance of the call. The call set-up time is an important consideration in using switched circuits for LAN interconnection.

call-user data

In X.25 packet-switching services, extra characters can sometimes be added to the host address to allow caller identification. These codes are transmitted in the preliminary packet that sets up the virtual links and they are used by the receiver to judge whether the connection will be accepted or not.

• User-data is a quite different ISDN D-channel function.

called party

This is the B-party. The A-party is the call originator.

caller

In a telephone conversation, the caller (the originator of the connection) is also known as the A-party, and the receiver of the call is the B-party.

• **Caller ID**: There is no Caller ID, the best a telephone network can do is to identify the name and number attached to the caller line, which might be owned by someone quite different to the caller. See CLID

CAM

• **Computer Aided Manufacture:** The control of machinery by computerization. Linked with CAD in CAD/CAM.

• **Computer Aided Management:** The use of computers for providing business information and forecasts. Now called MIS or EIS.

• **Computer Addressed Memory:** The main block of working RAM plus the system area in high-memory. It does not include bank-switched memory or virtual memory locations.

• **Content Addressable Memory:** See.

Cambridge ring

The first type of ring architecture for LANs developed by the Cambridge University in the UK. It was a token-passing LAN.

camcorder

Integrated video cameras and video recorders. Note that not all camcorders can replay the material they record. See Betacam.

camera movements

Camera movements in film and television production are described using a number of terms. Some are vague, some are specific, and some are used in different ways by different national industries. The most common terms would be:

— **Pan.** The camera pivots in the horizontal plane on a tripod. You generally pan at the same pace as the moving object you are 'following'. A Zip-pan is a very fast horizontal movement, so fast that objects blur completely: it is used as a transition.

— **Tilt.** The camera pivots in the vertical plane on a tripod.

— **Dolly.** The camera physically moves towards the subject or away. The term dolly-left/right is also used in America.

— **Zoom.** This is a 'virtual' movement. You zoom in when the subject enlarges on the screen, and zoom out when the subject's image size diminishes. Zooms and Dollies are not interchangeable; they create quite different effects — especially with the dominance of the background.

— **Track and truck.** These tend to be television terms which are sometimes substituted for dolly, but more often used for movements at right-angles to a dolly-in/out. In film, track and truck often means to move the camera with a walking subject, keeping at the same pace and usually slightly ahead.

— **Slew** is a TV term which means to arc around a subject with a pedestal or sometimes with a crane. Often it just means to shift the camera sideways a little. **Crab** can mean the same thing.

— **Pedestal** up/down is a TV studio term. The film version is Crane up/down.

camera shot sizes and transitions

See shot sizes and shot transitions.

camera-ready copy

CRC. In publishing, this is the final copy before the plate making and printing stage. It will have all the typeset material laid out with headlines; line graphics and artwork included. ·Halftone 'bromides' will be in place or spaces left for color separations to be added later.

With desk-top publishing (DTP) systems it is now possible to produce this stage directly from the computer output. However, since laserprinters traditionally only produce output at a resolution of 300dpi (while professional typesetting is 600dpi and up) it is best to produce the camera-ready copy as large as possible. At the printery the CRC will be photographed to produce a negative (if necessary by size-reduction).

Imposition (the arrangement of the pages on the master printing negative) takes place at this stage since multiples of eight pages are generally printed at once. Later the half-tone screens or color separations will be physically cut in to this 'forme' before the negative is sent for plate making.

camp-on

A PBX term which means that, if an A-party calls through a PBX, and the B-party is busy, the A-party can maintain a semi-connection on that line until the B-party finishes the conversation — and then the B-phone will ring automatically. It is a 'call-back-when-free' function.

This is a widely-used feature on digital PBXs and, in large corporations it will probably be available across the wide-area network joining many of the company's PBXs. See DPNSS and Q-SIG.

campus network

A network of networks, but within the same general area. It will usually consist of a relatively high-speed fiber or coaxial backbone joining many departmental networks together. This is not a wide-area network, although it can span several kilometers. These days it is usual for large campus networks to have a fiber-optic backbone running some protocol like FDDI or ATM.

CAN

• **ASCII No.5 — Cancel:** (Decimal 24, Hex $18, Control-X). This character indicates that the data which preceded it in a message or block, was in error and should be ignored. It usually tells the receiving station to dump the preceding text unit (block or message) and await a new transmission.

• **Customer Access Network:** The 'last mile' of cable from the local telephone exchange to you. The term is often used to refer just to the local loop but it may include the CPE and the linecards.

• **Controller Area Network:** The networks and protocols used in modern cars to integrate various control functions into a single network. The development of some CAN standards, and the arrival of cheaper chips, has allowed the standard to evolve into one that can be used for wider control functions — in factory automation, for instance.

candle

As a measurement of light brightness for photography, an ordinary wax candle generates one candle-power or candela. This is equivalent to 13 lumens. Many modern video CCDs can work at this level.

canned

One of those omnibus terms that will drive future context-sensitive AI interpretation programs crazy.

• Pre-written, preprepared, or finished — not to be changed. This is an old film production term which needs to be read in context. It can mean 'the job has been finished', but it can also mean that the project has been discarded, by-passed or abandoned.

• Also, stock-off-the-shelf or predictable. Processed to remove variation. Without innovation — as in canned laughter.

CAP

• **Carrierless Amplitude/Phase:** A multi-level form of modulation which can be used for high-speed networks, sometimes over low quality cable. In telephony it is a passband line-code technique using carrierless quadrature amplitude and phase modulation — closely related to QAM and PAM.

CAP is a two-dimensional line-code that is relatively simple to implement on DSP or VLSI chips. It has in-phase and quadrature phase components, and amplitude changes.

— CAP64 is the Pay TV system which handles 30 million symbols per second for a 150Mb/s theoretical data rate (without trellis coding). It transmits quinbits over normal twisted pair.

It has eight phase changes, each with a resolution of 8 amplitude levels, which creates a 8 x 8 constellation. Trellis coding is almost always added to reduce errors.

— CAP32 is a 32-level QAM modulation scheme which transmits quinbits also. It is used over UTP cable for local data. At 25 million symbols per second, it can theoretically transmit 125Mb/s.

— For ADSL and video-on-demand, CAP is a QAM/PAM modulation scheme (see) designed for longer distances and multiple use. The local loop's 1MHz passband spectrum is split to provide: 0 — 4kHz for POTS; 0 — 80kHz for BRA ISDN; 94 — 106kHz for upstream signaling (16-level CAP); and from 120 — 600kHz for downstream signaling (64-level CAP). They claim that it has a potential workable video-delivery rate of 2—6Mb/s over twisted pair in the streets. See DMT.

• **Code Address Plug:** The CAP address is the unique identifier used in paging systems. The name originated from a hardware 'plug'. This held the identification number, and it needed to be inserted into the early paging receivers. Now this is all programmed into EEPROM or battery-backed RAM. See GSC and POCSAG.

• **Competitive Access Providers:** In the US these are cable and wireless-in-the-local-loop companies that are offering an alternative to the local connection owned by the LEC. A CAP provides you with a way to bypass the local call company when making toll-calls. You need a connection from your own premises to the CAP's POP (which could be a switch co-located in your local exchange, or a multiplexer).

• **Communications Application Platform:** The term used by General Magic in Magic CAP — an operating system for the new PDAs.

capacitance

The property of static electrical-charge storage.

• Capacitors are condensers and they are used in electrical circuits to briefly store electrical charges. In the form of capacitors, capacitance is essential for many electrical circuits.

• Capacitors have the facility of completely stopping direct current (DC) along a line, while allowing alternating signals and current (AC) to pass through (provided the capacitor is large enough for the current to pass).

• In communications circuits, you will always have some capacitance-coupling when nearby conductors with different signals are separated by a dielectric. 'Coupling' here means the transference of signals across the gap without actual electron exchange. This can happen between adjacent wires separated by air.

• Capacitance creates problems in many communications systems especially over long distances and with higher data-rates.

capacitance disk

A form of analog videodisk technology use by RCA and also by the Japanese VHD system a few years ago. There were grooved and grooveless versions. The information is read from the disk by the measurement of the electrical capacitance at the point of a stylus. The Philips/Pioneer laser-read pit and land approach won out, and this later became the CD.

CAPI

Common API. This is for Euro-ISDN. It is a simple message-queuing interface for applications wanting to communicate over ISDN.

caps

Abbreviation for capital letters or upper case. A caps-lock key is often provided on a keyboard. Note that the action of caps-lock is not always the same as holding the shift key down — especially in its treatment of the number keys.

capstan

The main drive shaft on a tape recorder. The size of the capstan, and the rpm of the drive motor establishes the tape speed. The tape is usually held against the capstan by a pinch roller.

Captain

Character And Pattern Telephone Access Information Network system. A Japanese videotext system with very high resolution, but which requires an enormous bandwidth. It uses an alpha-pictorial approach.

Captain began operation commercially in November 1984 but it has been a bit of a failure. It was backwardly compatible with the CEPT and NAPLAP alpha-mosaic standards. NTT who developed the protocol and ran the system are now repositioning it as a communications tool for large corporations.

captioning

• The technique of using a teletex-type system to display text superimposed over television images as an aid to the deaf — or for foreign languages (although this is more often called subtitling). See closed captioning.

• The permanent addition of text material to camera images during video production.

• Adding explanatory text information to still pictures and illustrations in printed material.

capture file

A file automatically created by communications software, which holds in RAM (or puts on disk), everything that appears on the screen during a communications session. In many communications programs, the information can scroll off the screen and disappear for ever. So, to be sure of keeping everything, you should always turn capture on.

carbon mike

Edison's invention of the carbon microphone made Bell's telephone practical. Bell had used a 'wet' microphone in his original invention, and this was not practical or efficient. Later the Rev. H. Hunnings replaced the carbon powder with carbon granules and greatly increased the sensitivity.

Carbon microphones are moderate in quality but highly robust and maintenance-free although they require DC power. The vibrations of the diaphragm vary the compression forces on the carbon and change the resistance to current flows. Many phones still use carbon mikes, but see electret also.

card

• Originally this applied to the basic I/O device of a mainframe, which were standard cardboard 7 3/8 x 3 1/4 punched cards. These were fed into the computer in batches to provide both program and data.

• Now a card is probably a printed circuit board (PCB) with electronics, best know as an adaptor. These plug into expansion slots. (Why not 'adaptation' slots?)

• Increasingly, the term will apply to PCMCIA devices of various kinds, including modems and small hard-disk drives.

• It may also mean a smartcard.

Card Bus

See PC Card and PCMCIA.

cardinal

Numbers which supply a value — as distinct from the use of numbers to represent a position in a sequence. See ordinal.

cardioid mike

A microphone which has a heart-shaped polar response, (viewed from the side), and is omni when viewed head on. It has a strong rejection to signals arriving from behind (from the direction of the cable connection) and is therefore an excellent all-round mike for general recording.

caret

This is the ^ sign on most keyboards which is better called a circumflex, although neither is correct — a true caret looks like an upside-down italic Y. See circumflex.

Carin

An early name for the Carminot CD-ROM-based automobile navigational system.

Carminot

A European Eureka project for the development of an integrated car navigation, communications and diagnostic system. Philips, Renault and the Dutch and French governments are involved. It uses synthesized voice to provide directions, and includes automatic position correction at each corner to maintain its positional accuracy. Also called Carin.

Carpal Tunnel Syndrome.

Carpal Tunnel Syndrome (CTS) is the most common form of 'compression neuropathy' associated with RSI and muscular overuse. The problem is caused by pressure on the ligaments in the hand's carpel tunnel, due to prolonged friction and swelling.

When the median nerve is compressed in the tunnel, the motor and sensory signals to and from the thumb and first three fingers can't get through.

carriage return

CR. The old typewriter name still widely applied to the Return key, and to the control code which issues an ASCII No.5 control character (see CR). In screen-based systems this is interpreted as a command to return the cursor to the left margin on the same line of text (or sometimes on the next).

However, the CR will usually be coupled with a LF (line-feed) character to take the cursor to the next line — but this may be transparent to the user since it is provided automatically by the operating system.

CR is not automatically coupled with a LF in all WP programs. Daisy-wheel printers used a CR along the same line to underline or create bold print by overstriking (see auto line-feed).

Many computers now have a numerical keypad with its own Enter key also. This often acts like a CR, but it does not automatically provide the same ASCII value in every application.

carrier

In telecommunications this may mean:

- The public-owned or privately-owned supplier of the telephone/data service: the common carrier (see below).
- A continuous R/F frequency which can be modulated to carry an information signal using analog (AM, FM) or digital (ASK, OOK, QAM, COFDM, etc.) modulation techniques.
- Generated audio tone/s on a private line or public telephone network which are then modulated by the digital signals in a modem.
- The output of a continuous laser, which is carrying pulses along an optical fiber.

The terms 'carrier' and 'signal' are often confused — partly because a carrier is a signal in the physical sense. It is a signal which is provided for the purpose of modulation.

The distinction must be made between the physical transport and multiplexing mechanisms (relating to the carrier), and the logical information content (relating to the signal).

- If a steam whistle is switched on and off to signal a lunch break, the whistle's audio frequency is the carrier, while the on-off switching sequence is the signal.
- In FDM systems, each carrier will carry only one signal but there will be many of them at different frequencies, while in TDM systems, there will often be only one carrier which is carrying many signals, each occupying a separate time-slot.

• Local, long-distance, or international companies which own and operate a telephone network are also called common carriers. Resellers of carrier capacity or services are not carriers, although they may style themselves as such. In many countries the carriers are government monopolies, while in competitive countries they are usually privately owned but highly regulated monopolies.

• The use of the term 'carrier system' suggests that the baseband voice or data frequencies have been modulated to translate them to higher frequencies for long-distance transmission. See carrier-type pair cable.

• The frequency of the carrier sets an upper limit on the amount of information that can be carried — depending on the modulation scheme used. At present, the maximum that can be expected in this decade is about 8 to 10 bits of information for each Hertz — with either an audio carrier or radio-carrier. Most systems still only carry one or two bits per Hertz.

• **Printers:** In the early ink-jet printers a mixture of ink and a liquid carrier was sprayed through the jet on to paper. A fast-drying carrier could clog the jet. The new ink-jets use a melted solid ink. See bubble jet.

carrier detect

CD. With modems, this is better called Data-Carrier Detect. See DCD.

carrier sense

This is the approach of using a signal or 'event' which is injected into a cable system, to check whether the network is in use.

• In LANs — carrier-sense is a system of contention which allows the stations to grab space on the network on a first-come first-served basis. The nodes 'listen' for changes in the carrier transmission and await their opportunity to broadcast.

The term 'carrier sense' (aka 'bid') also applies to the signal sent on the channel to instruct other nodes that the channel is claimed. The distinction is with polling or token systems. See contention and CSMA.

• Carrier sense is also a signal which is sent (in the OSI stack) between the Physical layer and the Data-link layer to inform it that the network is in use.

carrier-type pair cable

These are special cable pairs used by telephone companies to carry relatively large numbers of voice calls. A single carrier-type pair can handle between 12 and 300 separate voice channels with the right multiplexing equipment.

carry bit

In decimal arithmetic, when we add two numbers which together make a figure greater than 9, a 1 is carried to the next column. In binary, whenever we add two ones, the result is 0 with one bit carried to the next numeric column in exactly the same way. See logical shift also.

cart

Short for cartridge. Cart machines are used in radio and television broadcasting — usually to hold short material such as commercials, stings, promos, etc. These machines are silo-like and can be automated. Usually they hold cassettes, not cartridges,

Carterfone decision

A 1968 legal judgement which allowed the customers of the US telephone carriers to connect their own equipment to the telephone network. Until this decision the Bell companies insisted on providing all devices connected to the network. Now the FCC regulates attached devices.

Cartesian coordinates

Referencing position on a page or screen by using X (horizontal from left margin) and Y (vertical) coordinates. Note that for programming purposes, some computer systems reference the Y coordinate from the top of the screen while others reference it from the bottom.

cartridge

• A plug-in ROM module often used in computer games machines, and increasingly in portables, laserprinters, etc.

• In disks and tape terminology, a cartridge is differentiated from a cassette by the fact that cartridges don't have a take-up spool. The cartridge has a tape-supply reel only, and the machine has a mechanism for seizing the end of the tape and threading it, so all take-up is done within the machine.

As a consequence, the tape must always be rewound before the cartridge can be removed. This terminology is not followed rigidly — often the terms are used indiscriminately.

• In CD-ROM disks, the term is also applied to the protective covering which must be used to insert and remove the disk in some machines (a caddy). This is a security measure and it also prevents disk damage. Don't confuse this with jewel cases used for storage.

CAS

• **Channel Associated Signaling:** An ITU standard also called R2D.

• **Communications Application Specification:** A communications protocol developed by Intel and DCA, but now in the public domain. This is a high-level interface specification that allows applications developers to embed a communications hook into their applications software which will allow the batch transmission of almost any formatted working files over LANs or phone lines. Versions of these CAS hooks were available for MS-DOS and OS/2 applications.

The main use appears to be in combined fax/modem boards; the standard allows personal computers and fax machines to easily exchange data. See OMI, Class 1 and 2.

cascadable

The ability of a multiplexed system to add or drop signals at intermediate points. Frequency division multiplexed (FDM) signals can provide these add–drop services relatively easily, but time division multiplexer (TDM) are much more complex and difficult. See add–drop multiplexers and SDH.

cascade

• In amplification, a cascade is a series of amplifying stages.

• In cable television networks, amplifiers are placed along the main trunk lines in series. The noise contributed by each amplifier is therefore additive; so noise on these systems is said to 'cascade' and the noise-floor will increase by 3dB every time the number of amplifiers doubles.

• In automatic re-transmission systems, this can refer to a series of redundant modules which will reconfigure if problems arise.

• In high-speed data networks, cascade implies that a small problem can generate large reactions and eventually disrupt the whole network. See packet-storm and black hole.

• With video production networks, different forms of compression are often used in series during the production and distribution phases. The original digital camera may use one compression system, then the editing system may use another, then the distribution may use another ... and so on. The cascading of different compression techniques produces all sorts of quality problems, since the compression algorithms aren't necessarily compatible.

cascaded star

A network topology where multiple hubs are configured for redundancy — to absorb any failure.

CASE

• **Computer-Aided (Assisted) Software Engineering:** CASE tools in computer programming are classed as 'applications generators'. They allow software developers to automate most of the most difficult parts of writing, testing and maintaining applications software.

A full CASE environment includes tools for programming, documentation, and debugging. The emphasis is on methodology, and the key resource is a good data dictionary.

The term is sometimes loosely applied to almost the entire range of 4GL software development tools, but CASE and 4GL languages should be considered as both separate and complementary.

CASE became a dirty word in the industry because the vendors of CASE tools couldn't deliver the productivity gains which they promised. CASE has since evolved into client-server development tools. See RIG, modules and SDLC also.

• **Common Application Service Elements:** In telecommunications, these provide common services for X.400 MHS (e-mail) and X.500 Directories. See ACSE.

case-based reasoning

CBR. This is an offshoot of AI, and something close to an expert system. Yale has been developed it since the 1980s.

The idea is that a CBR database holds a library of hundreds or thousands of short case-histories. When you come up against a problem, you access this database in a special way, and this allows you to adapt old solutions to new problems.

The CBR approach apparently works well with Help Desks — and in some areas of industrial problem solving. It differs from expert systems in not using explicit rules to reach a solution.

case-sensitive

The computer program will react to the upper-case alphabetic letters but not to the lower, or vice versa. This term is often found in Find and Replace utilities for changing an alphabetic word (a string).

Note that ASCII No.5 is fundamentally case-sensitive; it needs a small algorithm in the application to make it see 'c' and 'C' as the same letter.

• In computers, you should assume that systems are case-sensitive, unless you find otherwise.

CASS

Conditional Access Sub-System of a Pay TV operation. See subscriber management system.

Cassegrain reflector

A reflector dish (or astronomical mirror) having two reflecting surfaces. The main large dish focuses the beam onto a secondary dish (a convex hyperbolic subreflector), which then passes it back along the axis of the feed-horn to a LNA mounted behind the parabola.

CAT

Catalog or directory. See.

catalog

A list of all filenames on a disk. An old term for a directory.

Category # cable

These are the standard categories of Unshielded Twisted Pair (UTP) cable used for internal phone systems and local area networks:.

— **Category 1:** Used for basic voice (analog). No performance criteria were established. It is equivalent to UL Level 1.

— **Category 2:** Used for advanced voice applications (digital) and low-speed voice/data integration such as ISDN. This cable type is able to handle data-rates up to 4Mb/s, and it is very similar to the IBM Cable System Type 3 cable. It's the functional equivalent of UL Level 2.

— **Category 3:** Used for 10Base-T Ethernet cable (to 16Mb/s) for distances up to 100 meters. It will also handle 4Mb/s Token Ring; ARCNet, 100Base-VG and 100Base-T4. It's the functional equivalent of UL Level 3.

— **Category 4:** Used for 16Mb/s Token Ring and systems which are very sensitive to jitter and near-end crosstalk. It can handle up to 20MHz and is also used for low-lost 10Base-T. It is the same as UL Level 4.

— **Category 5:** Used for high-speed data-rates up to 100Mb/s (and later 155Mb/s) for multimedia applications. This is now being used for ATM over copper, and also for TP-PMD, CDDI, and 100Base-TX. The standard applies to UTP, ScTP or STP cables.

• These categories are set by the EIA and TIA under the 568 specification and Technical Systems Bulletins (TSBs), for both cables and connectors.

• The US-based Underwriters Laboratory's (UL) set of cable standards is now aligned with the EIA/TIA. Even the best quality cable can be compromised by bad installation practices, which is why the EIA and TIA have the 569 specification also.

catenet

The Internet is a catenet; a diverse network in which hosts are interconnected using routers.

cathode

In electron tubes and electrolytic reactions, the cathode is the electrode where positive ions are discharged and negative ions are created. It is the side where reducing reactions occur.

In a diode schematic, the cathode is represented by a bar. See anode.

CATV

Community Antenna (or Access) Television. This was the origin of Cable TV in the USA and the term is still widely used in electronic circles to mean Cable TV networks. They now translate it sometimes as 'Cable Access TV'.

CATV also became widely used for local television distribution, in the sense of within a campus — but here it is called CCTV (Closed Circuit TV) for the direct distribution of lectures (before VCRs).

• Public-access CATV systems were originally installed in locations with poor over-the-air TV reception. A large community antenna would be installed on top of a nearby hill, and the signals boosted and distributed to subscribers over a basic cable network using 75 ohm coaxial strung from power poles. These were often community cooperative projects, or sometimes they were installed by the local radio store owner under a local council franchise.

Later other sources of program were supplied by the CATV company (using VCRs, microwave relays, landlines, satellite, etc.), and this began true Cable TV operations (around 1948 in the USA). See Cable TV, Basic Cable and Premium TV.

CAU

Controlled Access Unit. This is a remotely managed MAU or a managed multiport wiring hub on Token-Ring nets. The control features consist simply of turning the ports on or off and bridging any destructive gaps in the ring. See MAU and MSAU.

Cauzin Softstrips

A technique of transferring small applications and data between computers using printed or photocopied 'barcodes' of a special type. These required a 'Softstrip' reader — a special optical scanner.

Magazines planned to print these 'barcodes' on a page as a way of distributing shareware to readers — but that was before software developers went on their multi-megaByte hike.

CAV

- **Constant Angular Velocity:** The system used for one type of interactive videodisk, and it applies to all your old 78 and LP records, where the rotational rate of the disk is fixed. The distinction in videodisk is with CLV where the 'linear velocity' (the speed of the head over the spiral track) remains the same.

CAV allows quicker access to data than CLV, because the player doesn't need to wait until the drive speed changes before data can be read or written.

The problem with CAV from a storage viewpoint, is that it doesn't use the disk's capacity to best advantage. The long outer tracks only hold as much information as the shorter inner tracks; with CLV they can hold much more.

- **Component Analog Video**: An analog video technique where the luminance signal and the color signals are kept separated. See component video.

CB

- **Citizen Band radio:** This covers unlicensed two-way personal communications. One band operates at 11 meters in the HF band (26.965—27.405kHz), and another operates in the UHF band (476.425—477.4MHz). Usually 40 channels are available with AM, lower and upper side bands, and another 10 reserved specifically for maritime use.

In the USA (the home of CB) the citizen's radio service has three classes of service:

— Class A is intended as a voice-only commercial service in the UHF.

— Class B is also voice-only, and this is the familiar 40 channel good-buddy trucker radio in the 27MHz range.

— Class C is for remote control devices, and it has allocations in both 27 and 72MHz bands but AM or FM is not permitted.

Also see DSRR and CBRS.

- **CB simulator:** On The Internet this is a special facility which allows keyboard 'chat' sessions. See IRC.

CBB

Computer Bulletin Board. Usually called a BBS. See.

CBDS

Connectionless Broadband Data Service. This is the name applied to the first European broadband (MAN) services which used the DQDB technology (802.6). Note that this is a 'service' name and not the name of a technology, so the same service may later migrate to a different technology (ATM, for instance). See DQDB and SMDS.

CBMS

Computer Based Messaging System. Just electronic mail!

CBR

- **Constant Bit Rate.** See
- **Case-Based Reasoning.** See.

CBRS

Citizen Band Radio Service. CB in the UHF region, between 476.425 and 477.4MHz. See CB (above) and DSRR also.

CBT

Computer Based Training. See CAI.

C&C

Computers and Communications. See convergence.

CCC

Clear Channel Capacity. See.

CCD

Charge-Coupled Device. Solid-state imaging chips used in modern video cameras to replace vidicon tubes.

The CCD was invented in 1970 and first used in about 1975. They are the reason why modern home video cameras are able to produce image quality of near professional standards at such a low cost.

Before the CCD, vidicon, plumbicon and saticon tubes were used in most small cameras and these were easily damaged, had low sensitivity, and high battery drain.

CCDs are now available up to 4096 x 4096 pixels. They are very small, have virtually no picture distortion, are not easily damaged by physical abuse or excessive light. They are also able to operate in very low light conditions down to the 5 lux level in many cases, and with low battery voltages.

- There are two main fabrication techniques: surface channel (SCCD) and buried channel (BCCD). The 'channel' is the area of the chip where the electrons are stored. BCCD devices are now used very widely. SCCD devices suffer from a tendency to trap electrons and thus produce an erratic output.

- There are two basic physical chip designs:
- — LIMs (linear imaging devices) are used in fax machines, bar-code readers, optical scanners. They are long and narrow.
- — AIDs (area imaging devices) are the more conventional square-rectangular type used in cameras and astronomical telescopes.

CCE

Cooperative Computing Environment. The integration of enterprise-wide resources in a company.

CCI

Co-Channel Interference. See.

CCIR

Comite Consultatif International des Radiocommunications. The French acronym for the International Radio Consultative Committee. It was the radio version of the CCITT, and it established standards for radio and ran the international WARC conferences. The old IFRB organization worked closely with CCIR and set the spectrum allocations. The CCIR is now called ITU-R. See ITU.

- CCIR 601 recommendation for digital television was very important. See 601 in numbers section.

CCIS

- Common Channel Interoffice Signaling. AT&T's version of CCS#6. See.
- NEC's CCIS#7 protocol, despite its name, is a proprietary PBX protocol which is based on a subset of the ITU CSS#7.

CCITT

Comite Consultatif International Telegraphique et Telephonique. The major standards-setting organization for telecommunications under the ITU. CCITT standards were only issued every four years, and were usually ratified by the ISO. The CCITT has now been replaced by the ITU-T group. See ITU.

CCP

- **Character-Count Protocol:** See.
- **Command Control Protocol:** This is a file of standard command sequences which can be called up and used like macros.

CCR

Commitment, Concurrency and Recovery. A CASE in OSI's application layer which helps establish the peer-to-peer communications links, and so ensure the reliable transfer of the data from user to user. See ASCE, ROSE, CASE and SASE also.

CCS#

Common Channel Signaling No.# The ITU-T signaling system which is now found in interexchange networks around the globe, and is the basis for all modern digital network signaling development. In some of the literature, CCS# is known as Signaling System # or SS#.

Seven signaling standards were set by the old CCITT (now ITU-T); some used common channel and some didn't. Only four are still in use.

- **No.4** is a European-only in-band signaling standard. It uses 2040Hz and 2400Hz audio tones in pulses which are 100, 150 and 350ms long, for both line and supervisory signals. It also has a four-element code of 2040Hz binary pulses, each of 35ms.
- **No.5** is in-band and used around the world. It uses 2400Hz and 2600Hz tones for line and supervisory signals. This is the international exchange signaling standard when CCS#7 isn't available. This was the phone phreaks delight, because the tones could be generated by simple plastic whistles (Capt. Crunch) or by the famous 'Blue Box' which funded the development of the Apple computer.
- **No.6** is still widely used around the world: AT&T call it CCIS. It is an out-of-band signaling system for SPC exchanges which transfer data over analog speech channels. They use four-phase modems operating at 2400b/s (each services 600 voice calls), and the signaling data is in units of 28 bits, of which 8-bits are for error detection.
- **No.7** is the international standard for all digital exchanges with integrated services. Signaling is by packetized data over 64kb/s digital bearers between exchanges, although lower-speed analog bearers can also be used. CCS#7 is now very widely used in all developed countries, and it links the new digital exchanges to the older SPC exchanges.

It is based on the OSI 7-layer model. At the Physical layer it is designed for 64kb/s digital links, although analog and satellite links are permitted. At Layer 2 it is similar to HDLC or LAP-B, while Layer 3 handles routing and relaying. Higher layers include facilities for route verification, testing and delay measurement.

CCS

- **Common Communications Support:** A component of IBM's SNA.
- **Common Channel Signaling** in digital telephony. See CCS#
- **Continuous Composite Servo** in optical disks. The disk is etched with a spiral of grooves for optical recording control. This is used with ISO 10089 standard Format A for both WORM and rewritable optical disks. An alternative technique is Sampled Servo which does not use grooved tracks.
- **Completed Calls per Second:** In old telecommunications switching terminology, CCS is often translated as Completed Calls per Second — but it more correctly refers to Centum Call Seconds ('C'=100). It is a measure of the number of simultaneous calls a switch can handle. In the Bell System, they used this measure instead of Erlangs for traffic flow.

With CCS, you would multiply the number of calls by the length of the call and divide it by 100. That calculation gives you a measure of traffic being carried on a line. The relationship between the two measures is: 1 Erlang = 36 CCS.

The distinction is also with the BHCA measures (Busy Hour Call Accesses) which grades switches on their maximum, rather than average, capacity. See Erlang also.

CCSA

Common Control Switching Arrangement. A special switching facility between the carrier and the private network which is designed to allow private control of some leased-line switching. See Centrex.

CCTV

Closed Circuit Television. This is usually a small internal cable network over which lecture or training material will be broadcast in real time.

CCU

- **Communications Control Unit.**
- **Card Controller Unit.**
- **Camera Control Unit:** A control unit in a television studio which remotely adjusts the iris of the camera lens, its contrast, and its color output. When three or more studio cameras are used through a SEG (switch) it is important that the density and the color of each match.

CCU operations is one of the most important roles in a production studio, and it is usually done by a skilled technician working in association with the lighting crew.

CD

- **(Data) Carrier Detect:** A control line in modems. See DCD
- **Carrier Detect** signal. In most modems this is a 2100Hz tone which is also used to disable echo suppression.
- **Compact Disk:** See below.
- **Change Directory** in MS-DOS.

CD (disk formats)

- **Compact Disk.** These 12-cm (4.7-inch) digital optical disks were designed by Philips originally for digital sound recording, but now provide data, video and multimedia through a number of different standards. See all of below:

— CD-DA for audio and small screen display information
— CD+ for audio with some limited text and image information.
— CD-Plus which is a later version of CD+.
— CD-ROM for all forms of data.
— CD-ROM/XA, which includes ADPCM sound techniques.
— CD-V for audio pop singles but with a few minutes of full-motion video.
— CD-G, a kareoke standard developed by JVC.
— CD-i for video interactive home/games/education use.
— CD-Photo for stills.
— CD-R for recordable (WORM) disks.
— CD-HD (HDCD), the new high-density standard which can hold a full movie compressed to MPEG-2 format.
— Multimedia CD is the latest Sony/Philips proposal. (see CD-Multimedia)

CD+

A proposed format for audio with very limited text and image output. See CD-Plus and CD-G below.

CD-DA

CD-Digital Audio aka CD-Audio. The digital audio compact disk — known as the Red Book (1981) standard. It was invented by Philips. CD-DA disks hold 72 minutes of sound sampled at a rate of 44.1kHz in stereo. The audio waveform is quantized as a 16-bit number for both stereo channels (a 32-bit number for each sample period), giving a primary data-rate of 1.4Mb/s.

Within the standard it is possible to also embed some character information which can be output to LCD screens. See Bitstream also.

CD-E

CD-Erasable. An early generic name applied to undefined standards for erasable/re-recordable versions of the CD-ROM standard.

• **1995:** CD-Erasable is now back in use for a proposed new standard for a recordable CD format which will produce read-only disks (WORM). This is totally confusing, since 'erasable' previously meant 'erasable and reusable'.

CD-G

CD & Graphics (aka CD+G). A non-interactive CD sound and image format which has graphic images and text displays coupled with CD-Audio. It was devised by JVC mainly for karoke (follow the bouncing ball) use.

— The early CD-G systems only provided 16 colors.
— Later CD-EG provided 256 colors, and instant switching between two buffered images.

Special decoders and disks are needed for CD+G, but the best this technology can display is the lyrics of the songs plus some simple images. More recently there has also been the development of a version coupled with MIDI codes to create music. These disks hold 16 interleaved tracks of music control (so these can handle a whole orchestra), and they can be used to drive synthesizers. See CD+ above and also CD-Plus below.

CD-HD

Now renamed Multimedia CD. See CD-Multimedia.

CD-i

Compact Disk-Interactive. A Philips/Sony standard intended primarily as a multimedia consumer product; announced first in Feb. 1986. This was a closed system with a built-in microprocessor and a joystick-operated pointing device. Cables hooked the unit to a standard TV set and stereo-sound unit.

Recently the emphasis has come off the TV-display only, and CD-i is available for personal computers also. The standard is now known as the 'Green Book', and this specifies both data formats and playback requirements.

In its basic form, CD-i provides text, graphics, and limited screen-size motion video only. It was intended for general educational, games playing, and home interactive use.

CD-i is an extension of the CD-ROM/XA standard and it draws on the XA Yellow Book extensions: Form 1 for computer data, and Form 2 for compressed audio and video. Most CD-i players will also playback CD-DA, CD-ROM/XA and Photo CD formats.

There are three audio quality levels called A, B and C:

— **Level A** is hi-fi with a 17kHz range and 90dB S/N ratio. It is equal to CD-DA, but it only uses 8-bits per sample with ADPCM.

— **Level B** has same frequency range, but with only 4-bits per sample it has a higher S/N of about 60dB.

— **Level C** uses the same 4-bit ADPCM sampling at half the rate, giving a frequency range only to 8.5kHz, which provides 19 hours of good speech. The standard doesn't currently support audio compression.

• CD-i units need a 68000 processor, 1MByte of memory and the OS-9 Real Time Operating System (RTOS) and real-time video and audio chips.

• There is now a Full Motion Video extension to CD-i called FMV which is added by means of a Digital Video (DV) cartridge. See below.

CD-i/FMV

Full Motion Video/CD-Interactive. Philips latest development which reproduces 72 minutes of full-screen motion video from a CD-ROM disk using extensions to the CD-Interactive system. This was Philips answer to the DVI development (now under the wing of Intel).

FMV digitizes every second image frame, then passes it through numerous data compression and motion analysis phases.

Philips has joined with Sony and Matsushita in the development of FMV and they have now agreed to use the video/audio MPEG-1 compression standard. The reduction in data rate can be as high as 140 : 1, resulting in a video image stream of just over 150kByte/s (1.2Mbit/s).

The MPEG video is then combined with a compressed audio stream of 0.2Mb/s, so the combined video/audio interleaved streams make up the CD limit rate of 1.4Mb/s. This is the maximum data rate that can be supplied by single-rate CD replay units — and even so, some image-quality losses occur when complex areas of the screen are rapidly changing. Also included, is stereo audio compression to below CD-Audio standards (but with 7:1 compression).

The FMV extension to CD-i (usually in the form of a plug-in cartridge) creates a universal disk standard which can run on both 50Hz (PAL) and 60Hz (NTSC) players. See MPEG.

CD forms and modes

• CD-ROM has two modes which allow the interleaving of dissimilar data and error correction:

— Mode 0 was the original CD-Audio mode.

— Mode 1 has only 2048 bytes per sector and is used for critical data that needs good error correction (specifically for computer data). A second layer of Reed-Solomon error correction is added to the auxiliary data field (the last 288 bytes in a sector).

— Mode 2 provides 2328 bytes of user data per sector and has no error detection or correction. It is used for compressed audio and video storage where errors are not particularly important since interpolation and other fill-in techniques can be used.

• In CD-i the modes were slightly modified and became known as forms.

— Forms 1 and 2 are the same as Modes 1 and 2 except that the form identity is included in the sub-header to permit interleaving of Form 1 and Form 2 sectors in real-time operations.

CD-Multimedia

The name for the new high-density standard created by Philips and Sony (plus 3M) for a true video and multimedia CD system capable of a continuous 100 minutes of good video. They have two new standards, and each is available in single-layer or dual-layer configurations. (Note: double 'layer' not double-side.)

• **Replay only:** The new single-layer Multimedia standard will store about 3.7GBytes of data, more than five times that of a standard CD-ROM, and enough to carry 135 minutes of MPEG-2 video and multiple sound/data channels. The transfer rate will be variable and from 1Mb/s to 10Mb/s (average 3Mb/s) which is potentially higher in quality than conventional TV.

This has been achieved using a 635 nanometer red laser by reducing the distances between the tracks (to 0.84 microns) and the size of the pits (minimum length 0.451 microns). New modulation (EFM plus) and error correction (CIRC plus) techniques have also been applied.

The disks can be produced in a standard CD manufacturing plant (with minor modifications), so production costs should be comparable with CD-ROMs.

The dual-layered version which will carry 7.4GBytes. This layered approach uses a disk pressed with two spiral tracks, one of which is deeper than the other. The laser can focus on one track or the other, and this doubles the replay time.

Philips says it has solved the problem of pressing disks with these storage densities, but they are still experiencing problems with mastering.

• **Recordable:** This is also to be available in single and double-layers, but the capacity is about one-third of the replay-only versions. The single-layer will hold 1.15GBytes, and the double-layer 2.3GBytes.

Apple, IBM, Compaq and Microsoft are working on ISO 9660 extensions to the volume and file standards for computer applications.

The Multimedia disks will only play on new drives. However these drives will be able to replay existing CD disks.

CD Photo

Also called Photo CD (see).

CD-Plus

A new 'Blue Book' specification for an enhanced music CD by Philips, Sony, Apple and Microsoft. It is based on the multi-session CD specification (recordable CD-ROM/XA) but with CD-DA audio tracks.

It is designed to be used fundamentally for music but in combination with video clips, lyrics, photos, animation and text. The audio will be backwardly compatible with CD-DA and most CD-ROMs.

CD-PROM

A term once used for a writeable (but not necessarily erasable) CD-ROM standard. It was to have used CLV techniques to be compatible with CD-ROM.

CD-R

CD-Recordable. Originally this term was applied to the recordable/erasable (and therefore reusable) version of the CD-ROM system. Now it just means the WORM version. This is the Orange Book standard.

There are single session recording standards (with only one table of contents) and multisession standards (where a new table of contents is added each time). Photo CD is a proprietary version of the multisession standard. See MPC, CD-WO and Photo CD also.

Since CD-R can be played on a conventional CD player, it can also be used to pirate audio disks, so the music industry is trying to block the commercialization of these products. See multisession format, WORM, Photo CD and audit trail. Also see ETM, PCR and magneto-optical disks.

CD-ROM

Compact Disk – Read Only Memory. This is a Philips/Sony standard for a computer peripheral. Currently it uses the Yellow Book standard of 1983-4, although there have been many later additions and extensions.

• The wider use of the term CD-ROM, also encompasses the various data formats such as CD-i, CD-V, CD-R, etc. Note that there is no single drive which can read all CD format types. See CD-ROM/XA.

• The specific term 'CD-ROM' is now more correctly known as ISO-9660/High Sierra (also ECMA-119).

The extraordinary digital capacity of CD disks means that CD-ROM is the cheapest form of text storage ever devised. A single disk costing only a few dollars to produce (in large production runs) can hold 700MB of usable data — the equivalent of 25 filing cabinets of letters, 400 novels, or 500 HD floppy disks.

The unformatted 'payload' capacity of CD-ROM is about 860MB, but it loses about 100MB in formatting. In fact, a full CD disk recorded right out to the edge can store over 1GB of information, but the standards published as the 'Yellow Book' call for very sophisticated error checking which consumes 16% of the space.

Note that CD disks read from the center out, but it is dangerous to record right to the edge of the disk because this part of the disk is subject to corrosion, dirt and scratching problems.

The original data rate which could be extracted from a CD-ROM disk drive was the same as CD-DA, which is fixed at only 150kBytes (1.2Mbits) a second. However, recently, double-, triple- and quad-rate disk drive units have become available. Triple-speed CD-ROM reads data at 450kB/s, which is often higher than the PC input port can handle. See Reed-Solomon, High Sierra, ISO 9660, and CD-ROM/XA.

— **Mode 1** is the conventional version used for data storage. It has a sector format with two levels of Reed-Solomon (see) error correction to ensure data integrity.

— **Mode 2** is the version typically used for games with less critical data and the interleaving of sound, image and data. The error correction is replaced by user-data (known as the auxiliary data field).

CD-ROM extensions

These are the extensions to the MS-DOS operating system that allow IBM-compatible PCs to use the full capacity of a CD-ROM disk. They appear as MSCDEX.EXE in the directory.

MS-DOS was originally only able to use hard disk partition sizes up to 32MB. With the 650MB capacity of a CD-ROM disk, this would normally require most of the letters of the alphabet to name the partitions (treating each as a separate volume) and would have made accessing information very difficult.

The installable extensions overcame this problem by providing a directory structure that resembles DOS, but with special features for the large disk capacity. The extensions make the CD appear to the OS as a network drive, and DOS doesn't place a limitation on network drives.

Some of the older CD-ROM applications still require this Microsoft TSR called MSCDEX.EXE. See High Sierra and ISO 9660 also.

CD-ROM layers

• These are protocol layers, as defined in different 'books':

— **Layer 0** is the bit structure of the disk.

— **Layer 1** is the arrangement of bits into frames and sectors.

• These first two layers constitute the original CD-ROM Yellow Book specifications.

— **Layer 2** is the sector organization into logical data blocks.

— **Layer 3** is the organization of the logical blocks into files.

These last two layers constitute the High Sierra logical format specifications.

CD-ROM/XA

EXtended Architecture. The XA version of CD-ROM was intended to bridge the gap between CD-ROM and CD-i, but it ended up as just another complication. The XA doesn't have it's own 'book'; it is defined by the Yellow Book's Form 1 for computer data, and Form 2 for compressed audio and video.

The key reason for the extended architecture was for sound to be interleaved with data/video to allow both information streams to appear simultaneously. It also allowed the actual retrieval software to be stored on disk.

XA also introduces ADPCM audio and, like CD-i, it included a 640 by 480 screen resolution-mode with 256 levels of color. Any XA disks will play on a CD-i drive but they are not compatible with CD-ROM ISO 9660/High Sierra formats; the XA-compatible drives need an additional decompression board.

The XA standard allows up to 16 parallel audio channels, and it supports DA audio quality, plus three lower audio levels that trade off space against quality, giving up to 19 hours of voice-quality audio on to a single disk. See CD-i for audio levels.

In April 1989, Microsoft and IBM announced a further extension to the XA extension. This was to allow the use of DVI video, then being developed by Intel. Indeo was later used instead.

CD-V

Compact Disk Video: The Philips/JVC White Book standard for a 12-cm disk which originally offered 5 minutes of video and 20 minutes of audio. It differs from the other CD standards in a number of ways and is mainly directed at the pop-music market, although it has other applications.

They also promised a smaller (8-cm/3-inch) version, and two larger versions — one, a 20-cm disk with 40 minutes of sound and pictures, and the other a two-sided 30-cm/12-inch disk (the old LaserVision size) with up to an hour each side. The video was to be compressed to the MPEG-1 standard. I haven't heard anything recently about these larger disk standards, so I assume they have been overtaken by later developments.

- CD-Video has now been revived by Philips and Sony for their new MPEG-based movie player. The 'White Book' standard is now at Version 2.0. It takes the interactive control of the software away from the recording medium itself (the Video CD) and into the controlling hardware. It also allows the recording of both high- and low-resolution stills, and allows the introduction of menu screens. However see CD-Multimedia .
- **Note:** for CDV (Compressed Digital Video) see below.

CD-WO

Compact Disk Write Once. This term was applied to the original recordable, WORM (write-once in sequence) disks that could be replayed on standard CD-ROM players. They conformed to the Philips/Sony Orange Book standard (now ISO 9660). CD-WO is also called CD-WORM and now CD-R.

See multisession formats, audit trails, and CD-R.

C&D bits

See C section. Page 97.

CDDI

Copper Distributed Data Interface. FDDI over copper. The maximum distance for either STP or UTP cables is about 100 meters. CDDI was also once known as TP-DDI (Twisted Pair Distributed Data Interface), but now this term seems to be restricted to unshielded twisted pair, while the term SDDI has come to mean shielded pair. It's a moving target.

The 125Mb/s data rate of raw FDDI corresponds, in Category 5 cable, to 62.5MHz (see TP-PMD also).

cdev

Control panel devices. A term used with Macintosh computers to indicate a small control program or desk accessory that is accessible through the control panel.

CDLC

Cellular Data Link Control. This is Vodafone's analog cellular phone (TACS) modem error protocol. It requires special modems. See ECT and MNP10.

CDM

Code Division Multiplexing. A modulation technique for cable systems. It employs spread-spectrum techniques. See CDMA.

CDMA

Code Division Multiple Access. A technique used for satellite and cellular digital communications. The generic technique is called Spread Spectrum Multiple Access (SSMA), and provides very valid alternatives to TDMA and FDMA techniques.

CDMA has long been used in the military and with VSAT systems, but recently it has migrated to terrestrial cellular and PCS systems. There are two main types of spread spectrum — direct sequence (DS), and frequency hopping (FH). CDMA is applied to the DS type, but there are various bandwidths (B-CDMA) and variations in the many standards.

With CDMA each terminal is assigned a unique encryption code (usually randomly generated), and the encrypted signals are transmitted over a very wide band of spectrum which is constantly shared with many other users. The receiver only listens for its own code — the other users are just noise — but the noise is minimized by the use of orthogonal Walsh codes. See Walsh codes and orthogonal.

The direct-sequence technique tolerates higher levels of noise and interference than TDMA and FDMA and has military security applications.

In cellular networks CDMA seems to have many advantages; capacity being the main one (10 x AMPS at least). CDMA cells can also cover larger areas (they need fewer basestations), and cell-planning and subdivision is comparatively easy. As an air-interface, it has many of the statistical advantages of packet and ATM networks — bandwidth-on-demand, variable rate voice coding, etc.

The cellular system designed by Qualcomm uses 1.25MHz of bandwidth (in each frequency division) to theoretically support up to 60 simultaneous users (possibly a max. of 20 in practice). New wideband systems will use 2.5MHz and InterDigital's system uses much wider bands again.

The capacity increase comes about mainly because the same frequencies can be reused in all cells, since the base-stations distinguish the mobiles (and vice versa) by the code, not by their frequency or a time-slot.

In practice CDMA's capacity is limited by unacceptable noise levels on the channel (too many users), not through 'blocking' where channels are unavailable. Variable bit-rate coders keep noise to the minimum — so capacity is proportional to actual usage, with no reserved/unused channel wastage.

Dual-mode handsets together with reuse of sections of the AMPS spectrum means that CDMA can be used in the cities, while analog remains for less populated areas of the country. See also Rake, Walsh codes, SSMA, spread spectrum and FD.

CDOS

Concurrent DOS, from Digital Research. The first true multitasking operating system for IBM PCs in the 1980s.

CDPD

Cellular Digital Packet Data. A system of public-access packet radio for mobile data applications which uses the spare time in normal cellular radio networks. It was developed by a consortium with IBM, Apple and a number of cellular operators, and has a transmission rate of 19.2 kbit/s, but only for bursts. The first CDPD commercial networks are on trial in the USA at the time of writing.

CDPD requires a separate wireline network but it uses the down-time on the R/F and transmission facilities of analog cellular phone system to transmit packetized data. It transmits its data packets in the gaps between normal voice calls. It is able to invisibly switch packets from one channel to another to maximize channel capacity without compromising normal voice service. Since there's no continuous online cost, transmission times are short and user pays only for delivered packets.

CDPD has enormous potential in providing a radio-linked short-messaging service because of the coverage area provided by the cellular phone system. It is designed to interwork with both X.400 and TCP/IP. CDPD modem uses the AT command set and has an IP or X.400 address which is assigned by the local cellular company. Applications should be able to interface to CDPD by way of telnet, SLIP, UDP, or TCP.

CDS

Call Data Standard. A US standard for call activity in telecommunications — specifically for communications software used by professionals who charge for their time. It specifies a group of file formats and guidelines on how to read and write to them. CDS keeps a 'call history file' which can be updated by modems and fax, and there's also a phone-rate file. These are both text files which can be read by a word processor.

C-DTE

Character-mode Data Terminal Equipment. A term used in communications to apply to dumb terminals, PC software/modem combinations and printer-terminals that use asynchronous signals. These devices all conform to the X.28 operating standard for asynchronous terminals.

CDTV

Commodore Dynamic Total Vision. The Commodore version of CD-i. It is a stand-alone device which allows motion pictures, sound and text to be stored on a standard CD-ROM disk and played back through a normal TV set and hi-fi. Currently the screen up-date rate for 'full-motion' video is below that of the normal TV standards (25 or 30 frames per second).

CDV

Compressed Digital Video. These days this means MPEG1 or MPEG2. It is a general term applied to all forms of digital video and television delivery.

CEBus

Consumer Electronics Bus. The US home appliance industry standard for control over lighting, home appliances, home security and entertainment systems. It can supposedly carry video images also.

CED

Capacitance Electronic Disk. The RCA SelectaVision grooved capacitance videodisk system that has now been discontinued.

cell (general)

• The storage space or screen area reserved for one unit of information.

In computer memory, this is one flip-flop or charged capacitor which stores a single bit in RAM or the equivalent in ROM. In a magnetic disk or tape it is a single spot which has an identifiable magnetic polarity.

• In spreadsheets, a cell is a location at the intersection of a row and a column. Cells are usually identified by a RxCx number.

• With batteries (or accumulators) it is one section producing the minimum voltage. Eight 1.4 volt cells, for instance, make up a 12 volt car battery.

• In general networking, a cell is a small 'fast-packet': it is a small, fixed-sized, information-bearing unit that provides the foundation for transporting and multiplexing. It is not a packet, but is usually carries part of a packet (packets are segmented and reassembled — see AAL). It does not have an address, but it has a VPI/VCI instead. See ATM.

The distinction between packet and cell here is two-fold: a packet contains a full address and is therefore self-contained, while a cell only contains a circuit identifier (which is why virtual circuits need to be established first), and the cell is fixed (and short) in length, while packets tend to be long and variable. There are other differences as well; packets usually have full error correction in each (cells only have some detection for header faults), and packets usually have sequencing numbers (cells don't — they rely on the virtual circuit to maintain order).

• It is important to realize that a long packet (which carries a full header, address, error checking, etc.) can still be segmented into a number of cells, each of which only carries a virtual circuit identifier — and so a connectionless packet, can move over stages, relayed between cell nodes along briefly established connections. The AAL defines how this is done.

• In ATM and DQDB, a cell consists of a 5 byte header field and a 48 byte information field. The bits are transmitted in decreasing order, and the byte numbers in increasing order.

• In animation a cell is a sheet of clear celluloid about normal letter page size. It has punched registration holes at the top, and different parts of the animation will be drawn onto different cells — and these will be stacked before photography so as to create a composite image. Non-moving parts will generally occupy lower cells in a stack. The background is generally not drawn on a cell, but on a strip of paper which can then be moved horizontally.

cell (mobile phone)

In cellular telephones, a cell is the geographic area served by one base-station installation and also a logical area serviced by co-located transmitters.

Originally, a cell-area was taken to mean the area covered by one base-station (using omnidirectional antenna) — but increasingly it is being used for sectors (directional beams from base-station antenna).

In omni-directional base-station transmissions, the cell is the whole area serviced by the base-station; with sectorized 'cells' it is a subdivision of this original area.

The real meaning of cell and sector is becoming increasingly confused in base-station facilities with smart antenna systems (which beam in a flexible way), and where channels can be dynamically allocated to various sectors.

• Except in CDMA mobile phones, each cell (analog or digital) requires a range of frequencies different to those used in all the cells adjacent to it — which is why the total bandwidth is allocated to cells in 'clusters'. Usually there are 7 or 9 cells to a cluster in AMPS, TACS and GSM. See co-channel interference.

• Neighboring cells don't like using neighboring frequency channels either, which makes bandwidth allocation even more difficult. See adjacent channel interference.

• In analog mobiles, cells cover an area usually 3 to 20 kilometers in radius (distance from the base-station). But TDMA/GSM digital systems often require higher antennas and smaller cell coverage areas (usually three to each analog cell) for equal drop-out-free coverage.

• PCS systems operating at 1.8—1.9GHz, generally need cells at least half, to a third, the size of current digital TDMA and GSM cellular mobile systems.

• CT-2 Telepoint cells are much smaller than mobile cells — they are generally no more than 200 meters wide, and they may not necessarily overlap. Since Telepoint operations don't generally involve hand-offs, cells coverage isn't necessarily seamless. You are expected to walk down the street until you register signal strength.

• There are micro, macro, pico cells and many other terms. See cellular telephones.

cell-based chips

A form of semi-customized chip with up to 25,000 gates. These are manufactured by using predefined (library) cells (layouts) that can range in complexity from only a few gates, to microprocessor-cores containing tens of thousands of gates.

cell-relay

A generic term for fast packet-switching systems using short cells. Cell-relay differs from frame-relay by working with fixed-length 'packets' — mainly ATM cells which consist of a 48-byte payload and a 5-byte header.

These cells are relayed in hardware at the OSI's Layer 1 (the Physical layer) in a manner similar to either a TDM multiplexer or an Ethernet switch. The term 'relay' indicates that the nodes don't store and examine the cells/packets as they pass through: they will send them on immediately, if possible.

Two cell-relay standards have been defined for broadband networks:

— IEEE 802.6, otherwise known as Distributed Queue Dual Bus (DQDB), for SMDS-type MANs; and

— ATM (asynchronous transfer mode) for WANs and general telephony.

cellular array processor

A type of parallel processing architecture pioneered by Fujitsu.

cellular radio

An old term for cellular mobile telephony. This covers a wide field, but it generally excludes the new PCS/PCN systems (despite the fact that they are cellular also) and cordless phones. See AMPS, TACS, NMT, GSM and CDMA.

cellular telephones

Also called cellular radios or mobile telephones. These are extremely popular forms of mobile telephony that work on the small-is-better principle. All the early systems used analog FM technology, but now digital systems using a range of technologies (FDMA, TDMA and CDMA) are being introduced — with not as much success as the promoters would wish.

With cellular phone systems, the city and more-populated country areas (particularly alongside highways) are dotted with transmitter sites connected via a special mobile switching center to the switched public telephone network. Transmission distances are deliberately limited by the choice of high radio frequencies and lower power so that communications is possible only within a relatively small cell area (between 0.5km and 35km in diameter).

As a mobile phone, engaged in a call, moves out of the base station's coverage range, a complex computer-control mechanism back in the mobile switching center hands off the call to the next cell — so cell coverage areas must overlap. There are various ways of controlling channel selections and hand-offs; special signaling channels are involved.

Credit for the first commercial cellular phone system should probably go to NAMTS in Tokyo in 1979. But the Scandinavians also claim to have invented cellular phones with the first NMT-450 standard.

The modern cellular systems were developed in AT&T's Bell Laboratories in the early 1970s. The AMPS standard (see) has been in operation (field and pre-commercial trials) in Chicago since 1977, but only in full commercial use since 1983. This early network used large cells with each of ten cells covering approximately 540 sq miles (1400 sq km), providing a total of 136 radio channels. The Chicago network had about 2000 paying customers in 1981. Later cell sizes were drastically reduced and the capacity increased.

- For history of cellular see NAMTS, IMTS, NMT-450 and MTS.
- Analog cellular in Europe used the following standards at various times in the 900MHz and 450MHz range: TACS 900, NMT 450 and NMT 900 (all used in various countries); Radiocom 2000-450 and Radiocom 2000-900 (both France), C 450 (Germany and Portugal), RTMS (Italy), and Comvic (Sweden).
- Digital cellular in Europe has had GSM 900 (most of Europe) and DCS 1800 (UK). Non-vehicular cellular systems include CT-2 (UK and France), BCT 900 (Netherlands), DECT 1900 (now being released)
- The USA and the rest of the world have used:

— **Analog:** AMPS (US and many others), J-TACS (Japan), and N-AMPS (US).

— **Digital:** NADC/TDMA (North America), JDC/PDC and PHP (Japan), CDMA (US and Asia), DCS1900 (PCN trials) and numerous other PCN developments.

- Cellular coverage areas are often categorized as being either pico-, micro-, or macrocell. But other terms sometimes creep in:

— Nodalcell, a single isolated cell up to 300m in diameter,

— Nanocell, a picocell up to 10m in diameter,

— Megacell (see macrocell).

CELP

Codebook Excited Linear Prediction. A digital voice compression (vocoder) technique invented by AT&T which provides good voice quality down to about 8kb/s and possibly to 4kb/s. The voice is first digitized as PCM and then passed through CELP circuits. Like all source-coder systems, CELP transmits a 'profile' of the voice characteristics, and then reconstructs a close synthesis of the voice at the other end.

The reference to a 'code-book' is analogous to a 'look-up table' which contains likely waveforms, which are based on statistical analysis of known signals. See linear predictive coding, source code, RPE-LPC, RELP, MBE, LPC and vocoders.

- LD-CELP is a form of low-delay or memoryless voice coding (see G.728).
- Q-CELP is a variation used by Qualcomm in CDMA systems at 8kb/s.

CEN

Comite Europeene de Normalisation. The European standards committee — the equivalent of the ISO.

- CEN/CENELEC: These two groups jointly look after information technology and the OSI functional standards. They are both independently members of the ISO and IEC, and with ETSI, they run the ITSTC (IT Steering Committee) which is closely involved with the EWOS's (European Workshop on Open Systems) OSI activities. See CENELEC.

CENELEC

Comite Europeene de Normalisation Electrotechnique. This translates as the European Committee for Electromechanical Standardization. The European equivalent of the IEC.

CENELEC is a member of ISO and IEC, and it has the European responsibility for setting EMI standards, and for researching potential radiation problems. See also CEN.

Cenpex

See Centrex.

Centel

See Centrex.

centi-

In the SI units, the prefix centi- represent 10^{-2}. The symbol is the small c.

central office

A US term for telephone exchange. The term encompasses the equipment (the switches) and the building itself.

- A central office code is the three-digit prefix in a seven digit local number. Don't confuse this with an area code which prepends the central office code.

centralized control signaling

Also known as Common Channel Signaling. See CCS#.

centralized processing

The processing is performed by a host computer at the logical central point of the network, although terminals may be located well away from the host. All communications between terminals must pass through the host. The contrast is with distributed processing where the computational power is distributed between devices connected to a network and they can communicate directly. This is often called 'host-centric'.

Centrex

This term has different uses in different parts of the world.

- Originally this was AT&T's name for a sophisticated PBX telephone exchange for customer premises. It could act as a sub-exchange the local exchange using DID.

- Small 'c' centrex: A US generic term for any modern form of digital PBX which can handle direct inward dialing (DID).
- Centrex: A 'virtual PBX' supplied by the local carrier to a company, by partitioning a section of the carrier's local exchange.

To all intents and purposes the company can use this partitioned section as if it were their own PBX; it will have direct inward dialing, direct distance dialing, and a console switchboard if necessary. Staff need not be all in the one building or site provided all sites are serviced by the same exchange (sometimes over long private links).

This, in fact, makes Centrex also into a form of virtual private network (VPN) even though it uses common public-access exchange equipment. Centrex can be extended to provide regional or national coverage by billing for exchange-to-exchange costs, usually at highly subsidized rates.

Centrex depends on intelligent networks with sophisticated features such as call-forwarding, and the ability to partition digital exchanges.

Centronics

A trade name which has been taken up generically to mean the particular 36-pin plug/socket used by standard parallel interfaces — and for the interface specifications itself. This is the parallel alternative to the serial RS-232 links on the back of most PCs for connecting peripherals like modems and printers. In the past few years, the Centronics parallel interface has become the standard in the PC world for printer connection.

Centronics is also sometimes also called 'ribbon' because of its use of a flat multi-wired cable (but many parallel cables are round). It is not designed to run any distance — 3 to 5 meters is generally taken as the maximum advisable.

— Pin No. 1 is the data-strobe.
— Pins 2 to 9 are used for the data-bits (least, lowest).
— Pin 10 is acknowledge.
— Pin 14 is supply ground.
— Pin 16 is logical ground.
— Pin 17 is chassis ground.
— Pin 18 supplies +5 volts.
— Pin 19 is return data strobe.
— Pins 20 to 27 are return data-bits (least lowest).
— Pin 28 is acknowledge.

CEPT

Conference of European Postal and Telecommunications administrations. A 27-country European telecommunications standardization committee which is equivalent to the US's ANSI T1 committee. It is the main governing body of the European PTTs. ETSI is a sub-group within CEPT which has some independence.

CEPT publishes NET and ETS standards. For telecommunications in all European countries under CEPT, Norme Europenne de Telecommunications (NET) standards are mandatory. However, outside the EU (but within CEPT) the ETS (ETSI) standards are only voluntary.

There are a few workable NETs: NET 1 is for X.21; NET 2 for X.25; NET 3 is for the Euro-ISDN user-interface, and NET 7 is for Euro-ISDN terminal adaptors.

CEPT-1

This is confusing, because the same term has been used in a number of different ways.

- For cordless telephony, see CT1.
- For digital transmission channels:
  — CEPT-1 is 2Mb/s; more often called an E1 channel.
  — CEPT-4 is a 140 Mb/s channel.

The US almost-equivalents are called DS1 and DS4.

CFDD

A receiver which can handle APCO's QPSK-C form of FM transmission (using a class C power amplifier) and the CQPSK form of AM modulation (using a linear class AB power amplifier). See also QPSK-C, APCO-25 and C4FM.

CG

Character Generator. A hardware unit used in television production for creating text on the screen. Cheap CGs are built into domestic camcorders these days, but the more elaborate devices hold thousands of fonts in many sizes and colors, and have the ability to distort and manipulate these and present them in rapid succession on the screen. Software on a PC can perform most of these functions in small studios. See DVE.

CG-VDI

The ANSI Computer Graphics Virtual Device Interface standard. See VDI and CGI.

CGA

Color Graphics Adaptor. IBM's first color monitor standard for the PC and the first monitor with graphic capabilities. You bought CGA as a plug-in adaptor card for your old monochrome system. It is no longer used but, fortunately, software written for CGA will generally run on EGA, VGA and MCGA.

CGA was designed mainly for video games, but it was able to display 40 or 80 characters of text in 25 lines, and in graphics mode it would handle monochrome with 640 x 200 pixels in high resolution mode, and four colors (from a 16-hue pallet) with 320 x 200 pixels. There were actually three graphic modes, but the lowest was only available by addressing the controller chip direct.

Text was displayed in only-just-readable 8 x 8 character cells, and text memory could be paged to create areas of foreground/background color on the screen. There was also a TV screen 40 character text mode.

CGA monitors were RGB and used TTL digital signals with a vertical scanning frequency of 60 Hz. You'd get a headache trying to run word processors with CGA, but if you liked pastel pink and cyan color combinations, you'd have loved this standard. CGA was once described as having 'all the charm and precision of a kindergarten finger painting.' See EGA and VGA.

• The CGA monitor connections were:

Pins 1&2 Ground; Pin 3 Red, Pin 4 Green; Pin 5 Blue; Pin 6 +ve Intensity; Pin 7 not used; Pin 8 +ve Horizontal drive; Pin 9 –ve Vertical drive.

CGI

Computer Graphics Interface. The ISO's draft graphics standard (modelled on CG-VDI) for writing graphics drivers which will be used for graphic devices such as screens, printers or plotters. ANSI's VDI standard is the current format for device drivers, and CGI is expected to replace it. See VDI.

CGM

Computer Graphics Metafile. A widely used intermediate graphic file format which has been developed from the earlier VDM (Virtual Device Metafile) format, and it also includes many elements of GDM (Graphics Device Metafile). The aim is to allow graphic information to be transferred from one program to another through the CGM common format. Implementations differ, so there are some problems with compatibility. It is an ANSI and ISO-draft standard. However, it is not intended to be used as a CAD transfer system.

chad

This is one for Trivial Pursuits: chad is the small 1 millimeter dot of paper punched out by Telex machines and other paper-tape storage devices. See perfory.

chaining

Linking in sequence.

• The ability of a program, during execution, to call another program, or to run one program after another, automatically.

• In SNA, messages (RU — Request/Response Units) are chained (grouped) for error recovery and for other purposes.

• Chain indexing is a hierarchical indexing system used in library classification. It is the indexing equivalent of hierarchical directory branching with pathnames.

• Chain searching means going from one hierarchical classification to the next level, and the next until the information is found.

• Chained lists are linked lists. See.

• In expert systems, backward chaining is the process of ascertaining the fact, then seeking the proof (by looking for a rule). Forward chaining is the normal progress of the expert system in moving through inferences to a conclusion. See forward and backward chaining.

• In telephony, chaining is joining one telephone circuit onto another in sequence through exchanges. You may phone long-distance using your private network, then receive dial-tone, and dial a local call through the company switchboard at the other end.

channel (general)

The ISO says that a channel is a single-direction path along which a signal can be sent. It also suggests that it is one part of a multiplexed bearer — as distinct from an individual signal on its own.

This may be strictly correct, but that's not how the term is widely (and casually) used. The literature is full of talk about two-way channels, and channels in unmultiplexed bearers. Very often people will refer to 'two-way channels' or duplex-channels, or simply assume this.

Channel is really a logical definition, while 'circuit' is a technical one — but the terms are constantly mixed up — which is understandable because it is confusing. With techniques like time-division multiplexing several logical channels occupy one physical bearer circuit.

Channel should mean the 'data path', while circuit should mean the route of electrical current along a medium, and the two are often not the same. In practice, treat channel as being almost synonymous with link, data-path, circuit.

• In telecommunications they talk about the channel as the generalized part of the system that joins the message 'source' to the message 'sink'. The size, or bandwidth of a channel determines how much information can flow through it, and is expressed in terms of Hertz (Hz) for analog and bits per second (b/s) for digital. In computer technology (internal communications, or with peripherals), they often measure bandwidth in Bytes per second (B/s).

• A channel in mobile phones is actually two channels separated by a duplex-offset. See duplex.

• In a time-divided system, a channel is represented by a single time-slot. Technicians sometimes distinguish between the 'physical' channel which carries the total signal, and the more numerous 'logical' channels which each occupy one slot.

• In satellite terminology, a channel is typically a one-way half-circuit which can carry one TV signal or 1200 voice telephone conversations — but one-way only. You would need another channel for the other side of the conversations. See forward and reverse channels; uplink and downlink.

• In storage devices, a channel is the logical information in concentric or spiral track which can be written to, or read from. CD-ROM uses the term channel bits and bytes to refer to the information as it exists at the disk level, before EFM conversion. See EFM and channel coding.

• IBM makes the distinction between what they call the ISA PC 'bus' (e.g. in the old AT-bus standard) and the 'channel' (as in MCA). By 'channel' they mean that an intelligent I/O device, such as a co-processor, is able to assume full control over both the system bus and memory. They say that with 'bus' architectures, the co-processor is always a slave to the main processor. So they are making the distinction on the basis of arbitration and control. Channels, in their terminology, provide bus arbitration between competing devices.

- An analog channel is often defined as one that will pass alternating current, but not direct current or baseband unipolar digital signals. However an analog channel can carry digital signals through modulation onto an AC carrier or with biphase line codes. This should be 'circuit'.
- A digital channel is defined as one that carries DC currents. This should be 'circuit' also.
- Channel logic is the defined interface functions between the Data link and the Physical layers. See OSI reference model.
- Channel identification: In TDM systems each logical channel is given an identification number. Also in modern telephone signaling systems, signals are transferred across the network in special packets, and each of these requires identification information to link it to the channel which is being controlled and switched at nodes. See VCI and VCC.

channel (processor I/O)

Channels are the quickest way to get a lot of information into and out of a CPU in large computers; they are often called data-pipes. This is specifically a terminology of mainframes, but it is migrating down. There are three main types:

— Channel–channel: From one host computer to another. These are mainly used for file transfers in bulk data applications. The channels will be high bandwidths, and handle large block sizes.

— Channel–peripheral: From the host to a major storage device such as a tape drive.

— Channel–LAN: A new role for data-pipes with the growing emphasis on the host becoming a high-speed server on networks.

- The range of channel interface definitions available include:

— Bus & Tag (FIPS-60) to 4.5MB/s for distances of 120 meters;

— SCSI-2 fast/wide, up to 40MB/s for 25 meters;

— ESCON (fiber) at 17MB/s for 20km;

— Fiber Channel 125MB/s for 10km;

— HPPI 100MB/s for 25 meters;

— FDDI (fiber) 12.5MB/s for 2km; and

— ATM in the 20– 80MB/s range for unlimited distances.

channel access

Also called media access. All stations on a LAN or any other form of shared media (including the radio spectrum), needs an access technique before multiple users can share bandwidth without chaos. They need some system of avoiding one transmission from over-riding another, and fair allocation.

In Ethernet the access technique is CSMA/CD; in Token Ring it is token passing; in GSM mobile phones it is TDMA; in CDMA phones it is CDMA.

Access to some schemes is contention-based and erratic (probabilistic), while others are deterministic. CDMA is neither since the technology is basically non-blocking.

channel aggregation

See aggregation and BONDing.

channel associated signaling

A form of inchannel but out-of-band signaling used for the control of digital traffic. This is quite different from conventional in-band telephony signaling of the type heard when we dial an analog phone, and it is also quite distinct from the packet-switching signaling of CCS#7 and ISDN.

Channel-associated is a class of signaling which can provide both telephony control, and a data control system. Within each digital frame on the network, some bits are reserved for network supervision or for flow-control and congestion-management. The control signals are carried either within the channel itself (by bit robbing), or by a special signaling channel directly associated with the traffic channel. See A&B bit signaling, T-1 and CCS# also.

channel-attached

Devices which are directly connected to a host computer, usually by a large bandwidth pipe. See channel (I/O) and channel extension.

channel bandwidth

In telecommunications, both digital and analog channels require that the appropriate bandwidth be provided over the whole chain of connections before the signal is carried. See bandwidth (guide).

channel bank

An old US term for a special form of telecommunications multiplexer and concentrator. Channel banks have been around since the 1930s under the name of carrier terminals. They began to be called banks in the early 1960s when T-type digital hardware began to be installed in telephone networks.

Digital Loop Carriers (DLCs) may use channel banks at the subscriber end. At the switch end, they may also terminate in a channel bank (LT — for Local Terminal) or be directly connected to a terminating card in the switch.

The name 'channel bank' is still used in T-1 discussions, and it now appears to refer to a relatively basic ('static') multiplexer or 'concentrator' which is only configurable by plugging in cards. Channel banks provide the multiplexing functions, but they also handle signaling information for each of the channels, and transmit framing bits.

- A VF channel bank (voice frequency), multiplexer 12 standard analog signals together into 48kHz of bandwidth (using single sideband AM). See group.
- D1 and D2 channel banks used 7-bit sampling, and they assigned channels to slots in a pseudo-random manner. See A&B signaling.
- D3 and D4 channel banks (from 1969) use 8-bit sampling, and fixed slot allocations. They are incompatible with D1 and D2. See T-1.

channel bits

See channel coding.

channel class

The media through which the channel transmits: coaxial cable, twisted pair, radio or microwave.

channel code/coding

This is the code actually used by the channel — whatever that channel may be. It is also known as 'constrained coding' and will be called 'line-code' on simple circuits. There is no hard and fast distinctions between channel codes and line-codes, although the first is done usually to increase data density (while maintaining reliability), while the second is done primarily to improve transmission reliability (while maintaining traffic density).

Channel coding is the broadcast and CD industry's equivalent of line-code, so it is the way in which the physical layer is handled in radio transmission or in a storage medium. Actually line—channel are two ends of a continuum with line-coding at the simple end, and channel coding involving more complex processes. Manchester coding sits between.

Channel codes are needed because simple unipolar codes (8-bit ASCII) of the kind used to process the data electronically, may not represent the data most efficiently when it travels along long copper lines, through optical fibers, or within the magnetic or optical domains on a storage disk.

Logical code is translated to a channel code before it is passed to the storage device or transmission system. These new substitute codes can better protect the data against errors, and they can increase the packing density of a disk, and provide self-clock synchronization.

— The term 'channel coding', emphasizes the mathematical alteration of the bit-sequences to prevent intersymbol interference — the interference caused by having adjacent 1s and 0s at too high a packing density. Often conversion will take place through look-up tables.

— The alternate term 'channel constraint', emphasizes the need to place a limit on the number of 0s that can occur between consecutive 1s. This is done to prevents loss of synchronization from long idle periods.

— The term 'line coding', emphasizes that the changes are usually made within the transmitting devices by voltage inversions, rather than bit changes (although they are both the same).

Each device and media has its own form of channel coding to suit the technology and the application. However AMI is very widely used, as is Manchester coding. See line-codes also.

• The 8-bit bytes being recorded on a CD-ROM or CD-Audio disk are translated to a 14-bit form, then to a 17-bit form, before becoming the pits and lands on the disk. The first stage from 8-bits to 14-bits is called 'Eight-to-Fourteen Modulation' (EFM), then a further 3 merging bits are added to each to create 17 channel bits. This 17-bit byte is the code actually recorded on the disk. This elaborate process actually increases the packing density (for the same error rate).

• With magnetic storage devices, IBM pioneered the approach known as MFM, which has now been upgraded to RLL. These techniques increase the capacities of high-end disk and tape storage systems. See MFM and RLL.

• The term 'channel coding' is often misused, so see also source code.

channel efficiency

The calculation of the total information that can be carried, taking into account the collisions, noise, overheads and all other problems.

channel-extension

Direct Access Storage Devices (DASD) are able to transfer data at extraordinary rates between a computer and a storage device, so special channel-extensions are used to carry the data to the storage site. These are now just called I/O processor channels. See channel (processor I/O).

channel group

See group.

channel logic

These are the logical functions between the transceiver and the data-link — between the Data link layer and the Physical layer.

channel rate

The primary measure of the data rate at which information is transmitted through the channel. This is not the throughput rate.

The channel rate is typically stated in bits per second (not baud) and it ignores the useful data throughput amplified by data compression, and the negative effects of overheads and retransmissions.

In packet systems it may sometimes exclude overheads associated with addressing, and with circuits it may not take into account framing, error checking and correction. So it is important to state whether this is the raw channel rate or the payload channel-rate (data only, excluding overhead).

• BRA ISDN has a raw channel-rate over twisted pair of 168kb/s, but after excluding framing, etc. the useful data-rate is 144kb/s, which divides into two B-channels of 64kb/s and one D-channel of 16kb/s (which is used for signaling). You'll find the channel rate of ISDN quoted variously as 64 (B), 128 (2B), 144 (2B+D) or 168kb/s (total). Take your pick; but I would chose 168kb/s.

channel separation

• Analog telephone networks always have near-end and far-end crosstalk; as a consequence of inadequate channel separation. Digital telephone systems don't have comprehensible crosstalk, but one channel can affect the bit-error rate of the other.

• Television stations (analog) cannot use adjacent frequencies in close proximity to each other because the sound of one intrudes into the picture of the other. See adjacent channels and taboo channels also.

• An audio playback system must maintain the distinction between the hi-fi left-and-right stereo signals. When the channel separation is inadequate you have crosstalk.

• In low and moderate-speed modems communicating in full-duplex, the problem is in keeping the go and return paths separate. There are three main techniques:
— Frequency division (using different audio carriers).
— Four-wire circuits (only practical in private networks).
— Echo cancellation (as used now in all the new high-speed modems).

• Cellular phone systems generally cannot use adjacent frequency bands in neighboring cells. This makes allocation of the various bands very complex.

channel spacing

The distance in bandwidth terms between two adjacent broadcasting channels. In cellular telephony this is called the channel separation. It represents the width of the band needed to carry the signal, together with any guard-bands.

In PAL TV, the center of the video will be 1.25MHz from the lower edge of the band, and then the audio is spaced about 5MHz above (depending on the type of PAL system being used). There is then a guard band between the top of the audio band and the next TV channel (except in NICAM, which actually intrudes into the next band).

channelize

The process of subdividing a channel into multiple channels with lesser bandwidth. This is what a multiplexer does.

CHAOSnet

A network protocol used by the artificial intelligence community. It was developed at MIT.

character

A very vague term for the visible result of a byte (the design of the font/typestyle), or for the interpretation of a byte in a standard look-up table such as ASCII No.5 or Unicode. A number of characters grouped together in sequence is known as a 'string'.

• In the widest sense character can mean any one of the 128 print or non-print 7-bit ASCII codes. More narrowly, it may refer only to the printable alphanumerics and punctuation codes in the ASCII No. 5 alphabet set such as A or [or 6 or =. A narrower application still would reject the punctuation marks, but some would also include the carriage return and possibly the line feed. The space is actually a 'printable character'.

Apple says a character is 'any symbol that has a widely understood meaning and thus can convey information.'
— Graphics characters, are actually graphical primitives (block shapes) which occupy a fixed character matrix space and which are treated as if they were normal characters, but in the upper range (the 8-bit) above the 7-bit standard character set.
— The IBM PCs had a special character set, quite separate from graphics characters. See IBM.

• With the recent emergence of double-byte WorldScript or Unicode code sets, it may now take two or more bytes to create a character.

• The character-error rate of a channel may bear little relationship to the bit-error rate. If the errors are evenly distributed the two measures should be in parallel, but if burst errors are the problem, then there will be a major discrepancy.

character control

• The use of control characters to initiate, modify, or stop an operation. The distinction here is with hardware handshaking systems where the control is exerted by raising or lowering the voltage on a specific pin at, say, the RS-232C port. XON/XOFF is a character-control system.

• The term may also include non-print control functions, such as line-feed and carriage return, used in word processing to create text documents.

• In the ASCII No.5 alphabet, control characters are the first 32 codes in the standard code table.

• Synonymous with 'function code' or 'function key'.

character-count protocol

One that specifies the number of characters in the information field. See Data link control.

character device

A device (usually a keyboard terminal) which sends and receives data one character (one byte) at a time. And any device which sends one character at a time is almost certain to be an asynchronous device. The distinction is with block devices. See C-DTE.

• A character printer receives information one character at a time, rather than a line or page at a time.

character editor

This is similar to a line editor (like EDLIN), but you can change one character at a time.

character framing

Asynchronous transmission systems which are using start and stop bits to 'frame' (book-end) each character.

character generator

• A ROM chip which converts the ASCII codes into the dot-matrix patterns that are projected onto the monitor screen or printed on a dot-matrix printer.

Not all computers now work this way. The Macintosh, for instance, handles its screen in a totally graphic way which is why the style of the characters can be changed instantly through installing software fonts, without changing a character generator chip.

• Software algorithms which take videotex codes from the page store, and create the characters on the screen.

• Special keyboard hardware devices used in television studios to generate captions, crawls, titles, credits, etc.

character graphics

A form of screen presentation used in conjunction with raster scan techniques where the graphical characters (block-like elements of a block-like image) are created within ROM, and sent to the video RAM as if they were alphanumeric characters. This is used in videotext and in some early PC programs. The distinction is with pixel graphics. See graphic characters.

character mapping

The process of converting character codes (say, in ASCII) to those in another format (say, EBCDIC). A direct one-to-one mapping can't be done in most cases, because not all characters are represented in both formats. For instance, 7-bit ASCII doesn't have the cent sign whereas EBCDIC does.

character-mode terminal

See character device and C-DTE.

character recognition

A number of techniques, the best known of which is optical character recognition (OCR). However the term also includes magnetic ink character recognition (MICR). See also optical mark recognition and ICR.

character set

The full 128 or 256 of characters that can either be shown on a computer screen or used to code instructions. Some of the character set will be 'printable' and others not. See ASCII No.5 and also EBCDIC

character signal

- The set of signal elements in text transmission.
- The quantized value of a sample in PCM audio.

character string

Any sequential group of characters (usually print characters) which the computer application treats as a single unit.

character terminal

In packet-switching terminology, this implies a terminal which cannot access the X.25 host directly, but must use an intermediate node with a PAD. Your PC with a modem and communications software constitutes a character terminal.

characteristic

In mathematics, see exponent.

chassis-based hubs

These were introduced in the 1980s to allow users to plug the appropriate network card into a standard chassis for complex networks. They followed the introduction of fixed-port hubs, and preceded stackable hubs.

With chassis-based hubs the user could add ports and mix cable media types and architectures (Ethernet, Token Ring, or FDDI). Network management, internetworking and terminal connections could also be added — and the system provided a degree of fault-tolerance. See fixed-port hubs and stackable hubs.

The problem with chassis-based hubs is that they proved to be very expensive. Stackable systems were much better and cheaper and the modules allowed incremental improvements.

chat

A real-time mode on many computer conference facilities and bulletin boards that allows you to keyboard 'chat' with other people on-line at the same time. See IRC.

Cheapernet

This form of Ethernet uses cheaper RG58AU 5mm 50 ohm coaxial cable and BNC connectors. The main cable connects directly into the transceivers in the PC, using a T-connector. The maximum number of stations permitted on a segment (max 200 meters) is thirty. See 10Base-2 and Ethernet.

cheat

In film production, this is the movement of something which has been seen in one shot, and will be seen again in a shot from a different angle. Strict continuity calls for all objects to remain exactly in the same place, but with a change in view angle, it is often possible to make a subtle repositioning of something — to 'cheat' it — so it is more convenient to the actor, or out of sightline for the camera operator.

check-bit

See parity bit formats and Hamming code.

check-character

This is most likely a checksum, but it could be a CRC. See.

checksum

The term has a general meaning that encompasses LRC and CRC techniques, and a more specific one which just means LRC additions.

A checksum is often just the binary (modulo 2) addition of all the bytes in a block being transmitted. This is LRC checking. The checksum is calculated and appended to each data block. The receiver then recalculates the checksum and checks it's addition against the received value to determine if errors have been introduced during transmission. This is commonly known as redundancy checking.

However in the wider sense, checksums aren't always just additions. They can be any calculated value which would most certainly be different if some of the data was lost or changed. A CRC is a checksum in this wider sense of the term — and an alternative to a checksum in the narrower sense. See XModem, LRC and CRC, and Reed-Solomon also.

Chernobyl effect

Packet storms where no information can get through on a LAN.

chevron

A symbol often used in computing. It is a double V usually printed on the side like «, where one fits into the other like a army sergeant's stripes.

- Chevrons are also physical devices used in chip memory.

child

• **Processes:** The term 'parent' and 'child' are applied to processes in software programming. The child process is one that runs while the parent process is in control.

• **Windows:** A child window must have a parent — they can't exist independently. If you attempt to close the parent window, the child will be closed and removed first.

CHILL

An ITU-T designed high-level language for stored program control (SPC) switching and other telecommunication applications. It has some outside applications also.

Chinese room

This is a reference to a thought experiment of philosopher John Searle when considering machine 'intelligence'. It hinges on the meaning of the term 'understanding'. He was suggesting a test for what could genuinely be called 'intelligent machine translation'.

Messages in Chinese are passed under a door, and returned in English. We don't know whether a Chinese-speaking translator or a translation machine is on the other side. If we can't tell from checking the returned messages, then the machine can legitimately be called an intelligent translator

The problem is whether a machine that mechanically responds (correctly) to questions can be said to be 'understanding' them. Is mechanical performance the same as mental processes?

• This is very close to the Turing test, and I'm not about to arbitrate on who had the idea first. See Turing test and ELISA

chip/CHIP

• An integrated circuit. See IC.

• In spread spectrum technology, a chip is the equivalent of a modulated bit. It varies slightly in form, depending on which type of spread spectrum is being used.
— In direct sequence: it is the equivalent of a bit which has been generated by multiplying the primary data by a orthogonal code. It is the length of the orthogonal code. This is the primary technique of 'spreading' the spectrum using Walsh functions.

In the narrowband Qualcomm system, a data stream of 9.6kb/s is multiplied 128 times (the Walsh code length) to a chip-rate of 1.2288MHz, to fit the 1.25MHz channel. The chip-rate is equivalent to the data rate times the spreading factor. The correct term is 'Walsh chip'.
— In frequency hopping, a chip is the length of time each transmitter spends on a single channel (hence chipping rate).

• **Clearing House Interbank Payments** A check clearance system between banks.

chip-card

See smartcard.

chip-set

The combination of various elements on to one or more VLSI chips. Even a single integrated chip is sometimes known as a chip-set. This consolidation of circuits constantly being done to reduce cost, lower power requirements, and improve performance. It is what George Gilder calls, the 'Law of the Microcosm'.

The term chip-set is most correctly used to mean a group of chips which form a single functioning unit, but it has been appropriated to also mean a number of library elements mounted on one piece of silicon.

chiplet

A two-layer chip fabrication technique which has a high-speed GaAs layer making connections to a lower silicon substrate. The technique is aimed at producing less expensive devices (particularly opto-electronics). See cubelet also.

chirp

• A change in the optical wavelength in an optical fiber system due to the pattern of the data pulses. Chirping occurs when a laser is turned on and off at high modulation frequencies. It causes transmission errors by making it difficult for the receiver to distinguish the pulses.

• A rapidly rising audio tone.

CHKDSK

A MS-DOS utility which checked the surface of the disk and marks out any damaged sectors. The new version is called ScanDisk, and it repairs crosslinks automatically, while CHKDSK only detected them.

choke packet

A control packet which instructs a transmitter to reduce its transmission rate because congestion exists down the line.

chop

• Rounding off numbers by truncating towards zero.

• Sorting: see binary chop.

chord/chordic

• A safety device where more than one key must be pressed simultaneously before a function will be performed.

• A type of one-handed keyboard for computer entry which has seven keys which need to be pressed two at a time.

Christensen's protocols

See XModem.

chroma bandwidth

The bandwidth of the color subcarrier on a 625-line PAL signal is 4.43361875MHz, give or take a Hertz or two. The bandwidth of a 525-line NTSC signal is 3.579545MHz.

The total bandwidth of the television carrier must include the sound sub-carrier and the separation between the two and some guard bands. NTSC does not have many variations, but PAL has dozens.

chromakey

A technique used in studio television production where all picture elements of one selected color (usually blue or green) are extracted from an image — creating an image-key into which a new background can be inserted. This is the way the games-show host is inserted magically into a postcard image of a Paris street scene.

Chromakey is usually an internal key process (see), with the key selected by color rather than tone. For this reason the background key color must be quite distinct from flesh-tones (or parts of Paris may appear in the face) which is why green is often chosen. (Blue-keying produces peculiar effects in some people's eyes!)

With good chroma-key equipment the tell-tale blue-green edging of the subject's outline (especially around wispy hair strands) almost disappears, and some very sophisticated new techniques allow ground shadows to be generated for added realism.

chrominance

Also called 'chroma'. The color information in a video signal. Both B&W and color images have luminance — or light intensity (usually designated as Y), but only color images have chrominance. There are three primary color components: Red, Green and Blue, and the 'chromaticity' of each is measured separately.

Note that there are pure and colour-difference forms of component color signals. RGB signals are additive to produce white (luminance), while other systems such as YUV use color-difference signals where a single primary color is subtractively combined with the luminance values (R–Y, or B–Y).

CHRP

Computer Hardware Reference Platform standard of Apple, IBM and Motorola with the PowerPC. This is supposedly a new standard for interoperability among PowerPC computers and PowerPC operating systems. They have defined cache, ASICs, controllers and other components; and they emphasize the use of an ISA and/or PCI bus. New CHRP chips will support PReP 1.0 from IBM, and Mac OS memory maps.

CHS

Cylinder-Head-Sector addressing. An extended version of this is being proposed as the best way of overcoming the 528MByte barrier on MS-DOS hard disk drives.

Two unused bits are borrowed from the head number (used in the INT13 interface) and added to the cylinder number. This now makes the cylinder number 12-bits and increases the address possibilities. This technique requires no modification to the OS or disk drive, and only slight changes to BIOS, so it is seen as a safer way to proceed than LBA.

chunky architecture/pixels

See packed-pixel architecture.

churn

The proportion of customers who cease subscribing to a service over (usually) a year. Churn is the measured rate of change in the industry and indicates either dissatisfaction with the supplier, or the success of the competition in some marketing campaign.

- Churning, is the process of persuading a client/subscriber to switch from one system to another — often by making claims which can't later be justified. It is a common practice in the mobile phone industry and in cable TV.

C/I

Carrier-to-Interference Ratio. See CIR.

cicero

See didot.

CICS

Customer Information and Control System. An IBM mainframe-based transaction processing environment (OS/2 utility software). It's a semi client-server subsystem which includes database management functions. It will communicate over a range of protocols (SNA, LU6.2, EDI, OSI and TCP/IP).

The idea is that transactions entered at remote terminals will be processed concurrently by the user-applications. CICS makes it easy to access the system without delays and to request data contained in mainframe files. On receipt of a request, the CICS program retrieves the data from the host files and sends it to the workstation's transaction program.

CICS was initially sold in a popular timesharing package which ran under the MVS operating system. It has now been ported to ESA, VSE, VM and OS/2. Most of the 30,000 CICS licences are for IBM VSE mainframes. See OLTP and transaction processing.

- CICS also provides facilities for building and maintaining complex databases.

CID

Caller ID. See CLID.

CIE

- **Computer Integrated Engineering.**
- **Commission Internationale de l' Eclairage:** A commercial color space standard for color printers and monitors. If this standard is adhered to for color alignment, colors should always look the same no matter what the process. Adobe uses this standard in PostScript 2.

CIF

- **Common Image Format** of HDTV. See.
- **Common Intermediate Format** of videoconferencing. See this and QCIF also.

CIG

Common Interest Group.

CIM

• **Computer Integrated Manufacturing:** A 1980s term for more advanced computerized production process control than CAD/CAM for linking company computers to factory automation devices. The 'integration' function here is between the original CAD application, numerical control, and the parts-storage files. See C4 also.

• **Computer Integrated Management.**

• **Computer Input from Microfilm:** Scanning microfilm images for computer storage.

cine

An old term for movie film and movie cameras — generally applied to the (so called) 'sub-broadcast standards' of 8mm, 9.5mm and 16mm.

Cinemac

A technique of archiving videotape images by converting them to the MAC component video format (with the audio, luminance and chrominance separated in time) then recording this as an image-trace of the wave-form on B&W 35-mm film. Well processed B&W film has proved to be the most time-resistant material available for motion picture archiving.

The problem with storing videotapes, is that the binder which holds the magnetic materials onto the plastic backing appears to lose flexibility with time (within ten years in many cases) and the magnetic surface then flakes off. We are seeing this also with computer tapes and disks.

Cinemascope format

The film industry's answer to television in the 1950s was to develop a wide range of exotic film formats, of which Cinemascope is the widest to survive. It has an aspect ratio of 2.35:1 which means that when shown full-height on a standard TV screen, only half the horizontal view is seen — so the film must either be letterboxed, or selected by 'Cut and Pan' — either at the time of showing, or during the pre-recording. See widescreen and Academy format.

CIO

Circular Intermediate Orbit; aka ICO. This is a narrower term than MEO (Medium Earth Orbit) because it makes the point that the orbit is circular, not elliptical. It could apply to LEO or GEO satellite which also take a circular path, but it seems only to be applied to middle distance satellites in the 10 —15,000km range, where satellite orbits can be circular or elliptical. See ICO and MEO.

cipher

See crypto systems.

CIR

• **Committed Information Rate** in frame relay. The guaranteed bandwidth which is available at each port on a FR network. At some periods you can exceed this. See EIR and BECN.

• **Carrier to Interference Ratio:** The power of the carrier signal to that of the interference signal. Both carrier and noise will be measured in dBmV, while the ratio is expressed in dB.

CIRC

Cross-Interleaved Reed-Solomon Code. This is two layers of Reed-Solomon error correction which have been interleaved to spread out any potential damage. CIRC can correct (not just detect) up to 4000 data bits in a damaged CD-ROM track. See Reed-Solomon code.

circuit

Any electronic architecture or part thereof — a path where electrons can flow. With digital data, there are both serial circuits and parallel circuits.

• In communications, the term originally applied to the physical resources of a telecommunications system — the lines and exchange connections used to create an end-to-end link for communications. It referred to the physical route dedicated to the traffic. It needed a power source (battery or generator) and a conductor (copper wire).

So 'circuit' once meant the physical/electrical connections of a network, while 'channel' meant the logical connection — but the distinction is now well and truly blurred.

Circuit's meaning has now come to be generalized as:

— The existence of an end-to-end channel (not necessarily a physical or electrical connection) in real-time. This generally implies a two-way link, so two analog phone channels = one circuit. These two channels are known as 'go' and 'return'.

— The physical linking component of a telecommunications system. For instance, the two wire twisted-pair that links your home telephone to the local exchange. With ISDN, this circuit can contain three quite separate two-way channels. Perhaps we should say here that the 'wire' contains three 'circuits', and each circuit is a two-way logical 'channel'?

• This physical definition of the word is further subdivided into:

— two-wire circuits (as with a normal telephone local loop),

— four-wire circuits (used between exchanges) See four-wire and hybrids.

And the circuits can be:

— switched circuits (public telephone/Telex services),

— leased circuits (private end-to-end network connections), aka 'tie-lines',

— shared circuits (LAN and other packet systems).

And they may be:

— narrowband (usually below 4kHz or 64kb/s),

— voiceband (4kHz or 64kb/s),

— wideband (from 64kb/s to 2Mb/s),

— broadband (from 2Mb/s up, but vague).

circuit group

Circuits in telecommunications are often grouped to be handled together through multiplexing equipment. See group and supergroup.

circuit parameters

The parameters which define the transmission performance of a physical circuit are:
— Attenuation distortion.
— Envelope delay (also called group delay).
— Signal-to-noise ratio.
— Non-linear distortion.
— Impulse noise.
— Phase jitter characteristics.

circuit switched

In telecommunications technology, circuit switching systems have a link extending from sender to receiver at the time of transmission, and one that is held open for the duration of the transmission. This term no longer implies an end-to-end electrical connection or physical link — simply that the dedicated channel/s exists before the message is transmitted.

The distinction is with message-switching and packet-switching, where packets are sent over the network, often without pre-arrangement, and purely local switching occurs at each node in the network during the transit of the message.
— Circuit-switched systems are always connection-oriented. A logical channel is always established before the data is transmitted.
— Packet-switched systems can be either connection-oriented, or connectionless.
— Message-switched systems are always connectionless.

• The distinction also needs to be made with LANs of the Switched Ethernet type where shared channels are effectively routed by a switching unit using packets. These are not circuit-switched systems but packet-switched.

circular file

• One in which the end of the file and the beginning of the file are joined. If the file is alphabetic in order, 'Aardvark' will be one record away from 'zymotic'.

• Slang for a waste-paper-basket.

circular queue

A data buffer where data is inserted at one end and removed from the other. The ends of the buffer are constantly changing so this is not necessarily the same as FIFO. Two pointers keep track of the current beginning and end of the queue.

circularity

A problem faced in computer applications, routers and a range of computerized devices where a job can't be finished, because it cycles back on itself infinitely.

• In a spread-sheet, the formula in one cell may call on information from another; but that other cell may have a formula calling on information from the first.

• In a bridge or router, messages may be sent along paths which circle back on themselves if the network layout hasn't been totally controlled, or if the correct routing algorithms aren't in place.

• Programming loops use circularity to function, but there will always be an achievable condition, at which time the circularity will be broken.

circumflex

This is a diacritic mark, an accent which is written above a vowel in many European languages. However the shift-6 key character ^ (aka 'hat' or caret) produces the same shape of character in isolation. This is an independent ASCII character whereas a true circumflex is an integral part of the Roman character vowels on a French keyboard.

CIS

CompuServe Information Service. This is often written CI$ for obvious reasons — although prices have dropped recently.

C-ISAM

The X/Open group's Unix standard system for indexed sequential file access. It is included in XVS but not in SVID.

CISC

Complex Instruction Set Computing — as distinct from RISC. This is the type of processing chip most widely used in current PCs — but probably not in the future. RISC seems to be winning the battle.

CISPR

Committee Internationale Speciale d' Pertuvations Radioelectrique (International Special Committee on Radio Interference). This group was originally set up in 1934 to establish test methods for radio interference, and to set emission and interference standards in Europe. Most national standards are based on CISPR recommendations.

They have a standard for 'unintentional' emissions (CISPR22) from IT equipment which is intended for electrical motors and radio receivers — but not as yet for 'deliberate' transmitters which pollute, like all the TDMA systems. They are currently producing special 'immunity' standards for ETSI — but there's no sign yet of an 'emission' standard — which tells us a lot about the companies which dominate this committee.

CENELEC usually produces these standards following CISPR and IEC guidelines. Some of the more important are:
— CISPR11 (1990): Measurement of radio disturbance characteristics of industrial, scientific and medical R/F equipment.
— CISPR13 (1975): Measurement of radio disturbance characteristics of broadcast receivers and associated equipment.
— CISPR14 (1985): Measurement of RFI characteristics of household electrical appliances, etc.
— CISPR15 (1985): Measurement of RFI characteristics of fluorescent lamps.
— CISPR20: Immunity from radio interference of broadcast receivers and associated equipment.
— CISPR22 (1985): Measurement of RFI of information technology equipment.

Note the one missing: ***Measurement of RFI from portable transmitting equipment — like cellular telephones!***

• Information Technology equipment standards are subdivided by CISPR into Class A and Class B. Class A is less strict and may not be allowed in all countries. For RFI, it is usually measured at a 30 meter distance, while Class B must be safe up to 10 meters.

CIT

Computer Integrated Telephony. The integration of telephony into computer applications. This gives a word processing application, for instance, the ability to dial and transmit a file from within.

But the term has wider applications than this: CIT also extends to systems used for automatic call dialing for marketing people, and allows voice and e-mail integration. This is a developing area where Microsoft and Novell are fighting a battle for control. See CITA, ACD, TAPI and TSAPI.

• The DEC/CIT and Ericsson MD110 Applications Link are now open standards.

CITA

CIT Application. This is the name applied to the ECMA interface standard for CIT.

citation

• In on-line databases, a citation is a record. It will inevitably contain title, author, journal, and accession number fields. But it will often also provide an abstract (say a 200 word synopsis), together with descriptor and identifier fields. See all.

• Citation can also mean only those references cited in a report or thesis. The bibliographic references.

CIX

Commercial Internet eXchange. The commercial Internet community association.

CL

Connectionless. Networks which do not create end-to-end connections (through circuits or virtual circuits) before dispatching data packets or messages. See message switching and datagram.

CL/1

Now called DAL. See.

cladding

In optical fibers, this is a lower-refractive-index plastic or glass which surrounds the central core, and (it is claimed) possibly limits the scattering. Don't confuse this with the buffer. Cladding does not 'keep the light in the core' as is often claimed; total internal reflection does that. This is a physical protector.

clamp

Also called line clamp or black-level clamp. A video processing circuit that removes low-frequency disturbances (e.g. energy dispersal) from the waveform.

clamp-on

See camp-on.

CLAN

Cordless Local Area Network. Also called LAWN. A generic term for all wireless LANs.

clapper

A small blackboard with a hinged arm on top, which is used in film production. The blackboard records the shot and take numbers, and the 'clap' of the clapper arm is used to synchronize the picture with the sound in the editing stage. The job is given to the most junior member of the camera crew, the 'clapper-loader'.

Clarinet

A subscription news service which travels around the Internet using the same transport mechanism as Usenet. It consists of many news groups.

Clarke's orbit

Named after sci-fi writer and futurist Arthur C. Clarke; he calculated the altitude which was needed for a satellite orbit which took 24 hours per cycle. Clarke estimated that three satellites in orbit placed at 30°, 150° East and 90° West at a distance of 35,881km over the equator, could provide coverage of the populated areas of the globe (ignoring some ocean areas and the poles).

Note: I have at least a dozen figures for the actual 'height' of the Clarke orbit over the land (presumably not including the Earth's radius). They all differ by unexplained amounts, and all claim a high degree of accuracy. The latest two altitudes I have is one at 35,784 and the other at 35,993km.

• The Clarke geostationary concept was further developed by John Pierce in 1955 at Bell Laboratories. See geostationary orbit.

• In Clarke orbit, C-band satellites needed 4° spacings, and Ku-band needed 3° (now back to 2°). With polarization, two or more satellites can be co-located (see DirecTV).

class/CLASS

• **Programming:** In object oriented programming, a class is a group of objects which share the same behavior and capabilities. Classes exist in hierarchies, and each layer of the hierarchy inherits characteristics from those above.

For instance: there is a class called Computer Systems, and a subclass of these called PCs, and a further subclass of these called Wintel PCs, and so on.

• **Exchanges:** In the hierarchy of a telephone network, see exchange class.

• **Fax modems:** See below and Group# facsimile.

• **Custom Local Area Signaling System:** See.

Class #

• **Fax:** These are new EIA/TIA standards for microcomputer communications with plug-in fax modems.

— Class 1 is little more than the definition of the basic hardware needed for sending a fax from a microcomputer.

— Class 2 adds a forty-command AT-command set. This is the standard needed for fax modems.

- **Clean rooms:** See clean rooms.
- **Telephone exchange switches:** See exchange classes.

class licence

This is a term used widely in radio telecommunications and broadcasting. In effect it means 'no licence'.

When a class licence is issued for part of the radio spectrum, any device (subject to technical and usage regulations) can operate in this band without needing its own license. CB-radios, radio microphones, emergency radio beacons, and ISM equipment. All use a class-licence.

In most countries, provided you use an 'authorized' piece of equipment in an 'authorized' way, you don't need your own licence.

A prerequisite of class licence, is that no specific frequency pre-assignments are required to operate the equipment. See CDMA and ISM band.

class of service

- In American telephony, this refers to the type of telephone service contracted by the customer — a specific billing plan. You may choose a class, say, where you pay extra for calls to nearby towns but have cheaper local calls. In some areas you may choose between classes which have free local calls, flat-rates or measured calls. Distinguish here between class of service and grade of service.
- In LANs and networking, this means 'how the packet should be handled'.

— TOS (type of service) is a class of service in IP.

— In SNA, this term covers a complex of requirements such as designated path control network characteristics, path security, bandwidth, transmission priority, etc. These classes will apply only to a specific session.

clean rooms

Special dust-free areas used for high-precision work such as assembling hard-disk drives, making chips, and coating videotape. Clean rooms are rated as 'Class 10,000' or 'Class 1000' or 'Class 100', with the smaller number representing the cleaner room. The filters in a Class 100 clean room will remove particles as small as 0.3 microns. Hard-disk drives require this level of cleanliness.

clear

- **Data security:** To send in plain language (as distinct from encipherment)
- **Telcoms:** In switched-circuit telecommunications, 'Clear' means to tear-down a circuit. Clear-forward or clear-back signals are sent through the network to force old inter-exchange connections to be dropped.

In packet switching, 'Clear' or 'CLR' means that the logical (virtual) circuit has been abandoned, but the physical circuits are intact. You will need to issue a new call set-up request and address to continue operating. The term can also apply to your command to break the logical circuit.

- A clear channel is one without any bit robbing. See A&B bit-signaling.

clear back/forward

If the called party (B) hangs up, it creates a clear-back signal which disconnects the elements in the circuit. If the calling party (A) hangs up, it creates a clear-forward signal. These procedures vary for different networks. This is important in modem operations and telephone exchange switching. See call clearing.

clear channel capacity

The ability of, say, an ISDN 64 kb/s B-channel or a T-1 channel to support unrestricted bit patterns end-to-end in the information stream. 'Clear' channels are transparent channels to 8-bit data.

This is possible in a public-switched digital network only when the signaling information is out-of-band (carried separately). Then all 8-bits and all 256 bit-patterns can be devoted to data. Other facilities use bit-robbing techniques because supervisory and control signals need to be carried in the data-stream. See A&B signaling and T1 also.

Clear Vision

A Japanese improvement to the normal 4:3 NTSC television receiver which improves the horizontal resolution, and provides line-doubling to 1050 lines for larger screens. Line doubling does not improve the vertical resolution of the image, but rather reduces the resolution of the scan lines (which is desirable in a big screen).

CLI

See CLID below

CLID

Calling Line Identification. (also called CND and CLI). This is an end-to-end service for the delivery of the caller's line number.

It is the delivery of the 'source' phone number [of the calling party (A-party)] over the local-loop in both PSTN and ISDN networks, to a computer or special display box attached to the B-party's phone line. Most home CLID units will store the last half-dozen numbers. Company PC's can be used to automatically store thousands and they can automatically recall their database information if the same number calls once again. Herein lies the main problem — privacy.

The Bellcore specification is GR-30; it has a 600Hz buzz preamble followed by a number of marks before the CLID is delivered by modem tones. The system is possible in modern telephone systems because calls are transferred and switched between digital exchanges using CCS#7 protocols and packet techniques — even if the end of the link is analog.

The base-standard CLID consists of up to 258 bytes of information delivered between the first and second ring signals using modem tones (although DTMF is also used), so it arrives at the customer's LCD display before the phone is picked up.

CLID originates at the first exchange in the network (the A-party exchange), and with Delux CLID, it can also carry name information along with the number. It can be blocked at the A-party exchange by a special dial-code preceding the dial sequence, or by a standing arrangement with the carrier.

- An enhanced protocol called ADSI allows the inclusion of real names and other messages. This is sold under the name Calling Line ID Delux. The display will show the name as it appears on a customer's telephone account (up to 15 characters) with the last name appearing first. See ADSI and ANI also.

client

An application that wants to receive or hold data from another application over a network. For some reason clients are also called 'containers'. Clients have only one user, while servers are constantly servicing many. The client software generates the data queries, and the server software processes it and returns the resulting data to the client.

client area

The screen area of a window available for use by the application. In Windows, it does not include the system menu box, title bar, size box, menu bar and scroll bars. The upper left corner of this area has the coordinates [0,0] and the unit of measurement is the pixel.

client layers

Telecommunications networks usually only provide the two lowest layers of the 7-layer ISO model. The remaining five layers are left to the client (user) to provide through hardware and software.

client/server

These were once called distributed systems. In many cases the software vendors took a mainframe application, slapped on a GUI front end, and supplied it with a PC instead of a dumb terminal.

So client-server covers a wide range of software sophistication. The only thing they all seem to have in common is that they are expensive and hard to implement.

The essence of the client-server concept is cooperative and distributed processing — but it is important to realise that the term is a catch-all for a wide range of technologies which include remote SQL, transaction processing, message-oriented groupware, and distributed objects (see CORBA).

In a client-server system, the user sends a query (or selection argument) to the server using a special language like SQL. Often the query is compressed to further speed up the process. The server then selects the records and sends back to the client only the required results.

Client-server refers to the fact that the front-end of applications will reside on the PC/workstation (to handle the user-interface) and the back-end on the server (often a mainframe) to do the work. The workstation does not download the application from the file server, but rather transmits and receives only the transaction queries and data it is working on. The aim is to reduce the load on the network by transmitting only the minimum essential.

The distinction being made is with host/slave computer/terminal relationships where virtually all of the processing is done by the host. Another distinction is with file-server operations where large files need to be transferred to and from the workstations for even a minor updating. The major advantage of using a client/server approach is that only processed information is sent across the network.

In a client-server system, the user sends a query (or selection argument) to the server using a special language like SQL. Often the query is compressed to further speed up the process. The server then selects the records and sends back to the client only the required results. See SQL.

The goal of client-server is to give desktop workstations seamless access to resources across an enterprises network, and to reduce the load on the network. However client-server requires special network management capabilities and a distributed relational database architecture. The concept hasn't been as successful as was first promised.

clip-art

Illustrations in printed or electronic form which are available for use without copyright restrictions, but on license — and the key license condition is that you have paid the publisher for these more general rights by purchasing the disk legitimately.

These disks are not 'copyright free' or 'in the public domain'. Generally when you legitimately buy disks of clip-art you can use them whenever you wish, within one organization, but not outside.

There's a lot of shareware clip-art, but beware of supposedly 'public domain' material since a lot has been illegally scanned in by someone who did not own the copyright.

clipboard

A Macintosh term for an area of memory (a temporary holding place) for material recently cut for transfer to another part of the file, or to another application file. It will then be 'pasted' into this document. The clipboard is a reserved area of memory that can be used by virtually all Macintoshes and Windows programs.

Clipper

See Skipjack.

clipping

In some long-distance wireline telephone connections, the DCME echo suppressors often cut in and out as they detect speech signals. Any delay in reversing the activation of suppression can result in the first part of the speech being clipped. You'll also notice this when two people speak at the same time. This can cause problems with data transmission, and this is yet another reason why echo suppression is usually disabled for two-way data transmission.

• In many of the modern digital phone systems, the output of the listener's transmitter is momentarily suppressed to reduce the interference in adjacent channels. This is known as DSI, and it results in speech clipping when it can't react fast enough.

• Another form of clipping is often deliberate in electronic circuits. For instance, the white signals of a video picture can reverse (solarize) if the circuits overload — which they will do in bright sunlight. So the dynamic range of these circuits must be clipped to provide a hard limit.

CLNP

ConnectionLess Network Protocol. This is part of the ISO 8473 standard — the OSI equivalent of IP. It is a packet-network protocol (also called ISO IP) which was considered as a possible IP replacement to extend the address space. It has now been merged into a new protocol known as IP version 6.

CLNP uses datagrams to route network messages. There are two OSI transport protocols: CLNP and CONS, and, of these, CLNP is the more efficient for LANs while CONS is better for WANs. See Internet protocol and CLNS below.

CLNS

ConnectionLess Network Service. As above. This is the key datagram service of the OSI.

CLNS segments larger packets of data into smaller packets and puts them on to the network. Since modern fiber links have relatively few errors, there's considerable advantage in handling data in this way. It means that any connection-oriented protocols can then be introduced at higher layers in the OSI stack, while the lower layers remain connectionless.

This reduces both the overhead involved in establishing a connection and also in handling data over the LAN. After a temporary network problem, the transmission of data can restart immediately if there's no connection to make. See CONS and CLNP.

• The CLNS address consists of the IDP

clock

• The central timing circuit of a computer. It generates a series of precise pulses which synchronize the operations of the chips and buses. All the calculations and processes within a computer occur in time to the clock pulse, so it is the heartbeat of a computer.

Clock-speeds are expressed in megaHertz (MHz) and they range from the oldest PCs at 4.77MHz up to 100MHz today. Generally, the faster the clock-rate, the quicker a computer can perform the various functions — but this is not universally true (and certainly only applies to the same type of chip with the same support). See wait states.

• There is often also a separate 'real-time' clock-chip which puts the time and date on the screen. See real-time clock.

• In telecommunications, the problems of clocks and timing are much more complex than in computers. A clock must supply the timing signal which guides the movement of data in synchronous systems, and clocks are needed to trigger the fraction of a micro-second when a waveform is to be analyzed (sampled) to detect the presence of a pulse (in a digital system) or the amplitude (in an ADC).

Clock pulse generation is by local crystal vibration for analog transmissions, but for synchronous digital networks, the clock signal will be extracted from a national (or service provider's) cesium clock. Clocks are synchronized by 'tracking, using phase-locking or frequency-locking systems.

Clocks can cause many problems. Jitter is an aberration in the timing of the signal pulse 'edge', and clock 'slip' (reading the edge of the wrong pulse) results in discrepancies between two clock sources. The main advantage of SDH/Sonet comes from the way it handles clock-timing nationwide.

• Modems and multiplexers which interwork on a one-to-one relationship with a similar device, will generally handle their own clock synchronization. One end will generate the clock signal and the other end will receive and lock to it, and often this will be nothing more than the phase change which carries the payload information.

• The importance of the start-bit in asynchronous communications is to provide a start-point for the clock. It triggers the points at which the waveform is sampled to identify the ones and zeros in the byte.

clock rate

The overall controlling speed of a computer. The clock rate is measured in Hertz (Hz), however a higher clock rate doesn't necessarily guarantee a faster computer; the number of 'wait states' needs to be taken into consideration also. There's also a major discrepancy between CISC and RISC in terms of their instruction-set complexity, and therefore in how many cycles they need to perform a single instruction. This is why MIPS is often used as a measure of likely CPU speed.

Other time-waste factors, such as hard disk access speed, and specialized design for particular applications, may be far more important than clock rate. Old-timers will remember that an Amiga running at 3MHz was much faster at graphics processing than an IBM AT running at 12MHz.

clone

An exact copy. This usually applied to copies of the various IBM personal computer architectures (although not MCA). Exact functional copies (100% compatible) weren't possible until the BIOS was successfully cloned in a way that didn't infringe IBM's copyright, but which functioned the same. Manufacturers don't like the term 'cloned' — they prefer 'compatible'.

CLOS

A type of matrix switch structure which is fast acting, and provides connection redundancy. The 'Two-sided CLOS' switching fabric may be used in future ATM systems. See banyan.

closed

• **Switches:** Switches are closed when they are on, and when electric current can pass. The term actually refers to 'circuits'. A circuit is 'open' when it is broken.

• **Files:** Files are closed when you cannot read or write to them. In practice, this means that they are no longer in RAM memory. In closing the file, any necessary changes have been made to the software pointers and controls, and the file has been stored on disk. It is actually the updating of all internal software pointers, addresses, offsets, and controls that is the action of 'closing' the file, rather than that of storing.

closed architecture

• A proprietary system where a third party manufacturer can only operate with special licenses and conditions. Usually there are some sections of the operating system or BIOS which will be patented, copyrighted, or just unknown (until reverse engineered).

• A computer design without slots. You are limited to being able to add only those peripheral units which can plug into the back-plane of the machine.

closed captioning

This is the use of teletext chips and special lines in the vertical blanking interval of the television image, to create special text-captions for the deaf. A special teletext decoder can overlay these on the picture.

closed circuit

• Switched on or active.

• Transmitted by cable to a specific group; not broadcast.

closed user group

CUGs are videotex, e-mail and computer conferencing systems on a public network where entry is strictly controlled. The entry restriction is usually imposed by passwords or callback.

cloud

The telecommunications concept of a 'cloud' is very like the computer concept of a 'black box'. It represents an important part of the network that doesn't need to be defined or understood by the user. A cloud indicates that the core connections between two points appears transparent to the user — but, of course, it only appears as a 'cloud' from the outside.

When the concept is a cloud, then obviously the user/s will have an access network to reach that cloud — and the access network entering the cloud will need a defined interface, or be a defined standard.

X.25 packet-switching provides a basic cloud interface for LAN-to-LAN connectivity. The X.25 protocol is only an 'interface protocol' — you won't ever need know how the data is transferred across the world. ISDN represents another access mechanism to a 'cloud' transport systems — but here we know that the signals are carried over the CCS#7 packet network, and the B-channels, on standard 64kb/s digital bearers.

A LAN manager can use these cloud core mechanisms locally, nationally, or internationally in the same way — you dump information into the network on one side using a standard protocol suite (like frame relay) and it comes out the other in frame relay format. But it could have been encapsulated in a dozen different ways between.

CLP

Constrained Logic Programming. These are new AI-like programming approaches which tackles difficult combinatorial problems, such as job scheduling, time-table developments, and routing. CLP is related to Prolog.

CLR

In packet-switching, this means 'clear' — the logical circuit has been broken, but the physical circuit is still intact.

cluster

• A point where a number of terminals can connect to a data channel. It is a concentration of physical devices at one convenient point.

• A group of disk sectors that form the basic unit in which disk space is allocated. The number of sectors in a cluster varies according to the operating system and the disk size — typically it will be around 4 sectors (each with 512 Bytes). The disk controller mechanism in MS-DOS allocates and uses sectors in clusters. Clusters are often 'lost' during an interrupted recording of a file, and can only be rediscovered by the disk controller after a CHKDSK/F command cleans up the disk.

If multiple platters (magnetic surfaces) are available with RAID, a cluster will be divided and recorded in all sectors corresponding to the same sector address. See RAID.

• A routing cluster is a group of routers on a LAN. Each cluster has a single virtual address to which all end systems send their off-LAN traffic. So if the primary router fails then another router in the cluster can detect the failure and take over, continuing to use the same address.

• In cellular telephony, a group of adjacent cells of which no two share the same radio spectrum. Ideally, a single cluster should contain cells which make available the total range of channels available, and this will then be repeated in adjacent clusters. Usually 7 or 9 cells will make up a cluster, although it can be as low as 3 in some systems, and one in CDMA. This is the N=3 or N=7 'cell-reuse' pattern.

• Database management people talk of VAX clusters, where a number of DEC minicomputers are grouped together to provide major communications- and database-server operations.

cluster controller

A computer or dedicated intelligent device that sits at the multipoint link between a group of terminals and their mainframe. It handles a range of communications functions on behalf of the terminals.

Cluster controllers are concentrators which have varying levels of intelligence, and they gathers messages and group them for more efficient transmission to a mainframe.

• Generally cluster controllers are IBM-compatible devices which allow attachment of 3270-class terminals. They will be channel-attached to the host, or use SDLC. The IBM 3174 and 3274s are most often used.

CLUT

Color Look-Up Table. See.

CLV

Constant Linear Velocity. A system of infinite rotational speed changes used in some magnetic disk drives, CD systems and videodisks. The aim is to keep the speed of the head constant relative to the track, and have this linear track-speed at the optimum for maximum data density storage. But since the track lengths vary across a disk's surface, the disks rotational speed must change.

In CD-DA and CD-ROM systems the change is from 500 rpm on the inner tracks to 200 rpm on the outer.

The problem with this approach is that the change of speed becomes a limitation when the head is required to dart around the disk extracting (or adding) isolated blocks of information to various tracks — it is only really a valid technique when the tracks are being used systematically and sequentially (replay of a movie). However, whenever access speed isn't important. especially in linear playback systems like music and entertainment videodisks, the CLV technique gives the highest packing density. See CAV

CM

Circuit Merit. A test and rating strategy for the quality of a delivered voice signal.

CMA

Consolidated Management API. This is a *de facto* standard for a range of open management systems and protocols which have links to CMIP, SNMP and X/Open.

CMAC

Satellites: See beginning of C section. Page 98.

CMC

• **Common Mezzanine Card:** See mezzanine card.

• **Common Mail Call:** This is the X.400 API Association's (XAPIA) mail server. See MAPI, MHS and VIM also.

• **Computer Mediated Conferencing.** Computer conferencing or keyboard conferencing — not in real time. In public access systems, these are called forums. CMCs may be more formal and run on internal networks within an organization.

• **Computer Mediated Communications:** Slightly wider, perhaps, than the above. It involves e-mail also.

• **Coherent Multi-Channel:** This is an upgraded version of SDH for Terabit/second rates.

CMI

Coded Mark Inversion. An ITU standard for telecom equipment. It is a form of AMI used on high-level multiplexers.

CMIP

Common Management Information Protocols. Pronounced 'SeeMip'. This is an open network management standard being written for OSI (ISO 9596) which has long been tipped to replace SNMP.

It has not been widely implemented, mainly because it requires a powerful microcontroller with associated hardware (including RAM and Ethernet controller). Most organizations have found it too complex and expensive.

The aim of CMIP is to define the various functions of network management software, and to standardize the way in which messages will be formatted and transmitted across the network between data collection programs and reporting devices.

There are five levels of management being defined:

— **Fault management** — responsible for detecting problems, isolating them and controlling them.

— **Configuration management** —involved in messaging to identify active connections and equipment and to determine the network's current state.

— **Performance management** — which collects statistics on packet throughput, error rates, and such things as disk access requests.

— **Security management** — creates alerts for unauthorized access attempts.

— **Accounting management** — which is responsible for collecting network data on the usage of resources.

CMIP is seen by some people as an extension of SNMP with emphasis on managing large heterogenous global networks. By others, it is seen as a rival to SNMP Version 2.

The key difference is that CMIP supports peer-to-peer interaction, while SNMP relies on polling — getting its information one frame at a time. CMIP also distributes its management intelligence out to the nodes and is thus able to handle more complex networks. CMIP is said to be more reliable in the guaranteed delivery of data. See SNMP, HML. CMIS, RMON, probe and agent also.

CMIS

Common Management Information Services. An OSI classification for standards which utilize both the underlying services of ROSE and ACSE and the lower layers, to manage heterogenous networks. See CMOT, SNMP, and of course CMIP.

CMOL

CMIP over Logical Link Control. This is an implementation of CMIP management protocols which was proposed by IBM and 3Com. It involves the creation of agents in a way that is more economical in terms of memory, than is normal found with CMIP or SNMP.

CMOS

Complementary Metal Oxide Semiconductor. This is a form of chip construction that uses very little power but is often slower than normal, and has a lower packing-density. These chips are widely used in portable computers, watches, and all sorts of battery-powered devices.

The term 'complementary' refers to the use of n- and p-type metal-oxide semiconductor field-effect transistors (MOSFETs) connected in series.

CMOS chips are clearly the technology of the 1990s, with advances leading to quarter-micron element separations, and clock speeds of 100MHz. They dissipate far less heat than TTL memory chips, and they can run on a wide range of voltages (from 3 to 15 V). See MOS and TTL.

CMOT

Common Management Over TCP/IP. This is the implementation of the OSI's CMIS network management protocols over the Internet's TCP/IP protocol stack. See CMIP.

CMS

Conversational Monitor System. An operating system used on IBM mainframes, usually in combination with VM/SP. It is then known as VM/CMS. It allows mainframes to run software such as DOS and OS/2.

CMT

Connection ManagemenT. This is an FDDI-II process defined by the X3T9.5 specification. It explains how transitions of the FDDI dual ring, through its various states (off, active, connect, etc.) will be handled for isochronous traffic.

CMTA

Concert Multi-Thread Architecture from DEC which is now used in DCE. Threads control the flow of programs and allow applications to process many commands simultaneously. Programmers can therefore use parallelism in a DCE environment. See DCE and thread.

CMTS

Cellular Mobile Telephone Services. See cellular telephones.

CMX

The acronym comes from the CBS-Memorex eXperimental group. In 1970 they produced the first computer controlled video-editing unit, and the name has since almost become generic for this type of machine.

CMX is a very commonly used videotape editing computer which controls the on-line stages of professional video editing after decisions have been made off-line. Most CMX-type machines use the same standard editing language.

CMY

Cyan, Magenta, Yellow. This is the subtractive version of RGB (Red, Green Blue). Subtractive colors are used primarily for color printing, while additive colors are used for screen displays. See subtractive color, RGB and CMYK.

CMYK

This is CMY (Cyan, Magenta, Yellow) with the addition of K — meaning 'black' (for some reason). It is a term widely used by printers and lately in desktop publishing. Conventional color printing is a four-plate process, with a negative made for each of these four 'colors' (black is a color to printers). See CMY and HSV also.

C/N

Carrier to Noise Ratio. See CNR.

CNC

Computer Numerical Controllers — of machine tools. The successor to the older NC (Numeric Control) techniques.

CNID

Call Number ID. In general use today, this is the same as Caller Line ID (CLID). However, the term seems to have been originally applied to the Enhanced-911 services (emergency) which US telephone companies provided to police and fire-departments. This displayed the ANI (before CLID was widely used), and the number could then be quickly processed against a reverse directory to find the name and address. See CLID.

CNP

Connectionless Network Protocol. This is a Layer 3 (Network) function in the ISO model. It is based on the ISO standard 8473, among others.

CNR

Carrier to Noise Ratio (aka C/N). The key measure (often taken at RF or IF levels) of signal quality; how much greater the signal carrier strength is than the system noise. Both carrier and noise will be measured in dBmV, but the ratio is expressed in dB. This is a more convenient measure than Signal to Noise (S/N) when dealing with radio frequency systems such as Cable TV.

- C/No is Carrier/Noise Density which is C/N per unit of bandwidth. See Eb/No also.
- In television systems the CNR threshold is the point where all snow disappears from the screen.

Cntl

- Another abbreviation for the Control Key. Aka Ctrl.

CO

- **Connection Oriented:** Networks that support virtual links. These are our conventional packet-switching networks (but not those involving datagrams) and the new ATM networks. See CL also.
- **Central Office** (see) the US term for telephone exchange.

CO sets

Character-oriented (cursor control) sets. These are the control codes used for text with ANSI standard terminals, They exert control to the X3.4 standard. See VT-# terminals.

- The graphics codes are known as GO sets. These were originally formulated in 1968 and updated in 1977.

co-channel interference

This is interference between the output of one transmitter, and that of another occupying the same frequency band, but usually broadcasting at some distance. Radio transmitters (especially TV stations and cellular phone systems) suffer from co-channel (same) and adjacent channel (neighboring) interference effects.

The variables are distance, power and directionality of the transmitter and receiver antenna. In TV, the special antenna used for reception in fringe areas are required to discriminate strongly in favor of signals coming from the front, as against co-channel signals from the sides or behind. This is not possible with cellular phone handsets which must be OMNI. See adjacent channel interference also.

- Co-channel interference in cellular phones causes crosstalk and drop-out. These problems are worse when the handset is well above the ground (in a high-rise building or on a hilltop).

- Not all modulation techniques have the same co-channel interference problems. Digital transmissions tend to be received in an all-or-none way. The digital receiver will either lock on to one station or the other, but not both. CDMA techniques ignore co-channel interference which is why the same frequencies can be used in adjacent cells.

CO-LAN

Central Office LAN. A form of small MAN (Metropolitan Area Network) offered by some American telephone companies using AT&T's Datakit Virtual Circuit Switch (VAC). Co-LANs are data and/or voice networking services within one local-calling area, and they are based on switching equipment located at the telephone exchange. The idea is to link Co-LANs via ISDN. SMDS has now taken over here. See SMDS.

co-operative

See cooperative.

coated paper

Paper which has been treated to provide a slick shiny surface. It is often used for color printing. Aka calendered paper.

coaxial cable

Also known as coax. This refers to a cable having one conductor enclosed within a conductive sheath and separated by a dielectric. Coaxial has at the center either a solid wire or a narrow single cylindrical tube of copper. The surrounding sheath is electrically conductive woven-metal or wrapped aluminum foil, which provides one path for the current and also shields the core against interference.

There are many types and qualities of coaxial. It seems to be economical and suitable for carrying signals with bandwidths from about 1.5MHz to about 50GHz. The bandwidth is tied theoretically (and closely in practice) to the diameter of the cable.

Coaxial is usually rated in terms of impedance, in ohms. Standard RG-58 LAN cable is rated at 52 ohms, and a lot of TV antenna cable is rated at 75 ohms (which matches the impedance of the dipole antenna).

In legacy Cable TV systems, the main feed cable is 0.75in. in diameter and the side street splits are 0.50in. The useful bandwidth (54—550MHz) of these cable networks depends mainly on the distance between trunk amplifiers, and the number of amplifiers in sequence at the furthest reaches. Modern hybrid fiber/coaxial systems, with no more than four amplifiers in sequence are capable of 1GHz in bandwidth, and of supporting interactive services.

- LANs coaxial cable is typically 50 ohm (used for digital signalling) or 75 ohm (used for analog and high-speed digital). LAN coaxial will carry data at 100Mb/s or more, as against a limit of about 10Mb/s for unshielded cable and it is relatively resistant to interference. The ECMA have standards for co-axial cable (ECMA/TC 24/82/54)

- In telephony, coaxial cable was widely used for long-distance and interexchange bearers until fiber became available. A coax cable would typically carry up to 3600 analog voice channels or two analog TV channels, depending on the repeater spacing.

- A coaxial channel using a 3mm cable wave-guide with repeaters at 4km intervals can handle frequencies up to about 16MHz. This cable would carry 3600 voice channels in four hypergroups. A 12MHz system will carry 2700 voice channels in three hypergroups.

COB

Chip-on-board. The distinction is with SMT (surface mounted technology). COB reduces mounting area and thermal resistance, and can produce lower cost PCBs in many cases.

COBOL

COmmon Business-Oriented Language. This language is quite old, but still very viable. It is widely used in commerce and business. There is an enormous amount of Cobol software available and Cobol programmers are always in demand.

cobweb site

An Internet site which hasn't been updated for some time.

COCOT

Customer Owned, Coin-Operated Telephone. In other words, a privately owned payphone in the US. It is a payphone not operated by the local carrier, and it will prohibit all sorts of different number patterns.

Codd's rules

These are the 12 rules which establish the credentials of a relational database management system according to Dr Edgar Codd the foremost theoretician on DBMS. His Twelve Rules are often treated as the computer equivalent of the Ten Commandments for anyone working with database management, but Codd himself says that they are just a way of rating various DBMS systems.

Ted Codd was a mathematician employed by IBM who established the theory of relational databases, and he created the 'relational data model' way back in 1970. Codd was also responsible for one of the early concepts (which now seems obvious) which was the need to separate the data itself from the database manager. So today, the storage and retrieval system are not inextricably linked as they were in the early years of computing.

Codd has since published another book with 333 database features, and 20 laws, rather than the original 12 rules. See RDBMS.

- The 12 rules for Relational Databases (RDs) are:

Rule 1. The Information: All information in a RD is represented explicitly at the logical level and in exactly the same way — by values in tables.

Rule 2. Guaranteed Access: Each and every atomic value (datum) in a RD must be logically accessible by resorting to a combination of table name, primary key value, and common name.

Rule 3. Systematic treatment of Null values: When representing missing or inapplicable information, nulls (as distinct from empty strings, blanks, zeros, etc.) are systematically supported in fully relational DBMS independent of data type.

Rule 4. Dynamic Online Catalog-based on the Relational Model: The database description and the ordinary data are logically represented in the same way, so that authorized users can apply the same relational language to queries.

Rule 5. Comprehensive Data Sub-language: A relational system may support more than one language and various modes of terminal use. However, at least one language must have statements which are expressible (using some well-defined syntax as character strings) to support all of the following: Data definition; View definition; Data manipulation (interactive and by program); Integrity constraints; Authorization, Transaction boundaries (begin, commit and rollback).

Rule 6. View Updating: All views that are theoretically updatable, can be updated.

Rule 7. High-level Insert, Update, and Delete: The ability to handle a base relation or a derived relation as a single operand applies both to the retrieval of data and also to data insertion, updating and deletion.

Rule 8. Physical Data Independence: Applications and terminal activities remain logically unimpaired whenever any changes are made in either the storage representation or access methods.

Rule 9. Logical Data Independence: Applications and terminal activities remain logically unimpaired when information-preserving changes, of any kind that theoretically permit unimpairment, are made to the base tables.

Rule 10. Integrity Independence: Integrity constraints specific to a particular RD must be definable in the RD sub-language and storable in the catalog, not in the applications.

Rule 11. Distribution Independence: A relational DBMS has distribution independence.

Rule 12. Nonsubversion: If a relational system has a single-record-at-a-time language, that low-level language cannot be used to subvert or bypass the integrity rules and constraints expressed in the higher level relational (multiple-records at a time) language.

code

- **Computer code:** This generally means those ASCII No.5 codes between 0 and 128 which are used for characters, punctuation, etc., however it can mean the instructions in a computer memory, as distinct from the data. Code in this sense means the applications (or perhaps the operating system).
- Code-cutters are computer programmers.
- More specifically, the term suggests low-level machine- or assembly-language programs.
- The result of encryption or encipherment.
- Any sequence of encoded signals, like Morse code or Baudot code, that needs to be translated back into English.

code-gain

A term used in telecommunications to indicate that the amount of information required to transmit a signal (usually voice) can often be greatly reduced by the application of new technologies. For instance standard PCM digitized voice requires 64kb/s of data bandwidth, but with new vocoder techniques very good voice can be transmitted at 4.8kb/s. This is a code-gain of 13.4.

code generator

See CASE.

code segment

A section of RAM memory which holds the application program/s — as distinct from that which holds data.

code violation

A break in the rules of a transmission system which is used to identify a section of code, or to carry a control message. See violation, AMI, HDB3 and B8ZS.

codec

Coder/Decoder (aka Compressor/Decompressor).

The term is applied to both hardware and software systems and the recent confusion reflects the fact that codec now refers both to both the digitization process, and to the digital-compression process.

A codec is the inverse of a modem. It takes an analog input, and digitizes it for carriage over a digital network. The input here is constantly varying voice/video signal, and it is carried over digital networks in digital form. So a codec takes analog data streams and digitizes them — transports them in digital — then converts them back to an analog output.

The term is applied to both voice and video digitization, but since the main value of digitization is in being able to compress the signal information, codecs have gradually assumed this function. However compression is secondary to the ADC—DAC conversion process.

Today codecs can be software based frameworks of algorithms which included, for instance, MPEG, H.261, QuickTime, and Video for Windows. These frameworks identify which decompression processes to use on the incoming data, and which to apply to the outgoing data.

- Delta and difference modulation techniques are used in high-quality voice codecs, and vocoder systems in lower quality ones. See ADPCM and vocoders.
- H.261, H.320, JPEG, DVI (Indeo) and MPEG codecs are all used for video.

coder

An analog-to-digital converter (ADC), or a programmer.

coding

- Coding and encoding, usually just means the conversion of binary signals, from any source, into electrical, radio, or light signals. There's no sense of encryption here.
- Programming. The term is applied to all forms of program production, from the high-level languages down to the lowest. At the machine level, 'code' is most certainly the correct term (machine language programming is different to machine coding).
- Encryption is quite often called 'encoding'.

coercivity

The resistance of magnetic materials (i.e. tape and disks) to change state during the recording process. This relates also to the likelihood of the signal degrading progressively after recording, and also to the thickness of the magnetic layers. Higher coercivity tapes lose signals at a lower rate and the layers can be thinner, which results in cleaner signals.

Coercivity is therefore an important factor in establishing the packing-density (the thinner the layer, the closer the domains can be together). It is also important in archiving tapes and disks. But the quality of the tape manufacture (the type of plastic and type of binder used) is more important.

- Coercivity force is magnetic flux density and it is measured in Oersteds (Oe). The Betacam SP professional recorder uses a high coercivity tape in the 1500 Oe region, while S-VHS uses a lower density tape at 850 Oe.
- The best high-density recording media, are probably tapes employing (respectively): metal evaporated, metal particle, thin-film alloys, chromium dioxide, and iron oxide techniques.

COFDM

Coded Orthogonal Frequency Division Multiplexing. This advanced broadcasting modulation system is used in multichannel DAB and the new digital television systems (satellite and terrestrial) to reduce multipath problems. Multiple data streams, often of only a few kilobits per second, are transmitted over a large number (up to 2000) of closely adjacent frequency bands. The bits are aggregated in the receiver, and converted back into a broadband stream.

The data bits are sent in bursts so that reflected (delayed) copies of the bits can be detected and rejected. Because of the low bit rate in each of the channels, it is possible to infill shadow reception areas with a repeater using the same frequencies, without causing serious interference. This is not possible with analog systems

- For DAB audio, each carrier is modulated by QPSK giving a 2-bits-per-Hertz data rate.
- For TV signals, an 8- or 16-level QAM modulation is first used to create the primary signals, then the COFDM modulation process is implemented by inverse FFT on each signal. The phases of adjacent carriers are orthogonal (at right-angles) to each other so as not to interfere. There's a heavy burden of processing — including high-speed FFT, Trellis Coding, and Viterbi decoding.

COFF

Common Object File Format — the Unix standard.

cognitive

The act of knowing, thinking and perceiving. In computer-aided learning theory it is important to make a distinction between cognitive (perception/knowledge), affective (emotional), and conative (endeavor, effort) processes.

- Cognitive overhead: An educationalist term which emphasizes the mental load that users often suffer in trying to use some supposedly 'user-friendly' computer systems.
- See combinational explosion.

coherence

- The opposite of random. Marching in step. Having waves (usually light waves) at the same frequency and in phase. This applies mainly to acoustics, radio and optical fiber systems. The term coherence also correlates with very narrow (single frequency) bandwidth sources, otherwise the waves would soon get out of step, even if they started together. See solitons.
- In disk caching, coherence refers to the way information in RAM memory matches the information on the disk. Unix systems refer to this as synchronization.

coherent light

In contrast to conventional light sources, laser light has highly coherent waves. The photons, which are emitted by the laser's individual atoms of light-producing material, are all aligned in phase by the internal reflection process.

This produces monochromatic light which can be sharply focused (which is why it is used with optical disks) and which suffers only slightly from scattering or absorption in most transparent materials. Coherent light (or something very closed to it) is produced by all lasers — but those with a resonance cavity are exceptionally coherent. See laser.

coherent optical system

An optical fiber transmission technique that uses coherent (very narrow bandwidth/in-phase) lasers. The data signals are merged with the transmitter lasers using either FSK or PSK techniques.

Coherence also allows a number of separate signals to be jointly projected down a fiber using wavelength-division multiplexing (WDM) within the narrow maximum-transparency band available in most fibers.

Data signals are extracted from the incoming complex of light by either heterodyne or homodyne techniques. These 'beat' the incoming frequencies with those of a similar laser, and extract a 'difference' signal which is then demodulated.

Coherence is probably the optical fiber technology of the future — especially for submarine cables. It increases detector sensitivity, increases the data-carrying capacity of the fiber, and works with optical amplification.

cold start/boot

See boot.

collaborative computing

A generic term now being applied to a range of computer network activities using software loosely known as 'groupware'. The activities include e-mail, computer teleconferencing, real-time teleconferencing and video conferencing. These systems also have applications in education.

At the present time this is little more than a marketing slogan which appeals to enterprises with mixed computer environments. It suggests that computer users should be able to interact with others anywhere, any time, and without regard to their location (home, office or hotel) or whether they are stationary or on the move.

So these directions involve the development of communications-rich applications, special service agents, inter-applications communications protocols, public-key encryption, the X.400 and X.500 protocols, and Lotus's Open Messaging Interface (OMI) for mail services — among others!

collapsed backbone

A sub-version of the Ethernet bus architecture, where the whole of the bus is confined within a wiring cabinet. Instead of a long coaxial bus stretching through the network area, the drop cables have been extended to produce what is essentially a star network.

Since UTP became popular in 1990 with 10Base-T, the design of Ethernet has radically changed. For the horizontal cabling, radiating twisted-pair wiring is often used from a wiring concentrator or hub. Each hub is connected via a LAN to a central multiport bridge, router, or switching hub, and these tie together the departmental LANs. This approach provides better performance, makes the network more manageable, and provides a simple way of by-passing faults when they occur. See wiring hubs and concentrators.

collate

To shuffle or stack pages (actually 'sheets') together in order to be ready for stapling or binding.

collating sequence

The sequence of 'shuffling' when a computer is asked to sort items into alphabetical order. Unless a special algorithm is run, the computer will view the alphabetical characters as numbers in the ASCII No.5 set. It will therefore see the capital A (65) and capital Z (90) as well ahead of the small a (97) and the small z (122). Numbers (48 to 57) will also precede the alphabetic characters, and various punctuation symbols will be before, between, and after, the alphabetical characters.

This is only the start of the problems. See alphabetic order.

collector

One of the three contacts on a standard bi-polar transistor.

collinear antenna

This is an antenna installation made up of dipoles which are stacked on top of each other. It is moderately high-gain .

collision detection

- **LANs:** CD is a necessary part of most contention schemes — although some depend on Collision Avoidance (CA) and others on Collision Elimination (CE). Ethernet's CDMA/CD is the best known collision detection scheme. Each station on the network is responsible for detecting interference caused by simultaneous transmissions of data which result in a collision on the shared media.

Collisions are often a consequence of the propagation delay when a packet crosses from one end of the network to the other. Two stations can be transmitting without either being aware of the other's attempts, until it is too late. This time of uncertainty depends on the length (actually the 'latency') of the network, and is known as the collision window.

Collision resolution is the way in which these problems are resolved; in the case of Ethernet without needing some central authority. See CSMA/CD.

- **Graphics:** Collision detection is a system of reporting when two images overlap.

color attribute

A bit pattern (a group of bits) which is held in the video buffer, and which sets the color of a pixel on the screen. This term was also applied to a very simple system used in the early character-based PCs, where attribute bits were attached to ASCII bytes.

color balance

- In video camera, this is the process of setting the sensitivity of the elements to achieve a realistic coloration on objects (mainly faces). Usually a white balance is used to adjust for the color temperature of the light sources.
- In film, it is the use of Wratten 80 or 85 filters to balance the color temperature of the various sources of light (some may be tungsten and some daylight) with the film type.

color burst

In color television a colorburst is a short signal provided as a reference for both hue and saturation of the color. It is sent in the horizontal blanking interval just before each line is drawn on the screen.

color coding

The way in which a computer signal represents the colors when sending information to a monitor or video recorder. The three most common color coding systems are:

– RGB (Red, Green Blue).

– HSV (Hue, Saturation and Value).

– YUV (Luminance, Red-difference, and Blue-difference).

In the first two, the bandwidth is equally shared between all three elements, while in the third the bandwidth is divided to provide half the bandwidth for luminance and a quarter each for the chrominance difference signals.

color correction

• In television studio production, this is the adjustment of the red, green and blue content of the electronic signal output, so as to provide natural flesh tones. Flesh tones are the most obvious reference for viewers.

• In film, color balance is achieved by selecting the brightness ('lights') of the light sources in the film printing machine. It is separately coded into the printing machine, and triggered by markers (often notches) on the film.

color difference

Color television signals consists of Red, Green and Blue. In combination (but not necessarily in equal amounts) these can provide both the chrominance and the luminance signals (R+B+G = White). Ideally, a color television signal would consist of three distinct and separate signals, one for each of the three primary colors and these would remain separate right through the production and post-production stages — even, if possible, during transmission.

However the old B&W television sets needed a single separate luminance signal and these had to be catered for. So the main image resolution was carried in the luminance rather than the color information.

Ideally, three color signals and one B&W signal would be handled in an analog component system — but each would consume large amounts of bandwidth in transmission and storage devices, and there'd be a lot of redundancy (duplication) in the information.

To avoid this problem, all three color signals are combined to create a (Y) luminance signal, and then two color-difference signals (R–Y and B–Y) are created by an extractive process.

This creates three baseband signals which are then properly called Y:(R–Y):(B–Y) — and since the eye's resolution of color is noticeably inferior to B&W images, these color-difference signals can each be handled at half the bandwidth of the Y. With digital equipment, this is expressed by the ratio 4:2:2.

The recommendation 601 for studio production equipment, says that this signal combination give the highest resolution and best color for the least consumption of bandwidth.

• 4:1:1 ratios are used for ENG, to reduce the bandwidth.

• 4:2:0 ratios are used in MPEG2. This means that the two color difference signals are transmitted in alternate lines.

• YUV and YIQ are two expressions of the above, when the signals are modulated onto a tape or transmission system. The U and the I are just different modulation techniques applied to (R–Y) and the V and the Q are the same for (B–Y). See component video, YUV, YIQ and Y/C.

color look-up table

A CLUT is a selection of colors which have each been assigned a digital value. To process a file for display, the computer will decode the image by looking up a value in the table for each pixel. It matches the code stored for the pixel, with those specified in the table — which then provides as an output, the full Red, Green and Blue bit-patterns (perhaps 8-bits each) required to generate the color.

CLUTs are needed to reduce memory requirements. If the full 16.7 million colors of a modern color monitor were to be stored in video-RAM, it would require 8-bits of information for each color (Red, Green, and Blue) for each pixel. This is a total of three bytes per pixel — so a single high-quality screen image of say, 1200 x 1600 pixels would need 5.76MBytes just for the bit-map. By using a CLUT and storing only 4 bits each, the memory requirements are manageable.

color separation

The process used in color printing for creating the negatives and then the printing plates. The printer creates four distinct screened monochrome film negatives, which correspond to the amount of Magenta, Cyan, Yellow and Black in a picture. In DTP this can now often be done in the computer. Each color separation must be a half-tone image with the dot-patterns of each set at different angles to avoid moiré patterns.

color space

The way in which color is described within a computer or for printing. There are many standards.

RGB	Red Green Blue.
CYMK	Cyan, Yellow, Magenta, and black.
CIE	An International standard.
HSV	Hue, Saturation, and Value.
HSL	Hue, Saturation and Luminance.
YIQ	US for luminance and color difference.
YUV	European for luminance and color difference.

See all of the above.

color temperature

A color scale which is measured in degrees Kelvins (K — from absolute zero). The degrees K is an measure of the redness or blueness of a (apparently) white source of light. Candles rate about 1000 K, and tungsten globes are on the lower end of the scale around 3000—3400 K. Full direct sunlight is in the region of 5000 K, and on a cloudy day the deep blue skylight that finds its way into shade areas can vary from 7000 to 14,000 K.

We need to measure for color temperature because the eye adapts to ambient changes, but color film or video cameras don't. So they need color correction filters or electronic 'white balancing' to recreate a 'natural' color balance to the scenes. See 80 filter, 85 filter, and white balance.

color under

Modern videotape recorders handle the luminance signal at a higher frequency than the chrominance information — the fine detail is in the Y, rather than the C component. Therefore the color is often modulated to a lower frequency — a sub-carrier. This is color under.

column move

A process performed in tables with rows and columns — particularly in spreadsheets, but also in databases and arrays. It is a technique of shifting the data, in one or more columns, to the left or right as a one-step process.

You see this in spreadsheets if you add another column to the left-hand side. This is a complex array process.

columnar transposition

A technique used in ciphers and error correction systems where the input is distributed horizontally across a matrix as a series of rows, but the output is read vertically down the columns — and vice versa for decipherment. See Reed-Solomon.

COM

- **File Extension:** As a file-extension, XXX.COM identifies a program which requires a fixed location in memory. The distinction is with EXE and BAT files. You execute a COM file by typing the name (without the extension). It is used for older programs inherited from CP/M. EXE is the modern version.
- **Continuation of Message:** See BOM.
- **Common Object Model:** See OLE.
- **Computer Output to Microfilm:** Also to microform or microfiche.

COM port

A serial communications port on a MS-DOS computer. The interrupts and addresses are typically:

COM1	3F8	IRQ4
COM2	2F8	IRQ3
COM3	3E8	IRQ4
COM4	2E8	IRQ3
COM3	(PS/2) 3220	IRQ3
COM4	(PS/2) 3228	IRQ3

Note that COM1 and COM3 use the same interrupt, and COM2 uses the same as COM4. See IRQ.

- **Physical and Electrical:** These are RS-232C serial ports which usually have D-shell 25-pin or 9-pin connectors.
- **Logical:** COM1 and COM2 were originally provided on the early PCs. Later COM3 and COM4 were added on some machines — but only two IRQs were provided for the four ports — so the same interrupt requests are often shared by two ports. Two different parts were added to PS/2 computers.

Be aware that in Windows, a plug-in card can be using a COM port (stealing its I/O address, and using its IRQ) although the underlying DOS may not recognize its existence.

comb

- **Television:** Comb filters were introduced to overcome the problems of cross-luminance and cross-chrominance in analog television systems. In composite television, colour and luma occupy parts of the same spread of bandwidth (at different ends) and tend to interfere with each other where they overlap. The old solution was to reduce luminance bandwidth, but this reduced picture quality. Comb filters are more selective.
- **Disk drives:** A set of magnetic heads that move across the spinning disk as a unit. This has multiple read and write heads.

combination/al

- **Combination:** In maths, this is a number of different elements which have been selected from a larger set, but without any constraint as to the order of selection or the order of arrangement.
- **Combinational logic:** The functions of an electronic circuit where the output depends only on the instantaneous discrete inputs at the time, not on any previous inputs. This is the opposite of sequential logic.
- **Combinational circuit:** A circuit where the output depends on the input values at the time of the clock pulse.
- **Combinational explosion:** This happens when the number of permutations reach the limit of a computer's capacity. It is also a cognitive science term which refers to the fact that doubling the difficulty of use of a computer system, might easily reduce the effectivity of that system to zero. The implications are Gestalt; the total problem can be much more than the sum of small difficulties.

combiner

A small passive electronic device (a 'Y' connector) that allows two sources to be fed into the one unit. For instance the TV antenna feed and the VCR output are combined and fed to the tuner of the TV set. These are also used on shared computer printers for multiple inputs without the need for switching.

COMDEX

COMputer Dealers EXpo. The major US computer show.

comma-defined

It is common in ASCII output from database files to have the fields separated by commas and the records by carriage returns. Another common approach is to use tabs.

- You often need to enter these in decimal or hex form: the comma is decimal 44, hex $2C. The carriage return is decimal 13, hex $0D. The tab is decimal 9, hex $09.

command driven

The former type of command/character interface which was used for personal computers. These are now called Character User Interfaces (CUI), and they were used in Apple DOS and MS-DOS machines before the Macintosh (1984) and Windows (1989) changed the way we communicate with PCs. See GUI.

command interpreter

A user-interface program which reads lines as they are typed at the keyboard (or detects mouse-clicks) and interprets them as requests to execute a program or perform some other function. Windows 3.x is a command interpreter, as is the Mac's Finder. See shell.

Command key

Aka 'Cloverleaf' or 'Snowflake'. This is the Macintosh special-process shift key directly alongside the space-bar. Don't confuse it with the Control key, or with the Option key (the Mac equivalent of Alt).

command language

A special-purpose language used for querying or job control. It will usually have a very limited range of commands. See SQL and DL1.

In on-line database searching, this is a specific syntax and a list of command terms which are used for defining the search. These search languages can usually be learned in an hour or two, but understanding the complexity of the database may take longer. The key search terms will be something like:
SEARCH or FIND, COMBINE, TYPE or DISPLAY, PRINT
The syntax will be based on Boolean combinations using operators, such as :

```
FIND x AND y NOT z
```

command line

- In smart modems using the Hayes AT command set, this is a line of controls beginning with an AT and ending with a carriage return. It can hold a number of different variable names and parameters.

- In MS-DOS computers this is a line which is typed to issue a command to the operating system. It is a video-screen line beginning with a prompt which is waiting to accept instructions.

command mode

Two-way communications systems need to have two distinct modes if they are dealing with an intermediate device which has a simple two-wire (one channel) connection on each side.

It will need a command mode and a data mode since the same channel must be shared. The first mode is used to set-up the call, transfer parameters and initiate the communications functions (setting parameters in the modem, issuing instructions). The second mode then transmits the data — so at this stage the intermediate device (usually a modem) must know to just pass the data through, and not attempt to act on it as if it were control signals.

We therefore need to have commands which switch the intermediate device from Command to Data mode (usually just a standard command), and another which switches from Data to Command mode (much more difficult since the device has been told to pass all ASCII characters through transparently). The Hayes smart modem solves the problem like this:

— In the **Command mode** the modem will accept commands to dial, make contact, break contact, change data rates, etc. Usually this is done by Hayes compatible codes beginning with 'AT'. When a carrier link is established (when CDC is enabled), the modem automatically switches to the Data mode.

— In the **Data mode** the modem is transparent (passes all ASCII characters through without action) to any flow of data, including the word 'AT'. However it will recognize an 'escape sequence' which is:

<one second pause> +++<one second pause>

This is a unique combination which is unlikely to be accidentally generated by data in a file. When a modem detects this, it will maintain the carrier connection, but suspend operations and switch to Command mode. 'ATO' will return you to the data mode again. See TIES also.

Common Air Interface

CAI. This is a general term for the protocols associated with the radio link in a cellular phone, PCN or Telepoint system. The distinction is being made between the radio links and the wireline network and the cellular base-station connections and switching system.

A CAI assumes that the handset from one manufacturer will work with the base-station of another. Perhaps not all the special features will be enabled, but the phone will communicate voice, and can make and receive calls.

- GSM is often said to be both a CAI and a wireline protocol suite since it includes special protocols for international messaging (see MAP and SMS). DCS-1800 uses similar protocols. AMPS, TACS, NMT, CDMA, DECT, NADC, etc. are mainly CAI standards, although they also specify how network switching for hand-off will work.

- The term CAI became well known a few years ago because of a specific problem with the CT-2 standard in the UK. CT-2 phones were introduced by four 'Telepoint' operators using three different and incompatible standards. Later in a vain attempt to salvage the mess, the UK government mandated a CAI for CT-2 telephones that permitted all makes and models of handset to communicate with any base-station.

- CAI is distinct from, but relevant to, Inter-System Roaming (ISR) which allowed call-charges to be authorized by any network provider, and so allows a user to make contact with any base-station. Naturally a CAI is essential before ISR can be implemented.

Common Applications Environment

The term CAE is usually applied nowadays to X/Open's move to develop a suite of standards for Unix. These include an operating system interface for Posix, networking services, database management, languages and the user interface.

- IBM also has a CAE for SAA.

common bus

See macroparallelism.

common carrier

A PTT, PTO, carrier, telco or telephone company. An organization which had been authorized to operate a public-access telephone or data communications service. This is largely a legal term which suggests that it is a public facility, available to all, and as much as possible it is available on an equal basis to all. In government legislation and regulations, the term common carrier often implies that Universal Service Obligations (USOs see) will be a condition of the license.

The term is applied to privately-owned and government-owned monopolies, and to the primary network owners in competitive environments (where resellers may also be present). It distinguishes service-resellers, who rent facilities from the carriers and resell capacity after adding some value, from the physical plant-owners (the 'primary carriers').

AT&T, MCI, Sprint, Telecom Australia, British Telecom, NHK, KDD and the seven Regional Bell companies (RBOCs), etc. are all common carriers.

common channel signaling

The use of an out-of-band signaling system in telecommunications — where the signaling and the data are carried over quite separate networks. This usually refers to the exchange-to-exchange signaling within the network, but it can equally refer to the local loop and to some private networks including LAN/-WANs.

Before ISDN, most telephony used in-band signaling (decadic/pulse or DTMF in the local loop) where the signals and the voice shared the same line, but at different times. CCS systems may use the same physical wires, but the signals will be in different channels to the voice/data. These common signaling channels and data-channels are then time-division multiplexed over the local loop back to the local exchange, where they are handled separately.

CCS allows signaling to be conducted while the voice or data circuits are in use. See ISDN and CCS#. The contrast is with 'channel associated' signaling or in-band signaling.

Common Channel Signaling system #7

Also known as SS#7, CCS#7 or Signaling System 7. See CCS#

common control

This refers to the modern signaling systems used on a telephone or data network which have become possible through the use of relay-based crossbar and later digital techniques — in particular stored-program control (computer) switching. The term differentiates this form of computerized 'centralized' switching from the old step-by-step switch method where, essentially, each access line had its own switch connection.

With these new techniques, all exchange lines may be switched under the control of a single control system — they share pooled control equipment, tone decoders and registers. Generally common control will also mean common channel signaling — but not necessarily so. See CCS# and in-band.

common interest group

- CIGs in general terms, are now known as forums or on-line open conferences. Don't confuse them with CUGs.
- In telecommunications regulation, CIGs are groups of individuals, professionals, or organizations, who are permitted to share phone or data lines which have been purchased on a group-lease basis. These restrictions are often tightly controlled to prevent reselling of bandwidth. From the carrier's viewpoint, CIGs must be tightly defined to prevent them competing with the normal public switched network.

Common Intermediate Format

In the CCITT's H.261 videoconferencing standard the two-way exchange of NTSC (525 line x 30 frame) and PAL (625 line x 25 frame) images is supported by way of this intermediate format. Images can be translated to the CIF format, which is based on 288 non-interlaced lines of picture repeated 30 frames per second, with a horizontal resolution of 352 pixels per line.

In a PAL system there are normally 576 active lines, or 288 per field, so a field's half-image (25Hz) is simply converted to the 30Hz frame rate by repeating one in every five. NTSC images already have the correct picture frequency, but must be converted from 240 active lines per field, by the addition of black image lines at top and bottom, or by duplicating one in five of the existing lines in the image. The CIF format is not mandatory within the H.261 standard, although it is most important for videoconferencing. Correctly, this standard should be known as Full CIF or FCIF. Also see QCIF.

Common User Access

The CUA interface is IBM's proposed standard for their own Mac-like, icon-based, GUI to programs — both those running under OS/2 and for SAA. The CUA defined the dialog rules between the human and the computer, and IBM says it is divided into:

— Physical: The QWERTY keyboards, screen, mouse.

— Syntactical: The sequence and appearance of the various displayed elements.

— Semantical: A uniform element definition.

communications controller/processor

- In large networks with mainframe hosts, it is usual to provide a special computer adjacent to the mainframe which has the special responsibility of looking after all communications; it does not run applications, and has no interest in the data-content of messages.

These computers are also known as front-end processors. The FEP will be connected to the mainframe by a high speed 'channel' link, and it will take responsibility for basic security, routing, dialog management, and any necessary protocol conversion.

• In SNA environments, this is a subarea node (usually an IBM 3745) that contains the Network Control Program. The NCP controls and routes the flow of data to outside network resources.

• At the chip level, communications processors are digital signal processors. This includes the new forms of specialized high-speed CPU chips being developed specifically for multimedia communications. The most important of these is probably a new media-processor from MicroUnity.

communications interrupt

This is a forced pause in transmission which has been created by the hardware of a serial adaptor in asynchronous communications.

Communications Link Control

CLC. This is an old term. In a layered model of a communications system, the CLC layer handles the actual movement of the bits and bytes onto and off the communications system. Both RS-232 and SDLC are at the CLC layer level.

communications port

Correctly, this is any port on a terminal or PC which permits the attachment of general peripheral devices. More specifically, this is usually a 'serial' port (most probably RS-232 or RS-422) for the attachment of serial printers, modems, etc. MS-DOS assigns names like COM1 and COM2 to these ports.

• The term isn't generally applied to parallel ports, although there's no real reason why not.

communications satellites

Most modern communications satellites are geostationary. They sit over the equator at 38,000km in the Clarke belt, and take 24 hours to complete an orbit — the same time as the Earth's rotation. However inclined orbits near to the Clarke belt are also used for communications purposes, and now many LEO and MEO satellites are being proposed for communications purposes.

The Molniya orbit has been extensively used by the USSR for 20 years, and the Tundra orbit is finding increasing applications — both of these are elliptical orbits. Motorola's Iridium constellation will use low-earth orbits as will Globalstar and Orbcomm. See geostationary satellites, Tundra orbits and Molniya orbits.

community

In SNMP network management, this is a logical group of devices and sub-management systems (NMSs), which are in the same administrative domain.

Community Access Television

CATV arose out of the need for better (usually higher) TV antenna in difficult reception areas. A community (or a local retailer) would install a good antenna mast on a local hill, boost the signals with a wide-band amplifier, and transfer them as a block via cables to the surrounding homes. A small charge would be made on a subscription basis for this 'Basic Cable' service.

Eventually, the operator would decide to add signals from further afield, perhaps via satellite, or even some of their own (from video tape recorders or a small studio), and before long the provision of these extra channels assumed greater importance in encouraging new homes to link up than just the original signal quality. This is the origins of the cable TV system in the USA, Canada and many other parts of the world.

• The term CATV is still used, but now they claim the acronym is for CAble TeleVision.

comp

The comprehensive artwork used by graphic artists and layout artists to present their concepts to an editor. These will show the general color and layout of the proposed page-spreads.

• In some printing shop the term comp means a copy of the final layout with all artwork and copy in place. It is used for checking and proofreading.

Compact Disk

See CD.

compacting

The word has slightly ambiguous meanings. It can apply to:

• The process of removing unnecessary disk or file clutter to reduce the file size, while still retaining the important information in the normal ASCII format.

• The use of various compression methods to substantially reduce the size of a text file. If this is the case, the file will need to be decompressed before it can be used.

companding

Compacting and Expanding again — or is it Compression and Expansion? Both terms are used. This is a technical term for a range of noise-reduction techniques involving non-linear forms of quantization. This produces data compression at the transmitter, with complementary non-linear expansion at the receiver.

• In analog radio systems this is a way of reducing the dynamic range of audio and video signals to fit the bandwidth available, while allowing expansion of the range again at the receiver. FM has a companding scheme.

• In the ADC process for digitizing voice to create PCM, the 8000 per second analog signal samples (PAM) are quantized by a non-linear scale. The non-linear scale is used because small amplitude signals are more likely to occur than large amplitude signals, and they are less likely to mask any noise.

So by using a non-linear approach, it takes a much smaller change in amplitude around the base-line to register one step of quantization than it does at the higher end of the amplitude range — the softer components are given a greater resolution. These coded binary steps are transmitted and, at the other end, this non-linear process is reversed. See A-Law and mu-Law.

This type of telephone device is called a syllabic compander, and generally, compression ranges of 2:1 are used. This technique also decreases the absolute level of noise when no signal is being transmitted. The characteristics of analog phone companders are subject to ITU recommendation G.162.

• To be strictly correct, the above use of non-linear quantization, is a substitute for companding rather than companding itself. Companding should be applied to the changes made to the analog wave-form before linear quantization is applied. The alternative is to make no change to the analog wave-form and then apply non-linear quantization. Both results are the same.

comparand

A value to be compared with another.

comparator

An arrangement of electronic gates that checks whether two signals (or codes) are the same, or decides whether one is greater than, or less than, the other.

compatibility

The degree to which two electronic systems or two products match — the degree to which they can work together. When they are compatible, the interface between the systems is transparent — everything passes through without any problems.

• However, 'compliant' suggests that something conforms strictly to all the requirements of a standard, while 'compatible' is often used when it only conforms to some. 'Supports' suggests an even lower degree of compatibility.

compatibility windows

Microsoft's OS/2 approach to running DOS software on 286 and 386-class PCs.

compatible

A term which often referred to IBM-PC clone. Compatibles, often weren't. A clone which was 99% compatible may be able to run 99% of the available software without problems. Alternately, it may almost (but not quite) run all of the software, all of the time! This second approach was more common in the earlier clone days than it is today.

compiler

The software which converts a program written in a high-level language (BASIC, Fortran, etc.) to binary machine code which the machine can understand on a one-off basis. A compiler converts source code to object code.

With compilation, the translation only needs to be performed once and it is then stored and used repeatedly at machine level. Since it is common to incorporate old routines already compiled, a second-stage process using a utility called a 'linker' is often used to process the compiled code to create the final application file.

With compliers and compilation, the distinction is with interpreters which perform the translation operations on-the-fly each time the program is run, and with assemblers which handle low-level programming in assembly language.

• The compilation of Forth programs is a quite different process to the above. It consists only of looking up the addresses of all Forth 'words'.

• See source code, object code, and preprocessor.

complement

A binary method of obtaining the negative of a number. It should not be confused with negation or inversion. The 'true' tens-complement of decimal 342 is 658 (you subtract the number from 999 then add 1) and the 'true' twos-complement of binary 1 1010 is 0 0110.

When a computer wants to subtract number B from A, it first calculates the complement of B, then adds this A. True complements of any number are found by subtracting each digit from one less than the radix (base), then adding 1 to the result.

• Subtraction in computers (A – B) is usually performed by calculating the complement of the number B (cB), then adding this to the original (A + cB).

complex bipolar

A chip memory device which is significantly faster than TTL technology. See bipolar.

complex numbers

Numbers comprising two parts — a 'real' number (the everyday integers) and an 'imaginary' number. The latter are figments of a mathematicians mind which allow him to solve equations involving the square root of a negative tier number. A complex number is therefore something like $Z = 3 + 2i$ where 3 is real and '2i' is the 'imaginary' 2 that can be squared to produce –4. This work is the basis of Mandelbrot numbers.

complex planes

Complex planes use complex numbers as distance values. This is a concept used in fractals and Mandelbrot sets.

compliance errors

Errors (usually in A-to-D conversion) caused by strict adherence to the rules of sampling. Often it is slightly better to have the samples averaged a little over the sample period, or to allow them to vary slightly in time. See DPCM and delta.

compliant

Compliant suggests that something conforms strictly to all the requirements of a standard, while compatible is often used when it only conforms to some. A device may be RS-232C compliant, but only RS-232D compatible (it may implement test mode and loopback procedures not in the C standard).

Component Beta

This is applied to the most common professional video recording format used today for location shooting of professional video. The term is applied both to the original Betacam and to Betacam SP.

component video

Component video treats the luminance signal separately from the color — and quite possibly, various color signals separately from each other. There are both analog component, and digital component systems.

The distinction is with composite video which is the way PAL and NTSC signals are now handled.

The most basic component system is that of S-VHS which keeps the luminance (Y) separate from the composite color (C). However, derived analog component video, using YUV (also called Y, R-Y, B-Y) is the now-standard way in Europe of handling video signals in the video production (and sometimes transmission) stages. The US also has a slightly different system called YIQ.

These are called 'derived component' systems since two color-difference signals must first be calculated from the standard RGB output of the camera. In YUV systems, the luminance information is the sum of RGB, and is kept separate. It is processed independently from the two color -difference signals. The MAC television standard is of this kind.

With both analog and digital TV, the most pure form of component video is the pure RGB (Red, Green, Blue) system used in computer monitors where these three primary colors are kept separate throughout the production system.

The world standard for professional digital video production is the CCIR's Recommendation 601 which standardizes a 4:2:2 derived-component, time-division approach for processing and storage — where for every 4-bits that the luminance (Y) has for resolution, the two color signals (UV) have 2 bits each.

With component editing and dubbing equipment, and with component studio and transmission equipment, the cabling is three times more complicated and expensive, since one cable is required for each color component and the electronics in the editing and mixing stages are all in triplicate. Studios also need to be completely rebuilt and re-equipped.

Component television was recently proposed for higher home quality using the MAC analog system. In component television transmission systems (and in some studio interconnection processes) the separate components are time-multiplexed. Within the equipment, this avoids the need for many separate cables. See MAC, YIQ, Y/C and YUV.

componentware

A proposed extension of object-oriented software design which allows the user to assemble the type of application he/she requires, using software modules which have been optimized for the task. This is not the same as object-oriented programming, but it is half-way there.

Visual Basic is a good example of componentware; it provides reusable 'off-the-shelf' software modules that you can use to build real applications.The Apple/IBM/HP development of the Taligent operating system is also to include 'Constellation Project' (from Metaphor Computer Systems) which allows component customization.

This modular control can be extended from the user interface to the core of the operating system itself — but generally it is restricted to applications at present.

Microsoft's OLE is similar to componentware, but Microsoft intends to treat major applications as modules; in Taligent and OpenDoc the intention is to have much smaller modules. See application builders and ALM.

• The distinction between an object and a component appears to be (I quote) 'that a component rests on the basis of encapsulation only, while an object has both inheritance and polymorphism'.

composite

Together — as one. A composite signal will consist of two or more signals multiplexed or merged in some manner.

• **Data links:** A composite data-link is a line or circuit connecting a pair of multiplexers or concentrators. This is the bearer circuit which carries the multiplexed data.

• **Protocols**: Composite protocols are used by statistical multiplexers to allow them to verify the proper receipt of the data. They carry information on: the destination channel for the data; the status of the multiplexer; and the channels connected. These composite protocols usually overlay any existing TDM protocols.

• **Monitors:** Composite monitors are those which accept composite video — with luminance, chrominance and synchronization combined in a single signal. Home television sets accept composite video signals after demodulation by the tuning circuits. See RGB 'component' monitors.

• **Video and TV:** Composite video is the type of video we are used to, where all of the components which make up the video signals (luminance, color and sync) are merged into one stream, either in analog or in digital form, after demodulation.

Baseband video can be composite or component. With composite, the system modulates the colour components to a higher frequency, and adds them to the luminance signals. So both components of a picture then occupy a single coherent band of frequencies, with luminance at the 'bass' end, and color at the 'treble' end. There is no clear separation between the two, which is why texture is sometimes interpreted as colour (moire patterns).

When composite video is modulated onto an RF carrier and telecast, it becomes composite television. PAL, SECAM and NTSC are composite TV standards. Audio is frequency-division multiplexed into a separate channel a few megaHertz above the video carrier.

• **Baseband:** Composite baseband is the term used for the raw demodulator output from a television tuner prior to filtering, clamping and (usually) de-emphasis. This signal contains both of the transmitted subcarriers.

• **Synchronization:** When raster-scan techniques are being used, some devices require separate vertical and horizontal sync feeds, while most require composite sync (which is part of the video signals — a complete square-wave reversal of the electron-gun drive voltage). This refers to a signal which has both the vertical blanking and the horizontal blanking.

composite/continuous format

One of the competing standards for 5.25-inch WORM drives. It is mainly promoted by Japanese companies in opposition to the sampled-servo format.

compositing operators

These are logical operators (like Booleans) which are used to define the way image elements interact when grouped together in the one graphic space. They define such things as whether one element is in front of, or behind, another. See alpha channel also.

composition

In printing terminology, this is the creation of the typeface which will be used for printing. In these days of desktop publishing, composition has been combined with layout.

• Composition coding was a way of transmitting two separate character codes which resulted in a one-character output. This was necessary to handle the placement of accents and diacritic marks above standard alphabetic characters in some Latin languages. See also Unicode.

compressed drive

See CVF.

compression

This is the process of compacting data and removing redundant material. It can also involve the intelligent selection of fine resolution (sound and picture) which can't normally be perceived, and marking it so that it can be abandoned.

The primary division of compression systems, therefore, is:

— Lossless compression. On decompression the original is restored in every detail. See entropy.

— Lossy compression. Judgements are made to abandon certain unnecessary detail (below the perception threshold).

Lossless systems are required for data/text files, and for high-quality editing of video and sound. Lossy systems are used in transmission of video and audio where slight degradations don't matter (they are ephemeral, or hidden behind larger features).

To be pedantic, compression in digital communications systems is usually Bit-Rate-Reduction (BRR) which involves both lossy and lossless techniques. In analog systems we also talk about companding (see) which is a way to reduce bandwidth requirements.

• The major digital data compression standards are: (lossy) JPEG, MPEG, DCT, DVI, CL-I, and (lossless) LZW, Huffman's, RLE, MNP and various CCITT and other ASCII-based systems.

• With databases, compression may largely involve the removal of gaps, empty fields, etc. although other techniques may also be applied. Very high compression rates are often possible because of the database's wasteful use of memory — but this is not data reduction.

• In the compression of raster-scanned still images, there are three main techniques:

— Background elimination. See Huffmans codes.

— Transform analysis (Fourier, wavelet or fractal).

— Pixel run-length encoding. See RLE.

• In graphics, fax and photographs, the numerous bits representing large areas of white or black may be replaced by codes which essentially say 'the next x bits are white (or black)' or 'this line is entirely white, move to next'. This is Run-Length Encoding (RLE).

• In transmission, some bytes or sequences of bits, will recur more often than others. Huffman's technique is similar to the Morse Code, where a shorter sequence of dots and dashes is assigned to the most commonly-used letters. Huffman's does the same with binary streams.

• In digital motion video, DCT frame-delta techniques may be applied which allow you to sort the more important details (large structures) of the image separately from the less important (fine-textures). You can therefore throw away some of the less important detail if necessary. In a moving image, near-enough may be good-enough. This is intraframe coding.

• Most motion picture compression comes from comparing one image with the next, and only sending the difference. This is interframe coding.

• Motion compensation is a technique also used in the more complex digital video systems.

• By summing the above three techniques, MPEG (see) achieves compression ratios of a high order. A 270Mb/s primary digital video stream can be reduced to about 1.5Mb/s with adequate processing time and average screen action, while high-action real-time sports may only reduce to 6Mb/s.

• In analog television (PAL, SECAM, NTSC) the 2:1 interlace scanning system is a basic compression system since it halves the bandwidth required for transmission.

• In audio, the term is often misused to mean reducing the dynamic range between the loudest passages of sound (music) and the softest, by boosting the soft, and attenuating the loud. See equalization and companding.

• The DCC audio compression technique used by Philips in digital cassettes is similar to the DCT video compression approach, where the 'find detail' of the audio (high-frequencies when they are blanketed by louder sounds) can often be discarded. Sony use a similar system on their MiniDisc.

• DCT is now a primary means of achieving compression in audio and video, but also see wavelet compression, DWT, and fractal compression.

• IBM and Kodak have announced an image compression scheme which can store digitally encoded identification pictures on low cost memory cards. See IVS.

CompuServe B

A very efficient file transfer protocol which is used by the popular online service CompuServe.

computer-aided design

See CAD/CAM.

computer-aided instruction

See CAI.

computer conferencing

A technique that allows a group of people, anywhere in the world, to exchange ideas in an orderly way. Computer conferences can go on for days, months, years — or never end. Each person is expected to make contribution to the discussion (sometimes anonymously), and it is often possible to backtrack through relevant material via 'trails' or 'threads'.

There are moderated and un-moderated conferences. Some forums have a coordinator (sysop) whose job may involve the condensation of past contributions and drafting decisions or conclusions. Sometimes the moderator acts as a gatekeeper by selecting which material is displayed.

Computer conferencing is probably the best technique yet designed for making logical group decisions. Some large organizations use these as a substitute for formal meetings.

• Murray Turoff and Starr Roxanne Hiltz wrote the seminal book *The Network Nation* many years ago and introduced most of the current ideas about computer-mediated communications. Turoff popularized the term 'computer conferencing' over 20 years ago, and Hiltz introduced the idea, that computer conferencing could form the basis of communities.

computer graphics

The term covers line drawings, pictures, charts and graphs created by computer.

Computer Graphics Metafile

See CGM.

computer-mediated communications

This refers collectively to electronic mail (e-mail), computer conferencing, and probably EDI. See all.

COMSAT

Communications Satellite Corporation — the US Intelsat organization.

concatenate

To merge two or more characters, words, sentences, or even files, together in sequence, without leaving a gap between. If we concatenate the sequence of single-characters: A B C D E, we get ABCDE; if we concatenate the words: 'The', 'quick', 'brown' and 'fox', we get 'Thequickbrownfox'.

• The space character is still a legitimate character when strings are concatenated; it is not a 'gap' (a null). So if I attempt to concatenate the typed phrase: 'The quick brown fox' I get no change since it is already concatenated — it is just that you can't see the character in the spaces.

• The opposite of concatenation is string slicing.

• There is a difference between concatenated characters and contiguous characters. Concatenated suggests logical proximity, while contiguous means that they are physically touching. Ligatures are contiguous.

concentrator

The one word covers a multitude of devices, some passive and some active; some specifically for telephone networks and some for LAN/WAN operations.

• **Telecommunications:** This is most probably a statistical time-division multiplexer which is used to concentrate traffic at a remote location and shunt it down a single twisted pair, coaxial cable, or fiber. If all the incoming lines were to attempt to use the multiplexer at the same time, then the system would overload. This is often what distinguishes a concentrator from a straight TDM.

Concentrators in the US are often called channel-banks, and they provide signaling and control circuits in addition to concentration. The new ITU specification for analog concentrators is the V.5.x interface standards. See pair-gain.

• **With LANs and WANs:** This could be anything from a wiring point, to a cluster controller, to a multiport repeater, multiplexer, switch or intelligent hub.

— At the lowest level, a passive concentrator will now be called a 'wiring concentrator' or 'wiring hub'. They were just solder points inside a wiring closet/center.

— With time-sharing systems and dumb terminals, concentrators were relatively simple devices that accepted feeds from many (usually 20) low-speed asynchronous sources, and linked them through high-speed (usually synchronous) channels.

— As Token Ring LANs evolved, concentrators became active devices which allowed a number of stations to be physically connected at the one point, and provided a by-pass mechanism for the ring if a unit failed. See MAU and MSAU.

— In Ethernet, the concentrator was originally a collapsed backbone, and later a multiport repeater into which the drop cables ran. It was a stand-alone box or sometimes a card in a network server which could often accept thin coaxial as well as twisted pair. With 10Base-T, where one device was on each line, the concentrator was a multiport UTP hub with management; often chassis cards, but later pileable or stackable hubs.

— The term has now evolved as a generic for centralized units which can include the functions of switching, repeating, bridging, routing, multiplexing, and management functions. However, it now tends to mean active hubs — especially stackable or pileable hubs with basic repeater functions.

• **In WANs:** With WANs there is a general use of the term to mean 'multiplexer' and a more specific use of the term to mean 'statistical multiplexer'; the difference being that the former applies to systems which have rigid channel allocations (usually TDM), while the latter dynamically allocate its channels on a demand basis.

• More recently, concentrator has been applied to the new Ethernet and Token Ring switches. Some so-called 'switches' have a bridging or routing function, while others are true switches.

concept map

A hypertext term which stresses the idea that the division of various areas of knowledge can be graphically illustrated by branching lines between concept modules. The range of knowledge and ideas that can be expressed in this graphical form is very limited.

concordance

In information retrieval, a concordance is an alphabetic listing of the key words contained in a file (or series of files), usually with details of the occurrence's locations and/or numbers; it is a form of index. In electronic systems the concordance can often be used in association with Boolean operators to specify a search path.

concurrent

Happening at the same time. In computing, this is stretched to mean 'appears to be happening at the same time'.

Concurrent processing is timesharing. For instance, when two processes for two quite separate tasks can be handled by the CPU, because it is rapidly switching attention from one to the other; each has a 'time-slice'.

All true multitasking operations are of this kind. The computer temporarily stores all essential details of each process in a process table (a special stack) while it works on the other/s.

condensed print

A feature of many printers where the width (but not height) of text characters can be reduced to about 60%.

condenser

• **Electronic devices:** See capacitance.

• **Microphones:** These are high-quality (for the price) electrostatic microphones which generally have a wide frequency range and low distortion. They always require a battery to drive the capacitance-diaphragm system, and there will almost always be a preamplifier built into the microphone.

conditional

Something that depends on the result of a test.: and so the test results must fall within pre-set boundaries which define the condition. The term is used in programming and CAI. The standard form of conditional statements is:

```
IF x occurs, THEN do y, ELSE do z
```

The condition may be: If x is greater than 100, or it could be the match of a name, or some internal match of two variables calculated within the program.

• Conditional expressions are the means used to test conditions in list-processing languages like LISP.

• Conditional access technologies in Pay TV systems are those encryption systems and devices used to prevent you watching these channels unless you have paid. Some new developments in conditional access also allow the customer to limit the use of the Pay system on a channel-by-channel basis to only those family members who are eligible. This permits 'moral levels' to be applied to programming, to keep the ankle-biters from watching porn. Smartcard systems are now being widely used in these systems.

conditioning

Adding equipment to analog leased lines to boost the quality enough for fast and reliable data transmission. The main techniques applied are amplitude and delay equalization.

Some of the circuit parameters which can be controlled or corrected are: attenuation distortion, envelope delay, S/N ratio, non-linear distortion, impulse noise and phase-jitter characteristics.

Conditioning by the carrier obviously can't be applied to switched circuits because the path through the network is constantly changing. However if the circuit is leased and dedicated to one connection, it doesn't pass through switches (or always through the same switch), and its transmission defects can then be measured and corrections applied. Generally this form of leased line conditioning has resulted in data rates which can be many times those possible through switched circuits — however, see below.

Carrier line conditioning is provided to various standard 'C' levels which emphasize attenuation and envelope delay distortions.

— C-1 is the next grade above unconditioned voice channels.
— C-2 is generally provided for a private-line service. It means that the line's attenuation and delay distortions have been corrected (by adding equalizers) enough for reliable terminal operations.
— C-3 is for access lines and trunks.
— C-4 (unknown).
— C-5 is the top grade theoretically possible.

The ITU has specifications for all of these grades.

— D-1 and D-2 conditioning standards were later introduced to handle 9.6kb/s operations by improving the S/N ratio and non-linear (harmonic) distortion.

• Most modern modems now perform elaborate handshaking rituals before beginning transmission. As part of these routines, they now test the line for many of the above parameters, and select correction techniques accordingly. For this reason, conditioned leased lines have tended to loose their advantage in terms of data-rate. In the latest V.34 modem specifications, a high bit-error rate in the process of transmission will trigger a new round of line-testing, and perhaps initiate a change in the transmission rate or correction techniques.

conference call

A connection with three or more callers involved in a simultaneous discussion. A multipoint connection. It applies both to voice and to video. See MCU.

confidence factor

In expert systems, heuristics are rules of thumb of the kind that we often use in everyday life. They are 'rules' not 'laws', and are rarely absolutely true in all circumstances. So in some expert systems the probability that these rules are true is assigned a confidence factor number. Typically, a confidence factor will be something like:

1.0 absolute truth.
0.5 reasonable likelihood of truth.
0.0 unknown.
–0.5 reasonable likelihood that it is false.
–1.0 absolute certainty that it is false.

There may be many other divisions, assigning intermediate values of truth or falsity.

CONFIG.SYS

A special file in MS-DOS systems that is the first to be run after booting. It is held in the root directory and contains information about the device drivers and general parameters of the system. This is where buffer size is set.

configuration control

In the management of large networks, there will constantly be a need to reshuffle devices and reallocate routes while maintenance and upgrading takes place in an orderly way. New nodes are constantly being attached, and new paths will be established which may alter the possibilities for all routing equipment on the network.

Configuration management requires a set of hardware and software facilities which are specially designed to keep track of these changes and to handle the problems without the need for a network expert to physically travel around the country. For instance, new software versions which are to be made available to all, may be mounted from a central server, or distributed to many servers throughout the network — rather than posted to every branch and copied onto every workstation.

configuration manager

• A module (a driver) in software with plug-and-play computer systems, which acts as the overall coordinator during the installation phase. It both configures the device, and informs all other device drives of the resource-requirements of all devices installed. See Plug and Play.

• Configuration management is part of CMIP, and one of the five defined requirements of network management. It involves messaging from the central console to identify active connections and equipment.

configure

To set it up to be used in a certain way. Software often needs to be configured to take into account the peculiarities of your system and this is usually done by the installation process. It may also require some input for user preferences. Plug and Play devices should eventually be able to inform installation programs about their requirements without human intervention.

• Adjusting a system so that all the elements work together. This is both a hardware and a software operation.

• Software configuration is usually a problem of telling the software what resources are available to it.

• Program configuring is also a very complex process used in parallel processing. It involves breaking down the progression of the code into parallel processes.

conflation

The technique of reducing the different forms of related words to the same root. This is used in library indexing. The words 'reduce' and 'reduction' would both appear under the same index heading.

conflicts

The PC often has problems where one card, application or TSR will conflict with the requirements of another. Check both the cards and the controllers when this happens.

The main sources of conflict are:

— Interrupt lines (IRQ),
— DMA channels,
— I/O addresses.

Some conflicts are resolved through software, some require DIP switch changes, and some require jumpers. This is the reason why Microsoft want to introduce Plug and Play.

congestion control

All forms of cell- and packet-switching network need some system of usage-control to handle overload conditions. Some rigidly allocate bandwidth limits (average available, and peak) at the time a call is set-up, and some rely on congestion notification during operations. Some use both methods.

In any packet technology, congestion control is the key to cost-efficiencies. Without such control during peak load period, the only solution is to provide extra bandwidth, and this has a low return on the capital outlay. Low error rates also help with congestion control by preventing retransmission queues from building.

The more common mechanisms used in frame-relay and ATM are FECN, BECN and ABR. See.

connect signal

A forward signal that sets up the call and secures the circuit.

connect time

A charging system for the usage of a computer or a connection which is made on a time basis — as distinct from charging on segments (number of characters) or CPU processing units. This is obvious with telephony, but not so obvious with some database access services.

connected speech

As distinct from discrete (paused) speech and continuous speech, in voice recognition systems. Connected speech usually refers to phrases spoken without unnatural pauses in a clear and precise manner, using a limited vocabulary. The limit is usually about 24 words in a sequence. See speech recognition.

connection-oriented

Any form of transmission where a connection is first established before the transmission takes place. The connection then reserves bandwidth, and may take care of error checking and recovery (which of course, imposes additional overhead).

Protocols taking this approach use a call-connect packet to initiate a session, and a connect-confirm response packet to complete the call sequence.

The requirements of 'connection-oriented' systems are that:

— The route for the packet is determined at call set-up time by allocating a virtual circuit between the two endpoints.
— At that time, all necessary resources on the virtual circuit are reserved and logical channels are allocated.
— During the data phase, fixed path routing is performed.
— The connection is then cleared.
— Then resources and logical channels are released for reuse.

• **LANs and WANs:** LANs are usually connectionless in the lower layers, and any connection required is imposed at the upper levels of the seven layer model (see CLNS). Usually LANs are connectionless.

• **Telecommunications:** Circuit-switched networks, are by their nature, connection-oriented. But we are in the process of migrating from circuit-switched to fast packet-switched (ATM) telecommunications.

— X.25: In conventional X.25 packet-switching systems, a special 'call-request' packet is first sent over the network to initiate a communications session. This first packet is connectionless, but it establishes the connection. Both the call-request packet and call-response packet carry a 12-bit LCI (logical channel identifier), together with the called and calling addresses (up to 60 bits each). From this point on, all data packets are directed on the basis of their LCI (not the calling address).

See CONS, CLNS, connectionless and datagram.

— ATM: With ATM networks a similar call set-up process is established using 'virtual paths' and 'virtual circuits'. Virtual paths have a number of channels sharing a single identifier.

Provision has been made in ATM for special connectionless services through AAL 5. See.

connectionist

A computer scientist involved in the development of neural networks.

connectionless

A form of packet-transmission suited to short messages or the transmission of limited numbers of packets, when there is no communications between the end devices before the first transmission of data. This is data transfer without the use of virtual circuits.

In simple bus or ring networks, there is no problem implementing connectionless systems because the path – choice is limited. However in meshed and complex networks, the problems are significantly different. Each router must have a large amount of intelligence to process the packet header, and the network requires an efficient mechanism to ensure that all routers or nodes have an up-to-date view of the overall topology.

However, the 'datagram' approach reduces network overhead since the network is not required to establish circuits or assume responsibility for error detection and correction.

• LANs are generally connectionless systems.

• Most telecommunications networks, however, require the establishment of a connection. Don't confuse this with the physical circuit-switched approach, however, we are talking here about logical connections (virtual circuits), and this applies to packet, frame and cell-relay networks.

• A LAN/WAN protocol stack can (confusingly) be connectionless at the lower layers, and connection-oriented at the higher layers. See CLNS and CONS.

• Telecommunications systems are said to be connectionless when they don't require the establishment of circuits or virtual circuits end-to-end before data is sent. In connectionless communications, datagrams can travel across the network without prior circuit establishment, so 'connectionless' implies that each data packet will carry a complete address (not just a virtual circuit identifier) and probably bits for error checking also. The DQDB MAN is a connectionless system, whereas ATM is connection-oriented. See datagrams, CLNS and CNP also.

• Store and forward messaging systems (e-mail) are 'connectionless' in one sense, but the term isn't generally applied in this way.

connectivity

The ability of devices to communicate and interwork with each other. Connections in computers are logical links (not just physical). So a connection is an 'association between functional units for conveying information'. This refers to both physical and higher protocols.

connector

• **Technical:** Engineers and the general public use the term differently. To the general public, the connector is the bit which makes the electrical contact — the interface — the plugs and sockets. But to an engineer, both connector and interface refer to the whole box and dice — the cable with all its wires and shielding, and the plugs/sockets — and particularly to the standards/protocols involved.

So connector is often a fancy engineering name for a cable — a bundle of wires with a plug or socket at each end. Connectors terminate with either male or female units — and the decision as to which is which, is made on the basis of the connecting pins, not on the outside shielding.

• **Databases:** In information retrieval, a connector is a word joining two or more search terms. These are Boolean expressions, AND, OR and NOT, as in the command:

```
Find Fruit AND Tomatoes NOT Oranges
```

CONP

Connection-Oriented Network Protocol. The OSI's recommendation for applying connection-oriented protocols to the upper-layer protocols in the stack. See below.

CONS

Connection Oriented Network Service. This operates at Layer 3 in the OSI model for connection-oriented X.25 services. Basically there are two OSI transport protocols: CLNP (connectionless) and CONS. CLNP is more efficient for LANs, while CONS is better for WANs. CONS allows the Transport layer to bypass CLNP when a single X.25 network is used.

console

• A terminal; nowadays usually a powerful computer or workstation equipped with a video display and keyboard. The term seems to be used with mainframes and minis to specifically indicate the supervisor's data terminal which issues commands to a host or to agents on the network.

• In net management, the console is the central management controller which runs the network management framework software, and to which the agents report. See MIB and RMON.

• In DOS terminology, console is used to refer to the keyboard, screen and motherboard functions only.

constant

In programming, a constant is a value that stays the same throughout the course of the program. Constants can be either strings or numbers: the number 10 is a constant, as is the word 'Hello'. If, during the course of running a program DISCOUNT is set and always remains equal to 20%, then this is a constant also. If it can be changed by the program during calculations, then it is a variable.

However, just to confuse the issue, in programming you put constants into variables to make them useful. So in the Basic statement: LET A$ = 'Hello', the A$ is the variable, into which the constant 'Hello' is being put.

If you wished, A$ could later contain the constant 'Goodbye', which is why it is called a variable.

Numerical constants can be positive or negative, and they may also be integer, fixed-point or floating-point. Sometimes they can also be hexadecimal or octal also.

Constant Bit Rate

Aka Fixed Bit Rate (FBR). Most telecommunications equipment is constant bit rate at present, however a lot of it would be more efficient in its use of bandwidth if it were Variable Bit Rate (VBR) and only transmitted the information deemed absolutely essential. CBR systems match well with time-division multiplexing and circuit switching, while VBR matches better with asynchronous packet and cell services where bandwidth can be provided 'on-demand'.

• Sometimes the rate may be' constant' (not-variable), but not 'fixed' according to some predefined standard. See ABR (Available Bit Rate) also.

• In ATM systems, the CBR way of controlling congestion is simply to reserve a fixed amount of bandwidth when the call is set up. If the required bandwidth is not available, the connection is denied. This is the simplest method, but it partly defeats the value of share media. See ABR, VBR, and leaky bucket.

constant length field/record

See fixed length field.

constant-voltage transformers

CVTs are used in power supplies of computers to deal with mains-power sags, surges and drop-outs. There are two types, square-wave and resonant. The first doesn't offer much protection against drop-outs (so be warned) and neither will protect circuits against hash or noise on the lines.

Line conditioners usually use resonant CVTs in combination with filters and spike suppressors. Generally you can only use one computer on one CVT. See power supply.

constellation

• **Distortion**: See group delay distortion.

• **Modulation:** When modem signals are modulated by both phase and amplitude, the combined signal can be illustrated by a phase and amplitude diagram along crossed X and Y coordinates. This is the constellation. Phase (or phase 'difference') is represented by the rotational angle, and amplitude by the distance from the center. Note that Trellis Coding introduces an extra level of complexity here.

• **Satellites:** Constellations are the combined satellite orbital planes when the satellites are acting in a coordinated way.

A constellation may consist entirely of circular polar orbits in which a number of LEO satellites will be orbiting in (say) eight different planes which parallel the lines of longitude. With inclined orbits the planes will cross the equator at an angle. Each plane will be separated from its neighbors by, perhaps, 30° or 45° around the equator. Station-keeping with this type of constellation involves both maintaining the even separation of the satellites around each plane/orbit, and keeping the planes at an even spacing around the globe.

Ellipso, however, has two entirely different constellations. One group of satellites is in circular orbit around the equator (the Concordia constellation), while another group (the Borealis constellation) follow highly elliptical (and 'inclined') orbits in a number of different planes which loop high over the northern hemisphere.

constrained coding

See channel coding.

constraints

In programming, these are arithmetical operators and expressions such as X > 0 or X + Y < 23. See CLP.

container

Containers are any structure into which data can be put. The data is usually then referenced by the container identification.

- A variable of any kind is sometimes known as a container.
- In reference to DDE or client-server operations. See client.
- The term is also used in SDH to define a frame structure into which plesiochronous data can be put. See virtual containers.
- Container objects in NetWare are branches of the directory tree used to organize other objects.

content-addressable memory

A new type of chip which provides flexibility when working with large data-sets. CAM is used for high performance applications which require look-up tables and cache searching engines. It provides a 1-bit processor built-in to each memory location, and this bit acts as a flag.

This is the basis of 'smart memory' technology claims. The flag allows a form of 'greater-than', 'less-than' and match-with-wildcards searching, and this can be performed on individual flagged bytes without the need for a sequential search through every address in the memory.

The technique has quick identification (through pattern matching) of information in memory, and currently it is used in LAN- bridging applications to match addresses on either side of the bridge.

This is an 'associative storage' technique generally used for databases of various kinds, but it is one that requires specially designed circuits.

contention

The result of two signals attempting to use the same resource at the same time. It is a term applied particularly to LANs which offer multiple access. Each party using an Ethernet LANs is free to butt in and fight for control whenever it detects a gap in the current transmission. Naturally, on occasions, two or more will begin to transmit at the same time, and this will result in data collisions. The stations can detect these collisions by listening on the network, as they transmit.

Multiple-access techniques of collision avoidance (CA), or collision detection (CD) must be employed to prevent overwriting data if two or more systems transmit at the same time. Ethernet and AppleTalk are contention systems. The distinction is between contention and deterministic (token passing and polling) systems.

- Contention systems are also called 'distributed link management procedures' since these networks don't need a central control. See CSMA and Aloha also.

context-sensitive

The ability to deal with ambiguities and idiom in the language.

- **Help:** Help files are usually available with most applications nowadays, but these will often need to be searched manually to find the explanation you need. Context-sensitive help systems are 'intelligent' enough to give you only information relevant to your immediate problem.
- **AI systems:** Info systems which understand the words,- within the broad area (the context) under consideration. Cyc is a major project which is an attempt to provide computers with common-sense. Before it can do that, it must understand the difference in 'pen' between the *pig is in the pen*, and *the pen is on the table*. It must know that pigs are unlikely to be in a pen on the table, so it can be said to be context sensitive.
- Any language translation program must be context-sensitive if it is not to make silly mistakes.

context switching

The ability of programs to switch from one to the other without closing files, while sharing memory. This is another name for pseudo-multitasking.

contiguous

Without gaps between: touching (sometimes): without barriers, divisions or segments.

- **Contiguous File**: A file which has its sectors stored as one continuous sequence in the tracks and sectors on a disk. This is not necessarily desirable since it means the blocks have not been recorded in the best possible way — the use of neighboring sectors on a disk isn't always the easiest and quickest way to access data. A disk controller requires processing time after each block of data has been read, so blocks need to be interleaved (separated) for the quickest access. See controller (disk) and RLL.
- **Graphics/text:** Graphics and text which have shapes that can touch each other. Some are deliberately contiguous. The distinction is made because text characters usually have a built-in one-pixel separation on the screen and in printers.
- See concatenate.

continuity

The vitally important job of a specialist assistant to the director of a film being produced on a shot-by-shot basis. Careful notes must be taken (also polaroids) at the beginning end of each shot to keep a record of the position of the subjects in the scene, where moveable objects were in relationship to the actors, how far down a cigarette had burned, etc.

There's many instances in feature films where hats suddenly jump on and off heads between shots, or even where the color/style of a shirt or dress changes completely.

continuous speech

The aim of voice recognition systems. It allows a normal conversational pace without pauses or limited word use. See connected speech, discrete speech and speech recognition.

continuous stationary

Fan-fold printer paper and rolls of paper rather than sheets. Continuous paper is usually tractor fed — and the torn-off edge called the perfory.

continuous tone

• In printing, this refers to the original photograph which has a gradient of tones from black through the various grays to white (both in terms of black ink, or in color). The key point is that it doesn't use a dot structure to produce the grays. These continuous tone photographs need to be converted to half-tones (using a screen) before they can be used in printing on any ink-based printing press.

• In desktop publishing, the term is often misused to mean large flat areas of tone across the page or behind text in a side-bar box. The distinction is being made between these areas and photo-images that require half-tone reproduction.

contour analysis

A technique used in optical character recognition which distinguishes letters by their outline. See OCR.

contrast enhancement

The various analog and digital techniques that exaggerate the difference in tones of an image. The extreme form of this is known as tonal separation. Here all the tones in a photograph may be converted to either black or white — with nothing between. This often then has the appearance of an ink-drawn portrait. See tonal separation.

control code

When the Control key on an PC keyboard is held down while another key is pressed, the keyboard generates a range of non-print characters from the standard 7-bit ASCII set. In use, these codes are relatively specific to each system or application, although some standards have been set.

For instance, <Control>G is the BEL character used to sound a bell or warning beep, <Control>J is the linefeed character. These are among the first 32 characters in the ASCII No.5 character set, which were used originally to control hardware.

• Note that ALT, ESC, and DEL are not treated as control codes. Alt generates a Null in association with another key; Esc and Del are part of the ASCII character set, but not control keys in the normal sense.

• The term is also applied to text format characters which may be embedded in data before it is sent to printers or modems. These are the format commands that center a word, or change the print font to bold, etc. They are more correctly called format commands.

control block

A group of variables set up in memory to hold information used by a device or by an operating system's utility functions. This is also called a parameter block.

control break

A way to stall MS-DOS and OS/2 applications by pressing Control and Break at the same time.

control key

A key on most computer keyboards. It needs to be held down while another alphanumeric key is struck, and it usually makes the computer perform some function. It creates a non-print character.

Note: control-key combinations are defined in ASCII for standardized telecommunications, however they are very often used for other purposes within applications. Each application will make use of these keys as it wishes. See control code.

control panel

All Macintosh, Windows and OS/2 machines have a control panel which allows the user to adjust such parameters as colors on the screen, rate of blinking of the cursor and insertion point, mouse-movement speed, date and clock time.

control track

On videotape, this is a linear edge track containing synchronization pulses.

controlled vocabulary

A fixed thesaurus of index terms used in some information retrieval systems. The contrast is usually with full-text indexing systems where every substantial word is indexed. Sometimes the controlled vocabulary is only used in creating descriptors. See identifiers and descriptors.

controller

There are both software entities, hardware circuits (or add-in devices), and full 'slave' computer systems that all qualify for the term controller.

Mostly it is a computer circuit which controls the flow of data between the computer and its I/O devices — usually disks, tapes, keyboards or screens.

• A disk controller is a specialized slave computer that takes over the housekeeping of the disk drives. It searches a disk for a file sector (usually 512 bytes), reads this to RAM, then transfers it to the CPU for processing. It then returns for the next sector. It may also be smart enough to cache information. See sector interleave and cache.

• Video controllers are the same as screen drivers or display drivers. Print controllers are print drivers.

• See also other types of controller: cache controller, cluster controllers, communications controllers, intelligent controller and network access controllers.

CONUS

CONtinental United States. A term used about US satellite coverage.The CONUS does not include Alaska or Hawaii.

conventional memory

The 640kByte of usable memory space in early MS-DOS machines. This gets very confusing because sometimes this section was also classed as low-memory (up to 640kB) while the space above was called high-memory (640k—1MB). But later, the term high-memory area (HMA) came to be used for a single 64kB block above the 1MB limit.

Below 1MB, the division of memory is now called 'conventional memory' for the lower 640kB, and upper memory area (with some upper memory blocks — see UMB) for the 640kB—1MB range.

Some conventional memory is used for storing vital system information, some is used for DOS (but see DOS 5.0 and above), and the remainder is available for applications and data storage. Conventional memory is managed by the built-in conventional memory manager. The distinction is with extended and expanded memory.

• In most recent Windows machines conventional memory is now given over to DOS. However MS-DOS can be run in the high-memory area in all computers later than the 286.

• Windows 95 allocates the lower 1MB of conventional memory exactly like the earlier versions. Device drivers and virtual drivers are installed in the conventional-memory area.

See EMS, extended memory, UMA and UMB also.

convergence

The coming together — of light rays, of non-parallel lines, and by analogy, of communications and computer technologies.

• The term is generally used today to refer to computers and communications, but also to the concepts of multimedia with text, graphics, sound and video all in one system. The three arms of convergence are computers, communications and broadcasting. A wider view of convergence would include another 'C', that of 'consumer electronics' — which includes home shopping facilities and video-arcade games.

• To the technician, convergence probably means ATM and what used to be called Broadband ISDN. The integration of all communications systems into a single ubiquitous network with a single transport protocol capable of providing all the multiplexing and switching functions over broadband links.

• LAN management: This is a quite different concept of convergence. Here it means the ability of an interworking group of devices on a network to adjust to a change in network topology. If a router is slow to converge, it can create problems with incrementing the hop count. See count to infinity.

conversation

A two-way exchange of information — either voice or data. The point being made is that both channels need to be open continuously, and be capable of symmetry (if need be).

• Conversation mode. See terminal mode.

conversational

Interactive: a two-way peer-to-peer exchange.

• Conversational mode is usually Terminal Emulation mode.

• Any connection which provides two-way traffic.

• Data conversations are usually in the nature of question and answer, request and response (transactions). The term is not generally used for file-transfers where, say, only ACKs are being conveyed on the back channel.

• Conversational systems must be 'almost' real-time operations. This generally means circuit switching, but packet-switching is used when short delays are acceptable. Fast packet-switching will allow conversational (isochronous) modes for voice and video as well as for data. The distinction here is with distributed (broadcast) services and store-and-forward messaging systems. See real-time.

• In SNA, a conversation is an LU6.2 session between two transaction programs.

conversions

Here are some of the more common conversions needed in computing and electronics:

Nanometer to Angstrom	multiply by 10
Inches to millimeters	multiply by 25.4
Feet to centimeters	multiply by 30.48
Yards to meters	multiply by 0.914
Sq. yards to sq. meters	multiply by 0.835
Miles to kilometers	multiply by 1.609
Sq.miles to sq.kms	multiply by 2.589
Ounces to grams	multiply by 28.35
Pounds to kilograms	multiply by 0.454
Pints to liters	multiply by 0.567
Gallons to liters	multiply by 4.54
Cubic feet to liters	multiply by 28.32
Electron volts to joules	multiply by 1.6021×10^{-19}
Horsepower to watts	multiply by 746

converter

• **DBS:** In direct broadcasting satellites, converters are needed to translate the C-band or Ku-band satellite frequencies down to the frequency range where they can be tuned by the television set or IRD. See LNA or LNB.

• **Cable TV:** This is a basic set-top box used in cable television. It provides tuning separation of the various channels (older sets may not be able to tune adjacent channels), and it may even have a descrambling function (some cable systems even scramble Basic services). Usually a more elaborate box with addressing functions will be needed for Pay TV. See IRD.

convolution

In the Queen's English it means coiling, twisting, folding back on itself. In information theory and error correction it means a checking system which looks forwards and back — one that turns back on itself.

• In computers and communications, convolution is closely related to the correlation process. Mathematically, it is what happens in the process of filtering. Often the same hardware is used to perform correlation and convolution processes.

• In graphics and image processing this is a process of 'averaging' where the value of a pixel in a selection is replaced by a weighted average of its neighbors. The process is used for smoothing, etc. Each pixel and its neighbors are accorded a certain amount of 'weight' (multiplied by a value) and the new output pixel value is the sum of these weights. The set of weights is called a kernel. When convoluted results are added to normal images, the effect is sharper edges and the appearance of greater resolution.

convolutional codes

These are error-checking systems which rely on checking past message blocks as well as current. The convolutional encoder has memory, and the values it calculates depends on calculations done with blocks already held in memory.

cooperative multitasking

• This now means task-switching. The user decides which process has priorities by bringing the application to the foreground. The Macintosh with System 7 is said to have 'cooperative' multitasking when what it actually has is cooperation in memory management, and only single tasking.

• Historically, cooperative multi-tasking was an idea for allowing essentially non-pre-emptive programs to act as if they were multitasking — they would share all idle moments of the CPU among themselves rather than having control exerted by the operating system. When you weren't typing, control would be handed back, say, to the database manager in the background, to continue with its sort — so this was a form of pre-emptive multitasking.

This cooperative idea depends on the applications sharing the use of the processor's time between themselves — and that meant they had to be rewritten. Once a schedular yields control of the CPU to an application, the schedular can't regain control until the application yields it, so this approach was only as good as the worst-written application. It was not a success. See multitasking.

cooperative processing

A revived technique where the execution of an application is shared between a number of processors operating under a single 'system image'. This extends to a number of distributed processors forming a single 'logical' machine, and running under a single operating system. See Plan 9.

Copenhagen interpretation

The standard theory of quantum physics. The two main principles of complementarity (wave/particle duality) and uncertainty (position and momentum can't be simultaneously measured) are counter-intuitive. See Debye's Law.

coprocessor

A separate processor chip that undertakes special tasks and thereby allows the main processor to concentrate on general operations. Co-processors are usually maths processors (number crunchers) or graphics processors, but they may also be used for video and sound processing.

This is not parallel processing — the main processor generally waits for the result from the co-processor (although there is some overlapping instruction execution). Often it can handle some other tasks after handing over to the co-processor, however, so there may be some parallel functions.

Co-processors extend the standard architecture [von Neumann] by taking over computational-intensive tasks and performing them more efficiently. The main processor still decodes the instruction, but the co-processor can access data directly. As a result, a co-processor can always be used without needing to rewrite the software.

Remember that in these days of outline fonts and vector graphics, many applications which aren't obviously math-orientated will benefit from a coprocessor.

• Maths co-processors are also called floating-point processors, maths chips, or numerical processing units.

• Co-processors often have a higher precision than CPU operations, also. It is not just the rate at which they perform.

copy

• **Publishing:** In desktop publishing, the written text of the article or report — as distinct from the graphics, pictures, headlines, etc. Body copy is the more correct term.

• **Computing:** To make a duplicate, without destroying the original.

copy-protection

A range of anti-theft techniques to prevent the pirating of software, which (thankfully) appears to be disappearing: the cure was worse than the disease. At one time almost all software had elaborate copy-protection mechanisms built in, so often you couldn't even make a backup copy of an application. If your word processor disk crashed (which it often did) you'd be left with a useless computer and hundreds of useless files.

When hard-disks became popular, copy-protection created too many problems because of the need to transfer these application programs to the hard-disk — and to be able to reinstall them again if something went wrong, or when you upgraded the system.

• Copy protection used a wide range of techniques, most of which used hidden bytes on the disk which would not be duplicated in a copy. Another technique involved physically burning spots onto the floppy disks, so that a copy would never have the same defect. An industry grew up around special copy programs which could defeat these systems, and the best known of these was Locksmith for the Apple II series.

Copyguard

An anti-copying technique developed to prevent video recording of movies over Pay TV channels in the USA. The system adds or deletes up to 16 lines from the 525-line NTSC standard.

Copyguard works because the electronics of a home TV set can rapidly adjust to missing or extra lines during the vertical blanking interval. However the mechanical servos which control head rotation in video-cassette recorders are not so forgiving, and so the picture breaks up with snow of distortion. This idea was not a universal success!

copyright

Copyright protects the expression of an idea, not the idea itself — and this is where it differs from patents most obviously. However, copyright will hold for translations of expression — from one language to another.

More than one copyright can exist on essentially the same work. A piece of music may have dozens of copyrights: on the lyrics, the music; the arrangement, the performance, etc. This is a legal minefield.

In most of the world, copyright is automatic at the time of creation. However the US only provides protection for American-created material if it contains the prominent notice:

```
Copyright 1995 John Doe
```

or

```
© 1995 John Doe
```

CORBA

Common Object Request Broker Architecture. This is the definition of a common set of reusable objects, and a set of APIs, used to create an Object Request Broker (ORB) which can accept requests from applications and match them to the mechanisms required for network management and information retrieval.

CORBA is middleware controlled by the Object Management Group (OMG) and supported by most of the IT industry. The OMG has specified a high-level set of services which permits 'objects' (encapsulated applications) to interoperate, while remaining shielded from many of the complexities usually associated with network distribution.

CORBA aims to bridge the gap between telecommunications and computers. The promoters of CORBA have lofty aims; they say that it will eventually change the way we think about networks, and that it possibly leads the way to the merging of the television, the PC and the telephone. It will conform to the ODP specification, and it may now incorporate OLE.

The cornerstone of CORBA is the Object Request broker (ORB). See ORB and TINA also.

cordless phones

These are not cellular mobile phones. Cordless phones talk to a local base-unit which is attached to a private line. Early cordless phones were FM analog and many of them used the same frequency, which created many problems. Anyone could walk down the street and make calls on someone else's phone, or listen to other conversations. Most later models are digital and have a wide range of security features.

The term has now been extended to include phones which can make contact with public networks when you are away from the home or office. See CT-2 and Telepoint.

• Wireless PBX technologies such as DECT and DCT-900 are also cordless, and the term is used here also.

cordless spectrum

• The various US domestic cordless spectrums are not contiguous.

• In the 43—49MHz region. Channel 1 is at 43.72MHz and has a 5.04MHz duplex offset. There are 25 channels (grouped in 10 + 15) with the 25th being at 46.76MHz. The frequencies are spaced at 20kHz.

• In the 900MHz band there are 60 channels (each 100kHz) between 902.1—903.84MHz and 926.1—927.84MHz, and these have a duplex offset of 24MHz. This is for both narrowband and spread spectrum phones up to 1000mW of power. The channels are 30kHz wide, with a 24MHz separation between send and receive.

• A wider range of frequency blocks have been used over the years in the USA and other countries for cordless phones: 1.7MHz, 27MHz, 30/39MHz, and 46/49MHz, in particular.

core

Central or essential. The kernel.

• A type of magnetic memory used in the early mainframes for system RAM. It used tiny doughnut-shaped (toroidal) magnetic ferrite rings supported on a grid of wires. This memory technology was invented by Dr Wang, who went on to found Wang Computers. Following from the above, the term is often now used to mean the main memory; the 'system RAM' or 'working memory' in a computer, even though the chips are probably DRAM or SRAM.

• A centralized sequence of programming instructions which act as a controller for other specialized routines. See kernel.

• The very narrow light-carrying center of a glass fiber. It is often surrounded by cladding (low R/I glass) and plastic buffer. The core size defines whether this is a single-mode or multimode system.

• The term is also applied to the slotted plastic 'holder' of the multiple fibers in a fiber cable. A small submarine cable will usually have a plastic core with six slots, and a fiber is inserted into each of these slots. Insulation, copper sheathing, etc. then surrounds this core.

core engine

The concept of writing software applications in such a way that most of the code for any program would be the same regardless of platform. Thus only a small part of the code would need to be rewritten for any particular system. See kernel.

core routers

On the Internet these are the primary routers maintained by the Internet Network Operations center.

corona

A bluish-purple glow surrounding high voltage systems, due to the ionization of the surrounding air.

correlation

The degree to which two things match — the degree of interrelation between two signals, etc.

- In ATM telecommunications, voice cells are said to be highly correlated because each cell in the communications link is closely connected logically to the last, and to the next. The cells can't arrive out of order as can connectionless packet systems (which therefore need a sequence number).
- See convolution also.

corruption

The destruction of data on a disk or in memory through a fault.

COS

Corporation for Open Systems. This is a non-profit R&D organization formed in the USA in early 1986 as a consortium to promote standards — mainly ISO and CCITT standards. It's members are international, and include vendors of equipment, service providers, and both large corporate and government users.

Their charter is to promote the use of OSI and ISDN as standardized technologies around the world. They provide conformance testing, certification, and they issue a COS Mark.

COS works closely with the ISO, ITU, and with the OSI Workshop. It has close ties to the European equivalent, SPAG, and also to the MAP/TOP group.

COSE

Common Open System Environment. Defined by HP, IBM, SCO, Sun and USL as a new unifying force in Unix. COSE doesn't seek to change existing Unix standards (already unified by OSF and USL) but rather advances the unification to higher levels in the computing environment — into networks, GUIs, object management and the like. The control of COSE is with X/Open. Its main role to-date has been in pushing OSF/Motif as the standard GUI, and also X Windows.

COSINE

COS Interconnection Networks in Europe. An EC project to build a large scientific and industrial network in Europe.

cosine transform coding

A video conferencing term. See transform coding

COST

Co-Operation for Scientific and Technical research group supported by 19 countries. This is a European group which is working on many projects under the EU.

— It has standardized a European 1/2-inch digital magnetic videotape standard along the lines of the CCIR 4:2:2.

— COST 211: In videoconferencing, the group's recommendations became the ITU's H120—H130 standards which were widely used in Europe before H.261.

— COST 244: Researching the biological effects of electro magnetic fields.

coulomb

The unit of electrical charge (C). It is a group of 6.25×10^{18} electrons. An amp is one coulomb per second.

count to infinity

When routing algorithms are slow to converge after a change in network topology, the routers can sequentially increment the hop count. This creates a problem unless some particular (arbitrary) limit is imposed on the number of hops to prevent them cycling for ever.

country code

In telephone systems, the formal numbering plan distinguishes between line codes (usually the last four digits), prefixes or exchange codes (usually three or four digits), area codes (one to three digits) and country codes (the first one, two or three digits) which are only used internationally. Convention is now to precede this international group with a + sign.

Country codes consist of two parts: a zone code, followed by one or two national identifiers:

- The zone code for the South Pacific is 6, and so Australia is +61 while New Zealand is +64. Smaller countries who gained telephone links later will normally have three digit country codes. So New Caledonia is +687, American Samoa is +684, and Western Samoa is +685. Note that there is no country with a +68 code; this two digit number had fortunately been reserved in time.
- The USA doesn't use a real country code, however the US, Canada, and some of the Caribbean share the +1 zone code. They make the next distinction the three-digit area codes. This system was in place before country codes were formally assigned.
- Europe has two zone codes (+3 and +4) and the old Soviet Union (zone +7) codes are now being redistributed to the independent nations. The problem has been in finding enough country codes to go around. Fortunately, West and East Germany combined and release a two digit code for subdivision.

coupled

- **Hardware – software:** Systems are often categorized as being 'closely coupled' or 'loosely coupled'. Mainframe and minicomputers and their associated applications are said to be 'closely coupled' because of the proprietary nature of both operating systems and applications — while PCs and their applications are said to be 'loosely coupled' when the applications and O/Ss work on a wide variety of hardware.

- **Electrical:** On electrical communications systems, some digital line codes can create DC-coupling which is undesirable. The ideal is AC-coupling with equal +ve and –ve pulses. See AMI.

couplers

- The devices used to join optical fiber. These are a constant problem with the technicians; they limit the frequency range and distance between repeaters of the optical fiber systems (through light loss) and are often very costly.
- Modems. See acoustic couplers.
- In electrical analog circuits, couplers are usually transformers which allow AC signals to pass across without direct electrical connection. Line-isolation transformers are couplers.
- In telephone networks, the term is sometimes used for hybrids.

coupling

- Physical joining, as in the use of optical fiber connectors.
- The close association of two data elements, such that a signal will pass across the boundary even though there is no electrical connection.
- To protect against lightning strikes, incoming telephone lines can be isolated from, yet connected to, the internal office or home network with iso-couplers.
- Relational links between database files constitutes a coupling.
- With business applications, hot-links produce couplings between different forms of data expression or view. If you change the figures in a spreadsheet application and the spreadsheet is coupled to a business graphics program, the graphic representation of those figures in a hot-linked bar-chart will also change automatically.

Courier

A typeface which has even spacing for the letters: it has a 'fixed pitch'. This makes it appear typewriter-like.

courseware

Computer-aided instruction programs. See CAI.

CP

Circular Polarization. Satellites use either horizontal or vertical polarization to separate the signals, or they may use left-hand, or right-hand circular polarization.

CP-FSK

Continuous Phase, Frequency Shift Keying. See FFSK also.

CPE

Customer Premises Equipment. A general name for anything connected to the carrier's lines within the customer's premises — telephones, telex, facsimile machines, modems, ACDs and PBXs, etc. Until the US Carterfone decision, the world carriers (not just in the USA) would not permit private equipment to be attached as CPE.

CPI communications

Aka CPI-C. An IBM API-layer which is part of SNA and SAA. It provides a very portable high-level function that allows applications running on a network to communicate with each other. See APPC and LU 6.2.

CP/M

Control Program/Microcomputers. Digital Research's widely used operating system (since 1973) which became the foundation for MS-DOS. CP/M in its various forms is still used in some older computers.

The family includes:

— CP/M-80 (the later version of the original),
— CP/M-3 and CP/M Plus (both the same),
— CP/M-86 (16-bit version),
— MP/M-80 and MP/M-86 (both pseudo-multitasking),
— Concurrent CP/M-86 (multi-tasking), and
— CP/NET for networking.

CPM

Critical Path Method. See.

CPN

- **Customer Premises Network:** The company's internal LAN or any telephone/data network connected to the PBX. Don't confuse this with the CAN or local loop.
- **Calling Party Number:** Another name/acronym for CLID (aka CLI).

cps

- Cycles per second. An old term now superseded by Hertz (Hz) which tells you less than cps.
- In printers and also in telecommunications — characters per second. This is usually the same as Bytes per second or B/s.

CPU

Central Processing Unit aka MicroComputer or Micro-Processing Unit (MCU — MPU) or 'processor'. This is the heart of a computer. Some would say this 'is' the computer; that everything else is peripheral.

In micros, the CPU is the microprocessing chip, in mainframes it may be a number of chips on a circuit card.

A CPU contains two main circuits: the control unit and the arithmetic/logic unit. It also contains various registers which are able to retain binary numbers, and all data in the CPU is processed through these registers. That's why, at the base, everything in a computer is a number (see ASCII and ALU).

A computer deals with only three types of information:

— **Control:** Information passed between the various parts of the computer to coordinate the processes. The language of control is the micro-code or micro-instruction.

— **Status:** Information which indicates the present condition of the CPU; it may be sending or receiving data, or in a control mode.

— **Data:** Information stored in memory at various locations and identified by an address.

See ALU, accumulator and wait states.

- **CPU bus:** This is the local bus. See.

CR

ASCII No.5 — Carriage Return (Decimal 13, Hex $0D, Control-M). A format effector that moves the printer head to the start of the line; on a monitor screen it moves the cursor to the left margin on the same line. It requires a line feed to advance it to the next line.

CR/LF

Some systems couple carriage-returns and line-feeds together; some treat them separately. The carriage-return should only move the cursor horizontally to the left, the line-feed should move it one line down. The distinction was important in the old days, when daisy-wheel printers needed to double-print some lines for bold or underlining. It is not so important now.

cradle

The on/off-hook switch for a telephone handset.

CRAP

Computer Re-Adjusted Photography. A most apt term for computer-enhanced, -altered, -adjusted photographs used in magazines and newspapers. It was suggested by US photographer Fred Picker.

crash

The close relative of the bomb — when a system stops working, forcing you to reboot. There are software crashes and hardware crashes.

- Disk crashes can cause physical damage. The read/write head of a hard disk normally flies over the surface without touching. When you power-down it is parked on one side — but when something unexpected occurs, it crashes into the disk surface doing a lot of damage along the way.

CRC

- **Camera-Ready Copy.** See.
- **Cyclical Redundancy Check.** See.

creator codes

In the Macintosh, four-character creator codes are used to identify the publisher and class of each file. For instance WNGZ is the code for WingZ files, SSIW for Word Perfect, 4D02 for 4th Dimension. A creator code, together with a type code, should uniquely identify a file for handling by the computer. The Mac operating system uses these codes to tell the computer which application to launch.

credit card

Bank credit card number sequences are reasonably standardized around the world, and codes to generate these are widely available on the Internet, so I'm not revealing any secrets here.

Cards have between 13 and 16 digits and an expiry date. The expiry date can't be generated from the card numbers, so it provides an extra layer of authentication protection.

The first group of numbers are bank codes, and the last group is a checksum used mainly with entry terminals to check that the previous digits were correct.

- Smartcards are being developed with both credit and debit functions. These are called 'relationship' cards.
- See SVC and STT also.

credits

In film and TV, these are the names of everyone involved in making the program, usually scrolled at the end of the program — although the half-dozen key actors, writers, directors often get special emphasis as part of the opening titles.

CRISC

A processing chip which combines both complex and reduced instruction set approaches (CISC and RISC).

critical error handler

This is an interrupt routine in DOS which is triggered when a serious device error has occurred.

critical path method

PERT and CPM are now interchangeable terms in program management. They are techniques used in the planning of projects (building construction and projects with similar logistics) to ensure that the resources are available to keep the job running smoothly, and to identify elements at each stage which are critical to maintaining the schedule.

The critical path is the route that takes the most time — it therefore marks the sequence of sub-projects which are most critical in performance.

CRMA

Cyclic-Reservation Multiple Access. A high-speed (gigabit) LAN protocol being developed by IBM. At these data-rates, problems arise in the access time (the time taken to place data onto the network) and with variable propagation times (time taken for data to cross the network). This protocol is seen as a potential rival to DQDB.

Cromalin

A common color proofing system used by printers to give customers a final check on their color separations before printing begins. It uses powdered pigments instead of ink — but the results are remarkably accurate.

crop

In printing and publishing, cropping is the elimination of parts of an illustration/photograph [Mark these on the back]. It also means to cut off physically. See also bleed.

crop marks

When preparing a page of 'camera-ready copy' it is usual to put crop marks (outside the actual area of the page) as a guide to where the paper should be guillotined. These marks are used for final trimming. They are also often called registration marks.

cross assembler/compiler

A compiler or assembler which is being emulated on another machine. The cross compiler or cross assembler will compile/-assemble code for a completely different machine. Some machines can't really handle their own development software, so this is done on a much larger machine.

cross color

In television, cross color problems are seen when viewing patterns such as the herring-bone weave in a sports jacket, or a fine pattern in a striped tie. The color moire patterns are the result of confusion between the chrominance and the luminance signals in conventional composite TV, but most of the effects can be removed by modern comb filters.

Cross color artefacts are intrinsic in composite (luminance + chroma) signals because both share the same bandwidth at the same time, and intrude into each other's territory. When these fine stripes or herring-bone patterns are scanned, the luminance frequencies generated fall into the frequency range used by the color subcarrier, and are therefore interpreted as color. See cross luminance.

cross compiler

A compiler which will generate code for an alien system. It allows programmers to design programs on a large, fast machine, to run on a small, slow one. Some small devices such as PDAs, don't have their own programming tools, libraries of routines, or compilers, so all programming is done on larger machines. See cross assembler.

cross-connects

A term which was derived from the old patch-panel way of connecting many lines (cross points) to many more, through patch-cords and sockets. Nowadays cross-connects are generally computer-based multiple switching devices. They can switch any input on one side of the 'frame' to any output on the other side, or they can switch any-to-any (with no reference to two sides). Multiple simultaneous connections can usually be made.

• Crossbar switches satisfy the requirements of a cross-connect.

• There are time-division cross-connects, and space-division; some use a shared bus, and some use banyan systems; and some cross-connects are 'blocking' and some are 'non-blocking'. See batcher-banyan and banyan.

• The present form of telephone network relies on being able to multiplex many 2Mb/s (or 1.55Mb/s in US) links. Unfortunately, these are clocked independently (PDH), which makes it difficult for any cross-connections to be made without demultiplexing all the signals. SDH overcomes this problem with STM cross-connect switches, and for the future broadband networks there are ATM-based cross-connects under development.

cross luminance

In the present composite TV standards, the color signal coexists in the same bandwidth as the luminance signal (the two actually overlap to a degree). It is therefore inevitable that each part of the signal will affect the other — despite the use of special comb filters. Luminance crosses into color, and color crosses into luminance.

Cross luminance is not as obvious as cross color. Because of the overlap in the composite signal, low-frequency color is interpreted as black/white detail, so the junction between saturated colors shows up on the screen as small crawling dots.

Component television systems and time-division multiplexing of the color and luminance, have been able to produce dramatic improvements in quality in the MAC analog component television techniques. See.

cross point

The switch/contact at the junction of the column and the row in an crossbar exchange matrix. This is the element that actually performs the switching function in many older telephone systems. The term is used virtually interchangeably with crossbar. See electronic crosspoints.

• With 1000 incoming lines and 1000 outgoing lines, an any-to-any switch would need 1 million crosspoints, if it weren't for the distributor.

• Originally this term was used for the mechanical switches in crossbar exchanges, and later for glass-enclosed reed-relays. Now these are being displaced by digital solid-state devices (logic gates).

crossbar switch

A widely-used mechanical telephone switch invented by Lars Ericsson in 1926. It was quicker and quieter than the Step switch and it was progressively introduced around the world in the 1940s and 1950s. This design was the last of the true electro-mechanical switches and it allowed STD (Subscriber Trunk Dialing) and IDD (International Direct Dialing).

These exchanges don't use linefinders or uniselectors. The early crossbar switches had ten vertical and ten horizontal bars (later 10 x 20) with no direct connection to each other. At the intersection of the 'cross bars' are small moveable contacts which respond to the appropriate dialed digit.

Since they can connect any vertical contact to any horizontal, they are classed as 'crosspoints', and a single switch can handle up to ten calls at the same time. The original design was modified later by the addition of reed-switches, and these have remained in use until today because they can be upgraded to use stored program (computer) control.

Control is exerted through 'markers' (a batch of relays) which connect the caller to a 'register' which is then able to absorb the dialing impulses and find a circuit. In crossbar operations, because of the high speed, only two markers and one register are provided for every 1000 subscribers.

The register converts the dialed pulses to tone codes (see MF) which are transmitted to code-receiver and markers in other exchanges. The main types of crossbar exchange are ARF (terminal exchange), ARK (small country terminal exchange), ARM (main trunk exchange) and ARE (crossbars with full computer-control).

• Crossbars with Stored-Program Control (SPC) use PC technology and software for control, but gradually they are being replaced by fully electronic, fully modular, digital switches. See crosspoint.

crossfire
Same as crosstalk.

crosshead
In page layout, these are the small 'sub-headlines' inserted in the copy of a long article to break up the sheer size of the columns. They are usually one or two keywords selected from the following sentence. They should be set in slightly larger bold type, and inserted into the copy-flow with space above.

crossover cable
See null modem.

crosstalk

• Unwanted intrusion of communications-energy from another channel — usually because two telephone lines run physically close and in parallel for a considerable distance. In microwave links (and cellular phones), crosstalk results from one system picking up and demodulating the radio signal from another nearby. More correctly this is called near-end crosstalk; written as NEXT. FEXT, or far-end crosstalk occurs at the other end of the line.

• For privacy reasons, telephone systems have a crosstalk index which varies from 0.1% to 1.0%. Shielding diminishes the effect of crosstalk.

• In stereo systems, crosstalk is a measure of the separation of the left and right channels.

• Crosstalk is a popular file-transfer program for PCs.

CRT
Cathode Ray Tube. An older name for a Visual Display Unit (VDU), or Visual Display Terminal (VDT) — which is an older name for a monitor — which, in layman's terms, is the computer or TV screen.

Video-type screens for computers were first introduced by General Electric with its first business computer in 1954. Before that, teletype printer terminals were the main I/O device, and before that, punched cards. Today, the distinction is probably between CRT/VDUs and LCD screens.

cryogenic propellants
With satellite launching, both liquid- and solid-fuel rockets are often used. Cryogenic propellants are the most common liquid fuels — these are the combination of liquid oxygen and liquid hydrogen. See hydrazene.

crypto systems
Systems that employ data encryption/decryption algorithms to provide end-to-end security for the data. These systems operate on files, electronic messages, EDI and EFT documents and make them meaningless without the decryption key.

— Cryptology is the design of secret codes.
— Cryptography is the science of secret writing.
— Cryptanalysis is the science/art of breaking codes.
— Encryption is the process of coding a file in secret form.

• Note a cipher is a means of secret communications that requires the use of a special code-book. Simple words or numbers convey pre-defined meanings: e.g. 'The Dog will bark in Autumn' means that a submarine will be in the bay at midnight. See MAC, DES, RSA, and public key encryption.

CSCW
Computer Supported Cooperative Work. See groupware.

CSDC
Circuit Switched Digital Capability. A form of high-speed telephone/data network in the USA.

CSG
Combinatory Solid Geometry. See graphics modeling.

CSMA/CA
Carrier-Sense Multiple Access with Collision Avoidance. A LANs type of which AppleTalk and Omninet are examples. The distinction is between collision detection and avoidance schemes.

An AppleTalk node waits for a break in communications on the LAN, then it waits a further 400ms plus a random period before claiming the line for its own use. This is not an OSI standard. See below also.

CSMA/CD
CSMA/Collision Detection. The IEEE 802.3 and ISO 8802.3 standard for LANs control. Examples are Ethernet and StarLAN.

A computer wishing to send a message on the LAN must listen on the cable to ensure that no other computer is sending a message. If the cable is quiescent, it sends the first 64 bytes of data message (the minimum message size possible) and then listens for a collision. If a collision is detected the computer takes a randomly generated number, and delays resending for this random period of time.

The problem with CSMA/CD systems is in the time wasted whenever collisions occur. When these systems come under load, collision numbers rise quickly to unacceptable levels.

The contention process works particularly well with keyboard input terminals where the data is delivered in short random bursts. Longer files will often create delays, and with client/server systems the likelihood of collisions increase.

CSMA/CE

CSMA/Collision Elimination. A LAN access method with pre-established priorities which was invented by NetWorth. When the computer wishes to send data, it waits for a quiescent state on the cable. But before beginning transmission, it then waits a further few microseconds (this is the DST or 'slot' time).

Each computer on the network has a unique DST, so the numerous stations waiting to transmit, won't all do it at once. Obviously, those with lower numbers have higher priority.

CSMA/DC

CSMA/Deterministic Collision. Intel's version of CSMA, which tries to have the best of both worlds. It operates as a normal Ethernet environment until a collision occurs. At this point deterministic control takes place where the entire network switches to a TDMA system; each station has a slot assignment, but Ethernet packets can fit between the allocated slots. After a time the system slips back to normal CSMA/CD.

CSO

Community Service Obligations. This is a term borrowed from social welfare and applied to telecommunications by the European Commission in 1969. It refers to those parts of a monopoly telecommunications carrier's activities that are deemed to be essential for all individuals in the community — even when the provision of that service is uneconomical (and certainly unprofitable) for the carrier.

It is generally assumed that CSOs will involve only the basic telephone service (although this is now being extended to data), and that the carrier will be compensated by the wider community in some form. This may mean that the carrier is allowed to overcharge the rest of the users, or perhaps the government provides a subsidy for remote-area users, pensioner, etc. The USA has a term 'universal service obligations' (USOs) to cover similar concepts.

CSTA

Computer Supported Telecommunications Architecture. A standard for the physical links between the telephone switch (PBX) and a LAN server which is being promoted by the ECMA and under consideration by ANSI. TSAPI on NT will support the CSTA protocol. See TAPI and TSAPI also.

CSU

Channel Service Unit, aka Customer Service Unit. This is a device which sits between a digital telephone line and the end-user's equipment. It is usually directly associated with a DSU.

The CSU interfaces with the user's multiplexers on one side, and the carrier's digital line-coding device (DSU) on the other. The main tasks of the CSU are:

— to perform equalization (line conditioning) and so maintain the signal performance,
— reshape and amplify the signal (reconditioning the pulses),
— check the bit stream for standard networking parameters,
— take care of the test signals put out by the phone company,
— perhaps perform loop-back testing.

• In ISDN, the CSU is the NT1 (Network Termination 1) unit. This is on the customer's side of the DSU and it provides a four-wire connection to the customer's equipment. See CSU/DSU and DSU.

CSU/DSU

Channel Service Unit/Digital Service Unit. The term covers a range of combined-function devices which provide an interface between the many digital circuit types from the carrier, and the customer's digital network. Normally only one device would be required here, but the USA has strict demarcation laws about who supplies CPE, so the functions needed to be split. Think of these in combination as a digital interface between the customer's digital system and the carrier's.

The 1.54Mb/s T-1 high-capacity links which are common in the US, require a CSU/DSU which costs in the $1500 range. A different type of CSU/DSU unit is also available for DDS services at about $500. These basic units offer only 56kb/s and V.32 (no RS-232) connections. Frame relay units cost more.

Many vendors are making CSU/DSUs which are SMDS and ATM-compatible because both use the same cell size, and this should bring the future price down for ATM.

The DSU is really a special high-speed line-coder which converts the terminal's (carrier's multiplexer) line-coded output into bipolar digital signals, and performs clocking and signal regeneration. The CSU then takes the data, and is responsible for line conditioning and signal reshaping.

In the new T-1 systems, the CSU also constantly monitors the 193rd bit in each frame for ESF information.

Also see DSU and CSU separately.

CT

• **Printing:** This means Continuous Tone.

• **Files:** Continuous Tone (CT) is also a file format used for exchanging high-level scan information.

CT#

Cordless Telephone Number #

— **CT0:** The undefined original FM cordless telephones for domestic use.

— **CT1:** These were the analog (one channel) cordless phones widely used in homes. There are CT1/UK and CT1/CEPT standards in Europe.

CT1/CEPT: The European version uses frequencies between 914—960MHz. It allocates channels dynamically.

CT1/UK: The British authorities (DTI) licensed 8 radio channels each 20kHz wide. The range is about 100 meters.

CT/USA In the USA the cordless frequencies are 1.7 MHz and 49 MHz. See cordless frequencies.

— **CT2:** A one-way, small-range, digital cellular telephone system originally designed by Ferranti wireless PBXs, and also for domestic cordless use. In this local mode it can both receive and originate calls. The extension of CT2 into public areas is known as Telepoint. You couldn't receive calls in the street, only make them, and the networks had no hand-off facilities for true mobile operations. Later versions have paging built-in.

CT2 uses FDMA digital communications in a ping-pong (TDD) fashion with 100kHz channels. Generally 40 channels fit into a 4MHz frequency allocation within the 864—868MHz band. About six channels are handled by each base-station.

Speech is encoded using 32-bit ADPCM. It is formed into 2ms bursts, and sent at 72kb/s over the 100kHz channel with both-way bursts interleaved into the same channel. Transmitter power is restricted to 10 milliwatts.

CT2 handsets signal via sync-bits contained within each voice/data frame. The value of the system for street use (Telepoint) is probably only realized when alphanumeric paging also provides a pseudo in-call facility, and this is being built into the newer handset models (but see CT2-Plus). True in-calling is available already with PBX operations.

- **CT2/CAI** is the official title for the later CT2 system which now sports a single common air interface (CAI). This was devised because the original four services used incompatible air interfaces, and this was the primary cause of most of the Telepoint business failures. The Canadians call this standard (with added features) CT2 Plus, Class I — for some reason!

Frequencies used in most of the world are 864—868MHz, but in Australia they are 861-865MHz.

- **CT2-Plus:** The Canadians have also developed a two-way version called CT2-Plus, Class II (known elsewhere as 'CT2-Plus') which consumes a full channel for signaling, handover, etc. It allows restricted two-way calling by permitting the user to 'register' his presence in a certain location. Canadian CT2-Plus is built for frequencies of 944—948MHz (the 4MHz version) and 944—952MHz (for the 8MHz version).

- **CT3** is sometimes used as a general term for a range of two-way microcell telephone systems designed around digital TDMA standards. However, recently Ericsson's registered this acronym in Canada for a version of their DCT-900 standard, and have therefore appropriated the term. DCT-900 is similar to DECT (but at different frequencies) and has no relationship with the CT-2 standards. See DECT, GSM, PCN, DCT-900 and CAI also.

CTE

Carrier Terminal Equipment. This makes the distinction between the multiplexers and switches that the carriers own, and the CPE (Customer Premises Equipment) that you own.

CTI

Computer Telephony Integration. This is a 25 year old dream. It implies integrated telephone and computer applications such as automated ordering systems. It also extends to screen-based telephony, where you click on a name, and the computer does the dialing. CLID extensions which tap into databases and extract information about callers are part of CTI.

Most network hosted CTI systems will run on NetWare and have servers running TSAPI. TAPI will probably be used at the PC level. See TAPI, TSAPI and NEST.

CTIA

Cellular Telephone Industry Association of the US. It is a subdivision of the TIA and it issues the IS-## cellular standards.

Ctrl

Another abbreviation for the Control key. Also Cntl.

CTS

Clear to Send. A response sent from a modem (or any DCE) to the terminal (DTE) to inform it that it can begin transmission. This is a form of hardware flow control (not handshaking) and it is always asserted in response to a request (RTS) initiated by the terminal (DTE).

In RS-232, CTS is asserted on Pin 5 as a raised voltage level of –12V on the connector. CTS will be inhibited by the modem when it is low on buffer space, and transmission must stop within a few bytes of this happening. When CTS is asserted again, the DTE can start transmitting.

- CTS is Circuit 106 (ITU) : in EIA terms it is CB. See RTS.

CUA

Common User Access. See.

cubelet

A form of chiplet which uses a 3-D stacking technology with between two and eight IC layers on a GaAs substrate.

CUG

Closed User Group. See.

CUI

Character-based User Interface. The standard MS-DOS, CP/M, etc. interface which is based on issuing commands by typed characters and selecting from basic menus by using characters. The distinction is with the Mac- and Window-like GUIs.

Curie point

The temperature at which magneto-optical materials can become magnetized. This is the point at which MO disks record in erasable/recordable optical disk drives. It is about 200° Celsius. See Kerr effect.

current

The bulk flow of electrical charge which is measured in amps (amperes). The distinction is with the electrical 'pressure' which is in volts.

current directory

The default directory name at the present time. All file operations are directed to this directory unless a path string is used.

current loop

The technique of physically interconnecting terminals where a binary 1 is represented by current on the line, and binary 0 is represented by the absence of current. This is the most basic form of digital communications.

cursive
A flowing hand-written style of typeface (font). The letters run together.

cursor
The small 'pointer' symbol on the screen which indicates where a command will be issued (over a button or icon) by clicking. Many people confuse the cursor with the insertion-point, because the cursor generates an insertion point when you click over text.

Another way to look at this is to say that there are different types of cursor, and often more than one will be on the screen at the same time: so we have mouse-cursors, and typing-cursors. See insertion point and I-beam.

• The change of the cursor shape is now used to indicate various conditions, especially with hypertext systems.

cursor-control keys
The arrow keys that cause the cursor to move. Sometimes these four keys are also coupled with other page-up, page-down, end, keys under the same common heading.

Custom Local Area Signaling System.
This is a later version of LASS which is standardized by Bellcore and has been progressively offered in US networks since the early 1985s. CLASS covers a range of technologies and protocols which extend elements of the CCS#7 signaling system over the local loop to the customer's premises — and thereby allow a range of intelligent network services to be offered by the carriers at a premium.
The main CLASS functions available are:
— Caller Line ID. This is a CMS (Call Management Services) protocol which delivers the caller's line number, time and date to a LCD while the phone rings. See CLID and ADSI.
— Call-forwarding. See hunt groups also.
— Call Waiting.
— Call-interrupts (hold one call while another is answered).
— Three-way calling.
Some more advanced versions of these services are now being offered under the 500 code system. See star codes.

customer-calling
Customer or customized calling features are those modern CLASS services such as call-waiting, call-forwarding, etc. The exchange must be computer-controlled, and therefore be ESS, ARE, AXE, SPC or a digital exchange of some type.

customer loop
The subscriber's access line. This is also called the local loop, local line, customer line, customer feed — and collectively, customer access network (or CAN) or subscriber access network (SAN), 'the last mile', and probably by about twenty other terms. It all amounts to the same thing — the wire between your home or office and the local exchange. The cost of installing this wire accounts for about 40% of the total capital cost of the telephone network.

customer premises equipment
See CPE.

cut/CUT

• Control Unit Terminal. An IBM 3270 terminal that is managed by the cluster controller. A CUT can only support one host communications session at a time.

• Cut and paste is a data transfer technique popularized in the Macintosh computers through their operating system. The cut transfers selected data to a Clipboard (a section of memory under O/S control) and the paste then reinserts it at the cursor point into another part of the same application, or into another application entirely. Once the paste has been completed the data has no connection with the previous application — there is no DDE-type hot-linking here.

cut-off frequency
The frequency above (or below) which signals are prevented from accessing the network (or a circuit). See high-pass, low-pass, or bandpass filters.

cut-through
Not buffered — allowed to power through without stopping.

Cut through mode, refers to switches which pass packets through on the fly – they just read the header and switch before the arrival of the full packet payload. This is fast packet switching, as distinct from store-and-forward switching. Most cut-through switches will revert to store-and forward when handling congested links.

CVD
Compact VideoDisk: A 12-cm videodisk standard with 20 minutes of full motion video using hybrid analog-digital encoding methods and a faster disk-rotation speed than normal. It was developed by SOCS Research, and needed a special player. Don't confuse this with the Philips CD-V.

CVF
Compressed Volume File. With DoubleSpace, a compressed drive is a 'virtual drive' which exists on the hard disk as a Compressed Volume File. This is how it appears to the uncompressed disk. It will have a name like DBLSPACE.000.

This file has Read-only, Hidden and System attributes. DoubleSpace assigns a drive letter (H:) to the CVF so that you can use it as a disk drive, and access any files it contains. Although it is a file, you treat it as if it were a separate hard disk.

CVSD
Continuous Variable Slope Delta modulation. A popular form of delta waveform modulation for voice. It is a type of delta modulation (see) which reduces the bandwidth requirement, but unfortunately introduces a lot of background noise. It encodes the 'differences' in magnitude of adjacent samples of the audio wave — and so it reduces the high degree of redundancy in sampling speech wave-forms. In telephone systems it

gives good results with voice at 32kb/s, and adequate results at 16kb/s, but it is not suitable for modems.

CVSD is also called Adaptive Delta Modulation or ADM. It is similar to ADPCM in that it codes differences not sample size. See delta.

CVT

Constant Voltage Transformers. See.

CW

Continuous wave as distinct from pulsed.

- **Historical:** The term will be used by any radio amateur over 50 to mean Morse Code transmissions. The US Coast Guard officially abandoned CW transmission on 1 April 1995.
- **Satellites:** A signal at a specific frequency that does not vary.
- **Medical:** Studies of the biological effects of radio waves take into consideration whether the radio waves are continuous waves (CW) or pulsed waves, and whether the CW radio signals were amplitude (AM) or frequency modulated (FM).

The current state of knowledge suggests that radio waves have the maximum biological effects when they are pulsed or AM, while very few effects are noticed with FM.

- **Laser:** You can operate many lasers in either continuous-wave (CW) or pulsed fashion. CW lasers put out a constant stream of photons and are rated in watts, milliwatts, or microwatts. Pulsed lasers are rated in joules, millijoules, or microjoules. See pulsed lasers.

cyan

This is a light blue colour (with a touch of green). It is the complement of red. CMY (Cyan, Magenta, Yellow) are the three subtractive primary colours, while RGB (Red, Green, Blue) are the three additive colours.

cybernetics

The study of control and communications in complex systems. The term has now widened to mean the associated areas of technological forecasting and assessment, policy analysis, AI and expert systems, industrial control, robotics and complex system modeling.

cyberpunk

This is a fashionable 'nerd' culture, with slightly anarchistic overtones — and with enough potential for 'naughtiness' to make teenage 'hackers' feel as if they are rebelling against authority. The concept was created in stories by sci-fi writers William Gibson and Bruce Sterling. It is currently a high-technology fashion based on virtual-reality 'experiences', video games, and Internet exchanges of trivia, raised to the level of a street culture.

cyberpunx

An art movement established around computer graphics, and more recently, using virtual reality systems.

cyberspace

Electronic or virtual space. There's both a general, and a more specific use of the term. Cyberspace was invented by science fiction writer William Gibson in his 1984 novel 'Neuromancer'.

- The general use of cyberspace is for any 'conceptual' area under control of computers (or by extension, electronics). So, a hacker, in this context, exists in cyberspace when he/she hacks into a company database, or when he/she communicates with someone by electronic mail, or purchases something through an electronic mall. Alexander Graham Bell supposedly invented cyberspace when he asked Watson to 'come here'. The key concept here is 'electronic networking'.
- More specifically, cyberspace means the conceptual three-dimensional multi-sensory environment into which you enter when using a virtual reality system. Within this environment you may be able to navigate, perceive, identify, evaluate, choose, modify and share information.

The rules of cyberspace do not necessarily follow those of normal reality— they are set by the programmer: for instance, water may flow up-hill, not down, and solid objects may metamorphose into new shapes whenever they are outside the field of view. The key concept here is 'perception'.

- Cyberspace has now come to mean (to the romantics), an 'amorphous, borderless world' — a new realm — an intellectual frontier. Time magazine used the term on its cover in this way and in so doing glamorized the concept of a *SuperHighway* which is probably paved with yellow-bricks of gold.
- Don't confuse cyberspace with hyperspace which is the illusory 'navigation' environment when you are within a hypermedia database. Hyperspace is an 'intellectual' environment while cyberspace is a perceptual one.

Cyc

A project of a US research consortium which has been financed by major computer makers to devise a knowledge-base of common sense. The idea is for future computer systems to use this common-sense database in connection with expert systems and other applications to avoid making those naive errors that often make computers appear to be stupid.

Cyc will know that 'cats sit on mats', and not vice versa and that a pig pen is something quite distinct from a drawing instrument. It will have a whole range of the basic concepts that we humans take for granted.

It is expected that eventually the Cyc base will have tens of millions of items, and know large numbers of 'micro-theories' which enable it to deal with differences and contradictions in its knowledge.

Cyc also uses the concept of 'inheritance'; dealing with similarity between properties of object classes. It is being programmed in a version of Prolog. A PC version should be ready to roll out onto a single gigaByte hard-disk in 1996.

cycle

• In analog waves this is one complete revolution of an action which has returned once again to its starting point.

• In discussions on video resolution, a cycle is a combination of a black and a white line. This is an attempt to overcome the confusions arising from the ambiguous use of the term 'line' in dealing with television resolution.

When you draw black parallel lines on a white card, different people will say that there are x black lines, or x white lines, or 2x 'lines' (white and black combined). Along the horizontal, TV engineers say there are 2x lines, but in the vertical they only count the scan line (x). But there will only ever be x 'cycles' of horizontal resolution.

In calculating vertical resolution, the problem is much more complex since scan lines and Kell factors enter into the equation also. The actual vertical resolution depends on a form of sampling artefact — the image is analog, but in the vertical plane the scan lines are integers (digital) and the Kell factor provides what amounts to a form of quantization. See image resolution.

cycles per second

The old measure of frequency. We now call these Hertz.

cyclic code

This is gray binary code.

cyclical redundancy check

CRC. The name is applied to both the error checking algorithms (not correction), and the result. The CRC is a special form of multiple modulo remainder which is transmitted as an attachment to the frame or packet (as a checksum). It is also known as polynomial code.

As each block of data is received, the attached CRC value is checked against one calculated by the receiver. Any mismatch will trigger a 'request for retransmission' (NAK) of that block.

The CRC approach is much more powerful than LRC in that it detects more errors. It is called a 'cyclic' check because it effectively uses every bit in the frame several times in computing the checksum character (LRC only uses each bit once).

To do this, the bits in the frame are shifted in a register one place at a time, and a recalculation made using a polynomial calculation.— hence the cyclic tag. There are a couple of variations on this.

The CRC frame detects the following errors.

— All single-bit errors.

— Almost all double-bit errors.

— Any odd number of errors.

— Any error when length of the burst is less than the FCS.

— Most errors with larger bursts.

The checksum is often known as the Frame Check Sequence (FCS) or Block Check Character (which is usually 16-bits long). V.41 is the CCITT's recommended CRC technique.

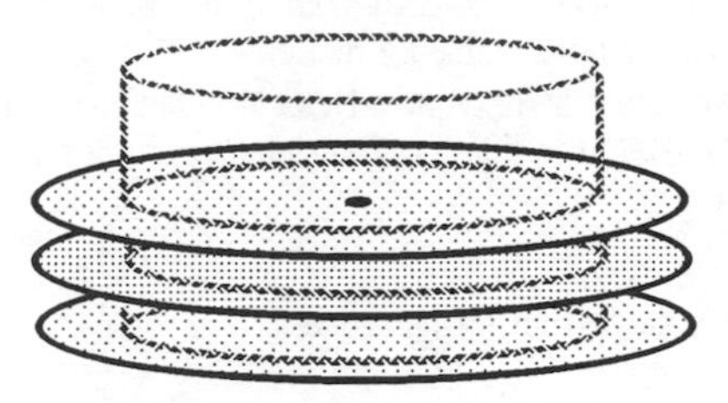

The theoretical cylinder through a multiple platter disk system.

cylinder

Cylinders are specified tracks on the various platters in a multi-plate, multi-layer, hard-disk unit. A cylinder is collectively the set of tracks which are at the same distance from the center of the disk on each plate, and each side of the plate.

The term is used with disk stacks where the multiple access arms move together across the surface of multiple platters. Rather than write a sequence of blocks of data to only one disk at a time, these systems write blocks to each of the disk surfaces without the access arms moving. These blocks are then recorded on a cylinder — and so the directory reference will be to the cylinder, rather than to a track. See disk.

• In Very High Density (VHD) floppy disks, a cylinder is a pair of tracks — one a servo track buried deep in the disk media, and the other a data track on the surface. This technology seems to have disappeared recently.

• The original Edison phonograph was a cylinder recorder. Unfortunately it proved to be impossible to mass-manufacture the cylinders, so this system was replaced by flat disks. Note that a cylinder is both constant angular velocity, and constant linear velocity.

cylinder density

In hard disk drives with multiple platters, any drive which has more platters (and heads), and more sectors on each track (probably through the use of RLL encoding), will have a greater cylinder density. It can access more data in less time.

cylinder write

In multi-platter disks, the computer will begin writing onto the track at the top of the cylinder, then switch instantly to the next platter in line, then to the next, and so on until the full cylinder has been written, without needing any physical movement of the heads.

Cyrillic languages

The written languages of central Europe (Russian and Bulgarian) which use special characters.

D

D (conditioning)

See conditioning.

D#

With T-1 transmission lines, D1, D1D, D2, D3, and D4, are all various stages of development in channel banks and framing formats. They all run at 1.54Mb/s.

D1 and D2 provided a framing format which could only deliver 56kb/s, even though the channels were 64kb/s wide; they always used AMI line coding. D3 and D4 later moved to better utilization of the frame and could provide 64kb/s channels using B8ZS line codes.

– D4 is probably still the most common T-1 framing format, but ESF is rapidly overtaking it. D4 helps the terminal locate payload channels, and it has a basic ability to signal the far end when problems arise, but ESF can send messages end-to-end and provide a level of line-quality reporting. See channel bank.

D-1

- **Video**: A professional component digital videotape recorder standard which conforms to CCIR 601 for 4:2:2 equipment and so is often called '4-2-2 video'. It is a YUV component system devised by Sony and it provides 10-bit transparent recording without bit-rate reduction.

D-1 is still classed as the IEC/ISO's international format for program origination and exchange in digital format. Because it is both digital and uncompressed, it can withstand multiple generation, editing, and general signal processing with no appreciable loss of quality.

This is the industry's benchmark video standard but it is too expensive to implement widely. The later D-2 digital composite machines from Ampex are more popular, and even later formats are making inroads into its traditional market. However D-1 is still the standard against which all are judged.

It is a helical scan system, and there are three cassette sizes: 13, 41 and 94 minutes, using gamma ferric oxide tape, 19mm wide. At the time the standard was defined (1984), metal particle tape wasn't available.

There are four digital audio tracks, and the video bit-rate is extremely high at 216Mb/s. Consequently there is high tape consumption which limits the applications because of cost.

D-1 wasn't designed for high-definition TV; however it has sufficient bandwidth to handle Extended PAL.

- **Multiplexers:** D1 was also the term applied to the early 7-bit channel-banks (multiplexers) used on the first US digital telephone link (T1) in 1962. The 8-bit version (1969) of these channel-banks is called the D2. See D# and channel bank.
- **Computers:** D-1 is also a digital tape recorder used by mainframes; this is actually the same machine as the D-1 recorder for digital video. An industry standard D-1 cassette using the ANSI ID-1 format, can record about 100GBs of data.

D-2

- **Video:** A digital videotape composite standard which is the equivalent of the current professional composite analog recorders, and is largely plug-compatible. It was developed between 1986 and 1988.

Ampex initially released details of a D-2 composite digital videotape recorder originally as a 'spoiler' to counter Sony's development of the CCIR 601 D-1 component standard, and they successfully threw the proposed world standard for digital production equipment into confusion. Eventually Ampex and Sony signed a peace treaty and co-developed D-2 as a new composite digital standard.

D-2 has since become widely used around the world, and it is likely that it will eventually completely replace 1-inch Type C equipment (if it is not overtaken by the newer formats). Unfortunately, D-2 has not proved to be a good specification for camcorders.

D-2 used the then-new metal-particle tape and the equipment was much cheaper than D-1. The tape is 19-mm wide and handles its composite images using the YIQ approach (Y = luminance, I = R–Y in-phase, Q = B–Y quadrature 90° out of phase). It has a very good Reed-Solomon error correction system.

- **Multiplexers:** D2 channel banks are telephone multiplexers. See D# and channel banks.
- **Computers:** The D-2 composite video format recorder is also used for computer recording. See D-1.

D2B

See Domestic Digital Bus.

D2-MAC

See D-MAC next page.

D-3

Panasonic's digital television camera/editing standard used at the Barcelona Olympic Games in 1992. This is a composite digital system with a reasonably low cost and relatively high quality. It was originally called the Dx, but it is actually a member of the M-I and M-II family, being built around the M-II tape transport mechanism.

This was designed to NHK specifications; it uses half-inch tape and competes directly with the D-2 video standard and with Betacam SP. It can be made smaller and lighter than D-2 and so is used as a camcorder.

The D-3 standard has a very effective error correction system (Reed-Solomon) and encodes the signal to the same audio and video parameters as D-2. It uses an improved metal particle tape and has multi-laminated heads. It samples the video at 143MHz and the information is stored in a YIQ format.

With PAL systems there are 8 tracks to each field, and with NTSC there are 6. The tape moves at 83.88mm/s (with a head-to-tape forward speed of 21.4cm/s) and the cassettes can hold up to 64 minutes, or 125 minutes in long-play.

D-3's limitations are in the use of 8-bit sampling (with EFM encoding) where the industry now wants true 10-bits, and also in the fact that it can't provide enough bandwidth for widescreen. See D-5.

D-5

A very high quality professional digital video standard from Panasonic (in 1993) which does not use compression. D-5 is a component system which has D-3 (composite) playback capability. It uses EFM channel-coding which turns an 8-bit sample number into 14-bits. The coding provides a lower error rate and less mistracking sensitivity.

D-5 is classed as a 10-bit recording system. The bit-rate requirements of the component D-5 are more than twice that of the composite D-3; it outputs a 270Mb/s data rate in a 4:2:2 component form. There is now an upgrade module which applies 4:1 compression for HDTV. See D-1 also.

- Note: there's no D-4: four is an unlucky number in Asia.

D-6

A professional studio HDTV 16:9 digital recorder standard just introduced by Philip's subsidiary BTS. It can record 64 minutes of HDTV, which is the equivalent of about 360,000 3.5-inch floppies. It has up to 12 channels of audio or data.

D-9

A type of plug used in linking computers to peripherals. It is a smaller version of the DB-25. D-9P is the plug; D-9S the socket.

D-channel

- Data channel: In ISDN telephony the D-channel is the local loop component of the packet-switched signaling network (as distinct from the B or 'Bearer channels').

The D-channel carries the signaling information between the subscriber's terminal and the local exchange, and it can also transport 'User Data' and telemetry. In BRA the D-channel operates at 16kb/s, and in PRA it is 64kb/s.

It is a datagram-based, packet-switching network connection which uses a variation on LAP-B (called LAP-D) protocols for error control. The various D-channel signals are converted at the exchange into Common Channel Signaling System #7 signals for inter-exchange communications.

- D-# channel banks are multiplexers. See channel banks.

D-MAC

D-MAC was a popular variation on the C-MAC satellite analog component TV system. In all MAC systems the video is the same, and only the treatment of audio and data changes.

- D-MAC was developed by the British in opposition to the European proposal for D2-MAC — but later the two were integrated into common chip-sets.

D-MAC had less flexible stereo channels, but at twice the voice and data-rate of D2-MAC. The European system which was specifically designed for narrowband (7—8MHz) cable systems, and to allow dynamic changes to be made to audio channelization. Both systems provided 8-channels of FM data, which allowed audio signals to be uplinked from separate locations — which was an important consideration in Europe with its different language regions.

- D-MAC was briefly used to provide Extended Definition TV (16:9) for BSB before it merged into BSkyB which uses PAL. D2-MAC was proposed for European HDTV (HD-MAC) before MPEG came along. MAC TV systems were overtaken by digital.

D region

The layer of the ionosphere at about 80 kilometers which attenuates HF radio signals in the daytime. It rapidly subsides at night to the point where it is hardly present after sunset.

D-shell connectors

Types of plugs and sockets used in serial computer communications. The 'D' refers to the shape of the connector (it's a very flat D). DB refers to the standard as a whole (followed by the number of pins) while DP is for the plug (male) and DS is for the socket (female). Note that the terms 'plug' and 'male' apply to the connectors (the pins) themselves, not to the outside shielding.

- RS-232 COM ports are usually DP-25 (aka DB-25P) on the back of a PC, but DB-9 is now also being used. On an AT, the COM ports use a DP-9 (aka DB-9P).

D-VHS

Digital VHS tape system. See DVHS at back of section.

DA

Desk Accessory. See.

DAB

Digital Audio Broadcasting. This applies to both terrestrial and satellite systems of broadcasting. The UK's first service is already working in London with five transmitters, each with 6 channels.

The European DAB is a development of the NICAM system used with PAL television. It achieves a drastic reduction in bit-rate by Musicam masking (eliminating sounds that are blanketed-out by other louder sounds), and is transmitted by a multiple channel multiplexing technique known as COFDM.

COFDM has a slow bit-rate on each channel and this means that the signal can be successfully received by mobiles. These would otherwise suffer from the multipath effects.

The current aim is to interleave at least 16 stereo CD-quality radio program channels together into 7MHz of the radio spectrum. About 4kb/s of LCD display data will be included with each channel. Musicam provides about 10:1 compression.

Since the signal is digital and COFDM modulated, the mobile can simultaneously accept signals from several transmitters, and handle multipath reflected signals as well.

Conventional tuners won't be used; the full wideband is demodulated and the required program is then extracted from the appropriate data stream by computer calculations. Simultaneous direct satellite and terrestrial broadcasting is possible on the same frequency band because of the robust and selective nature of the COFDM process.

The BBC and some US companies are planning to use the Eureka-147 digital radio protocol. This also has the backing of Canada, Mexico, Australia, Germany and generally other European countries — however, not all plan to use the same frequencies. The UK has set aside 217.5—230MHz in VHF, France is looking at the L-band (1500MHz) and German is looking at another band. See IBOC also.

DAC

- **Digital-to-Analog Converters:** The reverse of ADC. See digital to analog conversion.
- **Data Acquisition and Control:** This is where a central computer collects data from, and transmits data to, remote terminals. See GPIB also.

DACS

Digital Access and Cross-connect System. DACS are used in mesh network applications which require switching, grooming and routing capabilities.

DAD

Digital Audio Disk. An old abbreviation. See CD-DA.

daemons

Daemon means soul or spirit. These are sub-programs within programs that are triggered when certain values are found, or when a logical condition is generated.

They are small background processes which exist in a modular form so they can be extracted and replaced. Some of them act almost like independent agents.

One or more daemons may run on a Unix host and serve requests in queue order. Agents and daemons are similar; they say that 'an agent is not a process, but a set of executable machine instructions (pseudo) that overlay the daemon'.

Daemons are used, for instance, to control file locking on a network server and to process and order bulk electronic mail in mail servers. When a process needs the facilities of a daemon, it sends data to it. The daemon intercepts the data, actuates, performs its task, and then waits for the next event.

— In networking, a daemon can be system administration software that establishes lists, builds tables, and passes routing information to gateways. Examples of Internet daemons are finger, telnet and ftp. See.

— In software piracy, a daemon can be used to break protection schemes. It will intercept all disk requests from the protected program, and use this information to later bypass the software locks.

— In AI, a daemon is a co-routine.

daisy chain

A sequential link from one node to the next.

- Physical cable links: SCSI devices can 'daisy chain' from one to the other, but each device in the chain will need to have a distinguishing 'address' number (from 0 to 7).
- Processing functions: Technique for sequencing interrupts.

daisy-wheel printers

These were very popular a few years ago, but mercifully (for our ears) they have disappeared. The type characters were molded onto a springy-steel or plastic disk that looks like the flower head of a daisy. This disk rotated at high speed, and the particular petal with the required character, was hammered against a ribbon and onto the paper. Daisy-wheel printers were very noisy, but they produced quite high quality print for the time. The daisy-wheel disks could also be easily changed for different fonts or styles.

DAL

Data Access Language. Apple's new name for CL/1, which is a client/server protocol for remote data access. It is based on ANSI-standard SQL with some increased functionality.

DAL is actually a communications tool or connectivity language, which aims at providing painless global access to databases. It used a super-set of SQL-based commands, but extended its usefulness into hierarchical and flat-file databases.

DAL provided a uniform, simple way to access data on a mainframe from within any PC application, provided the mainframe database application has been DAL-enhanced, and it insulated both software developers and users from the host's operating system, database system or networks. See SQL.

DAMA

Demand Assigned Multiple Access. A satellite telecommunications technique that allows occasional-used, earth stations to be allocated channel space on a transponder only when needed. DAMA is widely used in VSATs.

When a remote VSAT transmitter receives data it sends a request to the hub specifying the length and number of packets it has for transmission. The hub then dynamically assigns TDMA time-slots in the incoming signal ('inroute'). It is more economical to assign communication channels only as they are needed.

In a satellite DAMA telephony system for under-developed countries, each remote outpost will have a satellite terminal (a weather-proof box of equipment and an antenna aimed at a GEO satellite).

When someone wants to talk to the outside world, this terminal sends a signal through a dedicated satellite signaling channel to the central control. The center then assigns a pair of one-way communications channels (which can be discrete RF carriers or time-slots) from the channel pool, and it then automatically instructs both the villager's terminal and the destination terminal to use these channels. Terminals listen constantly to the dedicated signaling channel, and each responds only to addressed instructions.

DAMA can also pool the available satellite power, and assign it to the terminals needing to overcome rain attenuation. Some demand-assigned systems automatically lengthened the bit-periods in those attenuated communications links, at one or both ends of the link. This maintains the required Eb/No for the demodulators and decoders to deliver acceptably low bit-error rate without wasting satellite capacity on the other links.

DAML

Digital AML. See AML

DAMPS

Digital AMPS. See NADC.

DAMQAM

Dynamically Adaptive Multi-carrier Quadrature Amplitude Modulation. A modulation system used by the Telebit/Trail-blazer high-speed modems. The technique depends on the intelligent assessment of the signal quality of a line by checking 512 separate carrier frequencies. 2-bit DPSK, 4-bit QAM or 6-bit QAM modulation schemes are then assigned to each usable frequency depending on this evaluation. This is the 'dynamic adaptive' aspect which is similar in principle to COFDM. See PEP and adaptive duplex.

dark fiber

This is a reference to fiber in cables which have been installed but which are not in use: quite probably the lasers or repeaters haven't even been installed. With the cost of the glass fibers themselves being such a small part of the total laid cable cost, it often pays telephone companies to put in place, say, 18 fiber pairs even when only one or two are needed. The remainder are left as dark fiber.

• The main question about dark fiber, is whether a large communications user should be able to lease dark fiber from the carrier, or whether it should only be permitted to lease capacity. The carriers prefer the capacity option.

• In 1990, the American FCC ruled that phone companies must offer dark fiber to all comers under the common carriage rules. Rather than accept this, the companies petitioned to withdraw from the business entirely.

DARPA

Defense Advanced Research Projects Agency (of the US Department of Defense). See ARPA.

DAS

• **Digital Acquisition Systems:** A general term for chips which perform A-to-D and D-to-A functions, together with some analog and digital circuit functions. These chips are generally used in monitoring devices, and most of the parameters they monitor will be analog in nature. The term is mainly used for computers and special equipment for logging data in scientific and technical research. See GPIB and SCPI.

• **Dual-Attached Station:** An FDDI term for devices (concentrators, bridges and routers) which attach to both of the counter-rotating fiber rings. This provides fault-tolerance. The distinction is with SAS (Single-Attached Station) which links to only one of the rings. However SAS is used for non-critical devices (workstations, etc.), and the attachment is less expensive than for DAS devices.

DASD

Direct Access Storage Devices. These are special hard-disk units which can be written to and/or read from, without the need for main CPU control. The emphasis is on random-access capability as distinct from the sequential-access of a magnetic tape system.

The ability to direct read from the disk at a high rate is the result of channel-extension support. IBM had a channel convention for this reason. Large DASD 'farms' (areas with many drives) are now very common in mainframe installations. Traditionally, DASD storage has been highly reliable, yet a failure can bring down the whole storage system. For this reason, RAID is becoming more popular.

dash

A dash — is not the same as a hyphen -, and the computer will see these as two totally different characters. In professional printing and computers the dash can be further subdivided into an en-dash – which is as long as a letter 'n', and an em-dash — which is as long as the 'm'.

In some word processors, you should use two hyphens '--' to indicate an em-dash, and this has long been the practice on typewriters where there was no dash character.

• Neither the dashes or the hyphen are related to the underline character, although all of these may be generated by the same key.

DASS

• **Direct Access Secondary Storage:** A term that is being applied to optical recording disks. The magnetic hard disk unit is seen as the 'working' or primary storage, and sequential tape systems have generally been used in the past for secondary back-up storage. But tape requires a slow sequential search to find data, so recordable optical disk units are becoming popular as a direct-access alternative.

• **Digital Access Signaling System:** A signaling standard used in the UK between PBXs and the public exchange. See DPNSS also.

DAT

Digital Audio Tape. A technology which uses the small (4mm) magnetic tape cassettes to store CD-quality digital audio. There were originally two proposals:

— **S-DAT:** Stationary DAT which uses a number of fixed heads and multiple linear tracks on the tape. This briefly became a professional recording standard, and a form of S-DAT now forms the basis of the Philips DCC system.

— **R-DAT:** Rotary DAT which uses a small helical-scan drum system like a VCR. This was designed as a high-end consumer standard, but the quality proved to be so good that it has been widely used professionally. Data versions are also popular.

For sound, the samples are taken at 48kHz with stereo 16-bit coding, giving a primary data-rate of 1.563Mb/s. When subcode and error-checking is added, this rises to 2.77Mb/s. Because the data is recorded in bursts, it must be time-compressed (the heads aren't constantly in contact with the tape), so the actual transfer speed of the data during record and playback is 7.5Mb/s. A two hour tape can store 1.3GBytes of information, the equivalent of over 1,000 floppy disks.

There are six DAT modes:

Mode I. The mandatory 48kHz sampling rate with 16-bit linear quantization.

Mode II. As above, but with 32kHz sampling frequency for satellite digital stereo broadcasting.

Mode III. A half-speed mode.

Mode IV. A mode with four channel recording.

Mode V. A mode used for playing pre-recorded tapes with the sample frequency at 44.1kHz (the same as the CD format). Mode V is mandatory in all DAT audio recorders. In the original specifications, this is only for playback to prevent CD copying, but this restriction has now been lifted.

Mode VI. An alternative pre-recorded tape format mode, with a wider track and faster tape-speed for even better quality. With barium ferrite tape, it is possible to physically duplicate these tapes using a contact process from a master 'mirror-image' metal tape.

• There are a number of systems in development for storing still images on DAT tape, and we now have both Digital Data Storage (DDS) and DataDAT logical file formats for computer data recording. See Digital Data Storage and Philips DCC also.

data

Anything with meaning on an electronic communications system — although often voice and video are excluded from the definition. The term is now most often applied only to digital forms of communications.

Unless you are a retired English schoolmaster with a passion for pedantry, data is both a singular and a plural term.

• Distinguish between data (symbols, or information in the form of a machine code) and the 'information' itself. Information is collated data in a form that humans can use. When you compress video, audio or text in a lossless way, you reduce the data, but retain the information.

data acquisition (capture)

• A general term which means the selection and entry of any type of data into a computer system.

• More specifically, it means the use of computers for instrumentation and monitoring. This area of the industry deals with connection to scientific or engineering equipment.

data casting

In the UK, television stations often transmit large amounts of general information in the vertical blanking interval of the TV signal (along with teletext). This is a way of distributing financial and stock-market information, for instance, and some supermarket chains use it as a way of updating price lists.

data circuit

The term is applied to both analog (with modems) and digital systems.

data circuit terminating equipment

See DCE.

data communications equipment

See DCE.

data compression

The techniques used to compress data fall into two categories:
— lossless systems (Huffman and LZW), and
— variable loss or 'lossy' systems (JPEG and MPEG, etc.).
See compression and redundancy also.

data dictionary

• The definition of a distributed database's architecture (field types, lengths, etc.) is kept in a central data dictionary. The client's workstation then doesn't need to know these details. The information will be held as metadata (data descriptions) and a single change in the dictionary will ripple through all database files in which these descriptions have been used.

Some data dictionaries include validity parameters, formatting information, form views, and indexing information also. This is a 'database' of the database's definitions. See distributed database.

• In CASE applications, see data repository.

Data Discman

Sony's 3 inch (8cm) 200MByte CD-ROM unit. It has a 15 character, 10 line LCD monitor screen which folds down on the keyboard. The disk is the same as the music disk format, and the player can play audio.

data elements

In EDI and general computer science, these are fields that comprise a data segment, e.g. the description of an item, a quantity, a unit of measure.

Data Encryption Standard

See DES.

data integrity

• Trustworthiness of the data. Error free.

• The term is used to stress the value of computerized record-keeping in simplifying the job of updating a database. If you change the record in one database to include a new address for a client, the system should be capable of ensuring that all address records (if more than one exists) are also changed. See file and record locking.

data-link

Data-links are bidirectional transmission paths with two data channels operating in opposite directions at the same rate.

• In telecommunications terminology, the data-link involves the total physical resources of the data exchange — the cables, lines, switching, and terminal equipment.

In telecommunications the protocols which handle a data link have two functions: a) to establish and terminate a connection, and b) to ensure message integrity during the passage of the data.

• On PC-based networks these functions are performed by firmware on the plug-in adaptor cards.

data-link control

DLC. Used in its widest sense in telecommunications, virtually all protocols are data-link controls. This is a generic term meaning a uniform discipline imposed for the transmission of data over a single link. So there are:

— Byte-control protocols. See BCP.
— Character-count protocols. See CCP.
— Bit-oriented protocols. See BOP.

There are many standardized complex protocols that incorporate these functions.

DLC protocols do such things as check frames (bit-groups) to see if individual bits have been lost, and they control the establishment and termination of the link. At the lowest layer these error checks can be performed by parity, but usually there are also higher-level data-link controls.

• DLCs follow well-established steps in managing channels:

— They establish the link.
— They perform 'handshaking' exchanges with the other station to ensure that both are ready to exchange data.
— They check data during the exchange.
— They send acknowledgment or retransmission requests.
— They terminate the link.

• Commonly used protocols are High-level Data Link Control (HDLC), Advanced Data Communications Control Procedures (ADCCP) and Bisync.

Data link layer

The second layer in the OSI's 7-layer model. It defines access and low-level network maintenance protocols — how electrical signals enter and leave the network cable.

According to the ISO this layer 'provides the transfer of data between directly connected systems, and detects any errors in the transfer.' It addresses (along with the Physical layer) the standard LAN types — CSMA/CD, Token Ring and Token Bus.

The Data link layer also defines the 'conversational' rules that are to be followed — who talks when, and when to remain listening. This layer deals exclusively with elements like bit-patterns, encoding methods and tokens. It is also the layer where data bits are grouped into 'frames', and where errors are detected and corrected by retransmission requests. Packet addressing, sequence numbering, and flow control is also performed at this layer, if required.

Controls at the Data link level are maintained through a variety of protocols which:

— set the signal level to be used by the network,
— define how data is to be handled,
— match speed in disparate ports
— determine the sequence of packets in a message.

• The Data link layer is divided into both upper and lower sections:

— **The Logic Link Control** (802.2): This upper section is common to all available physical media in the 802.x series. It deals with sending and receiving the data messages or packets, and it was designed to allow multiple stations to communicate over many links while using a single physical network. See LLC.
— **The Media Access Control:** The lower section, adjacent to the Physical layer, which is specified separately for each media type (twisted-pair, optical fiber, etc.). The MAC manages such things as token passing in a Token Ring system, or collision sensing in a CSMA system. See MAC.

data mining

This is the process of extracting information from a mass of data, shapes and patterns of data that we can recognize and use. This is the process of taking 'data' and using it to create 'information' in the human sense.

• The use is often judgemental; it suggests that if you dig through masses of data for long enough, you can always extract some samples that appear significant. There will always be a run of results which vary from average in a way that can be presented as 'significant' — and you can then use these to support or create an erroneous hypothesis.

data-mode

See command-mode.

data modem

The addition of 'data' in front of modem, is just to make a distinction between the stock-standard type of modem and fax modems (which are usually data modems as well).

Data modems come in three forms: a) built into the device as an integral part, b) plug-in cards (internal modems), or c) stand-alone units which plug into a port (external modems). Fax modems will tend to be plug-in cards.

data-over-voice

DOV. There are two different systems here:

• After digital data has been fed to a modem, it can be multiplexed and interleaved with digitized low-bit-rate speech (say, 4.8kb/s). This obviously requires a special modem at each end of the link and also special vocoders.

• Modemized data can also be multiplexed with normal speech over telephone circuits, carrying the data at frequencies above the 4kHz limit of analog telephony. This allows voice and low-speed data simultaneously within the same band — but only on analog circuits, and then only as far as the local exchange, or through a private network. Audio bandpass filters separate the two signals.

A new technique called Digital DOV appears to be capable of providing 19.2 kb/s data rates on a standard telephone circuit. These Digital DOV units are about the size of a standard modem. ISDN and ADSL actually offers much higher potential here. See PRC also.

data-pipes

See channel (processor I/O).

data processing

This is generally treated as synonymous for computing, but there is a subtle difference. Data processing originally referred to the use of computers for their ability to store and sort text information; at the time this was quite revolutionary. Until then computing was seen as mathematical operations in the sense of calculating or number crunching. Data processing now also involves image and sound processing.

data rate

A communications channel has a certain bandwidth (often expressed in Hz) and with a particular type of modulation and line coding; this will provide a certain maximum data rate (expressed in bits per second). However, this is not the throughput. To calculate useful throughput:

— From the data rate you must extract any overheads. These are redundant bit/bytes involved in addressing, error checking, supervision, etc.

— Allowance must then be made for any pre-compression of the data. So after decompression. the actual throughput may be much higher than the link rate.

data repository

A term for a library of reusable modules (algorithms) for programming. These are also called data dictionaries. The term is often applied to CASE products which are used for the 'definition, collection, manipulation and control of application-development information'.

Recently there has been a move towards the standardization of CASE repositories. IBM now has one called Repository Manager/MVS, and DEC has one called CDD/Plus. The DEC repository may become the ANSI standard (still under consideration) called ATIS. See data dictionary.

data segments

• In general communications, data segments are those parts of a packet which contain the message information — as distinct from the header (address) and the error checking and control information. The better term here would be packet payload (as distinct from packet overhead).

• In EDI, data segments are related strings of data, for example, an invoicing address.

data set

• An old name for a modem — hence the DSR pin on RS-232 which means 'Data Set Ready'.

• In a database, a data set is an index, or any logically-related collection of items which are organized in some set manner.

• In OS/360 on the IBM mainframes, a data set is a special file known by a special type of file-name.

data sink

In communications, the data sink is the end-destination of the message — the receiver or consumer of the information. It is where the signals finally disappear from the system. Data 'source' is the originator.

data source

See data sink.

data structure

The organization of the data in a file. A record in a database file is a data structure — it is organized into fields. The fields are also individual data-structures and have attributes of data-type and length. The FILO queue used in a stack and other structures are used in programming:

— **Scalar:** A single element data structure

— **Array:** A structure that holds many items of the same type, identified by one or more subscripts.

— **Record:** A structure that holds data of many different types. It may have names, numbers, and logical (Y/N) operators in its various fields.

— **Linked lists:** These are arrays where it is possible to insert items in the middle of the list, without needing to move data to make room. These use a location look-up table, rather than storing the items consecutively.

• Sometimes data structures are shared between users and sometimes between various applications, and it is then very important that mechanisms exist to maintain the data integrity. This is usually done through semaphores, the simplest of which is the lock. See record locking and file locking.

Data Transfer layer

In SNA this is the least defined of all the layers. It includes the CICS/IMS transaction processor (usually with database management functions), access methods VTAM or TCAM at the mainframe end, and emulation software at the PC end of the link. Both ends also have the 3270 screen format. See SNA.

data transparency

When flags are used to delimit frames, the pattern of bits in the flag byte must not be repeated accidentally in the data being carried as payload, so bit stuffing techniques are used. In HDLC, an extra 0 is always added after a sequence of five 1s in any payload data.

The HDLC flag sequence is 0111 1110. So the receiver examines the incoming payload data and, if after five 1s it detects a 0 in sixth place, then it will delete it. However if the sixth bit is a 1, it will look at the seventh. If this is a 0, the sequence is accepted as a flag.

This technique means that the data field can carry arbitrary bit patterns of any sort, and this is known as data transparency.

- Note, if the seventh bit is a 1, this is a rule violation — a signal to abort the transmission.

data type

Refers to the categorization of data in databases and in programming languages. Data structures usually need data of a certain type. The basic types of data are:
— integers,
— real numbers (with decimal places),
— alphanumeric strings,
— Boolean values.

In many databases, date–time is a data type, as are dollar values, etc. In some graphics programs two-dimensional (X/Y) coordinates will be a data type.

- In a programming language, data types are sometimes 'extensible' so you create your own to extend the capabilities.

- The specification of data types in database fields, is an important way to reduce operator errors. A name field should be restricted to alpha characters only (with no numbers permitted), while a zipcode data field is classed as an alphanumerics field, (some countries use alpha characters).

- In object-oriented languages the data and the function are combined into a single unit which is the object. The data is encapsulated (hidden).

data warehousing

One of the new buzzwords in enterprise computing. It refers to the use of large databases for decision-support. The data base is optimized for queries rather than for transactions.

database

Aka databank. This is a related group of data files which are themselves groups of records — not the program that manages (or manipulates) that data, which is the database management program. While the term is used very loosely today, it is still best to distinguish between the database and the database management system.

A database can be a single entity or it can be a number of distinct files which are linked by a relationship. Some people differentiate between flat-file systems and the more complex 'databases' (meaning relational databases) — but this often seems unnecessary. See DBMS, RDBMS and ODBMS.

- A flat file database is a two-dimensional table or array. It can be thought of as having 'rows' (records) and columns (fields), but no relationship with other files.

database management system

See DBMS.

database server

On a network, a database server is a more sophisticated technology than a file server; it is the key hardware for client/server operations so it may, in fact, be a mainframe. It must be able to deal with data-tables at the level of 'rows' (records, not just files) which greatly reduces the network traffic. Queries can then be optimized, and traffic over the network is reduced to a minimum.

The front-end (the client/PC) has the responsibility only of displaying the data and interacting with the user, while the back-end (the server) handles most of the processor intensive work — mainly data storage/retrieval and manipulation.

However database servers need to control all elements of the transaction and are responsible for preserving data integrity, so a failure anywhere on the network does not corrupt the data. They must therefore supply file and record locking, and be able to detect deadlocks in the system.

DataDAT

The computer storage version of the 4-mm Rotating Digital Audio Tape (R–DAT) audio cassette recorders. Originally each tiny cassette held about 1.3GByte and you could spool end-to-end in only 60 seconds and stop instantly along the way. Sony and Hewlett-Packard have devised the more popular DDS (Digital Data Storage) streaming tape format for backup from this format.

The tape only wraps 90 degrees around the drum. The rotary arm has only two heads at +/–20° azimuth angle, and it spins at 2000 rpm. The forward tape speed is 8.15mm/s.

The advantage of DataDAT was in allowing full-file updates to take place on the tape, although this was done at the expense of additional format overhead. It uses a random (rather than streaming) access method and the current capacity is 4GBytes without data compression.

The main competitor is DDS from HP and Sony; DataDAT has now largely been abandoned. See also DDS, Super-8 and 8mm tape.

datagloves

An automatic input device used in virtual reality to measure the movement of the hand and fingers. Most datagloves only recognize a very limited range of movements and gestures at the present time.

datagram

A packet sent over a connectionless packet-switched network. It is self-contained and requires no virtual circuit establishment. Most LANs use datagrams, but this is mainly because the route along which the packet can flow is predefined and limited.

In telecommunications, datagrams are packets on particular types of packet-switched networks which don't set up a virtual circuit first. The American Tymnet network was originally of this kind, and currently ATM is being defined with one AAL which is connectionless.

The datagram packets are all self-contained and the system doesn't use a preliminary call-request packet to establish the route, so each node in the network will switch the datagram towards the recipient using the best-available route at the time. And, since this will not be the same route for every packet, they can arrive out of sequence.

The value of the datagram approach (over virtual circuits) is that a circuit set-up process (usually involving special packets) is unnecessary and an imposition on overall capacity when only small amounts of data needs to be exchanged. There's also value in the fact that the failure or overload of one part of the network can automatically be avoided by rerouting the datagrams. See CLNS and CLNP.

• In X.25 the datagram identification is the first two bytes of the user-data field. X.25 has also recently introduced the concept of 'fast select' (see) which is a replacement for connectionless datagrams.

• A datagram is defined in the OSI model as an individual block of data delivered between lower layers.

Datagram Delivery Protocol

Apple's Network Layer in AppleTalk.

datapath

This is a technique which uses a normal two-wire telephone connection as if it were a four-wire connection to carry voice on one channel and data on another (or two data channels). It is a twin-channel system with a simple time-division multiplexer at each end. This is similar to BRA ISDN, but it pre-dates it by many years.

DataTAC

A radio packet system devised by IBM and Motorola, and used in the Ardis network in the USA, and elsewhere around the world. In packet-radio, it is the main competitor to Mobitex.

DataTAC offers data-rates (bursts) up to 19.6kb/s and it uses the very rugged RD-LAP (Radio Data LAP) air interface protocol. This applies 4-level FSK modulation for the 19.6kb/s rate on 25kHz channels, and achieves 9.6kb/s on 12.5kHz channels.

DataTAC uses Trellis coding, interleaving, error-detection and automatic retransmission. See Ardis and Mobitex also.

date

A number of ways of keeping track of dates have been used in computers over the years.

• In many of the early computer systems the date was referenced to 1 Jan. 1860, and a two-byte number was then used to calculate the current details. Now 1980 is used.

• Most other schemes are based on the Julian Day Number, which is the number of days which have passed since noon on 1 Jan 4713BC. The date is given in the form 2 449 958.5 which is days and hours (past noon in Greenwich). See Julian Date.

• The Modified Julian Date just truncates the Julian date by dropping the first two digits. Sometimes there is a three number addition. This is the number of days in the current year (since Jan 1). Sometimes a UTC 6 digit time-code is also added (hours, minutes, seconds — each of two digits) GMT.

• Many North Americans remain unaware that the order of date-numbers in many parts of the world is day-month-year, rather than the month-day-year sequence. So 12/10/95 is the 12th day of October, not the 10th of December. For this reason you should always use a date format with at least the first three alpha characters of the month: 23 Nov 1995 — and for future automation, it should always be in ascending or descending order (day-month-year or year-month-day).

• Many computer systems will produce erroneous results with the change of century from 31 Dec 1999 to 1 Jan 2000. Many pedantics will say this is not the change of the century or the millennium — that occurs on 1 Jan 2001, they say.

• Incidentally, in case you are writing date applications, the year 2000 is not a leap year, even though it is divisible by four. And don't use a period after the short-name for the month.

Datel

Data-telephony. A general term for carrier-provided private network services, which use the conditioned analog phone system for the transmission of data. Datel services need modems at both ends.

DATS

Digital Assisted Television System. See DATV.

DATV

Digitally Assisted (Augmented) Television. This is both a specific proposal by the BBC for analog HDTV, and a more general proposal for improving picture quality in any analog HDTV system.

• In the Eureka project's MAC HDTV plans, digital assistance (augmentation) reduced the bandwidth and retained maximum picture quality by transmitting additional digital information during the period corresponding to 40 lines in the vertical blanking interval. This ran in parallel with motion compensation to improved HD-MAC pictures. The cost of this approach is in the computational complexity needed.

• In the USA, for a time it was seriously contemplated that a separate channel would be set aside for digital augmentation information, and that home HDTV sets would then integrate the augmentation data with the analog NTSC channel to gain the wider, higher-resolution image.

DAVIC

Digital Audio/Visual Industry Committee/Council. A regrouping of the 300 carriers and equipment vendors who were behind MPEG, who have now formed this new organization under Leonardo Chiariglione (the father of MPEG). They are fast-tracking the development of the other key (end-to-end) standards in future digital video systems.

According to reports DAVIC is racing ahead with the development of a range of new standards which are all directed at standardizing the processes controlled by the set-top box. They had expected to report to the ITU by December 1995, but have recently announced delays due to the need to coordinate with the ATM Forum. See TINA also.

Daytona

The code name for Windows NT release 3.5.

D²B

Domestic Digital Bus. See.

dB

The decibel is a logarithmic means of expression ratios; it is a convenient way to compare large and small numbers on logarithmic scales. Decibels can be added and subtracted much more easily than actual numbers.

dB is generally used to compare electrical power levels — to measure signal intensity gain (usually sound) in terms of output signal power to input signal power. A change of 3dB in voltage is a gain factor of two, while 7dB is five times the volume gain, 10dB is 10 times the gain, and 20dB is 100 times the gain.

When used as an absolute measure it must be either a ratio to another measured valve (e.g. dBi, a ratio with isotropic radiation) or related to some fixed value (e.g. dBm, related to the fixed standard, the milliwatt).

- dBi is the antenna gain in dB relative to an isotropic source.
- dBm is the power in dB relative to one milliwatt. It is decibels above or below one milliwatt.
- dBW is the power in dB relative to one watt.
- dBmV is referenced to one millivolt. It is often used as a measure of carrier levels or the noise (usually) on a network.

When converting millivolts to dBmV, one millivolt is equal to 0dBmV, 2 millivolts is +6dBmV, 4 millivolts is +12dBmV and so on. Every 6dBmV doubles the voltage.

Also, 0.5 millivolts converts to –6dBmV, while 0.25 millivolts is –12dBmV. Copper wire has a noise floor of somewhere between –59 and –57dBmV in a standard TV cable system.

- In digital transmission, the ratio between the energy of each information bit (Eb) and the noise density (No) is called the Eb/No, and is usually expressed in dB also.

Note: the dB scale related to power is different from the dB scale relating to voltage. In power measurements, the power level doubles every 3dB (not 6dB as in voltage).

- See the European ratio measure called neper also.

DB2

A popular IBM mainframe RDBMS and procedural language which superseded IMS (the first commercially available DBMS). It runs on System 360 mainframes under VMS.

This is a high-end relational database which is very important in the mini/mainframe world, and it's apparently still used in about half of all large computer sites. It began to be used seriously only in 1985–6.

MVS, AS/400, AIX, HP/UX, Unix and PC versions of DB2 have been released, and the database manager in OS/2 EE is a scaled-down version of DB2. IBM has clearly indicated that DB2 database is of strategic importance to them, and that it will be brought into SAA.

DB-9

A popular connector used for serial connection. It is available in both male (m) and female (f)'genders'. It has two rows with 9 pins which are used in the RS-232 standard in a specific way (but not for synchronous communications — see DB-25). Note: the acronyms are slightly different in some literature.

Pin 1	CD	Data Carrier Detect
Pin 2	RD	Receive Data (Rx)
Pin 3	SD	Send Data (Tx)
Pin 4	DTR	Data Terminal Ready
Pin 5	SG	Signal ground (not earth)
Pin 6	DSR	Data-set (modem) ready
Pin 7	RTS	Request to Send
Pin 8	CTS	Clear to Send
Pin 9	RI	Ring Indicator

Any reasonable cable or connection should have all of these links between the DTE connector at one end and the DCE connector at the other.

When looking front-on at the pins in a DB-9P plug (male) the pins number from top left, in the normal way, to bottom right. The female socket numbers in reverse order, from right to left. See also null modem and RS-23.

DB-15

A standard plug used for monitors. It has three rows of pins.

DB-25

A popular connector used for serial connection. It is available in both male (m or P) and female (f or S) 'genders'. It has 25 pins which are used in the RS-232 standard in a specific way.

Pin 2	Tx	Transmission
Pin 3	Rx	Receive
Pin 4	RTS	Request to Send
Pin 5	CTS	Clear to Send
Pin 6	DSR	Data-set (modem) ready
Pin 7	SG	Common
Pin 8	DCD	Data Carrier Detect
Pin 20	DTR	Data Terminal Ready

Any reasonable cable or connection should have at lease these links between the DTE connector at one end and the DCE connector at the other.

When looking front-on at the pins in a DB-25P connector (male) the pins number from top left, in the normal way, to bottom right. The female socket numbers in reverse order, from right to left. See also null modem and RS-232.

DB-42

The standard connector used for RS-449 cables, and the one recommended for V.34 modems. However most manufacturers seem to be staying with DB-25 connectors. The DB-42 is too bulky.

DB-50

This is a large plug/socket used for SCSI connection. It has three rows of pins. Don't confuse this with the Centronics 50 or 68 which is also used for this purpose.

dBase

A brand name for a relational database management program developed in the very early days for microcomputers. It had a remarkably sophisticated procedural language which allowed some of the early PC hackers to do things that the mainframe jocks could only dream about.

The program was originally called Vulcan, and it came out of the Jet Propulsion Lab in the USA. The first version, dBase II was way ahead of its time. There was no dBase I, and the current most widely used version is dBase IV, although dBase III has retained its popularity. There is now dBase V for Windows which has GUI development tools.

So dBase it is still the micro database management program against which everything else is compared. The dBase language is not easy to learn, but the results are worth the effort.

• The American National Standards Institute has taken the dBase procedural language on-board as a standard under the name XBase.

DBMS

Data Base Management System. This is the correct term for a database application; the database is the data-files which have been created by entering information through the DBMS program.

DBMSs are more complex and flexible than paper filing systems, and they will often have a number of associated utilities — and perhaps links to other database files (RDBMS).

Management systems are needed to organize the files, establish the field categories, sort the records, store the records, lock the records (or files) while they are being updated, create indexes, retrieve the information, and publish reports.

Often a DBMS includes a built-in language for retrieving or manipulating data (see query languages) — and you can often design your own interface to the program, so data is entered via a screen 'form' with strict controls over the data-types (see). If the DBMS is 'relational (RDBMS) then you will be able to combine several different files with different record formats into one database. If it is distributed (D-DBMS) then all components don't necessarily reside at the same location.

There are three major types of data-base management:

- Hierarchical, where the files are related to each other like branches on a tree.
- Network, which allows the DBMS to establish many different relationships, since each file is not limited to only one higher hierarchical link.
- Relational, where each file is seen as a table called a relation. Key (shared) fields are required between relational files.

See Codd's rules and RDBMS.

• The main function of mainframes today is to run large corporate databases and management systems. At the mainframe level the most important DBMS was IMS, but now it is DB2. IBM also has AS/400 which incorporates a proprietary database into its operating system.

• DEC packages RdB with its VMS operating system.

• With Unix machines, Oracle is the leading DBMS, followed by Ingres, Sybase, Progress and Informix. These will also run under other operating systems.

• In microcomputers, the dBase family was once dominant and is still very important, and there are dBase clones, such as Foxbase. Paradox, Access and Q&A are also very important. In the Macintosh world, 4th Dimension is possibly the most sophisticated, but FileMaker is more popular. HyperCard is database-like, and very useful.

DBS

Direct Broadcasting Satellite. A geostationary satellite that rebroadcasts television signals from a transponder with sufficient power to be received by small privately-owned (household) receive-only earth stations (TVROs). The term applies to both analog and digital systems.

Typically these use dishes which are about 0.5 meters in diameter and they require a signal strength (EIRP) of between 55 and 60 dBW for good reception.

• For direct broadcasting in the USA, the down-link uses frequencies in the Ku-band between 12.2 and 12.7GHz which provide the standard 500MHz bandwidth. There are 32 licensed DBS frequencies; standard LNBs can handle the bandwidth.

The US now has its first commercial digital DBS with Hughes DirecTV. It offers 150 channels (shared between two programming companies) using two (later three) co-located satellites. See DirecTV.

• In Europe (Region 1) DBS satellites operate at frequencies between 11.7 and 12.5GHz . These satellites use both left- and right-hand circular polarization. They have orbital spacings of 6°, and provide 40 analog channels in 800 MHz of spectrum.

• In the UK, two incompatible systems were operating (Sky and BSB) until the two merged. Sky used the PAL standard, while BSB introduced D-MAC with a wide-screen format. This was later abandoned. See TVRO and SMATV also.

DC

• **Direct current:** The one-way flow of current from a battery, accumulator or rectified power supply; this is the main source of power in most computer circuits. Most computer chips require about 5 volts DC although 3 volts is now preferred for portable devices. The distinction is with AC.

• **DC-coupled** means that the devices are linked by square-wave pulses in DC voltage, rather than by modulated AC frequencies. DC coupling is undesirable because it requires direct physical attachment of devices; with AC-coupling, the link can be made through a transformer.

• **DC component**: In any binary transmission scheme where, say, pulses of x-volts are used to indicate 1s, and no voltage indicates 0s, the cables carrying the signal will have an accumulated DC component and therefore require DC coupling to attached transmission components. To overcome this problem, it is wise to use a positive voltage for half the 1s and an equal negative voltage for the other half (or some similar scheme), so that the DC components cancel each other out, and leave a form of square-wave AC.

This bipolar technique is better than unipolar transmission for long periods when the system may be idle. It will then be generating nothing but 1s or 0s. See AMI.

— The DC component of a video image is equivalent to the average illumination level of the image (the average by which the signal exceeds the reference level).

• **Dewey (Decimal) Classification** system for libraries. See DDC.

• **Device Control** in the ASCII No.5 character set. These are used for the control of ancillary devices or for special terminal features. How they act is specified by the application.

DC1	Device Control No.1	Decimal 17	hex $11	<Ctrl>Q
DC2	Device Control No.2	Decimal 18	hex $12	<Ctrl>R
DC3	Device Control No.3	Decimal 19	hex $13	<Ctrl>S
DC4	Device Control No.4	Decimal 20	hex $14	<Ctrl>T

DC-to-DC

Devices which convert one DC voltage supply into another. At one time this required generating an AC current, transforming it to a higher voltage, then full-wave rectifying it to produce DC again. These days, we have chips called 'charged-pump' DC-DC converters. They allow conventional 5 volt electronics to be driven by a single 1.5 volt battery, or permit an RS-232 serial link (which needs about 12 volts) to be driven by a conventional 6 volt portable battery pack.

DC600

The higher-capacity version of the DC2000 series of backup tape recorders. This is a streaming tape system which doesn't need to be formatted before use — but it can't overwrite existing data blocks for updates, all new back-up data must be appended. These tape units hold between 150 and 320MBytes.

DC2000

A standard small cartridge (1/4-inch) tape backup system. The tapes hold between 40 and 120MBytes, and need to be formatted before use. The system can find and overwrite particular data blocks — so it can make direct-access modifications, and doesn't need to append all back-up data to the end of the tape.

DCA

Document Content Architecture. IBM's SNA extension to give microcomputers access to host mainframes. Also see DCA/RFT.

DCA/RFT

Document Content Architecture/Revised Form Text. IBM's DCA/RFT provides an intermediate file format for the exchange of 'formatted' word processing (and other) documents between applications and platforms. It was designed for IBM systems, but is now more widely used.

However not all document features are supported: specifically there will be problems with some pitch changes, some types of character tabs, and 'hidden text'. And there can be problems with tables of contents, indexes, list entries and aligned columns.

Other problems that need to be addressed (and are by some conversion packages) are requirements for translation from American page lengths to European paper sizes; substitution of typestyles if the target system doesn't have the same fonts as the source; and multinational character sets.

Don't confuse this with Rich Text Format (RTF) which does much the same job.

DCC

Digital Compact Cassette. The DCC audio cassette system was developed by Philips and Matsushita as an alternative to R-DAT. But note that this is a (Stationary) S-DAT scheme.

The standard is backward compatible in cassette dimensions and replay with the current analog audio cassettes, so DCC players can handle both analog and digital stereo cassettes. But DCC recordings can only replay on a DCC player.

The audio is compressed by a selective culling process which ignores high-frequency sounds below the hearing threshold in certain channels. This reduces the digital bandwidth by about 4:1. PASC (Precision Adaptation Sub-band Coding) is used for modulation.

The cassettes have a play time of 90–120 minutes (half in each direction), and use chrome dioxide, low-coercivity materials, which pass the heads at the normal Philips cassette 4.76 cm/s tape speed. The tapes are auto-reversible, and the quick search mechanism will find any track within seconds (but DCC is slower than CD or DAT). Limited digital LCD display information can also be included with the audio. Special thin-film magneto-resistive heads were developed for the recorder.

The top sampling frequency (which is selectable), the error correction technique (Reed-Solomon), and the modulation scheme (8-to-10), closely match the original CD-Audio standard. PASC's bandpass filters divide the audio into 32 subchannels, each of 1.5kHz, and these are then compressed and multiplexed onto 8-tracks. DCC uses a stationary head linear recording system, so the mechanism is simpler than the helical scan of R-DAT or the laser optics of CD.

See DAT, S-DAT and MiniDisc also.

DCD

• **Data Carrier Detect:** In the RS-232 standards, this is a voltage change (enable is a raised voltage of –12V) carried on Pin 8 and generated by the modem (DCE). It indicates to the terminal that the modem has detected a valid carrier.

The primary purpose of 'carrier detect' is to prevent the DTE from responding to line noise. When DCD is on

(enabled), the DTE knows that valid data should be on the Rx line. When DCD is off (disabled) Tx ceases, and the call is dropped. Some printers require receiver-handshaking on Pin 8, so very often in these circumstance, DCD is bridged with Pin 20 (DTR) in the DTE.
— In ITU terminology, DCD is Circuit 109: in EIA terms it is CF.

• **Dynamically Configurable Device.** Another name for a PnP device.

DCE

• **Data Circuit-terminating Equipment:** Also incorrectly (but understandably) called Data Communications Equipment.

DCEs are the user's access interface to the phone network. They are generally modems or multiplexers or routers — although the term encompasses a range of other devices that exist between a terminal and the main network circuit.

A DTE (data terminal equipment) device is always designed to connect to the end of a communications chain, while DCE equipment is inserted somewhere along the chain. The DTE is responsible for high-level 'end' functions (size, sequence, format and error-checking of packets, etc.), while the DCE handles physical level functions, and the DSE (data switching equipment) handles switching.

With the RS-232 interface, remember that DCE equipment always transmits its data on Pin 3, while DTE equipment transmits on Pin 2. See also DTE and DSE.

• **DCE-Ready:** See DSR.

• **Distributed Computing Environment:** See client/server also.

DCI

Display Control Interface. This is the recent replacement for the old GDI Windows drawing engine. It is a low level interface specified by Intel and Microsoft which allows displays to handle video frames at a faster rate, and in bigger window sizes than GDI. It provides a direct connection from the software video drivers to the graphics display sub-system and frame buffers.

The use of DCI allows a graphics subsystem to convert the compressed YUV video into the RGB format needed by the monitor, without overloading the CPU.

DCME

Digital Circuit Multiplication Equipment (actually a trade name, but often used generically). See DCMS.

DCMS

Digital Circuit Multiplication System. This is the digital version of TASI which is now used to increase capacity of digital undersea fiber cable systems and also satellite systems.

It relies on the fact that in any two-way voice conversation, one person will always be talking and another listening (a minimum 50% capacity wastage). Time is also wasted during call set-ups, clear-downs, pauses, etc.

DCMS is a statistical technique which dynamically modifies the channels; it grabs these spaces, and allocates the bandwidth to other users. Usually it will be coupled with bit-stealing (see) techniques. It is claimed to give as much as a five-times capacity increase over the conventional two-way PCM channel allocation.

Note that with 16kb/s ADPCM (used on Atlantic cables) there's an automatic 4x increase in capacity over PCM. But the DCME figures are said to increase this code-gain 3-times — in which case, by coupling ADPCM and DCME they should get something like a 12x gain over two-way PCM (called 'nominal voice circuits').

DCS

• **Desktop Color Separation:** A format which creates up to five PostScript files for each color image. These are EPS files saved in five parts: a master, plus CYMK.

• **Digital Cross-connect System:** This provides multiplexing and switching capabilities in digital telephone exchanges.

• **Digital Cellular System** or DCS-1800. See below.

DCS-1800 (1900)

Digital Cellular System or Digital Communications Specification. (They change the name!) A digital cellular standard used by Mercury and Orange in the UK, and now promoted by European manufacturers for future PCN operations in the 1.8GHz and 1.9GHz bands. It allows two operators to share a network, and it relies on high-capacity small cells of about 1km diameter or less. This is a GSM derivative, but there are 11 additional 'Delta' recommendations added to the basic GSM standards.

At these gigaHertz frequencies the handsets are smaller and they should eventually be cheaper due to the mass production potential of PCN. However three times GSM's number of base-stations will be needed for the same coverage area, and this digital standard already needs more base stations than TACS.

Like GSM, the voice is coded by 8-to-13kb/s RPE-LTP coding. The standard frequencies are between 1710—1785MHz and 1805—1880MHz, with 374 8-slot carriers having 200kHz of separation. In the UK, two operators have been allocated 25MHz each, but have taken up just over half that bandwidth.

The Mercury service (called One-2-One) has been launched in London with 2 x 15MHz allocations and there is a wider-area service called 'Orange' (Hutchison) covering most of the UK.

In North America and possibly later in Asia, the DCS technology is being trialed for PCS, but in the 1.9GHz band. It is obviously then called DCS-1900.

DCT

• Discrete Cosine Transform. See.

• Digital Component Technology. See.

• Digital Cordless Telephone. See DCT-900.

DCT-900

Digital Cordless Telephone in the 900MHz frequency band (actually between 862—866MHz). This is Ericsson's version of what they originally called the 'CT3' phone standard.

DCT is similar to DECT (although at different frequencies) and is primarily intended for in-house use with radio links to a company PBX. It uses TDMA/TDD technology and currently has four carriers to a base-station, each carrier is 1MHz wide, and is subdivided by time into 12 half-duplex sub-channels.

A base station can therefore provide a total of 32 full-duplex communications channels within its cell. Voice is encoded at 32kb/s using ADPCM. DCT-900 has seamless hand-off and carries data also. Wireless LANs and a fax facility are said to be under development.

Most other companies have shifted over to DECT (at 1.8—1.9GHz) and Ericsson now seems to have followed.

See DECT, CT2, GSM, DCS-1800 also.

DD

Double-density floppy disks. It was a good name at the time, but these are now low-density.

DDB

See Domestic Digital Bus.

DDBMS

Distributed DataBase Management Systems. Ideally, distributed database modules on a network should collectively offer the same functionality as a database on a single host computer, and DDBMS aims to provide this. It results from the convergence of RDBMS and distributed network technologies.

Unfortunately issues such as data integrity, security, resilience and performance are proving difficult to implement across a network, and so most current products only supply a limited number of features.

However this is seen as the way of the future. Some tests of DDBMS performance are:

— Equal functionality to host computer systems.
— Transparency of location (you shouldn't need to know where data is located).
— Replication transparency; allowing the automatic duplication of often-used data at peripheral nodes to avoid network delays. This desirable function has updating problems. See two-phase commit.
— Query optimization.

Problems which still need addressing are decisions about the extent of data distribution, management of replicated data, data security and integrity, and a range of specific database design and administration issues.

• Note that IBM does not use the term DDBMS in this way with SNA and LU6.2.

DDC

Dewey Decimal Classification (often just DC) which is used in libraries and so provides a useful subject code for information.

DDE

• **Direct Data Entry.**

• **Dynamic Data Exchange** protocols. These are Inter-Application Communications (IAC) protocols designed by Microsoft for Excel, and later extended to Windows and OS/2. DDE allows communications pathways to be established between applications without user-involvement or monitoring.

You can use DDE to share data with any other program that supports it. However this is often a job for a programmer; before two applications can use DDE they must agree on an interchange format, and if the applications are from different vendors, then each vendor must publish detailed specifications.

DDE is actually a formatting and writing convention that describes the layout and meaning of messages passed through pipes and queues. These links can pass both data and messages, and they remain dynamically linked after the data has been passed. If the original information changes, you should not need to redo the data exchange.

With DDE, one program can use another's data as if it were its own, which gives us an alternative to cutting and pasting. There are five DDE mechanisms:

— **Hot links.** The server will automatically update the client application whenever the data changes.
— **Warm links.** The server will notify the client that the data has changed then the user can choose to request the update.
— **Requests.** This is a copy-and-paste operation which doesn't need an intermediate Clipboard stage.
— **Poke.** This is a back-channel transfer.
— **Execute.** One application controls another's execution.

See OLE, IAC, and hot-links also.

DDI

• Device Driver Interface.

• European term for DID. See.

DDIF

Digital Document Interchange Format. An Apple Mac-to-VAX data format that can work on both DEC and Apple systems. It allows certain 'complying' word and document processors to exchange formatted text, graphics and pictures. It was based on the ISO's ODA standard.

DDM

Distributed Data Management. IBM protocols which provide various ways to access a file, regardless of the file's physical location in the network. It is based on LU6.2.

DDN

Digital Data Network: Aka Dedicated Digital Network. A widely used term for a non-switched, commercial carrier service based on multiplexed digital links, usually at 48, 56 or 64kb/s data-rates, but sometimes offered at 1.54 or 2Mb/s and above.

DDP

• **Datagram Delivery Protocol:** See.

• **Distributed Data Processing:** The trend away from centralized data processing using mainframes in air-conditioned enclaves. It doesn't mean the mainframe host becomes superfluous, just that most of the processing is now taking place in the PCs on the network — and it suggests that a lot of data is being exchanged peer-to-peer rather than via the host.

DDS

• **Digital Data Service.** In telecommunications, this is often just a general term for leased lines which have been conditioned to carry digital data at rates of 48, 56 or 64kb/s. These are the services carried over DDN networks (aka Integrated Digital Networks or IDN).

• The specific transmission system known as DDS in the USA was developed by AT&T. DDS uses two pair of wires for local access to the exchange, one to transmit and one to receive. DSU/CSUs are used at the end of these lines. See.

Signals are sent in AMI format at a channel rate of 64kb/s and a data throughput of 56kb/s. DDS uses a primitive form of line-code which maintains synchronization by setting the 8th bit in each byte to a logical 1, so that long streams of zeros aren't possible. B8ZS is becoming popular because it doesn't waste these bits and allows a clear 64kb/s channel to be used. See T-1 and B8ZS.

• **Data Dictionary System**, in programming. See data repository, and data dictionary.

• **Digital Data Storage,** in streaming tape magnetic storage.

de facto

Standards which have just arisen through popular use.

de jure

Standards which have been officially established.

DEA

Data Encryption Algorithms. A generic term used in encryption techniques. It refers to a range of possible techniques. See DES, PGP and RSA.

deadlock

When two or more processes are each waiting for the other to finish, before they can proceed. Each holds resources the other needs. Systems need ways of handling this problem; usually by aborting one of the processes. See deadly embrace.

deadly embrace

This is a network 'mutual' lockout that occurs when two applications repeatedly attempt to access and secure the same resources. In a relational database (especially a distribute one) each contestant may need data from the other to complete the transaction. This is also called deadlock.

debugging

The process of locating and removing programming faults from a program, or minor design faults in a piece of hardware. Software publishers have two favorite methods of finding bugs:

— Give technical journalists and third-party developers the pre-release 'alpha' version but claim it as 'beta'. Then see what makes their systems crash.

— Release the program commercially as Version 1.0 and let the suckers who buy, find the bugs for you.

Unfortunately, it is cheaper to run a Help desk than employ professional debuggers.

• There are two major approaches to debugging: some products offer source-level debugging where you can run your program one line at a time. Others have the more complex machine-level debugging which allows you to watch how each instruction affects memory locations and check the microprocessor's status.

Debye limit

The idea that the size and spacing of elements on a chip are limited to gate-lengths of 0.1 microns with 0.5 volts of power. This is said by some to be the limit of our current planar chip technology.

• Hitachi are currently showing early prototype 1Gbit DRAM chips with 0.16 micron spacing, and using 1.6 volt power supplies.

DEC

Digital Equipment Corporation. The makers of minis. They don't like to be called DEC — but that's what happens when you use a generic term for your company name.

decadic dialing

The old telephony term for pulse dialing (ten position) — as distinct from DTMF or 'tone' dialing. See pulse dialing.

decay

• In music — the decay is the envelope characteristics of a fading sound, the shape of the sound envelope as the amplitude decreases. In hall acoustics, the term is also applied to the decrease in echo amplitude over time. See reverberation.

• In electronics the decay time is the time it takes for the signal to reach only 10% of the original level.

decibel

This is a measure of the ratios between any two power levels; it is expressed in a logarithmic scale based on Weber's law. A 3dB change corresponds to a halving or doubling of the power. Subjectively, with audio, it requires about a 10dB change to double or halve the apparent loudness of a signal. See dB for more technical information.

decile

The division of the total range of variations into ten equal parts. So the highest decile has all those above the 90% mark.

decimal

Decimal really means denary integers with fractions. The radix of both decimal and denary is 10 (the base). See denary.

decimate

Strictly, to reduce by one-tenth. In video compression you 'decimate' images by coding them to a lower bit-requirement, then later you will 'reconstruct' them when converting them back for display.

These are the terms used when performing both inter-frame (differences) and intra-frame code compressions. See DCT, MPEG and JPEG.

decimator

A type of digital filter (numeric selection process) used in ADC when oversampling results in extra digits; the decimator's job is to restore the bit rate to the required bandwidth. A decimator at the ADC end, is matched by an interpolator at the DAC end, and this restores the oversampled rate and so reduces quantization distortion.

Decimators can work simply by throwing away unwanted bits or by rounding. Obviously rounding is better.

decision feedback

A communications technique which uses ARQ (automatic repeat request) codes for error correction.

decision level

Any digital signal being communicated over a system will appear to the receiver as a variation in pulse amplitude. This amplitude can vary due to many factors including line attenuation. Therefore predetermined points must be established, above which the pulse is considered to be a 1, and below which it is treated as 0. These, obviously, are the decision levels or thresholds. Sometimes an area of ambiguity will exist between two thresholds.

decision tree/table

In expert systems, a decision tree consists of a sequence of questions (branches), each of which narrows down the range of possibilities. Ultimately one branch will terminate in a possibility that meets all the necessary conditions. The process is also called a decision table if it is laid out in this form.

declarative

To declare, or make a statement; to input a value.

- **Declarative language:** Languages like SQL which consist of declarations. This involves, according to the ISO, 'attaching an identifier, and allocating attributes, to the language object concerned'.

- **Declarative knowledge:** This consists only of statements which describe something. It underpins much of logic and mathematics because it forces a style of formal reasoning. Declarative statements do not tell us 'how' to do something — which is imperative knowledge — declarative knowledge exists in isolated, unconnected modules that just 'are'.

In expert systems the implication of the term is that the information in the knowledge-base is represented as isolated modules — that there is no sequence or order to that knowledge.

declarative programming

Programming by using heuristic rules — as in expert systems.

declare

In programming some languages require that variables should only contain one data type, which should be declared (established) before the variable is used and then never varied.

declination

The angle between an antenna beam and the equatorial plane (measured from the meridian) when a satellite dish is 'polar/equatorially mounted'. See azimuth and elevation also.

DECnet

Digital Equipment's (DEC's) networking system. They support PCs and host computers which can handle TCP/IP and OSI as well as their own proprietary protocols.

decode

Decoding is a wider and more general term than decrypting. In computing and communications we are coding and decoding all the time without encrypting.

The term is used for the translation of any form of code, from Morse Code to PCM. We decode it back to its original form of text, or analog voice.

The process of coding, as it is understood in computers and communications, is simply that of changing data from one form of representation to another, and the decoding changes it back again. For instance, a decoder is used in a computer to change machine numbers in binary, back to decimal numbers for display.

decrement

Counting down — usually by ones, unless otherwise stated. In programming this is usually done by a loop structure which only breaks out when a condition is satisfied. The opposite is increment.

decryption

The opposite of encryption. Distinguish this from decoding.

DECT

Digital European Cordless Telecommunications (previously 'Telephone'). DECT is a relatively new two-way DCT-900 derivative which can provide voice and data service using TDMA/TDD techniques in the 1.8GHz band. It was primarily designed for wireless PABX connections, but it could eventually have some public-switched application in city streets and slow-moving vehicles — but DCS-1800 appears to have taken over this role.

The design of DECT began under CEPT in 1988, and the standards were set by a special committee of ETSI in March 1992. Product became available in 1993/4 although some standards were still not compulsory — and this has produced compatibility problems. A common-air interface (called GAP) has also now been defined.

DECT uses 20MHz of the spectrum at 1880—1900MHz for ten R/F bearers (with wide separation), and it jams 12 full-duplex channels into each 1MHz of bandwidth. A total of 240 simultaneous calls can be supported.

DECT handsets typically have a power output of 250mW and a range of 200 meters. They transmit in 400μs bursts, once every 10ms (at 100Hz). The system doesn't attempt to cope with multiple-reflections or the Doppler effect of moving vehicles, as does GSM.

Voice coding is ADPCM with encryption, and there are now cordless LAN products under development. DECT is also seen by some people as the first stage of European PCN which will lead in turn to UMTS/UPT. It is specifically designed to interface with GSM and ISDN networks. See PCN, DCT-900, GSM, PCS and DCS-1800.

dedicated

A system/device that performs only the one function — or performs it only for one user.

• **Word Processors:** A few years ago, many second-rate personal computers with exotic and proprietary operating systems were sold as dedicated word processors until buyers woke up to the fact that they were paying more for less functionality.

• **Lines:** In telecommunications, a dedicated line is a leased or private line — as distinct from the ones available for public switched use. One advantage is that it can be conditioned.

• **LAN:** The term 'dedicated Ethernet' means that a single port on the device having the full 10Mb/s Ethernet capacity of a network segment, is being linking to a single device such as a server or high-performance workstation. Switched Ethernet systems make every workstation link into a dedicated link.

Dedicated 56

In the USA, 56kb/s data services are offered in either switched or dedicated form. Dedicated-56 is the hardwired form. See Switched-56 and ISDN.

dedicated servers

Servers on a network which are devoted to one task only. The distinction here is between those file servers on a LAN which are dedicated to the file-handling task (not used for any other purpose), and distributed file handling on a peer-to-peer basis.

It is also the practice with some of the smaller LANs to distribute the database functions between desktop computers on the network, so each can use the files of the others. In some cases a special 'dedicate' computer may perform the main file-server functions, but also be available for occasional keyboard input, etc. Normally, however, dedicated servers are special devices.

deductive learning

An AI term. Contrast this with inductive learning.

deep etch

In publishing, this is the process of cutting out the image of a person or object so that the image sits in isolation on a white page background. This is usually done by using stock-standard white-out (type-correction fluid) to paint carefully around the person/object before it is sent to the graphics camera for the production of the half-tone bromide.

default

A value or setting that the computer system assumes in the absence of explicit instructions.

— Factory defaults are values which have been set by the maker or vendor. They are usually non-erasable.

— User-defined defaults are modifications made by the user to suit his/her own requirements.

In most cases you can customize your software or hardware in such a way that any changes you make to these factory defaults are 'remembered' and reused on all future occasions — or until you later make other changes.

Factory defaults are usually stored in ROM, while user-defined defaults are held in EPROM (so they don't need to be reset every time). In modems, you'll find that the user-defined defaults in the EPROM take precedence over the factory-define ROM settings.

• In Hayes-compatible modems, the command 'AT&F' will recall all original factory defaults and make them active (overlaying your customized defaults). If you want to *permanently* replace your customized defaults in EPROM with the original factory ones, you will command 'AT&F&W'. The AT&F activates them and the &W writes them into the EPROM.

• The default directory is the one in which the operating system will look for files if no directory is specified. The 'ch' or chdir' command establishes the default file.

• The default drive is the disk drive on which the operating system will look for a file, unless an explicit drive designator is included in the file name.

defective media

A bad disk or fault on any other mass storage device.

deference

In some contention-type LANs, the data link controller will 'defer' by delaying its transmission to avoid data collisions.

definition

• The resolution of an image in subjective terms. The separation of image elements: features seen on a screen or printout, given good viewing conditions.

• The specification of a variable's data type in programming. See declare.

• The construction of a database by the selection and naming of fields, data types, field lengths, etc.

• A definition file in Windows is an essential part of each application file. It is used during the compilation and linking process, and it contains information about the application: the application's name, the name of the MS-DOS program that is to be run during execution, the stack size, heap size, etc.

defragging

Running a defragmentation program to bring the fragmented sectors of a file together on a disk where it can be quickly accessed, and recover disk space. See disk optimizers.

degauss
To demagnetize something. A degausser is a special demagnetizing coil with a high frequency AC current.

degeneration
Negative feedback. See.

degradation
A decreasing level of service or efficiency.

• In digital transmission, signal degradation is the gradual 'rounding off' of the square wave pulses that constitute the bits in the signal. Eventually the signal shape will deteriorate to the point where it is impossible to read accurately. In systems with this problem, regeneration is necessary, not just repetition or amplification. See dispersion.

• In analog systems, this usually refers to noise on the line. See SNR and distortion.

DEL
ASCII No.5 — Delete (Decimal 127, Hex $FE, no standard control equivalent). This signal is generated by a special keyboard key and it causes deletion of the last character.

• It is possible to change the setting of many applications so that the delete key deletes the last character, the one under the cursor, or the next character.

delay distortion
This is also known as envelope delay and as phase distortion. Different frequencies will travel at different speeds in communications circuits. This can cause signal distortion. The received signal is distorted by the variable delay of its frequency components. (The middle range frequencies usually arrive first.)

In analog circuits, this is generally not noticeable for voice because the main effect of these delays is to change the phase of the signal slightly. However for high-speed modems where 'symbols' are coded by phase-changes, it is a serious problem.

Equalizers can be constructed to compensate for this distortion by artificially delaying the faster frequencies, however there are other forms of delay distortion which complicate matters. See group delay distortion, amplitude group delay distortion, intersymbol interference and transmission impairments.

delay line
There are analog magnetic and acoustic delay systems, and digital delay systems. In composite analog TV, the color and luminance signals take different times to be processed in the receiver, so the color would normally arrive late, and be always seen to the right of the B&W image.

To solve this problem, the luminance is delayed by 400ns through a magnetic delay line. This is a coil of wire on a grounded coil form with capacitors. The combination of inductance and capacitance effects will slow the signal's progression through the coil quite significantly. See acoustic delay.

delay-sensitive
Traffic on a network which is more critical as to the time of delivery, than as to the data integrity: the information can arrive a bit damaged, provided it arrives on time. Voice and video come into this category, since it doesn't matter if a few bits get lost, provided there aren't delays (latency) across the network. The comparison is with data which is quite tolerant of delays but needs to have total integrity.

delimiter
A print or non-print character which is used to define the beginning, end, or boundary of a sequence (usually a character string). Inverted commas, for instance, are often used in programming to identify text strings. A prefixed full-stop has been used to delimit embedded commands in a word processing program.

Tabs, carriage-returns and commas are often used in database files to identify field and record limits. However, if the comma is used for a delimiter, it must obviously first be prohibited from use within the records as text (or converted to some other symbol first, using find-and-replace). See comma-defined and tab-defined.

Delphi technique
A way of forecasting future trends by combining the views of many experts. The value of this approach is disputable.

delta
Delta immediately suggests 'difference' techniques. However there are a few uses of the term, and in audio coding a distinction must be made between 'difference' techniques (involving 4-bit nibbles) and 'delta' techniques (with single bits only).

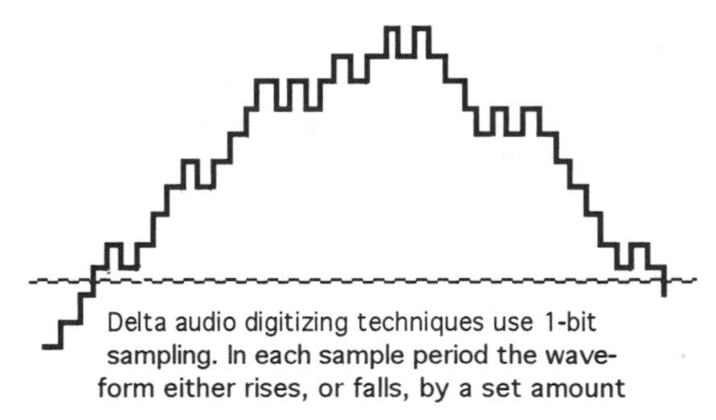
Delta audio digitizing techniques use 1-bit sampling. In each sample period the waveform either rises, or falls, by a set amount

• **Coding or modulation systems:** Digital delta techniques are 'polarity' techniques where a 0 represents a fall in value, and a 1 a rise in value — each bit indicating only the direction in which the waveform is changing (not by how much).

Therefore to indicate substantial falls, continuous streams of zeros are transmitted — and for substantial rises, a long series of ones. A steady state is indicated by an alternation of ones and zeros. These techniques are suitable only for voice.

Delta modulation systems put out pulse-trains (bit-streams) rather than identifiable bytes or samples. They often give a good approximation of voice where the waveform changes in a reasonably predictable way, but they don't handle extreme changes in waveform (such as phase changes in modem transmissions) at all well.

Delta modulation is a subclass of Differential PCM. Delta techniques need a sampling rate sufficiently high to ensure that no significant information is lost — usually 32,000 per second — but bits, not bytes are generated. If the slope of the signal varies steeply, the output of a delta system can easily be distorted, but adaptive schemes solve some of these problems.

— There are linear delta modulation schemes when the value of delta is fixed, and

— Adaptive delta modulation schemes (see) where the delta can be varied.

• **CVSD:** The most popular delta modulation system is called CVSD. See this and also DPCM, A-Law, Adaptive Differential Modulation and ADPCM.

• **Switches:** Delta network switches (actually cross-connect switches) are simple chains of switch units which have multiple inputs and outputs, but only in two rows. ATM cells entering either of the inputs will take one or other of the output paths, depending on whether the relevant bit in the address byte was a one or a zero. Each switch unit simply handles one bit of the address. The advantage of these over a more complex matrix structure, is in the speed, connection redundancy, and lower component count for a given number of connections. See Banyan, Omega and CLOS.

• **Color schemes**: Delta YUV (DYUV) are used in CD-i. There is a high correlation between adjacent pixels in any image (along a single line), so it is possible to encode only the differences between the absolute YU and YV values. Here delta is used to mean 'difference' and in practice it halves the sample dimensions — what was a byte, now becomes a nibble.

This color technique is also applied to full-motion TV. So rather than use 216Mb/s for normal uncompressed YUV 4:2:2 encoding, DYUV cuts the native bandwidth to 108Mb/s.

• **Channels:** For supervision or signaling. These are derived from a data stream by bit stealing. See A&B bit signaling.

• **Delta frames:** See frame delta.

• **Video compression:** See frame delta also.

• **Delta heads:** In old videotape machines it was impossible to 'freeze frame', but you could 'freeze field' (only the odd or even lines in the frame were ever shown). Newer videotape machines have heads which can wander across the tape sufficiently to find adjacent tracks and follow them. This is called 'dynamic tracking', and the heads are known as delta heads. With these heads it is also possible to vary speed during playback and to playback tape in reverse.

demand paging

This is a technique of bringing data into memory (usually from extended memory, but also from disk) a page at a time, but only when it is called for.

demarc

An American slang-technology term for the demarcation point, where the telephone company's network ends and your private network begins. This is usually a box or connection point of some kind. See network termination point.

democratic networks

One where each node has its own clock. The distinction is with despotic networks.

demodulation

Extracting the original (usually baseband) signal from a modulate carrier (audio for modems, or R/F for radio). This can be applied equally to analog or digital signal extraction.

Demodulation is the process of extracting 'information' from a carrier signal. We can have analog information piggybacked on an analog carrier (AM and FM radio) or digital data on an analog carrier (GAM) — or analog information carried on a digital signal (PCM) or digital information carried on a digital signal (CDMA). All of these processes involve modulation and demodulation of different types.

• In radio, the demodulator is the radio-frequency detection circuit which strips the information (the audio signal) from the radio-frequency carrier. The tuner selects and passes the signal through and then it is demodulated back to baseband audio.

• In modemized data communications, a demodulator takes the carrier audio frequency/s from the communications circuit and converts them into the bits and bytes that constitute the original signal. The signal can carry data by amplitude, frequency or phase changes.

• In video, a demodulator extracts the video, audio and sync signals from the composite modulated R/F signal of the broadcasting station.

• In CDMA, the demodulator is an Exclusive-OR process, which extracts the Walsh code from the chip sequence, leaving the original digital signal.

• In PCM, the demodulator is the D-to-A conversion process which outputs analog audio.

demon

See daemon.

demultiplexer

See demux.

demux

DEMUltipleXer — the secondary function of a full-duplex multiplexer. It separates the individual signals that arrive over the bearer — by time-division, (fixed or statistical) or frequency techniques. See TDM, FDM and statistical -multiplexer.

DEN

Document Enabled Networking. A document management architecture developed by Novel and Xerox from ODMA. It is designed to help developers create networked document-management applications. The DEN API opens the way to middleware services — including links to DEN-compliant back-ends.

denary

The correct name for decimal — base-10 numbers. Decimal should only really refer to denary 'fractions'. So with the number 134.567, the numbering system is denary, and the .567 is the decimal portion of it.

departmental LAN

A slightly larger version of a work-group LAN. It is a network which shares common local resources — applications, equipment and data. It is a vague term, but it suggests a network large enough to be reasonably self-sufficient. It will probably have its own laserprinter/s, fax server and file server, and perhaps one or more local CD-ROMs. It will be separated from other company networks by a bridge or router.

depth of field

The range of physical space within which objects are reasonably in focus with a camera lens. From the point of critical focus, it extends both closer to the camera, and behind that point.

Suppose a camera is critically focused on an object 5 meters away, using say an aperture of f-2.8. Objects closer than, say, about 4 meters and further away than 7 meters will be noticeably (and progressively) out of focus. We then say that the depth of field is 4—7 meters.

If you reduce this aperture to say f-4, you will increase the depth of field range to, perhaps, 3—9 meters. At f-8 (a smaller aperture still) it may be 2—20 meters. In all these examples we have left the critical point of focus on 5 meters.

Depth of field depends on:

- The **focal length** of the lens (F). Short focus (wider-angle) lenses have a greater depth of field.
- The **lens aperture** (f-stop). Smaller lens apertures (higher f-stops) produce a greater depth of field.
- The **distance** to the critical point of focus. A lens focused on a point at 5 meters may have a 3 meter depth of field (1m in front, and 2m behind), but if the point of critical focus is changed to 50 meters, the depth of field will likely be from 10 meters to infinity.

• Don't confuse this with depth of focus which is a similar concept. This takes place behind the lens at the film plane.

deque

A linear list in which insertions and deletions can be made at both ends of the list. It's a double-ended queue. The distinction is with FILO (stack) and FIFO.

derated

Satellite transponders rarely run at their nominal power for any length of time. A 64 watt transponder wouldn't last more than a few years if it was run constantly at full power. So when you rent a transponder, you always rent a specified 'derated' power output, and often it will be derated by 50% or more.

derived calls/circuits

The international carriers quote capacity in 'derived call' numbers when they mean the voice calls a cable can carry when it is used with digital speech interpolation techniques (DCME or TASI). Usually they calculate the number of 64kbps PCM calls that can be carried and multiply by 4 or 5 (in the Pacific area).

However in the Atlantic, voice is now being carried at 16kb/s (ADPCM) and DCME is then added, so the 'multiplication factor' may be higher. The number of 64kb/s channels is the 'nominal' circuit capacity.

derived channels

A name often given to a sub-channel in a frequency-division multiplex system; a subsection of some standard channel bandwidth. This is also called the data-band in digital FDMA.

DES

Data Encryption Standard. A rather old, but still widely used, ANSI-defined encryption system released in the US in 1975, and still probably the most widely used encryption system in the world. It uses a 56-bit or 64-bit cryptographic key code.

Many government departments insist that purchased software must support DES, and it is also widely used in EFTPOS, financial transactions, and similar sensitive communications areas. The US government supposedly tightly controls who can use the encoding/decoding chips but in practice these chips can be bought almost everywhere.

DES Types 1 and 2 are highly sophisticated, and are reserved for military and national security organizations. Two Israeli scientists at the Weizmann Institute, Dr Adi Shamir and Mr Eli Bihma have recently cracked DES. See RSA, PGP and EES also.

descender

In desktop publishing, the descender is the part of a lower-case letter which lies below the base line with characters like p, q, y and g. Some of the early computer screens couldn't handle descenders — which made them very difficult to read. See x-height and leading.

descriptors

• In computer software design it is a special binary code that aids in the location and identification of special data.

• Descriptors in programming identify how some other information should be interpreted or decoded.

• There are also descriptor registers on some CPUs.

• In an on-line database, special 'key-words' are often selected and stored in a descriptor field to aid in searching; these are terms which are fundamental to the subject at hand.

Descriptor lists are usually printed and supplied in a database 'thesaurus' (a controlled vocabulary) which searchers use to identify the most appropriate term. For instance, when applied to ocean mining, the terms marine, submarine, ocean, undersea and underwater may all be used interchangeably by the report writers, but only one will be chosen for use in the descriptor field and as a thesaurus term. See identifier and controlled vocabulary.

desk accessory.

In Macintosh terminology, a desk accessory (DA) is anything from a small utility to a large application which resides within the system and is accessed via the Apple menu — usually it is an alarm clock, calculator, control panels, etc. This provides a form of multitasking, since the DAs are accessible at any time even while another application is being used. It is not pre-emptive, but some memory is always reserved for DAs. See TSR.

desktop

The physical area on which you work and on which your computer sits with its own metaphoric desktop in cyberspace.

With the Macintosh and Windows, it is the imaginary space inside a computer in which you retain the applications and files on which you are currently working. The term refers both to the hand-disk space in which everything is stored (in volumes or folders on the desktop), and to the functional layer which provides the click-select GUI interface.

You could argue that the desktop is just the blank opening screen, before the GUI windows appears.

Desktop ATM25 Alliance

The charter of this group is to make 25Mb/s ATM products open, accessible and interoperable with other leading ATM products.

The members include IBM, Fujitsu, IPC, LSI Logic, Lexmark and many others. It was formed in mid-1994 and was strongly opposed for a time by other members of the ATM Forum. Now it, and its standards, are being accepted.

desktop conferencing

A term applied to low-rate videoconferencing systems, where the main emphasis is on the ability to share applications over the network — or at least point to words, spreadsheet cells, or graphics using cursor controls across the network, while retaining a reasonable voice connection. Usually you will have a small stutter-vision image of the other person on your screen.

desktop publishing

DTP followed the development of the Macintosh and the Apple laserprinter, and probably saved the Macintosh from a long and lingering demise. With page layout programs, and with a wide variety of special scanning, imaging, font manipulation programs which ideally suited the Mac's graphics interface, a whole new 'killer application' emerged.

DTP now means anything in the area of the graphics arts which outputs on paper. Multimedia is DTP carried into video (see DTV), and presentation graphics are DTP programs intended for lecture-projection.

despotic network

One in which the overall clock control is vested in a master clock. The distinction is with democratic networks.

destination code

In the US, this is the name applied to the full telephone number comprising a three digit area code, a three digit exchange (office) code and a four digit terminal number.

detection

The process of converting a modulated carrier to a useful 'message' by the process of demodulation. There are various forms of detection: the books list synchronous, envelope and coherent detection techniques. In my childhood it was a cat's whisker on a piece of coal; now they use diodes.

deterministic

A concept of network management where access to the network is controlled and distributed in some organized way — it is not 'opportunistic'. The distinction is with contention systems which grab space on the shared media whenever they can, but not according to any prearranged schedule.

Determinism is an important feature in the control of manufacturing systems, and wherever video and audio feeds (isochronous services) are passing through packet networks. The user needs to be confident that data-flows can be regularly supplied. Token systems are deterministic, while CSMA systems are contention.

deviation

- Variation from the normal.
- The level of modulation of an FM signal. The extent by which the baseband signal or subcarrier shifts the main frequency. It is essential on FM signals.

device control

A group of four ASCII characters DC1—DC4 which are primarily intended to turn subordinate devices on or off. See DC.

device driver

There are two types: some plug in to the operating system to allow it to handle disks, tape-drives or graphics displays. Others attach to the application to allow these programs to use a particular device.

Generally a device driver is a small systems-utility which attaches to the operating system via APIs, and which handles the communications between the computer and a peripheral (mouse, modem, printer, sound board, disk-drive).

A printer driver, for instance, translates the information from the computer application into a format which is usable by the printer; and since the requirements of printers vary dramatically, the device driver must be specific to the device. Often one manufacturer will dominate a niche so successfully that his device-drivers become the *de facto* standard.

Special driver routines may need to be written for devices that are not standard, and slight variations in driver routines create more problems than anything else when attempting to match peripherals and software in computer systems.

- In MS-DOS, device drivers are installed by referring to them in the CONFIG.SYS file.

• Windows 3.x, by being an application between MS-DOS and the working application, provided its own standardized device drivers, and so it was able to replace special drives for each application. This is probably a key reason why Windows was so successful.

• A device driver card is a circuit board (an adaptor) that fits into a slot and drives the device. The type of driver is usually specified — a video card drives the video screen, a printer card drives the printer, etc.

device ID

In Plug and Play, this is a code in the device's extension which indicates the type of device it is. These IDs, together with the vendor's ID (32 bits with manufacture, model and version of device), jointly create a unique identifier.

device independence

Device dependence was the norm in the MS-DOS days; each application needed its own special driver to work with almost every peripheral or adaptor card.

Windows provides device independence from MS-DOS. One device driver for Windows can handle the driver controls for any Windows application. Plug and Play, is the 'final frontier' for device independence on MS-DOS machines, but the Mac has been this way for years.

device partition

• The division of a large media (hard disk) space into smaller areas, each of which will be treated as a distinct 'volume' or logical entity. If your hard disk is partitioned, one section may become the C: drive, and the other can only be accessed as the E: drive. These are then known as logical drives.

• In the Macintosh, the term has a slightly different meaning. A partition can be a 'volume' that is part of a larger volume. The distinction here is between a 'hard' and 'soft partition'.

DFB

Distributed Feedback. In fiber optics, this is a type of laser with a very narrow bandwidth which is used for long-distance submarine cables and for coherent transmission systems. They are available for both the 1310nm and the 1550nm optical windows. They can be directly modulated, which makes them cheap and simple to control. See YAG.

DFC

Data Flow Control. A control and management element in X.25 packet switching networks. It is responsible for the direction and flow of packets.

DFSA

Dynamic Framelength Slotted Aloha. A contention access technique used in MPT 1327 trunked radio. See aloha.

DFT

Discrete Fourier Transform. Transforms analyze complex waveforms and produce a series of numbers representing the major-to-minor resolution elements (or harmonics) in a block of image or a sample block of sound. This is a practical implementation of Fast Fourier Transform techniques which is, itself, similar to Discrete Cosine Transform. See DCT and FFT.

DFWMAC

Distributed Foundation Wireless MAC. This is the proposal of the Wireless LAN group of IEEE (802.11) for a standardized wireless LAN technology. It standardized the MAC parts, while the PHY specification is different for each frequency bandwidth.

Access is controlled by CSMA/CA ('avoidance') which is similar to Ethernet's 'detection' scheme.

DGIS

Direct Graphics Interface Standard. A proposed specification for graphics processor chips. DGIS is widely used in CAD areas. It supports structured and bit-mapped graphics.

Dhrystones

An objective measure of speed and performance of a computer processor — measured in Dhrystones per second. This is now an outdated 'black box' benchmark; it was last revised in 1988. It measures overall performance by applying a weighted mix of the CPU's instructions, and seeing how long it takes for these to be performed, as compared with a VAX 11/780.

The benchmark includes weighted percentages of procedural calls, loops, integer assignments, integer arithmetic and logical operations. It evaluates not only microprocessor performance, but also compiler frequency.

• As a guide: 386-based computers run at about 8,000 Dhrystones a second, and fitted with the 486DX 33Hz chip, the PCs measure about 15,600 Dhrystones.

DIA

Document Interchange Architecture. IBM's SNA extension to give micros access to host mainframe data; it aids file transfers among dissimilar IBM devices and is associated with the DISOSS protocol.

Diablo printers

A trade name that, for a time, almost became generic for daisywheel printers.

diacritic

Aka diacritical mark. A range of small accent marks which are placed above or below some letters in certain European languages. Under this heading are acutes, graves, umlauts, circumflex, tilde and a range of other more exotic variations. These are known as national characteristics.

diagnostic programs

Many computers run diagnostic programs to check for possible hardware errors automatically on boot-up.

• Level 0 diagnostic are performed by PCs whenever they are booted. This is POST.

• Level 1 diagnostics are run from disk to check peripherals, memory and various controllers.

dial-in/out server
See ACS.

dial modifiers
Extra character/commands that are entered into a modem's dial-number sequence to modify the dialing behavior. These are part of the AT-command set, and they are included in the ATD dial string to modify the number-dialing sequence:

T or **P**	Tone or Pulse dialing.
, or ,,,	Pause before continuing (usually 2 sec. a comma).
W	Wait for a second dial tone.
$	Wait for a 'bong' tone (for calling card).
@	Wait for a quiet answer.
!	Flash the hook.
R	Reverse the originate/answer order.
;	Return to the command state after dialing.

dial-on-demand
A relatively new bridge/router feature where bridges and routers will dial up and establish a physical ISDN connection to a remote location on an as-needed basis. This may be the only connection, or it could be a back-up for overload.

dial tone
This is a tone of between 350 and 440Hz used by telephone companies to indicate that the phone is 'off-hook' and that the system is ready to accept dialing signals. This is not related to the choice of signaling technique — pulse or tone dialing.

• 'Offering dial tone', is another way of saying 'providing a conventional telephone service'. See video dial-tone also.

• 'Stutter dial tone' is used by North American telephone companies as an indication that a voice-message has been left in the exchange voice-mail facility, and not yet retrieved.

dial-up
Connections which are made on a temporary, rather than permanent basis by dialing up a phone line link. The significance here with Internet and packet-switching systems, is that with a dial-up link you can't have a permanent distinct address to which your messages will be automatically delivered. You need to deal through an intermediary host which will store incoming messages. Dial-up links to the Internet do still allow the use of SLIP and PPP.

dialect
A version of a computer language which varies slightly from the (*de facto*) standard. These variations are usually labelled 'enhancements'.

dialing parity
A regulator's rule which says that a dominant telephone company must allow the automatic transfer of local calls to the consumer's chosen carrier, without the need to dial extra digits. This is a similar concept to equal access, but for incoming calls.

dialog
• A small information window that appears on the screen to inform you of some mistake or error, or to force you to provide a parameter, or make a choice. Generally this is a 'child' window related to a higher window object.

• The Dialog information retrieval service was long a well-known commercial on-line database service provider.

• In telecommunications, dialog is a synonym for two-way communications of any kind. Dialogs proceed according to defined protocols. I can't find a clear distinction in the literature between dialog, conversation and sessions — except, perhaps in the length of time involved.

There is a basic hierarchy of dialogs:
— Single datagram packet, one way,
— Conversation/interactive exchange,
— Transaction,
— Sessions (multiple transactions).

• In transaction processing, dialogs are exchanges between two applications which take place across an 'association'. Either single or multiple (and sequentially) dialogs are usually supported.

diaphragm
See iris.

DIB
Device Independent Bitmap. This is Microsoft's Windows image file standard which is able to handle up to 24-bit color (16 million colors). It can do this independently of the computer's display card. DIB images also support monochrome, 16 color, 256 color, and 24-bit color. See BMP.

dibits
In some common communications systems (specifically multi-phase QAM modems) more than one bit is often sent with each baud (change of state). In PSK modems this baud is an angular phase change, often coupled with an amplitude change. If only two bits are transmitted with each change (each baud or symbol), then the device is transmitting dibits.

• This is why 'baud' (the number of events per second) and 'bits-per-second' aren't necessarily the same. See tribit, quadbit and baud also.

dice
The plural of 'die'. These are probably uncut chips, still in the large wafer form.

DID
Direct Inward Dialing. DID is a service offered by the telephone companies to allow large organizations to have numerous telephone lines and extension numbers attached to a relatively small number of trunk lines.

It works by treating the company PBX as a sub-exchange of the carrier's own local exchange and passing to it the dialled-number information. This means that the last three or four digits of a telephone number can also be the extension number of a PBX.

These systems are also used for paging, fax-servers and call-back systems so that the last few dialled numbers provide internal routing information after the message has been received over the trunks. Some US DID services are as cheap as $20 a month for 100 numbers.

• Most DID lines are receive-only. You dial out through a normal exchange link. The DID process is:
- The carrier signals on the line to the PBX by reversing the line voltage.
- The PBX accepts the call by reversing the voltage again (a wink).
- The carrier transmits the last 2—4 dialled digits.
- The PBX takes and uses these digits to reach an extension.

didot
A European typographic measurement of type size. One didot point = 0.148 inches; twelve didot points = 1 cicero.

die
An uncut chip on a wafer during the manufacturing process. The plural is dice.

dielectric
The non-conducting material separating a pair of conductors — the insulation between the conducting elements in a coaxial.

• It is the insulation/separation between the two conductive plates across which capacitance exists, in a condenser.

• In a hollow-tube waveguide the dielectric is air which has a very low charge-loss when dry.

• In coaxial cable it is a plastic material separating the inner core from the external shielding.

dielectric losses
The materials which form the insulation or which separate the conductors within a cable jacket always absorb some of the transmitted signals being propagated along a wire. Category 3 cable with PVC (which contains chlorine atoms) is rather notorious here in absorbing large amounts. Temperature rises exacerbate the problem, so the attenuation of the signal in these cables is related to a temperature change.

DIF
Data Interchange Format. DIF was the first standard for the interchange of spreadsheet files. It was the file-format of the original VisiCalc and every spreadsheet includes it as an option, so it became the way to transfer data from one spreadsheet to another. DIF carried values, but not formulas or calculations.

differential

• **Sampling Differential:** In digital coding, this is where the difference between two sample values is transmitted, rather than the absolute value of each. This is a way of reducing bandwidth requirements. Shorter 'binary-words' are used to represent difference-values than are needed for absolute-values.

Provided there is a reasonably close relationship between successive samples in digitized voice (which there always is) we can use 4-bit difference values (1-bit for direction, 3-bits for value), rather than the traditional 8-bit absolute-value samples and achieve the same voice quality as PCM. However with modem transmission, a 4-bit difference value often won't be able to follow an abrupt change in phase — the successive samples are greater than the system can handle.

Don't confuse differential with delta, which is another similar technique (see).

• **Angular Differential:** In phase modulation modem systems, differential means that the information is coded by the amount of phase-change between successive events, not by any absolute reference to a zero/standard phase. PSK systems are either of the absolute type (measured from some standard phase) or differential (measured from the last phase change). The differential type is easier.

• **Signaling:** Differential signaling is balanced circuit signaling where two wires are used for each circuit, and the voltage (direction of current) defines the signaling state. See RS-449 and balanced circuits.

• **Video:** In basic video coding, the differential is the difference between one frame of image and the next (usually calculated at a block level). Difference values constitute the main form of video compression. See differential prediction.

differential amplifier
A circuit which derives its output from the difference between two input signals.

differential echo suppression
Long-haul telephone connections use 4-wire circuits where each (go and return) channel is carried separately. Echo suppression (to reduce reflected signals) can be implemented by comparing the levels of signals on the different pairs. See hybrids, impedance, echo cancellation and echo suppression.

Differential Manchester
A line encoding technique which always has a voltage transition in the middle of the bit interval. If there is a transition at the beginning of the bit-interval, it signals a binary 0, and if there is no transition at the beginning of the interval period, it signals a binary 1. The mandatory transition in the middle of each bit interval solves synchronization and DC coupling problems. See Manchester coding.

Differential PCM
DPCM. (Note this can mean either Differential or Delta PCM.) This is a form of Pulse Code Modulation where the rate of change is measured, rather than the absolute amplitude of each sample.

The three major variants are: Predictive Differential PCM which is used in wideband TV where redundancy exists, Adaptive Differential PCM which is widely used in telecommunications, and Delta PCM which uses a one-digit codeword encoding technique and has some simple voice applications.

The term 'differential' here is used both specifically for 'difference-sampling' techniques, and also in a wider sense to encompass the 'delta' techniques.

Difference sampling is popular because the change in dimensions between successive samples is usually very much smaller than the magnitude of each sample itself — thus shorter binary numbers (usually 4-bit codes) can be used to reduce bandwidth — or alternately, the sample-rate can be increased to improve quality for the same bit-rate.

Distortion in DPCM systems comes about through compliance errors which can only be overcome by increasing the sampling frequency — hence 'Adaptive' DPCM.

Generic DPCM is subject to errors when a signal changes very significantly between samples; the reduced number of bits cannot satisfactorily code the change. See ADPCM.

differential phase modulation

Phase modulation techniques (for data transmission) involve shifting the cyclic stage of the carrier frequency by varying amounts (degrees) within one cycle of the main carrier wave. Each change is called a symbol, and each symbol carries meaning for one or more bits.

Phase changes are actually made by combining sine and cosine waves. In differential phase modulation, it is the change in phase following the last symbol (the last phase change) which signals the condition, not reference to any absolute phase value. See PSK, BPSK and QAM.

Differential Phase Shift Keying

DPSK is a variation on the normal PSK modem technique. PSK has a direct relationship between the angular phase difference and the di-, tri- or quad-bit being signalled. Differential phase modulation systems don't have this direct relationship. See.

differential prediction

MPEG-1 encoding is more complex than JPEG because it uses differential predictors and temporal redundancy (see). Differential prediction uses the information available in the decoder, to calculate a difference signal.

An image frame is coded and transmitted, and a copy is temporarily stored. A second frame is then introduced to a temporary frame-store. The first image is then decoded, and subtracted from the second — so only difference values remain. These are then subject to DCT before transmission.

In most modern digital video compression schemes, it is only this differential which is transmitted for most of the images, but occasionally a full-screen image will be inserted to ensure that possible corruption of the signal doesn't cascade down through a succession of images, since the differential values are always being referring to a past 'master' image.

The predicted frames (interpolated frames) can be either 'forward' predicted from the intraframe images (those coded as still images), or 'backward' predicted from previously predicted frames. All MPEG-1 encoders must also have a decoder to provide this predictive function.

• It is this predictive function which makes MPEG unsuited as a post-production compression system, since an edit could easily remove a reference frame from which the predictions are taken. See motion compensation, JPEG, MPEG and DCT.

diffraction grating

This is an optical spectrum-splitting device which relies on micron-spaced grooves etched or engraved into either optical-glass or a polished mirror surface. Gratings can work by either reflected or refracted light. The color separations arise because of interference effects from groove spacings close to the light wavelength. The angle of reflection or refraction will depend on the optical wavelength (color), so a diffraction grating can be used as a prism substitute to separate out different WDM laser-beams sharing a common fiber.

diffusion

To spread, scatter or disseminate — usually thinly through another medium.

• The manufacturing (or accidental) process where elements or ions leach into a material. Dopants are deliberately diffused into materials to alter their characteristics. See EDFA.

• Submarine fiber cables under enormous water pressure suffer from hydrogen ions in the water gradually passing through the copper-tube and insulating barriers, and penetrating the glass itself, gradually altering its optical characteristics.

Digicipher

A digital set-top TV unit with compression and encryption, designed by General Instruments. It uses DCT data compression techniques and motion compensation.

- **Digicipher I.** A system widely used in North America. A digital-PAL version is due in April 1995. The cost of each box is about $US1500. The video syntax is similar to MPEG but it uses smaller macro-block predictions and no B-frames.
- **Digicipher II.** A new version based on MPEG2 compression, but it remains a proprietary system. It is due for release in late 1995. The costs for a box are said to drop to $US600 through mass production.

Digicipher II supports both the GI Digicipher I standard and full MPEG2 at the Main Profile level. It uses a 120-pin monster-chip which does the signal processing. This is GI's digital set-top box standard for satellite delivery while DigiCable is the cable variation.

These systems are under development only for NTSC and the American HDTV system. GI claims that DigiCipher II can handle up to 10 digital 'entertainment-quality' image channels in the bandwidth of an old analog channel.

In MCPC applications it has a 27Mb/s data rate, using offset QPSK modulation and FEC. Audio is CD-quality Dolby AC-2 with 4 audio channels per video.

Renewable security cards are used, and the system includes a copy protection scheme.

digipeater

Digital repeaters. See packet radio.

digit

There is a distinction between a number, a figure, a numeral, and a digit. Decimal figures are 0 to 9 and decimal numbers are made up of one or more figures. These can be either integer numbers (whole) or real numbers (decimal). They represent a value on which mathematical functions can be performed.

Numeral applies to the symbol or symbols which represent numbers (such as the word 'ten'). A zip code is more correctly referred to as a numeral — as is first, second, third, etc.

Digit implies that there is positional value of each figure in the number. The number 234 has three digits, of which the first represents 200, the second 30, and the third only 4.

digital

Digital simply means 'of the fingers', and strictly should really be applied more to denary (base-10) counting than binary (base-2).

In fact the only real implication of the term is that the information is in discrete, countable, integer or 'quantized' form. The distinction here is with continuous changing (analog) states where progression is infinitely graded.

So we shouldn't confuse digital with binary. Binary is simply one form — the two-state (on–off) form of digital representation. There are also three-state (ternary), four-state (quaternary), five-state, etc. digital systems.

The clear advantage that binary has over other digital forms, is that on—off states are almost always unambiguous. With three or more voltage states on a line you can misread and make mistakes. But with only two (on—off or positive—negative), it is difficult to become confused.

With most modern information processing and recording systems, the value of the digital approach over analog for storage, reproduction and transmission of information is now obvious. It is accepted that there are always small losses of quality in transforming any analog signal to a digital form. But once the information has been converted to a chain of integer-numbers, the encoded signals will accumulate no additional noise (unless they are transcoded).

Digital symbols are virtually immune to interference or to distortions arising from signal losses. In this form, they can also be checked for errors, processed for compression, filtered for content, and can endure almost unlimited dubbing (copying), multi-layering, multiplexing and transferring over long distances with negligible loss of quality. See quantization and generation loss.

digital channel

A channel capable of carrying direct current and therefore of carrying two-state (at least) digital signals without modulation. You have digital channels between your computer and a local printer for instance. However digital channels are limited in the distance they will carry, unless special line-coding techniques such as AMI are used.

Even then, the distance is limited. Analog channels will not carry DC since they are designed for AC frequencies — which is why a modem is used to modulate the digital signals onto an analog audio carrier when the signal must be carried any substantial distance. Some ternary and quaternary channels will carry digital information without these restrictions. See Manchester coding.

Digital Compact Cassette

See DCC.

Digital Component Technology

DCT. A professional component digital videotape standard from Ampex developed in 1992. It offers Discrete Cosine Transform 2:1 compression and uses 19-mm videotape, making three hours of recording possible on the largest of their 'shell' cassettes.

The emphasis in this development was in maintaining video integrity through the use of error-correction; so it is seen to be the ideal mastering medium for program material originating on Betacam SP and film: DCT is claimed to maintain good picture quality over 30 generations.

digital component video

Digital component video exists in production studios to preserve the highest possible image quality through the post-production process. Recommendation 601 of the CCIR, sets the standard for a time-divided digital component system for television production, and the D-1 video recorder from Sony was the first to introduce a workable system. See 601 CCIR, D-1, 4:2:2, DVC, and analog component video also.

digital composite video

The introduction of digital component video to television production (see D-1) proved to be difficult and Ampex created a D-2 composite standard to overcome these problems. However it proved to be difficult to make light-weight camcorders with the D-2 standard. Panasonic's half-inch Dx standard (now called D-3) is also a digital composite format and so competes directly with D-2 and the older Betacam SP. See D-3, 601 CCIR and component also.

Digital Data Storage

DDS. This is a version of the DAT audio system used for information storage and computer backup — but it is a streaming-tape method (unlike DataDAT) using 4-mm tape. The system was established by Sony and Hewlett-Packard in 1989 and it has become the *de facto* data recording standard in the DAT format.

DDS performs random reads, but not random writes, so it isn't capable of updating files in place. About 60% of the tape surface is allocated to user-data and forward error correction, while the other 40% provides automatic track-finding information, save-set marks, and filemarks.

It uses a four-head rotating drum with two read- and two write-heads (normal audio DAT only has two) to allow immediate read after write (DOW). It also has an additional level of CIRC error detection and correction. The manufacturers claim an error rate of one in 10^{15} bits.

- **DDS-1** logical file format allows up to 2GBytes of uncompressed data on a standard 90 meter DAT tape cartridge, with a transfer rate of 183kB/s which takes about 2 hr to fill a tape.
- **DDS-2** which was specified in 1993 can handle 4GB of native data on a 120 meter tape using SCSI-II in burst mode. Some later cartridges take this up to 5GB.
- **DDS-3**, they say, will provide 12GB of native storage.
- **DDS-DC** is DDS-2 with data compression.

An alternative to DDS is the slightly slower DataDAT format. See Super 8 also.

digital exchange

Most telephone exchange equipment was analog until a decade ago, and the carriers generally used crossbar electro-mechanical switching. Later stored program control (SPC) systems (computers) were added to these exchanges — and sometimes these SPC analog switches are mistakenly called 'digital exchanges'. They are not.

Digital exchanges use computer-based switching equipment which has no electro-mechanics. All switching occurs within digital circuits. Some of the confusion comes from the fact that some major switches/architectures, such as the Ericsson AXE exchange, have both an analog and a digital version.

digital filling

In some synchronous communications systems it is often necessary to fill gaps in transmission with bits to maintain synchronization. These are identified and removed at the other end.

digital laser

See linear laser.

digital monitor

This was an IBM idea in the early days of PCs — and not a very good one. They decided to use a very cheap and simple monitor and drive the single (monochrome) electron gun with digital (on-off) signals. This was the MDA system.

Later they upgraded this to handle color, so both CGA and EGA had multiple digital channel feeds and special electronics switching the electron guns; the result was very low quality. Finally IBM gave up on digital monitors, and the conversion from digital to analog now takes place within the PC. Analog signals then drive an analog monitor. This approach has the flexibility we need for high-quality color.

digital multiplexer

Unless otherwise specified, this will be a time-division multiplexer or a statistical multiplexer. An analog multiplexer will be one using frequency-division. Of course, a FDM can handle either digital or analog, but with digital it makes more sense to use TDM or statistical multiplexing.

digital paper

A year or two ago ICI Electronics developed a new optical storage medium based on a flexible Melinix polyester-plastic film. It was a WORM system where the film is coated with a dye which is sensitive to infrared light. On the surface of this dye layer was a protective 'overcoat'. The 'paper' could be cut into reels or made into disks.

ICI claimed a shelf life of 15 years for the material and a storage cost of half a cent per million bytes. I haven't heard anything about it since.

digital radio

Microwave standards. See DR #-#.

Digital S

JVC's name for their new DVC video recorder standard.

digital signal processors

DSPs are super versions of RISC/maths coprocessors in VLSI chip form, which are optimized to perform spectral analysis and other signal processing functions. The discovery of applications for Fast Fourier Transform (FFT) resulted from a quantum leap in DSP technology.

No sector of the computer/communications industry has progressed as rapidly as DSP developments: DSPs have accelerated the pace of digital television dramatically. ATM switching and multiplexing, and MPEG compression, are all now possible because of DSP advancements.

DSPs differ from maths coprocessors in that they are independent of the host computer and can be built into a standalone unit. Most DSP chips have special floating-point multiplication and addition hardware, and most can complete instructions in a single clock cycle.

Like RISC, they depend on a small core of instructions which are optimized at the expense of a wider set. They are often capable of special addressing modes which are unique to the particular application.

- When processing video, the new DSP chips concern themselves with the enhancement of images (digital processing) and statistical analysis (mainly transform coding). The image-processing functions operate on the individual pixels within the video memory array. There are two main function-types: point operations and neighborhood operations.
- Speech recognition, robot vision, etc. all require specialized data processing at rates which are usually too fast to allow sharing of resources with a central processor. So special DSP chips are needed, together with large frame stores buffers and high bandwidth data paths.

These technologies are critical for future voice-activated systems, artificial vision, imaging applications, and to future filtering and control of massive data flows.

- Home TV sets are increasingly using digital technology, and future EDTV sets may require quite extensive digital image processing. Even current TV receivers can improve picture resolution and add extra features (anti-ghosting, multi-picture displays, picture-in-picture and special effects) using DSPs.

digital signatures

A technique of using private key encryption for authentication of the originator. The originator will add a short message which has been encrypted by his/her (secret) private key. This means that the receiver can check the source of the message by deciphering it, using the originator's public key, which he/she can obtain from a public directory database.

This is part of the X.400 and X.500 MHS standards. If the directory service is secure it is probably more difficult to forge a digital signature than a hand-signed contract. See public key encryption, PGP and RSA.

digital sound recording

The analog sound signal is sampled and quantized at a rate which must be at least twice the highest frequency to be reproduced. There are a number of different digital sound recording techniques, but the best known is Pulse Code Modulation and the differential variations on this. See ADPCM and CD-Audio also.

- DAT was the first digital sound recording system available for high-end public applications. It is now widely used for professional recording.
- DCC is likely to be the cheapest of the current digital sound recording systems, but it relies on a lossy form of compression which won't meet most professional requirements.
- CD-i and CD-ROM/XA systems offer a range of digital sound modes. The advantage of 19 hours of recorded sound on a single 12cm disk have not yet been fully exploited. Low-quality long-playing systems are likely to be just as important as high-quality systems with just one hour of recording.

digital speech interpolation

DSI. This is a way of conserving spectrum in radio and cable telephony systems. It assigns bandwidth only to the active voice channels (when you are listening, you aren't consuming bandwidth on your speech channel). This is a generic term for the digital version of TASI which was used on undersea analog cable systems, and also a way of reducing radio noise in cellular phones.

DSI is able to use high speed digital voice detection systems, and it can also be coupled with low-rate voice and bit-stealing techniques, so it can be many times more efficient than the best TASI systems. These days, some of the circuits will also be assigned to digital data rather than to voice — so data can have a lower priority, since a half-second delay is rarely serious. See TASI, DCME and DCMS also. See voice activity ratio also.

- GSM cellular phones use a suppression system based on the same technology. This is to reduce noise in the system when the radio channel isn't actively being used.

digital still camera

There are rapid advancements now in digital still camera, and the range of output devices needed to covert these to useful hardcopy or color separations.

- Canon's High Definition Digital Camera uses a BASIS CCD image sensor with 1.3 million pixels, and it stores the image on an IC memory card or hard disk. The CCD has a sensitivity equal to ISO 200 film, and it has a 10-bit ADC. See image stabilization also.
- Apple has a low-cost (under $1000) digital still camera which can hold a dozen images and put them directly into the Macintosh.
- Kodak has a US$300 digital camera under development, to be released in 1996. It already has a US$1000 digital camera for the professional market.

digital tape

A tape made specifically for digital recording. It has different characteristics to an analog recording tape. See coercivity.

digital termination systems

Digital radio links used to reach customers from a central satellite ground station.

digital to analog conversion

D-to-A conversion or DAC. This is the reverse of ADC. An analog output is needed by most electronic systems dealing with continuous variation waveforms. Both audio and video are inherently analog in nature.

- In D-to-A conversion, corrective approaches called 'oversampling' or 'bitstream' are widely used in CD-Audio to solve some of the quantization problems. See.

digital videodisk

Modern CD-based video disks (CD-HD and CD-Multimedia) are essentially digital in nature. However some of the older videodisk systems (despite using pits and lands in the same way) were fundamentally FM analog. However, these could be used to record digital information.

LV-ROM was a development of the LaserVision videodisk which could store digital text or numeric data — it was used in the UK Domesday project. The technique modulated the digital signals onto an FM analog carrier, and the LaserVision videodisk then stored the composite signal as analog FM. See LV-ROM.

digital zoom

A zoom-in narrows the angle of view by changing the focal length of the lens progressively. However the same image change (but not image resolution) can be achieved by selecting progressively smaller areas of the CCD surface, and performing mathematical functions on the individual picture elements to restore the correct number of scan-lines and pixels.

Resolution suffers because of interpolated 'averaged' pixels, and with doubled and tripled picture elements. See zoom.

digitizer

• A general term for any device which changes information into binary-digital form for use by a computer. Digitizers usually sample analog information, and therefore have quantization errors. The term covers a wide range of ADC-based input devices.

• A widely used term for paper-image scanners and their associated software. Scanners are only one form of digitizer. This may mean OCR as well.

• Digitizing tablets — see below and tablets.

• Video digitizers convert the output of a video camera to digital form for processing (editing, etc.).

digitizer puck

A mouse-like drawing tool for CAD work.

digitizing tablet

A graphics tablet, or just a 'tablet'. You draw on the plate using a special pen, and the movements are recorded in vector graphics form in the computer.

digrams

In some of the early data compression techniques, common two-letter combinations were sometimes given a non-print ASCII value. These two-letter combinations (th, in, er, re, and an) are digrams.

DIL

Dual In-line — see DIP.

dimensioning

• In the definition of an array, some programming languages and systems require that the maximum number of variables must be declared before they can be used. This is known as dimensioning.

• In CAD programs the process of dimensioning (adding arrows, measurements, etc.) should conform to the main ANSI, ISO, BSO and DIN standards. Options should exist for leaders, tolerances, units, arrow-styles, bearing and distance, etc.

DIN (connector)

Deutsche Industrie Normenausschus. A famous German standards organization. The term is applied to a number of products, but specifically to small round (about 8-mm) multi-pin connectors which are often used with keyboards and audio recording devices. There are numerous DIN plug and socket types, with many variations in the number of poles (pins).

• 6-pole DIN connectors are used for audio and video. The pin connectors are usually: 1 Switched voltage, 2 Composite Video, 3 Earth, 4 Audio (L), 5 Power, 6 Audio (R).

• 5-pole DIN connectors are often used for audio only. The pin connectors are usually: 1 Microphone (L), 2 Earth, 3 Audio (L), 5 Microphone (R), 6 Audio (R).

• The Macintosh keyboard and mouse connector is a special small DIN plug.

dinkus

In page layout, this is the small photo or caricature used to identify a newspaper or magazine columnist. It usually sits above his/her by-line.

diode

An electrical device which permits current to flow in one direction, but not in the other. It is a rectifier.

dip/DIP

• A drop in the mains power voltage. See sag.

• **Dual In-line Package**: A term applied to numerous electronic components designed to fit into two standardized rows of holes on a PCB. These devices are also called DIL (Dual In Line).

At one time most IC chips were a DIP standard. Today the term will more often refer to sets of small configuration switches — DIP switches. Be careful with DIP switches, there seems to be no standard for on and off being up or down.

A major source of confusion is the labelling of DIP switches:
— OPEN is the same as OFF.
— CLOSED means ON.
Switches are also sometimes numbered from right to left, rather than left to right, so check carefully.

diphase code

See biphase code and Manchester coding.

dipole

In radio systems, this is the simplest form of antenna — it has two arms, each one-quarter the wavelength in length. Modern TV-type antennas often don't distribute their power (or receive) in an omni-directional pattern (equal in all directions), but rather exhibit antenna-gain (emphasis at one angle of elevation, or one direction), and this is then compared to the dipole as a way of expressing the power gain. The measure is dBd, which is gain related to the dipole. Don't confuse this with dBi which is gain related to isotropic radiation.

• An 'omnidirectional' antenna at a cellular phone base-station may have a gain of 10dB. It radiates around the full 360^{0} but it is not isotropic (radiating skyward). A unidirectional antenna can have a gain as high as 16dB. Vehicle-mounted antenna usually have gains in the order of 3—4dB.

direct broadcast satellite

DBS. The transponder in the geostationary satellite needs to transmit sufficient power so that direct reception by small-dish antennas is possible. The transponder must therefore output more than average power (often in the 150 watt range).

It also often means that the transmitter power must be focused onto a smaller footprint (a spot beam) than normal, or that a special shaped beam must be used to concentrate 'spots' of power onto the most populated areas. The whole design of a DBS system is a trade-off between satellite power, antenna beam concentration, coverage area, size of home dishes, and efficiency of the LBN down converters.

Until recently most home satellite reception dishes were fixed in position, but now the dish often needs to be steerable to receive TV signals from a number of sources.

DBS was analog until late in 1994 when Hughes DSS launched the first digital system and Primestar followed soon after. See DirecTV and DVB.

direct call

A connection in a switched circuit that does not require a dialing operation. This is not a permanently leased line, but preprogrammed use of the switched network. When you lift up the handset, the call request is interpreted by the network and the circuit is switched to a predetermined address.

direct inward dialing

See DID.

direct mode

Some computer commands must (or can) be entered directly from the keyboard, rather than stored in a program on disk. These are direct mode commands — LOAD, to load a program; RUN to run it, etc.

Basic and other interpreted languages can often be used in either the direct or the indirect mode. In the direct mode, command statements are executed as soon as they are entered (when you hit the carriage return). The results of any arithmetical or logical operations can be displayed and stored for later use, but the statements themselves are lost immediately after execution. For this reason the indirect mode is generally used. Once the program has been entered, it can be executed in the indirect mode, usually by the RUN command, or something similar.

direct outward dialing

DOD. A PBX feature where outgoing calls can be placed directly from the extension by dialing an initial digit (the access digit — a 0 or 9). The distinction was with calls which needed to be dialled by the operator. See DID also.

direct read after write

See DRAW.

direct sequence

One of the two major subdivisions of spread spectrum. Direct sequence spread spectrum is best known as CDMA. Scc.

direct slot

A form of bus slot provide for one add-in card, used in some Macintoshes. It is not the same as the normal NuBus slot type, but it provides direct local bus access to the CPU.

directing code

In telephony — the collective term for numbers dialled before the normal local directory number when making long-distance calls. Directing codes are those used for zone, country, regional and area routing — as distinct from those used by the local exchange for local-call switching.

directional coupler/tap

On any broadband network (specifically a coaxial Cable TV network) the main trunk line will need to be split either to feed other trunk lines, or for distribution along feeder cables. These will need to be further split to supply the drop cables into the home.

There are two different passive devices used here:

- Splitters divided the signal power equally. So a four-way splitter would put 25% of the signal power into each distribution or drop cable.
- Directional couplers divert a set amount of the power into one circuit, and retain the remainder for the ongoing trunk. They will be limited to say 20% of the power being diverted to the distribution or drop cable, while 80% will continue on down the trunk.

directory

Any index of names and locations.

- Telephone directory systems are almost as important as the networks themselves — and the directory services (particularly the Yellow Pages) of a carrier can be the most profitable side of their whole operation (on a profit-for-revenue basis). In most of the larger RBOCs and dominant carriers, the Yellow Pages is a monopoly billion dollar operation.
- The new electronic directory standards are X.500. These are mainly for X.400 messaging, but the implications are much wider. See X.500.
- A computer's hard-disk directory consists of a single root directory and one or more sub-directories in a hierarchical fashion. Every file has one entry, and one entry only into a directory — but these files can be in either the root or in any sub-directory.

The full 'pathname' consists of the names of the directory and sub-directories higher in the hierarchy (along a single path) than the file, together with the filename.

- In practice, sub-directories are also known as 'directories'.
- In the world of the Macintosh, sub-directories are known as folders.

directory cache

In a large computer network, the directory on the server (directory tables and the FAT — which knows where the files are stored) is usually held in RAM, rather than just on the disk, to speed up access. Often these files will be duplicated on different areas of the hard disk also.

DirecTV

This is a Hughes direct-to-home digital satellite system which has proved to be highly innovative and successful. This, and competitor PrimeStar, are the first DTH digital video systems.

DirecTV uses the MPEG1 compression system at present, and MPEG2 will be introduced later. The proprietary transport standard for DirecTV is called DSS, and the 175 channels available are divided between two program providing groups: the Hughes DSS group and USSB.

The receiving dish is small (18-inch) and easily mounted and pointed by anyone, and it is purchased with a set top box for $US699. A million have been sold to mid 1995.

Currently the Hughes DBS1 and 2 satellites are co-located at 101° West, each with 16 transponders of 120 watts, and polarization is used to separate the signals.

The plans are now to co-locate another two satellites DBS-3 (1995) and DBS-4 (unspecified) with identical power and channels. This is not just to increase the power, but to allow each satellite to be reconfigured as 8 channels with 240 watts, rather than 16 channels of 120 watts. Each transponder channel can offer between 8 and 4 digital TV signals.

Hughes also plans to introduce statistical multiplexing for the channels along with MPEG2. They say that DSS receivers are already MPEG2 compatible; only the uplink changes.

DIS

- **Draft Initial Standard** of the ISO. So for DIS Ethernet, see Blue-book Ethernet.
- **Digital Image System**: Texas Instruments has a micro-mechanical video projection chip which utilizes thousands of tiny moveable aluminum mirrors, each only 16 microns square. These mirrors can be used to reflect light to form a very bright, video image.

  This chip has an active surface which is about 16mm square, with 442,368 of these mirrors in a grid. Each mirror can tilt through +/– 10° and they can be controlled very rapidly — fast enough to project motion video. TI says it can produce the chips using standard fabrication techniques.

DISA

Direct Inward System Access. A service provided by PBXs, where an employee can ring into the company exchange, and then make long-distance calls through the exchange which are charged to the company account. Usually access is controlled by the use of special phone numbers and access codes. It is a service usually provided only to company executives who travel, but it is a service that is much appreciated by phone phreaks.

disable interrupt

A computer needs to use interrupts fairly regularly in routine operations. For instance, the CPU in some PC architectures will constantly stall and suspend operations while it scans the keyboard to see if a key has been pressed. However, you don't want this break in work-flow happening when it is bringing large amounts of data into memory from a disk file, so a disable interrupt is used to prevent the process being disrupted. The screen may go blank or the keyboard apparently dead during these periods.

disabled

See enabled.

disabler

In long distance voice circuits which normally could exhibit echo (to which echo suppressors have been added), a disabler is used to deactivate the echo suppressor or echo canceller for the full-duplex transmission of data. An internationally standardized tone of either 2025Hz or 2100Hz is sent down the channel to activate the disabler. Any drop of carrier for more than 100 microseconds will reactivate the device.

For this reason, fax machines must have a 100ms break after the initial handshaking, since their tones activate the disabler. A fax line is half-duplex, so echo cancellation is required.

disassemble

A way of providing a hex dump (often with enhancements) to analyze machine-code programs. A disassembler program can produce a print-out of the code in assembly form. This is 'reverse engineering' if you are inspecting someone else's code.

A hex dump is simply a listing of the code in the memory locations, in sequence — but a disassembler attempts to provide some guidance as to the meaning of the code. It will attempt to group, say, an instruction together with the following byte/s of data or address to which it refers.

Unfortunately dissemblers often have great difficulty in distinguishing instructions and data — so they often get mixed up; their ability to unravel the logic behind a program's design is usually very limited.

disconnect

A disconnect in a telephone system is a series of active functions — not just the breaking of a circuit. The disconnect usually begins when one party signals by generating an on-hook condition, but it may also be initiated by the network management system. See call clearing.

Discovision

The US name for the Philips/MCA LaserVision optical videodisk. See LaserVision and videodisk.

Discrete Cosine Transform

DCT is the major intra-frame compression technique now being used with multimedia of all kinds — from videophones, to CD-ROM, to HDTV. To compress two-dimensional screen images, a mathematical transformation is applied to image-blocks (usually 8 x 8 or 16 x 16 pixels) and this process generates a series of numbers that progressively represent resolution factors within the block. Since the number series begins with the major image elements, and moves progressively towards finer (and less important) texture details, it is possible to abandon less important detail and so reduce the bandwidth when necessary.

Motion compensation and other inter-frame compression techniques can also be applied in motion video — however it is important to distinguish between the DCT intra-frame coding, and the inter-frame techniques.

DCT has been developed from Fast Fourier Transform theory and it is very similar to Discrete Fourier Transform. DCT is also applied to single still images in JPEG. See fundamental, FFT, MPEG, JPEG and H.261.

discrete speech

Aka discrete utterance: A voice recognition technology that requires an obvious pause between each word. Most systems are still of this kind, although connected and continuous speech systems are the aim of research. See speech recognition and parsing.

dish (antenna)

The dish is actually a reflector not an antenna — but the term is widely misused by everyone. The dish reflects the radio signals into the feedhorn for detection and amplification by the low-noise amplifier and converter. At this point, different signal-polarizations are also passed or blocked from entry.

The dish size needed for receiving data or TV broadcasts direct from a satellite is dependent on a number of factors:

- The wavelengths being used by the satellite (C-band needs larger dishes than Ku-band).
- The output power of the transponder.
- Position within the footprint of the dish (see boresight).
- The amount of expected interference from weather and electrical noise.
- The efficiency of the first stage, low-noise amplifiers on the dish. See LNB also.

However it is important to realise that dish size affects 'resolution' as well; a large dish can accurately point to a satellite, while ignoring another with only 1° of separation. A small dish can't do this.

But this also means a large dish will be much more critical with wandering satellites, so more thruster-fuel will need to be expended to keep the satellite in position unless the dish has the intelligence and mechanism to track the satellite.

Dishes also need a solid mount. In high wind they can tear away from roof-top mountings and the larger they are, the more stable they need to be to retain pointing accuracy.

disjunction

A Boolean, where the result is a value 0 if, and only if, each operand also has the value 0.

disk

- **Magnetic:** The primary form of secondary storage in a computer. For many years PCs only used 8-inch, then 5.25-inch, and now 3.5-inch floppy disks. These only spin when required to deliver or store information.

Hard-disks were previously called Winchesters, and they come in many forms — all of which rotate constantly. Most are single sided, but some are double sided, and some units come as enclosed stacked-platters which create a multi-layered system with perhaps as many as 12 recording surfaces (in large mainframe units).

Disks are logically subdivided into volumes (usually only one to a floppy, but there may be more on a hard-disk), and each volume will have its own root directory. This may then have subdirectories.

Data on a disk is stored in circular tracks, each of which is divided into sectors (a radial subdivision). The drive head will read one sector of information into memory as a single step.

In a multi-layered system, the selected track location on each surface, is known collectively as a 'cylinder'. See cylinder.

- **Optical:** See optical recording, CD, WORM, and MO.
- See Bernoulli box and hard-card also.

disk access time

One of the major measures of disk performance is the access time — the time it takes the disk's read/write head to find the block of data required. There is a random access time (for data anywhere on the disk), and a sequential access time (for blocks of data interleaved, but in sequence). See access time.

disk-based

As distinct from RAM resident; the file is kept mainly on disk, and only essential parts are shifted into RAM at any one time.

disk cache

Computers: Disk cache acts in a similar way to memory cache. It buffers data extracted from one or more of the disk sectors, following the one requested.

The cache controller is predicting that these sectors will be the ones next required. Caching works both with reading and writing . When writing to the disk, it allows the CPU to write data to the cache at full speed, then return to the other tasks.

Disk cache can be added to any system either by the installation of a cache disk controller, or by a TSR program that partitions part of conventional, expanded or extended memory.

There are three main types:

— Track buffering which uses a small amount of RAM. This approach is used by SCSI and ESDI disk controllers to hold only a few sectors.

— Caching controllers are more elaborate than simple track buffering systems. They have enough memory (usually a megaByte or more) to hold many sectors of data. They also use on-board processors to manage the cache.

— Software-based systems store data in system RAM and use the CPU to manage the data exchange.

The effectiveness of cache depends on the ratio of hits to misses — how often the controller guesses correctly. Disk caches can also slow down operations which are not disk intensive, especially in applications like word processors. See memory cache also.

- **Broadcasting:** A very large hard disk can be used as a buffer between radio and television station origination and transmission. Data flows onto the disk and is taken off simultaneously, so the hard disk becomes a temporary digital store with a dynamic input and output flow of program material. There may also be a disk which holds the commercials, etc. on a longer basis.

This technique makes it quick and easy to drop commercials into the program since it just requires a change in disk access. There is no cue time, and no problem with multiple VTRs. This approach also has applications in time-shifting for radio broadcasting in different time zones.

disk cartridge

These are large multiple-platter (usually six) disks for mainframes which can be removed and changed. They are inside their own container, but are removable.

disk controller

This is an essential part of a modern computer's hardware. It controls the flow of information to and from the hard-disk and floppies. The computer's overall functions are very often limited by the disk-access times, and to a large degree this depends on the interleave ratio of blocks of data. Adjusting this ratio can have a dramatic effect. See RLL.

disk farm

A large disk-drive center for storing masses of company information. There are also tape farms with removable cassettes which can be stored and loaded automatically. See DASD and RAID.

disk fax

A device which adds a floppy disk drive to a facsimile machine and allows the transfer of disk files over telephone lines using the facsimile's 9.6kb/s modem protocols. The Japanese consumer electronics industry is expected to support a new standard.

Given the complexity of computer data communications, this is a very good idea: you will treat a floppy disk just like a sheet of paper. Plug it into the fax, dial the number, and it should transfer the data.

disk geography

The physical layout of the disk— the positions of data segments, the location of the FAT and directory information, and the distribution of data files.

disk map

A graphic representation of the use of the disk by the disk operating system. A hard or floppy disk will be divided into concentric (not spiral) tracks, and each of these will have a certain number of sectors. Certain tracks and sectors are reserved for system use (especially the FAT), and others for the directory information.

disk mirror

The technique of automatically making a parallel copy of all changes to a file-server disk, also to a backup unit running at the same time. In the event of a server failure, the backup will switch into use transparently.

In critical services, different systems use different amounts of redundant hardware. Usually all disk drives will be duplicated, but often only some of the controller cards, and this creates a potential area of problems. Disk mirroring can be implemented as a software utility (driver) or in a hardware controller, or both.

disk operating system

DOS. Modern personal computers are dependent, to a large degree, on the operations of their hard and floppy disk drives, but this is only one of the concerns of an operating system.The first PCs had operating systems (not Disk OSs) because they stored their data on tape (usually an audio cassette).

The operating system of a computer is fundamentally concerned with the processor and memory functions: the disk drive is very secondary. However, having an integrated operating system with intelligent control of the disk units became important to make the early machines flexible and productive.

At this time the term DOS became synonymous with operating system in general. Unfortunately the term was appropriated by IBM and Microsoft to mean only PC-DOS and MS-DOS, and that's how many people use it now. Apple-DOS existed many years before MS-DOS was even a Quick-and-Dirty glimmer in the eye of the Seattle Computer Company.

disk optimizer

Aka defrag(mentation) program. These are software utilities which physically rearrange the allocation of file sectors on your hard disk to speed up performance. They will relocate file fragments while simultaneously updating the FAT to reflect the new cluster locations.

Ideally, every file will consist of a single contiguous run of consecutively accessible clusters (not necessarily consecutive sectors on a disk). A comprehensive defrag(mentation) program will put all files at the front of the disk, and move all free space to the back — but this doesn't necessarily give the best performance. See RLL and interleave.

disk servers

A basic LAN function where a hard-disk computer is added to the network to serve as a general magnetic store. Its function is to read and write data to the disk without any consideration of the contents; it is treated not so much as an intelligent device, but as a remote store.

The term is also often used when the hard disk is partitioned to provide private space for each user. The distinction is with file servers, where the computer's disk space is shared by files which the file server controls intelligently.

disk striping

See striping.

disk substrata

• The plastic poly-(anything) backing, to which the magnetic materials of a floppy disk are attached by means of a binder.

• In optical disk terms, this means the actual recording layer — the reflecting surface with the pits and lands embedded in it. In recordable/erasable disks the substrate can be either an amorphous material or magneto-optical. They are usually based on rare-earth metals.

diskcopy

This is a DOS, Windows NT and OS/2 command which make a complete copy of a disk.

diskette

The correct (but now quite quaint) name for a floppy disk. At one time it was applied only to 5.25-inch floppies (as against 8-inch), then later still, only to 3.5-inch floppies.

diskless PCs

Personal computers made specifically for operation on a LAN. The network allows them to access and use the hard disk of a file-server or disk-server, so individual floppy disk drives were not needed on each PC. The essential boot system is held on ROM in the network adaptor at each diskless workstation. The value is (theoretically) in lower costs, and greater data security since it is impossible to take a copy at a terminal. It is an idea that flourished briefly, and then died.

DISOSS

Distributed Office Supported System. A network package of electronic mail and document preparation programs. It was designed by IBM for mainframes and is based on LU6.2. Its associated protocol is DIA (Document Interchange Architecture).

dispatch radio

The two-way and trunked radio sector which services trucks, vans and commercial vehicles. Cellular phones are not generally used here because of cost, and also because the company often needs to make group-calls — to contact all its trucks or taxis, say, in one geographical area.

The category includes both voice, voice and data, and data-only dispatch systems. Traditionally dispatch radio has depended on half-duplex press-to-talk 'two-way' radio usually in the 450MHz band and below. More recently a shift has been made to 800MHz 'trunked radio' systems with both voice and data. See trunked radio, MPT 1327 and MIRS.

• There is also a subsection of dispatch radio used for emergency and police services. They often have different requirements. See APCO-25.

• At the other end of the scale are two-way paging systems which are now being used by couriers.

dispersion

The variation in signal velocity as a function of the signal frequency.

• In all telephone systems, lower frequency signals propagate at a reduced velocity when compared to higher frequencies. So the high frequencies will arrive first.

• In fiber optics this is the cause of pulse-spreading and therefore the limitation to the use of higher frequencies (and therefore of bandwidth). There are two types of fiber dispersion: mode and material.

— Modal dispersion occurs because different photons take slightly different paths along the fiber, and therefore they will arrive at slightly different times — thus blurring the edge of a square wave.

— Material dispersion is caused mainly by minor variations in refractive index through the glass in the fiber. The RI sets the speed of light, and therefore these variations have a similar effect to the modal dispersion.

• Don't confuse dispersion with 'dispersal'. See single-mode optical fibres and solitons.

displacement

• A computer can address a section of memory directly or relatively. In relative addressing, a displacement figure (also called an offset) is added to the current address in the register to arrive at the address of the new location.

• In graphics, displacement can be the x-y distance between one point and another.

display

Another name for the screen, monitor, LCD, CRT or VDT. Sometimes the reference will be to the normal video-screen functions of the computer, and sometimes it will refer to special motion-video functions.

• A display driver is also known as a video controller, a video driver or a screen driver.

• In video, a display adaptor holds the circuitry used to create the video signal, and is also called a video card or video adaptor. It will have one or more frame buffers to hold images during processing. Display adaptors need to be matched to the monitor type.

• Special video adaptors with special multiple functions (including overlay) are used for multimedia applications.

Display PostScript

A version of the page description language PostScript written as a screen driver. It is device independent; it allows the image to be processed to a much higher level of definition than that of the screen being used, in the same way that Printer PostScript can generate much higher resolutions for printing than the available printer might use.

display type

In desktop publishing, this is the style of type use for advertising and for headlines. It is usually more than 18 points in size, bold, and likely to be flamboyant or unusual.

dissolve

A film and video effect where one picture gradually fades out while the next simultaneously fades in. The distinction is with a fade-to-black or to a cut.

distance education

Teaching without direct face-to-face contact by using telephone, videophone or satellite connections. The term once meant postal correspondence.

distinctive ring

This is a CLASS service. Your single phone line can have two or more numbers. When these numbers are used the phone will ring in a distinctive cadence. This cadence pattern can be detected by an automatic switch and used to switch the line to, say, a fax or modem, or another extension. Provision has been made in most modern SPC-switched analog networks for four such numbers on each line, and with ISDN BRA it is possible to have eight (but these don't vary the ring cadence, since the number is carried on the D-channel).

distortion

In telecommunications this is any unwanted change to the signal waveform — it is the difference between the transmitted and the received waveforms. However a distinction is often made between distortion and noise. Noise is superimposed on the signal, while distortion is a change in the linearity, frequency or phase of the waveform.

• The general definition of distortion includes five distinct types:

— Attenuation.
— Intersymbol interference.
— NEXT.
— Noise pickup (ingress).
— Radiation (egress of power).

• Single sine-wave signal distortions appear as harmonics of the input frequency. The sum of all of these harmonics is then known as the Total Harmonic Distortion (THD). When more than one frequency is input, multiple cumulative distortions appear which are both sum and difference components, and this is then known as Intermodulated (IM) distortion.

distributed backbones

This is the classical design for a bus-type network architecture as used in the days of coaxial Ethernet. . Bridges and routers link departmental LANs to the high-speed backbone LAN, and so the routing functions were distributed geographically throughout the network.

This is the cheapest approach to building cabling. However, when 10Base-T became the dominant Ethernet standard, collapsed backbones and hubs became the order of the day.

Distributed Computing Environment

This is the group of fundamental technologies required for the interoperability of Unix and other computer platforms in a distributed processing environment. It provides a standard communications interface between network applications so that they can work together, regardless of location (or other variables). It was developed by the Open Software Foundation (OSF) and the current software release is version 1.1.

The main aim of DCE is to have a network which can transparently access all data without the user needing to know where it comes from — even if it is accessed on a different network, even on a different continent.

DCE is independent of the operating system and it can function in a heterogenous network environment. As it exists today, DCE is a group of basic services such as the distributed name service (network directory), remote procedure call (applications distribution), distributed time services (network sync) and concert multithread architecture (parallel processing).

• These are the main elements that have guided the design:

— Consistent user interface.
— Communications via OSI networking protocols, but also provide support for TCP/IP and X.25.
— RPCs to manage the communications between applications from HP and DEC.
— Name and directory services: based on DECdns and X.500.
— Time services (synchronization): DECdts.
— Threads service: CMTA from DECT.
— Distributed file system (AFS 4.0).
— Security, authentication and access: Kerberos with HP extensions.

• You can fit the services into two categories: Fundamental Distributed Services which contain the tools that software developers will use, and Data-Sharing Services which provide the end-user with functionality, without him needing extra programming. See also DME.

distributed data processing

This has now come to mean a range of different things:

• The use of interconnecting computers in a network, each sharing the resources of the others. This is peer-to-peer LAN without a file-server —where each user is able to share files on other computers around the network, but where there is no central store. Windows for Work Groups allows this. See IAC also.

• The use of a normal LAN with its file servers, but where a large part of the processing takes place at the PC and workstation end — as it does with most normal network operations. The distinction is with the old host-centric computer intelligence and storage which depended on large mainframes or minis.

• An extension of this is the client/server model. The point is that this approach divides the problem into separate tasks, so that it can be processed by the appropriate computer on the network. The PC application handles only the presentation and the application logic — while the RDBMS functions are handled (for example) by the main database server.

This is a two-tier model of software functions. It involves the use of distributed transaction processing or distributed data management at an interim stage for the application logic

• The term is also often applied to those techniques which enable multiple computers to cooperate in the completion of tasks (rather than just have access to each other's memory). This requires the task to be broken into sub-tasks and allocated to the various participants. The results are then compiled from the reports submitted from the sub-tasking computers.

The Unix International group is using OSF's DCE standard, but they have extended this and renamed it Atlas (see). See Distributed Computing Environment.

• See distributed transactions also.

distributed database

Where there are both many clients and many servers on a network, the databases are often remote from the user. However they can be treated as if there were a single resource, provided a data dictionary is available to map the data locations across the network. This is a necessary condition of client-server, enterprise-wide networks. See DDBMS.

distributed open systems

A general term for open systems (mainly OSI-based) which allow distributed processing. See OSI, TCP/IP, DCE and Atlas.

distributed services

• Services carried on a telecommunications network that are broadcast and therefore one-way, as distinct from interactive/-conversational two-way services.

• Services, in the sense of applications and resources on network servers, which are available across a network rather than being centralized in a mainframe or major host.

distributed switching

In a modern telephone exchange the switching units are usually on rack-mounted plug-in cards, and the exchange switch is modularized so that a section can be removed and replaced without disrupting operations. The switching intelligence of the system can also be distributed between the line-cards, rather than being centralized. Not all modern switches distribute the switch intelligence.

distributed transactions

Distribution of transaction processing is much more complex than normal distributed processing, because of the reliance of large centralized databases. However, increasingly, databases are also becoming distributed.

The first stage of transaction distribution is to split the functions between the front-end processor and the back-end processor in the mainframe installation. The FEP can perform the forms-processing functions while the back-end handles the database. This is client/server.

Efforts are still being made (largely unsuccessful) to distribute these functions even more, but generally large corporate/client databases and transaction processing still remains the stronghold of mainframe computing. See ODC and OLE.

distribution point

The street hub near to the end-user's premises where individual feeder cables radiate out to each home/office. This is the street pillar or the junction box on a telephone pole.

distribution services

As distinct from interactive services in B-ISDN terminology. It is applied to those asymmetrical or one-way services such as broadcast television, radio, teletext, etc.

distributor

In an analog telephone exchange, this is the general term applied to special equipment that switches incoming calls to outgoing circuits as part of a crossbar. In a modern analog exchange the distributor sits between a concentrator and an expander. The point where the incoming line and the outgoing line meet, is called the crosspoint.

DIT

Directory Information Tree. This is the hierarchy of names used in the X.500 Directory service.

dithering

The introduction of 'random noise' into a system to smooth out the artefacts, or approximate some condition. It is mainly applied to graphics.

• In bit-mapped graphics and DTP dithering is used to produce a pseudo gray-scale effect by varying the patterns of a group of pixels, rather than their size, as in printing. Small blocks of pixels (say 4 x 4) will be treated as a whole and various on–off patterns within those blocks will vary the overall density to produce a pseudo-gray. When all 16 print, the result is black, when only a few are printing, the result is light-gray.

Dithering reduces the image resolution to a substantial degree. It is an inadequate substitute for true halftones, which is why 300 dpi printers may be adequate for text and line drawings but clearly inadequate for quality photographic image reproduction. The distinction is with true halftones (see).

• In satellite television signals dithering is the addition of an energy-dispersal waveform to the video signal — usually a triangular wave of about 25 or 30Hz (at the frame rate), to produce a more uniform dispersal of the video-signal energy. Dithering is countered by video clamping.

• In digital audio, dithering is the process of adding a low level artificial hiss to mask the on–off switching of background (during quiet passages) — a consequence of the 20-bit to 16-bit conversion from tape to disk. See super bit mapping and EFM.

divestiture

The term which was applied in the US to the forced breakup of AT&T into seven RBOCs in 1984. See MFJ.

divide-by-zero errors

In maths, if you divide a number by zero, the result is always infinity. This creates problems in computing because most systems can't handle infinity as a number!

DIX

Digital, Intel and Xerox. DIX is a form of Ethernet. See.

DLC

- **Data Link Control.** See.

- **Digital Loop Carrier** — one form of SLC (Subscriber Loop Carrier). This is a general term referring to a system conveying digital information in the local telephone loop. DLCs come from many vendors and may use channel banks at the subscriber end. At the switch end, they may also terminate in a channel bank or be directly connected to a terminating card in the switch. Generally, a direct connect system is called an IDLC (Integrated Digital Loop Carrier).

DLCI

Data Line Connection Identifier. The frame-relay term for a virtual circuit identifier. See VCI.

DLE

ASCII No.5 — Data Link Escape (Decimal 16, Hex $10, Control-P). When used in the text of a message, it changes the meaning of one or more of the following characters, thus providing supplementary control information. It is an international control code.
— It also permits the sending of data characters having any bit combination, including those normally reserved for control.
— In X.25 packet-switching services, DLE will switch the PAD from the data mode back to the command mode.

DLL

Dynamic Linked Library. Windows and OS/2 allow modules of executable code to be loaded on-demand, and linked at run-time. DLL is coupled with Visual Basic Extensions (VBX) to provide Document/Dynamic Link and Load modules.

DLSN

Data Link Switching Net (Node). A standard encapsulation scheme for shunting SNA traffic over TCP/IP backbones.

DLSw

Data Link Switching. Despite the name, DLSw is a 'router' technique that encapsulates LLC Class II data into TCP/IP packets. It offers an alternative to source-bridge routing in SNA/APPN and NetBIOS traffic over TCP/IP networks. The IETS may ratify a DLSw standard to succeed TPC tunneling, but this is still open. The alternative is frame relay's RFC 1490.

DLSw technology was developed by IBM to carry data between mainframes and PC LAN and the LAN Server NOS. It routes (unrouteable) SNA traffic and maintains LLC Class II session integrity without needing modifications to the end-station software. It terminates LLC2 sessions locally, and spoofs all polls and ACKs when the WAN link is unavailable or congested. It appears to be ideal to use with IP, PPP and X.25.

DLT

Digital Linear Tape. A half-inch streaming tape drive and media type used for backup. DEC originally designed this system for mid-range and high-end computer systems, but it has caught on with LANs. It is a high capacity cartridge system which does fast backups of large quantities of information, compressed in hardware by DEC's proprietary standard.

It uses a dual channel read/write head and is designed to handle a transfer rate of 1.25MByte/sec (native mode) using the SCII-II burst mode — at a rate of about 3MB/s when compressed (DEC's data compression).

Like the QIC system, it also uses a linear serpentine track, but it runs the tape at 100 to 125 inches per second. 25% of the capacity is consumed by Reed-Solomon ECC and CRC.

- DEC makes a DLT2000 drive with a transfer rate of 2.5MB/s (9GB/hour) and a capacity of 20GBytes. There's also a mini-library system with 5 or 7 tape cartridges holding from 100 to 140GBytes respectively. This system may eventually be used for video-on-demand (VOD).

- Quantum's DLT recorder provides 16kB of Reed Solomon error-correction with each 64kB of payload, then there's another layer of protection with a 16-bit CRC and a 16-bit error detection code for each 4kB of data. There is a third layer of overlapping 16-bit CRC on each record, and they also perform DRAW (Direct Read After Write) just to be sure. This recorder has a transfer rate of 3MBytes/sec and a capacity of 40GBytes.

DM

Delta Modulation. See delta.

DMA

Direct Memory Addressing: Some peripheral devices (screens, disk drives, etc.) are able to gain direct access to computer memory locations (for storage or retrieval) without passing through the central processing unit. Normally the processor would need to process each bit of data, and pass it on to the appropriate memory cell one byte at a time.

By using this bypass process, storage devices can transfer data very quickly to memory; the processor is involved only in setting up the exchange. Most early DMA systems operated at single clock-cycle rates, but recently burst-mode DMA has become possible on PCs with MCA architectures.

Special IRQs and DMA buffers are allocated for this process. During the DMA transfer the CPU is suspended and it is usually disconnected from the bus. The DMAC (DMA controller) takes over. On completion, an interrupt instructs the CPU to read a status message which reactivates it. See intelligent I/O bus, PIO and Multiword DMA.
— Later PCs still only have a few DMA channels and buffers, and the same channel and buffer can't usually be used by two different cards.
— A new form of DMA, called Scatter/Gather, is available on PCI local bus systems. Numerous small blocks of data on a disk can be fetched with a single I/O request, and this improves performance in many of the new PC operating systems. See PIO.

- **Document Management Alliance:** This is the combined forces of Novell and Xerox (the DEN group), IBM, and the

Shamrock Document Management Coalition, under AIIM. They are releasing a new specification for the free exchange of documents across any network.

DMAC

- DMA controller. See DMA
- D-MAC television. See at front of D section

Dmax

Density-Maximum: The highest level of density possible on a film negative. It's a printing term.

DME

Distributed Management Environment. This is the OSF's protocol suite for the management of key routers, bridges, hubs and multiplexers across a wide area network. It is based, to a degree, on HP's OpenView but, over the years, it has been subject to many disputes and may never emerge as a final solution.

DME provide DCE-enabled management services, to provide network managers with a user-programmable mechanism in a distributed environment. See DCE, NetView and OpenView.

DMF

Microsoft's floppy format for Windows 95 which gives 1.7MB on a disk. It can't be copied by MS-DOS Diskcopy or Copy.

DMI

Desktop Management Interface. This is an important new specification for LAN management and it is related to SNMP. It adds to the normal network management systems the ability to manage individual NICs, mass storage devices, operating systems, and even applications. The companies behind DMI are Microsoft, IBM, Intel, DEC, HP, Novell and SynOptics. See DMTF.

DMS

- **Database Management System**, now usually referred to by the acronym DBMS. See.
- **Digital Multiplex Switching**. Northern Telecom's family of exchange switches. Variants are: DMS-100 (local POTS and ISDN), DMS-200 (trunk), DMS-300 (international gateway), DMS-MTX (cellular), DMS-STP (CCS7).

DMT

Discrete Multi-Tone. A modulation technique being promoted for video-over-copper, originally proposed by Stanford University and developed by Amanti. It is an alternative to CAP for ADSL (see both) for high-speed data (mainly video) over the local loop. It seems to be closely related to ODPM.

DMT requires 256 discrete subcarriers (at 4kHz spacings) from 20kHz to 1.024MHz, with each subcarrier being modulated by different levels of QAM. The spectrum below 4kHz is reserved for POTS, and some of the lower subcarriers are used for upstream data (possibly ISDN).

DMT dynamically adjusts its power spectrum around interference sources (impulse noise, crosstalk and echoes) and the sub-carriers with higher S/N ratios are allocated higher signal rates. See ADSL and CAP.

DMTF

Desktop Management Task Force. An industry group trying to define the tools needed for desktop workstation management. Intel, DEC, HP, IBM and Microsoft are involved, and the preliminary specifications for DMI (Desktop Management Interface) have been approved. See DMI and MIF.

The original offerings in this area were strictly point-to-point. There was no integration with other network management systems, and no standardization. Intel's LANDesk, which had a strong influence on DMI, currently controls the user PC configuration, backup systems and network virus protection.

- The vendors of desktop systems are also beginning to SNMP-enable their products, and a technical committee of the DMTF is charged with ensuring that both DMI and SNMP interoperate.

DMZ

De-Militarized Zone. See firewall.

DNA

Digital Network Architecture. This is the network architecture from DEC. DNA uses only one operating system (VMS) and one processing architecture (VAX) across its range of hardware.

DNC

Direct Numerical Control. The use of local area networks to control manufacturing machinery.

DNIC

Data Network Identification Code. See X.121.

DNIS

Dialled Number Information Service. This is one part of the ANI-DNIS (A-party number + B-party number) string which is forwarded from switch to switch when establishing a call.

DNS

Domain Name Service/System. This translates Internet host names (usually the Fully-Qualified Domain Name — using alpha characters) to IP addresses (numbers). DNS provides mapping and contact services for TCP/IP networks. See fully qualified domain name and IP address.

dockable

Integrated by sliding into a slot.

- A term applied to modern video cameras when a small portable recorder can be connected directly to the camera body to create a camcorder. The term is also applied to a camcorder which can be fitted into a mains-powered device for editing. It implies that no cables will be required since direct contact is made between the electronics, usually via a large and delicage plug.
- In notebook computers, dockable design allows the portable to slip into a tray in a desktop frame and immediately make electrical connection to a big screen, full-size keyboard and LAN. The advantages/disadvantages of having a dockable

portable as against using a separate stand-alone portable are:

— Security. There is only one copy of the sensitive files and you can take that home.
— Synchronization. You don't need to worry constantly about which version of a file you are using. With portables and separate desktop machines, you are always confused as to which is the most up-to-date file.
— The disadvantages are in having all your eggs in one basket.
— In order to keep dockable portables cheap and light, many don't have a floppy disk, but rely on the desktop floppy. This can be a problem if you are away from the desk for a long period of time.
— A very large complex contact/connector is needed between the portable and the desktop frame, and these can give you trouble.

document

In the widest sense, a document is any file created by an application program for the storage of data, however 'document' is usually used only for files which can be printed. File is a wider term which includes utilities, systems and applications.

• In common computer terminology, a document is a file composed of graphic or text material — usually from a word processor or drawing program.

• Recently the term has been extended to mean almost any kind of file. So, for instance, they will talk about voice documents (digitally recorded voice messages), image documents (photos, line art, graphics, fax, etc.), text documents (Telex, ASCII and W/P files), and I've also seen data documents (any executable program or machine readable data file).

• In the above context a composite document can be more than one of the above 'stapled' together — voice annotation on a text file, for instance.

• To document, means to record in text form — as in producing a manual or user-guide.

• A document reader is an optical character reader.

document-oriented interface

A new term being applied to object-oriented operating systems, where the emphasis is on the document rather than the application. The idea is that you don't actually need an application, but have instead a series of modules ('tools') which perform various tasks. You put together (customize) your own software to suit your needs. Taligent (IBM and Apple) and Microsoft are working on this approach to application building.

documentation

The manuals accompany new hardware or software, which, if you could read and understand them, would probably explain about half the important things you need to know. The Law of Documentation is that: 'The usefulness of the hardware or software is inversely proportional to the weight of the manuals.'

• Documentation, also extends to on-line assistance and tutorials.

DOD

• Direct Outward Dialing. See.

• Department Of Defense. The US department that involves itself in computer evaluation, especially of security measures. The DOD has levels A, B, C and D for software security. Most corporate software (with password protection) is classified as C sublevel 2. For instance:

— B1 is a relatively easy standard dealing mainly with access measures,

— C2 requires hidden passwords and auditing procedures. This is the standard needed for any LAN server.

doduc

A SPEC benchmark (see) based on a Monte Carlo simulation (a gambling simulation using random numbers) with high floating point precision.

DOI

Document Oriented Interface. See.

Dolby

Noise reduction mono and stereo systems used in television, audio tapes, films and some video recorders. There are a number of standards:

— **Dolby B** is a noise reduction system for tape recorders which boosts high frequencies during recording and reduces them during replay. The amount of boost is variable depending on the loudness of the music. This cuts the high-frequency hiss on soft music passages. This system is also used in FM radio.

— **Dolby C** has more boost over a wider frequency range, and thus eliminates hiss even better than B. However it can't be played on a non-Dolby C machine.

— **Dolby HX-Pro** is a amplitude-limiting technique that cuts the peak amplitude of the music energy to prevent distortion.

— **Dolby S** is a new noise-reduction system for analog recordings. It is claimed to improve sound quality dramatically, to the point where it is seen as a threat to DAT digital recording.

— **Dolby Stereo** (or Surround) is a technique used to provide three-dimensional (multi-channel) sound both in cinemas and with home VCR replay. It provides three channels; a central mixed channel, both left and right channels, and a rear surround channel which is 180° out of phase. The rear channel carries ambience and sound-effect information. All of this information can be stored and carried on only two audio tracks

— **Dolby ProLogic** is a more sophisticated version of Stereo, with which it remains compatible. (Both Stereo and ProLogic signals can be carried on two conventional audio tracks, and they can be transmitted over the air.) ProLogic is also compatible with non-Dolby stereo and mono signals.

ProLogic has four channels but requires five speakers. Special ProLogic amplifiers take the dual channel stereo inputs and processes them to create the five distinct outputs — front and rear (stereo L/R) + a mixed center channel. The amplifiers can also be programmed to add variable reverberation delays and responses to enhance the subjective response.

Note: Other similar cinema sound systems use George Lucas's (StarWars) THX, and Yamaha's Cinema-DSP.

— **Dolby AC-3** is a new 5-channel system which the American electronics industry is promoting as the future audio standard for MPEG2 compressed wide-screen television signals. See Grand Alliance.

dollar sign $

• In assembly language the dollar sign is often used to indicate hexadecimal numbering (hex). Hex code will often be written as $A7 or alternately as A7h.

• In Basic, the dollar sign may indicate that this is a string variable, not a numerical variable.

domain

A coherent area of influence. That part of any system under common control.

• In programming — a set associated with a variable.

• In material science — isolated regions in a medium that exhibits the same characteristics (e.g. RI variations across the width of glass fiber)

• Regions in a magnetic tape or disk where the direction of magnetization is uniform. Magnetic domains on tape or disk are equivalent to bits.

• In X.400 and X.500 Message Handling, a domain is the area of control of the organization which is providing the primary messaging service. Public domain, and private domain are two general types in X.500. Domains are logical not physical, and relate to management functions such as accounting, configuring, and error reporting, etc.

• In a LAN, a domain is one or more segments controlled as a single network. This is sometimes called an organizational, administrative or broadcast domain, and it will be managed by routers. In a company WAN, an administrative domain is bounded by a router (not a bridge).

• In communications generally, a domain is any sub-division of the system that uses the same protocols. To converse between domains requires a protocol converter.

• In the Internet (which is a heterogenous conglomeration of dissimilar networks) each network or sub-network, will comprise a 'domain'. So there is a hierarchy of 'domains' and these collectively constitute the international (IP) address of a host.

The full Internet address will consist of a person's account name, followed by the @ sign, followed by the host's fully-qualified domain name sequence. See fully-qualified domain name, DNS and IP address.

• In SNA networks, the SSCP is a focal point for managing the network, and it effectively controls and defines the domain. In SNA, a multiple-domain network is one with multiple SSCPs.

domain-specific knowledge

This is an AI term which emphasizes that, while an expert system must have a lot of information, the information must be closely related to one clearly-defined subject. The rules of chess are a good example of domain-specific knowledge. However see Cyc.

Domestic Digital Bus

Also called D2B or DDB or D^2B. This is a new system and a standards suite for home automation which was developed by Philips and Matsushita, and is now supported by Thomson and Sony. It is supposed to be compatible with previous Japanese and European proposals for a home bus.

D2B will combine control of domestic appliances, telephones, video-cassettes, television, hi-fi, central heating, lighting, kitchen equipment and security systems. This was Esprit II Project 2431 and is now an IEC standard. See NEST.

dongle

A small hardware device which needs to be added to a computer before it will work with a certain application. It is a form of security for access to the hardware, or alternatively a copy protection scheme for the software. Dongles are often added at the parallel or serial port, and the program will check for their presence fairly regularly — not just at start up.

Dongles sound like a good idea to stop software pirating, until you stop and think what your computer would look like if each application needed a dongle. What would you do with say, six applications on a multitasking operating system?

dopant

An intentional impurity. In the manufacture or transistors, lasers and optical fiber amplifiers, certain impurities are often deliberately allowed to diffuse into (to dope) the materials to provide them with specific electronic or photonic characteristics. These 'impurities' are dopants. See doped fibers and erbium amplifier.

doped fibers

Rare-earth doping (the introduction of rare-earth elements into the glass as an 'impurity') is performed on optical fiber to create optical amplifiers which do not require electrical connection at the point of amplification.

The rare-earth metals involved are mainly erbium and praseodymium, which are said to have 'restless' electrons which can be stirred up by an amplifying 'pump' laser (which is quite distinct from the signal laser). These pumped electrons rise to higher energy levels in their atoms, and when they fall back, they emit light (depending on the dopant element) in the frequencies used for the telecommunications signals.

The technique therefore amplifies the original signal without requiring direct power and without needing special electronic regeneration devices in submarine cables.

The practical limit of these systems appears to be about 5Gb/s without the use of solitons. See erbium amplifiers.

Doppler effect

The change in frequencies of a radio or audio signal due to the relative speed of the source to the sink. The whistle of a train coming towards you will sound high-pitched, but, as it passes and retreats away from you. the pitch will suddenly drop.

Similarly, in LEO satellite mobile communications, the motion of the satellite towards you as it rises over the horizon will cause a progressive upwards shift of the radio 'center' frequencies, perhaps by as much as 40kHz. There will be a corresponding drop in the frequencies when it passes overhead and is setting.

This means that the LEO mobile unit must be able to change frequencies to accommodate the Doppler shift. It also means that, when a mobile is first switched on and tries to acquire a satellite channel for communications, it must be able to search over a fairly wide frequency range to find the satellite — which may be rising, setting or overhead.

• The Doppler effect can be used by satellite receivers to calculate position, however this is much less accurate than GPS. It is, however, good enough for tracking shipping containers, etc. Usually it will provide a position within a kilometer or less.

DOS

• **Digital On Supergroup:** A telecommunications term used with high-speed (2Mb/s+) modems. A supergroup is 60 analog voice channels. See.

• **Disk Operating System:** A general term very often misapplied to MS-DOS only. The significance of the 'Disk' operating system, is that the early PCs used cassette-tape storage and therefore ran relatively simple OSs, rather than the more complex DOSs. Nowadays virtually all machines use hard-disks, so the term DOS has become almost synonymous with OS.

• Now that Microsoft has abandoned MS-DOS, PC-DOS 7.0 is now the current version. It requires less conventional memory and has the Stacker 4.0 disk compression and hypertext links.

• As a marketing ploy, IBM and Microsoft promoted the use of the term DOS as a synonym for both MS-DOS and PC-DOS. The gullible public accepted this use of the general term for the specific. However, there are other DOSs. The most important of these are:

– Apple DOS: probably the first to use this term.
– DR-DOS 5.0 and 6.0: from Digital Research. At one time DR-DOS looked likely to take the market away from MS-DOS 4, which is why MS-DOS 5.0 was hurriedly rolled out.
– NDOS 7.0: from Novell. Novell has recently bought Digital Research's DR-DOS in an attempt to give Microsoft a run for its money.

• **DOS Barrier** was the old limit of 640kBytes of working memory in the early PCs. See 640 kB Barrier.

DOS extenders/extensions

• To overcome the limitations of MS-DOS with only 640kB of working RAM memory, two techniques are employed — bank-switching, and DOS extensions. Bank-switching has memory overheads, but is excellent for those programs.

DOS-extenders give higher speed and performance by allowing direct access to expanded memory. The trick is to use the 'protected' mode of the 386 from within the DOS application. Real-mode procedural calls can be made to the DOS extension, which then deals either with DOS, or with the hardware direct. This is the role of the Extended Memory Manager feature. See.

• Special 'patches' that have been added to DOS to allow it to handle special conditions or peripherals. These allow easy access to the various hardware devices. Microsoft wrote special extenders to handle CD-ROM drives, for instance.

• DOS file-type extensions. See file-type extensions.

DOS-shell

A utility/application which provides a layer of menus or GUI functions around the normal MS-DOS, in an attempt to make it more 'user-friendly'.

• The original shell for MS-DOS 4.0 used a two-level menu structure which allowed selections to be made by keyboard or mouse. However, Norton Commander was probably the *de facto* standard for DOS-shells until Windows 3.x caught on.

• Windows 3.x is a DOS shell, but usually the term is reserved for the more character/menu oriented approaches. Windows is now seen as something above a shell, and from Windows 95 on, Windows is a true 32-bit operating system.

• A good DOS shell must be fast and intuitive, and provide excellent file management features, yet it should not force you to learn complicated commands. It should also allow you access to DOS direct.

The basic services provided by shells are: easier procedures for copying, moving, renaming, deleting and backing-up and archiving files. Extra services are: the ability to view directory structures in a tree format, applications menus, file views and simple text editing.

DOSKEY

A RAM-resident program that comes with MS-DOS 5.0. It allows you to re-assign any command to any other command. Doskey macros take priority over all commands. The function keys F7, F8, and F9 are all used by Doskey. The Editing Commands (Home, End, etc.) are also part of the Doskey suite.

dot address

In Internet addressing, the common notation of the IP addressing is to use four decimal bytes which are separated by dots (periods). See IP address.

dot-clock rate

This is the video bandwidth or horizontal resolution of a monitor or TV set (not to be confused with horizontal frequency). It is the rate at which the 'raster' (scanning beam) can change from one extreme of brightness to another (turn on or off), as it draws each line.

This maximum rate of change defines the horizontal resolution of the image (how close the different tonal elements can be together without merging into a flat gray), and for this reason horizontal resolution is often expressed in terms of video bandwidth or 'vertical lines/cycles'.

dot command

A command that was embedded in text files in some of the older word processors. The command is always in capitals and preceded by a full-stop. This makes it easy for the program to distinguish the format command from the text since a period followed by an alpha character is almost never found in normal English text. This approach was used for indenting, making characters bold, creating a new page, etc.

dot-matrix printers

There are two types of dot-matrix printer:

- Thermal dot-matrix which don't require impact.
- Impact dot-matrix, now the most common of all, which create their character through striking a number of 'pins' against a ribbon.

Dot-matrix types are graded by the number of pins; most of the earlier machines were 9-pin but now they have at least 24-pins. Some also use techniques which can improve the apparent quality by 'overstriking' (shifting the paper slightly then reprinting the character).

Good dot-matrix are usually called NLQ for 'near letter quality'. Dot-matrix printers usually offer a full international character set, and many can print medium-quality graphics.

dot pitch

In monitor resolution, this is the spacing between adjacent dot-groups within the pixels. In monochrome analog monitors a dot and a pixel are the same (they are theoretical concepts in B&W scanning), so dot-pitch is generally set by the diameter of the beam striking the phosphor surface (Kell factor) and the bandwidth of the signal.

In color monitors, dot-pitch depends on the fineness of the etched holes in the color mask (just behind the phosphors on the glass), and it represents the distance between any dots of the same color. There are three separate (R-G-B) dots (called a 'triad') which make one pixel. The pitch of even a large home TV set may be as coarse as 1mm, while the pitch of a smaller computer monitor will be down around 0.20mm or 0.25mm.

double-buffering

A computer display technique of storing video information in two sections of video memory. One is progressively filled while the other is being emptied. This approach is used for digital animation and similar image-intensive, real-time video applications. See video buffering.

double-byte code

These are ASCII variations, such as Unicode which can represent non-Roman languages. See Unicode.

double current

This is a type of DC circuit used for digital data transmission where the voltage varies from positive to negative, rather than positive to zero. It is usually called bipolar, and it can only be used for short distances up to a couple of kilometers. More often it is only used between DTEs and DCEs. See and AMI.

The technique has been used from the early days of Telex where the line voltage was swapped between +50V and –50V. The contrast was then with single current Telex which switched between +50V and earth potential.

double-density disks

Disks with a higher-quality magnetic surface capable of holding twice the data of normal disks. Obviously you need double-density disk-drives with a different track-spacing to use the extra capacity. Later improvements to floppy disk drives and coding systems have resulted in higher data densities than these so-called 'double-density' systems. See MFM recording techniques, and double-sided disks also.

double length maths.

See double precision.

double precision

A mode in some computers where the computer can handle twice as many digits as normal in floating-point arithmetic calculations. It uses a mathematical technique of allowing an operand to occupy two words. This provides greater accuracy in calculations, and is also called double-length arithmetic.

double-sided disks

The early floppy disk drives only recorded on one side of the disk — sometimes on the upper surface and sometimes on the lower. For this reason many or the early PC users cut a notch in the other side of 5.25-inch floppy covers and used both sides — one side was as good as the other. Later disk drives began to use both sides of the disk without the need to turn the disk over. These drives had both top and bottom read/write heads.

• Double-sided (DS) drives were followed later by double-density/double-sided (DD/DS) which was also called 'high-density' (HD) by some disk manufacturers.

These were then followed by quad-density, or high-density/double sided (HD/DS or 2HD) — with a write-notch in both sides and over 1MByte of storage on a 3.5-inch floppy.

• Some of the older control systems treat each side of a disk as a separate volume, while others treat both sides together as one volume.

double strike

Dot-matrix and daisy-wheel printers often used double-strike (printing each character twice with a slight shift of the page vertically) to produce bold type.

double-twist LCD

This is a version of the double supertwist LCD screen technology which uses a polarized plastic film instead of the compensating passive panel. This approach reduces power requirements since it doesn't need backlighting. See active matrix screens , Supertwist LCD and twisted-nematic also.

DoubleSpace

An automatic ('transparent') disk compression utility which originally gave trouble in MS-DOS 6.0 because many badly-behaved applications trampled on the DoubleSpace buffer in RAM and caused corrupted data to be written to the disk. The problem has now been solved, and the new versions have further improvements. DoubleSpace uses LZW compression. See.

• The original trade name of DoubleSpace was Stacker, and Microsoft paid dearly for patent infringements.

DOV

Data-Over-Voice. See.

DOW

Direct Over Write. Magneto-optical drives normally need a two-stage erase/rewrite process if they are to recover and then reuse the disk space. However a number of new optical techniques promise a single-stage direct-overwrite, where erasure and re-recording happens at the same time. See CD-R and MO.

down-converter

A device which accepts radio carriers in a wide block of frequencies and outputs them as a lower frequency block. Down-converters can be built just to handle a single carrier frequency, but in television they usually cover a block which is 500MHz wide (block down-converters).

They are used in satellite and MDS reception, and when receiving weak satellite signals they need both special low-noise components, and to be mounted near to the signal detection. Therefore 'low-noise block down-converters' (LNB) are usually mounted directly behind a dish antenna to convert the microwave signals to an intermediate frequency (IF).

These units usually use a heterodyne approach — a fixed-frequency local oscillator generates an intermediate 'beat frequency', and this IF signal can be handled well by coaxial cable feeds without significant loss or addition of extra noise. The set-top boxes are then standardized, and are able to tune the same range of IF channels.

down time

The time when systems are unavailable, either because of failure, or for maintenance. There is both scheduled and unscheduled down time.

downlink

This is the UK term for what the Americans call the forward link or forward channel in a radio system. The terminology is defined from the viewpoint of the service provider.

• In general communications, the downlink is the data channel, while the back-link or return channel is used for acknowledgments, supervision, etc.

• In cellular phone systems, the downlink is the one from the base-station to the mobile. The reverse channel or uplink is from the mobile back to base.

• In Cable TV, the down link carries the video from the head-end to the customers. Now that cable systems are becoming two-way, the back channel or reverse channel is created by a bandwidth split, and usually the frequencies from 5 to 45MHz are used for the back channel. The downlink will then occupy from 54MHz to, say, 550MHz, 750MHz or 1GHz. However, bandwidth splits vary now that digital is being introduced.

• In satellites, this term is applied to the signal transmitted from the satellite to the receiving ground/earth station or to the mobile or VSAT terminal. The distinction is with uplinks to the satellite. Note that the reference here is physical, not logical. Both go and return paths have an up- and downlink component. Downlink signals generally suffer heavy attenuation by the time they reach Earth, with an average loss of about 200dB. See EIRP.

• A downlink station is a satellite ground station. It is not necessarily a station in control of the satellite, however.

downloadable fonts

These are fonts that can be downloaded from the computer into the laser printer (or computerized typesetter), as distinct from those which are stored within the device on ROM (sometimes in a cartridge). This technique allows a large variety of different fonts to be used. The printer will need enough memory to store the outline characters, and it obviously must have sufficient processing power to generate the required bit-mapped images.

So there are three-ways that the font-characters can get from the computer onto the laser drum:

— A complete bit map of the page can be created in the computer, and then sent to the printer as an image bit-map.

— The ASCII characters for the page (plus formatting instructions) can be sent to the printer, and it will then use its internal/cartridge fonts to create the bit-mapped page.

— The downloadable font outlines (or bit-maps of the fonts), plus the ASCII characters and formatting instructions will be sent to the printer. It will then create the bit-mapped page.

• Many word processing and DTP programs have a downloading utility for this purpose.

downloading

• In telecommunications, this is always taken from the user's perspective. It is the process of retrieving information from a host computer, and storing it in your own.

• In PC to printer communications, downloading refers to sending fonts or data from the computer to the printer. Downloadable fonts are soft fonts (see).

downsizing

The business computing strategy of migrating applications onto a lower-level facility. Typically these days, downsizing is the rewriting of mainframe applications to allow them to work in a LAN client-server environment. About half the larger US companies have recently rewritten their key mainframe applications down to LAN-server operations, and apparently two-thirds are currently downsizing applications. Downsizing is also called rightsizing.

• Client-server applications have not proved to be quite as attractive as was once thought. The craze to downsize everything has slowed in recent times.

downstream PU

See split-stack architectures.

downward compatibility

See backward compatibility.

DP

Data processing. Now anything to do with computers.

DPAM

Demand Priority Access Method. The access technique being used in 100Base-VG—AnyLAN as a replacement for Ethernet's CSMA/CD. DPAM assigns to the hub the job of contention resolution. A node requests permission from the hub to transmit, and if multiple requests are received the hub will grant permission to each in turn. A node request will carry an indication as to whether the frame is high priority or data priority; this reduces latency problems for voice and video.

DPCM

• Differential PCM. See.

• Delta PCM. See Differential PCM.

DPE

Distributed Processing Environment of Bellcore. See ODP, CORBA, ROSA and TINA.

dpi

Dots per inch. A measure of the resolution of screens, printers, etc. The measure is linear — so 300 dpi will reproduce 90,000 dots in a square inch. Most laserprinters are 300 or 600 dpi machines, and ink-jet printers aren't too far behind. Monitor screens vary from about 72 dpi to 100 dpi.

The problem with this measure is that it doesn't specify what type of dots — so a laserprinter which only generates one size of dot will be markedly inferior to a new laserprinter which can vary the dot size, especially with tonal areas such as photographs. With type and line drawings, where all dots are reproduced full-black, there will be no appreciable difference.

Professional printing reproduces photos using a graphics camera system which produces highly variable dot sizes (called 'screens'), so newspapers can produce quite good pictures with 150 dpi (usually classified as 'a screen with 150 lines to the inch'), while a modern photo-typesetter may need 2450 dpi to reproduce good quality photos, because it lacks the ability to generate variable dot sizes. See dithering.

DPMI

DOS Protected Mode Interface. A specification which allows extended MS-DOS applications to run in a multi-tasking, protected environment on the later IBM-compatibles. This is now a widely agreed standard. It is basically a set of calls that the application can make while setting itself up in protected mode for managing the memory and interrupts. See DOS-extenders.

DPNSS

Digital Private Network Signaling Scheme protocols. The UK-devised European standard which allows two or more PBXs to communicate reasonably transparently. Currently DPNSS1 standardizes the link between digital (ISDN-based) PBXs over private networks. The most important part is the message-based signaling specification for the D-channel.

DPNSS allows fast call set up and network-wide supplementary services such as camp-on, but it is assumed that Q-SIG will eventually replace it. DPNSS has gone through about half-a-dozen revisions. It has a layered architecture which matches PRA ISDN at the lower layers. See APNSS, DASS, R2D and Q-SIG also.

DPP

Demand Priority Protocol. See DPAM and 100Base-VG.

DPPM

Differential Pulse Position Modulation.

DPSK

Differential Phase Shift Keying. See PSK.

DPUC

In the USA this means Department of Public Utility Control — the term for the state body which may control some aspect of telecommunications (along with the FCC). See PUC and PSC.

• The FCC controls inter-state telecommunications, while each state controls its own intra-state traffic.

DQDB

Distributed Queue Dual Bus (ex-QPSX). This is the IEEE 802.6 standard for Metropolitan Area Networks (MAN). It uses a twin optical fiber loop covering a geographical area of up to about 50km in diameter.

The data throughput is 140 Mb/s in each direction (on each fiber), with virtually no slow-down of the system at near full capacity. Individual users can expect between 2 and 45Mb/s links when needed.

The topology is 'ladder-like' with the nodes existing on the rungs of the ladder between the contra-direction fibers. Data frames are supplied on a first-come basis, although priorities can be given, and regular slots can be reserved for time-critical data. Some aspects of DQDB are very like ATM, however, this is a connectionless system. The cell size and data-rate has been brought into line with that proposed for ATM.

DQDB has applications in voice, high-volume data and video conferencing, but often the isochronous services in the MAN aren't implemented.

• The DQDB media access control layer is independent of the physical layer, and therefore a variety of DQDB networks can operate at different data rates over different transmission media and systems. The three which have been defined are:

- DS-3 (ANSI) at 44Mb/s over coaxial or fiber.
- Sonet/SDH (ANSI) at 155Mb/s (and up) over single mode fiber.
- G.703 (ITU) at 34Mb/s and 140Mb/s over coaxial.

• In the USA, DQDB is implemented as the access protocol for SMDS. See also connectionless, ATM and cell-relay.

DQPSK

Differential Quadrature PSK modulation. A four-phase system, but using the difference in phase from the last change (last 'symbol') as the measure rather than any absolute measurements of the change. All modern phase-shift-keying (PSK) techniques now use 'differential' phase angles, to the point where this is now assumed. This technique is used in NICAM. See QAM also.

DR #-#

Digital Radio. A series of microwave standards:

- **DR 6-30** operates in the 6GHz spectrum with channels 30MHz wide and is modulated to provide 4-bits per Hertz. Such a system can carry a maximum of 9408 PCM digital voice circuits over 7 channels, with hops of between 25 and 50 kilometers.
- **DR 11-40** operates at 11GHz with channels of 40MHz, with a capacity of 13,440 digital voice circuits in 10 bearer channels over shorter distances; usually 25km or less. These systems use 16 state quadrature-amplitude modulation (QAM).

drain

One of three contacts in a FET. See FET.

DRAM

Dynamic Random Access Memory. The normal type of semiconductor memory chip used in the main memory of most current computers. It is cheap and has a high packing-density. It was invented in 1970 by Intel (a 1kb chip).

Until 1995 DRAM was produced as 1Mb and 4Mb chips using 0.8μm line widths. The 4Mb chips are still widely used today; and either 8 or 9 of these are needed for 4 MBytes of computer memory. The ninth is for parity RAM. See.

16Mbit chips are now becoming the standard and 256Mbit chips are already being shown at trade shows, The 1Gb chips (which need 0.18μm elements) are predicted to be the current limit with a single-layer technology.

Dynamic RAM stores the data in small internal capacitors which require constant refreshing (about every 2 msecs) to retain the memory because the charge leaks away. For this reason the information disappears the instant the power is switched off. DRAM also creates problems with the new faster CPU processing rates because its read/write cycle is in the 80—120 nanosecond range, which is quite slow.

For installation, DRAM chips usually have a notch which allows you to orientate the chip. Don't rely on the printing on the back of the chip. Think of these chips as a cockroach with the notch as the head, Pin No.1 will be the left-hand front foot. See Static RAM, NVRAM and FRAM.

• Hitachi and NEC are showing prototype 1Gbit DRAM chips produced using 0.16 micron CMOS and powered at 1.6 volts, which they say are suitable for processors with a clock speed of 200MHz. They don't expect to be producing these before the end of the century.

• Toshiba in collaboration with IBM and Siemens, have recently shown commercial 256Mb chips. These are physically 13.25 x 21.55mm in size, with 26ns access times. We should see these soon.

DRAW

Direct Read After Write. A technique used in disk recording (optical and magnetic). DRAW systems read the disk constantly while writing to verify the material recorded. If an error is detected, the sector is immediately over-written (in erasable-recordable systems) or destroyed and rewritten (in WORM).

• Some optical DRAW machines use two lasers set 180° apart around the disk. One is a high-power laser which writes, while the other is a low-power laser which reads.

• Other systems use single read-write lasers which alternate between the two states — writing, then reading and verifying.

• Just to confuse matters, the term DRDW (Direct Read During Write) is also used for this second system.

draw programs

These are graphic programs which store elements (objects) of the image separately, and in a form that allows them to be later changed, shifted, or removed. Draw programs generally use vector images (lines and curves) and vector information, rather than bit-mapped (rasterized) information. However they still generate a screen/printer full bit-map for display — it is just that this isn't the primary source of the information, rather each element contributes.

This approach allows draw programs to use outline-fonts, and primitives (squares, circles, etc.) as distinct elements, and control the layering (shift to back, shift to front, make the foreground translucent,).

So you can shift all of the draw program's graphic elements around on the background, bring each to the front or send it behind other elements, or scale it to any size, and it will still remain an object in its own right.

Obviously this take much more memory than the simple paint programs. Another important aspect is that generally, with the vector generation of many image elements, objects (type, for instance) will be maintained at the maximum resolution of the display device. This will still be true if the object is scaled up or down.

The comparison is with paint programs.

DRCS

Dynamically Redefinable Character Set. See.

DRDW

Direct Read During Write — a variation on DRAW. See.

Drexler cards

A trade name for a form of optical card about the size of a credit card but capable of holding between 1 and 20 MBytes of information. It is a write-once, read many-times system. Drexler also developed a DRAW optical disk system.

driver

• Computer bus and telecommunications: A driver is a device which amplifies the power supplied by a line. See line-driver.

• Peripherals: A system utility which allows the PC to direct the operations of a peripheral device. See device driver.

drop and insert multiplexers

These are add-drop multiplexers, but drop-and-insert seems to be applied to smaller units. A drop-and-insert multiplexer will generally allow the network designer to extract or add one or more data channels (n x 64kb/s) to a bearer with many multiplexed channels en-route between major installations. The remaining channels are unaffected. See add-drop multiplexer.

drop cable

Jargon for a connection (cable) between a terminal and a major transmission line. Sometimes just called a 'drop'.

• In Ethernet, the drop cable links the MAU (transceiver) on the main cable to the AUI connector on the back of the PC. It is often called an AUI itself.

• In Cable TV, the drop cable brings the signal over the last fifty meters into your home from a four-way passive splitter on the feeder cable.

drop cap

A decorative style used in typesetting. This is a large character, the first in the column, which is inserted into the body type. Usually it will be three to five lines in height and it will look like the A here.

The alternative approach, is to use a raised cap, which aligns at the base with the first line, and stands high above the body copy. Both have their stylistic applications.

drop-out

A loss of signal, usually resulting from dirt coming between the read (or write) head and the recording medium, or possibly the actual loss of some of the magnetic surface.

• In video, it is caused by dirt or flakes of magnetic material on the videotape, and it appears on the screen as short horizontal white flashes. See drop-out compensation.

• In audio tape, drop-outs result in momentary losses of signal volume. They are usually caused by the tape transport mechanism not holding the tape tightly against the record or replay head, or by old worn or edited (cut) tape being reused.

• A sudden loss on a communications circuit due to noise or signal attenuation. A form of impulse noise.

drop-out compensation

The technique used in video-cassette and other video and audio recorders, of interpolating 'nearby' information into the signal gap left by dropouts. Usually the technique involves:

— A copy of each line of picture is fed to an acoustic delay.
— The acoustic delay unit buffers it for exactly one line-scan duration.
— A dropout is detected by the complete loss of signal, so
— the output from the acoustic delay is instantly switched in to fill the gap for the duration of the drop-out, thus filling the gap with almost identical picture information.

Without such a scheme, many cassettes would be almost unwatchable because of the white flashes on the screen.

dry-reed

A form of relay switch which is enclosed in a glass capsule. It is used in crossbar telephone switching.

DS

• **Directory Service** of OSI. This is one of the SASEs under development. It is the address directory for open systems.

• **Double-Sided** disks. See.

• **Data Signal** in the standard US transmission hierarchy. See DS# below.

DS#

There are a couple of these. The first is for a hierarchy of data rates in North America, and the second is for international videotex.

• **Digital Signal level #.** These are the common designations of North American signal transmission rates which are aligned with the ITU Rec. G.702 standard for asynchronous digital hierarchies. This is the accepted US multiplexing standards of the primary 64kb/s PCM data rate — used when multiplexing signals through the hierarchy from baseband to high-speed broadband trunks.

The DS-# hierarchy describes the protocols, framing formats and frequencies that various levels of transmission products use. They all relate to the carrier's T-channel facilities. The standard rates are:

DS-0	64kb/s	Sometimes 56kb/s (1 PCM channel).
DS-1	1. 544Mb/s	24 PCM channels for T-1.
DS-1C	3.152Mb/s	48 channels for T1C.
DS-1E	2.048Mb/s	The US designation for E-1 .
DS-2	6.312Mb/s	96 channels for T-2 — never used.
DS-3	44.736Mb/s	672 channels for F3 and T-3 [FT3].
DS-4	264.176Mb/s	4032 channels for T-4M.

For the ITU international equivalent standards see G.702.

• **Data Syntax I, II, and III:** These are the ITU-T categories for the different forms of videotex.

DS I alpha-mosaic (used in Europe, Asia and US);
DS II alpha-geometric (used in North America)
DS III alpha-photographic (used in Japan and the UK).

See all these terms.

DSA

Digital Signature Algorithm. A public key identification system. Unlike RSA, DSA does not encrypt the message, but rather appends to the message a number calculated from the message and the sender's private key. This appended number can be electronically verified by calculation, and so the process is well suited to computerization and smartcards. It can also be used to verify the identity of the sender. See digital signature.

DSAP

Digital Signal Array Processor.

DSC

Desktop Color Separation. This is a format which produces five separate PostScript files. The first is a positional image, and this is followed by four others — for the three process colors, plus black.

DSE

Data Switching Equipment. In a typical communications link over a public telephone network, the communications chain will consist of five distinct elements of which the DSE is one :

— A DTE (workstation or PC).
— A DCE (modem or multiplexer).
— A number of DSEs in the telephone exchange.
— Another DCE (modem or multiplexer).
— Another DTE (computer, printer).

DSG

Distributed Systems Gateway. A gateway (operating at OSI Layer 4) which allows connection of very dissimilar networks, particularly those with connectionless services having unusual protocols.

DSI

Digital Speech Interpolation. See.

DSL

Digital Subscriber Loop. The term applied to the 'local loop' (subscriber premises to exchange) connection in Basic Rate Access ISDN services.

DSP

- **Digital Signal Processor:** See.
- **Domain-Specific Part** — in addressing. See address.
- **Display System Protocol:** A *de facto* packet-switching standard which allows BiSync traffic to flow over a packet network.

DSR

Data Set Ready (old RS-232-C terminology, now 'DCE Ready' in Version D). This is a pin on the RS-232 connector (Pin 6 at the DTE end) which indicates to the attached terminal that the modem is now connected to the telephone system and is 'ready and rearing to go'.

Note that these handshakes are usually required to set up a link between a computer and its periphery (especially a dumb periphery), but overlying software controls may also be performing the same or similar functions across the telephone link. Commands often exist for exerting these signals through software controls in the event that the equipment doesn't provide them. See AT commands.

- In ITU terminology DSR is Circuit 107: in EIA terms it is CC.

DSRR

Digital Short Range Radio. A low-cost, simple-to-licence, public domain radio communications service — a modern replacement for HF and UHF CB radio.

DSRR is being devised in Europe by a consortium of manufacturers for personal radio services, but has lately been broadened in concept to include more business-oriented services (as a substitute for PMR). It is now an ETSI standard along with TETRA.

It uses digital techniques borrowed from GSM, and was expected to appear in 1992 but has been delayed. It is 'self-trunking' with individual subscriber identification.

DSRR operates in both single-frequency simplex and two-frequency 'talk through' repeater modes in the 933–935MHz and 888–890MHz bands. Each channel utilizes 24kHz of channel spacing to provide a 16kb/s data rate, and overall the service will have 77 traffic and 4 control channels. Australia and North America can only use the simplex version, since the 888–890 MHz band is reserved for AMPS cellular.

DSS

- **Decision Support Systems:** This is computer software which helps managers to analyze corporate data more easily — to support their decision-making ability. DSS software will generally run on mainframes or mid-range computers, and it will provide an extensive range of analytical capabilities.

  EIS is similar, but much more extensive and tactical. DSS uses 'decision trees' to help make strategic decisions.
- **Digital Satellite Service:** (generic). See DTH and DBS.
- **Digital Satellite System:** The name RCA, Thomson, Sony and Hughes give to the DirecTv technology. See DirecTv.

DSSD

Data Structured Systems Development. A complex programming methodology used with CASE. It is used to analyze requirements and create a diagrammatic representation of the proposed program. The analyst focuses on the transactions — while the DSSD mandates a range of specifications.

DST

Deference Slot Time (with CSMA/CE access systems). A way of providing fair access to a contention network while avoiding unnecessary collisions, by calculating a different time-delay for each node trying to access the network. The delay is determined by an algorithm which takes into account the speed of electron flow on a cable and keeps a history of the information content of all LAN transactions. Each computer on a network will calculate a unique DST, so access is always fair.

DSU

The Data Service Unit takes signals from the terminal/multiplexer (DTE) and translates them into bipolar digital signals for the CSU; it also provides clocking and signal regeneration on the channel. This is a general term for a number of different, but similar, devices.

It is a simplified high-speed 'digital modem' (there is no modulation — only code conversion for line-coding) for limited distance communications of a few miles in the local loop, and it will have been designed for a special protocol.

These days DSUs may also perform aggregation functions; they will take all of the bandwidth and make it into one fat data channel. Fractional-T1 DSUs are the obverse of that, they operate on a subset of the (p x 24) T1-channels.

• Essentially similar devices are called baseband modems, line drivers, or limited-distance modems.

DSU/CSU

See CSU/DSU.

DSVD

Digital Simultaneous Voice and Data — using V.34 modems (28.8kb/s) for the simultaneous carriage of compressed voice. It is being developed by AT&T, Hayes, Rockwell, US Robotics and Creative Labs. DSVD is an ANSI standard which will be put to ITU-T.

MultiTech appears now to have broken away from this group and established a sub-standard. Both are in competition with AT&T's ASVD.

• There's an alternative analog voice–data standard called QADM.

DSX #

Digital Signal Cross-connect (at levels 1 or 3). These are the electrical and physical interfaces to interconnect DS1 (T-1) and DS3 (T-3) signals.

DSZ

See GSZ.

DTE

Data Terminal Equipment. As distinct from a DCE. DTEs are usually computers these days, but printers are also configured as DTEs since they 'terminate' a link. Multiplexers and protocol translators are also often DTEs.

In a communications chain you should find DTEs at both ends, with two or more DCEs interposed down the line — and some DSEs in the exchanges. Sometimes it may appear that a DTE (such as a multiplexer) is not 'terminating' a connection, but in fact the multiplexer terminates links from both the line side and the trunk side. It is a store-and-forward device, so the link ends and begins here.

The importance of the terminology is mainly in the wiring of the plugs. When a standard like RS-232 is being used, a DTE device can only be cable-connected to another DTE device if a null modem is used to cross-over a number of the pairs. There is also a problem with plugs and socket shielding (and male vs female), but you'll find no coherent standard here.

There is also a difference in the primary functions between DTE and DCE, with the DTE handling high-level functions involving intelligence and digital data storage, and the DCE handling the physical functions.

This was once a clear distinction but, with modern modems having CPUs, memory, compression and error checking algorithms, it is impossible to make this distinction now. In truth, modems are now DTEs, but we still wire them as DCEs.

For serial connection, the key to identifying DTEs and DCEs is to know that, with RS-232, the DCE equipment transmits its data on Pin 3, while DTE equipment transmits on Pin 2.

DTE flow control

The point being made here is that the intelligence in the terminating equipment (the DTE) is serving as the control device for the flow of data over a certain link, rather than the modem (a DCE) or other intermediate device using hardware handshaking. DTE flow control is end-to-end flow control using message/code formats — from one terminal/PC to the other. See X-ON/X-OFF.

DTH

Direct to Home (US). A reference to satellite delivery of television via direct broadcasting satellites, rather than using (often secondary) transmissions through a cable or MDS system. The term DBS (Direct Broadcasting Satellite) means much the same. These will all be digital proposals today.

DVB (Direct Video Broadcasting) probably means the same today — it is the general name applied to European joint developments. But DVB was also used (as was DTV) for the old analog MAC systems. See DBS, DVB and DirectTV.

DTMF

Dual Tone Multi-Frequency. The touch-tone dialing system which uses a 4 by 4 matrix of audio 'sinusoidal' tone generators to create 16 distinct dual-tones for 'in-band' signaling over the local loop. Special chips are used to isolate the frequencies and decode the numbers in the exchange.

Most telephones only have keys for 12 of these dual-tones; the others (now known as A, B, C and D), are used for PBX or exchange controls.

The eight DTMF frequencies are:

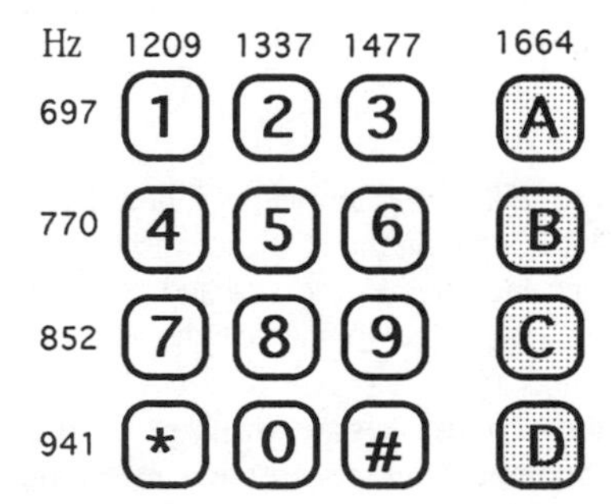

The frequencies are chosen so that neither their harmonics or their intermodulation products will fall in to another of the tone bands.

The distinction is with pulse-dialing and with packet-signaling which is now widely used in ISDN's D-channel.

• MF (MultiFrequency) is a different tone signaling system used between exchanges and between PBXs.

DTP

• **DeskTop Publishing:** See.

• **Distributed Transaction Processing:** An OSI application like MHS and FTAM. See distributed processing and distributed transactions.

DTR

• **Designated Transfer Rate:** The set rate for the transfer of information from the storage media to the system. The term is applied mainly to the replay of music and video from disk or tape.

• **Data Terminal Ready:** A signal which says 'The terminal is ready to operate'. It is sent by raising the voltage (–12V) on an RS-232 connection (Pin 20 at the DTE end) which indicates to the modem that the terminal is ready for transmission. It is controlled by the terminal (the PC), and is asserted and held high for the whole time the terminal is available.

Note that these handshakes are usually essential if you are to set up a link between a computer and its peripheral devices (especially a dumb peripheral), but overlying software controls may also be performing the same or similar functions. Control over DSR can be exerted through an S-register, but this is a stop-gap measure, not recommended. See flow controls, and AT commands.

• In ITU terminology DTR is Circuit 108; in EIA terms it is CD.

DTR/CTS

Hardware handshaking system where a connection between the RS-232's Pin 20 (Data Terminal Ready) in one device and Pin 5 (Clear To Send) in the other is used to control the flow of data between the two devices. You may find this with printers.

The DTR line is an output (from the viewpoint of the DTE, signaling to a DCE) and the CTS is an input. When the printer's DTR line output holds the computer's CTS line input at a logic 1 (–12v) it tells the computer that it is 'clear to send', while logic 0 means 'not clear to send'. See also handshaking.

DTT

Digital Terrestrial TV. This means UHF television with multiple digital channels per carrier.See MCPC.

DTV

• **DeskTop Video**. The general term for the multimedia successor to DTP (desktop publishing). it assumes that your PC will be able to handle more than 256 simultaneous colors, and be able to fuse (capture and process) moving images from broadcast, VCR, laserdisk, tape and video camera sources, and display these images in PC windows at normal scan rates.

• **Direct Television**. See DTH.

DTx

Discontinuous Transmission. This is an analog cellular phone term, where the mobile switches between two power levels depends on the amount of voice activity in its microphone. So when voice is not being used, the phone reduces transmitter power as a power-saving device.

This term is now being applied to a technique used in GSM mobiles to reduce the interference by detecting inactivity in the voice, and reducing transmission power. See DSI and VAD also.

dual-bus system

The approach to PC design that dedicates one bus to memory, and another to all other system enhancements. This allows one set of expansion slots to remain compatible with, say, the 8MHz requirements of many expansion cards, while allowing the memory add-ins to function at 20MHz.

dual cable

A broadband LANs topology in which two coaxial cables are used, one to transmit and one to receive. Normally a single cable is used with the total capacity divided. See Full Duplex Ethernet.

dual homing

A FDDI term for a way of linking cable concentrators and stations in a tree-branch configuration to provide a redundant path in case of failure.

Dual-ported DRAM

See VRAM.

dual processor

See coprocessor.

dual-spin

A type of spin-stabilized satellite where the main body spins to provide attitude stabilization, while the antenna assembly is de-spun by means of a motor so that it points continuously towards earth. Most of the early communications satellites were of this type. See spin-stabilization and body stabilized.

dual stepper servo

The technique of using two stepper motors to speed up hard disk access. A coarse stepper positions the read-head quickly and roughly, then the fine adjustment is made by the fine stepper motor.

dubbing

Copying. Usually applied to tapes.

dumb terminal

A terminal with no independent processing capability itself. The contrast is with smart or intelligent terminals. With communications software, the dumb terminal emulation mode is generally known as TTY or VT-100.

dump

To transfer data in a casual way. No attempt is made to format, or often to check for errors.

• To print out a listing of the contents of memory, or of the data on a disk. This is the first step in debugging a program.

• A screen dump prints out the image of the current screen, or transfers the image to a special graphics file.

duobinary

This seems to be the same as binary. It's a tautology.

duodecimal

A counting system using the base of 12.

duple

A pair: a slang term for duplex. It usually means full-duplex.

duplex

Doubling. Two in a pair.

• Duplicated equipment used to provide backup. Disk duplexing is the same as disk mirroring.

• Duplex circuits can transmit and receive simultaneously (such as a phone line).In European digital telecommunications this is short for full-duplex, as distinct from half-duplex or simplex. Full duplex allows data transmission simultaneously both ways on a circuit.

• Duplex spacing: In cellular phone systems the transmit frequency channel will always be a constant distance away from the associated receive frequency channel. This is duplex-spacing, and in GSM it is 45MHz. So a GSM mobile transmitting on 900MHz will be receiving on 945MHz.

duty cycle

A figure often quoted in printer brochures. It purports to be a rating, measured in pages per month, that indicates the type of workload a printer is designed to handle. A figure of about 3000 pages a month is about average for a small office.

DVB

• **Direct Video Broadcasting:** In older publications, this probably means an old analog satellite broadcasting (aka DTV) system like D2 MAC or one of the other MAC or PAL systems.

• **Digital Video Broadcasting:** The general term applied to joint European developments of digital TV technologies. This activity has now replaced the early-1990's European emphasis on the wide-screen and high-definition forms of MAC analog systems, and proposals such as HD-Divine.

DVB is now a group of 160 European equipment manufacturers and broadcasters with Canadian and Mexican support in North America. They have currently abandoned HDTV because of the cost. See COFDM and QAM.

DVC

Digital Video Cassette. A moderate-cost digital component videotape format (dating from 1993) which is aimed at both the consumer market and at professional users. Sony, Philips, Thomson, Matsushita (Panasonic) and six other companies are behind the standards development.

The standards call for DCT compression of about 5:1 ratio (almost lossless) and 4:1:1 component video output (4:2:2 in high-end equipment). The quarter-inch (or 6mm) camcorder cassettes will hold between 60 and 120 minutes of recording at rates up to 25Mb/s. Cassettes for home replay will be in two sizes: DAT-sized (66 x 48mm) and Large (109 x 76mm).

See D-5, D-VHS, component video and Quarter-inch.

• Matsushita released details of three new DVC camcorders in March 1995; The professional version (called DVCpro) will cost around US$30,000. The Compact domestic version will cost about $4000 initially and then drop to $1000. It will use a 1/3-inch CCD and have a 4:3 screen ratio.

The ENG version (4:1:1) uses 1/2-inch CCDs (4:3) and weighs only 5kg. It can be purchased as a complete system with laptop-computer editing, two very small tape drives and dual LCD monitors. The camera has 3 CCDs, uses 5:1 M-JPEG compression and produces pictures roughly equal to Betacam SP. The 6.53mm tape cassette has limited recording time.

The High-end production version will use a 2/3-inch CCD, and provide both 16:9 and 4:3 screen ratios.

DVCR

Digital Video Cassette Recorders. These are expected sometime in 1996. Currently a consortium of 50 companies (see above) have agreed on a world standard.
See Digital Video Cassette; DVC.

DVD

Digital Video Disk. This is a specific group of standards or rather 'requirements'. Don't confuse it with DVC (Digital Video Cassette) or DCC. These requirements are for digital TV, audio and data in a small (12cm) CD optical disk format.

The specifications were set by the Motion Picture Studio Advisory Committee. They wanted a new high-density disk (using MPEG2 compression) which would address the needs of both the entertainment industry and the PC industry.

The challenge was taken up independently by two groups, resulting in two different standards (MMCD and SD). They then came together (Sep '95) to create a single joint standard.

Philips and Sony lead one group which wanted to maintain backward compatibility with the present CD-DA and CD-ROM standards and provide recordable systems for computer data. They created Multimedia CD (MMCD), which was both a replay and a recording standard, suitable for both video and data.

Toshiba and Time Warner leads the other group which was more concerned with replay of movies for home entertainment. This group created Super Density CD (SD). This was primarily for replay, although provision was made for recording standards at a later date.

Finally, sanity prevailed and the two groups got together in September 1995 to create a single DVD disk format, based mainly on CD-Multimedia.

To retain compatibility with CD-ROM and CD-i formats, they have used Sony's 8 to 16 (EFM) channel modulation method, rather than Toshiba's 8 to 15 techniques.

This has reduced the disk capacity from 5GBytes to 4.7GBytes, which is equivalent (with MPEG compression) to a video reduction from 142 minutes to 130 minutes. However, this change makes the disks less susceptible to playback or encoding errors. The disks are made as a two-layer sandwich of 0.6mm poly-carbonate platters which can be coded on either or both layers.

It is now almost certain that this joint DVD format is the way in which future replay and recording systems will go.

See CD Multimedia (MMCD) and Super Density CD (SD).

DVE

Digital Video Effects. This refers both to the effects (everything, including morphing) and to the equipment used in modern day image manipulation.

DVE units can be low-cost computer equipment such as the Video Toaster (which runs on the Amiga) or the more modern versions such as Aladdin and Composium FX (under Windows) and Avid — now the industry standard.

High-end DVEs are made by Sony, Ampex, Abekas and Grass Valley. DVEs are usually integrated with digital switchers and paint programs. See also Henry and Paintbox.

DVHS

This is the new Digital VHS standard for home recording developed by JVC and Panasonic. It has some very innovative features to match the new digital television standards being developed in Europe.

In its HD mode, the recorder can store six interleaved TV signals at the same time (see MCPC). You can replay these later, isolating the channel you require and decoding it in the conventional way without needing to demultiplex or decode the other signals. So this will be used for Pay TV time-shift and possibly for Pay-per-view time-shift.

The same companies that formulated DVC are behind D-VHS (but with two years more experience), with JVC taking the lead. The advantage of this format, is that it is built around the mechanical components of the old VHS system, so it should be cheaper. The electronics of the system seem to conform to the basic DVC standards.

D-VHS will record at rates from 2 to 28.2Mb/s, and provide interleaved recording of a number of channels simultaneously (it can take them off-air in encrypted form). It will record both 4:3 ratio images and 16:9 widescreen images (called HDTV!), and store the information for a 3 hour widescreen movie on one tape.

There are three recording speeds:

— Long Play provides 2Mb/s rates for 49 hours on a tape.

— Standard mode gives rates of 14Mb/s and 5 hours on a tape.

— High-definition mode handles data rates of 28.2Mb/s and stores 2.5 hrs.

DVI

• **DeVice Independent:** A highly portable file-type.

• **Digital Video Interactive:** A technique developed by GE/RCA Laboratories (and since sold to Intel) for the compression and replay of full-motion video images using standard CD-ROM disks as the source. It is really a computer-based, pre-MPEG, peripheral standard, rather than a full system.

The technology relied on enormous computation power (but slow speed) to compress images. DVI could compress some images by a ratio of 120:1 — often taking minutes per frame — but it was lossy compression. It varied the pixel size whenever the data stream could not supply information at the necessary rate — and at this point scenes would look 'blocky'.

DVI differed from CD-I mainly in the fact that it was designed around standard PC technology which couldn't successfully incorporate the OS-9 real-time operating system. It had its own 'real-time executive' to handle audio/video disk-access concurrency. Intel later developed a special i750 video processing chipset for DVI compression and decompression, and the system then merged into Indeo. See MPEG.

Dvorak

A new design of keyboard that is 30% faster and more comfortable to operate. The problem is that no one who has learned to type on QWERTY wants to go through all the agony once again.

The design was by August Dvorak at the University of Washington DC, under a research program commissioned by the US Navy. The five primary vowels and the five most common consonants are grouped in the center of the keyboard where the fingers rest naturally.

It is said to be possible to type 3000 of the most common words without reaching out to another row, and 70% of all typing can be done without stretching a finger.

DVTR

Digital Video Tape Recorder(ing). This is a general term for a number of developments for professional use; there are two basic techniques — composite and component. The CCIR have produced recommendation 601 for component digital video. See 601 CCIR and 4:2:2 in numbers section.

dwell

The output pulse of a device may be shorter or longer than the input pulse. In many relay or logical amplification devices, the length of the output is adjustable. The difference is the dwell.

• Acoustic reverberation is also sometimes called dwell. See reverberation.

DWT

Discrete Wavelet Transform. An approach to video compression that has some advantages over DCT, but which arrived too late to challenge DCT's pre-eminence in JPEG and MPEG.

It is used in professional equipment where it gives almost lossless still-image compression at ratios from 12:1 to 30:1.

Wavelet Bitstream is the latest attempt to challenge DCT. It is a spacial transform process, which holds information about where a signal occurs, not just the frequency (as with DCT).

The claimed benefits of wavelet transform are:

- High-quality. The image is said to be easily scalable by resolution, with sufficient computational power.
- Software decoding.
- Transcoding abilities.

See transform coding also.

Dx

- Direction finding — radio location.
- Video standard — See D-3.
- Long-distance radio. A term used by hams (amateur radio operators) for those who specialize in long-distance.

DXC

Digital Cross-Connect. See cross-connect.

DXF

Document eXchange Format. A *de facto* standard CAD file format developed by Autodesk to translate AutoCAD drawings from one platform to another. See IGES also.

DXI

Data Exchange Interface of ATM. This is the third ATM implementation agreement (See UNI and B-ICI). It defines the protocols to be exchanged between the customer's router and a public or private ATM network. This allows existing LAN/WAN equipment to use a DSU to translate frames to cells, and therefore to use ATM networks.

Basically, it tells the routers how to format LAN data so that the DSUs can handle it. Two modes have been defined:

- **Mode 1:** for a maximum of 1024 virtual circuits, with AAL-5 support, and with the packet size limited to 8kB. This limit is set by a 16-bit CRC.
- **Mode 2:** for a maximum of 16 million virtual circuits. It supports AAL5 and AAL3/4, and handles 64kB packets (with a 32-bit CRC).

dyadic

Dual. A mathematical operation requiring two operands — addition, subtraction, multiplication and division. See monadic.

The term is also applied to operations where two processors or other system-components work together in a shared-parallel way.

dye laser

The use of organic dyes dissolved in viscous solvents to produce a tuneable laser (one that can be adjusted to a specific frequency). It is possible to design organic dye lasers which can operate over a wide range of continuously variable wavelengths. The dye molecules are 'pumped' by another laser (usually an argon) and then the tuned laser injects the signal information. The tuning matches the dye. This process may be used for optical disk "bubble" recording. See below and pulsed lasers also.

dye-polymer optical disks

This is one of the current WORM technologies — the others are ablative pit and magneto-optical. Dye-polymer technologies work by creating physical bubbles at the boundary between two bonded polymer plastic layers. The read-laser (at a lower power) reads these bubbles as 'pits' since they reflect less light back to the source.

- In some systems two lasers are used, each tuned to the dye-color of the layers. The dye in the bottom layer absorbs heat from its laser, and expands to form a bubble only when an area of the top layer directly above is also heated. To record, therefore, you use two lasers in synchrony; and to read, you use one reflecting off the surface of the bubbles. These disks are cheap and reversible, but still subject to a range of problems.

dye sublimation

A technique of computer color printing with photorealistic properties — it can produce continuous tones. It can also be used on plain paper or on transparent film for overhead projection. Costs of machines and paper, however, remains high.

This process is also known as dye diffusion, dye transfer, or 'sublimable' dye technology. It is similar to the thermal wax process. See.

Dynabook

A theoretical 'wish-list' portable lap-top computer idea which was popularized by Alan Kay in the 1970s. The name has since been stolen by a company making lap-tops.

dynamic

- In computing and communications this tends to mean 'change as needed' which is close to 'adaptive'. Dynamic buffering, will provide what ever buffer size is needed at the time it is needed.
- The other sense is the opposite of static. It means that constant attention is needed. This is how it is used with Dynamic RAM, which needs to be constantly refreshed.

dynamic address resolution

This suggests that a special address resolution protocol has been used to determine and store the address.

dynamic channel allocation

See adaptive.

Dynamic Data Exchange

See DDE.

dynamic equalization

See equalization.

dynamic execution

The term that Intel has applied to a system of forward instruction cache introduced with the P6 chip. The chip can execute several tasks simultaneously, and it can apparently think ahead.

It selects and readies tasks which are most likely to be needed next. It's a form of pipelining.

dynamic linking
See hot links.

dynamic microphone
A microphone which does not need batteries, but generates its own current using a moving coil. These are rugged, and reliable and usually have a cardioid acceptance pattern.

Dynamic RAM
See DRAM.

dynamic range
The ratio between the maximum level that can be handled by an analog electronic circuit, and the natural noise of the circuit in its quiescent state.

• In tape recording, it is the ratio between the highest level that can be recorded which results in only, say, 3% intermodulation distortion on playback, and the playback noise of the tape with no signal recorded.

• The term can also refer to the range of frequencies of an audio signal.

dynamic routing
Routing which adjusts automatically to changes in network topology, or to traffic loads. This is by far the most prevalent form of routing strategy (although see flooding). Dynamic routing strategies, adapt to changing conditions on a network.

There are three main categories (and many variations and hybrids):

– **Isolated dynamic routing.** The router adjusts itself to local conditions. For instance, it just sends the packets to the port with the shortest queue. This is almost a variation on random routing, and isn't much used.
– **Distributed dynamic routing**. The router receives information from adjacent nodes about delays and outages, and makes its decision locally about how to direct packets.
– **Centralized dynamic routing.** The router receives information about delays and outages from a central node, to which all nodes report.

dynamic tracking
In VCRs this is the technique of allowing video heads to flex minutely under the control of feedback circuits so as to follow accurately the path of a helical video track.

When the videotape is in motion, the 'resultant' angle of the track is set by both the forward motion of the tape transport, and the diagonal of the helical scan.

When a videotape is stationary, the angle of the prerecorded track to the direction of the sweeping head is substantially different from when the videotape is in motion.

There is no forward motion, so the spinning head tends to cross the helical scans, producing a screen image usually consisting of the parts of two to four or so images, divided by a horizontal band of noise (the guard band between the helical scans). Dynamic tracking was developed to overcome this and allow good quality freeze-frame images, and also to avoid the need for mechanical adjustment to the tracking control (necessary on many early VCRs).

These flexible head-mounts can absorb this angle change, even in still-frame; the heads are mounted on piezo-electric arms, which automatically adjust through electrical feed-back to follow the track with the maximum signal strength.

This is now the standard way to provide noise-free stop-motion images on most modern VCRs. The concept was developed by Ampex for the C-format, and it was then capable of reading two tracks (two fields or one full frame) for still-image reproduction. See delta heads.

dynamically redefinable character set
DRCS allows a terminal to show all forms of commonly-used text characters (English, French, German, etc.) used by non-ideographic languages. This was especially important with European videotex.

In true DRCS systems, the required character set could be automatically downloaded from a database when required; it would then be stored and used from RAM. This idea was also extended to include common non-text symbols.

The contrast is with terminals which can only display a limited range of characters stored in ROM. See Unicode also.

DYUV
Delta YUV. See delta.

E

e/E

- In maths, e is a constant with a value of approx 2.71828. The function e^x is its own derivative. In Basic it is computed by the function EXP(X).
- e = execution.
- In telecommunications E is Erlang. See.
- With large and small numbers, the capital E indicates exponential notation. See.

E#

- **E1** is a European digital data-rate service standard for a 2.048Mb/s connection. When used with standard DS-0 (64kb/s) multiplexing it can handle 32 channels on two twisted pair. In the framed version only 30 of the channels will carry voice. The ITU specification is Recommendation G.703.
- **E2** doesn't appear to be used.
- **E3** channels have a data rate of 34Mb/s.
- **E4** channels have a data rate of 140Mb/s.

These are the European versions of the American T# system based on T1 at 1.544Mb/s.

E.13B

The typefont used for Magnetic Ink Computer Reading (MICR). It was developed for the American Banking Association and adopted worldwide. It consists of the numbers 0 to 9 and four symbols which signify the meaning of fields.

E.163

A telephone numbering scheme which consists of a 'country code' (one to three digits) followed by a national number. The national number is fixed by the country's PTT administration. This is the current numbering technique used in PSTNs today. See also E.168, X.121 and F.69.

E.164

A form of international addressing used by ISDN. E.164 is a subset of E.163. These numbers are 15 decimal digits long and include country code, area (city) code, and local number. See also E.168 and X.121.

E.168

The new recommendation added to E.164 to extend the numbering scheme for eventual use by B-ISDN.

E&M

Ear & Mouth. An old interface for analog PBXs. The E&M interface uses some extra wires for signaling called E (ear) and M (mouth). This is battery-reverse or DC signaling.

There are several different types of E&M interface; and the terms 2-wire and 4-wire refer to the number of wires needed for the voice portion.

- A 2-wire interface which shares two wires to transmit and receive, and has 2 or 4 more wires for the E&M leads.
- A 4-wire interface will have two wires to transmit and another two to receive, and 2 or 4 more for E&M leads.
- Type I E&M uses 1 wire for an E-lead and 1 wire for an M-lead.
- Type II E&M uses 2 wires for both E- and M-leads.

An incoming seizure is signalled by a grounding of the E-lead (Type I) or a short across the E-lead pair (Type II). An outbound seizure is signalled by battery on the M-lead (Type I) or a short on the M-lead pair (Type II). E&M is converted to SF signaling for carrier systems.

E-channel

This was defined in ISDN (but not used). It is a 64kb/s circuit-switched control channel (not a bearer channel).

E/IDE

Enhanced IDE. See IDE and ATA.

e-mail

Electronic mail. E-mail is the generic name for the transfer of memo-type messages over telecommunications lines and networks: both globally, and within a company: publicly, or privately — but always between people. The generic term for this type of messaging is Message Handling Services, but MHS is a much wider concept including machine to machine.

Telex and telegrams were an early form of global e-mail, although the term isn't used in such an all-encompassing way. E-mail, as we now know it today, was probably invented at MIT in 1963.

An e-mail message is a packet of indefinite length. It has a source and destination address, followed by a message payload — and it may be protected by various error checking fields. Usually it is transferred as a single unit in a store-and-forward manner between nodes. 7-bit ASCII is preferred to ensure that the bytes can travel through any system which may be encountered.

Electronic mail messages are generally stored in computer 'mail boxes' until accessed by the receiver — although they may also be delivered directly to a LAN-server or to an on-line workstation. In the future it is expected that e-mail will be delivered to PDA-type terminals attached to home or business phones.

The old 'public' e-mail services of Telecom Gold, Telememo and Dialcom were a very narrow concept when compared to the new X.400 Message Handling Systems. These now have EFTPOS and EDI implications, and can interact with Telex, teletext, facsimile, postal and paging services.

The third generation e-mail architectures comprise four independent modules: the User Agent, Message Transfer Agent, Directory Services and the Mail Engine itself. See also X.400, UA, MTA, MHS, FTAM and X.500.

• The Internet is emerging as the world's most important e-mail service. The term e-mail on the Internet specifically means small files of a set format which can be transferred between servers using SMTP.

• Multimedia e-mail is sent over the Internet using MIME.

e-money

These are forms of electronic cash-payment systems which are usable over networks. The term covers a range of different proposals with varying degrees of security.

— Digicash is a way of producing what amounts to a tradeable electronic purse, into which sums can be entered, and sent around the network. These electronic purses are as good as banknotes (when backed by a bank) and they can be exchanged between buyers and sellers in the same way.

— Microsoft and Visa have a similar system which is backed by credit cards.
— Cybercash of Vienna (Virginía) and Wells Fargo Bank also have a system.
— Mastercard, Bank of America and Netscape have another,
— National Westminster and Midlands Bank have the Modex smartcard e-money system.
— Banks in Portugal, France and Belgium are running electronic cash schemes based on smartcards used as stored-value cards (SVCs — see).

• Some of these systems depend on you first having a credit card, and they then provide the normal credit card restrictions on the size of the purchase, and on the percentage taken by the credit provider. Others involve an exchange of 'electronic cash' but each transaction must be registered with the bank or service provider in order to prevent fraud (and to allow some percentage to be skimmed!). Others can be handled independently of the bank after the initial charging of the stored value.

• There are obvious privacy issues when anyone is tracking and recording minor expenditures.

E Region

The layer of the ionosphere at about 110 kilometers. It does reflect some HF radio signals during the day, but it dissipates at night after sunset, and this allows the reflective qualities of the F Region to operate more effectively. However on 'sporadic' occasions it remains as a reflector. See Sporadic E.

e-time

Execution time. See i-time also.

E-type

Any telephone device which can generate either tone (T-type) or pulse (R-type) signaling.

e-zine

An electronic magazine. Some of these are free, and some you must subscribe to.

EAPROM

Electrically Alterable Programmable ROM. Just EPROM.

early discrimination

In voice recognition systems the speech is first digitized, and then the first process of 'early discrimination' is applied to remove ambient sound — background noise, sneezes, door slams, etc. which will upset the recognition algorithms.

early token release

A modified token-passing scheme where a device can pass on its token before the communications packet has circled the network. Sluggishness in Token Ring systems is caused by the constant need to wait for data to flow around a large network.

earth

The earth lead on power cables and on many telecommunications links is (theoretically) set at zero potential. The voltage is usually referenced to the earth potential at the power source — and the neutral is also set at earth potential just outside the home or office (at power meter). In many household power supplies, the earth connection is made through the earth wire to a pipe buried in the ground near the switch board.

It is common practice, therefore, (but definitely not universal) to connect the frame or chassis of electronic equipment to earth at the consumer's end of the power feed, and to isolate both active and neutral wires from the chassis.

The idea is that if any 'hot' (active) element in the electrical path fails and makes contact with the metal chassis, it will instantly blow the fuse at the switchboard or trip the breaker.

Since the earth wire is nominally at earth potential, it doesn't normally give problems; however this can never be guaranteed without checking each wiring system individually. You must never assume. Some electronic equipment, especially old radio and television sets, were designed to have a 'hot chassis' with the metal at a high potential. These can kill.

Never work on any electrical device with its power supply plugged in to a power point. Don't believe popular computer myths about likely static damage. Always touch the case of the power supply with the back knuckles of your hand as a check (also to even out the static charge).

An unearthed chassis can be 'hot' for many reasons — and if you grasp it with the front of your hand with the power on, you will provide the earth connection, with potentially fatal results.

earth network

The term is used to make the distinction between those networks which rely on cable and point-to-point microwave/infrared links, and those that use satellites and short-wave radio with ionospheric ('sky-wave') bounce.

earth segment

In satellite communications, this is the side of operations involved in transmitting and receiving the signals between the earth station and the user or public network.

earth stations

In satellite terminology these are buildings with dish reflectors and the electronic equipment needed to transmit and receive satellite communications. They may be large or small, and some will also be equipped with telemetry and control systems (see TTC&M).

A professional earth station without control is often called a 'gateway', while one with control may be known as a Network Control Center (NCC — see).

At the other end of the scale, a home TV Receive Only (TVRO) dish system is not often classed as an earth station.

EATA

Enhanced AT Attachment. A new hardware interface standard for peripherals used with AT-type computers and the EISA bus.

The committee which designed EATA conceived it as an interface which would work with SCSI, ESDI, ST506 and other systems, so that special device drives wouldn't be needed.

Eb/No

Energy-per-Bit-to-Noise power density ratio. This is the digital version of C/No — which is Carrier-to-Noise Density (C/N per unit of bandwidth).

EBCDIC

Extended Binary Coded Decimal Interchange Code. IBM's unique 8-bit answer to the international use of the ASCII code system. It has 256 numbers, letters and symbols. This was widely used by IBM mainframes before the advent of the PC.

Fortunately IBM's PCs don't use EBCDIC (but 3270 terminals do). The character set includes values for control functions and graphics. See BCD also.

EBU

European Broadcasting Union. The EBU acts as a news and program distributor, not just an industry association.

ECC

Error Correction Code — as distinct from detection (EDC). See error checking and error control.

• Reference Technology's Layered ECC is used in CD-ROM. It adds further Reed-Solomon error checking layers to the basic CD format. The new layer of error checking, however, is performed by software within the PC, rather than in the CD player.

ECCO

This is Constellation Communication's joint Big LEO proposal, now in association with Bell Atlantic and Telebras of Brazil. It is a last-minute update of a previous proposal called Aries, which called for 54 satellites, and had an estimated budget of US$2.2 billion.

ECCO will now have 12 satellites (initially), and then only in a single ring around the Equator. It will provide mobile and fixed-site communications in the urban, rural and remote areas of more than 100 countries in Central and South America, South-East Asia, India, Africa and the Middle East.

ECCO is estimated to cost less than US$500 million, and to be flying by 1997 — later they will attempt to upgrade this to the more elaborate Aries. The Aries proposal was before the FCC but Constellation requested a deferral to draw up the ECCO proposal.

ECH

Echo Canceller Hybrid. One of two different techniques used in ISDN to provide basic rate (2B+D) over the local loop. In ECH, two hybrid circuits, one at the subscriber terminal and one at the network terminal, convert the two-wire loop to four-wire. The resulting reflection of the transmitted signal at these hybrids is removed by two echo cancellers.

The loop rate is the same as the transmission rate (144k/s) with ECH, but the hybrids make this approach more expensive. However, it can extend over a longer range. See TCM.

echo

• A reflection of a transmitted signal, sufficient to cause interference (or annoyance). The problem is caused by imperfectly terminated and (impedance) mismatched equipment (in the hybrids) at the telephone exchange.

There are two types of echo problems: long-haul and short-haul, and there are different echo cancellation and suppression techniques. Short-haul echoes come from the use of hybrids in both local modems and in local exchanges. If the impedance is mismatched, an echo results. These problems are usually handled by echo suppressors — which were older analog devices (see). Cancellers are generally more complex and expensive than suppressors, and are digital devices.

International telephone lines are most in need of echo cancellation to maintain voice quality, because of the distance involved. A delay of more than 30—45ms will generally cause problems — so generally any call over 300km requires suppression or cancellation.

- **Talker echoes** are those reflected back to the sender of the information.
- **Listener echoes** are those heard at a receiver point.

Most echo is generated at the point where the local-loop's two-wire circuits join the inter-exchange four-wire circuits at the hybrid. Others come from modems and phone handsets themselves (they have hybrids built-in). You can even get echo from the junction between two different gauges of wire.

• There is also an echo-like problem which arises in cellular phone systems, or any system using a vocoder, because of the time taken by the vocoder circuits to process the voice before transmission. Short distance links then still reflect these signals back with a long delay, and there are no suppressors in these circuits to reduce the echoes.

• Satellite transmission of voice and data is severely restricted by echo which restricts the data rate substantially. This is due to the signal propagation time to and from the satellite — and sometimes, in long distance calls, over two satellite hops. Despite this problem, for some forms of data transmission, echo cancellers must still be switched off because the cure is worse than the disease. See echo cancellation.

• The term echo also means the deliberate regeneration of a received character, and its retransmission along the back channel. Also called Echocheck and Echoplex (see).

• Local echo is when a direct feed exists between the keyboard strokes and the appearance of the characters on the screen. If these characters are doubled, it is generally due to the locally-generated characters being added to those returned (Echoplex) from a remote device.

• Echo is also a term used in on-line BBSs, especially with FidoNet. A message is echoed by being sent to all other connected bulletin boards; it is a form of keyboard conferencing. This is echomail (for everyone to read) as distinct from netmail (only for one person).

• In MS-DOS and OS/2 the echo command sends messages to the screen. ECHO OFF in the .BAT file will stop DOS from printing all commands to the screen as they are executed. Unix has an ECHO command also.

echo cancellation

An echo canceller is a circuit that removes echos, mainly for telephone networks. Cancellers work in a quite different way to echo suppressors — suppression is just an attenuation system, while cancellation uses digital convolution codes and intelligent timing devices to remove the unwanted signal. Echo cancellation is essential in long distance digital phone networks, and in two-way electronic circuits where there is a processing delay (as in the vocoder of a digital phone).

A reference signal is sent to the canceller, and this is replicated, inverted, and stored as a 'model'. This inverted signal is then reintroduced to the network with the appropriate delay (often more than once), and it subtracts an equal amount of energy from the returning signal. This is a digital computer process.

These techniques are useful for both voice and fax transmissions, but they can create problems with data — so a way is always provided to switch the cancellation circuits off. This will be a tone of either 2010Hz or 2040Hz, and of 400 ms duration.

• Long-haul echo is found in satellite and long-distance phone lines. Any distance over 3200km (and all non-LEO satellite communications travel further than this) produces a very obviously delayed echo.

Echo cancellation within modems enables them to work with the entire bandwidth of the circuit simultaneously in both directions (full-duplex). The older two-wire full-duplex modems at 1200 or 2400b/s split the bandwidth, with one party using the high frequency and the other the low. But now with echo cancellation in modems, both parties can use the same carrier frequency, and so use the full bandwidth. See V.34.

• The functions of an echo canceller are complex. There are many factors involved. The major ones are these:

– **Convergence time:** The time taken for the canceller to recognize and remove the echo. The best units take about 50ms, but the old ITU recommendation is still set at 500ms.
– **Phase roll:** This is a rapid variation in the echo caused by call-routing problems. The echo changes too rapidly for the model to ever be current.
– **PCM-offset:** Signal bias can affect a canceller's performance. This is directly linked to the number of units of PCM-offset with which it can cope.
– **Echo Return Loss (ERL):** This is an important measure of the attenuation across a hybrid, or across the ear and mouthpiece of a handset. It is ERL attenuation which enables systems to differentiate echo from local speech.

echo network

A bulletin board system where cooperating organizations automatically poll each other and share data. This is the Fidonet network system. Each network will have public 'echo conferences', so the range of the conference can extend internationally but often at a cost to the user of only a local phone call. See FIDOnet also.

echo suppressor

A analog transmission line device which reduces echo by attenuating the signals in one direction while the main signal is passing in the other. This happens at the hybrid point, where the two-wire splits to become four-wire half-duplex connection.

There are full echo suppressors which control both directions of transmission, and half ('split') echo suppressors which only control one.

Echo cancellers are used for longer delays with satellite circuits and with modern digital systems. See.

• In some echo suppression systems, the microphone of a transceiver will be turned off, while sound is on the receiving circuits — in effect, creating a half-duplex mode. This is done, for instance, with conference phones where the loudspeakers may feedback into the microphone after bouncing around the room. It was also common in early hands-free cellular phone systems.

• You disable echo suppressors by sending a tone of 2010Hz or 2040Hz along the wire for 400ms, and then retaining the carrier on the line with no more than a 100ms break (otherwise it will automatically switch back on).

echo suppressor (data/fax)

Long distance telephone networks need to have echo suppressors (older analog devices) or echo cancellers (modern digital devices) — the terms are often used interchangeably.

Echoes are common on any extended circuit wherever impedance mismatches are found (at a change from one media type to another) or where circuit hybrids are used.

Echo suppression is therefore provided automatically on long-distance circuits. These circuits cause many problems with data transmission — especially if Echoplex is to be used — since this depends on the deliberate creation of echoes.

For this reason the standard modem answer tones (2100Hz) are designed to switch echo suppression circuits off. The international standard calls for 400ms of a tone between 2000 and 2250Hz, to disable any suppressors or cancellers on the line. If the line isn't constantly in use, the suppressor will automatically switch back into the circuit again after a drop of carrier-tone for a duration of about 100ms.

• When data traffic is discontinuous and echo suppression is not required, a continuous signal must be placed on the line to keep suppressors disabled.

• However, fax devices (which are half-duplex) want echo suppression to remain on — so they deliberately leave a 100ms+ gap in transmission after the handshake to allow the echo-cancellation/suppression systems to reset.

However, if the transmitted audio signals are too strong and there is some residual echo in the system, the 100ms gap won't be detected as a gap (it will be filled by echo!), and therefore the suppressors won't come back into play. This is one of the main reasons why some fax machines experience constant errors on international calls (often experienced as the remote fax automatically resending page after page). See echo cancellation.

EchoNet

A basic EchoNet message exchange is used by many existing computer networks, including Usenet, Fidonet and Relaynet. FidoNet newsgroups are actually called 'echoes' and their electronic mail is called 'echomail'.

The basics of this approach are that when two machines make contact (usually by polling), they will exchange messages (e-mail) each way using EchoNet protocols. These are connectionless, non-switched contacts where any other EchoNet device may simply act as a relay. EchoNet message formats have a number of different header fields.

Echoplex

In full-duplex data operations it is possible to send messages simultaneously (each way) on a single circuit. If the characters you transmit are received, then simultaneously echoed back to you over this duplex circuit, you have a simple form of error detection — because you will see any transmission mistakes on the screen.

Echoplex works by cutting the direct link from keyboard to your local screen. The screen now only reproduces incoming data. The remote computer must then be set in Echoplex mode, so it automatically regenerates and 'echoes' back every character it receives.

So what you see on your screen is the characters which have been received and regenerated by the remote host. Generally you aren't aware of this two-way route.

This is often used for keyboard information retrieval, and so any transmission errors will appear on your screen as if they are typing errors; and you will automatically correct them by retyping.

• If local echo is inadvertently left ON in the communications program, you will see both the direct local feed, and the echoed characters, and therefore everything will be doubled:

```
lliikkee tthhiiss
```

ECL

Emitter-Coupled Logic. See.

eclipse

The period when a satellite passes through the Earth's (or moon's) shadow. During this time, satellite transmission power must be provided by batteries.

ECMA

European Computer Manufacturer's Association. This is a manufacturer's organization; the European equivalent of the US's EIA for standards formation. It consists of all major European equipment manufacturers plus a few from North America who operate in Europe.

Despite its name, it is not just a trade organization. It has headquarters in Geneva and since 1961 it has worked entirely as a standards and technical review group. Currently it acts as coordinator between the manufacturers and the international standards organizations, and has very close associations with the ISO and the ITU-T technical committees and study groups. ETSI has handed ISDN developments over to this group.

ECSA

Exchange Carriers Standards Association. A trade association of US communications carriers and suppliers. It is part of ANSI, and has set standards like T1. ECSA has the T1-Committee for telecommunications. See T1.

ECTEL

The Association of European Telecommunications and Professional Electronic Industry. ECTEL is a trade association which is the European voice of the manufacturers of telecoms systems and equipment. Its members are not individual companies, but the relevant national trade associations.

ED Beta

The 'final twitch of a dying format'. This is a video standard which Sony introduced to compete with S-VHS and possibly revive the dying Beta (Betamax) home half-inch VHS format. ED-Beta was a high-band Y/C analog format which was released as a small camcorder for semi-professional use. The price was generally held to be too high.

EDA

Electronic Design Association. EDA tools are used to design ASICs, FPGAs, and CPLD devices. Their speciality is in sub-micron IC design automation, and now multiple layer devices. See HDL.

EDAC

Error Detection And Correction.

EDC

Error Detection Code. Distinguish detection from ECC. See error detection.

EDFA

Erbium Doped Fiber Amplifiers. See FABM, erbium amplifiers and optical amplifier.

edge cards/connectors
The standard type of add-in card used in modern PC slots. The edge connector is just that part of the card which has been etched to provide contacts.

edge enhancement
A video technique for sharpening the edge of images to remove blur, and to enhance detail. In analog video systems the junction between tonal areas is represented by an abrupt change in voltage (brightness).

When the image passes through amplification and duplication this abrupt change is often blurred or flattened — and subjectively it appears as if resolution has been lost. Edge enhancement techniques detect signal level changes, and amplify the differences so as to boost contrast and create an illusion of greater definition. Unfortunately very often this results in a hard 'ringing' (a second negative edge) around objects on the screen.

edge-of-coverage
In satellite footprints, this is generally taken to be the limit of the defined service area. It is the 'isobar' where the signal is typically 3dB below that of the beam center (boresight). Reception is often possible well beyond this boundary by using larger dishes. You need twice the dish area for every 3dB reduction.

EDI
Electronic Data (or Document) Interchange. The process of transferring the information in trading documents (invoices, packing slips, statements, bills of lading, custom forms and shipping slips) from one computer to another — company to company. This is the first really practical implementation of the 'paperless' concept when applied to computer business systems.

EDI is more than just electronic mail, because it involves the use of standardized machine-readable forms coupled with security and tracking measures which are needed in the transfer of legal documentation.

There are two main regional standards, the old European GTDI, and the US ANSI X.12 — and there's a rapidly developing world standard called EDIFACT. See.

EDI-EFT
Electronic Document Interchange – Electronic Funds Transfer. This is the coupling of commercial document interchange and payment systems. EFT automates the payment process while EDI automates the trading process. The idea is to make reconciliation of payments, invoices, etc. possible immediately.

EDIFACT
Electronic Data Interchange for Administration, Commerce and Trade: the only true multi-industry, multi-national standard for EDI. This is part of the UNTDI (United Nations Trade Document Interchange) standards.

The standards specify forms and trading messages. There are also design guidelines, syntax rules, and specified segments and data elements. About 100 message formats are under development for various industries.

- In EDIFACT documentation you will find message format codes which translate as: 0 = draft only, 1 = draft formal trial, 2 = OK and in use, and P = agreed proposal. Only a few have reached stage 2.
- See also ANSI X12 and GTDI.

editor
A primitive type of word processor used by programmers to alter and correct source-code files. Nowadays, full word processors are more often used.

EDLIN
A small and inadequate text editor provided with MS-DOS. Some commands for EDLIN are:

4L	Lists the lines beginning at 4.
4D	Deletes line 4. (Subsequent lines renumbered)
4,9D	Deletes lines 4 to 9 inclusive.
4I	Inserts lines after line 4. (Ended with Ctrl-Break)
E	Saves the edit file and then quits.
Q	Quits without saving changes.

EDP
Electronic Data Processing. Aka DP. These days the EDP department is probably called MIS.

EDRAM
Enhanced Dynamic RAM. This is a new type of DRAM, which has read and write speeds as fast as SRAM, designed for use in high performance computer systems — including the new RISC and Pentium machines. They are moderately expensive.

EDTV
Extended Definition Television. The term applied to the half-way house between our present television systems and true HDTV. Now it generally means just digital, and widescreen.

The goal of EDTV was to provide a broadcast signal which could be received on existing home equipment as well as on advanced receivers with improved resolution, wider aspect ratios, and good stereo sound. EDTV is generally held to apply to a display system with 16:9 aspect ratio, 750, 625 or 525 lines of image (but the scan-lines may be doubled in the receiver), and greater horizontal resolution due to better bandwidth, clean satellite transmission, ghost reduction, video component separation, etc.

To confuse the issue, the European HD-MAC is really an EDTV system with extra motion-compensation added.

The term EDTV is now used to make a distinction with both SDTV (current 'Standard' 4:3) and HDTV (16:9 with 1000+ scan lines). See PALPlus, smart receivers.

edutainment
Education in an entertaining fashion — or that's the theory, anyway. 'Let's make it fun for kids to learn,' is the catchcry, and multimedia, is the supposedly the answer.

It fails to distinguish between learning being 'fun-fun-fun' and it being interesting or absorbing. It also fails to accept that satisfaction often comes from achieving something difficult. Neil Postman wrote a book about this trend called 'Amusing Ourselves to Death.' It's worth reading.

EE

Extended Edition — IBM's name for the built-in database manager that they added to the OS/2 operating system.

EECO

EECO numbers later became the SMPTE time-code numbers in film. See timecode.

EEMS

Enhanced, Expanded Memory Systems. (Often written E/EMS.) This was the superset of LIM EMS 3.2 which AST Research, QuadRAM and Ashton-Tate (see AQA/EMS) created in opposition to the LIM group. It allowed up to 64 pages of extra memory to be enabled simultaneously, and these could be mapped into the UMB via bank-switching. See EMS.

EEPROM

Electrically Erasable PROM. Aka E^2PROM. Erasure is made by electrical pulses rather than by UV light (as in EPROM) and the cells can be erased and reprogrammed at a bit level (not the whole chip as with EPROM — or in banks, as with Flash). These chips are often used in dumb terminals to store the customized set-up information.

The chips can't be used as a substitute for ordinary RAM because of their very slow write cycle (which is much longer than the read). They are reliable; it is claimed that EEPROM chips will allow 10,000 to a million cycles before failing, but experience has shown that the stored program can also leak away over a period of about six years.

The write cycle is done with 21 volt pulses, and the read with the standard TTY 5-volt systems. See flash memory also.

- There are both serial and parallel EEPROM chips.

EEROM

Electrically Erasable Read-Only Memory. This appears to be the same as EEPROM.

EES

Escrowed Encryption Standard. See also escrowed encryption and Skipjack.

EFF

Electronic Frontiers Foundation. A Washington-based lobby group which concentrates on the freedom of use of the open Internet and general computer communications.

EFM

Eight-to-Fourteen Modulation. A transcoding technique used on CD-audio disks to convert 8-bit bytes to 14-bit formats. EFM is needed by optical disks to avoid having two binary 1 digits separated by only one binary 0, since this would then mean that there was too little track distance along a pit or land for the discrimination circuits to react reliably. The logical 1 is created at a transition, while 0s are clocked non-changes — so they want bytes which never have single 0s between two 1s.

A look-up table is used for the 8-bit byte conversion. EFM allowed an increase in the density of data on the disk (through smaller bit spacings — even though it is adding bits). It also decreases the disk sensitivity to defects and damage since only certain patterns are legal. See channel bits and channel coding.

- EFM is also used in many of the new professional video systems. D-5, for instance, uses EFM channel-coding to solve a problem of DC bit-values interfering with adjacent numbers. This expanded coding also provides a lower error rate and less mistracking sensitivity.

EFP

Electronic Field Production. A rarely used television term for video used on location (outside the studio) for drama series and documentaries. Betacâm is the most popular standard used around the world for EFP use. A closely related term is ENG, for Electronic News Gathering.

EFT/S

Electronic Funds Transfer (Systems). A generic term for sending payment instructions over computer networks. It applies to a range of financial transactions by electronic means. There are four main areas where EFT is used: automated teller machines (ATMs), point-of-sale (POS) terminals, Electronic Document Interchange (EDI), and automated clearing houses. These are on-line transaction processing systems (OLTP), and security is obviously the main problem. See SWIFT, EFTPOS and EPOS.

EFTPOS

Electronic Funds Transfer at the Point of Sale. This is the use of POS terminals in retail stores to debit the customer's savings account immediately, using on-line communications with the bank's computer over the telephone lines.

In some cases the EFT and the POS functions are performed by distinct machines with no interconnection. In this case the EFT side will consist of a 'swipe' to capture the card's information, and perhaps a PIN-pad for authentication. The amount of the purchase must then be re-entered into the EFT terminal.

At the next stage the EFT and the POS units are connected by cable; and, at the last stage, (to provide maximum security) they are both integrated into the same machine.

- The EFT side can also be largely divorced from the terminals in large stores, and become a function of the retail chain's own central computer. With this arrangement, the POS terminal transmits both the sale and the financial information to this central computer which then switches the financial requests through to the card-issuing organization for verification and money transfer.

EGA

Enhanced Graphics Adaptor — IBM's second color monitor standard for PCs which appeared in 1984. It had a resolution of

640 x 350 pixels and 16 on-screen colors from a fixed hardware palette of 64 possibilities. Most MS-DOS and Windows 3.x applications (including the Windows interface itself) are still tied to this 16-color palette.

This mode displays its text in 8 by 14 character cells. It uses TTL digital signals which have a screen-scan rate of 60Hz. EGA has a resolution of 21.8kHz horizontally.

At the same time (1984), IBM announced the PGA standard for special purposes — however the next public standard for general use was VGA. See CGA and VGA.

- The pin standard for an EGA monitor connection are:

Pin 1 Ground; Pin 2 Red-2; Pin 3 Red; Pin 4 Green; Pin 5 Blue; Pin 6 Green-2/intensity; Pin 7 Blue-2/mono video; Pin 8 Horizontal drive; Pin 9 Vertical drive.

EGP

Exterior Gateway Protocol. An Internet protocol (RFC 904) for exchanging routing information between autonomous networks.

egress

The point of exit from a network. Where the CPE terminal equipment and the network interface. See ingress.

EHF

Extremely High Frequency. These are radio waves in the band between 30 and 300GHz. Apart from some special applications for satellites, we are really only just beginning to use the SHF band directly below EHF, and SHF has enormous capacity, so it will be some time before EHF comes into widespread use.

These are millimetric waves between 10 and 1 millimeter (between 10^{-2} and 10^{-3} meters). They will probably only ever be used over short distances, perhaps below 2km (except, perhaps, to satellites).

EIA

Electronic Industries Association. Founded back in 1924, this is the oldest of the computer-related standards associations. It publishes its own standards, but it also submits proposals to ANSI for accreditation.

The EIA is an hardware-dominated organization which is run by, and for, equipment vendors and parts manufacturers. It sets electronic and electrical interface standards, and it has been responsible for a number of important standards for equipment interconnection over the years (i.e. RS-232).

The TR-30 (Technical Committee: Data Transmission) is responsible for the Physical layer of the OSI reference model, and it meets regularly with ANSI X3.S3 to coordinate work on the Data-link and Network layers.

EIA###

For EIA number standards, see RS-###.

EIAJ

Electronics Industry Association of Japan. The EIAJ formulated some of the early small-gauge videotape standards which gave the Japanese industry a major advantage in the development of small reel-to-reel recorders, and later with video cassettes. These early reel-to-reel recorder standards were known as J-standards.

E/IDE

See IDE.

eigenface

These are the facial discriminating features used in Photobook security systems and other visual recognition systems. They are monochrome 'abstract primitives'. To identify a target image, the program compares its eigenface characteristics with those in the database. The code for a good face match can then be held in about 100 Bytes, and an excellent one in 500 Bytes of storage. It can handle hats, eyeglasses and facial expressions.

A new layer of discrimination recently added to the basic eigenface algorithm allows the specification of eigeneyes, eigennoses, etc. See thinning and IVS.

EII

Electronic Information Interchange. A fancy name for EDI with Open Document Architecture. It includes e-mail and voicemail, so it is a wider term.

EIR

Excess Information Rate. In frame relay, this is the maximum rate at which data can be transmitted from a port — and then only at times of low load on the network. See CIR.

EIRP

Effective (Equivalent) Isotropic Radiated Power. This is the measured power of a radio signal radiating from a transmitter antenna 'as if it were radiating from a point-source uniformly in all directions' (isotropic). So this figure represents the combined result of the transmitter's or transponder's RF power plus any antenna gain (any focusing of the power), as compared to a radiating antenna (with no reflector) in space.

- With satellite signals, EIRP is used as an indicator of the strength of the down-link signal measured on the ground — the terrestrial received signal strength. These signal strength calculations can be plotted on a contour map (like weather isobars) in terms of dBW (decibels referred to 1 watt).

The signal is strongest near the center of the footprint (the boresight) and it falls off regularly at the edges with a parabolic dish in the satellite, but it can be irregular with a 'shaped-beam dish'. The power distribution in each footprint will then depend on the satellite's shaped antenna design. See shaped beam.

- A difference of 3dBW between contours, indicates roughly that a dish with twice the signal collecting area (or better amplification) will be needed; all other factors being equal.

EIS

- **Executive Information Systems:** With computers, EIS refers to a suite of computer software which is designed to allow executives and managers to monitor and request information which is important to them. It aids the decision-maker in finding and assimilating information quickly, and helps identify problems and opportunities. EIS software will normally use a GUI, and reside on a personal computer, but it may then access information held in larger computers.

It is similar to decision support software (DSS), but for less technical users. EISs extend executives' ability to 'mine' corporate data to get a clear picture of activities and operations.

- **Electronic Image Stabilization:** In video, this technique of removing (or reducing) the instability of hand-held cameras by electronically processing the image from an over-sized CCD. The central portion of this image is normally selected for use, but when the camera shakes, the electronic image stabilizer system shifts the selection 'window' around to compensate — and it can do this at electronic rates — so even shaky shots appear stable (until the limit of the CCD area is reached).

Video cameras also use mechanical stabilization devices, ranging from gyro systems, to counter-balanced suspension (Steadicam) and to variable prism mechanisms which stabilize the light-path entering the lens.

EISA

Extended Industry Standard Architecture — a computer bus standard designed by the clone makers to compete with IBM's new MCA. Compaq led the 'Gang of Nine' in devising this standard. In a brilliant PR move the Nine promoted EISA as the 'extended industry-standard', which made it appear that IBM's MCA was not.

It is a direct descendant of the generic-AT (ISA) bus, and ISA cards can be used in EISA machines — the two connectors are jointly mounted in line. EISA handles larger cards than MCA (so manufacturers aren't limited only to surface-mounted technologies), which often made them cheaper.

EISA has intelligent I/O bus control which frees up the CPU to concentrate on computational activities. It is 32-bits wide and runs at 8–10MHz with a burst-rate throughput near to its theoretical maximum of about 33MB/sec. This is faster than MCA, and it is also more flexible and reliable than the older ISA bus. However complexity and costs have limited the application of this bus. See ISA also.

EL screens

Electro–luminescent screens. See ELD below.

elastic buffer

A function of an operating system that allows a data buffer to exist which is not of fixed size but which can grow to any dimension needed. Memory management in the operating system must ensure that the buffer doesn't destroy other information in RAM.

ELD

Electro–Luminescent Display. These are a form of gas plasma display using light emission from the junction of charged electrodes. These are etched at right-angles into the glass.

electret

A type of condenser microphone, now widely used in telephones to replace the older carbon-button microphones. The diaphragms are usually made of special plastics in which a dipole is formed in the manufacturing process. Electrets are also used for quality recordings in non-broadcast studios. They tend to be a bit fragile for location recording, and are often wind sensitive.

electrical noise

This is noise on a communications circuit generated by RF interference (from TV sets, radios, etc.), distant lightning, and electrical machinery. EMI from motors, heating units and similar devices, creates a constant background level for any unshielded communications circuit.

In twisted pair, there are two types of electrical noise:

— Common mode: between the line and the ground, and

— Transverse mode: across lines.

See transmission impairments.

electro-magnetic spectrum

The entire range of electrical signal frequencies, ranging from low frequency radio, through ultra high frequencies, into the infrared, visible, ultraviolet light range, then on to X-rays, etc. See radio spectrum.

electro-optical effect

The effect found in certain crystals where the effective refractive index (RI) changes when an electrical voltage is applied to the crystal.

The RI change, in effect, slows the passage of the light through the crystal, so this effect can be used to create a phase-change in coherent optical systems. This is a technique used both in switching and modulation of laser beams.

electro-static

See electrostatic on next page

electronic banking

This covers both personal finances, ATMs, company financial operations and electronic commerce where all exchanges take place over the phone lines between modems. See EFT.

electronic beam steering

See phased array.

electronic crosspoints

This refers to the modern type of totally-electronic switch used in a digital telephone exchange. The connection between one side of the switch and the other are made by logical AND gates which are set ON when a binary control signal is applied to the gate. The opposite binary control signal switches the gate OFF.

Both space switching and time switching techniques are used with these systems.

electronic ink

A term applied to the storage and transfer of handwritten notes as a 'facsimile' or image, as distinct from converting them with recognition technology to ASCII text, or holding them in some structured image form. Anything drawn on the screen in a paint program is really in 'electronic ink' form, as distinct from objects drawn in a Draw program — but the term is being applied to words and diagrams rather than to art. See Jot.

electronic mail
See e-mail at front.

electronic purse
See SVC and CAFE.

electronic whiteboards
A general name for a number of real-time conferencing techniques which transmit written information from one screen, across a link, to another.

The original design was a true, vertical, whiteboard of conventional size, but the surface of this board was a continuous white-plastic belt. You could draw and write on this surface, then press a button, and the belt moved vertically past scanning elements at the bottom of the board and creating, in effect, a fax image output. This is then transferred to a fax machine at the other end of the link.

The term has now been taken over for desktop video-conferencing, and here it means that images displayed on one computer can be transferred to another at the other end. Electronic chalk can be used by either party to point to words or graphic items, ring around items, or mark them in some way. See writing boards.

electrooptical effect
See previous page.

electrostatic printers
These printers operate on the principle of charged particles being attracted and transferred in some way to the paper. Ink jet printers are electrostatic, where the charged particles transfer directly to the paper. Photocopiers and laserprinters are also electrostatic, but they transfer via a drum, so they work in an entirely different way.

electrostatic screen
A wire net or similar shield around a computer or other instrument to prevent interference (ingress) or radiation (egress).

• On a PC board, an electrostatic shield can be a thin strip of conducting material that is earthed. It prevents one side of the circuit from influencing the other by capacitance coupling.

elemental areas
The abstract notion of a pixel in discussions of video and television design. It means the smallest area of a real image that can be scanned and transmitted in some discrete way so as to be resolvable by the viewer's eye at the other end of the system.

elevation
In satellite antenna, this is the angle between the antenna beam and the horizontal surface (measured as an angle in the vertical plane). See azimuth and declination also.

ELF
Extremely Low Frequency. These are the 'megametric' waves between 30Hz and 3000Hz (below the VLF radio band and within normal audio frequency range). Some references limit ELF to 30–300Hz, and others put it vaguely at below 100Hz.

ELF is being blamed for some possible health problems with computer monitors and power-lines. There is some evidence of increased cancer rates, and experimental evidence that biological material does suffer from the effects of both the near- and the far-fields.

Elisa
A famous MIT experiment aimed at proving the limitations of AI systems, but which went wrong and became a very interesting study in psychology. The idea was to mimic the way psychotherapists constantly throw a patient's answer back at them in the form of a question.

Elisa was programmed to do this by rephrasing the answer as a question, or asking a very general question if it couldn't parse an answer. It did it very well — so well, in fact, that it received rave reviews, and has been treated ever since as an example of machine intelligence.

In reality, it only demonstrated how little human intelligence was required to conduct a psychotherapy session.

Ellipso
A 'Big LEO', CDMA satellite communications system currently under development, and due to fly in 1997–8. It provides two sun-synchronous constellations with a total of 14 satellites (+ 2 spares): the first eight-satellite constellation (Borealis) uses four pairs in highly elliptical orbits (520—7800km) which stall at their zenith over the populated areas of North America and Europe. The second constellation, Concordia, has six satellites in almost circular orbits around the equator for coverage of the tropics and some southern hemisphere nations.

Ellipso will share CDMA spectrum with GlobalStar and the other spread spectrum proposals (see LEO spectrum). It has very simple spacecraft with bent-pipe transponders.

The antenna system projects a 19-cell structure on the earth. Data can be handled at rates to 9.6kb/s, and variable bit rate (average 4.8kb/s) voice coders will be used. Handset costs are said to be in the $1000 range.

Projected costs for the full satellite constellation is US$850m; and the backers are MCHI, Westinghouse, C&E, Harris, Fairchild, Arienespace, IBM and Israel Aircraft.

E&M
See beginning of the E section.

EM/Em
• ASCII No.5 — End of Medium character (Decimal 25 Hex $19 Control-Y).

A control character which is used to identify the physical end, the used amount, or the wanted portion of data recorded on a tape or disk.

• A printer's unit of measure the size of a lower-case m, which is now mainly used to specify indentation. It is actually a square of the type size — so a 10 point em is 10 points high and 10 points wide. It is equal to two Ens.

• Em-dash. This is the long — dash. See dash.

EMA

Enterprise Management Architecture. DECs distributed architecture for enterprise-wide network management. This is to be used with DECnet. EMA treats all components on the networks as managed objects. DEC has published open interface specifications.

e-mail

See at beginning of E section.

Embedded ADPCM

A packetized voice system offering bandwidth reduction for frame relay. This is a combination of ITU recommendation G.727 for the encoding system, and G.764 for packet control. It is a layered 'lossy' compression system which allows the least essential bits to be discarded whenever the bandwidth comes under pressure.

The 16msec samples of the audio are decomposed into layers by a transform process (See FFT). The core layer provides enough bits to provide clear speech (equal to about 16kb/s) while up to three higher 'enhancement' layers can be added to improve the quality for fax or modem transmissions, or to add higher voice quality when traffic loads are light. The third level of enhancement is needed to allow full V.29 fax transmission rates at 40kb/s.

The G.764 packet protocol allows the packetized datastream to wind back its bit-rate when the voice activity detection system detects silence. However a background 'comfort' noise is then added, so the phone doesn't sound dead.

embedded command

A formatting command (indent, center, boldface, etc.) which can be embedded in the text of a word processing program. Some embedded codes are trapped and used by the word processor itself, and other are passed through to the printer.

— For word processor applications, embedded commands are usually preceded by a period at the beginning of a line (following a CR).

— For printer commands the ASCII ESC (code 27) is usually the identifier.

embedded processing

The term implies 'hidden'. The use of computerization (specifically special microprocessors and memory) inside other common equipment (like copiers, telephones, television sets — down to controls for the hot water in a shower) to provide intelligence, but without obvious computer connections.

Embedded Systems Technology (EST) will also play a central part in multimedia systems, and it is inseparable from future network technologies.

In the office, the idea is to make smart copiers and smart fax machines possible.

— Microsoft At Work is an attempt to standardize a key software framework for these machines in the office so they will interwork. See At Work.

— Novell has just released NEST (Novell Embedded Systems Technology) which allows vendors to embed NetWare into any future intelligent device, including automobiles, microwave ovens, VCRs, television sets. Once embedded they can be managed through a file-server.

embedded SQL

The SQL query language used inside the processes of another application. This is used in client-server applications.

EMC

ElectroMagnetic Compatibility. The measure of the interference effect of one electronic system on its surroundings — the amount of EMI it generates and the amount it can tolerate. This is two way compatibility; a device is compatible if it doesn't interfere with others, and others don't interfere with it.

• 'EMC-designed' is a term applied to equipment that will function in an EMI environment without ill effects and will not emit enough EMI through its own operations to disturb other equipment.

• CENELEC is the European organization which controls EMC standards. Generally, however, when a popular piece of equipment (such as the GSM digital phone) puts out too much EMI, these organizations change the 'immunity standards' (the ability of other devices to suffer) rather than correct the problem.

• CENELEC also has strict standards for non-intentional interference emitters (such as electrical motors) but very lax interference standards for intentional emitters (such as mobile phones). See EMI and RFI also.

emergent behavior

A term which emphasizes the complexity of today's global information and communications systems — a complexity which is already beyond any single human capability to comprehend completely. Telephone networks, for instance, are well understood in terms of the hardware modules — we know what each switch and each exchange can do — but still, the network in New York crashed for six hours and no one could find out why.

This is because the combined effect was much more complex than a sum of the individual parts; each module worked 'correctly' (except for one line of code) but still the network crashed. This is 'emergent behavior'.

EMF

• Electro-Motive Force. The potential that exists between two ends of any 'unbalanced' circuit. It is measured in volts.

• Electro-Magnetic Field. The general term for radio and magnetic fields surrounding electrical and electronic equipment. See EMR also.

EMI

Electro-Magnetic Interference. This is R/F interference or RFI. EMI is the cause of TV's flashing white-line interference from electrical motors, switches, etc. and it creates unwanted noise on audio systems, telephones, etc. EMI reduces data integrity and increases error rates of digital transmission channels.

The study of EMI can be divided into:

— The study of transients, surges and RF-related phenomena,

— Electro-static discharge events (ESD).

See RFI, EMR and EMC also

• This term does not encompass the possible biological changes from non-ionising emissions. See EMR, CENELEC and ICNIRP.

emission

The propagation of non-ionizing energy — as distinct from radiation, which is the propagation of ionizing energy. Emissions are lower-frequency waves in the radio, TV and microwave spectrum. In practice, the terms radiation and emission are used interchangeably.

emitter-coupled logic

ECL is a variation of the most commonly-used TTL devices (particularly memories) which are the basis of modern microcomputers. ECL devices are faster than TTL but also more expensive, and have higher power consumption. They are based on bi-polar transistors, and are limited (by the heat generated) to mainframes and supercomputers at present. ECL based computers will often need water-cooling.

BiCMOS technology has recently been challenging the use of ECL in memory chips, and GaAs in gate arrays. See BiMOS, and GaAs also.

EMM

Extended Memory Manager. This was seen originally as a modification to the EMS standard, but in fact it is an emulation technique.

EMM386 is a utility which is loaded as a device driver at the start of DOS so as to control access to expanded memory. EMM allowed the 386-based machines to use the memory space above 1MByte without needing the clumsy bank-switching technique of EMS. See EMS, memory managers, extended- and expanded-memory, and XMS also.

emoticons

Used by bulletin board and Internet users in e-mail and forums to convey the 'spirit' of a typed piece of text. For instance the combination :-) is a smile and ;-) is the equivalent of a wink. Alternative expressions are to use <g> (meaning to grin). Dvorak's 'Guide to PC Telecommunications' lists of emoticons.

EMP

ElectroMagnetic Pulse. A sudden surge of induced current through the national power network in the event of a nuclear attack. Some experts believe this would destroy most of the computers in a country.

emphasis/de-emphasis

It is often common in audio and radio systems to apply emphasis (aka pre-emphasis) to certain signal frequency bands before modulating the signal, and then de-emphasis to restore the original balance (frequency response) after demodulating it.

Normally sound engineers are very keen to maintain a 'flat' response (no emphasis) through their equipment. However emphasis improves noise figures and greatly improves the frequency response (and therefore subjective quality) in some systems. Pre-emphasis and de-emphasis are built in to TV and FM standards. See also companding.

empirical

Based on experience or experiment — rather than theory.

• Empirical knowledge: see inductive learning and heuristic.

EMR

Electro-Magnetic Radiation. The term usually used when discussing potential biological effects of non-ionizing radio waves. In the near-field (for distances up to 1 wavelength) EMR must be considered as having two components:

— An oscillating electrical field measured in volts/meter (V/m).

— A magnetic field measured in amps/meter (A/m).

The intensity or power flux density is expressed in watts per square meter (W/m^2). See SAR.

• ANSI's guidelines for exposure to EMR (300MHz to 3GHz) is a permissible power density of 10mW/cm^2 for 6 minutes. This limit is now regarded as much too high; it was based on levels which produce a 'disruptive effect on behavior', and for 'acute exposures only' (not long-term cumulative).

In effect, if the rat back-flipped and expired instantly, the dosage was judged to be too high. If it didn't, then the exposure level was claimed to be 'safe'!

• An AMPS handset with a power output of 0.6 watts will produce an average SAR of 0.7 W/kg in the eye and nearly 1.8 W/kg in the ear.

• For many years the industry has maintained the fiction that the only biologically harmful effects of EMR comes from local heating of body tissues (the thermal effect). This has now been extensively and comprehensively disproved.

EMS

Expanded Memory Specification of Lotus, Intel and Microsoft (the LIM specifications). In 1985, Lotus, Intel and Microsoft devised a standardized way of expanding the 640kB working memory limit then available for MS-DOS users.

These LIM extensions opened the high-memory (HMA) space from 1MB to 16MB (later 32MB) for use by DOS programs (but originally only to hold data), by fooling DOS into believing that it was dealing only with a small block of memory in the upper-memory area (see UMB).

With MS-DOS, the 286 and 386 machines of the day could only address 1MB (in 'real mode'), so the computer needed to be fooled. They did this by using a vacant 64kB of spare memory between the normal video RAM and BIOS locations in the Upper Memory Block as a 'transfer- station'. This relies on the technical trick of paging or bank-switching.

As applied in EMS, this is a way of swapping 'pages' of memory in and out of four segments (each 16kB) of system memory space ($COOO to $CFFF) in the upper-memory (640k–1MByte). MS-DOS then sees this memory as lying within the normal addressable range.

Not every program can use EMS because it is a sophisticated form of bank-switching that requires hardware and software to work together. EMS 3.2 supported 8MB of expanded memory but only data could be stored in the high-memory area, and the modified Enhanced EMS (EEMS) could support even more. EMS 4.0 now supports up to 32MBytes, and handles both data and applications.

• See expanded memory, XMS, EMM, and also the variation called AQA/EMS and later AQA/EEMS.

• Make a distinction between this EMS 'expanded memory' approach and extended memory systems (even though both use the same actual memory locations above 1MByte). EMM deals with it direct, EMS only deals indirectly through the four 16kB pages.

• The general term 'expansion memory' may refer only to chips on an adaptor card, rather than to a mode of operations. See expanded- and extended-memory, EMM, XMS, EEMS, E/EMS and Virtual Memory.

emulation

To mimic or copy a function.

• Terminal emulation allows an intelligent PC or workstation to act as if it were a dumb terminal. See terminal emulation.

• Operating systems, and even chips are emulated. The new PowerPC RISC chip, running an emulation program, can look and act as if it were a Intel 80486, and therefore run DOS with Windows 3.x and Windows applications.

• The distinction should be made between emulation which mimics the function (but not necessarily the processes), and simulation which attempts to duplicate all the processes — to have all the attributes and problems of another system.

emulsion

The light-sensitive coating on a piece of film.

• 'Emulsion-down' specifies that the film image should be readable with the emulsion side facing away from the viewer. When contact printing one film onto another, the two emulsion surfaces must be in contact — so each generation reverses the emulsion side — from 'up' to 'down' to 'up' again.

In printing, the image negatives must match the requirements of the plate-maker, so your printer will specify whether the negatives needed for the plate making should be delivered as either emulsion down, or emulsion up.

En

A printer's measure. It is the size of a lower case n, which is half an Em. See Em.

• En-dash. The in-between dash size used for a minus (–6) and for connecting numbers such as 109–110MHz. See dash.

enabled

In telecommunications this means 'switched on and ready' (activated) or 'permitted to use'.

• Some systems are only enabled after identification and passwords have been entered and accepted as valid.

• In RS-232 serial connections, a control line is enabled when it has a negative voltage level (between –3 and –20v) applied and maintained. It is disabled when the voltage returns to zero (in practice, if it drops below –3 volts).

Encapsulated PostScript

See EPSF.

encapsulation

• **Packets:** There are two ways to use the term in packet switching:

— The process of wrapping data in a particular protocol header. So Ethernet data is 'encapsulated' in a specific Ethernet header before being transmitted. This term is widely used, but it doesn't make much sense.

— The process of packaging an existing protocol within the payload of a second protocol. For instance, if data streams are required to be carried for some intermediate distance over an X.25 packet-switching network, the original frames can be kept intact and encased within the X.25 packets.

At the other end, the X.25 wrapping is stripped off, so restoring the original frames.

This process of getting the 'native' protocols from one segment to another through an intermediary, is also known as 'tunneling' or 'bridging'.

In a similar way, larger packets can be segmented and carried within ATM cells, and then restored to the original packet form at the far end. This is really segmentation and reassembly rather than encapsulation; the terms are often confused.

• **Object-oriented programming:** You encapsulate objects to hide them. These are data structures with groups of member functions.

encode

The modification of information into a desired pattern or form required for processing, or for a specific method of transmission. Security is only a small part of encoding.

• To convert to machine readable format. When you hit a character key on your PC, the character is encoded into electrical binary digits.

• Analog voice is encoded into PCM to be carried over digital systems. It is transcoded when the 64kb/s PCM data stream is converted to 16kb/s ADPCM.

• Sometimes encoding is seen as the conversion of one form of digital information into another (often more compact) form — in data compression, for instance.

• The distinction should be made between encoding and encryption, although both terms are used loosely in discussions on security measures.

• We should distinguish these three:

— **Encoding:** The first stage of converting analog signals, images, words or ideas, into a symbolic form. Writing is an encoding process, so is typing into a word-processor. Transcoding in the translation of one encoded form to another — usually one form of line-code to another.

— **Encryption:** Encoding using secret keys and algorithms, with the express purpose of security. See crypto systems.

— **Encipherment:** The use of non-algorithmic techniques for creating secret documents (one-time code books).

encoder

• In computers, this is the term usually (incorrectly) applied to a device or algorithm which encrypts data for security reasons.

• In communications the term is often applied to the device which converts the analog signals to digital (D-to-A conversion).

• In cable systems, the term is often mis-applied to a black box which unscrambles an incoming video/audio signal in a Pay TV system. This is a decoder. The encoder is in the head-end.

• In video, it is an encoder which converts component video signals (RGB) to NTSC, SECAM or PAL composite format. This is more correctly, a transcoder. See.

encryption

The process of changing data so that it needs a special form of hardware, software or software configuration to read it. This involves encryption algorithms and the use of a key.

The distinction should be made between encryption and encoding, and between crypto-systems and ciphers.

— Encryption is the application of a special algorithm to the data so as to make it incomprehensible to those without the key. It is almost inevitably a mathematical process.

— Encoding is the use of ciphers (set tables, or protocols) known to both parties (code books, or ASCII tables) to change the structure according to some pre-arranged pattern. This is not necessarily done to camouflage meaning; most encoding is done for efficiency reasons. See line codes.

However, it must be admitted that the words encryption and encipherment are generally used as synonyms. See crypto systems, scrambling, DES, Clipper and public key.

end office/exchange

The end office is a North American term for the local exchange to which the subscriber's line connects. Within the carrier hierarchy, it is known as a Class 5 exchange. See exchange class.

end-of-file

The last data element in a file. This is usually a special character. Standards exist for End of Medium (EM), End of Text (ETX) and End of Message (EOM), End of Transmission (EOT) and End of Transmission Block (ETB).

end system

A host computer on the Internet. In OSI terminology the end system is the last device permanently connected to the network (as distinct from dial-up connections).

end-to-end

• **Circuits:** The term can apply both to physical or logical connections between two processes. If it is a physical circuit, then switching hardware will have been involved. If the circuit is logical, then no direct link necessarily exists at the time of input, the packets are switched as required, at nodes — but a virtual circuit still exists (with connection-oriented nets).

The distinction is with datagrams (connectionless packets) and messaging systems which don't need end-to-end circuits.

• **Protocols:** End-to-end protocols are those that manage the transfer of the data from the source to the sink, and have no effect on the functions of the intermediate nodes along the way. For instance, in frame relay, it is now considered that the channels are so error-free that it makes little sense to check frames for errors at each stage in the transmission. So frame relay relies on end-to-end error checking. Only the customer's equipment will know if a frame has been damaged, and therefore it alone can ask for retransmission.

• In a layered protocol structure like the OSI model, the control of the data flowing in the system can be seen in two ways:

— end-to-end control which the receiver imposes on the sender, and vice versa,

— congestion controls imposed by the system.

The contrast here is with node-to-network protocols.

endian issue

This is about the way in which memory is organized and how bytes are used in systems. There is no single *better* way, just a dispute about whether it should be done to a little-endian or a big-endian standard . See both.

Enet

• Ethernet. See

• Iso Enet. See Isochronous Ethernet.

ENG

Electronic News Gathering in television. The distinction is between the use of moderate-quality video systems and 16mm film. Until portable video equipment (ENG equipment) became available, all TV news was shot with film cameras.

The film took time to process, which was the major limitation for up-to-the-minute coverage. Now it can be shot and transmitted by microwave back to the studio, especially if real time coverage is required.

Most ENG is recorded on tape and then edited, however, it is still the Betacam SP 1/2 inch, Hi-Band U-matic 3/4 inch, and S-VHS video standards which are most widely used in ENG work. See SNG also.

engine

A very vague term which suggests some powerful 'behind-the-scene' function which exists as a self-contained module. It may mean hardware or software — it suggests the 'core' functions.

• In laserprinter terminology (and in a number of other new technologies), the central printing-drum unit together with the laser and driver electronics will be manufactured by one company and sold to others for further enhancement, design, marketing, etc. under numerous different trade names. The term 'laser engine' is applied to this central unit.

• In software, 'engine' can refer to a 'core' or 'kernel' around which customized applications can be built. This usually implies that APIs will be needed to link the engine to existing or special applications.

• We talk of LAN servers having 'mail or database engines'. These are the server's back-end applications which are dedicated to these tasks.

enhanced TV

The term is applied mainly to the wide-screen television system being promoted as an in-between stage to our present television systems and high-definition TV (1000+ lines). However the term is also applied to digitally enhanced standard-analog 4:3 television sets. These use DSPs to remove ghosting and improve the color overlay problems.

Enhanced Definition TV (EDTV) will probably have 16:9 aspect ratios, greater horizontal bandwidth (and therefore resolution) and about 750 scan lines. See EDTV.

Enhanced VGA

This is almost a 'Super VGA' or 8514/A standard for video graphic adaptors, but it only provides a resolution of 800 x 600 pixels, which is about 50% more than standard VGA. Enhanced VGA has 16 colors and an 8 x 8 character matrix for generated characters. This standard was adopted by the Video Electronics Association.

ENQ

ASCII No.5 — Enquiry (Decimal 5, Hex $05, Control-E). A transmission control character which is used as a request for identification ('Who are you?') from a remote terminal. It is used in a number of ways.

— In Telex systems it is a request for the receiver to send back an Answerback identification message.

— In other systems it asks whether the remote is willing to accept a message. The reply is often sent automatically.

— In BiSync, it is transmitted as part of an initialization sequence for point-to-point operations, and as the final character in polling sequences in multipoint operations.

It is often used as an ETX (end-of-text) control.

ENQ/ACK flow control

A technique used by Hewlett-Packard in printer-flow control. The ENQ (Control E) is sent to ask whether the remote is able to accept data, and an ACK (Control F) is sent in reply.

enter key

In some computers the enter key is the same as the return key, and it produces a carriage return and line feed. In the Mac, the enter key and return key are distinct keys, and they will sometimes act in a different way.

enterprise network

These are LANs and interLANs (WANs) covering the whole organization, as distinct from work-group or department interconnections. These are complexes of interconnected webs linked by LAN bridges, gateways and routers, and they carry the company's mission critical applications. Enterprise-wide networks are for large organizations and they give their users the freedom to access, transfer and share information across previously incompatible computing environments. These are complexes of interconnected webs linked by LAN bridges, gateways and routers, and they carry the company's 'mission critical' applications.

These environments are more complex than LANs because they tend to connect equipment from different vendors, use diverse communications architectures, and run over various transmission media. They will often need both centralized and distributed network management to isolate faults and control configuration, security, accounting and performance.

It is hard to find a distinction between 'enterprise network' and a 'WAN', except that 'enterprise-wide' stresses the size and complexity.

Enterprise Systems Architecture

IBM's later additions to SAA to make applications portable and easy to use. The new MVS/ESA operating system for 3090 computers has ESA. It extends the operating system by providing logical partitioning of data and Systems Management Storage.

entity

A single, self-contained unit or device. In network management, it refers to any unit which can be managed.

entropy

Entropy is the natural end state of the universe. Clocks run down, mountains turn to silt. The concept of entropy is used in different disciplines from engineering to philosophy, but in all cases it carries a sense of 'the degree of organizational complexity', the amount by which something departs from unorganized chaos.

Behind the term is the certainty that a system will regress from a condition with order, to a condition of random disorder. And, if there is order, there must be difference, while disorder is homogenous. The Second Law of Thermodynamics says that entropy is always increasing — that the ordered world is running down — that the universe is moving from a condition where energy and mass differences exist (and therefore there are flows of energy and gravitational forces) to a homogenous state without any energy flows.

By analogy, therefore:

— Total entropy is an inactive or static condition in a network.

— The term can also mean information or energy flows which are unavailable on a network.

• **Information Theory:** Shannon used the term 'entropy' first, as a measure of uncertainty or randomness in communications. The entropy of any data received in communications, is a measure of the uncertainty of the contents of the data stream. When entropy equals zero then we are certain of the data (it has a probability of 1). There are formulas to calculate entropy.

• **Coding:** Compression systems which get as close as possible to the basic information level — as distinct from our current use of symbols which have a high degree of redundancy. We can compress English language text because this redundancy exists, and these compressed files have much less order than the original. Highly compressed files look like random bytes. Entropy in a data stream, therefore, is related to the number of bits per symbol which are required to encode a message.

In practice, the term 'entropy coding' is applied to compression schemes such as Huffman which use shorter codes for the most used elements, and longer codes for the least. The code-lengths are assigned on the basis of the probability that they will be used.

• **Statistics:** Entropy is the mean value of the information which has been conveyed by any one of a finite number of exclusive events.

• **Video:** Entropy is now an important concept in compressed digital video because it affects how MPEG coders function. If, over a television system, you transmit a test card, then after the first frame has been sent, the second and subsequent frames are all the same. So the entropy drops to zero (provided the signal is noise free and there are no specially interpolated frames). However random noise on the screen would be totally unpredictable and so have maximum entropy.

Television pictures lie between these two extremes, depending on the amount of movement and detail in the scene. Hollywood movies have a lower entropy than video equivalents, because the film camera's reduced depth of field often leaves the background out of focus, and with less detail. So these images can compress the most.

E-NTSC

See NTSC.

envelope

• In music and sound — the low-rate characteristic shape of a sound represented graphically, with amplitude plotted against time (ignoring frequency). The envelope is divided into attack—sustain—decay—release phases. The envelope, therefore is a graphic representation of the way the sound starts, continues, stops and dies. With modern sound synthesis, it is often possible to alter the character of a sound's shape, by graphically manipulating the envelope on the computer screen.

• **Envelope Detection:** In demodulation of amplitude modulated signals, the envelope represents the original baseband signal. When the carrier is removed by 'detection', it is the envelope (the original audio or data signal) which remains. This is often known as 'envelope detection'.

• **Envelope distortion:** This is the same as group delay distortion. See delay distortion.

• Transmission of a complete spectrum envelope means that sidebands have not been suppressed.

• In communications an 'envelope' is a bundle of 'user data' that is required to manage the communications between hardware items in the network. The header on a packet is often likened to an envelope, because it carries the address information.

envelope sizes

There are a number of standard envelope and label sizes; some set by national authorities, and some by the ISO.

— Envelope sizes:

C5 is an ISO envelope 162 x 229mm.
DL is an ISO envelope 110 x 220mm.
Monarch envelopes are 3.875 x 7.5 inch.
Comm-10 envelopes are 4.125 x 9.5 inch.
A 'small' envelope is 3.67 x 6.50 inch.

— Envelope label sizes:

Executive label is 1.5 x 3.5 inch.
Small label is 1 x 3.5 inch.

• The standard envelope sizes in countries using A4 paper for letterhead, appear to be:

DL	110 x 220mm	Takes A4 paper folded in thirds.
C6	114 x 162mm	Takes A4 paper folded in fourths.
Banker	150 x 230mm	Takes A4 paper folded in half.
C4	229 x 324mm	Takes A4 paper unfolded.

EOA

End of Address. A unique set of characters separating the message header from the text. With ASCII, the EOA has been replaced by STX (Start of Text).

EOB

End Of Block. A non-print character often associated with ACK. It is used at the end of a line or a block of information. In the old tape-punch machines the EOB character was used as a carriage return.

EOD

End of Data. A character or code that marks the end of the data. Don't confuse this with the End of File marker.

EOF

End of File — or very rarely, End of Field. These characters are sometimes specific to the operating system. In CP/M, MS-DOS and OS/2 the End-of-File marker is Control Z (ASCII 26). Not all operating systems need an End-of-File marker.

EOL

End of Life. A term used when estimating the useful life of satellites and satellite transponders.

EOR

- Abbreviation for Exclusive OR in Boolean algebra (also called XOR).
- End of Record code in ASCII.

EOT

ASCII No.5 — End of Transmission (Decimal 4, Hex $04, Control-D). A transmission control character which is used to indicate the conclusion of the transmission. This transmission may have included a number of text messages identified (each) by STX—ETX (start test — end test).

EPHOS

UK version of GOSIP — a standard for government purchases of computer and communications equipment.

epitaxial

The process of building a foundation of semiconductor materials in successive layers. It is a class of techniques for growing single atomic layers of other materials on the surface of a silicon wafer (or similar material).

There are four techniques widely used:

- **LBE:** Liquid phase epitaxy, where a film is deposited from a melt of gallium, indium, etc.
- **MBE:** Molecular beam epitaxy where the materials are deposited under ultra-high-vacuum conditions (billionths of an atmosphere) from a flux of atoms beamed onto the substrata.
- **CBE:** Chemical beam epitaxy. This is the same process as above but the beam deposits metal-organic and inorganic atoms. This process is used for HBT manufacture.
- **MOCVD:** Metal-organic chemical vapor deposition. A low pressure system using organic and inorganic gasses, which replaces LBE and produces better yields.

EPLD

Erasable (Electrical) Programmable Logic Devices. These are semi-customized DSPS which are very useful because they have low power demands. EPLDs use erasable PROM technology, but have AND and OR gates. The users can program these chips many times. They are a more flexible alternative to an ASIC because if you make a mistake, you simply reprogram them. See GAL, FPGA, PLD and gate arrays also.

EPOS

Electronic Point of Sale equipment. See POS.

EPROM

Erasable and Programmable Read-Only Memory. A form of non-volatile memory chip in which the old data can be removed and new data implanted. These chips require exposure to UV light to erase, and you can usually recognize them easily in electronic equipment because they have a small window in the back (usually covered, of course). They are programmed through a special plug-in PC card called an EPROM burner.

EPROMs have slow access times, so the data they hold is usually off-loaded into RAM in order to maintain the speed of the computer. However some regular EPROM chips now have access times in the order of 100ns, and the new CMOS high-speed devices offer access times down to 35ns.

E^{2}PROM

See EEPROM.

EPS

Encapsulated PostScript. The key portable file-format for desktop publishing. It contains a set of PostScript instructions in modular form, and it is used when transferring PostScript image information from a desktop publishing system to laser-printers or typesetters. It spans both the PC and the Mac worlds, and is a common way of transferring everything from clip-art and layouts to top-quality color images.

PostScript printers can interpret EPS files, and a separate bit-mapped image is included in the file for the monitor-screen version. EPS files are known as EPSF, and they can contain text, paint and draw-type graphics.

EPSF

EPS Format. This file format includes both the PostScript information (high-resolution scalable object files) and the PIX information (low-resolution bit-map images for the computer screens). See EPS.

eqntott

A SPEC benchmark. It is integer intensive.

equal access

As competition began to be introduced into telephone networks, the dominant carrier often had the advantage of being the 'default' carrier. If you wanted to use another carrier for your long-distance calls, you were required to dial a special additional prefix code — and sometimes these codes were deliberately designed to be difficult to remember and annoying to dial.

Since local connections are generally still a monopoly, only long distance and international calls are competitive. Equal access is the term applied to ways of correcting this anti-competitive balance between the carriers.

- Generally, equal access means that the customer will nominate in some way, a preferred inter exchange carrier (PIC), and this selection will then be automatic within the local exchange whenever a long-distance number is dialed. Usually, an override code can also be dialed, if, for the one call, another carrier needs to be chosen.

The US FCC defines equal access as allowing customers to place long distance calls with the interexchange carrier of their choice, without the need to dial extra digits.

• In the USA, equal access meant that the BOCs (LECs) took control of the intraLATA calls, since they began screening the calls and handling any they were permitted to handle. Prior to this, intraLATA and interLATA calls were, in effect, indistinguishable, and were carried by the IXC.

• Equal access can also means that you need to dial a carrier-choice prefix with each long-distance call. This is a less desirable approach where you select your carrier on a call-by-call basis — but this applies equally to all carriers. This approach has all carriers at an equal disadvantage, but it is not liked by users.

• Sometimes in the USA, equal access means that a special dialing code (10xxx) can be used to select between several carriers, without needing to sign up with any of them beforehand. The dominate carrier will still be the default in this case, but a wide variety of other choices are available.

This stems from a historical problem in the US where small local exchanges would often only offer access to one long-distance carrier (who paid them for the privilege). Later under government pressure, they were forced to offer specially dialed access to many more long-haul carriers. It is certainly not 'equal access', since one carrier still has priority, but the term is often misused in this way in the US.

equalization

The restoration of original signal conditions by various means: correcting for its deficiencies. Equalization is the intentional departure from using 'flat response' or basic equipment to compensate for some characteristic introduced elsewhere in the system.

Equalization minimizes the affects of amplitude (loudness) or frequency (phase or time-delay) distortion by compensating for the variations in the data-channel quality.

— Frequency equalization restores the amplitude of different frequencies which have suffered non-linear distortion.

— Dynamic equalization counteracts delay distortion in analog circuits.

• In audio systems, equalization is the process of compensating for distortions and restoring the channel's original transmission characteristics by filtering and boosting selected frequencies. You can think of audio equalizers as being elaborate tone controls with a series of bandpass filters, each with a volume control on its narrowband channel.

• With group-delay distortion on cables, equalization is the process of reversing the deficiencies of a channel caused by some frequencies being delayed in the transmission more than others. This process requires selective filters with designed-in delays to restore the balance of the transmitted signals and to stop digital pulses from smearing. Adaptive and dynamic equalizers are used here.

• In mobile telephony and digital radio and TV, equalization refers to the correction of multi-path signal reception, where some signal components will have been reflected, and therefore arrive late and possibly create interference. TV ghosting is an extreme case, and some of the new anti-ghosting techniques are digital delay-equalization techniques.

• Echo cancellation is a digital delay equalization technique.

equalizer retrain sequences

The process of conditioning a telephone line in order to reduce the amount of interference and attenuation for data transmission. A training sequence, which is a known chain of 'symbols' or 'events', is transmitted over the line so any distortions or attenuations can be measured.

A negative filter is then constructed to compensate for the defects. It is applied to all data flowing over the line to correct the line deficiencies. This is used in some of the modern fast modems.

equatorial

• **Plane:** A slice through the earth which cuts clean around the equator — and extends out into space.

• **Orbit:** A satellite orbit above the equator. Geostationary satellites are all in equatorial orbit, but LEOs and other communications satellites tend to be in polar or inclined orbits.

equiband

In video, this is the name applied to the analog component technique of Y (luminance), R-Y and B-Y (color difference) signals. See YUV also.

equivalence

A logical concept. If three conditions A, B and C can all be either true or false then the equivalence of A/B/C can only be true if all three are individually true or false. They must all be equal in 'logical value' (either true or false) for the equivalence to be true.

equivalency lines

In a hunt group (call forwarding on busy), you will have one incoming number being distributed to many extensions according to the preset pattern. These extensions are equivalency lines.

equivalent noise temperature

In a communications satellite link, this is a measurement of the amount of 'noise' which is being produced by the receiving equipment. This noise degrades the signal's intelligibility.

equivalent voice circuits

A term often used in submarine cable installations where the carriers are attempting to camouflage the enormous capacity of the fiber in the cable. They will usually quote the capacity in terms of (say) '5 fiber pairs, each of 2.4Gb/s which is the equivalent of 187,500 standard (or 'nominal') voice circuits'.

What they don't say, is that voice is generally carried on trans-Atlantic cables at 16kb/s (not 64k/s) and that with DCME (digital interpolation) you can multiply this by a factor of five or more. Such a system would carry well over 1 million voice circuits at a reasonable quality, if required.

erbium amplifier

A form of optical amplification where the signal boost takes place in the fiber without opto-electronic conversion. The glass is doped with the rare earth erbium, and so the cable units are now often known as EDFA (Erbium Doped Fiber Amplifiers).

The erbium-booster technique was pioneered at the University of Southampton and later taken up by Bell Labs and NTT. Using this form of boost (these are not regenerators) it is now possible to transmit 5 Gb/s signals over a single fiber for thousands of kilometers without regeneration.

Erbium optical amplifiers are also capable of amplifying a number of different laser wavelengths simultaneously in the fiber (allowing WDM), and they are not fixed to a single datarate so these cables can be upgraded by changing terminal equipment only. This is a major breakthrough for undersea fiber cables.

However the limit of erbium doped amplification systems appears to be 5—10Gb/s, supposedly because of 'photonic noise' introduced by the amplifiers, but more likely because of long-distance pulse spread. Solitons will probably be the solution. See doped fibers, optical amplifier, and solitons.

ERC

European Radiocommunications Committee — a committee of CEPT which is responsible for licensing radio equipment in the European Union.

ergonomics

The study of computer keyboard height, key-positions, screen characteristics, table height, chair width/height, back-support, etc. which might give the user physical problems. About 10% of the claimed standards are supported by factual research. See RSI and Carpal-tunnel syndrome.

• It is now a marketing term designed to sell you a bendable keyboard with keys in totally new positions, an angled seat which destroys knee cartilages, or a screen with a scratchable plastic dull-gray overlay.

Erlang

A telecommunications measure of traffic flow. One Erlang is one voice channel continuously in use — or the equivalent number of channels used for a lesser time. So 15 four-minute calls, equal 1 Erlang of traffic.

Erlangs are used to indicate the number of calls in progress at any one time on a given group of circuits. If the 'instantaneous traffic' is given as 200 Erlangs, it means that 200 calls were taking place over the equipment at that time.

Erlang traffic loads are computed by taking the ratio of the call arrival rate and the call holding rate. See CCS also.

• Erlang Capacity Calculations:
— Erlang-B is used to predict blocking when a blocked-call is discarded.
— Erlang C is used when a blocked call is queued to wait for a free trunk.

Ermes

European Radio Messaging System from ETSI. A pan-European VHF multi-channel, wide-area alphanumeric paging network. It launched in 1995 across Europe (Denmark, Sweden and Switzerland first). Ermes devices may eventually be able to receive common frequencies across the US, Asia and the Pacific region.

The European commission allocated 400 kHz of spectrum between 169.4125 and 169.8125MHz for Ermes, and other frequencies around 800 MHz have been reserved in the VHF band. Ermes will use frequency-agile receivers which scan 16 possible frequencies.

The modulation format is 4-PAM/FM with a symbol rate of 3.125kbaud and a data rate of 6.25kb/s.

Ermes is designed to handle:
— Tone-alert (8-tones).
— Numeric signals of 20 to 16,000 digits.
— Alphanumeric messages with a high-speed, high-capacity mode of between 400 and 9000 characters (7-bit bytes).
— Transparent mode for files to 64 kilobits (8kB) in 8-bit bytes.

Ermes can handle reasonably long message-files downloaded to a PC, and it has a true throughput-rate of about 6.25kb/s on 25kHz channels.

It has had a slow start in Europe. The EuroMessage system is already well entrenched. The problem with Ermes appears to be the additional costs imposed by the need for agile receivers since normal pagers only use a single frequency.

Don't confuse Ermes with Hermes, which was once a communications satellite system, and now a European project to develop a space shuttle. Ermes competes with POCSAG, the new FLEX protocols, and TDP. There's also a pre-Ermes pan-European service (actually POCSAG) called Euromessage in the UHF band which is limited to the UK, Germany, Italy and France. See global paging also.

EROS

Emitter/Receiver for Optical Systems — a type of diode that can both send and receive light signals.

error

Distinguish between error detection, and error correction. Both are grouped together under error control. However, error detection can lead to correction by requesting a retransmission of a block, and some systems of error control have both detection and correction (FEC) characteristics. For instance, SEC-DEC (Single Error Correction — Double Error Detection). See ECC also.

• In binary, an error of one bit always reverses the logic. A zero bit becomes a logical 1, and vice versa.

error code/message

Some computer and communications systems generate specific diagnostic codes when a problem is detected. With some systems you will get even more sophisticated explanatory messages for particular forms of errors. These can be logged, displayed and printed out.

error control

There are literally hundreds of ways of checking for errors in transmitted or stored data. You should distinguish two types of error control :

— Error detection which may also involve the retransmission of detected faulty data, resulting in correction, and

— Error correction, which can actually calculate and restore missing or changed bits, from information included with the data (FEC).

The terms error-checking or error-control are casually used in a reasonably generic way for both of these techniques.We should distinguish a number of error-control approaches.

• **Duplex channels:** If the transmission or storage system is reliable and produces few errors, then we will tend to use an approach involving error-detection followed by retransmission of those blocks which have been corrupted.

In some communications systems, this form of checking is conducted at each node as the data blocks/packets pass through; in others, it is only performed at the end (end-to-end protocols).

This is the most common approach used in full- and half-duplex file transfers. See sliding windows also.

• **Simplex channels:** With simplex (one-way) systems, we often want error-detection but no correction — as in video signals and some other 'real-time' and broadcast applications. If we have no backward channel we have little choice (except FEC, see below). Generally with broadcasting and voice communications, small errors aren't fatal, and may even be unnoticeable.

Digital compressed TV has a much greater problem than analog, because, by virtue of the compression process, one lost piece of data can have very substantial affects on the image, and this could be passed down from screen to screen for a second or more (since only 'differences' are sent between 'interposed frames').

• **Storage media:** With digital data from a fixed storage device like a CD–ROM disk, it is essential to include the correction information in with the main data stream. This involves the supply of redundant (extra) information. The receiver will then be able to reconstruct damaged bit errors (to a finite limit), and therefore avoid the need to ask for retransmission. This is the class of error-correction protocols known as FEC (forward error correction).

• **Prediction and Interleaving:** Other techniques which are used to keep errors under control, involve predictive techniques (see Trellis coding) and interleaving. When more than one channel is being transmitted, the signal streams are interleaved (woven into each other). If a major problem (a burst error) then occurs, it will damage each stream a little, rather than one stream a lot. This makes it easy for the FEC systems to cope. Many systems will use FEC, but also fall back on ARQ as well.

• **Parity:** At the base level, error correction may involve the use of parity and Hamming codes to check each byte. Hamming codes can also provide correction, but only to a limited degree. Hamming code is a binary error-correcting code capable of correcting a single-bit error in a data word.

• **Echoplex:** In e-mail and information retrieval, the basic error checking system is often the human eye supported by full-duplex transmissions with Echoplex. (see) This provides an automatic check on both keystrike and transmission errors. The host will echo back the keyboard characters, as received, to the originator's screen. Any mistakes or transmission errors are immediately seen and corrected by the operator.

• **Checksums and CRC:** For file transfers, most detection systems depend on reasonably complex mathematical calculations resulting in the transmission of an appended checksum, or CRC.

When an error is detected, the receiver will request retransmission of the block in some way. The detection and the correction are quite separate operations. See error detection, checksum, CRC and ARQ.

• **Automatic Request for retransmission (ARQ):** With data, detection is almost always followed by a request for correction. This is a multistage process:

— The checksum or CRC is calculated and attached to the block, frame or packet being transmitted.

— The checksum of CRC is recalculated by the receiver and compared to that delivered.

— Often (but not always) a signal is sent on a return path (back channel) to ACKnowledge the correct receipt of each block. Some systems acknowledge every block, and others only send negative acknowledgements (NAKs).

— When a fault is found a request-for-retransmission signal (NAK) is sent.

— The replacement block is received, slotted in, and the checking process is repeated.

If the block/packet lacks a sequence number, then the retransmitted blocks must all be received in order. If sliding windows are used (where the sender continues to send, until a NAK is received) then a sequence number is essential. Even then, the receiver and sender may back-up and retransmit all blocks after the fault (not just correct the one block).

The most popular of the file-transfer protocols are Christensen's/XModem, Modem 7, MNP, Sealink, YModem (batch), YModem-G, ZModem and Kermit. See also CRC, LRC and ARQ.

• **Forward Error Correction:** FEC is essential whenever the originator can't be asked to retransmit. It is therefore used on storage media such as CD-ROM, and for deep-space probes where the request and reply might take days. FEC is also used on normal communications networks, provided the additional overhead (redundancy) is justified. See FEC, Reed-Solomon and Golay also.

error detection

Error detection systems do not always involve correction.

- In video, the detection of a drop-out error (producing a white horizontal slash on the screen) will be corrected in the receiver by the insertion of the matching piece of the line transmitted in the previous frame. See drop-out compensator.
- With voice, detected errors are usually ignored.

error peak

At certain times of the day the rate of errors on all communications systems will rise due to EMI creation — mainly from impulse noise (often created by old selectors in old exchanges) and to other factors in the radio environment.

error rate

The average ratio of the number of erroneous bits to the total number transmitted. The same term is applied to error detection (before correction), or to the data after error correction.

- Some error correction systems will be 100% perfect with single-bit errors but fail with multiple-bit errors.
- Errors in the bits which constitute the error-correction 'redundant' bits in FEC, can actually produce errors in the data. In ACK/NAK systems, such errors will just cause unnecessary retransmissions.
- A well engineered telephone data circuit should have long-term probability of errors of about 1 per 100,000 bits (10^5).
- CD-DA (audio) with two independent Reed-Solomon code levels has a claimed error rate of 1 per 1,000,000,000 bits (10^9).
- CD-ROM with layered ECC, has a claimed error rate of 1 per 10,000,000,000,000 bits (10^{13}). See BER also.

These CD rates are those after correction.

ES

End Systems in OSI language — meaning hosts.

ES–IS

End-System to Intermediate System. An OSI routing protocol. This is used as a way for hosts (end systems) and routers (intermediate systems) to find each other. The distinction is with IS–IS which is the router-to-router (intermediate-to-intermediate) version. See IS-IS.

ESA

- **Enterprise Systems Architecture.** IBM's mainframe range.
- **European Space Agency.** Apart from launching rockets and satellites, the ESA are also involved in running major on-line database services.

ESC

- ASCII No.5 — Escape (Decimal 27, Hex $1B, Control-[).

Originally the Escape code was used to halt a program prematurely, but it now has many different applications. The way it functions is highly variable, but its most common use is to extend command code possibilities by giving a specified number of following characters an alternate meaning.

For this reason it can appear much like a Control character, but in this case two or more sequential characters are transmitted, not one.

- The *Shift Out–Shift In* 'bookend' characters can perform the same functions, but the number of characters between these two controls is undefined. See SO.
- The *ESC character* is almost always followed by one, two or three control characters, but the number is always predefined in the application.

escape (sequence)

- **Program control:** The term escape, can also mean a control character sequence that caused the system to break. For instance the <Control C> was often used as an 'escape code' in early programs to break an application of an extended processing session (a loop) and return control to the keyboard.
- **Modems:** The standard escape code for modems is:

```
<1 sec pause> +++ <1 sec pause>
```

Note that the use of the +++ characters is set (and can be changed) by one of the S-registers.
Also see TIES.

- **Screen and Printer controls:** An old term for the standard printer and screen commands which were identified by the prefixed ESC ($1B or decimal 27) non-print character. These were multi-character command strings.

ANSI defined a set for controlling the screen, and Hewlett Packard defined another series which became the *de facto* standards for printers. They all begin with an ESC.

The printer command sequences will generally end with an uppercase letter or a special symbol such as @, and they are both upper and lowercase sensitive. Sometimes the sequences can be entered into an application in ASCII character format, and other times they need to be entered in hex.

ESCON

Enterprise Systems COnnectioN. This is a I/O processor channel architecture for IBM mainframes, used to transfer data. It's a 1992 standard for high performance computer connectivity using fiber optics which was designed to support devices such as terminal controllers, printers, tape drives and DASD. Some carriers are now offering ESCON tariffed services for banks, etc.

ESCON has a data-rate of 25MB/s (200Mb/s), which is a major step up on the older IBM Bus & Tag channel. It uses a proprietary frame structure with data frames up to 1000 Bytes in length.

The channel can reach out 20km without repeaters and 60km with repeaters, and an ESCON Director switch will allow a single channel to communicate with many different devices. ESCON uses the same channel command words as Bus & Tag, and ESCON DASD and tape units are both available.

escrowed encryption

Escrow is a legal term meaning to hold a bond or deed on behalf of another party under special written conditions. The objects held in escrow remain legally unavailable until certain conditions are met.

In encryption systems, this means that the essential parts of a decryption 'key' are to be held in secure repositories. The police or security service can (theoretically) only gain access to them after formal approval has been sought and received from a high-level judge.

The Clipper proposal was for RSA keys to be held in this way. This Clipper key was to be kept in two parts, with each held by an escrow agency, and the two parts could only be joined and used with the permission of the courts. However the Government itseif could gain access to a 'master key' which allows it to decrypt otherwise private messages if necessary for state security. See Skipjack.

ESD

Electro-Static Discharge. ESD is a threat to low-voltage electronic devices like chips, every time anyone replaces a cable or touches an I/O port or touches anything inside a computer. Damage caused by direct electrical discharge can occur during servicing and manufacture, and also from strong, near-field coupling (either conductive or radiated) effects. It is possible that ESD may result in more error and unexplained crashes than we realize.

However it also offers service departments an omnibus explanation; they can use ESD as a scapegoat whenever they have problems with equipment under warranty. What's more, it explains why the failure is actually your fault! See EMI.

ESD protection is now possible to a degree with computer chips, and the Europeans have set minimum levels of ESD performance (with tests specified in IEC 801-2). An EMC Act comes into effect on 1 Jan 1996. Equipment is also tested for ESD susceptibility using a human-body model or a machine-handling model (EIAJ).

- There are simple rules for reducing the likelihood of ESD damage to your computer, and mains-power damage to your health. Don't let anyone tell you that these are wrong.

Never touch the inside of the computer, the cards, or the chips while the machine is connected by a power cable to a wall plug.

It is important that you do not (repeat **NOT**) leave the chassis of the machine connected to the earth potential line of the power-point — this increases static potential difference.

So don't just switch off: **you must unplug**.

(The industry myth is the complete opposite of this).

Always touch the metal case surrounding the power supply with the back of your hand.

Do this before touching any electrical device or any component in the computer.

If you are moving around the room, keep touching the metal case regularly.

- Better still, of course, use a special wrist strap and ESD pad.

ESDI

Enhanced Small Device Interface, or Extended Storage Device Interface. A high-performance device-interface standard for serial computer communications (particularly for hard-disk drives) which IBM began using with the PS/2 series. ESDI has a claimed maximum data transfer rate of 10 to 15Mb/s (but 4 to 8Mb/s in practice) and has much more on-board intelligence than the ST-412/506 devices it replaced. However SCSI is even better.

It uses a 34-connector cable for control which can be daisy-chained from drive to drive, and a separate 20-conductor data cable for each drive (to 3 meters). ESDI controllers generally only support two drives (but the specification allows seven). IBM's ESDI hard disks used high-quality thin film technology, and voice-coil units for head placement. The standard can handle a 1:1 interleaving factor which means that the head can read every sector on a track in one revolution. See ST506, IPI and SCSI.

ESF

Extended Super Frame. ESF introduced a new way of handling the framing in T-1 systems using 'superframes'. The superframe idea was introduced with D2 channel banks, and the enhanced version came with D4 (see). D4 is probably still the most common T-1 framing format, but ESF is rapidly overtaking it.

ESF frames the data in the normal 193-bit way, but it also organizes groups of 12 frames into superframes with extra control bits. This frame structure then adds a data-link control channel which can provide a CRC, send operational messages end-to-end, and provide a level of line-quality reporting. With today's processing power, the line can stay framed even if half the framing bits are borrowed to form a message channel.

- There's both a proprietary AT&T version and an ANSI version (T1.403). The line code for both must be B8ZS.
- See ABCD, A&B signaling, T-1 and CSU also.

ESMR

Enhanced Specialized Mobile Radio. A digital form of trunked radio (for dispatch, emergence, etc.) which the FCC has licensed to use in cellular-like operations. The main distinctions between ESMR and cellular are:

- Cellular has higher voice quality, and less emphasis on data.
- ESMR offers one-way press-to-talk or two-way phones.
- Cellular can't make group calls to a fleet or to closed user group, ESMR can.

- ESMR pricing is usually on a per-unit monthly basis, with low per-call charges (if any)
- ESMR only secondarily connects to the phone network for normal dialed operations. Normally it links to the company base station. See MIRS and SMR.

ESN

Electronic Serial Number. A unique number assigned to every AMPS analog cellular phone as an identifier. It is plugged into the handset in an EPROM. At present, the ESN is an 11 digit number comprising:

- a two number manufacturer identification,
- a two number model identification,
- a two number country code,
- a five digit progressive number for each product.

The ESN identifies a piece of equipment, not the user. The other number in an analog mobile phone is the MIN. See MIN and white list also.

ESP

Enhanced Serial Port. See Hayes ESP.

espresso

A SPEC benchmark. It is an application which performs Boolean functions.

ESPRIT

European Strategic Program of Research in Information Technology. The European 10-year (1984–94; now complete) program of research into information technologies. The goal was to beat the US and Japan by 1992. See Alvey and RACE.

ESS

Electronic Switching System.

• A generic term for modern electronic analog and digital telephone switching systems used by a telephone company. There's nothing electro-mechanical in these switches, they all use solid-state devices and computer-type control equipment. They can offer custom-calling features, and often a wide range of CLASS services. See SPC also.

• AT&T's popular range of exchange switches are all labeled #ESS. The central office switch needs to be a 5ESS if it is to offer ISDN.

ETACS

Extended TACS. The original UK decision when introducing TACS, the analog cellular phone network, was for two companies (Cellnet and Vodafone) only to compete. They were each to have a fixed allocation of 300 channels in the 900MHz region of the spectrum.

The networks proved to be so popular that they soon ran out of spectrum in central London and a few other areas. Eventually the government granted a further 200 ETACS channels by taking back some military channels. Obviously only handsets with the new extended facilities can use these channels.

ETB

ASCII No.5 — End Transmission Block (Decimal 23, Hex $17, Control-W). A transmission control character which indicates the end of each transmitted block of data. It is used for blocking data, where the block structure is not necessarily related to the processing format.

ETC

Enhanced Throughput Cellular. A modem protocol which is designed to handle the noisy and erratic transmissions of cellular systems. It was developed by AT&T Paradyne to allow higher speed and better reliability of transmitting mobile data over the cellular network.

ETC gains its advantage by working both at the Physical (1) and the Data-link (2) layer. ETC equalizes the frequencies across a cellular connection (normally the central voice frequencies are boosted) and it monitors the cellular channel and performs auto-rating functions (monitoring phase jitter, harmonic distortion, SNR).

Normally it works end-to-end across the network with another ETC modem, but it can interwork with standard V.32bis/V.42 modems (both protocols are needed). Claimed data rates over AMPS networks are up to 14.4kb/s with stationary vehicles. See MNP-10 also.

E-TDMA

Expanded/Extended TDMA. The name Hughes Network Systems applies to a variation on the North American digital cellular phone (NADC) system which is said to give a 10-times capacity increase over NADC — or, a 15-times capacity gain over AMPS (note the discrepancy!). The voice quality is generally not very good, but recently they seem to have redesigned it more for use as a fixed wireless local loop technology. E-TDMA gains its capacity advantage by using a half-rate vocoder (all six slots in the TDMA frame are allocated to different users) and with digital speech interpolation. See DTx, VAD and DCME.

ether

This was a mystical/mythical substance that was believed to permeate the universe. It was postulated as the theoretical medium through which light and radio waves passed.

EtherFRAD

See FRAD.

Ethernet

Created in 1973 by Bob Metcalf's team at Xerox, from a radio system (hence 'Ether') using the Aloha protocols from the University of Hawaii. Metcalf was later assisted by DEC and Intel (about 1981), then later still, Ethernet was taken and modified by the IEEE to create standard 802.3.

The first standard became known as DIX (DEC, Intel, Xerox), which was later modified to DIX Version II. There are still Version II specs which create problems between Ethernet II and 802.3 — mostly, in the latter's use of a length-field vs. Ethernet's type-field in the frame.

The defining characteristic of Ethernet, according to Metcalf, is the use of the CSMA/CD access protocol.The other defining charactersistic was that it was both physically and logically a bus topology — but in the 1990s with the advent of 10Base-T, it evolved to a physical star network radiating out from a hub — but with logical bus segments.

Ethernet LANs are moderately fast, reliable and very flexible. The theoretical data-rate is 10Mb/s, but in practice the LAN throughput of data is about 4Mb/s. The standard address size is 48-bits (6-bytes).

However, some Ethernet systems run at 1Mb/s and others at 50Mb/s and now 100Mb/s — and some have 16-bit addresses.

Ethernet was designed for coaxial cables (Thinnet and Thicknet), but is now applied to twisted-pair wiring, and even to optical fiber. And while the original architecture was a bus topology which could be extended to tree-branch, it now includes work-groups in star configurations with dumb or intelligent hubs and even switched hubs. It is cheap, easy to install and implement, and it is reliable.

Ethernet uses Manchester line-coding and its CSMA/CD contention techniques work well up to 40% of theoretical capacity, particularly in an office environment. But there's some criticism about Ethernet in a manufacturing environment because of the contention nature of the access and the limited speed.

The DIX standard Ethernet frame (actually a packet) consists of an 8-Byte preamble, 6-Byte destination address, 6-Byte source address, 2-Byte type ID, between 46- and 1496-Bytes of data, and a 4-Byte frame check (CRC) sequence.

The IEEE 802.3 Ethernet frame consists of a 7-Byte preamble, 1-Byte start frame delimiter, destination and source addresses which can be between 2 and 6-Bytes, a 2-Byte length-of-data field, data up to 1500-Bytes (no lower limit but pad to fill the frame to a minimum length), and a 4-Byte CRC frame checksum. The shortest frame is 64-Bytes.

- Both flavors have a total of 1526-Bytes (including the 8-bit preamble). The maximum frame size of Ethernet is said to be 1518-Bytes, with a payload of 1024-Bytes of data.

- The terms 10Base-5 (thick 'yellow' coaxial cable), 10Base-2 (thin cable — also called Cheapnet), and 10Base-T (twisted-pair cable) are all Ethernet standards. (See 10Base-# in numbers section).

- The IEEE standard Ethernet is 10Base-5 which allows 500 meter segments of 10mm 50 ohm coaxial with up to 100 stations to each segment (a total overall of 1024 nodes). Up to four segments can be joined by repeaters. The nodes link to the main cable via a transceiver on a cable-tap, and with a drop cable (to the PC) of no more than 50 meters.

- 100Base-VG is not an Ethernet; it is a proprietary standard from HP and others.

- 10Base-T over UTP is now the most popular way to implement Ethernet. It requires each network node to have its own discrete twisted pair connection to a central wiring hub/concentrator. These concentrators can isolate a node in the event that they are causing trouble and most of them now actively repeat or switch the signals. See partition, wiring hubs and concentrators, Switched Ethernet and below.

- Ethernet TP is now called 10Base-T.

- Over optical fiber there is a closely related standard called FOIRL (see).

- There's two or more 100Mb/s standards being proposed. See Fast Ethernet, 100-Base? and 100VG-AnyLAN.

- There's also Full-duplex Ethernet at 20Mb/s. See.

Ethernet switches

The switching of Ethernet packets, either on a one-source-to one-sink basis, or the switching of shared-media segments, appears to be the current favored way to increase the overall data-rate of many networks without junking expensive NICs and software.

Kalpana's EtherSwitch started the trend; it is widely used to provide workgroups with their own 10Mb/s segment as an alternative to multiport routers and bridges. Later segments were reserved for power-users.

Multisegment switching with the 10Base-T radiating twisted pair links, allows many users to switch through a hub simultaneously. Both half- and full-duplex systems are available.

So Ethernet switching is a sub-category of intelligent switching hubs, but the term has been vaguely used over the years. The hardware variations are:

- Port switches. All the traffic from one port can be assigned to one of the separate backplanes via a software switch. These are used to segment a large network and provide permanent connections between segments. These switches are more like automatic patch-panels — they aren't true Ethernet switches in the sense of handling packets on a packet-by-packet basis. They can be remotely managed.

- Conventional switches. These switches are store-and-forward devices, but they can give power users a dedicated 10Mb/s rate on a single connection. The need for hub buffering slows these systems down to a degree, but being store-forward, they can handle rate-conversion provided they have the buffer capability and controls. Some are configured as bridges and some as routers.

  Most of these come with 24 or so 10Base-T ports, each of which can provide 10Mb/s to the user, and 100Base-T Fast Ethernet (or other) ports for backbone connections.

- Matrix or 'cut-through' switches tend to pass packets through on the fly, and are much faster. However if many segments are attempting to access a single server, the switch will respond no faster than a shared network.

  These are true Ethernet switches which stream data: the packet's front end leaves the switch before the back enters.

See multilayer LAN switching, intelligent switching hubs, virtual LANs, and Switched Ethernet.

EtherTalk

A relatively low-level (Layer 1 and 2) protocol for the Macintosh which allows direct connection to Ethernet at 10Mb/s. It complies with 802.3 and allows AppleTalk protocols to run on Ethernet. EtherTalk needs an add-in card on most Macs, although some of Apple's laserprinters and new high-end PCs come with the chips built-in.

ETM

Electron Trapping Material. A disk recording technology under development by US firm Optex. Europium and Samarium ions are held within the structure of the disk. When blue laser light hits the europium ion, and an electron is pushed into a higher energy state. When it falls back, the electron becomes trapped by a neighboring samarium ion. This is equivalent to storing a bit, because the recording can be read by light in the near-infra-red region.

Some of the excitement about ETM is because, instead of just recording single 0s and 1s, it is thought that multiple states can exist in the structure, holding dibits, tribits, and quadbits, thus increasing the capacity of the disk enormously.

ETOM

Electron Trapping Optical Memory. See ETM.

ETS

European Telecommunications Standards set by ETSI. These standards are only voluntary, while the NET standards are mandatory in Europe.

ETSI

European Telecommunications Standards Institute, the successor to CEPT, which began in 1988. It is the European standards organization in control of the development of ISDN, B-ISDN, MANs, DECT, GSM, PCNs and UMTS. It has a General Assembly, a Technical Assembly, 12 Technical Committees and 5 Special Committees. Most of the detailed work is actually done by 60 Technical Sub-Committees.

ETX

ASCII No.5 — End of Text (Decimal 3, Hex $03, Control-C).

A transmission control character often used when multiple messages are being sent. It terminates the text which began with STX and it tells the system when each one has finished.

- STX and ETX are book-ends around text messages.

ETX/ACK

A flow-control protocol used by the early Spinwriter printers and later for more general use. It works in a similar way to the more widely used XON/XOFF protocols. With ETX/ACK, a block of characters (usually 128) is sent to the printer followed by an ETX character (Control-C). After the printer has used these characters it will return an ACK character (Control-F), and the process will continue.

EU-#

See Eureka projects.

Eureka projects

These are joint national/company electronic research projects in the European Union. They become classified as Eureka projects only when the technology has been developed far enough for them to enter a product development phase.

France proposed Eureka in 1984 to counter the advantages US companies gained from being involved in SDI defense projects. A third of the current projects are environmental.

- Eureka 95 was the old High Definition TV proposal based on the MAC standard. The term is applied both to the committee and the proposed system.
- Eureka-147 is a digital radio protocol for audio broadcasting. See DAB.

Euroconnector

See Scart.

EuroMessage

A European paging system dating from 1988. EuroMessage is pure POCSAG and uses standard alpha pagers on a common frequency channel. Italy is linked into the network via SIP; France via France Telecoms and in Germany the paging company is Deutche Bundespost. In the UK, EuroMessage is offered by a couple of carriers. This was a stop-gap measure while Ermes was being developed, but it now appears to be Ermes major competitor.

Euronet

Currently this is a European packet-switching network, but the concept is to be extended by the EU as a European-wide fast data network.

Eutelsat

This organization has a constellation of seven GEO satellites. There were three in the first generation and four in the second. All were Ku-band, with about 20 transponders each in the most recent.

Eutelsat began in 1972 as a consortium of European telephone companies. It now has 41 members, and its latest satellites are called 'Hot Birds'.

EVE

European Videophone Electronics Assn. This group aims to develop a European market in videophones; it consists of the dominant carriers in the UK, Norway, Germany, France, Holland and Italy.

even parity

All bits in the original byte add up to an even number so no parity bit is added. So to be quite clear: even parity calls for a check 1-bit to be added to the byte, if the number of 1-bits is odd. This then makes the numbers of 1-bits even. See parity bit formats and parity.

event

- An event is any motion or action which the operating system or shell decides may be of interest to an application.

Mouse-movements, keystrokes, and the resizing of tiled windows are all events. Events can originate from the user, the operating system, the program or even another program.

- See event driven — computers and programming.
- In modem communications, an event is a symbol which may carry more than one bit. It is a change in conditions on the link. It is measured in bauds, or symbols per second (not bit/s).
- In LAN/WAN management, an event is a network message which indicates irregularities. This can be in the physical elements of the network, or in response to a request for information from the management console.

event driven

- **Computers:** The Macintosh is an 'event-driven' computer because it stores future actions in an 'event queue'. The CPU keeps a constant watch for user-initiated events from the mouse and keyboard, and gives these priority over other work in the queue. Windows is also event-driven. The distinction is with procedural systems in computers which constantly execute code unless deliberately interrupted. See IRQ.
- **Programming:** This means that the code in a program doesn't execute unless an event (either a system, or a user action — usually a mouse click) causes it to act. Objects are defined on the screen (buttons, forms of list boxes) and code is attached to them which will determine their behavior.

However the defined object will wait until a mouse clicks over its icon (or some other nominated event occurs) and, only then, is the code executed. This is how many of the popular new object-oriented PC programs and utilities now work.

event loop

A Macintosh program spends most of its time in an event loop waiting for something to happen. Mac programs are event-driven — they are always ready to respond to anything.

EWOS

European Workshop on Open Systems. This is the European version of the OSI Workshop in the USA. It is involved in conformance testing, e-mail standards (X.400 and X.500) and the defining elements of OSI.

exa-

In SI units the prefix exa- represents 10^{18}. The symbol is a capital E. An exaByte (EB) is a thousand pentaBytes or a million teraBytes.

ExCA

Exchangeable Card Architecture. Intel's extensions of PCMCIA Version 2.0. It is an open software standard which has been accepted by the PCMCIA standards group, but it's not universal. It allows hot-swapping of the cards.

exchange

A term applied to both the building which houses the telephone switches, and the switches themselves; however the term 'switch' is probably preferable for the machines. In the USA, a local exchange is called a 'central office' or an 'end office'.

There are a number of different kinds of exchanges; some are only involved in switching signals between other exchanges. See exchange class and tandem.

exchange area

This is a telephony term applied to billing. It means the area in which only a single local call charge applies. See LATA also.

exchange class

In North America, the telephone network is organized as a five-level hierarchy.

— **Class 1 exchanges:** These are regional exchanges which carry the intercity traffic.

— **Class 2 exchanges:** These are sectional exchanges.

— **Class 3 exchanges:** These are primary centers.

The above three levels are controlled and owned by the Interexchange carriers (IXCs). The next two come under the control of the Local Exchange Carrier (LEC).

— **Class 4 exchanges:** These are the intra-LATA toll exchanges which may also have some local switching functions, but which primarily provide gateway services to the IXCs: these exchanges provide the POP (Point of Presence). They are often called 'tandem' exchanges. See.

— **Class 5 exchanges:** The local exchanges which connect to customers over the local loop. Also known as the 'end office'.

exchange hierarchy

This deals with the topology, not the nominated class of the telephone exchange. Local exchanges are connected to primary trunk exchanges via a trunk line. These primary trunk exchanges are then linked to secondary trunk exchanges, and then to tertiary trunk exchanges, and so on. This is a branched tree structure. Routing is the process of switching calls through this structure. See exchange class.

Exclusive OR

In Boolean algebra this is a variation on the normal OR. If either of the binary input signals are one, then the output will be one (true). However if both are zero, or both are one, then the output will be zero.

EXE file

This is an executable file in MS-DOS, OS.2, the VAX/VMS which contains relocatable machine code utilities. To run these you type the name of the file without the .EXE extension.

executable code

The same as an object program or file. See.

executable files

In DOS, these are files ending in .COM, .EXE or .BAT. They are small files that make things happen to other files, or which control a sequence of events.

execute

To run a program/algorithm or perform some function.

executive
An old term for the operating system.

Executive Information Systems
EISs are software systems designed to gather and collate corporate financial information and make it available, in real-time, to executives. These programs supposedly allow the management to monitor a company's financial performance on an hour-to-hour, or day-to-day basis. They are information retrieval applications, rather than analytical tools.

The classic mainframe DP approach was to produce huge weekly paper reports for the executives, who were then expected to wade through the information and find the parts that were relevant. The EIS approach is to provide only the important pieces of information on-line for selection — so there is a certain amount of filtering and 'gatekeeping' involved. E-mail, automatic financial database retrieval, on-line calendars, project tracking, etc. are all EIS tools also.

executive programs
A program held (often permanently) in the internal memory of a processor to act as a master control.

expanded memory
IBM first used this term to apply to memory in the AT range between 1MB and the 16MB limit of the 80286 chip. Under older versions of MS-DOS this space could only be used for RAM disks, but it was directly addressable in the 286's or 386's protected mode (see VPAM) or by PC/IX or Xenix. This space is now called 'extended memory' when used by Windows and XMS-compliant applications, to distinguish from EMS (below).

• In 1985, Lotus, Intel and Microsoft (the LIM group) devised a standardized way of expanding the 640kB working memory limit. This 'LIM extension', set aside a 64kB block of spare memory in the upper-memory area (between 640kB and 1MB) for bank-switching applications — they provided a 'transfer window' through which programs could pass data, to use memory above 1MByte. The 4 x 16kB segments are known now as LIM/EMS (Expanded Memory Specification) page frames.

These blocks import and export pages of data from even higher memory locations (beyond 1MB) — but the operating system only sees them as coming from a memory location below the 1MB limit (the transfer page locations).

This technique is only used for DOS applications when they know how to use EMS (or EEMS), and it retained most of the upper-memory space for its normal use by video RAM and BIOS ROM, while allowing most current applications to function in conventional memory.

This technique was known originally as the LIM extender, but it is now known as LIM Expanded Memory Specification (EMS) — and it has since been extended and replaced itself. This is now a dying standard.
See EMS, EEMS, UMB and upper- and high-memory.

• When the 386 AT machine appeared, it was able to directly access 16MB of memory in protected mode. This allowed the software designers to fudge around the 1MB limit in MS-DOS without requiring the special EMS memory boards and bank-switching. All of the many DOS add-ons provide this new Expanded Memory Manager (EMM) feature. See DOS extenders and DPMI.

• A common misuse of the term 'expanded memory' is to refer to extra memory chips that you buy already mounted on a board that can plug into a memory slot in any DOS machine. What they really mean is 'memory expansion' (hardware) while 'expanded memory' refers to a logical function. Remember, that whenever memory expansion was used for expanded or extended-type operations, in early MS-DOS machines, it did still need special software — either memory managers (device drivers) or bank-switching utilities.

• In later editions of DOS for 386 machines, and above, extended memory can simulate expanded memory. EMM386 must be loaded by a device command in CONFIG.SYS, and it then simulates expanded memory.

• Windows 3.x can also simulate expanded memory if an application needs it.

Expanded Memory Manager
See EMM.

expansion bus/card/slots
Personal computers in the early days were either 'closed' or 'open' boxes in a hardware sense. The closed box approach (the early Macintosh) didn't allow for upgrades, or for third-party plug-ins. Open boxes, such as the Apple II, IBM PCs, and now Mac machines provide a series of expansion slots, into which cards can be added to expand the capabilities of the machine.

This can mean increasing the memory size (in some early machines), or it can mean introducing special devices to perform different functions, or special adaptor cards for running the monitor, printer or disk drives.

• The expansion bus is the bus into which you plug peripheral cards. It is designed to allow them to communicate with the rest of the computer usually via the CPU — but also using DMA.

The expansion bus is the bottleneck of PC design. It generally needs to be backwardly-compatible, which means that it had to remain very much the same design, while the speed of CPUs and memory chips have probably increased dramatically.

• The first IBM PC expansion bus ran at 4.77MHz — and it was only 8-bits wide to match the 8-bit data bus of the original 8088 processor. The slots were 8-bit slots.

• With the XT, the expansion bus slots had a new section added to one end. This added four more address lines, eight more data lines, and more control lines. These were needed by the 16-bit wide 8086 processor.

• The AT introduced the AT-bus which later became known as ISA (Industry Standard Architecture). The AT and the 386 were both 16-bit machines, so the slots were made 16-bits wide (but had many more contacts). However, to retain backward compatibility, the old 8-bit cards would still fit into these 16-bit slots.

• IBM then introduced the MCA bus (a closed approach) while the clone makers went with EISA (an open approach) — both 32-bit buses. Neither has succeeded.

• The MCA bus provides a 32-bit data pathway and is designed for multiprocessing. It also eliminates adaptor conflicts.

• When it became obvious that future PCs would need to handle data at video rates, there were numerous local buses designed. These were attempts to by-pass the motherboard bus, by providing a supplementary high-speed link between cards. The clone makers through their association, VESA, have attempted to standardize a video bus known as the VL-bus, but it has never been fully accepted, and the 'standards' are more than a bit chaotic.

• The *de facto* standard is now clearly the PCI bus developed by Intel. Apple and most Windows 95 machines will use this bus in the future for video. However Windows machines will retain ISA also, and the Mac will retain NuBus.

expansion memory

This is a very vague expression for memory chips on an adaptor card. It is add-in memory on a multi-function memory board, and it comes in 8-, 16-, and 32-bit banks.

Since all of the 80286 and later PCs came with a full 1MByte of memory on the motherboard, expansion cards were used to provide memory at locations above 1MByte, which we now call 'high-memory'. This can be used in three ways:

- By operating systems other than MS-DOS, directly in their 'native' mode. Unix and OS/2 use the memory in this way.
- By later MS-DOS versions and applications which could use it directly as 'extended memory'.
- By early MS-DOS versions and applications, which could only use it indirectly by bank-switching. This is EMS and expanded memory.

Anything that can plug into an expansion slot can be generally referred to as *expansion* memory — and that encompasses memory used for expanded (EMS), enhanced EMS (EEMS) and extended memory cards.

expert system

A practical form of AI which involves the use of inference engines and knowledge bases in a non-procedural manner. The program learns its expertise from human expert/s, and the program then allows non-experts to draw on that expertise in making judgements.

The expert system asks the user a series of questions, and the answers guide it on a unique path through the knowledge base. Both the rules needed for applying the expertise, and the knowledge itself, must be stored in the knowledge base.

All expert systems consist of three subsystems:

- The knowledge-base management for problem solving.
- The logical inference system to define what is both useful and relevant.
- A usable interface between machine and man.

See forward and backward chaining.

explicit

Directly expressed — spelled out.

explorer

A frame or packet which can be sent out by a device across a source-route bridging network, to discover the best route to another device.

exponent

Aka 'characteristic'. The small positive or negative integer to the right of, and above a base (e.g. 10^{18}) to indicate 'to the power of ....'. It says how many times the base (radix) is to be multiplied by itself. In printing this requires a superscript version of the font. See mantissa.

• In an ASCII file, the superscript is not provided, so the presence of the exponent is usually expressed by the ^, as in 17^3 = 17^3.

exponential projection

The staple diet of tabloid journalism when dealing with technology. The infinite projection of a short-term trend, in order to create a media sensation by arriving at totally erroneous conclusions. But, if you are among technophiles or technophobes, who will question your conclusions?

- The exponential projection in the potential speed of steam trains, if taken from statistics of the improvements made during the late 1800s, would put modern steam trains at about Mach 2.
- The improvement in bicycle speeds during 1900–20s, projected forward to today, would have the present world record at about 1000 kph.

All technological progression eventually plateaus. It either runs into finite limits for technical improvement or, more often, it satisfies the needs of the society and so development ceases, or turns in another direction for a time.

At certain stages, almost every technology begins to run up against social obstacles such as safe road speeds, carbon fuel consumption, traffic jams, smog, etc. With computers and communications these social limits may come from privacy fears, info-glut, or something else we haven't yet thought about.

However, infinite projection of developments or requirements is often mistaken for wisdom in this industry. To suggest that switched 5Gbit/s optical fiber feeds to every home is perhaps 'slight overkill' — when we can transfer wide-screen TV at 10Mb/s — is often scorned as a 'Luddite' reaction. In the guru business, it is always safer to project unlimited future requirements in bandwidth — and, no one can ever prove you wrong!

exponential notation

A way of expressing very large and very small numbers. This is scientific notation. The sign for this notation is the capital E in computer arithmetic.

• If you see the number 34.567E6 on your calculator, it can be translated to 34,567,000 in the following way:
– the terminating E6 here means Exponent 6 (with the decimal base of 10 being assumed).
– 10^6 is 1,000,000.
– The calculator expression therefore becomes 34.567 x 10^6 or 34,567,000.

• So 9E–7 meters, translates to 9 x 10^{-7} or 0.0000009m.

export

In computer terms you export text (or data of any kind) when you make it available (on disk, or over a cable, or across any telecommunications line) in a form that is usable by another computer. Exporting sometimes implies that the file will be transferred in a file-format which can be imported and immediately used by the other computer/software combination — but it can also equally be applied to ASCII.

extended ASCII

The partly-standardized upper 128 8-bit codes outside the ASCII No.5 set. The 7-bit ASCII standard created 128 characters which all modern computers recognize. At one time, this was all that was thought necessary for the alpha characters, numbers, punctuation, and some communications control codes.

However the modern use of the 8th bit adds another 128 characters which have never been fully standardized (except within MS-DOS or the Mac's proprietary applications). So this term refers to ASCII characters from Decimal 128 to 255.

extended control codes

Special codes used in transmission systems to provide cursor control, etc. These are codes over and above those found in the normal ASCII No.5 set. See VT–## terminals.

extended memory

This term has changed in meaning a couple of time. It has been used in different ways at different times by different people — and different dictionaries define it in different ways. So beware!

• Originally: This referred to 64kB of unused memory within the video display and BIOS area of the early MS-DOS machines. It was a section between 896kB and 960kB (below the 1MByte limit) which was later used for bank-switching. However some applications could make use of this spare space (making the overall working memory space in DOS 704kB), but this was mainly confined to utilities and disk cache.

• Later it came to mean memory space that lies above the 1MByte barrier of the early MS-DOS. Since memory was added on expansion cards, it was also known as 'expanded memory' — and the first application of these boards was with the bank-switching process known as EMS. This later became expanded (EMS) memory — which was a source of great confusion.

• Then the term came to mean any memory beyond 1MB which is addressed through the XMS specifications (see). Only 286 and later PCs can use this — and then only with a memory manager such as HIMEM.SYS.

• Windows 3.0 directly accesses this memory space in the 386 mode and above. This extension to the working RAM space could be made available as relatively straight forward additional memory space in 286 or 386 machines, but only in their protected mode — and not at all in the earlier PCs based on 8088 or 8086 processors. However it is widely used for disk-caching software and in some large DOS applications.

• All Windows and Windows-based applications now require extended memory.

• In current terminology, distinguish extended memory from expanded memory, by the fact that:
– Extended memory is 'normal' memory beyond the 1MB range which is addressed direct by software which conforms to the XMS specification. This is only possible for computers using the 80386 and later chips. It can be addressed without needing special software drivers. See DPMI also.
– Expanded memory is a way of tricking the old MS-DOS by bank-switching through an EMS page frame. This was only possible with computers using the 80286 and later chips, and is now a fading standard. It is now addressed through a software driver/emulator EMM386.
– Expansion memory is just memory on a card plugged into an expansion slot. This is a physical definition.

• This still sometimes gets confusing, because expanded memory emulators (notably EMM386) are able to use extra memory space above 1MB without bank-switching (see expanded memory). EMM386 is a memory manager which can work in either extended or expanded mode — but it actually only 'emulates' the expanded mode. It uses the extended mode.

• Also confusing is the fact that Expanded Memory Specification (EMS) system cards can be configured (to varying degrees) as extended (or rather 'extension') memory for OS/2 or Unix use. But generally EMS cards were for MS-DOS and then they used bank-switching via the EMS specification.

• We use the term High Memory now to mean the physical memory at locations above 1MByte, while 'extended memory' specifically refers to a DOS function which directly accesses this memory — as distinct from the 'expanded memory' function which can only access it indirectly via the pages in the Upper Memory Blocks.

Extended Memory Specification

See XMS (also EMM).

extensions

• **Codes:** See ASCII No.5.

• **Operating Systems:** These are sub-programs added to the operating system to give it additional capabilities. For instance, extensions must be added to MS-DOS before it can recognize a file bigger than 32 megaBytes. CD-ROM disks have files larger than this.

- **CD-ROM:** See High Sierra format.
- **MS-DOS Filenames:** See file-type extensions.

extent

In computer memory mapping, an extent is any region of directly-accessible space located between the defined upper and lower address limits.

exterior gateway protocol

A protocol used to exchange routing information between autonomous systems. This is a generic term which includes protocols such as EGP.

external key

In video production, a key is a matte or mask. In effect, this is the equivalent of cutting-out sections of an overlay picture, which then allows you to see parts of another picture behind. Some areas of the top picture are transparent and some are totally opaque, so a figure can be inserted into a different background, or strange effects can be created with one image appearing to peep through holes in the overlay from behind.

With video and film you have the problem that both images are always moving, so in film production the techniques are known as 'traveling matte'. In video, the secret is to use a very fast switch which acts on the co-ordinated output of two or more cameras, and selects just one. It can switch from one to the other in the time the scan takes to pass from one pixel to the next — and then switch back again, if it has to on the following pixel.

The key effect is the selection of the areas in the overlay picture which will become transparent (which trigger the switch to act). This can be made on the basis of:

— a cut-off tone or a specific color (usually green or blue),

— internal source (from tones or colors within the overlay picture itself) or external (from some outside third picture or graphic).

So keys are either triggered by color or luminance (levels), and they can be internal (one of the two images provides the trigger) or external (a third image, or outside source like a character generator supplies the trigger).

External key functions are built into most SEGs used in TV studios, and the technique is often used to insert moving images into a shape (map of Britain, say) or diagram on the screen. See internal key and chromakey also.

extraction indexing

The construction of indexes by automatic means. Usually the text files are first compared with a stop-word list of linking terms, pronouns, etc. and after these are rejected the remaining words are indexed according to some formula. In extraction indexing, the frequency of the words may be taken into account in deciding whether to keep or reject a word. See inverted indexes.

extrapolate

To make predictions about the future based on assumptions that the future will change in the same manner and speed as the past. It is a case of extending the plot-line of past changes into the unknown, and it requires no knowledge or intelligence.

The term applies particularly to graphic operations where the line on a statistical chart is extended beyond the bounds of the figures available. The idea assumes that change is evolutionary, predictable and continuous; without limits.

eye pattern

This has nothing to do with eyes or vision. It is a traditional graph type showing variations along one axis. The lines produce an 'eye'-like pattern (used with error detection).

- Eye height is a measure of the signal-to-noise ratio in a digital bitstream. The decoding margin is the extent to which this S/N ratio (or the Eb/No) exceeds that needed for an acceptable Bit Error Rate (BER).

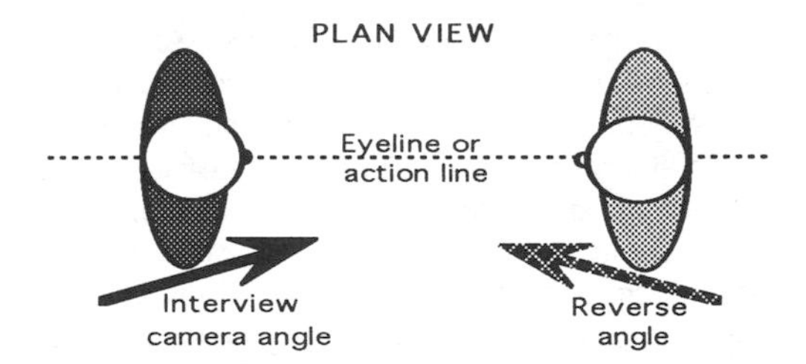

eyeline

In film and television production, it is important to orientate the audience by preserving 3-D concepts in what is just a 2-D screen. Rules of shooting are therefore essential to retain some sense of direction or the audience will become confused.

When two people talk (plan view above), they generally make eye-contact, and this line between the eyes (extended infinitely in space) defines the range in which the camera/s can move in that sequence. It/they must stay on one side of the eyeline only, if the shots are to retain a sense of continuity.

If a camera strays across this line, it can appear as if one person is looking at the back of the head of the other — or that they are both looking off-camera in the same direction, instead of at each other.

An almost identical concept (and one that is actually preferred) is to think of this as the 'action line' between the two geographic points of interest. This concept then also fits a car chase; the camera/s must always remain on the same side of the line of action otherwise one car may appear to be moving in the opposite direction to the other.

These are very powerful conventions. See reverse.

F

f/F (camera lens)

- The small f means 'aperture'. See f-stop below.
- The capital F is the focal length, which is always measured in millimeters.

Depending on the size of the image (the film in the camera gate) focal length relates to the 'angle-of-view'. It is classified by comparing it to the diagonal of the image frame.

When focal length roughly matches the diagonal, the lens produces a normal angle of view: when it is longer in focal length (a 'long-lens') than the diagonal, the view is more telescopic, and when it is shorter, the field of view is 'wide-angle'.

With a standard 35mm still camera, a lens of F=50mm is classed as normal and has much the same sense of perspective as the human eye. Lenses of 35mm or 28mm focal length are wide-angle, and those of 100mm or more are long- or so-called 'telephoto' lenses.

Zoom lenses are special multiple-element, variable focal-length lenses which are able to change the F value over a wide range. Their F is expressed as a ratio (10:1), or by the two limiting focal lengths (12—120mm).

Since the focal length of a camera is related to the size of the film/video image, an F=25mm lens, which is normal in a 16mm movie camera, is wide-angle on a 35mm still camera (and it will vignette the edges of the image, since it wasn't designed for this size of image) and telephoto on an 8mm camera.

Conversely, a 50mm normal lens or a 100mm portrait lens-from a 35mm camera, will become moderately 'telescopic' long-lenses when mounted on a 16mm or video camera.

A 16mm film camera with a 12—120mm zoom, covers the normal filming range (a 10:1 ratio lens) from wide-angle to good telephoto. Most professional video cameras require much the same lens focal lengths.

Video cameras vary in their lens focal length requirements, depending on the tube size or the physical size of the CCDs used.

F.69

The international standard for telex addressing. See X.121 (for X.25) and E.164—E.168 (ISDN) also.

F-connector

An F-connector is a well known coaxial cable and satellite antenna connector. It is often called an F-type connector.

F Region

The layer of the ionosphere (subdivided into F1 and F2) between 200 and 300km above the Earth. It reflects most radio waves that pass through the lower layers, but the plasma is very thin above 300km. Unfortunately, the F Region is subject to frequent changes in composition which is the cause of most of the variability in short wave communications. See D region, E region and Sporadic E.

f-stop

In lenses this is always written using the small f (capital F is the focal length.)

The f-stop is the theoretical calculation of aperture in a lens which relates light-acceptance to the size of the image being produced. This makes it the primary control of exposure, but it doesn't take into account any light-absorption from the glass or any lens surface losses (see t-stop).

The calculation of the f-number takes into account two factors: the size of the image, and the size of the aperture:

— The size of image (width x height) is related to the square of the distance from the center of the lens's theoretical optical back-plane to the film emulsion surface.
— The area of the lens surface which admits the light, is related to a square of the aperture (diameter).

So the calculation should be:

$$\text{f-stop} = \text{Focal-length}^2 / \text{aperture diameter}^2$$

But since both are squared, and the formula is reduced to:

$$\mathbf{f = F/d}$$

A lens of 50mm focal length, with its widest aperture being 25mm in diameter, will therefore have a maximum theoretical f-stop of f-2. When set at f-2, this lens should match the exposure of any other lens set at f-2, using the same film. A lens having a focal length of 100mm (a telephoto), with its iris diaphragm open to 50mm, will also be set at f-2.

The f-4 mark on each of these two lenses will correspond (respectively) to an iris diaphragm closed down to 12.5mm (on the 50mm lens) and 25mm (on the 100mm lens).

The theoretical f-scale begins at f-1 (which is a very low-light lens indeed) and it progresses by 1-stop amounts — each of which halves the light. Since f-1 represents 2^2 = 4-times (or two-stops) the image brightness of f-2, it was felt necessary to add the half-way division based on the square-root of two (1.414). So with a change in aperture from f-1 to f-1.4, the light entering the lens will halve, then from f-1.4 to f-2, it will halve again. From f-2 to f-2.8 it will halve once more. And these changes effect the depth of field as well as exposure.

The standard range of f-stops is from:

Large aperture: least depth-of-field to →

f1 – f1.4 – f2 – f2.8 – f4 – f5.6 – f8 – f11 – f16 – f22 – f32

→ *small aperture: most depth-of-field.*

Between each of these major full-stop marks, the standard sub-division is into thirds — so f-3.5 represents a one-third increase in aperture over f-4, and two-thirds of a stop below f-2.8.

As you move down the scale from high to low numbers, one full stop at a time, you double the brightness of the image each time (this is cumulative 2x, 4x, 8x, etc.). Moving up the scale halves the brightness in each step (50%, 25%, 12.5%, etc.).

• The f-stop also greatly affects the depth of field of a lens. The wider the aperture, the more limited the depth of field. Lenses set to f-22 and f-32 are almost pin-hole cameras, and so have a virtually unlimited depth of field (from a few centimeters to infinity). The focal length of the lens also influences depth of field, as does the distance the camera is away from the object on which the principal focus has been set.

F-type connector
The simple connector widely used on TV antenna ports with coaxial feed cable.

FABM
Fiber Amplifier Booster Modules. These are probably optical amplifier units used in submarine cables — although the term is wide enough to refer to the older signal regenerators. See EDFA and erbium amplifiers also.

fabric
As in 'switching fabric' (see). The distinction is being made between patch-panel-like cross-connects, and one-to-one or one-to-many switches.

Fabric is a vague term which signifies that there are many interconnects (multiple cross-connects) in a signal pathway, so there are many inputs and many possible outputs. It also implies that more than one signal may be passing across the 'fabric' at any one time.

The metaphor is of the warp and weft of woven fabric with many lines going in many directions. This is an extremely general term.

• This term is applied to the smaller LAN 'shared backbone', 'shared-memory' and 'delta' type switches (e.g. banyan) where the switch comprises numerous switching units, and many buffers, each quite distinct in its actions. Many people would dispute using the term 'fabric' for a shared backbone switch since only one signal can be passing over these switches at a time. See banyan, delta and cross-connect.

• The term can apply to any cross-connect — switching units that will switch any-port to any-port, or those which will only switch inputs on Side A to outputs on Side B. Some of these are not intended to switch at high-speed packet rates, but rather to make longer-term circuit changes.

• It is also applied to the total national telecommunications network, where the 'switching nodes' are the major regional telephone exchanges, and these make multiple interconnections.

fabrication
This usually refers to the manufacture of customized chips in a chip fabrication plant — which is also called a foundry.

facsimile
A faithful or exact copy. Today it is used for the technique by which document images are scanned, transmitted electronically, and reproduced.

• It applies to the current techniques and ITU standards of fax transmission. Group 3 is widely used around the world today, and fully-digital Group 4 will become popular if ISDN ever catches on.

Group 3 is the current world standard, however, and it is constantly being modified. Originally it was specified for 2.4 and 4.8kb/s, then later 9.6kb/s became the recognized standard, and now 14.4kb/s has been added. Fax machines work out between them which rate to use, during their initial handshaking period. See fax techniques and Group # facsimile.

• The term is also used for the paper copy produced by the machine. Most old fax machines use thermal paper which only retains its image for about 12 months (depending on the ambient temperature). Later machines use copy paper. One of the values of a fax card in a PC is that the images can be captured electronically, and then converted to a laser print-out if required.

• Some people also talk about WYSIWYG screens being a 'facsimile' of the final printed report — which is putting the cart somewhat before the horse! But you can understand what they mean. See fax card, fax modem, and Group # facsimile also.

facsimile telegraphy
Facsimile used by wire-services and newspaper groups to transmit photographic images which involve continuous tone.

factorial
The product of all integers, from 1 up to, and including the selected number. The factorial of 6 is the product of 1 x 2 x 3 x 4 x 5 x 6 = 720. It is written as 6! (with the exclamation mark).

factory defaults
These are settings which have been established in the device before it is shipped from the factory. Usually these defaults will be parameters stored in ROM or some other very stable form of memory so that they can always be recovered and reused.

You shouldn't ever be able to destroy factory defaults; when you make customized changes, the factory defaults should always continue to exist in the background, and there should be a mechanism for recalling them if necessary.

fade

• In telecommunications, a fade is a slow and long-lasting reduction in signal strength.

• In film and video, a fade can be a fade-out or a fade-in; it can be a fade-to-black, or fade-to-white (but usually to black). A fade signifies that a major 'chapter' has ended in the story, and quite possibly that time has passed. Don't confuse fades with dissolves which are two overlapping fades — a fade-in superimposed over a fade-out.

• Fade margin is the amount of signal absorption that is tolerable on a satellite or radio link for a small percentage of the time. Its a measure of the system's reliability in maintaining a certain S/N ratio on the link. This is one component of the link budget.

fallback

In facsimile and modem data transmissions, a high transmission rate (and associated protocol) is often initially chosen by the two communicating units in the expectation that the connection will remain constant and clear. But if line noise is experienced, the modems will detect an increase in errors, and they may then exchange commands which cause them to fallback to a lower data-rate and possibly use a different protocol.

Group 3 facsimile has a number of fallback standards and most of the new high speed modem standards specify fallback rates. MNP-10 and V.42 include special fallback protocols to handle these negotiations.

FAQ

• **Frequently Asked Questions:** Most newsgroups on the Internet will have a list of FAQs. It is considered polite for newcomers to check these first before asking naive (to regulars) questions.

• **Fair-Average Quality:** Commodity prices are usually set on the basis of FAQ.

far-end echo

Echo being returned from the far end of a long-haul connection. In telephony, this is usually from an impedance mismatch of the hybrids in the remote local exchange, or in the remote handset. See echo cancellation and near-end echo.

far-end frequency offset

The variation in the phase of the far-end echo in respect to the originating modem's output signal. It is a problem with high-speed modems.

far-field

In the investigation of potential biological effects from non-ionizing radiation (radio transmitters, VDUs, etc.) a distinction must be made between the near-field and the far-field effects.

— **Near-field** is at distances of less than one wavelength from the antenna, and here the magnetic and electrical components of the radiation are out of phase. This is generally held to be potentially more dangerous than in-phase radiation, and each component must be viewed as a separate entity having its own effects.

— **Far-field** is beyond one wavelength, and the two components are in-phase.

farad

This is the SI unit of capacitance. It represents one volt increase in potential when charged by one coulomb.

It is a large unit, so microfarad (10^{-6}) and picofarad (10^{-12}) are more commonly used.

• Standard off-the-shelf capacitors are generally in the range from 1 picofarad (10^{-12}) to 250,000 microfarads (0.25 farad). Capacitors in the 3–4 farad range are sometimes used to provide back-up power to CMOS memory for short periods.

farm

Mainframe managers speak of tape 'silos' (selectable banks of demountable tapes) and disk 'farms'. These farms are large areas of floor space, housing very many Winchester multiple-platter disk drive stacks. These days, they are generally DASD 'farms', and more recently the term has been applied to very large CD-ROM off-line storage devices.

FAST

• **Forecasting and Assessment in Science and Technology:** A research program funded by the European Union.

• **Frequency Agile Sharing Technology:** An intelligent base-station approach developed by APC for use with AMPS, CDMA and CT-2 PCS. It assigns frequencies to the basestations after first analyzing the spectrum available, checking for local interference.

Fast ATA

Confusingly there is now Fast ATA, ATA-2 and Fast ATA-2. These are various names for two related disk drive interface standards for PCs. Both Fast ATA (up to 11.1MByte/s) and Fast ATA-2 (up to 16.6MB/s) are now available. At least 25 major suppliers of hardware components have pledged support for the standards. The idea is to match the disk drive speed to the new performance capabilities of the current high-speed microprocessors, VESA, and the PCI bus. See IDE and ATA.

fast circuit switching

Special facilities available from some telephone carriers. It allows a call to be quickly connected and maintained (economically) for only a short period of time. This is done in many ways; the most popular is to have a single access number stored in the exchange, and this number dials instantly the phone is taken off-hook.

It is provided for EFTPOS-like transaction services where traffic on the network may only last seconds. This process still uses the standard CCS#7 public-switched network, so it is often little more than a special billing plan with very low up-front charges and timed calls.

Fast Ethernet

This applied originally to two new versions of Ethernet able to operate at 100Mb/s. Both were designed to operate over UTP cabling (Category 3 and 5, also STP and fiber).

The two technologies were:

• **100Base-T:** This is a shared-access technology which integrates with the existing 10Base-T. This technology is supported by the Fast Ethernet Alliance (most vendors) and has recently become an IEEE 802.3 draft standard. The main aim of 100Base-T has been to retain compatibility with the old Ethernet NICs. In more recent times this standard has assumed the 'Fast Ethernet' mantle alone, with VG now being called 'AnyLAN'. 100Base-X and 100Base-4T+ are signaling subsets of T.

While this is designed for 100Mb/s, like any CSMA system, at more than 50% (50Mb/s), the network runs into problems.

• **100VG-AnyLAN:** This is a superset of both Ethernet and Token Ring which is being promoted by the VG-AnyLAN Forum (mainly IBM and HP). It will be an IEEE 802.12 standard in 1995-6 and it will probably cost more than 100Base-T, but it should have many advantages. The main aim of VG-AnyLAN is to provide a fast system for multimedia. It uses a round-robin access method called DPAM.

• See Switched Ethernet also.

Fast Fourier Transform

A program/technique invented by Cooley and Tukey in 1965 which decreased the amount of computation needed to analyze a complex spectral waveform using Discrete Fourier Transform (DFT) analysis. Discrete Cosine Transform (DCT) is very similar.

FFT divides the waveform into a large number of equally spaced samples, and analyzes each sample separately. Special FFT chips handle the conversion of computerized data in digital form, and provide an efficient way of extracting the Fourier transforms from the signal elements.

FFT represented a major breakthrough which was applied in a number of computer/communications areas, including satellite reception, radar, telephones, HDTV, stereos and TV sets — wherever 'fluctuating phenomena' are found. The main application is now in the compression and decompression of digital images, although the DCT variation is now dominant here. See Fourier transform, DFT, DCT and DSP.

fast packet-switching

This is a general term applied to cell-relay and similar systems (frame-relay) which tend to have so little delay across the network that voice and video (isochronous services) can be sent along with data. The aim of FPS is to read the header information as the cell/packet arrives at the switch, and switch it without buffering it for any length of time. The cell may, however, be sent around a delay loop if it is likely to collide with another cell already being handled.

FPS covers a range of technological improvements, usually (but not exclusively) associated with optical fiber and lately with ATM and frame relay.

The distinction between 'fast' and 'normal' packet-switching is that the second is a store-and-forward system, while fast packets are 'cut-through' systems switched on the fly. This means that the switching must take place within hardware (special ASIC or gate-array chips) — and this generally needs fixed-length packets or cells. Routers, and similar packet-switching devices, tend to hold the information in memory, and then switch under software control.

See frame relay, ATM, and DQDB also.

fast select

This is a connectionless (datagram) packet-switching technique that allows credit-card/EFTPOS type transaction data to be transmitted very efficiently over an X.25 network. It provides a single, short, call set-up packet which also carries the data to the destination. The recipient responds in the same way. See X.25. This is not a true connectionless system, but rather a variation on the call set-up packet.

fast switching

This is a vague router function where special cache is provided to expedite the movement of the packets through the router.

FAT

• **File Allocation Table**. See.

• **FAT File System:** The term FAT File System was retrospectively applied by Microsoft to the old MS-DOS file system (and early OS/2), after the new HPFS was announced. The FAT File System is managed by two different structures: directories and file allocation tables.

• **Fat Bits:** In Mac Draw and Paint programs, are highly magnified areas of the screen which allow the user to modify the picture, one pixel at a time.

father

The middle copy of three generations of file normally kept for security — son, father and grandfather. The son is the copy currently in use, and it, in turn, will become the father, and then the grandfather. This cyclic method of keeping backups ensures that disks wear evenly, and that at least one of the two copies are always available — provided they are stored in different locations when the office burns down.

fault-tolerant

The ability to function correctly even with failed components, which usually means that parallel systems exist. For instance, a highly critical LAN-server disk drive may have an automatically-mirrored backup drive which will cut-in automatically if the first fails. RAID can also provide another form of fault-tolerance. Bridges gain some fault-tolerance from using the Spanning-tree algorithms.

• The term is applied to major financial and communications computer installations where redundancy is built in. Major computing facilities used for transactions often have duplicate sets of components (zones) housed in separate cabinets at separate locations, and linked by high-speed data cables or microwaves. They will have UPS, and probably dual-CPUs running synchronously.

• The term also applies to mirroring software which manages transactions, provides audit trails, etc.

• In LANs the main fault-tolerant techniques are RAID, shadow disk, mirror disk and remote duplicate storage

faulty sector

A common error message which tells you that a small part of the disk magnetic surface is defective. Some operating systems, and some utilities, will then mark this sector in such a way that it is never used — others will reject the disk. The Mac will often reject floppy disks that an MS-DOS will happily use. A question worth considering is which is the better approach?

fax card

Fax cards are not simply replacements for stand-alone fax machines. They can provide some sophisticated transmission services (personalized broadcast messages using mail-merge techniques) often not available on even the most sophisticated stand-alone machines.

However, common sense says that they are not an ideal 'first' machine for a small office, but they are excellent 'second' resources. These days, fax functions are often combined with data-modem functions in the same card so that they can handle e-mail and file transfers also.

Before you purchase, make sure the card can handle 9.6kb/s data rates at least for both fax and modem functions (a common trap), and check that the software is adequate. The Group 3 fax standard has now risen to a peak data rate of 14.4kb/s, and so a card which only supports a fax or data rate of 2.4kb/s is clearly inadequate, unless only used occasionally for local calls. See facsimile, fax techniques and fax modem also.

fax modem

In general terms this includes fax cards and stand-alone fax modems, but more often it means the stand-alone device only. It can also mean the modem chip which performs the fax function.

• Fax modems are external versions of a fax card which often doubles as a normal modem. They will link to the PC via a cable, usually running RS-232.

Many of these devices still have a maximum fax rate of 4.8kb/s, as against a modern Group 3 fax with 9.6kb/s, and this can be costly in terms of international phone charges.

Special software is needed to control the fax functions in a fax modem, and so some of these devices can cost in total as much as a stand-alone fax machine. However they do have added functions through their computer link — and, unlike a fax card, you can be reasonably sure that the fax modem will be useable with the next PC you own. See fax techniques and fax card.

• The term is also applied to the 'pump' chip which provides the central modem function in any piece of fax equipment: this is sometimes called a 'fax pump'. It should satisfy the ITU recommendations for V.29, V.27, V.21 (Channel 2 only), T.3, T.4 and the binary signaling requirements of T.30 for Group 3 fax machines.

fax router

A device used mainly within a company network. It can decode the normal fax tones, and extract the original digital data. This can then be sent as an 8-bit ASCII file over the standard corporate LAN/WAN internationally to another fax router where it is remodulated into fax signals and dials forward to the final fax destination.

Transmission is therefore over the national or international links at the original 9.6kb/s data rate (and on a private line) rather than in an expensive dial-up analog connection.

This has advantages in private and public networks. Normally companies don't fax across their own private voice lines because voice compression is used, and these channels don't have the bandwidth needed for standard fax.

fax techniques

The current highest-rate standard for Group 3 fax is now V.17. However with recent upgrades the image transmission can use any of the V.17, V.29, or V.27ter standards, which are all half-duplex-only modulation schemes.

Data-rate and other factors such as error checking, correction and page confirmation are negotiated during the handshaking phase before transmission begins. This is done at 300 baud (also 300b/s in this case) using V.21 modulation in the high channel. Other information (such as machine capabilities, page-width, etc.) is also transferred at this time using a protocol defined in T.30.

The ITU standard for fax machine transmissions are 2.4kb/s (for really bad lines), 4.8, 9.6, 12.0, and 14.4kb/s. But typically a fax machine which is required to print out hard copy will use 9.6kb/s. The latest fax modems more often use 12.0 and 14.4kb/s rates and increasingly the more expensive fax machines also support this V.17 rate.

• Some fax systems support error detection and correction, and some just leave a line blank when an error is discovered.

FBR

Fixed Bit Rate which is better termed Constant Bit Rate (CBR). The distinction is with Variable Bit Rate (VBR).

FC-AL

Fiber Channel Arbitrated Loop. A new serial data-storage interface standard which can handle transfers at up to 100Mb/s. It is designed for high-end, high-throughput applications and is used with mainframes and super computers as an Ultra-SCSI replacement.

FC-AL can handle up to 126 drives on a single interface. It is designed to support SCSI-2 fast/wide and to provide simple plug-and-play attachment. It is related to the Fiber Channel networking system. See Fiber Channel.

FCB

File Control Block. See.

FCC

Federal Communications Commission (USA). The agency which regulates and monitors US domestic use of the electromagnetic spectrum for communications and telecommunications. It also regulates wireline telecommunications, and cable systems. (Only to a very limited degree.)

It has a board of seven commissioners appointed by the President. Since 1934, the FCC has been the key authority which regulates all interstate and international telecommunications at the technical level — but it is not the only authority to regulate the phone service and the trading activities of the carriers. Most states have their own Public Utilities Commission, and the Department of Justice has had a direct and obvious influence since Judge H. Greene broke up AT&T in 1984.

• In 1965, the FCC set rules which limited the number of distant television signals a cable network could carry, and mandated some local signals as 'must-carry'. In 1985 the US Court of Appeal declared these rules unconstitutional, and new rules were adopted in 1987. These were later struck down again by the US Court of Appeal.

• Their 1972 rules established access channels and re-established some of the must-carry rules. See PEG channels, must-carry and A/B switch.

FCI

Forward Calling Information. A telecommunications industry standard developed for answering services. The Simple Message Desk Interface (SMDI) uses a four-wire analog circuit which allows it to carry information about the incoming trunk, called number, condition-for-forward, etc. from the switch to a voice mail unit in the Simple Message Desk Format (SMDF).

The FCI system can also carry commands which are to turn the extension's Message Waiting Indicator (MWI) on or off. See SMDI/VMI.

FCIF

Full CIF. See CIF.

FCS

Frame Check Sequence. The HDLC term for the redundant bytes used for error control. The term is also used for the CRC on Ethernet. See cyclical redundancy check.

FCSI

Fiber Channel Systems Initiative. IBM, HP and Sun's high-speed workstation interconnection standards. They are defining an FCSI standard for mass storage capacity on the desktop, and another for faster bridging and routing. See channel (processor I/O), Fiber Channel, and FC-AL.

FD

Frequency Division. All radio broadcast and radio telephony systems use frequency division as the primary means of subdividing the spectrum — so FD is often unspoken and assumed.

For instance, the GSM digital cellular phone system is said to be TDMA (time-division), but the full spectrum allocation is first divided into different frequency bands, and then TDMA techniques are used within each band to increase the carrying capacity — so strictly, it is an FD/TDMA/FDD system.

Similarly CDMA (code division) techniques rely first on being allocated 1.25 or 2.5MHz of radio bandwidth within a certain location, so this primary provision is 'frequency division' (or better 'frequency allocation').

Don't confuse FD with FDMA (frequency division multiple access). FDMA is an access technology (how disparate callers get to use a common set of channels) which uses FDM multiplexing. See FDD, FDM and FDMA also.

FDD

Frequency Division Duplex. Most cellular mobile telephone systems rely on FDD; using one frequency band to transmit and another to receive — these are called the 'forward link' (base to mobile) and the 'backward' or 'reverse link' (mobile to base). The difference between two related channels used in a single conversation, is the 'duplex offset' or duplex separation.

By contrast some pocket-portables (CT-2), wireless PBXs (DECT) and cordless telephone systems now use TDD (time-division duplex) which is a 'ping-pong' system having alternating send and receive time slots within a single frequency.

The term FDD is not widely used, simply because it is assumed with cellular phone systems.

FDDI

Fiber Distributed Data Interface. FDDI is a type of token ring LAN (a logical ring) which was designed for high-speed over dual counter-rotating fiber rings. There is a copper version also called CDDI.

FDDI has been under development since Sperry originated it in 1983, and it is now fully compliant with the OSI seven layer model. There are two major variations of the protocol, although the later FDDI-II has not proved to be a success.

This is a packet-based 100Mb/s backbone network protocol where the packet size can be chosen by the transmitting station to suit the application. The maximum frame payload is 4096-Bytes, and the variable packet size makes it easy to attach other LAN types. This allows FDDI to serve as a connectionless datagram service which is widely used for campus backbones.

It is deterministic, with multiple-tokens for high throughput — more than one packet of information or token can be on the ring at the same time. FDDI can use dual counter-rotating optical fiber rings for reliability, in which case the primary path is for the data, and the counter-rotating path provides a backup, and caters for serious cable breaks. The main advantage FDDI gets from the dual optical rings is fault tolerance, although fiber is also free from electrical interference.

FDDI has a distance limitation in the order of 2km between stations, with 500 dual-attached (Class A) stations, or 1000 single-attached (Class B) stations, before segmentation. (See below for Class A and B.) The total ring can be as large as 100km in length.

— FDDI-I is specified for data networking only. It is an ANSI (X3T9.5) and IEEE (802.5) standard. Token-passing rings are defined generically by IEEE 802.5, however FDDI deviates from this in several key application areas. While FDDI-I has been popular for high-speed LAN backbones, the costs for adaptors ($2–3000) made this an expensive desktop option until recently.

— FDDI-II is isochronous (it can carry voice and video) and since it offers subdivided capacity it can be used to support synchronous traffic. It is backwardly compatible with FDDI-I: it has the same 100Mb/s data-rate, but it carries non-packetized voice and slow-speed video through the implementation of a priority scheme.

The telephone companies once saw FDDI-II as a technology for future MANs, however DQDB (SMDS) has since displaced it. It can handle circuit-switched data synchronously in modules of 6Mb/s.

• Note that FDDI actually has a line code rate of 125Mb/s because one control bit is added for each 4-bits of throughput.

• FDDI extensions also now include:
- FDDI-II and Hybrid Ring Control (HRC) which integrates voice, data and video.
- Single Mode Fiber (SMF) extending FDDI into the MAN.
- Sonet/SDH Physical Layer Mapping, extending FDDI into the MAN.
- Shielded Twisted Pair (STP)using FDDI to the desktop.

See CDDI and TP-DDI also.

• The sub-sets of the FDDI standard which require compliance are:

PHY FDDI's physical layer.
MAC media access control.
PMD physical medium dependant facilities.
SMT station management (fault location etc.).

• There are two classes of device on an FDDI network:
- Class A devices are hubs, concentrators, servers and any other equipment with dual attachments to both the rings.
- Class B devices have a single attachment. These are workstations, printers, etc. They connect to only one of the rings through a wiring concentrator and a single duplex cable.

FDDI has a high degree of fault tolerance because the management interface can automatically reconfigure the ring after a break, within a few milliseconds. With a B-type device the concentrator activates an optical bypass.

FDHD

Floppy Disk, High Density. The term Apple applies to its very versatile 1.4MB floppy disk drive standard. The SuperDrive version of FDHD can read IBM PC (MS-DOS) and Apple IIGS disks also. See MFM also.

FDM

Frequency-Division Multiplexing. Multiple use of a media, depending on the sub-division of the available spectrum by frequency alone.

FDM is a multiplexing technique used in both analog and digital networks. Many narrowband signals can share the same wideband bearer using wire (mainly coaxial) or radio media. (FDM in optical systems is called WDM.) This is done by just allocating to each channel a different band of carrier frequencies separated by guard-bands. The baseband channels are then modulated onto each specific carrier frequency using FM techniques.

All radio systems make use of frequency-division (FD) as the primary way to sub-divide the spectrum — and in radio, different frequency bands are used for different purposes. However the term 'FDM' is generally only applied when multiple signals are to be carried over a narrow range of spectrum (say 100 channels of analog or digital data, each on a carrier with 30kHz of carrier separation within a 3MHz radio band) without other time-division or code-division techniques being applied.

The progressive stages required for a major FDM communications networks are:
- Transfer a group (12) of baseband channels individually to the primary multiplexer.
- Modulating each carrier onto a specific (and different) carrier frequency within the range 40—108kHz (a supergroup carrier).
- Transfer 5 of these 'supergroups' to a secondary multiplexer, and
- modulate each of these onto a higher carrier (and so on, up to five stages).
- Transmit the final output over the media (coax or the ether) using radio frequency technology.
- Use tuners (bandpass filters) in the demultiplexer chain to progressively isolate signals and carriers in a hierarchical fashion.
- Demodulate each of the final carriers to extract the original baseband signals.

• FDM handles analog and digital signals equally well. New multilevel QAM techniques will allow, say, 6-bits to each Hz of bandwidth.

• To prevent one carrier from interfering with an adjacent carrier (co-channel interference) a guard-band of unused frequencies must be left between. This is where most of the capacity inefficiencies arise. See FDMA, TDM and TDMA, and CDMA.

• Entertainment radio and television signals are transmitted in the spectrum and separated by frequency division. In this case, the 'ether' replaces the physical carrier, and a single tuner in each receiver just extracts one signal, and ignores the others. Otherwise it is the same process as FDM.

FDMA

Frequency Division Multiplex Access.

• **Mobile telephones:** In wireless telephony, the base-station is transmitting to numerous mobiles, while a number of quite independent and geographically separated mobiles are transmitting back to a single base-station.

Since the base is broadcasting from one-to-many while the mobiles are many-to-one, the problem of accessing the forward channel (base to mobile) is different from that of accessing the return channel (mobile to base). Channels need to be allocated by the base station.

The base station knows which channels are available for its own transmission, but the mobile must wait to be allocated frequencies. It must wait to be transferred to new frequencies when it moves into a new cell-coverage area. This is the 'access' problem, and in FDMA it is simply solved by giving each user exclusive use of one channel (one or two frequencies)

There are two fairly common ways of providing two-way channels in frequency division systems:

- **FDD:** Each conversation can be allocated exclusive use of two associated channels with a duplex spacing (see). This FDMA technique is also called FDMA/FDD. The 'access' problem is then one of allowing the mobile to issue requests for channels without interfering with the operation of other mobiles — and this is usually done over a separate control channel.
- **TDD:** Alternately a single allocated channel can be sub-divided by time (suited to digital transmission only) so that the channel rapidly alternates between being a forward channel and a reverse channel. This is FDMA/TDD which is known as 'ping-pong' or time division duplex. In reality, this is just a very rapid variation on the old 'press-to-talk' systems which were called 'half-duplex'.

• **Broadband LANs:** These have now almost disappeared but they used FDMA and FDM techniques to separate out the various signals. These were coaxial cable systems where many point-to-point links were established across a common cable using different radio frequencies.

• **Cable TV:** These networks (analog and digital) are FDM systems on the downlink, and FDMA on the uplink if they have any interactivity.

• **VSATs:** Single channel per carrier (SCPC) is a particular case of FDMA techniques. The bandwidth of the transponder is sub-divided into narrower bands, and each transmitting station can then beam its own channel at the satellite. Each will slot into an exclusive band within the transponder's frequency range (its bandwidth). See DAMA.

• **Fiber Optics:** Wavelength Division Multiplexing (WDM) which is just beginning to be used in optical fiber systems is a form of FDM, since the different wavelengths (colors) of the multiple lasers are used as the primary separation. With passive splitters this can be an FDMA system.

FDX

• **Full-duplex** (as distinct from half-duplex or simplex): The link is capable of carrying signals both ways simultaneously (and at roughly equal data-rates). Sometimes they are referring to the circuits, and sometimes to the functions provided by two circuits.

• Over time, FDX came to mean the use of a four-wire (or fiber-pair) circuit in telecommunications. Sometimes you'll see this translated as **Four-wire DupleX**, of **Fiber-pair DupleX**, but this is very confusing.

The point is, that in long-haul copper circuits, if the signals are to be amplified, the carriers need to use a matching pair of half-duplex circuits (one in each direction) since two-way amplification is very difficult in electronic circuits. So the parallel pair of half-duplex circuits together make a full-duplex communications link in telecommunications terms.

The junction between the true FDX circuit over the local loop, and the pair of HDX circuits for the long-haul, is in the hybrid at the exchange; and it is this hybrid that creates the echo problems. See half-duplex and echo-suppressors also.

FE

Format effector. See.

FEA

Fast Ethernet Alliance. The group of companies developing 100Base-T.

feathering

The vertical 'justification' of columns of text so that they reach their 'full-measure' (take up all the space allocated). This adjustment of line spacing is necessary only because without it, columns would often have problems with 'widows', (part-lines left over at the top of a column). If a widow is likely, a second line must be carried over to the new column, and the extra space will then need to be made-up in the first.

FEC

Forward Error Correction. FEC is always based on redundancy techniques (additional bits added to each frame or packet) — so an overhead is always imposed on the communications channel. The distinction must be made between:

- FEC, where the data already carries the information needed, both to check for errors and to make necessary corrections (to varying degrees).
- ARQ (Automatic Request), where systems rely on the errors being detected. The faulty blocks are corrected only through the cooperation of the sender in retransmitting the blocks.

ARQ can handle any degree of problem but needs the sender to cooperate, while FEC allows the receiving terminal to correct a finite number of errors only without retransmission.

FEC is always used when immediate interactive communications (a back channel) is absent. For instance, you can't easily ask a CD-ROM to retransmit corrupted data and nor, in practice, can you ask an exploratory satellite probe way past Jupiter — unless you are prepared to wait a few days for the reply.

FEC systems use any of the variety of detection/correction techniques such as convolution codes, Trellis codes, Viterbi codes, Reed-Solomon (RS), and Hamming bits within each block of information. If more than one channel is being transmitted in a shared bandwidth, interleaving is also used to improve the effectiveness of FEC by distributing the result of any burst errors between the channels — it is always better to have many small errors than one large block.

FEC systems impose a high overhead cost in terms of the number of redundant bits — in some cases actually doubling the number transmitted. However, this overhead can usually be justified.

The indicators for the use of FEC are that:

- Numerous transmission errors are to be expected and so a lower total overhead is achievable by automatically including error checking with the data.
- Retransmission is not possible.
- The delays inherent in retransmission would make post-correction untenable.

FECN

Forward Explicit Congestion Notification. (pronounced Fek-'n) In frame relay this is a bit which will be set in a frame as it passes through the network, to indicate that there is congestion in the network. It is transmitted in the header of a frame traveling from sender to receiver (on the forward channel).

It serves a similar purpose to BECN, which travels in the opposite direction. FECN warns the receiver, and BECN warns the sender. Terminals react to these signals in different ways, depending on the DTE. See CIR and EIR also.

FED

Field-Emission Display. This is a cold-cathode screen display which operates on the same basic principle as a CRT, but uses an array of low-voltage electron emitters to excite the screen phosphors. These screens are believed to have potential for lap-top portables since they produce brighter contrast screens while consuming less power. See active-matrix screens also.

FED/FIPS

Federal Government Standards; Federal Information Processing Standards. The US government's sub-sets of GOSIP (Government OSI Procurement) standards.

feed(er)

• **Feed horn or feed antenna:** In a satellite and microwave dish system this is the unit which gathers signals from the reflector and channels them to the detector and low-noise block amplifier. In transmission and reception, the feed-horn is said to 'illuminate' the reflector (even when it receives). See down-converter and LNB.

• **Feed line:** The 'feeder' is the coaxial cable which runs from the outside dish unit, indoors.

• In coaxial cable systems, the 'trunk' is a heavy coaxial cable (0.75 or 1-inch) which has amplifiers at regular intervals. Distribution or **feeder cables** (say 0.5-inch) split off from the trunk and run down side streets. These then have drop cables (0.3-in. or 7mm) which carry the signals into homes.

• A feeder is a set of wires (a cable) that supply signals to a number of peripheral units. Bus is sometimes a synonym.

• A feeder is the main trunk cable between telephone exchanges before it is broken down into individual lines through a multiplexer.

• In telephony, the feeder is the main arm-thick multiple twisted-pair cable in the local loop. It runs down streets, before breaking out into individual components at a street pillar.

feedback

Feedback is the return of some output to the input stage, usually as a control, or as a technique of amplification. Feedback is fundamental in control theory — from the first Watt steam engine to modern electronics.

• Control theory emphasizes the use of negative feedback loops. Watt's steam engine was only practicable because of the invention of the fly-ball governor (a negative feedback mechanism) which stopped the machine racing out of control, and most analog electronic devices are controlled in a similar way. Negative feedback is the transfer of energy from the output back to its input in such a way as to stabilize the conditions.

• Regenerative circuits use positive feedback as a way of increasing the effectiveness of the amplification — in effect, they are running the output power through the amplifier a couple of times.

• The term is also used in a non-technical sense, for information flows returning to the originator — usually comments and criticisms. The term is often used in CAI design.

• In programming, a looped For-Next, or Repeat structure is a feedback circuit.

• In studios and auditoriums, feedback is an annoying return of signal from the loudspeakers back to the microphone. If delayed, it sounds like reverberation or echo. It can cause the system to oscillate and 'squeal' but special delay circuits can reduce this effect.

female connector

These have holes, not pins. Always look at the pins, not at the outside shielding.

femto-

In the SI units, the prefix femto- represent $10^{-15.}$ The symbol is the small f.

• At the speed of light, a photon will only travel 0.0003 mm in a femtosecond, yet lasers that blink at 30 femtosecond rates have already been developed. However they are not yet able to provide continuous modulated bursts.

FEP

Front End Processor. A communications processor/computer (or these days, perhaps a plug in adapter card with processing power) that sits between the cluster controllers and the main host computer to concentrate the signals and handle the network I/O.

These are generally mainframe devices — typically an IBM 3745 which does nothing more than handle communications functions. In SNA, very often it is the front-end processors' bandwidth constraints which blocked network managers from introducing some of the more modern technologies. Routers are pushing FEPs into extinction.

ferreed array

The switching element used in a modern crossbar switch.

ferrite

An important compound of iron oxide and other oxides which has a high magnetic flux. The use of ferrite has been central to the development of many electronic-magnetic systems over the last 40 years.

ferro-electric RAM

A new type of RAM memory technology being developed by Ramtron, NMB, ITT and a number of other chip vendors. Initially the aim is to compete with EEPROMs and later with SRAMs and DRAMs.

FRAM retains the information within a memory cell after the power has been turned off. It does this by creating a polarized magnetic charge in a layer which is superimposed onto a basic 4-transistor SRAM memory cell. The same technique can also be applied to the 1-transistor DRAM architecture.

The cells retain their information by using polarized dielectric materials containing lead, zirconium and titanium (PZT) between capacitance elements. When power is cut to the RAM circuits, the FRAM layers are pulsed to create the polarized magnetic charge.

When power is switched on again, the memory is refreshed automatically, and so the memory chips hold their data without batteries. Current SRAM/FRAMs have a lifetime limit of about 10^{14} cycles, and an access time of about 25 nanoseconds, although the access speed is dropping quickly with later developments. With DRAM/FRAM, 4 megabit chips are on the drawing board. See flash memory also.

- The advantage of FRAM over flash memory, is that FRAM can be used within the memory map for conventional computer memory, and can freeze at any point — then resume later. So it can be primary memory. Flash is usually secondary.

FET

Field-Effect Transistor. A low-noise transistor used in amplifiers (LNAs) and other satellite and microwave equipment in the primary amplification stages. FETs are cheap and reliable and they have a very high packing density.

A FET has two islands of n-type material inside a p-type region. One island is the source, and the other is the drain. The third contact is an overlay of metal called the gate. See HEMT and MODFET.

FEXT

Far-End CrossTalk. Crosstalk caused by device interference at the remote end of the cable. From the receiver's viewpoint this is less of a problem than NEXT because the interference is countered by the power of the transmitter.

FF

ASCII No.5 — Form Feed (Decimal 12, Hex $0C, Control-L). The form feed signal was designed originally to advance the paper in a printer to the next page. On many video monitors this signal will clear the screen and put the cursor in the home position (the top left-hand corner). FF is a format effector.

FFS

FAT File System. See.

FFSK

Fast Frequency Shift Keying. Used in radio telephony and trunked radio systems. It is a variation on CP-FSK.

FFT

- Fast Fourier Transform. See.
- Fully Formatted Text. See DCA.

FHMA

Frequency Hopping Multiple Access. A form of spread spectrum. See spread spectrum and CDMA also.

fiber

In telecommunications, fiber usually refers to glass fiber 'waveguide', however plastic fiber is now available. There are three main forms of glass fiber used in communications:

— **Single-mode** (monomode) fiber is very thin and is widely used in public-switched networks where higher data-rates and longer distances are required. Usually this type of fiber is driven by pulsed lasers. See.

— **Multimode** fiber is often found in local area networks, where it is driven by LEDs. It is easier to handle than the thinner single-mode. See LED.

— **Graded or variable-mode** fiber is now usually classified as a multimode, although it acts more as if it were single mode.

Ideally the lasers used for monomode fibers will operate in the 1.5 micron band (usually 1550nm) while older lasers ran at 1.3 microns (1310nm). New glass fiber is more transparent to the 1550nm signals, however 1310nm fiber systems are still common (with lasers), and many local networks use a window of transparency at 850nm (with LEDs) for rates below 100Mb/s and distances of up to 2km.

- The distance between repeater/regenerators varies with single-mode fiber from about 40km to 150km. Generally shorter spacing is used on land.
- New low-loss fibers with a pure silica core have an average loss as low as 0.18dB per kilometer.
- Halide glass fibers have the potential to increase the span distances between repeaters/regenerators from the current 150km to an estimated 2000km using lasers in the 2000nm infrared range. See fluoride fibers.
- Direct optical amplification within the fiber is now possible. See optical amplification, erbium amplifiers and doped fiber.

fiber buffer

A thin coating which surrounds each light-carrying glass filament. See buffer.

fiber bundle

A group of unbuffered optical fibers which are act together to provide a single transmission channel.

fiber cables

These hold between 6 and 400 fibers, usually in multiples of six. Fibers are usually quoted in pairs also.

The large cables are used at a relatively low rate for city-wide inter-exchange carriage. A standard long-distance inter-exchange cable will be 36 fibers (18 pairs) running at between 560Mb/s and 2.4Gb/s.

- For Cable TV use in suburban HCF, the normal cable will be 6 or 12 fibers (perhaps with coaxial or twisted pairs), even if only one or two of the fibers are currently being used.
- Submarine cable usually has 6 or 12 fibers, but some may be dummies. Long-hop underwater cable requires copper sheathing to carry the 4–6000 volts of DC electricity fed down the cable to power the repeaters. This also helps block hydrogen diffusion but it needs thick PVC insulation, since the salt water provides the current return circuit.

Across the continental shelf submarine cable will require 'shark-bite' protection with a nickel-iron metal surround, and full armour-protection where it may be damaged by ships and trawlers. It is then usually buried about 1 meter.

Fiber Channel

Fiber Channel is a point-to-point optical link standard (also some copper for up to 10km) with data rates between 133Mb/s and 1.062Gb/s. It was originally a processor-to-peripheral channel extension interface for mainframes, where the storage devices needed to be in a remote location.

There is a Fiber Channel Association with 57 members, and ANSI is developing it in the A3T9.3 and X3T11 technical committees. They believe that it has the potential to be as cheap as Switched Ethernet in a few years.

Fiber Channel can be circuit- or packet-switched, and be used for ring or point-to-point configurations. Both switched-fabrics and an arbitrated-loop topology have been defined.

Currently it gets the lowest 100Mb/s rate from a line-code rate of 133Mbaud (there's a substantial line code overhead) using STP, but it can't run over UTP. At the higher end it provides switch-to-port and switch-to-switch bandwidths of 266Mb/s with no congestion. Adaptors cost $5–10,000, and switches cost $3–6000 per port.

The three main applications being targeted here are high-speed links from mainframes to: mass-storage devices, clusters of high-end workstations, and high-speed switched LAN interconnections. The downsizing of mass storage disk-arrays to LAN-servers is obviously a major future market.

Currently there are four classes of service:

- Class I: for hardwired circuit-switched connections
- Class II: for connectionless frame-switched links with guaranteed delivery.
- Class III: broadcast frame-switched links with no guaranteed delivery.
- Class IV: connection-based with guaranteed fractional bandwidths and latency levels.

HP and IBM are backing the protocol; Sun has an S-bus adaptor in the pipeline, and a Fiber Channel switch is being developed by at least two vendors. Also see FC-AL.

fiber optics

Fiber optics covers a much wider range of applications than just the transmission of data. Photonic and fiber techniques are now being used in optical computing and in detection devices. Fiber strain and pressure gauges are already widely in use in some industries, and soon optical switching will become widely available. Later, even optical memory may be possible.

- The major hardware units in the photonics side of fiber optics transmission are:
  - lasers and LEDs (used in multimode fibers),
  - modulators,
  - filters,
  - photodetectors,
  - and erbium-doped optical amplifiers and their pump-lasers.
- In the electronic side of fiber optic systems are:
  - multiplexers,
  - drivers,
  - amplifiers,
  - switches,
  - demultiplexers.

Fibonacci series

A numerical series where each number is equal to the sum of the two preceding numbers. 1, 2, 3, 5, 8, 13, 21, etc. is a Fibonacci series. This sequence is found in pine cones.

fiche

See microfiche.

Fidonet

An international cooperative electronic mail exchange service between many small independent bulletin boards. This was begun by programmer Tom Jennings. Fidonet members have their own newsgroups known as 'echomail', and there are gateways between the Internet and Fidonet.

There are over 4000 bulletin boards (nodes) on the system and these are polled automatically, nightly, in a hierarchical fashion for the local, regional and international 'gateway' exchange of electronic mail and other messages. Telephone bills are paid by the local system operator, who may, or may not make a small charge for access. Much of the international traffic now goes over Internet connections.

There is no Fido legal entity. It is a loose, cooperative system only, although there is an International Fidonet Association with 6000 members. Fidonet now carries about 500 echoed network conferences. See echo and echo network.

field

- **Database:** A field is the smallest data structure in a database. It is a data-unit (an 'object') which holds a single category of data within a record. In mailing lists, the Name, Address and

City are usually held in different fields — and one of these (usually the Name field or an Account Number field) may be classed as the 'key-field' to allow interconnection with other data files in relational database management.

Field types may include: alphanumeric, numeric integer, numeric decimal (with x number of places), time, date, logical (Yes/No), calculation fields, etc. and a number of other more exotic types.

In database management, columns in a table are called 'fields' and the rows are called 'records'. Fields are a consistent sub-division of records.

• **Television:** In current interlaced television systems, a field is a single video scan from the top to bottom of the screen. The odd-numbered lines are scanned first and the even-numbered lines then follow. The two interlaced fields then make a 'frame' or complete picture.

— PAL and SECAM systems have frames composed of two fields, each with 312.5 lines, with the frames repeated 25 times a second (the field-frequency is 50Hz).

— NTSC systems have frames of two fields, each with 262.5 lines, repeated 30 times a second (the field-frequency is 60Hz).

• **EMI/EMF:** Range of influence. See far field.

field dominance

Since any interlaced TV image frame is composed of two fields, it is often necessary to specify which field is dominant for still-image reproduction and for standards conversion.

This is especially a problem with live video recordings where every field can be quite different from the last. When film is scanned into a TV system, the pairs of fields (2:1 interlace) are taken from the same frame, but in live video they are independently encoded 1/50th or 1/60th sec. apart.

NTSC 60Hz systems also scan film at 24 fps in a 3:2 fashion to create 60 fields. This creates three fields from one film image, and two from the next. See interlace.

field effects

See FET.

field flyback

See vertical blanking interval.

FIFO

First In, First Out. A stack or register technique used in some types of buffers and shift registers in computer design. FIFO buffers are often used to compensate for varying data-flow rates, since they maintain the sequence of entry at the point of exit. The distinction here is with stacks which operate on the last in, first out principle.

• FIFO is also used as a form of shift register (see) but usually the control permits the input data to fall-through to the output stage, even though all intermediate stages are not filled. You can think of it as a variable-length shift register — its length is always the same as the data held in it.

• In Unix, a FIFO file is part of the file system that is created and owned by a user. It can be opened for reading and writing, with the same functions available for standard files. See queue, FILO and deque.

fifth generation

In general, the counting of generations in technology gets rather passe after about the third — so the terms fourth and fifth generation tend to become marketing slogans — and quickly lose credibility.

• **VCR copying:** The first generation is the original, so you count from there. VHS videotapes copied to the fifth generation are almost unwatchable; 16mm film copied to the fifth generation is about VCR quality; 35mm film copied to the fifth generation is still very good.

• **Language:** A term that has never really caught on. It is used sometimes to mean very high-level languages using AI techniques, but this now tends to be called 'natural language' input. See functional programming languages, LISP and formal language.

• **Computers:** Usually taken to mean the application of parallel processing (but sometimes neural networks) to create computers which will understand human language, and recognize shapes and scenes. Japan's MITI spent a fortune trying to gain a lead in fifth generation computing, but now they have admitted that very little came out of it.

• **PBXs:** These are the fully digitized ISDN-compatible units.

FIGS

Figures Shift. See.

figures shift

The old teletype machines which used the 5-bit Baudot code, could only provide 2^5 (32) character variations (ASCII No.2). This was only enough for capitals, but not lower case — which is why all telegrams and cables were caps only.

However even 32 characters didn't provide for 26 letters, 10 numerals, and 10 punctuation characters, so a physical shift of the terminal print-head was needed, as with manual typewriters. This allowed them to print numbers and symbols (but still no lower case).

They now had two-sets of 31 characters, plus the shift-up and shift-down character. This shift key, and the control character which initiated the up/down change, was known as FIGS.

file

For people just becoming involved in computing, the concept of what constitutes a file is always difficult to grasp — mainly, I suspect, because a paper 'file' from a filing cabinet is actually a folder in which documents are placed. It is also difficult to comprehend how the same word can be applied both to a tool (the application) and to the result of using that tool (the text file). But in computing, both are just sequences of numbers stored in a coherent way.

In computer and communications, a file is a single entity — a coherent data structure — which means one chunk of useful information with a beginning and an end.

The term simply means a collection of related information that is all structured in a similar way. The test of a file, is that it can be addressed by using a single filename. The term makes no distinctions between applications files, operating system files, or data files.

We can identify thousands of sub-categories, but probably four major categories will suffice:

— **The operating-system** is not strictly a single file, but I'm going to include it here. It is actually a collection of utilities and routines which work in a cooperative fashion. Utilities are small application files having very narrow functions. These do the computer's housekeeping, such as putting characters on the screen, or sending them to the printer.

— **Applications** are files that do something useful, such as process words, or data, or display data in spreadsheet form, or run communications networks.

— **Data files** are just stored data, often with formatting information which says how that data will appear on the screen. Without any formatting information, these are called ASCII or Text-only files. Data-files are those you will probably wish to manipulate, print out, send via e-mail, or just read for information.

— **Utilities and TSRs** are small applications that often just do one small job well. We could class them under applications if we wished.

• The key of a computer's universality, is that it handles all types of files in precisely the same way. It stores them all in the same large bank of working memory, and it moves through the files, location by location, in the same systematic way — but it is able to distinguish between instructions and information, because of the very careful design of the software.

• You should distinguish between a file and a database. A database is usually a collection of files which are jointly managed by a database manager.

file allocation table

The FAT keeps track of the location of file blocks in a particular 'volume' on a disk. A PC disk (especially a floppy) usually contains only one volume, but it can be partitioned into two or more volumes — especially large hard disks.

A volume is the primary division of a disk. When you address the C: drive, you are addressing a volume named 'C:'. And every volume has a disk directory, and every directory has, as part of each file reference, a pointer to a FAT which keeps track of which tracks/sectors have been used, and by which file.

This is the 'block reference' area on the disk, and the information is constantly being updated.

The FAT is best thought of as a series of pointers to the 'clusters' or blocks of stored information on the disk — and all of these change every time you save a file.

It is only by checking this reference that the computer system can keep track of those sectors on the disk which are free of data and can be reused, and those which have data that must be retained. Since these tracks are constantly written to, and read from, they are usually the first tracks on a floppy disk to wear. Since they are so important, there are often two copies of the FAT on a disk.

file compression

File compression of text or data is always lossless; the original can be restored in full in the decompression process. However with graphics and photographs, lossy compression is sometimes acceptable.

• The most common general file compression program in the MS-DOS world is Pkzip, and in the Mac world it is Stuffit. 'Zipped' files have the file extension .ZIP, while other compression systems create files with .ARC, .SEA and .LHA extensions. Stuffit files sometimes have an .SIT extension. GIF files, were, until recently, the favorite form of graphics compression, although JPEG is now close.

• Self-extracting files contain all the factors necessary to decompress themselves. You click on these, and they will expand to create an uncompressed document or utility.

• The three most widely used techniques in file compression are Lemple-Ziv (aka LZW), Huffman, and Run Length Encoding.

• Many data compression schemes work by making reference to dictionaries held at each end of a link. Instead of transmitting the data, they just point to a dictionary reference using a 'token'. MNP-5 uses LZW techniques and a fixed dictionary, while V.42bis uses a similar technique with a dynamic (built on the run) dictionary.

• In some forms of text and program-transmission, the compression ratio may be as high as 4:1 if the 'words' are repeated regularly (as with assembly language). Graphics may be compressed to even higher ratios. Motion video with its repetitive image sequences, can be very highly compressed (about 100:1) with very little quality loss. See MPEG

Generally with English language text, the average file compression will be 2:1 — or perhaps 3:1 in exceptional circumstances. See MPEG, JPEG, Huffman codes, RLE, and DCT.

file contention

On a local area network, two users may simultaneously attempt to access the same file on a file server. They may both wish to transfer it over the LAN to their workstation, and make modifications at the same time — and if both of these updated files were returned to the server, one of the changes would be lost. Special precautions need to be taken to handle this situation and to resolve the contention. See file locking and file lockout.

file control block

A structure in the memory of a PC that keeps track of the files currently in use.

file defragmentation
See disk optimizers.

file format
See logical format.

file grooming
A technique of file management which moves seldom-used files off a server when they are not active. Hierarchical Storage Management (HSM) processes categorize files according to the frequency of their use.

The main server, backup and parallel storage devices are also classed as being on-line, off-line or near-line (based on access latency and retrieval rates). Files which are not often used will then be migrated off the main server to one of the other storage devices, depending on the category and usage requirements. See HSM also.

file inversion
A method of indexing text material in a bibliographic database which indexes every word in the database, with the exception of simple connecting (stop) words like 'and', 'the', 'but', etc. This technique is the basis for full-text retrieval systems. The indexes are called 'inverted files'. See.

file locking
LANs must have ways to prevent people accessing and changing a file which is already in use by another computer on the network. File locking is one approach which can be applied generally to all types of files. Record locking is another approach which only applies to databases.

With file locking, the whole file is blocked to access by any second-user until the first transaction has been completed, while with record-locking, only access to the one particular record will be blocked.

The simple file-locking approach is only suited to small networks and to seldom-used files. In any large organization, where file sharing is common and essential for efficiency, more than one user must be able to access the database files simultaneously. These obviously require locking at the record level. However, note file lockout.

file lockout
On a LAN where a relational database management system is in use, two stations may simultaneously attempt to access two related files for updating. File-locking and record-locking may only deal with files on an individual basis — whereas this problem applies to different but related files, where both files will need to be accessed before an update can legally occur.

In rare circumstances each of the contenders will be granted access to one (but not both) of the files, and then a file lockout will occur. Special precautions need to be taken to ensure that this doesn't happen. See ACID test.

file maintenance
The process of updating a file by adding, changing, recalculation, or deleting data.

file manager
This is a very simple form of database software management, which can only handle single-level ('flat') files. These don't have any relational capabilities, but are ideal for PC address storage systems and other simple one-dimensional lists.

file-server
A general term for a range of hardware/software services which are mounted on a LAN for general use. File-servers can be a mini, a high-performance workstation, a dedicated server device, or any old PC (although inevitably with hard disk). They will always have special software which allows a large number of network users to store and retrieve files from their hard disks; usually they will have multitasking or multi-user functions. Unix is widely used here.

File-server has become both a generic term covering the wider range of server types, and also a more specific reference to a fairly basic level of file- and application-sharing functionality for small networks using relatively dumb terminals.

Larger networks would probably use the term network-server for a more powerful device, providing the same functions.

The first file-servers were high-spec 386 PCs. Now specialized file servers have much more RAM and are optimized for I/O operations. They will generally also have backup power supplies, other fault tolerant features, and network management capabilities.

- At a minimum, a file-server allows multiple users to access and share applications and files held on its storage devices. At this level, file-servers today are generally low-cost network servers which provide only a minimum of management, security and data-recovery functions.
- Many people find the term confusing because it often tends to refer to the software functionality, rather than to the hardware — and yet 'server' suggests hardware. In fact it is both.
- A further source of confusion is that there are two primary types of file server — dedicated and distributed. Dedicated servers need their own hardware, while distributed servers allow every PC and workstation to act as a server, supplying files to all others on the network. Control over these files stays with the user of the PC; he/she can usually choose to 'publish' the file for sharing or not.

— Dedicated file-servers were originally introduced to keep down the cost of computing by the use of 'diskless' workstations on small LANs. All applications and data would be on the server, with the workstations acting as basic keyboard I/O terminals.

— Later the dedicated file-servers would provide protocol conversion and central network facilities such as electronic mail applications, storage and telecommunications links. Gradually different names were applied to different servers. So, depending on the functions they perform, we now have mail-servers, fax-servers, modem-servers, etc.

• The key distinction today is probably between file-servers, database-servers, and network-servers.

- Client-server type database-servers use essentially the same hardware although they require more processing power and have much smarter software.
- Data-base servers inevitably offer 'record locking'.
- File-servers are generally fairly elementary and offer only file locking.
- Network-servers are a 'super-set' of the file-server category, which hold the primary network operating system, and provides security and management functions. These are now client-server devices.

• With file-servers (as distinct from database-servers), the network overhead is often very heavy as whole database files gets transmitted constantly instead of just records. The LAN then becomes congested and computing power lies idle. Some files may also be too big to fit the RAM memory of the end-user's PC. Special software is needed to look after these and other problems of security, access, etc. See semaphore.

file sharing

File sharing refers to the ability of a number of users on the network to access the same file at the same time.

file system

This is usually a reference to a flat-file manager, as distinct from more complex database management systems (DBMS) or relational database management systems (RDBMS). File systems are ideal for mailing list programs, address books, diaries, etc.

• This could also mean a format type: HPFS or HFS.

file-transfer protocols

The agreed sets of rules and procedures which govern how error checking, flow control, and sometimes file naming will be carried out on the blocks of data being transferred between two computers. Some will handle many files in a single batch while others won't; some can tolerate disruptions (Z-modem), and some (Kermit) can handle the peculiarities of mainframes.

Today, the use of the term also suggests that characters in the range above the normal 7-bit ASCII set (8-bits for graphics and binary files) can be exchanged using these protocols. Often PC software will specify 'ASCII file transfer' (which means 7-bit bytes) and 'Binary file transfer' (which means 8-bit bytes).

• Note that 8-bit ASCII files can still be transferred over 7-bit transport systems (old X.25 networks) if the right protocol is chosen; formatted file and binary transfers may need this. It is usually accomplished by splitting the standard 8-bit byte in half (two 4-bit nibbles) and transmitting each half separately on a 7-bit system. The protocol will pad out the 4-bits to 7-bits with zeros, or it will use the extra 3-bits for Hamming error checking code. Obviously, there is a 100% time penalty here.

• XModem is the lowest common denominator for file transfers. Almost everyone has it, but you can only send one file at a time.

• If in doubt for file transfers between PCs, use Z-modem (provided the other party can use it also).

file-type extensions

The three character 'decimal' behind the file name and period in MS-DOS and CP/M systems, which designates the type of file. Different word processors and spreadsheets will often generate their own specific extensions as an identifier for their own file formats.

These are the more common extensions:

.ARCFiles compressed by ARC.
.ARJ Files compressed by ARJ.
.ASM Assembly source program.
.EXE Executable program.
.BAK Backup.
.BAS BASIC source program.
.BAT Batch (both MS-DOS and OS/2.).
.BMP Bit-mapped graphics.
.BIN MacBinary files.
.C C language source program.
.CGM Common graphics.
.COM Executable program.
.DBF dBase Database.
.DBT dBase Text.
.DCA Text.
.DIB Device Independent Bit-map.
.DOC Document.
.EPS Encapsulated PostScript.
.EXE Executable program.
.GIF Compressed Graphics Interface Format.
.HQX Compressed by BinHex.
.LZH Files compressed by Lharc.
.PIC Lotus graphics.
.RFT Richly formatted text.
.SYS System (MS-DOS and OS/2).
.SIT Macintosh Stuffit files.
.TIF TIFF graphics.
.TMP Temporary.
.TXT ASCII text.
.WAV Sound files.
.WKS Lotus spreadsheet (V.1a.).
.WK1 Lotus spreadsheet (V.2).
.ZIP Files compressed by ZIP.
.$$$ Temporary.

filename

Files need a filename and, in some operating systems, they also need an extension to identify the file-type. Strictly, the file name is just the last level of the 'pathname' attached to the file itself. Unfortunately, the term is often used as a synonym for 'pathname' which is a series of hierarchical identifiers, in a tree-branch logical structure, beginning with the 'volume' name.

• A pathname in the MS-DOS world is the full directory-name\sub-directory-name\filename.extension sequence.

• A pathname in the Macintosh world is the full volume-name:foldername:filename sequence.

• The MS-DOS and CP/M wildcard * can be used to replace any filename, or any extension. So **copy A:*.*** will copy every file in volume A:.

filling

See padding.

film frame

A single image on a length of movie film or microfilm.

film-to-video conversion.

The techniques used for film-to-video are telecine or video scanners, where each film frame is held steady in a gate while it is scanned by what is essentially a video camera. There are also flying-spot scanners where the film moves continuously, while being scanned by an electron or laser beam in the horizontal plane only.

When film-to-video conversion is being performed for NTSC systems, the translation from the film standard of 24 frames per second to US television's 60Hz field rate, results in a 2/3 field-scanning sequence where every second frame is scanned three times. See field dominance.

In PAL and SECAM systems, the film is run at a slightly higher rate of 25 frames per second, with each frame being scanned twice. This means that a film will run for different times on a PAL channel from an NTSC channel. A 90 minute film on US TV becomes 86 min. 24 sec. in Europe — which gives the PAL countries time for an extra commercial or two.

FILO

First In, Last Out. The normal stack used by the CPU for suspending operations is a FILO store. The distinction is with FIFO and deque.

filter

A hardware or software entity that selectively removes unnecessary complexity or unwanted data — usually from carriers or signals traveling between two systems.

• In audio, a filter can alter the characteristics of the sound by modifying the composite waveform. There are three main categories:

– **low-pass** filters which remove all frequencies above a certain point,
– **high-pass** filters which remove all frequencies below a certain point,
– **bandpass** filters which both remove higher and lower frequencies — those on both sides of the set bandwidth.

• Most audio systems have a filter to remove the low-frequency 50/60Hz, mains 'hum', and often to remove the first harmonic at 100/120Hz also. Audio LP disk units usually had a high-frequency filter to remove the 'scratch' sound. Filters can be designed or selected to remove any selected band of frequencies.

• In telecommunications the analog telephone microphone output is filtered in the local exchange to limit the frequency range to between 200 and 3600Hz. The allocation per channel in multiplexed bearers is 4kHz which allows a little for the guard-bands. This is a bandpass filter function.

• Bandpass filters are called 'tuners' when they are selectable. They are devices which block all frequencies other than those of a relatively narrow range. A television or radio tuner is a bandpass filter which only permits one channel through.

• Traditionally filters have been analog electronic devices, but increasingly we are digitizing signals at the point of reception. We are also handling more signals in digital form (even if, in the end, we produce an analog output). Filtration in a digital system then becomes a computer-matching (selection/rejection) process. See Steinbrecher receiver.

• In programming a filter can be used to remove unnecessary documentation (such as REM) material.

• In Unix, a filter is a command which can handle I/O. Filters may be strung together into pipelines.

• In graphic image processing, filtering can be a very complex process where the value of certain pixels is checked against a 'look-up' table. DCT provides a way of filtering high-frequency (high resolution) image components, whenever the bandwidth can't sustain full rate transmission.

• In LANs, filtering is the process by which packets from certain sources and with certain destination addresses, are prevented from crossing a bridge or router into another part of the network. Improved filtering of broadcast or multicast packets can have a dramatic effect on the overloading of slower network segments. See firewall also.

• **Electronics:** See high-pass filters, low-pass filters, mains filters, bandpass filters, notch filters, and aliasing filters

• **Light:** These come in the form of optical glass for the front of lenses, 'gels' which are glued to the back of lenses, dichroics which fit across the surface of film lights, and celluloid rolls which are used in front of lights and windows. See ND (neutral density), color-temperature, 85 filter and 80 filter.

financial trading system

These are used in money markets by traders, brokers and dealers who need to make and receive calls from others very quickly. FTSs require a special operating console for each dealer, and are customized for the dealer's requirements — with hot-line buttons, hands-free operations, conferencing, etc. Barings Bros. Bank used one of these to lose $1 billion!

Finder

In the Macintosh, this is an application or 'shell' closely associated with the operating system. It is the point-and-click application that provides the user interface to the operating system — it creates the desktop and keeps track of the files.

Finder handles the icons, folder, etc., and it controls the way these act — it manages documents and applications.

finger

• A Unix/Internet command which gives further information about a person (usually phone number, postal address, etc.). You first need to know the name and location of the electronic mailbox. See Whois.

• In multiple carrier/frequency modulation schemes, a finger is a discrete band of frequencies modulated using one technique. See bean and COFDM.

finite-state grammar

In voice recognition systems, this is the process of using a syntactic model to predict the upcoming words, based on those already uttered. The system tries to guess ahead, which is one form of context evaluation. Finite-state grammars are usually highly specific, and are only used in computer programs with a narrow application (banking, airline information, etc.).

firewall

An impenetrable barrier to unwanted intrusion.

• A barrier for certain types of packets traveling across a LAN, internet or WAN. Routers can be used to set up 'firewalls', but not bridges; routers check packet addresses against a table, while bridges will transfer any packet that is not specifically required to stay on the source side of the network.

• Since companies all now want to connect to the Internet (usually with a full TCP/IP link), but few large companies want outsiders to connect to them, there's a lot of emphasis today on building network security barriers between internal communications systems and the Internet.

A firewall will usually consist of a router which is instructed to filter traffic. It will have special software and configuration techniques which can form a barrier. Even better, will be two firewall routers (an external and an internal router), with a computer between (often called a 'DMZ'). In combination, this is believed to be an impenetrable line of defense.

firmware

This refers to the program within the ROM chip. But it is often used for the hardware chips themselves. Important parts of the operating system are in firmware (the boot algorithms of a PC, for instance). Full operating systems and applications may be in firmware in dedicated devices — such as modems and disk drives. ROM, PROM, EPROM, EEPROM, or flash chips all hold firmware.

first-generation

• **Computers:** One of the room-sized mainframes from the 1940s and 1950s that used vacuum tubes.

• **Computer languages:** The original method of programming the first computers by directly inserting binary code via a bank of switches into each memory location. This is now called the machine-code level or machine-language (assembly). Both of these slightly dissimilar techniques are programmed using mnemonics and either hex or octal notation. Below this is the actual binary level which the computers use while functioning.

— Machine code directly creates object code (the machine binary).

— Machine language (assembly) programming creates a source code which must then be put through an assembler (a translator) to create object code. See language generations.

• **Film and videotape:** The original material, as recorded directly by the camera, from which all later copies originate.

first-order multiplexers

See PDH and plesiochronous transmission.

first-party clear

In some telephone systems, a call is only cleared (disconnected) when the calling party (the A-party) puts down his/her handset. Other systems clear the call when either party (A or B) hang-up. This difference has important implications in computer security systems where callback is to be used. First-party clear can allow hackers to defeat the callback-access defense. See call clearing.

FIT

Frame Interline Transfer. A new high-resolution form of CCD used in video cameras. Unfortunately FIT chips are low-yield and expensive, and an older version called IT (Interline Transfer) had smear problems. The distinction must be made between FIT and FT (Frame Transfer) CCD chips.

FITL

Fiber In The (Local) Loop. A general term covering fiber to the home (FTTH), fiber to the curb (FTTC), fiber to the business (FTTB) and all the other irritating acronyms they've come up with lately.

Most of these techniques use 1310nm or 1550nm lasers, and many of the proposals were aimed at the use of passive optical splitters because the carriers don't want to face the problem of providing power to active electronics in the streets.

The consensus now appears to be that future systems (for at least a decade) will split the signal at active corner-manhole hubs, and then provide shared coaxial cable links down the streets, rather than optical fiber. Each hub will service as many as 1000 homes, but the fiber service areas (FSA) can later be subdivided. These are usually called hybrid fiber/coaxial systems. See HFC, FSS and PONs.

five-level code

See ASCII No.2 and Baudot code. Don't confuse this with five-tone paging codes. See below and also the numbers section.

five-tone paging

See numbers section.

five-tone signaling

Used in radio transmission systems (mainly taxis and the early trunk systems) for addressing. Five different audio tones are used, and each receiver identifies a unique combination of these five tones. See 5/6 tone in numbers section also.

fixed-length field/records

Free-text databases are common today, but a few years ago the field and record size was strictly pre-determined and couldn't be altered. This rigidity still hangs over in some large database management systems.

Fixed-length fields are costly in terms of memory wastage, because a certain number of RAM locations must be allocated to each field, but then are often not used. The allocation must be for the maximum field-length anticipated.

However, fixed-length fields meant that the record length is also standardized and fixed, and this makes it easy for a computer to search large files by using very simple calculating algorithms.

It was this that made early file-searching appear to be 'random-access' rather than sequential, since the computer knew the length of each record, and could jump to a new location with a very small number of processing cycles. With modern computers this process happens so fast that fixed field-length is no longer such an important consideration.

See variable-length fields and records.

fixed-path protocol

A packet-switching technique where a virtual (logical) circuit is first established by a call-request packet, and all subsequent data packets will flow along this established route. The contrast is with path-independent protocol. See datagram and connectionless.

fixed point

The distinction is with floating point. Early PCs used fixed point arithmetic; it was easier because the computer didn't need to keep track of the shifting position of the decimal place marker.

Fixed point doesn't mean integer values only — it just implies that the position of the decimal (or radix) point is intrinsic in the way the numbers are used (fixed by predefined rules). Dollar/cent amounts are fixed-point numbers, because there will always be two, and no more than two, numbers to the right of the decimal point.

fixed-port hub

This was the first type of active hub used in small networks with relatively stable environments; they act as repeaters by regenerating the signals.

Fixed-port hubs are low cost and easy to install, but they only provide limited network-management capability. They can only support one media type, and every hub must be managed as a separate entity. See chassis-based and stackable hubs.

fkeys

- A Macintosh term for a small desk accessory program which is activated by a Command-Shift-Number key choice. Numbers 1 to 4 are already installed as part of the earlier systems, but new ones can be added.
- In the IBM system, f-keys are function keys: F1 brings up help, F2 invokes utilities, etc.

FL

Fuzzy Logic. See.

FLA

Fiber Laser Amplifier. See also erbium and optical amplifier.

flag

The term is used quite loosely to mean any condition which signals something specific to the system or application — it is a 'changed-status indicator'. A flag is generally either one bit, or a full byte, or, in communications, two bytes. The bit-flag will be either 'set' or not, and the byte-flag will be a specific and recognisable pattern. In one sense, the space character is a one-byte flag which indicates the end of a word in normal text.

- In CPU/ALU operations, a flag is usually one bit in a register which signals that some event has taken place (a status bit). In some cases a whole memory byte will be set aside and used as a series of eight 1-bit flags, each signaling something different.
- In synchronous bit-oriented TDM frames, the preamble is called the flag.
- In packet communications systems, flags are often used to indicate the beginning and end of a packet. They act as identifiable 'book-ends' to the packet.

Since the flag must be one of the 256 available ASCII characters, the use of one as a format-indicator then creates problems if the same character turns up within the data being transmitted. Rules must be created for handling these characters so they don't inadvertently trigger an end-of-packet in the middle.

- In programming, a flag is usually a Boolean variable or field-type which can be either true or false. This may be used to trigger some event in the program.
- In CD-ROM, a flag is a sequence of characters inserted in the text to identify information about a document's structure and display requirements.
- In one sense, the space character is a 1-Byte flag which indicates the end of a word-string in normal text.
- In film and video production, a flag is a large black piece of plywood used to block light from falling on some background or foreground object. You usually hold flags in C-stands.

flaming

Vigorously attacking someone with words on the Internet. Usually personal attacks.

flapping

A problem in routing when intermittent network failures cause the routers to constantly advertise two or more different routes. The nodes alternate (flap) between the two routes.

flash

- To switch off and on quickly. See hook flash.
- Flash EPROM. See flash memory.
- Flash ADCs are the fastest, but also the most expensive form of Analog-to-Digital Conversion. See ADC.

flash-memory

A form of EEPROM which is rapidly coming down in price and finding new applications. Developer Intel, named it 'flash' for the speed in which it can be electrically erased and reprogrammed.

Flash memory can be used to replace ROMs, RAM-disk and hard-disks (but not RAM), however it requires operating system extensions. In the form of PCMCIA cards, it is becoming especially important for portable computing.

Currently flash memory cards up to 60MB are under development, and they are projected to cost less than hard-disks on a byte-by-byte basis by 1996. However, at the time of writing (Aug '95) a 32MB flash card costs $2600.

The main problem with flash memory is that it can't be erased and reprogrammed on a byte-by-byte basis. Some chips require complete erasure, while others can be erased and re-recorded only a sector at a time (not a byte at a time). See FRAM also.

flash update

Normal route-table updating information is sent across networks at fixed intervals. Flash updates, are those sent asynchronously in response to a change of network topology.

flat-bed

• A scanner, fax machine, or photocopier where you lay the sheet to be copied on a flat sheet of glass, face down, and the scanning takes place through the glass. The distinction is with scanners and fax machines where you push the paper into a slot, or drag a hand-held scanner over a surface. The value is obvious; you can fax the pages of a book on a flat-bed scanner, but not on an insertion type.

• A plotter which draws on a flat surface. The distinction is with roller plotters.

flat-file database

A single, simple database (say, for names, addresses and telephone numbers) which does not have relational capabilities.

flat-panel display

This covers a range of different technologies, but most of these are versions of the LCD (liquid-crystal display) technology. However there are also flat CRT systems, EL screens, and plasma screens which are also flat-panel but not LCD. See active matrix screens also.

flat-rate

Services not charged on a usage or metered basis. You will make a single monthly payment for connection, and then use the system as required without volume charges.

• In the US, this means that your local phone calls will be free (not counting the monthly charge). The distinction is with charged local calls — on fixed-price per-call, and also on a timed basis. The term 'metered calls' is sometimes applied to the last two, but sometimes specifically only to calls with a timed component.

flat-response

This is the ideal characteristic of all parts of a communications system (especially analog audio). Every frequency being transmitted will hopefully be attenuated and/or amplified to exactly the same degree. The output should therefore have no emphasis in any part of the frequency range, and should not require equalization to restore its 'flatness'.

This is only possible in perfect electronics, with perfect interconnections — so in practice, flatness is constantly being restored by equalization as a signal passes through the system. See frequency response and equalization.

FLC

Ferro-electric Liquid Crystal. A type of high-quality flat display being developed by Canon. It is said to have a rapid response time, high contrast, high-resolution, wide viewing-angle, and large screen area. The base structure is LCD, with an overlay of a thin, high-molecular alignment film with the FLC molecules in neat arrays. The gap between them is 1.5 microns.

The same basic technology is being used in optical neurocomputers and other devices.

FLEX

A relatively new protocol for paging systems. It is supposed to provide a 10-times improvement in battery life (through the use of sleep periods) and may be suitable for wrist-watch pagers. It operates at three different speeds, 1.6, 3.2 and 6.4kb/s and offers a graceful migration path from today's 1.2kb/s POCSAG rates. It is said to be more robust than some of the earlier protocols, and missing messages can be flagged.

FLEX supports about 600,000 numeric pagers (5-Byte message) in the average paging channel, and the protocol is compatible with the majority of the paging systems worldwide. This makes installation simple and allows easy system growth.

It is a fully synchronous code and it wakes up the remote pager only when data is to be received, which greatly reduces pager power consumption

FLEX has both a two-level and a four-level (dibit) NRZ (non-return to zero) code format, and about 2/3 of the bits transmitted are actual information. See GSC and POCSAG also

flicker effect

On TV and monitor screens the scanning process will cause flicker if the raster scans at too slow a rate — and this rate is dependent on the type of phosphors used on the screen (the rate of decay of the image), and the method of scanning (progressive or interlace). When images cycle through the monitor at a high-enough rate, the eye interprets them as being continuous — this is the point of 'flicker-fusion'.

In US television, the field-scan rate is 60Hz with two interlaced fields creating a single frame every one-thirtieth of a second. In European countries, Asia and Australia the 2:1 interlace field-scan rate is 50Hz with frames every twenty-fifth of a second. Film projects 24 frames a second but the shutter opens twice to create 48 images a second.

These rates are borderline in flicker-fusion terms for normal screen brightness, but in low light conditions and with large screens (as in wide-screen TV) the flicker will become quite noticeable. For this reason computer screens use progressive scan and higher refresh rates. With future TV standards, it is intended to boost the field repeat rate to 90 or 100Hz. The same 30/25 frames per second will be delivered to the receiver's frame-store, but each frame will be scanned three or four times to create the screen image.

Sensitivity to flicker depends on a number of factors:

- Peripheral vision has less flicker tolerance than central vision (the flicker is more noticeable).
- Bright images tend to show up flicker more.
- Low ambient light conditions (high contrast between screen and surroundings) increases the flicker effect.
- Detail on the screen tends to camouflage flicker, so it becomes most noticeable in large flat areas of moderate color and tone (skies).
- Psychologically aware. You see flicker when you look for it.
- Some people are also more sensitive — especially those with epilepsy.

See phosphor time constant, large area flicker and fusion also.

Flight Simulator

A very popular PC games/simulation program which is often used as a benchmark for testing the speed and compatibility of various computer systems.

Flight Simulator was notoriously fickle on the early MS-DOS machines, and clones needed to be virtually 100% compatible with early IBM PC BIOS to run it.

flip-card animation

A process used in the preliminary stages of film animation (and in some CAD systems) to provide a sequence showing the major stages of the animation drawings, but at rates where the images don't merge into a flowing movement.

flip-flop

The basic building-block of static RAM computer memory. Flip-flops comprise two logic gates in a very simple circuit. This is a bi-stable device which stores a logical condition; it is either on or off. You need eight flip-flops to store a byte. See SRAM.

flipped cable

See null modem.

FLMTS

Future Land Mobile Telephone Services. See FPLMTS.

floating illustration

In desktop publishing, this is an illustration that isn't tightly linked to the text surrounding it.

So if changes are made to the copy, and the associated text moves to another column, the illustration may remain where it was previously located by the desktop publishing program. It is not 'anchored' to the text.

floating-point arithmetic

In contrast to fixed-point arithmetic, floating-point values have a variable number of digits both before and after the decimal point. Therefore the computer must keep track of the decimal point position.

Floating point numbers are held in a computer as a mantissa (significant digits with no zeros on the right) + an exponent (which gives the location of the decimal point).

Modern computers use floating-point arithmetic extensively since this gives them a higher degree of precision and saves memory space, but some of the early PCs could only work with integer and fixed-point numbers.

However, a standard CPU often needs to undertake many operations to perform a simple floating-point calculation. Integer multiplication of, say, 3 x 7 with a standard CPU, can be very many times faster than a floating-point calculation of the same numbers expressed in the form 3.0 x 7.0. For this reason special maths co-processors are often used to take over these floating-point calculations when number crunching or CAD graphics sizing, etc. is anticipated. These calculations can, of course, be done in hardware or software.

• Floating point numbers are stored as mantissa + exponent. So 123.45 becomes 12345 times 10 to the –2nd power, which is (12345×10^{-2}). The 12345 (mantissa) and the –2 (exponent) are stored in binary form, and all operations are, of course, done in binary.

Floating point division involves the use of look-up tables to find the closest binary equivalent. In most micros, these look-up tables are accurate to 16 digits (in the 'problem' Pentium, they were only reliable to 5 digits in some parts).

The IEEE has standards for formats and precision of floating-point numbers. See SANE, maths coprocessors also.

• When CPU's are compared ('benchmarked') the figures must specify whether they were for integer calculations or for floating point. Different CPUs are optimized for different calculations — which is why maths co-processors are often included for calculation-intensive floating-point tasks.

floating-point coprocessor

Aka maths coprocessor or numerical processing unit. See maths coprocessor.

flooding

A router receives new information, and immediately duplicates and retransmits that information through all its ports (except the one on which the data arrived). This results in flooding the network with the information.

This can be a legitimate alternative to conventional routing in small non-looped networks, but generally it is now only used as a way to distribute new routing information to routing nodes on the network.

Flooding can bring a looped network down through overload unless a hop count is kept in each packet, so that at a certain point they will self-destruct.

• The flooding technique has many advantages for delivering routing information:
- The information is guaranteed to reach every node with an active connection anywhere on the network.
- Most nodes will get multiple packets, but one will clearly be the favorite since it has the least number of hops from the source. It then defines minimum-hop routing.

FLOP

FLoating-point OPerations a second. A measure of computer number-crunching speed used mainly when comparing super-computers, where such a comparison matters. It provides a benchmark of floating point operations for number-crunchers.

The largest supercomputers today measure performance in terms of GigaFLOPs (1000 million FLOPs) but the new ones are measured in the TeraFLOPs (10^{12}) range.

floppy disk

This refers to the magnetic disk itself, not to the cover which is now hard plastic and about 3.5-inches square. The disk is flexible enough to be run in contact with the read-write head (unlike a hard disk which never touches the disk).

The standard 1.44MB floppy in most PCs will have 80 circular tracks with 18 sectors to each track, and it will use both sides of the disk. Each sector holds 512 bytes.

The first 33 sectors are used by MS-DOS to store information about the disk — the boot sector, the FAT and the root directory. The other 2847 sectors are available for data or applications.

Floppies must be formatted before they can be used This electronic process marks out the tracks and sectors and puts an initial null value into each byte.

• When the magnetic material manufacturers produce their 'jumbo rolls' of plastic, they test the surface thoroughly. The best material is retained for brand-name use as the highest-quality disks, while seconds and rejects are sold to cut-price disk makers.

So when you buy cheap disks, you may be getting identical quality in the material of an individual disk, but there will be a greater chance of any disk in the packet having a defect.

• The improvements in 3.5-inch floppy disks have followed this pattern:
- Single-sided (SS) floppy disks (140kB) were followed by
- Double-sided (DS) disks (380kB), and then by
- Double Density/Double Sided (DD/DS) (720kB). These were also called High Density (HD) by some manfacturers.
- Double Sided/Double Density/Double Track disks are the current standard. These are capable of handling 135 tracks per inch, which gives us 1 to 1.4MBytes of data. These are called MF/2DD disks with two windows (one locking).

• Be wary of some almost identical disks (with the same external characteristics) labeled MF2 HD.

FLOPS

FLoating-point Operations Per Second. See FLOP.

floptical

These are high-capacity magnetic recording floppies (in the 20 to 40MB range for 3.5-inch disks) which use an optical servo-mechanism and optical guide track for the magnetic recording head. The embedded disk information is optically-sensed and used to control a 'closed-loop' servo motor — this supposedly dramatically improves magnetic recording densities.

flow control

When a computer is receiving data over a communications link at a very high rate it needs to be able to send a message back to the originator to 'wait a bit!' so it isn't overwhelmed. Flow-control is exerted on the source either by software controls (from remote devices), or by hardware controls (from local devices).

Software flow-control is usually exerted by the transmission of the X-OFF character across the return channel. Later the signal X-ON is sent to restart the flow.

For local devices (such as PC-to-modems) which are connected by a multi-wire cable, flow-control is usually handled by dropping the voltage on the 'Clear To Send' line (hardware flow control) of an RS-232 connection. See XON/XOFF, RTS, CTS, ETX/ACK and ENQ/ACK also.

• Usually hardware flow-control only acts locally, while software flow-control acts over a distance. However many modems can be set to relay hardware flow-control; a modem whose RTS goes low can then pass that fact on to the remote modem by sending a X-OFF. Similarly, if a remote computer sends a X-OFF through its modem to your local modem, it can be set to lower CTS (this is often done if X-ON/X-OFF is disabled).

• Flow controls may be implemented at any level, from Application layer down, wherever there is the possibility of a sender sending the data faster than a receiver can handle.

• Flow control, which stops the other party from sending, or asks it to resend data blocks would usually be implemented as a Data-link (Layer 2) control. If this control is not in place, then it is usually a Transport layer (Layer 4) control. The session layer may also supply global synchronization through its sync points and resynchronization functions.

Every node in the network with a fixed buffer, where one entity is writing data and another is reading it, must have its own internal flow-control or semaphore mechanism to guarantee synchronization.

flowchart

A special form of diagram with lines, boxes, diamonds and circles, which is used to plan the basic structure of a proposed computer program. Some programmers swear by flowcharts; some never use them. The standard flowchart symbols are:
- Terminator (start and finish): flat oval.
- Instruction (what to do next): rectangle.
- Decision Point: rectangular wedge.
- Connector (pathway intersection): small circle.
- Input/Output: flat diamond.
- Manual Input: rhomboid.

fluoride fibers

The use of chemical halides in glass manufacture has been known for more than a decade to improve the transparency. Material-mixes of chemicals such as the fluorides of zirconium, barium, lanthanum, aluminum and sodium to make super-transparent glasses for optical fiber use, has not yet broken out of the laboratory — mainly because the glass has proved to be brittle and difficult to draw into the fine filaments needed. Theoretically it should be possible to make single mode fibers which can carry a laser pulse for 2000km without a repeater. Fluoride fibers should be more transparent at the current infrared wavelengths of 1310 to 1550nm, but their optimal performance is in the mid-infrared, from 2000 to 5000nm.

Fluorinert

A chloro-fluorocarbon-like fluid which has extremely low electrical conductivity, and can therefore be used to bathe electrical circuits and keep them cool.

flush

- **Cache:** If important information is stored in cache or in any form of temporary memory (buffers, etc.), the data will be lost if the computer is switched off — and quite often, if it changes applications during multi-tasking. Cache needs to be flushed (the information ejected, and put in storage), and well-behaved software does this automatically.
- **Printing:** Flush is the obverse of ragged. A justified column of text is both flush left and flush right.

A column or section which is set this way is called flush left, or ragged right —since the right-hand margin is not aligned, but is allowed to end wherever the line falls naturally.

A column which is set in this way is flush right, or ragged left, since the left-hand margin is not aligned. This is often used for captions and sometimes for poetry, or possibly just to create an effect.

flutter

- A fairly rapid distortion of mechanical turntables in sound recording and reproduction, which is caused by mechanical problems creating a variation in speed.
- Distortions of telephony caused by cross interference with other channels, resulting in a rapid variation in sound levels.

flux

- An unwanted change in signal levels over the medium term, usually in relationship to radio signals.
- The lines of force in a magnetic field.
- An anti-oxidation agent used in soldering to help the solder to flow. Residual flux creates problems with electrical contacts.

flux quanta

See magnetic flux quanta.

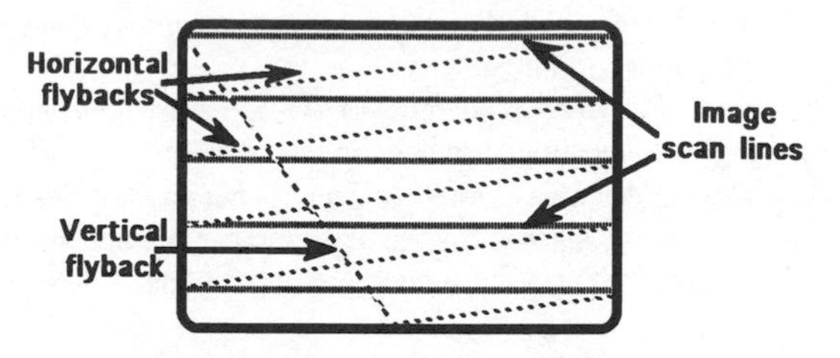

Horizontal and Vertical Flyback

flyback

In TV and monitors there are both horizontal and vertical flybacks. This is the phase where the raster scan returns to its start-point, diagonally at the end of a field, and horizontally at the end of each line.

The electron guns (RGB) are switched off during these flyback phases (by a negative voltage 'sync' pulse) so that the scan can reposition for the next line or next field. This is why the process is also known as 'blanking', and the time taken for the repositioning is called the 'blanking interval'.

These intervals represent wasted transmission time, and so they can be used for transmitting digital information. See horizontal frequency, horizontal retrace, vertical frequency, vertical blanking interval, and Teletext.

flying-spot scanner

A machine used for transferring film to video. The film moves continuously through the machine and it is scanned by a laser, light or electron beam (or beams) in the horizontal plane only. The film is in continuous motion, and this provides the vertical displacement. The distinction is with the older type of film 'kine' or telecine units. See film to video.

FM

Frequency Modulated/Frequency Modulation. The frequency of the carrier is modulated (modified) by the signal. The distinction is with AM which modifies the carrier amplitude, not the frequency.

- **Radio:** The FM process superimposes a signal on to a carrier in a way that sums the frequencies. An audio signal which varies from 200 to 20 000Hz may be superimposed on a carrier frequency of 2MHz. The audio frequencies will change the carrier's frequency within a defined band — here from 2MHz + 200Hz at the low end to 2MHz + 20 000Hz at the high.

Usually the frequency quoted for FM is the 'center frequency'. Carrier separation is the sum of the signal bandwidth and a guard-band.

— International FM radio station frequencies for broadcasting are those between 88 and 108MHz.

— AMPS, TACS and NMT analog cellular mobile phones all use FM modulation techniques. See FDD also.

Note: There is little to distinguish frequency and phase modulation. Both are essentially the same. Also FM side-bands exist at integer multiples of the modulating frequency on either side of the carrier. This is quite different to AM.

• **Magnetic Disk recording:** FM techniques really have very little relationship with radio FM. The disk technique relies on the use of dual one-zero pulse patterns (pulse then pause) being written to the disk, as a way of synchronizing the data. Ones are written in a one-one pattern. The first pulse is known as the clock bit, and the second is the data bit, and there's no pauses between pulses. This was the original technique used in 'single-density' floppy disks. See also MFM, GCR, and MMFM.

FM synthesis

Frequency Modulation synthesis is a way of generating the sounds of different musical instruments without using wavetables. It is an inexpensive way to produce sound cards. The voices (different sounds) are produced entirely electronically — and at the current state of the technology, this electronic-origin is generally obvious. See wavetable synthesis.

FMF

Fuzzy Membership Function. See ANN.

FMS

Flexible Manufacturing System. In computer aided manufacture (CAM), this is intelligent equipment which is capable of making short production runs (batches) as efficiently as long runs. FMS is the direction many manufacturing industries wish to move, because it reduces stock holding, warehousing, etc. and it is easier to make quicker model changes.

FMV

Full Motion Video. The term applied to CD-i disks used for video replay. The early systems could not maintain 24 frame motion video on the full monitor/TV screen — it was either small and smooth in motion, or large and jerky. The new FMV version is based on the MPEG1 compression standard.

See CD-i.

FNA

Financial Network Association. This is an international organization incorporated in Brussels in 1992 by a number of telephone carriers.

FO

Fiber Optics. See.

focus servo

The mechanism in an optical disk drive which keeps the laser beam focused accurately on the disk surface. The mechanism can adjust vertically to handle surface imperfections and external shock or vibrations. Other servos also maintain the laser beam horizontally on the track of pits and lands.

focusing coil

In a TV set or computer monitor, the electron beam passes through a focusing magnetic coil on its way to the screen. It also passes through horizontal and vertical deflection fields created by other magnetic coils (or sometimes by electro-static charges).

FOD

Fax on Demand. Interactive access to special databases where the information is delivered by facsimile. There are two approaches.

• You can phone the database, select the data using DTMF phone codes (following voice instructions), then transfer control to the fax machine. The pages will be delivered on-line.

• In other systems the information selection is separated from the delivery phase. The information is delivered only after the selection call has been completed and disconnected. The fax-server is able to extract the return phone number from the answer-back in the fax header, and it will then redial and transfer the information.

Many people with fax machines neglect setting the answer-back, and others forget to reset if after they've had a number change — which creates some problems here!

FOIRL

Fiber Optic Inter-Repeater Link. This is an IEEE standard for a signaling methodology when data is being transmitted over a fiber cable at 10Mb/s using 802.3 (Ethernet) framing and access techniques. It is used to link network wiring hubs, and distances up to 1km are common. With the addition of the 10Base-FL extension to the standard, the distance can be increased to 2km. See also 10Base-F.

foldback

In television studios, foldback loudspeakers return some of the sound being recorded to the studio to produce a richer 'reverb', and to allow the audience to hear what is going on.

folder

The Macintosh equivalent of a sub-directory. Other folders and/or files can be inserted into higher-level folders in a hierarchical fashion. When a folder is double-clicked with the mouse/cursor, it will expand into a new window and reveal its contents.

Foley studios

A standardized audio studio used in film production to generate sound effects. The floor of the studio is filled by meter-square boxes of sand, gravel, concrete slabs, etc. for footstep recording, and different doors and windows are provided for the appropriate opening and closing sounds. These are recorded in real time while watching a monitor.

folio

• A folio is actually a sheet of paper, folded into two 'leaves' which creates four pages.

• Book printer's jargon for the page number.

• Folio is also a large paper or book size. See foolscap.

font

The character-styles or typestyles used on the monitor screen and in printing. The term 'font' is used differently in computing, to professional printing. With computers, 'font' just means a collection of characters and symbols with a consistent style; it may include italic and bold versions, and the term refers to multiple type-sizes. So in computer terminology Bodini is a font, and 9 point italics is just a variation within the font.

A professional type-designer or a traditional printer would refer to the above as a 'typeface'. In their terminology the term 'font' refers to the complete set of characters, but only of one type design (say, Bodini) in one style (say, italics) and in one size only (say, 10 point).

See typeface, point, typestyle, pitch, proportional spacking, kerning, x-height, serif and sans serif.

In recent years there has been a complete revolution in computer fonts with a change from the common use of bit-mapped fonts, to the use of outline (vector) fonts. Most modern computer and printer equipment is capable of handling both. See TrueType, PostScript and multiple master.

With graphical user interfaces (GUIs) and laser/ink-jet printers we need to make a distinction between three sources of the font information:

• **Internal fonts:** A few of these are often built-in to the printer in ROM, and are therefore always instantly available. These are usually bit-mapped fonts in one or two sizes only in relatively dumb printers, but outlines can be added to the more intelligent printers (PostScript and others).

• **Cartridge fonts:** These are plug-in ROM font cartridges which extend the range of internal fonts in the printer; there may be dozens or hundreds of bit-mapped or outline fonts in a cartridge.

• **Soft fonts:** These are generated by the attached PC and font-generation software, and transferred into the printer's RAM before the text is sent across the cable. They are then used as if they were internal fonts to generate the page; only a single transfer needs to take place for one style and size.

In all of the above cases, the generation of the page is taking place in the printer by selecting font bit-mapped images, or by generation of an outline font. The alternative approach is to generate the pages entirely within the PC. The page is then transferred as an 'image' to the printer for printing. See PostScript, TrueType and outline fonts.

• Today printer and screen fonts are generally outline or vector fonts which are calculated and displayed as required from the same source but to different levels of resolution. Screens normally use only 72 dots per inch, while laserprinters require 300 dpi or more. With bit-mapped fonts in the past, it was often necessary to have different versions of the fonts for the two devices.

• Another primary distinction is between display and text fonts. Display fonts include brush-script, and special styles created purely for display purposes.

font cache

An area of RAM used by the font managers to store the bit-maps of characters generated by outline fonts.

font cartridge

These are ROM modules that plug into the laserprinter to provide it with a wider range of 'built-in' fonts. The alternatives are for the printer to:

- use its true-built-in fonts (outline and/or bit mapped);
- take downloadable fonts sent from the PC, and bit-map these in the printer;
- receive a complete bit-map of the page ready to print, from the PC.

font characters

These are the eight characteristics of a font in computing:

- Typeface: Courier, Times, or Helvetica, etc.
- Stroke weight: Light, medium or bold.
- Style: Upright (Roman) or italic.
- Point size: The character height (top of ascender, to bottom of descender). See x-height also.
- Spacing: Proportional or fixed/monospaced (usually associated with typewriters). See kerning also.
- Pitch: The number of characters printed per line cm or inch (only with fixed spacing).
- Condensed or expanded forms (in the horizontal) are sometimes possible.
- Symbol set: With normal Roman text; accents, grave, acute and circumflex (diacritics). There is also Extended ASCII; super- or sub-scripts; special numeric, punctuation or graphic characters; and scientific or mathematic notation.

font managers

This usually refers to utility applications such as Adobe Type Manager (ATM) and Bitstream's FaceLift which provide true device-independent printing of fonts. However font managers are now built into modern operating systems.

With a good font manager you are not tied to a printer command language since it will act as a filter/translator between the application and the printer driver.

Using these managers, you can print quality outline fonts on any laser or ink-jet printer, regardless of the device's characteristics. You can have PostScript quality on a laserprinter without a PostScript board or emulation program, for instance. However this doesn't necessarily mean that font managers can handle all the text and graphics images in page-layout.

Font managers generally create their screen and printer characters using outline fonts by a multi-stage process:

- The bit maps (font-maps) for each character are generated from outlines and stored in a font cache. Conversion only takes place once for each size, style and angle of font.
- As text is entered, the font-maps are used to build both a screen and printer bit-mapped image. The screen image is in video-RAM and is constantly scanned.
- The printer driver then sends the page bit-map to the printer when the page has been completed.

• There are many variations on the above, since some printer types (such as character printers, or dot-matrix) may take the text in pure ASCII form. In some cases, laserprinters will also take the text in ASCII and create the bit-map internally from its own outline fonts. The font-manager will then only handle the transfer of the outlines and the ASCII test.

foolscap

A paper size. The term foolscap is incorrectly (but almost universally) applied to the paper size called correctly 'folio' (although this term is also being misused). In printing terms, foolscap is 38 x 34.2cm (17 x 13.5in.) which is half a 'foolscap' broadsheet. However, countless typists have applied the term to the paper size which is 21.6 x 34.2cm (8.5 x 13.5in.) with dimensions equal to the half-fold of actual foolscap. This ancient misunderstanding came about because the foolscap broadsheet size was generally folded when used for legal printing. See paper sizes.

footer

A line at the bottom margin of a document page. A footer will appear at the base of every page (or every alternate page), and if it includes the same text it is generally known as a 'running footer'. Running footers are often chapter headings for quick reference. Footers can include text, images, dates and page numbers. Don't confuse this with a footnote.

footprint

The physical space occupied by something.

- **Computers:** The physical amount of desk space a computer or a peripheral occupies.

- **Chips:** The socket configuration for the connection of a semi-conductor chip. The socket's footprint must match the chip's pinout.

- **Satellite:** The theoretical area of good reception for a signal beamed back to earth by a satellite's transponder.

  Given the same transponder power: global, hemispherical and spot beams have different footprints and therefore different signal concentrations within the reception area. The center of an evenly-distributed footprint is known as the bore-site (bore-sight) and the outer edge beyond best reception, is called the penumbra.

  Not all satellite footprints have the signal distributed evenly across the area. Shaped or pencil beams are transmission schemes where most of the signal strength is concentrated in one or more very limited areas of the territory covered. Dish designers are now able to computer-model the satellite antenna shape to give peak power only to cities (where reception dishes must be small), and provide lower EIRP in country areas where larger dishes can be ground-mounted. See off-axis.

- The footprint is also the name applied to the paper printout of the contours within a satellite's coverage zone. It will be marked to show EIRP, PFD, G/T contours or antenna-size needed. A difference of –3dB in EIRP contours will mean that a dish-size must be doubled to provide the same signal strength.

foreground

A film and photography term which just distinguishes the objects of interest close to the camera from the general scene behind. Cameramen also talk about the depth of field being divided into foreground (that between the subject and the camera) and background (behind the subject).

foreground processing

There are two slightly different meanings.

- Multi-tasking operating systems with GUIs will, when you click on the window, select the application and data and bring it to the front of the screen. This is then the foreground task which is currently under operator control. On the screen the other windows will be overlayed, and all point and click operations will be referred to the foreground window.
  - True pre-emptive multitasking operating systems will allow the background tasks to keep running (sorting a database, or sending data to a printer). Usually the foreground window will automatically receive priority.
  - In pseudo multi-tasking OSs, background tasks will only run when brought to the foreground. They will be dormant in the background.

- Many of the true pre-emptive multi-tasking operating systems will also give priority to programs which are time-critical or which are selected to have a higher priority. These are said to be foreground tasks, while the others are less important and therefore background — although this form of priority may be done in a time-sliced way where the user doesn't notice — and where highest priority is not necessarily given to the front screen in the Window environment. This requires software to indicate to the O/S some form of priority claim, and this was the aim of cooperative multitasking. (It failed!)

foreign exchange

A telephone exchange other than your local exchange, to which you can connect directly. See FX.

fork

A subdivision of a file. In the Apple Mac, all files consist of two parts; a Data fork and the Resource fork. The Data Fork contains all the data (in ASCII form) just as with MS-DOS, while the Resource fork is an index to resources (formats) stored within the file. It is this division which gives the Mac its flexibility to create customized menus, and allows Mac-users to copy and transfer icons, etc.

form(e)

- In traditional printing, a form (aka forme) is the large film-negative in the plate-preparation process which comprises a number of pages (usually 4, 8, 12, 16, or 32). This is the first stage in which the pages have been 'imposed' (placed correctly in position) ready for the printing press.

- In CD-i: These are variations in the original CD modes. See CD forms and modes.

form factor

A rather pretentious term meaning little more than size and shape.

• It is applied in computing to the size of an add-in or built-in peripheral, or to the size and shape of the PC itself. Half-height and full-height disk drives are the two most common 'form factors' for disk drives, while PCs can be desktop, small-footprint, minitower or rack-mounted.

• Form factor is also used as a synonym for size. A 5.25-inch CD-ROM is said to have a different form factor to the 3.5-inch MiniDisk.

form feed

A command that shifts the paper up to begin afresh at the top of a new page. Some of the old printers had to move the paper one line at a time. Now a single command is issued. See FF.

formal

Formal implies provable, definable, and strictly in accordance with recognized scientific standards of analysis.

• **Formal Logic:** The study of structure and form of arguments and statements, without caring about the contents of the argument.

• **Formal Standards:** Those set by recognized standards-making bodies. The distinction is with proprietary standards, and *de facto* standards.

• **Formal Specification:** One that is able to prove mathematically the validity of an implementation.

formal language

A program written in a formal language may be considered as a string of program statements without need for a special order — although they will always obey syntax rules. This is an essential part of the formal method, and the languages themselves must be subject to formal methods in their specification. See below, and functional programming languages.

formal methods

Formal methods are programming techniques which divide into two parts:
— the formal language, which is usually a notation for describing the program in abstract (but very precise) terms, and in the correct syntax, and
— the proof apparatus. The translator just checks for syntax errors.

The importance of this approach is in specifying the total program structure in a way that allows it to be analyzed and examined for flaws, before the program is actually written. This is seen to be essential for safety-critical systems.

A formal specification uses a formal language to describe a set of properties that the system must eventually satisfy. Only after this stage has been verified is the implementation language used to program. See functional programming languages.

formant synthesis

One of three main approaches to speech synthesis. Formant synthesis is essentially a technique of modeling the natural resonances of the human vocal track.

Formants are resonant effects in all wind instruments including the voice. As the sound propagates along a tube, the various frequencies are selectively 'shaped' by reinforcement — and so peaks occur in the spectrum of the sound; these peaks are the formants.

Determining these formant frequencies is an essential part of speech analysis. Once formants have been calculated they can be used to predict values for the various character/sound combinations and so the voice can be coded in a very concise way. If these formants can then be effectively synthesized by the receiver, a close approximation of the original speech can be made by reproducing the formants, and applying numerical values to the strength of each.

• One variation of this technique is called phoneme synthesis. Each phoneme is assigned a code, and the creation of continuous speech is a matter of simply stringing the phonemes together. See linear-predictive coding (LPC) and waveform digitization.

format

The way something is laid-out or displayed. The standard style.

• **Disks:** To format a disk is to prepare it electronically to receive data. This is an electronic process in soft-sectored disks, performed by the operating system's disk drive controller in association with the disk drive. It marks out the tracks and sectors on the recording surface, and puts an initial value into each byte. Information is also added to the boot sector, the FAT and the directory sectors.

Usually the process will also test each byte location and either abandon a faulty sector, or reject the disk as a whole. Macs tend to do the latter.

Reformatting a disk erases all the data, and in most systems the data cannot be recovered because it has been overwritten. This is more than an undelete can handle.
— If the disk is to be used as a startup disk, the operating system will also need to be added after formatting. This was common in the early PCs without hard disks.
— On floppy and hard disks, the 'formatted space' is the area of surface reserved for data after the essential formatting processes have been completed. Formatting reduces the space available on a disk.

• **Text:** In desktop publishing and word processing, format effectors perform such jobs as shifting the printer head (or screen writing) to the next line (line feed), forcing it to begin at the left margin (carriage return), backspacing to make changes. Format effectors are part of the normal 7-bit ASCII set.

Format commands are added to these files to create indents, to make letters bold or italics, to justify the lines, etc. These are not a normal part of the standard 7-bit ASCII set, and are therefore removed when you save the file as 'Text Only'. They are usually single characters in the high (8-bit) extended ASCII range, but they can also be two and three character controls (usually, then preceded by Esc or SO).

format effector

A category of control characters (established in the ASCII No. 5 set) which are often used to control the layout and positioning of text characters on printers and display devices. They are back space, carriage return, form feed, horizontal tab, line feed and vertical tab. See BS, CR, FF, HT, LF and VT.

formatted files

Text files which include special control characters and commands to create a page-layout. A distinction is being made between the common formatted output of a word-processor, and the text-only (ASCII) output.

There are some 'format' effecting characters within the standard ASCII set, but these don't count (space-characters used for indents, carriage returns, line feeds and tabs). A file may have these, but still not be regarded as being formatted.

• Formatted files will have 'page layout' instructions (even of a very elementary nature) which are specific to the application, and not part of the ASCII text. The text itself may include bold and italic sections, larger headline fonts, and the page may have a designated page-width and/or length. All of these are controlled by non-ASCII format codes.

• There is some confusion with the use of the term in files where the width of each line has been set by a standard carriage return — rather than the carriage return only being used to signal the end of a paragraph. To call these files 'formatted' is an incorrect use of the term, but it is not uncommon; they are true unformatted files since they are totally 7-bit ASCII.

This problem comes about because the Mac and MS-DOS computers differ in the way they terminate a line of text on the screen (and store it in memory). The Mac only adds a CR at the end of a paragraph, while MS-DOS machines add one at the end of each screen line.

• Unformatted ASCII text files only use the 128 standard 7-bit ASCII set of characters. Any word processor (Mac, Unix or PC) should be able to read such a file, provided it can be transferred into the PC's memory (via a modem or disk, etc.).

formatter

• A program that formats (prepares) the disk for recording, and creates the sectors.

• An old term for a utility which was added onto a text editor in order to create a primitive word processor. The formatter controlled the output image of the text file — it supplied the page breaks, justification, etc.

formed-character printers

Printers of the older type which had the characters pre-formed, and which printed by pressing these pre-formed types against the paper through an ink ribbon, or some such similar system. 'Impact' printers — like the old mechanical typewriters are of this kind. See daisy-wheel and thimble printers.

Forth

A high-level programming language which is usually implemented as a 'threaded interpreted' language. This makes it highly interactive (like Basic) but, as the size of an application written in Forth is much smaller, it may be almost as fast as a compiled Basic program. Forth executes programs through threads of words or subroutines. The syntax is very simple; there are no syntax rules other than to separate each word by at least one space. It uses Reverse Polish Notation.

FORTRAN

FORmula TRANslator. A popular number-oriented scientific and engineering computer language, now reaching the age of retirement. Fortran is closely related to Basic.

forum

In keyboard conferencing, this is the 'electronic meeting place' where people belonging to a Special Interest Group (SIG) exchange information about their topic openly — they can all read and reply to (comment on) everything. It is a distributed discussion group of an electronic bulletin board which is devoted to one subject.

If you read forum material but don't contribute, you will be accused of 'lurking'. See UseNet. The plural of forum is forums, not fora — unless you are an ancient Roman.

forward chaining

In expert systems, forward chaining is the search technique which reasons forward in the knowledge base; it will reason from the user-specified knowledge towards a conclusion. The distinction is made between forward and backward chaining. In backward chaining, an expert system is asked to justify its conclusions, and so it reproduces the inference steps that led to that conclusion. With backward chaining we go to the end point, and see what factors were necessary to reach that point. See backward chaining.

forward channel/link/path

• In general communications, this is the data channel. The reverse or backward channel is used for acknowledgments and supervision, so the balance is often highly asymmetrical.

• In some cases the forward channel is defined by the call set-up; it is the path from the call initiator (the A-party) to the called party (the B-party). In telephone calls, these links are usually symmetrical.

• In cellular phone systems, the forward link is the one from the base-station to the mobile. The reverse channel is from mobile to base.

• In Cable TV, the forward channel is that carrying the video from the head-end to the homes.

• Over satellites, this is the outbound path from the earth-station to the satellite, and then down to the receiver. Don't confuse this with the up- and down-links, since a forward link involves both.

• It is also called the 'go' channel, as distinct from the 'return' channel.

forward error correction
FEC. See.

forward recovery
Sometimes when a system crashes and data is lost, the full range of changes that were made just prior to the crash can be recovered by taking an older version of the file, and adding in all the changes which were automatically recorded in isolation in a daily 'electronic diary'. This is forward recovery.

forwarding
To pass a cell/packet on. A bridge or router will filter some packets because it knows that they must remain on their own side of the network. It will forward all others from one LAN/-WAN segment to another. Bridges automatically all forward packets with addresses they don't know,while routers look up the address, and make an intelligent decision. See source routing and spanning tree.

FOTS
Fiber Optics Transmission Systems.

foundation services
In Cable TV terms, these are the four national US Pay networks which program current movies and big events: Home Box Office (HBO), Showtime, The Movie Channel (TMC) and Cinemax.

The term means that one of these channels is seen as the foundation of a 'Premium' package, and each package (usually of four to six channels) will include a few specialized pay channels such as a Family, Fine Arts, Adult or Sports channel along with a News channel. The distinction is between common-denominator foundation channels and the more specialized pay channels which service niche markets.

foundry
The manufacture of large numbers of complex ICs by companies like Intel, Motorola, etc. usually takes place in a chip foundry. Smaller runs of ASICs are usually made in a fabrication plant. However there's no real distinction between these two terms.

four-color process
See CMYK.

four-wire
The use of four-wire links between telephone exchanges and for special data lines, allows one pair to be reserved for single-direction (half-duplex) transmission each way. This makes signal amplification simple (it is extremely difficult to amplify signals in a full-duplex circuit) and it allows the signals to be fed into multiplexers. However, hybrids are needed to join four-wire systems to two-wire full-duplex links, and the mismatch of impedance here is a major cause of echo problems in telephone circuits. See hybrids, wire, and echo suppressors.

• Most telephone 'four wire' circuits these days are actually slots in time-division multiplexed signals traveling down a pair of optical fibers.

Fourier transform
Fourier transform is the mathematical application of Fourier theory. It is a theoretical approach to modeling (extracting the components of) complex waveforms. Fourier theory depends on the fact that any waveform, no matter how complex, can be recreated by the summation of a set number of sine waves. These will consist of a fundamental and any number of higher-frequency harmonics — so a complex waveform can be expressed as a series of numbers which represent the 'weights' of each harmonic.

Fourier analysis breaks down a cyclical function in either space or time into its sinusoidal components. These have varying frequencies, amplitudes and phases — and the resulting transform is a function that represents the amplitude and phase at each frequency.

Although this is basically an analog waveform analysis technique, it can be applied to number series which have a repetitive pattern — such as the pixels in a block of image (each scan line repeats roughly the pattern of the last).

By using Fourier transforms, it is possible to create a form of analog-to-digital video conversion and also digital compression. The key components here are the major features of the image (the fundamental) together with progressively lesser image components representing finer and finer resolution features (weighted values of harmonics). However the process is extremely intensive in mathematical terms.

• The Fourier approach reduces the number of bytes needed in analog sampling for audio also. The digital specification of waves by Fourier techniques will be much more efficient in bandwidth requirements than PCM or ADPCM techniques for the same reproduction quality.

• Fast Fourier Transform (FFT) is a program/technique that saves time by decreasing the number of multiplications needed to analyze a curve. Discrete Cosine Transform (DCT) is now the most commonly used variation. See FFT and DCT.

fourth-generation

• **Languages:** These were all the buzz of the industry in about 1985. They were an attempt to make complex programming easier by using even more English-like commands, and by removing much of the rigidity of structure found in 3GLs (Basic, Cobol, and Pascal).

4GLs are languages like dBase and HyperText and SQL — they tend to be narrower in their application than 3GLs. The first were really reporting and query tools, rather than full languages. Later these became 'application generators' on mainframes and evolved into CASE tools (although now 4GLs have progressed to PCs also). These languages are also sometimes called 'non-procedural' to emphasize their lesser dependence on structure — although this is highly variable.

Currently the term is mainly applied to tools that allow you to develop sophisticated client/server applications using visual programming techniques. 4GLs are said to require one-tenth the coding of a 3GL language and they use English-like syntax to generate code in such languages as C or C++.

— One characteristic of 4GLs is that they can generate code in proprietary languages (you'll need a special compiler). They will also generate standard code, of course.

— It is claimed that 4GLs will eventually facilitate the use of user-written programs. In general, they are quick and easy to learn, and have common applications built in — but they are also narrower in scope.

Many would regard SQL, for instance, as a query-language rather than a programming language. See language generations, CASE and visual programming.

• **Computers:** Fourth generation computers are probably the current form of computer using Very Large Scale Integrated (VLSI) circuits, but retaining the basic von Neumann architecture. This includes both CISC and RISC approaches, although one would suspect that this represents a generation gap.

FP

Flat-panel (screens). See LCD, supertwist LCD, double-twist LCD, active-matrix screens , EL and gas-plasma.

FPGA

Field-Programmable Gate Array. A form of erasable and programmable logic device which combines the attributes of standard gate arrays (high element density) with those of programmable logic devices (PLD — easy programming). It uses an 'anti-fuse' for gate connections — blowing it 'makes' the connection, rather than breaking it. The result is that about ten to twenty times the number of gates can be fitted onto the chip than with PLDs (usually 1000 only). See EPLD also.

The latest FPGAs have 22 000 gates and the numbers are climbing; sub-micron CMOS technology will take this to 40 000 by 1996. And the price now means that they are no longer just used for prototyping or bus-interface devices.

FPL

Functional Programming Languages. See.

FPLMTS

Future Public Land Mobile Telecommunications Systems. This is the CCIR/ITU acronym for what is called by ETSI, the Universal Mobile Telecommunications Systems (UMTS) which is part of RACE. The acronym is actually pronounced 'F-plumts' — I kid you not!

This is the general plan within the ITU to design for the future integration of cellular mobile, trunked radio, satellite cellular, Telepoint, cordless, wireless PBX, and even paging systems, which will be all jammed into one radio transmission standard. FPLMTS will offer all-digital voice, fax and data transmission, and even wireless computer networking. It will incorporate satellite and wireline links, and have small-cell cordless applications. Bandwidth will be on-demand.

These mobile services are to also be integrated into fixed networks, and all of this is to happen in the first decade of the next century. Some commentators think that FPLMTS is both unnecessary and unwise.

FPLMTS assumes:

— a single R/F front-end for all services,

— global roaming (location and billing) capability and agreements,

— the development of UPT and AIN also (see).

• A decision taken at WARC '92 requires 230MHz of bandwidth to be reserved in the 1885—2025 and 2110—2200MHz regions for FPLMTS. Further bandwidth would be needed for the associated satellite (1980—2010 and 2170—2200Mz) and paging requirements. Personal satellite services (Big LEOs) were allocated 1610—1626.5 and 2483.5—2500MHz with this in mind.

FPP

Fixed-Path Protocol. See.

fps/FPS

• frames per second.

• fields per second.

• Fast Packet-Switching. See this, ATM, and frame-relay also.

FPU

Floating Point Units. See floating-point arithmetic and maths coprocessor.

FR

Frame-Relay. See.

FR Access Device

See FRAD.

fractal

An element in a graphic design which follows the same pattern at higher and lower hierarchical levels. The term comes from the Latin 'fractus' which means broken or irregular.

There are two main applications: the first in graphics for interesting self-generation of exotic images, and the second in image and video compression.

The interest in fractals for image compression depends on its application of a simple algebraic formula which can be used and reused in micro, medium or macro dimensions. See fractal compression.

fractal compression

Fractal compression was invented by Iterated Systems. The technique analyzes images, and stores the various elements as fractal transform (see) codes. The potential claimed for fractal systems in achieving high data compression ratios does not appear to have been realized in recent years — but this could change with higher processing rates.

The 'generic' shape of a stretch of coastline when viewed from space is conceptually the same as the shape of a bay when seen from terrestrial tower heights, and these shapes are essentially the same as that of a short section of foreshore seen from eye height — especially when you stand on a rocky shore and look down. Similarly, a tree shape is roughly the same as a branch, which is roughly the shape of a twig.

There is nothing profound or significant about all of this, but fractals have the potential to make it simple for a computer to simulate mountains and trees and coastlines using only one 'primitive' shape for each, and reproducing this in different sizes to provide the textures and dimensions.

With more complex images, fractals also seem to provide a way of defining elements in a highly compressed form, and without the use of rigid pixel-blocks which DCT algorithms need (and which show up as blocks on the screen on occasions). Fractal 'blocks' are irregular and tend to make an image look like an oil painting. See DWT, DCT and fractal transform.

fractal geometry

This branch of geometry has provided us with a revolutionary way of describing complex mathematical computations as computer-generated patterns known as Mandelbrot or Julia sets. Fractal geometry has been closely associated with chaos theory mainly because fractal shapes characterize order within the apparent chaos.

Many natural shapes (trees, mountains, coast-lines) comprise many smaller pieces which are, themselves, scaled-down versions of the whole (fractals). So fractals appear to have wide applications in the computer synthesis of realistic images, and in music and art, as well as science.

fractal transform

A new technique of reducing the size of a digitized image file by converting it to various size fractal 'primitives'. This is an approach similar to DCT and Fast Fourier Transform, but in fractal transform, the computer scans the digitized image looking for parts that are the same as the library images held in memory.

When it finds a match, it transmits or stores the formula representing that pattern, not the pattern itself, and this formula can often be further compressed. This technique could eventually produce better quality images from smaller files than DCT compression. At present, the difference is not great.

fraction

In computing, fraction refers to decimal places, not just to the old quarter, half, three-quarters, etc.

fractional T1/E1

This refers to the supply by the teleco carriers of bandwidth in integer fractions below the standard T1 (1.54Mb/s) and E1 (2Mb/s) channel level. At one time the carriers would only offer the full T1 or E1 data-rates to customers on a take-it-or-leave-it basis, but recently they have been more prepared to offer bandwidth in lesser amounts.

In E1 (European) countries, Fractional E1 is sold in terms of p x 64kb/s (where p is any integer up to 30). With North American T1, the term 'Fractional T1' can mean p x 56kb/s channels but more commonly it means p x 64kb/s (below 23). For p x 64 you need also to specify B8ZS line coding (not AMI).

FRAD

Frame Relay Assembler/Dissembler. (aka FR Access Device) Frame relay requires a FRAD to packetize the data into frames and to put a header and trailer around the blocks of data. The FRAD also checks the data for errors both when it is sent and when it arrives. The PAD does the same job in X.25.

- An EtherFRAD acts as the DSU interface between an Ethernet and a frame-relay connection.

fragmentation

- Files which are constantly being updated on a hard disk drive very quickly become fragmented. Every time a file is saved it is completely rewritten ***before*** the old copy is disabled.

The save process tends to scatters segments of the file all over the disk, and this considerably slows the action of the drive when it tries to access the data. Special de-fragmentation utilities exist to overcome this problem — but always back-up your disk before de-fragging.

- Large packets of data on a network may need to be broken down (segmented) to fit the smaller frame or packet size of, say, an ATM network. This process is more correctly called segmentation and reassembly (S&R) but the term fragmentation is still widely used.

FRAM

Ferro-electric RAM. See.

frame (Ethernet and frame relay)

The common usage often fails to recognize a difference between the time-slot and the protocol-encapsulated data the frame may contain.

For any of the IEEE 802.x standards, a frame is treated as being the same as a packet. In Ethernet and 802.3 LANs, a frame is the basic packet of information which carries the addresses and data, so you will often find reference to Ethernet (Physical layer 1) 'frames' holding Ethernet (Logical layer 2) 'packets' of data.

In frame-relay, the frames are really packets, so the term is being misused in the Ethernet way.

To be strictly correct, Ethernet frames/packets are datagrams. Datagrams are packets which contain all the information necessary to travel through the network independently; they don't require a virtual circuit to be established first. In all simple networks (LANs), the packets are always datagrams.

frame (networking)

- **Wiring hub:** In its original context in analog telecommunications, the term frame meant the connection point for the local loops in telephone exchange wiring. In electronics and communications hardware, a frame is still the line-terminating panel where incoming and outgoing cables are joined.

• **Data structure:** In general, a frame is a standardized data structure into which information can be placed for storage, communications, etc. The frame provides a point of reference, so that the appropriate data-fields (both overhead and payload) can be identified.

The terms frame, packet, block and datagram, are sometimes used almost interchangeably.

'Frame' usually refers to a physical layer data-structure which surrounds slots in a time-division multiplexed system. However it is more than just flag bits, it will have its own control fields and perhaps error checking as well.

It is still correct to talk about an asynchronous byte being framed by a start-bit and a stop-bit. And with synchronous signals, the equivalent preliminary/trailing signals are called sync-bytes, flags or preambles/postambles — and this is framing in exactly the same way.

'Packet' usually means a self-contained data-structure with its own header containing addresses and sequencing information, and a trailer with error checking. It is usually either a fixed size (say, 128—2048 bytes) or it is strictly limited in length. However, note that a frame in frame-relay systems is actually a packet — so the term is used widely and vaguely.

• However, contrary to the above, some publications claim that a frame operates only at Layer 2 (Data-link), and that it contains its own control information for addressing and error checking in addition to the payload. This is how the term is used in HDLC.

• A frame may also be known as a 'block' in bit-multiplexing. It is a sequence of contiguous bits bracketed by, and including, beginning and end 'flag' bytes. Blocks are data-structures which usually contain payload for only a single user. These are the equivalent of frames on a single-user system.

• In ISDN, each frame consists of 38 data bits and 10 overhead bits, with a frame-rate of 4000 a second.

• In summary, don't take any of the above as prescriptive; the usage of all these terms is so casual that you must always interpret them in context.

frame (other)

• In **television** a frame is a complete picture.

— When 2:1 interlacing is being used, as in today's analog TV (PAL, NTSC and SECAM) systems, two fields comprise one frame.

— In progressive scan systems, a frame is one field.

• In **film** projection and in 30Hz NTSC television there are 24 frames of image per second. In PAL and SECAM television systems there are 25 frames per second.

• In **videotex** and **teletext**, a frame is the information needed to paint one screen. A number of frames (up to 26 in Prestel) may make up a 'page'. See video buffering.

• In **computer tape** systems a frame is the array of bits across the width of the magnetic tape.

• In **CD-ROM** a frame is synonymous with a disk sector.

• In **AI**, Marvin Minsky uses the term 'frame' to mean a 'mini-database' which collects related information together in a structured way (as used in the Cyc database). It is a data structure which represents a stereotype situation in machine translation.

frame alignment

The adjustment of timing by the receiving terminal, so that it is in synchronization with the sender. The receiver then knows which bits are to be used for what purpose — where the slots in a TDM frame begin and end, etc.

• A frame alignment byte or word, is the pattern which identifies the beginning of a frame.

frame buffer

See video buffering.

frame check sequence

FCS holds the error checking data for a frame or packet. It is most likely to be a CRC but it could be a simple checksum. Frame check sequences are generally tagged on to the end of each frame. See CRC and BCC also.

frame delta

An image storage and/or transmission technique that deals with information changes — only that which is different between this image and the last. The transmission of 'difference' information is the simplest way to reduce the amount of digital data needed to reproduce quality moving video images.

In modern video compression systems, frame delta techniques are the primary form of data reduction, and they are usually closely associated with DCT. See DCT and MPEG.

frame format

The format of the frame in data networking is the primary logical protocol. It is the definition which specifies the meaning associated with each bit, byte or block within the frame.

frame grabber

• In data and telecommunications, this is a temporary buffer which holds the frames of data passing along a channel only long enough for the address to be inspected and a decision made as to whether to accept, or pass it along.

• In teletext systems a frame grabber is used to take and store the screen page/s selected for viewing. These data frames are cycled rapidly, and are grabbed out of the TV signal's blanking interval when required. It may take a number of seconds for the required frame of information to become available.

• In television systems, a frame grabber is a large buffer which can hold the information of a single screen image frame. The process of frame-grabbing is one of digitizing the analog image and holding it in a very large (MegaByte) frame-buffer.

• In future home television sets, the full information for a frame will probably be held in computer memory, and the image can then be scanned to the display screen at whatever rate the display can handle (probably progressive scanning at 100Hz).

• In graphics, the frame grabber is also called video digitizer because it is responsible for the digitization of the analog images. It takes the analog video signal, frame by frame, and converts each to digital information — which is then stored in the frame buffer. Video frame grabbers are usually plug-in boards which store one field of the TV frame at a time, and which can transfer that image data to computer memory under DMA.

frame mode

See frame-relay.

frame rate

• The rate at which video frames are displayed on the screen. In NTSC systems, fields are repeated 60 times a second, but the frame rate is 30 fps. In PAL and SECAM which are 50Hz systems, the field rate is 50 times a second, and the frame rate is 25 fps. This is very close to the international film standard of 24 fps — but note that it reduces the playing time of films by 4%.

• In synchronous networks the frame rate is the rate of generation of packets or slots.

frame-relay

Once described as being 'X.25 on steroids', this is a cut-down, sped up version of X.25 packet switching for imaging networks and LAN interconnection — and possibly for voice.

It is popular, because many of the newer X.25 packet-switched networks can be upgraded to frame-relay by software changes — and frame relay is seen as lying on a direct evolutionary path to ATM. Generally it is more efficient than X.25 or ATM, provided the link does not have a high error rate. However, frame relay assumes that modern fiber networks will have low error rates, and it therefore treats the overhead normally provided for error correction as unnecessary.

Frame-relay is now a mature standard of the ITU and ANSI. It is both an access and end-to-end protocol, and in public networks it is used across the link between the user's devices (hosts and routers) and the network devices (switching nodes).

Frame relay differs substantially from cell-relay techniques in providing variable-length cells. It is intended for data traffic in the range from 56kb/s to 2Mb/s, but it is considered to be fast enough for low bit-rate voice using low-rate Embedded ADPCM and CELP vocoder.

The standard has an extensive range of congestion notification (not control) mechanisms — many of which are optional — but it relies on the end devices (terminals) detecting errors and demanding retransmission.

See BECN, FECN and FRAD.

frame-store

• The memory part of a frame grabber.

• Television receivers in the future will require digital signal processing to handle the combination of analog and digital signals, augmentation channels, data compression, anti-ghosting, etc. All of these ideas are incorporated in digital, wide-screen, and HDTV proposals.

So the incoming signals will need to be converted, stored temporarily, and processed, and this is done best with a digitized image. This digital image will be held in special Video RAM memory before being presented to the processing chips, and only after processing will it be scanned to drive the screen. To improve the flicker characteristics, it may be scanned progressively, two or four times. MPEG2 currently requires a frame-store of 8MBytes, and 16MB if it uses B-frames.

framed

This term is applied to data streams such as PDH and SDH, as distinct from asynchronous cell telecommunications systems such as ATM.

framework

Framework libraries are used in object-oriented software design to provide productivity benefits in building complex systems. They are an extension of the class-library concept, but at a higher level of abstraction. A framework may contain hundreds of objects.

framing

In communications this is the process of determining which groups of bits (within a frame or block) constitute a character, and which group of characters represents a message — in other words, which bits within a frame constitute a meaningful chunk of information. This process is handled according to the protocol (format and timing) definition.

framing bits

In some forms of time-division multiplexing it is necessary to regularly insert special control-information bits into the data stream. These provide points which reference the beginning of the slot sequence, and they can also serve to identify which terminal links are concerned with each of the message streams. These are framing bits.

Frankfurt specification

An upgrade on the Orange Book specification for CD-R drives; the original specification would permit only single-session recording. This is an ECMA development for a new hybrid CD-ROM file specification to work with Unix, Mac, OS/2 and Windows NT.

It is known as the 'Frankfurt' standard after the original formulation group, and it will remain part of the Orange Book standards for recordable CDs. It supports full multi-session incremental updating of the disk and directories, without current limitations. See multisession formats and CD-R.

free routing

The handling of message traffic (sometime packets) where the messages are forwarded over any available route — without a pre-determined channel being established. The contrast is with virtual circuits and virtual paths. This is also called random routing. See connectionless and datagram.

free space loss

In radio transmission, free space loss is the theoretical reduction in signal power in the path between the transmitter and the receiver, when treated as if it were in ideal conditions without atmospheric reflections and obstacle attenuation.

It is measured in decibels and is defined as 'the ratio of the power received by an isotropic antenna, to the power transmitted by an isotropic antenna' (e.g. no directionality to either antenna). The inverse square law operates here.

- Propagation models will also need to take into account:
  - ground specular reflection,
  - surface roughness,
  - line-of-sight obstacles,
  - buildings and trees,
  - through-structure and building losses.

Note that these are strongly influenced by the height of the antenna, and any directional antenna gain.

free-space transmission

Radio and television delivered over the air, rather than by 'guided media' (coaxial cable, fiber and video dial-tone systems). This is normally referred to as 'terrestrial transmission'. Generally the term does not include direct broadcasting satellites.

free-text search

The ability to search through a database for any name, date, word or number, held in any field. A search strategy without artificial limitations. This can be done by the computer just checking successively every word in the database (a lengthy job with any large database) or by the creation of an inverted index (without stop words). See inverted indexes and stop words.

Freenet

These are public access BBS systems which are usually run on a non-profit basis. The funding comes from many sources, but recently there has been a worldwide movement to place Freenet terminals in libraries and public spaces to give the average citizen access to the Internet and e-mail. Most of these services are funded by local governments. The US Cleveland Freenet has 35 000 registered users and handles up to 10 000 calls a day.

Often the local government authorities will make their own local information services available over the Freenet and they will also provide access to state and federal government databases. This is now a formal movement with an underlying philosophy, which is the 'empowerment' of the average citizen in the age of the superhighway.

frequency

The measure of cyclic phenomenon — the rate of repetition. Frequency is quoted in terms of the number of cycles in a certain time period (usually a second) and it is typically associated with audio and electro-magnetic (radio) waves. The old term 'cycles per second' has now been replaced by 'Hertz'.

If the propagation speed of a wave is constant (audio in air, or radio in free-space) then frequency and wavelength are inverse functions. See wavelength.

- Audio frequencies range from about 50Hz to 20 000Hz — although older listeners may only hear higher notes to about 15 000Hz. Below 50Hz you 'feel' the signal rather than hear it.
- Radio transmissions travel at 300 000km per second, so a signal which has a frequency of 300kHz will have a wavelength of 1km. A 3MHz signal will have a wavelength of 100 meters.
- For radio frequency bands, see radio spectrum.

frequency discrimination

Any tuner (a 'bandpass filter') must be able to select carrier signals within a certain range of frequencies (the carrier bandwidth) from a radio spectrum made up mostly of unwanted radio frequencies carrying other signals. How well it selects the desired band and rejects the unwanted, is a measure of its frequency discrimination.

frequency division multiplexing

See FDM.

frequency domain storage

See photon-gated.

frequency hopping

A technique of jumping around the frequencies in a set manner, either to avoid jamming (military), for security reasons, or as a technique for capacity and quality improvements. The idea was invented during World War II by actress Hedy Lamarr.

- In spread spectrum systems it is a basic coding technique. The receiver dedicates a large number of channels for the incoming signal and the signal hops around between these frequencies on a pre-established pattern. The technique requires good frequency-synthesizer devices to synchronize the transmitter and the receiver. Current limits on the hopping speed make frequency-hop spread-spectrum suitable for only limited applications, although recently claims have begun to emerge for PCS. See CDMA.
- In GSM and TDMA systems, frequency hopping is a technique which was added to the fundamental TDMA cellular system to average out bad channels. Each user gets a short burst of the bad channels along with the good. However, frequency hopping cannot be used whenever a beacon channel is being used on a frequency, and this limits its application seriously.

This bears little relationship to true spread-spectrum techniques since generally only two or three discrete channels will be available in a base station. In a single-carrier GSM base station, frequency hopping is obviously not possible.

frequency locking
See clock.

frequency modulation
The technique of modulating analog or digital signals onto a radio-frequency carrier by modifying its frequency in sympathy with the incoming baseband (analog or digital) stream. The distinction is with amplitude modulation techniques which modify the carrier's amplitude.

To be strictly correct the term FM should be applied to analog modulation techniques, and FSK to digital data modulation — especially of audio frequencies in modems. See FSK, FM, and modulation.

• Frequency and phase modulation are essentially the same.

• Narrowband FM is when the effective channel width doesn't exceed twice the bandwidth of the modulating signal. If it is to carry telephony at 4kHz, then the NFM signal will need a bandwidth of 8kHz.

• The sidebands of FM occur at integer multiples of the modulation frequency, on both sides of the carrier.

frequency offset

• Unwanted variations in the frequency of a carrier — typically in frequency-division multiplexed channels where offset can cause adjacent interference.

• An echo suppression scheme used in submarine cables and satellites. It creates problems for high-speed modems.

frequency response
The ability of a device or channel to handle the full range of the signal frequencies required. For hi-fi audio equipment, a 'flat' (equal amplitude) response is required for all frequencies from about 20Hz to 20,000Hz. Any electronics in the circuit are required to amplify to equal degrees every frequency within this range (within, say, +/–3db).

However, in practice, electronics always introduce emphasis at certain frequencies. To overcome these problems, pre-emphasis can be added before transmission, or equalization can be performed later.

frequency reuse
A term used in satellite communications, television and radio to emphasize the value of space diversity.

• The limited range of frequency bands that can be handled by the satellite transponders can be reused by polarizing one set horizontally, and the other vertically (or both in opposite circular directions — clockwise and anti-clockwise).

• Frequencies are also reused by virtue of the satellites being separated by angle in the sky. Provided the signals don't overlap geographically and the satellites aren't too close together, the same frequencies can be reused over and over again, provided the receiver dish has a high pointing accuracy and can discriminate between signals from different locations.

• In analog TV and radio broadcasting we need separation between broadcast channels, both in frequency (see adjacent channel interference) and in distance (see co-channel; allowing enough separation for best propagation conditions). These two factors establish the frequency reuse pattern.

Digital transmission systems generally allow much more frequency reuse, and CDMA digital techniques allow almost total reuse in adjacent cells — the limit is only in raised noise levels.

• In cellular mobile systems, a band of frequencies must be subdivided into clusters so that cells don't interfere with each other. Usually only one seventh (N=7) of the available channels can be allocated to each basestation.

With AMPS, TACS, NMT and GSM there will be seven or nine such subdivisions, with these cells forming a cluster (a grouping of cells within which the whole frequency allocation is used). With CDMA the frequency reuse is one (N=1; each cell can use all the frequencies).

• Sectorization complicates the above.

frequency shift

• For modems, see FSK.

• In telephone communications, frequency shifts are made to the incoming calls in an exchange so as to multiplex a number of them into a single channel for onward transmission over long distances. Each call is 'frequency shifted' to occupy a different frequency 'slot' in the FDM multiplexed signal. Eventually, after two or three multiplexing stages, there may be hundreds (or with coaxial, even thousands) of calls on the one carrier.

See frequency division multiplexing and coaxial cable.

• Translators are small radio transmitters, usually in remote places which can receive a television signal on one channel, frequency shift it to another, and retransmit it at a higher power. Beam benders retransmit it at the same frequency.

frequency splitting
The technique used by Telebit/Trailblazer, ADSL, COFDM and other ultra-fast modems of splitting the bandwidth of the carrier into different frequency sub-channels, modulating parts of the baseband signal on to each, and transmitting each using a selected modulation method often at a relatively slow-speed. The original signals are then reconstituted at the other end.

Theoretically, little should be gained from this approach, but in fact, intelligence can be applied to both the selection of channels (to avoid those suffering from interference) and the modulation techniques (applying the best method to each frequency range).

These techniques are being applied both to digital audio and video broadcasting.

• Telebit Trailblazer uses 512 separate channels. A powerful processing chip is needed in these modems to supply enough intelligence to supervise the transmission, check errors, and cancel the use of noisy channels. Note, however, that V.32 is even faster and more robust.

• DMT is the technique now being applied to ADSL video-over-copper technologies. This also uses a frequency splitting approach to achieve data rates in the order of 2–6Mb/s over the local loop for short distances from the local exchange (now 2–6km).

• COFDM techniques are being developed for digital television and radio transmission. See orthogonal also.

fricatives

In computerized speech coding, these are unvoiced sounds such as f, s, and sh which are all made without using the vocal chords. These sounds create special problems with speech synthesis, vocoders, and voice recognition.

front-end

• In personal computers, this is the user-interface; the man–machine junction of the keyboard, mouse and screen. The back-end is where the computer connects to the network or a printer. This alludes to the logical functions, not the physical placement.

Front-end applications are those which the operator uses to present, manipulate and display data.

• In client-server operations, the 'client' (the user's workstation) provides the front-end, while the server provides the back-end. So in LANs, front-end applications in PCs and workstations are used with back-end applications (mail or database engines) on the various network servers.

• In complex mainframe environments, a special communications processor (a FEP) may act as the front-end to the mainframe (in its connection to the network) to relieve the load on the CPU. It takes control of the communications protocols and isolates the mainframe and the application from problems. Note that here, front-end is paradoxically the network connection.

front-plane

See front-end.

front porch

See back porch.

Frox

A sound and video control system designed by Andy Hertzfeld (a major Macintosh designer) for the control of TV and video in the home, and for the provision of digital surround-sound channels. See THX also.

FRR

First Rake Read scanner. A barcode scanner which can read on one pass, no matter which direction.

FS

ASCII No.5 — File Separator (Decimal 28, Hex $1C, Control-\). A control character used to separate data logically into files. It is also known as IS4 — Information Separator 4.

• The four information separators (FS, GS, RS, US) are used in a variety of ways, but they should always be used in a hierarchical or nested manner with the FS being the most inclusive, and the US, the least inclusive.

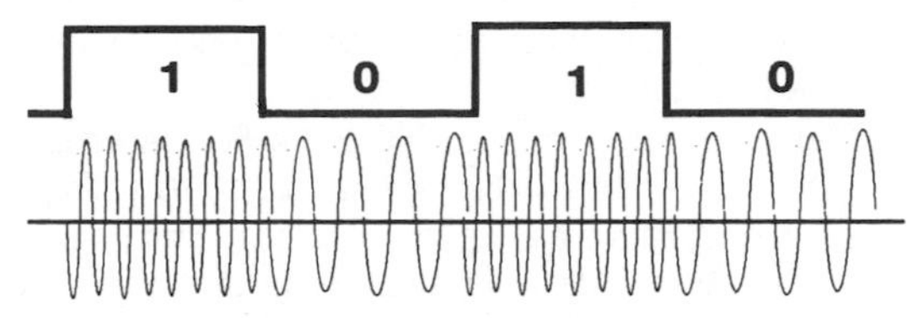

Frequency Shift Keying

FSK

Frequency Shift Keying. This is the technique used by older slow-speed modems (below 1200b/s). These modems transmit data bits by alternating between two audio frequencies, which is the 'frequency shift'. Center frequencies are usually quoted with frequency offsets, traditionally +/– 100Hz. By convention, higher frequency is called the 'mark' and is taken as binary 1.

So in full-duplex communications, two frequencies (the originate pair) will be used by the originator of the conversation, and another two (the answer pair) by the receiver — so four audio frequencies will be in use in total.

FSK techniques were easy to design into a chip, and they worked very well in modems over public-switched telephone networks for rates between 200b/s and 1200b/s. The US Bell modems use different frequencies to the CCITT standards — so these modems can't communicate.

FSL

Free Space Loss. See.

FSLA

Fiber Laser Sources and Amplifiers. A general name for fibers which have been doped (usually with erbium) and which can amplify a signal without the need for opto-electronic conversion. See erbium amplifier. Aka EDFA.

FSN

Full-Service Network. See.

FSS

Fixed Satellite Services. A wide term covering non-public telephony and DBS uses of satellites. These are generally direct-to-business services (aka BSS) where companies will purchase capacity to link, say, bank branches or remote mining camps. The ground stations will be at fixed locations. See VSAT.

FT3

A fiber-optical network standard used in US telephone systems. It consists of 12 ribbons, each of which contains 12 optical fibers. Each of these fibers carries 672 channels of PCM voice (or data) plus overhead. Overall this system can carry 144 data streams, each at a DS-3 rate of 44.736Mb/s.

FTAM

File Transfer, Access, and Management. An OSI (ISO 8571) vendor-independent data-exchange protocol which functions at the Application layer. This is one of a class of interface protocols known as SASE which allow local-style operations to be performed on remote files.

FTAM governs the transfer of files to and from different systems, and it is used for reading, writing, sharing, managing and storing files. It allows an application to create and delete a remote file, read the file, and transfer the file in binary or ASCII format.

It is a high-level interface which creates its own connections (associations) but you need FTAM on both ends of the link. See ACSE, MMS and MHS also.

ftp/FTP

• **File Transfer Protocol:** (The acronym, for some reason, is often written in lower case.) This is a vendor-independent industry standard. It is an Applications layer interface (a high-level protocol) for file transfers which is widely used on the Internet (with TCP/IP).

The protocol specification describes how one computer can host others for the purpose of transferring files in either direction; users can see directories on either and perform file-management functions. It's similar to Kermit, and is closely associated with Unix.

• An 'ftp site' is an archive of files. Some will only be available to password holding users, while others will be free for anyone to download and use. In general, the files are downloaded only to the nearest on-line host (with a full-time connection, not a dial up connection — although see SLIP and PPP).

• Anonymous ftp means that the library of files is freely available for use by anyone. You don't need to make a payment, or to establish a prior relationship with the organization. If asked, your Name is 'anonymous' (or guest) and your User ID/Password is your common e-mail address. See Anonymous ftp.

• The key commands are:

— ftp<remote.computer.name>
— open<remote.computer.name>
— cd change directory.
— ls list.
— get transfer the files to the ftp client.
— close

If you are using a dial-up connection, you will then need to transfer these files from your local host to your own PC in the normal way.

• **Future Techno-Polis:** A dreamer's term for a Utopian future city combined with technology park, where academic experts and high-tech industries would gather and blend to produce unimaginable benefits in R&D. No one has managed to build one yet. Also called Multi-Function Polis.

FTS

Financial Trading System/Services.

FTTC

Fiber To The Curb. These networks are being developed both for small business and residential use. They generally provide switched and non-switched narrowband services and one-way or interactive video distribution. The final section of the access network is implemented using copper cable — which is what distinguishes it from FTTH.

The specific use of the term was for a system which provided cable TV to about a dozen homes from each fiber. This has been replaced by much larger Fiber Serving Areas (FSS). This now also applies to a hybrid fiber-copper system where Switched Digital Video is transported over trunk fiber, and splits at a curbside passive hub to feed perhaps ten homes.

• The FTTC concept is now being challenged by HFC (Hybrid Fiber Coaxial). See FTTH, FSS and passive networks also.

FTTH

Fiber to the Home. The various techniques for carrying optical fiber in the local loop from the exchange (head-end) to the home or small business.

This has been described as 'Watering the garden with a fire hose,' which is an understatement! The use of very high data-rates (620Mb/s) to carry voice, data and video services right to the set-top box was felt necessary before video compression.

There are a variety of FTTH schemes. Some proposed using a single fiber between the local exchange and each home; others projected a single fiber 'trunk' which would then service a number of homes (typically 16 to 20) via a passive splitter at a local pillar box; others proposed dual fibers in a loop configuration. FTTH proposals were tied to Broadband-ISDN in the days when the established data rates were seen as 620Mb/s downstream and 155Mb/s upstream. Now this is no longer believed to be necessary.

You should distinguish between FTTH and FTTC systems.

FTTK

Fiber to the Kerb. See FTTC.

FU

Functional Units.

FUD

Fear, Uncertainty and Doubt — the FUD factor is a favorite marketing game played by the dominant hardware and software companies. They sow the seeds of FUD by leaking premature and misleading information to the press. See vaporware.

full adder

See half-adder. Full adders usually need three NAND and two XOR gates.

full-duplex

Both ways simultaneously.

• The situation where data travels simultaneously in opposite directions on the one cable. Your modem may be sending data to another computer, and at the same time the other computer can dispatch commands back along the circuit (the 'back channel') to prevent its buffers from over-flowing, etc.

The distinction is made with half-duplex usually, where the sender transmits a block of data then waits for confirmation before sending a second block.

Full-duplex is often confused with Echoplex (error checking – see), since the deliberate echoing of characters is often used within full-duplex systems as a form of error-control. But these are two quite different and distinct operations.

The alternatives to full-duplex are simplex (one-way only) and half-duplex (ping-pong) systems.

In the early days of modem communications, the problems of unwanted echo on the phone lines often made half-duplex more suited for higher rates, while Echoplex (which needs full-duplex) was very useful for keyboard low-rates. These distinctions no longer apply.

- In telecommunications the term is also applied to the use of two-wire circuits rather than four-wire (half-duplex) physical links. So two half-duplex links in parallel but opposite directions, constitute a full-duplex circuit.

- With voice telecommunications, the old press-to-talk radio systems were half-duplex ('Roger! Over and out!'), while the modern cellular phones are full-duplex (however they use two one-way channels). See duplex offset.

- Full-duplex in error control, refers to techniques such as HDLC and SDLC. See ARQ.

Full-duplex Ethernet

A recent version of Ethernet which can provide 10Mb/s both ways — for a total traffic rate of 20Mb/s on a single cable. The standard is supported by a limited number of vendors as a substitute for standard 10Mb/s Switched Ethernet, but many consider that the 2x improvement is not enough to justify the change (given that 100Base-X looks likely to be just as cheap).

Ethernet is inherently half-duplex. One wire pair in any 10Base-T connection is used to transmit and the other to receive. So if packets are ever seen in both pairs simultaneously, it indicates that a collision has taken place.

However with a single user on each segment of cable, a collision isn't possible — so full-duplex operations can be used on a switched star network. These can provide very high rates to isolated workstations attached to their own network segment.

In practice, this becomes possible only on an Ethernet switch which is capable of full-duplex operations, and obviously with only one user to each port. You also need to buy special full-duplex Ethernet workstation cards. This is for Category 3 and 5 cable. The IEEE 802.3 standard will have full-duplex capabilities in late 1995.

Full-duplex Token Ring

Unlike Ethernet, Token Ring is already a full-duplex scheme where one wire-pair carries data and the second carries the token and the error correction. However, if you dedicate a single segment to one user (using Token-Ring switching) then the token becomes unnecessary, and the second pair can be used to carry data. This provides total rates across the network of up to 32Mb/s.

Changes of this kind can often be made in software driver upgrades, and/or a new ROMs. Full-duplex operations will be added to IEEE 802.5 in 1995 for Category 5 and STP cable.

full-service network

The new idea of full-service communications, is a single monolithic service which would deliver phone calls, cable-TV, video-on-demand, and home shopping services. Usually the emphasis is one-way, from them, to us. It requires a hybrid fiber-coaxial network, and probably digital transmissions, and ATM switching and multiplexing. Time-Warner in Orlando, Florida, has the first trial.

full-text databases

Public on-line databases which contain the complete text of the information, rather than just an abstract or catalog card information. Full-text databases may be indexed by all the useful words (inverted or full-text indexing) or only by selected keywords. Don't confuse these with free-text databases.

full-text indexing

In a public on-line database the information is stored in fields within records. Some databases only index the few words typed into a 'keyword' field, some index on a hash value stored in one field, but others index every word in the database — with the exception of the small linking stop-words like 'and', 'but', 'the', etc. See inverted indexes.

fully-provided/provisioned

A term widely used to mean that the service is fully equipped and capable of handling any bandwidth required.

- A fully-provisioned fiber cable would be one where all of the fibers were equipped with lasers, detectors and repeaters. There would be no dark-fibers.

- A fully-provided route is a link which can handle everything required (including peak loads) without needing an alternate path or backup.

fully-qualified domain name

Don't confuse the naming system of the Internet with the IP address system, which consists entirely of numbers.

The naming system of the Internet has been undergoing progressive changes. To be a fully-qualified host (domain) name, the host address should have:

- a top-level country code on the far right (au=Australia, uk=Great Britain; None for most USA hosts);
- the common name of the computer immediately adjacent to the @ sign;
- one or more sub-domain names,
- often an organizational name,
- followed by an organizational type ID.

• The sub domain of the address will generally show one of the following to identify the type of organization:

com	a commercial organization,
edu	an educational institution,
gov	a government organization,
mil	military,
net	a gateway or network administrative host,
org	a private organization not in the above categories.

• Domain names must be translated to IP addresses; this is done by the DNS. See IP address also.

function

In programming. A function is a standard algorithm, subroutine or procedure, usually built-in to a programming language as a module which has been added to a library. It will return a result only when given an argument but it is always available to the programmer in the form of a command or call (other functions can often be added when needed).

Most programs are divided into functions (some written, some from libraries) in this wider sense of the term. A module is a group of related functions in structured programming.

• These are usually complex routines such as those that perform statistical operations like Average, Sin, Sum, etc. Trigonometry functions are sine, cosine, arcsine, etc; mathematical functions are integer, exponential, log, square root, modulus.

• The term is wide enough to encompass a whole range of built-in calculations and test operations which can be 'called' or used by the programmer.

Apple's HyperText language, for instance, has a clickLoc function which returns the x/y coordinates of the point on the screen where the user most recently clicked the mouse-button, and another which computes the present or future value of an ordinary annuity.

• String functions can be used to select, say, the first character, or the third character, or a non-alpha character in a text string.

• In communications, a function is a machine action such as a tab, line feed or carriage return.

• Function keys on modern computers usually trigger relatively-standardized macro functions which may be built into the operating system or added as utilities. Function keys generally created multiple-character messages or non-display characters.

• In the early days of PCs, function codes or function keys were often synonymous with character controls. The shift key, return key, delete, etc. were all considered to be functions. See character control and the ASCII No.5 alphabet.

function keys

The MS-DOS keyboard has 12 of these and the Mac has 15. They would be better called 'macro keys', since they can be assigned quite complex macro functions. F1 was traditionally used for Help. By including escape codes (F1 =59, F2=60 ... F10=68) in the PROMPT statement of AUTOEXEC.BAT you can assign keystrokes to function keys. Doskey in MS-DOS uses F7, F8, and F9, but the others can be reassigned.

functional group

In an EDI transaction, a functional group is a set of related electronic documents — the purchase orders, invoice, etc. associated with a transaction.

functional programming languages

FPLs: Specific types of programming languages of the so-called '5th generation' (5GL — a very vague term) which are derived from lambda calculus. LISP was a pure functional language originally, but it has since evolved into a procedural language.

The key distinction is that functional languages don't depend on the order in which statements are written or executed, while procedural languages do. See formal methods.

Procedural languages must define an order of execution because they use variables which hold different values at different times in the execution process. 'Functional', 'applicative' or 'single-assignment' languages don't have this problem, so an explicit order of operations is not essential. See language generations and referential transparency.

functional specification

One that describes what the device or software does. It will explain the relationship between inputs and outputs, but not necessarily what goes on inside the box.

functional unit

FUs are used in transaction processing to perform different and discrete operations. Some transactions (travel-bookings for instance) will involve a number of FUs — all of which need to be complete and committed before the overall transaction can take place. There's no point in having a hotel booked in Singapore if you subsequently can't get on the plane.

fundamental

The foundation.

• **Audio:** The basic frequency. This is the predominant note. If the sound appears to you as a Middle A (the international standard), then the fundamental will be 440 Hz, while the harmonics ('overtones') will provide this frequency with its tonal richness. It is the strength of each harmonic (which depends on the instrument used to produce the note) which makes a flute's Middle A sound different to a trumpet's Middle A. See Fourier transform.

• **Video:** When a video image is subdivided into blocks (say, of 8 x 8 pixels) the overall 'shape' of the scan output (within each block) will usually be roughly repeated over the eight vertical scans. Whether analog or digitized, the chaining of these eight short block-scans would generally take on the characteristics of a repetitive pattern wave.

This can be subject then to Fourier analysis to extract the fundamentals and the overtones. The dominant shapes and colors of the block can therefore be interpreted as a 'fundamental' and harmonics/overtones will progressively add finer and finer (higher-resolution) detail. This is the basis of Fast Fourier Transform and DCT.

fusion

To melt together into one mass.

• **Sound:** Two or more sources of sound fuse together to create a single composite wave. When you listen to an orchestra, you are only ever hearing one sound waveform (in each ear) at a time — and the shape of that waveform is a summation of the waves from all surrounding sources.

• **Sound stereo:** Waves arriving from different sources or directions will reach each of your two ears at slightly different times. However the brain fuses these into one single sound perception, and uses the time difference to calculate the position of the source. The fact that you only hear one orchestral sound, however, is the result of the brain's 'fusion' ability.

• **Vision:** There are three types of fusion in vision. One gives the sense of continuous motion from discontinuous images, the second merges the sense of flicker to create the impression of continuous light (not the same as the first) and the third is involved in creating stereopsis from slightly dissimilar images.

— **Motion fusion** is the process of interpreting a quick progression of still images as one continuous flowing motion — as in film and television. Generally 12 to 24 images per second will create a smooth motion fusion effect, but at these rates the image will appear to flicker in brightness (see below). Cartoons are generally produced at the 12 fps rate (each is doubled).

— **Flicker fusion** requires that we show each image two or four times in bursts with black space between. For this reason, movie film is shot and projected at 24 frames per second, but each image is projected twice in the cinema to create a 48 flicker rate. In TV, 25 frames are converted to 50 fields for the same reason.

— **3D vision** of stereopsis comes about because each eye sees close objects from a slightly different position, and therefore the image in each eye is slightly different. When the brain fuses these two images together, the difference between the two, creates an additional feeling of 'depth' — the stereoscopic effect. This effect normally functions only over distances of a few meters. With stereoscopic photography the effect is exaggerated, which is why these images tend to have a 'cardboard cut-out' appearance.

• **Flicker:** The flicker-fusion frequency (the point at which flicker becomes unnoticeable) depends on the brightness of the image and brightness of the background, the amount of peripheral vision involved, and rate of change of the images. Also, to a slight degree, the flicker depends on the time-ratio between the bright images and the dark gap between.

Flicker is more noticeable in the peripheral regions of the eye, and across large areas of relatively flat, even color.

Futurebus(+)

The tentative name applied to a proposed new computer architecture (bus standard) being developed by a US consortium of 150 chip and computer manufacturers; DEC, Motorola, Sun and Unisys are involved.

The Futurebus architecture is planned to be about 100 times faster than current PCs, with a top speed of 3.2GBytes/s when shifting data internally. The IEEE is now coordinating this standard. Note that some companies are adding the plus sign to the name which signifies something.

fuzzy

A way of dealing with imprecise or ambiguous subject matter by mimicking common-sense 'logical' reasoning which allows the incorporation of imprecise data such as warm, near or approximate.

• **Fuzzy theory:** This is a branch of AI (overlapping with expert systems) in which a program uses concepts like 'large' or 'old', rather than the precise information it would normally require. Fuzzy theory programs will use a database of heuristic rules much like the knowledge base of expert systems (but often at a lower level) in the interpretation of words rather than phrases. There's a lot of cynicism about fuzzy logic. The Japanese have taken fuzzy logic up as a marketing tool, however it is doubtful how much true 'fuzzy' is involved.

• **Font:** See anti-aliasing.

• **Search:** When you look for the word 'war', the computer application checks its synonym dictionary, and so it will also turn up matches with 'conflict' and 'battle'.

See fuzzy logic, and fuzzy set also.

fuzzy logic

The common explanation is that conventional computers only deal in Yes and No (logical 0 and 1) and so don't handle shades of grey or shades of meaning — but fuzzy does.

This explanation of fuzzy logic is too simplistic to be useful. How does a digital TV screen or a color monitor screen work if the computer is only capable of Black and White?

Obviously binary logic is capable of infinite gradations to the limit of floating-point precision, and weightings can be provided to ranges of data. See fuzzy sets for a better explanation.

The Japanese have become the major players in fuzzy logic, but it appears to be used in that country as little more than marketing hype. You can buy fuzzy logic refrigerators and washing machines.

• Hardware manufacturers are now producing coprocessors and other devices which can automatically provide 'weighted' actions. Thomsons, for instance, have a WARP (Weigh Associated Rule Processor) which is a dedicated high-speed fuzzy logic coprocessor or 'inference engine'. It works on input variables called 'membership functions' and uses rules-processing as with an expert system.

• One explanation of fuzzy logic: 'The computer's ability to develop multi-valued logic for simulating human responses to continuous choices.'

- See ANN and FMF.

fuzzy sets

In expert systems, many terms are often applied in a vague way. Does 'middle-age' apply to someone who is 35? The answer can't be determined by a simple yes/no answer.

However we can apply various confidence factors to a distribution of ages, so that age 50 has a confidence factor of, say, 1.0 that it is 'middle-aged', while 40 may only have a confidence factor of 0.95, and 35 may have a confidence factor of 0.65. This table of confidence factors will produce a fuzzy set which lets us deal with vaguely defined concepts. See confidence factor.

FX

Foreign Exchange line. An American term for a special line which goes to an exchange (central office) other than their local exchange. Such a line may cost $1000 to install plus monthly and usage (both way call) fees but it may by-pass high local call charges and provide special features. The alternative now is often an 800 number. FX was devised decades ago for businesses with high constant volumes of traffic.

fyi

- Internet shorthand for 'for your information'.

- FYIs are also a more formal resource on the Internet as a way of distributing information. They are a subseries of RFCs. These FYI files will have general information about various topics associated with the operations of the Internet.

G

G

- G — giga (thousand million)
- G — gravity. Acceleration equal to Earth's gravity.

G#

- Fax: See Group # facsimile.
- Videotex code set.
- **G1:** The set of graphic characters used in alphamosaic videotex graphics.
- **G2:** In European videotex these are a set of characters which provide special symbols (such as diacritical marks, etc.) for different languages.

G.702

This ITU specification is for telecommunications multiplexing hierarchies. It is the international equivalent of the US DS-# standards. See E1 — E4

E1	2.048Mb/s	30 channels	(PCM)
E2	8.448Mb/s	120 channels	
E3	34.368Mb/s	480 channels	
E4	139.264Mb/s	1920 channels	
E5	65.148Mb/s	7680 channels	

G.703

An ITU electrical and mechanical interface standard for connecting telco equipment to a DTE — mainly for PBXs. This is a raw 2Mb/s interface, while the G.704 standard is framed. G.703 uses either double coaxial lines or a 4-wire twisted-pair on the link side. At 2Mb/s, you can either divide the channel into thirty 64kb/s channels, or keep it as one. See HDB3.

G.704

G.704 is the ITU's standardized way to frame a 2Mb/s bearer for multiplexing 64kb/s channels. It will be broken into 32 distinct time-slots, 31 of which are available for data communications: the other is used for synchronization.

- With ISDN, only 30 channels are available; one of the remainder is for timing, and the other for control. See HDB3.

G.711

The ITU standards for PCM. See.

G.721

The ITU-T's ADPCM voice coder standard used by most PCS systems. It is an international standard and is used in CT-2 and DECT mobile phones. The standard is also widely used in US and Canadian local-loops, whenever pair-gain systems are in place. It doesn't like modems.

G.723

The ITU's new voice compression standard for voice at 6.3 and 5.3kb/s. It is designed to be used with a 20kb/s video stream on 28.8kb/s modems.

The voice coder is a variation on the TrueSpeech algorithm already used by Microsoft and Intel. It is based on LPC. The 6.3kb/s version is called MP-MLQ (MultiPulse, Maximum Likelihood Quantization) and it uses tenth-order LPC. This can be implemented on a 16-bit fixed point DSP running at 18MIPS. Overall system delay is 97.5ms.

G.726 & G.727

This is another of the ITU-T's recommendations for ADPCM, but this time at rates of 40, 32, 24 and 16kb/s. It also defines the encoding and decoding algorithms for packet voice. These voice coders produce a compressed voice standard suitable for packet transmission using G.764. See ADPCM and Embedded ADPCM also.

G.728

The ITU standard for 16kb/s Low-Delay, Code-Excited Linear Predictive voice coding (LD-CELP), designed for use when capacity is scarce in wireline systems or with radio networks.

G.764

The ITU standard which, when associated with the G.727 protocol for packet voice, makes it possible to lose some of the packets and still have an acceptable voice quality level.

G-code

See VideoPlus.

G-line

A lightly-insulated wire used to channel microwave energy.

GaAs

Gallium Arsenide (pronounced ˈgas) is a compound used as a substitute for silicon in integrated circuits. Faster chips can be created because the atomic structure of the material allows electrons to travel up to 7-times the speed (although, typically 4 to 5-times). These transistors also have less resistance, produce less heat, and so IC elements can be more closely packed.

However GaAs for chip manufacture is costly and difficult to work with. It is extremely brittle and so it's often layered onto a silicon foundation to make fabrication easier. Currently the largest GaAs wafers are only the size of silicon wafers of a decade or so ago — which means lower yields and higher costs.

At the present time GaAs accounts for only 1% of the semiconductor market. They are widely used in low-noise amplifiers (for satellite receivers) and wherever chip speed is important.

- A low-temperature superconducting form of GaAs has been discovered at the University of California at Berkeley, which may be used in super computers.
- GaAs chips become faster with higher temperatures (the opposite of silicon).

GaAsFET
Gallium Arsenide Field Effect Transistors. Along with bipolar transistors, these are the basic chips used in low noise amplifiers for satellite signal reception systems. See GaAs and FET.

gain
To increase in magnitude. This term is applied to both current and voltage levels. It usually means the boost in signal strength through an amplification system, measured as the difference between the received signal power-level and the output signal level. In communications, gain is usually expressed in dB (as a ratio). The opposite to gain is attenuation.

• In amplifiers, both the required signal and any introduced noise will be amplified. All amplifiers also introduce noise and distortions of their own. Regenerators for digital signals don't amplify the noise component, however.

• Gain-hits are some unwanted surge of power on a communications line caused by lightning strike, switched power, or something similar.

GAL
Generic Array Logic chips. The simplest and fastest of the Erasable (Electrical) Programmable Logic Devices (EPLDs).

galley
In magazine and book printing, the galley (proof) is the first copy of the typesetting. It is usually sent for proofreading, and given to the author for final checking.

gallium arsenide
See GaAs.

gamma
In television and photography, gamma is a measurement of the distribution of tones between black and white — and therefore how much contrast the image exhibits. In film, gamma is the 'tangent' of the exposed image-density plotted against development time, while in electronic systems gamma is defined by voltage levels.

Think of it as a measurement of the evenness of the distribution of tones along a scale from black to white — how 'compressed' the light and shade tones are, or how 'expanded'. Tones can be compressed in dark areas (hence the picture will lack detail in shadow areas) and expanded in light areas (which destroys the texture on white and off-white surfaces) — or vice versa (the whole picture will look flat and lack sparkle and density). The first are 'high-contrast' images and the second are 'low-contrast'.

Gantt chart
A type of bar chart used in project management to depict scheduling deadlines and milestones (the points where each phase is complete).

GAP
Generic Access Profile. The Common Air Interface (CAI) of DECT when used as a wireless PBX. Many DECT units were sold before the common radio-link standard was devised. PAP (Public Access Profile) is a wider term; a GAP is a PAP without some options.

garbage collection
Running through the memory or disk space to reclaim all space not now in use. See defragging also.

gas arrester
These are protective devices against voltage transients, which act by creating an arc in a gas-filled tube when the 'spark-over' voltage is exceeded. Gas arresters don't handle mains voltages, but work well on telecommunications lines.

gas plasma

• **Display:** A type of flat-panel monitor screen which uses an inert gas sandwiched between two glass panels — so it is more like a type of neon light. One panel has horizontal current-carrying elements ('wires') and the other, vertical 'wires'. When an x/y coordinate is selected, the gas at the junction point ionizes and glows. These displays may become popular in laptops because they are easier to read in low-light conditions.

• **Highly ionized gas:** A plasma is electrically neutral because it contains an equal number of positive ions and electrons. In this state it is also highly conductive.

gate

• A logic circuit which usually has one or two inputs and only one output — it is the fundamental building block of computers. There are AND, OR, NOT, NAND, NOR, XOR and a few others which are defined by the Boolean rules they obey.

• One of three connections in a field-effect transistor (FET).

gate arrays
These are pre-programmable, semi-customized chips used for small-run production as a substitute for programmable logic devices (PLDs). They consist of arrays of transistors (logic gates) which are prefabricated on to the silicon.

The distinction between gate arrays and PLDs is that, with gate arrays, the connections and non-connections between array elements must be established at the time of construction. A PLD has every connection made at the time of manufacture, and those not wanted are 'fused' or broken later in a programming process. See also FPGA.

gateways
A term that has been generalized so much that it now means any device or operation which interconnects one system to another. Usually, it means any form of hardware (including the necessary software) that converts protocols between two communications formats — or perhaps, between dissimilar networks or dissimilar network devices.

• In the telephone network, a gateway exchange is the primary contact point for other networks. Each country will have at least one international gateway where submarine cables or satellite feeds terminate. See ISC.

• Satellite systems will have gateways which are 'access points' for signals, and which usually don't provide telemetry or controls. A country the size of the USA, may have one control center, and a half-dozen gateways making connections to the regional telephone networks. See TTC&M station.

• Communications servers (front-end processors) are gateways, and routers are often mis-named as gateways.

• A gateway is generalized as any hardware–software combination that allows otherwise incompatible protocols to communicate. Typically, a gateway will be used to link a workgroup LAN using, say, the XNS protocols to another using TCP/IP. Here it acts as a protocol translator.

The term implies a hardware+software device (usually PC-based) that provides a link between a LAN and a larger information resource, such as a large packet-switching network. You can identify three main types of gateway in general data networking: LAN-to-LAN, LAN-to-WAN and LAN-to-host. See DSG.

• In OSI terminology, gateways are a level up on routers because they use Layers 4 and above. They provide both a physical connection between the LANs and some kind of application link. Most gateways use all (or most) of the ISO's seven layers, but the major functions are at Layers 4 to 7 (Session, Presentation and Application) with protocol conversion at the Transport layer.

• In the IP community, a router was once considered to be a gateway — but now the term is reserved for more specialized devices which perform conversion or translation functions (especially those involving most of the layers in the OSI stack).

• A database gateway will be a dedicated file-server which has been programmed to look like, and act through, its LAN interface as if it were a database of the type that the client's application can access.

• SNA Gateways are hardware/software devices that connect LANs to an SNA mainframe. They handle the translation between the PC and the mainframe by making the PC appear to be a 3270 terminal. The PC can then access applications, files and printers on the SNA system.

• In videotex, a gateway is simply a connection between one videotex system and another — usually between a public system and a private. It implies that the videotex supplier will bill you for the service, and pay the other service supplier in bulk.

GATSOL

Garbage At The Speed Of Light. A commentary on the Information SuperHighway's tendency towards info-glut. It refers to the commercialization and trivialization of the Internet which overloads the interconnecting networks by broadcasting unwanted and useless information.

gauss

A unit of magnetic flux (induction) used, for instance, in the calibration of magnetic tape. The symbol is capital G.

Gaussian

• **Noise:** The 'white' or random noise (background hiss) heard over a telephone line or radio channel. This is noise which is distributed equally across the bandwidth.

In many systems this noise, at a very low level, is caused by the natural movements of electrons in the cables and electronics and is therefore dependent on the temperature of the circuit and components.

• **Distribution:** The voltage of Gaussian signal components is distributed in a normal probability curve — hence Gaussian distribution means 'normal' distribution.

This curve establishes Gauss's probability-for-error distribution when making measurements. In communications, it is used to determine the likelihood that the information signal will exceed the random noise-voltage on the network.

Gaussian Minimum Shift Keying

See GMSK.

GCR

Group Code Recording. A digital tape recording technique that Steve Wozniac translated and used to run the original Apple floppy disk drives. GCR is still used on 6250-bpi half-inch 9-track magnetic tape systems and for some double-density disk drives, although the more-efficient MFM techniques later became more popular.

GCR is block-oriented, and has algebraic error-correction codes which deal well with burst errors. It encodes four-bits as five-bit groups on the tape or disk, and the coding rules state that it must never have more than two zeros in a group of five bits; a process which allows synchronization without clock pulses.

GDDM

Graphical Data Display Manager. IBM's proprietary graphical interface which was incorporated into OS/2's Presentation Manager.

GDF

Graphical Display Format. An IBM image format.

GDI

Graphic Device Interface. Microsoft's screen display imaging model or 'drawing-engine' for Windows. This is quite different to the Graphic Programming Interface (GPI).

GDI is a page-description language in the same category as HP's PCL, Apple's QuickTime, and Adobe's PostScript. GDI printer-drives use the Windows software to create the page in the computer, and then send the final results to the printer as an image. This means cheaper printers, but slower printing. Unfortunately, there are serious quality and flexibility limitations to GDI: it doesn't compare with PostScript.

• GDI has recently been replaced by Display Control Interface for video. See DCI.

GDM

Graphics Device Metafile. See CGM.

GDMO

Guidelines for the Definition of Managed Objects. Specifications for agents and objects in a managed network.

GDSS

Group Decision Support System. See groupware and collaborative computing.

GE

- Greater than or Equal to.
- General Electric Company.

GEM

An early GUI interface program from Digital Research which attempted to make an IBM PC look and act like a Macintosh. See WIMP.

Gemstar

See VideoPlus.

gender benders

These are small plug-in devices for RS-232 cables, similar to null modems but different in function. They look like two identical plugs (or sockets) back to back . They swap the 'gender' of a cable interface so that the cable terminating in a 'male' plug, may be changed to a 'female' socket (or vice versa). However they don't swap the cable connections.

generation loss

The loss of quality, experienced in analog systems (mainly) when copies are made. Each copy process will introduce noise and frequency changes (or in film, resolution degradation), and reduce discrimination. Copies of copies therefore get progressively worse as noise and other defects accumulate. 'First generation' is the term applied to the original recording.

The loss of resolution is proportional to the general 'area' or 'packing density'. So 16mm film copied, say, four generations becomes distinctly blurred, while 35mm film can be copied to six generations with no noticeable degradation.

Analog component video systems also copy better than composite video, and generally the wider the tape format, the better it will copy. VHS low-band tapes will generally only take three or four generations before the image begins to break up.

Digital systems don't suffer these generational defects since digital copies are 'all-or-none'. With good error protection, we can guarantee that each generation of a digital file is identical to the original. This is probably the main reason why the world is progressively migrating from analog to digital media and communications systems.

generic

Applying to a wider class of objects — not specific to one.

genlock

Synchronization signal-generator lock. A television term for the hardware device (often built-in to equipment) which allows the synchronization of two video images. Synchronization is essential if images are to be modified or 'processed' in any way; and it is essential for them to be 'gen-locked' if one image is to be superimposed on another, or 'keyed' into it.

Genlock was originally a proprietary term, but it is now being applied generically to the synchronization units in TV studio equipment (in many semi-professional video cameras), and to plug-in cards that allow TV images and computer graphics to co-exist on a computer monitor.

There are two parts to a genlock system. A signal generator will provide the 'master' (sync creator) signal, and the genlock device is then the 'slave' which follows the master.

In a professional TV studio all cameras and switches slave to a special sync-generator or to the Special Effects Generator (SEG). In non-professional camera equipment, the genlock output signal is likely to be supplied by one of the cameras.

GEO

Geosynchronous or Geostationary Earth Orbit — also called the Clarke orbit. GEO satellites are geosynchronous because they take 24 hours to make one orbit, and they are also geostationary because that orbit is directly over the equator. See geosynchronous and GEOS below.

geometry

The layout of a chip or a logic board. This is also the computer industries' slang term for the width between the circuit tracks on a chip, usually measured in microns.

Geoport

A plug-and-play serial interface which is backward compatible with the UART in most personal computers, but which offers 200-times the bandwidth. Geoport was developed by Apple and is supported by AT&T, IBM, Motorola, Siemens and Zilog among others. Recently there's been a decided lack of interest.

The serial communications architecture of Geoport is optimized for computer-telephony integration. They claim that:

- Any telephone line (up to T1/E1 rates) can be connected to any computer, in any country in the world.
- Telephony APIs such as TSAPI, IBM's CallPath, Microsoft's TAPI, or Apple's Telephone Manager are supported.

Supposedly, with Geoport, any phone can take full advantage of the services provided by the computer, and vice versa. It allows multiple simultaneous streams of information, including real time (voice and video), to pass through the telephone connection and be processed by the computer.

GEOS

- **Geostationary Earth Orbital Satellite:** See GEO.
- **Geostationary Operational Environmental Satellite:** The first in this US series was launched in 1975. These satellites monitor weather, solar flares, etc.

geostable

This is a geostationary satellite which is not in inclined orbit, and therefore requires no tracking mechanism on the dish. It flies within its 'box' of about 130 sq km.

To be geostable, the TTC&M earth station will need to make slight adjustments to position (trigger short thruster bursts) about every second day. Any satellite needs this level of regular adjustment to its position, to maintain this degree of accuracy. See inclined orbits.

Geostar

An American GEO satellite communications project which proposed to transmit short text messages in very quick bursts from a small handheld terminal, using the L-band frequencies. The first transponder failed within six weeks, and the second satellite was lost when a launch failed in 1986. Orbcomm's LEO system is the natural successor here.

geostationary satellite

GEOs. A communications satellite which is in an equatorial orbit at a height of 22,438 miles (35,788km) — in the Clarke orbit. In this circular orbit, a satellite has a period of revolution which coincides with that of the earth so these satellites appear to remain stationary. But note: to appear stationary they must always be over the equator — and therefore they cannot provide services to the polar regions.

GEO satellites need constant correction from a TTC&M earth station or they will drift under the pressure of sunlight and through constant changes in the earth's magnetic flux. The thruster fuel consumed in correcting this drift, ultimately limits the lifetime of a geostationary satellite. (See inclined orbit)

Geostationary satellites are allocated a slot (a degree of longitude) by the ITU, with an angular separation between slots of 2° to 4°. Ku-band satellites can be at 2° while C-band needs 4° of separation. Each degree represents about 700 miles in actual distance at this altitude and satellites are often allowed to drift out of position by nearly half this amount.

Satellites can be 'co-located' (within the same slot) provided they use different frequencies and/or different signal polarizations. See DirecTV.

In the Clarke orbit, a satellite will have a velocity through space of 11,070 kph (6880 mph) and theoretically it will be able to contact 42% of the earth's surface with its global beam (but the poles are always out of sight). In practice, a higher sighting angle is needed for good reception, and so the coverage is much less. But this was the basis for Clarke's projection that the world would need three GEOs for global coverage.

Signals suffer an 0.24 sec transmission (propagation) delay over the 72,000km round link, and this is increasingly a problem with high-speed data — hence the recent emphasis on LEOs for data communications.

• See inclined orbits, spin-stabilization, 3-axis stabilized, and TTC&M station.

geosynchronous

A satellite having a prograde orbit period equal to that of the Earth's daily rotation. Inclined orbital satellites can still be geosynchronous, so geosynchronous is not exactly the same as geostationary although people use them interchangeably. The two are only the same for a satellite held in position over the equator. See geostationary satellites and geostable.

gesture

The ability to mark, tick or point to something to identify it, or to cause some action by a movement on a computer screen.

This is not a trivial matter in modern communications — gesture is likely to play a central part in the I/O systems of all new PDAs (especially those providing handwriting recognition).

Gesture is also used with 'electronic blackboard' communications, and with some remote groupware applications, so gesture now takes its place as one form of direct or remote access control in computer systems, and as a specific form of network communications. It must now rank alongside voice, video, image and data in the basket of multimedia technologies.

GFLOP

GigaFLOP — see FLOP.

GGP

Gateway-to-Gateway Protocol. A ex-military networking protocol (used widely) that specifies how routers should exchange information. It uses distributed shortest-path algorithms.

ghost

A duplicate but offset image on the TV screen caused by a late arrival of one component of the signal. This is the most obvious example of multipath, and it's also a good example of radar.

Ghosting is the product of the original 'direct-path' signal interacting with a secondary signal which arrives later because it has been reflected from some object, aircraft, or structure. This secondary signal has therefore traveled via a longer route, and because of the direction of scan the offset is always to the right of the main picture (unless you normally watch a reflected image). The delay can be calculated and used to identify the source of the reflection.

• The term is also sometimes incorrectly applied to image 'burn' or 'stick' on video camera systems which have been left viewing a stationary scene for too long.

• Modern TV sets and monitors with digital signal processing (DSP) circuits can create an inverted numerical filter which removes ghosts almost entirely.

• Ghosts can often be reduced or removed by the use of highly directional antenna systems which accept the direct signal, and reject any coming from reflective surfaces at an angle on either side.

GIF

Graphics Interchange Format (aka General Interchange Format). GIF is a public domain graphics format (but see below) devised by CompuServe in 1987 to provide a common file format suitable for many types of computer. It is now very widely used. GIF allows the exchange of bit-mapped files up to a very high resolution. It uses LZW compression algorithms, and supports MS-DOS (VGA and EGA), Amiga, Mac and Apple II. It is used widely on the Internet, WWW, etc.

GIF supports color images with palettes ranging from 2 to 256 different colors, and there are current two subsets (87a and 89a — see LZ#).

• Recently Unisys revealed that it had a submarine patent on the LZW algorithm in GIF, and began to charge for its use.

giga

The SI prefix for one-thousand million or 10^9. Abbreviated by the capital G.

gigaByte

A thousand million bytes of information which is almost twice the full capacity of a CD-ROM disk or about 800 of the high-density 1.4MByte floppies. This is 170 million common English words, or about 1000 average 'airport' novels. It is also roughly six times the amount a fast typist could type in a lifetime.

GIGO

Garbage In: Garbage Out! The point of this humorous acronym is that a computer program can't produce valid results unless the input data is correct and sufficient. Don't blame the computer for mistakes in data-entry, or for poor design in the data-capture phase.

• This has also been translated to Garbage In: Gospel Out — a cynical observation which is intended to point out that some people tend to rely overmuch on the results of computerized output (especially spreadsheets). They assume that because the calculation has been done by computer, it is in some way more trustworthy. See GATSOL.

GII

Global Information Infrastructure. Vice President Al Gore coined this phrase for the Rio conference in 1994. He was suggesting that the US National Information Infrastructure (NII) should be extended to cover the world. It is his 'Information SuperHighway' concept on a global scale.

Giro

A European bill-payment system which is used in EFTPOS. It is a complex system where a single payment-order can implement money transfers from the customer's account to a number of other accounts. It differs from the American check system by using credit-transfer rather than debit-transfer.

GKS

Graphics Kernel System. GKS is possibly the most important CAD graphics application standard today. It is a DIN, ANSI and ISO (ratified in 1984) device-independent standard which describes formats for both image-geometry and text in graphical files. It can create 2D, 3D and raster-scan graphic images.

GKS is the preferred international standard for CAD; it is only slowly being implemented in many areas. The original version did not cover 3D graphics, for instance, so it needed to be extended. See PHIGS, X Windows, and Motif.

• GKS 3D is an ISO superset of GKS.

glare

You lift up the handset to begin to dial, and at exactly that second an incoming call arrives. This is 'glare' and it gives some PBXs and exchanges real headaches.

• Glare spoofing, is a way hackers defeat callback mechanism. I'm not telling you how.

global

Universal in application. International in scope.

• **Satellite Beam:** Relating to the whole system, or the area under consideration. In satellite communications, a global beam covers the entire visible surface of the earth as seen from the satellite, which can be about 40% of the earth's surface (but the low angles aren't really usable). The distinction between this and a hemispherical beam, is that the hemi-beam covers only half of the observable surface area but about 70% of the usable area.

• **Address:** A global network address consists of a network address followed by a MAC address. See NSAP address.

• **Links:** In enterprise networks, global signifies a WAN which has international links. It also carries the implication that there will be problems dealing with multiple languages, cultures, standards, regulators and carriers.

• **Variable:** In object-oriented languages and some procedural languages, variables can be used either globally or locally (within that 'handler' only). Global refers to the reservation ('declaration') of that variable-name for use throughout the program. In some cases a variable may need to be declared for global use, while it can be used locally without declaration.

— In mainframe applications a global variable is one whose name is accessible by both the main program and by all its subroutines.

global paging

Ermes and an international Inmarsat development, both of which aim to provide global paging services.

• Inmarsat will use Standard-M satellite technology and their system will carry paging and short-messaging (initially one-way, but later two-way) services anywhere in the world. See Inmarsat.

• Ermes is a European project to set a multiple channel terrestrial paging standard which gets over the problems of different countries using different frequencies for paging. They use 'agile' receivers which can jump around and use a wide range of frequencies. See Ermes.

• Orbcomm is a Little LEO project which will provide direct paging-like (but two-way) messaging. See Orbcomm.

Globalstar

A satellite mobile voice/data system proposed jointly by Qualcomm and Loral in the USA. This will be a low-earth orbit (a 'Big LEO') system using 24 to 48 satellites at 1400km from the Earth's surface in circular orbits. The project levers off Qualcomm's development of CDMA cellular phones and Loral's military satellite expertise (ex Ford Aerospace).

The main funding (total US$1.5 billion) is already at hand, and the FCC has allocated spectrum (which will be shared with other CDMA satellites). See LEO spectrum.

Globalstar proposes to use very simple satellites with bent-pipe transponders. Voice and data links will be provided at rates up to 9.6kb/s, and handsets, they say, will be priced eventually in the $750 range. These will probably be dual handsets with CDMA cellular, and eventually tri-mode with AMPS.

The Globalstar satellites will be launched in 1997 from Cape Canaveral on a two-stage version of the Delta II rocket which will place four satellites in a transfer orbit at a time.

• Other backers are Vodafone, Airtouch, Alcatel, Deutsche Aerospace, Hyundai, and Dacom.

Glonass

GLObal NAvigation System. The Russian version of the GPS. See also GNSS.

glossary

A library of regularly-used ('boilerplate') text material which can be selected and inserted into word processing or spreadsheet documents. You would keep your own name, address and company logo in a glossary.

glyph

The exact shape of a text character — what it actually looks like. In modern printing terminology, the term 'shape' is generally applied to a generic character (e.g. all capital As of any font), while different 'glyphs' (shape variations) define whether the character is italic, bold, etc. 'Font' is generally used for different type styles.

• The term is also used specifically for printed symbols which contain information — arrows, pictographs, and hieroglyphs.

GMRS

General Mobile Radio Service is a UHF personal radio service in the 460MHz band. Only private individuals may obtain licenses. It is a US variation on CB radio.

GMSK

Gaussian Minimum Shift Keying. This is the form of phase modulation for digital radio communications which is used in GSM mobile phones. In GSM, the primary data rate of 270kb/s is modulated onto the carrier using GMSK with a bandwidth data rate product of 0.3.

The term 'minimum shift' denotes that the phase of the carrier doesn't contain discontinuities when plotted against time. The modulating waveform is filtered using a Gaussian filter where 'the product of the 3dB bandwidth, times the duration of one bit period, equals 0.3.' (I quote!)

GMT

Greenwich Mean Time. The English standard used almost universally (until recently) to avoid ambiguities; it is based on solar time. GMT is also called UT for 'Universal Time', and it has recently been replaced by UTC (Universal Time Coordinate) which is based on atomic clocks. See UT and UTC.

G/N or GNR

Gain over Noise ratio. A measure of amplification in low noise conditions. See gain.

GNSS

Global Navigation Satellite System. This is an Inmarsat proposal to fly 6 GEOs and 15 ICO satellites to augment the US GPS system. The motivation appears to be a suggestion that the US may begin to charge for its GPS services. Inmarsat also quotes the need for greater accuracy and reliability.

If either GPS or the Russian Glonass became unavailable, the number of satellites in the GNSS constellation could quickly be increased to provide full global coverage. See GPS and Glonass.

GNU

Gnu's Not Unix. A Unix-compatible operating system developed by the Free Software Foundation.

go path

The channel for outward-bound signals in telecommunications. The distinction is with the return path. These can also be called forward and backward channels, however this tends to suggest asymmetrical transmissions, while go-return is a symmetrical channel between two equals, but viewed from a single viewpoint.

If the two channels aren't symmetrical, then the go-path will carry the signals and messages, and the return path will have the supervisory and error information.

GO sets

Graphics Oriented sets of controls for terminals. See CO sets.

Golay

• **Error Checking:** A system of matrix (row and column) error checking which usually has a 100% overhead (i.e. it doubles the data transmission requirement), but is very effective. It has now been largely superseded by Reed-Solomon codes except, perhaps, in transmission from deep space probes.

• **Paging codes:** See GSC.

golden numbers

Easy-to-remember telephone numbers that are usually sold at a premium by telephone carriers.

golden rectangle/ratio

The classic aspect ratio of width to height, used in Greek architecture. It is said by many to be the perfect aesthetic ratio for film, video and graphic forms.

The golden rectangle is based on approximations from the Fibonacci series (1,2,3,5,8,13,21...) of numbers. It is calculated by dividing all numbers in the series into the one immediately above them. So 8/5 = 1.6, 13/8 = 1.625, 21/13 = 1.625. On average, this makes the golden ratio approximately 1.62:1.

This ratio, expressed as a rectangle, has the characteristic that if you extract a square equal to the shorter side, the area remaining will also be a golden rectangle.

Golden Scart

See Scart.

googol

A very large number, often used in a humorous way. It is supposedly 10^{99} — a 1 followed by one hundred zeros. It is a yotta squared, squared. See SI-units.

• There is also supposedly, a googolplex which is 10 to the power of a googol.

GOP

Group of Pictures. In MPEG, because the successive frames of image are transmitted as 'differences' rather than full pictures, it is difficult to edit on a frame-by-frame basis. Normally you can only edit at every 8th frame (1/3rd of a second) where a full frame of image is transmitted. However different systems have different requirements, and this is their group of pictures. For this reason M-JPEG is used for editing.

• It is possible to overcome this problem if the editing system decodes and holds all the information in a way that allows full image reconstruction. For this reason M-JPEG is used for editing.

Gopher

An Internet service which allows you to browse Internet resources using lists and menus: it groups information resources by category. This is a tree-branch approach to finding information. You will navigate down through a hierarchy of categories and sub-categories until you find the information.

Gopher is usually very fast, and it is often the best way to find information resources. It runs as a client on your own PC, and you can either connect direct to a Gopher server, or you can Telnet to a Gopher site.

You can navigate the Internet by selecting items from Gopher lists. However, some Gopher servers will go further than just provide you with their list information, they will also search for information-categories through the Internet on your behalf. You decide on the information you want, and the Gopher server will get it for you.

Anyone can set up a Gopher server and offer it over the Internet, so there are numerous special-interest Gopher sites run by enthusiasts who are willing to help others find information. See WAIS, WWW, Archie and Veronica also.

• Gopherspace is like cyberspace, but more restricted. However, as one Gopher site can deal with another transparently, this creates an enormously wide information environment.

GOS

Grade of Service. See.

GOSIP

Government Open Systems Interconnect (OSI) Profile. Government standards (and various standards groups) for the purchase of computer and communications equipment.

From 1990 all US government agencies were told to buy only equipment which conformed to GOSIP standards. In Europe, any government IT purchase which exceeds 134,000 to 200,000 ECUs must be GOSIP-based. Procurement policies are in place also in the UK, Australia, Sweden, Norway, Denmark, Finland and are pending in other countries.

The US and European versions of GOSIP differ, mainly because of the European emphasis on packet-switching. GOSIP is also gradually loosing its attractions. See OSI and TCP/IP.

GOSUB

In Basic, this is a 'call' to a subroutine. It is a module of a program which is constantly being reused, so the control of the main program passes via the GOSUB command to the subroutine, and then passes automatically back to the point where it left off. See call.

GPC

Graphics Performance Characterization committee. This group has a benchmark for graphics performance.

GPF

General Protection Failure. A GPF message in Windows indicates that a memory conflict has occurred. Windows 3.x can detect these conflicts but can't block them or fix them, so a full reboot is necessary.

GPG

Group Portability Guide of the X/Open group. The X/Open group's specification to ensure portability of Unix-based software. See X/Open.

GPI

Graphic Programming Interface. Microsoft's screen display imaging model for the Presentation Manager in OS/2. This is quite different from GDI.

GPIB

General Purpose Interface Bus. This is an early parallel instrumentation bus and printer interface —it is 8-bit parallel ('byte-serial') and asynchronous. GPIB is considerably more complicated and versatile than the Centronic interface, and is mainly used for data collection in laboratories where it has some advantages over normal serial connections. Hewlett Packard (they invented it) uses the same standard, but calls it HPIB.

A number of peripherals can be attached to one GPIB port because each device contains electronics that effectively give it an address. Each of the output devices can control the rate and timing of information in and out of the output port, rather than having the rate controlled by the computer.

GPIB later became the IEEE-488 interface connection and communications standard, which has recently been updated to the new ANSI/IEEE standard 488.2. This version 2 removed the ambiguities of the earlier standard by defining data formats, status reporting, error handling, controller routines and a common set of configuration commands to which all instruments must respond in a precise manner.

GPIB is used with a host of electronic equipment, and particularly with test and measurement instruments. It can also be linked to a SCSI port through a translator. See SCPI.

GPRS

GSM Packet Radio Service. A new project of the ETSI GSM group which might provide a multiple-access scheme more suited to short messaging and transactions. Currently data over GSM and other digital phone networks is circuit-switched (except for CDPD).

GPRS will provide a data-rate of about 12kb/s in short bursts, and it is said that there will be low probability that packets will be lost. It will interwork with X.25.

GPS

Global Positioning System. GPS follows earlier Loran and Decca systems, and it is based on the fact that if you know the exact distance from three known positions, you can plot your position in three-dimensional space. (The surface of the globe is three-dimensional.)

The range is calculated by the propagation delay of signals arriving from the satellites at the speed of light, so it presupposes very accurate timing capabilities in the receiver. The signals from a fourth satellite will supplement the key three, for maximum accuracy.

Each satellite transmits a coded signal which gives the receiver the time of the transmission and the satellite's position. The passive receiver then extracts from this information, three coordinates and one time-difference.

GPS transmissions use a CDMA (DS/SS). A signal of 100Hz is multiplied by a 'Gold-code sequence' at a chip-rate of 1.023Mb/s, resulting in a signal with a bandwidth of about 2MHz (broadcast at both 1575.42MHz and 1227.6MHz). The first frequency carries signals for civilian and military use, while the second carries the precise (P) military code only.

The claimed accuracy of 'clean' (P) GPS is well under a meter, but the signals are deliberately 'dithered' (varied in random fashion) to create unreliability, resulting in a civilian accuracy of a theoretical 10 meters — but more like 100 meters.

Even without the P-code, it is possible to still calculate the discrepancies, and local authorities can broadcast local dither corrections if they wish, using a fixed transmitter.

The Navstar satellites which provide GPS aren't geostationary; they have inclined orbits which take 12 hours to circle the earth at a distance of 1500km. The first workable GPS satellite constellation depended on 6 satellites, but now there are 24 in the US constellation, and almost as many in the Russian Glonass (GLObal NAvigation System).

GPS will eventually open up the international radio frequencies by freeing up large sections of the spectrum used for military and civilian aviation position location and navigation.

- The use of GPS is free since it is a one-way system only.
- A new PCMCIA card has recently been promised to provide GPS to notebooks. This could be vaporware.
- Small GPS devices for hikers now sell for $300. They provide the latitude, longitude and height above sea level on a small LCD screen.
- Inmarsat is now planning a similar system. See GNSS.

graceful degradation

- **Computers:** Graceful degradation refers to a redundancy technique used in 'mission critical' enterprise computer applications where it is essential to maintain all the computing resources online while they are still functioning. These systems are redundant and modularized; when modules fail they are bypassed and the total computer power reduces 'gracefully'. Error recovery techniques hide the failures from the users.

There are a number of different approaches. With on-line transaction processing, for instance, computer modules are often run in parallel. If one fails, the load shifts to the remaining units, reducing the response time but still maintaining a working system. See also triple module redundancy.

- **Cellular phones:** In CDMA cellular systems there is a theoretical limit, but no (practical) finite limit to the number of calls that can be handled by a cell before blocking begins. However at a certain load-level, the quality of the voice begins to degrade for everyone. At this point, those with less urgent needs will discontinue their conversation and free up the spectrum, but at no time are new connections actually blocked.

The distinction is being made with TDMA systems where there is a certain number of frequencies and slots available, and when these are filled, all other calls are blocked.

- **Television:** In digital television, the term also applies to specially engineered systems which might transmit the key elements of an image with a high level of error-correction, while less important details in the image receive a lesser degree of protection. There may be three or four steps of 'quality', so during difficult periods (where the bandwidth available can't cope) the image quality can drop — but this is still much better than having the picture freeze on the screen.

gradation

In printing, this is the smooth transition from black to white, or between colors. In desktop publishing, it is important that banding doesn't take place. See.

grade of service

GOS Don't confuse this with QoS (Quality of Service). GOS refers to the blocking potential of public-switched telephone networks. It is the probability that a call attempt will fail due to inherent design factors in the network. The phone network is designed on the assumption that not everyone will attempt to call at the same time.

• When contracting with a carrier for special services, GOS is also a measure of the service levels which the carrier contracts to provide to the customer.

• In public network design it is the percentage of attempted calls blocked by the network. Most phone networks are engineered to provide a required GOS during the busy hour. For instance, the P.01 grade of service has a 1% blocking probability (this seems to be the world standard for PSTN lines); there is a 1% chance that any user will need to wait the average call-duration time (or longer) to gain access to the network. A cellular phone system will have a much higher (worse) GOS in inner city areas.

• The term-variation 'service-grade' once had a more general meaning.

• Don't confuse grade of service with terms like 'voice-grade'. Telecommunications services are roughly divided into bandwidths; narrowband, voiceband, wideband, broadband. A 'voice-grade service' is one that provides sufficient bandwidth for voice only. When used with modems, the term 'voice-grade' suggests throughput rates of about 28.8kb/s can be expected.

graded-index fiber

A type of optical fiber where the refractive index (RI) decreases the further it is from the core center. Think of it as tree-rings, with the dense glass at the core and then progressively less dense towards the outer cladding. This is a form of multimode fiber, but it is intermediate in function between mono- and multimode.

Light tends to travel down the high-RI core and, when the cable bends, it automatically refracts back towards the core through the various outer layers, rather than reflecting from the outer fiber surface. Over distance this helps avoid signal-spread (dispersion) so these fibers tend to act as if they were monomodal (see mode).

This approach has many technical advantages over multimode fiber. These are large fibers, so they can be similarly driven by LEDs and they are easily spliced. Unfortunately graded-fiber is expensive and difficult to manufacture. See step-index fiber.

gradient

The degree of inclination; the angle of an up- or down-hill slope measured from the level. And, by analogy, the angular direction of the graphical representation of some function.

• At any point along a curve, a line has direction and/or gradient relative to some base. Gradients are usually measured in degrees, radians or ratios.

grammar

The formal rules of a computer language. See syntax.

Grand Alliance

The consortium of US equipment manufacturers who are working together to formulate the next generation of digital HDTV systems. They are producing a hierarchical system where a standard 4:3 ratio picture is embedded in a HD wide-screen signal (16:9), so that either can be chosen. This is likely to be extremely inefficient in its use of bandwidth.

They will use MPEG2 video and the associated systems syntax (including B-pictures). For high-definition images, they will provide both interlaced (1440 x 960 at 30Hz) and progressive scan (1280 x 720 at 60Hz) modes.

They still haven't finalized the selection of:
- The modulation technique (QAM, VSB, OFDM). VSB was tentatively selected while COFDM is favoured by many,
- The convolution code (MS or Viterbi), or
- The error correction (RSPC or RSFC — both based on Reed-Solomon).

Zenith currently has commercial 'beta' test equipment in production, but final FCC approval will not come until late 1995 — or possibly later. They've settled on Dolby AC for the sound.

The Japanese and European manufacturers have their own plans for digital HDTV, although both Philips and Thomson are involved in Europe and the Grand Alliance also. See also DVB.

grandfather

A backup file once removed — the backup before the current backup was made. This is in reference to the grandfather/father/son methodology of always keeping at least two previous backups, and rotating them in a cycle.

The two previous version/s of the file (the grandfather and the father) should be kept (hopefully off the premises) for comparison as well as for safety. Don't ever backup onto the same disk over and over again, See backup.

• Grandfathering is the process of providing something with a legal cut-off date. A communications technology may be licensed until a certain date, at which point it must be reviewed and relicensed.

granularity

The ability to detect structure in what appears, at a further distance, to be an even (amorphous) body.

• In networking, it refers to the degree by which the original larger network segments have been broken down into smaller segments by the use of bridges and routers (not repeaters) in order to increase the capacity of each. The rule is to keep segments at a balance, with 80% of their traffic being local, and only 20% crossing the bridge.

• In telecommunications, granularity can refer to the smallest part of a data-stream which can be handled as an independent unit. For instance, in a highly multiplexed data-stream (with several layers of multiplexing in a MUX hierarchy) granularity refers to the degree that it is possible (at any one level) to switch the signals.

At the base or primary level of de-multiplexing, it is obviously possible to switch each signal independently. But at the secondary level, it may only be possible to handle traffic in 'groups'. These are twelve 4kHz signals which have already been multiplexed together by a primary-multiplexer. So the granularity at this level is more coarse. In digital systems where millions of voice circuits are being modulated along optical fiber cables, SDH/Sonet is being introduced to allow a single circuit to be added or dropped without full de- and re-multiplexing: this is the ultimate in granularity.

• In computing, granularity refers to the fineness of the divisions between modules (handlers) in object-oriented programming. The smaller the average object (module) in a program, the greater (or the 'finer') the granularity. Really this is trying to make a distinction between component software (encapsulation only) and true object-oriented software. See VBX and componentware.

graphic

There are four levels to this term in computing:

- In the widest sense, all images on the screen are graphics.
- More specifically, we mean non-text elements (we divide the image forms into text and/or graphics).
- At the third level we divide graphics into drawn or human-created images (graphic illustrations) and photos/video images (pictures).
- At the fourth level we make a distinction between simple scanned bit-mapped graphics, and mathematically-defined vector graphics.

Usually we mean the second level — those screen images that aren't alphanumeric characters, punctuation, or standard keyboard symbols. If it is on the screen, yet isn't part of the 8-bit extended ASCII set, then it is a 'graphic element'.

Modern screen displays are Graphic User Interfaces (GUIs), so there's both a generic and a specific use of the term here.

Early computers dealt with drawings and images on the screen in an entirely different way to text characters and numbers (except for some graphic 'primitive' characters in the 8-bit ASCII set). These early PCs displayed alphabetic text and numbers in a very rigid form. Each character's image was generated by a character chip (a ROM) and displayed within a standard character matrix. These were Character User Interfaces (CUIs) and sometimes they also had a graphics mode — but these two tasks were quite distinct and they handled the video output in entirely different ways.

Modern graphically-oriented computers now handle text and graphics in essentially the same way; they create and treat their text characters on the screen as graphic elements in an overall graphic window image. These GUIs are therefore far more flexible, and this has allowed us to control font, style and size changes, kerning, line-space, and DTP layouts.

There are many different architectural approaches to handling graphics. Some computer architectures transfer graphics processing to a separate processor (a graphics co-processor), but others still use the CPU for normal graphic tasks.

The two main categories of graphics are bit-mapped 'Paint' graphics (raster graphics), and object-oriented 'Draw' graphics (vector graphics). With the second type you can, at any time, extract, shift, or manipulate any graphic element on the screen.

In the first type, any new element added becomes an intrinsic part of the image and can't later be shifted or changed (only erased): it always leaves a hole.

• When discussing bit-mapped and vector graphics, you must keep in mind the distinction between what is happening in the application, and what is happening in the video-RAM of the operating system. If you have a normal raster display (as 99.9% of us do) then the image you see is the result of a raster-scan from the video-RAM's bit map. But this will probably be a secondary 'image file', generated by the application's vector graphics driver and the application file in the main memory.

• Graphics are stored as bit-mapped vector files, or held in a range of other special hybrid or compressed graphic file-types. Without compression, bit-mapped files are extraordinarily wasteful in memory requirements. See GUI, TIFF, PICT, GKS.

graphic characters

These are relatively primitive blockish shapes of different configurations which are pre-defined in a set screen matrix as if they were characters. In Prestel videotex, for instance, there are 64 pre-defined graphical characters with each occupying a 2 x 3 matrix of pixels.

These are selected and displayed in the same way as alphanumerical characters. Usually they are assigned some of the upper 127 bytes in the 8-bit ASCII character set.

The IBM PCs have graphical characters defined in the 8-bit 'extended ASCII' set also, and these each fit into the standard text-character matrix. So in CGA displays with 8 x 8 character cells, a graphic character would occupy one cell. EGA had character cells which were 8 x 14 pixels.

• Don't confuse this with IBM's special character set. See IBM and MDA.

graphical display

A monitor which can handle graphics. All modern monitors are of this type — they handle graphics and text. There are very many graphical display standards; in the IBM PC range we have CGA, EGA, VGA, 8145/A and XGA. The Mac has Color QuickDraw. For videotex see alpha-mosaic, alpha-photographic and alpha-geometric and NAPLPS. See text display, vector graphics and raster scan also.

• A graphics card is a video card which can handle graphics as well as text. Some of the early PCs had outputs which were text only. However, these could show a primitive form of graphics using pseudo-characters defined in the 'extended ASCII' — in the 8-bit range. See graphical characters.

Graphical User Interface
See GUI.

graphics modeling
In graphics terminology, this means the ability to create a three-dimensional real-life appearance in a graphical image. The first stage is wire-frame representation, and the second is rendering. Advanced modeling systems can then provide reflectance (matte or mirror), transparency (opaque to invisible) and smoothness (from fuzzy to polished surfaces).
There are two main advanced modeling techniques:
1. Boundary Representation (B-rep) which uses interlinking (basically 2 dimensional) primitives (usually polygons) to define the surface.
2. Combinatory Solid Geometry (CSG) where simple primitives like spheres, cones and cubes are added to, or subtracted from, each other to create the more complex 3D shape.
- **Choreography,** is the term applied to the design of motion in the animation of these models.
- **Lambert shading** is a simple cosine calculation to define how the light reflects from a matte-surface 3D object.
- **Phong shading** is a later and better calculation which allows a degree of transparency and reflectiveness.
- **Ray tracing** is the brute-force method of shading, and creating reflections, etc. It is computer intensive.
- **Radiosity** is a similar modern rendering technique, but less computer intensive.

See wire-frame and rendering.

graphics tablet
See tablet.

grating
Gratings are optically-flat glass sheets which have been microscopically engraved with thousands of parallel grooves. They are a substitute for prisms in separating light into its component frequencies (the rainbow), using interference between the photons passing through the grooves.

• In optical fiber detection systems, a grating is used to deflect the light on to different detectors when a number of wavelengths are being used. This is a technique of Wavelength Division Multiplexing (WDM) which increases the carrying capacity of fiber — it is the equivalent of frequency division multiplexing in optical fibers.

• Gratings can also be used as optical focusing devices.

gravure
A ink-printing technique which uses the etched areas on the plate to hold the ink. It is a high quality process, only used with very long print runs — mainly for calendars, wall-prints, etc.

gray binary
A form of 4-bit binary notation which involves only one bit-change for every unit increase. This is an alternative to the normal method of binary counting (which was modeled on the old decimal system we use in everyday life).

For instance, in the regular binary sequence, when numbering from zero to four, we make changes in this way: 0000, 0001, 0010, 0011. Note that the progression from 1 to 2 (0010) involves changing both the far right-hand and its adjacent: the two least significant digits. This creates problems in some computer systems. Gray binary offers a way around these problems.

With gray or 'reflected' binary, the same sequence would be: 0000, 0001, 0011, 0010. Note that now, in this new sequence of number progression, only one digit at a time has ever changed. Obviously, you also need to perform calculations (additions, etc.) in a different way, using different 'carry' rules.

Gray binary is said to have a 'Hamming distance' of one, whereas regular binary has a variable Hamming distance. See symmetrical binary also.

gray market
The selling of hardware and software, outside the 'authorized dealer' channels. This has copyright implications in software. The software has usually been purchased wholesale outside the regular distribution channels and it often won't be supported by the authorized distributor. The term doesn't imply that the software fell off the back of a truck.

• There are conflicting opinions about the rights and wrongs of this. One argument is that licensed dealer channels become quasi-monopolies able to profit too much. The opposite argument says that these dealers offer customer support and updating services on a regional and national basis, and shouldn't be expected to do so for customers of the gray marketeer.

gray scale
A strip of tone-blocks printed on a card, and used to align and test video and film cameras. The tone-blocks vary from full-black to full-white in incremental steps — usually with a logarithmic change of reflectance. These are used to test the linearity of a video system (the gamma), or to check the exposure and processing quality of film. The strip provides a standardized test of tonal-range and contrast.

• When used for exposure, film cameramen often use a 7-point gray-scale with approximately one stop of exposure between each step: Black, Near Black, Dark Gray, Mid Gray, Light Gray, Off White and White.

Mid Gray (subjectively half-way between black and white) reflects about 18% of the light. Each step on either side reflects either twice or half this amount.

• Television studios tend to use a ten-point gray scale with finer gradations.

• In computers, gray-scale refers to the range of binary values that can be assigned to a pixel to indicate shades of gray. With a B&W monitor the computer may be able to assign 8-bits per pixel just to levels of gray (a range of 256 values) but this would be beyond the tonal discrimination limits of the eye. Usually 5-bits with 32 levels is enough for the eye to see the tonal range as continuous.

• Printers also use a special gray-scale for checking camera exposure during plate-making.

• With computers, the number of steps possible depends on the number of bits allocated to each pixel (pixel depth). A 5-bit pixel-depth can produce 32 levels of gray (white, black and 30 levels of gray). With color monitors, the tonal variation is more complex since each of the colors (R, G and B) has, in effect, its own 'gray scale' — and green makes much more contribution to tonal levels than either red or blue.

• At best, a laserprinter is probably only capable of doing even moderate justice to a 4-bit (16-level) gray-scale image. Some techniques, such as 'dithering' improve the subjective appearance of half-tones, and some of the new laserprinters can also vary the size of their dots in one dimension.

• A gray-scale monitor is able to depict tones between black and white, but not colors.

grazing

In information jargon this means absorbing (or claiming to absorb) information in an indiscriminate and casual manner — neither hunting or browsing. The example usually given is of a vidiot watching television all night without switching channels — he will stumble across some useful information.

However serendipity (making desirable, but unsought-for discoveries) is a very important part of learning; we graze newspapers every morning.

greek

• **Scientific:** A fancy form of printing on some daisy-wheel machines. It was used to approximate Greek letters for scientific notation, by a process of overstriking (double printing).

• **DTP:** Small text-like characters which, during page resizing on a computer monitor, have become too small to be resolved as readable characters. The position of characters is usually indicated by small random black blobs. This still gives you the overall impression of the completed page layout, but in a highly reduced size.

• **Printing:** In printing, Greeking was the use of 'gobbledegook' (a sort of pseudo-Latin) which was run into a page-layout to give you the impression of readable copy.

Green Book/Paper

This is a widely used term as is White paper. It usually means a large volume of standards or proposals in the process of being defined — a preliminary or intermediate stage.

• However, various bodies like the ITU also print their yearly determinations in book series, where they use different colored covers each year for easy differentiation. So this can be confusing; a green book standard may mean nothing more than a standard finalized in a certain year.

• For CD and CD-Interactive, this is the current standard for this particular sub-standard of CD-ROM. Other CD standards are distinguished by other colors.

green numbers

In Europe, this refers to freephone numbers.

greenfield

A popular telco and general business term which means completely new installations, as distinct from updating or replacing old 'plant' — which in telephone terms means the wires in the ground. Greenfield projects are not hampered by 'legacy' systems or devices.

GREP

Global Regular Expression Parser. A very powerful string-search macro 'language' found in Unix, Nisus, QUED/M, etc. GREP is a rather complex system for searching a file to find matching patterns of characters using powerful and versatile conditional commands.

grid

In many graphics and layout programs the computer can generate a grid of various dimensions as an overlay. This can later be switched off, or on. When grid-lock (aka 'snap') is active, it is only possible for lines to be drawn from set grid junctions, or along grid lines.

• A grid of sorts is also available in layout programs to ensure that related pages in a book, magazine, or newspaper, look the same. When you define 'stationary' or 'template' layouts, you are fixing a grid within which the copy will be confined.

grid fitting

With outline fonts. See hint.

GRINSCH-SL-MQW

Graded Index Separate Confinement Heterostructure Strained Layer Multi-Quantum Well Distributed Feedback Laser. This is Nortel's new commercial strained-1550nm laser. For some strange reason the DFL was truncated. I just had to put this one in for the acronym alone!

grooming

Grooming means cleaning up, organizing, or optimizing.

• In multiplexed telecommunications, grooming is the process by which individual channels are sorted according to their destination, so that they can be multiplexed into higher-order groups, all heading in a common direction.

• In disk management, grooming involves optimizing the use of disk space, the migration of some material to other disks or other storage devices, and applying load management techniques. It is much more than just de-fragmentation.

Grosch's Law

'The performance and productivity of a computer rises as a square of its price.'

I've included this 'law' just to show how wrong these laws can be. In the 1950s, this principle was both correct and widely accepted — because mainframes were then being used for time-sharing. The economies of scale with time-shared 'big-iron', favoured the biggest and most expensive mainframes.

ground

There are two quite distinct uses of the term.

— As a synonym for 'earth'; the connector at earth potential.

— As a common reference point created by electrically connecting the major metal shieldings and (usually) chassis of the equipment, to provide a standard voltage reference.

This second is not necessarily connected to the earth, and it may exist at any voltage level above earth. So this can be any conductor considered to be of zero potential — but not necessarily earth voltage.

• However, most 'ground potential' electrical contacts for devices using mains current will be connected to the earth at some point, usually through a metal pipe buried about two meters into damp earth. Dampen around the pipe periodically.

• The metal frame of a piece of electronic equipment is often assumed to be connected by an earthwire to earth. However, be warned! There are many pieces of electronic equipment which have a 'live-chassis' — where touching the metal frame while the power is on, can kill you. Don't assume the metal frame of any piece of equipment is earthed.

• In normal domestic (110V or 240V AC) mains power, one wire will be the active, and another the neutral. If a third wire exists, this will usually be at earth potential or ground. In a power plug, this earth contact should be clearly separated from the other two, of a different shape, and probably have a longer prong than the other two so that it makes contact first. Green has traditionally been the color of earth potential wiring.

• The shielding around communications cables should be grounded, via the plug shielding at each end, to the EMI protection case around the terminal equipment — but not necessarily to the chassis or external casing of the power supply.

• In RS-232 don't confuse Common Return (Pin 7) which is the reference point for all voltages, with protective ground (Pin 1), or with an earth point. The shielding around the cable and the connectors should be earthed, but not Pin 7. Protective Ground would better be called 'Chassis Ground'.

ground loop

A transient current which can be created through incorrectly linked equipment. It occurs when two or more conducting parts of equipment which should be at earth (zero) potential but aren't, are connected by a conductor — so stray electrical currents migrate along these links.

Ground loops often make equipment behave erratically, and they can disturb or destroy computer or communications operations. One of the values of optical fiber in communications is the absence of any ground loops. With RS-232 connections, if you suspect ground loops try disconnecting Pin 1 (optional).

ground plane

The area under a radio transmitter's quarter-wave antenna. It should be a good conductor (damp or metal) and at least a quarter-wave in distance from the antenna.

ground start

A method of signaling between the telephone and the local exchange, used originally in SWER (Single Wire, Earth Return) telephone local loops, but now in special phones and many analog PBXs. The ring wire carries –48v battery power, and when you take the phone off-hook, the wire is 'grounded' (shorted to earth). This voltage change is detected at the exchange which then switches in its 'signal ground' connection on the tip wire, thus completing a communications circuit.

This system was widely used in telephone call-boxes at one time. The ground's electrical resistance between the subscriber and the exchange is unpredictable at best, and twisted pairs are vastly superior to single copper wires. See loop start and wink.

• Ground start is apparently still used as a standard for analog PBX lines, and some callback security devices need a ground start line.

• In some countries ground start is called earth recall.

ground station

In satellite terminology, this applies to any communications unit which has both up- and down-link capability. It can both send and receive. However, even this concept isn't strictly maintained.

• A receive-only VSAT terminal is still often called a ground station. The term TVRO is used for receive-only television ground stations.

• Handheld or mobile two-way communications devices using satellite hops are not usually classed as ground stations.

• The term is most often applied to data-only satellite or radio transmitting and receiving stations, and these may also be called gateways if they make connection to other networks.

• The larger ground stations tend to be known as 'earth stations' which suggests some control and monitoring functions as well. The primary controlling earth stations are called TT&C or TTC&M.

ground wave

Radio waves are broadly classified as ionospheric waves, tropospheric waves and ground waves. Ground waves are that part of the radiation which is directly affected by the earth and its surface features.

They have two components: a surface wave which is 'earth-guided' (curves with the earth), and the 'space-wave' (not to be confused with the ionospheric sky wave). The space wave has two components also: the direct wave (line of sight), and often a ground-reflected wave. See propagation and space waves.

group

In analog telecommunications this is twelve telephone channels each with a 4kHz separation (this includes guard bands). A group therefore requires 48kHz of bandwidth with FDM, and it can be carried in a single twisted-pair if necessary, for a few kilometers. This is the level of first-order multiplexing.

For higher orders of multiplexing, the first group would normally be modulated to occupy the spectrum between 60 and 108kHz. Five of these groups would subsequently be multiplexed to produce a supergroup band from 420 to 612kHz in increments of 48kHz. This is second-order multiplexing.

The idea is to have a hierarchy of standards for communications equipment. A supergroup is 5 groups or 60 voice channels (this bandwidth needs coaxial), and a hypergroup is 960 channels. The US and ITU differ in their use of the terms mastergroup, supermaster group and jumbogroup. See.

Group # facsimile

— G1 fax: This is the name applied to a standard that existed very briefly in 1968. The ITU's Recommendation T2 for 'Standardization of Black and White Facsimile Apparatus' covers these early machines. Group 1 fax machines took between 4 and 6 minutes to transmit a standard A4 page with a good local telephone connection. Since the Americans had a previous standard which was partly incorporated into G1, faxes could be sent to North America from Europe, but not received.

— G2 fax: This 1976 analog standard was very complicated and expensive in its day, and it took about 3 minutes to scan an A4 page. Most Group 3 faxes have a Group 2 function as well.

— G3 fax: Since 1980 this has been the current digital (but with analog modem transmission) fax standard used worldwide. The success of fax is due mainly to the development of data compression systems (see RLE, Huffman) which cut phone costs.

G3 fax has been progressively up-dated to transmit a single A4 page in about 1 minute at the 9.6kb/s rate but, if the line is noisy, the machine pairs will fall-back to a lower rate. The most recent additions to this ITU-T standard have increased the rate again to 14.4kb/s. At this rate (after handshaking) pages are transmitted in 10 to 12 seconds.

Progressive improvements to the G3 technology have been possible because the machines always exchange parameter information at 300b/s (V.21) during a handshaking period before transmitting the data. They decide which data rate to use, and check for common features. Some provide error correction, for instance, while most don't.

Group 3 fax applies a One-dimensional Modified Huffmans code for initial compression, and then a Two-dimensional Modified READ. And G3 scanners have a fixed horizontal resolution of 203 dots per inch (pels or dpi). They also have selectable vertical resolutions of 96 lines/in. (only in early machines), but later, 192 and 384 lines/in.

The modems conform to the ITU standards V.27ter (2.4 and 4.8kb/s) and V.29 (4.8, 7.2 and 9.6kb/s) using dibit and quadbit PSK. The new 14.4Kb/s rate uses V.17, and the handshaking is in FSK using V.21. See fax modems also.

— G4 fax: This 1984 fully-digital fax standard was designed for the ISDN telephone system. It will run on an ISDN's B-channel at 64kb/s and transmits a typical page in 3 seconds. The resolution is 400 x 400 dots per inch which is higher than many laser-printers. Error checking will be the norm, and some models will have a selectable 64-level gray scale. Most will print on plain paper. This standard has not been wildly successful.

— G5 fax: This proposed standard is for B-ISDN and it will provide greater than 300 dot/pixels per inch resolution, and have color reproduction at a 24-bit (3 x 8-bit) level.

group address

This is the multicast address used for broadcasting to multiple devices on a network.

Group Code Recording.

See GCR.

group delay distortion

In a communications network, not all frequencies propagate along wires or through circuits at exactly the same rate.

So, over a reasonable distance, a phase difference will occur because some frequencies of the group arrive before others. This creates problems, especially with multiplexed channels. Group delay defines the time it takes all signals (of all frequencies) to pass through a device or network.

In normal voice conversation, group delay is unnoticeable, but it is highly significant with high-speed data over long distances. Delay equalization is necessary to overcome the problem. Group delay distortion is also called envelope distortion or constellation distortion.

group-enabled

Software which has been adapted for groupware use. Some applications are more oriented towards a messaging-enabled view, while others use a shared database approach. In Notes, Lotus uses a replicating document store which is coupled with an e-mail system.

group velocity

- In conventional wire-line data transmission, different carrier frequencies are propagated at different rates. This can produce group delays in data, which then require delay-equalizers to slow down the faster signals. See group delay distortion.

- In optical fiber systems the 'group velocity' is the speed of the communications along the fiber. This is in contrast to the 'phase velocity' or 'wave velocity' which is the speed of the light along the fiber. It may not sound logical, but apparently these two are not necessarily the same.

All measurements of the velocity of light are done by interrupting a beam of light so as to form groups of waves. So the speed measured is always group velocity.

In air, the difference between the group and phase velocities is only 1 part in 50,000, but, in glass, the difference is a ratio of 1.5 to 1 — about equal to the refractive index.

groupware

This is an extremely wide term. Groupware is also known as CSCW software or 'Computer Supported Cooperative Work' applications, or just 'workgroup' applications.

The terms all refer to networking software which is designed specifically to permit a number of people (in a workgroup) to communicate while simultaneously and jointly working together. As one magazine put it: 'to integrate an organization's culture, work practices and applications whilst improving productivity and business performance!'

At a much more realistic level, workgroup computing can apply relatively standard applications. These are enabled by e-mail, DDE and OLE. The 'groupware' term is generally applied to word-processing (for group reports), chat programs (real-time group decision-making) and spread-sheets. However it also includes the management of joint calendars and diaries, and keyboard conferences.

The term seems to exclude databases, despite the fact that multiple-user access has been a reality in database transactions for many years. The file- and record-locking techniques used to protect database records when they are being accessed and changed by a number of different people, are not possible with spreadsheets or word-processing documents which must be treated as a whole. So special groupware/software is required to handle the simultaneous use of these files.

Probably the archetypal groupware product is Lotus Notes — although generic e-mail could arguably qualify. A useful categorization of groupware is:

— Communications: e-mail, keyboard discussion groups.

— Personal tools: scheduling, diary, calendars.

— Workflow tools: routing, retrieving and processing documents automatically.

• The vague use of the term also covers management information systems where teams of people can be united with sets of computing resources to provide answers to business problems in a more efficient way. This goes under the catch-all phrase 'business process re-engineering' (changing the way a business operates). See EIS.

• An even more specific form of groupware, in this sense, is that of group decision-support systems.

• Another definition limits groupware to discrete applications, such as calendaring, scheduling or forms-routing workflow applications.

• Another definition ties groupware specifically to shared LAN software which allow teams to work on joint projects.

• One definition of groupware is just 'virtual office'. This assumes that people will work from home, but retain links to a workgroup. The term 'meetingware' often creeps in here also.

• Some groupware will always be on-line, but other types of groupware can be off-line — where people add their comments, then pass on the document, even via disk exchanges.

These concepts are similar to workflow management software, and to special smart-forms for the entry of group data, and to modern e-mail-enabled applications.

Grove's Law

'Only the paranoid survive.' Andrew Grove is the 'Titan of Intel', and he should know!

GS

ASCII No.5 — A data Group Separator (Decimal 29, Hex $1D, Control-]). A control character used to separate data logically into groups. It is also called IS3 or Information Separator 3.

• The four information separators (FS, GS, RS, US) are used in a variety of ways, but they should always be used in a hierarchical or nested manner with the FS being the most inclusive, and the US the least inclusive).

GSC

Golay Sequential Coding. A Motorola standard for paging systems which provides for tone, numeric and alphanumeric pagers. It uses FSK modulation, and operates at both 512b/s and 1.2kb/s This is a dying standard, but only because paging carriers want to get faster systems operating; Golay protocols are very rugged and ideal from a user's viewpoint — especially with short-message numeric paging.

When the GSC coding system was introduced, tone-pager controls were included in the standard, and the possible variations in tone-patterns became even more elaborate than those that were available with the older 5/6-tone systems. See POCSAG, ERMES, TDP, TAP and FLEX.

GSI

Giga-Scale Integration. The next stage of chip design (past VLSI and ULSI) with more than 1 billion transistors on a chip. Many people doubt that this is possible. Component distances of less than 0.25 microns are likely to give trouble, and beyond 0.1 micron separations quite different laws of physics operate. Giga-scale designs may be approaching the fundamental limits of even MOSFET technologies. See Mead's and Debye limit.

gsm/GSM

• **Grams per square meter:** The measurement of paper thickness.

• **Groupe Speciale Mobile:** Originally the French term 'GSM' applied both to the CEPT committee (really a consortium of PTTs in Europe) which was working on the new pan-European digital cellular telephone system from 1982, and to the technology itself. CEPT handed over part-control of GSM to ETSI in 1988, and total control in March 1991. The use of the English version of the GSM acronym (below) was for general marketing purposes. The old committee is now known as the Special Mobile Group (SMG).

• **Global System for Mobiles.** The pretentious new name for the European digital cellular system — especially since neither America or Japan use the standard.

GSM is intended to be a European-manufactured, globally sold, digital standard for two-way mobile cellular telephones (and data). It uses TDMA technology to cram 8 (later 16) channels into a 200kHz wide channel in the 900MHz band.

The system is designed around duplex 25MHz bands set aside in the spectrum to provide 124 (+4 signaling) radio carriers and 992 full-rate voice channels in total — although these need to then be divided among the five potential carriers.

GSM uses a RPE-LPC 13kb/s voice-coder (8kb/s plus FEC) and another 3kb/s is added for in-band signaling, monitoring and control (a total of 16 kb/s). The channels have a transmission rate of 271kb/s, and in large cells at this rate, dispersion occurs. Modulation is π/4 GMSK.

When the cellular switch has the facilities, and the user has the required equipment, data and fax can be transmitted over the GSM channels at rates up to 9.6kb/s by bypassing the vocoder in the mobile and the basestation, and making links to standard modems in the switch premises. A Short Messaging Service (SMS) can also be implemented if the basestation has the modification and all networks are interconnected (and have the necessary agreements in place).

The standard spectrum allocation for GSM is between 890.2—914.8MHz for mobile transmit, and 935.2—959.8MHz for mobile receive, with a 45MHz duplex spacing. Each of these 25MHz bands is divided into five 5MHz chunks for competing carriers. A healthy guard band (at least 1MHz) must be left on both sides of this bandwidth, since GSM is a notorious generator of EMI.

The first European networks began in June 1991. Germany now has over 1.6 million users (nearly half of the world total). Elsewhere it has been less successful. Analog AMPS and TACS have proved to be just as popular and provide better coverage, and GSM has been widely criticized for interfering with other radio and non-radio electronic devices.

It is said that GSM will later (1996) be provided with the new half-rate vocoder in 1996, and the bandwidth halved for more capacity. This is called GSM-II or Phase II, and there is now a further stage called Phase II+.

A higher frequency version of GSM in the 1.8GHz range (called DCS 1800) has been designed for PCN. See air interfaces, DCT-900, CT2, PCN, CTIA, DCS-1800, DAMPS, N-AMPS and CDMA also.

GSM power classes

GSM has five power classes, although the 20 Watt class has generally been discontinued because of interference problems.

- Class 1 vehicles and portables at 20 watts (now 13 watts).
- Class 2 portables at 8 watts.
- Class 3 hand-helds at 5 watts.
- Class 4 hand-helds at 2 watts.
- Class 5 hand-helds at 0.6 watts.

Since the signal is pulsed, all power is quoted at 'nominal peak output power'. Penetration is eight times that of the equivalent continuous output device.

GSNW

Gateway Services for NetWare. Aka 'Gates Sees NetWare in his sights'. This is the key element in Microsoft's NT networking strategy. GSNW is built into the 'applications server' operating system to convert requests from any Microsoft-based clients (SMB commands over NetBEUI) into IPX-based NCP requests that a NetWare file server can handle.

This is a very clever way of insinuating large numbers of Windows NT servers into existing Novell networks, and then gradually taking over the networking environment.

GSO

Geo-Stationary Orbit. Now called the Clarke Orbit or GEO. See also ICO and LEO.

GSV

Global Videophone Standard — developed by AT&T and submitted to the CCITT as a possible world standard. It died.

GSZ

This is an implementation of ZModem that uses graphical characters — an update of DSZ. Both of these were created by Chuck Fosberg. GSZ is available as shareware.

GT

Greater Than. Usually the > sign is used.

G/T

'G over T' means gain-over-temperature — the ratio of antenna gain over noise temperature, expressed in decibels.

It is a measure commonly used in satellite signal reception, and is the single most important performance parameter of an earth station, since it describes the system's ability to pick up and amplify the extremely weak signals from the satellite.

Ideally there would be no noise in low-noise amplifiers (LNAs) but, in practice, the figure of interest is the gain which is possible above any inherent Gaussian noise (from the movement of molecules) and this depends on temperature. The bigger the antenna and the better the LNA or LNB, the higher this G/T figure will be. See footprint.

GTDI

The European standard for EDI. See EDIFACT.

guard band

- In telecommunications, an unused frequency band which separates two traffic-carrying frequencies to prevent mutual interference. Most TV channels have a guard band of 0.5MHz on each side, but even then, their antennas can't be co-located if they use adjacent bands unless signal power is matched and receivers are 'channel-ready'. See taboo channels.
- In videotape, the guard band is a width of unused magnetic surface between the recorded tracks. Not all videotape systems need guard bands; it is often possible to reverse the phase or the head-angle (azimuth) and overlap neighboring tracks.

GUI

Graphic User Interface — pronounced 'gooey'. GUI is a general term which is applied to all the modern Macintosh/Windows-like screen interfaces being used today on personal computers.

These ideas were developed at Xerox's PARC research lab, but Apple made them work as a coherent 'intuitive' system for modern computer control. After many years of rejecting the idea as being 'wimpish', Microsoft, IBM and the other MS-DOS users suddenly had a Paulian religious conversion, and discovered Windows.

The distinction is with the older text-based Character User Interfaces (CUIs) which simply put characters of one fixed font size and style up on the screen. See also WIMP, Windows, Presentation Manager (OS/2), Motif and Open Look (Unix).

guided or guiding media

- Guided Media: Meaning the use of copper or fiber cable for the delivery of media services which would usually be broadcast by ground waves or satellites using the radio spectrum. This is radio and television delivered by cable.
- Guiding Media means almost the same, except that the term media now applies to the glass itself, or to the copper in the cable. This is a general communications term.

guided wave

A wave (acoustic or electro-magnetic) which travels down a tube — a waveguide or pipe or fiber — as distinct from one that runs through a solid wire. With copper cables, radio guided waves travel across the surface of the copper, not through electrical conductivity. See waveguide.

H.120 & H.130

Two associated 1983 ITU interim standards for videoconferencing which were widely used in Europe but largely ignored in the USA and Japan. The H.120 standard was for video encoding and compression to enable video codecs to use 2Mb/s links, and H.130 defined related variations, including split-screen. See the later H.261 and Rembrandt.

H.221, H.230 and H.240

See H.261.

H.261

The current ITU standard for visual telephony and videoconferencing. This is also called 'p x 64' because it allows a range of data rates in 64kb/s steps (where p is an integer).

H.261 uses DPCM modulation algorithms and DCT techniques to allow substantial image compression. Motion compensation is also applied, along with run-length encoding.

Video luminance is sampled at twice the rate of the color-difference signals, and both are digitized at 8-bits per pixel. Frames are then divided into 16 x 16 macro-blocks with 8 x 8 color-difference blocks mapped into the same frame. Inter- and intra-frame coding, and motion compensation is then applied before DCT transformation, normalization and quantization. Finally, run-length encoding (RLE) reduces the bit numbers even more.

The two main resolution standards currently supported for videoconferencing are the CIF and the QCIF standards, of which only the last is mandatory (see).

- Associated standards are:

H.221	frame structure for 64kb/s to 1.92Mb/s channels.
H.230	frame sync controls and indication signals.
H.242	communications protocols.
G.722	7kHz audio coding.

- H.320 is the desktop videoconferencing and videophone standard (MPEG, and later MPEG4, will play a role here).

H.320

This is a group of conformance standards for multimedia, and now the major standard for desktop videoconferencing. It controls the transport of audio, video and data. It is also the new ITU standard for digital videophones suitable for use over ISDN. Interoperability is now much less of a problem.

H-channels

An ITU designation for wideband digital channels designed for ISDN and B-ISDN. With ISDN they originally specified B-, D- and H-channels — however few ISDN systems provide the last. The idea was that H-channels could be used for user-information at higher bit rates than those of the B- and D-channels.

These H-channels were originally intended to be aggregates of 6, 24 and 30 B-channels, but the ITU has now added variations on these basics. If the type of H-channel is unspecified, a H0-channel is probably assumed; H0 is used for videoconferencing.

H0	386kb/s	equal to six DS-0 channels.
H1	2.048Mb/s	
H11	1.536Mb/s	interpret this as H1.1.
H12	1.920Mb/s	interpret this as H1.2.
H21	32.768Mb/s	
H22	45Mb/s	
H4	139.264Mb/s	

- The ISDN Primary Rate Interface (PRI) can therefore consist of either four H0, three H0 + D, or one H11 within the North American standard T1 rate of 1.54Mb/s. The European 2Mb/s E1 standard can carry five H0 + D or one H1.

- Instead of using H0-channels for videoconferencing, ISDN carriers have tended to use inverse multiplexers to create channel bandwidths in excess of 64kb/s. See BONDing.

hacker

This term has had four quite distinct meanings:

- Originally a hacker was a term of praise for an exceptionally skilled computer programmer.

- Later this became a slight sneer at a keen computer 'nerd' who devoted a large part of his/her life and energies to the use and understanding of computers. This was seen as involvement in work, taken to the level of 'obsession'.

- PCs and modem communications allowed these enthusiasts to extend their interests into other people's computer systems using modems. So then the term began to be applied to anyone who maliciously entered and/or altered material in a remote computer. It now carries the implication of vandalism, theft, and possibly spying: and it certainly makes sensational tabloid newspaper headlines.

- And, just as with 'punk', this media-generated 'bad-boy' image has become a cult among teenagers who now style themselves as hackers. They now inhabit 'cyberspace' and create 'handles' (nicknames for themselves) and hold endlessly long blow-sessions over the Internet and on bulletin boards boasting about fictitious exploits (sexual and electronic).

HAL

- **Hardware Abstraction Layer:** In Windows NT this isolates the NT kernel ('Executive') from specific hardware components, thus providing something like an open architecture, which allows vendors to support different drivers on different I/O subsystems.

Apple has also given the new Copland O/S a HAL so as to support hardware diversity in PCs and allow the new Macs to become open machines. HAL hides the hardware from the system software — it decouples the two layers.

• Arthur Clarke's famous computer from 2001, which is said to have been named after IBM. Subtract one place from each letter and see what you get.

half-adder

A part of the arithmetical functions of a processor chip. A half-adder can generate a carry bit, but it cannot take a carry bit from the preceding half-adder; you need a full-adder to do this. A half adder is two NAND gates with an XOR gate.

half-duplex

• In point-to-point data transmission the link between the transmitter and the receiver can be:

— Simplex (where all the traffic is one way),

— Half-duplex (where each device takes a turn at using the circuit in alternating directions), or

— Full-duplex (where both devices simultaneously share the circuit; both sending and receiving).

Half-duplex circuits still carry information in both directions, but only ever in one direction at a time. At its most basic, this is the technique commonly used in press-to-talk radio communications systems where both transmitters share the same band of radio frequencies.

In computing communications, half-duplex was ideal for downloading a substantial amount of data from one machine to another, where the required transmission path is predominantly one-way (asymmetrical). Early modems had to subdivide and 'share' the full bandwidth of the circuits, so half-duplex systems were faster than full-duplex — but not any more.

In half-duplex operations, after the transmission of each block of data, the computer will pause for a fraction of a second and check that there are no error (ACK or NAC) messages being returned along the line. This pause-time is called 'turn-around', and fast turn-around is essential.

If you are using half-duplex and you wish to see your transmitted data on the screen, you will need to set your computer/terminal to Local Echo. You can't use Echoplex.

• The RTS and the CTS connectors in RS-232 are primarily designed to control half-duplex operations.

• In telecommunications, half-duplex can refer to either the physical or the functional aspects. Telcos sometimes apply the term to four-wire (two-way) connections between exchanges, as distinct from the two-wire (two-way) connections in the local loop. The point is that, between exchanges, calls are carried independently by two different one-way channels so that the signals can be amplified and multiplexed.

A device called a hybrid at the local exchange is used to join the two independent half-duplex channels into the single full-duplex local loop.

Optical fibers are also 'half-duplex' devices (you generally need a pair to carry a conversation), although it is possible to have two-way fiber systems.

• Half-duplex error-control protocols are just another name for ARQ.

half-gateway

LAN gateways are often broken into two units to simplify the design and maintenance. These systems may use an intermediate transfer protocol which neither matches the source protocol or the sink protocol. The two halves of the gateway may be thousands of miles apart. See gateway.

half-height drives

Disk drives come in pretty standard packages since they are usually made by one or two major manufacturers and are designed to be interchangeable in the various computer boxes. Half-height drives are made to a size that allows two drives to occupy the space of an old standard single drive. So a half-height 5.25-inch disk drive is approximately 5.75-inches wide, by 1.6-inches high (150 x 40mm).

half-inch tape

• IBM is about to announce a new computer storage system based on half-inch magnetic tape. This is IBM's 3490 cartridge system which can hold 10 gigaBytes.

• DEC are also ready to release their 100GByte digital linear tape in the DLT cartridge. See DLT.

half-repeaters (LANs)

These are devices which connect two halves of a LAN or two different LAN segments, using some quite different medium (usually dial-up circuits) between.

The two halves can be any distance apart. As with full repeaters, they extend the LAN in terms of physical distance but not in the number of stations which can be added — they don't segment the LAN. Half and full repeaters transfer all packets on a LAN segment to the other segment — they don't filter at all.

The term 'half' repeater just refers to the need to use two of these devices (with the dial-up link in between) instead of a single normal repeater between two segments.

half-transponder

A way of transmitting two television signals over a single satellite transponder by reducing the deviation and power allocated to each. Half-transponder television carriers reduce the cost of distribution signals to cable head-ends. They are also widely used for international and national delivery of TV news, etc. A half-transponder will generally operate at a level 4—7dB below the normal single-carrier power.

• Direct broadcasting satellites generally require the full power of the transponder to be applied to the one signal (even though it is using only part of the available bandwidth) — however, with digital TV, more than one channel can be multiplexed into a single stream.

Don't confuse half-transponder operations with vidiplex.

halftone

An illustration or photograph that appears to the unaided eye to have continuous tonal areas, but which, under magnification, consists of a regular pattern of dots — and the dots are of various sizes, creating the appearance of tones.

In the printing process, each dot is printed full-black (or full-color in the case of color printing), but the size and separation of the dots determines the area's subjective tonal value.

At present, laserprinters don't make halftones, but rely instead on 'dithering': a technique using varying patterns of fixed-sized dots to create pseudo-gray tones. However, some of the more advanced laserprinters have a hybrid system which uses, in part, a dot-size variation and part dithering.

Halftone reproduction is the basic technique used in book and newspaper printing for handling 'continuous tone' photos. A 'bromide' is made of the photograph using a graphics camera and a screen which has been chosen to match the quality of the paper, and therefore of the printing.

The distinction is with 'bitonal' line art, where the illustration consists of fine lines of black only — there are no gray areas and also with continuous tone (as in the original photograph) where true grays exist.

• In DTP, a standard 300 dpi laserprinter will have more than adequate resolution for text and line art, but it will clearly be inadequate for photographs because laserprinters can't reproduce true halftones. So if good quality variable-tone images are require, a true halftone reproduction technique must be used, and the image cut into the camera-ready copy. This will be physically inserted at the final make-up stage or into the plate negative itself. See dithering and dpi.

• A halftone screen is a sheet of transparent film on which is etched a fine matrix pattern. This screen is sandwiched in the graphics camera underneath and against the plate or 'bromide' during exposure, and it produces an image consisting of halftone dots of varying size — the dot's grow in size during processing, depending on the exposure each 'pixel' receives. The screen's granularity is measured in lines per inch (linear). The coarser the paper-quality to be used in the printing (say, newsprint), the more it requires a coarse screen. See screen.

halide fibers

See fluoride fibers.

Hamming code

A forward error correction (FEC) system related to parity. Hamming bits are calculated by adding the binary values of a byte or nibble using modulo-2 arithmetic.

If you have a four-bit nibble, Hamming will sum the bits in the nibble in three stages: 1st + 2nd + 3rd, then 2nd + 3rd + 4th, and lastly 1st + 2nd +4th. Summing in modulo-2, is equivalent in this case to calculating three sets of 'odd parity'.

These three Hamming code results will be appended to the nibble during transmission, and they provide enough information to restore any single lost bits of the original four.

However, note that the Hamming bits are almost as vulnerable as the original nibble (3:4 ratio). So, as a simple error-check, the system has a high overhead, is vulnerable itself, and it breaks down with multiple bit errors. Hamming still has some value as part of more complex error-correction systems. See Reed-Solomon code and parity.

Hamming distance

In some numbering notations it is important to note the number of digit changes that have happened between two numbers. The Hamming distance ('distance' here means 'number') is a measure of how many digits have been changed.

For instance, the Hamming distance between 12345 and 12846, is two. The 3 became 8, and the 5 was changed to 6. The Hamming distance between 1000 0101 and 0000 1100 is three (three of the binary places have changed sign). This is important in some forms of binary notation. See gray binary.

hand-helds

• Small mobile radio (phone) transceivers of any kind.

• Small portable computers in the one pound weight-range which can fit into a coat pocket. Usually they require one-finger keying. Most hand-helds are used as personal organizers, but some are true MS-DOS machines with the ability to handle full-scale computer applications and files.

The term is used mostly for keyboard-driven small computers, but it increasingly is used as a generic term to mean tablets (used in factories and warehouses) and PDAs, some of which have touch-screen keyboards, and/or handwriting recognition.

hand-off/over

In cellular mobile phone systems there must be a mechanism in place at the basestation which can constantly monitor the signal level of the mobile. When the signal drops below a predetermined limit, the remote computerized switching center will immediately request transmitters in the immediate vicinity to check their reception power of the mobile.

The best of these will then be selected, and an almost-instantaneous switch-over will occur (the hand-off) to the new station, and a change to a different pair of frequencies since the same frequencies can't be used in adjacent cells, except in CDMA.

This hand-off problem (with its momentary break in transmission) disrupts cellular systems when transmitting data — so they need a robust error scheme. See ETC, MNP-10 and also MAHO and soft handoff.

handle

On the Internet: your name. This may also be a pseudonym you use in forums, MUDs, or chat/CB sessions. It was the old CB radio term for nickname.

• In drawing programs (as distinct from paint programs), each shape created on the screen exists as a separate 'object' which is resizable. These objects generally have handles (either directly attached, or on a surrounding marquette) which the cursor can 'grab' and stretch to the desired size.

handler

• In object-oriented languages, a handler is a self-contained module of script (program) which performs a distinct function.

• In conventional operating systems, this is a device driver which also deals with interrupts.

handshake

The formalized exchange of special signals (usually operating parameters) between two devices which are attempting to establish communications.

For instance, a computer and its printer will conduct a handshake before data is sent down the line. Fax machines handshake with each other to establish transmission rates, and check if they can use error correction. Modems exchange information about their capabilities and jointly select the highest data rate that will run on the connecting link via handshakes, and they may also choose a common data-compression system and error correction mechanism (V.42 or MNP4).

Handshakes like this are an agreed sequence of binary codes which are an essential part of each communications protocol.

• In local cable links over multi-connector cables (such as RS-232), the handshake signals can be in the form of substantial DC voltage changes on one or more of the connecting lines and pins. The line is said to be asserted (voltage raised) or inhibited (voltage dropped) — and the 'asserting' voltage can be either +ve or –ve (negative in RS-232).

• In RS-232 connections, the hardware handshaking of DTR/DSR establishes that the equipment is connected and ready to go. Don't confuse this with the two lines involved in the control and regulation of the flow of data across the link (CTS/RTS). These were for controlling half-duplex operations.

However, the term 'handshaking' often incorporates flow-control, and in some local cable systems the primary go/no-go and flow-control circuits are linked (see DTR/CTS). See also hardware handshaking, flow control, DSR and RTS.

• SNA networks engage in almost continuous handshaking between host and remote computer systems. See spoofing.

hanging indent

This would more correctly be called an 'outdent'. It is where the first line sticks out from the column like this, rather than being inset. It looks OK in the right circumstances.

hard boot

A computer reset, which is equivalent to turning it off then on again. A hard boot is accomplished by hitting the reset button. This causes the operating system to reload, and the computer then re-runs the POST diagnostic tools. See warm boot.

hard-card

A small hard-disk unit (usually 3.2 inch or 2.5 inch) mounted on a PC card which plugs directly into an expansion slot inside the computer. The main advantage of this approach is, ease of installation. You can also take your hard disk with you when traveling, and you can take all your private records home at night — so there are security implications also. Some hard-cards are now being made to fit PCMCIA Type 3 slots (42.8MB) and some are being made for Type 1 slots.

hard-copy

Printed output.

hard disk

Usually a fixed disk unit (although there are removable hard disks in PCs as well as in mainframes) with a multi-megabyte capacity. Large hard disk units may have more than one platter (disks in a stack) and they vary in the type of media (plated or spluttered), the number of heads, and the control systems used. The early hard disks used simple FM recording techniques, while later ones used MFM, then GCR, then MMFM modulation techniques. Later still, RLL became popular.

The distinguishing feature of a hard disk is that the magnetic read-write heads don't actually touch the disk surface but fly a micron or so above on a thin layer of air. For this reason, the disk must be enclosed in a totally sealed chamber to prevent dust settling on the disk — which would cause a head to crash.

A typical hard disk spins at about 3000 rpm (about ten times as fast as a floppy) and has a single head with a magnetic-gap of about 0.18 microns. The same head handles both read and write functions.

To avoid the head ever making contact with the surface, there's a shipping/landing zone towards the center of the disk-assembly, where the heads rest whenever the power is off. The heads are then said to be 'parked'.

The surface of each disk is divided into thin circular bands of data known as tracks, and if a number of disk surfaces are available (in a disk stack) then all tracks at the same location (distance from the center) are collectively known as a cylinder. Some hard disks will have 1000 cylinders which, with (say) six recording surfaces in a stack, would translate to 6000 tracks.

Each track is further segmented into sectors of a uniform size (usually 512 bytes). The tracks towards the margins of the platters may hold more sectors (say 200) than those towards the center (say 150). Sectors have a small number of bytes (usually 6 or 8) allocated for error checking.

Each sector is identified by a number hierarchy which consists of a cylinder or track number (say from 0 to 999), a head number (say from 0 to 5 if multiple heads are used), and a sector number (say, 0 to 199). See voice coil, and stepper.

hard hyphen

An ASCII hyphen which must remain between two words; not one introduced by the application to break a word at the end of the line. Hard hyphens and spaces are also called 'required'.

hard return

A (usually) two-byte control code entered into a word processing document when you press the return key (to mark the end of a paragraph). In most MS-DOS applications, this key-press will enter both a Carriage Return (CR — ASCII Decimal 13) and a Line Feed (LF — ASCII Decimal 10).

However, not all word processors use the same form of hard return. WordPerfect and Unix W/Ps, for instance, only use LF. When sending files to printer, the printer-driver in the WordPerfect application will make any necessary adjustments.

• Soft returns are those entered into the text at the end of each line to force word-wrap and to move the cursor down the screen: these are added by the program itself and should not normally be stored as part of the file. Soft returns vary; some use the CR with the eighth bit set (Extended ASCII).

• It is the inclusion of unwanted screen-control characters in ASCII files, and the interpretation of soft returns as hard returns, which create problems when a Macintosh attempts to read an ASCII file generated in MS-DOS. Small utilities can be used to strip these extra CRs out.

hard-sector

This refers to the positioning of information on a disk against a mechanical reference point. The early floppy disks had reference holes punched on the disk near the spindle to provide the fixed/hard-sector reference point, and a photocell in the disk drive detected the edge of these holes and used them as a pointer to the beginning of the sector.

Soft-sectored disks are referenced only to a point magnetically recorded during the formatting process.

With hard disk drives, the hard-sectored approach has advantages in terms of the density of data that can be stored in each track. See soft-sector.

hard space

An apparent space in text which is usually created by a combination of keys strokes, rather than by the space bar alone. The idea behind providing a hard space, is that the word-processing program will see it as a pseudo-print character, and treat the two word-strings on each side as a whole. It prevents the word-wrap function from breaking the line at this point.

You would use a hard space, between the two strings '1000 mph.' to ensure that the number '1000' doesn't end up on one line, and 'mph' on the next (and maybe over the page).

One other solution, which is becoming increasingly popular (and used in this book) is to set all contractions, immediately behind the number: as in '3000km' rather than '3000 km'.

hard wired

• **Favorable:** This means what it says. That the logic of the system is built into the electronic connections on the silicon, or by wires across a PC board. That you can't modify or change it by software, and it works very well (fast) in a limited capacity.

In computing, very often certain functions (video processing, for example) can be hard-wired into a specially-designed chip, or they can be performed under software control in a CPU + memory processor. Hard wired will inevitably be faster, by orders of magnitude.

By extension, a dedicated tie-line is hard wired, while a dial-up switched connection is not.

• **Unfavorable:** This also suggests a lack of flexibility. It refers to the old way of connecting items or equipment electrically by wire and soldered connections. The distinction is partly between electronic and electrical equipment (which is generally hard-wired) and computer driven equipment, which is configured and controlled by software. Hard-wiring makes a TV-set a TV-set, while it is the software control of a computer which makes it a universal machine — able to act like a TV-set with the right software (and admittedly, some extra hardware).

hardware

The physical parts of a computer system, as distinct from the programs and data in memory which are software.

Removable magnetic disk/tapes and optical disks could be classed under either the hardware (the medium) or software (the information stored) definitions, but they are always known as software — because it is the software which has the value. The internal non-removable hard-disk itself, however, is classed as hardware.

• There are also some forms of software which are stored in non-volatile memory (ROM) and then known as 'firmware'. These are hardware devices (chips) holding specific software.

hardware addresses

In communications, these are MAC layer or physical addresses (although they are in the Datalink layer). They are directly associated with a particular network device. The distinction is between the MAC device address, and the network address; a bridge uses the former, and a router uses the latter.

hardware handshaking

Local connections between a PC and a printer, or a PC and a modem, can use baseband communications and multiline cables because they only carry signals for a few meters at most.

The most commonly used is the serial connection known as RS-232. It is used between computers and printers or computers and modems, and it is connected to the device at each end through a UART. It requires (generally) 8 wires. See Big 8.

Some connections are used for handshaking (to inform the other device of status), others for flow control, and others transfer the data.

The two main hardware handshaking connections in RS-232 are Data Terminal Ready (DTR) and Data-Set Ready (DSR). They are asserted (activated) by raising the negative voltage on the appropriate connector pins and maintaining this voltage (called a 'logic level') for the duration of the session. Printers, modems and PCs will often monitor other connections also, in order to check whether buffers are full (see RS232), but this is not strictly 'handshaking'. However the term is now often used vaguely enough to include all of these signaling changes.

Hardware handshaking acts on the UART chip, and it usually initiates action by way of ARQs.

• In reference to flow control systems, the distinction being made here is between this hardware approach and the use of binary code systems (software handshaking) like XON/XOFF to indicate ready states. These are initiated by full-buffers, and are transmitted on the back channel. All software codes need to be interpreted by the computer before they can initiate action. See DTR, DSR, ETX/ACK, DTR/CTS and XON/XOFF.

hardware tree
A database of configuration information which keeps records of plug-and-play device requirements. From Windows 95 on, new PC operating systems will all have a hardware tree.

harmonic
In any physical (audio) or electro-magnetic (radio or light) wave, a harmonic is an integer multiple of the fundamental frequency. The first harmonic is twice the frequency of the fundamental, the second harmonic is three-times, etc. The summation of all harmonics generated, produces the identifiable tone of a musical instrument. See Fourier transform.

harmonic distortion
This is caused by electronic devices in the communications system which produce unwanted additional signals (artifacts) which are integer multiples of the fundamental frequency of the original signal. The effect is also known as non-linear distortion. There are various kinds of harmonic distortion. See transmission impairments and distortion.

Harvard architecture
A processing chip (DSP) layout that separates data and program memory, and provides each of these memory areas with its own independent bus.

hash
- **The # character:** (aka the 'pound') Pound is confusing because of the alternate £ sign on English keyboards; the correct name for this symbol is octothorpe.
  - The hash is used as a substitute for the English pound sign on any American keyboards without this symbol.
  - A key with the # (hash) sign is widely used on DTMF keypads in both touch-tone telephones and videotex.
  - The hash is often used as a variable symbol for 'number'.
  - In typesetting it can mean 'add space' or 'end'.
- **Hash values:** Since ASCII characters in a PC have a set decimal and binary number equivalent, it is possible to sum 'words' (alphabetic strings) and produce a highly identifiable numeric value. This can be used for authentication, error checking, or for indexing.

  The simplest way to generate a useful hash value is to take a bit-by-bit exclusive-OR (modulo-2 addition) of every byte in a block or message. A CRC in data communications is a form of hash value.
- **Hash functions**: Algorithms which compute some value based on the contents of a message. Usually the purpose of a hash function is to create a 'fingerprint' of a file or a block of data, so that the file or block can later be authenticated. Hash functions produce fixed length outputs, and they are reasonably easy to implement.
- **In database retrieval:** To hash a word is to convert a key alphabetic string (say, company name) into a single number identifier. Words can be hashed in various ways, and this process is widely used for indexing. See hashing.
- **In security:** A large hash function is a non-reversible function — or rather, it is extremely difficult to change the original message secretly if the original hash value is known. You would need to know how the hash value was computed before you could modify the message and match the hash value. This would be extraordinarily difficult with complex calculations.

hash table/index
The index created by using a hashing technique.

hashing
- This is an text-indexing method used for very large files, where a search value is calculated to define the location where the data will be found. There are a few different very simple hash approaches. For instance, the first three characters of a company name will be converted to ASCII decimal equivalents, then summed. This number will be used to define the storage location for all the data about that company.

  One apparent problem is that the same hash value can be generated by another company with a similar name. However, within this location, company names will be stored alphabetically. Hashing provides a quick search technique for large databases and with large storage media, particularly with CD-ROMs.
- Intermixing elements — a communications term.

hat
An alternate name for the ^ sign. It is also incorrectly called a caret or circumflex. I assume it doesn't have a correct name, but it is most commonly called a caret in computing.

Hayes compatible
Hayes Microcomputer Products have long been the leading manufacturer of modems in the USA to the point where their computer-to-modem 'AT-command language' has become the *de facto* standard for most communications software and modem hardware.

Modems are mostly Hayes compatible these days. What makes them so is the use of the V.# standards (see) of the ITU, and the control of the system from PC communications software using the AT-command set. See AT command set.

Hayes ESP
Enhanced Serial Port. This is a Hayes-developed plug-in card to overcome the bottleneck of slow UARTs driving the new V.34 modems (with V.42bis compression).

The card has an on-board processor and RAM; it provides a 1kByte buffer for both sending and receiving. It can emulate the standard 16450 or 16550 UART in normal operation, but with its own special driver, it can also scream along at the much higher rates necessary to feed the new modems.

Remember that the UART is handling raw data while the phone line is transferring compressed data.

HBT

Heterojunction Bipolar Transistor. These use a blend of silicon and germanium, and IBM's are currently the world's fastest silicon-based transistors, running at 75GHz. HBTs are used to modulate communications signals beyond 100Gb/s data rates.

HCC

High Capacity Cellular telephone technology. This is an old term for what is now the current form of analog mobile telephony. The distinction is with the Low Capacity Cellular systems that were phased out in most countries many years ago (but not all). See IMTS and AMPS.

HD

- Hard Disk — aka HDD.
- High Density (floppy and optical disks). See double-sided.
- High Definition (television).

HD-Divine

A European project for high-definition terrestrial digital television. It has since been merged into the common European digital-video broadcasting standards projects. See DVB.

HD-MAC

High Definition, Multiple Analog Component. A television transmission system that almost became the European standard in 1993. It was not true high-definition, but it provided 16:9 aspect ratios and 625 lines of image, and these lines could be doubled in the receiver to provide 1250 lines. This technique did not improve the vertical image resolution much (there was some augmentation), but allowed the use of large screen sizes without the scan lines becoming so noticeable.

The image quality was supplemented, by three levels of motion compensation which was delivered by a digital augmentation channel in the HD-MAC's transmissions.

All production and post-production needed to be done with a true full-bandwidth HD Video format (1250 scan lines) and then coded down to the HD-MAC format for transmission. See component, D2-MAC and Eureka projects.

HDB3

High Density Bipolar, Three zeros coding. A form of AMI line-coding which is used in Europe and Japan for 2Mb/s circuits (B8ZS is used in the USA). After three consecutive zeros, the mark is always inverted, so an input string of four zero-bits will be replaced by three zeros and a code violation (see).

The aim of HDB3 is to provide a clocking signal which can retain synchronization during long delays in transmission. This is part of the G.703 recommendation, and is widely used in transmission systems (E.1 and Primary Rate ISDN).

It can be framed or unframed. E1 has 32 time slots (each of 8-bits), and the frame format is defined in G.704. Time slot 0 is reserved for frame alignment and alarms. Time slot signaling is used (if required) for channel associated signaling. The voice channels are numbered from 1 to 30 (with 0 and 16 frame numbers ignored). See scrambling, violation and B8ZS also.

HDCD

High Density CD. The Philips/Sanyo code name for the new generation of double-density video CDs. See CD-Multimedia.

- Toshiba and Time Warner also have a rival high-density CD standard under development. See Super Density.
- There is also a HDCD-R (recordable) standard under development in Japan, and both of the above have plans for recordable versions. They are currently holding talks and may merge. .

HDD

Hard Disk Drive. As distinct from the floppy. The average HDD in a modern PC sold for business is now 350MBs, and it is projected that it will be about 1GB by 1996. Graphics-intensive software and 32-bit operating systems are the main culprits, although the prospect of video must play a part.

- Word for Windows already requires 24MB of disk storage, and Wordperfect for Windows requires 33MB.

HDEP

High Definition Electronic Production. The use of special digital high-definition video equipment for feature film production. The aim is to produce electronic images as good as, if not better than, 35mm film. This is more than HDTV.

In general, it is expected that HDEP systems will save approximately 15% of production costs for single camera crews on a normal feature film shoot involving both studio and locations. Some HDEP proposals suggest that 1575 scan lines (3 x 525) will be enough, together with 16:9 horizontal aspect ratios. However, good 35mm film has a vertical resolution equivalent to about 3000 lines. In the past, the production version of MUSE has been used for feature film production mainly in the USA by Francis Ford Coppola. See MUSE and HD-MAC.

HDH

HDLC Distance Host. This uses the LAP-B packet protocols to carry 1822-bit data packets over synchronous links.

HDL

Hardware Description Language. This refers to analysis and simulation tools for designing hardware (mainly chips). The IEEE 1364 standard is based on Verilog HDL. See VHDL.

HDLC

High-level Data-Link Control. A bit-oriented, Datalink layer, communications protocol defined by ISO 3309 and 4335 for packet-switching. It has become very popular.

HDLC is the ISO's version of IBM's original SDLC (see for technical explanation). It is designed to detect errors, provide flow-control, and handle retransmission of data on point-to-point links. Unfortunately the standard has not been tightly controlled and so every vendor implements the process in a slightly different way.

HDLC was defined for high-level synchronous connections on X.25 packet networks. Variations are used in ISDN, V.42, frame relay and, more recently, in PPP. LAP-B is a subset, and ADCCP is almost identical.

The HDLC standard defines three types of station (primary, secondary and combined); two link configurations (unbalanced and balanced); and three data transfer modes (normal response, asynchronous response, asynchronous balanced modes). The frame structure defines how the data is to be encapsulated (framed) on synchronous networks. Redundant bits are added to each frame for addressing and error detection, and when the checksum doesn't compute, the frame is retransmitted. See SDLC, Data Link Protocols, LAPB and LAPD.

- The so-called 'HDLC 7e' protocol, are such well-known names as X.25, SDLC and SNA. They all use variants of HDLC.
- HDLC defines three types of frames:

— I-frames are information frames which carry user data.
— S-frames are supervisory frames which provide the ARQ
— U-frames are unnumbered frames which provide supplemental link control functions.

The first bit/s in the control field identify the frame type.

- The 'standard' frame for HDLC provides the following fields:

— 8-bit flag (0111 1110),
— one or more bytes of address,
— 8 or 16 bits of control information,
— variable length information,
— 16 or 32-bit frame check sequence,
— 8-bit flag again.

Bit stuffing is used to ensure that a character equivalent to the flag-byte isn't accidentally transmitted in the data. An extra 0 will always be inserted after five 1s in a row.

HDMM

High Density MultiMedia. This is the old name for the CD-Multimedia standard. See.

HDP

High Definition Progressive. The proposed digital studio production standard to be used in Europe for TV and feature film production. It is a progressive scan system with 1250 lines, and it has 1920 pixels per active line (horizontal resolution). Color differences are sampled only on every second line and at half the horizontal rate.

With 2.4 million pixels and a 50Hz refresh rate, sampling is therefore at 120MHz for the luminance and 30MHz for the two color-difference signals. With a 10-bit (luminance) and 8-bit (color) pixel depth, this translates to an uncompressed data-rate of about 1.728Gb/s. See HDEP

HDSL

High-speed Digital Subscriber Link (Loop). This is a parallel proposal to ADSL video-over-copper to homes. Both have their ancestry in VLSI developments for ISDN.

Generally HDSL is more expensive than ADSL and it calls for two (or three) copper twisted pair to the office/home (ADSL only uses one). HDSL is bi-directional, and it has the data rate needed to carry a compressed MPEG image for videoconferencing both ways. There are a few variations of HDSL using different line-coding techniques:

- ANSI originally standardized on a three-pair 2B1Q transmission system (rather than CAP or DMT) for HDSL at 1.54Mbit/s, but it now appears certain that they will endorse DMT which has proved its superiority. More recently, two-pairs with 2B1Q line-coding has been used, however CAP is favored by AT&T who tend to dominate here.
- With the CAP line coding techniques (with two-pair), each cable is used bi-directionally at 768kb/s for a total of 1.54Mb/s full-duplex. HDSL-CAP has sophisticated digital filters for 'integral echo cancellation, interference de-correlation, and decision feedback'. To combat NEXT and impulse noise, a two-dimensional (8-state) Trellis code has been included in the CAP specification.
- In Europe, ETSI is examining a proposed HDSL standard which provides 2Mb/s data rates (each way) over distances of up to 3.6km using a single 24 gauge copper pair (or 2.7km with 26 gauge) without repeaters. A variation on the ISDN (2B1Q) line coding, called 1B1Q, is proposed.
- The development of HDSL repeaters will extend the range of all of these systems to about 6km.
- An HDSL frame also contains overhead-bits, such as frame-sync bits, CRC bits, and embedded operation-channel data bits. The CRC is used to calculate the BER-feedback of each sub-channel for CAP and DMT.

HDTV

High Definition Television — a term that is very vaguely used. Don't confuse High Definition with extended or widescreen TV systems — and HDTV is not tied to either analog or digital techniques. HDTV implies only that there will be an approximate doubling of the vertical image resolution because there will be at least 1000 scan lines.

In the interest of retaining 'square pixels' they will also need at least a doubling of the horizontal resolution over that of present 4:3 PAL/SECAM and NTSC. An unspoken assumption is also made that the number of frames transmitted every second will be enough to avoid flicker and artifact problems — both of which are more obvious with larger screen areas.

Note that the HDTV production standard (more correctly called HD Video since they often differ from what is finally transmitted) often has twice the line resolution of the transmission standard, since the video production equipment is also designed for making conventional feature films for celluloid cinema projection. In most studios, HD Video will be based probably on the CCIR-601 standard called 4:2:2 (see 601).

The latest generation of professional video camcorders is capable of producing widescreen images which are nearly HDTV standards — usually as an option; most have wide-screen capabilities. The old Sony D-1 professional 4:2:2 recorder standard has now been upgraded to handle HDTV images.

• Most industry organizations apply the term HDTV only to 1000-or-more active 'image' (not doubled) scan lines, with an 16:9 aspect ratio, and multiple channels of high-quality sound.

• Another definition, once widely used, was that HDTV should be 'at least equivalent to 35mm film' — but that would mean about 2000 scan lines and 3000 or more lines of horizontal resolution.

• Full 25 frame per second HDTV systems, on the most optimistic of current estimates, need bandwidths for transmission of about 12Mb/s with MPEG2 compression, and 20Mb/s for real-time sports. The digital data-rates during production (before compression) will be about 1Gb/s. See HDP.

• It is expected that home HDTV sets will need to have a diagonal screen size of 1.4 meters if the full resolution is to be seen in an average small living room. With conventional CRT technology, this is a TV set about the size of a small refrigerator on its side. Even then, the living rooms in most homes will not provide good viewing or audio conditions, which is why many people question whether home viewers need more than 750 lines of vertical resolution coupled with a wide (16:9) screen.

• Many HDTV monitor developments are driven by the desire to create a single standard for both TV sets and for computer monitors. Good workstation monitors already exceed HDTV's vertical image specification (although they aren't widescreen).

See Grand Alliance, DVB, 4:2:2 and 4:2:0. Also see augmentation, simulcast, widescreen, enhanced TV, PALplus, MAC and D2-MAC, HD-Device, ATV, agile receivers, B-ISDN.

• MPEG2+ has defined two levels of the HDTV signal image format. The low level is 1440 x 1080 pixels at 30Hz, and at 25Hz it is 1440 x 1152. The high level is 1920 x 1080 pixels at 30Hz, and at 25Hz it is 1920 x 1152.

HDWDM

High Density Wavelength-Division Multiplexing. A new BT (UK) development for passive-optical telephone networks that seeks to move exchange switching and bandwidth allocation functions to the customer's premises and equipment (CPE).

Bandwidth can be used as a replacement for switching, and this idea aims to supply a vast bandwidth by the use of fiber-to-the-home.

BT describes the idea as 'optical ether' (as in radio). They propose providing multiple channels using lasers at different wavelengths — as many as 50 are projected. Your laser would tune to an unused wavelength, just as your cellular phone re-tunes automatically to select an unused channel.

They expect to increase bandwidth available over the FTTH, eventually it would be in the order of 10 or 20Gb/s per home in about 10 to 15 years. See quantum cascade laser and WDM.

HDX

This is an old telecom engineering term derived from 'Half-Duplex'. It means that a single two-wire circuit is being used (as in the local loop). The distinction is with FDX, which is a four-wire connection between exchanges. However this term is totally confusing — they are calling the local loop twisted-pair an HDX circuit even though it carries two-way voice.

head

• In computers, communications and media, the head is the beginning, and the tail is the end. In film and videotape, the terms 'head' and 'tail' are applied to the beginning and end of a shot or sequence. So the head end is the beginning — the first frame or first shot in a sequence.

• Short also for magnetic-head, which is the tiny electro-magnet which makes contact (or near contact) with the magnetic material, and causes it to flip magnetic orientation. The head-gap size and its proximity to the magnetic surface, and the thickness, coercivity and granularity of that surface, all contribute to the packing density.

head-end

The source of signals in a broadband network.

• In Cable TV systems this is the central amplification and distribution point for the signals. The head-end collects all the signals (off-air, via satellite, coaxial or optical cable distribution, or simply off tape), modulates them each to the required carrier frequencies, scrambles the 'Premium' (Pay) channels, then combines the signals for transmission over the network. Cable head-ends are relatively small buildings with banks of amplifiers, and R/F modulators.

• In broadband (FDM) LANs all nodes transmit towards the headend, and the headend then re-transmits the signals to the destination station/s. In point-to-point cable and broadband systems, the headend is a form of transponder/ translator device; it receives data that has been transmitted at one set of frequencies, and transmits it at another. See high-split.

head tracking

• In recording systems, this is the technique used to keep the read head directly over the track laid down during the write cycle. In audio recorders, if the tape weaves slightly from side to side (usually through wear), the head will mistrack and may read part of the adjacent track, but it is rarely noticed.

In video recorders, head tracking is a major problem and requires special controls including piezo-distortions of the head position within the head drum. Mistracking results in a very noisy picture. Early video recorders always had a track control, which provided the broad adjustment.

• In virtual reality, head tracking is the measurement of movements of the head. This must feed back to the visual image system and displace the projected image horizontally and vertically by precise amounts.

Head tracking is a very costly coordination problem for VR developers. Usually the measurements are based on transmitters and receivers that triangulate to determine the helmet's position in the real world, and sometimes they use acceleration detectors within the helmet.

head unit

- In a computer disk drive, see access arm and voice-coil.
- In a satellite antenna, this is the Low Noise Amplifier (LNA) or Low Noise Block Converter (LNB or LNC). It is the wide-band linear amplifier directly behind the antenna's feed horn which boosts the signal (and translates it to an intermediate frequency) before noise is introduced by the cable.
- In audio and video tape recorders this is the magnetic record/replay head (in professional equipment these are separate heads). It may possibly include the erase head and control head (in a video recorder) also.
- In a video or DAT recorder, this is usually the whole head-drum assembly.

header

- The first identifying elements (routing, destination, segmentation and error-control information) in a message/-frame/packet of information on a public packet-switching system or a LAN/WAN. The header will usually include source and destination address, sequence number, message length and type. Cell-switched systems (ATM) have much shorter headers with cryptic address information (VCI and VPI).
- In SNA, the packet payload is known as an RU (Request/Response Unit) and the headers attached to it would include: response header, which identifies the packet to the data-link controls; transmission header, which identifies the routing path needed, and link header, which contains information relevant to line-error checks and controls.
- In formal documents, the header is the embedded controls at the beginning of a file before the text or data begins. This section of the file holds coded information about the document/record structure, but not text or data. In database files, the first record may contain information such as the name of the file, the number of records, and the last update, etc.
- In printing, a header is a line of text (often a chapter heading) which is repeated on the top of each page. See footer.
- A head (sometimes called a header) is the common term for the headline or title on a newspaper report or article.
- In a CD-ROM, a header is a disk sector which contains the absolute sector addresses and mode bytes.

header prediction

Internet LANs protocols like TCP/IP were designed to utilize complex routing possibilities. However when routes aren't fixed, packets can travel over different routes, take different times, and therefore arrive out of order.

However, within the local area net, there is generally only one possible route from one station to the other, and so the sequence of packets always remained in order across these networks. These WAN packet headers therefore carry a lot of redundant information on a LAN, which slows down the network. Header prediction is a way of reducing the overhead in these circumstances, by discarding the redundant information and allowing the router to replace the header on WAN traffic.

heap

- In computer memory systems, the heap is an area of RAM used for the allocation of data structures. This is similar to a stack, but there's no restriction placed on the modes of access (like FILO).
- In Windows, the data segment allocated to each application is divided into a number of segments. The heap is a variable size area of memory which is used to hold dynamically-allocated local memory objects.

heartbeat

A signal which an Ethernet transceiver sends back to the network controller, to advise that the collision circuitry is working.

Heaviside layer

The E region of the ionosphere. See both and ground wave.

HEC

Header Error Control. A parity-bit or checksum used in the header of a frame- or cell-relay system to improve the integrity of the cell/frame header. This type of error control is used in the header when a packet protocol does not provide total error checking of both header and payload. This is an ATM term.

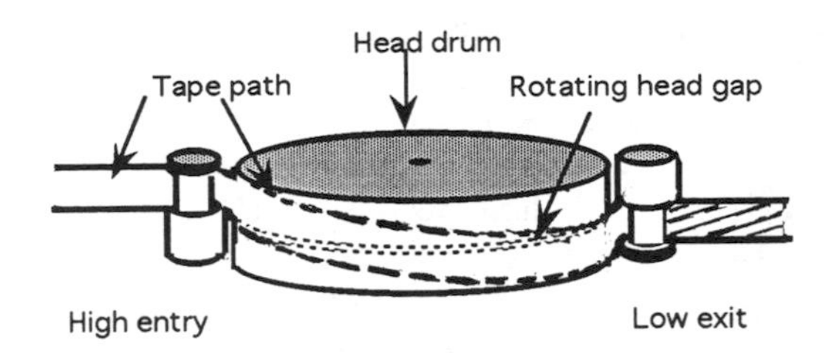

Helical Scan

helical scan

The video recorder technique (also DAT) which lays slightly diagonal tracks along/across the tape by angling the tape path along a helical (vertical spiral) track. The tape enters the head-drum, at a tape-width higher than it leaves the drum (usually after traversing half of the circumference) about 6° from horizontal, as it passes around a rotating head-drum. Each field/-frame recorded then exists as a diagonal and parallel track, separated from the last by the forward movement of the tape.

Helical tracks have a relatively long, diagonal-but-finite length which corresponds to one picture image (field or frame) in most systems. Conventional linear tracks on the edge of the tape provide control tracks for synchronization, one or more audio tracks (although audio can be embedded in the helical scans), and time-coding information.

Helical scan is the most common video recording technique used these days in both amateur and professional equipment, and it is also used in some audio and data recording systems. See R-DAT and DataDAT.

The distinction is with longitudinal recording (see S-DAT) which lays parallel track down the length of the tape, and also with the old Quad video recorders which striped the recorded images in four tracks across two-inch wide tape.

helios noise

The electronic noise created by the sun when the orbit of a communications satellite passes across the face of the sun, or near to it. This can cause a sun outage.

Hello

There are a few types of Hello protocols; one is used by NSFnet nodes, and another by OSPF systems — and they do slightly different jobs.

- NSFnet: Hello is a routing protocol which helps the packet switches to find the route with the least delay.
- OSPF: Hello establishes and maintains neighbor relationships in a multi-access network.

Don't confuse these with the Aloha protocols.

help

Help is offered in many ways in computer software. These days on-screen help is often 'context-sensitive' which roughly means that it should only give you information relevant to your predicament. Usually it doesn't.

The second level of help is to read the manuals. The third level is the company's hot-line.

- The F1 function key is often used to provide Help.

hemi-beam

Hemispherical beam from a GEO satellite.

- It really means a shaped beam which covers about half of the visible surface of the Earth as seen from the satellite. This is normally the widest coverage (and weakest) down-link signal provided by a communication satellite's transponder. It is shaped so as to provide less signal strength in the footprint center (where the ground is at right-angles to the beam) and more to the periphery (where the angle is acute) — so, compensating for additional signal attenuation at these look-angles.
- Many of the earlier hemi-beams were not shaped, so the EIRP fell-off sharply towards the edge of the coverage area.
- This term is being widely misused. Don't confuse it with a global beam. A hemi-beam has less coverage and a higher EIRP.

HEMT

High Electron Mobility Transistor. Aka MOdulation-Doped Field Effects Transistor (MODFET). A new type of low-noise amplifier (invented by Fujitsu and made by Toshiba) which appears to be ideal for satellite TV reception as a substitute for GaAs technology. HEMTs are also being used in super high frequency (SHF) terrestrial equipment — they operate at between 5- and 10-times the speed of silicon devices.

Fujitsu has announce HEMT logic and memory chips — in gate arrays and SRAM. This is a relatively recent, and very important development.

Henry

- The unit of inductance.
- A Quantel-designed hard-disk unit for storing video images during the post-production editing and effects-creation processes of professional video production. Henry is usually used in association with Paintbox, the popular professional digital effects unit. See Abekas also.

HEO satellites

Highly Elliptical Orbit satellites. The term is used for non-geostationary satellites used for communications. See Ellipso, Tundra and Molniya orbits.

Hercules Graphic Card

HGC (also called MGA). This was a non-IBM proprietary monochrome graphics standard (named after the adaptor card) for MS-DOS computers. It became highly successful because IBM's own CGA text-standard was so poor, and the later MDA standard didn't have a graphics mode. So for a time in the early 1980s, the HGC became the *de facto* standard for MS-DOS machines whenever graphics were needed.

HGC has a monochrome resolution of only 720 x 348 pixels, but it had good text resolution because the characters were displayed in a 9 x 14 dot matrix which was easier on the eye than the IBM alternatives. CGA and HGC graphics were not compatible, and quite a few of the later graphics software packages wouldn't drive Hercules boards in the graphics mode.

Hermes

- The European space shuttle program. Don't confuse this with the European paging system Ermes.
- Microsoft's network management/installation 'collection of technologies' intended to ease the installation and support of software on an enterprise network. The correct term is now Systems Management Server. See SMS.
- Recently, a major consortium of railway interests and some funding corporations have attempted to establish a pan-European communications network using the railway rights-of-way.

Hertz

The SI unit for cycles per second, written Hz. This is the measure of frequency these days. It is now incorrect to talk about cycles-per-second or cps. Thousands (kHz) and millions (MHz) are widely used measures also.

Hertzian waves

Anything in the radio frequency spectrum including visible light, infrared, ultraviolet light, and X-rays.

heterodyne

Heterodyning is the process of combining two frequencies to obtain a 'difference' or 'intermediate' frequency (IF). This radio technique has now been taken up in optical fiber transmission systems; it is used in signal detection with coherent optical lasers.

With satellite reception the Ku-band frequencies are too high and too weak to travel any distance down a cable, so the first stage transformation takes place immediately behind the dish. This is a heterodyne stage which produces an IF.

With lasers the heterodyne technique relies on creating 'beats' between two nominally similar base light-beam frequencies (usually highly coherent lasers) to extracting a 'difference' signal. See beat.

Heterodyne is used for tuning (bandpass filtering) and for amplification.

• Superheterodyne techniques were developed in the early days of radio, and were an important breakthrough. These techniques are still used in systems such as cellular phones. The 'super' referred to the fact that the beat frequency was 'super-sonic', which in those days meant 'ultrasonic' (above hearing frequencies).

The heterodyne frequency is now known as the intermediate frequency (IF). Set-top boxes are designed to work with standard IF frequencies.

heterogeneity

The term is applied to large mainframe systems and networks — especially those involved in distributed processing. The network operating system must be able to support competing technologies at the various 'layers' — different protocols, different equipment, etc. This is the opposite to 'proprietary systems' where everything is homogenous — provided by one manufacturer to one standard.

heterogenous network

A network composed of elements of different kinds — usually different vendor equipment and probably different protocols. The point is, that the variety of equipment and protocols will certainly create problems in a large network because there is probably no common protocol. Some parts of the network may be running Ethernet, other TCP/IP — and there may be a mix of old and new PCs, some high-powered workstations, and some Unix hosts. You often need multi-protocol routers on heterogenous networks.

So this is a judgemental term; it carries the implication of an overblown, uncoordinated network structure that will be expensive and difficult to maintain.

However the term sometimes carries the opposite implication. It suggests that the network elements have been designed to suit the user's needs (by incorporating the best equipment available) rather than by being forced into conformity with only one manufacturer's equipment.

• With modern equipment, managing a heterogenous network is not as difficult as it was a few years ago. With all the international take-overs, and buy-outs, it is inevitable that large heterogenous networks will emerge.

heterojunction

This is a future chip technology which uses logic-gates made from two or more special materials which change state at a very fast rate. This is said to be a couple of orders faster than silicon.

heuristic

A heuristic is a rule-of-thumb which has been demonstrated to produce valid results — even though the rules may be inexact and incomplete, and have no logical experimental base. These are 'criteria-methods' — principles for deciding which among several courses of action promises to be the most effective in order to achieve some goal.

Heuristics depend on inductive reasoning and are important in expert systems and AI programs. A chess-master uses heuristics because it is impossible for him to calculate all the possible moves and outcomes. The contrast is with structured programs using pre-defined algorithms. See fuzzy logic also.

• In OCR, a heuristic system is one where the application can be taught to recognize a certain typeface (style). Non-heuristic OCR programs already hold the type-shape information in internal files — so they can't adapt or 'learn'.

heuristic learning

Learning from experience. Not planned or systematic. Note that it means intelligent learning, not random blundering.

hex

Hexadecimal. See.

hex dump

The listing to a screen or printer of sequential codes (or data) from a range of memory locations. The listing is always in hex format, and its often used by assembly programmers as part of the debugging or disassembly process. See disassemble.

hexadecimal

Hex is a notation system based on the number 16 which is widely used in assembly-level programming. In print, a prepended $-sign (i.e. $D9), or an h-suffix (D9h) is often used to identify hex notation.

In practice, the counting system uses the numerals 0 to 9 followed by the letters A to F to represent the zero-plus-fifteen symbols required.

Hex provides us with a relatively easy way to deal with the computer binary digits as an alternative to trying to recognize machine-code (binary notation). For instance, the normal binary notation 1111 becomes $F (or Fh) in hex, which we would recognize in denary as decimal 15. The 9-bit binary byte 1111 1111 similarly becomes $FF (FFh) in hex, which is decimal 255. Assembly language programming is always done in hex or octal, using mnemonics commands as well.

HF

High Frequency — the part of the radio spectrum between 3 and 30MHz. These are 'decametric' waves from 10 to 100 meters in wavelength — also called 'short-wave'.

HF is more limited in range than MF in normal circumstances when ground waves are being used (see). However the band is strongly (but erratically) influenced by the ionosphere's F region at altitudes which vary from 80 to 300km. This is why HF (aka 'short-wave') is widely used for long distance communications.

During the day the F region is blanketed by two lower radio-absorber layers, the D and E regions, but at night it is highly reflective. See F region and Sporadic-E also.

The main applications in this HF band are:

- Defense voice and low-rate data communications.
- Emergency services (2, 4 and 6MHz)
- Short-wave radio (3 to 26MHz)
- Citizen's band radio (27MHz)
- Aircraft, marine and land mobile (3 to 30MHz)

HFC

Hybrid Fiber-Coaxial. This refers to modern cable television networks which will eventually become the 'data superhighway' in suburban streets. It is now fairly widely accepted that FTTH (fiber to the home) is not viable in the near future, especially since HFC can provide the likely bandwidths needed domestically for the next decade or more.

HFCs are a combination of fiber-trunks with an active hub which splits the signals into say, 12 or so radiating coaxial cables. Each of these coaxial street feeder cables will have no more than four amplifiers in series, and so they can carry data at rates of about 1GHz in a relatively noise-free environment.

Using such systems, a single fiber would feed perhaps 200 to 2000 homes at a much higher signal quality than provided by the present full-coaxial cable networks (which can have as many as 40 amplifiers in series).

A recent tendency has been to limit these coaxial splits to the servicing of only 500 homes in what is called a 'Fiber Serving Area' or 'Customer Access Area'. This makes bandwidth-provision for future interactive services. This is important when the cable network may, in future, carry two-way telephony and substantial data-rates. ATM switches may eventually replace these hubs.

• Often dark fibers are left in the trunk cable feeds for a possible later sub-division of the Fiber Serving Area if it becomes necessary.

HFS

• **Macintosh:** Hierarchical File System. The current file system used by the Macintosh. The early Macintosh used a very simple one-level Macintosh Filing System (MFS) which was acceptable with this rather-limited machine. HFS is a multilevel (hierarchical) filing system, which helps organize the information within a volume. It looks much the same as MFS, but new implementations act much faster. It uses the metaphor of folder, and folders-within-folders rather than sub-directories.

• **MS-DOS:** See HPFS.

HGC

Aka HGA and HMA. Hercules Graphic Card. See.

Hi-8 video

Sony's high-band 8-millimeter videocassette standard. This is a Y/C component analog format which uses metal-evaporated tape. The earlier Video 8 format used metal-particle tape.

Hi-8 is now being used for ENG and professional business production but the quality is only marginal and it won't stand much editing (unless transferred to a higher standard).

With the metal-evaporated tape, a horizontal resolution of about 400 lines can be expected, as compared to 230 lines for standard Video 8. This is still not enough resolution to sustain the generation losses in professional editing and dubbing, so for professional purposes Hi-8 is generally seen as an 'acquisition' format only.

hi-band

In analog video recording systems, lo-band systems are generally introduced first, then hi-band comes in later with modifications which improve performance. Hi-band video recorders are used at the low end of the professional ranks for news gathering, and in businesses and universities for low-cost production.

The term hi-band actually refers to a technique of shifting the luminance FM carrier to a higher frequency and increasing the bandwidth stored on the tape, but this is only possible with radical improvements to the physical, magnetic and electronic components. Image resolution information is carried more in the (Y) luminance part of the signal than in the (C) chrominance component/s — and this process increases the separation, and provides a wider bandwidth for each.

Hi-band U-matic was the first of these developments and it was very successful around the world. S-VHS and Hi-8 video are also hi-band professional versions of a lo-band domestic recording standard. These are not true component systems, but they are called Y/C because the Y (luminance) can be separated and processed apart from the C (chroma) signals.

Hi-band equipment often has other special features which are needed in small studios and with editing equipment — such as genlock, time-coding, etc.

hi-fi

High Fidelity: in almost perfect conformance with the original.

• In general it just means quality above that experienced in cheap recordings or radio. This can, of course, be used to mean almost anything better than telephone quality — but generally it is applied to amplifiers, speakers and playback equipment which has an almost 'flat-response' (equal energy across all significant frequencies) from about 100Hz to 20,000Hz.

• In CD-i terminology, the term Hi-Fi is specifically applied to the second level below CD-Audio. This audio mode is available on CD-ROM XA and CD-i disks, and it provides a frequency range only up to 17kHz — the equivalent of an old LP record. With this level of quality, a CD-i music disk can hold 288 minutes of mono-recording, or 144 minutes of stereo.

Hi-Vision

The marketing brandname of the Japanese MUSE-E standard for HDTV. Unfortunately, when true MUSE-E television sets proved to be so expensive, some of the members of the Japanese consortium began to produce widescreen TV sets which used only basic widescreen technology, and still sold it under the Hi-Vision label. This devalued the name's currency, since many of these were inferior to the standard NTSC TVs in image resolution. See MUSE, HDP, HD-MAC and Eureka project.

hidden lines removal

In CAD this is the first stage in modifying wire-frame models before rendering — which involves converting those planes nearest to the imaginary camera into a solid appearance by adding reflective surfaces. The lines on the 'backplanes' of the wire-frame image must be removed or hidden by the front-plane surfaces, as they would in normal vision. See graphics modeling and rendering.

hierarchical

The generic term for the concept of a branching tree structure, with different levels (root, trunks, branches, twigs, etc.).

• **A hierarchical directory** will begin with a single 'root' directory, which will then branch into any number of sub-directories at the second level. Each of these will then branch into any number of sub-sub-directories (possibly including files) at the third level — and so on.

In a hierarchical system there is only ever one pathway to a file, and the full 'pathname' will begin with the root and include sub-directory levels until the filename is reached.

• In networks, the implication of having a hierarchical organization is that some nodes will be superior to others in exerting control. One 'root' node will be a master node.

• **Hierarchical addressing** is of the type used in IP. The address comprises network numbers, host numbers, and then subnet numbers in descending order of coverage.

• **Hierarchical routing:** Routing based on hierarchical addressing systems.

Hierarchical File System

See HFS.

hierarchical source coding

HSC. In packet-switched systems some of the cells (small packets) are likely to be discarded because they have low priority when they meet congestion. For instance, cells carrying voice and video traveling over ATM data networks may be dropped because of network overload. However it is possible to reduce the impact of the discard by using HSC in large networks.

Cells sharing a VPI will be temporarily stored in a two-dimensional matrix and the information will be input row-by-row — but it will be read out and transmitted column-by-column. Each row will be 53 x 8 bits in length — enough to contain one standard fixed-sized cell. Obviously the reverse happens at the other end of the link.

This produces pseudo-cells in which one cell carries all the most significant bits, the next carries the next-most, and so on. Priority can be given to the most-significant cells — and the least can then be dumped if necessary. With video and voice cells, 'near-enough is good-enough' for the occasional burst.

Something similar to this is normally done in Reed-Solomon error correction using bytes, but hierarchical source coding uses the matrix in a different way. HSC has no value for data, but it is ideal for voice and video.

high-band

See hi-band.

high definition television

See HDTV.

high density

• **Magnetic disks:** The techniques used to increase the density of modern floppy disks include the use of different formats, codes, narrower track widths, absence of guard bands, and better magnetic media. These factors have combined to allow double-density and quad-density disk drives to be manufactured and used with reasonable reliability.

Note that some manufacturers call double-density/double-sided floppy disks (720kByte) 'high-density', where others reserve this name for the 'quad-density' 3.5 inch disks with over 1MByte. See double-sided disks.

• **Optical disks:** See CD-HD.

high duty cycle

A term applied to non-bursty data traffic. Long file transfers have a high duty cycle.

high frequency

See HF.

high frequency effect

As data rates increase along cables, the signals become more susceptible to voltage loss due to capacitance and inductance effects. These problems increase roughly in proportion to the cable length. So the maximum data rate possible through, say, an RS-232C cable depends on numerous factors, including the cable's length. Intersymbol distortion also increases. See distortion and inductance.

high-level controls

As a general term, this applies to packet-switching and similar systems where control is exerted at a higher level than that needed for just transporting streams of data in the Physical layer. Devices (computers) at each end of the network act in pairs on the information contained in the packet headers to provide:

— Data flow controls (dialog and session management).
— Access path controls (addressing, packetizing and routing).
— Data link controls (HDLC error checking).

Control information is exchanged end-to-end in the form of bit patterns. See HDLC.

high-level languages

This generally refers to 3rd generation languages which need to be compiled or interpreted before they can be run, but it includes 4th generation also.

• Third generation languages used words and syntax which are English-like and therefore easier to program than assembly. Basic, Cobol, APL are all seen as high-level languages, and Fortran was probably the first.

• Fourth generation languages are even more user-friendly. They are non-procedural — which, in effect, means modular and object-oriented to varying degrees.

Whenever a high-level language is used, the programmer writes source code. This is then translated and linked (compiled or interpreted) to create object code which is linked with library modules to create a machine code that the computer can run. See language generations.

high-line

Over-voltage. See surge.

high-memory

In MS-DOS terminology this originally referred to memory locations above 640kB and below 1MB — now called Upper Memory. Later it meant the memory locations above 1MByte. These memory locations are used by both expanded (indirectly) and extended (directly) memory specifications, and also by other operating systems running in native modes.

• **Historical:** The old 'high-memory' (640k—1MByte) address space was reserved for video bit-maps and the BIOS ROM — but these didn't occupy all the 360kB of (now called) upper memory. Some gaps in the memory space can be detected by third-party memory managers, and these could then be assigned to utilities, TSRs, or used for bank-switching. The spare memory locations here are now known as the Upper Memory Blocks (UMB — see). See expanded and extended memory also.

What sometimes makes this confusing, is that a block of 64kBytes of memory in the UMB range could be used by DOS 5.0 for small DOS programs. And, at the same time, DOS 5.0 could load itself into the new 'high memory area' (HMA) above 1MByte, which left the main 640kB of working memory almost entirely free for applications and data.

• Extended memory is managed by HIMEM.SYS. Expanded memory can be emulated by the memory manager called EMM386. Both will then use high-memory locations. Remember that a device driver must be loaded by a device command in CONFIG.SYS.

high memory area/blocks

See HMA or HMB.

high-order multiplexers

In multiplexing over digital telephone systems the concept of analog 'groups' with has given way to high and low 'orders'.

A Primary (low-order) multiplexer will generally take thirty 64kb/s (PCM voice) data streams and multiplex them into a single 2Mb/s bit stream.

These 2Mb/s data flows can then be multiplexed into higher-order 'data rivers' still for onward transmission between the various telephone exchanges. The standard 'orders' of multiplexing in European countries are:

First order	30 x 64kb/s	into	2Mb/s.	(E1).
Second order	4 x 2Mb/s	into	8Mb/s.	
Third order	4 x 8Mb/s	into	34Mb/s.	
Fourth order	4 x 34Mb/s	into	140Mb/s.	
Fifth order	4 x 140Mb/s	into	565Mb/s.	
Sixth order	4 x 565Mb/s	into	2.4Gb/s.	

high-pass filters

These are filters which pass frequencies higher than a preset cutoff frequency, and attenuate all lower frequencies. See low-pass and bandpass filters also.

high-power satellite

One having transponder power output of more than 100 watts. DBS will use high-power transponders so that home reception is possible over a wide coverage area with a small dish.

high-resolution

• **Monitors:** This is an old term from a time when monitors were very low in resolution, so it can be applied to any monitor which reproduces a readable 80 columns of text across the screen. Any monitors capable of providing at least 600 x 800 pixels can probably be correctly termed 'high-resolution' at the present time — and that includes them all (or the vast majority). See Super VGA.

• **High resolution display** in CD-i usually means a mode with 560 x 768 pixels.

• In **CAD**, a high resolution monitor probably has a capability of more than 2000 pixels in both directions.

• **Laserprinters** characteristically print 300 dots to the inch, but some now output 600 dpi and professional typesetting machines can handle 1200 dpi.

• **TV:** See HDTV

High Sierra format

The HSG format is now known as ISO 9660. The High Sierra Group (HSG) of computer companies standardized the logical format for CD-ROMs in 1985 at a hotel on Lake Tahoe in the High Sierra mountains. Their purpose was to specify how the information was to be organized, and where the directories were to be located on the disk.

These standards were added to the existing physical CD-ROM (Yellow Book) standards so that consistent retrieval software could be devised to suit all computer platforms. However they didn't solve all the problems.

Disks in the High Sierra format are common across all systems, but the information retrieval program you need to get at that information will be specific to your computer's operating system. Since control software is usually kept on the disk, you will still find a distinction between the Mac, Unix and MS-DOS CD-ROMs, even when they all conform to ISO 9660.

high-split

In a Cable TV network the frequencies between 5 and 54MHz are traditionally avoided because this area of the spectrum has the most impulse noise and interference. Now that cable systems are becoming interactive, this low-end of the spectrum is being reserved for return-channel data for Pay-per-view, home shopping, etc. (and telephony in some systems).

By placing this 'low-split at 54MHz, TV signals from off-air sources can be directly mapped into the cable without remodulation (Channel 1 begins at 54MHz in US VHF).

Generally the spectrum between 54 and 550MHz (and sometimes up to 750MHz) is used for analog signals, and since these HFC cable networks are now capable of carrying good quality signals as high as 1GHz, the region between 550MHz (or 750MHz) and 1GHz is reserved for future digital transmissions. This is the high-split.

• In some of the early multipurpose broadband-data cable networks the uplink frequencies were between 6 and 180MHz, and the downlink (from the headend) were only between 220 and 400MHz (25 channels). There was a guard band between 180 and 220MHz. With the split occurring at 220MHz, these were sometimes called high-split systems. See headend.

high-tier

The concept of 'tiers' seems to have crept into fashion in the USA telecommunications industry, but it is not always clear what they mean by high- and low-tier.

A term increasingly applied to cellular and radio paging systems in radio spectrum auctions and regulation. High-tier signifies technologies which offer the best quality and widest range of services: including voice, data, and messaging. In PCS, high-tier services are those requiring high mobility. These are the higher-cost cellular-like services for vehicles, using technologies like DCS 1900 and CDMA 1900.

The distinction is with low-tier paging and e-mail type radio technologies which may get away with much less bandwidth. Low-tier services are the pedestrian pocket systems using technologies like PHS, PACS and CT2, with range limited to a few hundred meters at most. Paging systems are low tier also.

• Note, however, that with CDMA's ability to cram many more users into the same spectrum, some of the PCS bidders were bidding for low-tier bandwidths, while intending to use them for high-tier services.

highlight (hi-lite)

• A video monitor effect which usually reverses blacks and whites on the screen (or uses color) to indicate that an area of the screen or a line of characters has been selected.

• In photography and printing, this is the lightest area in a photograph or illustration (the darkest part of a negative) — usually a spot of light reflected from some surface and reproduced as pure white. It is the reference point used for control of exposure (or with video; voltage) during subsequent stages.

hint/HINT

• **Fonts:** Outline fonts need to calculate the shape and size of every character on the computer screen (or in the printer) from a common mathematical description. These need to take into account the granularity of the screen/printer.

If, at 9 point size, a descender is 3 pixels wide — what should it be at 10 points? Should it be 3 or 4 pixels wide?

Hints are rules to tell the computer what to do in making these compromises, both for the screen, and for the printer.

Hints are also called 'encryption rules' and this process is known as grid-fitting. It is essential with all outline-font descriptions such as PostScript and TrueType. The hints also detail the approach to be used in the adjustment and rasterization processes, and therefore they help define how the final words will look and act (how they fit together) — especially with extreme size reductions. See outline fonts, jaggies, alias.

— Hinting program: This is the program/language/technique used to convert drawn typographic characters into full outline font specifications.

• **Games:** Fanatical game players also exchange 'hint files'.

• **HINT:** (Hierarchical INTegration) is a benchmark developed by Iowa State University. It is a cross-platform tool, suited to any architecture, and designed to provide useful metrics on reasonably simple mathematical calculations. The benchmark results are called QUIPS.

Hiperlan

The ETSI alternative to the IEEE's 802.11 specification for wireless LAN. This is a new system which will have higher data rates and operate in the 5—17GHz band.

HIPO

Hierarchy plus Input-Process-Output — a form of flowchart that illustrates block program functions. IBM developed it as a documentation tool. You can delve down through layers and see progressively more detail.

HIPPI

See HPPI.

histogram

A bar chart.

hit

• A successful search in a database.

• In the World Wide Web, this is an on-line request to view a page of information — from the viewpoint of the information provider. Organizations count the number of 'hits' to see how well their messages are being read.

• In cache terms, it is a successful attempt to bring into the cache, items of data/instructions that will be required next by the CPU. See cache hits.

• In audio recording, it is a momentary impulse noise. Not, as you might think, a best selling record!

• In telecommunications it is a transient of any kind. A 'gain hit' is a current spike on the line which is enough to stop a modem from functioning. See transient.

• In messaging services, a hit is the loss of a character. An alpha pager, for instance, is more likely to suffer hits than a numeric pager, because of the length of the message. But the hits are less of a problem with an alpha pager since you can usually still read the message. A numeric pager uses short messages, but every bit needs to be correct.

HLLAPI

High Level Language API. A set of tools developed by IBM for writing applications that conform to SAA.

HLS

• Higher Level Service. In telecommunications: see BCS.

• Hue, Luminance and Saturation. See the preferred HSL.

HMA

• **High Memory Area:** This refers to memory locations directly above the old MS-DOS 1MB limit. The HMA has addresses above 1MB, and it should not be confused with the older MS-DOS term 'high memory' (now called Upper Memory) which was between 640kB and 1MB and reserved for system functions. MS-DOS 5.0 and above can use the HMA if there is a statement: DOS=HIGH in CONFIG.SYS.

• **Hercules Monochrome Adaptor:** See Hercules Graphics

HML

Heterogenous LAN Management. A standard for PC management over networks, which was devised by IBM and 3Com. It was a sub-set of CMIP, and it only includes the lower layers. It seems to have been put on the shelf until CMIP makes more impact.

HMOS

See NMOS.

hold-downs

Some routing protocols have a way of preventing regular updates in cases where the update may attempt to reinstate a route that is still giving problems. These are hold-downs.

holding time

The time in which a device making a call locks up a public switched telecommunications channel. It is the total time from call set-up to call clear.

hole

In semiconductors, a hole is a vacancy for an electron. It is a space in the semiconductor's atomic structure into which an electron can migrate. When this happens, the electron will then also leave a hole behind, into which another electron can migrate, and so on. This is how the current flows through semiconducting materials.

holographic memory

A developing technology using lasers to store information in crystals. Bellcore is the leader in this research, and it has recently produced an array of 1000 semiconductor lasers which solves the problem of lenses and prisms for retrieving the holographic images and information. They claim that a single crystal measuring $1cm^3$ (the size of a dice) can store about 1 trillion bits of information. They say that they can retrieve the data at a rate of a 100kbit 'page' every nanosecond!

holography

The creation of reproducible images by the use of diffraction patterns. The technique was invented by Denis Gabor in 1948.

Conventional holograms are produced by recombining laser light that has been split down two paths, reflected from a surface, and then recombined to produce interference patterns. The two beams are known as the 'object' and the 'reference' beam. The hologram is the interference pattern formed where the two beams meet.

• A holograph, strangely enough, is nothing to do with the above. It is a document written entirely in the author's hand.

home

• An old hacker term for the top-left corner of the screen. Home, usually also implies that the screen is cleared. In the days of non-scrolling screens this clearing and repositioning of the cursor was essential.

• The **Home key**, on most modern keyboards, positions the cursor as above at the top left.

• The home position on a line is the beginning of that line.

• In HyperCard, the **Home stack** is virtually the root-directory but it holds other user-defined information as well.

• **Home Brew** is a term applied to the culture of the computer 'hot-solder' jockeys, who built their own computers in the early days. Steve Wozniac designed the first Apple as a home brew project for a local club.

• **Home Bus:** See Domestic Digital Bus and NEST.

• **Home page:** On the WWW this is part electronic shop window and part an index of goods and services for sale by a person or a company. Home pages are set up through WWW servers (most people rent space on a commercial server).

homodyne

This is an old radio technique, recently taken up again by researchers working on coherent optical fiber systems. It is similar to the heterodyne technique, but is much more sensitive to misalignment. However it can detect much more subtle phase changes, and is therefore potentially capable of handling greater data rates.

• For those who understand — 'the electrical signal from the photo-diode is held at baseband in optical homodyne systems. This is in contrast to it being centered around a non-zero frequency (IF) with heterodyne systems.'

homogeneous networks
Everything is from one vendor, and uses matching or compatible protocols and standards. See heterogenous network.

homonyms
Words that sound the same — as in 'two' and 'too', 'know' and 'no'. These words present major problems in voice recognition systems — especially when voice is being used for input to a text output system. To get around this problem, the system must have some intelligent notion of the context of the words.

hook

• **Programming:** Points (memory locations) in a program which can be used to make connection to 'patches', or to 'add-in functions' which are to be inserted at a later date. Hooks have an address, which often points to another address.

For instance, some routines in the operating system can often be replaced or modified for special purposes — usually by device drivers. The address of these routines will be stored in a special part of RAM, and if you know these memory locations, you can 'hook' your new driver routine into the operating system by changing this address.

- Hooks may be deliberately built into an operating system to be used later for running a CD-ROM, etc.
- Hooks are also built into applications.
- While they are part of the complex of functions involved in an API, hooks are not quite the same. APIs are system calls.

• **Telephone:** The switch which disconnects the telephone handset from the line — the one that signals to the exchange that you wish to make a call, and later signals that you've finished the call. In the early phones this was physically a hook, and you hung the earpiece on this to disconnect the call. To go off-hook means to initiate the sequence which will provide dial-tone. To tell a modem to go 'on-hook' means to tell it to drop the carrier and switch off.

hook flash

• Hanging up a telephone connection for a very short period, only as a signal. Some PBXs will respond to a half-second hang-up as a signal to transfer a call.

• A hook-flash is often used to swap between calls in Call Waiting (provided by the local carrier).

• 'Flashing the hook' was a make-do way of signaling over any old-style pulse dialing system. Since the rotary dial just makes and breaks this connection to create pulses, you could signal numbers by jiggling the hook the required number of times. This was an old way of by-passing the need for coins in some payphones; it was relatively easy when the numbers are low, but difficult when they are in the nines.

hop
A single stage between two routers on a WAN, or two packet-switching nodes on a packet network. On a LAN it probably means the number of segments (sub-divisions of a LAN).

• Hop count is a measure of the number of individual links (hops) a packet must make between the source and sink. Note that a link will terminate in any store-and-forward device such as a router — so to use the term 'segment' is sometimes misleading. So a hop count gives an indication of the efficiency of transfer between routers — or of the complex of transit across a network. It is counted end-to-end.

• Ethernet, for instance, limits each network to four hops (here segments of the same LAN joined by repeaters) because of time delay problems, but some later twisted-pair Ethernet implementations have broken this barrier.

horizontal application
Programs that most PC owners will require: word processors, spreadsheets, database managers, communications, etc. See vertical applications.

horizontal cavity
Aka 'edge-emitting' lasers. Horizontal cavity lasers are also called edge-emitting lasers because these units beam the signal out from the edge of the chip in the plane between the two active layers. Vertical cavity lasers beam at right-angles to the flat plane-surface of the chip. See semiconductor lasers.

horizontal frequency
A misleading term when applied to monitors because it can be confused with the bandwidth which sets resolution. Horizontal frequency usually refers to the horizontal scan rate (the line frequency), and is the number of scan lines (including those in the vertical blanking interval) that can be drawn every second (as distinct from every frame). It is usually measured in kHz, and varies between about 15.75kHz for NTSC/CGA monitors, to 38kHz for SuperVGA. See below and dot-clock rate.

horizontal line frequency

• NTSC television, at 525 lines and 30 frames per second, must draw 15,750 lines per second. This is an allowance of 63.5μs per line. About one-sixth is lost in the horizontal retrace, leaving 52.5 μs to trace each active part of the active lines (create the image). The rapidity with which the signal can change from black to white (a 'cycle') within this 52.5μs, sets the upper limit of horizontal resolution.

• PAL and SECAM television at 625 lines and 25 frames per second, must draw 15,625 lines per second. So the horizontal drawing time of each line is almost identical — which means that the horizontal resolution is almost identical. However more bandwidth is consumed because of the higher vertical resolution (more scan lines).

horizontal resolution
The horizontal resolution of a video screen is measured:

- **Theoretically:** By the bandwidth of the beam drivers (e.g. the number of cycles it can handle, switching rapidly from black to white, in the time the beam scans each line of image). TV traditionally provides about 5MHz of total bandwidth, while computer monitors are in the 18MHz region.

– **Practically:** By the number of vertical black/white lines (as in the test pattern) that can successfully be resolved across the width of the screen by the human eye. When they can't be resolved (when the scan beam can't alternate quickly enough), the vertical stripes merge into a uniform gray.

In a PAL studio video monitor, the horizontal resolution will be about 572 lines, and with NTSC it is about 522. This quality is degraded in transmission, and so 300–350 lines is better than average for home reception. Consumer receivers are generally made to this low standard — which is why TV sets don't make good computer monitors. Home VCRs generally can only reproduce about 200–250 lines on the horizontal.

• Cycle (one black line followed by one white line) is a less confusing measure of horizontal resolution than 'lines' (at half the figure), because the term 'lines' (of resolution) and the scan lines are constantly being confused.

horizontal retrace time

The time it takes for the scanning beam of a TV set or raster monitor to return from the right-hand end of a scan line, to begin a new line on the left. Also called horizontal fly-back or horizontal blanking interval. It occupies about 17.5% of the line trace period. NTSC lines take about 52.5μs to trace the video, and about 11μs for the retrace time. PAL is about the same.

horizontal wiring

In a structured wiring system, this is the copper that connects the workstation to the hub. These days, this will probably be UTP rather than coaxial. The image is of a high-rise office, with a vertical fiber backbone linking the floors at active hubs, and with the horizontal wiring reaching out to the workstations on each floor.

horn antenna

A satellite or microwave signal collector. A large horn can sometimes be used independently, but generally it acts as a collector for the signals reflected from a dish.

host

This means master in a master–slave relationship: A warehouse in terms of a storage facility; or an on-line service supplier.

In on-line systems, this is the master computer which holds the primary database or the main applications. The location of the main applications (rather than the network operating system) probably defines the host — and usually the main application in large corporations is a database management function.

This database is accessible from remote locations by terminals, PCs or other computers using on-line information retrieval techniques, remote terminal (emulation) operations, or client/server functions.

The term host is now a bit vague but it suggests a tree-branch network structure and master–multiple-slave relationships, with the host controlling the trunk and the key resources. The distinction is with distributed processing environments where each workstation deals with the other on a relatively equal basis (peer-to-peer).

• A bulletin-board's service computer is a 'host' in this sense, even though it may only be a PC itself.

• For Internet dial-up connections, the host will be the first computer you contact via the dial-up phone line which is permanently connected to the Internet (discounting its front-end communications processor).

• The host is the primary or controlling computer in any multiple computer operation — the one on which the other smaller computers depend for most of their main resources. There is a master–slave relationship here. While peer-to-peer networking and client-server systems are becoming important, host-centric operations are still very common. See distributed data processing also.

• Host is often used just as a casual name for any mini or mainframe. However the term host is best reserved for a computer system running useful applications, whereas words like 'device' or 'node' will generally refer to terminal servers and routers on the network.

host-server

A device on a LAN which allows a large computer, of a type that could not normally be connected, to connect to the network. The host-server provides all the necessary LAN support.

Hot Birds

This originally just meant satellites which were active and functioning. But the name has been appropriated by Eutelsat for three DBS satellites which will broadcast digital TV and analog TV signals, at a higher power than Astra.

Each transponder will have two half-channels. One of these will be 27MHz wide for analog signals, and the other will be 9MHz wide for digital data (video) coded at 8Mb/s.

hot-key

• The command key combination that is sometimes used to boot up an application.

• Any key (or key combination) which can stall running applications, and cause another event to happen. They can interrupt the other program and take control. So this is another name for macro keys that call up a TSR/DA program.

• Hot-key programs are TSR 'pop-up' programs.

• In PC-to-mainframe communications, the hot-key lets you switch from PC functions to local terminal functions.

hot-link

This is a form of association-linking used in some integrated programs (and now some operating systems like Windows and OS/2) to update data in, say, a graphic chart which has been derived from some spreadsheet figures.

After the data has been transferred, a dynamic link still remains. If you later change some of these figures in the spreadsheet, the graphics chart will automatically re-draw to reflect the changed information — even though it is in an entirely different application, or module.

• Warm-links are where you need to take some action (issue a command) to create the dynamic change. Usually the system will advise you that the data has changed, but it will not automatically update the data. See queue, IAC, DDE and OLE.

hot-swap

The ability to change plug-in cards in a device without blowing them, or the device, up. Some devices can, most can't.

Hot-swappable modules are becoming increasingly important as more and more computers and communications devices become central to the operations of more and more companies. A lot of equipment is now 'mission critical' so you can't switch it off while you change one component.

• Some so-called hot-swappable devices have dual power supplies and fully redundant circuits — so one set can remain powered and active, while the swap is made on the other.

• Microsoft is adding hot-swappable features to Windows 95. See Plug and Play.

Houdini board

A plug-in processing card for the Macintosh which allows it to run MS-DOS and Windows. The original version was a NuBus card with an Intel 486SX at 25MHz. Later versions are for the PDS slot, and provide a 66MHz Intel 486DX2. This is not the same as the emulation of MS-DOS — it is a true secondary system just sharing peripherals and resources.

house style

Most publishing houses have a preferred form of spelling, punctuation, hyphenation and formatting. This is the house style. It is generally fifty years out of date, and its application is rigidly enforced by the most pedantic, SOB sub-editor in the place.

HP-PCL

Hewlett-Packard's page description language for laser printers. See PCL.

HPFS

High-Performance Filing System. The file structure and disk management system in later versions of OS/2. It replaces the old File Allocation Table (FAT), and it increases the maximum disk space available to PC users. HPFC makes disk access faster, adds security features (passwords) and longer file names (to 254 characters) and allows more descriptive information to be kept about each file.

HPGL

Hewlett-Packard Graphics Language. A graphics file format used by many CAD systems with HP plotters. It has become a *de facto* standard for plotters.

HPIB

Hewlett Packard Interface Bus. See GPIB.

HPPI

High Performance Parallel (Peripheral) Interface (aka HIPPI). A very mature, 32-bit high-speed parallel interface for supercomputers and workstations wishing to transmit over their own channels (to disk drives, etc.) at about 100Mb/s.

This is a development of the older Cray Channel (1980) standard used for supercomputers and it is now defined by ANSI as X3T9.3/88-023. The recent interest in the HPPI standard follows the rise of extremely high performance computing super-networks on optical fiber, and it is now being used for connecting large hosts computers to RAID disks as a substitute for multiple-SCSI connections.

The ANSI X3T9.3 standard called HPPI-PH (HPPI Physical), also defines dual 100-pin connectors and parallel copper cabling for rates up to 800Mb/s. This rate will later be doubled by doubling the number of path connections (but 25 meters will remain the maximum distance).

There is also an ANSI specification for HPPI switch control, which will enable a single host to access multiple peripheral databanks. See also Channel (processor I/O), Fiber Channel, ESCON and FDDI.

HRC

Horizontal Redundancy Checking. HRC refers to techniques of error checking where the 'redundant' (check) data is included along with the data block (the byte) for validation. This term covers the various forms of parity, including Hamming code. See LRC and CRC also.

HSB

Hue, Saturation and Brightness. See HSL.

HSG standards

High Sierra Group standards. See.

HSL

Hue, Saturation and Luminance. An alternate specification for color in some computer graphics (as against RGB, CMYK, HSV, etc.). In Paint and Draw programs it is often easier to control the colors by applying incremental changes to each of the three HSL factors, than to those of RGB.

You'll also find references to HLS since both versions of the acronym seem to be in use, and there's also HSB (Brightness), as well as an HSV color space where the V stands for Value. This last appears to be slightly different to L (luminance).

HSLink

High Speed Link. This is a relatively new file transfer program which has a popular following in some circles because it handles data flows bidirectionally. With the latest in high-speed, full-duplex modems, data can be moving each way simultaneously down a phone line. HSLink can support V.32, V.32bis and V.34 modems.

HSM

Hierarchical Storage Management. A form of file categorization which selects files, not just by date or name, but also by the file's activity rate. This is important because it allows us to match the cost of the storage medium to the performance requirements. Lower-cost, lower-performance storage media can then be used for data which is not frequency accessed, and higher-cost, higher-performance media is reserved for time-sensitive information and information needed regularly.

HSM systems are often incorporated into automatic archiving, disk-volume management, capacity-management and backup procedures as a way of identifying which files should be instantly available on-line, which should be off-line, and which should be near-line. See file grooming.

- NetWare 4.0 offers a rather basic form of HSM with SMS.

HSSI

High Speed Serial Interface. This is a relatively new and popular WAN serial link standard which will eventually replace V.35 and RS-422 for computers and multiprotocol routers. It is a high performance, flexible Physical layer specification for an interface which can work with different wide-area network circuits and services via an external data service unit (DSU).

HSSI was designed jointly by Cisco and T3plus and it is now with the IEEE. It will carry data at rates up to 52Mb/s for routers and multiplexers, and currently it is being designed to support T3, E3 and HDC/Sonet OC-1 circuits.

HST

High Speed Technology. A proprietary modem standard (US Robotics) for a hybrid rate of 14.4kb/s with 300b/s or 450b/s on the back-channel. The standard is actually half-duplex with rapid turn-around. HST uses Trellis-coded multilevel-QAM, and it encodes 7-bits per baud with one bit being used for parity.

HSV

Hue, Saturation and Value. A color monitor system control used in some graphics programs. HSV must be converted to another model for printing (usually CYMK), or to be used on a monitor screen (RGB). This appears to be different to HSL.

HT

ASCII No.5 — Horizontal tab (Decimal 9, Hex $09, Control-I). This is a format effector that works like a typewriter's tab. The cursor jumps along the line to the next tab location. These tabs are usually preset to every eight columns, but the position can be altered on most terminals.

HTML

HyperText Mark-up Language. A document format used on-line for systems offering hypertext links; it is designed for screen-based information retrieval rather than as a printing control. HTML is now being used in WWW communications and it is increasingly important on the Internet — however it lacks some (almost) essential features (such as tabs) and badly needs upgrading.

This is not a programming language, but a way of tagging text so that a browser knows how that text should be presented, and how documents should be linked. It is a type of (an 'instance' of a) SGML. The elements consist of URLs, the Common Gateway Interface (CGI); tools, editors and document translators; browsers, and servers.

HTML tags are delimited by the greater-than less-than signs <>, and they enclose a simple command. For instance, s <P> indicates a paragraph break, <B> and </B> would force the word 'and' to be printed bold. To use the European character e with a grave accent, you would type è. Other tags provide links.

There are now four levels:

— **Level 0** was the original basic command list

— **Level 1** is mandatory in WWW browsers; it is essentially the same as Level 0 with some updates for images.

— **Level 2** is currently being finalised. This is a catch-up with the level currently used on most WWW browsers.

— **Level 3** is also called HTML+, and it includes markup tags for tables, figures and equations.

- HTML+ seems to be giving way to HTML 3.0, which may also be Level 3. See VRML and Acrobat also.

http/HTTP

HyperText Transfer Protocol is the WWW standard for the construction and use of documents which can apply URLs to link with other items on the Internet. Mosaic is an HTTP browser.

There are both secure (RSA) and unsecure versions. HTTP makes it possible to follow pre-established links which have already been established between documents, even though they might reside on different computers on a global network.

hub

A point of concentration, for network wiring. The center of rotation, action, etc. A very vague term!

- **LAN:** This is a general term used to describe anything that sits at the center point of a star-wired network. There are passive hubs (wiring concentrators) which split but do not repeat any signals sent through them, and active hubs which are multi-port repeaters. There are also more intelligent devices such as LAN switches, which can be true switches or fast-acting bridges or routers. See wiring hubs and concentrator; LAN hubs; fixed-port, chassis-based and stackable hubs.

- **Disks:** The central part of a magnetic or optical disk. Hub wear can cause disk problems, especially in old floppies.

- **Network:** A telecommunications term for the large centralized switching centers (public or private exchanges), based in primary regional locations, often operated by international carriers (telcos) and some resellers.

Large multinationals and international VAN suppliers may have a half-dozen private hubs around the world which are interconnected by leased circuits using undersea optical fiber cables and satellite earth stations. These hubs would then distribute the signals in their regions through smaller feeder cables or domestic satellites.

With the growth of very large international private networks, hubbing facilities are being competitively offered by many regional PTTs in order to bring business into the country — usually with government assistance, or subsidy.

• Satellite communications systems (VSATs) have a large earth station at the hub. Very often all communications will pass through the hub, even though the end-to-end connection may be from one VSAT to another.

hubbing

This is where a domestic carrier in one country may offer a number of overseas destinations via their own country network at a special price. This allows one-stop discount shopping of a sort, for an international corporation dealing in a region.

hue

Hue is the dominant color — the major visual wavelength of a 'color' in its pure state (without black or white added). Red is a saturated hue — pink is an unsaturated red hue. See HSL, HSV and saturation.

Huffman codes

In data compression this is a technique of reducing the number of bits used to transmit or store alphanumeric bit-streams and image-files, by using variable length codes instead of the standard 8-bit bytes. Huffman's codes are created at the end-point, just before storage or transmission.

This is called a form of entropy coding, analogous to Morse Code where the most frequently-used characters (T and E) have the shortest codes (dash and dot) while lesser-used characters (Q) have longer codes. Huffman does much the same by transcoding to variable length strings of bit-patterns — each of which is quite unique.

With Huffman's, short binary codes are used for the most frequently found values/sequences, and longer codes for the less frequent. You can only do this if all of the code-combinations aren't used and so many combinations of 0 and 1 patterns are prohibited — but the overall result is a highly efficient transmission coding system.

Huffman's codes are sent as a continuous bit-stream without breaks or frames between the 'bytes'; since each binary sequence is unique it can therefore be identified within a bit-stream. Huffman is encoded and decoded through the use of a 'binary tree' structure creating a look-up table.

• Arithmetic coding can apparently improve on Huffman's compression ratios.

• Run Length Encoding is often used as a primary stage, as is Lemple-Ziv-Welsh. See RLE and LZW also.

• Group 3 fax applies a One-dimensional Modified Huffman code, and then a two-dimensional Modified READ compression system to the data.

hung

A computer that has 'hung' has discovered some error in a program which stops it from proceeding. You should try to close down the application before taking the fatal step of switching the machine off.

In a Windows application try holding down the Control key and then pressing Alt and Delete simultaneously; this may return you to Program Manager. In MS-DOS the most likely effect will be to perform a warm-boot.

• These days most applications have built-in ways to overcome simple bugs, so you only get to see major disasters. For this reason when a modern PC hangs while running a well-known application, the result is usually fatal.

hunt group

Hunting is a call diversion process. Sometimes it can be done by the carrier in the exchange, but more often it is done within a company PBX for call distribution.

It is applied to the process of finding a free extension/line within a group of extension lines, and diverting an incoming call to this line. The PBX is said to 'hunt' for this line (in the old days, by using a rotary switch). The group available for selection is the 'hunt group' and the equipment is called a 'rotary'.

Many telephone companies offer hunt groups at the telephone exchange. You may have between two and a half-dozen independent lines coming into your home or business, but if an incoming call finds the line busy, it will divert to the next in the pre-established hunt-group sequence. This is similar to call-forwarding, but it operates at a more basic level.

Hunt groups may be accessed in a variety of ways:

- The basic pattern is to always begin looking for the first vacant line, from a predefined point. This puts most load on the low extension numbers.
- Alternately, they may average out the load by constantly cycling around the hunt group, always beginning from the next unused line in the sequence. See ACD.
- However with modern computerized PBX systems, an incoming call can be diverted to a free line in the hunt-group according to some predefined rules, and these rules can be quite complex.

  For instance, you can 'hunt-on-busy' or hunt on other criteria, say, if not answered by the third ring. Time of day, and day of week changes can also be included.
- Jump-hunting is when the number used for overflow calls is not immediately in sequence with the line which is busy.

hunting

• In information jargon this means specifically seeking some information, as distinct from grazing or browsing.

• Call forwarding on busy: See hunt group.

hybrid chips

In chip manufacture, hybrid refers to the use of two or more different component-manufacturing technologies — particularly, the combination of monolithic, thin-film and discrete-component techniques.

• Note: I also have a definition which says that hybrid ICs are those 'where the active elements are attached to the surface of a passive (e.g. ceramic) substrate.' This doesn't sound right.

hybrid coding

With voice coding, this is a combination of waveform coding and source coding. RELP is an example of this dual approach. It uses linear prediction to model the vocal tract but it also transmits the voice after passing it through a bandpass filter which restricts frequencies to those between 300 and 800Hz only. Multipulse Coding and Adaptive Predictive Coding are also hybrid coders.

hybrid disks

In optical disks, this refers to the Photo CD or CD-WO disk formats.

hybrid languages

Hybrid languages are conventional languages such as Pascal, Fortran and C which have been retrofitted to accommodate object-oriented programming techniques. C++, for instance, is the hybrid form of the C language.

These hybrid OOPS languages reduce the learning curve, because programmers can apply OOPS techniques within a familiar programming environment.

hybrid networks

There are three slightly different uses here:

- Networks which use different media. The new hybrid fiber/-coaxial networks being laid for the future data supertollway to your house will use optical fiber links for the trunks running to a local hub, and coaxial for the local feeders. See HFC.
- Networks with equipment from many vendors. This term is applied only to small networks, because it is almost inevitable in the larger networks. It suggests that everything might not always work perfectly with everything else. See heterogenous network.
- Networks involving different protocols. These are more properly called heterogenous networks. See.

hybrid satellites

These are satellites which have both C-band and Ku-band transponders.

hybrids

Hybrids are small audio-frequency transformers (couplers) with a single coil on one side, and dual coils on the other.

In telephone exchanges, hybrids exist at the junction between the two-wire (full-duplex — single pair) local loop circuit and the four-wire (half-duplex — one pair for each direction) circuits used between exchanges or to make a connection to the optical fiber multiplexers.

There is also a hybrid in each telephone handset which is used to couple the microphone and earpiece to the twisted-pair telephone line.

Hybrids can create echo through impedance mismatching, but these devices allow amplification of the signals because they convert a full-duplex circuit to two half-duplex circuits — and half-duplex amplification is easy.

A fiber is also a half-duplex pathway, so hybrids are needed before fiber pairs. Multiplexing also requires the data to only be traveling one way. See four-wire and echo cancellation also.

hydrazine

A type of satellite physical-control thruster fuel which is widely used to keep satellites in their orbital station. Hydrazine is a hyperbolic (spontaneous combustion) propellant, and is therefore extremely reliable for long periods in orbit when electrical ignition systems can fail. The bi-propellants used in hydrazine systems are nitrogen tetroxide and monomethyl hydrazine. Some smaller systems use bottled nitrogen as an alternative, which is released as a jet to make minor corrections.

Hyper-G

A development of the hypertext linking in Mosaic which is being used in the Internet WWW. It is capable of identifying and displaying document links which may have been created by other users.

hyper-realist

A hyper-realist is a philosophizing networking fanatic who believes that the real world will eventually be transformed into cyberspace.

They say that, to a degree, we already live in a world of 'heightened reality'. 'When a movie actor can become the US President; politics and the fictional environments have already merged.' This is the 'society of the spectacle'.

I get the point: I just don't see anything profound in the message!

Are they commenting on the permeation of the trivia and the rise in power of intellectual light-weights, or promoting it?

hyperband

The ability to tune a TV 'cable-ready' set well beyond the current UHF/VHF limits. Hyperband tuning is a vague term for higher than the conventional analog VHF. The assumption is that the cables in the future will probably carry more than 100 channels. However it is also probable that most of these channels will be selectable/tunable by some device before the signal is fed to the TV set, and most probable that they will be digital and switched. See IRD.

hypercomputer

A form of parallel computing, achieved by connecting together a number of PCs with a LAN or WAN, and programming them to focus on one particular problem during their idle time. If you could eventually link thousands of PCs during the night-time and weekend hours, and if you had a way of dividing the tasks at hand, and then recombining the results — you may have a low-cost supercomputer. This was the idea. See macroparallelism also.

hypercube

A way of effectively creating a pseudo-parallel computer by combining individual computing elements (often transputers) into a cube-like architecture. Each node is interconnected by a direct-channel, and is linked to every other node in the neighborhood. The 'cube' distinction is with other possible parallel architectures such as ring or rectangular grid topologies.

hypergroup

In telecommunications this is a combination of 16 supergroups (each with 60 analog voice channels) making a total of 960 voice channels, which are then frequency division multiplexed down a coaxial cable. See group and supergroup. The ITU uses the term Supermaster group to refer to 900 voice channels.

hypermedia

An over-hyped term which has two quite distinct, but confusing, meanings:

- The term is applied to hypertext systems that also contain vision and sound files. The term hypertext is now used in an almost generic way to cover all these hypermedia extensions.

The point is that you always have some form of active control over navigating through the data. By clicking on a button you will branch (jump) to another part of the material following a pre-defined path — and this can be done for text, images, sound and video. The term implies that the material is not constructed essentially in a linear way. See hypertext.

- It is also used to mean the combination of new text/sound/-image technologies which are now becoming available — but the term multimedia is preferred. However, some people use 'hypermedia' to suggest a stage beyond multimedia (which they often take to be something based on CD-ROM). So, it can include High Definition Television, MIDI sound systems, CD-Audio, DVI, videophones and conferencing, and the like also.

hyperspace

This is the illusory 'navigation' environment which surrounds your current position in a hypermedia or hypertext database or programmed environment.

The designer controls the user's access and movements through the database by providing 'hyper' links, but this can quickly get confusing because there's no hierarchy or logic to the progression. Often the program will provide navigational aids such as link markers and backtracking mechanisms. It is important that users have signposts or visual clues for the user's hypermedia travels.

Don't confuse hyperspace with cyberspace, which refers to generic perception environments rather than intellectual ones.

hypertext

This now generic term applied to the storage/access of data in non-linear, non-structured modules (as distinct from records in strict hierarchies). The concept includes free-formed links which can be created between those modules, and these can be author-generated, or user-generated.

The key to hypertext is that the structure of the material in the database is not defined by the computer, but by the user (at least, after a while).

Early hypertext modules only contained text, but now sound and video image files can be linked in the same way. Hypertext uses concepts of browsing and navigating, together with some general information-retrieval techniques. Certainly the most successful and famous of the hypertext applications is Apple's HyperCard with its HyperTalk language — but this is only marginally 'hypertext'.

There are many similarities between what can be achieved by linking hypertext documents, and what is being done with AI and expert systems using a totally different approach.

See HTML and non-linear also.

hysteresis

When an electric current passes through a coil, it produces a magnetic field (flux). When the current is switched off or reverses in phase, the flux collapses, and in its collapse it generates a reverse current in the wire. The difference in time between the cause (the current change) and the effect (the electrical current generate from the collapse) is the hysteresis.

- The term has a wider application than this also; it is a reference to the lag between an electrical event, and the response to (or effect of) that change.

Hz

Hertz or cycles per second.

I

i### chips

Intel attempted to add an 'i' before the chip numbers a few years ago. This would have allowed them to brand-name their chips and prevent the clone makers from using the same number designation. But they failed, everyone ignored the i.

So the chip which Intel labeled 'i486' is popularly known as the '486', and its real designation is 80486. For information about chips from i386 to i950 see the related number in numbers section (before A). For i586 see Pentium.

I.#00

These are the ITU-T's broad series of recommendations for ISDN.

- I.100: The general introduction to, and outline of ISDN.
- I.200: The recommendations for ISDN services.
- I.300: Networking aspects of ISDN.
- I.400: User-network interfaces. These describe ISDN's physical and protocol-related aspects of the user interfaces, terminal adaptors, and multiplexing methods — both PRA and BRA.

I-beam

A term for the insertion point 'cursor'. In a word processing document, this will remain blinking at the point at which the next typed character will be inserted or added. It is shifted to another location by using the mouse or 'pointing' cursor, and clicking at another screen location.

I-frames

In MPEG, these are the 'inter-frames' used in high-speed compression and decompression. IBM makes an MPEG2 I-Frame Encoder and Decoder which are a matched set of chips that compress video data at one end and decompress it at the receiving end. Higher orders of compression are achieved by using Interframe Predictive 'IP-frame' and 'B-frame' (IPB) techniques.

i-time

Instruction time. A cycle of CPU processing consists of i-time and execution time (e-time).

IA#

- IA2: International Alphabet No.2 — the old Telex standard. See ASCII No.2
- IA5: International Alphabet No. 5 — the current ASCII standard code used by most computers. See ASCII No.5.

IAB

Internet Activities Board. The main coordination committee for the Internet which sets many of the policy recommendations followed around the world. It has a number of major (and many subsidiary) committees:

- Internet Engineering Task Force (IETF) which specified protocols and recommends Internet standards.
- Internet Research Steering Group (IRSG) This appears to be the oversight group.
- The Internet Research Task Force (IRTF) conducts the research on new technologies under the IRSG.

IAC

Inter-Applications Communications. See DDE and OLE also.

IACK

A VME term meaning Interrupt Acknowledge.

IACS

Integrated Access and Cross-Connect System. A wideband packet-switching (frame relay) system which is based on a special fast-packet switch. IACS switches can sense the difference between (time-critical) voice and (non-critical) data and allocate different priorities to them. Voice is digitized using ADPCM with 4:1 compression. See frame relay.

IAV

InterActive Video. This can mean videodisks used in an interactive way, or it can refer to new two-way cable TV networks — or video games and virtual reality.

IBCN

Integrated Broadband Communications Network. This is the term used by the RACE group to cover the general integration of terrestrial Broadband ISDN (B-ISDN) and the various mobile and satellite communications systems. It is a wider term than B-ISDN, but is often used to mean essentially the same thing. See ATM and B-ISDN also.

IBM

- **Terminals:** IBM has many types of terminals, cluster controllers, etc. in the market place. See 3270 and 7171 in the numbers section.
- **Special characters:** In IBM machines, but not in clones, you can type a range of special characters by holding down the Alt and typing a three-digit number on the numeric keypad.

For instance:

- The a, c, e, i, n, o, u and y variations used in European languages with different diacritic marks, are numbered from 128 to 165.
- The yen sign is 157.
- Greek alphabetic characters are from 224 to 239.
- Maths symbols are from 240 to 247.

Don't confuse these with the original graphic characters which were once in the upper range of the 8-bit ASCII code (the extended set).

IBM Cabling System

IBM introduced their Cabling System in an attempt to establish one common cabling standard among IBM users. It is based on

150 ohm shielded twisted pair. The system has nine cable specifications, but not all are used:

- **Type 1** contains two individual shielded twisted-pairs of 22 AWG with solid conductor and braided shielding. It is capable of multiple Mb/s rates over distances up to 300 meters.
- **Type 2** is the above cable grouped together with four telephone-grade twisted-pairs (six in all) used for building wiring.
- **Type 3** is the international specification for twisted-pair telephone wire of 22 or 24 AWG. It is used with data for short runs and slow speeds.
- **Type 5** is 125 micron fiber-optic media used for Token Ring repeaters.
- **Type 6** is a single pair of flexible stranded 26 AWG copper wires for patch panels, etc.
- **Type 8** is a flat cable designed to run under carpets. It has two pair of 26 AWG shielded cable without twists.
- **Type 9** is a thinner version of Type 2 with a limit of 200 meters.

IBMBIO

Another name for IBM's BIOS (Basic Input/Output System). See BIOS.

IBMDOS

An old IBM name for the MS-DOS routines that handle the reading and writing to disk. See MS-DOS.

IBNS

IP Backbone Network Service. This is a very high speed project by the NSF and MCI to link US supercomputers. It will use SDH carrying ATM cells at 155 and 622Mb/s, and an IP switching matrix.

IBOC

In-Band, On-Channel. A term applied to the version of Digital Audio Broadcasting being promoted in the US in opposition to the Eureka 147 project of Europe. The US FCC plans to allow IBOC to coexist with current AM and FM in the same frequency spectrum. New receivers could incorporate both analog and digital functions.

IBS

Intelsat Business Services. A range of business services provided by the Intelsat organization to large corporate customers, but only with the approval of the regional or national carrier of each country. You don't by-pass the carrier's controls by using Intelsat!

IC

Integrated Circuit. An IC is a circuit whose connections and components have all been fabricated onto one integrated structure — usually a postage-stamp size wafer of silicon.

This is now a very general term that can be applied to a wide range of devices.

The main distinction is probably with discrete components which need to be joined by soldered wires and mounted on PCB boards. ICs are mounted on PCBs also, but they reduce the solder-contact points enormously, and make equipment both more reliable and much faster, because of the reduced distance the signals travel.

ICs were made possible by the invention of the planar transistor; and the first were used in 1963 in a hearing aid. Early ICs were based on bipolar-junction transistors which proved to be ideal for TTL circuits. See LSI, VLSI also.

- The main type of IC chips are:
  - Microprocessors, which are entire processors on a single chip.
  - Transputers or Computer-on-chip. Devices with entire microprocessors, memory, clocks and often other functions on a single chip.
  - Logic chips, which can perform some or most of the functions of a processor.
  - Memory chips: RAM, ROM, PROM, EPROM, EEPROM, flash, FRAM, etc.
  - A-to-D and D-to-A conversion chips.
  - Logic and Gate arrays. See PDL.
  - Digital signal processors.
  - ASICs.

ICC

Integrated Circuit Cards. A smartcard. See.

ICD

International Code Designator. A global network numbering scheme. See X.121, E.163, E.164, E.168 and NSAP also.

ICE

In-Circuit Emulation/Emulators. These are expensive software applications which are used by software writers to check and debug programs for future microprocessors and/or operating systems before they are manufactured. Emulators make one machine look and act like another.

Iceberg system

The Iceberg system uses numerous hard-disk drives, each with 1.6GByte capacity, to offer a total high-capacity on-line storage system for large enterprises. These disk-farms range in capacity from 100 to 400GBytes. This is a RAID approach which is said to be cheaper than the more traditional DASD. However, very complex software is required to manage the data so that it can be found and recovered when needed.

ICMP

Internet Control Message Protocol. A Network-layer Internet protocol specified in RFC 792. It is a set of messages which are exchanged in IP communications to report errors and to provide other information relevant to IP packet processing.

ICNIRP

International Commission for Non-Ionizing Radiation Protection. This is a committee of the International Radiation Protection Association (IRPA) which is examining the biological effects of radio.

ICO

• **Intermediate Circular Orbits** for satellites lie at distances of 10,000 to 15,000km above the Earth. At this height, communications satellites move slowly across the sky and are easy to track. However a constellation of between 12 and 15 is needed to provide global coverage.

The main problem with ICO constellations is that the designers need to take into account the problems of van Allen ionization, so the space-craft needs special protection.

The distinction is with LEOs and GEOs. Note that there are also intermediate elliptical orbits (see Ellipso, HEO, Molniya and Tundra). The general term for all these circular and elliptical orbits is MEO (Medium Earth Orbit).

• **ICO** n: The .ICO format is Windows icon representation. These are always 16-color bitmaps measuring 32 x 32 pixels.

iCOMP

A benchmark, devised by Intel to compare its range of microprocessing chips internally. This is a single-number index which is totally processor-based.

• The Pentium 120MHz chip is the first to break the 1000 iCOMP barrier. Previous chip ratings are: Pentium 100MHz (815 iCOMP), Pentium 75MHz (610) and 486SX25 (100).

icon

Graphic symbol used to represent an application file, object, or a process. All GUIs use icons for identification (some types are also 'buttons'). There are passive and active icons. Generally in GUIs, icons are active ('hot') so when you click on them, something happens. See GUI and WIMP.

ICR

Intelligent Character Recognition. A variation on OCR which doesn't use matrix-matching techniques but rather reviews the words in context. ICR relies on sophisticated AI tools such as context analysis, page analysis and lexicons. This is probably the future of OCR — but it is also dangerous because the ICR application is sometimes taking 'educated guesses' — and it often isn't very well educated.

The phrase 'This is no~ essential' within a legal document (where the tilde is a missing letter), can be interpreted as 'not' or 'now' with equal context-weighting.

ICS

Intelligent Communications System. IBM's new AIN architecture which relies on user agents. It is designed around e-mail and intelligent router technology.

ID

Identification. Usually your membership number — but sometimes your password.

IDA

Integrated Digital Access. A cut-down form of ISDN which British Telecom offers in the UK. It has one B-channel (at 64kb/s), a half-rate packetized D-channel (at 8kb/s) for data, and another 8kb/s D-channel for signaling. The total data rate is 80kb/s. There's also a Multi-line IDA with 30B + D.

IDE

• **Integrated Digital Exchange.** See.

• **Intelligent (or Integrated) Drive Electronics:** A relatively high-level disk drive interface for IBM compatibles, which is also known as the AT-bus interface. It is an ANSI standard, and is fast becoming the dominant interface on new 3.5-inch drives. IDE interfaces are often built into the motherboards, which saves an expansion slot.

This is a very simple interface which relies on an intelligent disk controller being integrated into the circuit board of the disk drive. Since most of the new electronics are in the drive, often only one chip needs to be changed in older motherboards.

IDE provides drives that are as fast as the ESDI drive, and as intelligent as a SCSI (so they say!). It certainly has the simplicity of EDSI and ST506 (its ancestor), but test results show that it can't match SCSI rates, and it can only link two similar devices as against SCSI's eight.

Currently IDE is faster than the ISA bus, but MCA, EISA and PCI can handle it. Cables are limited in length to 18 inches and data rates to below 4Mb/s. Most IDE drives will have 34 or more sectors per track and run at 1-to-1 interleave. It emulates the standard ST506, so no BIOS modifications are required.

— The new Enhanced IDE needs a new controller card, and can't be installed in many older DOS-based computers. Enhanced IDE can handle up to four drives at rates to about 8Mb/s. However this is not the same as Fast ATA. See ATA and Fast ATA.

— There were problems getting new IDE disk drives to work with older models: the standard took a long time to stabilize.

identifier

• A label or name for a string or numeric variable, and sometimes for a constant such as pi (π).

• A MAC address on a network.

• In an on-line database, special types of 'key-word' fields are often created to aid in searching. Some of these will use a controlled vocabulary (thesaurus terms) while others will hold important, but undefined terms. The most common terms for these are Descriptors and Identifiers

— Descriptors are key-words widely recognized across the whole subject area, or Dewey-decimal type subject codes. A database provider will often publish a list of these terms in a 'thesaurus' along with synonyms — so that you'd know not to use 'under-sea' when looking for telephone cable information, but rather 'submarine'.

— Identifiers are those terms which may not be subject-specific, but which are record-specific and important. They will therefore not be found in a thesaurus.

For instance, a report on Italian volcanoes may include in its Identifier-field the terms Sicily and Mount Etna, while the descriptor-field would hold terms like volcanos and lava. See descriptors and controlled vocabulary.

IDI

Initial Domain Identifier.

IDLC

Integrated Digital Loop Carrier. About 30% of US telephone lines are now carried in the local loop by an IDLC system; while most of the rest of the world uses single-carrier twisted pair. IDLC relies on a line concentrator in a local street box which is equipped with the appropriate interface cards (for the type of services being provided). The box will link to the exchange via a TDM optical fiber pair or coaxial cable. Without introducing high-priced components, up to 240 subscribers can be serviced by a 2Mb/s link — although not all telephone subscribers can use their phones at the same time.

Homes or businesses will be linked to the street boxes by copper pairs, coaxial or optical fiber. This is a fiber-to-the-curb telephone-only system that also has applications in more remote areas. See RIM.

idle bytes/characters

Redundant data which is often sent on a synchronous communications link simply to maintain synchronization. This is usually a stream of zeros, but with some form of timing inversion. See AMI.

IDN

Integrated Digital Network. Interexchange and long-distance networks in which both switching and data transmission is digital. Most developed countries are at this stage in development of telecommunications now — although the more far-flung regional links will still be analog.

There seems to be two common applications of the term.

- Usually IDN is specifically applied to a carrier's 'internal' core network which, in advanced countries, has evolved over the years from being a purely analog service to now being almost totally digital on fiber. This is the network that ties all the carrier's exchanges together (the IntereXchange Network — IXN).

It will probably transfer voice calls using PCM 64kb/s links — even if the calls originate and terminate in an analog local loop; ADC and DAC is performed at the first digital exchange. Signalling is carried separately to voice traffic by using a message-oriented packet network (see CCS#7) between exchanges.

The term is also used for any kind of integrated digital network. If you accept this terminology, then ISDN is a specific form of IDN providing local access. The confusion also happens in reverse; many people mistakenly talk of ISDN as if it were an end-to-end network with its own switches and interexchange circuits, but ISDN is only a local loop technology, which uses IDN for on-carriage past the local exchange.

IDP

Initial Domain Part. Part of the CLNS address which holds the format identifier, domain identifier, and the authority.

IDRC

International Digital Recording Compression. An IBM text-compression system developed for tape storage about ten years ago and still widely used on backup systems.

IDRP

InterDomain Routing Policy. This OSI protocol for IS-IS specifies how routers communicate in different domains.

IDTV

Improved Definition Television. This means conventional analog 4:3 NTSC and PAL television sets with improve quality, mainly through upgrading the signal processing rather than any modification of the transmission system.

The concept depends mainly on proposals to use internal digital frame-storage techniques, the application of DSP to process the signals and remove ghosts, and perhaps faster scan rates and line-doubling at the receiver to remove flicker and reduce line visibility. IDTV also sometimes implies wide-screen (16:9) aspect ratios, but this is generally now referred to as extended definition (EDTV).

IEC

International Electrotechnical Committee (of ISO). See ISO.

IEEE

Institute of Electrical and Electronic Engineers. Founded in 1963, the IEEE is one of the oldest and most respected professional societies in the USA. It is a 300 000-strong membership organization of engineers, scientists and students, and an accredited standards committee of ANSI.

The computer and communications societies within the IEEE have concentrated recently on defining standards in electronics and information technology with particular emphasis on LAN and WAN standards. The best-known of these are the IEEE 488 instrument bus, and the '802-dot' series for LAN/-WAN/MAN. See ITU and 802.# also.

IETF

Internet Engineering Task Force. This is the group which standardizes the Internet, and therefore SNMP and TCP/IP. It is a task force within IAB, and it has 40 subgroups working on engineering problems. See IAB.

IF

See intermediate frequencies.

IFF

Interchange File Format. This is the standard Amiga bit-mapped image.

IFRB

International Frequency Registration Board — which manages the allocation of satellite and radio frequencies. It is now called ITU-R. See ITU.

IGES

Initial Graphics Exchange Specification. An ANSI-standard CAD file format for graphic images. It enables CAD drawings to be converted from one system to another by providing a neutral intermediate format between two proprietary CAD file standards, through which the data can be exchanged using translators (pre- and post-processors). IGES is an SASE under development for OSI. It involves the use of GKS. See DXF also.

IGP

Interior Gateway Protocol. A family of protocols used in the Internet to exchange routing information within an autonomous system. RIP and OSPF are both IGPs.

IHQ

Intelligent High Quality. A term that Akai are using for an enhancement to VHS video standards to 'greatly improve' the picture quality. The machine records a standard signal onto the tape, then replays the tape to test the characteristics. It then adjusts the recording parameters to get the best results. Nokia have a similar system called ASO.

ikon

See icon.

ILD

Injection Laser Diode. See.

illuminate

In satellite terminology, a feed horn 'illuminates' the antenna dish whether it is sending or receiving. So illuminates just means that it 'gathers signals from' or 'sends signals to' the dish.

IM distortion

Intermodulation distortion. See distortion.

image/IMAGE

- When a radio signal is produced or detected using heterodyne systems, an intermediate frequency is produced, but there is also an unwanted 'image frequency'. Unless this image is filtered out, there will be two bands on a receiver's tuner where the signal will be found. The one not wanted is known as the image.
- A graphics file format used in the IP suite on the Internet.

image orthogon

An older type of camera tube about the size of your forearm. It was widely used in professional studio cameras before vidicons and CCDs came along.

image processing

Analog images suffer from degradation whenever they are processed or duplicated. The copy of a film or videotape, or the transmission of a TV signal, always results in an image that is inferior to the original. See generation loss.

The conversion of images to digital form creates an initial slight loss of quality (depending on the sampling and quantization processes) but it can prevent any further degradation in the image through generational copying or transmission, and it also allows us to use intelligent devices to improve or modify the images.

The problem is that, in digital electronic form, the amount of data involved in any photographic-quality image is substantial. A color TV picture of home PAL quality (equivalent to 450 x 350 pixels, each of 16-bit pixel depth) requires 2.5 million bits per frame, or 62Mb/s, and a typical color postcard-sized photo of high quality requires about 3MBytes of storage.

Without compression, these figures are too high for most forms of current storage device. So data reduction (compression) is the only way to practically handle most digital images today. However some of the techniques used do not allow the video images to be subsequently edited (MPEG2 for instance).

There are many different compression algorithms available and usually more than one will be applied for best results. Recently, there has been almost universal acceptance of transform coding techniques using DCT for the last-stage reduction in bit numbers. DCT combines the advantages of significant compression with impressive image quality, and it has a reasonable cost of implementation. Where moving pictures are being transmitted, transform coding is augmented by prediction (comparing one image with the next) and motion compensation (looking for blocks of image which have shifted between frames) to further improve the data reduction factor.

- DCT is used in JPEG for still-frame single image compression, and in Video or Motion JPEG which is used with video in the post-production stages (MPEG is not easily edited). It is central to MPEG1 (CD-ROM) and MPEG2 (digital television) which are now the main video transmission standard, and it is also used in H.261 and H.320 for videoconferencing.

See DCT, motion compensation, JPEG and MPEG.

image resolution

The number of screen pixels which are being used to produce a picture do not provide a measure of image resolution — however they do give us the upper limit possible. If a screen has 1 million pixels, then the best image shown can never exceed that of the screen. However if a low-quality picture is shown, then the image resolution will be that of the picture, not the screen pixels.

Image quality is measured in terms of 'lines', or 'cycles' in analog systems, and pixels in digital systems (although analog color monitors are often quoted in terms of pixels). There are problems with both these measures:

- These 'lines' of resolution in the horizontal plane, have nothing to do with the scan lines of a video image. In the vertical plane there is an odd relationship between scan lines and image resolution lines. We usually multiply the active scan lines by 0.6, to create a rough approximation of image resolution 'lines'. A TV picture with 400 lines of horizontal resolution (300 vertically) will be exceptionally good.

• You need to distinguish between screen pixels (physical dot-groups on the screen) and image pixels (groups of these dot-groups) that may be switching on and off together. See resolution, and video monitor resolution.

image scaling

In TV this usually means reducing the window size. A picture requiring 640 x 480 pixels, is reduced by subsampling to an image of 230 x 240 pixels.

In computers, it is more complex, and is generally only applied to object-oriented 'Draw' programs or font outlines, where screen objects can be enlarged or reduced in both dimensions.

image setter

An electronic/optical device which takes its output from a computer image and transfers it at high-resolution onto photographic film or paper. This is similar to a laserprinter, except that it is specifically designed for printing shops to produce much higher quality images (in the 1200 dpi region).

image stabilization

Hand-held cameras or cameras in motion have shake and blur problems. Both professional and amateur cameras, both movies and still, need techniques to overcome these problems. There are a number of techniques:

— **Gyro-stabilizers:** These were mounted under the camera, and driven at high-speed by battery-powered electric motors. They were noisy sometimes, and the batteries didn't last.

— **Steadicam:** The counter-balance method has been used for many years in movies. Steadicam is a body-harness with springs and balancing counter-weights which tends to leave the camera upright and moving relatively smoothly. If was first used for the training sequence in Rocky I.

— **Lens-prism systems:** These are very successful, and are finding their way into good amateur cameras. The front element of the lens is a variable angle prism driven by electronic actuators, and it is able to control the angle of refraction of light entering the lens, and correct for image movement. It relies on movement sensors being built into the camera body.

— **Image-selective methods:** A CCD is used which is larger than that required. However the camera intelligence is able to select which area of the CCD it will use, and this is under dynamic control from movement sensors in the camera body.

imaging

The term is now widely applied to the use of scanners to produce images of documents for business archiving — and to the problems of transmitting large X-rays, or the full negative of a newspaper page across the nation for instant printing.

Imaging processes usually include most of these five functions:

— Scanning,
— Processing (compression/decompression),
— Facsimile transmission,
— Copying,
— Laserprinting.

• Image storage has become practical only with large-scale storage devices — particularly recordable optical disks. The problem with imaging of this kind is that an indexing and retrieval program often needs to be run in parallel if the stored images are to be usefully employed. With complex multi-paged documents, the software must handle the relationships between images.

• Image transmission is becoming increasingly possible due to two factors: the use of fiber optics has given us cheaper and more reliable high-bandwidth networks, but even more significantly, digital signal processing has given us lossy and lossless video compression. See JPEG.

IMAP

Interactive Mail Access Protocol. This replaces the POP standards for Internet e-mail access. Currently IMAP 3 is in use, but 4 is coming soon. This is a client-server system.

IMAX

A projection system showing super-wide, top-quality images in specially equipped film theatres. See Showscan and Iwerks.

IMBE

Improved Multi-Band Excitation. A technique of low bit-rate voice encoding based on Sinusoidal Transform Coding. See STC and MBE also.

IMDS

Image Data Stream. A file format produced by IBM's ImageEdit.

imho

Internet shorthand for 'in my humble (or honest) opinion'.

imo

A more aggressive version of imho.

IMP

Interface Message Processor. A name once used in Internet literature for packet-switching nodes.

impact printers

A generic term which encompasses dot-matrix and formed-character (daisy-wheel, golf-ball) printers. It applies to any printer where a ribbon is hammered against the paper — although now it is often used as a synonym for dot-matrix.

impedance

This is the apparent resistance of a circuit to alternating currents, as distinct from the true resistance to direct current. Both are measured in ohms — but you can't measure impedance with just an ohm meter.

— Transformers will have high AC impedance, but low DC resistance.
— A capacitor will have infinite DC resistance, but low AC impedance.

Impedance affects the networks propagation delay, and it attenuates signal strength. All LAN line-coding systems, and all physical topologies, have their own impedance values which must be adhered to if the network is to function well.

The coaxial used for Broadband LANs and CATV has an impedance of 75 ohms, while standard Ethernet uses 50 ohm cable.

• Impedance matching is important in balancing network devices for maximum throughput.

• Impedance mismatching causes echos. Echoes occur in telecommunications whenever signals encounter an irregularity in impedance. This is the major cause of problems in high-speed data transmission. See hybrids.

imperative

See declarative.

imposition

The process of deciding where pages should be placed (in what order, position and orientation) to produce a forme for the printing press which will then be correctly sequenced when the sheets are folded into a signature, and trimmed.

In desktop publishing, camera-ready copy will often be sent to the printer as a series of single pages. But a large printing press always prints many pages at a time (usually 8, 16 or 32) and often on both sides of the paper at once.

The printer will arrange the pages in a way that the large sheets, when folded, place pages the correct way up and in their required order. You will want page 2 to be printed on the back of page 1, not page 3; and you may want any spot-color or four-color pages to be only on one side of the sheet (to save costs) — and this will only happen if imposition has been worked out beforehand with the printer.

improv

A type of IRC in the Internet. It is an on-line version of theatre sports where the actors make words up on the fly. You are given a situation and a character by a Master of Ceremonies, and told to improvise.

impulse noise

Short-duration, sporadic, high bursts of energy which disturb telephone signals and are often disastrous for data. They are usually caused by nearby electrical equipment, remote electrical storms, power sources, and switching or signaling equipment. You probably see a flash of impulse noise on your television screen when you switch a light on or off.

IMS

• **Information Management Systems** (a generic term).

• **Information Management System** — an IBM mainframe-based transaction processor, released in 1968, which includes database management functions. This was the first real database management system, and it was tied to the DL/1 language. IMS was very successful and is still in use today, however it lacked true relational capability and was widely replaced by DB2. See Codd's rules and RDBMS.

— IMS/VS (Virtual Storage) is an 'operating environment' used with MVS for transaction processing.

IMTS

Improved Mobile Telephone System. The mobile phone systems in use before true cellular came along. The base station towers were high, and placed on hills for a coverage range of 20—30 miles. They typically radiated 200—300 Watts, and the mobiles 40 Watts.

Up to 60 analog channels were available from one base-station and there was no cellular frequency reuse. In Chicago (the center of developments), it was possible to service only about 2000 customers on 23 channels and provide good service. At its maximum a city the size of New York would only be able to support 60 simultaneous conversations.

With IMTS, when you lifted the handset in the vehicle, a channel would be allocated, and you'd be presented with dial-tone if one was available. You would then dial yourself, using in-band signaling tones from 1300Hz to 2200Hz.

Incoming calls were received in the normal manner, provided you were within your home location area (serviced by the city-wide base-station). Outside this area, the mobile was a 'roamer' and advance arrangements needed to be made. Calls were made to a roaming number through a roaming operator who kept check on the roamer's location and manually patched calls through.

IMTS is still being used and it is often cheaper than cellular. Modern versions have integrated paging. See Radiocom, MTS and NAMTS.

IMTV

Interactive Multimedia TV.

IN

Intelligent networks. See, and also AIN.

in-band signals

In-band and in-channel signaling systems seem to be virtually the same — with in-band being used for analog, and in-channel for digital.

The term in-band refers to the use of a single voice/data frequency band for both signaling and analog voice or data traffic. The same bandwidth is used, but usually separated in time. However there are two types of in-band:

— Sequential use of the voice/signaling channel. Our present analog telephone systems uses in-band signaling because the dialed tones/pulses are carried through the network by the circuit that will later be used for voice.

— Simultaneous use of the voice/signaling channel. In the US, an external tone generator called an SF (single frequency) unit typically convert pulsed dialing signals into 2600Hz on-off tones. This is within the voiceband (300—3400Hz), but special notch filters block these signals from callers.

In-band signaling is easy to implement, but it is very limited in terms of the information it can convey during a conversation. However it is very widely used still in the local loop.

See also out-of-band signaling and CCS#7.

in-channel signaling

In circuit-switched analog channels the signaling has always been within the channel, until recently. This means that the same channel is used to set up the call, as will later carry the call. The signaling begins with the originator of the call, and follows the path of the call to its end.

In-channel signaling seems to be the digital version of in-band; but here bits are 'borrowed' or 'stolen' from the data channel to carry the supervisory and signaling information.

The early digital (T1) systems used a form of in-band signaling by robbing bits from the voice-bytes. This is known as channel associated signaling — and this is why US 64kb/s channels often only have a 'clear' 56kb/s of bandwidth.

The alternative approach is Common Channel Signaling which has a separate packet-switched network for signal where signals may take an entirely different path between exchanges.

• One reasonably reliable reference says that in-channel signaling has two types:

— **In-band signaling:** The type most often used over the local loop with analog phone systems.

— **Out-of-band:** Channel-associated signaling.

This makes it appear as if the term is used generically in some parts of the industry. Either that, or the author is confused.

• Note: ISDN local-access links have a separate D-channel which uses out-of-channel packet-switched signaling (even though it is carried over the same wires as the B-channels).

Similarly all digital interexchange networks now use packet-switching to carry signaling information as messages. The value of this approach is that signaling and message exchanges can continue while the data is being transferred, or while the voice channel is in use.

in-circuit emulation

A program developer's productivity tool that provides a simple, yet effective way of testing hardware and software. It is used in the development, integration and debugging stages of hardware and software. The microprocessor is removed from the unit under test (the target) and connections are made to a second computer system (host). The emulator may allow the software engineer to experiment with different clock speeds, etc. to detect subtle timing problems that may be experienced later with the actual processor.

in-slot signals

These are in-band or channel-associated signals which occupy a fixed time slot (or bit position) in the frame associated with each channel in time-division and packet-switching systems. See channel associated signaling.

• When the term is applied to analog phones it means standard in-band pulse or DTMF signaling.

INA

Information Networking Architecture is Bellcore's framework, closely associated with TINA. Both TINA and INA include the Distributed Processing Environment (DPE) software backplane for plug-in applications. These are all based on distributed object technologies such as CORBA and DCE.

inclination

The angle between the orbital plane that a satellite is taking as it passes around the earth and the equator. Geostationary satellites would have zero inclination, but many medium and low-earth orbiting satellites are deliberately inclined to provide maximum coverage over specific geographical regions.

As satellites age, they are often allowed to slip slightly out of true equatorial orbit to save fuel, and this is also called an inclined orbit. From the ground this appears to the observer, to be a GEO performing slow figure-8s in a north–south direction. See inclined orbits.

inclined orbits

There are two distinct uses for this term:

• When the thruster fuel load of a geostationary satellite is nearing its end, the TTC&M controllers will allow the satellite to drift further from its geostationary slot than is normally felt to be acceptable.

From the ground, these geriatric satellites appear to wobble around the sky in a figure-8 pattern and, past a certain point (depending on dish size), this requires constant tracking by moving the dish. However this is often allowed to happen to old satellites to prolong their life, but only over a limited range of deviation — no more than +/–5° N/S and very little E/W. The technique conserves the remaining thruster fuel in the satellite, but the satellite can only then be used with earth stations capable of tracking.

Inclined orbits can often add three of four years to the useful life of a satellite, and so the owners can afford to lease transponders on these satellites for a lower rate.

If a satellite is wandering only by +/– 1°, a 3 meter dish can probably receive it without tracking — but it will need to track a +/–3° movement. However a small dish will be able to receive signals over +/–3° of movement without tracking. See station-keeping and beacon.

• LEO and MEO communications satellites need to provide world-wide coverage at high (vertical) angles to their users, so they are typically put in either polar orbits (Iridium) with orbital-planes spaced at equal distances around the globe, or in orbits which are inclined to the equator (Globalstar, etc.) usually in a criss-cross pattern of planes.

These satellites will be orbiting in numerous planes inclined at different and opposite angles to each other, to produce a 'basket-weave' coverage of the earth — and they may be supplemented by one or more polar planes and/or equatorial planes. The choice depends on the coverage areas required, and most of the world's population live in the Northern Hemisphere and around the equator.

• Elliptical orbits, such as those used by the Russian Tundra or Molniya satellites which service the polar regions, are also both inclined and elliptical. Ellipso uses such an inclined elliptical orbit for some of its satellites. See all.

increment(al)

To raise the value, usually by one step at a time (unless otherwise specified). Decrement is to lower the value.

- Incremental backup is when you only copy those files which have been changed, or not backed up before.
- Incremental compilation is an approach where the application/language compiles the source code as it is typed into the computer, rather than as a batch-job later.

IND$FILE

An editing utility for mainframes. It is often used in PC-to-mainframe file transfers.

Indeo

Intel's video compression and decompression system which evolved out of DVI. Intel has licensed it to Microsoft for Video for Windows. Indeo video can be replayed using a special board with decompression hardware, or it can be handled in software — Intel created special i750 chips which speed up the process.

Indeo files can be recorded in 8-, 16- or 24-bit color form. Currently, with software, Indeo can decompress (but not compress) the full color at 15fps image at a resolution of 160 x 120 pixels with a 486DX running at 33MHz. The new Indeo boards will also provide the MPEG standard. This now seems to be called ProShare.

independent logical unit

This is a LU that doesn't rely on SSCP to establish its sessions. It must be LU6.2 with PU2.1, be capable of sending bind requests and XID, and of supporting parallel sessions.

indirect addressing

In computer programming, indirect addressing allows programs to be loaded at variable locations. This is essential now that multi-tasking computers keep more than one application in memory, and so programs can't be guaranteed to load at the same memory locations each time.

Indirect addressing is a way of referring (forwarding) the CPU to the address required, by providing an offset.

An instruction always requires data from a location, and previously the instruction would directly access the address of that data. With indirect addressing, the location provided will be that of the address at which an offset amount will be found. This is added to the base to find the data. See relative address.

indirect mode

In programming. See direct mode.

inductance

A circuit carrying any electrical current will be surrounded by a magnetic flux (field), and this magnetic flux passing through any nearby electrical circuit will generate a current. So any adjacent circuit will be influenced by changes in the flux of a signal-carrying circuit, and the second will generate a low-level signal in sympathy with the first. This is inductance.

When wires are wrapped so as to produce a coil (an inductor), the magnetic flux of each wire reinforces that of its neighbors, which is why coils produce electro-magnets, the regular coil-windings act in concert to produce substantial magnetic fields.

The microsecond delay in the collapse of such magnetic fields, and the reverse current generated by the collapse, is also the reason why inductance effects can be used to change the oscillation rate of circuits, and therefore why coils find use in radio tuning circuits (bandpass filters).

- An inductor exhibits only low resistance to DC and to very specific AC signals, but very high resistance to any AC signal not alternating in an integer relationship to the coil's collapse rate.
- Inductance effects also produce near-end crosstalk (NEXT). To overcome these problems, telephone wire pairs are constantly twisted around each other, and around other pairs in the bundle in a very controlled manner, so that two wires associated with different circuits will never lie close alongside each other for any distance.
- The unit of inductance is the Henry.

inductive coupling

A form of interference in a cable system caused by inductance effects from electrical systems nearby. The most common example is fluorescent light ballasts which can create problems with unshielded LAN cabling if the two are in close proximity. See NEXT also.

inductive learning

An education and AI term used in expert systems; learning from experience rather than systematic via theory — empirical knowledge. Most of our knowledge of the world is by inductance; much of what we profess to be 'deductive knowledge' (systematic and rational) is actually ***post-hoc*** rationalization. Contrast this with deductive learning. See heuristics.

industry-standard

- Any system/technique used by more than one vendor (If you believe the marketing men promoting the product which uses that particular system/technique!) If a standard is accepted by three or more companies, it will then be known as an 'international standard'!
- A general term that is now used by the clone-makers to mean the PC's ISA bus — and often EISA — as distinct from IBM's MCA. It's been a very effective marketing put-down.

inference engine

In expert systems, the inference engine is a section of software which maintains a list of all the assertions that must be proved true in order to come to a conclusion. It holds the rules and applies them in an IF/THEN form. The rules may be weighted by certain graded factors (fuzzy logic), or they may be a fixed set of values (as in classical logic). See backward chaining.

infix notation

This is normal algebraic notation. 1 + 1 = 2 is a simple sum in infix notation, as is X = (3 * Y) +5. The distinction is with prefix (or Reverse Polish) notation. See.

infobahn/strada

Catchy names given to the information superhighway by Wired magazine and others. These are roads through cyberspace which service a Multi-Function Polis. They carry truckloads of hype. Others prefer to call them 'I-ways' or 'Supertollways'.

• The term probably means little more than hybrid fiber-coaxial (HFC) cable networks with ATM cell switching and multiplexing for interactive applications, and using MPEG2 compression for video and audio. Direct satellite, SHF terrestrial broadcasting and HDS over twisted pair will supplement this HCF network.

informatics

The all-encompassing study of computing, communications, and broadcasting; information storage and retrieval; artificial intelligence and artificial reality — but possibly not robotics.

The area of study includes the hardware and software technologies, but the emphasis is more on the social impact, practical value and the problems that may develop from the use of these technologies.

Informatics deals with cognitive sciences, perceptions, interface design, etc. but it recognizes the importance of understanding the conventional hardware/software aspects of technology.

information

This is often used in the computer and communications scene to mean data. It is a bit like saying that the characters of the alphabet are the same as words, and that these in turn are the same as the development of ideas in a book.

Information is a high-order term which implies logical construction incorporating something meaningful to humans — not just to machines. But the term is used very vaguely in the IT industries.

information bits

Those bits being transmitted on a communications system that convey the useful parts of the message — as distinct from synchronization bits, check bits, framing bits, etc.

information provider

• In videotex, this is the organization or group which publishes information in the system. They are responsible for inputting and updating the information; they therefore need special terminals and special access rights to enable them to enter, edit, and amend frames in the videotex database.

• In the large public-access database utility, the distinction is made between the information provider who owns copyright on the database, and the utility-supply company which looks after the computers and the charging.

information retrieval

• The generic term for selecting and extracting the required information from any database.

• On the Internet, this probably means ftp or Telnet, or the use of a WWW browser.

• With on-line use of professional database utilities, where the information is held on host mainframes, you will need to register first and establish credit. You can access most of these professional services nationally or internationally via X.25 packet-switching networks, but you'll need to learn the basics of their specific search language (only a half-dozen cryptic commands usually, to get started).

Most of these services hold abstracts of articles only, but they are associated with research organizations which will take your on-line order, and mail to you a copy of the original paper. Reports and articles cost between $10 and $100.

information separators

Control codes which are used to maintain separation between fields of information in applications. These are:

FS File Separator.
GS Group Separator.
RS Record Separator.
US Unit Separator.

information superhighway

See superhighway.

information theory

• As applied to humans this means cognitive science, semantics, linguistics and a few associated disciplines.

• Librarians see information theory in terms of categorization.

• As applied to electronics, this means the theories about circuit noise, channel capacity, data rates/bandwidth, etc.

There are two classical problems in information theory:

— How to code a message in the best possible way (least bits).
— How to send a message in the presence of noise.

Claude Shannon concluded that 70–80% of the letters in English text is redundant — which means that the optimal compression system would allow us to store each byte, using about 2-bits on average.

• Informatics would include all of the above.

information utilities

A general name given to on-line public database, electronic mail, and general information (videotex) service-providers.

infotainment

Information in an entertaining fashion. Most light-weight TV current affairs programs which feature children with disabilities and puppy-dogs fit this category. Many of the latest CD-ROM encyclopedias also have a lot of pictures along with little text.

infrared

The part of the light spectrum between 0.77 microns (770nm) and about 1mm (or 1000 microns). IR lies between microwaves and visible light.

The infrared spectrum is divided into four sections:

- Near IR $773 - 1.3 \times 10^3$nm
- Middle IR $1.3 \times 10^3 - 6 \times 10^3$nm
- Far IR $6 \times 10^3 - 4 \times 10^4$nm
- Far-far IR $4 \times 10^4 - 10^6$nm

For fiber optics, the near infrared wavelengths between 0.75 and 1.3 microns (750—1300nm) were once generally used, but now the preferred wavelengths have shifted to 1550nm.

Infrared radiation is readily absorbed by most kinds of matter and, in the process, the energy agitates their electrons — which is why we treat IR as synonymous with radiant heat. In fact, all frequencies in the electromagnetic spectrum produce heat when they are absorbed, but the absorption is much greater at the IR wavelengths.

However IR waves also penetrate some translucent forms of matter more easily than visible light: the atmosphere, for instance, is almost completely transparent to all IR in the neighborhood of 10 microns.

The various ways of defining IR are:

- Wavelengths between 10^6nm (1mm) and 770nm (0.77 microns).
- Frequencies between 3×10^{11} and 4×10^{14} Hz.
- Photon energies of 0.4 eV.

• Infrared in the free-space of an office environment can be used to carry data, and is therefore a wireless technology. It is now being used for short-distance exchanges of data between PDAs and desk-top computers and peripherals. There is a new standard under development. A group known as IrDA are developing the Versit standard.

ingress

A point of entry to a network. The access point where the node or CPE interfaces with the network. This term can refer to wanted or unwanted signals; when a Cable TV network gets old and the shielding breaks down, the cables get unwanted ingress from CB radios. See egress.

inheritance

In any hierarchy there is always a relationship between the individual, and the layers above and below. We inherit certain characteristics from our parents, and pass certain characteristics on to our children. Inheritance is the way certain properties or attributes are passed from one level to the other.

In object-oriented programming the concept applies to classes of objects. We have classes, then sub-classes, then sub-sub-classes. The lower classes share characteristics with the ones above. There are base classes (root) and derived classes (branches), and the derived classes are 'constituents' of the root class. Inheritance allows properties and attributes to be passed from a class to its constituents.

So through Inheritance, new general-purpose object classes can be refined and stored in libraries for future use, and this leads to the concept in OOP of reusability.

Init

A small Macintosh utility that is copied into the System folder for housekeeping functions. Inits activate themselves automatically on boot-up and perform their functions in the background. These are not TSRs or F-keys.

initialization

To set-up something ready for action: Setting up hardware or software so that it will function with a particular computer.

• The formatting of a disk, and the creation of FAT sectors, etc. so that the disk is prepared to store data or applications. Initialization involves two or three steps — formatting and organization (directories, etc.). Usually a verification process will follow (depending on the formatting utility and the commands issued).

• If you reinitialize a disk with verification, you clear the disk of all information (this is more destructive than just an erase or a simple reinitialization), and reformat it for further use.

• Setting up hardware or software so that it will function with a particular computer.

• The activity of a computer during boot-up, when it tests and clears its memory, then reads in data from ROM, etc. During this process all computer counters will be reset, and switches and addresses set to zero.

injection

Geostationary satellites reach their required orbit in two distinct phases. In the first stage, a multi-stage rocket (or the shuttle) boosts one or more satellites into a low earth orbit where they are 'deployed' or 'parked'. This usually marks the end of the contractual obligation of the launching organization.

Control then passes to the TTC&M control center for the satellites, and after a few days of checking, the controllers will fire a booster stage to transfer the satellite to its higher permanent orbit. This preliminary boost from the low 'parked' orbit to the higher one is called the perigee-thrust or transfer stage. When it reaches the outer orbit, reverse rockets are fired to slow it down and circularize the orbit — to 'inject' the satellite into its final orbital slot. See hydrazene.

injection laser diode

ILDs are semiconductor devices that emit a beam of light when a voltage is applied. These are more costly than LEDs for optical fiber communications, but they can sustain higher data rates.

ink-jet printers

A type of quiet, high-quality printer pioneered by HP and later by Canon, which creates an image on paper by ejecting minute drops of ink and electrostatically driving them against the paper — in the same way a serving tennis player throws up a ball and then hits it with a racket. The problem of ink-clog has been overcome with these printers, and the quality is now approaching that of the best laserprinters. Generally, however, they cost a bit more to run on a per-page basis.

ink standard

See Jot.

Inmarsat

INternational MARitime SATellite organization. This is a consortium of carriers from 75 countries that operate a network of satellites for maritime, aeronautical and land-mobile communications and navigation.

Be sure to make the distinction with Intelsat, which operates international communications satellite services to fixed ground stations. Inmarsat operates mobile services to ships, aircraft, cars, trucks, pagers, hand-helds, etc.

Inmarsat offers a range of different mobile communications:

— **Inmarsat-A** is a maritime voice, data and facsimile service using analog techniques, which was developed by Comsat. About 10,000 ships around the world use this system via eight satellites. Terminals are manufactured by 15 different makers around the world, and cost in the region of US$50—$60k.

They need a gyro-mounted 1m (3ft) wide dish, usually in a radome. The receivers need half to one kilowatt of power, and there is a land-transportable mains-powered version which fits into a couple of suitcases. Voice, fax, telex and data are available at rates up to 64kb/s. The costs are $5—$9 per minute.

— **Inmarsat-B** is a maritime digital system that provides voice, data, telex. It is the digital version of A. It has a peak data rate of 64kb/s and costs about $6—$12 a minute. Terminal costs are about the same as A.

— **Inmarsat-C** is a messaging and location system using Standard-C technology. These are small briefcase-sized units which can be fitted to a car, boat or aircraft. They consume about 100 Watts DC, and use a small 30cm (12 inch) dish or a phased-array panel to communicate data at about 600b/s. This is basically a store and forward messaging system which interfaces with telex and X.25 networks.

— **Inmarsat Aero** is a data-only system under development for commercial and private aircraft. There are two versions: Aero-L is for low speed data at 600b/s, and Aero-H is for high-speed data to 10.5kb/s. Aero-H is for voice, facsimile and data on commercial airliners and business aircraft, and it will provide in-flight telephone links also.

— **Inmarsat-M** is a low-cost portable voice, data, fax and paging terminal for ships and land-transport. The unit is briefcase sized. Maritime versions use tracking antennas, while portables are fixed. The messaging rate is 2.4kb/s and voice will be available over these systems at a digitized rate of 4.8kb/s (however, with full error correction, the rate is 6.4kb/s) Terminals are priced at about $10—$15k and usage costs about $6 per min.

— **Inmarsat-P** is the future hand-held voice-data communicator proposal using ICO satellites. It is being developed by a consortium within Inmarsat, and the hand-set is being developed by Ericsson. See Project 21.

• Inmarsat is now deploying its third generation of satellite, and is examining the feasibility of a possible global positioning system. See GNSS.

• Inmarsat systems are also known as Marisat. Comsat is the US representative. Worldwide there are probably 25,000 users.

Inode

A numbering system that is used in Unix (in addition to the file names) to keep track of files.

input mode

The data mode in a modem where all input is interpreted as bits which need to be transferred across the phone link. The distinction is with the command mode. In this input mode the modem intelligence will only detect and use as a command, the escape sequence (usually <1 sec>+++<1 sec>).

INS

Intelligent Network Services which are more often called IN (Intelligent Networking). This usually means call-forwarding and similar basic services of this kind. See IN, AIN and CLASS.

insert

To interpose between existing material.

• In word processing, you insert when you add extra characters between characters already in memory (and on the screen). The insert operation forces a repositioning of the other characters to new memory locations.You don't notice the memory location updates these days because it all happens so fast.The distinction here is with overwrite, where the new characters destroy old ones.

• In analog videotape, insert editing is used to add a shot into a sequence (or into an over-long shot) already in place in the edited roll. Obviously, this is a different problem to assemble editing, where you just add shots in sequence to the end.

Insert editing is easy with film, where you just cut in the extra length of film, and splice it at each end once again — but video is a different matter.

Even if a shot is to be replaced or shortened, it is unlikely that the inserted shot will exactly fit the gap left. However the main problem is that the synchronization of the images may drift between the insert and exit points, causing the picture to misalign at the end of the inserted shot (the picture will roll).

So true insert editing requires that the old control track remains in place, and only the picture and sound can change. This means that the pre-erasing of old picture must be perfectly controlled — and this can only be done by editing machines with flying-erase heads (erase heads on the rotating head arm), not by stationary erase.

Video editors, therefore, have special controls and circuits for insert editing to ensure that the new image and sound is locked (synchronized) to the existing control track and that the insert exactly matches the erased area of tape.

insertion

• **Loss:** Each device or circuit in a telecommunications system contributes in some way to the erosion of signal power (except those providing amplification). The 'insertion loss' of a transmission system is simply its overall attenuation — from the source to the sink. It is measured in dB for audio signals and, unless otherwise stated, the measurement is made using a 1004 Hz (US) or 800 Hz (ITU) audio tone.

• **Point:** The point (some call it a cursor) at which new typed symbols will be inserted into text. See I-beam.

• **Sort:** An insertion-sort algorithm places the elements in an array with either ascending or descending order. After examining a pair of elements, and deciding to shift an element to a higher location, it will check to see whether the gap left can't be filled by something lower in the line. It creates 'holes' and attempts to fill them logically, rather than just shifting everything by one place. This is quicker than simple bubble sorts.

in-slot signaling

See Page 340, associated channel, and in-band signaling.

instruction

• One individual command at the machine code level (in binary code) of a computer's operations. This is usually a four stage process: data and control information is brought into the CPU; an arithmetical process is performed according to the control information; data is transferred back to memory; then the computer moves to the next valid memory location. See wait states.

• The instruction-fetch is the one made to extract data from the location point (the current memory location) which has been established by the CPU's program counter. After the fetch, with linear programming, the program counter then automatically increments one location at a time unless a jump (branch, etc.) occurs.

• MIPS is a measure of millions of instructions per second. Don't confuse this with MOPS which measures operations, not instructions.

instruction-bursting

In RISC operations, instruction-bursting frees up the address bus for data access by placing the first instruction address into RAM. The RAM controller hardware then assumes the task of incrementing the address.

instruction set

The commands which are built-in to a microprocessing chip at the time of manufacture. These control the functions of the CPU and they are manipulated directly by assembly language coding. With assembly, instructions are usually entered by the use of mnemonics — but there's a one-to-one relationship with an instruction function. With higher-level languages the relationships are much more complex, which is why assemblers and compilers are necessary.

If a compiled program is to run on two different CPUs, then the basic instruction set of each CPU must be the same. If the instruction sets are different then cross compilation must take place, or alternately, the program must run on a 'virtual machine' (using emulation). See cross compiler.

• Instruction sets are the 'atoms' of processing. They include assembly mnemonics commands such as ADD — which just adds two binary numbers in a register, or JMP — which causes the computer to jump to a location stored in a location register.

• CISC processors which use complex instruction sets, try to solve the speed problem by having hundreds of different instructions to perform more and more complex tasks. RISC takes the opposite approach and keeps the number of instructions available to a minimum, and is prepared to use two or three in sequence to perform complex operations.

CISC clearly won, while the main job of the processor was highly variable — while it did everything itself; RISC is now winning because more of the CPU's work is being offloaded to specialized coprocessors and controllers, which run the screen and disk drives and perform the graphics processing. Now the CPU is almost totally involved only in handling memory I/O and the basic instruction processing . Here RISC excels.

insulator

The opposite of conductor. A material in which electrons don't easily flow.

integer

A whole number without any fractions or decimal places. The distinction is with real numbers (floating-point) which have variable-length decimal 'fractions', and with fixed-point numbers which have fixed decimal points. Don't confuse fixed-point with integer. An integer is a floating point number with no decimal fractions. See.

Some computers use binary-coded decimals (see) to carry out integer maths operations, and the computer stores each number as two bytes.

integrated

• Matched, and designed to work together.

• All in the same pipe — such as Integrated Services: i.e. the one network serves all requirements. Note that the 'integration' of ISDN was originally from the viewpoint of the carrier — the idea was that the IDN digital network would carry everything, and so ISDN would simplify the service provision by carrying this out over the local loop to the customer. It was sold to the customer as a way to integrate his requirements: to talk and send data at the same time. These are two entirely different views of integration; integration hasn't sold ISDN, but the higher data-rate and fast set-up time has.

• Incorporated into something — rather than an add-on. For hardware, the implication is that no cables will be needed.

• Embedded is now sometimes used as a substitute.

Integrated Broadband Communications Services
IBCS. See B-ISDN.

integrated circuit
See IC.

Integrated Digital Exchange
Don't confuse this with ISDN, although ISDN can only function with IDE exchanges and IDN (networks). IDEs are modern exchange switches which handle voice and data in digital form. Analog voice is translated to digital PCM (or perhaps ADPCM) at the exchange, and everything is then treated the same as digital data, using TDM multiplexing and (usually) fiber connections between exchanges.

These exchanges are integrated because they are intelligent enough, and flexible enough to handle voice, data, Telex, videotex, etc. with the same equipment. The contrast is with the earlier SPC exchanges which had digital computer controls, but were basically electro-mechanical switches. See IDN.

integrated optics
This is a new telecommunications technology where the numerous optical components of an optical fiber subsystem are enclosed in one discrete miniature package.

integrated processor
This is the concept of the basic functions of a chip-set (currently excluding main memory) and the CPU of a microcomputer, being etched onto a single piece of silicon. They are effectively making a computer on a chip. The first example was probably the AM286LX/ZX IP which includes BIOS, clocks, memory and keyboard controllers with a basic 80286 microprocessor.

Integrated Services Digital Network
See ISDN.

integrated software
A packaged collection of applications/functions which share a common user-interface and have a mechanism for sharing data. This may be nothing more than cut-and-paste, but some have hot-links. Usually integrated software will provide a word processor, spreadsheet, database manager and perhaps communications all in the one package. Quite often, all these applications are cut-down versions of larger packages. See IAC, DDE and OLE also.

- Integrated software is sometimes called 'bundled' software, but bundled more often means quite distinct software packs, wrapped together and sold with hardware.

integrity
The reliability of the data or of a function. 'Data-integrity' is the opposite of 'data corruption', rather than 'data loss'.

intelligent controllers
A computer/cluster-device used to handle network control operations for a number of remote terminals. It may be programmed to act as a server for specific tasks.

intelligent switching hubs
There seems to be two levels:

- Some of the early devices labeled 'intelligent' switching hubs could only turn the ports on or off in a slow and fairly basic way. This is port-switching and it was used for balancing loads on network segments. Intelligent versions of these switches were reconfigured from a central console.

- The latest are high-level Ethernet, FDDI and/or Token Ring switching hubs which are often able to provide full bridging and routing functions. Usually an Ethernet switch which only connects Ethernet segments will be called an 'Ethernet switch', while 'Intelligent hub' or Intelligent switch' will be reserved for the more sophisticated forms able to handle multiple protocols. In fact the first is likely to be a true switch, while the others are little more than bridges and routers with RISC functions added to speed them up.

Intelligent hubs will also provide access to management functions, and, like standard hubs, they can provide wiring flexibility. They will have port-level management and they also provide network segmentation and in-built bridging and routing.

Multiple dedicated or shared segments can be handled by the hub, as can multiple media types (thick, thin, UTP and fiber) in most. See Switched Ethernet, FDDI and Token Ring.

intelligent I/O bus
In all early PCs the CPU was responsible for transferring data between main memory and the peripherals, in addition to performing its more important computational activities. However with intelligent I/O buses (EISA and MCA) these transfer processes are performed within the bus architecture itself. This frees the CPU for more important tasks.

intelligent networks
In telephony, this term is applied to special computerized database facilities which are checked by the switch whenever a call-request is received. Adding intelligence to the network, is the way of the future for telephone companies.

The database holds information which allows call diversions, call-waiting signals, and a range of other special functions. See CLASS. The term IN has now been raised a stage and called 'Advanced Intelligent Networks' (AIN) but it is hard to determine the division between the two.

Intelligent networks need powerful centralized computers, large databases and high-speed CCS#7 packet networks. Some simple CLASS functions (called LASS) were possible without CCS#7 signaling, but not much.

- IN allows the carriers to add sophisticated new services for a fraction of the cost (and time) that it now takes, because most of these will depend only on software upgrades. The telcos see this as their financial future since the carriage functions are fast becoming basic 'pipeline' services. And, with IN, these new services are implemented at a single point rather than needing to load and maintain databases at hundreds of switches around the country.

• IN is relatively vague. It means something more than was possible with basic stored-program controlled switches, but something less than we will have tomorrow. CLID and Call-forwarding are IN functions.

• AIN involves concepts such as the Universal Personal Telephone (UPT) number, portable numbers, PCN, worldwide call-diversion, credit-card charging, enhanced information services, virtual private networking etc., and it exists as an overlay on both ISDN and POTS services.

Between IN and AIN are some services which could be categorized either way. These give the subscriber access to a range of call features currently only available on highly sophisticated private networks: closed-user groups, Centrex, ACD, call routing, origin-dependent routing, and virtual private networking, to name but a few.

The architecture of the intelligent components of a telephone network are now relatively standardized. The main components are:

— SCP service control points.
— SSP service switching points.
— SMS service management system.
— INAP the intelligent network application protocol.

These all have network connections which overlay a standard CSS#7 signaling system. The ITU-T has released a set of recommendations known as IN Capability Set 1 (CS1) giving a framework for the introduction of these services. See 800 service, SIM and AIN.

• There are two major trends in IN developments at the moment:

— ITU and ETSI-based research projects are defining an IN architecture (and service logic distribution) for the integration of intelligence in broadband and mobile networks.

— Various groups, such as the TINA consortium, RACE, and ACTS program, are creating a migration path that will extend the development of IN. There's also a move towards a more open platforms for functional entities like SSP, SCP, and SDF.

intelligent player

A videodisk or CD player which has extra built-in computer control features (memory and intelligence) for interactive educational/training use. The term is also applied to a dumb videodisk player under the control of a personal computer.

CD-i uses the intelligent player concept by incorporating both a processor and 1MB of RAM within the box, while CD-ROM is simply a dumb peripheral to a computer system.

intelligent receivers

• In television: See smart receivers.

• In paging: See agile receiver.

intelligent TDMs

Time-division multiplexers which dynamically allocate bandwidth according to usage requirements. Dumb TDMs (the most common kind) will continue to provide channel time-slots even when the channels are idle. Intelligent TDMs optimize the available bandwidth. If they do this dynamically they are better called statistical multiplexers. See.

intelligent terminal

Data entry terminals which provide variable amounts of local computing power and usually have their own CPU and memory. The term is vague (and has changed over the years), but an intelligent terminal will at least allow the insertion and deletion of lines and characters — even if it isn't a full PC.

You can't use high-level control functions end-to-end unless the terminal is intelligent enough to handle the error correction and cursor controls. Fortunately, these days most end terminals are PCs or workstations running client-server or emulation software.

Some people don't make a distinction between terminals and peripherals, especially in the mainframe literature. So an intelligent terminal is likely to be an intelligent peripheral which may be a laserprinter with its own buffer/spooler, or a modem with V.42bis compression.

Intelsat

International Telecommunications Satellite consortium. A 25 year old consortium of the major PTTs which was set up to run international communications satellites in 1964. Originally Intelsat had 11 members and it now has 170 'signatories'.

This is a private club of carriers with a UN-type charter. A country must be a signatory to the charter before a carrier can join, and the dominant carrier is usually the local representative, and it therefore controls access and sets the local fees. This avoids price-cutting through competition.

Intelsat has flown more than 30 satellites. It currently (1995) has 14 in orbit (some are geriatrics in inclined orbit, however). You can rent a transponder from them for about $1 million a year, or you can rent circuits through your dominant carrier.

Intelsat communications satellites are mainly used for fixed-base communications and require large satellite dishes.

Don't confuse Intelsat with Inmarsat. See Inmarsat also.

• Comsat is the US representative of Intelsat.

Intelsat Business Services

In the international communications satellite business, these are private (non-switching) commercial digital services which include voice, high and low speed data, telex, facsimile and videoconferencing offered over the Intelsat satellites. IBSs use small and medium-sized earth stations which can be located on or near the end-user premises.

• The cost of leasing an Intelsat voice-grade circuit (any distance, one hop) in 1965 was $US2667 per month, but by 1993 it was down to $US390 a month, a drop in nominal terms of 85%.

With competition now coming from PanAmSat, it is probably already much lower in many parts of the world.

Intelsat satellites

These satellites have played an important part in international communications since 1965 over all oceans. They are owned by essentially the same group which own the submarine cables, which is why the cost of circuits is comparable. Recently, however, a couple of private companies (mainly PanAmSat) have challenged their dominance.

— **1965:** Intelsat I (aka 'Early Bird') had 50MHz of bandwidth, and could carry 240 telephone circuits *or* 1 TV signal. It weighted 36kg.

— **1967:** Intelsat II had 130MHz of bandwidth, and could carry 240 voice circuits and 1 TV signal. This was the first GEO.

— **1969:** Intelsat III had 300MHz of bandwidth, and could carry 1200 voice circuits and 4 TV signals.

— **1971:** Intelsat IV had 430MHz of bandwidth, and could carry 4000 voice circuits and 2 TV signals. This was a series of eight satellites launched between Jan 1971 and May 1975 for all ocean regions. Each weighed 722kg.

— **1975:** Intelsat IV-A had 720MHz of bandwidth, and could carry 6000 voice circuits and 2 TV signals. This series was launched from Sep 1975. Weight was 826 kg. With the right terminal equipment (16Kb/s ADPCM) these satellites could handle 33,000 telephone channels overall.

— **1981:** Intelsat V had 2144MHz of bandwidth, and could carry 12,000—15,000 voice circuits and 2 TV signals.

— **1985:** Intelsat V-A had 2250MHz of bandwidth, and could carry 15,000 voice circuits and 2 TV signals.

— **1989:** Intelsat VI had 3300MHz of bandwidth, and could carry 24,000 PCM voice circuits and 3 TV signals. With compression and interpolation this was actually 120,000 voice circuits overall — which made it even larger in capacity than VII. They have 48 transponders; 10 in the Ku-band and 38 in the C-band.

— **1992:** Intelsat VII: This series is higher and power for low-cost earth stations, but lower in capacity. It can carry 18,000 two-way PCM circuits and 3 TV channels. With DCME and compression, telephone capacity can be boosted to 90,000 circuits.

inter-applications communications

IAC refers to the process by which the operating system in a multi-tasking computer allows the exchange of information between applications. Cut and paste techniques are an early form of an IAC. The new IAC protocols in the Macintosh's System 7.0 (called Apple Events) aims to support application-to-application ('smart clipboard') interprocess links at a number of levels, and the next versions will be much more dynamic. Microsoft calls its version of an IAB, a DDE. See also OLE and hot-links.

inter-exchange

- Between exchanges.

- Inter-exchange carriers (a misnomer; they should be called inter-LATA carriers) are known as IXCs in the USA. See.

inter-frame coding

This refers to the technique used in video compression codecs, where the image frame next-in-line to be transmitted, is compared with the last (or with some previous 'reference' frame). Only the 'difference' between the two images is then coded and transmitted. See differential prediction.

As the text-books put it, these codecs 'remove redundant information created by the repetitive nature of movie film/-video images'. The distinction is with intra-frame coding which compresses by comparing repeated elements within the same image frame. Most of the compression in video systems comes from removing inter-frame redundancy.

However, when the video image stream is a series of 'difference' values (not full images) then the stream can't be easily edited. Feature films need to be cut to an accuracy of one-frame, and this isn't possible unless each frame is self-contained. For this reason Video JPEG is used for editing video, rather than MPEG2; JPEG doesn't apply inter-frame coding.

See DCT, JPEG, MPEG, and motion compensation also.

inter-layer interface

In the OSI 7-layer model and in other similar complex layered models of communications protocol structures, the aim has been to modularize the various functions. When this is done, changes can be made progressively and in a modular fashion to allow for different requirements of different systems, and to accommodate new developments.

The major function categories of protocol are therefore treated as a hierarchy of 'layers', with the boundaries between each layer then requiring an interface specification so the two faces match. If the rules are strictly adhered to, a layer can then be removed and replace by another without disrupting the remainder of the stack, provided the two inter-layer interface specifications (above and below) remain the same.

Inter-Personal Messaging

In the terminology of electronic message services, IPM now refers specifically to the central e-mail 'memo-sending' function. It distinguishes the human-to-human e-mail/IPM functions from the machine-to-machine functions such as EDI, EFT, facsimile, and telex, which are now part of any Message Handling Service (HMHS). See X.400.

inter-process communications

IPCs are sophisticated internal techniques used in multitasking operating systems like OS/2. They ensure data integrity, and handle the shared memory, semaphores, pipes, queues and the Dynamic Data Exchange (DDE) functions. See IAC also.

interactive

- The use of computers, terminals and/or computer-based equipment where user and machine carry on some form of dialog — usually in the nature of a Q&A session. The distinction is with file-transfer (essentially one-way) and batch processing systems. Interactive is a more general term than transaction —

but both suggest that the operation happens in real-time.

• In the data superhighway context, interactive services are often seen as two-way applications on existing networks (Cable TV used for telephony, for requesting VOD or PPV movies, for interactive games, home-shopping, etc.). The distinction is being made here only with the old one-way broadcasting of television programs.

• The term 'centralized interactive services' is a new euphemism for highly asymmetrical interactivity; you send brief requests or acknowledgments back to a centralized computer, and it sends broadband streams of video to you. This can be done on a network with most of the bandwidth devoted to downstream delivery — or, indeed, all of it: the requests may travel upstream by telephone. However, most people wouldn't then accept that this was 'interactive' in any meaningful way.

• An alternate view of the interactivity on the data superhighway is that it means providing fully distributed, symmetrical, two-way links worldwide. These links will allow anything up to full-motion video images to flow both ways — so everyone can be a service provider. In the short term, this is probably blue-skies, but eventually with ATM it may be feasible.

Lower quality videophones and videoconferencing will be interactive and symmetrical, and layered video architectures are intended to allow them to be both interactive and asymmetrical — but at the receiver's choice.

• We think of interactivity as something associated with broadband, but telephones are our primary communications tool and they are interactive. Interactivity is also the reason why keyboard forums have been so successful on the Internet and other text-based systems — but this does not necessarily mean that video interactivity will prove to be just as popular. In all forms of communication, there are often many advantages to reducing the amount of information transferred or revealed.

• Note that interactive does not necessarily mean that both forward and return channels must necessarily be over the same network — just that the two channels must exist. Video may come to you via terrestrial microwave, MDS, or GEO satellite, but your request back to the head-end could go via the telephone line, a cellular phone link, or even via a LEO satellite.

interactive languages

Basic and Forth are 'threaded interpreted languages' which means they are highly interactive. You can use them to instruct a computer direct — on a line-by-line basis — and watch the response, without needing to write the whole of the code first. The original Basic was interactive when written without line numbers, and an interpreted language when used with line numbers. SQL is another type of interactive language.

interactive media

• Information storage systems where the user exchanges information with the machine in such a way as to systematically and progressively choose a unique path through the information modules. This could be a hypertext or hypermedia system, but the term is usually applied to branched programming of the kind used in video kiosks for information, for programmed learning, and of course for videogames.

The suggestion is that the user is in control of selecting the pathway, but this is often illusory. Often pathways diverge, but then converge once again to keep the number of branching possibilities to a minimum.

The comparison is with linear media structures where the path is fixed. Interactive media structures usually branch in a hierarchical way, but they can also be meshed or randomly linked as a form of hypertext. See branch and hypertext.

Some interactive media is undefined, so the possibilities are infinite because the pathways are being generated by pseudo-random factors within the program. Games often have elements of this.

• The term interactive media was used for branched video programs on tape and the early laserdisks, many years before interactive 'multimedia' became popular.

interactive TV

This applies to dozens of different techniques at many levels of sophistication. Some depend on the telephone line for the return path, and others will require wide-band connections over the new Hybrid Fiber Coaxial cable system.

The main systems in current use are:

— Videoway. A 6 year old system developed in Canada and used in Montreal (also in US under the name ActivCard). BSkyB are considering it in the UK. The new version has a smartcard.

— TV Answer. Also called EON. This is a US wireless system.

— VEIL (Video Encoded Invisible Light). Used successfully in a couple of countries, and still under development.

The term does not only mean games playing. It includes the ability to control video-on-demand servers (stop the movie when every you want), home shopping selection, and virtual reality feedback.

It also includes selection from multiple choice signals — for instance, instead of Near VOD broadcasting with multiple channels for one movie, the station transmits multiple image streams from cameras at a football game. You choose yourself, which to watch. Your choice could be the output of two or more cameras, and perhaps of a stop-frame recording disks. With digital video there are likely to be hundreds of channels which can be used in this way.

This is not a futuristic idea, it is already in operation in a couple of places around the world.

• With HFC and 1GHz of cable bandwidth, a cable can theoretically carry 150 analog or about 500 digital channels. So the multiple channels needed for delivery of multiple images from a local head-end should not be a problem. This is already common for near video on demand (NVOD) movies at nights, and with sports, it is likely to be just as popular.

• Ultra wide-screen signals (not screens) are also being trialed in the laboratories. Normally you will only see part of the digital wide-screen picture, but you can control the pan and scan of your image across the larger image frame. With layered video architectures, you may be able to select more than one sub-image from a single wide-screen image.

• The term also applies to Video on Demand.

interactive videotex

The term videotex is being used here in its widest sense, as a generic term covering broadcast data systems (teletext) as well as wireline (videotext). The term 'interactive' here just emphasizes that the normal telephone-connected wireline videotext (with a 't') service is being discussed, as distinct from the non-interactive broadcast teletext.

interconnect

In a competitive telecommunications environment, and where local carriers and long-distance carriers interact, the technical problems of interconnection are largely solved by ITU and national standards.

However, the billing problem is still a matter for individual negotiation in most cases; the long-distance carrier will need to pay some of its income to the local carrier for use of the local interconnection services.

Any interconnect agreement in a telephone system will be either line-side interconnect (before the switch), or trunk-side (after the switch).

interconnectivity

A catch-all buzzword which holds as its ideal, that all information processing equipment in an organization should be able to communicate without needing special translation equipment, gateways, etc. In their ideal world, all should be able to share data and resources transparently. It shouldn't matter where the information resides, what type of machine (Mac, PC or mainframe) it is on, or in what format (text file, database, graphics, ASCII or EBCDIC) it will be found.

Unfortunately the world isn't like this, so interconnectivity is a constant problem. Usually it relies on intermediate systems and protocol translation to handle the differences. For this reason X.25 was very highly successful because it was an intermediate standard able to perform rate adaptation. ATM will perform the same function in the future.

interface

The point of contact of any two dissimilar systems. This term is so wide that it can be (and is!) applied almost everywhere. Interface standards are either specifications (hardware interfaces) or protocols (software agreements on the logic).

• In hardware terms, an interface is any boundary shared between two components that wouldn't otherwise be able to interact with each other. The interface protocols (or specifications) define how signals are exchanged between the two. Standards often change at the interface.

• The cable linking two systems is also an interface, as are the plugs and sockets. The plugs and sockets, pinouts, and the various voltage changes on the lines, these are all interface standards or specifications.

• With computer peripherals, the term interface card is often applied to the circuit or plug-in board that passes control signals and data to the peripheral. You could also call these adaptors or extension cards.

• In routing terminology, an interface is a network connection — a port.

• In software, the term is usually applied to the functional link between two systems, or between the system and the user. The API in an operating system provides an interface (a series of 'calls') so that applications can work with the OS.

• In the OSI's 7-layer model of communications, each layer is modular and variable, but the interfaces between the layers are strictly defined. The stack can therefore always pass information up or down through the layers.

• The man–machine interface is the keyboard and screen.

• Digital communications interfaces for hardware have four characteristics:

— **Mechanical:** The physical connection of the DTE and DCE; the plugs and sockets.

— **Electrical (or photonic):** The voltages, brightness, and timing of changes.

— **Functional:** The assignment of meaning to the various interchange circuits, such as data (Tx and Rx), control (RTS, DTR etc.), timing and common-return (voltage ground).

— **Procedural:** The sequence of events that are needed.

The electrical and mechanical specifications are generally said to require 'functional compatibility'.

We should also add:

— **Software interfaces:** These are high-level specifications which define how the information will interact with the application.

— **APIs:** These allow applications and operating systems to interact.

— **Man–machine interfaces:** Rules covering I/O devices (keyboards, touch-screens); how human should act.

interlace

In television terms, 2:1 interlace is the common process of writing all odd lines (one field) onto the screen, and then all even lines (the second field). Taken together, these two fields create one frame of image.

Ideally every line would be drawn on the screen every 1/50th of a second (1/60th in USA) but this would double the width of the radio spectrum consumed by each television channel. By only writing half the number of lines each time, bandwidth is reduced, yet each area of the screen is 'refreshed' each 1/50th of a second (it would flicker at any lower rate).

So interlace is a form of analog video compression and it is not lossless — some vertical resolution is lost because interlacing lines 'wander' as they run across the screen. During the time delay between the painting of successive lines, there is a greater likelihood of time-base errors. See TBC.

Interlace TV systems usually trace half the number of lines (one field) in each mains-current cycle.

— In 50Hz PAL and SECAM television, 312.5 horizontal lines are scanned (from top-left to bottom-right; odd lines first) on the screen every 1/50th of a second. The next 312.5 lines (even lines, for a total of 625) are scanned beginning with a half-line in the middle of the screen, in the next 1/50th second.

— In 60Hz NTSC, 262.5 lines are scanned in each of two fields to create a frame of 525 lines. Actually the scan rate in NTSC is misaligned with the mains power deliberately by using a 59.97Hz scan rate. See also field dominance.

With TV, the 2:1 field interlace system was chosen in the early days as the best way to compromise on the need to reduce flicker within the limitations of bandwidth. A TV image must be able to change relatively quickly in action sequences, so TV screen phosphors must not be so long-glowing that the images 'stick'. Yet, at a refresh rate of only 1/25th of a second, quick-fading phosphors at the top of the screen would noticeably be losing image before the bottom 'painting' was finished. So the use of interlace scanning allowed them to double the rate of the scans, without increasing the bandwidth.

• In progressive scanning the lines are all drawn in sequence from top-left to bottom-right. Computer screens use progressive scanning at rates of 60—70Hz or more, both to improve the resolution and reduce the flicker.

Future digital TV systems will probably all have progressive scan, with frame-rates in the order of 100Hz. When a digital image is held in a frame-store, it can be scanned as many times as necessary, so the scanning is no longer tied to the bandwidth.

interlace factor

In television systems using 2:1 interlace there is always some reduction in the vertical image resolution as a consequence of the interlace process. This is generally taken to be about 10% of the best results of progressive-scan video systems with the same number of lines.

When attempts are made to relate scan lines to 'lines of image resolution' the number of active scan lines needs to be multiplied by about 0.6, while progressive systems have a factor of about 0.7.

This degradation is often incorrectly credited to the Kell factor but it is actually an artifact, the result of a clash between vertical 'digital' screen components (scan lines in integer increments) and vertical 'analog' image components (the actual lines of image we are attempting to reproduce). Even though this is an analog television system, the scanning produces a digital sampling artifact with image 'lines'.

interLATA

Between one Local Area Transport Area and another in the USA. LATAs service a geographical area with a number of different exchanges. The size of the LATA was chosen to represent a fair division of the old AT&T business, with some reference to state boundaries — there are 200 of them in all. New York state has six LATAs, while Wyoming state is entirely within one.

Under current regulations, telephony between LATAs must be carried by long-distance (IXC) and not by local carriers (LECs) even if the boundaries served by the LECs abut.

This is now subject to a number of regulation changes, legal challenges, and intercompany agreements. A local call in the USA is usually within a very restricted area. So even within a LATA , a call can be classed as local or toll, depending on the distance and whether it passes between two companies or not.

interleave

To weave two or more data streams together.

• With TDM telecommunications, interleaving is used to send blocks of data from the different channels over a single bearer in a strict rotational order. Short bursts of data are inserted into slots in a frame. These blocks of data are generally either one bit, or one byte long.

• On hard disks, blocks of data are interleaved to speed up the access process. See interleave ratio/factor, RLL, disk controller and sector interleave.

• On CD disks the above applies also, although the track (singular) is spiral rather than circular. By interleaving two different files (say audio and video) you can play them back simultaneously without the head needing to jump around sampling different tracks.

• In error checking systems, interleaving is a way of preventing a disk scratch or a transmission glitch from destroying whole blocks of data in a single file. By interleaving data from two files, or from different parts of the same file, the damage is spread. The wider it is spread the easier it is to correct fully, using sophisticated error correction codes. See Reed-Solomon.

• In CPU operations, cache interleaving speeds up the processor by using two banks of on-chip cache memory instead of one. The first bank can be supplying data to the CPU while the second is fetching the next batch.

• In communications or storage devices, interleaving can be between bits, bytes, blocks, or even messages. Error correction systems need bit or byte interleaving, although short blocks are acceptable. With hard disks it is almost always block-interleaving. In high-speed communications, bit-interleaving is becoming more popular with the new high-speed switching chips. ATM systems really use a form of block interleaving where the block is a 53-byte ATM cell.

interleave ratio/factor

On a hard disk, the delivery of information blocks from the read-head to the disk controller must be separated by a finite time. The disk controller needs this time to process a block before the next arrives.

If all file blocks on a disk are contiguous (in successive sectors along a track), the head would be only be able to read one block in each disk rotation because more than one would overload the controller. Blocks are therefore interleaved in a way which provides this processing time, yet keeps the flow of blocks at the highest rate possible. The head reads one block, then skips the next one or two, before reading the next.

The ratio of blocks skipped to blocks read, is the interleave ratio (it is usually 4:1 in ST-506 drives). This ratio can be changed by the user, and often this change can result in a dramatic improvement in disk access times, especially with RLL encoding. However, if the ratio is set too high or too low, disk-access slows. With a 4:1 interleave factor, the disk will read (on a single track) sectors 1, 5, 9, and 13 during the first revolution, then 2, 6, 10, 14, then 3, 7, 11, . . ., etc. on the next. It will take four revolutions of the disk to read all sectors on one track.

With a 1:1 interleave factor (possible on some ESDI drives) all sectors can be read in a single disk revolution which cuts access time to nearly a quarter. See RRL, ESDI, ST-506.

interline flicker

Also known as 'twitter' on analog television sets. This is the flickering effect often seen on horizontal lines (tennis court base lines) or at the border of objects on the screen. It is a type of flicker caused by the 2:1 interlace scan system, and is especially noticeable with modern high resolution cameras. Progressive scanning overcomes these problems.

interline spacing

See leading.

interlock

In a CPU, this is the problem caused by the processor needing to wait for the result of one operation before undertaking the next stage.

intermediate block character

A transmission control character that terminates an intermediate block of data on a communications system. It is usually sent immediately before a Block Check Character (BCC) and it is provided to allow error checking using smaller block sizes. See error checking.

intermediate code

A special form of object code which must be interpreted before it can be run.

intermediate frequencies

IF is a half-stage often used in electronics when translating very high-frequency signals down to relatively low levels (or vice versa). This is often done as a two-stage process because very high frequencies can't be carried more than a few inches in cables without problems. In the distance between the roof antenna and the set-top box tuner, the signal would degrade very substantially if an intermediate frequency translation hadn't occurred. This can be done either for one selected channel only (see LNA) or for a wide block of frequencies (see LNB).

With satellite reception, a Ku-band dish which is receiving signals, say at 12—12.5GHz, will feed the signal to a LNB converter immediately behind the feed horn. This will convert the 500MHz wide-band of received signals into a similarly wide-band, but now in the spectrum between 950 and 1450MHz. Within this intermediate band the signal can then be carried over the coaxial cable to the tuner where the individual channel will be selected.

Often there will be two IF stages with the first being in the range 960—2050MHz and a second from 300—680MHz. It is then passed to the tuner and detector stage.

- IF is generally the product of a heterodyne operation. Two signals are 'beat' together to produce the sum and difference frequencies, so there is an oscillator which produces IF radio sine-waves. These frequencies radiate out from the tuner or LNB amplifier, and can be detected. This is how the early TV licence-detection systems worked in Europe and many other parts of the world — a vehicle would drive down the street listening for the IF frequencies, and the operator would then know that a house had a TV set. They would check against their list of licence holders.
- IF stages are used in the tuners of most radio and television sets to maintain the linearity of the various amplifying stages.

intermediate system

In Internet terminology, this is a router. See IS-IS.

intermodulation

IM distortion. See distortion.

intermodulation noise

This is carrier noise caused by slight linearity problems in multiplexing systems. On voice circuits, it sounds like distant grumbling.

internal key

In video, this is a 'keying', 'matte' or 'mask' technique used in program production to insert one moving or still image into selected parts of another. With internal key inserts, you select a certain tone as a cut-off point, and tones below that point will be replaced by the second picture. Subjectively, this appears as if the darker tones are made transparent: you can see through them to another moving image behind. Anything lighter in tone than the cut-off will remain totally opaque (normal).

This effect is used to insert a figure or object into another scene, or to produce unusual optical effects.

Technically, internal keying requires a fast-acting switch which takes the scan-feed from one of two available sources (usually two cameras). It switches instantly on the basis of the voltage level of the internal key source (which is always one of the two pictures involved) and it can switch in the time the scan takes to pass between adjacent pixels on the screen.

The distinction is with external keys (which switch between two pictures in the same way, but use a third image to drive the switch) and chroma-key (which switches on the basis of color, rather than tone). Linear or luminance keys, are either internal or external keys. See external key also.

International Alphabet

See ASCII No.2 and ASCII No.5.

international fanfold

A continuous stationary size which measures 305 x 210mm (12.25 x 8.5in). There is a difference of about one centimetre between the US fanfold and international fanfold, and this discrepancy is responsible for more people throwing perfectly good printers out the window, than any other cause. This is why you constantly get 'creep' on a fanfold printer.

International Standards Organization

See ISO.

international units

Systeme International units. See SI units.

internet

There is both a general (lower-case i) and a specific (upper-case I) use of the term. See Internet for 'The Internet'.

An internet is a linking of local area networks (LANs) into a larger network. The components are joined by switches, bridges, routers and/or gateways (usually routers). We would generally now call this a WAN (Wide-Area Network) if it were in one company or organization, or perhaps, an 'internetwork' if it were shared. It is an interconnecting network of networks (where 'network' is defined by segments using one protocol and having a single network address).

Internet (The)

Since it doesn't exist, I don't know why we capitalize it — but then we go even further and give it a pretentious 'The'. 'The Internet' has no real governing body, no real shape, and almost no rules. It is an internetwork of networks — an anarchistic organization of independent host computers and their international connections.

- In the USA, 'The Internet' is the name sometimes given to a conglomeration of interconnecting American research and university backbones, together with some large-scale regional and site-data networks in the research/government area. This is now more correctly called 'The Matrix'. The real Internet is now very much a global phenomena — and it usually now loses the pretentious 'The' prefix.
- The main Internet backbone consists of The Matrix, MCI Mail, Bitnet, Fidonet, etc. and thousands of others. You can argue endlessly about whether CompuServe, Prodigy and other commercial services are part of the Internet.
- Internet connections are global, and the USA networks are now only part of the whole (although an important part). Worldwide it comprised more than 60,000 networks in 150 countries (late 1994 figures) and probably has more than 2 million permanently-connected computers.
- No one really knows how quickly Internet usage is growing or how far it has expanded each year. 1995 figures suggest that another million hosts were connected in 1994, and that there were over 30 million individual users in late 1994 — but another claim is that there are only 20 million regular users in mid 1995. However, 1995 figures are no more likely to be true than 1994 figures. What constitutes a 'user'? How do we incorporate and measure the LAN users accessing through a server?
- The best definition is that the Internet is a common language, and that language is TCP/IP (mainly IP), and the usage of the language is spelled out in many RFCs. TCP/IP was implemented at the University of Berkeley along with BSD Unix, and this became the reference implementation.

Today the Internet rests more on specialized routers than general-purpose computers, but the major Web servers and FTP sites (file repositories) are still usually Unix machines.

- Another definition is that the Internet is 'ham radio for the computer generation'. It certainly has many of the same characteristics including international free exchange of ideas, lack of regulation or organization, etc. It can also be seen as a renaissance of the humble written letter.
- The Internet has no relationship with such popular concepts such as 'broadband' or 'information superhighway' — except, perhaps, as a model for how anarchistic networks can thrive and foster. The Internet today is essentially a narrowband, text-based, e-mail network, and it gets by with surprisingly little bandwidth. Obviously more is better in bandwidth when so many people are sharing, and highly compressed voice would be nice, but the provision of broadband video services isn't seen as a priority.
- The Internet provides a number of different defined 'services':

— **E-mail** is the most basic. Store-and-forward access is provided by many organizations which have hosts connected directly to the Internet, or via other private mail services with a gateway to the Internet.

— **Newsgroups and News:** Messages sent to a newsgroup are re-broadcast to those who have identified their interest in obtaining news about these subjects. The information is also stored for further reference. This is a realtime interactive exchange. You 'belong' to a newsgroup, and you obtain the information via e-mail.

— **Gopher** is a menu-based access service to help you navigate around the world and access databases with information. You may need to use key-word searching. You will requires specific gopher software for your host. Archie and Veronica are similar navigation tools.

— **World Wide Web and Mosaic:** WWW is a multimedia capability for graphics sound and video. Mosaic is the software, and is available for Windows, Mac and Unix machines.

— **Hyper-G and Harmony:** This is a series of extended capabilities beyond WWW. They have hypermedia links. Harmony software is available for Windows, Mac and Unix machines.

— **Usenets** are circulating forums, where messages may be posted to several thousand news groups. Usenet is simply an organised collection of newsgroups that is distributed around the various networks; it is not really a network itself.

— **Internet Relay Chat** is a real-time chat section. See IRC.

• Internet addresses are IP addresses. See.

Internet Protocol

IP is the main communications standards now used for datagrams in wide-area networks. It is a Network layer (Layer 3) protocol closely related to TCP, and it governs the way packets are segmented and forwarded in a connectionless way across WANs. Its job is to connect two or more networks into an internetwork.

IP software keeps track of the node's internet addresses, and it handles the routing of the outgoing messages and recognizes incoming messages. It is similar to the ISO CLNP standard (see) with which it is merging. Currently you will find it documented in RFC 791. IP was originally used on Unix but it now has much wider applications. See IP address and ULP also.

• IP version 6 is the proposed convergence of IP protocols and CLNP. The address has now been extended to 16 Bytes.

internetworking

Internetworking refers to a specialized telecommunications industry sector which has grown up around solving the problem of interconnection and interoperability of different equipment and protocols on a national and global scale. See VPN.

interoffice

An American term for between telephone exchanges. Interoffice signals are those transferred between exchanges when setting up or clearing a call. See CCS#7.

• An interoffice trunk is a junction circuit linking two local exchanges.

interoperability

The current buzz word for a Utopian ideal. It depends on the ability of the computers, networks and applications to exchange and use data. 'Transport independence' is another similar term.

Interoperability is the ability of one company's computer equipment to operate in parallel with, and exchange information with another dissimilar computer over a network.

The current most important issues in the quest for interoperability are:

— The OSI standards for open systems. (but lately TCP/IP).
— The ability to translate management and control standards.
— Remote Procedure Calls (RPCs).
— Interprocess communications (IAC and DDE).
— E-mail. Specifically X.400 and X.500 and Internet standards.

Recent emphasis has been on using Internet protocols rather than OSI. The key to interoperability in future heterogenous networks is possibly DCE or ONC. See.

interpolated frames

In video compression schemes such as MPEG2, inter-frame coding is the process of calculating the difference between the current and the past frame/s. However, in order to take compression losses into account, the compressed reference frame is decompressed and used to 'predict' what the next frame is like (it assumes that the frame will be identical). The second frame is then compared with this prediction, and the differences are calculated and coded. This process is then repeated using the second frame, predicting the third, and so on.

However, to prevent cascading problems if an error occurs, full-images that have undergone intra-frame coding only (no predictive process) are regularly inserted into the video stream.

Between these regular insertions of compressed full frame, the predictive 'difference' process is used. These are the 'interpolative frames'.

interpolator

In ADC and oversampling — see decimator.

interpretive language

A high-order language like Basic which can be run directly through an interpreter. It doesn't need to be first compiled. However interpretive language programs are slow when compared to a compiled program. These days it is often possible to write and trial a program in the interpretive form of the language, and then compile it later. You can now do this in Basic.

interpreter

Any program written in a high-level language must be converted to machine level micro-instructions before it will run on a computer.

Compilers do this permanently by creating object code in a single 9once-only) operation, while interpreters (often held in ROM) make the translation on a line-by-line basis each time the high-level program is run.

The advantage of the interpretive approach, is that during the programming stages, an interpreted program can be test-run, debugged, changed, and run again without problems. This is time consuming with a program that needs to be compiled.

However once the programming has been completed the interpreter makes program execution slow, which is why compilation is used for more permanent conversion.

In compilation, the translation and checking process is quite distinct from the execution, while in interpretation, the translation and execution is effectively combined. Commercial software is usually compiled to create fast-running object code, although some new high level languages allow the best of both approaches — you can use an interpreter during the code-writing stages, and then compile the final product. See C-language.

interrupt

To break the flow of the system. These signals tells the processor to stop what it is doing and await further instructions.

The interrupt 'flag' is usually checked by the microprocessor during every instruction cycle, and if any device has asserted (activated) the interrupt line (sent an IRQ), the microprocessor will suspend its current processing and branch to a specified memory address (specified by an interrupt table) to check for instructions.

Generally, the cause of an interrupt is an adaptor card or a device controller notifying the operation system that something has happened in their neck of the woods (perhaps a key has been struck, or data has arrived at a communications port). The identity of the interrupt (the IRQ's number) then tells the processor to pick up a command byte from a location corresponding to that port (or rather to that IRQ#).

Many of the older adaptor cards used a character-level interrupt across the control bus, but some newer adaptors will use a message-level interrupt through the DMA channel.

Not all computers use interrupts in the same way. The Macintosh, for instance, checks constantly to see if a key has been struck, while the IBM-type machines only check for a keystroke when told to do so by the IRQ.

• Interrupts are often used in operating systems to change the order of execution of tasks. Most PC DOS and BIOS routines are called through interrupts rather than through ordinary subroutine instructions.

In the PC they are particularly clumsy because I/O boards can't share an interrupt line (unless only one is active at a time). So multiple IRQs must be established for different uses. However this does then save the processor from constantly needing to check memory locations to see if information has been added.

• There are two main types of interrupt:

— Software interrupts are generated by the program executing a machine instruction and they act through the operating system.

— Hardware interrupts work in exactly the same way, but they are generated by hardware events from outside the microprocessor.

• Interrupt types are numbered both according to a standard which is set within the operating system, and also by the hardware manufacturer's own standards. IBM defined the interrupts for the PC, while Microsoft defined those for MS-DOS. Windows has its own interrupts, and so it intercepts and checks these signals before passing them on to the underlying DOS and hardware. See IRQ# also.

interrupt driven

A form of computer architecture use by IBM PCs and compatibles. Any device (keyboard, disk drives, communications ports) that has a task to perform must first gain the CPU's attention by transmitting an interrupt.

The distinction is with the Mac-type architecture where the CPU checks for input constantly.

intersymbol interference

ISI. All twisted pair cables show non-linear responses to low frequencies (usually, below 20kHz), and this introduces intersymbol interference.

In digital transmissions, ISI is the distortion of one pulse by an earlier pulse: the line has not settled to a zero state before the next bit arrived.

Digital data usually travels on network wiring as a series of DC analog pulses (each is a bit). When viewed on an oscilloscope, the pulse will have a sharp rise time, but a slow decay (the 'tail'). This is the nature of copper-wire transmissions, so one symbol (pulse) can interfere with another if they are too close together. Distance is also a factor.

• ISDN has significant problems with ISI because of the high rate on unshielded twisted pair.

interworking

This doesn't imply that two devices, or applications, should just work together — but that they should work together efficiently. Interworking is not just an issue of compatibility but of 'productive compatibility'.

intra-frame coding

Motion video compression involves both intra-frame and inter-frame coding. Still image compression involves only intra-frame coding since, with one frame only, there can be no inter-frame prediction.

There are various techniques used for intra-frame compression, but generally the image is broken up into blocks (say 16 x 16 pixels) and the blocks are processed independently. Sometimes block sizes are variable but generally they are fixed: DCT coding is generally used within the blocks. Both still (JPEG) and movie (MPEG) systems use intraframe coding.

The output of intra-frame compression will possibly also be run through lossless RLE and/or Huffman's coding systems to further reduce the bit-rate.

• In video-conferencing codecs, the term is often applied specifically to a special high-resolution mode which is used with a graphics camera to send detailed drawings or maps. Often in video conferences it is desirable to be able to send these high-quality graphics images (architectural plans, for instance,) and, since they are static, they will be processed in a different way.

• When a video-conferencing or general video compression codec decides that there is a substantial difference between two successive frames of image, it will switch from its normal inter-frame mode pattern, to coding the image using intra-frame coding entirely.

• This break in the normal pattern also occurs at the junction between two scenes (a cut) in motion video. Following a cut, and every half second or so, it is important to have one full-frame image which is totally self-contained, and therefore has intra-frame coding only: no prediction. This is done regularly in the MPEG standards. See JPEG.

intraLATA

In the USA, any telephone call made within a LATA may be carried by local companies only. LATAs define the boundaries across which a local carrier (a LEC) must pass the call only through an interexchange carrier (an IXC). There can be more than one carrier operating within a LATA area, however.

Local calls are always within a LATA, however this does not necessarily mean that all calls within a LATA are classed as local calls. Some intraLATA calls will be local, some will be intraLATA toll calls; some will remain within the bounds of one carrier, and some will be passed between two carriers within the same LATA. Different charges apply to each, and different dialing patterns may be necessary between carriers.

intranode addressing

The LAN equivalent of a local call. It is a transfer of data from one station to another station attached to the same node.

inverse multiplexing

The function of aggregating channels and providing the essential synchronization to allow the multiple channels to act as one. ISDN's B-channels (each 64kbit/s) are often aggregated into wider channels for videoconferencing by using an inverse multiplexer (which may also be called an access controller). See BONDing.

inverse video

Where black becomes white and white becomes black on the screen. Used for highlighting. Aka reverse video.

inversion

To reverse something — to turn it upside down or change left for right. In binary, this is changing all zeros to ones and all ones to zeros. See negation.

- In filtering, it is the process of constructing a reverse-match of the direct signal (the first to arrive) and using this to counter any multipath signals or delayed echos.
- In echo control on two-way circuits it is the process of constructing a reverse-match of the outgoing signal, holding it in delay signals, then applying it to any incoming signals to negate the echo.
- See convolution.

inverted backbone

A network architecture where a concentration of hubs or routers becomes the center of the network, with all of the sub-network linked back to this concentration. It is a collapsed backbone architecture — as distinct from using an elongated campus backbone, and having all sub-nets linking into this.

This may appear physically as a star network, but it may still operate logically as bus or ring.

inverted files/indexes

An alphabetically ordered list which consists of every word (except stopwords) in the record (sometimes restricted to titles or specific fields), together with the record location. This is a way of assuring that every word of any importance (not just selected keywords) in the original records of a database are searchable in the index. Inverted files are naturally very large indexes (often almost as big as the original). See file inversion.

inverter

- **Logic:** A NOT gate is an inverter.
- **Power supplies:** These are driven by DC low-voltage current from batteries or accumulators, and create high-voltage AC current. They are often used to run mains powered electronic equipment out in the field where car batteries or solar cells are the only source.

Some inverters produce a square-wave 50 Hz 240V (or 60Hz 120V) output which is not suitable for many applications where the load is highly reactive. Microcomputers generally don't like this type of supply; it may need to be conditioned first.

- Resonant inverters modify a square wave through a resonant tank load circuit which has a frequency identical to that of the mains. This produces a stable sine-wave output. However, these devices can be noisy.
- PWM (pulse width modulation) inverters produce a sine-wave output, and these are used in many modern UPSs.
- Linear Amplifiers can provide good stable sine-wave power supplies up to about 1kVA. The technology is basically the same as a high-powered audio amplifier. See UPS also.

invocation

To 'call' (command the use of) a subroutine or a special procedure.

Inwatts

The US term for the free-phone telephone service provided by the telco for retailers, large companies, etc. that lets you ring them from any country area or state, at a cost to you of only a local call. This term seems to be falling out of use. These are now known as 1-800 numbers (or just 800 numbers) and a very large number of US private citizens have them because they provide a single-number/nation-wide link which doesn't need to be changed if you shift residence. See 800 service in numbers section, and also 500.

I/O channel memory

The IBM term for additional memory on an adaptor or expansion card. The distinction is with 'planar' memory which is the memory on the motherboard.

I/O device

Input-Output device. Any device which bridges across the gap of the man–machine or machine–machine interface. The classic I/O devices for PCs are keyboards, screens, disk drives, modems and printers.

You should also remember the less usual I/O devices like touch-screens, handwriting PDA screens, drafting tablets, audio microphones, telephone DTMF devices, etc.

From the viewpoint of the CPU, the problems with any I/O device is that it may demand attention at any time, often while the CPU is busy processing other data. For this reason, the core-system must be designed to look at various I/O port locations (memory locations into which I/O data flows) regularly, which is a bit of a time-waster. Alternately I/O devices will use an interrupt technique which signals to the CPU that something is happening, requiring its attention. See IRQ# and BIOS.

I/O processors

These are not coprocessors. An I/O processor is a complete CPU which executes input/output routines alongside the main processor. The NeXT machine uses both a DSP and a customized I/O processor (called an ICP) to speed up disk operations. Se DMA.

IOC

Independent Operating Company. In the US this means telephone companies (LECs) which aren't tied to the seven RBOCs — GTE, ITN, Contel and United are all IOCs.

ionosphere

Multiple layers of the upper atmosphere (above the stratosphere and troposphere), between 50 and 1000 kilometers above the Earth, that is ionized by ultraviolet radiation and high energy particles from the Sun. It is usually considered to begin at the Heaviside layer (at about 100km). The height and thickness of the ionosphere varies from day to day and from day to night. It is affected by the pressure of sunlight.

The layers acts like a reflector to some radio frequencies, and sometimes they also act like a wave-guide. Either way this carries signals over long distances apparently around the curve of the earth (although signals always travel in straight lines). Sunspot activity affects this layer.

The ionosphere reflects 'sky-waves'. This is the term applied to high-frequency (HF) 'short-wave' radio (3 to 30 MHz) signals and also medium-frequency (MF) AM broadcast signals which can often skip long distances when conditions are right. However sky-waves are subject to fading, distortion and periodic blackouts. Contrast with ground waves. See D Region, E Region, F Region, Sporadic-E.

• Note that LEOs generally orbit within the ionosphere, whereas the signals to and from a GEO need to pass through the ionosphere.

At an angle, radio waves incidental to the ionosphere will either refract, reflect, or be channeled as will light passing though variable density glass. The low frequency (3.5–7MHz) radio waves reflect before high and are more reliable for long-distance communications.

The lowest useful layer of the ionosphere at night is the E layer (or region); in the daytime there is an even lower D region (highest at noon). Above the E layer is the F layer which has the best reflection characteristics just before sunrise.

• The ionosphere's reflective properties depend on geographical position, time of day, season, transmission frequency and the periodic and capricious nature of the Sun (spots). This is why many alternative frequencies are needed for short-wave broadcasting.

IOR

Indian Ocean (operating) Region — for satellites. See POR and AOR.

IOSS

Integrated Office Switching System — extended PBXs used for voice and data.

IP

• **Internet Protocol.** See.

• **Information Provider.** See.

• **Integrated Processor** — new chip designs. See.

• **Intelligent Peripheral** — in discussions about Intelligent Network design. See intelligent networks and intelligent terminal also.

• **Interframe Prediction** in MPEG2. See MPEG2.

• **Intellectual Property:** Patents, Copyrights, etc.

IP address

A form of packet address used for wide-area interconnection on the Internet — as distinct from the Ethernet address. The IP address is actually a 32-bit (4-byte) binary number, but it is most commonly represented by four decimal numbers joined by periods, such as 143.31.345.672

In practical use, these numbers are replaced by character strings and, within any domain, there is a Domain Name Service (DNS) which is responsible for making the translation. Names are not assigned arbitrarily, and each address is filed with the Internet's Network Information Center (NIC).

The alpha-string e-mail address of an individual will consist of a personal name (usually their name + initial, concatenated), the @ sign, then the domain name with its organizational type and country code. But from the Internet viewpoint, it will still only be four numbers connected by periods.

• IP addresses have two parts, a network number and a host number. There are three classes of network numbers. **Class A** networks can have up to 16 million hosts; **Class B** can have 65,000 (and they are running out) and **Class C** can have a maximum of 254 hosts.

• Recently it has been decided to increase the address size from 4-bytes to 16-bytes. See CLNP. See DNS, fully qualified domain name and Internet Protocol also.

IPB

Interframe Prediction with B-frames. See MPEG2.

IPC

Inter-Process Communications. See and also multitasking.

IPI

Intelligent Peripheral Interface. A proposed device-interface standard for communications between a CPU and its peripherals (mainly disk drives). The IPI handles up to 10MByte/s rates. See SCSI, ESDI, ST506 and SASI.

IPM

Inter-Personal Messaging. See MHS and X.400 also.

IPSIT

International Procurement Standards for Information Technology. See EPHOS, GOSIP.

IPSO

IP Security Option. The part of the Internet protocol which defines the security level.

IPSS

International Packet-Switching System. A world-wide network of X.25 systems run by national PTTs.

IPX

Inter-network Protocol (Packet) eXchange. A protocol for the exchange of message packets on an internet. This is Novell's implementation of the popular XNS protocol in NetWare, and it corresponds roughly to the protocols in the Datalink and Network layers of OSI. It is most like a Network layer (Layer 3) protocol and it is very similar to IP.

IPX is NetWare's native LANs communications protocol, and it is used to move data between the servers and workstation applications running on different network nodes. It assembles the packets and provides routing control which governs packet-forwarding across internets. IPX packets can be encapsulated within Ethernet packets or Token Ring frames.

IQ

See YIQ.

IR

InfraRed. See.

IRC

Internet Relay Chat. CB radio in cyberspace; you type instead of talk. IRC is a party-line where people converse in real time. It has more than 1200 channels on-line in peak periods. There are both public and private channels. Some are constant, others are ephemeral.

This is a loose federation of about 200 interconnected computers which function in a client-server relationship. Behind the scenes it is a network of real-time servers which accept connections from client programs (one per user).

Through the use of specially-written software, they allow users to establish 'channels' in forums where people go to discuss specific topics. You create a channel with the command:

```
/join channel-name
```

The command jargon of these systems is the `? help` and `/join xxxxx.` If xxxxx doesn't exist as a channel, it will create one. Other useful IRC commands are `/nick xxxxx` to establish a nickname, and `/whois xxxxx` to find details of the other person.

- Anyone can set up a forum on IRC. See virtual community.

IRD

Integrated Receiver/Decoders. The term applies to any form of set-top box which provides both the tuning and selection system, as well as a way to unscramble programs. This is the correct name for the box on top of your TV set which allows you to watch Pay TV.

These units will become the communications controller for the future 'information superhighway', and therefore handle much more than just video. The DAVIC group is attempting to standardize the various functions.

There are five main standards-based components in the new digital boxes:

- The demodulation and channel selection functions (depending on whether it is satellite, cable, or terrestrial). See VSB and COFDM.
- The transport mechanism. Which may extract one signal from interleaved digital channels (See SCPC and MCPC).
- The decompression system (now fixed on MPEG2)
- The conditional access system, probably using a standardized box with standardized scrambling techniques, but based on smartcards holding the keys.
- The home-bus control and distribution standard. See NEST.

See DAVIC also.

IrDA

Infrared Developers Association. See Versit.

IRE units

International Radio Engineers units. The peak signal voltage of a TV signal is usually set a 1 volt. This represents 100 IRE units. However the scale goes from –40 IRE units, with zero representing the division between the image signals and the sync signals. This places 0 IRE at about 0.3 volts.

In practice, the black level is set slightly higher than zero (usually at 7.5 IRE units).

Iridium

A metallic element with 77 electrons — hence, Motorola's name for a mobile communications system that originally relied on 77 small, low earth-orbit (LEO) communications satellites in polar orbits. Unfortunately they now only have 66 satellites. (Before you ask, the element is Dysprosium!)

The satellites will have three phased-array antennas, each with 16 spot beams — a total of 48 channels. Each of these can be shared four ways using time division. This is a TDMA-TDD proposal, which will use some of the GSM chips and protocols — but it is not a GSM system.

Iridium wasn't the first Big LEO to be proposed, but it has been the most widely promoted. However it is also one of the most electronically complicated and expensive, with a projected budget of $4 billion. It is not due to fly until 1999.

The main distinction between Iridium and the other Big LEO proposals (apart from using TDMA and therefore having different frequencies) is that Iridium's satellites can switch messages between themselves without returning the signal to earth. Theoretically, Iridium can drop the calls to any ground station connected to the international telephone network — and this may be outside the territory where the calls were made. This adds complexity to the satellite and has created some political problems with local PTTs and governments. See Big LEO and Little LEO also.

• Iridium's backers include Verba, BCE, DDI, Great Wall Industries, Raytheon, Sprint, STET, and United Communications.

iris

The diaphragm in a camera lens which progressively closes to reduce the light entering the lens, or opens to increase the brightness of the image. The hole remaining at the center of the iris is known as the aperture.

Iris diaphragm ('aperture') settings are not graded by direct physical measurement of a diameter of the aperture, but by a calculation which takes into account the focal length (F) of the lens also. This is done so that the same aperture (f-stop) on any lens will provide the same density of light to the film plane, in identical circumstances.

There are actually two different aperture scales — the f-stop and the t-stop. See both.

• To take wear of the iris mechanism into account, you should always set the iris diaphragm by opening the iris to its fullest extent, then close it down to the desired aperture (f- or t-stop). Always set to the scale by a physical closure of the lens.

Irma

A plug-in 3270-emulation board produced by DCA for the IBM PC range. It has become a *de facto* standard.

IRPA

International Radiation Protection Association. See ICNIRP.

IRQ#

Interrupt ReQuest. An I/O device (keyboard, modems, mouse, scanner, etc) needs a way of signaling to the processor that it wants the CPU to pay attention — which is the purpose of IRQs and interrupts. An IRQ says to the CPU, 'stop what you are doing, and check my memory location'; the number of the IRQ tells the CPU which location to check.

Unfortunately, every bit of hardware (and a lot of software) that can be added to an IBM/clone PC demands its own IRQ# so that it can command the attention of the processor. And these systems can't share IRQs, as can other computer architectures. This is where interrupt conflicts arise.

IRQs are used in a seemingly random way by serial and parallel ports, floppy and hard drives, keyboards, coprocessors, system clocks, timers and VGA cards. Some 'standards' have arisen just by common consent. The key user-set I/O IRQs are numbered (from 1 to 16), and are usually only set when configuring external devices (most are from IRQ# 1 to 4). New devices need an unused IRQ if they are to be able to claim the attention of the processor (except if it uses a DMA channel).

Don't assume that the IRQ number coincides with the COM or LPT port number. Normally IRQ4 is used with COM1, for instance, but you'll find that IRQs in a PC are used in a seemingly random order by both the serial ports and the parallel ports, and also by the keyboard, floppy drives, hard disk, coprocessor, system clock, and video cards.

• From the AT on, there were 16 of these interrupt request channels. As a rough guide only, the ten which appear to have *de facto* standards are:

IRQ1 for the keyboard,
IRQ2 for the video card,
IRQ3 for COM2 and COM4,
IRQ4 for COM1 and COM3,
IRQ5 for LPT2,
IRQ6 for the floppy drive,
IRQ7 for LPT1,
IRQ9 identical to IRQ2,
IRQ13 for the maths coprocessor,
IRQ14 for the hard drive.

You don't need an IRQ to be reserved exclusively if you are using a DMA channel. See interrupt.

IRTF

Internet Research Task Force. This research group is governed by the IRSG. See IAB.

IS

• **Information Separators:** A category of control codes used to separate data. The IS characters in ASCII No.5 are US (IS1), RS (IS2), GS (IS3) and FS (IS4). See all.

• **Intermediate System:** OSI for router.

• **Information Services/Systems:**

— In companies, this is same as IT or DP. The services that a centralized computer processing department provides within a corporation.

— This is also a general term for anything to do with computers and communications. It is often synonymous with Informatics.

— In libraries and paper-based research divisions this term covers everything they do.

IS-41

IS-41 is an automatic roaming standard for cellular phones, but it has very little to do with digital cellular technology (it is more administration). However, it describes the way different networks should hand-off mobile calls across system boundaries when the call has been made on a digital phone (either TDMA or CDMA).

There are currently a few variations:

- IS-41-0 provides for both the hand-off between switches, and for validation before calls are made. Version 0 is incompatible with the later A and B versions.
- IS-41-A has the above plus call delivery, subscriber data transfer, roamer validation and registration facilities. It uses CCS#7 signaling.
- IS-41-B and C, both under development, will add further features.

IS-54

The North American (TIA) standard for the TDMA digital mobile telephone system, also known as NADC (North American Digital Cellular) and Digital AMPS. The later version (once called IS-54B) of this standard is now incorporated into IS-136. IS-54-C was never published, nor were the IS-7X series of standards. Instead, IS-136, IS-137, AND IS-138 were created. See NADC.

IS-88

The TIA's Narrowband AMPS standard. See N-AMPS.

IS-95

The North American standard (TIA) for CDMA digital mobile phones based on Qualcomm's technology. The associated substandards appear to be: IS-96 for the vocoder, IS-97 for the mobile phone itself, and IS-98 for the base station.

IS-99

The PCS version of Qualcomm's CDMA digital cellular phone system operating in the 1.9MHz band. Apart from the frequency change, it is essentially the same as IS-95.

IS-136

The TIA's IS-136 standard is essentially the same as TDMA IS-54 on the voice channel, but has an entirely new digital control-channel structure. The standard was published Dec 1994. See IS-54.

IS-IS

Intermediate-System-to-Intermediate-System protocol. This is a dynamic routing method used in OSI. It is similar to TCP/IP's OSPF and IBM's APPN. The 'intermediate system' here refers to nothing more complex than a router. See also ES-IS.

- Integrated IS-IS adds support for IP and other networks. The implementations of this variation results in the system sending only one set of routing updates, rather than two.
- Dual IS-IS is the same as Integrated IS-IS.

IS-PBX

Integrated Services PBX. A PBX designed for ISDN; it is digital throughout and uses D-channel signaling.

ISA

Industry Standard Architecture. A marketing term which should only really be applied to the 16-bit AT-bus developed by IBM for AT machines using the Intel 286 CPU. However it often includes the earlier 8-bit bus architecture. The term ISA was used originally to suggest that any computer not using this bus was 'not industry standard' and therefore somehow 'naturally' inferior. Rubbish!

The 8-bit bus was developed originally by IBM, and used for the original PC and XT, and its confusion with ISA arises because most AT machines had both 8- and 16-bit slots, with the smaller slot taking 'half-sized' add-in cards.

The 16-bit ISA standard was originally called the 'AT-bus' and it was an important development. With the AT-bus, for the first time in PCs, data could be transferred across a 16-bit wide path at 8MHz (the theoretical maximum is about 10Mb/s). Improvements in the ability of the bus to provide DMA access, and in the number of IRQ channels, were also made at the same time.

The theoretical transfer rate with some later versions of ISA is said to be 16Mb/s but, in practice, it is probably less than half of this. The ISA bus is now widely used in combination with the VL-bus and especially with the PCI-bus. See these, EISA and MCA also.

The fact that ISA remains the most widely used standard today, shows that 16-bit bus architectures satisfy most of the needs of PC users.

isarithmic flow control

A flow control technique which uses a system like token-passing. 'Transmit permits' must be held before a unit can transmit.

ISC

International Switching Centers. The special telephone exchanges responsible for international connections via submarine cable or satellite. Usually they will be called 'international gateways'.

ISD

International Subscriber Dialing Aka IDD. The provision of direct-dialed international services using standard prefix numbers which identify the zone and the country. National telephone switching networks must now be able to handle 15 or 16-digit numbers in order to carry these codes. See zone code and country code.

ISDB

Integrated Services Digital Broadcasting. The translation of some B-ISDN concepts to radio. It is an attempt to standardize and harmonize broadcasting, computers and telecommunications as much as possible. MPEG, with its new ability to compress multiplexed video and sound is part of this concept. The ITU has been studying ISDB since 1985, but has not made much progress.

ISDN

Integrated Services Digital Network. The decade-old world standard for a digital telephony 'access' technology which aims to merge all telephone and low-speed data into one system by providing digital techniques end-to-end from the customer premises. But since the inter-exchange network is already digital in ISDN countries, these protocols only deal with the last mile — the 'local loop'.

The first specifications were released in 1981. It is important to note that ISDN depends on digital exchange switches and on digital links between exchanges and CCS#7 packet signaling. ISDN is the local-access part and the rest is IDN.

The standard digital channels in ISDN are 64kb/s PCM voice and data channels (B- or Bearer channels) and also 16 and 64kb/s packet-switched D- or Data channels. There is a higher-order system using H-channels (see).

- ISDN is provided either as:

— **Primary Rate Access** (PRA) having 30 (or in the US system 23) B-channels and one D-signaling channel for PABXs and large companies. These are carried over E1 (2Mb/s) or T1 (1.54Mb/s) links.

— **Basic Rate Access** (BRA) has two B- and one D-channel for homes and small businesses. The total data rate (including framing and overhead) on the twisted-pair is 192kb/s.

Both ECH and TCM line-coding techniques are used for ISDN.

- The following channel types have been officially defined by the ITU-T for ISDN:

— **B** (for Bearer) channels at 64kb/s.

— **D** (for Data) channels at 16kb/s (BRA) or 64kb/s (PRA).

— **H** (aggregated) channels at 384kb/s to 135Mb/s (not always provided).

- D-channel signaling is not identical to CCS#7, but translation between the two is relatively easy. For this reason true out-of-band signaling and messaging services can easily be provided over the local loop.

The standard US transport protocol for ISDN is 2B1Q (see) although 4B3T line-coding is now standard in Europe.

- Basic rate ISDN in the US is normally carried on a single pair and it uses a 2B1Q modulation scheme with 160kb/s data rate. The US system transmits 80k quad-bits per second and uses adaptive echo cancellation at both ends to negate its own transmit signals.

- Each ISDN frame consists of 38 data bits and 10 overhead bits, with a frame-rate of 4000 a second each way.

- The ITU has defined the functions of the TA, TE and NT and four reference points in the ISDN network, the U, T, S and R. See TA, TE#, NT#, and T-interface, S-interface and R-interface.

- **ISDN2** is the term often applied in Europe to Basic Rate Access (2B+D) which can be carried over a standard twisted-pair telephone cable in the local loop.

- The UK also offers a cut-down version of ISDN called IDA. See.

- **ISDN30** is the European name for Primary Rate Access (30B+D) which can be carried over a standard 2Mb/s link (E1).

- **Euro-ISDN** is the name now applied to the trans-European standards for PRA and BRA.

- Many of the early US ISDN implementations only provided clear B-channels of 56kb/s bandwidth, which makes it impossible for the US system to interface with other global ISDN systems except through a special rate-adaptation gateway. This service is now more correctly called 'Switched 56' and is not now seen as an ISDN offering. A B-channel is now always a 'clear' 64kb/s channel with no bit-robbing.

- In the USA, standards incompatibility between implementations by the switch manufacturers has limited the success of ISDN; also for long distance calls many American intermediate data backbones are often only available in 1.54Mb/s (T1) channel dimensions because of bit-robbing. Some US carriers have also modified their CCS#7 signaling standard, leading to further incompatibilities.

- Bellcore has now devised the US National ISDN 1, 2 and 3 standards which should eventually solve the problem. A recent upgrade has resulted in National ISDN 2. The US PRI provides only 23B+D but at the standard 64kb/s channel bandwidths, it can fit into a standard T1 circuit with ESF.

- Despite the fact that ISDN is postulated on twin 64kb/s channels, variations are widely offered within the carrier network. For instance voice is often identified and carried across the network (between exchanges) at 32kb/s ADPCM in the IDN.

In the USA this is often called '3.1kHz audio' (which is also used for voice-band data and fax). End-to-end, the full data rate available on a PCM channel can be either 56 or 64kb/s with the intermediate carriers in the US (depending on the type of T1 circuit used between). The carriers sometimes provide multi-rate (i.e. n x 64kb/s) services also.

Also in the USA, some carriers offer different rates for a more rigid form of BRA service where each channel must be pre-specified either for voice or data — rather than used as the customer requires for either. See IDA also.

- The ITU standards allow X.25 data calls to be placed over ISDN (both D- and B-channels) and to use ISDN phone numbers (E.164) instead of the X.25 addresses.

ISDN modem

This is an ISDN terminal adaptor. See below.

ISDN terminal adaptor

TAs. A plug-in PC board or an external box (a digital 'modem') which allows your computer to connect with the ISDN telephone service, usually through an NT1. Plug-in adaptors can utilize the full features and speed of ISDN while most external boxes may be limited to 19.2kb/s (because of the UART).

TAs are a wide variety of devices which allow pre-ISDN equipment to exist on the digital 64kb/s B-channels and handle the D-channel signaling. For larger installations there are special TAs which will usually also handle connections to a bridge, router or PBX. The adaptor converts digital signals to the ISDN-compatible format.

TAs are both sophisticated speed converters which pad-up the data rate with null bits, and they are the ISDN equivalent of a modem. Some can handle only one channel at a time, others have two BRA links, and others can handle the full 30 channels of PRA. They also vary in the way they handle the D-channel packets. See TE# also.

ISDN-UP

ISDN User Part — another name for ISDN's 'User Data'. See User Data.

ISDNBIOS

Hayes' application program interface (API) which may become a standard way of using an ISDN terminal adaptor on an ISDN telephone network. IBM has one with the same name which supports data but not voice.

ISI

Inter-Symbol Interference. See intersymbol interference and distortion.

ISL

Inter-Satellite Link. An ITU term for space-segment links between satellites. Iridium and Orbcomm both uses these.

ISLN

Integrated Services Local Network. A high-speed LAN designed to carry data, voice, video and graphics.

ISM band

The Industrial, Scientific and Medical bands of the radio spectrum. These are three frequency bands set aside in the USA for unlicensed use (mainly for spread spectrum) under Part 15 of the FCC rules.

The bands are 902—928MHz (26MHz of bandwidth), 2.4—2.4835GHz (83.5MHz) and 5.725—5.850GHz (125MHz). The spread spectrum (CDMA) devices using these bands must emit one watt of power or less.

• Note that the lower band (902—928MHz) is generally used for cellular phones around the world.

iso-*

• **Iso-Ethernet** (aka Iso Enet, Isonet or Isochronous Ethernet) is a modified Ethernet LAN protocol which, with good voice compression and faster data rates, is now able to handle voice as well. See Isochronous Ethernet.

• **Iso-FDDI.** This means Isochronous (able to carry real-time voice and video) FDDI. Iso-FDDI is actually FDDI-II. See.

ISO

International Organization for Standardization (as they pedantically insist) or International Standards Organization in common parlance.

The IOS/ISO reversal comes about because they insist that ISO is not an acronym (even when capitalized) but they say it refers to the Greek term for the 'same' or 'equal for everyone'. The ISO is said to be a voluntary meeting of like-minded national organizations only — with no formal status. But in practice it functions like every other standards-making body.

The ISO works in conjunction with the ITU (ex CCITT and CCIR), but it is not in the United Nations hierarchy. It is a voluntary organization consisting only of national standards organizations from each member country.

In 1984 the ISO was requested by the US, France and UK to direct attention to international telecommunications standards, and since then more than 200 have been defined. These are cooperatively developed and published by three international standards-making bodies: the ISO, the International Electrotechnical Commission (IEC) and the ITU-T.

• Differentiate between the ISO (the organization) and OSI (the layered standards). The ISO defines standards in many fields, not just electronics.

• ISO defines film sensitivity ratings, for instance, however it is best known in electronic communications areas for the OSI 7-layer reference model which it developed in 1978.

The joint ISO/IEC committee responsibility for terminal-based communications and computer issues is called JTC1 (Joint Technical Committee #1).

ISO 646

See ASCII No.5 which, with some slight modifications, became the current ISO 646 standard.

ISO 9000

This is a series of international standards for quality certification The ISO developed these back in 1986, with the help of 90 other countries. They address business and manufacturing processes used to produce products in general, and they involve rigorous certification as well as regular, independent third-party audits to prove continued compliance.

The main substandards of interest in IT are 9001 (design, development and servicing), 9002 (production), and 9003 (inspection).

ISO 10646

This is the ISO proposal for a standardized code capable of handling most of the world's languages. It is a 4-Byte (32-bit) code capable of storing Japanese, Chinese, Korean, and Burmese, etc. character and ideographs. However may people feel that the allocation of four bytes for each character is excessive. See Unicode.

ISO IP

See CLNP.

ISO reference model

See OSI reference model.

ISOC

The liaison process between ISO and the Internet Society.

isochronous

Strictly, this means 'equal in time', but a better explanation is 'predictable' in the sense of needing regular delivery. When referring to telecommunications, signals are said to be isochronous 'if the time interval separating any two significant events is an integer multiple of the unit interval'. The term has also been jargonized in physics to mean 'constant wavelength'.

The reason why this term is so confusing in communications, is because it is used in two distinct and almost diametrically opposed ways.

— For data in synchronous circuits (note: not asynchronous), isochronous and asynchronous are often treated as synonymous. They are end-user controlled forms of access to a synchronous data link — not under any central control of timing. (There is another form of asynchronous transmission which does not operate on synchronous channel!)

— **For voice**, the terms isochronous and asynchronous are almost opposites. Isochronous implies constant bit-rate sampling and real-time transmission without random delays, while asynchronous suggests erratic and occasional transmission — whenever an opportunity presents.

• **ATM and Frame Relay:** For voice and video today, an isochronous communications service must be able to guarantee a certain bandwidth, and guarantee deliver of the data with very low delays (low network latency), and at regular (predictable) intervals. So isochronous traffic has a 'real time' requirement of regular delivery. It can tolerate some delay — but it can't tolerate variable delay. Voice and video are 'isochronous services' — while data and electronic mail can be stored and forwarded seconds or minutes later.

• There seems to be at least a couple of vaguely different applications of the term in telecommunications.

— An isochronous channel is a 'synchronous' channel which is capable of carrying the clock information along with the data, 'in-band'. This is a confusing use of the term because it mixes two concepts. It is best to call this a 'synchronous' channel.

— Isochronous traffic is a term often used to specify non-packetized transmission systems. This usually (but not exclusively) means that they are talking about the time-division multiplexing over a circuit-switched network.

The term is being used here to distinguish between the store-and-forward aspects of packet systems vs. the real-time needs of services like voice telephony which are time-dependent and need 'predictable' delivery. However, see ATM.

— In discussions on packet-switching, isochronous communications usually refers to equal-time interval, fixed-bandwidth, 'virtual circuits' — perhaps within a prioritized queuing arrangement for the real-time data. The term is used here to distinguish ATM cell-relay and frame relay systems from the older and slower packet technologies.

— Within ATM, isochronous is often used to distinguish the normal connection-oriented use of the system from the connectionless protocols.

• ATM, despite being a 'packet-based' network system, can provide isochronous services to voice and video because it is based on the use of short cells, high data rates, fast packet-switching and low network latencies.

Delivery of 'time-dependent' data can therefore be guaranteed within the perceptual limits of the eye and ear. Fast-packet switching technology contributes by not requiring the packets to be stored at intermediate switching nodes while packets are checked for errors.

• Isochronous packet schemes move prioritized data in a predictable manner and with a pre-determined interval between each data packet, and they will often let the non-prioritized data look after itself. To maintain these services in high rate fiber-based systems, a number of isochronous channels can share a single 'virtual TDM circuit' (time-slot) within the packet system — and these isochronous frames/packets will always have priority. DQDB handles voice this way. Each voice channel on the network may only have a few bytes within each 48-byte cell, so a dozen or so different voice calls may be interleaved into the one cell structure.

• Isochronous communications in modems is sometimes said to be 'a mix between synchronous and asynchronous data transmissions'. What they mean is that the individual characters are still framed by start and stop bits between the PC's UART and the modem, but the modem strips start and stop bits transmit the data across the telephone network between the modems in a synchronous way. If you accept this rather odd use of 'isochronous', almost all modems running at more than 2400b/s are isochronous these days. This usage seems very odd to me. See asynchronous devices also.

• Digital circuits always provide a clock of some sort along with the data (even if they are self-clocking) — there must be some clock timing to provide the sample interval for the pulses to be read. So all digital circuits are 'isochronous' in one sense, but the services provided over them may not be. Think of this as an escalator — the regularity of the steps moving up is fixed, but when you get onto the escalator it can be a user-defined casual choice.

Isochronous Ethernet

IsoEnet: A form of high-speed Ethernet which separates voice and video from data traffic and carries both on asynchronous channels using a different part of the spectrum on the same cable. This new technology was designed by National Semiconductor and is supported strongly by IBM. It is now a proposal before the IEEE's 802.9 committee.

IsoEnet provides two different data pipes: it adds 6Mbit/s to the normal 10Mb/s Ethernet bandwidth (total data rate is 16.144Mb/s) and this extra 6Mb/s is subdivided into 96 separate 64kb/s ISDN-like B-channels. These channels handle fixed-length packets at constant rates for voice, and these can be aggregated for videoconferencing. The isochronous services won't affect the normal Ethernet data services or the use of normal Ethernet NICs.

Apple has demonstrated H.320 videoconferencing calls over the IsoEnet using QuickTime Conferencing. Support for the standard came from Microsoft, IBM, AT&T and many others.

ISODE

ISO Development Environment. An implementation of the upper layers of the OSI stack on an TCP/IP protocol stack.

isoEnet

Isochronous Ethernet. See.

isometric

At equal angles or measures.

• In computer graphics an isometric projection of a cube is a pseudo-perspective drawing, where the vertical lines remain vertical, but horizontal sides are at 30° or 45° to the horizontal and appear to have equal depth on both left and right of the near-point. This is an orthographic projection, but don't confuse it with oblique projection.

isopoint

A mouse-like device for laptops. It is physically a roller-bar along the bottom of the keyboard under the space-bar. Vertical cursor movements are made by rolling a sleeve on the bar, and horizontal movements by a sideways sliding motion.

isotropic

Having properties that are the same in all directions. Non-directional. Without 'planes' (layers) or polarity.

• Isotropic materials are those which are uniform throughout.

• A light-bulb in space would radiate in an isotropic manner; it becomes anisotropic if you put a reflector behind it. This is why isotropic is chosen as the reference when measuring antenna gain — they are measuring how much better (in terms of power concentration) is the antenna's directionality, than a free-space source. See EIRP.

• Isotropic mapping of graphics. In this mode, when you change the size of objects on the screen you preserve the previous aspect ratio — the x and y axes of the graphic remain in proportion. With some drawing tools, you do this when you hold down a command key at the time you drag one of the object's handles. This is the way you force the drawing application to create circles rather than ovals, and squares rather than rectangles. See anamorphic also.

ISP

• **International Service Provider:** A company or that may be willing to rent you a 64kb/s connection across the Atlantic, either via satellite or cable. It is probably reselling a bulk-rental itself. See IVAN.

• **Independent Service Providers:** These are small commercial companies (or sometimes cooperative organizations) which provide private access to the Internet, usually for a fee.

The organisation will have a computer/server with TCP/IP protocols. It is therefore constantly addressable over the Internet, and it can act as a store-forward device for phone-in users.

Various levels of access are possible, from simple e-mail, up to full-access (SLIP or PPP) status. Some only have a character interface, while others have special graphic user interface facilities (WWW, etc.).

The ISPs are able to do this because they will have a leased-line permanent connection to an Internet node. Their main computer (your 'host') is always on-line, and can therefore have a true Internet Class 2 address. See IP address.

ISPBX

ISDN Private Branch Exchange. These are private automatic switches which can handle analog or digital lines. See DPNSS and Q-SIG.

ISR

• **Interrupt Service Routines:** These are routines which are added to (or included in) application software. They will intercept an interrupt when it occurs, call a certain correction or signal element of the program, then return control to the system — thus ensuring that the normal interrupt-vectors are restored.

• **International Simple Resale:** A licence where a foreign telephone company is offering private telephone links over an international leased line, with break-out links to the public-switched networks at both ends. It is a cheap way for large companies to make international calls.

ISUP

Integrated Signaling User Part. The signaling system of ISDN. It is closely related to CCS#7. See CCS#.

IT

• **Information Technology:** A popular UK generic term for computing that also goes by the names IS (Information Services) and DP (Data Processing) in large corporations.

At one time the term referred to data processing (mainframe computing) only. In recent years it has included all forms of information processing (office, information retrieval, transaction processing, manufacturing control) and also the communications and telecommunications side. It is not always clear whether robotics, as such, are included in this definition.

• **Interline Transfer** — in video. See FIT.

italics

A sloping, cursive (written) style of typeface (font) invented back in 1423. Italic is slightly different to oblique. See font and typeface.

ITAR

International Trade in Arms Regulation. Used by the US government to classify encryption systems as weapons of war, and so prohibit the legitimate sale of these overseas.

ITB

InTermediate Block character.

ITDM

Intelligent TDMs. Time division multiplexers are generally transparent to the information they convey, so they are unable to react to requests to switch to another similar resource or recover from some forms of errors, or determine service-utilization statistics.

Intelligent TDMs using microprocessors can provide these facilities and are therefore used as a network management tool. They can divert connections around a problem area, selectively control (by priorities) channels under maximum load conditions, and provide adaptive voice compression, etc.

iteration

Solving a problem by repetition or repeated recalculation — the process of constantly feeding data back (looping) in a cyclical form, until some end is reached, or of carrying out a process time and time again. The term can also mean 'expression'.

- In computer programming, iteration implies that the program is repeating a sequence of steps under the control of a condition. DO WHILE and FOR-NEXT instructions perform iteration functions.
- Often iteration is simply a synonym for repetition, which is how it is most commonly used in English. See loop.
- Sometimes it carries the suggestion of trial and error, or a brute-force approach to solving problems (running through all possibilities until one matches — rather than a more considered approach). To find the word 'iteration' in a dictionary by iteration you would begin at Aardvark and work systematically through, a line at a time, until you eventually got to the 'I's. More intelligent people would jump immediately to the 'I's, then to the 'IT's, and so on.

ITU

International Telecommunications Union. Founded in 1865, the ITU is now a UN body with about 160 member countries. Until 1993 it had four divisions: a General Secretariat, the IFRB, CCIR and CCITT. It has now been restructured:

- **ITU-T:** This (the ex-CCITT) committee makes recommendations (not laws) and it works through a hierarchy of Study Groups (i.e. SG2) which consider 'Questions' (i.e. Q.5/2 = Question 5 of SG2).

They generally set four year 'Study Periods' in which a number of questions should be resolved — although more recently some of these questions are being reviewed annually. The home base is in Geneva.

Note: The interim acronyms ITU-TS and ITU-TSS are no longer used.

- **ITU-R:** This group looks after radio. It is the old CCIR with some FIRB functions also.
- **TSB:** This is the ITU's Telecommunication Standardization Bureau which looks after publications. It currently has the fax number: + 41 22 730 58 53 if you need anything. You could also Telnet to ties.itu.ch or info.itu.ch (login name 'gopher'), or e-mail (with message 'HELP') to itudoc@itu.ch for documentation.

ITV

Interactive TV. This means anything from ring-up-and-pay for a movie over cable, to full-motion video, and two-way virtual reality (if and when it comes). It is a very vague term.

IVAN

International Value Added Network (Services). See VANS.

IVD

Integrated voice/data. The IEEE now has a draft standard (802.9) for the support of IVD systems on LANs, and for the integration of these IVD LANs systems with ISDN telephone systems.

IVDS

Interactive Video and Data Services. In July 1994 the FCC conducted auctions for 594 IVDS licenses in 279 markets across the USA.

IVHS

Intelligent Vehicle/Highway System. This is a general term applying to the following functions:

- Commercial vehicle location and weight (measured while in motion),
- Electronic toll collecting,
- TRaffic Information and Management Systems (TRIMS),
- Advanced Driver Information System (ADIS),
- Automatic vehicle control,
- Stolen vehicle trace.

TRIMS involves congestion notification, and the automatic re-routing of cars around trouble spots. ADIS includes navigation.

IVPN

International Virtual Private Network. See VPN.

IVR

Interactive Voice Response. This is a technology still in its early development stages. The AT&T systems are probably the most advanced in the world, but they still won't recognize many of their standard words in Australian- or English-English, because they've been programmed to recognize American accents.

IVS

Image Verification System. IBM and Kodak have recently announced this image compression scheme which can store digitally encoded identification pictures on low cost memory cards. They claim to use 'hypercompression' to store a recognizable face-shot on a magnetic card or smartcard in just 400 bits or 50 bytes.

They won't say how they do this, but the only feasible way is for these bits and bytes to register with a library of stored image primitives of facial characteristics. See eigenface.

Iwerks

A very wide-screen super-definition film projection system for special theatres. It uses 70mm film shot at 30 frames per second. See IMAX and Showscan also.

IWU

Internet Working Unit — another name for a LANs router or gateway.

IXC

IntereXchange Carriers (IXC). These are the American long-distance carriers like MCI, AT&T and Sprint. The distinction is with the LECs (ex-BOCs and independent local companies). Inter-exchange carriers, until late 1994, had the exclusive right to carry telephone signals between LATAs, while the LECs had the right to carry signals within a LATA.

With recent deregulation moves, all these distinctions are disappearing. See intra-LATA and inter-LATA.

Note that under the US system, a toll-call can be either inter-LATA or intra-LATA, and more than one local carrier may inhabit a LATA — but each as a regulated monopoly over local territory for wireline services. This is changing with telephony over cable .

IXO

A proprietary paging protocol which is almost identical to PET and TAP. See TAP and ETC also.

J

J-standard
See EIAJ.

j-type defects
Microscopic defects in microfilm. They are often only tens of microns in diameter.

jabber
• A stream of frames longer than the legal limit on an Ethernet LAN — or sometimes just junk on other networks. This is a problem usually caused by faulty cabling, but it can come from electronics.

jabbering node
• A node on a network which, through malfunction or software problems, constantly inserts junk data. NICs which are jabbering without listening, are one of the main causes of network management problems. The problem is usually one of faulty cabling.

• In Ethernet, a jabbering node is one which places packets longer than the accepted standard on the network. The frames will be longer than the 1518-Byte limit.

jacket
The outside plastic cover around a floppy disk.

jaggies
The rough-edged, stepped appearance of graphics and fonts which results from scaling up bit-mapped images. This is also called aliasing. See object-oriented graphics.

jam signal
• **In radio:** A jamming signal is one which is deliberately transmitted at high power to prevent the normal use of the channel. Spread spectrum techniques were specifically designed to overcome these problems.

• **In Ethernet:** When a collision is detected by a node, it immediately transmits a jam signal to ensure the other nodes on the network are aware of the problem. The jam is a warning to others that a collision has occurred, and it initiates the recovery sequence.

JDC
Japanese Digital Cellular. A TDMA cellular phone system used in Japan, but not exported. Attempts were made to sell it in Asia under the name Pacific Digital Cellular (PDC — see) but with little success. Now it has been renamed Personal Digital Cellular. See PDC.

Don't confuse this with the later 1.9MHz Personal Handi-Phone Service (PHP or PHS).

JEDEC
Joint Electronic Device Engineering Council. This is an American organization that standardizes the external dimensions, pin arrangements, etc. on electronic devices like RAM and EEPROM chips. It also has other standardization functions.

—JEDEC 'files' are for programmable array logic (PAL) devices.

—JEDEC memory modules use a standardized 88-pin socket.

JEDI
Joint Electronic Data Interchange committee. The UN committee which devised EDIFACT in 1987.

JEIDA
Japan Electronic Industry Development Association. JEIDA has a number of standards including a popular memory-card standard which was the precursor of the new PCMCIA standard. JEIDA 4.1 appears to be the direct equivalent of PCMCIA 2.0.

jewel case
The flat plastic case that holds a CD-type disk for general shipment and storage. This is not a caddy or a cartridge, but a storage and shipping container.

JIT
Just in Time — a Japanese manufacturing philosophy that is being widely adopted in association with EDI. JIT is inventory management where supplies are order, to be delivered only when needed by the production line or manufacturing facility. Computer databases and management information systems play a vital part in JIT.

jitter
Very short-term variations (relative to the allocated time-slots) in the arrival times of digital signals in a communications system. Jitter is a major cause of errors in synchronous systems, especially data loss at the higher data rates and over longer distances.

Long distance digital links usually derive their clock signals from the service-provider's cesium clock, and these clock pulses are read from the leading edge of the signal. Jitter occurs when there is a slight displacement of the edge relative to the time when it should occur and it is becoming an increasing problem with today's higher speed data circuits.

Jitter exists on virtually all synchronous transport mechanisms, whether they are over copper, fiber or microwave. The significance of jitter as a problem depends on the combination of bit-rate, repeater design and bit-detection mechanisms.

There's actually amplitude-, phase- and time-jitter (and more recently 'delay-jitter'). Amplitude jitter is unwanted variations in the amplitude of a signal, but this problem doesn't generally worry data circuits. See phase-jitter and timing-jitter.

• **Telephony:** In long distance telephone circuits additional jitter is produced in each repeater stage, but some types of regenerator (not repeater) will suppress jitter. But not all — it depends on from where their timing is derived.

The jitter that exists in PDH telephone networks is sinusoidal in nature. With PDH tributaries now combining into wider SDH streams, other sources of jitter are now being introduced.

• **In LANs**, jitter is essentially variations in phase or frequency (time) of the incoming data. If the jitter level is high enough, a node will misread the incoming frame and then transmit incorrect data to the next node. In Token Ring systems, each node on the network will add to this problem.

• **In ATM:** Delay jitter is a type peculiar to ATM systems. It is caused by delays in cell assembly, cell transmission, cell queuing, or just by load on the network.

• **Video:** A quite different use of the term jitter is for video artifact problems caused by vertical synchronization faults in video and television. Jitter in video recording is often caused by the drag of the tape around small video head-drums.

Josephson junction

A form of switching technology that requires near zero electrical resistance. It can therefore only be used in association with ultra-low-temperature devices and superconductors. This technology has been the subject of much experimental work ever since the superconductor rush began. Theoretically it is possible to achieve one billion MIPS rates with Josephson junction computing devices.

The junctions are a sandwich consisting of a very thin insulating layer between two superconducting layers. Electrons can tunnel through the insulator to produce a current, and this current is extraordinarily sensitive to magnetic fields, so the current can be switched on and off using a minute magnetic field. Alternatively, the junction can act as a magnetic field detector. See SQUID and quantum effect transistors.

Jot

An 'electronic ink' specification which is designed to help applications share handwritten notes, sketches and other pen-generated data, in the form in which they were drawn or written. It works with a range of pen- and non-pen-based systems, and the initial application will be for electronic mail.

Jot is being developed by Slate Corporation, Lotus and Microsoft, and has the backing of 22 companies including Apple, Go and General Magic. See electronic ink.

journal

• A daily record of messages and transactions conducted by a computer, or over a network link. In transaction processing it is essential that a daily record of all changes made to the database be kept in case of database problems.

• While it may mean 'daily', in library terminology, a journal is ant magazine or periodical which is regularly published. Newspapers are journals ('daily record'), but these days the term journal is also applied to monthly, quarterly or annual publications — also called 'periodicals'. The distinction is also with monographs, which are a casual (non-regular) publications.

joystick

A mouse substitute which is mainly used for video games but which has some application in business-type programs. A joystick combines the compactness of a trackball with some of the feedback characteristics of a mouse.

JPEG

Joint Photographic Experts Group. This collaborative group (ITU and ISO/IEC) has established JPEG as the new international standard for natural still-image compression. JPEG is a intra-frame compression scheme only, although a form called Video JPEG is also used in movie and video post-production where MPEG has editing problems.

A JPEG image can be compressed about 10:1 for a gray-scale image, with no noticeable degradation. With color images, it is possible to go to 20:1 with only very little degradation.

A standard image of 640 x 480 pixels, with a 24-bit pixel depth, will normally require 921.6kBytes of storage in native form. With JPEG compression it will occupy about 50kB (20:1).

The latest upgrading of JPEG is now strictly called JPEG2. It provides compression ratios of about 24:1 for still colour images before degradation becomes detectable.

JPEG supports a wide range of image quality ratings. The user can determine the required trade-off between compression and quality. There are four JPEG functional modes:

- **Sequential encoding** which is the baseline lossy compression algorithm. The image is encoded in a single overall scan.
- **Progressive encoding** which builds the image more slowly in a series of scans and retains a higher quality per megaByte.
- **Lossless encoding.**
- **Hierarchical encoding.** The file has multiple resolution levels, stored in a way that allows the lower quality images to be extracted quickly and easily, without needing a top-down decompression approach. You can have a thumbnail version that can be used for indexing, and then various higher quality levels — but all are built into the same file.

• The JPEG compression processes are:
  - Digitization.
  - Transform.
  - Quantization.
  - Huffman encoding.

Initially, the JPEG process divides the screen area into 8 x 8 pixel blocks and then uses the DCT algorithms for compression (except in the lossless mode).

During the quantization process the algorithm can be set to either retain or discard the very high-frequency image components which the eye often can't resolve — so this stage can increase the apparent compression ratio even further by dumping bits. A default Huffman-tree is used for lossless compression.

• For motion video production, Motion JPEG (M-JPEG) is

often used successively on each frame to provide an almost lossless picture quality while reducing the digitized files to a manageable size.

Compression ratios of 4:1 and 8:1 are achieved with no noticeable degradation of the image. M-JPEG is preferred over MPEG in motion video post-production, because interframe coding techniques like MPEG, rely on updating the image from previous frames, and they therefore can't handle the editing processes. Sony says it will have editable MPEG2 in 1995.

Once the video has been edited in M-JPEG, very much higher compression ratios are achieved by passing it over for MPEG2 compression. See MPEG, DCT, DWT, fractal and wavelet compression also.

JTACS

The Japanese version of the TACS analog cellular FM phone standard. A variation on AMPS, it was introduced in 1988.

J-TACS uses frequencies in the 860—870MHz band (Base Tx) and 915—925MHz band (Mobile Tx). The most noticeable feature of this standard is that they split the 25kHz carrier in two, and created a half-width version in very much the same way that Motorola did with N-AMPS (which was split into three). However J-TACS was never a success — mainly for cultural reasons rather than technical.

JTC1

Joint Technical Committee 1. This is the ISO/IEC committee responsible for IT standardization.

JTM

Job Transfer and Manipulation. An ISO 8831 and 8832 standard related to VT/VTP and FTAM. This is an SASE which is still under development. It will provide a way of controlling a distributed processing service over a network, between a number of servers. Each completes a specific aspect of that service.

judder

A television image which is being refreshed at a rate of 50 or 60Hz will suffer judder when objects move rapidly across the screen. The image of the moving object will appear to jump in discrete stages, rather than move smoothly and easily. This is a problem produced by the low refresh rate, and it is common even at refresh rates which appear to be sufficient for image fusion (flicker-free).

• In movie cameras, the film will judder vertically when it has not been loaded correctly. This is a totally different problem caused by a lack of free-loop in the camera's film gate.

jukebox

Disk libraries. A machine which can hold a number of video disks, CD-ROM disks, or cassette tapes. It feeds them, when required, to a replay (or recording) device. With the advent of CD-ROM, MO and CD-R there are three levels of essentially similar devices:

— **Disk changers** usually have no more than six disks, which can rotate into position on a single replay unit. These are always mounted and ready to go.
— **Jukeboxes**, where the changer must grasp the disk caddy, and physically mount it on the spindle. These typically hold 5 to 150 disks (usually in caddies) and often have more than one replay device attached to the network to avoid delays.
— **Silos or farms,** where thousands of disks may be available, and the automatic selection device may be handling a large number of different replay units. These tend to be very slow in access and used for tapes rather than disks.

Julia sets

Closely related to Mandelbrot sets but corresponding only to one single point on those sets. As the point changes so does the Julia set, but always in a totally predictable way.

Julian Date/Day

Julian Date and Julian Day Number. The start of the Julian calendar is figured from noon on January 1, 4713 BC. This is an early date in the Egyptian chronology, of which the experts are reasonably certain.
The present day number is expressed in the form 2 449 958.5 (10am Tues 29th August 1995 in Sydney, Australia). This is the number of days since the Julian start date. The decimal fraction (here 0.5) shows that it is midnight in Greenwich in the UK.

The *Modified* Julian Day Number is the above with the two first digits truncated — 49 958.5. This is the form most often used in computer calculations. Both forms take into account days lost in calendar changes over the years. There's a gap of ten days in October 1582.

• In GPS systems, the date/time is often expressed by the Modified Julian Date, followed by a day count since the beginning of the current year, plus a six-figure time number (two digits each — hour, minute, second).

• What is often misnamed the Julian Date on IBM-PC computers, is a memory register which keeps track of the days elapsed since Jan 1 1980. The Basic command MOD 7 with this number will show the day of the week on which any date falls.

• Another form of date often called the 'Julian Date' is a five digit numeral. The first two representing the year, and the last three being the cumulative days since Jan 1. So 95087 would be the 87th day of the year 1995.

• Another usually reliable source says that sometimes the Julian date is represented as a five-bit number, with the first two being the year (presumably in the 20th Century), and with the last three being the day-number of the year (1-to-366) from January 1. Obviously, the term is being used for a wide variety of different methods of date calculation.

jumbo

• **Chip:** A full wafer with many IC circuits before they are broken down into individual circuits.

• **Roll:** The large roll of magnetic plastic material used for making magnetic tape or floppy disks. These rolls are about 1 meter in diameter and 1.5 meters long, and the surface is strictly checked for defects before the tapes and disks are cut.

Second quality rolls are sold to small companies who use them to make cheap disks and tapes. There are only a half-dozen companies in the world capable of making first-quality jumbo rolls.

• **Group:** In US terminology this is 3600 analog voice channels which have been frequency multiplexed on to 16,984MHz of bandwidth in the spectrum between 0.564 and 17.548MHz.

A jumbo multiplexed group is 10,800 voice channels which are occupying 57MHz.

jump

In newspaper and magazine printing, this is the process of 'spilling' or carrying-over material, by transferring it to a later page.

jumper

A short length of wire (or a small plug), used to create an electrical bridge around some obstruction, or across a gap in a circuit. Jumpers are often used for test purposes, but in some cases they make semi-permanent modifications.

junction

The boundary region between two semi-conductor materials having different electrical properties. It can also be between a metal contact and a semi-conductor layer.

It is this boundary region that controls the flow of electrical current through a semi-conductor device.

junction circuit

This is an interoffice trunk in North America. It is a circuit between two local exchanges which by-passes the trunk exchange. In other parts of the world it can be any link between a local exchange and any other exchange.

junk diallers

This is telemarketing equipment which dials numbers in sequence. Some immediately hand-over to the human operator, while others deliver a message and record responses.

justify

• **Programming:** To shift the contents of a register to a specific position.

• **Publishing:** A publishing and printing term for the expansion of the space between words (and sometimes between letters) so that each line has a 'full measure' and aligns on both sides of the column. Both left and right margins are straight or 'flush' (except for the last line in each paragraph). The lines in this text column are justified.

The distinction is between 'justified' text, and copy which is set 'ragged right' [unjustified] text. Ragged-left text is less often used; except possibly for captions. Text can also be 'centered' (both ragged right and left). See flush also.

Most inexperienced layout designers over-use ragged right, ragged left, and the centering of text in the belief that it looks modern. Sometimes it does, but more often it just looks messy.

• Vertical justification is a feature in some desktop publishing programs. Here the text's line separation (leading) is automatically expanded until the text fills the available column length. This is usually called 'feathering'.

• In communications, justify can mean the forced-alignment of data bit-rates between two channels (one feeding into another). This is usually done by bit-stuffing in time-division multiplexers. Justification both fills the slots and adjusts the timing.

K (or k)

- **SI units:** The small k is the SI standard for kilo, or one thousand. That's how the general public use it, and it is often used in computing in the same way: for instance with 'kHz'.

However both the small k and the capital K are interchangeably used to mean 1024 (2^{16}), with 'k' or 'K', 'kB' or 'Kb', 'kb' or 'KB' — all of which I've seen used to mean kiloBytes and also to mean kilobits. This is hopelessly confusing.

Many magazines will express computer memory sizes in 'Ks of RAM', which most readers would understand to mean kiloBytes when referring to the computer's memory as a whole, and kilobits when referring to the individual chips — eight of which are needed to create a byte-wide memory location.

But this usage is sloppy and too casual in communications, so it is essential to specify bits or Bytes. I don't think it matters whether the k stands for 1000 or 1024, so I intend to stay with the small k and permit Kelvin to retain the capital K as part of the international SI standard.

- **Color:** See Kelvin.
- **Temperature:** See Kelvin.
- **Electronic Noise:** See Kelvin.

K-9

An American expression for the education market from kindergarten to Grade 9. This was, at one time, the traditional market for Apple II and GS computers.

K-12

This is the US education system's K-9 market extended up to the 12th grade — the year before college or university. This market is more sophisticated in its technical requirements than the K-9 market.

The computers used are: Apple IIs 46%; DOS compatibles 32%; Macs 15%. New purchases in 1993–4 school year were: Macs 61%, IBM PCs 18%, and DOS-comps 18%. In late-1994, CD-ROMs were in 37% of these schools and in 50% of high-schools.

K-band

The range of satellite frequencies from 10.7 — 36GHz.
— Ka-band is the 20—30 GHz band.
— Ku-band is 10.7—18GHz. See both.

Ka-band

A (mainly) satellite frequency band which operates between 20—30GHz: it is a sub-set of the K-band. At these high frequencies, the Ka-band is direct line-of-sight, but it has enormous potential capacity.

Ka-band only refers to the satellite and radar use of the spectrum. In fact, this is part of the Super High Frequency (SHF) band which was thought at one time to have few terrestrial applications. Now it has many.

This area of the spectrum is still under development, so at present there are few mass-produced products working this band to bring the component costs down. However it looks very promising, and despite the short terrestrial range (line of sight only) it is finding many new applications.

Satellites, however, will be the most important users of the Ka-band this decade. Ka will be initially used for super small aperture terminals (SATs) which are called VVSATs (very, very small), MSATs (micro), and USATs (Ultra).

- The new NASA ACTS satellites use transponders in the Ka-band, and they promise to provide '20-times the data transfer rates now being tested on the Ku- band'.
- The Japanese are developing Ka-band digital satellite television systems using transponders with 100MHz of bandwidth at 21GHz

See SHF and the K-, L-, X-, C-, Ku-, S-bands and also radar bands.

Kansas City standard

This is an old standard for holding data on an audio-cassette tape.

Karn algorithm

An algorithm which helps the Transport layer protocols to distinguish between good and bad time-samples when measuring round-trip time estimates. This is important for building router tables.

kb/kB

- kb = kilobit which is 1024 bits. Also written as kbit.
- kB = kiloByte or kByte, which is 1024 Bytes or 8192 bits.
- The capital K has often been used here to emphasize the difference between k = 1000 and K = 1024, but no one seems to worry much about the extra 24. Far more confusing is the bit vs. Byte problem, and editors and sub-editors are often irresponsibly casual here.

keepalive

Network devices often need to send a keepalive message to another device to inform it that it should remain on line, that the session has not yet been completed, so the virtual circuit is still active. These keepalive messages must be sent at regular intervals.

Kell factor

A video image is painted on the screen by a scanning electron beam which has a finite screen dimension. This spot-diameter established the width of the image line in the B&W sets, and this painted line was always narrower (about 0.7) than the space available. The Kell factor is a ratio of the width of the line to the interline space.

• The Kell factor was previously (and incorrectly) used to determine the theoretical vertical resolution of a TV set. Video engineers have traditionally multiplied this factor by the number of active lines on the screen, and arrived at a figure which they call vertical image resolution. But this is a fallacy.

The Kell factor does have some impact on the vertical resolution, but not much — and then mainly on how clearly you see the lines, not the resolution of the image.

• Kell factors must be kept low for interlace systems (scans wander a little during interlace), but they can be increased to almost 1.0 (lines touching) with progressive scan systems, and this makes it more difficult to see the line structure.

• The old mistaken concepts have been rationalized by TV engineers. They now mean by Kell factor a constant of about 0.6, and they use this to calculate an approximate vertical image resolution expressed as:

```
active lines x 0.6 = perceived image lines
```

This 0.6 constant only matches the true Kell factor by coincidence. It was derived from statistical tests of human eye resolution on TV screens.

This makes the subjective vertical 'image resolution' (with a test card set at 90° to normal) of a PAL screen at about 350 lines (175 cycles), and it doesn't degrade much during transmission. Subjective vertical resolution improves dramatically if objects move slightly on the screen (especially vertically).

Kelvin

Aka absolute temperature. The unit of absolute temperature which is also used in electronic noise measurement and for the color of light sources.

• The freezing point of water (zero Celsius) is 273 Kelvin. When 1 calorie is absorbed by 1 gram of water, the temperature rises 1 degree K. Absolute zero is –273.15°C.

The reddishness of artificial light and the bluishness of sunlight is on a continuum and it can be measured by reference to the color output of a piece of 'red-hot' or 'white-hot' tungsten.

Since all early light bulbs used a tungsten filament, this is how color and temperature are related.

Often the light-source color must be corrected in color photography before it will look natural (since the eye automatically corrects for ambient color), and this is done by the addition of filters in front of the light, or in the lens. So Kelvin provides a measure of both the light's coloration, and the filters needed to correct it. This is called the color temperature measurement and the rough standards for average lighting conditions are:

Candle flames	1850K
Quartz lights	3200K
Photoflood lamps	3400K
Normal daylight	4300—5600K (direct sun)
Overcast sky	6000K
Summer shade	8000K (Skylight fill only)

• Movie and still film is generally manufactured to be color-correct in daylight ('daylight film') or under tungsten film lights ('indoor' or 'tungsten' film). To use a tungsten film outdoors (the standard for most documentary production) you will need to add a Wratten 85 orange filter to the front of the camera lens.

See colour temperature and 80- and 85-filters.

Kermit

A specialized file transfer protocol which is named after the frog 'who would talk to anybody'. However Kermit is less efficient than XModem, and it is way behind ZModem in throughput and ease of use — provided these are viable alternatives in the circumstances.

Kermit places emphasis on multiple file-transfer (batches) between micros, minis and mainframes, and it also provides terminal emulation. It's probably the only popular file transfer program that can handle 7-bit files (still widely used by mainframes) in a robust way, and it has some special features for remote searching — for instance, the ability to specify filenames using 'wild-cards', and to handle them in batch.

Kermit has a sophisticated handshaking system which negotiates on-line for a graded choice of speeds, block sizes, and error-check protocols. Some versions have compression.

• Kermit variations can use 6-bit checksums, 12-bit checksums and 16-bit CRC. The main variations are:

— **The original Kermit** transmits data packets from 10-Bytes to a maximum size of 94-Bytes. This is still the default mode, but see below.

— **Long-packet Kermit** increases the packet size limit to 9024-Bytes.

— **Sliding-windows Kermit** tolerates networks with periodic response delays. The 'sliding window' reference means that the sender will keep transmitting information for a while, and wait for ACKs which arrive much later. This allows the system to tolerate satellite delays.

— **Super Kermit** (same as sliding-window).

• Kermit has also been released as a full communications package which now includes communications and other file-transfer protocols. The latest version supports X.25 and TCP/IP and has a scripting language.

The main problem with Kermit is that there are simply too many options; you need a college course just to understand the choices.

• To speed up Kermit, the on-net advice is to change the packet size from 94-Bytes to something in the 5000—9024 range. Further rate-increases can be had from setting the sliding window to between 3 and 5 packets.

kernel

This term is often used generally, just to mean 'core'. It usually means the central core of a program — more often of an operating system. The term is mainly applied to Unix, AI, expert systems, and some modular programs.

Operating systems are best viewed as tree-rings; at the center there is the core or kernel; around that exists a ring of critical utilities; then surrounding those are rings of other utilities. Procedures in one ring, call procedures in another.

With object-oriented operating systems, this rigid hierarchy is not quite the same — but the more critical objects constitute the (micro)kernel.

• Kernel is often used vaguely to distinguish the core of an operating system from its shell — which generally refers just to the user-interface and device-drivers. Windows 3.x was a shell around MS-DOS — but you wouldn't really call MS-DOS a kernel because it was a full OS in its own right.

• The kernel of an application supplies the interface between the hardware and the more fundamental parts of the operating system, so the linkage is never seen by the user. This type of kernel is responsible for the API 'calls'.

• In CPU operations the kernel-mode is also the supervisor mode, and should be distinguished from the user-mode. All CPU processes have both a kernel and a user mode, and while in the kernel mode, all CPU instructions are permitted.

kerning

A typesetting term for adjusting the space between text characters — the 'letter space'. With typewritten text there is always a fixed amount of space provided for each character no matter how narrow or wide the characters. However in the proportional-spacing used in printing, the gap between letters varies. An 'i' requires much less linear distance than 'w', and in the word 'To', you actually want the end of the 'T' to hang slightly over the 'o'. But not in 'Th', where the 'T' and 'h' must be separated slightly by at least one pixel.

Kerning therefore requires rules and tables which look at the two adjacent characters and decide how much space to leave. Automatic kerning is built into an outline-font's hints; while manual kerning requires the physical adjustment of these distances — and is not available in many applications.

• In desktop publishing, it is often necessary to adjust the kerning of a headline (the large type exaggerates any imbalances), and sometimes you need to compress or expand the letter spacing of a whole line of characters. Some 'display fonts' have these controls. See tracking and ligature.

Kerr effect

In magneto-optical recording disk-drives, the recording surface is heated by the writing laser and simultaneously that area of disk surface is subject to a strong magnetic field.

The area under the spot then changes under the magentic influence, and while the change (the 'domain') is microscopic, the surface then reflects the read-laser's light in a slightly polarized way — the plane of polarization of the surface has been rotated slightly by the magnetic flux. This is the Kerr effect.

During the read phase, the reflected (or, in some systems, transmitted) light passes through polarizing filters to exaggerate the difference, and the drive controller can then detect these (so-called) 'pits' and the original 'lands' through brightness changes. A pit is a point of change, and a land is a track-length of no change.

The temperature at which the polarization/magnetic change occurs in these special coating materials is known as their 'Curie point'.

key

Ignoring the obvious, this also means:

• In a relational database manager, the key field is the one chosen to create the relational links. The same field must exist in both of the files being 'related'.

• For data security and authentication systems, see encryption, public key, RSA and DES.

• Key disks are special disks which must be held securely, and sometimes inserted into a computer to keep an application working. This is a technique used at one time to prevent piracy.

• In telephony, a key service is a small-business multi-line telephone system comprising a number of special handsets. These are generally wired with multi-wire cabling so calls can be forwarded from one to another, and they function without a dedicated switchboard. See key station.

• In television there are a number of different types of video-key. The most widely used today is chromakey which allows the operator to select a saturated color in a scene (usually a blue or green) as the key-color. The special effects generator (SEG) will then subtract from the image any area with this color, and insert instead the image from another source (camera or videotape). This is a matte process, where the image supplies its own matte. See chromakey.

TV luminance (also called linear) keys of various kinds were widely used in B&W days of television and are still used today. These functioned on the tonal levels of the scene rather than on the color. See internal key and external key.

key management

A term used in the security side of EFT, EFTPOS and data security to mean the creation, storage, protection, distribution and exchange of cryptographic keys. The principles followed are that keys must always be:

— generated by a random process,

— exchanged in a secure manner, and

— securely stored in the issuer's computer.

And keys must:

— never be extracted or displayed on an EFTPOS terminal.

Traditional key management uses two keys to allow the host to check the authenticity of the terminal before the transaction.

• There is a master key (for each terminal), and either a session key (used for multiple transactions) or a transaction key (for single transactions). The transaction key approach is more secure since it changes each time; a session key might last the whole day and give the thieves time to crack the system.

• Damien Doligez recently broke the Secure Socket Layer (an RSA variation) of Netscape. This was a 40-bit key (shorter than would ideally be used) but it was broken by brute force (trying each combination), in only 8 days on 120 workstations.

key service
See key systems.

key sets
See key stations and key systems.

key systems
Also called Small Business Systems (or NSUs) which are groups of identical telephone handsets without a central switchboard. They usually require more than two wires for interconnection, but they can carry multiple incoming lines, and switch calls between terminals.

This approach is widely used in small offices. It avoids the need for a central PBX-type switch, but the signaling process used usually means that data can't be sent over the lines. Also called SBX, SBS or key services. The individual units are called key sets, or key stations. See KSU also.

• Apparently some of the new digital key systems can handle data.

keyboard
Keyboard, as distinct from keypad, suggests that full-text messages can be input.

• Computers: The standard keyboard layout is called QWERTY (although many of the smaller pocket organizers use an ABCDE layout) and the French use AZERTY. Other computer keyboard layouts have been tried, but none have ever succeeded. The best-known is Dvorak which has constantly proved to be faster than QWERTY, but, after laboriously learning to touch type, it is hard to get enthusiastic about relearning such a hard-won skill.

• Keyboards can sometimes be changed by software controls. There are different layouts for different languages.

• Soft keyboards are virtual keyboards. They are created on a touch-sensitive screen and usually used with a stylus (PDAs) or, in a kiosk, by finger.

keying
Aka digital modulation. This refers to modulation techniques where digital signals are superimposed on analog carriers — and the distinction should be made between the digital 'keying' techniques and the parallel analog version.

So we have AM (amplitude modulation) radio broadcasting where an analog signal is superimposed on an analog R/F carrier, and ASK (amplitude shift keying) where a digital/binary signal is superimposed on an analog (audio frequency or R/F) carrier. The 'keying' is the process of varying the signal output between two amplitude levels.

FM (frequency modulation) and FSK (frequency shift keying) are obviously similar. FM superimposes analog on analog by modifying the carrier frequencies in a continuous way, while FSK superimposes digital on analogy by flipping the output between two audio or R/F frequencies under binary input control. See Phase Shift Keying (PSK) and modems also.

keypad
A vague term for a small I/O keyboard with limited functions — typically the numeric (calculator) type, or a teletext/videotex-type controller or the number-buttons on a touch-tone telephone. The telephone and videotex types of keypad will have the ten numerals plus the * (star-asterisk) and the # (octothorpe-hash-pound) keys.

• A small self-contained keyboard device used for the control of machine functions (TV remotes). Some keypads use infra-red control systems.

• A soft keypad is a 'virtual keypad' which exists only on an LCD screen. It is used by touching the various key-icons with a finger or stylus.

keystation
A single handset on a key-system.

keystroke-stacking
A process akin to batch-files (and similar to macro recording) where a program stores a series of keystrokes and replays these in sequence whenever the application is booted or run.

keywords
• In any high-level programming language, these constitute the 'vocabulary' of English-like words which the language can interpret. So SAVE and GOSUB are keywords in Basic.

The distinction is with other symbols also used in the language — numbers, multipliers, etc. So the syntax (rules) of a language comprises keywords, symbols, and the ordering of these into lines, sequences, etc.

• Specially selected ('substantive') words in a database record (or sometimes in a special field) which are indexed to allow easy searching of the database for relevant information. See descriptors, identifier, stop words and inverted file.

kill
To kill a file is to destroy it; to remove it from the directory.

• A kill file is a personal file used by Usenet 'clients'. Every user keeps his/her own kill file; if you add a name/address into that file, it will filter and reject any message written by that person. This is widely known as a bozo filter.

killer application
An application which redefines the computing or communications industry at the time of its release. VisiCalc was the first killer-app. When it appeared, overnight the Apple II changed from an enthusiast's toy to being a serious business machine.

The second killer-app was probably dBase II which created a whole industry around the CP/M operating system, and introduced most of the world to relational databases and sophisticated procedural search languages.

Wordstar was probably the killer-app that helped the early IBM PCs to establish a market. For a long time the PC was seen mainly as a word processing device. You didn't do anything serious on a PC.

Desktop publishing and the laserprinter certainly gave the Macintosh the boost it needed in the early days. Note, that with the exception of DTP for the Macintosh, all killer-aps have been horizontal market applications. See horizontal applications.

kilo-

The SI prefix for a thousand, or 10^3. Always abbreviate kilo with the small k, even though all other SI units above the fundamental use capital letters.

The capital K is now sometimes being used to represent the computer 'kilo' which is 2^{16}, or 1024, but this is also used for Kelvin.

kiloByte

A thousand bytes (strictly 1024 bytes). In average English text this is about 170 words — about one-third of a column on this page.

kiosk/kiosque

The two versions are becoming confused, so are best treated together.

- Kiosk information systems are stand-alone public information booths, often with touch-screens and menus. You find these at modern amusement parks, in museums, etc.
- Kiosk/kiosque services over electronic networks are systems of money collection, which are run by the telephone carrier or network operator, on behalf of information providers. The carrier will bill you for small amounts for casual usage through your normal phone bill.

KISS-rule

Keep It Simple, Stupid!

KLIC

Key Letter In Context. An indexing system where all terms are sorted by every letter in every term. The remainder of the term is also displayed. This is a permuted-list indexing system based on letters not words. See also KWIC, KWAC and KWOC.

kludge

(Pronounced 'klooge' — meaning smart, in German) To get around a bug or hardware problem by a make-shift solution. This is jury-rigging. This means that the problem is probably deep-seated and can't easily be fixed at source. There's also the implication that a kludge solution is 'inelegant' — so there's a professional sneer hidden in this term also.

Klystron

A type of high-power beam tube used in some microwave and satellite systems and in television broadcast transmitters. For TV transmission these are available in output powers of 30kW and 60kW.

knowbot

This is an intelligent agent (software) which is sent over the network and given the job of discovering and filtering information for you, or of performing some other simple task.

The ultimate idea is that, eventually, agents consisting of many knowbots (aka 'bot' — see) will roam the world's networks, collecting and customising information in your fields of interests, and send it back to you.

So knowbots and agents are a new form of I/O interface — but unfortunately they lack the important element of 'common-sense'. That will come with Cyc. See.

knowledge base

In expert systems, a knowledge base is a set of hypotheses, observations and rules usually gathered together by using 'knowledge elicitation' techniques (asking experts). This is heuristics — picking the brains of an expert (or two).

Unlike a database which is highly structured, a knowledge base is a set of unstructured facts. The paths by which these facts are related are determined on the fly (as needed).

Korn shell

A standard user-interface shell for Unix. C and Bourne shells are two others.

KSR

Keyboard Send/Receive. This acronym is also used to mean a dumb terminal, because all it can do is send and receive.

KSU

Key Service Unit. This is a demarcation device used in key systems (see) which handles the connection to the phone line.

Ku-band

A widely used frequency band for satellite communications between 10.7 and 18 GHz. The Ku-band was established for international satellites by the WARC in 1979. It is a subset of the K-band (see).

When Ku is used, ground stations with small dishes in the 0.5—4 meter range can be used, so this band is particularly suited for VSAT operations. However a Ku-band transponder will require greater receiver sensitivity and it will need to generate higher radiated power than the equivalent C-band satellites.

Generally the Ku-band satellites transmit on bands centered on 11GHz (the downlinks) and receive on 14 GHz (the uplinks) but there are many variations. In the US, uplink frequencies are restricted to 14—14.5GHz, and downlink frequencies to 11.7—12.2GHz (called the 12/14 band). In other parts of the world frequencies from as low as 10.95GHz, and as high as 12.75GHz, are used for downlinks.

Ku frequency bands are not well suited to the tropics because monsoon rains can blanket the earth stations and prevent reception of the signals. See also the L-, X-, C-, S- and Ka-bands, and also radar bands.

Kurzweil

Usually this suggests reading machines for the blind, or speciall scanner-and-OCR devices. Kurzweil has long been the leader in this field, but recently others have surpassed them.

KWAC

Key Word And Context. A type of indexing system where the titles of documents are permuted (changed in order) to bring (successively) every key word to the fore. So 'The Informatics Handbook' would be indexed under 'Informatics Handbook (The)' and also under 'Handbook: (The) Informatics'. People looking for 'The Handbook of Informatics' would then have a fair chance of finding it. See KWIC, KWOC and KLIC also.

KWIC

Key Word In Context.This is a form of automatic library indexing. As entries are made to the database, keywords are extracted from the title and often from the abstract, for indexing. Stop words are eliminated.

A KWIC index usually consists of four columns. The second column contains the indexed keyword. Columns to the left and the right hold associated words from left and right respectively, and the fourth column holds the reference number.

KWOC

Key Word Out of Context. The index is by keywords, and these are chosen by the indexer. They are not necessarily only those taken from the title.

L-band

The L-band is used for satellite-to-mobile communications. It uses microwave (UHF) frequencies from 1646.5—1660MHz for the uplinks, and 1545—1558.5MHz for the downlinks.

It is mostly used for mobile services only, mainly because it only requires a small frisbee-sized antenna or a large helical aerial pole with vehicular mounting.

The output from a LNB amplifier behind any receiving dish, is generally in the standard L-band range of 950—1450MHz (a 500MHz wide band) for either C-band or Ku-band receivers.

See Inmarsat-M, the X-, C-, Ku-, Ka- and S-bands, and radar bands also.

L-carrier

A hierarchy of coaxial cable rates used in analog telephone systems. The first coaxial telephone transmission system in the early 1940s had a capacity of about 600 voice circuits. This was L1. The series has continued to evolve for long-haul use. Type L5 has a capacity of 10,800 frequency multiplexed voice circuits. See N carrier and T1.

label swapping

The routing technique used in the original APPN. In a network, each router independently determines the best path to the next router — when a packet arrives at a router it is given the next segment address and sent once more on its way. This approach limits the network's ability to find alternate paths when individual routers failed, so in the later APPN+ this technique of label swapping was replaced by source routing (see).

Lamba-PONs

A UK passive optical network (fiber to the home) which uses multiple wavelength lasers (WDM) to create an almost switchless system.

LAN

Local Area Network.

• An internal company computer communications network, not involving a telephone-type PBX switch. The term can be applied to a single link over distances of one meter between a PC and a dedicated printer terminal using LAN communications techniques, or to a network of 10,000 personal computers, minis and mainframes joined together over optical fiber backbones stretching a couple of hundred meters.

WANs are larger inter-networks — chains of LANs — and these LANs may not all be of the same type.

The LANs remain the primary network units. LANs are defined as the range of nodes (computers, workstations, servers, etc.) on different segments which can be addressed by using a local addressing system (the MAC address). Another way of defining a LAN is as a single 'domain' which shares a common network address.

Bridges can join LANs by filtering packets from one side, and only transferring the few intended for the other network across the gap. This can only happen with a bridge if both networks are of the same kind.

Wider-area connections, and those between different types of network protocol, need to make connection through a router or gateway. This needs an additional network address. So this is the distinction between a bridge (using MAC addresses) and a router or gateway (using network addresses). In the OSI model, the MAC address is called generically the NSAP. See NSAP address.

• The two primary types of small-scale LANs are those which use peer-to-peer network operating systems, and file-server-based networks. See access protocols.

• Work-group LAN, is a term applied to a relatively small network which only links those people who constantly need to share data on a daily basis. This is the smallest practical unit in a large corporation, but it may constitute one or more network segments.

• Local Area Networks are often made up of segments, and all segments must share a common protocol. There is a limit on the number of segments the system can support in a series. Segments are joined by repeaters.

• See Ethernet, 10Base-#, 802.#, Token Ring, LocalTalk, CSMA, LAWN, contention and deterministic.

LAN analyzers

These are small hand-held network-testing devices which don't have the full functionality of a Protocol analyzer. However, some LAN analyzers are available as plug-in cards for network workstations.

Generally, they will identify network protocols and incompatibilities, gather statistics, and isolate which components are generating errors. They may also be able to generate test messages, and create high traffic loads on the network to place it under stress.

LAN classification scheme

- **Modulation**
  - Baseband.
  - Broadband.
- **Access**
  - Dedicated links.
  - Switched circuits:
    - Data switch.
    - PBX or PABX.
  - Multiple Access techniques:
    - Contention.
    - Token passing.
    - Polling.

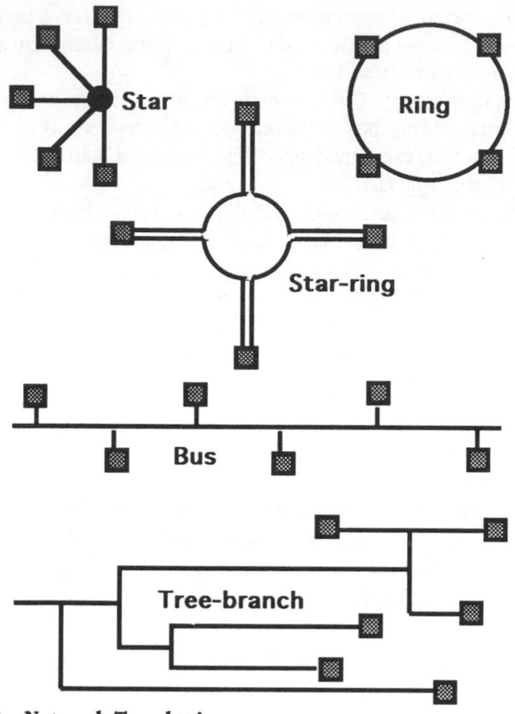

- **Network Topologies**
- **Transmission medium**

  Fiber Optic cables:
  - Monomodal.
  - Multimodal.

  Coaxial cable.
  Wire:
  - Unshielded (Twisted Pair).
  - Shielded.
  - Ribbon.

  Terrestrial wireless:
  - Radio, microwave.
  - Infrared.

  Satellite wireless:
  - UHF.
  - C-band.
  - Ku- and Ka-band.
  - L-band.

LAN emulation

ATM systems are cell-switching systems while LANs are shared-media networks. However ATM systems can emulate LAN functions. See LES, LEC and LE-ARP.

LAN gateway

A PC or special server on a local network which has been reserved for the task of protocol-translation between two networks, or between the network and a mainframe or mini.

LAN hubs

A concentration point for the physical network wiring or cabling. This can refer to a passive concentrator, a MAU or MSAU, a multi-port repeater or an active router /bridge. The term also encompasses network management devices.

The earliest active LAN hubs were repeaters following the 1985 IEEE 802.3 specification. While the original network repeaters were simple devices with two ports, thin-Ethernet vendors began making multiport repeaters as network hubs. They were unmanaged and could be expanded only by linking port to port (segment to segment). There was no rate adaptation possible.

With 10Base-T and twisted pair, came the concept of star networks with one device on each cable, and this required multiport UTP hubs. Each station was connected to the hub by two twisted pair (transmit and receive) and the hub repeated the signals received on each one of its outgoing lines. Although this is physically a star, it is logically still a bus.

Multiple levels of hubs can be cascaded in a hierarchical fashion, and multiple stations can be attached to each cable run (or a mix of stations and hubs).

The first of this kind were chassis hubs; large racks with numerous line cards and management hardware. Individual ports could then be managed from a remote console. Later, smaller units were made for smaller networks, and from these evolved the 'pileable' and later the 'stackable' hub.

Pileable hubs are individual repeaters which interconnect by standard port connections (max. of four for Ethernet), while stackable hubs are managed as a single repeater, and can handle different media (thin Ethernet, fiber, etc.).

Hubs then also took on a true switching function in some cases, and pseudo switching using store-forward principles.

These were fast bridges and routers which could also provide other functions — rate adaptation, and translation to backbone protocols such as FDDI.

- A recent development is Switched Ethernet (switched, bridged and routed), Switched Token Ring and Switched FDDI. See wiring hubs and concentrators, stackable hubs also.
- In Ethernet, Token Ring and FDDI networks, hubs now incorporate SNMP management software controls which can sometimes monitor LANs down to the port level.
- Some networks also use hub polling, where the controller progressively contacts the peripheral nodes and asks if they wish to use the system, or commands them to download data.
- People are now beginning to talk about generations of hubs.

— **The first generation** were simply wiring points — junction boxes with perhaps by-pass functions (MAU) — or multiport repeaters that supported only a single type of LAN.

— **The second generation** were multi-port repeaters that supported only a single type of LAN, and later smart hubs which featured some network management and accommodated multiple LAN types.

– The third generation hub devices, are switched units which have multiple high-speed buses. They can support different types of LAN as well as many different media, and in multiple domains. They come with integrated bridging and routing capability, and offer advanced network management.

LAN Manager

Microsoft and 3Com's extension to OS/2 which provides the core functions of a network operating system. It sits at the level of OSI's Presentation layer and supports several protocol stacks, so it is similar to NetWare. LAN Manager uses NetBEUI or TCP/IP network protocols.

Although LAN Manager was written for OS/2, it is not dependent on this operating system. Both HP and AT&T offer Unix versions, and DOS, OS/2, Windows, Macintosh and Unix clients and servers can coexist and communicate on the same network.

LAN Manager has been defined to have a common core (file, print, administrative and security services) on to which different companies can hang enhancements. It removes the distinction between workstations and file-servers and replaces this with a distinction between 'requester' and 'server'. Each device on the network will then be either a requester or a server.

LAN Manager has not been successful, especially when compared to Novell's NetWare 3.x. IBM, which was involved in the original design, later broke away from Microsoft and 3Com to develop a fast version which was renamed LAN Server.

LAN Network Manager

This is IBM's source-bridge and Token Ring management package. It can pass alerts to NetView.

LAN Server

IBM's network 'pseudo-operating system' and distributed file system which was derived from LAN Manager, and is therefore based on OS/2 and NetBIOS. It supports DOS, Windows, OS/2 and Macintosh clients. LAN Server actually runs as an application under OS/2, so it is not a true operating system like NetWare — although it may run almost as fast. See LAN Manager.

LAN Switching

See LAN hubs and Ethernet Switching.

Landmarks

Landmark Speed is a small CPU/memory-intensive benchmark. It compares the computer with an IBM PC or XT running at 4.77MHz. There is no disk I/O in this test, and the presence of a maths coprocessor won't make any difference. The major influence on this benchmark is the number of wait-states.

lands

On an optical disk, information is stored in the pattern of 'pits' and 'lands'. Lands are the residual flat areas of the original disk surface which has been anodized to make it reflective. The pits are the 1.6 micron depressions in the surface created by a 'stamper' or holes/bubbles/magnetic domain changes in the reflective layer created by the recording process. What defines a pit is merely a change of optical reflection from the land surface. A typical pit on a CD disk is about the size of a bacterium (0.5 microns wide by 2 microns long).

Optical disk systems react to the 'change in state' between a land and a pit — not to the presence of pits and lands themselves. In CD systems, the change from pit to land or land to the, expresses a binary 1. The steady state, whether the laser is riding over a pit or a land area, expresses only binary 0s, depending on the number of clock pulses.

landscape

- **Printing:** This is page printing with the lines of text along the page's longest axis. The distinction is with the more usual 'portrait' printing. In some laser printers, a special set of landscape fonts must be held in memory, although standard font rotation is possible in others.

- **Screen:** The standard monitor is a landscape monitor. One which has the vertical axis greater than the horizontal axis is called a portrait monitor. Some monitors can be rotated into either position.

LANE

LAN Emulation standard. This is the virtual LAN standard of the ATM Forum, which is intended to emulate Ethernet, Token Ring, or FDDI — and for bridging these protocols across an ATM network. The ATM backbone then appears transparent to the user. The interface is known as the LAN Emulation Client (LEC) interface.

language

The term is applied to coding systems (like assembly language or machine language), but strictly it should only be applied to the high-level programming tools. There are three main categories of computer language:

— Procedural languages like Cobol.
— Logical programming techniques like Prolog.
— Object-oriented languages like SmallTalk.

The term is often used interchangeably with 'code'. So a code generator, may actually produce a program written in a high-level language, which will then need to be interpreted or compiled to machine or assembly 'language' (code).

language generations

The recognized generations of computer 'languages' are:

- **First:** Machine languages, where just 1s and 0s are entered directly into computer memory or via some simple storage device, originally by rows of switches but now by a simple translator. This is the level at which the computer actually functions.

- **Second:** Where simple mnemonics are used in assembly language coding. Each instruction consists of a mnemonic and it is usually followed by one or more hex operands such as 'LDA ADh 00h 03h' or 'CLC 18h'. These are difficult for humans to read, but the machine can translate them easily to machine binary because of the one-to-one relationship of instructions and mnemonic input.

• **Third:** These are the so-called 'High Level' languages which use more English-like instructions. Languages such as Basic, Pascal, Cobol and Fortran, and more recently C. They require either interpretation or compilation before they can be run. Usually 3GL programs are readily transferable between hardware platforms.

• **Fourth:** These are non-procedural modern languages. There are two types; production-oriented 4GLs and user-oriented 4GLs. This category includes languages like HyperText and dBase, and also query languages like SQL.

• **Fifth:** These are functional programming languages (see) and AI-type languages. Functional languages don't depend on the order in which statements are written or executed.

LAP

Link Access Procedure. The protocol used at OSI's Datalink layer for framing, synchronization, error detection. The varieties of LAP all use variants of the HDLC frame format.
— LAP-B is widely used in packet-switching systems.
— LAP-D is for ISDN's D-channels.
— LAP-M is for modems.

LAP-B

Link-Access Procedure, Balanced — for packet-switching. LAP-B is a subset of HDLC that has been standardized by the ITU for X.25. It is a Datalink access control technique, using a frame which is variable in length, delimited by flag bits, and which has a two-byte CRC.

The 'procedures' are those which are responsible for framing, synchronization, error detection, and maintaining the link between a terminal and its network node. LAP-B utilizes ARQ-retransmissions for the correction of data errors. See HDLC.

LAP-D

The Link Access Procedure used in Basic Rate ISDN's D-channel and the frame format used in frame relay. It was derived from LAP-B and designed to carry the signaling part of ISDN (ITU Q.920 and Q.921).

LAP-D specifies a frame format made up of several elements, most of which are identical to LAP-B. The frame is variable in length and delimited by flag bits, and it has a two-byte CRC. It supports multiple parallel logical links to allow several terminals to communicate concurrently.

A new field in LAP-D is the two-byte header, made up of an 11-bit data-link connection identifier (DLCI), and some bits which indicate congestion. The DLCI denotes a connection address (not destination) and it has local significance only, so it can be reused in other data-links.

Because LAP-D is such a small modification to HDLC, many products can be upgraded for use with frame-relay with no more than a change in the software.

LAP-M

Link Access Procedure for Modems. This is a variation on the ISDN LAP-D error control protocols which allow a smooth migration from the use of modems and the analog telephone system to ISDN. See V.42.

Lapack

See Linpack.

laptop portables

A small fold-up computer usually with an LCD screen, powered by batteries. It has a full-size (all fingers) keyboard and it weighs under 8lb. This is now a compulsory item of apparel for anyone wishing to fly Business Class.

I can't find any obvious difference now between laptop and notebook computers, but I suppose someone will always claim that there is. See also notebook.

large area flicker

The eye is able to see flicker on large uncluttered areas of the screen when screen refresh rates are about 50Hz (depending on how long the phosphors allow the image to 'stick'), so the PAL refresh rate is marginal. The NTSC 60Hz rate is substantially better, and a higher rate of 80–100Hz would be better still, especially when screens get larger with wide-screen TV.

The refresh rate at which flicker is not perceived, is known as the 'flicker-fusion frequency'. Flicker-fusion frequency is a quite different problem from the perception of the motion-continuity frequency, which generally requires 24 frames of image per second for a feeling of smooth motion (see fusion).

Flicker decreases at lower brightness levels, and is more noticeable where lighting contrasts are high. It is also worse in the peripheral (rather than central) area of vision, which means that it is exaggerated by large screens and closer viewing distances. It also varies slightly from individual to individual. See phosphor time-constants.

laser

Light Amplification by Simulated Emission of Radiation. Lasers were first developed in 1960 and grew out of the earlier work on masers.

Lasers were widely characterized in the 1960s as a 'solution looking for a problem' because no one could find a use for them. The theory of coherent waves had been known for many years, but it had no practical implications at the time.

Lasers come in many varied models; the three primary types are Gas, Semiconductor (diodes), and Solid State. They produce light frequencies that cover the visual and invisible spectrum, and are generally characterized by their output wavelengths (expressed in nanometers, microns or Angstroms), deviation from a single frequency, and power.

They all produce highly coherent light (none are ever perfect), but some are fitted with a resonance 'cavity' which allows them to be tuned to even narrower bandwidths. There are now horizontal cavity lasers, and vertical cavity lasers. See these and pulsed laser, z-laser, CW, and LED also.

• Ideally the lasers used for monomode fibers will operate in the 1.5 micron band (usually 1550nm) while most of the older lasers ran at 1.3 microns (1310nm).

• The wavelength of the laser used in CD and videodisk systems establishes the packing density of data on the disk. Current systems use red lasers of 750nm wavelength, but claims are being made that shortly, blue-green lasers of 523nm will be commercially available. This will increase data density 2.3-times.

In the interim, the technique of Second Harmonic Generation (SHG) has surfaced which produces a blue light from a red laser. See. Another development is a form of doped fiber which also changes the wavelength.

laser font

The term usually means an outline font such as those provided in PostScript. See TrueType and outline font also.

laser line-width

This is the 'bandwidth' of a single spike on a laser's spectral output. 'Narrow line-width' lasers are necessary for coherent optical fiber transmission systems. Technically, the line-width is judged as being the frequency interval between the two points at which the laser power is half its peak level. Graphically translated, this is the width (in Hertz) half-way up the spike on the graphed output of a single-mode laser.

laser recorders

In film and video production, laser recorders provide a high-quality way of transferring electronic images to film for cinema or television distribution. The system is essentially the reverse of a flying-spot scanner. The unexposed film moves constantly through the machine while three colored lasers (R, G, B) scan across the width of the film and write the film image onto a film negative under electronic control.

laserdisk

A generic term which should encompass CD-Audio and CD-ROM, but which is generally reserved for the earlier large-disk units which stored their video and audio information in FM analog form. Note that the laserdisk was only one form of videodisk, which is a wider term encompassing technologies which include capacitance and physical (needle) tracking. See CD-V, CD-Multimedia, Super Density CD, LaserVision and LV-ROM.

laserprinter

Laserprinter technology is based on the photocopier principle of exposing an electrostatic metal drum to scanning by a strong light beam. The photons knock out electrons, leaving the surface of the drum electrically charged. This charge attracts and picks up dry-powder toner, which is impressed onto the paper by the rolling action, and then heated ('cooked') to fix it.

There are dozens of different factors involved in choosing a laserprinter — everything from the use of parallel connections, for instance, which are much faster than serial . . . to paper handling. Paper handling is very important, especially for printers mounted on LANs, and you will need at least 1MByte of memory in most printers.

Some laserprinters are designed to provide good general, fast, silent printing in an office environment, and handle dozens of users (who need to be able to switch paper/letterhead remotely) — while others are single-user printers with a minimum of features.

• Some printers are PostScript-compatible, for use in desktop publishing — although with TrueType, most laserprinters will do a pretty good job for DTP. These pages have been generated using a normal 300 dpi, base-standard Hewlett Packard IIP laserprinter, using TrueType font-generation from a Macintosh. The camera-ready copy has been photo-reduced to about 60% of the original.

• The often quoted throughput figures for Laserprinter printing times (x pages per minute) do not take into account the time needed to establish the page as a full bit-map in the printer memory. So the first copy of a page may take minutes to be generated, and only then will multiple copies flow at the quoted rate. If you print one copy only of each page, then the 'x pages per minute' figure is meaningless.

• The time taken to generate a page is variable, and it often depends more on the computer application and the drivers, than on the type of link and the capabilities of the printer.

• The current state-of-the-art for low-cost (under $10,000) DTP laser printers, is probably the Xerox ColorLaser 4900 which can provide output of 300, 600 or 1200 dpi along the horizontal, and 300 dpi in the vertical, in full color. QMS also makes one with a 600 x 600 dpi output.

LaserVision

The Philips/MCA trade-name for their analog optical videodisk which later became the basis of CD (digital) technology. It first appeared about 1970, although it only came into common use in 1978.

This was a 12-inch (and later 8-inch) standard which held either 30 minutes (CAV) or 60 minutes (CLV) of PAL-quality frequency modulated video with sound on each side. For feature length movies, variable rotation-speed (CLV) disks were used with both sides holding information for feature-length movies.

The pit length on the disk represented a single time-parameter of a composite (sound and picture) FM waveform. This is an analog recording (the pit length was not in integer steps) and it could be easily reconstructed as an analog FM signal. See PFM.

Disks were either constant linear velocity (CLV — variable rotation speed) for linear program material like movies, or constant angular velocity (CAV — fixed rotation speed) for interactive material where the head had to skim quickly across the disk to find an image frame or sequence. These disks were often used for single-frame image storage also.

CAV NTSC disks spun at a constant 1800 rpm (30 revs per second) and PAL at 1500 rpm, so each revolution stored two fields or one frame of an image, at a quality level slightly above the best PAL image. There were 54,000 tracks on each side of a disk — and each track/frame could be accessed directly. Each track was numbered, and the control program simply addressed the track.

LaserVision is still widely used for legacy interactive video applications, mainly in education. With some adaptations, the disks can hold 1GByte of data modulated into the FM carrier. See interactive (video), CD-i, VHD and LV-ROM also.

LASS

Local Area Signaling Services. Call-waiting, call-forwarding, etc. This was the original AT&T designations, the 'C' was added by Bellcore after divesture. See CLASS.

last mile

Slang for the local loop. See.

LAT

Local Area Transport. DEC's protocol for terminal-servers.

LATA

Local Access Transport Area. There are 161 of these in the USA, and they define areas of exclusivity — the areas within which local telephone companies (LECs) are permitted to complete calls without reference to the IXCs (long distance carriers).

Within LATA areas the wireline companies effectively have a regulated monopoly. Outside and between LATAs the area is defined as competitive. So the LATA represents the old division between the LECs and the IXCs which were designed to give equal advantage to the regional companies during the break-up of AT&T. However LATAs are not necessarily 'owned' by one RBOC and certainly not by any one LEC, it just means that the LATA boundaries define the areas in which they operate.

Some LATAs are physically very large: one in Utah ranges 500 miles north–south. However, in the southern states some are very small; there are 35 LATAs alone in the state of Georgia — while the state of Wyoming only has one.

The 'mean average' LATA has 700,000 telephone lines and 100 local telephone exchanges. However some LATAs range down to a size with as few as 3000 lines, and others extend up to 5 million.

When a telephone call has to cross a LATA boundary, the LEC (local exchange carrier) must hand this 'inter-LATA 'traffic to a long distance carrier (IXC) via the carrier's POP (point of presence). The customer chooses which IXC gets his/her business, but often only from a very restricted list which will always include AT&T. This boundary requirement is now subject to change, but generally the signal must be carried by an IXC between LATAs — even if the originating and terminating LEC share a common boundary.

A call made within a LATA may be handled by one company, or it may be passed directly between two LECs servicing different parts of the same LATA. This intra-LATA traffic may be charged as either a local or toll call (usually as an intra-LATA toll). So while calls within a LATA are 'local calls' for regulatory purposes, this does not mean that they are necessarily 'local calls' for billing purposes. They can be intra-LATA toll calls — which, in fact, are often more costly than national or international calls.

latch

To grab and hold temporarily. The suggestion with 'latch' is that the system will grab one element from a stream, but hold it only temporarily. Latches are used whenever it is necessary to preserve or freeze information. In PCs there are generally internal latches in any I/O interface chip.

- Latch is a general term for flip-flops made from NAND gates or NOR gates.
- In computer terminology, latch sometimes implies that some unusual external event has triggered the holding action.
- In broadcast teletex systems, the image frames circulate past the receiver at hundreds of pages a minute until the required page number is recognized by the receiver. The frame store then 'latches' onto this image and holds it temporarily in memory for screen display. This is also called frame grabbing.

latency

This word has two quite distinct meanings:

- Not apparent, hidden, concealed. In film, an image is latent until the film is processed.
- Delayed. The delay in disk-access time while the system waits for the disk to make a part-rotation before the next sector can be written or read. See sector interleave, RLL and interleave ratios.

— **Network latency:** The propagation delay, end-to-end across a network, is the measure of the network's latency.

— **Router/Switch latency:** Typical figures for network's packet latency is about 2000 microseconds for a bridge and slightly longer for a router. For Ethernet switches you might expect delays of between 40 and 800 microseconds, depending on whether the switch is a cut-through or a store-forward type.

— **Ring latency** is the time it takes a signal to propagate around a ring. In Token Ring LANs it is measured in terms of bit-periods.

— **Rotational latency** is the time it takes for a magnetic disk to rotate until the next useful sector comes under the head of the access arm. See RLL.

- **LAN:** This is the average time between a station seeking to access the LAN, and when that access is granted (or possible).
- **LAN Switch:** This is the time it takes for a packet to pass across the switch. In store-forward switches it must stop and be processed, while in cut-through switches it powers on through.
- **Network:** For isochronous services, the network latency is vitally important. This is the average (or maximum expected) end-to-end time of any packet or cell on the network.

lateral inversion

The mirror reversal of an image. Left is exchanged for right, and vice versa. Top and bottom remain the same.

lathing

In CAD, lathing is the ability to rotate a 3-D object realistically on the screen and see all sides.

launch vehicles

In satellites, this is a general term covering either multi-stage rockets or the shuttle — the propulsion system needed to get satellites into low-earth orbit. The cost of the launch phase (including insurance) will be about one-third of the overall cost of the satellite.

• Orbital Technologies now make the Pegasus, which is a second-stage launch vehicle much like a cruise missile, which is launched from the Lockheed Tristar aircraft.

Lavelier

A neck mike. A small microphone hung on a cord around the neck, or attached to the tie or clothes by a small clip. These mikes have bass de-emphasis to allow for directly transmitted vibration from the chest.

Law of Large Numbers

One of the key 'laws' in telecommunications. It says variability is more pronounced in small groups than in large.

For instance, the law suggests that between any two cities, the packetized voice and data-streams from a half-dozen large companies may vary by an order of magnitude (say, 10-times) during the day, however the multiplexed data-streams from 6000 small companies won't vary to the same degree in percentage terms.

With many more (but smaller) users, both the low and the high-point will randomize, 'average out' and regress to the mean. So the variation experienced by the network between low and peak-high loads with thousands of companies may only be half as much as that created by the larger companies.

So the law can be interpreted to mean that it is more economical to have many small users of bandwidth on any network, than it is to have large users. Networks must be designed and provisioned, not for 'average' loads, but rather to handle 'peak' loads — and it is large users which create the need for uneconomical bandwidth safety-margins.

The consequence of the Law of Large Numbers, is that network economies demonstrate features which are alien to people trained in classical economics, who often look upon Economies of Scale as natural law. Taken from a purely engineering point of view, carriers should not provide special discounts to large users, but rather charge them a premium.

The ideal customer is one with small bandwidth requirements and a randomized time of access (or off-peak access). Discounts should be given on a time basis to average-out the load, not on volume which encourages over-engineering. Of course there are other complications here, such as switching, billing, and local loops that lie idle.

The same principle often applies with other networks — such as road systems, power supplies, etc. Big is not better.

LAWN

Local Area Wireless Network. The IEEE 802.11 standard will set a new wireless protocol for line-of-sight IR systems and spread spectrum devices. It will allow more than 1000 nodes, links up to several hundred meters; and transmission rates up to 20Mb/s.

There are a number of wireless LAN network systems now on the market. Some of the more notable are:

- NCR's WaveLAN is a spread-spectrum system released in 1990 for 2Mb/s links using the 2.4MHz ISM bands.
- Motorola's Altair which connects with Ethernet and can support 32 users of a single microcell of approximately 500 square meters. It uses frequencies in the 18 GHz range, and handles data up to 15Mb/s.
- AT&T's WavePoint bridge operates in the 2.4MHz ISM band at 2Mb/s.
- Airlan uses ISM CDMA frequencies for 2Mb/s links up to 10 miles, for LAN interconnection.
- There's also a range of CDMA techniques being developed by Nynex and a number of other US companies.

• Note that the term can be applied to both infrared (IR) and radio links.

layered

• **Protocol:** This usually refers to complex protocol suites which conform to the OSI reference model, but SNA and TCP/IP are also layered protocols. Protocols can be either monolithic or structured.

It is clear that communications is now too complex in most networks to be handled by a one-piece protocol. So suites are structured, using layers. The main advantages of layering these protocol suites, is that those users who require only simple network functions can utilize only the lower layer protocols, while those wanting more complex functions can add progressively higher layers.

Another advantage is that some layers are network-specific and some are application-specific, so by layering the protocols and tightly specifying the layer interfaces, one part can be substituted by another without affecting the overall functions. See OSI reference model and interface.

• **Video:** Video is layered so as to provide different standards of quality, and/or different bandwidths, to different receivers. If you want the highest possible quality you will take all the layers (probably delivered in ATM cells), but if you only have a small portable, you may only take the lowest layer. See UVC.

• **CD Multimedia** and **Super Density CD** optical disks are layered in an entirely different way. This is a physical sandwich. See CD Multimedia and Super Density CD.

• **ECC/EDC:** Layered ECC refers to an additional layer of error detection and correction which is used in Mode 1 of CD-ROM. This is a second level of Reed-Solomon correction which is added to the CIRC (which is enough for CD-Audio) to provide the level of reliability needed in data delivery.

• **In CAD:** This is where the software has the ability to develop a drawing as a series of separate layers which can be viewed or printed individually or in combination.

layout

In desktop publishing, the layout is the rough overall design of the various pages showing where the illustrations will be, and how the headline will fit. The term is technically correct only for the rough design phase, while the final product should be called camera-ready copy. However the term layout is now often used for both. See comp. also.

LBA

Logical Block Addressing. A way of overcoming the DOS 528MB barrier on hard disk drives. It requires major software changes. The standard addressing convention is replaced, so it requires changes to the OS, BIOS and disk drives. See CHS.

LBV

Local Bus Video. See local bus.

LBX

Low Bandwidth X. See X-windows.

LCC

Low Capacity Cellular telephone technology. The intermediate mobile radio stage before the introduction of the present analog cellular telephone systems.

LCD

Liquid Crystal Display. Liquid crystals were first discovered in 1888. They combine the properties of a liquid and a crystalline substance, and they can change polarized states under control of a specific voltage, so they are now used in computer, watch, and many other electronic systems for flat-panel display. However, LCDs are only one type of flat-panel device (see EL and plasma also).

There are many variations on the basic LCD system, however all the various types have a similar physical construction. The liquid crystal material is sandwiched between two very flat glass sheets into which a cross-hatch of transparent conductors have been etched. This sandwich is, in turn, sandwiched between two polarizing filters and the screen is constructed so that light either passes through this sandwich ('backlit LCD') or is reflected from it.

There are three basic types of liquid crystals: smectic, nematic and cholesteric. Nematic crystals are generally used for most LCD displays today.

The term 'twisted' or 'supertwisted' refers to the degree of polarizing rotation between the layers of active material.

There are also passive screen technologies (most of the old LCD screens) and active (the new LCD screens).

- The main problem with passive LCD driver technology is the need to individually control many conducting intersections simultaneously; active-matrix techniques overcome this.
- With active screens, the displays are virtually very large-scale chips with the switching elements integrated into the screen surface. These go under the name active-matrix, thin-film transistor (TFT) or ferro-electronic (FLCD).

• Twisted nematic screens have a higher contrast than conventional LCDs, and respond in about half the time (30–40ms). Response time is important for flat-panel TV sets where images are constantly moving. However twisted nematic screens have very narrow viewing angles for TV.

• Color LCDs for computers usually use the twisted nematic crystals, with thin-film transistor switches. The light is twisted as it passes between successive polarizing layers with the various color filters.

• Matsushita now has a color LCD display screen of 264mm square, now on sale, for about US$3500. This is the largest size commercially available.

See active-matrix screens and twisted-nematic also.

LCP

Link Control Protocol of PPP. See.

LCR

Least-Cost Routing. See.

LDCELP

See CELP and G.728.

LDM

Limited Distance Modems. Aka line-drivers. Special 'modems' which don't use conventional modulation techniques. There are both synchronous and asynchronous LDMs, and various forms go under the names 'Data Service Unit' (DSU), 'line adaptor' or 'baseband modem'. At the top end of the scale LDMs can have data rates up to 19kb/s for twisted pair cables (up to 15km) and 230kb/s for coaxial. See line-drivers and DSU.

LE-ARP

LAN Emulation Address Resolution Protocol. See LES.

lead-acid batteries

The type of accumulator used in cars. There are both sealed and unsealed types, and naturally only sealed types are used in portable computers. The advantage of a lead-acid battery over ni-cads is that you can monitor the state of charge, since the voltage drops significantly over the drain period. The disadvantage is that they are heavy in relationship to the power they can deliver. See ni-cad and lithium ion.

leading

(Pronounced 'ledding'.) A typesetting term meaning the amount of space (measured in points) between the lines of text in a column. These columns have one point.

Leading is actually the distance between the lowest descender of one line of characters and the highest ascender of the next. With zero leading, the bottom of a 'g' and the top of a 'd' directly below it on the next line, would normally touch.

• Point size and line-space are generally expressed as, say, Times Roman 10/11 ('ten-on-eleven') which indicates that there is 1 point of leading between the ten-point characters.

leading edge

• **Pulse:** In electronics, when dealing with square waves (digital transmissions), the leading edge of the wave is the point at which it changes the state of the circuit instantly from a low to a high voltage, or vice versa. Most synchronization and timing uses this leading edge, since the trailing edge is more easily distorted.

No circuit can switch instantly from one voltage state to another, so a perfectly 'square' wave is only a theoretical construct and, in practice, the leading edge is never perfectly vertical but rather takes on the shape of a high-frequency sine wave. It is this frequency which, in fact, defines the upper limits of bandwidth in a digital circuit.

• **Hype:** In marketing all technologies are 'leading edge', so the words are now virtually meaningless. It is naive to buy leading edge technologies — only tyros buy Version 1.0.

The term is intended to suggest that the designers of the system are well advanced in their research over those of every other manufacturer. But since virtually all manufacturers of useful products source their chips from only a few chip makers, and all need to conform to industry-wide standards, these claims can rarely be justified.

leaky bucket

A concept applied in ATM communications networks to refer to a form of network traffic congestion control. At any node on the network, a counter (the 'bucket') fills as cells arrive, but it empties at a fixed rate. When the bucket reaches its pre-set limit (the 'overflow'), no more cells are accepted. Cells are then dumped. These systems rely on end detection of cell loss, and retransmission for replacement of lost data. With video and voice, the lost cells are usually just abandoned.

learning bridge

A bridge which can automatically learn where addresses are located. See spanning tree.

leased line

The term simply means a standard link over the telephone network that is rented for permanent assignment from A to B. The term covers everything from analog to digital, and 32kb/s to (say) E1/T1 lines. Although the old term 'lines' is being used here, these are most likely to be standard telephone channels.

The distinction is being made with dial-up lines, and with 'virtual private networks' and 'semi-permanent' links where a channel appears to be exclusively assigned for use, but is actually part of the switched network.

Leased lines are usually connected through standard switching exchanges but on a non- or permanently-switched basis — they are 'tied up' connections. Apart from instant availability, one of the main values of a leased line is that its connections are always made through the same hardware. This means that these lines can be 'conditioned' to compensate for transmission-line distortions, and so noise-induced errors can be significantly reduced.

You can't do this permanently on a switched circuit, because different electronics and paths are used every time — although modern modems now condition lines each time before they begin transmission. With these modems much of the advantage of conditioned/leased lines has been lost.

There appears to be a subtle difference between leased lines and private tie-lines, in some countries. The first term is often applied to guaranteed 'clean' 56kb/s or 64kb/s connections suitable for data, while tie-line is more specifically used for analog voice. In some digital public networks, tie-line voice may be carried over 32kb/s ADPCM channels.

The most common leased line sizes are:

— Standard 4kHz analog telephone tie-lines.
— 64kb/s digital links for data (in the USA, 56kb/s).
— T1 and E1 lines (1.54Mb/s and 2Mb/s).
— Fractional T1 or E1 (n x 64kb/s).

least-cost routing

• **LAN/WAN:** Virtually all packet-switched networks use some form of 'least-cost' (LCR) calculations as the basis for their routing decisions outside the immediate local environment.

— Dijksra's algorithm finds the shortest path from a given source node to all other nodes.

— Bellman-Ford algorithm also finds the shortest path, but with the added constraint of requiring no more than one, two, or three links.

— Other systems are dynamic, taking into account load on the network, network failures, etc.

Circuit-switched systems use different least-cost algorithms.

• **PBX Switching:** PBXs often have enough intelligence to work out which carrier should be given the job of carrying a particular long-distance call, based on the table of changes and discounts. The more sophisticated of these can take into account, day of week, time of day, holidays, etc. These tables are a constant source of frustration to communications managers because the long-distance carriers constantly change prices. See ARS also.

• For LANs and WANs. See link-state routing.

least significant bit

In any byte (or series of binary digits), the bit at one end of the string will be the most significant (if changed it will create the largest effect on the number) and at the other end it will be the least significant. In binary, by convention, the least significant is written at the right, as it is with denary/decimal notation.

The significance of this terminology is that often communications systems don't handle the data-bits in the normal order of left-to-right. Sometimes they will transmit the least-significant bit first ('little-endian'), and sometimes it will be transmitted last ('big-endian').

• In quality loss-tolerant application such as the transport or manipulation of voice and video, you can often afford to dump the least-significant bits in a byte, if necessary. But not the most-significant bits.

• When bits in a byte are numbered, the least significant bit is No.0, so the numbering proceeds from right to left with the most significant bit in an 8-bit byte being bit No.7.

LEC

• **Local Exchange Carrier:** An American term for the local telephone company. There are perhaps still 2000 of these; mainly small companies which service only one or two towns in a small region. Some are 'mom-and-pop' companies owning one small switch and having just a few hundred customers.

In the larger cities and more developed areas, the LECs have millions of customers. The largest were the original 22 Bell Operating Companies (BOC) which became the key components of the seven Regional BOCs (RBOCs) after the break up of AT&T in 1984.

The RBOCs are often treated as if they are LECs because they are seen as identical to their component parts. See POP, LATA, RHC and IXC also.

• **LAN Emulation Client:** In ATM networking, this is the entry point to the emulated LAN. The LAN emulation server issues a two-byte code called a LECID (LAN Emulation Client ID) which is unique to each LEC. See LES and LANEW.

LECID

See LEC above.

LED

Light Emitting Diode. Small semiconductor light sources used for indicator lamps in all sorts of electronic devices, and as drivers (laser substitutes) for multimode optical fiber over short distances. Both visible and infrared LEDs are available.

They are now also being used as light sources for some printers. LEDs do not burn out; they require very little voltage, and they switch almost instantly between the on and off state.

• If you wish to use a two-contact LED as an indicator in a circuit, the longer of the two contacts is the anode. You should mount a resister in series with the LED: at 6V use 220 ohms, at 12V use 560 ohms.

• In three-contact, dual color LEDs, the longest pin is the common cathode.

• LED printers are very similar to laserprinters, except that the light source will be very many LEDs (a 'LED array') rather than a single laser sweeping across the printer drum surface.

legacy

Systems, data, devices or protocols which are old, but which can't be discarded. This usually refers to equipment remaining after a revolutionary upgrade to a new standard or new technology. Often legacy equipment is too valuable (in book terms) to discard, but too difficult to use. See greenfields.

• The major problem that network managers face is often how to mix LAN traffic with legacy traffic — usually SNA/SDLC, async, Bisync, STDM and X.25.

• IBM prefers to call its own older systems 'heritage systems'. They don't sound quite so dead.

• Legacy data usually refers to flat-file databases on a mainframe (as distinct from relational databases).

legend

A list explaining the meanings of symbols, colors, etc. on a map or drawing.

Lempel-Ziv

See LZW.

LEO

Low Earth Orbit satellites. These are usually orbiting from a few hundred kilometers above the Earth to about 2000km at the most. They are just outside the atmosphere.

• Big LEOs are those which offer voice and data, and which have frequency allocations above 1GHz. The FCC has said that these systems will operate in the 1.6GHz and 2.5GHz frequency band. Their plan will assign CDMA systems 11.35MHz of shared bandwidth, and TDMA/TDD (Iridium) systems will have 5.15MHz of dedicated bandwidth.

• Little LEOs are systems which provide non-voice, store-and-forward mobile satellite services at frequencies below 1GHz in the UHF band. At these frequencies they are sharing the spectrum with terrestrial users on paging and two-way radio systems around the world. See Big LEO and Little LEO.

LEO spectrum

The FCC has allocated the range from 1610 to 1626.5MHz (uplink) and 2483.5 to 2500MHz (downlink) for Big Leo operations.

The primary division is between CDMA and TDMA systems on the uplink:

- 1610 to 1621.35MHz is for CDMA (shared bandwidth).
- 1621.35 to 1622.6MHz has been allocated to TDMA, which CDMA may use, if necessary.
- 1622.6 to 1626.5MHz is for TDMA/TDD exclusively.

CDMA may use the whole of the downlink band from 2483.5 to 2500MHz.

Iridium, Globalstar and Odyssey have all been given FCC licenses at the time of writing, although Ellipso has been rejected and has resubmitted its application. Only Iridium plans to use TDMA/TDD; the others will use CDMA.

LES

LAN Emulation Server. This uses the LAN Emulation Address Resolution Protocol (LE-ARP) to map a LAN's MAC address to an ATM address so as to establish an end-to-end virtual circuit connection (VCC) for ATM internetworking.

letter box

• This refers generally to the image dimensions of Cinemascope when shown on a conventional 4:3 TV screen. The full image of a wide-screen or Cinemascope movie is transmitted, and black bars are left at the bottom and top of the screen.

The alternative is for the TV station to 'pan and scan' the image before transmission. Selecting those parts to transmit.

• The term letterbox is also being applied to a specific form of interim HDTV standard proposed for PALPlus in Europe. They are also showing a wide-screen image, and they will need to keep it viewable on the remaining standard PAL sets.

When transmitted, the conventional PAL receiver will receive a 'letterbox' image which uses only 432 active image lines, with (apparently normal) black lines top and bottom. But the black bars actually contain 143 lines of information which have been modulated at a signal level between the black-level and sync.

With a special 16:9 television set, each of these extra 143 lines can be demodulated and inserted back into the image after every third line of the letterbox scan to produce a full 575 active lines of image. The restoration will provide a wide-screen, full-height, image with a 16:9 aspect ratio. See PALPlus.

letter shift

The receipt of this character caused a physical shift in old teletype terminals using Baudot code to enable the printing of alphabetic characters (caps only). This alpha-mode remained in force until the figure-shift code reversed the process and allowed numbers and special symbols to be sent.

letter space

The process of adjusting the space between letters is known as kerning. The adjustment of this space is often important when large sizes and capital letters are used for display or headlines. See kerning.

LF

• ASCII No.5 — Line Feed (Decimal 10, Hex $0A, Control-J). This moves the paper up in a printer by one line. On a monitor it scrolls the display up one line (or moves the cursor down one). Some systems always couple CR and LF together; some often treat them separately. See hard return.

• Low Frequency (also called LW or Long Wave). In the radio spectrum these are frequencies between 30 and 300kHz with kilometric wavelengths from 1km to 10km. These propagate as surface waves and therefore follow around the curvature of the earth. The main application of this band is for aircraft and maritime radio navigation beacons (200 to 400kHz).

library

Any store of pre-defined objects, modules, text, images.

• **Programming:** A stored series of routines, images, data modules, or whatever, which can be taken and used by a computer programmer in creating a program. Programmers and CASE tools use libraries of special algorithms and routines for programming.

In some programming environments, libraries are segments of code meant to be shared. The contents of a library file can often only be accessed by an application through an already existing task.

• **Chip design:** Complex chips, especially ASICs, are usually designed by taking standard library elements, laying them out on the surface, and then interconnecting them.

licenses (software)

Mainframe software licenses were relatively easy to control, but when PCs came along the question of licensing became very confused. LANs complicated matters even further.

A software license is a contract granting the buyer the right to use software on a limited basis. There can be a wide variety of different agreements, but this is a guide:

— The **basic license** is for a single machine/single user. This has recently been varied to allow the same license to be transferred to a copy also used on the customer's laptop.

— A **site license** is a volume-discount contract with an organization for a specific number of copies to be used on a specific network installation.

— **Corporate licenses** are an enterprise-wide variation on site licences.

— A **LAN license** is usually priced according to the number of users on a LAN. A variation on this is the 'concurrent' license which is priced according to the number of users who can access the software at any one time.

ligature

A stroke or bar connecting two letters — and by extension, the two letters treated as a single character. You find this often in printing if you examine print closely. The letters 'ff' and 'ft' for instance, are often joined together to produce a more pleasing effect. Without ligatures, the kerning control will often insert too much space. The & sign was once 'et'. See kerning.

light

See visible light, infrared and light spectrum.

light-based

Color systems using additive primaries. See.

light-emitting diode

See LED.

light-gate

An approach to TV projection systems which doesn't rely on a high-output being generated by the image source but by filtering. This approach to the projection of video images, is almost identical to the projection of film images — where the film acts as a block or filter, with the source of the light behind. These techniques are also known as 'light-valve' methods.

Most video projectors available today use three high-output tubes to generate very bright (R,G,B) image sources, and a wide-aperture lens system then throws these images onto a highly reflective screen. These systems are limited in size and application by the brightness that can be generated by the video tubes.

By contrast, the light-gate is a subtractive system. It just blocks part of the light, so it can project images of almost any size. The brightness of the image depends only on the output of the projection lamp. See passive matrix and TSTN also.

• The GE Talaria projection system uses a light-gate composed of an oil-film; others often use transparent LCD displays.

• One recent light-gate with distinct possibilities for home use is the Projectavision. The gate is an active matrix, thin film transistor LCD unit in the form of a small-glass plate sandwich.

light-pen

A hand-held 'pen' with a photo-diode as a 'nib'. The pen is linked back to the computer by a cable and it sends a simple digital pulse when it sees a bright light — and it only sees the light when the electron beam scan goes flashing past the end. So when the pen is pressed against the screen, the computer is able to detect its x–y coordinated position on the screen by its own knowledge of the scan-line position at the time the diode 'saw' the sweep of the VDT's electron gun. This information can be fed back into cursor control, or into a drawing program, or used to find map coordinates, etc.

light spectrum

The light spectrum includes the visible spectrum, and both infrared (IR) and ultraviolet (UV) light. IR is at the lower frequency (longer wavelength) end of this range up against the high end of microwaves (actually the EHF spectrum).

Light waves are usually defined in terms of wavelength. IR is from 1mm to 740nm (740×10^{-9} meters): visible light is 700nm (red) to 400nm (violet), and UV is from 400nm to 5nm. UV then butts up against X-rays, then gamma, and lastly cosmic rays at the end of the electro-magnetic spectrum.

light-valve

This can probably apply to a number of photo-optic devices these days. In image projection, see light-gate.

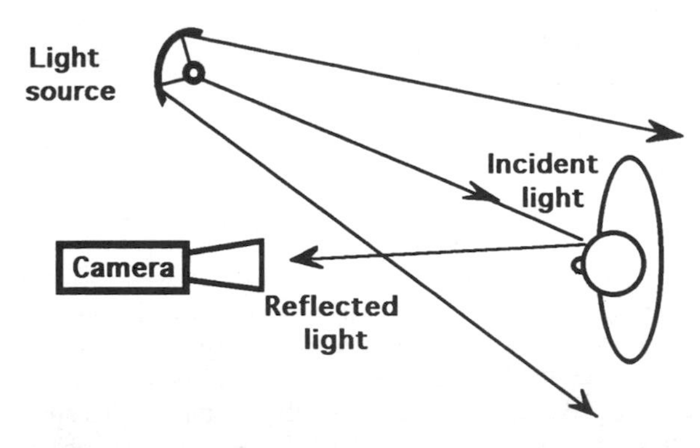

lightmeter

There are three basic types of lightmeter used in film and television production.

• **Reflected lightmeters** measure the light after it has reflected from the objects to be photographed. These meters are always held at a position near to the camera lens and pointed at the object. They are therefore making a complex measurement that depends on:

– the brightness of the light source/s, and
– the reflectivity (gray-scale variations) of the objects (and background) within the light-meter's field of view (its acceptance angle).

Spot meters are reflected meters with a very narrow angle of view. They are used to judge tones after basic exposure has been set by the cameraman.

Incident lightmeters are used as a substitute for the object to be filmed. They usually have a white dome, and are held directly in front of the object, and face back towards the camera lens. They only measure one factor — the brightness of the light source/s, and do not take into account whether the object or background is predominantly white, gray or black.

Good exposure depends on taking both light-brightness and object reflectance into account, so top cameramen use both meter types, and calculate a result to produce the desired effect.

• **Color temperature meters:** These are incident meters which measure the balance between source light at the blue and red ends of the spectrum. This allows filter grades to be calculated.

lightwave/guide

Another name for optical fiber and fiber cable. Note that the term 'cable' applies to the whole structural unit which will normally include a number of fibers, plus strength members and power feeds (in submarine cable), usually sheathed in dense PVC plastic, and possibly steel wound armor plating for crossing the continental shelf. See single mode and multimode optical fibers.

LIM

Lotus, Intel, Microsoft.

• LIM Emulators: The above group jointly devised the EMS standard which allowed bank-switching on MS-DOS machines to (apparently) use memory above the 640kB RAM limit of the system. This is an 'expanded memory' system which uses locations in high memory. LIM emulators make extended memory (above the 1 Megabyte DOS limit) appear to the EMS system extensions as expanded memory so that older EMS-dependent programs can run. The most common of these is EMM386 which comes with DOS. See EMS and high memory also.

• LIM/EMS: See EMS.

• LIM memory. This is expanded (EMS or bank-switched) memory above the 1MByte limit.

limited distance modem

A line adaptor or baseband modem. See line-driver and DSU.

line

This originally referred to a single connection, which was then an open wire using earth return (now called SWER — Single Wire, Earth Return). Line later came to mean a single connection via twisted-pair. Now it seems to be synonymous with channel or circuit — whatever is necessary to carry a conversation end to end.

• Line (singular) is now applied collectively to both 2-wire and 4-wire telephone circuits.

• A single trace of the electron beam as it sweeps across a television or monitor screen. This is known as the 'raster scan'.

• As a measure of vertical resolution of a video screen, the number of scan lines provided is fixed by the television standard — but this isn't a measure of the resolution (it just sets the upper limits). The actual resolution of 'lines' is about 2/3rds of the number of active scan lines. This has nothing to do with the Kell factor. See active lines and bezel.

• As a measure of horizontal resolution of a video screen, the number of lines of an image which can be resolved is directly related to the bandwidth of the signal (often limited by the quality of the driver circuits). This figure represents how fast the electron scan can switch from black to white as it draws each scan line — the limit of the system's ability to handle the quick changes in contrast which are necessary to provide fine resolution. So to measure horizontal image resolution, the test-card will have vertical lines. It is better to use the term cycles here; a white line plus a black line equals a cycle. (A 2:1 ratio.)

line access

The physical point at which a handset, facsimile or computer terminal is connected to the telephone line. The end of the customer's circuit, and the beginning of the carrier's. This is often called a demarc(ation).

line adaptor

This can be a number of different devices, but it is usually a limited-distance modem. See DSU

line art

Illustrative material that is just in black and white form — not halftones (with shades of gray). See halftone.

line booster

This can be an amplifier, limited distance modem, repeater — almost anything. See line driver and DSU.

line-card

The modules in a telephone exchange rack that act as the interface between the subscriber's line and the central switch unit. It is the first electronics that the signals from the local loop meet when they enter the telephone exchange. The decoding of the signaling (pulse or tone) may be done at this level.

Line-cards are service-specific. An exchange will have analog line-cards on an analog line, ISDN linecards, DDS cards, etc. A moderately large exchange switch may have up to 100,000 line-cards — one for each connected customer service line. On some systems, one line-card will often handle a number of incoming lines (usually about eight) and in some distributed intelligent switches, each line-card will have its own processor.

line-codes

The physical-layer techniques used to transmit the digital data over cables and telephone lines. There are literally hundreds of these line-codes but only about a dozen are widely used. There's no real difference between line-codes, channel-codes or constraint-codes, although the term 'line-code' is best used for techniques at the simple end of the spectrum.

Line codes need to work over different cable types, for different lengths, at different rates, with different tolerances for error and interference, and at different cost structures. Some are baseband, and some are passband.

The simplest baseband line-code is a 'unipolar' technique that applies a DC voltage to the line, and switches it on for a 1 and off for a 0. But this can only work effectively over very short distances and it needs parallel control circuits.

The most popular line-coding formats are:

— NRZ and NRZI.
— AMI and Pseudoternary (bipolar).
— Manchester and Differential Manchester.
— B8ZS and HDB3.

See all of the above, plus channel-code, 2B1Q, CAP and DMT.

line conditioning

See conditioning.

line discipline

Control procedures for orderly communications with many stations on a data-link — this refers to access control techniques and protocols. See poll, contention, token and CSMA.

line displacement (rotation)

A video scrambling technique. See scrambled video.

line doubling

A technique used with large television screens to give the appearance of high-definition without requiring the bandwidth. It is pseudo high-definition. Each image is held in a memory store, and scanned twice. So an image with 575 active scan lines (PAL), appears on the screen as 1150 lines.

• Be careful with the claims about line doubling; this technique has no effect on image resolution at all, but it halves the resolution of the scan-lines (which is good).

Most companies promoting line doubling systems multiply the pixel count by two on the basis that there are twice as many lines — but this is not correct. In discussing image resolution, pixel should refer to the number of image elements, not to the number of scan-line elements. See pixel.

• While line doubling doesn't improve the vertical resolution of an image (unless it is coupled with motion compensation, or some other technique which augments the information — see MAC), however it does improve the subjective feel of picture quality. It reduces the eye's ability to discriminate the scan lines (they are smaller and closer together) when the screen is viewed from a close distance. So it appears to improve picture quality by reducing the scan-line resolution!

line-driver

A general term applied to limited-distance 'modems' (where there is no audio or R/F modulation function). The main operation of line-drivers is to translate line-coding techniques, although more intelligent functions may also be involved. These are the variations:

— A simple form of driver, also called a **line adaptor** or **data service unit (DSU)**, is used only for local connections. These are serial line-drivers which typically operate at speeds up to 19.2kb/s (but some go up to 250kb/s) for distances between 30 meters and several kilometers using (usually) two pair of normal telephone twisted pair cables (4 wires). See DSU.

— There are also **multi-drop**, **multipoint** (polling), async and sync line-drivers.

— Line boosters are signal converters which condition (reshape/regenerate) a digital signal to ensure reliable transmission over extended distances when a line driver is being used. A line driver/booster can also be a power amplifier on a baseband system (analog and digital).

line editor

A simple text editor used with ASCII (usually for programming) which edits one line at a time. There's one that comes with the MS-DOS utilities, called EDLIN.

line feed

LF. An ASCII No.5 character (Decimal 10, Hex $0A, Control-J) which forces the cursor to reposition onto the next line. See LF and hard return.

line flyback

The horizontal blanking interval in a TV or monitor scan, when the electron gun (raster) is returning from the right side of the screen to begin a new line on the left. The beam is turned off during this period of about 11μ (actually the driver voltage is reversed). See horizontal retrace time.

line frequency

In television and monitors. This is the number of scan lines written on the screen in every second. In the NTSC system with 525 lines being scanned every frame and 30 frames a second, this is a line frequency of 15,750 per second (interlace scanning). In PAL with 625 lines and 25 frames per second the line frequency is 15,625 per second. These figures are usually written as 15,750Hz or 15,656 Hz.

• A good computer monitor may have 1000 lines which are being scanned progressively (not interlaced) 70 times a second — which requires a line frequency of 70,000 Hz. Even if the horizontal resolution of your TV set and your computer monitor were the same (which they are not) it's obvious that the bandwidth of the monitor must be much higher than the TV standard to provide this rate of line writing.

• The line period is the time taken for each horizontal scan, in total. This includes both the raster scan time (the 'active' time), and the horizontal flyback (HBI) time.

line hit

These are sudden and abrupt changes on a telephone line. Various kinds of hits appear in phone circuits without any obvious cause:

— Dropouts. Sudden losses of signal.
— Gain hits. Sudden unexplained increases in amplitude.
— Phase hits. Sudden shifts in the phase.

These can all cause distortions and disturbances of various kinds. See transient.

line isolation unit

A circuit/unit used with modems and other gear which may be attached to telephone lines. It electrically isolates them from the phone lines so as to avoid problems in the event of failure — either the modem's failure or the phone company's.

This is not done to protect your computer, but rather to ensure that the telephone technicians aren't electrocuted back at the exchange. Line isolators generally don't protect the computer from surges or lightning strikes.

• Some of the best line isolation units depend on optical signals passing over a short glass-fiber gap. Others are line-isolation transformers: Modems will have one of these built-in.

line level

In audio equipment this is a high-level input or output circuit. Normally this input can handle up to 1 volt (the approximate level expected on, say, a loudspeaker lead) while a microphone input can only handle millivolts. See AUX.

In an audio or TV studio, the line level for all feeds will be highly standardized in impedance, and usually requires pre-amplification.

line noise

A general term for all sorts of electronic interference introduced by external factors and by the communications system itself. Every system has line noise

Claude Shannon calculated the data rates that were theoretically possible over various circuits of various bandwidths, taking line noise into account. at various levels. His calculations establish the absolute highest rate at which your modem can shunt data over the line. But note: Shannon's calculation is for the whole line connection (end-to-end) not just for your local loop.

line-of-sight

Certain types of transmission carrier — specifically lasers, IR transmitters and microwaves — travel only in straight lines, and therefore need line-of-sight between sender and receiver. Line-of-sight isn't the best expression, because reflectors, prisms, and beam-benders can often be used to bend these signals around obstacles (such as another building) to reach points that you can't see. But let's not get pedantic!

VHF and UHF, however, are scattered by atmospheric particles and sharp hard edges, and can therefore appear to bend over the curve of the earth or around buildings. These frequencies propagate by 'ground waves' (they don't reflect off the ionosphere like sky waves).

• After line-of-sight is lost with the UHF/VHF transmitter, the ground wave signal will drop off in strength substantially, but from that point on, the remaining signal strength can often be detected for quite a distance.

• Radio waves also bend over sharp hill crests using an effect known as knife-edge refraction. The hill crest acts as a secondary transmitter. This is an interference effect.

• Another phenomenon known as 'surface waves' allows very long wave (VLF and LF) signals to 'wrap-around' the curve of the earth. The wave-lengths being used are in the order of the curvature of the earth.

line period

The time taken for a TV raster to scan across the screen and fly-back. In PAL 50Hz systems this is 64 microseconds (μs) and NTSC is almost the same. This is reduced by 'overscan' (the area of image wasted behind the bezel) and the fly-back period, so only about 52—53 μs are actually available to draw the image on the screen (the 'active' display time).

line printer

This has two quite different meanings:

• A printer which prints one line at a time, as do most bi-directional dot-matrix printers. They will begin printing when the first line or two are in their buffers, and they will continue to print as long as characters are delivered.

The contrast here is with page printers which need to generate the image of a whole page before printing begins, and with the early impact printers that only handled one character at a time, and were totally controlled by the driver in the PC.

Some line printers are uni-directional and some are bi-directional.

• Line printers for early mainframes, however, were special high-speed printers which produced a full-line of characters with each 'impact'. These were multi-pin dot-matrix devices which produced low-quality output on fan-fold paper. Fast line printers could produce 1500 lines of text a minute.

line protocol

An OSI Data-link control protocol. See Data link.

line shuffling

This is the most common device used in the scrambling of digital video. It relies on transmitting the image lines out of order, in a constantly changing pattern. The correct order is restored in the frame store.

line spacing

• The measurement of the space between text lines when printing. This will always be the sum of the point-size of the text copy, plus the point measurement of any leading. See leading. It is usual to specify the size of the type and the line-spacing in points, rather than the leading.

• The term can also refer both to increasing or decreasing the 'leading' (space) between normal printed lines to make copy fit the space available on a page. You may decide to change from 10/11 copy (10 point type on 11 point line space) to 10/12 copy, just to extend the depth of each column, and make it fit the space available.

• In office-work, line spacing refers to the use of double or triple spacing. Double spaced type was widely used in draft typewritten copy to provide room on the printout for corrections. This was also done in professional printing to allow for editing and marking up the copy for typesetting.

line switching

The same as circuit switching.

line transient

A general term for impulse noise, dropouts and phase- and gain-hits in a telephone circuit. See impulse noise and dropout.

line turnaround

The reversing of the transmission direction from sender to receiver (or vice versa) when using half-duplex operations. Some systems of half-duplex communications have line turnaround which is so fast that the system is treated as being full-duplex. In others, it is painfully slow.

linear

Sequential. The term is often used in computers and communications to mean the opposite of random-access. In linear systems, the elements are read, stored (or whatever) in strict sequence — beginning at 1, then 2, then 3 .... to x. Tape storage devices are linear, while disks are random access. The distinction is important, because it has wide implications in the way the information can be indexed, accessed and used — and therefore in how it is stored and handled.

Random access isn't necessarily 'better' than sequential access, however: it's horses for courses. A feature film will probably be watched in a linear fashion, while a video encyclopaedia would probably need to be randomly accessed.

• Linear programming is structured programming. Object-oriented programming is close to 'random-access' with modules that are linked by a message passing system.

linear amplifier

An electronic amplifier which provides an output arithmetically proportional to the input over the range of frequencies being handled. It doesn't introduce frequency distortions, and it is not deliberately boosted to provide equalization.

This is usually a Class A amplifier. Don't confuse this with a linearized amplifier, which is usually a Class AB amplifier.

• Repeaters are often just linear amplifiers.

linear instruction

An educationist's term for 'non-branching' instructional programs that take the student through a strictly-ordered sequence of learning modules. Film or television programs are 'linear' media, because you will watch an instructional film from beginning to end in an orderly sequence defined by the film-maker. Dictionaries are random access media where you need to dip in at any point.

Half-way between these extremes is conditional branching where you proceed in a linear way, but via tests or choice, you will be directed along new pathways through the maze of information modules in a root, branch, twig structure. A distinction should also be made between these traditional learning structures and the freedom of hypertext (mesh-networked) structures for both instructional material and information.

linear keys

In video, these are either internal keys or external keys, but not chromakey.

linear laser

A laser which can be amplitude-modulated. It is capable of a smooth range of amplitude outputs, which are tightly coupled to the level of input signal, and it can therefore transmit analog signals over optical fiber. The contrast is with digital lasers which are only capable of on-off signaling.

However, be wary here; an on-off laser can still transmit analog signals if the timing of the switching is not clocked. This is how analog FM modulated video is stored on a LaserDisk.

See PFM.

linear post-production

See non-linear.

linear predictive coding

LPC is a digitized voice compression technology used in the first vocoders. It is really an analysis and synthesis technique.

It encodes pitch, energy level and other voice parameters and allows reasonable conversation at an extremely low data rate. Voice patterns are highly predictable, so the LPC coder can sample the speech, and predict ahead as to what should follow. It then transmits only the differences between its prediction, and the actual.

LPC was originally very expensive to implement, and it provided poor quality with the female voice. It is possibly being displaced by MBE techniques. However it is the voice analysis/-synthesis technique used in Texas Instrument's Speak & Spell, and is widely used in speech recognition. See source code, CELP and vocoder.

• TI's Speak & Spell uses hardware modeling of the vocal track which is similar to formant synthesis. By storing the LPC parameters for both filter coefficients and gain and excitation frequencies, adequate speech quality can be created with only 1200 to 2400 b/s data rates. See also formant synthesis and waveform digitization.

linear programming

• A technique of breaking down problems in a way that they can be solved by computers. The assumption is that the program will be basically linear from a single start-point to a conclusion. However most programs of this type will have recursive elements, and GOTO statements.

• Structured programming should exhibit a direct line from beginning to end without the use of loops or branching instructions (such as GOTO statements). This was once seen as an ideal, but it is very rarely realized.

linearity

Arranged along a line with just one dimension. Linearity implies uniformity, equality and regularity.

• **Amplification:** The ability of an amplifier or other electronics to process a signal without changing the characteristics of the input signal. Sometimes in analog electronics, the input will be pre-corrected (pre-emphasis) to compensate for known losses of linearity.

• **Informatics:** In database searching or information retrieval, linearity is the use of systematic one-dimensional procedures of working towards a goal, as distinct from hypertext or random browsing.

linefinders

In the earlier telephone switches, linefinders were used to concentrate the traffic from a number of subscribers onto subsequent ranks of switches.

In the older rotary exchanges, step-by-step linefinders search automatically for a line when the subscriber picks up the handset. The later crossbar switches had concentration functions without linefinders.

• Linefinders are also called uniselectors.

lineside

The side of the telephone exchange that connects to the local loop. The distinction is with 'trunkside'.

link

• This is the total communications channel/circuit. It should not be confused with line or channel, which are service definitions. The link may contain radio or satellite segments, and the term is often used vaguely to include devices, media, and software intelligence. It will usually refer to full-duplex or two-way operations (channel often means just one-way).

• In programming, a link is a 'call' on a subroutine or on another program.

• In compilation, the last stage before generating machine code which the computer can run, is the linking stage. Links are established between the generated object code and library, or other code modules. See linker.

• In database management, a link is a pointer that refers to data in another record, or in another file.

• Hypertext and hypermedia systems establish links between modules of information at the command of the user. These can be temporary or permanent, and are used for navigation through information modules in a non-linear way.

link budget

This is the calculation of the expected receive-site levels of radio signal power being transmitted from a base-station, broadcast transmitter or a satellite. It may or may not allow for expected rain attenuation. A link budget is calculated when planning any radio or satellite system and it must obviously take into account the sensitivity of the receiver, the electronic noise of the environment, and problems like multipath.

- In satellites, the 'link budget' is the calculation of the signal power needed by the ground receivers, taking into account permissible noise levels of the amplifiers. This is calculated over both the uplink and downlink to judge the size of dish. This varies between the EIRP contours of the satellite footprint.
- In cellular mobile phones, the link budget calculations depend on the sensitivity of the handsets as well as the power of the base-station and the coverage area (and type of topography). These calculations tell the network designer how large the cell coverage areas can be for certain antenna locations, heights, and power outputs, and therefore how costly the network will be to construct.

link control

Link controls are all the procedures that are added to basic data communications to ensure the data is delivered error-free.

Most reliability controls depend on adding 'redundant' bits of control information to the data in the frames/packets. These redundant bits will be automatically checked by the receiver, and if they fail to compute they will initiate recovery or retransmission procedures. Low-level link controls are simple error-checking functions used in most file transmission protocols. High-level data link controls are HDLC, ADCCP and similar complex protocols. These are also known as Data-link Controls (DLC). See DLC, ARQ, FEC, LAP, LAP-M and LAP-B.

link layer

A short form of Data-link layer which is Layer 2 in the OSI protocol stack. See Data-link.

link management

The many varied communications procedures concerned with channel allocation and contention resolution in various types of network.

link margins

For all forms of terrestrial radio or satellite systems, this is the amount (expressed in dB) by which the carrier-to-noise ratio (in normal working conditions) exceeds the threshold C/N ratio. See link budget.

It is the amount of signal power that can be made available above the minimum required for reception in the open, with line-of-sight back to the transmitter (best possible conditions for that location). It is the extent (in dB) of power which is to be held in reserve.

Cellular phones must have good link margins for in-building penetration, or to allow blanketing by a passing bus, etc.

link-state routing

These are routing algorithms where each router broadcasts information about the cost of reaching its neighbors. These costs are sent to all other nodes on the internetwork to be stored in tables, and they will then be summed and used in calculating the best path through the network for packets. The contrast is with distance-vector routing algorithms.

This is a form of least-cost routing for LANs and WANs.

linkage editor

See linker.

linked lists

These are data structures, or ways of organizing data so that it can be retrieved in a particular order. This is not necessarily the order in which it is entered. Each item of data has with it, an associated 'field' which gives the location of the next item.

To read a linked list, you only have to know how to find the first record at the head of the list. The advantage of this approach is that it is relatively easy to add or remove items; you just need to change the field which provides the location information — you don't need to reshuffle every record.

It is also possible to have more than one path through the records, by using two or more linkage fields.

linker

A linker or linkage editor is a utility in the operating system which is responsible for processing the object-code routines generated by a compiler to create a single application machine language file (object-code program). This is a two-stage process. Linkage editors allow the programmer to incorporate common library sub-routines into an application without laboriously reprogramming them. The output is sometimes called an 'executable program', or 'load module' and it is ready to run.

Linpack

This is benchmark which consists of a set of Fortran routines designed to simulate the number-crunching requirements of average scientific and engineering applications. It was ported from the Unix environment. Linpack measures CPU speed with a wide variety of machines, up to super-computers, and is highly intensive in floating-point operations. The results are expressed in MFLOPS. Linpack and the associated Lapack were developed by Jack Dongarra.

LIPS

Logical Inferences Per Second. The MIPS of AI. It is a measure of the relative speed of AI machines. Despite the acronym it has nothing to do with LISP.

liquid crystal display

See LCD.

LISP

LISt Processing. LISP was developed at MIT way back in 1965. It is a high-level programming language which, along with Prolog, is central to artificial intelligence developments — especially in expert systems and in natural language parsing. It is a symbol manipulation language which makes no explicit distinction between the program and the data.

list

A data-set which is stored in a way that allows accession in a particular order. This can mean arrays, or linked lists.

listservs

These are electronic mailing (distribution) lists which are automated by a server as a way of managing the contact and information requirements of special interest groups on the Internet.

There are thousands of these, and people manually subscribe to them via simple e-mail messages. You send a message to a centralized list and it distributes identical messages to all on the list.

literal

In computer programming, this is a symbol which stands for a particular value. In the Basic statement:

```
30 LET Z$ = "THE TIME IS "
```

the Z$ is a string variable, and THE TIME IS is a literal. It can't be varied, but only replaced by another literal. It is also a parameter and a string — which makes this term rather superfluous.

lithium

Lithium batteries are widely used in portables. There are both non-rechargeable batteries, and rechargeable lithium-ion accumulators. They provide excellent power for their size (about 3 to 4 times that of ni-cads) but they are expensive.

For a while it was thought that NiMH accumulators were better, but there's recently been a return to lithium-ion for portables. See NiMH, ni-cad and lead-acid.

lithium ion

Lithium-ion accumulators (batteries) are solid polymer batteries which have energy densities about twice that of the Nickel-metal hydride accumulators they are replacing in mobile phones and other battery-powered electronic equipment. Up to 125Watt-hour/kg is possible. This is about three times that of Ni-Cad.

The lithium batteries use a compound lithium oxide as the positive electrode and carbon for the negative. Charging takes place because the ions move in and out of each electrode through a micro-porous film.

One advantage of Li-ion technology, is that a single cell produces a voltage of +3.6 volts, which matches the new IC power requirements.

The problem is that Li-ion costs about twice as much as Ni-metal hydride.

- There is also a non-rechargeable lithium battery.

little-end(ian)

See big-endian and least-significant.

Little LEO

Small data-only Low Earth Orbiting satellites using frequencies in the UHF band. The major player (the only player, at present) is Orbcomm. Orbcomm's system is satellite e-mail with links into the Internet, using IP and X.400 addressing. Little LEOs will also provide position-location and monitoring functions. See Orbcomm and Big LEO.

live links

See hot-link.

liveboards

Electronic whiteboards. See.

liveware

Humans.

LLC

Logic Link Control. This is a subsection of the Datalink layer (Layer 2) in the OSI seven-layer model (the other subdivision is the MAC). The basic LLC protocol is modeled after HDLC and it handles error control, acknowledgment and flow control.

The LLC provides queuing services for applications running on each network node, and it frames, sends and receives packets and message. It exists above the MAC sub-layer to define how the packets should be treated by the upper layer protocols.

The LLC is common to all IEEE LANs standards in the 802.x series while the MAC will change with each to match the physical media. The LLC protocol defined in IEEE 802.2 has three distinct classes of operation:

— **Class or Type I** is a form of connectionless communications which therefore requires no link establishment. It does not acknowledge packets (PDUs), or have flow control or error recovery; it is very simple. It is widely used in modern networks which have low errors. In this Class, the connection control will reside in the Transport layer, and higher layers will often provide the necessary reliability and flow-control mechanisms.

— **Class or Type II** is connection oriented. It provides both flow-control and error recovery and is therefore needed for reliability with simple devices such as terminal controllers. These connections require constant acknowledgments to stay alive, and any congestion, no matter how brief, may cause sessions to crash. See DLSw also.

— **Class or Type III** is a hybrid version of Classes I and II. These use datagrams which need to be acknowledged.

LLC2 tunneling

Logical Link Control Class 2 tunneling (encapsulation). 3Com's proprietary technique for encapsulating SNA data for routing, and also for spoofing session pairs. See encapsulation.

LLI

Logical Link Identifier. All terminals on a network have a unique LLI of some kind.

LMDS

Local Microwave Distribution System. This is Super High Frequency (SHF) MDS, which is called 'local' in contrast to MMDS ('Medium/Multicast' MDS). Both systems are now known as 'wireless cable'.

The term LMDS is now being applied to the new one- and two-way, analog and digital television services operating in the 28GHz SHF range. These have the bandwidth and the range of possibilities of hybrid fiber coaxial (HFC) cable with which they might directly compete in a few years.

The transmitters average only 25 W in power, so a large city would need many transmitters. In a typical flat city, it would require 30 cells to cover 4000 sq km, but this makes equipment investment higher than MMDS.

Currently CellularVision is operating a trial SHF system in New York using 20MHz wide analog channels, and trials are proceeding in the UK.

Two-way communications is possible because of the short range and fixed antenna. There is some suggestion that two-way telephony could also be offered over these services.

In effect, these systems are taking the lessons learned from cellular mobile (the smaller the cell size, the more the bandwidth reuse) and applying it to broadband fixed services.

LNA

Low Noise Amplification. The primary signal pre-amplifiers with satellite earth stations (TVROs) in the mid-1970s. Now most people use block converters (See LNB).

The signals transmitted by any satellite transponder are extremely weak by the time they reach earth, although large collecting dishes help to concentrate the signal. However the signals need to be boosted at the dish before they begin to collect noise from terrestrial sources via the antenna cable.

LNAs therefore sit right behind the feed horn at the focus of the dish and amplify the signal at radio frequencies. LNAs have special gallium arsenide electronics and a very high signal-to-noise ratio, and they can boost a channel (by as much as 60dB) before sending it down the cable.

With LNAs there is no intermediate frequency conversion here, so the antenna cable must be thick coaxial cable (22mm) to handle the C-band satellite frequencies (3.7—4.2GHz). This approach proved to be impractical with Ku-band frequencies — it was necessary to convert Ku frequencies at the masthead to intermediate frequencies before putting them down the cable. See LNC and LNB.

• At such low signal levels, the noise from molecular motion in a transistor can be a limiting factor, so the quality of a LNA is expressed as the temperature equivalent (in Kelvin) of a perfect system. At 0 K (absolute zero) there would be absolutely no molecular noise, so 80 K is obviously better than 150 K.

Super low-noise amps for large earth stations are cryogenically cooled (liquid hydrogen), but for domestic systems most are based on GaAsFET amplifiers.

• Don't confuse the R/F amplification function with the block down-conversion function which results in an intermediate frequency. See LNB and LNC.

LNB and LNBC

Low Noise Block-Converters for satellite reception systems. LNAs were mainly used with the early C-band satellites. Later LNCs added a conversion stage to intermediate frequencies, but only for the signals from one transponder. LNBs were later developed to cover the full satellite output from many transponders — they are wideband amplifiers and wideband block down converters.

These devices are typically used with Ku-band direct broadcasting satellite systems. LNBs take a 500MHz block of signals directly at the focus point of the antenna (in the outdoor unit) and amplify and convert them to an FM signal in what is called the first IF stage. Originally this was between 900—1400MHz, but it later became 950—1450MHz, and it can now lie anywhere in the 950—2050MHz range.

The IF signals then travel easily down the antenna cable to the set-top box tuner/selector where they are further down-converted to UHF frequencies the tuners can handle (480MHz). Satellite receivers capable of handling both C-band and Ku-band are now manufactured to a standard which accepts and tunes intermediate frequencies from 950—1450MHz.

• LNBs use a heterodyned oscillator to beat with the original signal and produce the intermediate frequencies. If this oscillator is higher in frequency than the satellite band (as it is with C-band) then this is called high-side injection; if it is lower in frequency (as it is with Ku-band) then it is low-side injection

• LNBs are graded by noise figures in dB (as distinct from the C-band grading in Kelvin). Common noise figures range from 2.5—3.8dB, which is the Kelvin equivalent of225—405 K. See block down-converter, LNC and LNA.

LNC

Low Noise Converter. LNCs added a frequency conversion stage to an LNA (amplifier), but only for one transponder at a time. LNCs were fitted behind the early C-band dishes to perform the translation of a narrow band of signals to an intermediate frequency which could be carried down standard antenna coaxial. They were popular in the 1980s before it was possible to build the wide-band (500MHz) LNBs.

LNCs would convert the signal to an intermediate frequency of 70MHz which the set-top box could then handle direct. The problem with LNCs was in getting them to handle multiple channels. See LNB.

load balancing

Load balancing is important to increase the utilization of almost every network, and make more effective use of bandwidth in large networks. In any large, highly segmented client/server networks, load balancing between the segments and the resources available, is a constant challenge.

Intelligent load-balancing routers will spread traffic between different output ports whenever both have roughly equidistant paths to the destination address. A load balancing algorithm in the router will hopefully take into account both the line-speed and reliability of each connection.

High-performance servers can saturate narrow bandwidth channels; a server running at 100 MIPS and providing services for hundreds of clients, is unable to perform at its designed rate across a 10Mb/s Ethernet network. So load-balancing will also involve adding extra 10Mb/s links between the hub and server, or upgrading a hub-server link to FDDI, etc.

load module

See linker.

loaded coil

The ideal antenna on any transmitter is one that is half the wavelength. However, is often impossible with cordless and cellular phones because the whip antenna would be too long. A loaded coil is therefore added to the antenna system to simulate a (pseudo) antenna of the required length.

loaded line

A telephone line which has been equipped with loaded coils to add induction. This was very necessary in the early days of telephony to counter-act the signal losses caused by capacitance over any substantial distance.

Line loading techniques were used to minimize amplitude loss and amplitude distortion. Coils were added to all analog telephone lines longer than about four kilometers, and they dramatically improved the voice levels over long distances.

However, loading coils also acted as 'bandpass filters' and in the 1990s they are unnecessary since cheap amplification became possible, and they will generally make it impossible to transmit high-speed digital signals over the cable. Most loading coils have been removed, but older networks will still have some lingering somewhere.

Over any distance, telephone signals are now multiplexed so the problems that coils solved are no longer an issue. However some reports say that probably a quarter of the local loop connections in many developed countries still have loading coils somewhere along their length. These now need to be removed for modem transmissions.

loading

• In many electronic circuits it is essential to test equipment under circumstances where the expected electrical resistance will be experienced. These systems are loaded by adding resistors and impedance-matching devices to simulate large loudspeakers, etc.

• Parallel wires (specifically telephone wires) running for long distances experience substantial power loss due to the effects of capacitance. To reduce this loss, many of the older telephone networks used inductance coils which counteracted the capacitance effects. See loaded line.

LOC

Lead on Chip — a chip design format employed by IBM on 1Mb DRAM chips, and by Hitachi and TI on 16Mb (SOJ) DRAM.

local area network

See LAN.

local bus

There are two slightly different ways this term is used.

• In the original PC architecture the bus linking the CPU and memory was also the I/O bus, and all I/O was controlled by the central processor. Later the CPU began to exceed the speed of the peripherals, and special I/O processors also came into use to take the load off the main CPU.

A local bus is the one which now ties the CPU and memory together, and also services a few other high-speed devices near to the processor (maths coprocessor, for instance). It is also called the CPU bus — but don't confuse this with the buses within the CPU chip itself.

This type of local bus will now be running at a much higher rate than the I/O bus, and the I/O bus will work independently with the peripheral devices.

• The term has recently been appropriated by card manufacturers, so now it is used to mean a bus that doesn't send data for any distance over the motherboard (or connect normal expansion slots) but provides a special high-speed pathway between the CPU and special peripherals on some adaptor cards.

Today's computers often handle 32-bit wide data paths at rates of up to 50MHz, but only between the CPU, cache and main memory. When they are coupled with the much slower ISA bus, bottlenecks occur: this will often happen, for instance, when data is being sent to a memory card or a peripheral.

MCA and EISA were two attempts to create faster general-purpose I/O buses, but neither has been wildly successful.

VESA decided, therefore, to link the CPUs private local bus back into the I/O system by adding in-line connectors on the motherboard. These were to supplement the ISA slots.

So for high-speed memory cards or video cards, a secondary local-bus connection can now be used to provide the full 32-bit wide high-speed path to and from the processor. This reduces the load on the I/O bus, and keeps the CPU working on speed-critical functions to handle the new disk controllers, network interfaces and video.

• Sometimes a local bus is provided on (and entirely within) a plug-in card which has its own memory and special processor. With video, this technique is often called Local Bus Video or LBV.

• There are a number of proprietary local-bus standards based on ISA. VESA standard (called VL-bus) isn't well standardized.

• Intel's PCI (which is only marginally a local bus) seems to be the clear winner. 60% of new PCs have both PCI and ISA.

local echo

Local echo is the normal condition of a PC. When you type, the computer also creates a screen display of the characters, instantly raster scanned from its video RAM.

The term applies only when using a computer for communications. The distinction is being made with the signals which are echoed back to your screen from a remote node (Echoplex). The value of Echoplex is that any keying or transmission errors will appear on your screen immediately as you type. You will probably assume that all of these are typing errors, and you will therefore correct them immediately.

With local echo only, the screen characters are generated directly by your keystrokes, and simultaneously they are being transmitted over the line (to the modem). You see what you type instantly (not a millionth of a second later), but with local echo only your own typing errors appear on the screen as mistakes — so it gives you no check over any transmission problems.

The main point is that you should use either Echoplex or local echo — but never both. If both are functioning, every letter typed will be duplicated on the screen lliikkee tthhiiss. In such a case, just switch local echo off.

• Not all remote systems will provide Echoplex, and it often isn't given this name even when it is provided. It may be called 'full-duplex' or just accepted as the default condition.

• Don't confuse local echo with near-end echo which is an unwanted signal bounced back from the hybrid in your local exchange to the hybrid in your local modem (and vice versa). See Echoplex.

local loop

The cable connecting your domestic or business telephone directly to your local exchange. The total cost of the local loops represents almost 40% of the cost of the whole telephone network since capital and installation costs are high. Yet it is also the least productive use of capital. In domestic telephones, the cable is generally only in use for a small part of each day.

The local loop is the 'customer access network' (sometimes called the CAN) and it is exclusively used by one customer (except in party lines), while the rest of the telephone network (the 'backbone' and switches) is a shared network.

With the introduction of PCN, the term local loop could also mean a radio link, and with telephony over the new cable networks the local loop could be a contended channel over shared coaxial using ATM.

With BRA ISDN, we get two B-channels and a D-channel over the local loop, and with ADSL the local loop may be carrying up to 6Mb/s, but only as far as the nearest exchange.

• The local loop is generally true twisted pair until the last pillar, and from there to the home/business it is often plastic covered quad. In the USA, much of the local loop is multiplexed coaxial to local IDLCs. See.

• In conventional modem use, the problems created by the local loop are mainly to do with mismatched segments, each often having different wire gauges (and different electrical characteristics) which are solder connected at pillars, and often poorly insulated. Part of this network will also consist of large multi-pair cables where individual pairs are in close proximity over long distances, so slight near-end crosstalk is common. These multi-pair bundles are often paper-insulated, and in winter the insulation can break down.

Because of the crosstalk, impulse noise from rotary dialing is common, and these cables suffer from high-frequency attenuation, signal reflections, longitudinal imbalance and phase distortion.

local loopback

In modem testing there are:

• Local analog loopback tests where the analog transmitter (carrier) of the local modem is routed directly to its own receiver. A test 'pattern' is sent from the transmitter to the receiver and automatically checked. There is no actual transmission of data over the phone line, so this test just checks the transmitter against the receiver to ensure that both are functioning correctly. This is known in the ITU as a Loop 1 test.

• Local digital loopback tests include both the local and remote mode, and the telephone line between the two. Both A- and B-parties must therefore co-operate, and the remote modem must be capable of reflecting back data sent to it. This test causes all characters to be looped back to the originating terminal and checked. This is known in the ITU as a Loop 3 test. See loopback and remote loopback.

local mode

The use of communications equipment for local functions without linking it over a circuit. A printer-terminal may be used for local printing, or alternately for on-line (remote) services.

local terminal

LT. In telecommunications, an RT (remote terminal) is at the subscriber's end and a LT (local terminal) is at the switch end.

local variable

A local variable is one used only in the particular subroutine, function or handler; it may be reused in different parts of the program — and the two uses will have no connection. The distinction is with global variables which are declared and used throughout the program. Often you will need to pre-define a global variable, but you shouldn't need to do this for local variables.

LocalTalk

The original physical layer of the Macintosh AppleTalk protocol stack. LocalTalk uses a bus topology, with shielded twisted-pair cables (there is also an unshielded version called PhoneNet). It operates at 230kb/s, over a network segment length of 300 meters, and uses CSMA/CA for access control. Apple provides alternate physical layer protocols for AppleTalk also.

location

You contact locations by using addresses or parameters.

- An addressable area of memory.
- An addressable node (station) on a network.
- The physical position of data on a disk which can be addressed by track and sector parameters.

lock

The action taken to prevent a file or record from being accidentally or deliberately altered and/or accessed.

- User-activated locks divide into physical locks (i.e. sliding tabs on some disks), software locks which can be changed by anyone and password-type secret locks.
- On a network, the file-server must have a way of locking records or whole files to ensure that the material isn't inadvertently altered by a second user while the first user has the file open and is possibly making changes. See file- and record-locking.
- Software publishing houses sometimes build-in locks of many kinds to prevent unauthorized copying of the disks. See below, copy protection and dongle.

lockout

The term is used in two ways — as a synonym for locking, and to mean permanent exclusion from something.

- File locking/lockout procedures are those used to prevent a file or record being updated by one user while it is simultaneously being used by another. You can't afford to have updates being made by two people at the same time on the same record. Database files can be locked at either the file level or the record level.

In the early days a whole file was locked while just one record was being changed, but now records are locked independently. See file and recording locking — and also file lockout.

Lockout facilities are those provided to deal with contention when two users try to update the same record.

- Lockout may also mean that the file has been damaged in some unspecified way, to the point where no one can now gain access to it without some remedial action being taken.

In relational databases there may be two or more quite different files with related material in various records. With simple file-locking, it would be possible for one employee to be changing a record in file A, and another to be changing the related record in B, and both may lock the other out. But as part of the updating of A, it may be necessary to automatically change B — and vice versa. So each may lockout the other, and mess the files up significantly in the process.

LOG

Logarithm. But beware, there's more than one type:

— In most forms of Basic, LOG (X) is a function that will calculate the natural logarithm of X.

— In other forms of Basic, LOG (X) will calculate the common logarithm, and the function LN(X) calculates the natural log value.

- Base-10 logarithms are called common logarithms.
- Base-e logarithms are called natural logarithms. See e.

log-PCM

All PCM voice sampling systems use a logarithmic scale for quantization because a linear scale is inefficient. See A-Law and mu-Law.

logic board

The motherboard — the main printed circuit board in the computer which usually holds the main CPU, ROM and system RAM, the bus, and the expansion slots.

You can change the motherboard as a way to upgrade an old computer, but remember that your old plug-in cards may not fit into the new expansion slots. Also, if you take this upgrade approach, you are retaining old disk drives and the old keyboard — all mechanical devices which are more subject to age and wear problems than the motherboard. You are upgrading the most reliable part of the system, and retaining the least reliable.

logic bomb

This is a form of malicious program similar to a Trojan Horse and a virus. It can be triggered by a key-stroke sequence or it may have a predetermined trigger date — often Friday the 13th or something similar.

Logic bombs have usually been left in operating systems by disgruntled IT employees of very large organizations. So if you are going to fire a DP professional, always wait until they're on holiday!

logic chip

Slang for a microprocessor's central processing unit.

logic gate

An electronic switch which may have more than one input signal but only ever one output signal. These switches perform according to the rules of Boolean logic.

Logic Unit 6.2

See LU 6.2.

logical channel

The term sounds more profound than it is. It just means the appearance of a continuous circuit between two terminals using circuits, packet-switching, frame-relay or cell-switching. This is also called a 'virtual channel'.

- In packet-switching networks, a logical channel is established by the header address information in each packet.

To the user it appears to be a dedicated connection between two devices, but in reality it is a shared, non-dedicated path way. Packets will use either full addresses, or virtual circuit identifiers to travel along this channel. Many logical channels will simultaneously share a single physical circuit.

• Some packet-switching services offer a connectionless mode, in which case a logical channel would not exist. The packets find their way through the network in a step-by-step way using best-path algorithms in each node.

• In cell-relay systems such as ATM, the logical channels are called virtual circuits and have a Virtual Circuit Identifier (VCI). Channels going to the same destination are further grouped into virtual paths (VPI).

• In a TDM connection, a logical channel will generally occupy one slot in the time-divided frame. Sometimes slots can be aggregated to provide more bandwidth for some services, in which case the logical channel may occupy more than one slot.

• Logical channels in a time-division multiplexed system are often assigned logical channel numbers.

logical errors

The distinction is being made between syntax errors and logical errors. Logical errors are created when you tell a computer to do something that it can't do: the syntax might be correct, but the logic isn't.

For instance if you tell the computer to PRINT X when you don't have a variable X (perhaps you meant to say PRINT X$) then this is a logical error.

logical format

This is synonymous with file format. It defines such things as the file size, the directory structure and the location.

logical interface

This is how two devices interact by sending data messages of some sort. The distinction is with the electrical or mechanical interfaces which interact by changes in voltage or position.

The specifications of a logical interface will include details of the meaning of all signals between the devices, and the response that one device should expect to a signal from another and, under what conditions.

logical ring

On a token-passing bus network, the token must circulate in a regular manner, as it does in a Token Ring.

So the network is configured logically as a ring (even if it is not physically a ring), where each device is given a sequential address and the token is relayed to the next device address (no matter where it is physically on the network), thus ensuring circulation of the token.

When the token is passed back from the highest address, it is immediately transferred to the lowest.

logical shift

In binary arithmetic, a logical shift occurs when you add a bit to one end of a byte in an 8-bit register. This process forces each of the other bits to shift one place to the left or right (depending on the end added) and one bit will fall off the register at the other end. If you add your bit to the least-significant end, the most-significant bit will fall off the register.

logical unit

This is a very confusing term since, in one sense, every frame, packet, block, datagram or ATM cell is a logical unit.

• In SNA it was originally applied to the end-points of a communications link — say, a mainframe application and a 3270 terminal. These would be called Network Addressable Units (NAUs) to everyone else, but to IBM they were LUs.

• Later in SNA, these LUs became associated with software constructs (say, terminal emulation programs) that act as a bridge between the end-user and the SNA network. LUs can communicate with both the host system and the applications, or with other LUs of the same type.

The idea now is that LUs will always present a uniform interface to the network — to a LAN they will all look alike. SNA establishes communications among LUs, so that the application doesn't need to be concerned with the physical location or characteristics of the logical unit. Each LU has a network name (used by the application) that SNA translates into a network address. If a circuit fails, SNA will switch to a new one automatically.

• LUs range from LU 0 to LU 7 (with decimal-divisions there are currently nine types), and each provides a different service and requires a different protocol.

— LUs 1, 2, 3, and 7 are designed to control peripherals.

— LU 6.0, 6.1 and 6.2 communicate only with LUs of the same type.

— LU 6.2 is general purpose for APPC and allows applications to communicate. See LU#.

logical units

In computing terms, these are electronic switches which perform according to the rules of Boolean logic..

• IBM uses the term in a different way. See above.

Logical Unit Convergence

See LU 6.2.

login/-on/-off

The log-in, or log-on procedure is the sequence of actions that a user needs to perform before he/she can get access to a file or a system. You log-out, or log-off to leave the system later.

The log-in process will usually involve a password and user-identity number/code. Log-offs may also involve a special sequence that must be followed.

• The time you are connected to the system is known as a 'session'.

• When trying to contact a new system which permits guests, try these names: anonymous, guest, newuser, logon, login, trouble or help.

• When exiting from an unknown system, you should try any of the following:

```
q, quit, logoff, off, end, close, or bye
```

Only as a last resort would you switch off your modem.

long-space disconnect
This is a continuous break signal which is transmitted by the local modem for 4 seconds. It will do this after receiving a break signal from the remote modem (prior to it disconnecting). The long-space disconnect is used as a system check to ensure that an unattended remote system has actually received and recognized a break.

long-term storage
This often means that the data will be stored on half-inch magnetic tape or something similar, but it could now mean any of the optical or magnetic archiving systems.

• At one time this meant any form of memory that was not volatile. So RAM was short-term storage, and any magnetic medium was classed as long-term storage.

long-word
See word.

longitudinal recording
Audio tape recorders use longitudinal recording. Currently, all video recorders (and DAT) use helical recording where the recording track slants across the tape at a shallow angle (and therefore each has a beginning and an end). Helical recording needs moving heads, while longitudinal uses stationary heads.

Attempts have been made over the years to build video recorders with multiple stationary heads, but so far with little success. See S-DAT and helical scan.

longitudinal redundancy check
See LRC.

look-ahead
MS-DOS 4.0 and later versions can read more than one disk sector at a time. When the disk controller is commanded to read a disk sector, it also grabs as many as seven following sectors and stores the data in a buffer ready for use. This is look ahead. See disk cache and look-aside cache below.

• The new Intel P6 chips have a similar look-ahead function in bringing instructions into the CPU in the expectation that they will be used next.

look-and-feel
The vaguest of vague terms invented by corporate lawyers to enable them to make big bucks from lawsuits; and used by computing companies to try to stall the opposition's marketing efforts. Apple was a culprit here, they wanted other computers to stop looking like their Macs, even though the Mac's look-and-feel was largely borrowed (licensed) from Xerox.

• Of course, the interest of users is in having all computers look and feel alike — just as we like to have all rented cars with the same look-and-feel.

look angle
The degree above the horizon which provides line-of-sight to a communications satellite (usually to a GEO above the equator). Low look angles mean that you will need to penetrate a lot of the earth's atmosphere, so high look angles are better. At the equator you get the shortest and most direct path. Near the arctic circle, you will not be able to see a satellite at all, unless it is in polar, inclined, Tundra or Molniya orbit.

look-aside cache
The look-ahead cache in a CPU can have one of two basic architectures. It can be:
— serial, or look-through cache, or
— parallel, or look-aside cache.
With parallel cache techniques, the memory-access requests are always sent to both the main memory, and to a parallel block of memory simultaneously. The parallel 'cache' block has been 'set-aside' and is seen as not part of the main memory. If the block of data required is in this cache, then it is delivered to the CPU and an abort signal is sent to main memory. The advantage of this approach is that the parallel cache can be removed or changed without disrupting normal operations.

With serial cache techniques, the request is made first to the cache, and only if it doesn't respond with a 'hit' will the request be passed to main memory. This increases delays when the request fails, however if cache is well designed and a high proportion of requests are cache hits, then the result is a faster operation. There is less load on the main memory bus.

look-through cache
See look-aside cache.

look-up table
One way for a computer to handle complex information quickly is for it to hold a table with all relevant calculations already made. The computer will then treat the process as a database search and look up the results, rather than calculate them (even if it could) itself. This technique is also used for non-calculation type operations. See color look-up table (CLUT) also.

loop

• In computer programming, this is a recycling process which repeats a section of the program a set number of times or until a condition is met. Algorithms are nested within these loops so that the algorithmic process will be performed the required number of times. See iteration.

Loops may just rely on a counter (do x times) or they may be executed an unknown number of times until a particular condition is met (do until x=y). Loops can also be nested — with one loop inside another.

• In analog telephony, this is the completion of the electrical circuit between the exchange and the subscriber when a handset is lifted (off-hook). It is also a shortening of the term 'local' loop. See local loop.

• Loop disconnect, is an old term for pulse signaling on a decadic telephone system. The digits are signaled to the exchange by the rotary dialler making and breaking the local loop's DC current. DTMF is slowly replacing this technique.

- In network topology a loop is either:
- — A ring which is bridged at one point by a control system. This is usually logically a bus. DQDB uses a dual fiber loop for MAN operations and a bus for inter-MAN links.
- — A broken-ring. The control devices (MAUs) will usually bridge the break, or
- — A route where packets cycle endlessly through a series of network nodes and never reach their destination. See looping (WAN).
- Loop # tests — see local loopback and loopback.

loop start

A DC voltage of 48—50 volts is applied at the exchange to one of the wire pairs of a standard telephone pair. The other wire is at earth potential, but from the telephone exchange.

When the phone is taken off-hook, a loop closure is made allowing current to flow between the 'tip' and 'ring' wires. This is a 'supervision' signal to the switch, which instructs it to put dial-tone on the loop pair; it is called a 'line seizure'.

The user can then signal the required number, by either rotary (pulse) dialing or DTMF tones. Loop start is the standard way most phones work. The distinction is with ground start. See ground start and wink.

loopback

A technique used to test full-duplex communications equipment (mainly modems) where the device generates a test signal at an output, which is redirected back into the same equipment for checking. Loopback tests may be performed internally in a device, or they may be directed along the communications channel and reflected back from the other end. The standard ITU loopback tests are:

- — **Loop 1** tests are internal loopbacks within the modem.
- — **Loop 2** tests are remote digital tests. The signals are reflected by the remote modem, but they do not include electronic regeneration of the signals (the remote modem acts passively).
- — **Loop 3** tests are local analog or digital loopbacks.
- — **Loop 4** tests are remote digital tests which actively include the remote modem.

See local and remote loopback.

looping (WAN)

Bridges don't expect to encounter loops — they expect tree-branch pathways between network nodes and segments. So if a loop is accidentally formed, packets can travel forever around this 'circular' circuit, being constantly regenerated and handed on by the bridges and repeaters.

Accidental looping like this can generate traffic load which can bring down an enterprise-wide network. The spanning-tree protocol overcomes this problem, but many remote bridges can't support this protocol. Routers can generally deal with loops.

LORAN

LOng-RAnge Navigation. A way of obtaining navigational fixes by using low-frequency radio signals. The principle is based on the time-difference for pulsed radio signals arriving from a pair of synchronized transmitters. There are terrestrial as well as satellite Loran systems. See GPS also.

- Loran-C is widely used around the world for location fixes and accurate time signals in UTC form.
- Because of terrestrial transmission, certain areas are unable to use Loran transmissions and are known as 'Loran Holes'.

lossy/lossless compression

Lossless compression systems are used when the reconstructed material must be identical to the original. Data text and applications files must be stored and transmitted in a lossless way, while video or audio can afford to lose some selected detail (particularly high-frequency components beyond the limit of human resolution) without the audience noticing.

Some compression systems are designed to restore the original file in full, while others need only be close to the original.

Huffmans and Lempel-Ziv-Welch compression systems are of the lossless type. Usually 2:1 or 3:1 compression is the best that can be expected (depends on the original material) with lossless text, but more with programs and general files where repetition is common.

Lossy compression is used with video, audio and imaging where the output needs only be 'near-enough' to the original. MPEG is an example of lossy compression techniques. Compression ratios in the order of 150:1 are being achieved with satisfactory results on motion video, and 20:1 on still images using JPEG. Motion video offers the highest compression ratios because each frame of image in a sequence will be almost identical (especially with videoconferencing).

low noise amplifier

See LNA, LNB, and LNBC.

low noise converter

See LNC and LNBC.

low-pass filters

These pass lower frequencies and attenuate the higher frequencies. See high-pass filters and bandpass filters.

low-power satellite

One with the peak transponders output power less than 30W.

low tier

See high tier.

LP

Long Play.

- **Audio:** Used once to describe vinyl audio disks of 12-inch diameter, played at 33 rpm (also called 'albums' to distinguish them from the smaller 45 rpm 'singles').
- **Video:** Applied to VHS machines which can record and replay with the tape motion at half standard forward speed.

This causes the video tracks to overwrite each other, and decreases video quality, but it results in tapes which can store twice the normal program material.

— LP is also applied to extra-thin very long standard tapes played at normal speed.

LPC

- Linear Predictive Coding or Codec. See.
- Longitudinal Parity Checking. See LRC.

LPTV

Low Power TV. A relatively new analog UHF television transmission license issued in the USA, generally to community or non-profit organizations. These are classed as non-commercial, but they are allowed to raise funds through local advertising and sponsorship. Don't confuse this with LMDS.

LRC

Longitudinal Redundancy Check. A generic term for a range of relatively basic error-checking protocols where the checksum consists (usually) of 8-bits. It is effectively a form of 'longitudinal' parity checking. All the least-significant bits in a block are summed and parity calculated — then all the other bit-positions have parity checks made independently, until the most-significant bits are reached. So the checksum bit sequence consists of 8 parity bits, using one bit from each byte.

If LRC is combined with individual parity-checking of each character it can be fairly powerful, but generally CRC is preferred. The distinction is with VRC, HRC, CRC and FEC. See modulo-2 also.

LRU

Least Recently Used. A term applied to both the algorithms and the data controlled by those algorithms. The idea is to keep check of data held in disk cache or in virtual memory systems.

In disk cache, the LRU blocks may be required again, so they are kept at hand while, with virtual memory, the LRU blocks will be the ones written back to the disk to make room in RAM for incoming data blocks.

LSB

Least Significant Bit — the right-most bit.

LSI

Large Scale Integration. A term that seemed right at the time, but it now applies to chips that only have about 500 circuits or up to 10,000 gates (transistors). The first true microprocessor, the 4004, is from this generation.

Originally it meant an IC with more than 100 gates. The key breakthrough which made LSI so important at the time, was that this one chip was able to both carry and connect transistors, resistors and capacitors. This was a major achievement. VLSI and ULSI are the terminological progressions.

LT

Local Terminal. A terminal at the switch end of a digital loop in telephony.

LTC

Linear Time Code. The form of video frame measurement for editing, where a time code is recorded on a linear track. Early systems used one of the linear audio tracks for the time code, while later systems reserved a special track. See timecode also.

LTP

Long-Term Prediction. Used in RPE-LTP codecs. See.

LU#

Logical Units. These are network addressable units (NAUs) which are a primary component of SNA because they manage the data exchanges between end-users. LUs allow end-users to communicate with each other and to gain access to SNA network resources: they are more protocol suites than simple protocols.

With IBM networks, those numbered below LU 6.2 are now supposedly obsolete, but LU 2.0 for 3270 terminal emulation is still the most widely used of all. (However it doesn't have APPN support).

Communications between LUs can occur only between those of the same type. The main functions are:

- **LU 0** uses non-SNA protocols (implemented by the LU itself) and provides program-to-program links.
- **LU 1** is for interactive, batch transfers, or distributed data processing on SNA. It provides control of host sessions.
- **LU 2** provides services for 3270 data streams, and control of host sessions.
- **LU 3** supports 3270 connection to printers, and provides control of host sessions.
- **LU 4** supports SNA character string and LUs in peripherals. It supports both host-to-device and peer-to-peer communications.
- **LU 6** provides program-to-program links:
  - **LU 6.1** supports application subsystems in a distributed environment.
  - **LU 6.2** (see below).
- **LU 7** supports sessions on IBM mid-range computers. It is similar to LU 2.

LU 6.2

Logical Unit 6.2 is IBM's peer-to-peer, half-duplex, transaction protocol for a distributed processing environment. It is independent of hardware, operating systems and underlying data-link controls, and was previously called Logical Unit Convergence.

LU 6.2 is not an application or a file server, but rather a protocol suite — and it is often confused with APPC which it supports (and of which it is a part). Without making too fine a distinction: LU 6.2 is the implementation of the protocol (the type), while APPC is the API (the collection of commands that the applications can use to communicate).

The importance of LU6.2 is that it frees up SNA from being entirely a tree-structured hierarchical network dependent entirely on a host, and it therefore paves the way for cooperative processing. However, it is a memory hog and is difficult to work with. The half-duplex decision was a substantial mistake in terms of performance.

The seven logic unit types that preceded LU6.2 are now becoming obsolete. See LU, NAU, SNA, APPC and APPN also.

luggable

A large portable computer. The term often refers to a machine that requires mains power, but is otherwise self-contained and portable. The comparison is with laptops and notebooks.

lumen

A unit of illumination. It is defined as the luminous flux (the brightness) of a standard candle within a 'solid angle' of one steradian (the square of a radian).

luminance

Strictly, this is the measurement of the radiance of a light source using a photometer — as distinct from the subjective sense of 'brightness'. But in practice, we use the term interchangeably with both subjective and objective concepts.

So luminance is the characteristic of brightness or intensity in a video signal or of a light source. In television, it is the brightness of the image, as distinct from the color.

Luminance signals create black and white images; chrominance signals create color. However a luminance signal can be generated by mixing Red, Green and Blue in the correct proportions (not exactly equal amounts), and Green has the highest weighting in this blend because it is central in the spectrum.

- In video, luminance signals are identified as Y.
- The color green contributes most to the luminance of an image since it lies in the middle of the spectrum, at the eye's most sensitive range. For this reason R, G, and B are not summed in equal amounts to make white, but are weighted with green having the highest value.
- B–Y or R–Y signals are color-difference signals created by subtracting the luminance value from the specific color value. See YUV and YIQ.
- Luminance keys in video are either internal keys or external keys, but not chromakeys.

luminosity

A value which corresponds to the brightness of a color or tone.

lurking

A slang term applied to people who read the information on a computer conferencing system (a forum), but don't contribute to the discussion.

LUT

Look-Up Table.

lux

A measure of illumination. It is actually one lumen per square millimeter — a measure of how brightly illuminated an object is across its surface — and therefore an indication of the exposure needed to photograph the object.

It is used to express a camera's ability to shoot pictures in low light conditions. Normally 10 lux is an acceptable figure for most cameras, and a camera in the 3–5 lux range is excellent.

This depends on four factors a) the maximum lens aperture b) the type of imaging device, c) the quality of the electronics and d) the acceptable amount of electronic boost provided specifically for low light conditions.

LV

LaserVision videodisk from Philips/MCA, Pioneer and DiscoVision. Don't confuse LaserVision with videodisk (many people do); LaserVision was only one form of videodisk. See VHD and LaserVision.

LV-ROM

A form of digital information storage based on the analog LaserVision videodisk. This was a hybrid system designed by Philips for the BBC's Domesday project. It used CAV rotation for faster access times, and held close to 1GByte of information on a 12-inch disk. Some hybrid schemes developed at about the same time had storage capacities up to 4GBytes. No real standards existed in this area.

LW

Long Wave — see LF.

Lynx

A text tool for WWW. See.

LZ##

Variations on the Lempel-Ziv compression systems are often known by an LZ number identification (presumably the date), such as LZ77, LZ78, etc. More recently they've used letters.

- LZ77 and LZ78 are dictionary-based compression systems where common sequences of bytes are referenced in a data dictionary.
- LZSS is similar to the above. As in most LZ systems, only the last 4 to 8kB of data are used in the dictionary reference. Instead, a pointer is stored (which references the first byte of the previous matched string), along with a length indicator. See LZW.
- LZHUF is an open version of the LZW standard.

LZS

Lempel-Ziv Stack. A variation on the standard LZ and LZW compression technique which is widely used in routers.

LZW

Lempel, Ziv & Welch. A very sophisticated series of algorithms used in data compression which were developed from the theoretical work of mathematicians Abraham Lempel, Jacob Ziv with later additions by Terry Welch.

LZW operates slightly differently to the original LZ coding; it builds a dictionary from string and bytes in the whole file as it is transmitted (not just the last 4–8kBytes). This is happening continuously, so the dictionary being used at the beginning of a file will not be identical to that in use at the end.

LZW works very well for graphics also, because it treats bytes and bytes, and just looks for repetition. It works well wherever the computer comes across a repetitive sequence of bytes — whenever the algorithms can discern a pattern in the file.

It doesn't work at all on random (encrypted) files. Unisys has patented the LZW algorithm and the company is now charging software developers for using it — although most people thought until recently that it was in the public domain.

LZW uses an encoding table (or 'tree') and encodes repeated strings (words, prefixes, suffixes, letter pairs) as fixed length 'tokens', originally up to 10-or-more-bits in length. The current version is a 14-bit code.

Its pointers point back to a previous occurrence of a string. Microsoft gives this example of the way LZW is used in DoubleSpace:

```
the rain [3,3]Sp[9,4]falls
m[11,3]ly on [34,4]pl[15,3]
```

The [3,3] tells you to step back three characters (include the space) and reuse the next three [in]. The [9,4] tells you to step back 9 and use 4, and so on.

There are a number of popular applications:

- In MNP data compression using LZW, the modems have a standard dictionary stored in ROM in the modems. The tokens point to these indexed sequences of bytes in the dictionary.
- In V.42bis data compression a dynamic data dictionary is constructed as the compression/communications program works through the file. This dictionary is duplicated at both ends of the modem-to-modem connection.
- Long files, after compression, will consist almost entirely of short pointers to earlier instances of the text string tokens. See Huffman codes and RLE also.

M

m/M

In the SI units this is confusing, because:

- m = milli — thousandths of a unit. Millionths is micro — mu or μ. (aka micron).
- M = Mega — million. MHz is megaHertz or millions of cycles per second. Mb is megabit, MB is megaByte (but this standard is not always adhered to. See Mb).

M-I and M-II

Professional videotape standards. The 'M' refers to the shape of the lacing path.

- **M-I** is the RCA/Matsushita professional video standard known as 'Hawkeye'. It is a half-inch component analog video standard, somewhat similar to U-matic. It uses YIQ, where the IQ color signals are recorded on a separate track, and it preceded Sony's Betacam in 1980.

  It also provided two program audio tracks, a time-code linear track, and a control track. It used the standard VHS cassette, but with the tape moving at about 10 times the normal VHS speed.
- **M-II** is another component analog video system which was introduced in 1986 to compete with Sony's Betacam. It uses 'equiband' (Y, R-Y, B-Y) recording on to metal-particle tape, and it has had only limited success.

M-II's advantage is that the camcorder can be modified for widescreen, and they say that the professional equipment costs will drop substantially in the next few years, although Panasonic has now released DVCpro. NBC adopted M-II in 1986, however, the standard was expensive and it soon came under pressure from low-cost S-VHS and the new quarter-inch digital systems at the low end of the price spectrum. It made little inroad into the Betacam market at the high end. See DVC.

M44 specification

This specification is for the use of 32kb/s ADPCM over standard T1 links to extend the network's voice circuit capability. The 24 standard PCM voice channels are now increased to 44 ADPCM channels but, if required, some conventional 64kb/s channels can also be integrated with a lesser number of ADPCM channels.

This approach actually subdivides channel bytes into 4-bit nibbles, and then 12 nibbles are grouped together into 48-bit 'bundles' before transmission. A delta channel (the 12th nibble) in each bundle will contain the supervisory and status information.

M-JPEG

Motion JPEG. See page 429.

M-PSK

See page 440.

M-QAM

See page 440.

M&E

Music and Effects. When a film or video program is made for wide distribution, it must take into account the need to provide translation of any narration or commentary track. This is especially so in documentaries, current-affairs and news.

Therefore when the sound tracks are being produced (mixed) it is common to create a separate track which will have music and effects, but no narration or commentary.

The term 'effects', here, also includes any natural voice seen on-camera, including interviews.

M&E tracks are also created for drama films, where the intention is to dub any speech. In that case on-camera voice would obviously not be included.

MAC (instructions)

Multiply ACcumulator. An instruction in a CPU or DSP. Two numbers are to be multiplied together then added to the accumulator register.

MAC (ISO networking)

Medium Access Control. A sub-layer of the Data-link layer in the OSI model. The MAC is one of two sub-divisions here, and it sits between the LLC and the Physical layer to regulate the way packets can access the shared physical media.

It handles the token passing on Token Ring networks, and senses any collisions in CSMA systems, so it is specific to the type of LAN or WAN being used. MACs differ between the different LAN types, while the LLC sub-layer (directly above) remains the same. It is the MAC which supplies the logic on CSMA/CD to manage the packets, and it is the MAC which determines whether a LAN can use tokens.

The MAC is also responsible for packaging the data and sending it out over the network, and then checking that the packages are received.

MAC (satellites)

Multiplexed Analog Component. An analog component television standard for satellites. Until 1993 it was believed in Europe that this was the next evolutionary stage in television developments.

It is a hybrid which uses digital techniques for audio and data, and analog component video (e.g. where the luminance and chrominance are transmitted sequentially) for the image.

There are A, B, C, D and D2-MAC variations which differ only in the way the sound and data streams are modulated; the video handling is identical. Since MAC systems differ so little, it is possible to design a single receiver chip which handles C-, D- and D2-MAC.

B-MAC was proprietary to Scientific Atlanta. More recently there has also been the development of N-MAC (see).

Because they provide at least one digital control channel, MAC systems can handle teletext, banking and retail transactions, Pay TV, cable networking, etc. They usually have multiple sound channels which can be varied in bandwidth and dynamically subdivided. This allows them to have multiple low-quality narration tracks in different languages for one program, and switch to stereo high-quality music channels in the next.

MAC systems generally have very sophisticated in-built encryption mechanisms which enable individual receivers to be given or denied access 'on-the-fly'. D2-MAC was once selected as the pre-High Definition TV standard in Europe (to be followed by HD-MAC), but was later superseded by digital projects, particularly MPEG. See DVB and component video.

MAC (security)

Message Authentication Code. An encryption code based on a secret key which generates a small block of data (the MAC) which can be used as a form of electronic signature. MACs are used, for instance, between an ATM and the bank's host computer to protect the transactions.

Modern EFTPOS standards often use a MAC which is based on block cipher techniques and the last few bits of a DES cipher text. The ANSI-standard MAC is now fairly old. See hash.

MAC addresses

MAC layer addresses are also called physical, data-link, or hardware addresses. They are directly associated with a particular network device. The distinction is with 'protocol' or network addresses.

The first 12 bytes in an Ethernet frame are the MAC address. The first six are the destination address, and the second six are the source address.

MAC bridge

These are used to tie two different LAN cabling schemes together. They are 'media translators' which forward frames of data from one physical cable network type to the other. MAC bridges can, for instance, link a Token Ring network to an Ethernet network. Obviously only data addressed from one network to the other passes through the bridge.

MAC protocols

Media Access Control protocols are the way any shared transmission system has of sharing out the transmission opportunities equally. See MAC (ISO networking) below.

There are a number of categories of MAC techniques:

— **Round Robin:** Each station on the network takes its turn. There are two subcategories here: Centralized control systems are called polling systems, and distributed control systems are token-passing systems.

These are most efficient when all stations wish to transmit fairly regularly.

— **Reservation:** DQDB has a dual-fiber approach where a station wishing to transmit places a request on the frame-header which makes a reservation in all queues further up-stream.

— **Contention:** Each station tries to transmit, but also listens. If it hears a data collision, it will take some action defined by the protocol (usually to wait a random period of time) then try again. These systems have distributed controls and are simple to implement, but performance tends to collapse under moderate to heavy load conditions.

MacBinary

This is a file format used in Macintosh communications software. It transmits the two Mac sub-files called the data and resource forks. The transmitted files will then retain the standard Mac icons and have standard screen formats. MacBinary transfers also include Finder information.

Mach

A Unix-like operating system kernel developed at the Carnegie-Mellon University from the Berkeley 4.3 extensions, which are themselves based on AT&T's Unix Version V. Mach is used by NeXT, and it will be integrated into AIX. It has been adopted by IBM and the OSF, and is included in Microsoft's Windows NT.

machine code

Since, at the base level, computers function on electrical charges stored in memory, the fundamental machine code is a binary representation of instruction sequences and data, and is most accurately represented on paper in a format like '0010 1011'. With the first PCs, programs were entered into memory by flicking eight switches to set the bits, then pressing a button to enter the byte into a memory location; it was a slow and laborious process.

Binary codes are difficult for humans to remember, so hex (often signified by the $ sign) representation is generally used. Instructions and data can still be introduced into some computers in the form 'B6' (1011 0110) or 'FF' (0000 0000).

This is often done through a program known as a monitor; assembly programming is at a slightly higher level again, because it allows the use of mnemonics for a wide range of operations. However these are all really very close to machine-level coding techniques.

Machine code programs use mnemonics as do assembly language programs (but different mnemonics). Machine code mnemonics can be two, four or five letters long, while assembly is rigid on three, and machine code memory locations will usually be specified by a fixed address.

Each register also has a letter code: e.g. **LD Y, X** will be interpreted as 'Load into register Y, the contents of register X', and the mnemonic **JPNZ** is interpreted as 'Jump, if not zero'.

There are four main groups of machine code instructions:

— Arithmetic and logical operations.

— Transfer of data.

— Conditional instructions.

— Unit arithmetic, I/O and general commands.

Machine code programs are stored as single bytes in consecutive memory locations, and they require a loading program before they can function.

• Today we write most programs in a higher form of language (3rd generation compilable languages), then use a compiler or interpreter to perform the translation to the machine code. So there are three ways to get the machine code that the computer needs without writing it direct:

— **Compilers** translate a high-level program to the machine code version only once for each program; it is then saved in this form. Usually a linker is needed also.

— **Interpreters** translate high-level programs to machine code instructions on a line-by-line basis, and they do this every time the program is run.

— **Assemblers** translate from low-level assembly language programs to machine code.

Programmers never write in machine code these days, although the term is often incorrectly used as a synonym for assembly — which is close, but not correct.

machine language

Just to confuse the issue: the term machine code is most correctly used to mean a specific type of program which runs at the true object code or binary level, while the term machine language should only mean assembly language. The language is that of the mnemonics.

machine-readable

This can mean any possible form of computer input from any type of information or storage (tape or disk), but it usually refers to:

— Punched cards and paper tapes (mainly in bygone years).

— Magnetic print on bank-checks.

— Barcodes.

— Causzin softstrips.

— Mark-sensed forms.

— Magnetic-striped cards (aperture cards also).

It could also apply to handwriting recognition and scanned OCR. Possibly also to Drexler-optical and smartcards.

macro

• **Assembly:** A user-defined abbreviation for one or more lines of code. A macro assembler is one that allows the programmer to define macro instructions.

• **Applications:** A set of stored sequences which are reproduced by striking a selected key or key-combination. These can be:

— Instructions: Traditionally macros can repeat a sequence of commands (key commands or menu selections) called a macro-script.

— Actions: Some macros can record and reproduce a sequence of more complex commands ('go to top of page, write "Joe Bloggs", then down two lines, then tab twice').

— Data: They can reproduce typed sequences of data (say, name and address). These may be called Glossary items.

Macros can be written in a form of scripting language, but they are often recorded by: turning on the record function, performing the required task, switching off the recorder and assigning a macro key. This sequence is stored in a macro file and can be later recalled for use by a keystroke combination or F-key.

In spreadsheets, you often define macros by typing a formula into a cell and then giving that cell a name.

• **Lenses:** A lens which physically extends forward to provide extreme close-up functions, outside its normal focus range. This is usually also a zoom lens, and the zoom barrel function then becomes the focus control.

macrocell

In GSM digital mobile telephones, this is a coverage area with a diameter (from the base-station) of between 1 and 5km. The distinction is with micro and pico cells, which are much smaller.

Normal analog mobile telephone cells are often between 3 and 35km in cellular systems, and satellite cells can be larger than 500km in diameter. The term megacell is sometimes used for cells in the 20 to 100km range.

• Not everyone agrees on the definition of macrocell.

macroparallelism

A technique of tying together a number of microprocessors of the same type into a parallel processing configuration. There are two different ways of connecting these processors — with a common bus, and with point-to-point message passing.

• Common bus architectures are used to create shared-memory machines, because the processors can communicate with each other through the shared-memory. This approach makes the system easy to program, but performance doesn't increase proportionally as more processors are added — they spend too much time contending for the memory bus. Common-bus architectures quickly reach an effective limit.

• Message passing. Here the processors each have their own memory and communicate via messages. This improves the bus-arbitration but it introduces problems with synchronization. Some processors will always be waiting for data to arrive from another processor, so overall through-put is reduced. However, this architecture has a linear relationship between the performance and the processor numbers.

• See multi-processing also.

magnetic flux quanta

The smallest possible amount of magnetic energy that can exist in a superconductor. The quanta size is related to the magnetic field strength. In some ways these quanta are the magnetic equivalent of an electron, and they can be used to carry signals around a superconducting circuit.

Single-flux quanta-switching devices which are under development may be ten times as fast as the best electronic switching devices (currently 30 to 300GHz). Researchers have produced AND and OR gates, adders and A-to-D converters, using single-flux quanta technology, where one flux-quanta represents a logical 1.

magnetic recording

The first magnetic recorders were wire recorders built for audio and telegraphy recording by German engineers just before the Second World War. Later plastic (polyester) tape was used, and these materials evolved into today's hard and floppy magnetic disks (8-inch, 5.25-inch, and now 3.5-inch) and videotapes.

Magnetic data storage is the process of magnetizing regions of the medium in one of two opposite directions, using an electromagnetic head with a very narrow gap. Reading just reverses the recording techniques; the magnetic domains (regions) on the tape generate a current in the head as they pass the head gap and head coils.

High coecivity (resistance to change) is necessary in the magnetic medium to keep the layers thin and resilient, and the thinner the magnetic layer the higher the bit-density. Iron oxide was the first magnetic material, but later cobalt or chromium dioxide was added to increase the coercivity. The magnetic materials are bonded to the backing plastic material by a 'binder' which is the main problem when archiving magnetic recordings (it dries out and releases the particles).

Later pure metals were used but this required the development of new deposition techniques. See metal tape, hard-sector and octave.

magneto-optical disks

MO. This is a form of optical recording disk where the laser heats an area of the disk to which a strong magnetic field has been applied. This reorientates the magnetic spot which subsequently produces polarizing changes when a laser is reflected from (or passes through) the surface. The detection (read) is by shining a laser onto the spot and measuring the reflected brightness variations (+/–20%).

To erase an MO disk, you bias the magnetic field strongly and reheat the area with the laser. This means that the disk must first be erased, and then recorded — a two-step process.

MO disks are polycarbonate with pressed guide-grooves and address pits (laid down at the beginning of each sector). The magnetic alloy is splutter-plated onto the substrata.

- MO systems are erasable and rewritable but, by the use of special disks, they can also be write-once (WORM).
- Early MO disks were 5.25-inch and held 650MB; lately the track spacing has been reduced, and the zone/sector definition has been modified to give 2GB.
- The 3.5-inch 230MB version is also growing in popularity, and could eventually dominate in the computer market.
- For professional video and for the video-server (VOD) market, Sony is developing a standardized 8-inch platter system that offers 3.5GB per platter. Several platters are combined to create a server, and four heads are used to access each platter. A single server installation can have up to 100 of the MO player units, and offer up to 30 hours of MPEG2 at 2Mb/s, or for post-production in a TV studio, 3.3 hours at 18Mb/s.

See CD-R, WORM, MiniDisk and MD-Data.

MAHO

Mobile Assisted Hand-Off. Some digital systems allow the mobile unit to take a more active part in deciding when the hand-off between two base-stations should occur. Without MAHO, the mobile can occasionally alternate rapidly between two cellular base-stations, handing off backwards and forwards as radio conditions change.

These handsets monitor the various control channels, looking for the one with greatest reception power and then advise the current base-station.

mail-enabling

Applications software (usually word processors, spreadsheets or database managers) which incorporate e-mail functions.

main memory

In the PC world, it is sometimes not exactly clear what is meant by main memory today. In MS-DOS, some would say that it is only the 640kB of RAM below the video and BIOS RAM, while others would include extended memory.

Outside MS-DOS, generally, main memory would refer to the memory stack from $0000 to the upper addressable limit, and other terms such as working-, high-, upper-, would apply to the subdivisions.

mainframe

There's no accepted definition of a mainframe — but there's little doubt in most people's minds what it means. It is anything between a mini- (midrange) and a super-computer.

In earlier times, a mainframe was a multi-user computer with a CPU which could be reprogrammed with different microcode (unlike a microprocessor, where the microcode is built-in).

Many mainframes now use PC processing chips, most can now run Unix (which also runs on a PC), and some are using parallel techniques. Some are also adopting new programming techniques, such as object-oriented methodologies.

- The main mainframe companies are IBM, Amdahl and Tandem.
- IBM doesn't sell mainframes now, it sells 'enterprise servers'!

mains filters

Filters used on the mains power supply to reduce voltage spikes and hash. Mains filters don't have any significant effect on slow variations such as surges, sags and brownouts. See power supply.

mains peak voltage

AC power is a sine wave, and the quoted voltage of these systems is not the peak electromotive force, but rather an average (called RMS). The top of the peak in US mains power is higher than 110 or 120 volts — about 160—170 volts.

In the more common higher-volt world systems, the nominal voltage of mains power is 240 volts, but this is an 'average'. At the top of the AC voltage cycle (every fiftieth of a second) the EMF reaches a peak of 340 volts — so this is the mains peak voltage. In all sinusoidal waves, the peak is the average amplitude, multiplied by 1.414 ($\sqrt{2}$). See RMS.

mains power

North and South America, and ex-colonies of the USA, have tended to use 110—120V mains power at 60Hz. Most of the rest of the world has standardized on power at higher voltages, between 220V and 250V, usually at 50Hz (although 40Hz was also popular at one time).

The higher voltages gave less problems when carrying power over long distances, and required less copper in the wires. Unfortunately these voltages also kill you easier.

The European Union has now (IEC 38 /HD 472-S1) standardized on 230V to be strictly enforced by 2003, However variations of up to +6% or –10% will be permitted for a time.

• Despite the regulations, most of Europe is still on 220 volts, 50Hz, while the UK still uses 240 volts, 50Hz.

• Generally you can assume that mains power will be maintained within +/– 6% of the nominated voltage (depending on your location and distance from the last transformer) and that it will have long-term variations of no more than 2% over time. Mains frequency (Hz) is normally maintained to within 0.1Hz and corrected over time. If a slow-down has been recorded, the generation plant will speed up the cycles to compensate (for time-clocks, etc.). See power supplies.

majordomo

A listserv function which automatically extracts your name and address from your subscribed message.

make-up

In desktop publishing, this is the combination of text and graphic elements pasted down as 'camera-ready copy'. It may not have color separations in place, but it will have an indication of exactly where they are to go.

male connectors

These have pins or jacks, not holes or sockets. Look at the contact pins, not at the outside shielding.

MAN

Metropolitan Area Network. There are two quite different concepts involved in MANs.

- A MAN is a telecommunications network service — a common-carrier tariffed service provided by the telco for LAN-interconnection, imaging, and other high-rate data applications.
- A MAN is the same as any wide-area network (WAN) but only for private networks covering a single metropolitan area. These are larger than LANs, but smaller than national or international WANs. The term WAN is then used for national or international coverage.

This second use of the term is going out of fashion, luckily — it was unnecessary and confusing.

• More specifically, in the IEEE the term MAN is a service definition applied to various forms of multiuser fast packet-switched networks, which allow the exchange of data (and possibly voice also) along a LAN-like loop covering a substantial geographical area. The IEEE definition is for a dual-fiber loop with optimal diameter of 50km. However, a number of interlinked MANs can provide city-wide or nation-wide coverage.

The IEEE has recommended the 802.6 DQDB standard as the main technology specification for MANs, but FDDI is also used. In fact, FDDI-II was, at one time, the favored MAN technology. The service definition used by the American RBOCs is usually SMDS, which is not necessarily now tied to DQDB.

• The best way to think about a MAN is as a large metropolitan-wide LAN which has facilities which allow it to be shared among many users (billable) and where the company attachments are treated as nodes. The main problem with this type of public-access shared network, is in providing equal and fair access to all users. This was the value of the queuing system invented for DQDB.

• The most common application for MANs is to provide LAN-to-LAN links but, being a connectionless service, a MAN can easily add new access points and can be used as a public-switched broadband data service and carry isochronous traffic.

In the USA and Europe, these services now go by the name Switched Multimegabit Data Services (SMDS). See WAN, DQDB, SMDS and FDDI.

managed networks

This doesn't just mean services supplied from outside an organization, but includes those run by the company itself. It refers to the provision of basic voice and data links, together with such essential features as control over network configuration, capacity-distribution, network diversity and redundancy, tailored billing arrangements, network management reports and planning.

• VANS and IVANs are now national and international managed networks.

managed object

Any device on the network which can be managed by a network management protocol. Usually these will be SNMP-based, and the management will be via an SNMP agent in the device.

Management Information System

See MIS.

Management Resource Planning

See MRP.

Manchester coding

Also called diphase or digital biphase code. This is a very popular line-coding technique for today's communications systems; it is the basic line-code of Ethernet. Manchester is seen as an alternative to NRZ, Bipolar AMI and CMI techniques and also used with some disk recording techniques.

When used in networking, Manchester is backed by a complex standard which specifies the voltages and voltage-reverses which are required to specify a space and mark. These techniques are also used today in digital magnetic tape recording, optical fiber, and coaxial links.

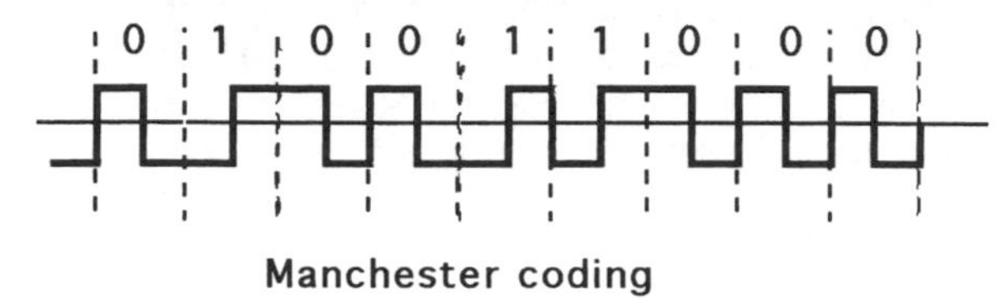

Manchester coding

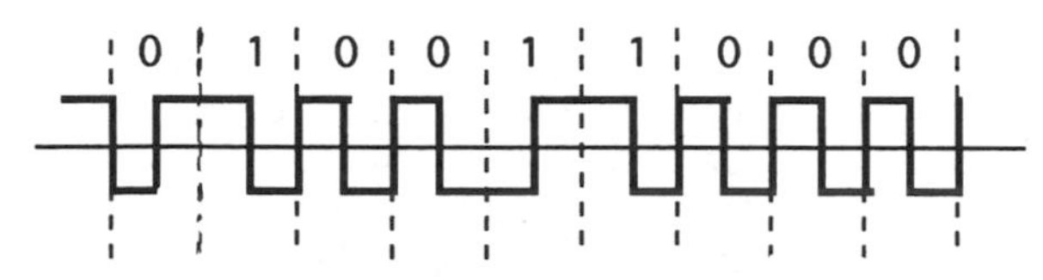

Differential Manchester

Manchester is a two-level code with a mid-bit transition for each bit-interval — in other words, it has two symbols (baud or events) for every bit. In the middle of each signaling interval (the bit-slot or bit-cell) there is either a transition from a high voltage to a low, or from a low voltage to a high. It is the direction of this middle transition that is important, not what happens at the beginning of the bit-slot.

Unfortunately there is no official standard for Manchester.

— In Ethernet LANs: Logical 0 is represented by the change high-low in the middle of the interval, and logical 1 by the transition low-high.

— In older telecommunications networks: Logical 1 is high-low, and logical 0 is low-high.

Voltage levels can also change at the juncture between bit-slots, but these don't count — only those of the mid-bit transition, and this is also used for clocking.

By only recognizing signal change within bit slots, Manchester coding solves the synchronization and idle-bit problems encountered with AMI and similar line-coding techniques — it is impossible to have long periods without any change.

However, it is also correct to say that Manchester's use of two-states to represent one bit, is extremely wasteful in network capacity. In effect, Manchester systems have a 100% redundant overhead when compared to the potential signaling rate.

• **Differential Manchester** coding is used in Token-Ring. It is a variation on the standard code with more forced transitions to help maintain synchronization. It registers a logical 0 if there was a voltage change at the beginning of the bit period, and a logical 1 if there was none. It is not concerned with the direction of the change. See Differential Manchester.

• The advantages of Manchester systems are:
— Synchronization; the system is self-clocking,
— There is no DC component, and
— A substantial degree of error detection and signaling is incorporated into the line code itself. See violation.

Mandelbrot set

A mathematical formula discovered in 1980 by Bernoit Mandelbrot which produces extremely beautiful fractal images — the delight of graphic artists and spaced-out video freaks. This is part of Chaos theory. The significance is that simple algorithms can produce very complex shapes and images. See Julia sets.

mantissa

In scientific notation, the number 2 followed by 6 zeros is written 2×10^6. The power to which 10 is raised is called the 'characteristic' or the 'exponent', while the number itself is the 'mantissa'. An example: consider the value 123.4. This can be expressed as 0.1234×10^3. The mantissa is 0.1234 and the exponent is +3. See radix.

map/MAP

• Memory manipulation. See mapping.

• RAM allocations. See memory map.

• **Manufacturing Automation Protocol:** A LAN architecture created by General Motors in 1983 as a sophisticated broadband communications blueprint for its automated-equipment vendors. The physical foundations are in a complex layering of software (taken from the OSI model) to guarantee the delivery of data to machines, and to allow many different types of computer to speak a common language. MAP uses special ISO protocols sitting on top of the IEEE 802.4 Token-bus network running at 10Mb/s. See Token-bus.

• **Mobile Applications Part:** In GSM. This is the key messaging protocol between base-stations and switches. It will need to be completely upgraded and replaced for the 1996 GSM Phase 2 introduction.

• **Multiple Applications Platform:** This is a generic term, but it is used for small mass-produced hardware devices which can be reprogrammed to provide a wide variety of functions or applications. The idea is that MAP hardware will eventually become commodity objects or modules which can be re-used for many tasks in computing and telecommunications.

MAPI

Messaging Application Program Interface. Microsoft's specification which allows e-mail messages to be sent directly from applications. It comes in two parts: the first defines interfaces for the mail server and the second, those for the mail client.

MAPI is an attempt to deliver a standardized interface for different host-based, LAN-based and electronic messaging systems. With this API, the developer can add e-mail messaging to Windows applications, even though the program remains quite independent from the messaging and directory services.

Microsoft have integrated MAPI into their SQL Server for instance, and this means that e-mail can be used for updating a database. See VIM, CMC, MHS and WOSA.

mapping

Mapping can be the act of mathematically converting one set of binary numbers into a different set of matrix values, or the act of translating the bits in memory locations into the images on a screen (or laserprinter), or of taking one table of values in a router and converting them into another table. The term suggests the transfer of values or information, in a very tightly controlled way, from one set of pigeon-holes to another set of (perhaps slightly different) pigeon-holes.

• Usually the term is applied to bit-mapping where the elements to be displayed on the screen (or on a printer) are first inserted into memory cells (in video RAM) according to a calculated pattern. The way this happens has changed a lot over the years, but the term used hasn't; we still talk of mapping the word-processor's text into the video bit-map, for instance.

The computer is converting from logic-values, to screen-matrices at a certain coordinate — from a byte which represents, say, the letter T, to a series of bits set in upper memory which, when raster-scanned, will produce the letter T in the correct size, shape, weight and position on the screen.

Mapping of text is generally done under the guidance of multi-dimensional look-up tables or by using complex algorithms (See outline fonts). Mapping of graphics may involve a wide range of different techniques — however Paint-type programs are fairly close to being in final screen bit-map form from the start.

However the data has been mapped into video RAM, after the whole bit-map has been constructed, the video RAM locations are scanned sequentially and used to modify the screen. This is a strict relationship — each screen pixel lights up or changes color by virtue of how certain bits are set for each location in this memory matrix. The video RAM bits are then being 'mapped' to the screen.

• The above process is even more complex in some cases: see pixel depth, packed-pixel and planar architectures.

• The video memory map represents the screen as a two-dimensional matrix. This allows the rapid control of the raster-scans which are constantly refreshing the screens. One memory location (at the bit level, the byte level or the multiple-byte level) can hold the full representation of one screen pixel.

So, when using an outline font like TrueType, the algorithms will calculate the shape for a character of the correct size and weight, and store it as a character bit-map in a look-up table. This will be mapped into the appropriate area/s of video memory (the video map) which represents the locations of that letter on the screen. The screen driver then constantly extracts the information and sends it to the monitor's electron guns.

• With object-oriented drawing programs, each object on the screen will have its own bit-map, and these will be transferred as a secondary process to the main video RAM corresponding to the screen. Here the system must control which objects lie in front of the others. When you move an object, the coordinates of the object in the new position must first be transferred to the new locations in the object's video map, and then to the screen.

Paint programs won't have this two-stage approach, so you can't just move elements of the image around.

• Different screen-mapping modes can be provided by the same software. Windows has eight: a pixel-based mode, five fixed character modes, one isotropic mode and one anisotropic mode.

• A quite different example of the term 'mapping' would be the translation of application commands intended for one type of database into those that can be used by another (say, at a database gateway). The individual requests would need to be mapped into requests suited for the new application. This is a cross-table operation.

• A memory map (showing the areas or memory allocated for various usage) is quite different from memory mapping; so don't confuse. See memory map.

Marisat

An ancestor of Inmarsat. The early international maritime organization launched its first geostationary satellite over the Pacific in 1976 for ship to shore communications. This was the beginning of mobile satellite communications. Later the same year, two more satellites were launched over the Atlantic and Indian oceans. Marisats operated in the L-band and the ships needed a 40W transmitter and 1.2 meter dish antenna. The Marisat satellites were handed over to Inmarsat in 1982.

mark

A telegraphy term for the physical condition on a line which signifies a binary 1 as distinct from binary zero which is known as 'space'. It is still a useful term because it is unambiguous.

Some systems use a low voltage to indicate a 1, while others use a high voltage. Some will use a high–to–low transition to signal a 1, while others use a low–to–high transition. So the use of the term 'mark' makes a lot of sense.

mark parity

A form of parity that tests the accuracy of transmitted data by checking whether the defined parity bit (an eighth bit in a 7-bit character byte) is always set to a logical 1. Space parity is the obverse — the parity bit is always set to zero. These techniques are not often used because they will only detect gross byte-sized errors — but there have been systems in the past which required mark parity to act as a 'marker' at the end of a byte.

mark-sensing.

See optical mark recognition.

marquette

A Macintosh GUI term for the tool which surrounds an area on the screen with a dotted rectangular selection box. This is used to surround and activate a number of icons, or select an area of a graphic image. It is used to identify objects or regions on the desktop or in a drawing or CAD program on which some process will be performed. The Mac also has a vague-shaped marquette called a lasso.

MARS

Microfiche Automatic Retrieval System. This is widely used in libraries.

MASCAM

Masking-pattern Adaptive Sub-band Coding and Multiplexing. A digital audio signal bit-rate reduction technology developed by a German radio research institute. The masking process reduces the coding rate by a factor of about 10 without seriously degrading the subjective quality of the audio. In this case, masking refers to discarding high-resolution elements that can't be heard because of louder low-resolution sounds. This is the forerunner of DAB and NICAM. See PASC and DCC also.

maser

Microwave Amplification by Simulated Emission of Radiation. The radio equivalent of the laser, invented back in 1954.

MASH

See Bitstream.

MASIC

Memory-intensive ASICs. An ASIC chip which is customized for a specific application, and which contains a large amount of memory. FRAM will probably be important in this market.

mask

A general term meaning to protect an area in order to selectively (or sometimes accidentally) block some action.

• In a bitmapped image, a masked area is an inactive area which will not respond to change. In a color 'paint' program you may wish to prevent changes being made to certain areas or certain colors — so you mask them.

• During the input of data into a database, each field may have a 'mask' which prohibits data not of the right type (or outside certain parameters) from being entered.

• In the production of chips, this is the 'negative' image used in the photo-lithography process. In ROM manufacture a mask is used to 'program' the chip. It blocks off certain areas of the chip when the ROMs are burned. See masked ROM.

• In the manufacture of PC boards, the areas which are not to be etched will be masked before the photosensitive material is exposed to the activating light source. In this case, the mask is more a stencil.

• In film production, people can be inserted into previously filmed scenes by creating moving dense-black masks (also called mattes) which are silhouettes of the moving figure. These are later sandwiched with the print-stock in the printing process. The mask blocks the light from penetrating through to the print-stock and leaves the appropriate area un-exposed. The image of the moving figure is then 'inserted' (overprinted) into this 'reserved' area. This is generally called matte or 'traveling matte'. See chromakey — the video form.

• In video a mask is an electronic insertion of a shape generated by image tones or colors, used to 'cut-out' part of one image, into which an exact match from another image is inserted — it is an electronic version of the traveling matte film technique. So an actor can appear to walk through the streets of London (which may only be a transparency or video) because the mask he generates (via the video camera) is used to reserve (switch out) part of the London scene — and then his original image is inserted into this.

This involves a very fast switch, which can take the individual lines of image from either of two sources — and the switch is triggered by (usually) the color as it sweeps across the scene. Blue and Green are preferred as background colors (those that trigger the switch) because they are well away from flesh-tones. See chromakey.

• In video, tones (within the picture, or external) can also be used for a key effect, similar to chromakey. See key, and internal and external key.

• In digital audio, masking in human hearing means that some louder sounds will blanket-out softer ones. When these can be identified, it is possible to ignore many of the quieter sounds which will not be heard. This then reduces the number of bits needed to code the sound at the same subjective quality. DSS uses this technique, as does the Sony MiniDisk.

masked ROM

This refers to the process of creating a ROM (non-volatile memory) chip by masking the surface of the chip during manufacture. The chip is said to be mask programmed. This type of manufacturing set-up requires a large investment, so the process is only economical for large quantities. The distinction is with PROMs and EPROMs and other forms of non-volatile memory which are programmed electrically through a link to a computer and can be programmed one at a time. Virtually all chips referred to simply as 'ROM' are actually MROM.

masked words

See stop word.

masked write

A process of changing one pixel in a video memory location without changing the whole byte. See VRAM.

mass storage

Refers to large disk and tape storage devices, and now to optical recording disk systems.

master

• The original, from which copies are made. The host that controls the peripheral terminals. The key system upon which slave systems depend.

• The final post-production format of a disk/tape/film, etc.

which is being created for mass replication. The master is used to make the printing negative or the stamper, and is then put away securely in a safe. All subsequent copies are made from these negs, or stampers.

- The key unit in a distributed system — the central point. There can only ever be one master station in a network. In a polling system the master will control the polling sequence.
- The dominant unit in terms of control, as distinct from the 'slave'. The terminology master–slave was widely used at one time for multi-user systems, but now the terms 'host–node' are preferred. The distinction here is often with client–server.
- Fonts: see multiple master.

mastergroup

In North American terminology this is 600 voice channels which, in FDM need 2.53MHz of bandwidth, and which occupy the spectrum between 564 and 3084kHz. A mastergroup is then ten supergroups.

- In ITU terminology this is 300 voice channels which in FDM need 1.232MHz of bandwidth, and occupy the spectrum between 812 and 2044kHz.

masthead amp

These are amplifiers (usually coupled with intermediate frequency converters) which are mounted near to the antenna to boost the signal before it passes down the antenna cable where it may be seriously attenuated or pick up excessive noise. If the cable run is long, a line-amplifier might also be used at some intermediate stage down the cable. See LNB and LNBC.

maths coprocessor

A specialized processing chip which plugs in alongside the main CPU and takes over many computational-intensive tasks. This chip is often called a floating-point unit or numerical processing unit. This is not a parallel processing operation; either the normal CPU or the coprocessor will be active, but not both.

The value of having the maths chip, therefore, is that it is especially designed to handle floating-point arithmetic. It will have a number of internal registers (usually 8 to 32) each capable of holding a long floating-point number (often 80 — 90 bits long). The chip is therefore able to perform complex calculations without constantly accessing memory. See coprocessor.

matrix

A table or array with two or more dimensions. To be strictly correct, arrays have a single dimension while matrices have two or more dimensions. The term 'table' is also applied to the two dimensional (row and column) form.

- Matrix matching. See OCR.
- Matrix printers. See dot matrix and impact printers.

matrix manipulation

The type of data manipulation done in spreadsheets. This also applies to other large tables not in the spreadsheet form.

matte

See mask.

MATV

Master Antenna TeleVision system. The more limited form is also called SMATV or 'Satellite' MATV. Master antenna systems are small coaxial-cable distribution networks usually in large blocks of flats or which run through a campus or industrial site.

This is a small, self-contained CATV system which possibly receives both terrestrial and satellite signals and distributes them through its own short run of cables to its own residents; it will usually rebroadcast the signals down the cable at their received frequencies, so they will be tuned by the various TV sets in the normal way. For Premium Pay TV, each home will need its own descrambler.

MATV systems often have problems when they are receiving off-air signals via an antenna on the roof, satellite signals via a dish, and have a cable connection in the basement, all trying to use the same cable and often the same frequencies. Often two or three independent MATV networks are needed, or many of the channels transposed in frequency to fit into slots available.

MAU

The same acronym is used for dissimilar devices which perform much the same functions, but which are distinguished by different names in different LAN specifications.

- In Token Ring (802.5) this acronym means Multistation Access Unit. To avoid confusion, many people now refer to this as an MultiStation Access Unit or MSAU. See.
- In Ethernet (802.3), MAU is translated to mean Medium Attachment Unit. In general terminology, however, it is often called Media or Medium Access Unit. and in the Ethernet literature these devices are often just known as transceivers. Transceivers perform the basic Layer 1 functions of injecting bits into the network, and of collision detection. They provide the electrical and physical connections.

maxdata

The maximum data size for a frame on a network link.

Mb/MB

- **Mb: Megabit** or Mbit. This is 1,048,576 bits.
- **MB: MegaByte** or MByte. This is 8,388,608 bits.

Note: in some computer dictionaries these are presented the other way around with Mb = byte, and MB = bit. This doesn't make sense to me, but there is no recognized standard.

- **MB: Motherboard.**

MBE

- **Molecular Beam Epitaxial:** A technology for depositing materials such as compounds of gallium arsenide with a single atom precision. MBE was developed at Bell Labs in the mid-1960s by John Arthur and Al Cho.

It began as a technique for growing crystals, but it proved also to be capable of producing ultra-thin layers of different (and doped) semiconducting compounds. This is the key process in the modern development of opto-electronic devices — lasers, modulators, switches and filters. See quantum.

• **Multi-Band Excitation:** MIT's voice codec technology for low bit-rate cellular telephones. It is based on STC. See.

MBE is replacing LPC and CELP as the preferred technology for very low-rate voice reproduction (but it is terrible with music) because good telephone quality can be preserved down to 4.2kb/s and comprehensible speech can be encoded at 2.8kb/s. MBE also provides superior reproduction of the female voice which is always a problem with LPC. The term IMBE (Improved) is also used — I guess it is the later version.

MBone

Multicasting Backbone of The Internet. This is an overlay multicasting virtual network which spoofs routers into handling wider-band streams than they would normally carry (for videoconferencing).

MBS

Mobile Broadband System. A project in the European RACE program for the development of broadband radio links in the 60GHz range. It is one of two related RACE telecommunications projects — the other is UMTS. MBS aims to provide broadband services from 1Mb/s to 70Mb/s for mobile users, while UMTS is for narrowband services.

MCA

• **Micro Channel Architecture:** IBM's attempt to regain control over the PC clone makers by implementing a new proprietary bus standard for their new PS/2 line. The idea was that MCA would then be licensed to other PC manufacturers at a high cost. These companies were also tied, in fact, to a penalty payment for having previously cloned IBM technology. It was not one of IBM's most intelligent moves. It effectively lost them control of the PC architecture, because everyone rejected their conditions.

While there was no dispute that MCA was a major improvement, the MCA bus was not backwardly compatible with the 16-bit ISA (AT) bus, which annoyed potential customers for a start.

The reaction of clone makers was to create an alternative standard called EISA which was about half as fast again. In the final wash-out, neither MCA or EISA has made much of an impact; the old ISA standard survived and is now commonly coupled with VL-bus and/or PCI.

MCA is a 32-bit bus with expansion slots said to be capable of rates of 80MHz. The term 'channel' here refers to an intelligent bus, the theoretical throughput of which is about 20Mb/s.

The boards are smaller and use surface mounted components on each side.

• **Media Control Architecture:** An Apple Mac specification for computer control of videoplayers and CDs.

MCD

Multimedia CD. See Super Density CD also.

MCGA

Multi-Color Graphics Array. An IBM color monitor standard introduced with the Model 30 (8086-based) PS/2 machine. It had all the features of MDA and CGA, and some additional modes of its own. However it didn't have some of the modes of EGA and VGA.

When IBM upgraded the Model 30 to a 286 chip, they also dumped the MCGA monitor standard and replaced it officially with VGA. The MCGA modes had a rather coarse 320 x 200 pixel resolution, but with 256 colors.

MCM

Multi-Chip Modules. A chip package type used for multilayer devices.

MCN

Micro Cellular Network. Vodafone's micro-cellular GSM mobile phone network. They designed their network to have a small (micro)cellular structure but with a macrocellular overlay, and they connected them via microwaves. Different channel frequencies are used for the micro- and macrocells, and the network will switch users between micro- and macrocells dynamically when required. This network design seems to offer a few advantages.

MCPC

Multiple Channels Per Carrier. In satellite broadcasting, terrestrial broadcasting and general communications systems, it is common for a number of compressed digital channels to be interleaved into a single wider-band radio channel.

In digital direct-to-home satellite TV with the new MPEG compression standard, for instance, between four and six channels can occupy the bandwidth previously allocated to one analog TV channel. This can be done either by assigning a specific frequency band to each digital channel (SCPC — subdividing the old analog bandwidth and allocating each separately), or by multiplexing and interleaving the four or six channels into the one wide-band carrier.

The MCPC approach makes the channels more robust and flexible. Burst errors are spread among the channels, and individual allocations of bandwidth can be varied, on the fly, to match requirements. But obviously the MCPC approach can only be used when all interleaved channels originate from the same earth station.

If it is essential that the various signals originate from different earth stations at different geographic locations, then each needs its own carrier bandwidth — which is the SCPC (single channel per carrier) approach.

MCU

• **Micro-Controller Unit:** These are devices used in the control of disk drives. These control units will often now have digital signal processing (DSP) functions as well.

• **Multipoint Control Units:** A term in videoconferencing which refers to what is essentially a video 'bridge'. It is the point in the network where all signals are controlled, duplicated and transmitted when multiple codecs are involved. Standards conversion may be performed at this point also.

This is a carrier-supplied function. As an example, the Bell Labs' MCU offers:

- Three control modes. auto, chair control, broadcast/presentation.
- Dial-in or dial-out (people can join conference either way).
- BONDing, for codecs greater than 128kb/s.
- Audio only involvement.

MD-Audio

MiniDisk. See.

MD-Data

Sony's magneto-optical (MO) format derived from the 3.5-inch audio MiniDisk (MD-Audio) and is used for computer data storage. The capacity is 140MB, and the disk's data transfer rate is 150kB/s, the same as basic CD-ROM.

There are three variations on MD-Data:

- Writeable disks which can be erased and rewritten any number of times.
- Read-only (also called MD-ROM).
- Hybrid, which are partly writeable, partly read-only.

The MD-Data disk has its own file system and is therefore interchangeable between the PC and Mac.

The technology differs from conventional MO, in that it can erase and rewrite data in a single pass (DOW). They achieved this by having the laser and the magnetic heads on opposite sides of the disk and using a new layered magnetic material called terbium-ferrite-cobalt which changes state very easily.

MD-ROM

See MD-Data.

MDA

Monochrome Display Adapter. This is IBM's monitor standard for the original PC. It provided 25 lines of character cells in 80 columns, and each character cell has a fixed size with 9 x 14 monochrome dots. The theoretical graphics resolution was 720 x 350 pixels — but there were no true graphics.

MDA was designed for text only, and it could only use TTL digital signals. It could, however, use graphic characters.

The monitor screens generally came in white or 'cat-sick green', and the early-model monitors were once described as '... brutish, nasty things, designed by microcephalic troglodytes in the secret pay of opticians'. See HGC, CGA and EGA.

- The connections for an MDA monitor are:

Pins 1 &2 ground; Pins 3, 4 & 5 not used; Pin 6 +ve intensity; Pin 7 +ve video; Pin 8 +ve horizontal drive; Pin 9 –ve vertical drive.

MDB

Modified Duo-Binary. A communications technique using four voltage states to transmit dibits. See quaternary.

MDS

Multipoint (or Microwave) Distribution Service. Originally this was called 'Metropolitan Distribution Service'.

MDS is the use of microwave frequencies (higher UHF, with true line-of-sight transmissions) for the local distribution of data, television or radio broadcasting — often for niche or 'narrowcast' programming.

The transmitter power is often restricted (100 watts seems to be common) and so MDS transmitters generally cover areas up to 50km in radius. Different frequencies are used for MDS in different parts of the world, but generally 100 to 150MHz of bandwidth is made available in the spectrum around 2GHz or 2.5GHz.

The 'multipoint' term refers to the fact that if many TV signals are transmitted at the same power from the same antenna, then problems with adjacent-channel interference disappear, and so every channel can be made available. Usually about 20 are allocated in a single band in the upper end of the UHF spectrum, which is why the term 'microwave' is sometimes used. These are all analog systems at present, but digital will arrive in the 1997 time frame and double channel capacity.

Traditionally the MDS approach to TV transmission is associated with 'narrowcasting' of data and information to specific users (education, tourist hotels, medical centers, etc.). More recently it has begun to be used for Pay TV or Pay Radio services, which can be classed as 'narrow' or 'broad'-casting at the whim of politicians and lawyers.

When the same company transmits multiple channels from a single masthead, it is more correctly called Multipoint MDS (aka 'Wireless Cable' — a contradiction in terms) to emphasize the fact that a number of programming channels are grouped together. But MDS can deliver, say, a dozen channels cheaper than cable in many situations.

With microwave frequencies, there is always a problem with signal shadowing behind buildings, and in valleys. So, unlike normal VHF and UHF, a number of transmitters will be needed to provided metropolitan coverage, and 'beam benders' (repeaters) will often be used to push signals (of the same frequency) into shadow areas. Transponders will be used to carry the signal between transmitter sites.

Transmission power will be in the range from 10—100 watts for small-area coverage, and up to 1000 watts for city-wide coverage.

MDS offers a number of advantages over conventional UHF/VHF systems of broadcasting. Because of the line-of-sight limitation, the MDS frequencies can be reused in other nearby 'cells' in the same way that a cellular structure allows reuse of mobile phone bandwidth. So, with adjacent channel use, and cellular structures, MDS often provides a very efficient way to use spectrum bandwidth. See MMDS, LMDS and wireless cable.

MDX

Multiplexer/demultiplexer. Also called a MUX.

ME

Metal Evaporated tape. See metal tape.

Mead's limit

Carver Mead, the inventor of the gallium arsenide MESFET transistor, has predicted that the behavior of transistors on a chip will deteriorate when the feature size gets below 0.2 microns. MicroUnity's developments have thrown doubt on this. See GSI and Debye limit.

• In Feb 1995 NEC showed a 1 gigabit DRAM memory chip using 0.25 micron spacing (about the width of a bacterial cell). They say they expect to bring line-width down to 0.15 or 0.18 microns in the next few years.

measured lines

In the USA, the local telephone carrier will often offer a range of different local-call billing plans. Customers can choose a higher monthly rental and free local calls (unmeasured), or a lower rental fee and be charged for calls (measured).

It is not always clear which way Americans use the term, because there are three basic forms of call charging:
— free local calls, but high rentals,
— flagfall charges with no timed component,
— timed calls (or combined flagfall and a timed).
Timed calls are clearly measured, but the second category can be classed as either measured or unmeasured depending on what you define as being measured: call time or call connections. Virtually all long-distance or toll calls are measured. These lines/calls are also known as 'metered', which clearly indicates time-charging.

mechanical

• In desktop publishing, this is the term for camera-ready (paste-up) artwork which is ready for offset printing.

• Mechanical interface: This is the physical construction of the connector — the pin placement, sockets and shielding.

media

• The physical material through which the signals pass. At the base physical level this can mean copper, glass or the ether. But, in fact, the term is usually used in a much more specific way than this. It will refer specifically to a type of copper (unshielded twisted-pair, shielded twisted-pair, or coaxial cable), or to a type of glass fiber (monomode or multimode fiber), or to the VHF radio band as distinct from the MF band.

Pedantics will tell you that media is the plural of medium — but don't believe a word of it. It is still media, whether you have one fiber or two. (Ask the pedant whether the ether is 'the media' or 'a medium'. Is a universal imaginary environment, singular or plural?)

• A general term for print and broadcast communications.

Media Lab

Usually this refers to Nicholas Negroponte's MIT Media Lab.

media processors

Very high-speed CPUs and DSPs (but also including massively parallel systems) which are designed specifically for encoding and decoding compressed video. These will be used to handle the high bandwidths required in new radio/TV transmission systems. See communications processor.

medium capacity systems

In telephony — this means a system that will handle up to 300 telephone extensions in analog form, or data at a rate of between 2Mb/s and 34Mb/s.

medium-power satellite

One with RF output power from its transponders between 30 and 100 watts.

Medium Wave band

MW. This is a specific section of the MF band used for AM broadcasting. It is at the longer wavelength end of the spectrum. The band extends from 530kHz — 1610kHz.

mega-

In the SI system this means a million or 10^6. The abbreviation must always be capital M to avoid confusion with milli- units.

Mega-chip

A 1 megabit DRAM. You need eight of these (nine with parity) in a computer to hold a megaByte.

• NEC has just produced a 1 gigabit DRAM.

megaByte

A million-plus bytes (strictly it is 1024 x 1024 = 1,048,576). In terms of common English text this is about 170,000 words or about the length of two average novels.

megapixel

A million picture elements. A measure of the quality of a single high-definition digitized video image, and also of workstation monitor resolution. Good workstations will always have more than one megapixel of resolution. And for PCs, some portrait (vertical format) screens now provide 1280 x 1024 pixels in a two-color mode — which is 1.3 megapixels.

A full-frame scan of a 35mm motion picture image can yield about 12 megapixels at the limit of the film's resolution (much less by the time the film is edited and printed), and HDTV systems need about 3Mpixels per frame. Note that this measure does not take into account pixel depth — it can equally apply to a B&W (1-bit) pixel, or to a color pixel with 24- or 32-bit depth.

memes

Thoughts, ideas, concepts; as part of human communications.

• Richard Dawkins uses the term to mean ideas that infect your mind, just as viruses infect a biological host.

Memex

• A theoretical ideal source of information proposed by Vannebar Bush in the 1930s. It included some early ideas about hypertext linking.

• Also memex is a hardware/software unit which holds text in a very compressed state, and has lightning retrieval despite not using an index. It allocates 128 non-ASCII characters to common words, then by using a two-byte system it creates a special vocabulary of over 16,000 words. All words in a book then need only one or two bytes to identify them. This approach is especially useful for fixed reference material such as the Bible where the vocabulary is limited. See substring replacement, Lempel Ziv.

memory

• **Computers:** In computers, the term is usually applied to the RAM (working memory), although strictly it should include ROM and EEPROM memory types. However, it is often used vaguely to refer to the whole memory-map, which may include V-RAM and ROM or EEPROM.

Sometimes memory is extended to long-term storage such as hard-disks, floppies, and tape systems. This is confusing — they are better called 'secondary storage' or just 'disks and tapes'.

Non-volatile memory is a general term applied either to battery-backed RAM, to flash, or to EEPROM. The term non-volatile is sometimes incorrectly applied to Static RAM also.

• **Chips:** A new development is 'smart memory'. See content-addressable memory.

• **Batteries:** With batteries, memory causes a gradual reduction in capacity. It is common when ni-cad batteries have been charged regularly without fully discharging them first — as a result, the total charge capacity of the cell drops. The cells fail to return fully to zero and there's a growth of crystals inside the cells which needs to be broken down. To do this connect a length of wire between anode and cathode (in short bursts) and flatten each battery completely before recharging. Intelligent chargers perform this discharge cycle before recharging.

memory cache

Modern processors often have internal or external cache (pockets of memory available for temporary storage of work in progress) to speed up operations. The Intel 486 chip, for instance, has 8kB of internal cache on the chip. Other systems may set aside (say) 64kBytes of SRAM in a special section of the main memory.

Memory cache is designed to improve CPU performance in a few different ways:

- By holding a copy of the last data accessed from memory, on the assumption that it may soon be required again.
- By storing closely-related information in fast memory.

In some systems, when a memory location is accessed both the contents of the location and the contents of its neighbors are automatically moved to the cache, and the new addresses are cross-referenced and stored in the cache controller. The CPU will check the cache controller before accessing normal memory. Cache requires special controls, now usually held in ROM. There are other forms of cache also. See disk cache.

memory card

• **Expansion:** A memory-expansion plug-in board which occupies a computer slot and expands the amount of RAM available. This usually provides high-memory locations (above the old 1MByte DOS limit) which can be used for expanded or extended memory applications — or just as normal memory with any of the modern operating systems. Older cards were often more specific: some were called EMS cards, for instance. See EMS and EMM.

• **Smartcard:** A general term for the lower-level smartcards which hold personal and financial information in a secure way. They are basically just memory-chip cards using EEPROM. Smartcards usually have a PIN and some in-built intelligence to control access; and these systems can be used by some POS terminals and/or computers through peripheral card-readers. See SVC.

• **Drexler optical cards:** These have information added to the surface by laser, and they are not erasable. However modern optical cards can hold an enormous amount of information, and are now being used for hospital records. They don't like being scratched, and they are WORM — they can be updated and the information can be erased (destroyed).

• **Plug-in:** A form of ROM, EEPROM or flash memory (and sometimes battery-backed RAM) in a credit-card sized memory unit. These are nearly always flash cards now — however I've also seen the term used for the small PCMCIA Type 3 plug in hard-disk drive. See flash and PCMCIA.

memory managers

Don't confuse software memory management with memory management in the sense of multitasking.

• In modern computers, a section of the processor (or a separate MMU chip) has the job of providing protection for the software against being overwritten, and this is absolutely essential when the computer is multitasking.

The MMU provides memory functions critical to applications when they are associated in multitasking or multi-user situations, where a memory manager must keep one application from interfering with another.

• Quite different software techniques were designed to provide PC users with more effective ways to utilize the working memory in the early machines and overcome the old DOS limitation of 640kBytes. These were add-on utilities.

To use extended memory, expanded memory, or upper memory area, your CONFIG.SYS file must contain a command that loads a memory manager. There are three types:

Expanded memory managers: Expanded memory cards were originally managed by special expanded memory managers (they came with the EMS card). These days, EMM386 will manage programs requiring expanded memory by emulation, using extended (sic) memory. MS-DOS doesn't actually have an expanded memory manager.

Expanded memory has certain limitations on use, and it can be slower and more cumbersome than extended. It was a stop-gap measure and it is mercifully dying, however some older programs still require EMM386 emulation.

Extended memory managers: These all provide the first, and at least some of the following functions:

- They can temporarily borrow the EMS page-frame to initialize large device drivers, and allow them to load into the extended area of memory.
- They may be able to discover unused addresses within system ROM and utilize this in Upper-RAM.
- They can possibly remap ROM addresses to provide contiguous blocks of Upper-RAM (in the 640kB—1MB range).
- They may be able to hide some of the ROM code and allow DOS to use the entire Upper-RAM area.
- They may push video display memory and BIOS ROMs as high as possible in the Upper-RAM area..
- They can load the DOS kernel, command processor, buffers and stacks into high memory (above 1MByte).

DOS Protected Memory Interface managers: (DPMI) which is a new way to bypass the 640kB limit for some programs.

• All versions of MS- or PC-DOS will manage the conventional memory (below 640kB) themselves with their own built-in memory managers.

• MS-DOS 5.0 performs some of the higher level functions itself (using Memory Control Program) as does any version of Windows. However third-party memory managers are still used.

• MS-DOS 6.00 and IBM DOS 6.01 both offer full memory management.

- HIMEM.SYS provides access to extended memory. It ensures that different programs don't try to use the same part of extended memory at the same time. It conforms to the LIM (+AST) extensions.
- EMM386 is dual purpose for 386 and 486 computers with extended memory. It provides access to the upper memory area and allows you to run device drivers and TSR programs in this space. It also uses extended memory to simulate expanded memory.

• Windows 3.x provides its own memory management including allocating EMS space to DOS applications.

• OS/2 doesn't bother with high and low memory concepts and deals with memory up to 4GB as a block. However if the application requires a memory extender, OS/2 will emulate the operations of the memory manager so that the application doesn't need to be rewritten.

In the 640kB space with OS/2, there is now no DOS or Windows code, only a small block (about 10kB) of active OS/2 system redirecting calls.

memory map

A graphic representation of the standard allocation of the various locations/addresses (usually in pages or segments) of memory within a computer. The memory map includes both RAM and ROM, and it will begin at the zero address location $0000. The first page of memory is called the zero page. See.

• Memory mapped video is the type of screen control we use today. A section of the memory (called Video RAM or VRAM) is set aside as a bit-map and the full screen display (including all information surrounding the actual windows) is included in this display.

• Elements affecting this bit-map may come from the Finder or Windows shell (providing the desktop, and background window), the operating system (cursor), the application (menu, etc.), and the document itself.

In earlier machines (DEC Rainbow), the monitor screen was often mapped directly by a combination of stored character bitmaps and the document file. These were all character-based user interfaces.

memory-resident

This means the whole file is constantly in memory and available for use. Memory-resident utilities these days are usually called TSR (Terminate and Stay Resident). These utilities earned their name because they became popular in days when multitasking wasn't common and used a pseudo form of multitasking called task-switching.

Data files and applications are usually kept fully in RAM while in use, however, if they are too large for the memory allocation they may also be shifted in and out to disk. This is called a virtual memory operation.

memory segmentation

The 1MByte memory space addressable by the original 20-bit address bus was built into MS-DOS, and to further complicate the process, this space was subdivided into a number of 64kByte segments. Conventional memory occupied the first ten segments (up to 640kB) numbered from Seg 0 to Seg 9. BIOS and Video RAM occupied most of the space in Seg A to Seg F (640kB —1MByte limit).

memoryless coding

See LPC.

mentalities

Ways of reasoning, understanding and responding to the world. Jerome Bruner says there are three categories:

- **Enactive:** involved in object manipulation.
- **Iconic:** understanding through observation.
- **Symbolic:** abstract reasoning.

menu-driven

The use of menu selections for controlling an application, as distinct from text commands and command keys. Both CUIs and GUIs use menus.

MEO

Medium Earth Orbit. These satellites will be above the earth's surface at about 10,000km. MEO designates the distance, but these orbits can be circular (ICO for Intermediate Circular Orbit) or elliptical (HEO for Highly Elliptical Orbit) — and they can be equatorial, polar, or inclined.

At this intermediate distance, the satellites move relatively slowly across the sky but require more transmission power from the mobile unit than with a LEO. If the satellites are in elliptical orbit, they may also have problems passing through the van Allen radiation belt. The highly elliptical Tundra and Molniya orbits are 'medium orbits'. See LEO, GEO, Ellipsat, Molniya orbit and Tundra orbit also.

meridian

An imaginary plane which passes through a geographic location, the poles, and along the earth's axis. A slice through the earth which cuts both North and South Pole.

A line of longitude is along a meridian, and LEOs in a polar orbit will fly along a meridian line (however the earth will be revolving under these orbital planes).

MES

Mobile Earth Station. These are two-way briefcase sized mobile earth stations of the type used by Inmarsat-A and -M, rather than the handheld units used with LEOs.

mesh network

A mesh network is one in which any node can communicate with other nodes without the need for a strict hierarchy or sequence (order) of tree-branch connections. This is now seen as the ideal for public networks since it provides fully redundant pathways.

The distinction is with tree-branch, bus and star topologies which all have restrictions on linking different devices because they often need to avoid creating loops (although routers can be used to prevent feedback errors).

A meshed network, however, is always a series of loops — and this is the main problem in using this topology at the present time. However loops also provide redundant paths, so it is expected that in the future, meshed networks will become commonplace in private and public networks.

ATM protocols can handle meshed networks while SDH/-Sonet requires rings.

message

• A message is an arbitrary number of characters. It is undefined, but both the beginning and the end are clearly identified, and a message will always have a header with address information (and usually length and CRC data also).

• The term 'message' is sometimes used as a synonym for a block.

• When talking about message-, frame-, and datagram-handling within the OSI protocol stack, messages are Application layer entities.

message channel

A voice channel (VF channel) in telephony.

message digest

A message digest is a hash function. See.

Message Handling Service

See MHS.

message-oriented signaling

This usually means Common Channel Signaling, D-channel signaling, or sometimes packet-network signaling as used in X.25. GSM mobile phones also use message-oriented signaling. See MAP.

Signaling in telephone networks can be carried by any change of state from changed voltages on lines, to pulses, dual-tones, stolen bits in frames, to packets and full-blown independent messages. In this case they are carried as self-contained message packets. See CCS#7 and D-channel.

message passing

• Object-oriented programs use message passing to provide links between program modules ('handlers').

• A form of parallel processing where the processors each have their own memory. This is a supercomputer architecture. The problem is in synchronizing the activities of each processing/memory unit, since there is no theoretical limit to the number of parallel processors that can be interconnected. See macroparallelism, multi-processing and shared-memory.

Message Store

In a X.400 Message Handling System, the Message Store (MS) computer is the one that holds the mail awaiting the receiver's call — it is where the electronic mail-box is located. With X.400/X.500 protocols, the MS may also be instructed to deliver the message to a fax or Telex machine, or to beep a pager as a warning that a message has arrived.

message switching

This is best considered as an extreme form of packet-switching where the packet is of undefined length. Most electronic mail or MHSs will store and forward the complete message through the system as a single packet. The message has an address and most likely some error checking, and so it can be treated as a very, very long packet.

This can only be done when the nodes in the network are capable of holding messages of indefinite length — which, in practice, means being relayed from one computer's hard disk to another. See Inter-Personal Messaging and Fidonet.

Message Transfer Agent

This is an intermediate node in an X.400 network. It is responsible for reading the address in the message header and passing the message on through the network.

messaging channel

An old name for a voice channel (VF channel) in telephony.

messaging services

These are often seen as e-mail only, but the term covers:

— Electronic mail.
— EDI, EFT and EFTPOS.
— Electronic directories.
— Paging (one and two-way).
— Radio e-mail.
— Radio monitoring and control.
— Cellular short messaging services.
— Voice mail services.
— Voice response services.
— Facsimile, and fax store-and-forward.
— Fax downloading services.

messenger

In coaxial cable systems this is a strong steel supporting wire which is either strung above the coaxial (held on by loops of wire) or incorporated into the cable. This allows the cable to be strung from poles. (Tarzan cable — it swings from pole to pole down the street!) The alternative is to duct the cables.

meta

The prefix actually means 'among' or 'behind' (as in 'stand-behind'). There is also a use of the prefix in 'metaphysical' which implies pure thoughts, beliefs and abstract concepts — in the sense of 'not of the real world'. In computer use it often seems to mean 'at-a-higher-level' in the sense of 'master' or 'supervisory'.

• **Meta-data** — data about data, data descriptions, as in a data dictionary system. This can be used in multiple databases and shared by applications. If you later make a change in the dictionary, then the change ripples through all the databases — you don't need to change each one.

• **Meta-language** — a type of macro language which is translated into computer instructions. These language extensions have features which lie outside the range of a given programming language. See Backus Naur.

• **Meta-knowledge** — knowledge about knowledge

• **Meta-signaling** channel in B-ISDN — a predefined virtual channel which is always available for use to establish other specific signaling channels.

• **Metafile**. See.

metafile

A metafile is a hybrid of the two main categories of two-dimensional image files: bit-maps (raster graphics) and vector graphics. Metafiles were developed to overcome the problems of translating a vector graphic image to another vector or bit-mapped image format. They include both bit-mapped data and mathematical information.

metal tape

Video and audio tape traditionally depend on an active layer of magnetized metal compounds in the form of oxides, etc. These are usually 'glued' to the plastic tape backing with a 'binder'.

But in an attempt to pack more information into each square millimeter of tape, the manufacturers of video and audio equipment are now beginning to use pure metal particles, and some direct application techniques. There are two forms:

• **Metal Particle tape** (MP) is made by a reasonably conventional process, using a binder to stick the metal particles onto the backing. However the very fine metal particles are able to hold more information. Some of the early media appeared to be corrosive; after a time rust appeared on the tape.

• **Metal Evaporated tape** (ME) is made by 'sputtering' minute metal particles directly onto the tape surface in a vacuum — a similar technique to that used to manufacture anodized materials. ME tape is usually more expensive than MP tape, but it will produce better quality if the machine has been designed for its use. The binder has always been the problem with tape-ageing, so ME tape may be better for archiving.

metaphor

The design of Graphic User Interfaces (GUIs) depends, to a large degree, on the use of familiar metaphors to simplify learning and make computer use more 'intuitive'. Both the Macintosh and Windows use the image of a desktop with folders as a metaphor, because it mimics the way people normally work with paper. The trash-can is another metaphor.

• Unfortunately, the idea of hierarchical directories is not part of the metaphor, and is not intuitive to first-time computer users. Nor is the idea of storing both applications and data at the same level in the same directories. Metaphors only work to a limited degree with new concepts.

Metcalfe's Law

The value of any network increases geometrically with the number of people who use it.

meteor(ite) bounce

See micro-meteor burst.

metered

Generally, this just means that something is measured, usually with the intention of charging on a usage basis.

• **Metered calls**. See measured lines.

• **Metered pulse:** A way of charging for telephone services which is widely used in Europe. A fixed charge is made for every pulse unit, but the rate of the pulses varies according to time-of-day, distance, and other factors. This is a 'unit' charging method, but there can be a fixed up-front call set-up charge as well. The distinction is with direct time-distance charging systems used in the USA and elsewhere.

The pulse tone is generally at 12kHz which is well above the 4kHz of the conventional bandwidth of a telephone, but it does generate some audible sounds. European modems are said to need a notch-filter to block this.

Pulses can be, say, 3 minutes apart for local calls, and between 60 seconds and 10 seconds apart for long-distance.

The significance of unit charging is that for a call of 3.01 minutes, two units of cost are incurred — so when comparing international costs of telephony, this should be taken into consideration. However, see unit cost.

method

In OOP this has a specific meaning. Each class has a function of procedure defined for it, and this is the method. Most methods are associated with objects. For instance, a method defines what happens when a user clicks on a button or icon.

Metropolitan Area Network.

- See MAN, SMDS , DQDB and FDDI-II.
- In the USA the term is often mis-applied to point-to-point private links over optical fiber connections across a city.

mezzanine cards

These are plug-in cards which are inserted parallel to the main logic circuit board on a standard bus-based computer. They bypass the main backplane in order to provide faster data transfer rates.

Mezzanine cards are widely used on VME-bus systems. The IEEE is attempting to standardize these with the P1386.1 (PCI) and the P1386.2 (S-Bus) recommendations. There's also a Common Mezzanine Card (CMC) standard which is a superset of these.

MF

- **Medium Frequency:** (also called Medium Wave or MW) That part of the radio spectrum between 300kHz—3MHz which is used internationally for the AM radio broadcast band, because it propagates locally by diffraction over the ground. These are surface waves which tend to follow around the curve of the earth. Don't confuse medium frequency with intermediate frequency (IF) which is a quite different concept.

These are 'hectometric' waves from 1000 to 100 meters in wavelength and, in ground-wave conditions, they will provide good coverage for general AM broadcasts throughout the day without ionospheric reflection.

During the day, MF radio signals are heavily absorbed by the D-Region of the ionosphere. But this rapidly retreats at night, and so the MF waves are then able to bounce off the F Region, making long-distance reception possible.

MF radio waves are also used through wires and coaxial cables in telecommunications.

The main domestic/civilian applications in this band are:

— AM radio (500kHz to 1.6MHz)

— Cordless telephones (1.7MHz)

— Radio telephony and fixed links (above 2MHz)

- **Multi-Frequency:** (telephony) These are different to DTMF tones, and they are used between exchanges for inband signaling with T1 carriers. This is an early form of DTMF which is used between PBXs and exchange equipment (toll facilities). See MFC below.
- **Microfiche and Microfilm:** See microfiche.

MFC

Multi-Frequency Codes/Controls. These are dual-tone codes used for telephone exchange interswitch signaling and also on T1 circuits. Older PBXs also use these for in-band signaling because this form of signaling can be handled by the voice amplifiers in the long-distance circuits.

MF differs from DTMF in a couple of ways; MF doesn't support the # or the * keys, and the system is based on the transmission of tone pairs within the old reliable central voice frequency range from 540Hz to 1860Hz.

There are six frequencies which are grouped into pairs to provide the number sequence 0—9. The frequencies are 700Hz, 900Hz, 1100Hz, 1300Hz, 1500Hz and 1700Hz.

Usually, each dialed digit is acknowledged by a 'send next digit' signal on the back channel.

MFE

Media File Exchange. A file-compatibility framework for multimedia developed by ImMIX.

MFJ

Modification of Final Judgement. The document which spells out the way in which AT&T was broken up into seven RBOCs in 1984. The MFJ defines what both the new BOCs (and RBOCs) and AT&T can, and cannot, do.

It basically gave the 22 BOCs the right to become independent regulated monopolies within their own territories but allowed them to operate only within a LATA. It also created the RBOCs by grouping these 22 BOCs into seven larger regional companies (also monopolies), and it created Bellcore.

Not all US local telephone companies were involved in these arrangements, however. Many thousands of small independent companies and cooperatives had not been swallowed up by AT&T, and these remained independent. The largest of these was GE.

In the MFJ, long distance was made competitive, and AT&T was permitted to operate here to carry calls between LATAs and internationally. MCI and Sprint and hundreds of smaller IXCs now operate in this area also.

The MFJ was a Justice's Department document under the control of Judge Harold Greene. See divestiture.

MFLOP

Millions of FLoating-point Operations Per Second. A measure of the number-crunching speed of a computer. These days we are talking about GFLOP (Giga = 1000 million) and TeraFLOP (1000 GFLOP). See processor speed, FLOP and MIPS

MFM

Modified Frequency Modulation. The MFM format recording is a widely used method of magnetic disk-storage; the original PC/XT, the PC/AT machines, and the Macintosh SuperDrives all use it, as do many hard disks. The famous ST-506 drives used MFM recording methods and IDE electrical interfaces.

The term MFM refers to the way in which data is converted into magnetic fluctuations suitable for magnetic storage; it combines clock and data information. MFM takes the FM recording technique, and modifies it by doubling the pulse-rate (halving the 'cell' time like Manchester coding) — it then follows elaborate rules. It writes clock bits to the disk only if no data-bit has been written in the previous cell, and then only if no data-bit will be written in the present cell. This makes the coding complex, and rather wasteful:

— A single one-zero pulse pattern then represents a logical one.
— The logical zero is represented by an x-zero pattern with the x alternating — depending on the previous value.

• With hard disks, the replacement of MFM by RLL increased the data density on the disk by 50% without increasing the actual 'packing density'. An off-the-shelf MFM disk drive could be combined with an RLL controller, but later special RLL drives were introduced.

• MFM drives have 17 sectors per track. These drives are relatively dumb with all the intelligence residing on the control card in the computer. They use a standard connector known as the ST-402/516 with 34-way ribbon cable which can be daisy chained to four drives in a row (each drive has 20 connectors, some shared).

• Apple's conversion to MFM on the Macintosh (replacing GCR) has enabled Apple SuperDrives to read/write/format PC disks. See MF, MMFM and RLL, and also FM and GCR.

MFP

Multi-Function Polis. An Australian and Japanese term for a FTP. See.

MFS

Macintosh File System. This old Macintosh disk file system was actually a 'flat' architecture with all file information held at the 'root' level. When the Macintosh didn't have hard disks, and the 3.5-inch disks had limited storage capacity, this was acceptable — but not in the modern day. See HFS.

MGA

Monochrome Graphics Adapter. See Hercules Graphic Card.

MHEG

Multimedia and Hypermedia Experts Group. This is an ISO activity concerned with the coordination and indexing of information for multimedia applications.

MHS

Message Handling Service. This has much wider implications than just e-mail. MHS encompasses EDI, EFT, telemetering systems, and the type of computer-to-computer reporting and querying that are necessary for international secure messaging and directory services.

In the ITU definition, MHS is an applications interface at the top of the OSI stack. It does not need a connection to be established before transmission (FTAM does).

In more down-to-earth terms, MHS is a store-and-forward service which encompasses the X.400 Mail Messaging Service and Motis. It consists of a User Agent (UA) through which the users interact; Message Transfer Agents (MTA) through which messages are exchanged; and Message Stores (MS) where messages reside until read by the recipient. See X.400 and FTAM.

• The true e-mail part of the MHS is actually called Inter-Personal Messaging (IPM). See.

• Note that many LAN vendors (especially Novell) use the term MHS in a proprietary way. Novell calls its popular e-mail protocol 'Message Handling System'.

MHSnet

Message Handling Service Network: A more advanced form of UUCP connection to the Internet. It is still a dial-up system, so no real-time TCP/IP connections are possible. See UUCP.

MIB

Management Information Base. The database of information about equipment on a LAN. This is the repository for the characteristics and the set parameters of devices which will be managed (both managed objects and data variables), and it stores information about recent events on the network. Agents in remote devices report back their findings to the MIB.

If centralized management of a WAN is to be effective, the MIB information (on the traffic operations, alarms and faults) must be stored in a standardized way, since the MIB acts as an access point for the agent of the centralized management console applications.

The management station monitors the network by retrieving the values of MIB objects. Both SNMP and CMIP rely on MIBs to hold the attributes of their managed systems.

• The way to conceptualize the relationship of MIB, agent and resources is:

— The parameters of the various managed 'objects' are stored in a conceptual repository called the MIB.
— The MIB is made visible to managers (management stations or applications programs) via an agent.
— The agent will also both react to manager requests and initiate communications with the resource's interface card, to retrieve information.

• There is a Remote Network Monitoring MIB requirement also. See RMON.

MIC

• French acronym for PCM — Pulse Code Modulation.

• **Media Interface Connector:** The *de facto* standard for a FDDI connector (a NIC).

MICR

Magnetic Ink Character Recognition. The system of representing a number or alphabetic letters in a way that makes it easy for a computer to read by magnetic scanning. You see this type of character as a strange line of space-age blobby numerals on the bottom of bank-checks.

micro-

• The SI prefix for one-millionth or 10^{-6}. Abbreviated by the Greek letter mu (μ).

• Micro also just means 'small', without necessarily carrying the idea of a millionth.

Micro Channel Architecture

See MCA.

micro-floppy

An old term for the 3.5-inch (and the defunct 3-inch) floppy disks.

micro-meteor burst

A technique of sending point-to-point messages over a medium distances (a few hundred miles) by radio using the ionized atmospheric particles as a reflector.

Focused VHF radio beams are directed in a way that causes them to bounce off the ionized trails left by micro-meteors hitting the Earth's upper atmosphere. Usually dish antennas are used, directed like searchlights at a point in the upper atmosphere.

The process is this:

— The *receiving* station directs a beacon signal at the calculated bounce-point in the upper atmosphere.

— The proposed sender listens until the beacon signal is detected (ie. the atmosphere is temporarily ionized by a micro-meteorite).

— The sender then immediately transmits a burst of information in reply, back along the same path.

— The transmission continues as a series of erratic bursts.

The technique has potential for mountain hopping to remote locations, and low (100b/s) but reliable bit rates are achieved. The highly directional beam has reasonable coverage while still providing space diversity — in a hub configuration, the same frequencies can be reused to other locations.

micro-ops

In modern semi-parallel processors like the Pentium and the P6, a number of instructions can be processed simultaneously.

In the P6, as many as 20 micro-op(eration)s can be executed out of order — some of these, running ahead of the processor's instructions. This is called 'speculative execution'. The required instructions are being predicted and conducted ahead of time, and then the results are either sequenced for use, or abandoned.

microcell

In mobile telephony, the term microcell has a couple of different meanings:

• With cellular mobiles for vehicles, this just means a base-station which has a coverage range of between, say, 200 meters and 2km (radius) — but others would use the term only for cells in the 250 meter range. It is a vague term for a low-cost, low-power basestation.

Sometimes these microcell transmitters work as repeaters ('dumb-transmitters'), and backhaul all their signals to a conventional base station. At other times these are intelligent base-stations, just with low power. There are problems with multiple hand-offs for such small cells, unless they are being used for pedestrian mobiles.

• For pedestrian Telepoint and PCS operations, the term microcell is generally taken to mean a cell radius of 20 to 300 meters. These generally have no hand-off capability (see CT2).

microcode

The embedded control instructions built in to a microprocessor chip, which interprets the binary machine-code instructions. Collectively, these form the microprogram.

A modern PC is called a 'micro', not because it is small, but because it has in-built microcode which can't be changed — the microcode is built in to the microprocessor chip.

The function of the microcode is to carry out the particular data-manipulation sequence defined by an instruction, and there may be between 50 (RISC) and 300 (CISC) of these. The instruction decode-logic interprets the data-word in the instruction register and converts it into a timed pattern of control signals to the other processor elements. It will perform three main types of function:

— Moving data around in the computer.

— Making arithmetical calculations.

— Making comparisons between values.

• The term microprogram is sometimes used as synonymous for microcode. See ALU also.

• In older and larger computer systems, a 'microcoded' computer was originally one that had another computer within — one with its own microcode instruction set. By writing new microcode, these assembly-level machines could be changed over to a completely new instruction set.

Microcom Network Protocols

See MNP.

microcomputer

An old term which doesn't refer to the lesser physical size of the micro when compared to the mini or the mainframe, but rather to the use of microcode embedded in the silicon of the microprocessor. At the time these terms were being formulated, only PCs used microprocessors.

microcosm

Guru George Gilder's term for the future world which is being created by the increasing sophistication of computer chip real-estate. More and more elements are being added each year and processing is getting cheaper and cheaper. The Law of the Microcosm states that linking any number ('n') of transistors on a chip leads to 'n' squared gains in computer efficiency.

Gilder believes that this exponential increase in chip densities is likely to continue through the next two decades. See Moore's Law, Debye limit and telecosm also.

microfiche/form/film

A photographic process for storing the images of documents in very much reduced form. Computers can output their data to microfiche through special peripherals.

Microfilm comes in reels, while microfiche is small envelope-sized cut-film, which can hold perhaps 250 pages of readable text. A special camera is needed to expose the film. Some forms of microfiche also carry coded information to provide a machine-readable index. See MARS and COM.

micrographics

The generic term for microfiche and microfilm.

microkernel

The core of a modular (object-oriented) operating system. The microkernel is a small privileged core consisting of the essential services. It is surrounded by utilities/sub-systems which deal with the outside world.

Mach, the OS for the Next computers, was the first operating system to use a microkernel architecture, but now all seem to be moving in the same direction. There's a clear trend away from monolithic OS architectures towards modular designs. Extensibility is one of the key aims of this approach; ease of porting is another.

micromechanics

This is quite different from nano-mechanics which have been in all the newspapers and TV talk shows recently. This form of mechanical device operates at a chip level as part of telecommunications — not at the bacterial level, using micro-machines to drill out clogged arteries. Recent developments have proved the potential of micromechanics in large video projection systems. See DIS.

micrometer

One millionth of a meter — μm. The old term for a micron, or 10^{-6} meters. People often confuse micrometer with millimeter (which is one thousandth of a meter) — these were very badly chosen terms, which is why micron continues to be used.

micrometeroids

These are small particles in space which threaten to damage satellites and space vehicles. To be classified as micrometeroids, they must weigh less than 1 gram (1/30th oz.) — however when the relative impact rate is thousands of kilometers an hour, the damage can still be considerable.

- They can also be useful. See micro-meteor burst.

micron

The widely used and preferred term for micrometer which is 10^{-6} meters. A micron is about 1/80th the diameter of a human hair, and the very best ultrafine wool ($1 million a bale) has a fiber thickness of 13.8 microns.

- The Pentium chips use 0.35 micron technology (350nm).

- A Harvard University team say they can litho-etch circuit lines only 0.05 microns wide (50nm) through a gold layer.

microparallelism

A form of macroparallelism (see), but one which is confined to the design of a processing chip. This technique provides multiple execution pathways within the same processor.

The Intel 860 chip is of this type; it contains separate pathways for integer, floating-point addition, and floating-point multiplication. Each pathway has its own pipeline, and they can all execute simultaneously. See pipelining, superscalar architecture and superpipelining (a similar technique) also.

microprocessor

See MPU and CPU.

microprogram

The microprogram in a computer is a ROM-based interpreter which takes the control instructions to the CPU and supervises performance at the machine level. It will typically have between 50 and 300 different instructions See microcode, RISC and CISC.

microsecond

A millionth of a second. A musec or μsec.

microsegmentation

Segmentation is the subdivision of overloaded LANs to improve performance. These subdivisions will be linked by switches, bridges or routers. With microsegmentation, the subdivision is often taken to the point where there is only one or two workstations on most of the LAN segments. With high-end workstations, each will now have its own switched LAN segment. See Switched Ethernet.

Microsoft At Work

See At Work.

microvalves

- An old form of vacuum tube used before the transistor was invented; it was often called a peanut valve, and they were about the size of your thumb. The valve technique of amplification still has some inherent advantages, and by decreasing the size (beyond the limits thought possible a few years ago) these advantages have been recently enhanced.

- The term 'microvalve' is now being applied to very small experimental devices which are much faster at switching than transistors and more resistant to electro-magnetic noise.

microwave

Micro here means 'small', not 'one-millionth'. The term has changed in meaning over the years.

Marconi (the company) first used the term for the 600MHz (0.5 meter) band, but this is now considered far too low. The term was then applied to the use of highly-directional radio waves above 890MHz, but more often to those above 1.5GHz (0.2 meter wavelength) in the UHF band.

Many people still apply the term to mean any frequency above 1GHz. Others will only accept that microwaves are above 3GHz (SHF, not UHF). However, some place the upper end of microwaves as high as the beginning of the infrared spectrum (3000GHz). These people divide the spectrum very simply into only three parts: radio waves, microwaves, and lightwaves.

• Microwave ovens operate at 2.45GHz, which is the point of maximum energy absorption by water molecules.

• The best way to think about microwaves is to assume that they are half-way between light and conventional radio waves — and have many characteristics of both (line-of-sight transmission, 'searchlight' focused beams, etc.)

• Generally the spectrum between 1 and 15GHz is used for microwave telephony and, at these frequencies, the relay towers need to be spaced about 30 to 50km apart. They don't suffer much from atmospheric changes or conditions.

In general, the use of microwave frequencies requires a clear line-of-sight for good reception, although the signals can bounce off some structures and will pass through light foliage.

These waves pass directly through the ionosphere without deflection, which makes them valuable for satellite communications. However certain bands within this range suffer badly from fog and rain attenuation (above about 10GHz), and also from oxygen molecule absorption at the higher end.

There is now a problem with satellite vs. terrestrial microwaves interference. Many cities are criss-crossed with microwave telephony beams and these frequencies are also widely used for satellite broadcasting.

• Currently TN-1 microwave systems are used in the US for telephony, with 1000MHz of available bandwidth (10.7—11.7GHz). These systems can be configured as 12 two-way RF bearers using a common antenna with 1800 voice channels each, or as 23 one-way RF bearers (needing two antenna) with 1200 voice channels each. A total of between 6000 and 25,200 voice channels over distances of 5 to 23 miles.

• Long-haul TM-2 microwave systems operate in the 4—6GHz range, with stations 3—5km apart. They provide sixteen RF bearers, each with 1200—1800 voice channels. A total of between 8400 and 12,600 voice channels per system.

• Microwave developments came out of the invention of radar during the war. See Klystron. For communications, these systems were first used in the USA to link networked TV stations in about 1969. MCI later used the same basic approach to carry long-distance telephone calls in competition to the AT&T monopoly at that time — which was the beginning of the commercial war which eventually led to the AT&T's monopoly judgement. (See MFJ)

The first microwave telephony systems carried 1,800 two-way voice channels, and later analog systems carried 2700 voice channels or one TV channel. Microwave is also now used by the carriers for lesser capacity bearers in the 300, 600 and 960 channel range where cables aren't economical.

• See also Microwave Distribution Services (MDS).

microwave transmission

In the USA, microwave relay for public telephone networks operate in several bands: 2GHz, 6GHz, 8GHz, 10/11GHz, and 18GHz. Many of these frequencies are also used for private relay systems, however, private users are being pushed into the new higher bands of 18GHz, 22GHz, 23GHz, and 25GHz.

Microwave links are often installed and maintained by carriers or by turnkey service providers. They are generally for relatively short range, typically for a single hop only — and often as a way to by-pass a local carrier which is making unreasonable charges.

• The **2GHz band** is being reassigned to PCS. It provided relatively narrow bandwidths (typically 7MHz) and low data-rates (typically 12Mb/s). It has been widely used in the past to provide links to cellular basestations.

• The **4GHz, 6GHz** and **8GHz bands** typically provides 30MHz bandwidths which allows digital transmission rates in the order of 90Mb/s. Telephone carriers tend to use these lower microwave frequencies for long-haul communications, with spans up to 80km.

• The **10GHz** and **11GHz bands** are coming into use by carriers and cable TV operators (12GHz) as lower frequencies become congested. These links typically provide 40MHz of bandwidth and data rates in the order of 100Mb/s. Rain attenuation becomes noticeable over distances above 10—50km above 10GHz. In the USA the FCC has reserved the 10GHz band for local digital data distribution. 11GHz is used for medium capacity, medium-distance hauls.

• The **18GHz** and **22GHz bands** are also used for television and radio studio-to-transmitter links (STLs) and for cellular phone links in the range under 25km (often limited to 12km in metropolitan areas). Bandwidths are in the order of 220MHz with digital data rates of 274Mb/s. 22GHz is also widely used for short point-to-point links between buildings. These frequencies (and higher) are only suited to short-hauls in areas with moderate to low rainfall.

• The **23GHz band** is used for data, voice or mixed traffic over ranges up to 16km — but mainly for inner city, inter-building and campus connections. This hardware is relatively compact and inexpensive and it will typically carry four T1/E1 signals, two videoconferencing channels, or a single broadcast-quality video channel.

microwriter

A small text-input device designed to be portable and usable with one hand.

mid-range

In computer systems, this is the non-desktop arena where Unix is firmly entrenched. It originally meant minicomputer, and the DEC VAX was/is the archetypal 'midrange' mini. Now it means anything smaller than a mainframe, and that encompasses the larger super-workstations and LAN servers as well.

This is the market that Microsoft is now entering with Windows NT. The mid-range arena is also the major database market (Oracle, Sybase, Informix, and Ingres), and the server side of client-server operations.

mid-split

In a broadband LANs, the carrying capacity needs to be roughly symmetrical. So the frequencies are simply divided in the middle with one half being used for transmission towards the head-end and the other for reception from the head-end. See high-split also.

middleware

This is the key to distributed computing. Ovum has described it as 'off-the-shelf connectivity software which supports distributed processing at run-time, and which is used by developers to build distributed software.'

They identify six categories:

- database connectivity products (with 73% of the market),
- message-oriented middleware,
- distributed computing environment middleware,
- remote procedure calls,
- object request brokers, and
- distributed transaction process monitors.

• **Middleware architectures** are DCE (see) and ONC (SunSoft's Open Network Computing). These are designed to allow interoperability of heterogenous networks. Each provides a standard communications interface between network applications so that they can work together, regardless of location (or other variables).

ONC is reasonably well established because of its relationship with NFS. It has been implemented on PCs and mainframes. DCE is supported more by the proprietary database vendors.

• **Middleware software:** This is translation software; these applications will exist in the network simply to translate one standard to another.

• **Middleware tools and libraries** help developers to create complex applications. Here the term is not directly related to translation applications.

• **Middleware technologies** in future broadband 'superhighways' are such things as directory services, digital signatures, encryption, or search agents that protect intellectual property, privacy and security.

MIDI

Musical Instrument Digital Interface. A computer standard for the control of musical instruments and synthesizers.

MIDI was first defined in 1983. It enables the musician to record the note struck (or played on an electronic keyboard) in a complex way. The recording will define the velocity of the strike, the duration of the note, the attack and fade, etc. These parameters are stored and can be reproduced later. See ADSR.

A number of intelligent MIDI peripherals, such as synthesizers, rhythm machines, sequencers, audio mixers, sound processors and even lighting controllers can be combined (orchestrated) and run remotely under MIDI control.

MIDI can handle 16 concurrent links at 31.25kb/s. The channels are 8-bits, 1 stop-bit, and no parity (8N1).

Each manufacturer of MIDI equipment is identified by a special 'system exclusive' (SysEx) code, and two different bodies oversee MIDI implementations — the MIDI Manufacturers of America (MMA) and the Japanese MIDI Standards Committee (JMSC).

• GM (General MIDI standard) exists at various levels. Level 1 requires support for all 16 channels, with 24 voice polyphony (simultaneous sounds) and 128 preset sounds. This is the lowest common denominator.

• GS (General Standard) adds some extra capabilities. Eventually this will become GM Level 2.

• MIDI is a serial interface with opto-isolation and a current-loop connection. No current represents a logical 1, and a current of 5mA represents a logical 0.

MIF

Management Information Format. The Desktop Management Task Force (DMTF) has three new DMI (Desktop Management Interface) specifications. See DMTF.

• **Software MIF:** This offers LAN managers a way to have software report the version ID and audit date automatically across the network to the management tools and MIB. It allows external audits to check for the presence of software on the network.

• **LMO (Large Mailing Operation) MIF:** This defines classes and attributes of mass printing jobs.

• **Printer MIF:** This allows printers to communicate their status to the desktop user. It will also support printer management over the network.

• **Servers and Modems:** MIF specifications for servers and modems are still being completed.

mil

Thousandth of an inch.

millard

The UK name for the American billion, or 10^9. The British now appear to have accepted the inevitable, and most UK computer magazines now use billion to mean 1000 million. However some old die-hards still cling to the 10^{12} (million-million) definition of billion.

milli-

The SI prefix for one thousandth or 10^{-3}. Always abbreviated with the small m. So 300 millivolts is 0.3 volts.

millimetric waves

See milliwaves.

millimicron

The old term for a nanometer or 10^{-9} of a meter. It's a thousand-millionths of a meter, or one millionth of a millimeter, or one thousandth of a micron.

milliwave

Radio waves in the region between 30GHz and 300GHz — the EHF band. At 300GHz, radio waves are exactly 1 millimeter in wavelength. These waves only travel very short distances (in the region 2 to 20km max.) and are easily affected by rain, so they will tend to be used in pico cell applications and within office environments.

Milnet

A Military Network.

Milspec

Equipment made to the specifications required by the Military. This is synonymous with 'expensive as all hell'. Most defense departments around the world are moving away from Milspec to conventional equipment, because they can get better equipment for the same money. They may then add their own little touches, like encryption.

MIMD

Multiple-Instruction, Multiple-Data. This is one of the two popular parallel computer architectures (the other is SIMD). This is the current approach favored by the massively parallel supercomputer industry where up to 100,000 processors may be working on a single task.

MIMD processors carry out different operations on many pieces of data at the same time. The main problem with the MIMD parallel architecture approach is that it increases the programmer's problems because of the need to break down the application code into multiple, parallel streams of logic. When each of the processors has its own local memory, coordination can be extremely difficult.

- MIMD vs. SIMD claims:
  - The MIMD approach is better for problem solving.
  - SIMD is better for image processing.
  - MIMD is said to be more flexible.
  - SIMD is easier to use.
- New developments may combine both architectures, allowing the bulk of the tasks to be performed as MIMB, but providing a control network which can broadcast the same instructions to all processors and synchronize their execution whenever the SIMD approach makes sense. When the execution gets to the point where each processor needs to act independently, it will begin to fetch instructions from its own local cache memory.

MIME

Multipurpose Internet Mail Extension. A message encoding system used for transmitting files (not e-mail) over the Internet. MIME has been built on top of SMTP which provides the mail service (transport protocol), and it allows you to attach computer files to a text message.

MIME can handle multi-part and multimedia messages, so it can transfer non-text data, graphics, fax, international character sets, PostScript files, binary, audio messages, and even digital video. It defines a structure for encoding diverse types of data, from spreadsheets to multimedia.

These files can be sent over any Internet e-mail-capable link and be handled by any e-mail server, so nothing in the network needs to change. However the client in your PC must be upgraded to handle the new file-types. The other common standard for the Internet is the (incompatible) UUEncode.

MIN

Mobile Identification Number. The phone number of an AMPS cellular phone. This is a number that uses area codes and number patterns that are part of the national numbering system; it is provided by the carrier. Mobile phones need a MIN because the customer may change phones over the years but want to keep a phone number. See ESN also.

mini-computer

A medium sized computer; there is no strict definition. Minis are now merging with the top end of the workstation range and with the bottom end of the mainframe range.

Generally, mini-computers were designed as multi-user, multitasking central computing systems (often clustered), but they are now increasingly being used as general network servers. They have lost their primary position in medium-sized companies, and tend to be seen now as major peripheral resources on the corporate network.

Early minis were 16-bit machines (at a time when PCs were 8-bit), and they had perhaps 64kB of RAM and a few dozen communications ports. Now they are usually 64-bit machines with many megaBytes of RAM, and a few hundred communications ports.

MiniDisk

MD aka MiniDisc. Sony's portable optical-disk recorder and player audio standard, with CD-like quality. There is also a data version called MD-Data. It is both a play-back standard and a standard for magneto-optical (MO) recording, so it has uses for both domestic and office applications.

For audio, it can record 74 minutes of high quality digital sound onto a 64-mm (2.5-inch) disk using magneto-optical technology and a 5:1 lossy data compression scheme. The sound is compressed using ATRAC algorithms which take a similar masking approach as the Philips DCC tape. SCMS anti-copying was supposed to be used with the MD standard, but this appears to have been forgotten.

The MD-Audio is the audio version and the MD-Data (140MByte) is the data version (this is a DOW system). See MD-Audio, MD-Data, DCC and PASC also.

- In the 1980s, a minidisk was a 5.25-inch floppy!

Minitel

Also known as a Teletel terminal. A small keyboard-and-screen terminal and database access service supplied by the French PTT to subscribers in lieu of phone books. There are now 6.5 million Minitels and 0.5 million PC emulation programs working Minitel.

This is probably the most successful on-line public information service in existence. Currently 20 million French telephone users can access electronic directories and over 23,200 information services, via Minitel.

But while the network provides gateways to a wide range of non-PTT services, the most common use of Minitel is still as an electronic yellow pages. The French spend 23 million hours each year searching for phone numbers. The second most common use is for matching freight carriers to freight; then thirdly for purchasing tickets; and fourthly for sex-talk and romance (called 'pink' services). These sex services generate $US1.4 billion in revenue.

• 'Messageries' are general chat sessions, and 'messageries rosa' are specific sex chats (billed).

• Minitel is still standardized on the old videotex 1200/75 b/s rates, although this is now changing. They have now launched a 9.6kb/s version which costs more, but supports multimedia.

MIPS

Millions of Instructions Per Second. A very primitive and often misleading way of measuring the processing speed of a computer. It only really makes sense to use MIPS to indicate a microprocessing chip's raw performance, and it is totally misleading to quote it when comparing computers without taking into consideration the comparative 'wait states'.

Other versions of the acronym are:

— 'Misleading Information to Promote Sales'.

— 'Meaningless Indicator of Performance for Suckers'.

Basically there are two MIPS measures — the normal one which was designed for IBM architectures and the 'VAX MIPS' one for DEC architectures. And there's an enormous difference between using MIPS for a central processing unit and also for digital signal processing (DSP). DSPs are specially designed for a task and therefore function better with a lesser number of instructions.

As a rule of thumb, a microprocessor's MIP rating must be divided by 5 to get a DSP MIP rating.

See processor speed and MFLOPs also.

• The standard for VAX-MIPS is that set in the Stanford integer benchmark suite. A VAX-MIP is the equivalent of the old VAX 11/780 minicomputer which was once the king of the midrange systems.

• MIPS Technologies is also a company (a subsidiary of Silicon Graphics) and a processor brand.

• There is also a benchmark test based on Dhrystones (see) which is called the MIPS test.

mirror

In computer terminology this can either mean a direct exact copy or a copy with a left-right inversion.

• **Disk mirroring** is the process of having a second drive running in parallel, thereby making an identical and simultaneous backup copy. This second drive is usually able to cut-in automatically if the first fails, so this is a fault-tolerant system.

• **Mirrored networks** are those which have exact duplicates of all data (usually at a distant site), so that an up-to-the-second duplicate is available in the event of earthquake or fire damage to the primary storage center. Mirrored networks are said to maintain 'real-time redundancy'. They usually need expensive high-speed data links. Fiber Channel and some of the similar new high-speed channel protocols were designed specifically for this purpose.

• **Mirrored databases** are often just duplicates — not necessarily done in real-time and not always done for safety reasons. On the Internet you may find all RFCs mirrored at a local site. This is done just for convenience so that unnecessary loads aren't placed on long-haul links.

• **Mirror sites** on the Internet are computer sites which keep copies of the most commonly requested files from anon.ftp sites around the world, to save the high cost of constant international traffic. You should always first check your local mirrored site first for a current file.

• **Page mirroring** in desktop publishing is when the format for the left page is repeated (but inverted) for the right.

MIRS

Motorola Integrated Radio System. Motorola's new digital communications technology, which includes trunked voice, data, dispatch and messaging. It is based on low-rate voice coding (4.2kb/s) and it uses time-divided channels (TDMA).

It is said to provide high clarity and good range, but reports about the Nextel service (the first user in the USA) aren't all that good — although Craig McCaw (Cellular One) has just bought in. Currently MIRS is designed to work in the 800MHz trunking band and it may prove to be cheap for the many organizations. See DSRR, TETRA and MPT 1327 also.

MIS

Management Information System (or Services). MIS shouldn't be confused with EIS (Executive Information Services).

MIS is a general term which is applied to both financial and general-operations applications and data, and which are generally carried on a company's centralized 'host' database computers. The emphasis in EIS's is on the productivity side, rather than on the provision of real-time management information.

MIS implies that these services are centralized on a large computer or computer cluster, run by a full-time computer department. So MIS has now become a trendy title for what used to be called the DP (Data Processing) or the EDP (Electronic DP) department, but it suggests a wider role.

MISC

Minimum Instruction Set Computing. This is a variation on RISC.

mission-critical

The buzzword of 'enterprise computing'. Apparently executives get an ego-boost out of imagining that corporate greed and market dominance is a soul-cleansing 'mission' on a level with taking religious vows of chastity, altruism and poverty.

This term refers to key applications, networks and databases which are essential to keep a large corporation functioning. Almost everything is now classed as 'mission-critical' if it runs on a mainframe or large mini-server.

MIT

Massachusetts Institute of Technology. A lot of new ideas have come from here, especially from the Media Lab and its director, Nicholas Negroponte.

MITI

Ministry for International Trade and Industry. The Japanese trade promotion and industry organization which directs a lot of activities in the private sector. It has immense power (it is often known as 'Japan Inc.') partly because it allocates research grants in addition to its other activities. MITI also promotes cooperation between companies on some development projects, and promotes competition in others. It is evidence of a strange coupling of bureaucratic central-planning with free-enterprise and democratic rhetoric.

MITV

Microsoft Interactive TV platform. This is a proposed operating system and broadband network architecture (+ management software) for servers, set-top boxes, and network switches.

M-JPEG

Motion JPEG. This is a version of the JPEG compression scheme which is used in the post-production of video. Unlike MPEG, it does not depend on images previously displayed, so it can be edited on a frame-by-frame basis. Later it will be converted to MPEG for release. See JPEG.

MMAC

Multi-Media Access Control. An intelligent wiring hub capable of handling a number of different forms of media. See MAC (networking).

MMCD

See Multimedia CD.

MMDS

Multichannel Multipoint Distribution System or Medium Microwave Distribution Service (or any variation between) — aka Multichannel Television (MCTV) or Wireless Cable. These are just the latest in a whole range of acronym redefinitions and names which include Multiple Microwave Distribution Services also, and more recently the term 'Medium' has now begun to be used with this acronym to distinguish it from LMDS 'Local' MDS. In Europe it is often called MVDS.

This is a type of MDS (dating from 1983) which is licensed for multiple channels of television — in the USA between four and ten are sold as a subscription package. This techniques is now being used in 60 countries as an alternative to cable.

The key to using adjacent channels (up to 30 or more are possible, in sequence) in any one location is that they are all transmitted from the same antenna at the same power. Adjacent channels don't then interfere with each other, and so all can be used, provided TV sets are 'Channel Ready'.

The term 'microwave' is used here because the station sends out its video and/or data programming on a quite low-power (10—100 watt) microwave carrier, usually at the high end of UHF around 2—3GHz. Some are licensed at 12GHz also.

Low power transmitters are reasonably cheap, only costing in the $60—$120k region. The subscriber needs a special small rooftop antenna and a down-converter (about $250 total) to translate the signals to a form that a normal TV set can use. When MMDS is being used for Pay-TV, an IRD (decoder set-top box) will be needed as well.

In the USA, the original 'wireless cable' systems were licensed to carry only four channels, and in much of the world the limit is set at about 20 channels. However the technology is improving rapidly, and more spectrum is being allocated at the high end of the UHF band and some in the SHF (called LMDS). Digital carriers will double the number of channels available again, but there are still some problems to be solved with digital transmission. See MDS and MMMDS also.

• In the US, MMDS currently operates in the 2.5—2.686GHz band (31 channels of analog NTSC) with an ancillary band 2.15—2.162GHz which provides another two channels. Output power is generally 100 watts from a 100 meter antenna, and this is able to cover about 7500 sq km of a flat city. Each signal is 30MHz wide. The average city can be covered by six 10 watt transmitters, but the signal suffers from rain attenuation.

MMFM

Also known as 'M-squared FM'. This is the Miller Modified Frequency Modulation technique for double-density disk recording. It differs only slightly from conventional MFM. It writes clock-bits only if neither a data-bit nor a clock-bit has been written in the previous cell, provided that no data-bit will be written in the present cell. See MFM.

MMFS

Manufacturing Message Format Standard. An OSI Application Layer standard for MAP networks. See MMS.

MMI

Man–Machine Interface. A general term for keyboards, touch screens, voice-recognition, handwriting recognition, and eventually retinal-scan and alpha-brainwave controls of computerized equipment.

MMIC

Monolithic Microwave Integrated Circuit. A relatively new Japanese technology used in small satellite (VSAT) transceivers and in cellular phones.

MML

Man–Machine Language, as specified by Bellcore.

MMMDS

Millimetric Microwave Multipoint Distribution Service. If this wasn't used so dead-pan, you'd be tempted to think this is a joke. This is actually called the M (cubed) system and also Local MDS (LMDS). It refers to a new MDS broadcast system which uses frequencies in the 'millimetric' region around 28GHz (in the SHF band).

These frequencies can only transmit over relatively short distances (a few kilometers), but the short distances will eventually allow them to also provide a radio return path for interactive services. See MMDS and LMDS.

MMR

Modified Modified Read. A modified version of the already modified version of the READ algorithm used in run-length encoding of Group 3 facsimile files. See READ and Huffman.

MMS

• Manufacturing Messaging Service — now Manufacturing Message Format Standard (MMFS) or Specification. This is part of the OSI suite (9506) — an ODA/ODIF for factory automation. It is used for machine-to-machine communications and it includes the concept of Virtual Manufacturing Devices (VMD).

MMS is a high level applications interface (a SASE) which is the EIA RS–511 standard for MAP. It is a powerful language with a defined syntax which is used to control manufacturing processes in CNCs and PLCs. It was specifically designed to control industrial robots. See SASE.

MMU

Memory Management Unit. PCs needed an MMU chip to become multitasking. Generally, without hardware specifically designed to keep track of memory use and allocate blocks of memory locations to the various programs and data-files on the desktop, memory conflicts would be inevitable. See memory managers.

mnemonic

• An easy-to-remember word or letter-string chosen to represent a computer command or instruction. For instance, the name given to an address field in a database (a variable) could be called Z32, but it is easier to remember if it is called 'ADDR' — which is a mnemonic for Address.

• Mnemonic code is the form of computer instructions used in assembly language. An instruction to store something may be 'STR' and an instruction to jump to a new location might be 'JMP'. These simple codes are easily translated to binary machine code by the assembler. See op-code also.

MNP

Microcom Network Protocols. A series of error correction and data compression protocols which are now becoming widely accepted standards in modem telecommunications. Different classes of MNP are available.

The MNP protocol family is an expandable and layered system which conforms to the OSI's model. It aims to provide reliable flow-controlled and transparent data transfer, both as a stream of bytes and as files. It requires the modem at each end of the channel to recognize the appropriate MNP class protocol so it requires end-to-end coordination.

RS-232 is used at the Physical layer, but at the Data link layer, MNP provides data transparency, flow, and error control. Byte stuffing is used on byte-oriented machines, while SDLC framing is used for bit-synchronous hardware and sync modems.

MNP requires positive acknowledgment, and has 16-bit CRC with retransmission when necessary. The main standards are:

• **MNP-1** was an old asynchronous half-duplex protocol which is rarely used today.

• **MNP-2** is asynchronous, byte-oriented (with stop and start bits) and full-duplex. It is a good error checking system, and is the fall-back protocol for all MNP products.

• **MNP-3** is a popular PC modem standard that uses synchronous communications between the modems to reduce overhead. It is fast, bit-oriented, and has excellent error-checking. Note that the link between each modem and its PC is asynchronous (10-bit bytes) while between modems it is synchronous.

• **MNP-4** is the current standard for error correction — the upgrade of MNP-3. This protocol has the ability to adapt to line conditions. It sends data in variable-sized packets (set by testing line quality), and control information is constantly being exchanged to keep the link running at the highest possible rates. It is synchronous, taking its timing from the data stream.

• **MNP-5** provides a low level of data compression (about 2:1 for text), password protection and security call-back checks. It uses Huffman's and RLE techniques of compression.

• **MNP-6** allows one modem to 'train' the other by gradually raising the modem rate and changing the modulation techniques. It works between 300 and 9600b/s. It is a fast turnaround half-duplex standard.

• **MNP-7** is an improved (3:1) data compression standard, but it is still inferior to V.42bis (by about 25%). It uses in-built data dictionaries of various sizes, and engages in extensive negotiations with the other modem to determine the best possible use of dictionaries and standards.

• **MNP-8** (no details available).

• **MNP-9** is the top of the range with data compression of 4:1. It runs very fast (claimed to be 38kb/s on an average telephone line) with special modems, but it has never been widely accepted. It allows a sliding-window type operation which can indicate which of the last few blocks to resend.

• **MNP-10** is a fall-back/fall-forward protocol which specifies the triggers which initiate the rate changes. The modems will fall-back in data rate if problems arise, and also fall-forward to higher speeds if line conditions improve.

Data rates are changed mainly by varying the packet size: the modems begin by transmitting small packets and progressively increase the packet size as conditions permit. The two modems constantly exchange information about line conditions using Link Management Idle packets which are exchanged during pauses in the transmission of the user's data.

• **MNP-10 EC** (Enhanced Cellular) is a recent enhancement which is backwardly compatible with MNP 10 and LAP-M. It is an attempt to match the ETC protocols for radio modems (Rockwell helped in the design). See ECT.

• Note: The CCITT/ITU V.42 standard modems always include MNP as a fall-back.

MO

Magneto-Optical disks. See MiniDisk, Curie point, Kerr effect, and PROM and OROM also.

mobilesat

A general name for mobile satellite communications services. These provide voice and data direct via satellite to moving cars, aircraft, ships and trucks.

• A specific name for an Inmarsat-M derivative developed by Aussat in Australia.

Mobitex

Swedish Telecom's packet-switched radio data service — which is now controlled by Ericsson in association with a user's group (the Mobitex Operators Association). Mobitex is available to public subscribers through telecommunications service operators in North America (US and Canada), Europe (UK, Sweden, Finland and Norway), Australia, and S.E. Asia. DEC has now become involved in a cooperative agreement.

Mobitex was designed to provide a packet-radio transmission facility with high data integrity and features such as store-and-forward. It uses a fixed packet length of 532 bytes (512 byte payload) and it can encapsulate Ethernet or X.25 if required. Error correction adds 100% loading, which means that the current latest version of the system has a data rate of 9.6kb/s, but a throughput of only 4.8kb/s.

There's a three-level system of cellular nodes in the fixed service network. All exchanges are intelligent, and data is routed through the lowest common exchange in the network hierarchy. Intelligence is distributed, so that the network will continue to function should an exchange or link die.

mod

• A modification or update.

• In programming or mathematics, modulo. A type of maths which deals with remainders.

mode

A specific method, manner, or way of doing something. An identifiable aspect or kind of operation. It is one of those 'use-anywhere' words that we all know and love!

• **Fiber optics:** Mode here refers to the variety of ways a photon of light can travel down a fiber. It can pass predominantly down the center of the fiber (the shortest route), or it can constantly ricochet off the reflective walls of the fiber (taking a longer path). If the fiber is thick, multimode transmission is inevitable, but if it is thin, the light is confined to one 'mode' (monomode).

Thick fibers and multimode transmissions have the advantage that: LEDs can be used instead of lasers; the receiver will be activated by more light down the thicker fiber; and splicing will be less critical. The disadvantages are: the cable will carry only moderate bit-rates; and carry them for shorter distances without the pulses smearing (intermodulation distortion).

So for higher data rates over telecommunications distances, thin fibers and monomode transmission is always preferred. With only a single mode (actually a very restricted range — rather than a 'single' mode), pulse smearing is much less at the far end of a link (a pulse can travel 3000km at 5GHz rates and still be identifiable). But this type of fiber needs to be driven by a laser and amplified by laser-based amplifiers (regenerators or fiber amplifiers). See multimode and monomode.

• **Lasers:** Semiconductor lasers generally emit light in several wavelength 'modes' and the output power rapidly fluctuates between these modes. This terminology has nothing to do with the modes of thick and thin fiber. However, coherent optical systems need single-mode (totally one color/frequency) lasers. Distributed Feedback (DFB) lasers will suppress the side-modes so that the output spectrum can be said to be single-mode.

• **CD-ROM** — See CD forms and modes.

modec

A special form of ISDN terminal adapter which translates between analog and digital data protocols. These adaptors are designed to link baseband outputs and inputs from a remote terminal to an ISDN network's local loop.

A modec is an exchange terminal device, which can be connected to one of the ISDN local loop B-channels. It generates modulated audio signals which mimic those produced by conventional V.32 (or other) modems, and it works over a link to a remote modem user through the interexchange network.

On the other side, it deals with the customer using standard B-channel digital signals, with the D-channel for control.

Modecs are necessary if remote analog subscribers wish to use modems to communicate with an ISDN subscriber who only has digital equipment.

model

In computer terminology, modeling is the creation of an (always imperfect) representation of some real event or a real construct. Much of our thinking, organizing and planning involves model building, if only in the abstract.

Computer models are generally analyzed mathematically and used in simulation. They are attempts to recreate the complexity of the real world in order to predict future behavior: so sometimes they prove to be useful, and sometimes they are little more than mathematical masturbation.

As one guru put it: 'models are precisely-calculated, computer-generated, logically-constructed abstractions of what is, in reality, some real-life, fuzzy-logic, random input conditions in an unpredictable environment.'

• Model is a very general term when used in computing. For instance a weather chart is a model of the weather — and the mathematical program in the computer that draws the weather chart is a model of the weather chart.

modem

MOdulator/DEModulator. An electronic device that converts digital baseband DC signals to digital variations in electronic acoustic frequencies (and vice versa) — not, as many writers continue to claim, 'convert digital to analog signals'.

Twisted-pair cables don't handle DC digital signals well, but analog AC frequencies up to 1MHz travel very long distances in standard telephone twisted pair (up to 10km). However, within the telephone network the bandwidth is deliberately limited to 4kHz within the exchange, so that many 4kHz circuits can be multiplexed onto megaHertz-wide interexchange bearers. This is where the bandwidth limitations lie in the telephone network — at the line-card and switch.

So the modem's job is to superimpose digital signal variations onto an analog audio carrier limited to 4kHz, which can then travel through the global network. There are three different ways of modifying audio signals to carry digital data: change the frequency, amplitude or phase (or two of the three simultaneously).

You must distinguish between the signal and the carrier here — just as you distinguish cars from roads — one provides a way for the other to travel. The audio tone is the carrier, and it is modified by the digital input in the modulation process.

Modems conform to agreed standards. Originally this was Bell in the USA, but the ITU's V.## standards are now used in most of the world. Each standard uses different modulation techniques (ASK, FSK, PSK and QAM) and runs at different data rates — however most modems have three or four standards built-in, and can choose between them automatically.

When establishing a connection, the modems at both ends of a link will briefly exchange information on their capabilities (the handshake) before agreeing on a standard to use (the highest common-denominator available), and then begin transmission. Sometimes if the line gets noisy they will 'fallback' to a lower rate, and if it later gets better they may be able to 'fall forward' to the higher rate again.

Modems are used for simplex, half-duplex and full-duplex (mostly) exchanges of data, and they can be asynchronous or synchronous — although almost all of the newer modems are hybrids: they are synchronous across the phone lines and asynchronous back to their PC.

You need special communications software to drive a modem, usually software that is 'Hayes compatible' and uses the AT-command set. The new modem standards have software built into the modem which performs compression-decompression and error-correction functions.

See the various V.## modem standards, Bell standards, MNP, and the V.42 and V.42bis compression and error checking standards. See also AT-command set, fax modem, soft modem, LDM, line driver and DSU.

modem eliminator

A device that allows the direct connection of two DTE devices (PCs) over a meter or two, for transferring data from one to the other. In practice, this is either a normal RS-232 cable with a null modem, or a special null-modem cable (with Pins 2 and 3 swapped, plus other control lines). See null modem for details.

• You will use these with PC communications software to control the transfer — usually in terminal emulation mode.

modem interfaces

• **Wireline:** The output of a modem interfaces with a standard POTS phone line on one side (and over it to another modem), and with a UART connection on a PC (or other communications device) on the other.

The most common links across this local gap are the RS-232C and RS-422 standards, but there are others. Parallel links (Centronics and GPIB) and SCSI have also been used, but with limited success.

Some other high speed interfaces combinations used with modems are V.35, V.24/V.28/ISO2593 (from the ITU-T) and EIA/TIA 530-A. There's also the ITU-T's V.10/V.11/V.24/ISO 2110 combination.

• **Cellular:** See modem pool below.

• **Wireless LAN:** See DECT, Wavelan, Altair, CDMA, and LAWN.

• **Packet radio:** See Mobitex, Ardis, DataTAC, and CDPD.

modem pool

• A modem pool allows a large number of LAN users to share a small number of modems, without each PC needing to have its own modem or modem card.

• Modem pools are now been supplied by some mobile phone carriers at their cellular basestations for data over digital cellular phones. The rates are generally fixed to 9.6kb/s by the digital standard. See GSM.

Analog cellular phones generally pass the modem tones through directly as audio tones to the remote wire-attached modem. They will use ETC or MNP-10EC protocols for maximum data rates (now 28.8kb/s).

• For ISDN see modec.

modem server

See modem pool and ACS.

modem standards

American modems originally used a couple of standards, of which the Bell series was the most popular. The rest of the world used the CCITT's V.standards (now called ITU V.## standards).

More recently, the USA and European standards have merged, and so almost all high-speed modems (above 9.6kb/s) use the ITU V.## standards at these higher rates.

• Modern modems will generally support a range of new and old modulation and data rate standards. You'll find most of them listed in the V.## section.

Part of the confusion comes from multiple V.## standards which can apply simultaneously; for instance, a V.34 modem may also have V.42 and V.42bis operating at the same time. The first V is for a modulation scheme and data rate, the second is for error correction and the third is for data compression. The same modem may have fall-backs to V.32bis, V.32, V.23, etc. all built in.

These modems are now full-blown computers with their own CPU, RAM, ROM and operating system, so a link between a PC and a modem should be considered as a full link, equal to that between a PC and a printer. The modem is now not a 'pass-through' device, but an independent terminal in its own right. It is actually a store-and-forward processing and translating 'bridge'.

• The early modem standards were:

200-300b/s	V.21 or Bell 103
1200b/s	V.22 or Bell 212A
1200/75b/s	V.23 (No Bell equivalent)
2400b/s	V.22bis for both.

(Note: in V.22bis, the US version had fall-back to Bell 212, while the European version had fall-back to V.22.)

Modem chips were often made with the dual CCITT and Bell protocols so either could be used. In the AT-command set, this selection was made by the B-command.

• The modern modem standards are V.34, V.32bis, V.32, V.23, V.22bis, V.22 and V.21. Also the compression (V.42bis) and error correction (V.42). See all.

moderator

A person who manages (chairs) a 'moderated' keyboard discussion forum or newsgroup. In bulletin boards the System Operator (Sysop) sometimes plays this role, but generally not. Many people prefer moderated groups, while others like the open-access anarchy of unmoderated discussion groups where anything goes.

MODFET

Modulation-Doped Field Effects Transistor. See HEMT.

Modified Modified Read

See MMR.

modifiers

On a computer keyboard there are a range of keys which do not generate ASCII characters or commands themselves, but which modify the output of other keys. The Shift key is an obvious example, as is the Caps Lock. Note that the Caps Lock does not necessarily function the same as Shift; it modifies the number keys in a different way in most keyboards. Different keyboards work in different ways, and have different names for essentially the same function.

Alternate, Command, Option, 'Snowflake', Control, and ESC keys are also modifiers (IBM and Mac). The Alt key, for instance, works by adding Nulls in front of the character and transmitting a multi-byte command.

Note that the modifiers can also modify the actions of a mouse/cursor movement. Holding the Shift key down while drawing a straight line ensures that the program will draw strictly horizontal or vertical lines in some drawing programs.

modular hardware

Anything capable of being unplugged, removed and replaced — usually without needing technical assistance. Some of the best new PC designs are being constructed in a plug-in modular way — with the power supply, hard-disk unit, etc. completely removable and replaceable by the user.

modular software

Software is always modular — but the modules are of varying sizes up to a full application in one piece. There is a software hierarchy between the computer hardware and the user which consists of at least four modules — the operating system, the programming language interpreter, the application, and the user-interface as a shell.

However, the term 'modular' when applied to software usually means that the application can be customized by adding or removing functions. Don't confuse modular software with object-oriented. Modular software doesn't satisfy the full requirements of an object (inheritance, etc.) that object-oriented requires — but it is rather a practical compromise.

Future operating systems will be object-oriented and hence totally modular — and they will probably incorporate many of the functions that we now find only in applications (text entry and editing, file management, etc). Currently the direction of modular software is towards 'build-your-own-applications' from a library of modules with different functions.

modulation

The technique of modifying a carrier's characteristics so that it carries a signal. This applies to both analog and digital signals.

It is important first to distinguish the 'carrier' (usually a radio or audio frequency) from the 'content' (the information-carrying signal). Modulation techniques modify one or more parameters of the carrier — amplitude, frequency, or phase.

Amplitude and frequency modulation techniques are widely used in radio, television and modem communications.

Frequency modulation is used in disk drives. Phase + amplitude modulation is used by high-speed modems.

Amplitude and phase, or amplitude and frequency can be used together, but not phase and frequency since a phase change is, in effect, equivalent to a very brief high-frequency change.

• Modulation can have many layers — each one modulated onto another. A PC's data can be modulated by a modem onto an audio carrier, then that analog audio signal can be digitized to Pulse Code Modulation standards for carriage over the network.

• Both analog signals and digital signals are modulated onto radio-frequency carriers.

— AM radio is analog on analog. An amplitude-modulated baseband audio signal, is modulated onto a single radio frequency sine wave (analog) carrier, by using amplitude summation techniques.

— FM radio is also analog on analog, except that the baseband signal modulates the frequency of the carrier rather than its amplitude. This create a wider band of radio frequencies.

— ASK is digital modulation of an analog carrier, by either switching the carrier on and off, or by modifying its amplitude greatly. Morse Code radio transmissions were a form of ASK.

— Code Division (CDMA) is the digital modulation of a digital code (an XOR process) which itself is modulated (BPSK) onto an analog carrier. A multistage modulation process.

• Modems modulate digital data in various ways onto analog audio sine waves, which carry voiceband AC analog signals as far as the exchange.

— There they may be frequency-modulated onto analog bearers, or suffer an analog-digital conversion, to be carried over digital circuits.

— Along this digital path, the highly multiplexed digital signals will be modulated onto a laser light beam or onto a microwave analog radio carrier.

The point is that multiple layers of unnecessary and inefficient modulation exist in today's phone networks, which is why there is so much interest in a ubiquitous system like ATM.

• The carrier for an information system can be a DC current, an AC analog sine wave (audio or R/F), or a digital code (which, will itself be modulated on a carrier). All of these can carry signals, but you generally wouldn't refer to the baseband case as 'modulation'.

— A DC current, when modulated using any of the normal digital line-code systems, creates baseband signals which carry for a limited distance. A telephone voicelink over the local loop is a DC current 'modulated' by analog resistance changes in the microphone.

— An analog carrier can be modulated with either an analog signal or a digital signal. See FM, AM, ASK, FSK, PSK, QAM and COFDM.

— A digital code, when modulated by another digital code, becomes a direct sequence CDMA system. The modulation technique used here is an XOR code-substitution process.

• Be careful to distinguish between the 'modulating signal' which is the baseband signal, and the 'modulated signal' which is the signal + carrier.

modulation rate

The modulation rate is the baud, event or symbol rate, not the bit-rate. The two will only coincide when each baud produces only one-bit, and that only happens with modems in the 300—1200b/s range. See baud, dibit and quadbit.

module

A reasonably self-contained unit which adds some capability to the whole. This implies a building-block approach to design — modules are sub-assemblies or sub-programs or sub-routines. They are related groups of functions, which become modules when they are put together into a larger entity.

• In software — a utility or function which can be added to (or removed from) the overall application or operating system.

• Object-oriented software consists of numerous 'modules' which exchange information through messages.

• In programming — a set of logically related program statements. In structured programming it applies to the smallest manageable size of sub-function.

• In hardware — a discrete unit which can usually be plugged in to provide some specific service. Increasingly, hardware is being designed for modular construction to make manufacture and maintenance easy.

• Modern IC design consists largely of selecting modules from 'libraries' and then joining them together in a unique way to perform a specific function.

modulo

Don't confuse modulo and modulo-2 with either Modula or Modular-2 which are programming languages.

Modulo refers to a family of arithmetical concepts which are concerned with the values remaining after a calculation. Modulo-2 arithmetic has only two digits, a zero and a one, so it applies to binary numbers. Modulo-10 has ten and applies to decimal numbers, but you can have modulos for any number base.

• Modulo arithmetic — an arithmetical operation in which the modulo integer is the remainder after a base-division has taken place. You divide the result by the 'base-n' and keep the remainder. For instance with modulo-5 (base 5), these two sums are correct:

2 + 4 = 1 i.e. 2 + 4 = 6 6 divided by 5 goes once (ignored) and this leaves 1 remaining. So 1 is the modulo-5 integer result.

2 x 4 = 3 i.e. 2 x 4 = 8 8 divided by 5 goes once and leaves 3 remaining.

This strange arithmetic has important applications in computer number crunching (parity checking, check-sums, etc.). The value of modulo-2 arithmetic is that it can be performed with simple logical circuits using exclusive OR gates and serial shift registers.

• Modulo 2 arithmetic uses binary addition with no carries. This is just an exclusive-or (XOR) operation, and it is used in calculating CRCs. It means that you can perform complex calculations on any series of numbers, of any size, and still come up with a checksum of known size which will perform just as well as a full check-sum (simple addition) would.

MOF

Male or Female? A question on the Internet. It is OK to ask — but don't be so naive as to expect a 'straight' answer!

moire

An artefact, alias or unwanted pattern caused by interference effects.

• Color moire patterns usually appear as fine rainbow bands following contours in some background. B&W moire looks like fine venetian-blind or contour striping over what should be plain-flat objects.

• You see moire patterns on analog TV when someone wears a stripped shirt or a herringbone suit. See cross color.

• In professional printing, moire is an undesirable effect caused when the halftone screen patterns of the three color separations are only slightly misaligned, and so the junction of the dots of the half-tones become visible. To prevent moire in printing, each of the color separations must be screened at a totally different angle.

molecular size

The size of individual molecules varies, but currently the largest molecule of one element, is the fullerene molecule of benzene rings. These can be either 60 carbon atoms in a spherical 'soccer' shaped ball, 'buckyballs' measuring 0.7 nanometers in diameter, with a spacing between molecules of 1.04nm, or buckytubes (made from buckyballs) can be 15nm wide and 200nm long.

• Infrared light has wavelengths in the 600—800 nm region.

• Microwave ovens (at 2.45GHz) are effective because the wavelength is a harmonic of the water molecule size.

Molniya orbit

An inclined elliptical orbit used by Russian communications satellites. It has a 12 hour cycle and is inclined at 63.4° to the equator.

At it's nearest point (perigee) the satellites fly 1000km above the earth, and at its highest (apogee) they are out 39,000km. This means that, in each orbit the satellites must pass through the Van Allen belt, which is a high radiation environment. Relative movement of the satellites also causes a Doppler shift in the frequencies. However, at the apogee (there are two of them a day — 180° apart on the earth's surface) there's roughly a four hour period when the satellite is virtually stationary and high in the sky over the northern part of Russia. It then appears to move very quickly to the next apogee, by flying low and fast through the perigee. It recharges its batteries at these times.

In effect, a satellite in this orbit traces a large double-looping path over the surface of the earth. But if three Molniya satellites are provided (separated around the globe by 120°) full-time coverage can be provided over two areas of the globe's surface, separated by 180°. This is a way of providing 'almost geostationary' coverage of Russia's regions to the north, which can't be seen by equatorial GEOs.

The orbit was named after Molniya 1, Russia's first communications satellite, and it means 'lightning'. See Tundra also.

MOM

Manager-on-Manager products. The integration of enterprise-wide network management. AT&T's OneVision architecture integrates OpenView (HP) for managing the infrastructure, statistical data collection from NetLabs, and their own BaseWorX telecoms manager.

monadic

A mathematical operation requiring only one operand. The negation of a number is a monadic. See dyadic.

monitor

• The VDT screen unit used by a computer, in a television production studio, or one driven directly by a video cassette recorder outputting signals at baseband. Monitors do not usually have radio-frequency tuners but require signals to be fed to them either as 'composite' video or as RGB baseband.

The average bandwidth for a monitor is about 18—20MHz, which is three or more times that of a TV set, so you should expect to see images at a much higher horizontal resolution on a monitor than on a home TV set. See digital monitor, analog monitors, and video-DAC.

• Any device used to check the status of a system. A monitor program checks and supervises the activities of other programs.

• A collection of computer routines usually stored in ROM which are strictly part of the operating system. In some computers you use monitor to read or write directly to RAM memory locations. It presents the binary code in hex form. See monitor program.

• On IBM's Token Ring, the monitor is the workstation on the network, with control of the token generation and monitoring.

monitor program

Part of the operating system that has the routines which create the interface between the hardware, user, and the application. This is usually the first part of the system loaded from hard disk during the boot phase. The monitor initializes the system for start up, controls the application program execution, and prioritizes the input and output operations.

In some systems you can inspect changes to the contents of registers and memory locations directly from monitor.

monochrome

This is often taken to mean Black and White (B&W) as distinct from color. In fact, it doesn't signify 'white', but only 'one-color'. The old green or amber computer screens were still 'monochrome' since green or amber simply replaced white.

The significance here is that the information needed to control a pixel can be stored in one bit — it is either on, or off (logical 0 or 1). The video image (in those days almost entirely text or line graphics) then had a pixel-depth of one.

However more sophisticated monochrome monitors can also show shades of gray/tone and reproduce half-tone images. So 'monochrome' is used in four ways. It can mean:

— Black-and-white only, (1-bit pixel depth).
— Black and one other colour only (1-bit pixel depth).
— Black, white and grays only, (multiple bits per pixel).
— Black and one range of color tones (multiple bits per pixel).

monograph

A librarian's term for a single 'paper' or thesis on a subject. It is not part of a serial (regular) publication, which is called a journal or periodical. The term monograph includes theses, pamphlets and leaflets. It signifies that the paper must be cataloged separately from the regular publications.

monolithic

Literally 'one-stone'. In archaeology, the term is generally applied to large stone carvings. But in electronics it is applied to creating a single circuit with many elements on a single piece of silicon — it means all-on-one chip. This is not necessarily the same as planar (see).

monomode fiber

See single-mode optical fiber.

MOO

MUD Object-Oriented. An Internet term for a cyber-environment in which people meet, discuss and play games. This is a Multi-User Domain which has objects, zones and structures. It is more intelligent and extensible than a normal MUD. See MUD.

Moore's Law

Formulated by Gordon Moore (co-founder of Intel) in 1975: 'The number of transistors which the current technology can put on a microprocessor chip will double every 18 months'.

His law was not widely accepted at the time, but it later proved to be a good predictor of the rate of development for the last decade or so. But as a predictor of future chip potential it now only has limited application.

Moore later revised his law downwards in 1975 to a doubling every two years (double the component count, and double the power). It is now apparent that this can remain true only for another five or six years — two to three generations of ICs at the most. The limit of photolithography is reached at about 0.18 microns.

• Moore has now created another law: 'The cost of a new fabrication doubles for each new generation of microprocessors'.

Currently Intel are spending nearly $2 billion in designing and tooling up for each new processor (and then they get a faulty Pentium!). See Debye limit also.

morphing

The translation of one video image progressively into another. It comes from the idea of metamorphism, which is the process by which a pupa changes into a butterfly.

Don't confuse morphing with a dissolve — where one image progressively overlays the other; the new fades in, as the old image fades out. Morphing uses computers to establish intermediate positions between the key points in both images — usually the eyes, mouth and general face contour. It also creates a transition for colors, hues and textures for key elements in the before and after image.

morphological variants

A term used in spelling checkers. 'Write' is a word, but writes, writ, wrote, written, writing, writer, are morphological variants.

Morse code

An early form of quaternary digital communications. It used four signaling states; a long and a short tone, and a long and a short pause. The long-states were supposed to be three-times the length of the short. Character codes were variable in length and the long-pause was used to frame the character-byte.

You can transcode Morse beeps as an unambiguous binary stream by clocking to produce 111 for a dash, 1 for a dot, 000 for a letter space, and 0 for an inter-element space. So the Morse characters 'ask' (dit dah: dit dit dit: dah dit dah) come out as a bit stream of 10111000101010001110101111. Huffman's coding would probably reduce this substantially.

MOS

• **Metal-Oxide Semiconductor:** One of two basic designs of logic and memory chips — the other is bipolar. There are many varieties of MOS devices which are characterized by the type of channels they employ.

— **NMOS** devices (negative-type) employ the movement of electrons.

— **PMOS** devices (positive-type) employ the movement of positive holes.

— **CMOS** (Complementary) combines both types in the same circuit and it is able to match the requirements of circuits based on TTL.

— **HMOS** is a high-performance form of NMOS.

Other MOS devices are VMOS and DMOS. See MOSFET, NMOS and CMOS.

• **Mean Opinion Score:** A measure of subjective voice quality used in telephony. See QOS.

mosaic

In a regularly 'tiled' fashion — laid out in a regular pattern.

- In alpha-mosaic videotex systems this refers to the use of 2 x 3 unit matrixes, which are 'tiled' on the screen. Each matrix will hold either a pre-set graphic image or a character. There are 128 in all.
- In the Internet, Mosaic is an excellent graphical user interface and communications program which is in the public domain. It is a World Wide Web 'browser' which was written by the University of Illinios, and requires SLIP or PPP connection.

MOSFET

Metal-Oxide Semiconductor (MOS) integrated circuit technology using Field Effect Transistors (FET). MOSFETs have a high input impedance and a much slower switching speed than comparable bipolar devices. But they also consume less power and dissipate much less heat, and can therefore be packed onto the chip surface at much greater densities.

motherboard

The main logic board of a computer to which all the major components and expansion slot connectors are attached..

Motif

Aka OSF/Motif. An alternative graphical interface for Unix which was developed by the Open Software Foundation from the GUIs of H-P, Microsoft and Digital. It is part-based on X-Windows, and resembles Microsoft's Presentation Manager. Motif has now become the IEEE 1295.1 standard. It provides a graphical-user toolkit, windows manager, style guide and user interface language. See Open Look also.

motion compensation

A technique used by digital and high definition television systems to permit greater compression; MPEG1 and MPEG2 both have it. The encoding system measures the direction and speed of moving images — or rather, of changes between successive images. It is able to identify, say, a flying football, or code the screen movement of the whole scene if the camera is panning.

It holds the image from one block in memory and checks it against later frames — comparing a number of image blocks in each, starting in the original position and then working out in a set pattern until it finds a near match. It then supplies this vector information to the encoder in coded form. The algorithm is only interested in low-resolution color and shape matches, which can be supplied by the DCT coding.

The receiver then uses this vector data to create intermediate images by shifting blocks of image to new positions, instead of needing to totally decode and decompress image blocks which have changed position. This technique is vitally important when the camera itself has moved. Most of the scene is then in motion, and there are differences between almost all the blocks in the image frame. Without motion compensation, the decoding process would overload with every camera pan.

This is called both an 'inter-frame' and a 'spacial interpolation technique' (adding details for in-between stages). Although there is obvious picture degradation in just selecting the low-resolution components, the problem is not serious because we perceive less detail in moving objects anyway.

The block-matching process is called motion estimation or prediction, and the vector which identifies the best match between an old block and the new block, is called the motion vector. This is the vector transmitted along with the final difference image, created by applying DCT and then RLE.

- John Watkinson has come up with a slanderous interpretation of the MPEG acronym: 'Motion Predicted by Educated Guesswork'.

motion estimation/prediction

See motion compensation.

MOTIS

Message-Oriented Text Interchange Standard. An ISO standard which is a more primitive version of MHS. It will eventually be merged with X.400. See also SMTP.

mounted

Attached or installed. In software, this often means that the program is installed on a server and is ready for use.

mouse

Graphical User Interfaces generally use a mouse (with one, two or three buttons) to control the cursor — and the whole lot is then called a WIMP interface (see). Other alternatives for cursor control include trackballs, pen interfaces, digitizers, joysticks, etc. Voice control is also being developed, but mainly for the handicapped.

Most mice get their position information from a rolling ball, but some use an optical sensor to read information from a grid printed on a mouse pad, and there are some new radio- and infrared-linked developments. Optical mice tend to be more precise than mechanical mice.

If you've got an RSI problem, change your use of the mouse to the other hand. It will only take a few hours to learn to use it on the other side.

- You should regularly clean the mouse ball, the mouse pad, and the small rollers surrounding the ball. They all accumulate hand grease.

MP

Metal Particle tape. See metal tape.

MPC

Multimedia PC. A standard (of the Multimedia PC Marketing Council) for CD-ROMs. This is the closest we have to a true multimedia standard for CD playback devices which can support the high-bandwidth requirements of multimedia. There is both MPC-1 and MPC-2.

— **MPC-1:** The Class 1 (1993) standard calls for a drive to have an average seek time of less than one second, and to provide a sustained transfer rate of 150MByte/sec into a buffer (minimum of 32kBytes) while using no more than 40% of the available CPU cycles. The standard needs 2MB of RAM, and can be run from single-speed drives. It requires 8-bit sound capabilities only.

— **MPC-2:** The Class 2 (1994) standard is now the minimum rate of data handling for CD-ROM drives, and all SCSI drives should conform to this specification when playing video from a CD (parallel-port drives will often drop more frames than this specification requires). Class 2 requires access times in the order of a fifth of a second, sustained transfer rates of 300MByte/sec to a buffer of 0.5 MByte. It sets the sound standard at 16-bits. This standard needs 4MB of RAM and at least a double-speed drive.

You will need an MPC-2 compliant sound card (at least 16-bits, supporting mono and stereo) if you want to play today's MPC CD titles. New Classes will add different requirements.

MPEG

Motion Picture Expert Group. This is an ISO group which has selected a range of algorithms for image and audio compression — for both storage and retrieval. MPEG is not a standard suite so much as a framework or toolkit of standards, and is therefore very flexible.

The MPEG definition group now comprises computer, video and television companies. Currently they have 11 sub-standards (profiles + levels) which are accepted internationally. The MPEG standards have replaced almost all other video compression techniques (except H.261 for videoconferencing) and they are being extend to HDTV (originally called MPEG2++).

This group reformed in early 1995 (and expanded) to standardize the other problem areas surrounding MPEG which are needed before a true world-standard television system can exist. See DAVIC.

MPEG uses DCT transforms which are applied to 8 x 8 pixel blocks in essentially the same way as H.261, but it adds many other layers of compression including Huffman's coding. Most of the compression comes from calculating differences between successive frames of image, and some from motion compensation techniques. The broadcast versions will provide motion picture compression ratios in the order of 150:1, and this could rise to 200:1.

These versions (MPEG2) use B-frames and motion prediction (optional in MPEG1). They are highly asymmetrical: a lot of processing power is needed to compress the images, while much less will decompress them. This means that movies can be highly compressed 'off-line' (not in real time) while live sports need high-powered expensive compression systems and achieve much less in the time available. However home set-top boxes can decompress either with cheap chips and much less processing power.

MPEG is a 'generic' standard in the sense that it provides information which allows the selection of a range of different algorithms — so it acts as an index of compression techniques. However there are now some incompatibilities within the MPEG framework. Currently MPEG4 is outside the framework.

• The MPEG framework (matrix) is divided into Levels and Profiles; currently there are four Levels (Low; Main, High-1440, and High 1920) and five Profiles (Simple, Main, S/N Scalable, Spacial Scalable, and High) which makes 20 pigeon-holes in all — of which 11 are defined. As at April 1994, the Main-Level & Main-Profile (ML&MP) was defined for 2–15Mb/s video and audio over cable, satellite or terrestrial broadcast channels and for digital storage.

• The bandwidth needed for MPEG systems is flexible, varying from about 1.5Mb/s for VCR quality of pre-processed TV, to 20Mbit/s for wide-screen HDTV — depending on the profile and level used. It also allows for VBR and CBR data streams.

• With the exception of the Low-level (NTSC-to-PAL) interchange format (60Hz with 352 x 388 pixels), each of other MPEG levels are defined for both 50Hz and 60Hz, and for 525- and 625-line structures. To equal the analog system quality (which has a vertical blanking interval) the MPEG standards for screens are 480 or 576 lines with 720 horizontal pixels (NTSC and PAL respectively).

• The two high definition/wide-screen levels (1920 and 1440) have both 1080 and 1152 line versions.

• American and European committees are currently trying to settle differences over the use of Dolby-AC or Musicam sound systems.

• There's now an Open PC-MPEG consortium which is creating APIs for application software.

See MPEG1, 2, 3 and 4 and also JPEG.

MPEG levels and profiles

• **Levels:** In MPEG frameworks, the levels are used to establish screen aspect ratios and resolutions. They are:

— **Low:** For video-telephony 60Hz (only) at 352 x 288 pixels.

— **Main:** For general TV: 50Hz @ 720 x 576 pixels, 60Hz @ 720 x 480 pixels.

— **High 1440:** For high-definition 4:3 screens: 50Hz @ 1440 x 1152 pixels, 60Hz @ 1440 x 1080 pixels.

— **High 1920:** For high-definition 16:9 screens: 50Hz @ 1920 x 1152 pixels, 60Hz @ 1920 x 1080 pixels.

• **Profiles:** Variable sets of algorithms may need to be applied to the compression-decompression process. All profiles will use DCT to transform the image, but only some will use motion-compensation techniques. The different profiles are:

— **Simple:** Requires minimal processing.

— **Main:** Requires 'average' processing and is used for most general TV.

— **S/N Scalable:** Uses a layered approach to provide different image qualities where different conditions apply.

- **Spacial Scalable:** Uses as layered approach to provide different amounts of processing where different screen resolution requirements apply,
- **High Profile:** Is the future system where substantial processing is possible for high-definition.

Each Profile can be used with a different Level which sets the resolution of the screen (for both 50Hz and 60Hz systems).

MPEG1

MPEG1 is a non-interlaced (progressive scan) system which was devised mainly for video and audio storage systems where delays in making the initial compression aren't a problem.

However it is being used for direct satellite broadcasting by the Hughes DirecTV system (later to upgrade to MPEG2). MPEG1 has been incorporated into the new framework of MPEG2, and many of the older MPEG1 decoder chips were designed to allow some of MPEG2's advanced characteristics.

Philips CD-i-FMV uses the early MPEG1 (Format 1) compression scheme to reduce the standard digital video rate to match the 1.5 Mb/s delivery capacity of a single-speed CD disk (1.15Mb/s for video and 0.25Mb/s for audio) — a compression ratio of 140:1.

MPEG1 now uses the Source Input Format (SIF) resolution of 352 x 240 pixels, which is about VHS-replay quality, as the standard image size for Format 1. The new Format 2 is for double-speed CD-ROM drives (3Mb/s), for video CDs, karaoke machines, PCs and workstations, games units and TV set-top boxes.

• The standardization of audio has proved to be a problem: Audio levels 1 and 2 have been finalized, but Audio level 3 (the highest CD-like quality) is still in dispute because of the likely expense in implementing it.

MPEG2

MPEG2 is an extensive suite of interlaced standards for real-time broadcast applications. It was 'signed off' in March 1993, but then modified and upgraded again in June 1994. Profiles and Levels have been defined regularly in the intervening period, and the process appears to be continuing.

The broadcasters in the committee want the real-time PAL-quality image-compression to be 4–6Mb/s, but it is proving to be difficult and costly to make these real-time encoders ($300k each). Fortunately, the decoders are relatively simple, and cheap enough for a mass market.

In these compression systems, the quality of the final image is traded off against processing power and processing time during the encoding process. When ample pre-processing is possible (films and pre-recorded programs) MPEG2 offers good VCR-level pictures at 2Mb/s and possibly down to 1.5Mb/s. However, for real-time transmissions of high-activity sports, it is generally assumed that about 5–6Mb/s is needed for good NTSC quality, and at least 6Mb/s for PAL. Wider-screen images will need proportionally more, in the 10–12Mb/s range.

• MPEG2 is not suitable for video post-production, since it is impossible to clean-edit a frame which only exists in terms of difference values. For this reason Video-JPEG is used for video post-production, and MPEG is then applied only to the finished product. Sony has a system able to edit MPEG using Motion JPEG.

• The most common of the MPEG2 frameworks is Main-Profile & Main-Level (MP&ML). There are both PAL and NTSC versions of this. See MPEG.

• The NTSC-equivalent MPEG2 (MP&ML) has an image format of 720 x 480 pixels. The standard PAL image format is 720 x 576 pixels. However some chip-makers cheat and only provide 600 pixels or less in width.

• IP (Interframe Prediction) in the decoding process comes with or without B-frames. A decoder with simple IP requires about 1MB of memory, while one with B-frames needs 2–4MB, and possibly much more.

• MPEG2 with B-frames, sometimes called MPEG2B, provides better picture quality and higher compression ratios by allowing the decoder to look ahead five or six frames. Not all MPEG2 systems use this predictive approach.

• The standard consists of three parts: the MPEG2 System part deals with synchronization and multiplexing issues; the MPEG2 Video part deals with the video compression techniques, and the MPEG2 Audio part deals with the sound. The current standard audio coding will provide five full-bandwidth channels plus additional low-frequency enhancement channels. You can apparently use seven commentary or multilingual channels in total.

• It is now felt that 6:1 compression can be expected from MPEG2 with almost total recovery of the original image. In other words, at this level MPEG2 is virtually lossless.

• MPEG3 was merged into the MPEG2++ standards which are now treated as part of the MPEG framework. The Grand Alliance is defining the Main Profile, High Level (MP&HL which is 1920 x 1080 pixels at 60Hz) version for high definition use. Currently this requires a data rate of 80Mb/s, but it will later be reduced.

• The ATM Forum is now working on the standardization of MPEG2 over ATM networks. The transport stream is to be made up of interleaved 188-Byte MPEG2 packets of video and audio, which will be broken down in some way into the ATM standard 53-Byte cells.

MPEG3

MPEG3 was merged into the MPEG2++ standards which are now treated as part of the MPEG framework.

MPEG4

MPEG4 is a very low-level coding system for AV and videophones at PCM rates. It is intended for mobile applications and for use over ISDN B-channels. See scalable.

• The Main Profile/Low Level standard of MPEG2 is designed for videoconferencing and video-telephony also. It requires a decoder rate of 4Mb/s and substantial buffers.

MP/MLQ

MultiPulse, Maximum Likelihood Quantization. See G.723

MPP

Massively Parallel Processing. A cheaper technique (if you think in millions of dollars) for building super-computers with GFLOP and possibly TFLOP rates. Oracle's new n-Cube is an example of this form of computer architecture. A single n-Cube machine has more than 65,000 processors and is capable, as a video server for VOD, of delivering MPEG compressed video simultaneously to 150,000 users, so they say.

There are a number of techniques which differ mainly in how memory is allocated to the multiple processing chips, and how the coordination of tasks is handled

• Some MPP systems eliminate the shared-memory bottleneck by using a distributed memory structure, with each of the processors having local memory and communicating over very fast local CPU buses. See MIMD and SIMD.

• 'Shared-nothing' architectures are linked through the disk-access subsystem. See SMP.

MPR#

MPR is the Swedish acronym for the National Institute of Measurement and Testing, and it has now published MPR2 guidelines for computer ergonomics. This includes a lot of things including keyboards, but by far the most significant is the standard for magnetic and low-frequency radiation from VDUs. This is now the *de facto* standard used around the world.

M-PSK

Multi-level Phase Shift Keying. A general term which refers to both 8-level PSK and 16-level QAM and similar phase-shift + amplitude modulation methods. These all use increasingly complex constellations of differential phase-shifts, with finer and finer angles (and amplitude levels) of distinction. See QAM and Trellis code modulation also.

• The term is also applied to systems like COFDM which use multiple channels, each of which is modulated with a multi-level QAM technique.

MPT 1327

Mobile Protocol Trunking 1327. A non-proprietary signaling standard and air-interface specification for analog trunked radio, which has been widely adopted in Europe. It is currently the *de facto* European standard for public access mobile radio (PAMR) and for general dispatch purposes. It was devised by the UK's Department of Trade and Industry.

MPT defines services such as call queuing, priority calling, group calling, call diversion, PSTN connections, etc. which are used for fleet control — and it permits links at the basestation into the telephone network.

It uses 1200 baud FFSK signaling and it appears to be suitable for any of the common mobile radio-frequency bands. The contention problems have been solved by modified Slotted Aloha (DFSA). The key protocols are numbered 1327 and 1343. ETSI is formulating a digital trunking standard. See TETRA, MIRS and DSRR.

MPU

Micro-Processor Unit. A more restricted term than CPU (which covers microprocessors and other larger processing units) but one that you don't often see used. Early microprocessors were called 'microcomputers', which was also confusing. See CPU.

• The microprocessor was born in about Oct 1969. Intel were quite late with the 8080 in 1973 (originally costing $400 each, now about $1). There are now about 10 billion microprocessors in operation around the world (two for every man, woman and child) and the design of each new chip costs tens of millions of dollars. See 4004, 8008, and 8086, CIRC and RISC, Debye limit, and Moore's Law.

M-QAM

Multilevel QAM. QAM involved both phase shift keying and amplitude changes. Phase-shifts of 180° only provide a single binary digit, while 90°, 45°, 22° respectively allow dibits, tribits and quadbits to be sent. Coupled with two levels of amplitude these modems can yield quinbits.

Multilevels are only really limited by the ability of the systems to discriminate between minute changes of phase-angle, and small differences in amplitude.

• Nortel says it is using 515-level QAM for carrying Sonet/SDH signals over microwave links. This doesn't seem possible, but it could mean that they are using some system like COFDM or DMT with multilevel QAM on a large number of 'fingers'. See M-PSK and constellation.

MR

Magneto-resistive. A new head design for hard disk drives, which employs a magnetically sensitive thin-film resistor element to detect data bits.

This type of head can handle data bits just 0.4 microns apart and 4 microns wide, and it does this relatively independent of disk speed.

MR is replacing the conventional inductive thin-film read-write heads which have trouble distinguishing signal from background noise when disk rates are high. See PRML.

MRM

Mobile Radio Modem.

MROM

Masked ROM. The standard type of ROM used in all large-scale production runs. This is normally just called ROM. See masked ROM and ROM.

MRP

Management Resource Planning. Formal software that allows modeling, forecasting, simulation, etc., in a business.

MRP II runs on company mainframes in manufacturing corporations as part of MIS. It encompasses a philosophy for integrating computers in manufacturing, production scheduling, automation and education.

ms/MS

- Millisecond. One thousandth of a second.
- Microsoft — the Bill Gates-owned diamond mine.
- Message Store.

MS-DOS

The world's most popular operating system which was acquired by Microsoft Inc. for use on the first IBM PCs. At that stage it was called QDOS ('Quick and Dirty OS'). Although Microsoft quickly improved it beyond recognition IBM didn't purchase an exclusive contract, which enabled Microsoft to keep control.

There were a number of early versions: Lifeboat Associates began distributing the system under the name SB-86, IBM called it PC-DOS, and the general-purpose version released by Microsoft was called MS-DOS. It has an obvious but superficial resemblance to CP/M, however it also has a close relationship with Xenix (Microsoft's Unix).

MS-DOS underlies Windows 3.x, but it has been replaced in the new versions of Windows 95.

- MS-DOS 6.0 had problems with the DoubleSpace data compression.
- PC-DOS 6.1 is the IBM attempt to steal the market away from Microsoft while they had problems with data compression. (PC-DOS 6.1 doesn't have any.) It has most of the same utilities, but other features are different.
- MS-DOS 6.2 is Microsoft's upgrade to the MS-DOS 6.0 which gave them such problems.

MSAT

Micro-Small Aperture Terminals. See VVSAT.

MSAU

MultiStation Access Unit. This is a Token Ring term which is used for MAUs to make a distinction between those used for Token Ring and those used for Ethernet. However, MSAU and MAU seen to mean exactly the same thing in the Token Ring literature.

These are the wiring concentrators and up to eight terminals can be connected to one MSAU/MAU concentrator. The MSAU is more than just a wiring point, it is an integral part of the ring and it is responsible for handling the token and the packets on behalf of the terminals.

This approach allows the development of a star-wired subnetwork consisting of relatively dumb terminals. These can each be linked into a logical ring at a point fitted with a relay-switch (activated via power from the attached stations, or remote) which can heal the ring (by-pass the terminals) if a device disconnects or fails.

This is important because the Token Ring LAN architecture is not naturally resilient. If a device failed, without this by-pass mechanism the ring would remain broken, and the whole network would cease to function.

- If the MSAU is under remote management, it is called a CAU. See.

MSB

Most Significant Bit — the left-most bit. Binary notation has the most significant bit at the left, and the least significant bit at the right — as does normal 10-based decimal (denary)notation.

Mscdex

Microsoft CD Extensions. See CD-ROM Extensions.

msec

Millisecond. A thousandth of a second. This is usually shortened to ms.

MSN

- **Multi-Service Network.** This is a virtual private network operated on behalf of some large corporation by a number of service providers cooperating.
- **MicroSoft Network.** The new Internet connected personal computer network, forum and e-mail system, with links into Windows 95.

MSnet

A communications package which is an add-on for Microsoft Windows 95. It offers a totally graphics interface with special menus for e-mail, networking, etc.

MSO

Multiple System Operator.

- This was originally a US term for a cable network operator which ran a number of franchised cable networks which were linked together by fiber or satellite. Time-Warner and TCI are both multiple cable owners, and so they earned the name MSOs when they began linking these and other networks into Premium (Pay) services. Eventually the term MSO came to mean companies which 'wholesale' these Premium channels to single-system cable, satellite, and wireless cable operators. The are the distributors of numerous channels.
- In the present US context, the term is limited to the ten major cable programming companies which control at least 40% of the total cable market. Time-Warner, for instance, owns HBO, the USA Network and Cinemax. Other MSOs are Viacom (Showtime, Movie Channel, MTV, Nickelodeon and VH-1 Music) and Group W (Nashville Network, The Home Theatre Network).

MSS

Mobile Satellite Services: the subcategories are Big LEO and Little LEOs, MEO/CIOs and GEOs. MEOs and CIOs are generally grouped with Big LEOs. The term MSS can be used for GPS applications but it doesn't generally refer to the GPS constellation itself, and it may also include the Inmarsat maritime systems. MSSs are used for voice, e-mail, data and location, however some LEO services (such as Orbcomm and the Teledesic proposal) are not primarily aimed at mobile users.

MTA

- **Message Transfer Agent:** This is the mail server in an X.400 MHS network. It is actually the software in a message switch, responsible for routing the message across a X.400 backbone. The MTA's job is simply to read the address of the message package and pass it on towards the appropriate Message Store computer.
- **Major Trading Areas**: The Rand McNally's geographic divisions widely used in US telecoms. There are 47 of these across the USA.

  BTA (Basic Trading Areas) are less significant, and there are 483 of these. PCS spectrum licensing is on the basis of MTAs and BTAs, and priced according to POP (population within each of these areas).

MTBF

Mean Time Between Failures. A measure of the reliability of computer equipment, usually expressed in hours. These figures are supposedly arrived at statistically by subjecting a number of models to prolonged testing — however a cynic would feel rather uncomfortable with the way some of the figures are calculated with 'accelerated testing procedures'.

MTP

Message Transfer Part of the CCS#7 signaling system.

MTS

- **Multichannel Television Sound:** A stereo format.
- **Message Telephone Service:** Yet another US name for long distance or 'toll' services.
- **Message Transfer Service (System):** The part of X.400 MHS (e-mail) which handles the transfers between Message Transfer Agents (MTAs).
- **Mobile Telephone System:** The oldest mobile telephone system, which was really only two-way radio with a telephone connection. It was one stage before IMTS, which came before cellular. It used inband signaling tones in the 600Hz to 1500Hz range.
- **Mobile Telephone Switch:** A cellular phone network switch. This controls the hand-offs. There will usually be one or possibly two to each city.

MTSO

Mobile Telephone Switching Office. The central computer controller for a cellular telephone network.

MTU

- **Magnetic Tape Unit.**
- **Message Transfer Unit:** The payload of a message (excluding the header and other redundancy/overheads) in packets, cells and messages.
- **Maximum Transmission Unit:** The maximum size, measured in bytes, permissible for a packet in a particular network or with a certain protocol implementation. For instance, ATM uses a 53-Byte cell, but with only a 48-Byte data payload. Therefore the MTU of the ATM network will be 48-Bytes (despite the fact that this payload often carries packet headers as data). Any larger packet must be fragmented when carried over ATM networks, and they will be segmented and be fitted into this MTU area — then the packets are reassembled at the other end. See PMTU

mu-255

See A-Law.

mu-Law

(Actually μ-Law). This is the North American standard for pulse code modulation (PCM). Like the European A-Law it uses a non-linear quantization of each sample, so that the sample numbers reflect a subjective scale of sensitivity to high and low variations in the audio waveform.

American PCM systems also sample 8000 times a second with 8-bit quantization. However the logarithmic approach makes 64kb/s μ-Law roughly equivalent to 128kb/s linear-sampled PCM — so it can be viewed as a bit-rate reduction technique, or a quality-improvement one.

It is said that μ-Law is slightly superior to A-Law in terms of low idle-channel noise, but inferior in S/N ratios. It uses slightly different companding techniques to compress the amplitude range and reduce quantization errors. See A-Law for details.

- US mu-Law PCM can be transcoded to European A-Law PCM with only a slight degradation of S/N with a converter, but a direct feed makes them almost unintelligible.

MUD

Multi-User Dungeon (aka Domain or Dialog) This is an Internet term for a forum in which participants meet for the purpose of on-line games, exchange of information, or just to talk and perhaps learn — but it can also be seen as a fantasy game played by many players. MUDs are computer-generated environments where the participants take on a character role. You need to Telnet to a MUD and ask to be admitted.

muldem/muldex

Two old terms for a mux (multiplexer/demultiplexer). The demultiplexing function is now just assumed. See multiplexing.

multi-access

Multi-access computer. A type of multi-user computer system where the main computer can handle only one application, although a number of user-terminals can access it. This is a form of time-share, still used for mass input/retrieval of text.

multiband

See broadband.

multicast

The transmission of messages to a number of specific users. The distinction is being made between point-to-point connections, narrowcast and broadcast. See multipoint distribution.

- A multicast address is a group address.

multidrop

See multipoint distribution. There is a vague distinction between multipoint and multidrop. Multipoint is usually applied to star-network configurations with a central hub linking to peripheral devices.

Multidrop usually implies that the connection is via a bus-type architecture with the peripheral devices making contact along the cable length. Multidrop can be a one-way broadcasting system. However, two-way interactive systems will probably use polling and be in the control of a primary head-end node.

multi-format

Television receivers and monitors which are multi-format, are capable of handling signal from 50Hz PAL, 50Hz SECAM or 60Hz NTSC or perhaps from a range of computer devices.

multiframe

This is a group of frames of data which are considered as an entity. It requires a multiframe alignment signal.

multihomed host

A host which is attached to a number of different network segments, but each segment sees the host as on its own network.

multilayer LAN switching

Remember that LAN switching is often not a true switching function at all; it is actually a bridging or routing function which has been beefed up by special RISC processors.

The early LAN 'switches' tended to work only at OSI Layer 2 (Data link), so they were *de facto* bridges. However further developments in switch-based internetworking requires the addition of routing functions, which were at Layer 3. This in turn, requires routing tables and intelligence. In effect, these became very fast brouters.

To avoid unnecessary complexity in the switches, the routing calculations can be performed by dedicated routing processors which are resident in other devices. Best-path information is then distributed to any number of multilayer LAN switches from a central site.

Multilayer LAN switches offer excellent throughput and provide dedicated LAN segments to individual users. They also ease network administration problems by decoupling the logical address from the physical port. See Switched Ethernet and LAN switching.

multilevel code

Transmission schemes where more than one voltage (amplitude as well as polarity) must be detected by the receiver. These overcome some of the deficiencies of NRZ and NRZ-I codes. See AMI, pseudoternary and 2B1Q.

multilink

See BONDing.

multimedia

The use of computers and digital techniques to blend text, music and video into 'a rich interactive experience'. Multimedia files will, by definition, be large, and be delivered from local storage (mainly optical disks) or over broadband networks (mainly coaxial cable). Video servers will eventually be the most common form of storage, and ATM over HFC the most popular form of delivery technique.

At the present stage of development, the term multimedia assumes that computers will control videodisk players, CD-ROMs, VCRs, or large hard-disks. Later it is assumed that we will have access to on-line multimedia databases via the 'Information Superhighway'.

Multimedia will usually involve the manipulation of text and computerized graphics along with video and sound — with each of these elements being treated separately when required. The term probably extends to virtual reality, although the two terms often seem to be used in distinctive ways.

Don't confuse multimedia with hypermedia which deals with the way multimedia modules may be linked together. See MPEG and JPEG.

- See CD Multimedia and Super Density CD.

Multimedia CD

See CD Multimedia with the other CD-format listings.

multimode optical fiber

A form of thick optical fiber which allows the use of light emitting diodes (LEDs) and makes for easier coupling and splicing because the fibers don't need to be so perfectly aligned.

Multimode refers to the fact that the photons of light will travel along slightly different paths through these fibers, and therefore travel over varying distances. Some signal dispersion (blurring) can be expected at higher rates and longer distances. However, multimode fibers are often ideal for local area telephony and LANs. See single mode optical fibre.

multi-multipoint

A form of multipoint operation where only the primary station operates in a full-duplex mode, while the secondary stations are all half-duplex.

multipath effect

Also called alternative path propagation. This is the cause of ghosts in TV sets, and distortion and lost calls in cellular phone systems. It results from the signal traveling via two or more different paths (because of reflections) from the transmitter to the receiver, and therefore arriving at slightly different times.

Multipath signals can therefore arrive out of phase and tend to cancel each other (see Raleigh fading) or be well out of sync and produce complex overload and signal cancellation effects. In analog cellular phones, multipath causes sound fading and noise; in a car radio, multipath creates a fluttering of the sound as the car moves across different paths of reflection (the picket-fence effect); in digital cellular phones it causes all sorts of intermodulation distortion problems (but see CDMA and rake receiver); and in TV it creates positive and negative images which are always offset to the right (in the direction of the scan) of the primary image.

Directional antenna can reduce the multipath effect, but this is not practical in mobile communications.

multiple master

In outline fonts, these are typefaces which have only one source of definition, but which allow the generation of many variations in weight and width. Before this technique was developed, outline typefaces could either be normal or bold in weight, but nothing between (although sometimes extra-bold faces were available for display purposes).

Multiple master techniques allow different weights and widths to be defined from a single master outline, so extremely heavy faces can be created, along with condensed or expanded print, all from the same source. This is now incorporated into ATM 3.0 and is available in the Macintosh.

multi-platform

This means that a service will support MS-DOS, MacOS, Unix or OS/2 or, alternately, it may support more than one network operating system. The term is usually used for services over a LAN. Note that it generally applies to 'services' not software.

multiplex

A cinema complex with several screens. It is the way of the future because they only need employ one projectionist and staff one box-office but can run several movies.

multiplexing

• **Chips:** At the chip level, a multiplexer is a logic circuit that can accept several inputs, but it only allows one to get through to the output at a time. The selection of inputs is controlled by a Select (aka Address) input.

• **Networks:** A system of carrying a number of distinct channels on a single physical bearer. At the practical communications level, these are expensive devices used by the carriers between exchanges, or by private network operators on rented lines in order to make better use of very expensive leased circuit bandwidth.

To save renting a large number of tie-lines, a company will multiplex many signals into one bearer at one end, and demultiplex them at the other — using essentially the same two-way communications equipment at each end.

Multiplexers come in many different sizes, costs, form factors — and they use a variety of different techniques. Some are highly intelligent and others are extraordinarily dumb. Generally they are boxes or racks with numerous slots into which port cards and trunk cards are added. The box provides both the allocation intelligence and a backplane for transferring the data between buffers.

There are a few main types:

— **Frequency-division multiplexers (FDM):** Either analog or digital signals can be multiplexed in this way. There is one physical connection, but along this link are hundreds of distinct channels which use different carrier frequencies. The data streams are each modulated onto one of these carriers. The separation process is equivalent to having many TV tuners, each of which isolate one channel from the many transported.

In the past, all telephone signals were transported between exchanges using FDM multiplexing over coaxial cables, microwaves, or twisted pair, and the technique is still used widely for analog signals. It is cheap and simple.

— **Time division multiplexers (TDM):** Time-division techniques are only suitable for digital data because the process requires the temporary buffering of incoming data in digital memory devices. There will be one buffer for each channel.

Only one block (a bit or byte at a time) is released into the bearer from each buffer during the multiplexer's cycle, but all input channels get their turn in strict order of rotation. If the bearer can operate at a high-enough data-rate, dozens or even hundreds of channels can be transmitted across a single link without data loss and without noticeable delays (although there always is some). With fiber optics, thousands of voice channels can be carried simultaneously in a single fiber.

At the remote end of the bearer, a demultiplexer cycles in synchronization and shunts the right bits or bytes into the correct output buffers. These then output the data as a steady stream along individual channels. Obviously the data rate of the bearer must be at least as fast as the total of all individual input channels.

TDM is used in optical fiber transmission systems carrying enormous numbers of channels. It is also used in coaxial and twisted pair cables, and in microwave systems. See bit-oriented protocols and byte-multiplexing.

— **Code Division Multiplexing (CMA):** See CDMA.

— **Statistical multiplexing:** This is best treated as an adaptive form of TDM, but packet-switching systems are actually a form of statistical multiplexing. However it is best to make a primary distinction between circuit-switched and packet-switched, and then the term statistical multiplexing really only applies to circuit systems. See statistical multiplexers.

• The multiplexing of analog signal in the telephone network (always FDM) is usually handled in voice-channel groups and super-groups (see). The stages are known as 'primary', 'secondary', etc. multiplexing, or by their official designations. (See channel-banks and T1.)

Digital TDMs have a similar sequence of hierarchical orders, however, the purpose of Sonet/SDH is to avoid these multi-layered stages of processing for long-distance bearers.

The μ-Law PCM carriers and multiplexers have a hierarchy of data-rates:

1st order 24 channels or 1.544Mb/s (T1).
2nd order 96 channels or 6.312Mb/s (T2).
3rd order 672 channels or 44.736Mb/s (T3).
4th order 4032 channels or 274.176Mb/s.

• See statistical multiplexing, high-order multiplexer, plesiochronous transmission, add-drop multiplexers, PDH and SDH.

• The term multiplexing and demultiplexing is also being wrongly applied to the provision of 'distinctive ring' on home phones. With distinctive ring, you have one line and two or more phone numbers — and you can detect which of the numbers has been rung by the distinctive ring pattern. Some of the literature claims that this identification service allows you to 'demultiplex' the line automatically. You can see what they mean, but they are stretching the idea a bit. However a computerized switch can detect the distinctive ring and connect the normal phone line to a fax — so I suppose this is multiplexing of a sort — but only one call at a time can be on each incoming line.

multiplicand

One of the numbers (factors) used in multiplication — it is the number which is to be 'multiplied by x' The other factor is known as the 'multiplier'.

multiplier

See multiplicand.

multipoint distribution

The term usually implies the distribution of signals from a central point to a finite number of peripherals. For instance in a polling environment, a single centrally-controlled communications system can support many devices — so this is a multipoint distribution of control.

The links can be by microwave, cable with multiple taps, satellites, or normal radio broadcast. However 'broadcast' suggests that anyone can receive the signals, while multipoint indicates that the distribution is tightly controlled (by scrambling).

Multipoint channels may be full or half-duplex. If the primary station operates using full-duplex and the secondary stations are half-duplex, the system is known as having 'multi-multipointoperations'.

With either multipoint of multidrop operations, one network station must always be the control or primary station. Other stations are called 'tributary' or 'secondary'.

multi-processing

This is applied to a chain of stand-alone CPUs linked together by a connection of some sort, and running an operating system (or networking extension) which allows them to share a processing task. It is a form of parallelism. Don't confuse this with multi-programming or multitasking.

Sometimes these multi-processing systems will all be in the one box, with both a central processor and a number of independent processors on accelerator cards, all plugged into the main bus. Each processor acts independently, so this is not a true parallel processing operation — but it is part-way towards parallelism.

There are two main multi-processing architectures: Symmetrical (SMP) and Asymmetrical (ASMP). See both.

• There are two main applications for this approach:

— **Fault-tolerant computing:** This is where the purpose of the configuration is to ensure that in the event of one system failing the other will take over the task. If three processors are used, the triple-mode redundancy approach can be used.

— **Shared processing:** For instance in ray-tracing applications for photo-realistic images, three computers can be linked to share out the processing tasks. A separate computer will handle the Red, Green and Blue components of the final image.

• Special computers, such as array processors, also use a multi-processing approach to provide concurrent maths processing on the data sets. See vector processors.

• See macroparallelism.

multi-programming

An older term for a computer system able to run a number of different applications concurrently. We would now call this multitasking if the control was totally in the hands of one user, and multi-user if many terminals were able to access and control the computer (seemingly) simultaneously.

• In the original use of the term, it simply meant a system that was the opposite of mono-programming computer.

• These were the early PCs; if a disk drive was operating, the CPU would just sit idle until it was asked to perform a task, while in multi-programming computers, when the CPU didn't have anything to do on a process, it could swap to another process and utilize the idle time.

• The term later came to mean systems that could split up a program into two or more different processes, and this later led to the use in reference to time-sharing and multi-user operations.

multi-protocol routers

LAN routers which are capable of handling different interconnection protocols simultaneously. These are special intelligent routers. The term is often misapplied to brouters, which also have bridging capabilities.

multiscan monitors

Aka multisync monitors that can handle a number of different feed types — EGA, VGA, etc. They can automatically detect the scan frequencies and set themselves to match. Aka multisync.

multisession formats

Several recording CD-ROM specifications have recently been developed. The Orange Book specification for a CD recording system, originally only dealt with recording taking place in a single session, because it only allowed for a single table of contents. Later multisession standards were added and now data can be written to the disk in numerous sessions. There are several multisession formats already in use and more under development. See Photo CD.

multi-sync

See multiscan monitors.

multitap

Broadband taps. See tap.

multitasking

The sharing of routines, data space, and files to execute several processes at the same time. This is an operating system function, which is possible only when the OS has been designed to handle a number of different application programs simultaneously, and when facilities exist to manage computer memory.

All modern computers are now able to keep more than one application in working memory (provided the space exists) at the same time, but only one of these is often active. This is called the foreground task, and the others lie in the background. Usually these background tasks are dormant, but they can, with true multitasking, be databases performing sorts, or word-processors sending novel-length files to a printer, while you are working on your spreadsheet.

The three basic forms of multitasking are:

— **Parallel or Pre-emptive:** This is true multitasking, where background applications can continue to use time-slices from the CPU. This is not really a parallel operation, but rather a quick interleaving of the various processes.

— **True Co-operative:** Applications needed to be specially written for true co-operative multitasking, so this technique fell out of favor and is no longer used. However see co-operative multitasking.

— **Serial or Task-switching:** Aka context-switching. This is often called cooperative multitasking, but it is actually pseudo-multitasking where only the foreground application is active. The background applications stay in memory, but they are stalled until the foreground application gives up its primary position. There is nothing co-operating here except the sharing of memory. See cooperative multitasking.

All modern computer hardware is multi-tasking since they can all handle more than one process at a time through very fast interleaving: even single applications running alone require a number of processes to be functioning. But both pre-emptive multi-tasking and task-switching introduce a higher order of complexity by allowing more than one application to be resident in memory and share time of the CPU. They therefore require sophisticated memory management. See MMU.

• Multi-tasking is not quite the same as multi-programming which is a wider term. Multi-processing is something quite different again (it is a form of parallel computing).

• A computer can't actually run processes, tasks, or applications simultaneously, but it can appear to be doing so if the hardware acts fast enough. A pre-emptive multitasking system actually uses the microseconds of delay during the input and output processes of the active (foreground) program, and applies this processing time to background tasks. The tasks are handled by a schedular.

• For pre-emptive or task-switching, all data-spaces, stacks, process tables, etc. for the different applications must be isolated from each other. This is process management and it also requires good memory management.

• Note that in the PC world the terminology has changed in recent years. Multitasking in PCs originally meant that only one application was active, but that a number of documents/files for the same application could be maintained on the desktop and quickly be bought into use. This was one application, many data files.

Later the term came to mean systems which maintained a number of applications on the desktop, although only the foreground one was ever active. This was many applications (one active), many files.

With OS/2, true multitasking came to the desktop. This was many applications (many active), many files.

• Lately and unfortunately, the term multitasking has also been applied to a form of primitive parallel processing. See multi-programming also.

• Note that some add-in accelerator cards can function independently of the main CPU and therefore appear to perform in a multitasking or multi-processing (parallel) way. In fact this is really simultaneous tasking; there are two complete CPUs operating quite independently and performing tasks that are not really related in any way.

multi-threading

This is a more generic term than multi-tasking, but it seems to be used for the same type of operations as a multi-access or multi-user computer (see). This term can be applied to time-sharing computers, or to one user where the computer can juggle several running application programs and simultaneously perform such tasks as printing or copying a file.

• One definition says a multi-threading computer system will run only one application, but that it will have many simultaneous users or application — each requiring access to the CPU for only short periods of time.

• With 'threading', the processing of one transaction is not necessarily completed before the next one begins. It is possible to suspend the transaction operation at some intermediate stage, and resume it later (on a microsecond basis). This seems to me to be identical to 'process'. See re-entrancy.

multi-user

• A single processor that makes provision for a number of simultaneous terminals and a number of simultaneous users. Traditionally this was mainframe territory, but now with modern 386 and 486 PCs it is possible to use multiple terminals with a single PC 'host'.

Some multi-user systems should strictly be called 'multi-access' since they require all users to be using the same program and accessing the same files; others are multi-user and/or multi-tasking. To all intents and purposes, each person has the attributes of a stand-alone computer, but they can also share information as if they were on a LAN.

• The term is also used to refer to the obvious with LANs and LAN-servers. The point is that file and record locking is essential with databases in a 'multi-user environment' — meaning other people may be trying to access them.

multi-user DOS

This was once promoted as an alternative to building small-scale LANs. Multi-user DOS was said to link dumb terminals and personal computers to a central processor at a fraction of the cost and trouble of the early LANs, but they were suitable only for relatively small work-groups. This idea only really caught on for some tedious data-input installations.

The emphasis was on text-based applications, and the idea gained momentum once again with the release of the Intel 386 and 486 chips which were seen as unnecessarily powerful for a single user.

Multi-user DOS systems use DOS conventions and commands and run the OS by remote. File access to the shared disk and file system was done directly over the bus — not through additional layers of network hardware and software.

multi-vendor network

Since standards are often loosely applied, it is natural that networks comprising equipment from many manufacturers will give more problems than networks with equipment which all comes from one. However, this point is often overplayed by marketeers.

The up-side to multivendor networks is that equipment can more easily be chosen to fit the requirements and satisfy special needs. No one vendor makes everything for everybody.

Multiword DMA

Multiword DMA is an alternative data transfer technique to PIO (Programmed Input/Output. It boosts a disk drive's transfer rate by handling multiple words of data with one set of commands, and so is more efficient with large blocks of data.

— **DMA Mode 1** protocols are designed to transfer data between system memory and disk drives at rates in the region of 13MBytes/s. Fast-ATA devices can handle this rate by direct connection to the CPU bus.

— **DMA Mode 2** will increase these rates to 16.6MB/s. Fast-ATA2 can provide these rates.

See ATA, PIO and DMA.

MUMPS

Massachusetts (General Hospital) Utility Multi-Programming System. This is both an operating system (computer!) used by medical health care institutions, and a multi-purpose, multi-user (interpreted or compiled) high-level language. It also has strong database support features. IBM, DEC and ANSI have promoted the standard.

Murray code

See Baudot code.

MUSE

Multiple Sub-Nyquist Sampling Encoding system. A bandwidth compression technique which was the basis for NHK's analog HDTV broadcast system.

There are a number of versions of MUSE, with the current domestic system being MUSE-E. It has 1125 active lines with a 2:1 interlace on 60Hz, and a transmission bandwidth of 8.1MHz. However the true frame rate is only 15 frames per second — enough for static images, but requiring motion compensation for all moving elements. Bandwidth reduction was also achieved by halving the line-rate to sample only every second horizontal pixel of the production MUSE standard image. See quincunx.

MUSE was quite revolutionary in its approach — which also created problems since it didn't have evolutionary paths to aid development. The Japanese planned to run it in parallel with existing TV systems (using DBS) and so it required completely new production, transmission and reception equipment.

MUSE sets went into production and were sold under the HiVision identification — however they were too expensive, and later some of the Japanese manufacturers began to sell cheap cut-down MUSE also under the HiVision brand. See HiVision.

MUSE has now virtually been abandoned (but not officially). The Japanese, like everyone else, have now moved on to full digital TV projects.

music software

You can isolate five major categories of music software:

— **Sequencers,** which provide control over synthesizers, etc. See MIDI.

— **Editors/librarians**, provide cut-paste, insert and delete functions to a control sequence. More sophisticated ones allow graphic manipulation of the envelope, etc.

— **Notation/publishing,** allow manuscript printing from the devised sequence.

— **Pattern generators** and **Composition Aids**. By applying random pattern algorithms to thematic source material they can trigger new ideas.

— **Film Score utilities**, for sync locking a sequencer's timer to the audio or video recorder. They use time-code.

Musicam

Masking-pattern-adapted Universal Subband Integrated Coder And Multiplexer. The sub-band coding technique chosen for

Digital Audio Broadcasting (DAB) in Europe and also for MPEG. The alternative technique long considered was ASPEC, which had a slightly more efficient transform coding system, but other problems.

With Musicam, a CD-quality data-stream can be reduced to about 128kb/s with no noticeable degradation of subjective quality — even with classical music. The US Grand Alliance still intends to use Dolby AC.

must-carry

An FCC rule for US cable companies. It has been modified a few times, and subjected to attacks under the Constitutional right (of the cable companies) to 'free speech'. At present it is not legally functioning, but in practice the FCC still wields it over the head of cable operators.

The 1987 version of the rule says that a cable network must carry as 'Basic cable' the signal of any broadcast station within 50 miles of the cable head-end which has at least 2% of the audience. In effect, this means the major four national TV networks, and a few local independent syndicated stations. These stations get the benefit of this rule by having wider exposure to their advertisements. See A/B switch.

MVDS

Microwave (or Multichannel) Video Distribution System. A general term for terrestrial television transmissions in the microwave frequencies around 2GHz — but extending up to 40GHz. This is the European term for 'Wireless Cable' systems. See MMDS, MDS and LMDS.

MVP

A multimedia processing chip designed by Texas Instruments for Vtel and Sony. It operates at rates up to 50MHz, performing 2—3 billion processing steps a second — a rate roughly equal to about 1000 or 1500 DSP MIPS. This is a MIMD chip with four 64-bit DSPs, a 32-bit RISC CPU, a FPU, two video controllers, 64k of SRAM cache, and a DMA controller.

The chip can simultaneously encode and decode video using MPEG or H.261 (for videoconferencing), and handle audio as well.

MVS

Multiple Virtual Storage. The name given to the flagship IBM mainframe operating system. It is the mainframe equivalent of MS-DOS, and is based on OS/1: it is almost completely compatible with OS/360.

It is a multiple-user system which runs mainly on the IBM 360 mainframes. There are three current versions: MVS/SP, MVS/XA and MVS/ESA — and a new version with a higher degree of open system compliance is due for release at the time of writing. MVS uses the newer SNA and Network Job Entry protocols. See CMS and VM/SP also.

MW

Medium Wave. See MF.

MWI

Message Waiting Indicator. This is part of a voice mail system. The use of 'stutter tone' is common in some American public voice-mail services; when you next pick up your phone to dial you hear the distinct pattern of the dial-tone. Hospitals, hotels and private PBX networks will usually have a warning light on the phone as an indicator. The new ADSI protocols for Caller-Line ID also include message indicators. See SMDI.

mylar

A form of polyethylene used as a base for floppy disks and magnetic tapes. It is resistive to stretching.

N

N#-carrier

This is a hierarchy of standards for twisted-pair analog telephone carrier systems introduced in the early 1950s. The standards are designed for medium and short-haul applications.

— N1 carriers have a single voice-circuit capacity.

— N2 will carry 12 circuits using FDM.

— N3 and N4 will carry 24 circuits using FDM.

See group and supergroup.

N-AMPS

See page 450.

N-ISDN

See page 458.

N-MAC

See page 458.

N0X/N1X

An Americanism for the pattern of numbers permissible in area codes. Here N= any number between 2 and 9 (0 and 1 are used for access codes), and X= any number between 0 and 9.

Until 1989, all area codes were limited to the N0X/N1X form with a zero or one as the second digit, and they were getting in short supply. The pressure has been relieved by allowing area codes to be in the NXX form.

However this has created enormous problems with public and private exchanges which are coded to reject numbers other than 0 or 1 for the second area code digit — these switches 'believe' that any number in the NNX form (with second digit between 2 and 9) is a local exchange code, and refuse to recognize it as an area code. For this reason a prepended 1 is used in many areas of the country to identify the area code. See NXX and NNX also.

N-type

A type of semiconductor used in transistors. The N-type material has an excess of electrons. See transistor.

N-way

A Fast Ethernet standard which gives adaptors the ability to detect whether a switch, hub, or NIC is using normal 10Base-T or Fast 100Base-T Ethernet, and switch accordingly. It will always default to 10Base-T.

NA

- **Not Available.**
- **Numerical Aperture:** In optical fiber — a measure of the light acceptance of a fiber cable.

NAB

National Association of Broadcasters. (USA) This is a very powerful lobby group. In setting new standards, the cable equivalent is probably CableLabs. See ATSC also.

NAC

- Network Access Controllers. See
- Network Adapter Cards. See

NACK

Negative Acknowledgment, aka NAK. See NAK and ARQ also.

NACN

North American Cellular Network. This is McCaw Cellular One's transparent roaming and message-delivery network. Some other independent operators linked with Cellular One also own part of the NACN. It provides for a very rapid sign-on procedure when roaming with AMPS phones. The authentication system uses centralized tandem fault-tolerant computers, and the signaling is based on CSS#7.

NADC

North American Digital Cellular. This is probably the least confusing name for the IS-54 cellular phone standard also called Digital AMPS in much of the world or, in America itself, 'TDMA' (which is a generic term for the access method).

Digital-AMPS suggests that this is the digital evolutionary form of the AMPS cellular mobile system, but in fact it bears no relationship to the older technology, other than to use the same 30kHz channels in the same part of the spectrum, and the same control channels. (This was an FCC requirement.)

NADC is being defined for both 8kb/s and for 4kb/s vocoders. In the standard 8kb/s version each mobile uses two of six time slots in the 30kHz wide channels. This gives them a theoretical 3:1 increase in capacity through time-division, but in practice it is nothing like this. The later 'half-rate' version will use the 4kb/s coder and only one of the six slots for a theoretical 6:1 capacity gain.

Ericsson invented the TDMA technique used here, and this company is the main proponent. It has been less than wildly successful to date, although McCaw are pushing it in the USA. Some of the AMPS network providers who used Ericsson switches have little alternative than to stay with this system.

- The frequencies reserved for NADC are those of the AMPS network (824—849MHz and 869—894MHz). The duplex separation is 45MHz and the RF carrier spacing is 30kHz (the same as AMPS). There are three traffic channels per carrier, and the transmission rate is 48.6kb/s using $\pi/4$ shifted QPSK. It uses a 13kb/s VSLEP vocoder.
- Two other standards are competing with NADC; the Narrowband AMPS (see N-AMPS) which is an analog system which subdivides the channels and provides lower quality voice, and CDMA which takes a 1.25MHz swathe of AMPS spectrum for a high-quality, high-capacity digital system.
- Digital data is only possible on NADC with changes to the

control channels (implement in analog to match AMPS). The current revision IS-54C permits data. Also see GSM, TDMA and air interface.

nadir face

The side of a body-stabilized satellite which always faces Earth.

Nagle algorithms

Two algorithms which are used on TCP networks for congestion control. The first reduces the sending window, and the second limits small datagrams.

nailed-up channels

These are 'permanent' links although they may pass through a number of switches (PABXs or whatever). They are nailed up usually via entries made in the routing tables of those switches, to permanently maintain the connection.

NAK

ASCII No.5 — Negative ACK (Decimal 21, Hex $15, Control-U). A control character sent by a receiver as a negative response to the sender.

- In communications, it indicates that a packet was received with some form of corruption, and that it should be re-sent.
- In polling systems it indicates a negative response.

name resolution–server

The process of associating a name with a network address. A name server is provided for this function.

named pipes

An inter-process communications (IPC) technique (an API) used in Microsoft's version of OS/2 LAN Manager to provide a buffered bi-directional channel between two processes (not the machines themselves). In OS/2, named pipes are treated very much as you would files.

This is the way LAN Manager handles the setting up of sessions, and security checking. It may be considered a replacement for NetBIOS — but it is at a much higher level because a single named pipe function call equates to many NetBIOS calls.

Named pipe processes can send data to each other (from a parent process to a child process) by referring to the identity of the pipe — and these pipes are named according to the Uniform Naming Convention (UNC) format. They are created and destroyed by APIs, and are widely used in client-server links. See NetBIOS.

nameserv

On the Internet, this is a computer than handles Internet names and IP addresses.

naming

Addressing systematically. Naming is essential in any distributed system. It is the process by which objects on the network are identified, located and accessed. Host names and network addresses are both included here.

• **Naming Services:** There are a variety of naming services which help users find the location of network resources in much the same way as a phone directory. It is a naming service which allows you to access applications and printers using English language names. In file-server systems, naming services become very important.

• **The X.500 directory** is a 'naming' standard although it includes additional functions such as descriptive searching. Naming services are generally lower in level than X.500, but they are gradually migrating upwards.

• See named pipes and IP addresses.

N-AMPS

Narrowband AMPS. A digitally-enhanced FM-analog cellular phone technique invented by Motorola, which packs three analog channels into the 30kHz space of a carrier in the older AMPS band. It uses a digital control system and provides digital enhancement, but the transmission is analog. Handsets are dual-mode and very small (5.9 oz). This is the IS-88 standard.

Typically a few channels at one end of the AMPS spectrum band will be taken over for N-AMPS use. Most American cellular carriers see N-AMPS as a transition technology to lighten the load on their AMPS spectrum during a transition to CDMA. See AMPS for spectrum details.

Since N-AMPS is a relatively late design, the handsets are as small as the smallest AMPS portables, and offer about 30% more battery life. They also have automatic notification of voice-mail and use the digital control channel for alphanumeric messaging. Motorola claims reduced noise and less susceptibility to dropped calls also.

NAMTS

Possibly the first true cellular system which began in Tokyo in 1979. It had a capacity of 4000 subscribers, which was later raised to 8000. Like NMT450, it operated in the 400MHz band and used 180 frequencies (but no frequency reuse). See IMTS.

NAN

Neighborhood Area Node. In modern hybrid fiber-coaxial (HFC) networks this is the hub point where the fiber-trunks connect the coaxial feeder cables to each street. Usually a hub will service 500 homes (although some service up to 2000).

NAND

In Boolean logic and algebra, this means NOT AND.

• NAND gates are the primary building blocks of computers. They are electronic elements which obey the NAND operation of Boolean algebra. When its two input binary signals are different, a NAND-gate will produce a binary 1 output. When both input signals are 1, then the output will be 0. When both input signals are 0, the output will be 1.

nano

The SI prefix for a one thousand-millionth or 10^{-9}. It is always abbreviated as the small n.

• **Nanosecond** — one thousand-millionth of a second. In a nanosecond, light travels only 30cm, about 1 foot.

• **Nanometer** — one thousandth of a micron or micrometer. One billionth of a meter. The largest of the carbon molecules (a fullerene with 70 carbon atoms in a ball) is about one nanometer across, and this is about three times the average size of an atom.

NANP

North American Numbering Plan. Administrated by Bellcore. This plan covers World Zone 1, which is the USA (including Hawaii and Alaska), Canada and parts of the Caribbean. This numbering plan is a real mess for historical reasons. There are no distinct country codes, and area codes have no regional relationship or geographical significance. Different forms of access code need to be dialed to reach different areas of the country. See LATA, LEC and NPA.

NAPLPS

North American Presentation Level Protocol Syntax. Designed by AT&T and standardized by ANSI and also closely associated with the Canadian Telidon (videotex and teletext) technology.

This is a presentation system for video screens. It is an alpha-geometric standard which can handle a high level of graphic detail (display characters in a 2 x 3 pixel matrix) in a reasonably compressed form. It was primarily designed for videotex, but it had much wider applications. However the coding/decoding units are very complex. See alpha-mosaic, alpha-geometric.

narrowband

• In LANs, this means baseband (unmodulated).

• In telecommunications it means voiceband, although there has been a tendency recently to use it to mean low speed data, at rates below 200b/s (Telex mainly). See broadband.

narrowcasting

This term is very vague and appears to be progressively changing in meaning:

• Originally it was used to designate the range of possible services which could be made available via Cable TV, as distinct from terrestrial broadcast. Only those connected could receive the signals — not everyone.

• Later it was applied to specialized radio and TV transmissions intended only for a very narrow group. For instance, the transmission of medical TV to doctors, or of legal programs to lawyers. It is assumed that users will have specified their particular area of interest, and that they are equipped with special decoding devices to receive the narrowcast signals.

• More recently it has been applied to Pay TV services which don't aim to capture the mass market, but which specialize in 'narrower' markets like opera-lovers, foreign-film buffs, etc.

• Sometimes it is incorrectly applied to point-to-point interactive communications systems.

• Now it is sometimes being applied to Video-on-demand services, which are true point-to-point.

NAS

• **Network Applications Support:** This is another attempt to define a global blueprint for programming interfaces across distributed, multivendor environments. It is a related standard to Applications Integration Architecture (AIA). AIA sets the software framework standards that will allow interoperability and portability for applications, while NAS provides the toolkit and libraries.

• **Benchmark:** NAS is also used for a 'parallel benchmark' suite developed by NASA's Ames Lab, which works with many parallel computer architectures.

National Host

The USA National Host is a newly formed consortium consisting of MIT's Research Program on Communications Policy, Penn Uni's Distributed Systems Lab, the University of Illinois' National Center for Supercomputing Applications, and the National Media Lab. They are involved in global planning for the Global Information Infrastructure (GII).

National ISDN

An attempt by the COS (computer vendors) to force some much needed standardization on the ITU definition of ISDN in North America. There are now National ISDN-1 and -2, standards and -3 is coming. Europe has its own Euro-ISDN. See ISDN

native

In the original form. The mode for which it was designed.

• **Mode:** Without emulation. The CPU distinction is with processing operations which need to be backwardly compatible with previous members of the chip family, and which therefore need to emulate past modes. Native mode is without emulation, and is therefore the fastest and most flexible.

• **Code:** This is another name for machine binary code at the level in which a computer functions.

• **Storage:** Native storage statistics are those of uncompressed data storage.

natural language

The languages humans speak.

• **Programming:** The application of natural (usually English) words and phrases to computer instruction programming (don't confuse this with processing — below). This is part of 5GL and AI/expert-system research.

Natural language programming shares some similarities with processing, in that it concentrates on problems of the comprehension of human language. The main problems are categorized broadly as:

— syntax (sentence structure),
— semantics (meaning),
— phonetics (speech recognition and synthesis).

• **Processing:** Natural language processing is applied to written, spoken or scanned language after it has become normal text. It aims to create applications which can perform language translation, information retrieval, text summarization, and even auto-plagiarization applications.

There are two distinct approaches:

— syntax, which examines the mechanics of language (see parsing), and

— semantics, which examines meaning.

This is an ambitious approach which relies on grammatical rules and reference information about the world (see Cyc). It is looking for patterns, not for meanings, however. In text-summarization, for instance, a statistical examination of the file will find phrases that occur most frequently. The program then 'stems' the words (reducing them to roots), weights them, and finally reprocesses them to produce a summary. This involves natural-language creation in addition to the text analysis.

NAU

Network Addressable Unit. A general term for an SNA addressable entity (a 'logical port') which can send or receive traffic; they are logical mechanisms for providing open access in SNA.

There are three types: Logical units (LU), Physical Units (PU) and System Service Control Points (SSCP). An SNA node is a collection of NAUs which always contain at least one PU and one or more LUs, and, if it is involved in managing the network (as a router or gateway), it will also contain an SSCP.

LUs, PUs and SSCPs handle the middle layers in SNA protocols (Presentation Services, Data Flow Control, Transmission Control, and Path Control) which correspond to the Presentation, Session, Transport and Network layers of the OSI model.

NAUN

Nearest Active Upstream Neighbor. In Token Ring and 802.5 networks this is the closest upstream device still active.

navigation

• GPS is now being regularly used to provide navigation fixes for ships, aircraft and mobile land vehicles. Car navigation systems for city driving have already been announced. Aircraft systems include the concept called '4D-control' which means that automatic reports are generated and transmitted for latitude, longitude, altitude and time. Older systems were Omega and Loran. See Carin also.

• A hypertext term meaning the deliberate choice of paths through the hypertext maze to a goal. Hypertext navigation usually requires some form of module map or the user quickly gets lost in the complexity. See browse.

NBS

National Bureau of Standards. The NBS is now called NIST (National Institute of Standards and Technology).

• NBS/DES is the NBS's Data Encryption Standard. See DES.

NCC

Network Control Center. This term is widely used in satellite circles. Don't confuse it with gateways; you may have four or more gateways communicating with the satellite, but only one NCC. The NCC is not necessarily a TTC&M center: with LEOs you may have an NCC in each region, but only one TTC&M globally.

• In LANs the term NOC (Network Operations Center) is now preferred.

NCP

• **Network Control Program:** A program which resides in SNA communications controllers. It controls the flow of data between controller and network devices, and it routes data.

• **Network Control Processor:** A predecessor to TCP/IP (it grew out of Arpanet). This is a special IBM program for front-end processors. It works with VTAM to link the host computer and terminal controllers. It is now used in PPP.

• **Novell/NetWare Core Protocol.**

ND

Neutral Density. In film and video these are filters which reduce the light entering the lens, but do not modify the color.

— ND3 reduces the light by 50% (one-stop).

— ND6 reduces by two-stops (to 25% of original).

— ND9 reduces by three-stops (to 12% of original).

A standard polaroid filter also has a neutral density function plus a specific reduction of any cross-polarized light. The general reduction is about ND4 (1.3 stops), while the specific reflected light (polarized) reduction may be ten or more stops.

NDIS

Networks Driver Interface Specification. The generic device-driver specifications for adaptor cards used by LAN Manager and others. Microsoft devised this as a hardware- and protocol-independent device driver for NICs.

NDT

Net Data Throughput. This is a measure of the rate at which data is received at the end of a communications system. It is calculated by deducting from the total data received all overheads; allowing for call-connect times, retransmissions, start and stop bits, control and signaling, etc. In many radio systems the NDT will be less than half the claimed throughput.

near-end echo

Echo being returned to the source, usually by the impedance mismatch of local hybrids and the presence of old unterminated tails. The hybrids are usually in devices at the local exchange. Normally this echo has such a short delay that it is not noticeable in voice, but it can limit the data rate of modems. See far-end echo. Don't confuse this with NEXT.

near-field

See far-field.

near-line

This refers to CD-ROM jukeboxes or cassette silos which are not on-line, and take a few seconds to swing into action.

negation

In binary, negation means to reverse each digit in the numeric string. For instance, a binary number 0110 becomes 1001. You should distinguish this from complementation which calculates the actual negative of the number. Negation is also known as 'inversion', 'NOT operation' or 'Boolean complementation'.

negative feedback

Systems with negative feedback are generally highly stable: this is the primary method used for controlling systems of all kinds; Watt used negative feedback controllers on his steam engines.

Negative feedback means that the system reacts to changes in such a way as to reduce the effect of those changes. When they try to accelerate, it slows them; when they go too slow, it speeds them up; when they are too loud it softens them.

negentropy

The delivery of information. The structure of things. The opposite of entropy. I can't think of an uglier word!

Negroponte Switch

Nicholas Negroponte, the head of MIT's Media Labs once made the perceptive remark that broadcast services which had traditionally been radio-broadcast, were now swapping to cable delivery — while traditional cable services such as telephone and data, were swapping to radio delivery. This claimed 'switch' remark was widely perceived at the time to be deep, profound and predictive. In fact it turned out to be superficial and largely incorrect. We must be wary of over-simplifications.

— Cellular phone and PCS radio links don't replace wireline services, they just extend them.

— Cable only replaces radio for the delivery of broadcast entertainment when the radio spectrum is regulated and restricted, or unsuitable (in high rise areas). Direct-to-home satellites, could replace cable. And terrestrial SHF, MDS and the UHF/VHF with digital compression, could provide thousands of TV channels without billion dollar HFC expenditures.

— Data services are 99.9% cable, and show no signs of swapping to radio.

— Interactivity does not need an 'integrated' service. The return channel for a radio link can be wireline telephony. ATM over cable and telephone networks will probably make these distinctions academic.

neighborhood operations

• In image processing, point operations work with individual pixels, while neighborhood operations perform with interrelated groups of pixels.

• In routing, neighbourhood operators are those dealing with a nearby device on the network. See NAUN and Hello.

neper

A European measure of amplitude used with voltages and currents. It is similar to a decibel in that it registers ratios.

One dB = 8.686 Nepers. John Neper was the inventor of logarithms and probably of the slide rule.

NEST

Novell Embedded Systems Technology. This is a portable version of NetWare which can be used to link up a small office or home to control appliances, such as ovens, washing machines, faxes, PCs and telephones, etc. T

NEST is fighting it out with Microsoft's At Work as the software architecture for future offices. Microsoft has been in the lead since June 1993, and now has over 60 vendors supporting the concept. Microsoft has a problem with At Works, which requires the At Works OS, and then only on Intel processors.

NEST came late on the scene in Feb 1994, but it appears to be more flexible; it can work with a selection of microprocessors and real-time operating systems.

• Novell is designing the Portable Operating System Environment (POSE) interface with APIs for various OSs.

See SuperNOS.

nesting

A structured hierarchy which involves enclosing one structure within another like Russian dolls. For instance, (the use of the curved parentheses [in this example] allows us to nest the square-bracketed phrase within the phrase defined by the curved parentheses) which is, itself, nesting within the main phrase.

• Nested folders in a Macintosh are the equivalent of subdirectories in a hierarchy.

net/NET

• Any network.

• 'The Net' means the Internet.

• **Norme Europeane de Telecommunications:** These are mandatory standards for telecommunications in Europe. The distinction is with ETSs (ETSI standards) which are only voluntary. There are a few workable NETs:

— NET 1 is for X.21.

— NET 2 is for X.25.

— NET 3 and 5 are for ISDN user-interface type approvals.

— NET 7 is for ISDN terminal adaptors.

• **Network Entity Title:** The network address as defined by the ISO and used in CLNS.

NetBEUI

NetBIOS Extended User Interface. Microsoft's version of NetBIOS. It seems to have all the defects of NetBIOS, and it is used in Windows for Workgroups.

NetBIOS

Network Basic Input/Output System. The API originally offered by IBM and Microsoft with PC Network for IBM PCs and compatibles running MS-DOS. It allows peer-to-peer communications at the Transport and Session layers; which means in practice that it provides a set of commands that the application program can use to send and receive data across the network.

NetBIOS has since been replaced to a degree by the later APPC and APPC/PC, and recently by Named Pipes (OS/2).

Like NetWare, NetBIOS was an attempt to provide a common software interface for a variety of connected hardware. It governed data exchange and network access, but it lacked a Network layer and so couldn't be routed across the network. This made the creation of large internetworks based on this protocol very difficult.

NetBIOS operates at the Session layer, and it specifies the way to 'talk' to other computers regardless of the network hardware. There are implementations for Ethernet, TCP/IP and the OSI standards. It provides a programmable entry into the network, and supports communications by using a generic networking API that can run over multiple transports or media. See NetBEUI, named pipes and LAN Manager.

Netfind

A way of searching for Internet addresses. Usually you will Telnet to an Archie server and address it in the form:

```
Netfind pname cname
```

where pname is a single name for the person you wish to find, and cname is the name of the computer you wish to search

nethead

A networking fanatic.

netiquette

The etiquette of networking. Dos and Don'ts.

NetView

IBM's network management architecture and a suite of related applications. The correct name for this network management platform is NetView for AIX.

IBM began developing a Unix network management system in 1992, using (under licence) HP's OpenView as their base.

NetView now runs on large mainframes (it is host-centric) in an SNA environment, and it has evolved into a suite of management tools with several monitoring and control programs.

Many device suppliers are now porting their management applications to NetView. See network management, DME, SunNet, and OpenView also.

- NetView/PC interconnects NetView with Token Ring nets.

NetWare

Novell's NetWare operating systems (a family of them) are now the *de facto* standard for Ethernets. The original NetWare 2.x range was not highly successful. The real breakthrough came in 1989 with NetWare 386 which was later renamed NetWare 3.0.

This is Novell's suite of interfaces for applications: it is treated as a LAN protocol because it provides workstations, PCs, etc. with access to a network file-server. It intercepts the commands issued by the workstation, and decides whether they are to be handled locally, or passed on to the network. NetWare can operate on a very wide variety of LAN types (Ethernet, Token Ring, ARCnet and StarLAN), and it is very widely supported by other software companies. The NetWare environment can also use IPX/PSX, TCP/IP or NetBIOS, and it supports DOS, Windows, OS/2, Macintosh and Unix clients.

— NetWare version 2.2 is a 16-bit operating system, and it was relatively inflexible.

— NetWare versions 3.x are 32-bit systems. These versions still dominate networking.

— NetWare version 4.0 is also a 32-bit system, but it uses a 'domain' architecture. Most of the new features are to help developers. It has been moderately successful.

- **NetWare Loadable Modules** are applications which run on a NetWare service in parallel to the operating system. They are usually written in the C language (using standard C libraries and tools) and then compiled. NLMs provide superior performance to other applications which may be running at some distance away from the operating system core.

- Novell's AppWare and UnixWare are becoming of equal importance to NetWare, so the company says. See SuperNOS.

network

A general term which applies to Local Area Networks (LANs) , Wide Area Networks (WANs), Metropolitan Area Networks (MANs), Value-Added Networks (VANs), cooperating PABXs, the Internet, and the global telephone system.

In computer terminology, networking usually implies the use of a LAN, but these days the term will often mean a group of LANs (a WAN) which exchange messages through bridges and routers and requires a degree of management.

- In telecommunications, network often implies the whole global telecommunications system. However there are analog networks, and digital networks, Telex, voice and data networks. Most phone networks today are hybrid, with the core digital, and the local loop analog. ISDN is an attempt to integrate the local loop and digitize it, while IDN is the integration of the inter-exchange network.

- Distinguish between the segments and the network itself: LAN networks connected by a router are physically and logically separated, while LAN segments connected by a bridge are physically separated but logically the same, since they share the same network address. A repeater in a segment does nothing except extend its length by amplifying the signals.

- Private inter-networks (WANs) mostly consist of leased lines (from various carriers) which join together the offices of a company. To make these more cost-effective, the company will probably multiplex a large number of different channels (voice and data) down these lines, often using voice compression.

- Networks differ in many ways. See LAN topologies

Network Access Controllers

NACs are hardware boxes used in large packet-switching (and other) networks to provide a range of services to the user's links. They are concentrators for multiple terminals, and they provide PAD services, switching, protocol conversion, and some network management.

Network Adaptor Card

Aka NICs. The card you insert into your PC's expansion slot to enable it to communicate with the LAN. Not all computers need these cards since some have the hardware built-in. This card provides the transceiver functions.

network address

Confusingly, this term is used for two quite different addresses, even though a distinction must often be made between them.

- In a generic sense 'network address' can mean the unique address of a computer or node on the network — the device's individual MAC address or 'physical address'. These are used by bridges (but not by routers) to transfer packets.
- However when a distinction is being made between bridges and routers, 'network addresses' are specifically those used by routers (not bridges). This is the address of each LAN as a whole, and these are assigned to each LAN by the network administrator who follows a set of rules within the network operating system. By following these rules a router can then make decisions as to the best route a data-packet should follow to reach its final destination.

The 'network address' directs each packet on a WAN to the right network, and the MAC address then takes over and directs it to the right computer. See MAC address and SAP also.

network administration

The tools and systems that are used to manage the network. These consist of diagnostics, reporting methods, assignment of user-priorities, capacity management, management of MIBs and agents. See RMON, SNMP, and CMIP.

network-aware

Applications that offer file- and record-locking features to guarantee data integrity in a multi-user environment. They have the basic functions needed to run on a network.

network file system

See NFS.

network interface card

See NIC.

network-intrinsic

As distinct from network-aware applications, this is an application which will take substantially more advantage from being networked by fully utilizing the distributed resources. Client-server applications are of this type.

Network layer

The third layer in the OSI's 7-layer model — sometimes called the Packet layer.

This is the layer responsible for addressing and delivering messages — it is where control of switching and routing takes place. The ISO says that it is involved in 'routing and relaying [packets] through intermediate systems'. IP and IPX are examples of protocols operating at this level.

The Network layer defines network routing methods for the packets of information, and it provides high-level network protocols such as congestion and flow-control for maximum efficiency. It decides which physical path the packets should follow using the software switches within the network itself, and it establishes, maintains, and terminates connections, and manages the billing.

In smaller LANs, this layer has little value because there will be only one path between the nodes, but it is important in WANs and internetworks where it aims to guarantee end-to-end delivery of the packets.

Both connection and connectionless protocols have been specified; however the Network layer is usually specified as connectionless (using CLNS to reduce the error-handling overhead) to make it easier to restart a session after a failure.

The network layer has been subsequently subdivided into: Inter-network, Convergence, and Intra-network sub-layers.

network management

Networks need to be managed to:

- add and subtract nodes (users);
- find, isolate and fix faults;
- gather traffic information for future planning; and
- to establish security and priority conditions.

The ISO has defined five categories of the essential management functions. See CMIP and SNMP.

Recently there's been a shift away from proprietary device-level management applications to generalized network management platforms. These are integrated frameworks, able to gather data from a wide variety of devices and present it in a standard GUI form on a centralized console. Agents, probes, and MIBs are the key here. See also RMON.

Many of the new network management systems are attempts to automate the administration of large, heterogenous networks. The major platform developers are HP with OpenView, Sun with SunNet Manager and IBM with NetView for AIX.

network operating system

See NOS.

network relay

A form of gateway. It operates at Layer 4 and above in the OSI model, and interconnects dissimilar networks. See DSG.

network termination

See NT#.

network termination point

The point at which the responsibility of the general telecommunications carrier ends, and that of the private citizen or company begins — also called the 'demarc'. This is usually at a plug or point within the building, but in some cases it is being interpreted to be at the land-boundary — leaving the customer with the responsibility of bringing the connection into the premises.

Ideally the point of termination for a domestic line should be in an RJ-11 socket, and all further extension lines should run out of here using an RJ-11 plug for connection. If problems are then ever experienced, it is a simple matter to test which side of the demarc the problem exists, by simply plugging in a phone, known to work.

NETZ C-450

An early cellular phone standard used in Germany. It was analog in the 450MHz band. The handheld phones cost between $US14,000 and $US16,000 — and they wondered why the system wasn't popular!

neural networks

Hardware/software combinations of various kinds which are set up, supposedly, to 'imitate the biological structure of the brain in processing information — that is, through the use of interconnected neurons'. These ideas have been developed only since about 1982, and they are based on theories of 'connectionism'.

A software neural network acquires its knowledge through training, and it is this ability to learn from example that distinguishes the neural network from an expert system. Many people suspect that neural networks may be the way of the future for all computing, but currently the potential appears mainly limited to pattern recognition — both spacial and temporal.

There are, however, numerous practical applications. The sale of neural network products is already at the US$500 million level and will double by 1997.

These aren't called neural 'computers', because most of the neural-connectionist functions are being simulated on traditional computers. The 'virtual neural' software has outstripped the hardware — but it is the use of conventional hardware that limits these systems.

Unfortunately, neurons with their masses of connections, take up a lot of chip real-estate. With just a few hundred neurons, you've got a chip the size of a Pentium.

Neural networks work on layers; at the first each 'neuron' has a single input path and many paths leading out, one is connected to every middle layer neuron. The output from these have connections to all neurons in the next layer, and so on.

At the final layer, each neuron has only one output but, as you can see, the problem is in making all these connections. The learning process is one of varying the voltages across the paths — changing the 'strength' of the connections — and this is a cumulative process.

A human has 100 billion neurons, and even a cockroach has 100,000, so currently computer neural networks are still well behind the cockroach, and very limited in scope. They tend to be application-specific and applied only to simple recognition tasks.

Neural networks, fuzzy logic and chaos theory seem to be related in the development of pattern recognition. There's also a relationship with the control of engineering tolerance imprecision, and prediction of fluctuating time series. See ANN.

New Sync

A later addition in hardware control for communicating with direct-connect devices, available when using an existing protocol like RS-232. It often doesn't figure in the specification because it was added later. However, 'New Sync' is provided by some DTE devices on Pin 14. Some multipoint master modems require this control circuit to be in place.

Newton

Apple's name for a new RISC-based 'personal digital assistant' (PDA) computer architecture which emphasizes recognition technologies (printing, then cursive writing; later voice) for data input. It was released too early and proved to be a bit of a disaster — but it obviously has enormous potential.

The Newton is based on the ARM chip but it may need a PowerPC chip to work at the speed necessary to handle cursive input. It takes a modular approach to file storage and used IR for direct communications with a Mac or PC.

NEXT

- A computer company started by Steve Jobs, ex-owner of Apple. They make high-end workstations and software.

- Near-end Cross Talk. This is the normal type of 'crosstalk' experienced by people using telephones — the transfer of the signal down one pair of wires to another pair of wires in close proximity. See FEXT also.

In data systems, NEXT is the main limiting factor in UTP performance. With twisted pair, the tighter the twist the better the cancellation of NEXT, and also the higher the data rate which can be supported. However in bundles of twisted pair (telephone cables in the streets) the twisting and positioning of all pairs within the cable must be carefully calculated and controlled. Next is a complex phenomena with both frequency and phase characteristics, but generally, the higher the frequency, the more the crosstalk.

- Another definition says that NEXT is 'the crosstalk from the transmit channel getting into the receive channel' which seems to confuse it with feedback or sidetone.

NFS

Network File System. Sun Microsystem's distributed file- and peripheral-sharing protocols which have become the *de facto* networking standard. It's been licensed to most workstation manufacturers.

NFS refers to both a suite of protocols, and one specific protocol in the suite. The suite includes XDR (External Data Representation), RPC (Remote Procedure Call) and others.

NFS is based on IP and it has become widely used for distributed file and print-sharing on TCP/IP networks. The software is available for many computers and operating systems (there are more than 90 commercial implementations) and it is found on virtually every Unix network — you'd say it is the Unix equivalent of Novell NetWare.

It uses the upper levels of the OSI model for remote file operations, and it is both machine and operating system independent. See RFS.

NI-#

National ISDN. (US) They are up to their third version of the national standard so you'll get NI-1, NI-2 and NI-3. See N-ISDN and ISDN.

ni-cad

Nickel cadmium accumulators (batteries) are the dominant form of power supply for portables because they are relatively light in weight for their capacity. However it is almost impossible to monitor them to read the state of charge, and battery power can fall quickly past a certain discharge level which makes them less than ideal.

Ni-cads also suffer from a memory problem, which means that they must be fully discharged by shorting across the contacts fairly regularly — preferably every charge. See memory, lead-acid batteries and lithium ion.

• If ni-cad are charged in parallel or series (not individually) they can develop voltage reversal on the weaker cells.

• Modern ni-cad batteries have a very efficient overcharge mechanism and can be left on trickle charge indefinitely. This is not true with the rapid rechargers, however.

NI-DPCM

Near-Instantaneously (companded) Differential PCM. A audio coding system used in the MUSE HDTV standard. See MUSE and HiVision.

NIB

Non-Interference Basis. This is a term used in radio spectrum discussions. Originally, various types of equipment had exclusive use of parts of the spectrum, but eventually the spectrum began to get overloaded. The British then made a pragmatic suggestion that often two services could co-exist in the same band and not interfere with each other by virtue of the way in which they function. For instance, pocket-portable phones (with very low power and small cells) could use the same frequencies as broadcast-radio directional links which had high-power, but used dish reflectors.

Therefore a secondary service can be licensed on a Non-Interference Basis — with the assumption that it will be de-licensed if it does interfere with the primary service.

nibble

A half a byte — four bits. An alternate spelling for nybble.

NIC

• **Network Information Center:** Where you find address information, on a large network. On the Internet, the NIC provides user assistance, training, access to RFC documents, and other information.

• **Network Interface Controller:** This seems to be the same as below.

• **Network Interface Card:** The plug-in card used to physically link a workstation, server, bridge or router to the LAN's media: it is the transceiver. The NIC supplies both the hardware and the software needed to communicate over networks.

Most NICs can transmit data at a rate which is fast enough to saturate the LAN, and they can also receive at a rate that ensures that all packets at the network addressed to their workstation, will be received.

This is essential because shared media provides the full bandwidth, but only for short periods. The fundamental problem in NIC design, then, is in making low-cost devices which can work effectively at these high rates.

There are three main architectures:

— **DMA Master:** Low-cost NICs use system memory for storage rather than having a large buffer in the card. The NIC then needs a fast path to offload the data to the device. Direct Memory Access must be prioritized in the PC so packets aren't lost.

— **Programmed I/O:** This more expensive approach relies on large memory buffers on the NICs, but fast memory is expensive. These are store-forward systems between the PC and the NIC. This approach has the CPU copying the network data by using normal I/O.

— **Shared memory:** This approach also relies on large memory buffers on the NICs, and so it is also a store-forward systems between the PC and the NIC. This approach has the CPU copying the network data by using memory instructions.

NICAM

Near-Instantaneous Companded Audio Multiplex. A form of digital 'CD-quality' sound broadcasting for PAL television, pioneered by ITV and BBC in the UK in the mid-1980s. It is now finding application terrestrially and in Direct Broadcasting Satellites (DBS) and Direct Audio Broadcasting (DAB).

NICAM uses a 728kb/s signal which is scrambled then modulated (using DQPSK) on to a carrier spaced 6.552MHz (9 times the bit-rate) above the vision carrier. The digital stereo signal supplements the traditional FM mono analog sound signal. It provides a usable bit-rate roughly only half that of conventional CD audio (using 10-bit samples), but the DQPSK modulation compensates for this reduced bandwidth.

• A Scandinavian version has the audio carrier 5.85MHz above the video. See DAB and COFDM also.

nickel-cadmium

See ni-cad above.

NII

National Information Infrastructure. A proposal put forward by the Clinton Administration, and championed by Al Gore. It is setting an agenda which includes research into high-speed links (NREN), but more importantly it looks at the more general provision of information to the public. See universal service and CSO.

• The national Interagency Infrastructure Task Force (IITF) is the US Federal Government's agency in charge. It has a Technology Policy Working Group (TPWG).

• NIIAC (NII Advisory Council). Industry representatives from both private and public sectors, who advise the IITF.

• The NII 2000 Steering Committee is a group of high-level executives and academics who are working with the TPWG to develop a baseline understanding of what technologies are to be deployed when, where, and by whom.

NiMH

Nickel-metal-hydride. A new form of battery (accumulator) used with portables and notebooks. This technology has not been a success; the lithium-ion batteries appear to be better.

Nine-track

Tape recorder: See 9-track in numbers section.

NIS

Network Information Services (of Unix). A global naming support system developed by Sunsoft and used by the Unix International group.

N-ISDN

• **Narrowband ISDN** — which is our present ISDN service with the B-channels supplying 64k/s data rates. The distinction is being made here with B-ISDN. See ISDN.

• The acronym has also been usurped by Bellcore for the North American **National-ISDN.** They have made three attempts (N-ISDN 1 and 2, with version 3 in process) to standardize ISDN. See ISDN and IDA also.

NISO

National Information Standards Organization. A US organization which establishes standards for libraries, publishing and the information sciences.

NIST

National Institute of Standards and Technology. (US) This was previously the NBS (National Bureau of Standards).

NIU

Network Interface Unit. See NIC.

NIU-F

North American ISDN User's Forum.

NLM

NetWare Loadable Module. Small applications. See NetWare.

NLQ

Near Letter Quality. A marketing term applied to dot matrix (and other) printers which are marginally above average in print quality. Another way of saying this, would be 'Less Than Letter Quality', but this doesn't seem to find favor for some reason! See dot-matrix printers.

NLSP

NetWare Link Services Protocol. Novell's version of OSPF.

N-MAC

Narrow MAC. A new form of Multiple Analog Component TV (a variation on D2-MAC) developed by the British Telecom Research Laboratories. It boosts channel capacity and reduces the antenna size.

NMAC has a 12MHz mode and an 18MHz mode. With 12MHz bandwidths, up to six signals can be transmitted on one 72MHz satellite transponder and each of these channel feeds can be uplinked from a separate site (see SCPC). Up to four stereo audio channels can be carried. Alternately the narrowband channels can carry encryption or data services at 64kb/s rates. See MAC, D-MAC and TMTV also.

NMC

Network Management Center. See NMS also.

NMOS

An n-type Metal Oxide Semiconductor. The distinction is between p-types and n-types. NMOS transistors are fast, dissipate little power, and can be crammed together to produce high-density memory chips. This type of IC can operate from a single 5-volt power supply, although the early n-type devices needed several supply voltages. A high-speed version of NMOS is called HMOS. See MOS.

NMS

• **Network Management Systems:** The wide term for all the network control, fault-finding, traffic-flow management, and billing functions, etc. All of the functions that enable, support, or enhance the communications network. See DME.

• **Network Management Station:** A large and powerful workstation which is responsible for communicating with agents in remote devices, and of holding the network databases on errors, traffic flow and resources. This refers to the main console, or to one which is only managing part of a network.

NMT

Nordic Mobile Telephone system. The Scandinavian cooperative approach to analog cellular mobile systems which offered the first (though limited) international roaming.

— **NMT-450** was the first in Europe (1981) and was also sold into the Middle and Far East. It provided a variable number of channels (around 200), and was designed for vehicles only.

The maximum capacity in a city was said to be about 30,000 users (180 channels) with 25kHz carrier spacings, although 20kHz spacings (225 channels) were also used.

The main NMT-450 frequencies are 453—457.5MHz, although other frequencies are used in some areas.

NMT-470 is a popular variation used in some countries at a slightly higher frequency.

— **NMT-900** was the second generation system (1987), originally called CMS 8900. It is currently popular in Scandinavia and operates in the 890—960MHz band which was specified by the council of European PTTs (actually 872—905MHz).

NMT-900 has both 25kHz and 12.5kHz versions which provide either 1000 or 2000 channels in the available bandwidth. The duplex separation is 45MHz.

• NMT-450 and NMT-900 systems still operate in Europe.

See AMPS, TACS, and NTT.

NMVT

Network Management Vector Transport. IBM's connection-oriented network management protocol for SNA.

NNI

• **Network to Network Interface:** The term is mostly used these days for the specification of devices attaching low-speed network segments to high-speed networks. You may have an Ethernet store-forward switch which takes 10Mb/s of Ethernet on one side (the UNI) and outputs 100Mb/s FDDI or ATM on the WAN side (the NNI). See UNI.

• **Network Node Interface:** This is probably just a variation on the above.

nntp/NNTP

Network News Transfer Protocol of the Internet. This is used by the Internet to distribute newsgroup information and discussions (most Usenets are distributed this way today).

It operates in a client/server fashion usually with an NNTP server in an organization. These are often just old PC AT's with hard disk storage. The client, called a 'newsreader', talks to a news-server over TCP/IP using the NNTP protocols. See UUCP also.

NNX

In the USA, the use of certain telephone number-combinations were prohibited in the numbering plan. This is now changing. NNX refers to the old system where the letters represent the first, second and third digit of a 7-digit local telephone number. The Ns could only be a number between 2 and 9 (never 0 or 1), while the X could be any number from 0 to 9. See NXX.

The point is that, in the 10-digit version of these numbers (used outside the local area) there was a three-digit exchange prefix. To identify this as being exchange-related, the second digit was always a 0 or a 1 under the old system. So these combinations were prohibited in NNX local numbers to avoid confusion. Now they avoid confusion by prefixing all long distance dialing with a 1. See NXX.

NOC

Network Operations Center. See NCC.

nodal processor

This is like a fancy channel bank with software configurability, multi-T1 networking, etc.

node

An autonomous unit within a communications network where two or more links meet. This just translates to mean an active device of any kind which is attached to a network, through which data passes. Switches, routers, workstations, etc.

• In telecommunications the word node seems to imply a point in the circuit where switching or a signal ingress or egress takes place. Usually the node is just the telephone exchange — but it could also be a remote multiplexer in an outdoor cabinet which acts to concentrate local signals.

• In LAN terminology a node is any point of attachment of a station or a server. The term 'end-node' seems to be synonymous with 'terminal' (as in DTE). Any active device attached to a LAN is called a node. However the NIC (transceiver) is probably correctly the 'node', not the workstation or PC, but the term is flexible.

• The term is also applied to packet-switching systems, and here the nodes are usually full-blown computers or routers capable of storing packets and redirecting them. So a node here is a major switching or control point. On a dial-up circuit to an X.25 network, you deal directly with a primary node called a PAD. See.

• In SNA, node is a vague term for Physical Unit. The term 'embodies the rules that govern physical attachment to the network'.

node-to-network protocols

In this type of network the access-path is managed one step at a time. The management is from one node to the next adjacent node, which is in contrast to end-to-end packet protocols.

noise

Any (usually electrical) disturbance which affects the characteristics of a signal. Originally the term applied to audio signals, but now it applies to all audio, video, and data.

• In the audio sense, noise was any unwanted signal component which wasn't related to the original signal, and which tended to obscure the original. In audio hi-fi systems there are two main types:

— random noise which produces a hiss (see Gaussian),

— powerline frequency (harmonics) which produce a hum.

• An electronic signal picks up noise every time it passes through an electric circuit. This is manifested as 'static' on telephone lines, 'snow' on television images, and there's even optical noise on optical fiber systems.

Technicians identify a number of different types of noise:

— White noise. Broad spectrum 'hiss' or Gaussian noise.

— Impulse noise.

— Intermodulation noise. This is found in multiplexed channels through intrusion of artifacts from another channel.

— Quantization distortion noise. In PCM systems.

noise-floor

All electrical circuits generate noise. A cable network or electrical circuit has a lower noise limit (the noise-floor) which is set by the ambient temperature, the bandwidth and the impedance of the system.

This noise-floor is measured in terms of dBmV and it is calculated by comparing it with the noise generated by the electrons within a straight piece of copper wire (which is assumed to be the lowest possible noise generate in the same circumstances).

- The noise-floor of the average coaxial cable (as used in the US for Cable TV) is about –59dBmV.
- In PAL cable systems (with a wider bandwidth) the noise-floor is about –57dBmV.

Cable amplifiers generate noise above the ideal minimum (which would be the noise floor of their cable). So an NTSC coaxial amplifier with a noise figure of 10dB would have a noise-floor of 10 minus 59, or –49dBmV.

noise words
See stopwords.

nomadic computing
The new term for laptop and portable users who need to have full access to company information resources via radio links, cellular phones, etc. This used to be called portable computing — but nomadic suggests people who rarely go into the office.

nominal voltage
For AC current, see mains power voltage and RMS.

nominal voice circuit
This usually means the number of circuits that a cable would carry if voice was being carried at the old PCM rate of 64kb/s.

It also suggests that the voice is actually being carried at a much lower rate (probably ADPCM at 32kb/s or 16kb/s) and that DCME digital interpolation is probably being used.

Carriers only use the term 'nominal' when this is happening, and they don't want to reveal the actual capacity figures.

non-blocking
See blocking.

non-deterministic
Unable to control access, and therefore unable to accurately predict the performance of any one point-to-point transfer of data. This is applied to network designs where collisions or conflicts (and therefore delays) are possible.

CSMA systems (Ethernet) are non-deterministic because there is no central control which allocates 'turns', while polling systems and token systems are deterministic.

non-heuristic
In OCR — see heuristic.

non-interlace
See progressive scanning.

non-isochronous traffic
In telephony, this generally means packet-switched data. They are making an old distinction that really doesn't apply any more; that only circuit-switched systems can carry real-time voice and video — that packet-systems are too slow and erratic. This is no longer the case. See frame relay and ATM.

non-linear
Not evenly distributed over the spectrum, or, not in a direct line of progression.

- **Distortion:** This is mainly harmonic distortion (see). But it is also the modification of a signal due to different frequencies suffering variable attenuation through the communications link. Frequency division multiplexers have problems of this kind, and television signals require special pre-emphasis to overcome these problems. Non-linear distortion can be overcome to a large degree at the receiver end by equalization.
- **Video Post-production:** Non-linear editing in video is the use of the new PC-based editing systems with the shots stored on disk. Scenes can be 'cut-and-pasted' into a video sequence (in fact, only the identity and location is being stored), so changes can be made very easily.

These systems depend on modern computers being able to dodge around in random, and buffer video streams for long enough to give the appearance of there being a continuous sequential stream of shots. But they are not actually editing these shots in the direct lineage between the camera and the transmission — only marking them for replay.

Non-linear editors either use no compression (and cost a fortune) or use high compression (often Motion JPEG) and sacrifice some quality. Don't confuse this with on- or off-line editing — non-linear editing can be either. It is on-line if the final output is used as the master copy, and off-line if it is only used as a guide for an on-line editing machine (using time-code correspondence).

non-preemptive multitasking
See multitasking

non-print (code or character)
A code or character sent as part of the message in communications, which does not result in a visible character-image on the screen. However a space is classed as a 'print' character.

So non-print generally means one of the ASCII set (the first 32) involved in control signals from NULL ($00) to US ($1F).

non-procedural languages
Fourth and fifth generation languages: those languages which are event-driven (modular or object-oriented) rather than straight-line procedural. Non-procedural languages send messages between the modules to initiate actions. See object oriented language and functional language also.

Non-Return to Zero/Inverted
See NRZ and NRZI.

non-synchronous

- See asynchronous
- Demodulation: In a coherent optical fiber system using heterodyne detection, non-synchronous demodulation is used to extract the data signal without requiring an electrical carrier at the intermediate frequency.

non-volatile

• **Memory:** There's both a general and a specific use of the term.

The general use is any form of chip device where the memory is retained even when the power is switched off. The specific use is usually referred to as NVRAM.

There are three different concepts here:

— **DRAM and SRAM:** From the user's point of view DRAM loses its memory instantly and is classed as volatile. SRAM requires less refreshing and so is classed as semi-volatile, but it still loses contents when the current is off, so it is actually volatile.

To be non-volatile 'computer-memory', the memory must retain data when the computer power is off. Current Non-Volatile RAM (NVRAM) has its own power supply, usually a lithium battery, and it will often use SRAM so it doesn't need to be refreshed constantly.

— **FRAM:** At the 'chip-memory' level, the view is different. Some new forms of SRAM are now being made using a FRAM overlay, and FRAM is definitely non-volatile since it holds information indefinitely without a battery.

FRAM has a built-in mechanism which creates a magnetic polarity change above each cell if power is lost, so later the computer can be switched on and the working memory is restored exactly as it was when the power failed.

— **Flash:** Flash memory is also non-volatile, except that it must be changed in blocks and can't therefore be used as working memory. It is used as secondary storage only. Disks are secondary-storage devices also but you normally wouldn't stretch the term 'non-volatile memory' this far.

— **Other:** The term non-volatile can also be used to mean all the different forms of ROM as distinct from RAM — it depends on the context. This covers PROM, EPROM and EEPROM.

• **Non-Volatile Storage**. See NVS.

• **Non-Volatile Log:** In network management, information from agents in the peripheral devices will generally be transmitted over the network (using SNMP) to a non-volatile log (on disk usually, but it could be in flash memory). If the network fails, then the information about the cause of failure isn't lost.

non-von computers

Computers which do not have a von Neumann architecture. This usually means parallel processing or neural networks.

NOR

NOT-OR in Boolean Algebra.

nor-gate (circuit)

NOR is a logical operator in Boolean Algebra, and a nor-gate is an electronic unit which follows the NOR rules. It will produce an output digit only when the two input signals are absent at the time of the clock pulse. NOR means 'NOT-OR' and is the negation of OR.

normalization

To re-establish the normal or 'unstressed' conditions.

• In large database systems, normalization is the process of breaking down data structures to their most rudimentary level and storing the components on separate tables — thereby creating a relational database. The aim is to reduce data duplication.

Norton

• **Utilities:** A series of excellent disk-recovery, optimization, and general utility programs for IBM-compatibles (and later Macs) written by Peter Norton.

• **SI rating:** Peter Norton also created a 'System Index' which is measured by a special program. It gives an index of PC performance relative to the original IBM PC. An SI rating of 5.0 means that, on average, the system runs five times the speed of the original IBM PC.

NOS

Network Operating System. System software that is the logical side of the network and controls who can use it, how, and when. There are two fundamental types: network operating systems can be server-based or peer-to-peer. In the latter case, they use every node as a *de facto* server (see TOPS). The term NOS is generally reserved for the server-based forms suited to large networks.

NOSs provide the core functionality needed to maintain the most basic network operations. There are three main components:

— Server operating systems.

— Client-server applications.

— Workstation connectivity software.

These all work in combination to provide the network's communications services. Early NOSs ran on file-servers and basically just governed access to the files and other resources on a single server. Now NOSs run on database- or network-servers and provide network file systems, memory management and the scheduling of processing tasks across very wide networks.

The reigning champion of NOSs at present is Novell's NetWare with LAN Manager (generally with OS/2), NFS and Banyan VINES well behind. The others of importance include AppleTalk, Windows NT and SCO Unix.

NOT

A specific term used in keyword selection systems and in operations involving Booleans. It is also a database retrieval command modifier. You would use NOT to remove some items from consideration; for instance in selecting records containing 'A NOT B' you specify that records containing A should be selected only when B is not present. The distinction is with pure Boolean AND and OR operators.

• NOT-AND. This is also known as NAND

• NOT-OR. This is also known as NOR.

• A NOT operation is a negation. See.

notation

A number can be written in different ways depending upon the base you choose. The number 137 uses a 10-base notation which consists of ten numerical characters with the least-significant to the right. Decimal notation is denary with fractions.

The Roman form of the decimal seven, is VII, which uses a base of 5 and complex rules as to the placement of characters. And 0000 0111 is binary notation using the base of 2.

Computers use binary notation at the machine level. But in order to make it easier for programmers, some operations are conducted using octal notation (base 8), and more often hexadecimal or 'hex' (base 16).

In all of the higher-level languages however, the data is usually presented to the user in his/her 'native' form which is denary/decimal (base 10). See hex.

notational

Simply any standard way of recording information. It can mean written text or a music score.

• When applied to numbers, it means the standard way of counting using the base 10.

notch filter

The opposite of a band-pass filter. A notch filter acts to block only a very narrow range of frequencies — and they are sometimes used to remove specific harmonics. Notch filters are needed in European modems because a 12kHz tone is used for metering (billing). With the old in-band signaling systems they were needed to block the 2700Hz tones from being heard by the callers.

notebooks

Small portable computers of the A4-page size with a keyboard which is just large enough for a touch-typist. The first were the 286 notebooks introduced in 1989. The form-fact was 8.5 x 11 x 2 inches.

High-end notebooks are now down in the 7-pound and less range, and notebook sales represent about 25% of all portable sales. Notebook screens are usually 10 inch active-matrix LCD (color and mono) displays, and the systems are powered by the new 3.3V Pentium chips which allows more battery life and lighter batteries.

In the new notebooks you can have a choice of many new features including: color screens, PCMCIA cards, data/fax modems, 800MByte hard drives, integrated CD-ROM drives, infrared I/O, sound and video I/O for demonstrations. Many now have removable components, so they are flexible. Screens can usually be ordered as active-matrix or dual-scan passive-matrix; in color or mono.

• Dockable notebooks are coming back into fashion. In the office the notebook 'docks' into a unit which has a full-size keyboard and screen, mains-power supply, mouse, and subsidiary floppy and hard disk drives.

• See portable computers.

NPA

• **Network Printing Alliance:** Both a group of vendors and a standard for a bi-directional high-speed parallel printer port which will be backward compatible with existing parallel printer ports. It will be independent of any operating system or print-data stream. The IEEE has designated this as the P1284 standard and is working on draft specifications.

• **Number Plan Area:** In American telephone numbering terminology this is an area code under the new NANP. The Los Angeles metropolitan area has six NPAs (213, 310, 714, 805, 818, and 909), some only a block or two from each other; you virtually need to include the area code to dial a premises on the other side of the street.

Sometimes you'll need to add a 1 before the 3-digit NPA area code, and sometimes you won't. Life wasn't meant to be easy!

— When they run out of numbers in an area code, they can either use 'overlay codes' (extra codes covering the same area) or perform code-splits where half the premises get a new number.

NPT

Non-Programmable Terminals. Dumb terminals and very cheap mainframe workstations.

NREN

National Research and Education Network. NREN was originally conceived as a university and government tool for scientists, but now it is being viewed as a public utility to which commercial interests will have access (for such services as video conferencing). However, it must first include an accounting mechanism to allow charging for copyrighted material and the use of the network.

The US Congress passed legislation in Dec 1991 to establish this as a national high-speed computer network, and as the US Internet backbone.

The Bill originally provided NREN with $US3 billion over a period of five years. As envisaged, this new network will have a backbone bandwidth of over 1Gb/s, and feature large-scale parallel computers operating at 1000 GFLOPS. It will also have nationally distributed libraries of software and data.

The main immediate aim is to link government laboratories, universities and research centers across the US by the end of 1996. The National Science Foundation (NFS) is funding the test beds.

• The main argument is about the role the present common carriers (phone companies) will play in all this, and also how the NREN backbone fits into the National Information Infrastructure plans for data superhighways.

NRM

Normal Response Mode. In this mode, HDLC secondaries can only transmit over the link if they first receive a poll from the primary.

NRZ (NRZ-L)

Non-Return to Zero (level); aka NRZ-L. The most common way of transmitting data, where a positive voltage represents one value (logical 0), and a negative represents the other (logical 1), so zero voltage is never used. This is bipolar communications, used to overcome the problems of DC -coupling (see).

This is a very popular method of line coding used between computers and their peripherals. It is also directly related to the way of storing data on a magnetic surface — creating north pole magnetic domains, say, for a 1, and south pole domains for a 0. See NRZI also.

- A similar system is used on CD disks. See pits and lands.

NRZ-I

Non-Return to Zero — Inverted. A technique of 'differential' encoding used in SDLC protocols. A 0-bit is identified by a change of state on the line, while the 1-bit is a steady state with no change. So the data is encoded by the presence or absence of a transition — not by the voltage states themselves. This is more reliable in the presence of line noise than basic NRZ (level) techniques.

NRZ and NRZI are both usually associated with a higher-level protocol which prevents a long string of bits from being grouped in such a way as to lose synchronization. See AMI, Manchester coding and HDLC also.

NSAP address

Network Service Access Point. The network address format which was first defined by the ITU in 1984, and which is now ISO 8348/Ad2. These addresses are for computer/router use and have a maximum length of 40 decimal digits or 20 bytes.

Each node/station/terminal on a network must have an NSAP (usually a standard MAC address) and this must be unique. However a node can have more than one NSAP.

- When the NSAP for the device is combined with the address of the network, we have a global network address. The router uses the network address to shunt the packets to the correct network, and the MAC address is then identified by bridges and devices on the network.

NSF

- **Non Standard Facilities:** General. But in fax machines these are proprietary protocols.

- **National Science Foundation:** Their network was, until recently, the backbone of the Internet. See NREN.

NSF/CSI/DIS

In the opening (handshaking) sequence of our present range of G3 fax machines, this is the sequence which follows the introductory Carrier Detect 2100Hz signal. It stands for three primary parameters which need to be agreed and exchanged:

— Non-Standard Facilities,
— Call Subscriber Identification, and
— Digital Identification Signal.

NSU

- **Non-Switch Units:** Another name for key systems or SBS (small business system) telephones. The distinction is with small PBXs.

- **Network Service Units.**

NT-#

Network Termination type #. This is usually the equipment which provides the 'interface point' between the carrier's network and the customer premises equipment (CPE) and cabling system. However the term can also apply to private networks.

- **In ISDN:**

— **NT1:** The NT1 box terminates the main transmission line from the exchange (it is often the demarc) at the physical level only. The main job of NT1 is to convert the four-wire T-interface into a two-wire U-interface extending back to the exchange. It multiplexes the CPE's individual 2B + D connections into a single bit-multiplexed channel using TDM across the local loop.

With BRA, the NT1 may also need to handle contention resolution within the premises when a number of terminals attempt to seize the line. In the USA the basic small-business or residential NT1 unit (costing about $200) must be supplied by the customer; in most other countries it is supplied by the carrier.

— **NT2:** This is a far more intelligent device than NT1 (it is a switch or multiplexer), and it performs switching and concentration functions within the customer's premises. A digital PBX, terminal controller or LAN can all provide these NT2 functions.

— **NT1/2:** Both of the above functions can be in a single box. In many countries the carrier will own the NT1/2 and thus provide the full ISDN and PBX-like services.

- The T-interface exists between NT1 and NT2. See TE# also.

NTFS

Windows NT's file system.

NTIS

National Technical Information Service of the US government. This is an amazing library of information on technology which can be accessed through most international databases utility services.

NTSC

National Television Standards Committee. The television system used in the Americas, by Japan, and by ex-US colonies like the Philippines. The format uses 525 lines and is usually associated with 60Hz mains power.

The original video standard was called RS-170 in the 1940s and it was designed with the idea of saving valuable bandwidth. It was essential that the same channels be viewable by owners of both the new color sets and the old B&W.

By sheer genius, the design team was able to incorporate color information into the old B&W transmission processes in a way that the color tones continued to be represented on

B&W screens in a realistic way. However WTSC was always a compromise between maintaining backward compatibility, and providing best quality.

The image is displayed in 486 active lines of information and the other 39 lines are wasted in the vertical blanking interval. (Some references say 483 — the difference is probably bezel.)

In digital terms, the NTSC signal is related to a digital sampling of about 640 pixels per line, for 480 lines of image (chosen for easier standards conversion to PAL). This creates 307,200 pixels per frame of image which are then repeated 30 times a second (actually 29.97Hz).

- The horizontal scan time of a line lasts about 64 microsecs, and if you remove the horizontal blanking interval, you are left with an active line time of 53 microseconds.
- NTSC video signal requires about 4.2MHz of bandwidth including blanking, but excluding guards. The composite TV signal has two carriers. The visual carrier is 1.25MHz above the lower limit of the band, and the aural carrier is 4.5MHz above that. The upper limit of the band is 0.25MHz above the aural carrier.
- E-NTSC or 'Enhanced' NTSC attempts to improve the analog quality with digital signal processing in the receiver.
- Super-NTSC uses adaptive digital filtering techniques to avoid artefacts. This is probably the same as E-NTSC.

NTT

Nippon Telephone and Telegraphy. This is the now-privatized Japanese telephone monopoly. The term is also applied to an old mobile telephone technology used by the Japanese.

NTU

Network Terminating Units. The junction between the incoming carrier's data lines and the Customer Premises Equipment (CPE). See NT-#.

NUA

Network User Address. The address of true X.25 packet-switched terminals (not character terminals working through a PAD) on a packet network. The distinction is with the NUI address for PAD users.

NuBus

A standard form of slotted computer address and data (expansion) bus now used in the Macintosh line of computers. It is distinguished particularly by the fact that it is plug-and-play, making the installation of add-in cards very simple.

- Apple is now officially abandoning NuBus because it is becoming a bottleneck with the new PowerPC chips. They will replace it with PCI which is three-times faster.

NUI

Network User Identification. For dial-up X.25 networks, this is the public identification number/string which identifies the user for accounting purposes. The packet-switch host (where the PAD resides) has a NUA, which is a true address, and the user has an NUI, which is a local ID. Don't confuse this with a password or PIN — which only the user should know.

NUL

ASCII No.5 — Null character (Decimal 0 Hex $00 Control-@). The null is generated by the keyboard as the first of two characters when you press 'Alt' and any key. It is also used in many other ways in communications — often to pad out transmissions and to fill space in data records and in storage systems.

Don't confuse this with the zero character in ASCII, which is Decimal 48, Hex $30.

null

Nothing — zero. You put zeros into any memory location to destroy previous character bytes.

- **Null character:** Make the distinction between the null character in ASCII ($00), the space character ($20), and the zero character ($30). Null leaves a variable location effectively empty (or filled with zeros, which is the same), while the space and zero characters fill it with a bit-pattern in exactly the same way as any of the visible print characters.
- **Null string:** This is a string in a data structure which contains no characters (in fact, full of null characters — since if the bit-cells aren't set to logical 1 then they must all be at logical 0).

Null strings are often used to pad out a transmission, or to fill a block on a disk, or within some storage space. The null characters are also the default characters — if nothing exists in a location, it is the equivalent of a null.

- **Null suppression:** This is the process of ignoring all those null characters that may be filling null strings when transmitting or storing a file. It reduces the file size by removing all the padding.

null modem

A small double-connector unit used to deliberately swap circuits in a computer cable connection from one pin to another — it is a form of physical 'spoofing'.

A null modem is usually needed on an RS-232 cable when it is being used to join two computers to transfer data, since both PCs are generally soldered up as DTE devices — and so they both want to transmit and receive on the same pins. Normally with a straight-through serial cable (the standard) a DTE will only transmit to a DCE, so the DTE transmits on Pin 2, while the DCE normally transmits on Pin 3.

With a conventional cable the pin numbers on one end are directly connected to the same pin numbers on the other — a send, connects to a receive — but one end has a male plug and the other has a female. The null modem converts the DTE's pin assignments effectively to those of a DCE (or vice versa).

Special 'crossover' or 'flipped' cables are used for the same purpose — these are cables with the null-modem functions built-in. The wires are soldered so as to perform the connection of outputs on one end to inputs on the other. Be careful: some of these cross-over cable also have jumper connections.

- The main crossovers for an RS-232 null modem are:

Pin 2	connects to Pin 3	Tx to Rx.
Pin 3	connects to Pin 2	Rx to Tx.
Pin 4	connects to Pin 5	RTS to CTS.
Pin 5	connects to Pin 4	CTS to RTS.
Pin 6	connects to Pin 20	DSR to DTR.
Pin 20	connects to Pin 6	DTR to DSR.
Pin 7	connects to Pin 7	Mandatory common return.

With printers, Pin 8 (DCD) is often bridged with Pin 6 (DSR) within the connectors at the both ends of the link.

- With DB9 plugs the null modem has:

Pin 3	connects to Pin 2	Tx to Rx.
Pin 2	connects to Pin 3	Rx to Tx.
Pin 4	connects to Pin 1 & 6	DTR to DCD & DSR.
Pin 1 & 6	connect to Pin 4	DCD & DSR to DTR.
Pin 7	connects to Pin 8	RTS to CTS.
Pin 8	connects to Pin 7	CTS to RTS.
Pin 5	connects to Pin 5	Common return.
Pin 9	The Ring Indicator isn't generally used.	

number-cruncher

Slang term for machines whose design and software is primarily devoted to arithmetical computation, as against the modern PC's emphasis on text handling, spreadsheets, databases, etc. Supercomputers are generally giant number-crunchers.

number theory

The branch of pure mathematics that is suddenly proving to be invaluable in computing and communications. The major development in this discipline in recent times has been the solution to Fermat's Last Theorem by Dr Andrew Wiles, of Princeton University (Nov 1994).

Number theory is the purest form of mathematics, and it gives us understanding of prime numbers and encryption technique, and allows us to develop efficient encoding systems. Wavelet and Fourier transform coding techniques are pure maths solutions; queuing theory is central to many developments in communications and even in CPU control; RSA encryption comes from prime numbers; and now supersymmetry and chaos seem likely to produce useful developments.

numeric control

The control of machines (lathes, drill presses, etc.) used for industrial production by computer systems. This is the ultimate aim of CAD-CAM; to both design on a computer and then run the whole of the manufacturing processes through numeric control. If any design changes later need to be made, they can be done from the desktop. Eventually this could lead to full product customization on assembly lines.

numeric coprocessor

A floating-point or maths coprocessor. See coprocessor.

numeric keypad

The calculator-type keypad to one side of the main computer keyboard, or as a stand-alone unit.

Don't assume that the numbers on this pad will necessarily be the same as numbers typed using the top row of numerical keys, or that the Entry key on the keypad is the same as the Return key on the keyboard. It depends both on software and hardware how these act.

numeric word

A string of numeric characters (not necessarily numbers). Classification numbers are often numeric words, and they may contain spaces and slash characters, etc. Postcodes (zipcodes) are numeric words.

Numeris

French term for ISDN.

Numlock

A key on IBM-type keyboards used to toggle between the numeric key pad's use for numbers and counting, and the alternate use of the same pad for cursor control.

NURBS

Non Uniform Rational B-Splines. See patch.

NVM

Non Volatile Memory. This could refer to disk and tape storage, but usually to chips which don't lose information when they lose mains power. See ROM, PROM, EPROM and non-volatile.

NVOD

Near Video On Demand. This is a cable and satellite technique of showing the same movie on different channels at different delay intervals. Usually they will begin on different channels every 15—20—30 minutes. The more popular the film, the shorter the interval and the greater the number of channels needed.

NVOD can be supplied over conventional terrestrial (UHF and MDS) networks if there are spare channels available. It is usually offered as a Pay-per-view service, but it need not be. The same approach could be applied to normal Premium/Pay channels, or even to free-to-air channels if the program provider so wished. All it needs is enough channels.

The distinction is being made with true VOD, where each user has a one-to-one relationship with a channel, and the video source is under individual control. VOD is a true 'server' operation.

NVRAM

Non-volatile RAM. This is usually static RAM (SRAM) with its own power supply — usually a miniature lithium battery which keeps the memory fired up while the PC is off. It could also use FRAM which doesn't need a battery, and possibly flash memory in the future.

The term NVRAM is more specific than just 'non-volatile memory' and it is applied to chips on the motherboard which store real-time information for clock and calendars, and sometimes for the computer's setup program etc. This is all working memory, rather than secondary memory.

PROM, EPROM and EEPROM are also chips that are part of working memory, so the term can apply to these devices as well — although, more often you would spell out the type of ROM.

NVS

Non-Volatile Storage. See NVM.

NWN

Nationwide Wireless Network. The $150 million, Microsoft and Mobile Telecommunications Technologies (Mtel) joint development of a two-way, nationwide personal communications services (PCS) network in the US. By the second half of 1995 they expect customers in the nation's top 300 markets to be able to receive and send wireless text messages over 50kHz channels. This will become an extension to the Microsoft Network: it will have outbound messages at 27.2kb/s, and receive at 9.6kb/s.

NXX

In the North American Numbering Plan, some area code number sequences were not permitted. This was essential in these early systems because the first three digits could also be mistaken for the first three in a local number sequence.

The older area numbers are known as N0X or N1X where the N is any number between 2—9, while X is any number in the full-range 0—9. (Note: 0 and 1 are always prohibited in the first digit., but were mandatory in the second)

Originally, the presence of a 0 or a 1 as the second digit in a three-digit sequence (N0X or N1X) signaled to the exchange that this was an area code, while NNX which used numbers between 2—9 as the second digit, signaled that this was the first three digits of a local number.

NXX is the later version of the numbering rules and it represents the designated form for new area codes where the middle digit can be any number, not just a 0 or a 1.

In this new scheme, a 1 is pre-dialed for long-distance (before the area code), and this 'access code' tells the exchange that the next three digits are an area code. Therefore the prohibition on using anything other than 0 and 1 no longer applies.

So in carriers using this new scheme, the switch can recognize any three-digit sequence as an area code without needing to check the status of the middle digit. This now opens up many more area codes to be used.

- The vast majority of the US carriers now intend to use the 1-NXX–NXX–XXXX form for all long distance calls, whether they are intra-NPA or inter-NPA.

nybble

Aka nibble. A half-byte. In computer terminology, a nibble is four bits. The basis of hex notation is a nibble (four bits can provide numbers between 0 and 15). To make them readable in binary notation, bytes should always be written as two nibbles with a space between (1011 1001).

Nyquist sampling

A well established ADC relationship which H. Nyquist first described (although Shannon also has his supporters). It establishes the relationship necessary between the sampling rate, and the theoretical ability to reconstruct the original analog waveforms reliably. It is applied to A-to-D conversion.

The Nyquist rule simply states that the sampling rate must be at least twice the rate of the highest frequency which the system is required to handle. It almost states the obvious.

With any waveform you will need to sample the waveform once above the line and once for that part below, if you are to detect the rate of alternation.

This means that a sample rate of twice the peak Hertz is essential in analog to digital conversion (ADC).

Since complex analog audio or video waveforms can be considered as a blend of the fundamental and its harmonics (see Fourier), with the fine video detail or the richness in the audio quality being provided by those frequencies at the upper end of the frequency scale, then Nyquist's rule defines the quality limits with a given rate of ADC sampling.

If we measure an audio signal waveform at a rate of 20,000 samples per second, then the best audio reproduction we can expect to reproduce will have frequencies to only 10,000Hz; this will be the frequency cut-off. So Nyquist establishes the upper limit of quality — but don't forget that you also need to take into account the way (fineness of gradation, and type of scale) each sample is measured (the quantization).

- Digital telephone systems need to pass audio frequencies up to 4kHz to be compatible with analog systems. Thus PCM was designed to use 8kHz sampling. The potential quantization of each sample was 8-bits, so 8000 samples of 8-bits is the basis for the 64kb/s channels. See companding also.

- Note that the Nyquist limit says nothing about the quality of the signals (only the frequency limits). Quality will also be set by how fine the gradations are in the measurement of the samples, and whether that quantization was linear or logarithmic.

By using logarithmic companding techniques and equalization, it is possible to get much better quality across a long-distance digital circuit at 64kb/s than an analog circuit at 4kHz. Then by applying 'difference' principles to the quantization value (using ADPCM) rather than just transmitting the measured value, even higher audio quality can be achieved for the same data-rate — or excellent quality at a lower rate.

Nyquist only calculates the upper frequency limit — not how faithfully the restored frequencies will follow the original.

O&M

- Organization and Methods.
- Operations and Maintenance.

O&U problems

Over-promised and Under-delivery Problems. This is an industry-wide disease of epidemic proportions. It has resulted in such spectacular marketing failures as pen-based computers (Go and Newton), which will effectively destroy their market segments for many years, and inhibit development of the technologies.

OA

Office Automation.

OAM&P

Overheads, Administration, Maintenance and Provisioning. Often used without the P.

Oasis

Open And Secure Information Systems. A Eureka project to establish security rules for computers and communications.

objects

The term has been taken up by the computing industry as a buzz word, and used in a variety of ways, all hinging on the idea of 'being self-contained' or 'cell-like' or 'modular'.

• **General:** Computer systems have hardware objects (CPUs, disk drives, printers, memory segments, terminals) which need to be identified (addressed by a unique name) and protected. They also have software objects (processes, files, databases and semaphores) which also need protection — which need to run in a 'protected domain'.

In an operating system, objects are classed as 'abstract data types'. Processes must also be policed so that they don't access an object without authorization.

A process must be given the right of access to a certain 'domain' for a 'legal operation' (it may be permitted to read, but not write to a memory segment).

• **Programming environments:** In all forms of programming, objects are sub-sections of the program which can be manipulated and tested as a single unit. Everyone agrees that having a standardized library of reusable objects would be a good thing, since these could be tested once, then used over and over again.

'True' objects are reusable and have three qualities that are of interest to programmers: they have inheritance, classes, and encapsulation. Modules are almost objects, but of the three essential object qualities, they only have encapsulation.

With both objects and modules, the program will consist of discrete sections of active code which are linked in some way, usually by messaging.

Microsoft has its object environment description called OLE, and Apple and IBM are promoting OpenDoc — and both of these are more modular than object-oriented if you want to make this distinction. Most of the mainframe vendors and carriers have thrown their support behind CORBA for larger systems. (See TINA also)

There are many different views of objects in programming:

— To some, objects are just self-contained modules of code (ALMs) from a library. In the newer versions of COBOL, for instance, objects are modules, which include data and procedures (methods).

— In OOP, objects are both 'the data and the procedures that revolve around that data'. They are classified according to the properties they may have in common. Objects communicate with each other by message passing — and these messages are either queries or commands. See object-oriented language.

— To most PC users, objects are identifiable screen-based images (screen icon, buttons, lists, forms, or even words) to which some program code can be (or has been) attached. You click on the object, and it does something. However a screen object is more than just its icon; it is also the active program structure behind. See event-driven and Visual Basic.

• **Classes of objects:** In OOP there are many classes of objects, but the typical categories are: physical objects; elements of user-environment (menus, windows, mouse, keyboard); logical elements (graphic primitives, like lines, circles, rectangles); programming constructs (stacks, arrays, lists, binary trees); databases (inventory records, a dictionary); and user-defined data-types.

• **Directories:** In X.500 directory terminology, objects can mean persons, job titles/positions, computer peripherals or telex machines — anything which can be addressed individually by a message.

object code/module

Distinguish between object code and object program, and between object program and object-oriented program — these are all quite different concepts. 'Code' is sometimes used as a synonym for 'program' but it shouldn't be, here.

• **Object Code:** The core machine language commands in a program, based on the micro-instructions of the CPU. The distinction is with source code.

Object code results from compiling source code, or from running an assembly language program through an assembler. Later object code will be linked to create an object program which the computer can run.

Many contract software companies will only provide object code to their customers — they want to keep the source code to themselves so all future modifications need to be made through them.

• **Object modules:** An object code file does not necessarily contain the micro-instruction needed to make a program run. The source file is first run through an compiler/assembler to create an object module file. Object modules are not executable. See preprocessor also.

This object code file is in an intermediate format which contains machine language instructions and tables with external reference information — often to other modules in a library, and so it will generally need to be linked with several other modules to form the final machine-language program. It is the job of the linker to resolve external references (called 'binding') and to create the machine-language instructions.

object graphics

These are structured or vector graphics, as distinct from bit-mapped (raster) graphics. All shapes are described as algorithms which provide a mathematical description of an object rather than just the matrix position of individual dots. Some are standard library structures known as primitives (circles, squares, lines, etc.).

Entire objects can therefore be manipulated — stretched in one or two dimensions, scaled up or down, shifted as a whole, lines made darker, changed fill patterns, and grouped together. There has been some success recently in combining the attributes of bit-mapped (raster graphics) and object graphics.

• Paint programs use raster graphics and have single bit-maps corresponding to the whole image. Object graphics are Draw programs where each object has its own bit-map, and each is layered onto the screen independently. See primitive.

• CAD programs are always based on object graphics.

• Note: It is incorrect to call these 'object-oriented' graphics. Object graphics are a data-type in procedural language programs.

object language

The output translation language of a cross assembler or cross compiler (see), not to be confused with object-oriented language. Sometimes object-language programs produce an intermediate stage which requires further translation before a computer can use them at the machine level.

Also, compilers don't necessarily produce a direct machine-readable code — rather they produce object modules, which can only be executed by computers after the linking process.

object-oriented interface

This is the type of graphical user interface (GUI) invented by Xerox and refined by Apple on the Macintosh. Later, after many years of decrying the 'WIMP'ish approach ('Real men don't use GUIs!'), Microsoft also introduced Windows. It also uses icons to represent 'objects' (files, folders), windows, and manipulates them with the mouse-controlled cursor.

This is a misapplication of the term 'object-oriented' rather than just 'object' (It was the trend in the 1980s to make everything 'object-oriented'!), but I guess it is close enough.

• We should probably distinguish more between techniques which are object-oriented, and those which are object-based.

Many systems now allow you to create a custom object, such as an icon or button, that will inherit the features and functionality of existing objects — yet they aren't true object-oriented systems.

object-oriented language

A computer programming language where specific functions are performed by program modules called objects. An object is a single unit which combines both data and functions (specifically functions that operate on that data).

The only way to access the data in an object (aka 'instance variable') is through the object's functions (aka 'method') — the data is hidden from other functions, and is therefore said to be 'encapsulated'. It is this isolation of the data that makes modules easy to write, debug and maintain because each is then predictable.

The distinction is with a normal linear type program where progression is predominantly from beginning to end (both structured and unstructured languages). These programs have both global and local variables which are being dynamically changed throughout the program flow

With OOP (object-oriented programming) each module/object is treated alone, and messages ('calls') are relayed between modules as required. Since the objects exist in isolation they can be stored in a library and reused; for this reason the OO approach includes the concept of putting together custom programs from off-the-shelf parts.

• There are five features in OOPs: objects, class, encapsulation, inheritance and reusable code.

• The main languages are still probably Smalltalk, Turbo C++ and Objective-C.

object-oriented programming

See OOP.

object program

One that has been compiled or assembled and then link-edited. It is the final stage of linked object code or machine code, and is ready to run. This is also called 'executable code' or an 'executable file'. See object code/module.

• Note: OOPs (Object-Oriented Programming) should never be called object programming and vice versa — they are two entirely different things.

oblique

A font-style which is similar to, but not the same as, italics.

• Oblique projection is used in computer graphics. One face of a square will always appear to be parallel with the screen surface, while the depth lines are usually angled at 45° or 60°. Don't confuse with isometric projection, or with perspective.

OC

Optical Computer. There are some of these now working in the labs. They can perform basic computing functions using light and photonic switching, but there's no satisfactory form of memory for photon-bits, at the present time.

OC#

North American standard data rates used in Sonet, which is the US version of the international SDH standard.

- OC-1 (Optical Carrier 1) operates at 51Mb/s and serves as the optical carrier for copper-originated DS-3 (Digital Services 3) transmission. The grade of service provided at this rate is known as STS-1.
- OC-3 is 155Mb/s.
- OC-12 is 622Mb/s.
- OC-24 is 1.2Gb/s.
- OC-48 is 2.4Gb/s.

These are used internally in telephone networks. See STM, Sonet, STS and SDH.

• Note that the data-rates given for the OC and DS standards will vary slightly because of overhead.

Occam

A parallel processing language developed by Inmos for use with their transputers. It is a 'concurrent control-structure language'.

OCE

Open Collaborative Environment. A sophisticated proposal for file transfers and networking in an open environment. It is constructed on an e-mail specification introduced by Apple, and built into their operating system for the Mac. It is very like MAPI, with which it will work. OCE is the natural consequence of extending the Mac's Apple Events and Microsoft's DDE into a common command system.

OCR

Optical Character Recognition. This is a general term applied to a number of different approaches to scanning pages, creating graphic images and then deriving the ASCII text characters from this graphic image.

Some of the old OCR systems required the characters to be in a special font which was designed not to be ambiguous to the reader. You purchased special type for your IBM golf-ball typewriter, and then the pages, so created, could be scanned and read into the office computer.

Nowadays, OCR means hardware and software combinations which can select elements on a page, and identify a wide range of written or printed characters by the use of a scanner (or video camera) and computer. The images are stored in memory (usually as a TIFF file) and the computer then applies (usually) matrix-matching or contour-analysis techniques to this image — although there are other approaches.

The OCR program is attempting to convert essentially a bit-mapped file to a text-character (ASCII) file, but it is effectively limited to fonts that it can recognize (or learn to recognize) and therefore it has problems with smudged photocopies, new type-styles etc. — and for reasons never fully explained (to me!) it is sometimes completely flummoxed by first-quality print of a standard font (Times Roman), on top-quality paper, laid out in simple columns. Shiny paper is sometimes the problem — in which case, try a photocopy.

• There are two basic types of OCR software: the 'teachable' and 'non-teachable'. The way to test an OCR program is to phone a professional typesetter and ask them to send you some sample pages of their typestyles, and use photocopies of these as a standard check. You'll be amazed at how well some OCR programs do on some difficult fonts, and how badly they perform on other easy fonts.

• As a preliminary stage, the OCR software will usually also be required to identify columns and flows of type. This is not really an OCR function, but a good program will have this ability. Test the program out on newspaper cuttings.

• Modern OCR must deal both with headline size/width, and the column size, width and flow — and often with the columns also broken by sub-heads which are of a different size and style.

• If the output ASCII text is only to be used for reference, OCR can prove to be useful if it gets only 95% of the characters right. However, if the data is intended as copy in an important report or publication, then OCR must get at least 99% right. It is almost as quick to have material retyped, as it is to find and correct 1% of errors. See also ICR.

• OCR-A and OCR-B were standard typefaces devised by ANSI to make life easier.

OCR-A/-B

Special typefaces for golf-ball typewriters, which were to have allowed old IBM typewriters to become input devices for computers. Unfortunately, it wasn't very reliable.

octal

Octal is half-way to hex. It is a base-8 notation, and therefore uses eight numerical characters (0 to 7) only; decimal eight is therefore 10 and nine is 11.

Octal is usually printed in number groups of three, with leading 0s, to differentiate it from decimal or hex, so octal 017 = decimal 15; octal 106 = decimal 70.

So, although it may look like decimal, it is not. Octal has some advantages in machine-level programming, but hex is usually better. Don't confuse octal with octet.

octave

In music an octave change is a doubling of the frequency. If you can play a tune within the range of one octave, and you then move up the scale and use the next octave, the frequency of every note played will be doubled. So octave is logarithmic.

In magnetic video recording, the octave is used as a measure of the dynamic range of the material. Typically a magnetic medium can handle frequencies over a span of about ten octaves (from the lowest to the highest Hz).

Unfortunately, the range of frequencies produced by a good video camera (say, a S-VHS with Y/C output) will be from 20Hz to 5.5MHz, which is a span of 19 octaves — well beyond the limit of the magnetic material.

So octave compression is necessary, and this is done by frequency modulating the luminance signal (the Y) onto an HF carrier where it will occupy a reduced octave range (but at a higher frequency), and recording it on a helical scan which can handle these higher frequencies. During replay, this signal is 'down-converted' to restore it to baseband.

octet

The communications term for 8-bits or one byte. Don't confuse this with octal. The term is used instead of 'byte' because not all communications systems/equipment use 8-bit bytes — most of the old systems were 7-bit, and so octet stresses that these bytes are 8-bit. The old Baudot telegraphy used 5-bit bytes, and later 7-bit bytes were used in most packet-switched systems.

octothorpe

The correct name for the 'hash' or 'pound' sign (#).

OCX

OLE Control eXtensions. This is a component extension of Microsoft's open environment, OLE.

ODA

Office Document Architecture. ISO's common interchange standard for text, formatted documents, graphics and pictures. It is an SASE, which is concerned with the representation of mixed-content, structured, office documents — from memos up to desktop publishing.

ODA allows office workers to use the same standard sets of software to manipulate documents on many types of computers, and to interchange documents electronically between these systems. The documents can include character-text, graphics, and fax-quality images. There's an ODA Consortium which was founded in 1991 by IBM, DEC, ICL, Bull, Unisys and Siemens-Nixdorf. See ODIF and SASE.

ODA/ODIF

This has now become ISO 8113. It is providing the main competition to Microsoft's OLE.

ODBC

Open DataBase Connectivity. These are APIs which provide reasonably transparent access to data-diverse DBMS from within Windows applications. See WOSA.

ODBMS

Object-oriented DataBase Management System (aka OODBM). These are becoming very important now that object-oriented programming is increasingly being used for mission-critical applications. ODBMS involves the 'persistent storage' of objects, and this could be the next generation of database management applications after Relational DBMS.

In early forms of OOP, the storage of objects onto a disk or tape involved the conversion of the whole program in memory into a flat-file, or later into a relational series of files. This was a complex problem.

ODBMS differ from other forms of persistent storage by its lack of conversion of objects; objects are stored as objects on the hard disk or tape. This technique also provides all the other essential features of any object-oriented methodology.

- ODBMS are reputedly having much success in the CAD field and are now migrating to MIS. They often store complex data files — particularly annotated text files, voice, and graphics that normal mainframe relational databases don't handle well.

- ODBMS are also cheaper to program than relational databases: some can provide a gateway through to RDBMS.

ODC

Open Database Connectivity. This is a database access system from Microsoft, codenamed Nile. This is to extend the OLE to remote database access and to allow distributed transaction processing. See OLE and OR also.

odd-even checking

Parity checking. See parity bit formats.

odd parity

When all the bits in the original byte add up to an odd number, no parity is added. Let's be clear about this because it is often totally confusing: Odd parity calls for a check 1-bit to be added to a byte only when the original number of 1-bits is even, and this then makes the total number odd. It is the same for 7- or 8-bit bytes. See parity bit formats.

ODI

Open Data-link Interface. A specification for generic NIC device drivers from Novell. ODI lets you load multiple protocol stacks (such as IPX and IP) simultaneously, and it provides a standardized way to access networks.

ODIF

Office Document Interchange Formats — an OSI ODA standard for electronic document interchange. This is a standard which allows documents which include both text and graphics to be transmitted in editable form with all their format-attributes intact.

- There's also a neutral form for archiving text and graphics, which can be read by many applications.

ODMA

Open Document Management API. This is an attempt to standardize document management architecture, by standardizing desktop access to document-management clients. Through ODMA, applications (word-processors/spreadsheets) can talk via the Windows Dynamic Link Library (WDLL) to a local document manager. This manager will then talk, in turn, directly (or through middleware layers) to document stores. ODMA has led to Document Enabled Networking (DEN). See.

ODP

Open Distributed Processing. An ISO project which aims to specify an ODP reference model along the lines of OSI. It will specify the generic functions and the associated interfaces, and define guidelines for the 'open' distribution of processing between disparate multi-vendor applications.

The ODP reference model (based on object-oriented technology) will eventually provide a complete conceptual framework for specifying interoperability and portable distributed systems. See CORBA.

ODP

OverDrive Processor. See OverDrive.

ODS

OverDrive Support. Most of the new Intel chips allow a faster CPU to be installed on the motherboard in a special socket. See OverDrive.

Odyssey

This is a Big MEO (circular orbits at 10,500km) proposed by TRW in association with Teleglobe and Mantra/Marconi. It is being licensed by the FCC under the Big LEO category, and it shares CDMA bandwidth with Globalstar and Ellipso (see LEO spectrum) for both voice and data.

Odyssey will use reasonably-standard TRW communications satellites built originally for the US Navy (they need only six to provide coverage — but are planning 12) but the unique feature is that these satellites are fitted with steerable antennas so that each can direct 32 cell-beams on the populated areas of earth. Odyssey will use dual-mode CDMA handsets with an output of 0.5 watts. The cost of this project is put at $2 billion and launch-date is in the 1999 time-frame.

O/E

Optical to electronic junctions in fiber optics.

OEIC

Opto-Electronic Integrated Circuit. At the present time this hybrid approach has begun to yield useful optical fiber devices. Both light and electricity are used within the chip to transmit signals and both optical signal-detection and electrical signal-amplification have been incorporated within the one chip using surface-emitting lasers.

• The term is also used in a slightly different (more generic) way to mean chips which have optical devices such as diode lasers and photodetectors on the same chip as the electronic circuits.

OEM

Original Equipment Manufacturer. A term which, half the time, now means roughly the opposite of what it says.

• It really means a manufacturer who sells all of his hardware to vendors. They then rebadge it, add some value, and on-sell it under another name. However:

• By some strange inversion of logic, the OEM has become the buyer as well as the seller. Very often you'll hear that OEM's are companies which buy equipment manufactured by large corporations and put together their own systems using these elements — which is putting the cart where the horse used to be. They'll often source this equipment from many suppliers, and write their own software for the combination. In this case OEM and VAR are used interchangeably.

OEMI channel

See BMC.

OFDM

Orthogonal Frequency Division Multiplexing. This was developed for Digital Audio Broadcasting, and is now being used with digital TV under the name Coded OFDM. For DAB they are currently proposing to use QPSK with a 2 bit/Hz data-rate on each channel. See COFDM.

off-axis

Satellite reception outside the intended footprint. There is always a bit of signal scattered over a wider area than the contours show, and with a big enough dish and good enough low noise amplification, TV reception may be possible well outside the intended area. It is very hard to predict the quality off-axis reception without a trial. To a large degree it depends on the polarization and the power, but much remains unexplained by the theory.

off-hook

The condition that indicates that the customer's telephone circuit is in use — either ready to dial, or sending voice or data. The distinction is with on-hook. This applies to modems and fax also. In a normal POTS line, the off-hook condition signals to the exchange by shorting across the ring and tip lines (the two in the pair). See loop start and ground start.

off-line

Not on the equipment used for day-to-day operations. Not in real-time.

• In video editing, this is the process of making all editing decisions by using a 'window-dub' or 'work-print' of the original material. It may be viewed on any quality screen or replay unit, and within the image window will be the time-code which allows each frame to be identified unambiguously.

After all decisions are made, an EDL (Editing Decision List) is compiled, listing by time-code all 'in-points', 'out-points' effects (fades, dissolves) and audio dubbing — and this is then loaded into a CMX computerized editor (or something similar) for the final assemble editing process. This last stage is a very expensive process, which is why decisions are taken on cheap off-line equipment.

• This is not the same as linear and non-linear post production. See non-linear.

• In communications, off-line devices are any which are not physically connected to and using, a communications medium.

off-mike

Microphones have different patterns of reaction for different frequencies. They are designed to produce the best results when the sound is in the center of the mike's reception pattern (cardioid, omni, etc.). If a person is speaking off-mike (to one side of this pattern), the sound will be both attenuated differently at different frequencies and sound 'thin' (it will have few low frequencies).

office

In US telecommunications terminology this means a telephone exchange.

- Small office, home office is SOHO.

offset

The amount that must be added to a computer memory address, relative to a reference point, to reach a required location. The distance from here to there. Relative addressing always deals with offsets.

- The original PC (using the 8088) always gave memory locations as two numbers. The first was an absolute number of a segment, and the second was an offset within that segment. So to find the address, the second was added to the first (using a shift to the left).

- In real mode addressing, Intel chips from 8086 to the 80286 deal with memory by segments (up to 64k each) which are addressed relative to their lowest location. To this address number, the offset is added to find the required location.

- The location of an item of data in the database relative to the beginning of a file or record. In fixed length records this was always a fixed amount which made it easy for the old computers to step through the records in a file looking for a particular item.

- Offset is also a type of printing process developed from lithography, which is now the most widely used professional process for printing almost everything. It has replaced letterpress as the most common printing technology for business stationery. The inked image on the plate is transferred first to a rubber roller, and then to the paper — this is the offset process.

- In cellular phone systems the duplex offset is the distance in frequency between a channel on the forward link and its matching channel on the reverse link. The difference between the transmit, and its matching receive channel.

Oftel

Office of Telecommunications. The UK's telecommunications regulator.

ohm

The unit of measurement of both resistance (DC) and impedance (AC).

- Ohms law is I=V/R where I is the current in amps, V is the EMF in volts, and R is the resistance in ohms.

OIM

OSI Internet Management. The group responsible for integrating the OSI network management protocols into TCP/IP nets.

OIRT

French for International Radio and Television Organization which sets radio and TV broadcast standards.

OIU

Office Interface Unit. An American telecommunications term where 'office' means central-office or telephone exchange. The distinction is with SIUs or Subscriber Interface Units. These make a pair which they propose to fit at each end of a future fiber -to-the-home connection.

OLE

Object Linking and Embedding. Microsoft's open environment for object-oriented software components for data-sharing. It is a way to integrate applications across a network; but it hasn't yet taken the world by storm. It's been around for a while but it only works with Windows and NT. It has limited Mac support.

OLE is based on Common Object Model and written in 32-bit code. The term 'OLE controls' includes objects with clearly defined properties and I/O interfaces. Both ODC and OR are being developed to extend it. See both.

OLE specifies an object-environment for use under Windows with DDE; however it works at a layer above DDE and it insulates the programmer from many of DDE's problems. It standardizes a rich file structure for documents, and it provides APIs through which OLE-enabled applications can cooperate for the display and editing of a document.

The distinction between OLE and DDE is this: before two applications can use DDE they must agree on an interchange format, and if the applications are from different vendors, then each vendor must publish detailed specifications. However with OLE, one application places the data into a 'container' in the other application, and it can then be easily extracted from this container and displayed. If the data needs to be changed, however, the second application can invoke the first through a special interface. This approach is also used in New Wave.

OLE is currently outstripping OpenDoc which is from IBM and Apple, and the main challenger. OLE has probably reached critical-mass with the number of OLE-enabled applications especially as it has support from Novell. However CORBA from the Object Management Group is now also a serious challenger since it has become incorporated into TINA.

- OLE 1.0 required the user to step through several menu commands to insert an object from one application into another. However, OLE 2.0 now allows the user to drag and drop a selected object between applications. See DDE and OpenDoc.

OLTP

On-Line Transaction Processing/Processor. The processing of real-time information, triggered by 'events'. OLTP systems are traditionally mainframe hierarchical database operations used for banking, share-market trading, airline reservations, etc. They perform credit and debit operations, manage personal records, and perform scheduling tasks.

The distinctions being made are usually with 'batch' processing (non-real-time) and also with real-time on-line information transfer (extended sessions of interactivity) services. More recently OLTP has been migrating to client-server applications and distributed databases (the last with limited success).

Transactions are characteristically short exchanges triggered by events such as the entry of a card and PIN number into an ATM machine at a bank, or the booking of an airline seat. See transaction processing and ACID test also.

— If we are talking about hardware, the OLTP service is usually provided by the front-end communications processor for a bank's mainframe storage system. Both CICS and X/Open's XA monitor standards are important here.
— If we are referring to the process (such as ATM transactions, funds transfer, or airline reservations), then this is ideally a parallel-application environment where many concurrent users or clients can simultaneously execute independent requests to large shared databases.

This niche was, until recently, the preserve of massive host-centric mainframes and dumb terminals, but these have now been replaced to a degree by client-server architectures in an open systems environment. However, fault-tolerant conventional mainframe systems still dominate and will probably continue to dominate here. Massively parallel computers are having some impact however.

OM-1

A consortium of MPEG users who are trying to standardize the development and deployment of appropriate tools and technologies. The organization will test and certify product — chips, adaptors, software, and system platforms.

omega wrap

In helical scan tape recorders, these are machines where the tape forms an 'omega' shape when in contact with the head drum. The tape does not quite wrap around the full 360 degrees of the drum surface. See C-format video.

OMF

Open Media Framework. A file-compatibility framework developed by Avid for digital video. See MFE.

OMI

- **Open Microprocessor Initiative:** A European R&D program which is pushing developments in ARM and the Inmos Transputer, as well as Sun's SPARC and Mips' Rx000. See ANDF.
- **Open Messaging Interface:** Now known as VIM. See.

omni-directional antenna

This simply means that the antenna is not directional (except that the energy is probably projected in a horizontal plane, rather than up or down). It can send to, or receive from any place in the hemisphere surrounding it.

Handheld communicators, particularly cellular phones, need to have omni-directional antenna, simply because no one can guarantee the direction in which they will be pointing. Mobile navigational aids using GPS or LEO need to be omni-directional so that they can receive from a range of satellites in any position above them.

OMR

Optical Mark Recognition. See.

on-hook

To hang up a telephone or disconnect a modem. The circuit between the ring and the tip wires on the local loop has been broken. The term signifies that the phone is now ready to receive an incoming call.

on-line

The common use of this term now has wider implications. It suggests:

— A terminal connected over an active link of any kind. For instance, the real-time use of a remote computer facility over the telephone line or by digital links.
— Doing something in real-time, rather than by batch. In video editing, the term refers to the time spent with full-quality editing machines (usually under CMX computer control) as distinct from the time spent with sub-standard (usually VHS) machines viewing material and making decisions. Don't confuse this with linear editing.
— Able to make direct decisions which affect the final product, or only indirectly affect them through a secondary stage.

With modern digital editing in computer workstations, the distinction between on-line and off-line editing is whether the images being manipulated are in a direct line between the original and the final product. When they are to be included in the final output, the computer system must be manipulating high-resolution image files with only M–JPEG compression (not MPEG). See window dub.

on-line mode

The on-line mode of a modem is more correctly called the data-mode. The distinction is with the command mode.

on-line transaction processing

See OLTP

on-ramps

This means the commercial Internet service providers (and possibly some of the non-commercial ones as well). These are 'on-ramps' to the 'information superhighway' (said in a slightly cynical tone!).

ONA

Open Network Architecture. This is an FCC mandated requirement in the USA by which the RBOCs must deliver equal-access service to a variety of carriers and information providers.

It depends on having common standardized interfaces. See also the European ONP.

ONC

Open Network Computing. This is a middleware distributed-applications architecture from SunSoft originally, but the standard is now controlled by a consortium. It provides network applications with a standard communications interface. This allows them to work together regardless of location or any other variables. ONC is popular in some areas because of its relationship to NFS which is widely implemented. NFS is actually a part of ONC. See DCE also.

ONP

Open Network Provision. An attempt by the European telecommunications groups to standardize, among other things, leased lines, packet switching, voice telephony, ISDN and intelligent networking protocols. The various parts of ONP will be issued progressively as a series of EC directives. The Americans are developing a similar open standard known as ONA (Open Network Architecture).

ONU

Optical Network Unit. The junction between an optical fiber network and a coaxial or twisted pair feed. In Cable TV hybrid networks these are also known as 'hubs' since they feed TV signals out along many coaxial distribution lines. See HFC.

OODBM

Object-Oriented Database Manager. See ODBMS.

OOK

On-Off Keying. This is the extreme form of amplitude shift keying (see ASK) which was once used for modem communications over phone lines, and is now used in light-guide (optical fiber) systems. This means what it says; the light is off to signify a 0, and on to signify a 1. It is a monopolar line coding system.

OOP

Object-Oriented Programming. In the last few years this radical change of approach to applications programming language has moved solidly into the mainstream, mainly because it can use and reuse tested programming modules, and because these programs seem likely to be responsive to future change.

Simula (by OJ Dahl) was probably the first OOP language, but Alan Kay's SmallTalk is the best known of the earlier forms, and Apple's HyperTalk is probably the widest used (at a less substantial level). A couple of versions of C++ are now very important.

The success of these languages is due to a combination of modular programming (which allows easy debugging and experimentation) and the use of English-like words and syntax. This is making OOPs usable by an increasing number of non-professional programmers.

Objects are entities where some data is bound together with its associated functions and procedures into one isolated module. Objects within a program interact and make use of each other by sending messages. They respond to a message by executing certain internal procedures and accessing internal data.

• The key concept of an 'object' is that it provides:

— **Abstraction.** This is the process of characterizing the common behavior and properties exhibited by a collection of similar objects into a 'class definition'. Object Class, establishes the relationships between modules.

— **Encapsulation,** in which data is segregated within an object and protected from outside access except through the object's functions. Encapsulation hides the internal implementation of various objects from the user; they therefore become 'black boxes' — independent modules which can be developed and tested outside the application.

— **Inheritance,** which is the ability to pass on certain characteristics, and so create more 'generic' modules which can be reused in other programs. It is therefore the expression of commonality between class definitions, enabling one class to be declared (say) a sub-class of another. The new will then inherit the properties of the higher-class.

— **Polymorphism,** which allows objects to respond in their own way to the same message. Each object can be designed to interpret and respond to a message in a different way to the other objects. And, from the above comes **reusable code**. See hybrid languages.

OOT

Object-Oriented Technology. This generally applies to off-the-shelf applications development software which can be customized to a particular application. The 'object' here, is to sell a software package that contains some data and provides a set of procedures that manipulate or use that data. The claims made about OOT are: reduced development time, increased quality, reduced maintenance costs, enhanced modifiability.

op-code

Operation Code. This is seen as a two-digit hexadecimal representation of the instruction — an assembly language instruction. Each different assembly instruction is encoded into a unique symbol known as an op-code. This code has a one-to-one relationship with machine level functions.

A mnemonic, is a three alpha-character version of the op-code which is easier to remember. So, 'BRK' is the assembly mnemonic which would be keyed in during assembly language programming. This would then create the hex $00 op-code, which then functions at the machine level as the binary instruction code 0000 0000. See instruction set.

open

A general term that means 'not proprietary' or 'not fixed in concrete'. It can also mean 'available for use'. In fact it is almost an omnibus word that can mean anything to anyone.

• **Open files:** We open files in a computer in order to work on them. With Wintel or Mac machines, you will open the file by clicking on its icon — or by choosing it from the Open File menu within the application.

• **Open connection:** A broken wire, or a switch that is off.

• **Open architecture:** A computer with expansion slots. See.

• **Open standards:** In the definition of computer and communications standards, 'open standards' means that anyone can use them. See open system/standard.

• **Open marketing:** Often, the marketing people mean by 'open standards', nothing more than a proprietary standard which is available for other competitors to follow (which they usually won't), or available for third party vendors who can add their own peripherals/features.

• **Open Standard Interconnect:** See OSI.

open architecture

A computer hardware design (architecture) where public-domain specifications exist, and which is available for use by any third-party developer. It also usually means a computer box with expansion slots, into which you can plug device driver cards, or other processors, or memory expansion cards, etc. See S100. The computer is not 'closed' at the time of manufacture, as was the early Macintosh.

The old Apple II was regarded as an 'open architecture' machine even though everything within the box was proprietary. For instance, the Apple II became the world's most successful CP/M machine because of a plug-in adaptor built by Microsoft.

With the later IBM PCs, open architecture meant the combination of hardware design and the operating system which is available to any developer in the early days of MS-DOS. This came about because Microsoft owned the rights to the O/S and was keen to sell it to anyone, and the hardware was cobbled together from bits and pieces freely available to any clone maker. However, the IBM PC wasn't a true open system until the BIOS was successfully cloned — for years they could only get 95—98% compatibility.

• These systems were then 'open' by virtue of the fact that different adaptor cards (boards) could be added to:

— upgrade the CPU (accelerators),
— change the CPU (Apple IIs ran CP/M for many years),
— upgrade resources (extended and expanded memory),
— customize operations (various I/O devices and drivers).

• See open system/standards also.

Open Look

A graphic user interface supported by Sun, AT&T and Unix International in opposition to Motif. Make a distinction between 'Open Look' compliance, and the ABI compliance for Unix; both are needed.

open loop

A programming term which means a section of the code will continue to be processed until a human intervenes. A closed loop will take corrective action automatically.

Open Software Foundation

An international non-profit organization developing an open, portable environment to which vendors and users have equal access. It is a member-sponsored R&D organization funded by a group of 200-odd companies led by IBM, DEC, Apollo and Hewlett-Packard.

The aim was to develop an alternative form of Unix to AT&T's standard, and they later developed some more general 'open system' aims. These were to be specified as the Application Environment Specifications (AES).

The OSF group was originally reacting to a decision by AT&T and Sun to use the SPARC chip as a binary standard and to merge Berkeley and System V without consulting other leading players in the Unix field.

The OSF's interface is called Motif and it is based on IBM's AIX; they have also rewritten the Unix source-code to completely free themselves from dependence on AT&T while conforming to the international Posix standard. This is OSF/1 (aka Digital Unix), and its user-interface is called Motif (or OSF/Motif). It is in direct conflict with AT&T's Unix interface called Open Look. See USL, Unix International and X-Open as well.

• The Unix Wars are now over. In Apr 1993, USL and OSF publicly agreed on common standards, and also agreed to conform to each other's published interface specifications.

open system/standard

This is an amazingly vague term. Basically it means computers and devices which rely on public-domain industry-standard technologies, rather than proprietary ones. This is then supposed to provide interoperability between different vendor equipment, and so protect a company's long-term investment in applications from indefinable, but unavoidable, changes in technology.

'Vendor independence' was the jargon term. However it didn't always mean that everyone stayed strictly within the standards — which often made open systems and open standards incompatible. Unfortunately, life isn't this easy in a business where everyone wants to get an edge.

With open systems, applications are supposedly developed in an open hardware environment, so they are not so dependent on the underlying technology, and so are easy to port from one system to another. This is the theory!

However, open systems in the enterprise environment (where the term is mainly used today) would be expected to support an industry standard like SQL, and they should be scalable (able to handle large and small implementations).

• The term open system was used at one time to mean open architecture. It was the term applied to any personal computer which provides expansion slots.

• The OSI (bless their heart!) define an open system as one that conforms to their standards when communicating with other systems. Don't imagine that only OSI standards and systems are open. TCP/IP and a range of others are also open.

Open Systems Interconnect(ion)

See OSI and OSI reference model.

open wire

This is telephone wire strung between the cross-bars of telephone poles (almost unknown these days) with each wire being insulated from its partner by an 8-inch air-gap. For any distance, open wire systems carried more than one voice call on a wire pair. Between 3 and 12 voice channels was the norm, but 28 pair was possible using frequency modulation to 143kHz. With these systems repeaters could be as infrequent as every 240 miles in a three-channel system. These were full-duplex systems using different frequency bands. See carrier-type pair cable.

OpenDoc

This is the rival protocol to Microsoft's OLE. It was created by CIL (Component Integration Laboratories) which is owned by Apple, IBM, Oracle, Sun and Xerox. WordPerfect is now in the group also, but Novell has left to join the OLE camp.

They wanted to take the initiative away from Microsoft's OLE, but currently they are seriously lagging behind, although recently IBM has shown OpenDoc for OS/2.

OpenDoc is a component software architecture that actually does away with the idea of the application; instead it provides a container of user-selectable modules (componentware). AppWare could be the glue that binds it all together.

The original aim of the CIL group was to create a standard which:

— was vendor-neutral and cross-platform,

— had application-to-human interfaces which are consistent, and

— could modularize and standardize the support features (such as spell checkers) so that the same one can be used by many applications.

OpenDoc provides object-oriented functionality across diverse platforms from a PDA to a supercomputer, and it supports multiple operating systems. They list OS/2, Windows, MacOS 7, Taligent, AIX and several other forms of Unix.

The main components of OpenDoc are:

— **SOM:** System Object Model (object-linking) from IBM. This is a 'language neutral' development technology and it underlies OpenDoc.

— **SLM:** Shared Library Manager from Apple.

— **Bento:** the component storage technology.

— **Taligent:** a fully object-oriented operating environment. (Apple, IBM and HP).

• OpenDoc has class-libraries which are compatible with OLE, and OLE 2.0 could run in the OpenDoc environment.

By comparison, OLE retains the idea of separate applications, but tries to make them interoperate. See OLE and Taligent also.

opening a file

Computer files generally need to be opened before they can be read or changed, and the files also need to be closed after use. Usually these operations are performed by the database management application and so they remain unnoticed by the operating system. But if a computer system fails, a file may inadvertently be left 'open' and it may later be difficult to retrieve the information, even though it is intact and is in a known position on the disk.

OpenView

This is HP's network management platform, and potentially the most successful. It arrived in 1990, and it has been highly popular ever since. NetView from IBM is a sub-set of OpenView, and much of the OSI's DME has its base in OpenView. See DME, SunNet manager and NetView also.

operand

• **In maths:** This is the number which is to be the subject of a mathematical operation. In the calculation: 36 x 2, the 36 and the 2 are both operands and the x is the operator. 72 is the result.

• **In computers:** It is a fancy term for an item of data which has been fetched from a memory location, and is now subject to some form of manipulation by a program instruction. In the example 4 x 5 = 20, both the 4 and the 5 are operands.

• **In SNA:** A parameter used to make the meaning of a macro instruction more specific.

operating environment

A general term applied to the operating system and the hardware on which it is running. Also called the platform.

• It could also mean whether the computer was being used in a cold, dry, hot or humid climate.

operating system

The program (or set of programs and/or sub-programs) which controls the processes of the machine itself and manage all the resources. It will take care of all I/O — tasks like reading the keyboard and displaying text on the screen. It will look after the internal management of the computer itself.

Don't confuse operating system and system software. The operating system is the key element (the kernel) of system software, but the latter is a wider term (see).

The four key components of OSs are:

— Management of the various processes (including multitasking control).

— I/O management.

— Memory management.

— File system control.

With modern OSs, the supervisory elements which call up and release utilities and handle peripherals are the vital elements — hence the common term 'Disk' operating system. DOS functions are part of the I/O management and they operate through an intelligent disk controller. See also NOS (network operating system).

operation code
See op-code.

operational research
A field of study developed in the Second World War which stresses qualitative techniques of management for repetitive activities — traffic flows, queues, assembly lines, etc. It has a philosophy which centers on the use of statistical methods and models to find solutions to business problems — and therefore it is ideally suited to computerization and simulation. It is very popular in some business circles. The problem seems to be the old one of 'drawing precise answers to vaguely worded questions'.

optical amplifier
A device that amplifies the light signal in an optical fiber system without first converting it back to electrical form. There are two main types — semiconductor and doped fiber.

Optical amplifiers use a process of stimulated emission, as do lasers — in fact the special length of doped fiber can be considered as an elongated laser cavity.

The way a doped-fiber amplifier works is roughly as follows: the optical signal along a standard glass fiber cable eventually passes through a spliced-on few meters of special glass fiber doped with erbium. Light from a laser 'pump' (high energy but different wavelength to the signal laser) is also coupled into the optical system. This 'pumped' light (ideally at a wavelength of 0.98 microns) excites the atomic levels of the erbium atoms.

The incoming signal-carrying light now reaches the excited erbium fibers, and each photon acts in a way which discharges the excited state of part of the fiber. This re-energizes the pulse, so the fiber emits additional light at 1.5 microns in synchrony with the signal.

The doped fiber can be manufactured to match the normal fiber exactly, which eliminates coupling problems (such as reflection at the junctions). See EDFA.

These amplifiers can work with a number of different light frequencies simultaneously, which suggests the use of Wavelength Division Multiplexing (WDM) and solitons in the next generation of fiber systems.

• Unlike an optical fiber regenerator, the optical amplifier is not rigidly fixed at one data rate. New undersea cables will operate at 5—8Gb/s using optical amplifiers instead of regenerators. Within reason, a 3000km undersea fiber system using optical amplification can begin with low-speed terminal equipment, and later upgrade when higher data rates are needed.

• The latest in the development of this new technology is the Pirelli T-31 Bi-Directional Line Amplifiers (BDLA) which can boost two opposite-direction light signals instead of one, by routing them using WDM. Light waves are received at 1557nm and 1533nm from opposite directions. A single amplifier therefore handles two signals over the same fiber which will allow the carrier to halve the number of optical line amplifiers it needs, and also increase the distance between optical regenerators from 140km to 400km.

optical character recognition
See OCR.

optical compensator
These are rotating polygons of glass which use internal reflection to maintain a reasonably steady image on a screen, while the film moves past steadily.

This is a film viewing (and projection) system, which doesn't require the stop-go motion of a conventional film gate. And since the images overlap (they appear to dissolve into each other), it is possible to use optical compensator telecine machines to run old 16 fps silent film at any variable rate (usually at 24 or 25 fps) to restore the original movement. The other way, is to step print, duplicating every second image.

optical drive
• A disk drive which reads information from a reflective disk by means of a laser. The term can actually be applied to drives which read-only, or which read and write. We can distinguish three main categories of optical disk drives:

— **Read only**. This includes the CD-Audio, CD-ROM, CD-i and others, including the laserdisc.

— **Write-Once, Read Manytimes**. See WORM.

— **Erasable.** These include different base technologies such as magneto-optical (MO), ablative pit and dye-polymer, and also different techniques of access, such as single-session and multi-session. See PhotoCD and CD-R also.

optical fiber
A thread of glass (or sometimes plastic) which can pipe pulsed light flashes over long distances. Optical fiber only became practical in the 1970s when pure glass was made by Corning in the USA. The first practical fiber systems had losses of 20dB per km; today they have losses in the order of 0.5db per km.

Optical fiber is immune to electrical interference (lightning), difficult to tap (secure), but most importantly it can carry enormous amounts of information over long distances at very high rates.

There are three main types:

— **Multimode fiber** is mainly used for short-hauls, for local area networks etc, and it can be powered by cheaper LEDs.

— **Graded index fiber,** which attempts to provide a compromise between the thicker multimode and the faster monomode fibers. See.

— **Monomode fiber** (single mode) is for long-hauls and high data rates and it requires special diodes and lasers.

Optical fibers can carry both analog and digital signals. An analog signal is transmitted via continuous linear lasers, or by pulsed lasers using a form of FM modulation. But most optical systems are digital.

There are two main ways of sending multiple channels of digital signals down a fiber:

— **Time Division Multiplexing** (TDM — where each channel has its own time slot), or

— **Wavelength Division Multiplexing** (WDM — a form of frequency division modulation using different 'colors' of the laser light).

The first is by far the most common, providing thousands of channels over fibers with rates in the 5Gb/s range today.

The second form (WDM) is only now becoming practical. Until the invention of optical amplification it was difficult and expensive to regenerate signals from different lasers — now it is relatively easy. See fiber.

• Because of the enormous bandwidth, optical fiber cable is an extraordinarily cheap and reliable transmission medium when used in telecommunications. The standard bundle of fibers used today has 6, 12, or 18 fibers for undersea use (copper wrapped and insulated), and for long-haul terrestrial communications the cables have between 12 and 36 fibers with repeaters every 60 to 100km. For the shorter interexchange links the cables often have up to 400 fibers at a low rate. This is cheaper than using high order multiplexers and a few very fast cables.

Data rates in common use around the world today are from 140Mb/s and 560Mb/s per fiber (fibers almost always work in pairs), to 2.4Gb/s (standard long distance) and more recently to 5Gb/s and 8Gb/s (the new generation with erbium amplification).

When you calculate the capacity of these cables in terms of 64kb/s PCM channels (for data), or 32kb/s and 16kb/s ADPCM for voice (often with DCME as well), the amortised cost of distance in telephony becomes almost nothing.

• Optical pulse amplification can now take place within the fiber. See optical amplifier, and erbium amplifier also.

optical frequencies

Optical fibers act as waveguides for light frequencies in the region of 10^{14} and 10^{15} Hz which is the 3000—300nm wavelength band. These frequencies are more than 10,000 times those of microwave, and therefore they have a very high bandwidth.

However, unlike microwaves, the modulation schemes used to modulate light are quite primitive. At present it is exclusively OOK — on-off keying — which treats the light as if it were a DC baseband source. There is then no relationship between the number of bits carried and the frequency of the light (you will have thousands of cycles per pulse).

The modulation techniques now being applied to microwaves are complex modifiers of waveform (both phase and amplitude) which produces many bits of output for each cycle of the carrier. About 6-bits per Hertz is now common.

This is only possible with direct modulation of the waveform of the carrier.

• Ideally the lasers used for monomode fibers will operate in the 1.5 micron band (usually 1550nm) where the glass is most transparent, while many still run at 1.3 microns (1300nm). Older systems ran at 850nm.

• For optical disk drives; the shorter the wavelength, the higher the packing density. Hence the emphasis on developing blue lasers and harmonic systems which generate blue light from red lasers. Current CD systems use red lasers of 750nm wavelength, but claims are being made that shortly blue-green lasers of 523nm will be commercially available.

optical mark recognition

OMR. This is used mainly in tests, surveys, etc. for statistical purposes which are scanned in and computer-read. It usually requires the application to identify boxes (usually surrounded by a color which can be easily identified) which are pencil marked.

This is used in education to mark multiple-choice exam papers, and also for survey and census work. It is also called mark-sensing.

optical printer

In film post-production, an optical printer is used to add titles and special effects. It is virtually a camera, which is rephotographing the image in a projector — but it allows multiple layers of film to be sampled, to create mattes, etc.

For mass production of the final release prints (copies that go to theatres) a continuous printer is used.

optical receiver sensitivity

A measure of the optical power coming into a receiver which will result in a specific probability of error. The error level in these systems is usually expected to be one incorrect bit in 100 million (10^9).

optical scanners

See scanner.

optical sound track

Film can carry 'married' sound either as a magnetic track on one edge, or more often (especially in older films) as an optical sound track which is read by a light and detector system.

This track runs on the edge of the film between the images and sprocket holes, and it is advanced (sound ahead of picture) by 26 frames. There are two types:

— Variable area tracks: The track is exposed and developed to the full black density of the film, but the width of the track varies, and this modulates the light beam.

— Variable density tracks: The track is always the full width, but it varies in density between white and black — this looks like a series of very find bar-codes.

Both can be read by a standard projector.

optical transceivers

Transmitters and receivers on an optical fiber.

option cards
Add-in cards with accelerators and/or coprocessors.

opto-electronic devices
Lasers, modulators, switches and filters used in optical fiber communications systems, and in the experimental optical computing systems. It is difficult to know whether this term means only devices that convert between optical and electrical systems, or whether it includes pure optical-optical devices (optical erbium amplifiers, or optical switches) when they are driven or controlled by electronic circuits.

opto-coupling
The process of joining optical transmission systems to electrical. Opto-coupling is part of optical fiber transmission systems of course, but the term seems to mainly be used with systems which protect computer equipment and telephone exchanges from power surges, accidental shorting, etc. These are small devices that use an optical path to isolate electronic CPE connected to telecommunications lines.

OQL
Object-oriented Query Language. This is SQL for OODBs. It is now being standardised.

OR

• **Open Repository:** Microsoft's new add-on to OLE and ODC which will provide a library and a gateway for executable OLE components.

• **A Boolean operator** used in keyword selection systems and in database retrieval. The command to select records containing 'A OR B' specifies that one or the other of A or B must be present in each record selected. If both are present then the condition is satisfied also. The distinction here is with a variation called an Exclusive OR, where the condition is satisfied by one being present only.
— **OR-gate:** In Boolean algebra and logical gates, OR will output 1 when either of two inputs has a binary 1. If both inputs have binary 1, then the output will be 1 also. If both inputs have binary 0, the output will be 0.

• **Operational Research.** See.

Orange

• **CD Book:** The Philips/Sony CD-ROM specification for CD-WO (Write Once), CD-MO (Magnetic-Optical) or CD-R.

• **Security Book:** This is a set of criteria for security in computer systems designed by the US Department of Defense.
— Division A deals with 'verified protection'. Formal systems to protect sensitive information which is being processed and stored by the computer.
— Division B is concerned with mandatory protection measures. It has three classes relating to labeling security.
— Division C is discretionary protection.
— Division D is for minimal protection.

• **Card:** For the Macintosh. This is a DOS coprocessor for the PowerPC. Don't confuse it with DOS emulation. It has a 486DX4 running at 100MHz, but uses the 286 instruction set.

ORB
Object Request Broker. This defines the platform upon which object-based interactions will occur. In network management, ORB accepts requests from various management applications and matches them with the appropriate mechanism which can retrieve the necessary information. Many types of ORBs will need to exchange data. See CORBA and DME.

Orbcomm
A Little LEO, data-only, satellite project promoted by Orbital Technologies and due to have the first working satellites in orbit early 1996 and most of the full constellation up by 1997. Orbcomm uses small meter-wide 'banjo-shaped' satellites which can be stacked, six at a time into a Pegasus launch vehicle for launch from a Lockheed Tristar aircraft.

There will eventually be 26 to 36 satellites in orbit in six planes at a height of 785km. The frequencies set aside by the FCC are 148—149.9MHz for uplink, and downlink bands at 137—138MHz and 400.5—400.15MHz (all in VHF). Some of these frequencies are already in use on the ground for paging, so Orbcomm relies on agile transmitters and receivers which check a range of frequencies and chose which to use when they are over each continent.

This is for e-mail and monitoring only (including location). Packetized data is carried on downlink at 4.8kb/s, and on the uplink at 2.4kb/s. Some space segment cross-links will be provided for relaying signals. Handset and monitor devices should be in the $100 to $400 range.

The total cost of the first working constellation with four US ground stations is only $180 million.

orbital velocity

• LEOS at 200 km height have an orbital velocity of 29,000 kph and an orbital period of about 90 minutes.

• MEOS at 1730km have a velocity of 25,000 kph and a period of 2 hrs.

• GEOS at 35,700km have a velocity of 11,300 kph and a period of 24 hrs.

• The moon at 386,000km has a velocity of 3700 kph and a period of 28 days.

orders

• **Orders of multiplexing:** See high-order multiplexers.

• **Orders of magnitude:** An increase of 'one order of magnitude' in common terms, is a multiplication of ten times. Two orders of magnitude means 100 times.

ordinal
Originally this just referred to sequence. So if you had a series of chapters (as this book does) which are 'numbered' A, B, C,. . ., Z then these were ordinals. Now the term seems to be restricted to numbers only.

Ordinal numbers are those which indicate sequence as distinct from value — so first, second, and third are ordinal numbers, and one, two, and three are cardinal numbers.

• In maths and in data-structures, ordinal means 'unsigned' integers (neither plus or minus). They will usually be in the range 0—255 for 8-bit systems, and 0 —65,535 for 16-bit computer systems.

• In computers, however, we will speak of record number #23 in a database as being an 'ordinal' 23, since we are only concerned with its sequence in relationship to others.

• The contrast is with cardinal numbers where you are concerned with values: e.g. a box holds +23 apples and you can remove 2 (–2).

• The similar distinction is often made between a number (a value) and a numeral (a numeral is a symbol often used for numbers, but also for ordinals).

In a variable, numerals (say, for a postcode) should be held as an alphanumeric string, not as a number.

ordinate

This is the vertical axis in a two-dimensional graph. It is synonymous with the Y-axis in an X/Y set of coordinates.

Organization & Methods

O&M implies the systematic examination of the way an organization functions. This is to distinguish the process from the company's normal preoccupation with administrative and financial duties.

originate/answer

Low-speed modems transmit binary signals by alternating between two audio frequencies (FSK), and so in full-duplex operations four different (two pairs) frequencies will be needed. The originator has two standard frequencies, and the answerer has another two.

Most off-the-shelf modems switch between the answer and originate modes automatically by choosing the originate frequencies if they originate the call. However there are some old modems which need a hardware switch, and some are switched by software commands.

Some old host computer modems are designed to work in originate mode only, whether they receive or originate the call. So if you initiate the call to these hosts, you will need to switch your modem to 'answer' immediately after the connection has been made. Normally this is done by ending the dial sequence with an R (for Reverse).

Higher-speed modern modems share the same single transmission frequency (singular) between both ends and use phase and amplitude modulation, so this originate/answer problem doesn't exist. See AT command set.

OROM

Optical Read Only Memory. This is a standard very like CD-ROM but with CAV which gives shorter access times. It was developed by 3M and is now an ISO standard. The latest generation of multi-function magneto-optical disk drives is designed to use both write-once and rewritable optical disk drives in the OROM format.

There is also a pre-recorded form which can store 122MBytes of data, and be read by standard magneto-optical drives. With these the information is stamped into the disk in the factory. There's also a Partial (P-)ROM version which leaves some space on the disk for the recipients to store their own information alongside the pre-recorded information. See MO and P-ROM also.

orphan

In desk-top publishing this is a single line, the first in a new paragraph, which is left all alone at the bottom of a column. The remainder of the paragraph is at the top of the next column. Most good publishing programs will identify orphans, shift them to the top of the next column, and feather-out (increase the line space) on the first column to fill out the column depth.

The term is loosely used, so now an orphan has also come to mean a single part-line left over at the bottom of a page or column. The point is, that it looks messy. See widow.

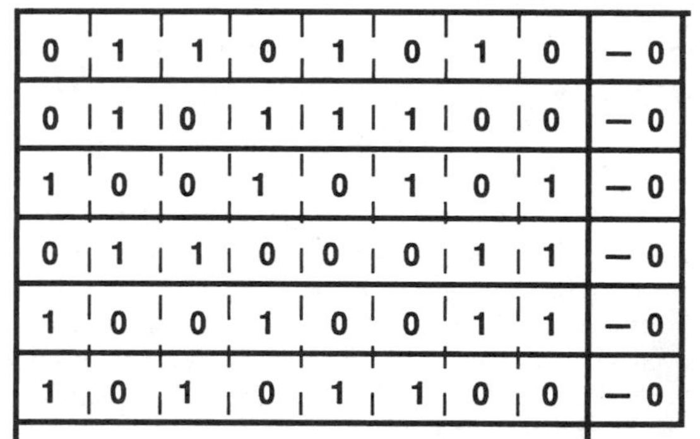

0	1	1	0	1	0	1	0	– 0
0	1	0	1	1	1	0	0	– 0
1	0	0	1	0	1	0	1	– 0
0	1	1	0	0	0	1	1	– 0
1	0	0	1	0	0	1	1	– 0
1	0	1	0	1	1	0	0	– 0

The sum of the codes in the horizontal and in the vertical columns are equal to zero. On average, one code then cancels the other.

Orthogonal codes

orthogonal

This word is widely used in computers and communications, and it is often difficult to grasp the implications. It often means balanced or counter-balanced.

There's a common concept here. When adjacent objects or 'things' are orthogonal, the unwanted effects of one cancel out the unwanted effects of the other. There's also a sense of even-handedness between opposites — that the two will balance or counter-act each other.

• Aligned at right angles — rectangular. This is said to be a sampling technique using a rectilinearly-aligned sampling structure.

• In satellite and terrestrial transmission, mutually at right-angles (eg. between horizontal and vertical polarization) or counter-related (between right and left-handed circular polarization). The point is that adjacent frequency channels can overlap slightly if the carriers are orthogonal, without causing interference.

- In CDMA digital phone systems, the Walsh codes used to transmit each chip are orthogonal, as in the illustration above. These are long-strings of unique bit-patterns which are chosen specifically because the number of 1s and the number of 0s in each pattern is equal. So in any cumulative transmission system (where everyone is transmitting in the same channel) they will, on average, cancel each other out.

For this reason, many CDMA users can share a single wide-band channel, since at any time, the BPSK pulses (180° out of phase with each other) will generally cancel out, and consequently the channel will look only slightly noisy to each user.

orthographic

Right angled. A form of computer graphics where 'perspective' lines don't converge on one point, but remain parallel.

OS

Operating system. This is generally called a Disk Operating System these days — on PCs, anyway! See operating system, kernel and DOS.

OS/2

Operating System 2. An operating system originally designed back in 1987–88 for 80286 machines, and at that time jointly controlled by IBM and Microsoft. This was a Microsoft product which IBM later took over and extended after Microsoft left the partnership to concentrate on Windows NT and 95.

IBM's basic version was called OS/2 Standard Edition, and all versions from 1.3 have been entirely IBM. Presentation Manager is the WIMP interface for OS/2, which is much like the early Windows.

The first OS/2 was a 16-bit, true single-user, multitasking operating system which could address only up to 16 megaBytes of RAM. Version 1.1 had a Graphic User Interface (GUI) for the first time running on an IBM PC platform. Multi-user applications were only possible on a LAN running LAN Manager. Later versions added many more features (windows, database management, etc.).

OS/2 has been described as Windows/DOS on steroids. It can run Windows/DOS programs without needing to have DOS loaded, and it is a true multitasking environment.

- OS/2 Version 2.0 released in 1992 is a 32-bit system for 486 and Pentium systems, and this introduced the first object-oriented user shell. Presentation Manager for multimedia was an option. Version 2.0 is a multi-threading, multitasking, single-user operating system which can run native applications, as well as those created for DOS and Windows. It uses paged memory management and HPFS files.
- OS/2 Version 2.1 has had few changes. Options are now rolled into the main product.
- OS/2 for Windows allows OS/2 to run as a task over Windows 3.1, on MS-DOS machines. That sounds like a desperate cry to be noticed.
- **Warp**: Warp is a cut-down and refined version which is proving to be very popular. It is IBM's counter to Windows 95.
- **WarpPower** is OS/2 for the PowerPC.

OS/2 EE

The Extended Edition of OS/2 which uses IBM proprietary code. This extension bundles a full range of communications functions and a relational database engine with the OS/2 operating system. The database manager is a scaled down version of IBM's very popular mainframe database software, DB2. The second phase of the EE developments will allow PCs to use IBM mainframes and AS/400 minis as servers, and vice versa.

OS-9

The real-time operating system used in CD-i and in many Motorola-based systems. It has a shell and a scripting language similar to the Bourne shell in Unix. Files are in a tree structure with separate user areas, and everything, including devices, are treated as a file. The later version which was written in C is known as OS-9000 and is available for the new CIRC Motorola and Intel chips.

OS-9000

See OS-9.

OSF

Open Software Foundation. See.

OSF/1

A Unix operating system designed by the OSF and released in Nov 1990. It incorporates elements from Berkeley's Version 4.3, Carnegie Melon's Mach 2.5 (the kernel), IBM's AIX (commands and libraries) and a few other sources. It complies with the XPG3 specification.

OSF/Motif

See Motif.

OSI

- **Open Source Information:** A movement in the USA to free up copyright limitations for electronic systems.
- **Open Systems Interconnect:** This is a major development program for the international standardization of networking and equipment interoperability. It began under IEEE and is now promoted by both the ITU and the ISO.

— **OSI/NMF.** This is the Network Management Forum of the OSI which is trying to speed up the OSI's bureaucratic processes.

— **OSI/RM.** OSI reference model.

— **OSI protocol stack.** OSI reference model.

— **OSI TP.** An OSI standard protocol for transaction processing which could become the basis for many on-line (specifically bank and finance) transaction processing implementations in the future. It can be used by any application requiring inter-process or reliable communications. The IBM equivalent is LU6.2 plus its API, APPC. See Transaction Processing.

OSI reference model

Aka the OSI protocol stack. This is a seven-layered architecture of 'interface definitions' which aim to provide us with a standardized way of interworking with all types of computer equipment and all network operations. (For a basic understanding, see Three-layer model.)

The design of the OSI model stack (the structure) was first released in 1984. By dividing it into seven layers and carefully specifying the interfaces between layers, the OSI approach allows each layer to be insulated from its neighbors (and therefore replaceable) while still interacting with the layers on both sides through carefully-defined interfaces at the boundaries. See SAP.

The stack is a hierarchy from the 'user-services/general-application' facilities at the highest layer (Application), to the 'specific network' facilities at the lowest Physical layer.

Each layer must conform to clearly defined protocols at the interface between both its higher and lower neighbors, and each has a specific job to do. Each layer will use the services of the next lowermost layer, and provide services to the next highest layer — but they are entirely ignorant about all other layers in the stack.

The functions these layers perform are: addressing, flow control, error control, encapsulation, reliable message transfer, and many others. One of the main advantages of layering protocols is that those users who require only simple network functions can utilize only the lower layers (those network-specific protocols), while those wanting more complex functions can progressively add higher layers.

The layers are defined as: (lowest, in contact with the media) **1) Physical, 2) Data-link, 3) Network, 4) Transport, 5) Session, 6) Presentation, 7) Application.** (highest, in contact with the user).

They are grouped into two-sets of three: User-services (Session, Presentation and Application), and Network Sublayer (Physical, Data-link and Network), with the Transport layer between. Layers 1 to 4 of the model deal with hardware, and Layers 5 to 7 involve the software user-interface.

- The main problem with the OSI approach is that it is not as efficient as competitive proprietary systems — there's always an added complexity inherent in taking a modular approach like this, because additional work needs to be performed at each layer boundary to makes sure the information can transfer across a pre-defined interface, and this tends to slow the system down.

- See the individual layer-names for further information.

OSN

Open Systems Network.

OSPF

Open Shortest-Path First protocol. A well-known and popular open routing protocol (an Interior Gateway Protocol — IGP) standard which is part of TCP/IP. It provides for the dynamic exchange of information between routers. This uses a 'link-state' routing technique, rather than the 'distance-vector' techniques of RIP.

The aim of OSPF is for the router to take into account network load and the bandwidth of the available links when routing packets over the network. It can handle least-cost routing, multipath routing and load-balancing.

OSFP replaces the older RIP. It only takes about 10 seconds to define a new route whereas RIP often took minutes. Also it doesn't overload WAN with constant table updates (once every 2 hours only, if no change). Previously OSPF was called OSPFIGP (OSPF Interior Gateway Protocol). See RIP, SAP, IS-IS and APPN also.

- Three types of OSPF networks can be defined:
- Point-to-point (PTP) directly between routers.
- Broadcast Multi-Access (BMA) on a LAN.
- Non-Broadcast Multi-Access (NBMA) for frame relay.

OSS

- **Optical Storage Systems:** This includes WORM and erasable/recordable disks as well as CD-ROM (generally). See optical drive.

- **Open System Security:** The security aspects of public communications networks like EDI, e-mail, etc.

- **Operational Support System:** A suite of computer-based tools used to make the construction, maintenance and operation of telephone networks (business and general) both efficient and easy. The term is usually used for multi-million dollar software suites which the major telephone carriers write (or purchase) to run their networks and control their value-added services. These also supply management services for the front-desk, sales force, and ordering system, etc.

In large carriers, the systems themselves may comprise a half-million terminals, PCs or workstations; thousands of larger computers; and millions of lines of code. Together they create the business, operations and maintenance nerve center of a telephone network.

OSS software interfaces with both the management-information system and the commercial applications. They provide the 'basic care and feeding of the telephone network' — ranging from billing and provisioning to surveillance and testing.

There are hundreds of different OSSs being used around the world, and the problem is how to integrate them. See TL-1.

- Older OSSs were quite distinct from the administrative support systems (which deal with financial or general information). Today they tend to be integrated.

OTDR

Optical Time Division Reflectometer. This is a fiber cable tester. See TDR.

OTLP

This is probably a mistake in typesetting. See 0TLP in numbers section.

OTPROM

One-time Programmable ROM. Since all PROMs are one-time programmable, this just stresses that it is a PROM, not an EPROM or an EEPROM.

out-of-band signaling

In-channel signaling means that the signals are carried in the same circuit as the conversation — as distinct from CCS#7 and ISDN where separate packet-switched channels are provided.

There are two types of in-channel signaling: in-band (within the voice frequencies) and out-of-band (above the range of the normal voice frequencies).

In the local loop it is possible to use a wide range of frequencies above those carried by normal telephone switches and multiplexing equipment (up to 4kHz), but generally local loop signaling uses in-band (current pulses, current loops, and DTMF).

However, between telephone exchanges, a type N1 carrier system will convert the pulse-dialed signals into a 3700 Hz on-off tone signal which is above the 3500Hz limit of the voice-band. Filters in the channel units block these signaling tones from the callers. See in-band signals.

Multiplexed digital networks have quite complex methods of signaling on-hook, off-hook, dial and other essential information. Some of the old systems were limited to only 56kb/s of usable data on a 64kb/s link, because some of the bits are reserved for signaling. This is a digital in-channel system called channel associated signaling. See channel associated signaling and A&B bit signaling.

See also CCS#7, ISDN and D-channel.

outframe

The number of outstanding frames which are allowed in a PU2 servicer, on SNA.

outline fonts

These are the type of fonts which are defined mathematically, rather than through bit-mapped images. Adobe pioneered this technique in PostScript. The font file just records the character shape as a series of mathematical formula for curves and lines.

Using these formulas, the shape of any character of any size can be generated, then stored temporarily as a bit-map, and used over and over again on the word-processor screen, or sent to a printer for printing.

This process requires far less memory than holding multiple bit-maps, one for each size of type, and different ones again for light, bold, and italics. The system is even more flexible now because condensed and expanded versions can also be generated. See multiple master.

When you select a font, style and size, the outline is drawn to the appropriate size, and then the enclosed areas are filled in with black. This requires substantial amounts of processor power and time, although with modern CPUs it is hardly noticeable. And, since the basic shape of an outline font is mathematical, the font can be infinitely scalable without leaving jaggies (to the limit of the display resolution).

However, both printing and screen displays use a rasterization (scanning) process and individual finite-sized pixels, so there are still some difficulties in deciding on the width (weight) of fine lines. For this reason a further process of grid-fitting is used to adjust the outline and improve the final image.

The encryption rules which specify how this grid-fitting is to be performed are called 'hints', and it is this process of hinting which is the legally copyrightable part of outline fonts. This made it difficult to clone PostScript before TrueType came along and forced Adobe to do a deal.

It is therefore the hinting techniques more than anything that distinguishes one outline font technique from another. Outline fonts are also called vector fonts or scalable fonts. See TrueType, PostScript, PCL; also hinting and multiple master.

outline processor

A word-processing function which allows you to establish a hierarchy of headings and sub-headings, and sub-sub-headings, etc. These can each be 'folded' into a previous level to create lists at the various levels — and when required, unfolded to add more detail. These were often called 'ideas processors' because they helped you create a framework for writing a report, a major article or a book.

outsourcing

To contract an outside company to maintain and/or run (either or both) a company's computer and networking facilities, rather than perform these tasks in-house. The theory is that this allows the company to get on with what it does best — making widgets. The term is usually used for mainframe computer operations, but also for private networking. VPN is sometimes seen as a form of outsourcing, as is Centrex. See.

over-voltage

See surge.

OverDrive

Intel's single-chip processor upgrade for 486 machines (and later). Some PCs come with an empty OverDrive socket on the motherboard. With some computationally-intensive functions you may get a 60–70% speed increase (as measured by iCOMP) with an OverDrive chip in place.

overhead

• The cost, in terms of extra characters/bits or time, in providing addresses, error checking, frame structures, header and tail flags, etc. on an information system. Overhead bytes are the protocol-bytes, and the term is distinguishing between those 'redundant' bytes and the useful 'payload' data-bytes.

• Overhead bits. These are bits other than data-bits which are used as check bits, flag bits, signaling bits, framing bits and parity bits.

overlay code

In the new US numbering system, the area serviced by a cellular phone network may exist at a layer above that serviced by the wireline carriers, and cover quite different territory. Some of the new NPA (area code) numbers are being reserved for cellular use, so these are 'overlay codes'. They don't match any existing exchange area.

oversampling

This is the process of using a sampling rate for ADC and DAC, which is higher than the Nyquist rate (twice the highest frequency required). By doing this, it is possible to use anti-aliasing and special reconstructive filters which have a much more gentle cut-off slope, so there is less loss of phase linearity.

In a digital-to-analog conversion process, oversampling converts the input data frequency to a much higher output data frequency, but it has an interpolating stage between. In effect it smooths out the steps inherent in digital sampling systems (quantization steps).

It is also suggested that often, the least-significant bits are thrown away if the processing load is too heavy, and this results in coarse quantization.

The main advantage of oversampling is to 'smooth-out' the jumps in quantization. This then allows low-order analog filters to be used following the DAC, to reduce non-linearities.

See Bitstream and decimator

• In CD-Audio there are two-, four- and eight-time oversampling, and each improves the sound quality marginally.
See Bitstream.

overwrite

• To add information to a memory location that already is storing data, so destroying the old data.

• In word processing, the definition remains the same, but the emphasis is on the visual replacement of characters on the screen as you type. The distinction here is between overwriting the existing information and inserting the new information at the cursor point.

• In the old daisy-wheel printers, if you wanted to create bold type or underline, you could send the spinning printer head back across the same line of type and 'overstrike' the letters on the page, or add the underline. This is one reason why carriage return (CR) and line feed (LF) needed to be kept as separate functions.

P

P# chip

• **P5:** The P5 chip from Intel became the Pentium or 80586. See Pentium.

• **P6:** Pentium Pro. The P6 chip is Intel's latest microprocessor chip capable of operating at 200+ MIPS, which is twice the speed of the Pentium. P6 is based on a four-processor architecture and has 5 million transistors. It will be 'completely binary compatible' with previous generations.

The P6 runs at 133MHz (SPECint92 rating of 100) and needs 256k of Level 2 SRAM synchronous cache, and has its own 16kB of on-chip (Level 1) cache. Although the P6 is a 32-bit processor, it will have a 64-bit bus between the processor and the Level 2 cache.

It moves into a new era with a technique called 'dynamic execution' where the processor appears to think ahead, to select and ready the tasks it is most likely to perform next. Power dissipation will be between 14—20 watts. (4.5 times the Pentium). It competes with the PowerPC, DEC's Alpha and Sun's Ultrasparc.

• **P7.** This is to be Intel's first true RISC chip. It is 64-bit and it is due in late 1997.

p x 64

The old name for the H.261 videoconferencing and videophone standard. It refers to the use of an integer number (p) of 64kbit/s channels (PCM) in aggregation. See H.261 and H.320.

P-channel

A packet channel in the 802.9 specification. A MAC frame for the P-channel consists of two flags and five fields: flag; service identification number; frame control; destination and source address; then the data-unit; frame check sequence; and final flag. See B-, C-, D- and H-channels also.

P-code

See P-system below.

P-EDI

See page 502.

P-parity symbols

A form of error detection and correction used in CD systems.

P-ROM

See PROM on page 530.

P-system

Pseudocode System. An operating system which can run on a wide range of mini and microcomputers. It was developed by the University of California in San Diego (UCSD) and written in a language called P-code. All languages supported by P-system are compiled (Pascal, Fortran and Basic). The programs are initially translated into a series of universal P-code (pseudocode) instructions instead of directly into the machine language. A second step, using the machine-specific 'P-machine emulator', then translates P-code into machine instructions.

P-type

A form of semiconductor used in transistors. See.

PABX

Private Access Branch eXchange (aka Private Automatic Branch Exchange). The European term for a company's internal telephone exchange. Also called a PBX in the USA.

• American terminology refers to the PABX (*sic*) specifically as a 'first-generation' electro-mechanical switch (from about 1920), which uses analog signaling. These were the first to replace manual switchboards.

Now the term PBX is used in the USA to emphasize the difference between these PABXs and 'second-generation' (from 1970) PBX electronic-technology switches which use internal digital signaling, and also from the latest PBX 'third-generation' integrated voice/data switch systems.

PAC

Personal Authentication Code. See PIN.

PACE

Priority Access Control Enable. A new technology for Ethernet switching which allows real time (isochronous) and multimedia services to run over what is basically a 10Base-T Ethernet network. This protocol upgrade is supported by 3Com, Apple Novell, Silicon Graphics and Sun.

It can only be used on star-wired switched systems, but it integrates 10Base-T and 100Base-T technologies. It requires special switches with PACE incorporated.

pacing

Flow control. See XON-XOFF.

package

• **Packaged Media:** This distinguishes CD-ROM and disk delivery of electronic entertainment from the other five paths to the home: terrestrial broadcast (one-way), SHF LMDS (possibly both two-way and cellular), direct satellite, HFC cable, and telephony cable. All will be digital and use MPEG compression.

• **Computer software:** In terms of 'applications software package', the term implies that there are a number of separate applications (word processor, spelling checker, thesaurus, etc.) which will probably not all be integrated into one major multipurpose application. The distinction is with 'integrated software' where each module should be part of the whole, not a separate item isolated within memory.

• **Cable and Pay TV systems:** The subscription channels which are grouped together and promoted as a whole for a single charge are called a 'package. This is an all-or-none offer.

With modern set-top boxes, there is no credible reason why the video service market, like every other form of service, can't be sold on a per-use, per-channel basis rather than as a package. Pay-per-view is a non-packaged service, as is Video-on-demand — although they may only be offered as part of a subscription package.

packed-pixel architecture

This is a technique used in handling video-RAM, especially in some of the more advanced video cards. The image is stored in an area of very fast 'frame buffer' memory with each pixel represented by a series of bits — from 1-bit for monochrome, to 5- or 8-bits (each) for RGB color.

The pixel's information is packed together in 'chunks' (It is sometimes called 'chunky' architecture!) in the video memory and stored sequentially, with memory locations representing pixels from the top-left to the bottom-right of the screen.

When the screen is being painted, each pixel receives its 'triad' information in full (RGB) before the next pixel is accessed. With the alternate technique called planar architecture, the Red triads would all be painted before the Green, and then finally the Blue.

This architecture is the standard form used on the Mac and IBM PCs for video. See planar.

packet

The unit of information (a logical unit) by which a LAN or packet-network communicates. The primary distinction is with circuit-switched communications where the data flows in a constant steady stream, but only after a call has been set up.

'Packet' is both a general term and a specific. As a general term it covers frames (Ethernet), frames (as in frame relay), and cells (as in ATM) — but not blocks or segments which don't have a header. An Ethernet packet, for instance, is a MAC-layer data structure.

Packets are distinguished from TDM-type multiplexing frames by virtue of the protocol information contained in the packet header and trailer. Each packet contains the address of the sending and receiving stations, perhaps sequence numbers, error control information, and other service information, plus any data that is to be transferred (called the 'payload'). Packets are groups of bytes (often 128- or 1024-Bytes, but highly variable) which are treated as a whole throughout the network. However packets are also occasionally fragmented to be carried over a network segment using a different protocol ('encapsulated') and then re-assembled into the original packets again at the other end.

Some packet protocols strictly set the size of the packets, while others allow variable sizes and provide for a size field in the header. See cell and frame.

• The term is also very widely used to mean a specific packet-protocol, X.25, which has dominated both the public and private WAN business for a decade. X.25 is a store-forward system widely used for LAN interconnection, file transfers, transaction processing, and information retrieval.

• In X.25-type packet-switching systems there are two major types of packet: the call-request packet which sets up the call and establishes the route, and the data packet which is used interactively. See Tymnet, PAD and frame-relay.

• Packets differ from messages, in the fact that the message is totally self-contained; the whole 'file' is in the one bundle. So message switching nodes need to be able to store substantial amounts of data usually on disk, and they generally aren't expected to forward the message instantly.

• Fast packet-switching introduces another complication here. When it is said that ATM is a 'packet' system, this has meaning at two different levels:

– The first is that packets (such as Ethernet) can be carried across ATM networks by segmenting the packet into 40+ byte segments, and reassembling the packets at the other end. This is done at the AAL level.
– It can also mean that, at the cell level, the cells travel through the network using 'fast packet-switching', where the cells aren't delayed by being checked at each node.

• 'Fast-packets' don't exist. The term is 'fast packet-switching', and this generally means ATM cell-switching in silicon rather than software.

packet-mode hosts

Special computers with built-in X.25 ports.

packet radio

A technique once only used by ham radio operators to send data in packets over the airwaves, but now an important part of commercial mobile data communications. It gives portable computers with special packet radio modems, the same facilities as, say, an X.25 network. It is ideal for e-mail and transactions.

The Swedish Mobitex was the first commercial system to gain wide acceptance, but Motorola's DataTAC/Ardis system is now challenging them. The latest, however, is CDPD which reuses AMPS cellular facilities. GSM and CDMA systems will also have packet-radio facilities eventually (they only have circuit switched data at present). See all.

Packet radio has enormous advantages over circuit-switched radio data channels when it can be combined with a cellular infrastructure, especially when the location of the receiver is known (or can be tracked).

It provides a two-way paging system, in effect, which can be used for mobile e-mail, for navigation systems, for monitoring, for credit card transactions on buses or trains, for making changes to electronic signs, for controlling traffic lights, for emergency beacon transmissions, for fleet and taxi dispatch, etc. Circuit-switching is probably better for long-file transfer.

• The ham-radio packet system works in amateur bands, and include relayed transmission by other ham transceivers (called 'digipeaters') as onward switching nodes. With these systems there are three standardized types of frames: information (for data), supervisory (controls) and unnumbered (onward transmission).

packet storms

Aka broadcast storms. Packet storms have many causes on a LAN — a malfunctioning adaptor card may begin to send out thousands of malformed packets, or a packet with an incomplete address may confuse other devices and trigger thousands of queries from the other devices — resulting in the system flooding.

On very large networks packet storms happen regularly, and they create an overhead load on the network. Routers and repeaters are then generally used to break the network down into smaller segments, and so lessen the spread of disruption.

packet-switching

The technique of sending data in segments of a convenient size with a 'protocol header' attached. This is store-forward transmission.

Long files are broken down into packets (usually of a fixed length) and switched through the network individually using the header to identify the end-address. The switches are actually specialized routers, which are computers with CPU, RAM memory, an application, and multiple input-output ports connected to other nodes.

With conventional X.25 packet-switching, the packet is temporarily stored at each node and a full check-sum calculation is performed. If the packet has been damaged, the node will request a retransmission from the previous node. So the packet crosses the network in a series of hops.

X.25 packets always have two addresses (source and destination) in their header, and the destination address is read by each node in the network, which then makes a decision as to which direction the packet should be sent. In connectionless services, this is done independently at each node for each packet. Each packet then takes the best available path available to it at the time.

However, it is now more common for packets to be dispatched across the network only after a 'virtual circuit' has been established. All intermediate nodes will be advised as to which path to forward the later data packets, and the packets will be identified by a virtual connection identifier (VCI).

When the packets finally reach the recipient, a note is taken of the source address (within the header) and the packets are reassembled into a file.

This packet approach creates a quite different type of network to a normal circuit-switched network. In packet-switched networks all devices on the network remain physically connected at all times, and the networks can be configured in mesh-like topologies rather than tree-branch architectures.

The main problems in packet-switching networks are those caused by variable delays (which have, in the past, stopped them from being used for voice), problems of congestion (how much overload can each node handle), and what to do about damaged packets (ignore them, throw them away, and rely on the end nodes to request retransmission — or fix the damage at each stage).

• There are five major subdivisions of generic packet-switching systems:

— **LANs and WANs** in their wide variety, are all packet-based systems. PBXs are usually circuit-switching systems.

— **X.25-type packet systems** are the best known and are found almost everywhere around the world in private and public-access networks. X.25 has packets from 128- to 4096-Bytes which move step-by-step between switching nodes (routers), and each packet is error checked at each node. These are full store-forward systems, and they were designed to be highly reliable over the old copper networks.

— **Frame-relay** is a faster form of X.25 designed to operate in the more reliable environment of today with optical fibers.

Frame-relay systems pass the packets through the network without checking or asking for retransmission. If a packet is corrupt they will simply dump it and expect the end-node to detect and request retransmission. — they just 'relay' the packets. Because of the faster handling of the packets, these systems can carry private networked (low bit-rate) voice.

— **Cell-relay systems.** These are very short packets with virtual circuit-type 'addressing' in the header. These are very fast systems which can handle voice and video — and are the way of the future for telecommunications. See ATM and fast-packet switching also.

— **Message type systems** which send each message as a whole. A message can be seen as a packet of indefinite length. See X.400.

• Generally an X.25 packet on a national network will find its way from one end of a network to the other in less than 1/3rd of a second, however, under load conditions it may take a second. Most delays experienced are caused by the host computers outside the network, not the network itself

• One of the key values of packet systems has also been their rate translation and moderation. They are able to handle different input-output rates through large buffers and thus provide rate-translation, and they often act as the common denominator link between two otherwise incompatible systems.

• Packet systems are designed to be connection-oriented or connectionless (datagrams) — and sometimes both. If they are connection-oriented, a call set-up packet will first be sent through the network to establish the optimal path between sender and receiver, and this path will be identified by a virtual channel identifier and virtual path identifier, which will then be carried as the 'address' at the head of the packet.

Datagram services do not establish an optimal path, and therefore carry the full source and destination address. In fact datagrams may take completely independent paths to their destination, and can therefore arrive out of sequence — so they need a sequence number for reordering.

• Some of the newer packet protocols are classed as being 'fast packet-switching' where the packet passes through a node like an express-train through a switch junction — without stopping. This is the aim, anyway. Currently the switches are a bit slower.

The first element (the first virtual address bit) in the header address will be read, and the switch thrown to select a circuit between two outputs (0 or 1) before the next element reaches the junction. The rest of the cell then follows. See batcher-banyan switch and ATM.

• See also LANs, X.25, X.75, Tymnet, frame-relay, DQDB, FDDI, PAD, ISDN and D-channel.

packet terminals

A terminal capable of interfacing directly with the X.25 or Tymnet system using a leased line. It doesn't need a PAD because it handles the packetizing itself. The distinction is with character terminals which use dial-up to connect to a PAD. See also X.32.

packing density

A measure of the number of information units (pixels, magnetic domains, pits-lands, etc.) that can be stored in a certain area of physical space, or along the length of a recording track.

This is always a complex matter involving many trade-offs. The quality of the recording surface, the type of channel code used, the modulation method, the physical width of the track, the physical dimensions of the magnetic head gap, the wavelength of the laser light, etc.

• The term is also used in chip fabrication. Modern memory chips, for instance, need less than 1 micron of separation between the different elements.

Intel are currently fabricating chips using 0.35 micron technology and they are achieving 0.18 micron separation in the laboratories. Remember that packing density increases as the square of the separation change, so 0.33 is nine-times better than 1.0 micron separation.

Harvard University has announced a fabrication technique that etches circuits that are only 50nm wide (0.05 microns), well beyond the possibilities of conventional chip lithography. The process, however, may not be suited to mass production.

See Moore's Law and Debye limit.

PACS

Personal Access Communications System. This is a new PCS (1.9GHz) cellular radio design developed by Motorola. It combines the Bellcore WACs approach with the Japanese PHS/PHP Personal Handiphone. The T1 committee and the TAI have both recommended this as the PCS standard.

The wireless system is intended to interface over ISDN to the existing PSTN (Class 5) exchange rather than to a special cellular network. The wireline network will also provide Advanced Intelligent Network (AIN) capabilities.

Bellcore's WACS requirements covers the air interface, the radio system-to-network switch interface, and the radio system elements. There are two separate proposals before ANSI for Unlicensed PACS, dubbed UA and UB, for PHS- and WACS-based approaches respectively.

PACT

Public Access Cordless Telephony. The distinction is being made between Telepoint, CT-2, DECT (generally low-power and with limited, or no hand-off capability) and the cellular mobile systems made for use in moving vehicles (PMTS — Public Mobile Telephony Services).

PCS/PCN sits half way between these. It was originally seen as a super-cordless phone that you could take down to the local shops, but now it is hard to distinguish from full mobile cellular — except that it uses a higher frequency band.

PAD

Packet Assembler/Dissembler. That part of a packet-switching node that provides the interface between a terminal (usually a PC or dumb terminal connected via a modem) and the packet-switching network. It allows devices which don't themselves generate packets to connect to the network.

The full functions of a PAD node include setting up the calls, assembling the characters into packets, adding addresses, and providing buffering, check-sum calculations, rate-conversion and flow-control. You dial in to a PAD via a modem.

PADs operate using a number of pre-set 'profiles' and 'parameters', most of which can be changed by the user — but you need to know what you are doing.

To be compliant with the X.23 specifications, PADs must support X.3 (packet assembly and disassembly), X.28 (interface with asynchronous terminal), and X.29 (methods for controlling a PAD). See packet terminals, X.32, and X.25 also.

• The main error-checking used between the dial-up terminal and the PAD is Echoplex, so set Local Echo off in your communications program.

padding

Additional blanks, nulls, or zeros are often added to the beginning or end of data to make it up to some pre-defined size — usually to fill out to the end of a frame. In multiplexing, this may be used for synchronization purposes, and it is then called 'bit stuffing'.

• Many of the older databases required each field to be a fixed size, so padding characters are added at the end of the typed characters to make up this length. See null.

page

• **Computer memory:** A page in computer memory was usually 256-Bytes in the early 8-bit computers and 16kBytes in the later 16-bit PCs. It was the amount of memory that could be addressed by a single transfer of (respectively) an 8-bit or 16-bit address over the bus.

• **Videotex:** A videotex page is a specified small block of coherent information, consisting of one or more frames.

• **Publishing:** Don't confuse page (which is one side only) with sheets (2 or 4 pages), or with signatures (a folded forme ready for binding). You have a 'folio' (in this case meaning a page number) to each page.

page description language

A language like Adobe's PostScript or HP's PCL which can describe a page graphically for the transfer of a layout and all its text and graphics to a laserprinter or photo typesetter.

They are able to take all the elements entered into them by a desktop publishing program, order and layer all the elements (illustrations over text, etc.), handle the creation of the outline fonts, and build a full-page bit-map which is mapped straight to the laserprinter's memory and then scanned to the print drum.

So PDLs are both an intermediate file-type and a printer language. The later PDLs provide a page-imaging model with features for creating line art, halftones and color images. See PostScript and TrueType.

page imaging models

These are page description languages. See PDL.

page make-up

In desktop publishing, this is the same as camera-ready copy. The full layout with all elements, ready to be printed. It will usually have all text and graphics included, but it may need half-tone or color illustrations to be physically inserted at the pre-plate negative stage.

page printers

These are printers such as laserprinters, inkjet and LED printers, which first create the image of a whole page in memory (as a bit map) before the printing process starts. The distinction is being made with line or character printers.

Page printers always have substantial delays before printing begins, because the whole bit-map must first be generated (either by the printer or the PC) and stored in memory, before it is transferred to the printing drum as a continuous flowing stream. However once the bit-map has been established, numerous copies of the same page can be printed quickly.

For this reason, the oft-quoted figure of 'x-pages per minute' is almost meaningless with a page printer if you generally only print one copy of each page. You might be able to print pages at a rate of one a second, but it may take a minute to create the first copy in each case.

• A full A4 page, stored in monochrome bit-mapped form with a resolution of about 300 dots per inch, requires about 1MByte of printer memory.

paged virtual memory

The 80386 and 80486 chips allow programs to access more memory than is actually in the machine (using disk space as a 'virtual RAM') by swapping 4kByte 'pages' in and out as necessary. This is the same as the expanded memory process using bank-switching, but it draws its data from disk rather than high-memory. Segmented virtual memory is something different. See expanded memory, and segmented virtual memory also.

paging (messaging)

Paging is a one-way (simplex) messaging system, although people now talk about two-way paging (probably best called 'short messaging' or SMS).

There are a variety of receiver technologies:

• **Tone:** Originally pagers were radio systems which simply activated a buzzer or beeper tone. These were later modified to allow varying tones to indicate a pre-arranged message or to signal a degree of urgency. Usually four tones are now provided on tone-alert pagers.

• **Voice:** Some of the earliest pagers transmitted voice. They are now only used within hospitals or in very restricted locations, and some of these operated by induction rather than radio. There are many very funny stories about pagers making inappropriate announcements without warning.

• **Numeric:** Later numeric pagers were introduced, and these provided you with a coded message (using 4-bit 'bytes') which could be interpreted as 'Ring Office' or something similar (with pre-defined codes). The first just provided numbers, but later forms included codes which could be interpreted by the unit. Typical message length is now 5-Bytes, often with a maximum of 15 denary digits. Callers can send messages themselves, they phone direct to the transmitter and code the message by DTMF dial tones and/or voice responses.

• **Alphanumeric:** Later still, full alphanumeric paging came into being using 7-bit bytes. These can now hold a long series of messages in memory and replay them on command. Some take data automatically from broadcast newsflashes and stock-market sources. The typical message length is about 80 characters but some devices can handle up to 240 characters.

— **Transparent** or **value-added** pagers can handle long message files using 8-bit bytes — but not necessarily providing the automatic message-checking, etc. of normal paging. The maximum length is set by the service provider, and is usually in the range of a few kiloBytes — these files hog his transmitter.

• **Group:** To send a message to a pager, some numeric services require that you phone the special number associated with each pager and do-it-yourself with a tone-keypad. Others will have the call answered by a full-time pager operator, who will then type the information into the system — this is more common with alphanumeric systems.

Most paging services also provide the larger customers with PC software which allows them to send their messages to the paging transmitter direct (this is usually much cheaper for volume). Some of these services can only contact individuals, while others can contact a pre-defined group.

• **Protocols:** The major protocols used in paging are 5/6 tone, RDS, GSC, POCSAG, FLEX, Ermes, and TDP (see all). These protocols operate at rates from 512b/s to about 6.8kb/s, and some have high (100%) error correction overheads. Some also provide ways for the pager to check for missing messages. POCSAG is by far the most popular protocol around the world, although GSC is very popular in the US.

One of the characteristics of paging is that a number of different protocols can co-exist on the same broadcast frequency and be transmitted in sequence from the same transmitter site if the data rates are the same. Pagers are now designed to follow a strictly-timed schedule, and they only listen to their own time-schedule, and then only 'wake-up' when they recognize a preamble: this is a battery-saving technique.

There's nothing secure about paging. Each pager has a unique address, but it can receive every signal from the local transmitter. Generally it will only take note of a message carrying its own address.

• About half the world's pagers are still tone-alert, but this is declining. France and Australia both have a high percentage of alphanumeric pagers (about 70%) while most other countries favour numeric and tone. Numeric and tone paging is cheap because many more customers can be supported on a 5-Byte (average message) number system, than on one that allows 200-Byte alpha messages.

• Paging systems are divided into local (on-site) and wide-area pagers which cover a region or a nation. Local pagers may use an induction system rather than radio waves. A normal paging area will cover a large city. If you travel interstate you will advise the paging company to redirect messages to that new location (assuming it has coverage).

paging (general)

• Computers. See bank switching and expanded memory. Also virtual memory and paged virtual memory. Paging in programming terminology is similar to segmentation.

• In cellular mobile phone systems, paging happens when the base-station is actively seeking a mobile phone. It is either alerting it, to check its location (done about every half-hour), or to set up channels for an incoming call. Separate paging channels are assigned for this purpose.

The network will begin by paging the last cell in which the mobile's location was registered. If it doesn't reply, the reference area is widened to page all cells covered by a particular switch, and if this fails, then it pages all the network.

• In Telepoint systems (see CT2) the small one-way pocket phones were coupled with a one-way paging system to provide a call-in facility without needing location tracking. You would call and leave your number which would appear almost immediately on the pocket-phone. The owner would then reply.

• GSM has the Short Messaging Service (SMS) which is a very good two-way paging system. Unfortunately, at the time of writing the GSM MoU still haven't got their act together to provide universal roaming with this service.

paint programs

In a generic sense, these are graphic programs which treat the image in a single bit-mapped form — as a grid of pixels. Paint programs are simple to write and easy to use, but they have distinct limitations. You can't shift elements in a paint program or resize them, because they are not 'objects' but rather an integral part of the whole image.

If you remove an element, you leave a hole in the picture since the element wasn't layered over the background, but part of it. The distinction is with draw programs which are object-oriented, use vectors, and have multiple layered objects. There are also some hybrid forms.

• If you attempt to scale a paint program to a larger size, you are left with the fundamental problem of aliasing. If you double the dimensions (four times the size) then one pixel will simply be replaced by four, all set to the same color and density as the original. If you scale by less than double, however, the computer is faced with making decisions involving which pixels will now be the same and which will be different. See alias, hinting, and the jaggies.

pair

• **Twisted pair:** The twisted pair of a telephone local loop, and the extension of this into your house, probably using quad. Both Shielded (STP) and Unshielded (UTP) twisted pair is used for LANs. See both and Category #.

• **Fiber pair:** Optical fibers generally work in pairs because it is easy to regenerate the pulses traveling only one way down a fiber. However, see optical amplifier.

pair gain

A term used in telephony where basic multiplexing is involved. Pair gain is used in the local loop when the telephone company runs short of twisted-pair. It will then carry two or more distinct phone connections down one twisted pair using a small two-channel multiplexer on each end. See AML.

If 30 subscriber phone lines can be concentrated through multiplexing to travel over only two-pair (one outbound and one inbound pair), then you have a pair-gain ratio of 15:1. This is very common in older areas of US cities.

paired disparity code

This is Alternate Mark Inversion. See AMI.

PAL

Programmable Array Logic: A form of gate array or PDL. This is an integrated circuit whose exact logical function is programmable at the time of manufacture. These are PLDs in VLSI form where only the AND gates can be programmed.

Don't confuse this with PLA which is Programmable Logic Arrays (aka Programmable Gate Array). See Programmable Logic Devices.

• **Phase Alternation (by) Line:** The German-designed, pan-European television system which is based on 25 images a second and 625 lines, with a 2:1 interlace (50 fields a second).

It came into Europe in the 1960s (along with SECAM)

Phase Alternation refers to techniques developed to overcome the old NTSC problems of constant drifting and blurring colors ('Never Twice the Same Color').

PAL overcame the US problem of differential phase distortion which was affecting color fidelity by reversing the phase of the R–Y color subcarrier on alternating lines. Differential phase errors will then cancel out in the decoder.

This is possible with the quadrature phase modulation of NTSC and PAL subcarriers, where the two color difference signals are carried simultaneously, but not with SECAM.

PAL is the most widely used television standard around the world but there's a range of minor and major variations on the standard which are identified by added letters (i.e. PAL-B).The most extreme variation found in PAL standards is a version used in South America with 60Hz/525-line (PAL-M). However, there are many subtle versions of 50Hz/625-line PAL which have different separations between the sound carrier and the vision frequencies. Some TV channels are 7MHz wide and others are 8MHz (needed for NICAM).

In digital terms, a PAL video frame requires about 780 samples horizontally and 576 vertically (active lines only), or 449,280 pixels overall (at 25 frames per second). Each pixel is expressed in three colors at a pixel depth of two or three bytes (16 or 24 bits overall). When converted to the 4:2:2 format, and quantized at the 10-bit level, this is the basis for the standard 216Mb/s digital data rate (before any compression).

• PAL is recognized as requiring 4.43MHz of bandwidth for video (including blanking), and this would translate to about 159Mb/s in uncompressed digital form. Generally, the combined sound and vision is reduced to 140Mb/s, or even with compression to 34Mb/s, with little loss in quality.

PALcode

Privileged Architecture Library code. This is a layer in DEC's Alpha RISC chip which provides a BIOS-like library of routines. It substitutes for both BIOS and microcode, and is loaded into main memory from a FlashPROM. This approach is very flexible and provides the Alpha chip with a way to run more than one operating system.

palette

• In digital video and computers, this is the range of colors available for pictorial presentation. This is limited by the number of bits needed to create each pixel — which is why you have 2-bit, 8-bit, 16-bit video cards in computers. Note that the number of palette colors 'available' (say, 256) may be much greater than the number which can be held in the bit-map, or 'usable' at the one time (often only 16) on a digital color monitor. Analog color monitors have infinite variation.

• In modern computer graphics terms, the palette is usually 16.8 million colors. This is based on the number of bits of pixel-depth — it is 2^{24}, where each of the three colors (RGB) has a pixel depth of 8-bits. Professional video production equipment requires 10-bit pixel depth — but this is distributed unevenly, with more going to green and luminance, than to red and blue.

If, in a computer's video RAM, three full bytes were to be set aside for each pixel, and the screen had say 1 million pixels (1200 x 900), then the memory needed for a single screen bit-map would be 3MBytes.

So computer screen systems are often designed to handle only 256 colors at any one time — and these are then related to the 'possible' palette by reference to a CLUT. Then only a single 'reference' byte needs to be stored for each pixel — and that will reference 256 colors on the palette. The full range of colors can't then be offered 'at the one time' but a selection can be made from the full range by choosing the palette — and this can be changed very quickly by the program, when required.

• By extension, the word palette is also used to mean a series of icons which constituted a menu on a graphic-oriented application. You can have a 'tool palette' or a 'view palette'. These are really just menus in an icon form.

Palo Alto

A city on the outskirts of San Francisco. It is synonymous with Silicon Valley.

PALplus

An upgraded version of the standard analog PAL developed in West Germany (by Philips, Nokia, Thomson and Grundig) as an alternative to the MAC proposal for widescreen (16:9) TV.

It requires modified TV sets for the full-widescreen, but these are backward compatible with existing PAL sets to the degree that old sets can still view PALplus programs in letter-box format, and the new sets can view 4:3 screen images with side bars.

• Conventional PAL sets viewing PAL-plus signals handle it this way: of the normal 625 PAL lines, only 432 are used for the PALplus image (instead of the normal 575 for 4:3 formats). The extra 143 lines appear to be black (they are actually a negative voltage) — so producing the letterbox format.

• A PAL-plus 16:9 television set viewing a PAL-plus signal: These extra 143 lines are inverted in voltage and demodulated to produce additional image lines. One new line is inserted after every third line of conventional image, thus restoring the full 575 active lines of resolution — but also producing a 16:9 wide screen aspect ratio.

This technique requires a multiple frame store in solid state memory, and some extra digital augmentation codes are also transmitted in the vertical blanking interval. Ghost reduction techniques are also included.

This appears to be an excellent EDTV system which is backwardly compatible with PAL. It was seen as an interim standard in some parts of Europe until full digital comes along — but digital is coming fast. Even so, there are some arguments which say that the digital processing will mostly be done in the set-top IRD box — and the output will be analog.

So PALplus allows widescreen analog monitors to be introduced now, and digital transmission to be later handled through the set-top IRDs.

PAM

Pulse Amplitude Modulation. This was an analog telephone technique, trialed long before the development of PCM, which was intended to allow analog signals to be multiplexed together in a time-divided way. The idea was to transmit short pulses of signal, each pulse at an amplitude which corresponded to the signal's sampled amplitude at the time. This was an analog waveform sampling technique.

The standard PAM approach was to provide 8000 clock pulses a second with each pulse being amplitude-modulated. But since each pulse only occupies a time period of 5.2 microseconds (and there are 125 microseconds between pulses) then a number of PAM channels (usually 24) could be 'interleaved'.

This technique is very similar to digital time-division multiplexing except that there is no digital quantization of the pulses — they merely have varying amplitudes (in any fractional amount).

When it was used to transmit analog voice, however, the varying amplitudes of the pulses suffered badly from noise. So PAM proved to be as noisy as the normal analog transmission techniques it was intended to replace. However it was (and is) used in PBX systems, and it is the foundation of PCM coding.

- **Digital PAM:** With line-coding systems such as 2B1Q, the line voltage varies between pulses which are, say +5V or +15V, down to –5V or –15V (the four quaternary states) which transmit dibits at each 'event' (pulse). In effect, these are a digital version of PAM since the voltage states are integer values and not infinitely variable. It would be possible to use six or eight voltage states and send more than 2-bits with each pulse.

PAMR

Public Access Mobile Radio. A form of wide-area trunked radio which lies half-way between private mobile radio (PMR — including trunked private systems) and public cellular mobile telephony. This is not the normal CB (Citizen's Band) radio but a commercial trunking operation by the carriers. See MIRS and MPT 1327.

pan

A rotation of the movie or video camera in the horizontal plane. Tilt is in the vertical plane.

pan & scan

This is the alternative to letterboxing for Cinemascope and wide-screen movies on conventional 4:3 television screens without losing essential action at the edges. Pan & scan requires a controller to sit with a joy-stick viewing the film, and to pan & scan the 4:3 transmission area across the Cinemascope image to follow the essential action.

With broadcasting systems this process can be pre-recorded at the source in a video laboratory if all viewers are watching 4:3 screens. However when both 4:3 and 16:9 screens are being used by viewers, then the job of panning over the wider image is best done by the Pay TV set-top decoder box using control signals issued from the transmitter site.

The secondary value of this is that, for sports, home viewers can take control of pan, scan and even the zoom on their own screens via controls to their set-top box.

panoramic video

This is the only new form of image technique to be invented this century; it is a computer process which puts you at the center of a real-world environment at photo realistic quality. The current systems are driven by Apple's QuickTime VR.

Some of these images are computer created, while others are taken from real photographs which have been joined in a totally seamless manner. You can pan the display around 360 degrees, and tilt up and down, and often choose a new viewpoint location.

PANS

Since the humorous acronym POTS ('Plain Old Telephone Service') began to be widely used to distinguish the old analog system of PSTN from the ISDN variety, everyone has been trying to devise a version of PANS ('Pots and Pans' — Get it?).

There's also one called CUPS. However, the term PANS is now being used for some of the new advanced services. So far I have recorded these variations on the acronym:

- Pretentious, Although Necessary Services.
- Progressive And New Services.
- Picture And News System.
- Peculiar and Novel Services.
- Pretty Advanced Network Stuff.

Pantone

See PMS.

PAP

Public Access Profile. This seems to be a generic term for a Common Air Interface for PCN radio systems and wireless PBX services.

paper sizes

There are dozens of different paper sizes still in use. However, international paper sizes are now the A- and B-series (See A# and B# paper sizes).

These are the ones that matter for desktop publishing. For book publishing, see book sizes.

Many sheet-fed offset printing presses handle Medium size broadsheets (see) of 23 x 18 in., which provide a standard quarto-fold with pages of 9 x 11.5 in. or an octavo-fold of 5.75 x 9 in. (before trim) for book and booklet printing.

In many applications we are required to set accurate paper sizes for printer control.

• Since most software is of US origin and therefore uses inches, here are the equivalents:

A4 Letter	8.25 x 11.66 in.
US Letter	8.5 x 11 in.
US Legal	8.5 x 14 in.
Executive	7.25 x 10.5 in.
Tabloid	11 x 17 in.
A3 Tabloid	11.7 x 16.5 in.
A5 Letter	5.83 x 8.27 in.
B5 Letter	6.9 x 9.8 in.

• Some other useful older sizes of paper are:

Common Foolscap 8 x 13.25 in. or 203 x 330mm.
Quarto — see below

American Foolscap	8.5 x 13 in.	or 216 x 330mm.
Intl. Fanfold	8.25 x 1.0 in.	or 210 x 350mm.
SixMo is generally	6.5 x 8 in.	or 165 x 203mm.
Octavo is usually	5 x 8 in.	or 127 x 203mm.
Draft	10 x 16 in.	or 254 x 406mm.

• The so-called 'computer paper' which is widely used, is also highly variable. These are the main sizes.

279 x 356mm or 11 x 14 in.
328 x 279mm or 12.93 x 11 in.
368 x 279mm or 14.5 x 11 in.
389 x 279mm or 15.3 x 11 in.
452 x 279mm or 17.8 x 11 in.

• Quarto is not strictly a paper size by a fold, and so it differs in size in different countries. The UK quarto is considerably larger than the US quarto. Here are some variations:

Common Quarto	8 x 10 in.	or 203 x 254mm.
UK Quarto	8.25 x 10.5 in.	or 210 x 265mm.
American Quarto	8.5 x 11 in.	or 216 x 279mm.
It is also:	7.5 x 10 in.	or 190 x 254mm.

• Fan-fold paper differences give the most problems with page advancement. American applications use US-Letter fan-fold which is 8.5 x 11 in., while European systems generally use A-4 which is a metric size equivalent to 8.25 x 11.66 in.

Printers set to the wrong paper setting will constantly advance the page too much or too little by about a half-inch (18mm). This may need to be corrected in the control panel window, and also (for manual advance) in the printer itself, sometimes via DIP switches.

• In book, magazine and report publishing, allow the best part of a centimeter for cut-off from three sides (not the fold). See A#, B#, foolscap, quarto and book sizes also. The entry for envelope sizes includes standard label sizes.

paper tape

In the early days of computing, perforated paper tape was an effective means of storing programs and data — it was the primary I/O technique for a time. The system was taken over from the old Telex machines where it was originally used as a repeater device. See punch-cards also.

paper weight

The thickness of paper is measured in terms of its weight. In the metric system, paper is measured in grams per square meter (gsm or just gm). This is known as the 'grammage'.

English printers often still measure in units of the pound, and here the term 'basis weight' refers to the weight of 500 sheets (a ream) cut to a standard size (the basis). It therefore varies with the use of the paper.

For most normal page printing, paper should be in the order of 60 to 105 gm in weight; copy paper is usually 80 gm. Normal letterhead and leaflet paper is about 100 gm, while 140 gm is used for post cards and business cards. 60 gm is the lightest that can safely be run through a laserprinter, but 80 gm is better.

P/AR

Peak to Average Ratio.

paradigm

An old English/Greek term which has been appropriated and constantly debased by pretentious people in the computer industry, to the point where it has almost ceased to have meaning. The strict definition is 'pattern' or 'example' — but often it is used to mean 'model' or something similar. In computing we use it mainly to mean 'organizing principle'.

• Joel Baker's book, ***Future Edge***, defines a paradigm as a set of rules and regulations (written or otherwise) that establish or define boundaries. The paradigm tells you how to behave inside the boundaries to be successful.

• 'Paradigm shift' is a phrase coined by science historian and philosopher Thomas Kuhn in 1962 to describe what happens in scientific revolutions. He believes that a revolution takes place when those involved begin to see the world in a fundamentally different way — and this is the paradigm shift.

paragraph

• In the MS-DOS world, a paragraph in a text-based application is signaled by a hard-return (as distinct from soft-returns at the end of each screen line). These involve both a Carriage Return and a Line Feed character, in most word processors — but there are no real standards.

• In Macintosh files, the paragraph is signaled only by the Carriage Return, and nothing is inserted into the file at the end of each screen line. See hard-return.

• In the Intel 286 and 386 microprocessors, a paragraph is a unit of 16-bits. The segment address in these chips is one paragraph in length.

parallel

• **Peripheral connections:** See parallel interfaces below.

• **Computer buses:** The internal addressing and data buses within a computer are always parallel. The UART at the serial port has the job of converting the parallel signals into a single serial stream.

- **Processing:** See parallel processing
- **Video:** In video, see parallel processing and component video.
- **Coprocessors:** Don't confuse the use of 'parallel' maths coprocessors (& etc.) with parallelism. These processors effectively stall the operations of the main CPU whenever they take over the task, so this is not true parallelism.
- **Cache:** See look-aside cache.

parallel interface

Parallel interfaces and cables (i.e. for computer-to-printer links) have multiple lines, and transmit all eight bits in a byte 'abreast' (simultaneously). Usually, these eight data-lines are supplemented by a couple of lines of control information. Parallel interfaces of this sort are often known as 'Centronics', IEEE-488, or GPIB interfaces.

The value of the parallel connection is in the rate at which these peripherals can be supplied with data. A byte of information (which creates a character) is transferred roughly in the time it takes a serial interface to transfer only one bit.

Most parallel printers follow the Centronics protocols which are highly standardized from early PC days, so they are easy to set up and use.

Serial (RS-232) connections are much more complex and involve a range of controls which often make them difficult to set up — however, the parallel interface is often severely limited in length. Furthermore, the serial interface proved to be more versatile.

parallel processing

Parallel computing implies that a task will be broken down into a number of component parts, and each will be handled independently (and simultaneously) by a separate processor.

Computation using parallel processors appears to be the way of the future for many processor-intensive applications. There are three problem areas with parallel processing:

— communications between the processors,
— synchronization of the processors,
— scheduling the work to maintain orderly sequences.

Relational databases are inherently suited for parallel processing because they can be readily partitioned and distributed among the multiple processors. On-line transaction processing is also often inherently parallel with multiple tasks to be performed before a transaction is completed.

The main distinction here is with von Neumann machine architectures.

- There are three major topologies in parallel computing:

— pipelines,
— tree structures, and
— hypercube.

See SIMD and MIMD, transputer, Occam, microparallelism, superscalar architectures, superpipelining and pipelining also.

- In video the term is applied to the handling of separate components of the video signal — the luminance (Y), the chroma (C) or the color-difference (UV) or basic color (RGB).

Devices such as switchers, SEGs, and recorders will need to use parallel processing internally, unless the components are combined into a 'composite' form (PAL, NTSC, of SECAM) as is currently done in many studios with legacy equipment.

parallelism

See parallel processing, pipelining, hypercube, SIMD, MIMD and MPP.

parameter

Variable information — usually data which you provide to the computer as a form of control — such as dimensions on a CAD drawing, phone numbers for a modem, character strings for searching, etc. When measuring widgets, the length, breadth, weight, and color are the widget's parameters.

A parameter in a computer procedure, is a symbol or variable name. Later it will be replaced by a value — but only when the procedure is called. For instance:

```
PUT X INTO THE BOX
```

Here X is the parameter. Later it can be changed to 'Toy' or 'Book'. There are both formal parameters and actual parameters — but I can't find a definition of the difference. My guess is, that the difference is between the X and the 'Toy' (above).

- When we talk about vector graphics using x/y coordinates, then x and y are parameters.
- The term is often incorrectly used to mean 'boundary'. This confusion comes from the casual use of 'dimension' to mean limit, and/or confusion with the word 'perimeter'.
- In the PAD control of X.25 packet-switching systems, a number of 'Parameters' jointly make up a 'Profile'. Each profile (a list of parameters) will hold different values for at least some of the parameters. A profile is just an easy way of handling a related group of parameters.
- A parameter block is a control block. See.

PARC

Palo Alto Research Center — of Xerox. Where many of the best brains worked on many modern aspects of computers. The modern microcomputer was virtually invented here.

Pareto Principle

The Pareto Principle says that 80% of the result comes from 20% of the resources. This is called the 80/20 rule. It has also been applied to say 80% of the problems will come from 20% of the network (or whatever!).

It is commonly quoted as if it were natural law, but it has no statistical or common-sense validity whatsoever — other than that exhibited by any function with a normal bell-curve distribution.

It is often conjured up to add pseudo-scientific support to untested observations. You could equally make up a 70:30 rule, or a 90:10 rule, or a 60:40 rule.

parity

The idea is simple; add up the number of 1-bits in the byte (usually a 7-bit byte), and then add an extra 1-bit or not, depending on whether the count was odd or even. This gives you a check if one of the bits in the original byte becomes accidentally changed during transmission or storage.

- Parity is an Exclusive-OR function.
- It is used in a number of ways:

— Parity is added to bytes in the ALU of a microprocessor to protect the data during the CPU's operation.

— Parity is often added to computer memory in the form of a 9th bit to keep a check on memory. See Parity RAM.

— Parity is used in communications today, but almost only when 7-bit bytes are being transmitted. The parity-bit then becomes the 7th (most significant) bit since the bits are numbered from 0 to 7 (R to L). Both sender and receiver need to know that parity has been set to odd or to even — or to none. See odd parity and even parity.

- Today, when the full 8-bits are normally transmitted, parity is usually set off (to 'None'). However there are some strange mainframe systems which use 8-bits and add parity to create a 9th-bit. With start and stop bits added, these then creat an 11-bit format in modem transmissions. This is rare.

- The concept of parity extends well past the use of odd/even and no parity. It is more widely used than as just single bits added to the end of 7-bit bytes in data communications.

For instance: consider the problem of sending 8-bit bytes over a 7-bit system (graphics or binary files over many X.25 networks). The 8-bit byte can be broken down into two 4-bit nibbles. But if these 4-bit nibbles are to be transmitted over a 7-bit system, we have room to to append to each, a 3-bit parity-check.

The first will be the parity of the nibble's 1st, 2nd and 3rd digits; the second is the parity of the nibble's 1st, 2nd and 4th digits; and the third is of 1st, 3rd and 4th. This is called Hamming code; it provides error correction, not just detection.

- Parity checking is also used internally in most PCs as 'Parity RAM'. The 8-bit bytes will normally be stored in eight separate chips, but a 9th-bit can be added to store the parity value of each byte. If a fault is detected, you'll get a 'parity error' message, that means you have a faulty memory chip (or a faulty parity chip). US government regulations require many PCs to have this internal parity bit.

parity bit formats

In 7-bit modem communications, the parity bit will be the 8th bit (this is bit No.7, the most significant, and it is generally the last to be sent — since these are little-endian transmission systems).

In 8-bit communications, parity is generally not used — but bytes of data flowing in 8-bit buses within a PC may have a 9th bit added as far as the UART. If parity is applied to an 8-bit byte in communications, it becomes the 9th bit (bit No.8).

Generally in computers or communications networks using 7-bit ASCII, the parity bit becomes the 8th. With 8-bit EBCDIC, the parity becomes the 9th. Some systems send parity first, and some send it last.

You'll find these parity check formats in modem communications software:

- **Odd Parity** — if the sum of the one-bits is even, the parity bit is turned on (the total is made odd).
- **Even Parity** — if the sum of the one-bits is odd, the parity bit is turned on (the total is made even).
- **Mark Parity** — the parity bit is always on.
- **Space Parity** — the parity bit is always off.
- **No parity** — the parity bit is ignored (and it may be removed, as superfluous).
- **Ignore Parity** — the eight bit is not a parity bit, but part of the extended (256) character set, so it may **not** be removed.

Parity RAM

Most PCs (but not all) provide a ninth parity check in RAM. This means that nine chips will be required instead of eight for main memory, and this adds 12.5% to the cost of memory. For this reason some clone makers have dropped this ninth chip.

The main value is in detecting the occasional bit in memory which may have been corrupted by cosmic rays. Note, however, that it can only detect an error, not correct it.

park

The read/write head of a hard disk flies just above the surface of the disk on a layer of fast-moving air, it is not permitted to touch the disk surface. So when you turn the power off, the disk must fly the head over to the center near the spindle and 'park' it in a safe area. Be careful when you move old computers if the disk is spinning.

PARS

Private Advanced Radio System. An 80 channel personal two-way radio system using trunking technology at present under development in Europe.

parsed text files

Spreadsheets.

parsing

The process of identifying scientifically a word (or phrase) by nominating the part of speech, the inflectional form, syntax relationship, etc.

- In voice recognition, parsing is the process of isolating words (identifying a beginning and an end) before passing the digitized sound on for template matching.
- In program languages, parsing is a procedure the computer must perform to identify the various parts of an expression in higher level language programs. In dBase, for instance, the command SUM PRICES FOR ORDERS = "CURRENT" must be parsed to identify SUM as the basic command, PRICES as the

argument/field name, FOR as the conditional request for a search, and ORDERS as the conditional field name — which must be matched to the data, CURRENT.

Parsing requires a grammar which defines possible structures. It can be done top-down, or bottom-up.

Part 15

A section of the FCC regulations which permits cordless phones and low power radio devices to be used without a specific license. It is mainly for CDMA systems up to 1 watt, or FDMA at 100mW and below. See ISM band.

partitioning

- **Hard disks:** The division of a large hard disk into sections — each of which will be treated as if it were a separate hard disk (a 'volume'). Partitioning allows multiple users each to have their own 'virtual' hard disk, and it can also be used to provide multiple operating systems on a single disk.

It is also commonly used these days to separate volumes so that only one has automatic data compression (DoubleSpace) applied, or to make it easy to perform full-volume searches. To partition a hard disk in MS-DOS machines you will use the Fdisk utility. But remember that partitioning destroys any existing data on the disk.

- **Trunked Radio:** In trunked radio, when a large number of mobiles are contending for access to the base station, the signaling system could easily experience overload. In the MPT 1327 protocol, therefore, the base will partition the mobiles and only permit half of them to contend. If overload persists it will sub-partition, so only a quarter are able to contend, and so on.

- **LANs:** In a 10Base-T network the wiring center acts to partition malfunctioning nodes. This is just a fancy term for 'disconnects'. Token Ring MAUs also do this job.

PAS

Publicly Available Specifications. In ISO-speak, this just means *de facto* standards.

PASC

Precision Adaptive Sub-band Coding. The compression technique used by Philips in the DCC digital tape technology. It was derived from DAB techniques, and it gives compact cassettes the subjective quality of a CD audio disk at about one-quarter of the bit-rate. It is a masking technique which throws away high-frequencies if it judges that these will be blanketed by louder low-frequency sounds. See DCC, Mini-Disk, NICAM and MASCAM also.

Pascal

- A unit of pressure (pascal).
- A high-level computer language.
- The inventor of the first mass-produced working mechanical calculator.

passband

The range of frequencies that can pass over the network, within the current limitations.

- So, for an average analog telephone network, end-to-end connections pass through local loops at each end, and the interexchange network and electronics between. When passing through switches and multiplexers, the passband is only a guaranteed 300 to 3200Hz (up to 3800Hz in some more modern networks). This sets the rate limit for modems.

- However the local loop (from the CPE to the local exchange) has a passband in the order of 1MHz — provided the signal does not need to carry further through the network. For this reason ISDN and ADSL (access technologies only) can carry data at much higher rates than modems.

passive devices

A passive device is one that doesn't require power for amplification, switching, etc. Most passive devices will consist of transformers, resistors, condensers, beam-splitting prisms, etc. to perform filtering or bandwidth-splitting functions.

The distinction is an important one in telecommunications, because active devices in a street cabinet or manhole require mains power, and this creates problems. However most of the experiments with passive optical splitting techniques for fiber-to-the-home, have been replaced by active-hubs which now split and amplify the signal, and feed it to coaxial cable.

- In chip fabrication and electronic circuits, distinction is made between active devices (mainly transistors and gates) and passive components such as resistors and capacitors.

passive matrix

An LCD screen which uses a 'subtractive' approach to producing color — as distinct from the active-matrix light generation approach. These systems were previously called 'light gates'. See TSTN and light gate.

passive nodes

All nodes (computers, or LAN cards) on a network can read every packet that passes, but some nodes repeat (regenerate) every item of data on the network (Token Ring). With other systems the nodes are passive, and they stand aside and just watch (Ethernet, AppleTalk).

patch

- **Programming and hardware:** A 'quick-fix' on a programming problem without getting at the fundamental cause of the problem. This suggests that further bugs may be generated. Patches can be made in many ways; some intercept error messages and then automatically intrude and fix the immediate problem, then set the program back on its way. This sort of patch may work perfectly 99% of the time.

Patches can be hardware or software. In the Pentium floating-point inaccuracy problem, a temporary fix was provided by a couple of software patches which corrected a hardware problem with divisional look-up tables.

These patches were in the nature of interrupt/utilities. They checked to see whether a certain type of calculation was being performed, and if it was, they would provide a corrective multiplication using a look-up table.

Patches can also be system instruction of a more permanent nature that override or modify the way hardware works. These will usually be used to update an older machine.

• **Graphics:** In graphics, patches are curved surfaces whose shape is controlled by numerous control points. Smooth curves can be created and changed by moving these control points. There are a number of algorithms for controlling patches. See B-Spline and NURBS.

patch panel

A plug-board. The old hand-operated manual telephone exchanges were patch panel operations. Patch panels are still used in audio studios or wherever many different cables need to be interconnected in various ways at various times. Patch panels are a 'cross-point' or a basic form of 'switch matrix' (but with very slow manual switching!).

Note that some will only allow an entry point (say on the left) to be switched to an exit point (on the right), while some are any-to-any devices — and this is the same with modern matrix switches.

path-control

• Path-control layer: The level 3 layer in the SNA architecture. This is the layer that routes packets through an internetwork.

• Network path control: The contrast here is with NAUs which provide upper level service in SNA networks. Path-control networks consist of low-level components that control only the routing and data flows, and which handle the physical data transmission between SNA nodes.

path independent protocol

See fixed path protocol.

pathname

Every file on a disk volume has a name — and sometimes a name-extension. When the file is held at low level in the disk directory hierarchy, it will have as its unique pathname the filename + extension, and all the sub-directory and directory names of the levels above.

For instance a text file named TEST, will have a filename TEST.TXT. If this file is stored in the LETTERS sub-directory of the MAIN directory on the volume C: of the hard disk, the full pathname would be:

```
C:\MAIN\LETTERS\TEST.TXT
```

• In the Macintosh, the same principle applies. However the directories and subdirectories are known as folders, and the colon is used as a separator rather than the slash. The Mac doesn't use extensions or cryptic volume names either. So the Mac path-name for the above would probably be:

```
MacintoshHD:Main:Letters:Test
```

Pay TV

A variety of cable or radiated (terrestrial and satellite) broadcast systems where the user pays to see a program on either a subscription basis or on an individual viewing basis. Note that the payment is for the program, not for the supply of an electrical connection which delivers advertising-supported programming material (probably stripped off-air). This is not generally understood in countries which haven't developed a cable system along US lines.

Pay TV services in the US are 'Premium' services which are 'extras' added to (and charged for) above the Basic Cable subscription service. This is not a technical necessity, but rather the consequence of how the system has developed.

These services will scramble the image of the Pay channels (but not necessarily the Basic channels), so that a special decoder is needed to see the pictures. This is the only way the provider has of forcing payment for the premium, as distinct from the basic services (base on connection).

Modern set-top boxes are very sophisticated, and the scrambling key can now be stored in a plug-in smartcard, rather than in the box itself. This means one standard box can be defined which all program providers can use (see DAVIC).

It also means that a variety of payment systems can be implemented which don't depend on monthly prepayments, or programming channels only sold in multi-channel packages. The program provider can address thousands of set-top boxes over the air in a few seconds, and authorize them to receive signals on a program-by-program basis.

So Pay TV is now a vague and generic term in some parts of the world, and it can be used to cover:

— **Subscription Premium Package services** (usually 4–6 channels of movies, sports, news and entertainment on a monthly basis).

— **Pay-per-view** (individual payments for each program, bookings made in advance, usually by telephone).

— **Impulse Pay-per-view** (same as above, but paid instantly by local electronic purse or smartcard, or via direct interactive connection over the cable).

— **Near Video-On-Demand** (multiple channels of Pay each running the same program but staggered in time).

— **Video-on-demand** (the channel shows what you want, when you want it, and remains under your control for the duration — you can stop, rewind, and pause).

• See premium TV, NVOD and VOD, and DAVIC.

payload

The payload of a system is usually referenced to the system as a whole, and this can create confusion.

• For instance, the payload of a launch-rocket or Shuttle is the satellite it is carrying. But the payload of the satellite is the electronic devices it carries — not including the thruster and control systems needed to control it in orbit.

• In packet-switching, the payload is that part of the packet not required for use by the transmission system (the overhead). In other words it is the packet, less the header (source and destination address plus other) and the tail (error checking bytes).

However, even this is still a bit vague, especially with ATM. The new fast packet-switching systems segment Ethernet packets, for instance, and carry them as 'payload' (within the data segment of each cell). But some of that space is occupied by the packet header, and the data-segment of an ATM cell is now normally subdivided in formal ways which constitute additional 'overhead' (and therefore reduced payload). See AAL.

PBX

Private Branch Exchange. A company's own internal telephone switching unit which is also called a PABX in many parts of the world. The term is applied to both analog and digital automatic switches, but not to the old manual exchanges.

— **1st generation** (1920) private switches were analog electromechanical. These were known, at the time, as PABXs.

— **2nd generation** (1970) switches were electronic with internal digital switching. The voice was still handled in analog. In North America these became known as PBXs, while most of the world continued to use the PABX acronym.

— **3rd generation** (late 1980) switches are integrated digital voice/data systems.

PBX was an old generic term in most of the world which applied to all types of customer switching gear — automatic, or not; the old manual exchanges were often called PBXs. See ACD and Centrex also.

PBXer

A PBXer is a phone phreak who uses company call-diverters to place long distance calls. See call diverter.

PC

• **Personal Communicators:** As in PCNs (Personal Communications Networks). This term encompasses pocket cellular phones, two-way pagers, and PDAs.

• **Personal Computer:** A general term applied to all microcomputers used as stand-alone desktop units until Sun came along and began to call their high-end powerful machines 'workstations'. At a later stage there were some disk-less versions on LANs which became known as 'diskless workstations' but these were at the opposite end of the power scale.

The distinction is still made between PCs at the more mundane office-work (word processors and spreadsheets) end of the complexity scale and 'workstations' at the upper end.

It is hard to accurately define a point which separates these two categories, although workstations obviously have much more specific and power-hungry applications. They will probably also have larger full-color screens, and run graphics-intensive programs for CAD and similar applications.

• The term PC was also arrogantly appropriated by the IBM camp to refer to the original IBM personal computer which used the dreadful 8088 chip. They were trying to suggest that the IBM PC was the only 'real' personal computer and that Apple IIs and Amigas (etc.) weren't. At the time the Apple IIs and a half-dozen other machines were probably more powerful and flexible than the IBM PC.

By extension, and through heavy promotion by IBM, the term 'PC' gradually came to mean the IBM-PC, XT, AT, 386 and all the compatibles — and to largely exclude the Apple Mac, Amiga and others which had pioneered this industry.

PC Card

It appears that all PCMCIA cards after Version 2.1 were supposed to have been called 'PC Cards' — but this didn't happen. This new PC Card standard was designed to support 32-bit interfaces to the new PCMCIA Card Bus, which can operate at multimedia rates up to 132MB/s. It has DMA and can operate on the lower 3.3 volts. The card standard also has excellent built-in power management. See PCMCIA.

PC-DOS

The first version of MS-DOS was a rewrite of QDOS, and it was leased, on a non-exclusive basis, by Microsoft to IBM for the first IBM PC. IBM itself owned the rights to that essential part of the operating system called BIOS (Basic I/O), but this part was in ROM, so it didn't affect Microsoft's rights.

However BIOS, both by being in ROM and being machine specific, successfully stopped the clone-makers from making machines that were 100% compatible for many years — and being 99% compatible wasn't good enough. Later IBM's BIOS was reverse-engineered and re-created by the clone makers — which then opened the market to Microsoft for the wholesaling of MS-DOS to the clone makers.

Microsoft has since made its operating system available directly to anyone who wished to buy, while IBM packaged their version for their own machines under the PC-DOS title.

To all intents and purposes they were and are the same, however there are now enough incompatibility with some versions to create slight problems. The main difference was in the utilities and drivers included in the disk packs.

• PC-DOS 7 is the current release (Aug 1995). We don't know whether Microsoft will continue to bring out new versions of MS-DOS, now that Windows 95 no longer needs it as a base.

PCB

• **Printed Circuit Boards:** Modern boards use multi-layered techniques and surface mounted components.

• **PolyChlorinated Biphenyl:** This is a very dangerous chemical used in some old transformers and fluorescent ballasts.

PCI

• **Protocol Control Information:** This is OSI-talk for the header on any packet of data.

• **Peripheral Component Interconnect (Interface):** The PCI-bus. See.

PCL

Printer Command Language. Hewlett-Packard's page description language invented for the first commercial laserprinters. PCL is now up to Version 5 and is very widely used by many laserprinter manufacturers; it is really the *de facto* standard.

It provides scalable fonts and page layouts without the expense of PostScript cartridges. PCL has gone through many evolutionary phases. Make the distinction between PCL and a PDL (a generic term for all page description languages).

See PostScript, GDI and PDL also.

PCLP

A framing protocol used with ATM and DQDB fast-packets. The common PLCP standards are 622Mbit/s (STM-4), 155Mbit/s (STM-1), 140Mbit/s (E4), 45Mbit/s (SD3), and 34Mbit/s (E3).

PCM

- **Pulse Code Modulation.** See.
- **Plug-Compatible Mainframe:** Vendors like Amdahl and Hitachi which make mainframes that can run the IBM mainframe operating systems (like MVS/ESA).
- **Plug Compatible Manufacturer:** A company which makes peripheral equipment that can be plugged into standard computer devices.

PCMCIA

Personal Computer Memory Card International Association. This group standardizes memory cards (The 'PC Card'; see).

The current PCMCIA Version 2.1 standard in memory cards is for a semiconductor-based card, the same size and about four-times the thickness of a common credit card. It has 68 contacts on one end.

The major improvement of this version over 1.0 (Aug 1990) is the ability to execute programs in card memory. Version 1.0 cards were secondary-storage memory cards only.

Version 2.1 allows the computer to treat the card RAM as an extension of computer memory — a process known as XIP. These cards also have enhancements to support I/O devices and dual-voltages (5 and 3 volts).

So with Version 2.1 and XIP, the PCMCIA interface specification is now closer to a system bus than just a physical connection to a memory card. It can also support a wide range of computer-peripheral functions, such as plug-in modems, LAN controllers and small hard-disk drives.

Be warned! There are many inconsistencies in the PCMCIA standards — some in physical dimensions and others in electrical incompatibilities. Always try before you buy.

- There are three card thicknesses:
- — Type I are 3.3mm thick,
- — Type II are 5mm thick in the center, but with 3.3mm edges.
- — Type III is 10mm thick in the center and cannot fit into Type I or II slots even though it still has edges of 3.3mm. It is specifically for small plug-in hard-disk drives. Type III cards take two normal slots in many laptops.
- After Version 2.1, PCMCIA cards were supposed to have been called PC Cards — but this marketing ploy didn't work. This new PC Card standard supports 32-bit interfaces to the new Card Bus, which can operate at multimedia rates up to 132MByte/s.
- It's an oldie — but a goodie! I still like this interpretation of PCMCIA: 'People Can't Memorize Computer Industry Acronyms.'

PCN

Personal Communications Network. A term which has both a generic and two specific meanings — and the meanings are changing constantly.

- In the generic sense, PCN is the UK version of PCS. This refers to the future pocket-sized communications devices (phones, PDAs, two-way pagers, etc.) which use a cellular mobile network. Both TDMA and CDMA are competing for territory here, with DCS-1800 (TDMA) winning in Europe, and CDMA winning in the USA.

PCN will rely on intelligent networks with facilities to locate and track users, and methods for automatic billing, identification of roamers, etc. The theoretical concepts projected for PCS include the distribution of intelligence across the public network, so that wireline and mobile services merge and special mobile-cellular exchanges will no longer be necessary. This becomes UPT.

It is claimed that 'PCN has only a little to do with the air-interface'. This may be true in theory, but, in fact, the whole battleground is to do with the control of the air-interface. So DCS-1800 is winning in Europe, CDMA is winning in North America and parts of Asia, and PHS is winning in Japan. See PCS also.

- The term is used in Europe for DCS-1800 (see) which is based on GSM and being used as a cellular system operating at 1.8GHz. In the USA the same standard is called DCS-1900. DECT is also a possibility in Europe, but they haven't yet defined an acceptable public air-interface standard. See GAP.
- The US FCC has defined the PCN terminal as an inexpensive pocket-sized terminal by which 'it may soon be possible to reach individuals at anytime in any place using a single telephone number'.

They say it is a microcellular, digital, low-powered, wireless system for neighborhood, residential cordless, small-business Centrex and Wireless PBX use. See PCS, DCS-1800, CDMA, DECT, CT2 and Telepoint.

PCR

Phase-Change Rewritable disks. A name applied in 1994 to a new optical disk process developed by Matsushita and Toshiba for high-density (1.5GByte) CD writable disks. In the laboratory, it was apparently possible to get 500,000 write cycles, and the claim is that new techniques can achieve a 5–6GByte disk storage capacity with new short-wavelength (blue) lasers. See PD, CD-R,CD-WO and Super Density CD.

PCS

• **Personal Communications Services:** Those cellular-type phone services provided by a PCN, but also the US general term for any personal communications device licensed to use a set series of frequency bands in the 1.9GHz region.

The term was originally applied to low power pedestrian-oriented services with a range of only two-to-four city blocks, using simple basestations.

The US FCC has now auctioned a number of frequencies in the 1850—1895 and 1910—1975MHz bands for national and local PCS services — which will range from two-way paging, to PDAs and cellular phones. The FCC says that PCS devices will: 'help meet our ever more mobile society's rapidly growing demands for on-the-go communications. PCS will include small, lightweight multi-function portable phones, portable facsimile and other imaging devices, new types of multi-channel cordless phones, and advanced paging devices with two-way data capabilities.'

Some base stations will be part of two-way HFC cable networks which will supply telephony services and video. See FSN.

The wider philosophy of PCS, however, is much the same as PCN. When coupled with UPT, you would use your pocket-size phone at home via a home base-station; take it with you when you leave home in the morning to use in a car or train/bus while commuting; and at work it would dial and receive calls through wireless links to the company PBX.

In the view of AT&T, Apple, IBM, HP and a few other companies, the PCS concept equally extends to palmtop, notebook and pen-based computers with built-in radio links, and short-messaging capabilities.
See PCN, DCS-1800 and DECT.

• **Personal Conference Specification:** This is a cross-platform, PC conferencing standard being devised for RISC/CISC, cross-platform standards, by Intel, AT&T, Lotus, HP and DEC, etc. It will use MPEG compression.

PCS bands and regions

The FCC auctioned the PCS spectrum as two main bands: the narrowband and broadband PCS (some minor bands are due to be auctioned later). The distinction between these two is in the amount of spectrum and the frequencies allocated to each.

• Narrowband PCS is in the 900MHz band with 26 channels. Some are 50kHz unpaired channels, and others are 50kHz duplex pairs. The unpaired channels are probably best used for paging, while the pairs could be for packet radio.

• Broadband PCS is in the 2GHz band from 1850—1990MHz. These come as either 30MHz or 10MHz duplex-pair allocations per license. The two technologies most favored here are DCS-1900 (TDMA) which require 30MHz of bandwidth for a substantial service, and CDMA which can work well in either 10MHz or 30MHz bandwidths.

The licenses are also being awarded in terms of coverage area in the following classes: Nationwide, Regional, Major Trading Areas (MTAs), and Basic Trading Areas (BTAs) based on Rand McNally. In effect, a BTA is a group of counties, and there are 493 in total. MTAs are much, much larger, and there are 51 of them covering the country. Lastly, there are six narrowband PCS regions, with each of these being a collection of MTAs.

• The FCC has also released 50MHz of spectrum for unlicensed PCS use for a range of devices including wireless LANs, from 2390—2400MHz, and 2402—2417MHz (Part 15 — for data, cordless phones security, utility monitoring) and 4660—4685MHz for fixed mobile services. The band 1910—1930MHz was released previously.

PCTE

Portable Common Tool Environment. A standard environment for CASE developers which was proposed by the European PTTs and developed under the Esprite program. It has now been accepted by many US telecommunications equipment and mainframe vendors.

PCX

PC Paintbrush files. An image file format developed by Zsoft Corp. It is used also by Microsoft in Word for Windows and Excel and in a number of desktop publishing applications (along with TIFF). It supports monochrome, 16 colors, 256 colors with gray palette, 256 colors, and 24-bit color.

PD

• **Public Domain:** This means that it can be used without copyright release. It doesn't necessarily mean that copyright doesn't exist — just that the originator has issued a blanket license for anyone to use it non-commercially. Copyright can be exerted at any time if the originator/owner so desires, and it may be conditional on only allowing non-commercial use.

Also, don't confuse Public Domain with Shareware rights, which certainly remain under the control of the copyright owner.

Note that the term public domain is often used incorrectly on the Internet, just to mean that something is freely available.

• **Par Disk:** Matsushita's marketing name for a rewritable optical disk which can read data four-times as fast as conventional CD-ROMs. See below.

• **Phase-change Dual:** The Matsushita PD technology itself. The recording medium is a crystalline layer of germanium, tellurium and antimony. This is an amorphous recording system where the difference in reflectivity between minute spots on the disk surface which are crystalline (reflective) or amorphous (unorganized and less reflective) can be read by a low-power laser as pits and lands.

The recording process is one of heating specific areas of the disk with a high-power laser, and the erasure process is one of reheating the whole disk and quickly cooling it to produce a general crystalline surface once again.

Matsushita has a half-height drive with one-sided disks holding 650MBytes, and having a read-rate of 870kb/s. The tracks are circular, rather than spiral like the CD's — although they still call this drive a writable CD-ROM because it can sense which type of disk is in the unit, and change accordingly.

• The original PD unit was known as 'Phase-change Drive' and the technology was then called PCR (see) for 'Phase Change Rewritable'. This system is also sold under the PD label, but it requires a higher-powered laser and so is incompatible. NEC is producing a PD drive as part of its multimedia line, and several US manufacturers are tooling up to this standard. Toshiba is producing a 3.5 in. PD optical disk drive unit.

PDA

Personal Digital Assistant. Apple's term for the Newton MessagePad and other similar handheld computers which tend to use handwriting recognition or soft-keyboards for data entry. Apple coined the term for the small pocket-sized devices in their own Newton line which they also plan to extend to larger tablet devices for EDI warehouse use, and for messaging storage devices on wireline telephones.

PDA has now caught on as the generic term for similar devices made by other companies, such as Thinkpad, Zoomer, and Presario. See PIC also.

PDC

• **Program Delivery Control (Code):** A video-recorder control service devised by the European Broadcasting Union (EBU) which signals the start and end of a television broadcast for automatic control of a VCR. This is a variation on teletext control using a bar-code for input.

It is a development of the German VPS idea which sends small packets of coded information during a television program. PDC requires a special decoder in the VCR to identify and select the timing and channel codes. See also VideoPlus.

• **Pacific Digital Cellular:** The renamed JDC (Japanese Digital Cellular) technology which they tried to flog to Asian countries. It has since been renamed again — see below.

• **Personal Digital Cellular:** The PDC standard is now the Japanese digital cellular standard which is similar to American TDMA (NADC). It operates at two frequencies:

— In the 800MHz band between 810—826MHz and 940—956MHz,

— In two 12MHz wide 1.5GHz sub-bands between 1429—1441/1477—1489MHz and 1453—1465/1501—1513MHz.

The duplex separation of PDC is 130MHz in the 800MHz band, and 48MHz in the 1.5GHz band. The channel spacing is 25kHz, (actually, interleaved 50kHz).

There are three traffic channels to each carrier and 1600 carrier bands in all three frequency ranges. PDC is said to be highly spectrally efficient — they claim 2.2-times the capacity of GSM per megaHertz.

The vocoder is VSLEP at 11.2kb/s, and the overall transmission rate is 42kb/s using $\pi/4$ shifted DQPSK. A half-rate coder is under development.

PDC/JDC has been less than a raging success in Japan and it doesn't look like going anywhere outside Japan. It is far too expensive. See PHP also.

PDE

Portable Data Entry devices. These are small handheld data-entry units used for stock-checking, etc. in warehouses and supermarkets. The term covers a wide range of equipment, from notebook computers to A4-sized tablets with handwriting recognition. Many will have a bar-code reader attached, some may use soft keypads or keyboards for additional data entry, while others will have a form of electronic mark-sensing. See PDA.

PDF

Portable Document Format. Both PostScript and non-PostScript files can be translated into PDF format. See Acrobat.

PDH

Plesiochronous Digital Hierarchy. PDH is the current way in which digital data streams are multiplexed in a standardized way into wider-bandwidth bearers for carriage across a network. Plesiochronous means 'almost synchronous' — which refers to the fact that different streams of data arriving at a multiplexer will possibly each have different timing points — and need padding for an SDH bearer.

The terms used with the various PDH data rates are First, Second and Third-order multiplexing.

The highest current data rate carried on PDH bearers is 565Mb/s, but multiplexing up to 1.2Gbit/s is available. Generally, at these higher data-rates, however, the SDH standard will be used, and any lower-rate PDH signals will be mapped into SDH containers.

• The American PDH standards are based on T-1 channels of 1.5Mb/s data rates and the European on E-1 channels of 2Mb/s rates. See SDH also.

PDI

• **Picture Description Instructions:** These are codes used in alpha-geometric videotext systems. They define primitive components of the screen image, such as lines, arcs, points or polygons, and the screen image is drawn using PDI/primitives.

• **Product Data Interchange:** This is the manufacturer's version of EDI which uses different formats and documents from EDIFACT, and is production oriented. EDIFACT is trade-oriented. There are two major standards within PDI:

— PDES: Product Data Exchange Standard.

— STEP: STandard for the Exchange of Product marketing information.

PDL

• **Page Description Language:** the generic term. This is a category of languages used to describe a fill page of image and text for printing on a laserprinter (or any other page printer). It includes Adobe's PostScript, Microsoft's GDI and HP's PCL (the first, and *de facto* standard). PDLs are often now closely associated with outline-font technology (but not necessarily by any means), and they are more than just handlers of text.

Graphics are also an essential part of PDLs. See PCL, PostScript and GDI.

• **Project Description Language** from Smartware II.

PDM

Pulse Density Modulation. This is a form of 'delta' modulation used sometimes as an interim stage in translating analog wave forms into PCM, but also to create bitstreams.

• **CD-Audio.** PDM is the basis for the new 'bitstream' CD-Audio players using one-bit digital-to-analog conversion (instead of 16-bit DACs normally used). The music is still sampled and encoded in 16-bit form (actually EFM), but bitstream is said to be a superior way of reproducing digital CD sound.

There are two main bitstream techniques: Matsushita's technique is called PWM (Pulse-Width Modulation) or MASH and it differs from the Philips PDM system (Pulse-Density Modulation).

PDM converts the stereo 44.1kHz samples into an 11Mb/s stream of binary data by multiplying the 1s and 0s according to how 'significant' they are in the 16-bit word. The output has 8-bits for every 1-bit of input.

As an example of the idea, take the 4-bit nibble 1011. The right-hand bit (least significant) is only worth a single 1, while the next-from-right is worth two, the zeros are worth four, and the left-hand bit is worth 8. So this nibble would produce a bitstream of 1111 1111 0000 11 1, which could be interleaved with the next nibble to provide an appropriate average of 'delta'-stream bits — still representing the sample.

The nibble 1011 becomes 111011101110110.

This continuous stream of bits then passes to a single-bit DAC which directs the shape of the wave-form in a delta fashion. Obviously the weighting of 1s and 0s in the bit stream reflects the original sample numbers, and it determines the output audio wave shape but without the abrupt quantization changes of direct samples. See bitstream.

PDN

Public Data Networks. Any data-carriage network operated by a carrier or a private service provider, which is offered to the public for a fee. Since circuit-switched data networks usually need to be private and ISDN is both a voice and a data network, this term usually means packet, frame or cell-switched services.

Usually these are X.25 packet-switched networks or recently frame relay, although SNA must be included. ATM will be the next stage, and a few are already on trial.

TDM circuit-switched data networks could also be included under the definition but, generally, unless otherwise stated, PDNs are assumed to be packet-switched. They are mainly used for WAN interconnection, for information retrieval, and to supplement leased-lines (where the cost isn't justified by the usage). Transaction services generally suit packet-switched networks of this kind also. See PSDN.

PDP

The PDP 8 and PDP 11 were two very famous models of the DEC mini-computer. Unix was written for, and on, these machines.

PDS

• **Processor Direct Slot:** This is a slot in some Macintosh computers which gives special adaptor cards direct access to the CPU. This is quite different from the standard NuBus slots. PDS slots provide local bus functions.

• **Premises Distribution System:** AT&T's wiring system for buildings.

PDU

Protocol Data Unit. These are nothing more than the standard connectionless (datagram) packets in an OSI network.

In common use, PDU often also refers to packets which are exchanged across a data-link during the handshake period to determine which file-exchange, data rate or error-checking protocols to use.

pedestal

• The local out-doors raised 'pillar' box which provides a connection point for underground telephone cables.

• The blanking level in a TV picture. It is the lowest section of every signal, and it corresponds to the black level. To 'set the pedestal' is to make blacks black.

• In a TV studio, this is also the hydraulic platform on which the common studio cameras sit. They are highly mobile, smooth in action, can change direction in an instant, and rise smoothly from knee-level to well above head-height.

P-EDI

An ITU recommendation for an X.400-based messaging service to carry EDI traffic.

peek

A programming command in Basic which allows the direct inspection of the contents of the computer memory. This was a way of incorporating machine-level data into a high-level language. You could peek into an addressed location, and add the number found there into a Basic variable for further action in the high-level language. See poke.

peer

Most people think peer-to-peer means from one PC to another (rather than to a host). It can have that meaning, but 'peer' is a specific term in telecommunications and it is often misunderstood.

In the OSI seven-layer model, communications take place between matching layers. So the Physical layer of the sender, communicates with the Physical layer of the receiver (often by way of intermediate Physical layers of nodes along the path).

Similarly the Data-link layer of the sender communicates using higher-level protocols with the Data-link layer of the receiver. This happens for all the layers — each communicates with its matching layer at the link's other end using protocols specific to that layer. Peer is a logical construct, not a physical device.

peer-to-peer

• Strictly, this means communications from one level of functionality to another at the same level (in the OSI reference model). See peer.

• In general use, two different 'nodes' (devices) on the network can communicate directly with each other, without the need for an intermediary. This can be a client-server relationship, or two PCs talking to each other as equals. In peer-to-peer links, each party can act as the client, and each can be the server, because 'peer' refers here to a protocol layer not to an overall device function.

The term is almost always used vaguely in IBM circles to mean communications between PCs on a network where the network resources also include minis or mainframes. It comes from IBM's gradual recognition that mainframes no longer sit at the center of corporate networks. At one stage the only way for PCs to communicate between each other on such a network was for them to pass their messages through the processor of the mainframe. Both could talk to the host, but not directly to each other. This is 'host-centric'.

So with this terminology, the distinction is being made here between symmetrical and asymmetrical relationships. Peer-to-peer is being contrasted with the old mainframe-based, host-centric networks with their master-slave/host-terminal relationships, and also with some of the newer client-server relationships (even on peer-to-peer hardware). See APPC and APPN.

peer-to-peer network

Medium and large corporate networks are generally based around a dedicated computer known as a file or network server. However, many smaller networks now function without any dedicated servers, by using the resources of each PC on the LAN, on a shared basis. The network operating system for this type of network will usually be held in RAM at every station on the network. Windows for Workgroups and TOPS allow this type of low-level networking.

PEG channels

Public, Education and Government channels. Under the 1972 rules of the American FCC, larger cable systems are required to make free channels available for PEG use. The rule applies to cable networks with more than 35 channels.

pel

In television or in IBM computers this is a picture element. Everyone else calls it a pixel. However, it is only ever used for a screen element, while pixel is sometimes also extended in definition to mean a memory location which controls a screen element.

PEM

Privacy Enhanced Mail. This is an encryption system which is freely available on the Internet but not yet widely used. It provides confidentiality, authentication of messages, and message integrity.

PEN

Public Electronic Network. This is a general name for community networks, often involving local government authorities, and sometime providing Internet access to local citizens. The term seems to suggest 'non-profit' but it is hard to be sure.

pen interfaces

Many companies are experimenting with the idea of written (rather than typed) input, especially for executive-type notebook and hand-held computers and PDAs.

These are also called 'slate' devices, and they are generally based on RISC CPUs — but they don't necessarily use handwriting recognition. There are cursive forms and non-cursive forms. And some use soft keyboards and soft keypads for data entry. Others use electronic ink. See all and JOT also.

• Pen interfaces are a good example of how this industry becomes carried away by its own hype. In the early 1990s, you couldn't open a computer magazine without being inundated by articles claiming that handwriting recognition was about to wipe keyboards off the face of the Earth.

• This, of course, has now been replaced by speech recognition as the way of the future!

pencil beams

See shaped beams.

penta-

This means five. Don't confuse it with 'peta' as in petaBytes, which is 10^{15} or a million gigaBytes.

Pentium

Aka 80586, i586 and P5. This is Intel's 64-bit superscalar microprocessing chip which has 3.1 million transistors and uses 0.8, 0.6, and now 0.35 micron technology.

The number of transistors is three to six times that of the 486DX, and the increased packing density provides roughly four times the performance at 112MIPS (vs.486DX at 27MIPS — although the 486DX2 ran at 54MIPS). Initially the Pentium clock speed was 60MHz, but 66, 75, 90, 100 and 120MHz versions are now available.

The chip uses a combined CIRC/RISC approach. A pair of RISC integer execution units run in parallel, and these perform all the instructions in hardware. So there are two instructions processed in a single clock cycle. There is no microcode for the RISC, but both instruction and data cache exist on the chip. Floating-point functions are also integrated.

Another section of the chip is CIRC. This does have microcode in a separate execution unit, and it is entirely compatible with the 386.

There are separate 8kByte cache for data and code write-back (with multiprocessor support), and branch-prediction boosts performance by about one quarter.

• The early Pentium chips were discovered to have a flaw in the look-up table used for floating-point division. In the original design a few transistors had been accidentally omitted. This required a software patch for accuracy, and later, under public pressure Intel offered full replacement with a modified chip. Intel says the chip stumbled about once every 27,000 years, while IBM says it does so about every 24 days.

• In Mar 1995 Intel announced a Pentium using 0.35 micron technology and running at 120MHz. This delivers 140 SPECint92 and 103SPECfp92 performance, so they claim. It is a 3.3 volt BiCMOS chip built in four layers.

PEP

Packetized Ensemble Protocol. A packetizing technique used in DAMQAM modems. Since these modems are sending data in parallel over many beans (audio frequencies — there's usually about 400), the data must first be assembled into packets with a header containing a sequence number.

The packets are all of the same size in any one bean, but they vary according to the modulation techniques being employed (DPSK or QAM 4/6-bits). PEP technology enables the modem to take advantage of the full bandwidth of the lines. Both sync and async techniques are used. See DAMQAM.

perceptron

An old term for neural networks. See connectionist.

Perfect

A supercomputer benchmark of the Perfect Club of the University of Illinois. It is similar to SPEC.

perfect-bound

The process of binding magazines or booklets by use of adhesives, rather than staples. This gives a magazine or book the 'square-backed' look. But with magazines, it is more costly and requires different imposition to stapling. See saddle-stitch also.

performance management

This is one of the five key issues in network management, as defined by the OSI in CMIP. It involves the gathering of statistics on packet throughput, error rates, and such things as disk-access requests. This information will be sent to the central management console.

perfory

The torn off edge of tractor-fed continuous fan-fold stationary. A major environmental hazard of the computerized world. This term won a 'name-something-useless' competition. See chad.

perigee

In a non-circular orbit, the apogee is the highest point and the perigee the lowest point. In most geostationary satellite launches they will aim for a perigee of about 200km and an apogee of 35,993km.

• After a communications satellite has been released from the shuttle ('parked') the perigee kick motor is fired to boost the satellite into an elliptical transfer orbit which takes it up to the Clarke orbit. The process is called 'transfer' and it requires the subsequent firing of reverse thrusters at the apogee in order to circularize the orbit.

• The preliminary boost from the low 'parked' orbit to the higher one is called (confusingly) either the perigee-thrust stage or the apogee-boost stage.

period

• The time between successive points in a repetitive signal — say, between the top of successive waves in a waveform. It is the time taken for two successive waves to pass a single point.

• In satellites, the time taken for a complete orbit.

peripheral

Any unit which is attached to the central computer by cables and is controlled by the computer while remaining physically independent of it. Strictly, the keyboard and the video monitor screen are peripherals — but the term is usually applied to disk drives, printers, plotters, modems, etc. The term is not usually applied to plug-in cards, but it could be.

• A peripheral node in SNA is a node that uses local addresses, not network addresses. These nodes then require boundary function assistance from an adjacent subarea node.

Peripheral Component Interconnect

PCI is Intel's new standard for an input-output local bus which is capable of full-motion video transfers. This bus is to be used to transfer the digital video signal to/from A-to-D converters, codecs and display controllers.

It is actually only an 'almost-local-bus' architecture (some call it a 'virtual bus') and it is now an open, non-proprietary standard being established by more than 160 companies. This is the immediate future in broadband computing; PCI is now on 60% of all new PCs being sold.

PCI is more complex than VL-Bus and more expensive, but it has considerable throughput advantages and it decouples the peripherals from the processor. PCI is usually found in combination with an ISA bus in normal computer use, because the ISA is required for most normal computer applications, and companies already have huge investments in ISA cards.

The PCI bus interface acts as a buffer (the clock rate of the CPU is independent of the clock rate of the bus), so that high-speed peripheral cards can feed their data onto the bus without creating contention problems — even though the CPU may be engaged in other duties.

It is processor-independent, and it runs at a 33MHz clock speed. PCI has a 32-bit architecture and boasts a peak bandwidth of 133Mb/s (ISA has only 5—10Mb/s). This top rate is only ever achieved in burst mode; the sustained throughput is generally about 80Mb/s, which is still more than enough for full-motion PAL-quality video. However the rate achieved is still 200 times that of the PC's network connection.

By including support for future 64-bit data, the designers have given us an extensible system with enough capacity to handle future high-definition TV systems and intensive requirements like real-time 3D CAD rendering and virtual reality.

PCI cards are designed to be self-configuring. This is a 'plug-and-play' specification which provides automatic configuration for memory, I/O space, DMA channels and interrupt levels. Lower voltage (5—3.3 volt) versions are also proposed. See local bus, Nu-Bus, SCSI-2 and VL-Bus also.

• PCI's theoretical throughput of 133Mb/s. By comparison the EISA is 30Mb/s, MCA is 20Mb/s, and ISA is 10Mb/s.

• PCI has also been married to SCSI-2 in new adaptors.

• PCI has (since Feb 1995) new specifications for 64-bits, and also there is another proposal for a 66MHz bus. Eventually these two specifications will merge into one.

peripheral nodes

The SNA communications network is formed by sub-area nodes, each of which can be a host or a communications controller. Peripheral nodes (terminals and terminal concentrators) attach to sub-area nodes.

peripheral sharing

At one time devices like laserprinters were much too expensive for companies to afford more than one or two. The same applies to large-capacity hard-disk units.

So peripheral sharing was the old idea of using a LAN to interconnect a number of workstations or PCs together with a few printers and hard-disk units, so that many computers could share a limited number of expensive peripherals.

Our concepts have changed, sharing hardware resources has become less of a requirement, while sharing software and data has probably become more important.

Peritel

This is an audio-visual connector used in Europe. See SCART.

permissive dialing

When a carrier makes changes to its numbering system it is usual, for a period, for it to decode both the old way of dialing and the new way. This is the 'permissive dialing' period.

This is then followed by a period where specific dialing instructions are given if the old number is used but the number is not connected.

persistence of vision

The eye retains an image for about 1/15th of a second under average room lighting conditions. This makes the illusion of television and film screen motion possible, but at 15 images a second the flow of the image will have a cartoon (silent movie) jerkiness and it will appear to flicker.

This is the persistence of vision effect: don't confuse it with the flicker-fusion effect. Images will appear to flicker because the image-retention has been partially lost before the next image arrives — and this requires image refreshment every 1/50th of a second. To solve these associated problems we show the same image two or more times before we change it. Cartoon images are shown four times (each frame is double exposed), motion picture frames twice (screen as well as TV), and video once only (each field is different).

You need about 50 images a second to remove flicker, but this is highly variable. Flicker will be most noticeable when a) the peripheral vision is involved, b) at lower the levels of ambient illumination, c) when the image is excessively bright relative to the ambient levels, and d) when the image has large areas of flat, even, color.

For the above reasons, large wide-screen television sets have a different order of flicker problem to conventional TVs.

persistent storage

This is just a general term which applies to various ways of storing corporate information with object-oriented programming. It is the collective storage techniques used for mission-critical objects (OOP objects) on a hard disk.

Early OOPs used a flat-file approach where each object needed to be converted to and from a file by the program that created it, before it could be stored for later reuse.

Another approach was to store objects in a relational database. This involves constant conversion also.

• The latest approach is ODBMS which stores the objects as objects. See.

Personal Identification Number

Your numeric 'password' used at a Point-of-sale (POS) terminal or ATM (Automatic Teller Machine) to authenticate you as the owner of the card, and therefore the legitimate user of the bank account.

Personal Information Managers

A new category of business software (vaguely defined) which lets you store and retrieve almost any type of information in an unstructured database. Agenda from Lotus was the first major product of this kind. PIMs are generally (but certainly not always) used on portable computers.

• The four most often used PIM modules are: calendars, contact lists, to-do lists, phone-books and notepads. However, other modules such as expenses and graphical time-scheduling may also be added. Some portables and organizers have integrated modules, while others are just stand-alone. PIM modules need synchronization with desktop systems.

PERT

Project Evaluation and Review Technique. A form of critical path analysis (method) developed as part of operational research. PERT is a common chart form used in project management software.

PET

Page Entry Terminal. This is an old paging protocol still widely used. It is virtually the same as TAP and IXO. See TAP and ETC.

peta-
The SI prefix for 10^{15}. The abbreviation for peta is the uppercase P. A petaByte is a thousand teraBytes, or a million gigaBytes. See exa also.

Petri Net
Petri Nets and performance models are techniques for increasing the performance of networks (or concurrent systems). These techniques are based on theory developed by Dr Carl Adam Petri in the 1960s. They are used in fields as diverse as computer systems, communications networks and protocols, neural networks, manufacturing, library, banking and even legal systems.

PEX
PHIGS Extensions to X Windows. These are recent 3D graphic extensions to the X Windows system. Interoperability is one of the major PEX features. See PHIGS and X Windows.

PEXlib
An API for 3D PEX graphics on X Windows-based system. It was jointly developed by DEC, HP and Sun, and is an extension of PHIGS.

P/F bit
Poll/Final bit. This is a protocol used in bit-in-bit systems to indicate the function of a particular frame. When set to 1 in a command-frame, it indicates that the frame is a poll-frame. When it is set to 1 in a response-frame, however, it indicates that the current frame is the last in the response sequence.

PFD
Power-Flux Density of a satellite beam. This is a measure used to provide beam footprint contours. It is the 'illumination level' — the measure of power received per unit area. See EIRP also.

PFM
Pulse Frequency Modulation. An old analog approach which was used for a time on optical fiber. PFM is sometimes claimed to be 'quasi-digital' rather than analog, but this is incorrect since one of the dimensions (time) is always infinitely variable.

In PFM systems the duration or time the laser is on or off is the variable, while the light level is discrete (on or off). LaserVision laserdisks from Philips used PFM as the modulation technique to hold the composite analog television signal as a series of pits and lands in an optical disk.

PGA

- **Programmable Gate Arrays:** See PDL and gate arrays.

- **Professional Graphics Adaptor:** IBM's CAD-type analog color monitor standard for PCs which was announced in 1984. It had a screen resolution of 640 x 480, and gave 256 colors out of a possible palette of 4096. The vertical frequency of PGA monitors were set at 60Hz. This was a very expensive monitor standard at the time and it died from a lack of support. See EGA, MCGA, VGA, XGA.

PGP
Pretty Good Privacy. An encryption software package written by Philip Zimmermann in an attempt to by-pass strict US defense laws. It uses RSA public key algorithms. Full acceptance of PGP has been hampered by some incompatibilities between the various versions, and there have been legal difficulties caused by the US Defense Department's weapon-sale laws (encryption is a 'weapon'!).

- There are two flavors of PGP: the first is freeware from MIT, and the second is ViaCrypt's commercial version. PGP version 6.2 is now available on an MIT server.

See DES, RSA and PEM also.

phantom circuit
A third circuit which has been derived from two existing physical circuits by using a repeating device installed at the ends. Usually this is applied to two pairs of telephone wires which are already carrying a circuit. A third can be added ***between*** the two pairs, without the other users knowing.

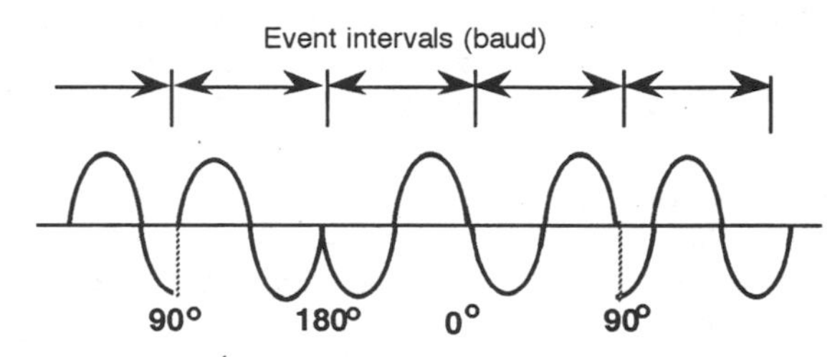

Quadrature phase changes

phase
With sine waves, this is a measure of the relative timing of the wave. How far it has progressed through its cycle.

- **Phase shift:** A technique used in PSK and QAM modems and radio modulation systems. A phase 'shift' or change measures the angle of discontinuity of the sine wave. This is measured in degrees — either from some absolute reference signal (difficult and unnecessary), or as an angular difference from the preceding phase change.
  - A phase change of 180° reverses the expected polarity of the cycle. This radical change can be used to signal one bit for each event, and it provides a form of end-to-end self-clocking also. This is BPSK.
  - A phase change of 90° increments (by 0°, 90°, 180° or 270° degrees) can be used to signal any of four dibit pairs (00, 01, 10, 11). So each 'event' or 'symbol' signals two bits. This is QPSK.
  - Phase changes of increasingly finer resolution (45°, 22.5°, etc) can progressively signal tribits or quadbits. This, together with amplitude changes, is QAM.

It is here that the relationship between baud and bit-per-second breaks down. Baud is the number of 'events' (here phase-changes) each second, but each event can signal 2, 3, 4 or more bits per second.

• **Phase delay:** The time or angle by which a signal is delayed, with respect to some reference position. See sine wave.

• **Phase hits:** Unwanted change in phase of a signal, caused by some electrical impact on the line.

phase-change optical disks

In optical disk technology, phase change is where the recording surface (usually rare-earth metals) changes from a crystalline to an anamorphous state. This happens with local heating and also with magneto-optical systems.

The reflection properties of some exotic materials used in write/read optical disks can change, depending on whether the material is in an amorphous (non-structured) or a crystalline (structured) state. These substances are known as chalcogenide alloys and if you heat them quickly with a laser to just above melting point (about 5 milliwatts for 100 nanoseconds), they change from the crystalline (low-reflection) to the amorphous (higher-reflection) state. The difference in reflectivity is only about 30%, but enough to create pits and lands for optical recording. A half-power laser reverses the state. See MD.

phase jitter/wobble

Phase jitter is a variation in the phase of a signal transmitted over a communications link; usually caused by multiplexing. It isn't noticeable with voice but is potentially disastrous with high-speed data. Like all jitter it results from timing problems.

It appears in communications networks as angular modulation changes, and it causes errors in high-speed communications — especially those carried at the higher frequencies in multiplexed channels. Most phase jitter is below 300Hz, but phase wobble is between 3Hz and 20Hz. See transmission impairments and distortion.

phase lock

When, in a synchronous communications system, the clock signals are in an absolutely fixed and consistent position relative to each other, then they are phase-locked. They may also be frequency-locked if both clocks generate the same frequency — but they can be phase-locked without being frequency-locked (although there will necessarily be an integer relationship between the two). See clock.

phase modulation

PM. The technique of modulating information onto a sine wave by changing the phase of the carrier in sympathy with the incoming data stream. When this takes place in each cycle of the carrier wave, phase modulation is almost identical to frequency modulation, and with analog modem signals they are effectively identical.

Phase change detection can be handled in frequency systems by passing the signal through a simple differential circuit before the FM stage. See Phase Shift Keying.

• The relationship between phase and frequency are important. It is possible to combine either phase or frequency shift modulation systems with amplitude changes to increase the data-rate, but you cannot combine both phase and frequency.

Phase Shift Keying

See PSK, phase, BPSK, QPSK and QAM.

phase velocity

In optical fiber systems the phase velocity is the speed of the light along the fiber. This is in contrast to the group velocity, which is the speed of the communications along the fiber. These are not necessarily the same. Don't ask me why.

phased array

A type of transmitting and receiving antenna used to replace the traditional dish in satellite and similar communications. Phase arrays are flat panels which can direct their beam in certain directions (or receive from certain directions) using electronics and special signal-based controls.

PHIGS

Programmers Hierarchical Interactive Graphics Standard. An ANSI/ISO graphics standard designed to remedy the lack of hierarchy in GKS, and to cope with 3D images. It appeared in 1984 and mostly replaced the original GKS standard. PHIGS has since been extended to include functions which allow it to remove hidden lines and hidden surfaces. See GKS, X-Windows, Motif, and PEX also.

Phoenix BIOS

This is a version of IBM's BIOS written by Phoenix Technologies to skate around the copyright laws. It mimics IBM's standard BIOS very well, but was different enough to escape the long arm of the law.

For many years clones made use of this and similar BIOS copies (mainly those of Award Software and American Megatrends International). More recently, in late 1993, BIOS was 'standardized' by a wider group of clone-makers led by Compaq and Phoenix Technologies, with Intel and Microsoft also.

phoneme encoding

A vocoder system of digitally encoding and reproducing voice using, essentially, a total synthesizing technique for the sound production. This is a lower level of vocoder than normal, and the technique requires only about 100 bits per second for intelligible speech. However it has a very poor machine-like voice quality.

• Phoneme encoding has little application in public-switched telephony but it may be used in internal networks.

• English is said to have only 45 different phonemes.

phosphor time-constant

A TV screen or monitor needs to hold the image for a fraction of a second until the next screen refresh occurs. This is the phosphor time-constant. If it is too long, then the image of moving objects will tend to smear on the screen; if it is too short, the image will tend to flicker at low refresh rates (below 1/60th sec). See flicker effect, large-area flicker and fusion.

Photo CD

Kodak and Philips have combined to produce this form of still-image display. You take your color film to be processed, and you'll have the option of getting regular prints, or images on a compact disk. Each disk will hold about 100 prints. These are WORM-type disks; you keep taking a disk back to have more images added until it is full. Your CD Photo player will display them on a television or computer screen, and they can be printed out on a high-resolution Kodak printer.

The Kodak/Philip's Photo CD digital disk format is actually a recordable extension of CD-ROM/XA. These disks can be replayed on any CD-ROM/XA and CD-I player to show photographic images stored in computer digital graphic form, and they can also be used to store data other than picture images.

The Photo CD image is coded in a 24-bit YCC form (called Photo YCC — see) with 8-bits each for luminance and two 'difference' color components. The file structure is quite unique; files are stored in a layered hierarchical fashion so that various quality options are available — from the low resolution image of 128 x 192 pixels, to 256 x 384, to 512 x 768, to 1024 x 1536, and on to the highest resolution of 2048 x 3072 pixels. Only the lowest resolution files are stored uncompressed.

In the initial scanning process, 12-bits are allocated to each of the RGB components in the 2048 x 3072 pixel.scan. This is then color corrected and converted to YCC form (in three stages) with each of the three components now being allocated 8-bits. This composite file is then compressed to provide an Image Pac file of 3—6MBytes.

This file technique gives an 600MByte 12cm disk an average storage capacity of about 100—150 images. And, since this is a standard CD-ROM file format, it can be read by a computer. However extra microcontroller readable sectors are included for the special Photo CD and CD-I players as well.

• Portfolio format has now been added to the above with up to 800 images of 768 x 512 pixels, for still-video presentation.

photo resist

A solution applied to conductive boards (PCB) or glass mastering disks to make them photo-sensitive.

• In CD disk mastering, the information is then burned in by a modulated laser. After development, the glass is etched and then electro-plated to create a stamper which makes the disks by direct pressure.

• For PCB manufacture, a light source is projected through a mask (outlining the tracks required for electronic connections), and the material is then developed and etched.

There are both negative (insoluble) and positive (soluble) photo resist materials.

photo typesetters

Photo typesetters are used to produce high-grade master 'bromides' which are used in offset printing. This is a special type of laserprinter which can reproduce about 1000 dots per inch (dpi); you need this for high-quality printing on glossy paper.

The Linotronic 300, for instance, produces 2540 dpi, as against a standard laserprinter with 300 dpi. The average monitor screen only has about 72 dpi. The high resolution is necessary because these machines do not produce variable sized dots (as with a camera-screen system), but use dithering for half-tones.

Phototypesetting has come to the desk-top publishing world because Linotype machines now handle PostScript, so pages can be set on a PC and then output to a Linotronic 100 or 300 setter.

Photo YCC

Kodak's color 'space' specification. This is now a proposed industry standard way of representing color in digital form for computer, broadcast and print media applications.

The Y is the luminance, and the two Cs are for color-difference signals. The image is coded in 24-bit form with 8-bits devoted to the Y and each C component.

The Photo YCC format is derived from a RGB scanned image. In the first stage, 12-bits are allocated to each component in a 2048 x 3072 pixel scan. The image is then color corrected, converted in three stages to the YCC form, and compressed to provide an Image Pac file of 3—6MBytes. The sub-files are stored in a hierarchical fashion which makes the various resolution options directly available.

photodetectors

The general name for electronic devices which produce an electrical output when struck by photons of light.

photodiode

A type of photodetector based on transistor-diode technology.

photon

The basic 'atom' of light. See boson.

photon-gated

These are compounds (both organic and inorganic) which react in a complex way to light wavelengths (often only at very low temperatures). There's a lot of research going on into the use of these materials.

The prospect is that an optical disk made from photon-gated compounds may be able to store a broad spectrum of wavelength information at each spot on the disk. This is also called the 'rainbow disk' concept. Theoretically, photon-gate disks could have a million-times capacity increase in storage over current technologies: in practice, they hope to get one hundred times. This is also called 'frequency-domain' data storage.

photonic switching

The direct switching of light signals without the need to convert them first to electrical form. These devices currently use a lithium niobate crystal substrata on which are formed two optical waveguides. The guides are almost touching, and if an electrical potential of, say, 20 to 30 volts is applied across the gap, some of the light is transferred from one side to the other. The switch must be preceded by a polarizing filter.

photonics

The study of light and lightwave communications and computing. The optical equivalent of 'electronics'.

PHP

Personal HandiPhone aka PHS (service). The very successful Japanese attempt to build a low-cost digital PCS standard. It is used in Japan and Hong Kong.

PHP is designed for the 1.9GHz frequency band and for pedestrian rather than vehicle use, although handoffs can be handled at vehicle rates.

The frequency band allocation is 1895—1918MHz (23MHz), and there are 40 carriers with a carrier separation of 300kHz. PHP provides four TDMA/TDD channels per carrier (eight slots) with a carrier transmission rate of 384kb/s. It packs 12 of these channels into 1MHz, and claims a cell-reuse factor of 4.

The vocoder is of the 32kb/s ADPCM type and the modulation is $\pi/4$-shifted QPSK. Handsets have an output power of 10mW.

The standard was devised by the R&D Center for Radio Standards in Japan, and it is known as CRC-27. I have had good reports about this technology, although it is virtually unknown outside Japan (and recently, in Hong Kong). It will mainly be used for wireless PBX, but many of its elements are being incorporated into the PACS standard in the USA. See DECT, CT2, DCS-1800.

phreaking

Aka freaking. The term originally meant stealing free long distance calls from telephone companies by providing them with fake tone codes which caused the billing systems to switch off. Now it appears to mean any theft of phone services, even from private individuals. See blue box and CCS#5.

PHS

Personal Handiphone Service. Also called PHP. See.

physical address

Aka MAC-layer address. The distinction is being made between the physical address for the device itself which is held at the MAC layer, and the network address. Bridges will use the MAC address, and routers the network address. See OSI reference model.

physical channel

There's not necessarily anything 'physical' about these channels at all these days. Originally this meant a bit of wire.

In time division systems, each slot is effectively considered to be a physical channel, and the name continues to be applied to the link established in a TDMA communications system, even when the system is frequency-hopping and constantly changing radio channels.

- Don't confuse this with physical media — which means the wires, fibers, etc.

physical control layer

This refers to the bottom layer in the SNA model — the equivalent of the OSI's bottom layer, the Physical layer. Both define the electrical, mechanical and physical interfaces to their networks.

Physical layer

The lowest (No. 1) of the OSI's seven-layer model architecture — the hardware layer — it deals with the communications medium. The ISO says that the Physical layer 'provides the transparent transmission of bit-streams between systems, including relaying through different media'.

It defines the connectors, cables and other hardware, the signaling method and the bit-rates. Repeaters, which are completely unintelligent signal regenerators, exist at the physical level because they don't change or manipulate any signals, just amplify and repeat whatever they are given.

This level is involved with modulation methods, line-codes, multiplexing and signal generation. Some of the RS-232 standard exists at this level — which pin does what, the voltage levels used, and all physical protocols.

The four sub-sections of the Physical layer are:

- **Mechanical** — connector types and pin allocations.
- **Electrical/Light** — voltages, impedances, laser brightness and attenuation, etc.
- **Functions** — the meaning of physical changes, say, of voltages on a pin in the RS-232 cable connector.
- **Procedural** — the rules applying to the above: the sequences of events, etc. The key factors set in Physical layer attributes are: serial or parallel; full-duplex or half-duplex; synchronous or asynchronous; hardware handshaking connections; point-to-point or multipoint.

physical media dependent

A subsection of the FDDI optical fiber standards that deal with pinouts and electro-optical characteristics.

Physical Unit

- In SNA, see PU.
- The physical (what the device is and how it connects), as distinct from the logical (how it functions and what it does.) unit in any communications system.

PI

Program Identification. This is associated with RDS data transmissions over the new European FM radio systems. The PI data allows the radio receiver to identify and lock onto the appropriate transmission frequency. See PS and RDS.

pi/4 GMSK

See GSM.

PIC

- **Preferred InterLATA Carrier:** (aka Primary Interexchange Carrier) In the USA, this is your choice (your 'default service') of which company (IXC) will handle your long-distance calls. You can choose from a number offering links at your local exchange.

• **PIC file:** (See PICT below) In MS-DOS, a .PIC file is a standard PICT image file format.

• **Personal Intelligent Communicators:** Another name for PDAs or small handheld computers with (generally) a pen-based input. However PICs place more emphasis on the communications functions, so they may include a microphone, earphone and cellular phone connections. This is the term being promoted by General Magic for PDAs. See Telescript

pica

• A printing measure used to express overall width, or the depth and length of a line. However type is usually measured in points, while line length is sometimes measured in picas, and line indentation is expressed in ems. For the linear measure of type, there are 12 points to a pica, and 6 picas to an inch.

• A type size and style used in impact typewriters and printers. Pica has 10 characters to the inch. This is an Anglo-American term which is called cicero in Europe. See pitch also.

PICK

A portable, multi-user operating system (similar to rival Unix) developed by Dick Pick about 1970. It has proved to be flexible and it supports a wide range of applications — especially in the relational database area. More recently Pick has become 'seamlessly integrated' with Unix.

pico-

The SI prefix for 10^{-12}. The abbreviation is the small p.

picocell

In mobile telephony, this is a cell which only provides coverage for a few meters around the base station. However the term is more generally used to mean 'within an office' area, or the equivalent. A typical picocell would provide coverage of, say, an office corridor or a work-area. See microcell.

picon

A term for an 'icon' or 'thumbnail' image. These are used to represent a still image or a piece of video footage in a video-controller or editing application.

The term often comes into general use as a replacement for 'icon'. In hypertext and hyper media picon can also refer to special areas of the picture which have been hot-linked to object handlers, and so perform some function when clicked.

picosecond

Often written 'psec'. One millionth of a millionth of a second. 10^{-12} secs.

PICT

A graphics file format; the native mode for the Macintosh.

— **PICT (1)** was the first object-oriented image file format for the Macintosh. It could also handle bit-mapped black and white images.

— **PICT2** is a version that has been adapted to include color information. Color bit-maps are called PIX maps and may contain up to 24-bits of color information per pixel (16.7 million colors). However this is called 32-bit color, with the extra 8-bits used as an alpha channel.

PICT2 file may contain objects, bit-maps and PIX maps. PICT and associated PICT2 documents can't be separated.

— **PICT2+** is a version of the standard that allows 'chunky pixels' operations (RGB pixel painting).

— **Object PICT** is for CAD. It contains vector primitives.

• Common file extensions are .PCT and .PIC

piezo-electric

Ceramic devices which generate a voltage when a mechanical force is applied are piezo-electric. The reverse also happens; when a voltage is applied, piezo-electric material distorts physically. So these modern materials are widely used to make transducers — specifically medium-quality microphones and earphones for telephones, etc. They are also used to create sparks in your gas-cooker lighter.

PIF

Program Information File. A file which is used to tell Windows how much memory a non-Windows application requires. It carries information about whether the application writes direct to the screen, whether it runs in the background, etc.

piggyback

In an ACK-NAC error checking system, piggybacking is when the acknowledgments are carried within a packet of data traveling the other way. It is used with bi-directional transmissions to save on bandwidth.

pigment-based

Coloring systems that use the subtractive process in printing. This usually refers to the CYMK system. See additive primaries.

pileable hubs

These look like stackable hubs, however high speed links can't be made between the backplanes of the various hub modules, but only via daisy-chaining repeater ports. This limits the number of hubs that can be connected to an Ethernet (usually to four) because each in the chain introduces a repeater delay.

Pileable hubs will also appear to the management system as a number of different repeater segments rather than as a single unit. See stackable hubs.

pillar

This is 'street furniture' used in telephone networks. There's usually one on every second street corner in the suburbs.

There are active pillars (with mains-powered electronics and sometimes with air conditioning) and passive pillars (by far the most common) which are solder points. See pedestal.

PIM

Personal Information Managers. See

PIN

- Personal Identification Number. See.
- Positive Intrinsic Negative. See
- The male contacts on a multiple connector.

pin diodes

Cheap replacement for avalanche photo-diodes in optical fiber systems. They can operate at speeds of about 100Mb/s. See Positive Intrinsic Negative.

pinch roller

See capstan.

PING

Packet InterNet Gopher. To ping the host on the Internet is to send it a wake-up message. This is a process involving an ICMP message (Internet Control Message Protocol) and its reply. The Ping checks that the host is active and available, and helps establish the link.

Many Mosaic interfaces have a Ping icon. You should use only the host-name (the DNS). If you are successful, the link should then be logically established and you should receive back, as an echo, whatever you sent. You can then try to make contact normally using the full WWW address.

- In a more general sense, a Ping is a test message sent either during LAN testing or to keep a link alive. TCP/IP and IPX constantly send packets from three different layers to maintain their presence on a network. So when there are many devices pinging on a network, they can generate a lot of traffic.

ping-pong

- **Modem protocols:** In some 4.8kb/s and 9.6kb/s modems running in the half-duplex mode, special synchronization techniques are used for rapid and regular line-turnaround. This simulates full-duplex operations, while still being half-duplex.
- **LAN loop:** In LANs this is the unwanted action of packets in a two-node routing loop; they keep re-sending to each other the same packet.
- **PCS:** In CT2 telephones, the channel bandwidth is divided in time so that the incoming and outgoing voice signals share the same bandwidth. Within the same channel frequency, you both send and receive. DECT and PHP also use this two-slot approach in a single carrier. See TDD.
- **Cellular:** The term 'ping-pong' is also used for a problem in cellular phone handoff when the mobile is at the junction between two cell areas. Often as it moves it will be subject to changing signal paths back to alternative basestations and therefore to wildly fluctuating power levels. The basestations may agree to handoff the signal, but then an instant later they hand it back. Special algorithms are needed to ensure that ping-pong handoffs don't occur.

pinout

- A chip has a 'pinout' and the socket has a 'footprint'.
- This also means, to test systematically that the correct cable connections have been made in the connector.
- It can mean the overall configuration of electrical signals in a connector. In RS-232, for instance, only eight of the 24 pins are essential, but these eight are vital.

PIO

Processor Input/Output. A system of bringing data from a peripheral (a disk drive) into memory. The distinction is often being made here with Direct Memory Access (DMA) with which PIO is closely aligned. See Multiword DMA also.

PIO depends on the CPU handling the data transfer, while DMA allows the disk-drive controller to handle the transfer after the processor performs an initial setup.

PIO was initially faster than DMA, but since faster data buses have been introduced DMA is probably faster.

- PIO/DMA come in modes.

— Mode 1 for DMA transfers up to 13.3MByte/s.
— Mode 2 for DMA transfers up to 16.6MByte/s.
— Mode 3 for data rates of 11.1MByte/s (ATA).
— Mode 4 for data rates of 16.6MByte/s (Fast ATA-2).

pioneer's preference

In recognition of the R&D effort that some companies put into the development of telecommunications technologies and services, the FCC has granted a few special licences. These are usually radio spectrum license, and originally it was suggested that the pioneers would get these free.

However, in the recent spectrum auctions for PCS, the pioneers were to have been granted a guaranteed license priced at the going rate. Since there was no bidding on this license allocation, this represents a healthy discount — especially since many people in the industry would question the 'pioneering' nature of the companies' achievements.

PIP

- **Peripheral Interchange Program:** An old CP/M term for a copy program.
- **Path Independent Protocol:** In packet-switching this is a connectionless protocol.

pipe

- A pipe is used to connect two processes together. It is a unidirectional communications path, usually from a parent process to a child process (or vice versa), but since pipes are one-way you usually acquire them in pairs.

This was originally a Unix term but OS/2 and MS-DOS now also use pipes. A pipe is the simplest form of inter-process communications (IPC) structure. Once you create a pipe, you can read and/or write to it as if it were a file. See process.
In OS/2, pipes are either anonymous, or named:

— **Anonymous pipes** are used internally for multitasking.
— **Named pipes** are used between applications on separate PCs (they use the standard file naming conventions). These

connect unrelated processes and operate in a client-server fashion. They can be inbound, outbound or duplex.

See named pipes, pipelining and queue.

• A pipe also exists when two or more commands are chained together into a sequence: the output of the first becomes the input of the second, and so on.

• The term data 'pipe' in telecoms emphasizes that the interconnection is playing a completely passive role. Data is pumped into one end of the pipe, and it comes out the other — it is not checked or processed in between. So the carrier shouldn't be charging premium fees for 'value added' services.

pipelined mode

See scalar mode.

pipelining

Pipelining is used to speed up the delivery of data to a CPU. A pipeline is a linear flow of processing elements that are able to work concurrently on a stream of data. These elements pass values from one to another in the same way that cars are constructed on a production line as a series of progressive stages — which then finally deliver the data to the CPU.

• This is a technique used in mainframe, mini and now microprocessors to improve CPU performance. This technique is especially useful in RISC chips, and it is found in Motorola's 68030 and Intel's 80368 CIRC chips and in all their later chips.

Pipelining is effective because it attacks the problems of lost CPU time. A very large part of each instruction cycle is normally wasted accessing the next memory location. With pipelining, while the processor is executing the first instruction, new instructions and data are being pre-fetched to ensure the CPU is never idle. This is possible because the CPU can usually generate the address of the next memory transfer before the current transfer is complete.

In effect, pipelining allows the main processor to handle two, three or more instructions at once, since the data access time of the chips can be almost as long as the processor-execution cycle.

In CPUs, pipelines are applied in different ways, however. The Intel 485 has a five-stage instruction pipeline, while the Motorola 68040 maintains separate pipelines for data and instructions, and has separate cache for each.

RISC computers often have two or more processing units, each with its own pipeline, and are therefore able to execute multiple instructions simultaneously. This is a microparallel technique.

pirates

The name publishers give to those users who rip off commercial software without paying for it — and the name users give to publishers who charge $1500 for second-rate software that hasn't been debugged; and then charge another $500 for a version upgrade that may just work (provided you purchase another few megaBytes of memory).

pit

See lands.

pitch

• The pitch of a typewriter font is defined by the number of characters to an inch. The 12 pitch Elite font has 12 characters to the inch, and Pica has 10. Typewriters had 'fixed pitch' letter spacing, while modern computers usually have proportional spacing. See courier.

• In a TV or monitor screen the pitch is the horizontal distance between the picture elements (in color monitors, between the triads). This is often called 'phosphor pitch'.

— With most medium-sized home TV screens (4:3) made today, the phosphor pitch is between 0.7 and 0.8mm.

— High-quality monitors for TV studios cost about ten times as much as a TV set, but they can have a pitch as low as 0.4mm (Note: this affects the horizontal resolution only).

— Computer monitors often have a dot-pitch as small as 0.3 – 0.25mm.

Note that there's no direct relationship between pitch and resolution, except that the pitch sets the maximum horizontal resolution possible. If you have a low-resolution image showing on a high-resolution monitor, you gain nothing.

• Vertical pitch is set by the number of scan lines. It is therefore less on computer monitors because these will have from 1000 to 2000 lines on a relatively small screen. See active lines.

PIX

Color bit-maps. See PICT.

pix-maps

The pixel equivalent of bit-maps. Pixels are logical structures. Groups of, say, six screen 'spots' may be constantly switched together in some graphics operations — and these together constitute a pixel.

A pix-map would usually only hold one bit for each pixel (B&W) and this would then control the larger bit-map at the maximum screen resolution — or the conversion to the raster scan would be done on the fly. This use of multiple-dot pixels was common in the early days of PCs, but it is rare now.

So in the past few years pixel has come to correspond to a memory location (the bit-map). See pixel-depth, packed-pixel and planar architecture also.

Pixar graphics

See Renderman.

pixel

Picture Element — aka a 'pel' in television. The smallest dot you can draw in a video display with a complete range of colors (which is only one, if B&W). The term is sometimes used ambiguously, but it shouldn't be.

• When discussing a screen's potential resolution, a screen-pixel is a single dot on a B&W screen, or a groups of three dots (RG B — called a triad) on a color screen. These are controlled and switched on and off together, under control of a specific location (holding from 1-bit to 32-bits) in video RAM.

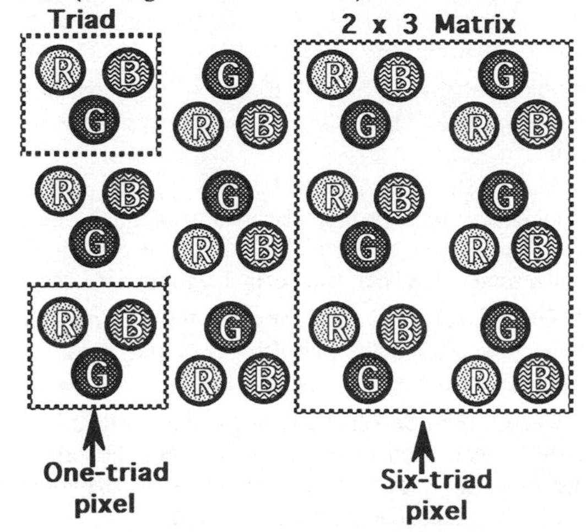

• When discussing image resolution, a logical-pixel is any number of dots or triads which are switched as a group, under the control of a single location in video RAM. Line doubling, for instance, switches all of the triads on the twin lines in parallel, but still under the control of single memory locations. This doesn't double the pixels from the viewpoint of increasing resolution.

Chunky videotext graphics switch a matrix of 3 x 2 or 6 x 4 triads, under single memory location control. This was common in videotex because picture quality was traded off against modem data rate. Each of these matrix blocks is then a pixel.

• Most of the confusion about pixels comes about because people don't appreciate the distinction between the hardware pixel (which in color monitors is better called a 'triad') and the resolution (logical) pixel which deals with images. See pitch also.

• There was little confusion in the early B&W monitors. A pixel was related to one bit of memory, and this almost always corresponded to one dot on the screen. These devices always worked at the highest resolution possible in computer graphics and text.

• When videotext and teletext came along, however, computer-controlled screens sometimes functioned at a lower resolution than their full capability. One bit (or one byte) in memory could be used to switch two, three or more adjacent dots (still classed as a pixel).

Most videotext systems used a basic matrix of 3 x 2 triads on the screen, and, when producing graphics, these were all switched together. (They were very blocky graphics!)

• The minimum video RAM memory needed for a color monitor which can reproduce only eight colors (including white and black as two colors), is 3-bits for each pixel. This corresponds to the possibilities when switching a single R, G and B triad (one pixel) either on or off in various combinations; it can produce white, black, pure red, green and blue, and the additive blends of R+B, R+G, G+B.

• For full-color high-resolution workstation monitors, there will still be a single triad to each pixel, but up to 8-bits may be controlling the intensity of each R, G, and B dot in the triad, giving $2^8 \times 2^8 \times 2^8$ color possibilities — which is where the claimed 16.7 million color variations come from. Most of us can only detect differences in the order of 2^5 in each of the three primaries.

• In pseudo-HDTV systems (like HD-MAC) where line-doubling is used to create '1250 lines of picture', the manufacturers often 'double' the normal vertical pixel count in their claims. With line-doubling a 'logical' pixel simply becomes two triads rather than one; the image resolution doesn't improve at all.

However line-doubling does *reduce* the resolution of the scan lines (they are thinner and therefore more difficult to see — which is good). You don't want to see the lines, you want to see the image. The screen looks much better from a few feet away, but no better from a few meters away.

• I've also seen a sales brochure where the term 'pixel' is used to refer only to each phosphor dot on a color screen. The company then calculated the total pixel count for their monitor by adding the red-pixels, the green-pixels, and the blue-pixels — in other words they claim that three pixels make up a triad. It's certainly a novel way to have better specifications than your competitors!

• Pixel is often also used when referring to any location in video memory, when that location controls (corresponds to) a point on the graphics screen. This is incorrect, but it is a very widespread (mis)use of the term and quite understandable.

— In this case, a B&W monochrome program will have a 'memory pixel' which is only one bit. Early PC screens were like this and so they were very economical on video RAM. They could hold the information for eight pixels in a single byte of video memory.

— If a gray scale was to be reproduced, more than one bit would need to be allocated to each pixel (usually up to 8, which would then give 256 levels of gray).

— With color images, it will take between 3 and 24-bits to store one pixel depending on the color variation required. Normally 16-bits (2-Bytes) per pixel (3 x 5-bits + 1) is enough for good quality color reproduction with quite subtle tones and hues, with the extra bit sometimes being used for transparency.

However 24-bits (3-Bytes) per pixel (3 x 8-bits) is considered to be optimal for photo-realistic images, and suitable for even the best still photos which are always more critical than movies. When 32-bits are provided, the extras 8-bits are usually used for another purpose. See alpha channel and pel.

• Professional video equipment now has an 8- or 10-bit pixel depth. Eight-bits is used for news, and 10-bits for production.

pixel-depth

A simple monochrome monitor will show black or white pixels only. It therefore needs each pixel to be represented in the video-RAM bit-map by only a 0 or a 1. This graphic representation therefore has a pixel-depth of one.

A pixel-depth of two will allow four tone-levels to exist on the screen (black, white and two grays). It could be used to provide some color, but the minimum generally needed for color is one bit for each of Red, Green and Blue, and obviously more bits for each primary, if we are to have subtleties in the hues and tones.

Pixel-depth can either be applied to monochrome B&W, mono gray-scale, the provision of a few saturated colors or to full tone/color quality of a photo-realistic image. Each improvement in the range of possibilities, must be matched by an increased need for more bits to be controlling each pixel. With computers the limits are probably 8-bits per pixel per color (24-bits in all), but often 32-bits are used and the extra 8-bits used for an alpha channel. See.

pixel graphics

A form of screen presentation used in conjunction with raster scan techniques where the screen image is created by individual pixels (as in a paint or draw program) generated by an application.

The distinction is with character graphics, which were videotex-type blocky images created from a library of 'primitives' attached to the extended ASCII set, and vector graphics based on co-ordinates and algorithms.

pizza boxes

Low profile stackable hubs and routers. See pileable hubs also.

PKE

Public Key Encryption. See PGP, RSA and contrast with DES.

PL/1

Programming Language 1 is a modular language which combines some of the features of COBOL and some of Algol.

PLA

Programmable Logic Arrays. See Programmable Logic Devices.

Plan 9

The operating system being developed at Bell Laboratories by the original Unix authors (Richie and Thompson). It is said to be revolutionary, and intended for research and education.

It is actually a small shell program around Unix which allows transparent distributed processing over networks. Bell Labs have released it on CD-ROM.

Their aim is to support tens of thousands of users, terabytes of data, and distributed cooperative processing over telephone networks. A single file protocol will be used to control all local and remote sources.

planar

In a single flat plane. Two dimensional only: everything at one level in a hierarchy or stack.

- **Planar transistor:** The current design of transistor used in most chips. Fairchild invented the process which allowed all three electrical connections to reside on one surface of the device. This made the IC possible.

- **Planar chips:** In the early days of IC development, complexity was added by building up multiple layers of active elements.

  Fairchild, however, realized that this approach was wrong, and it built its reputation on planar chips with only one active layer. This became the dominant architecture for chip design until recently. Now layering is beginning to appear, once again.

- **Planar board:** The motherboard — the main multilayered PC board in a computer, to which most of the major components are attached.

- **Planar memory:** IBM's term for the standard 'conventional' memory on the motherboard of a PC. The term distinguishes these chips from memory chips on adaptor cards, which IBM refers to as 'I/O channel' memory.

- **Planar video architecture:** In planar architecture the screen image is physically organized in video memory as a number of bit planes (see), each of which will be stored in a different area of memory. This approach is often better for image processing. The distinction is with packed-pixel architecture.

plant

This can mean almost anything in this industry. However it is most often applied to the cabling system.

- Cable TV networks and their amplifiers are often collectively called 'plant'. The engineers will talk about the 'optical plant' (laser drivers, fibers and hub) and the 'copper plant' (coaxial and trunk amplifiers and drop cables).

- In-plant equipment, just means internal — as against that supplied by an external carrier.

- The telcos often refer to 'plant and switches' which suggests that they just use it to refer to street cables.

plasma display

Plasma displays appear to be ready to leave the laboratory and go into mass production. These may be our best chance for large-sized wall-panel screens capable of handling television quality colour images.

Present indications are that they don't appear likely to have the resolution necessary for computer screens, however. There is also a question about their long-term reliability. They may fade with age.

- Thomson has shown a 56cm screen, but it requires 350 volts AC and has only moderate resolution. It is UV sensitive also. Fujitsu and NEC also have plasma screens with high AC voltages.

- Matsushita has developed a DC system in association with Du Pont, while Sony has opted for a hybrid (see below) of a plasma and LCD. See gas plasma.

Plasmatron

Sony and Tektronix have designed this combination LCD/-Plasma system which they believe is capable of producing cheap, large-area (50-inch diagonal) flat-panel displays. It dispenses with the normal switches behind pixels (these give problems in manufacture) and uses Plasma discharge tubes.

plated media

These are platters (disks) in a disk drive that have higher magnetic packing densities because of improved magnetic materials and manufacturing techniques. The distinction is with the older techniques of 'sputtered media', where the magnetic material was sprayed onto the disk surface.

platen

The black roller on a typewriter or printer that provides the backing behind the paper during the impact printing process. The term has come over to computing, and so platen-width defines the maximum paper width that the printer can handle.

platform

A vague term that means a computer system plus its operating system environment. It sometimes appears to be applied only to proprietary hardware, but the term usually encompasses the operating system as well. It is the total environment in which an application will run.

• You 'port to other platforms' when you translate an application written for a Mac, to run in Windows or on Unix.

platter

The magnetic disks in a hard-disk unit. There will be one, two, three or four platters in most conventional mainframe hard-disk units, and in some cases both sides will be in use. In some of the large-capacity hard disk stacks, one side of one platter is often used as a 'serving platter' (with a read-only head), and this generates the servo-signals that define the track numbers for each location.

Multiple platter disk units (stacks) have not become as important with PCs as was once assumed, because excellent results have been obtained by refining single platter systems. However they are still widely used with the larger mainframes and minis. See cylinder and RAID.

PLC

Programmable Logic Controllers. PLCs are hardware devices used to monitor and control some technical processes (mainly for data acquisition: temperature, pressure, etc.). These are often intelligent two-way devices which can receive commands and act on them, and can simultaneously report on processes, monitor flows, and activate valves, taps, and switches, etc.

Some are linked to the central controls by relatively standard external communication modules which support both wire and radio modems; some can phone up a paging system. Fault tolerance is obviously important here.

• Don't confuse these with Programmable Logic Devices.

PLD

Programmable Logic Devices. See.

plesiochronous transmission

Not strictly synchronized transmission ('plesios' means nearly or almost). Our present telecommunications systems are generally of this type. Two synchronous signals might be carrying data at exactly the same rate, but it is unlikely that they will be perfectly synchronized, and in phase.

When a number of PCM channels are multiplexed together, each with its own independent clock-timing (which translates to very slight differences), the result is plesiochronous transmission.

To compensate, padding bits are often added at each multiplex stage — but this creates problems with add-drop multiplexing in accessing a single channel, since the switch can't distinguish the bit-padding from the data itself. Currently with high-level multiplexing the carriers still need to demultiplex the whole supergroup or hypergroup of signals (through perhaps three stages) to get access to one channel — then reverse the process through the re-multiplexing stages for onwards carriage. The solution is SDH. See PDH and SDH.

Plex

A language used in telecommunications for Stored Program Control (SPC) exchanges.

PLL

• Private Leased Lines.

• Phase-Locked Loop. This is a common method of maintaining synchronization on a network, using an oscillator which is controlled by a phase detector. It measures differences between the local clock signal, and one derived from the incoming signal.

PLMN

Public Land Mobile Networks. A general term which includes cellular telephone systems, Personal Communications Networks (PCN) and Personal Communications Services (PCS). It matches PSTN as a general term for wire-line networks.

plotter

A peripheral output device which draws graphs and graphics. Most plotters also have character generators, and can therefore reproduce text. The distinction between plotters and printers is that the plotter draws the images on paper usually using a pen or a series of pens. The plotter is also a vector device, while the printer is a raster device.

PLP

Packet Layer Protocol. See X.25.

Plug and Play

• The term originally referred to peripherals and add-in cards which didn't need configuration. The NuBus used in the Macintosh didn't require you to set DIP switches — this was all done automatically.

This couldn't be done with MS-DOS machines because there were so many non-standard variations, and therefore many incompatible standards. The PC is also very restricted in the way it uses its interrupt lines, and this was the main cause of problems with other devices.

• Microsoft is now using Plug and Play (developed by NEC for VESA) to mean a future collection of standards (incorporated initially into Windows 95) which allows hot-pluggable boards (inserted and removed while the machine is functioning) and automatic configuration of peripherals and expansion cards.

PnP requires cooperation on three fronts: the operating system, the hardware, and the applications. You'll need a PnP BIOS which can't be supported on today's ISA bus — and most of your current applications will not be PnP-aware.

PnP devices don't install automatically; rather they identify themselves and the resources they require (IRQs, DMA and I/O channels) to the operating system. The computer then builds a central registry. MCA systems have had this registry for years, but they haven't had the operating system or applications.

In Windows 3.1, most of this work is done by Configuration Manager, but Windows 95 takes the automation of the process much further. However, while PnP is important, it is fair to say that it actually seems to provide little more than Apple had with NuBus in 1987.

• PnP's work is done by a software configuration manager which draws information from:

— the hardware tree (a central database of configuration information),
— resource arbitrator (information as to how a resource works),
— drivers (aka Bus enumerators),
— and also from unique identifiers burned into ROMs in the devices.

plug-compatible manufacturers

Manufacturers of peripherals and plug-ins, whose products are so standardized that they can be readily interchanged.

plug-pack transformers

The standard off-the-shelf 'power-savers' which you can buy. These plug into mains-power and produce a low-current DC voltage for battery-operated equipment.

PM

Phase Modulation. See PSK.

PMBX

Private Manual Branch Exchange — a company switchboard of the older type. It needs a full-time operator.

PMD

Physical Media Dependent. This refers to the use of communications which must be carried on copper wire or optical fiber — not by radio. It cab also carry the suggestion that the particular protocol will work with one media (say, copper) but not with another (say, fiber). It is used vaguely.

PML

Programmable Macro Logic. These are high-density (2500-gate) CMOS gate array devices used for microprocessor interface, peripheral bus control, memory/cache control, etc. Both NAND gates and flip-flops will be provided on the chip. See PAL, PLD and gate arrays also.

PMMU

Page Memory Management Unit. A chip controller used in the Macintosh to handle the allocation of memory during (pseudo) multitasking operations.

PMOS

A p-type Metal Oxide Semiconductor. The distinction is between p-types and n-types. See MOS.

PMR

Private Mobile Radio. The term is vaguely used:

• Some distinguish between PMR and 'public-access' PMR — which is then called PAMR. Others don't make this distinction.

• In the wide (non-public access) use of the term, some include 'all private land-based systems excluding paging and telephony-based mobile networks'.

• Most people use PMR as a pretentious name for the once-ubiquitous 'two-way' radio.

• Traditionally PMR has meant a private basestation with a few mobiles in a company fleet; they will have the licensed exclusive use of a frequency (or maybe two) within a clearly defined area.

The early systems were all push-to-talk and were not allowed to be linked into a telephone network. Recently, single exclusive-channel two-way PMR systems are giving way to private and public trunked radio and cellular telephones.

We can therefore subdivide the PMR category into:

— Conventional PMR (two-way).
— Trunked PMR (see MIRS, MPT 1327, Tetra).
— Trunked PAMR.
— Personal Radio (see CB and DSRR).

Many people would argue with the last category, but the technology is much the same. See also SMR, ESMR, MIRS, Tetra, trunked radio and cellular telephones.

PMS

Pantone Matching System. A standardized way of ensuring that 'spot color' printing inks match the chosen colors. Every shade, tone and hue will have a distinctive PMS number. Printers can purchase simple comparison charts which provide the necessary ink mixing information to match that color. Standard PMS charts are obtained from art-supply houses, and are an essential item for anyone dealing with color printers.

PMTS

Public Mobile Telephony Services. Cellular telephone services as distinct from 'pocket portables' or PACT systems (CT-2, DECT, etc.). US PCS probably comes under this definition.

PMTU

Path Maximum Transmission Unit. When packets are passing through complex (or unknown) network links where different protocols may be in use, the payload size of the most restricted segment becomes the bottleneck. The MTU of this section of the 'path' needs to be discovered before messages are sent.

Without knowing this 'minimum' MTU, the packet payloads may be subject to a large number of sequential segmentations and fragmentations as they travel over the various links; it is better to send only the size that will get through without further segmentation.

Path MTU 'discovery' provides a way to determine the minimum MTU along a path. This then allows the end systems (and only the end systems) to perform fragmentation or adjust segment-sizes, as necessary. See MTU.

PN

Personal Numbering. See UPT.

PN-3384

The ANSI designation of the 1.9GHz version of CDMA used for PCS.

PNA

Parallel Network Architecture. A technique associated with DCS-1800 networks which allows two independent suppliers to share the same basic network structure. See PCN also.

PNNI

Private Network Node Interface. A protocol series adopted by the ATM Forum in July 1994. It is an interoperability standard for ATM switches which allows users to construct multi-switch, multi-vendor networks, and to then set up switched circuits (SVCs) across those networks without using proprietary protocols. Phase 0 is the initial standard, and Phase 1 (a more scalable version) is due by 3Q-95.

PnP

Plug and Play. See

pocket computer

A miniaturized computer which has been reduced to pocket size. The keyboard is, of course, too small for more than one-finger typing, which limits its usefulness. These are often just called personal organizers. The more modern approach is probably a PDA. See.

POCSAG

Post Office Code Standardization Advisory Group. An international open standard for paging systems. It is also known as CCIR Radio-Paging Code No.1 (RPC1) Recommendation 584-1. About three-quarters of the world's paging systems now use POCSAG.

POCSAG was designed by a committee under the auspices of the British Post Office. It is designed to handle tone, numeric and alphanumeric paging. It provides basic forward error correction, uses CAP addresses (see), and provides for a 'sleep' mode in the receivers to save battery power.

POCSAG uses FSK modulation, and operates at the 'robust' 512b/s (POCSAG 512), 1.2kb/s (POCSAG 1200), and now at the 2.4kb/s (POCSAG 2400) data rates.

- When the POCSAG coding system was introduced, tone-pager controls were included in the standard, and the possible variations in tone-patterns became even more elaborate than were possible with the older 5/6 tone systems.
- The other commonly used paging standards are Motorola's GSC (aka Golay), Ermes, TAP, TDP and FLEX, and these standards can often be mixed together on a common broadcast channel (usually transmitted in batches of the same type, at the same rate). See all protocols mentioned.

POH

Power-On Hours. Used in the measurement of MTBF with computer equipment.

POHM

Page-Oriented Holographic Memory. An optical-computing version of RAM which is now in the early laboratory stages. It uses a holographic technique and it can theoretically store one million hologram 'pages', each with about one million binary bits. The information extraction rate for these gigabits of information is very, very high since you can often read and write bits into the POHM in parallel.

point

A measure of printing sizes abbreviated as pt. One point is 0.01383 (or about 1/72) of an inch, or a third of a millimeter. Ten point type is usually taken as a general standard for body copy in newspapers and magazines, although 9.5 pt type is also widely used.

In traditional typography there are 72.307 points to an inch, but this is usually rounded off in PostScript and other systems.

— Note body copy is normally expressed in terms of 10/11 ('ten-on-eleven') which means ten point type with one additional point of leading between each line. See leading and line spacing.

— Not all fonts of the same point size are equally readable. See x-height.

Point-of-sale

See POS and EFT/S.

point operations

In image processing, 'point operations' work with individual pixels, while neighborhood operations perform tasks on connected groups of pixels. See neighborhood operations.

point-to-point

The connection of two (and only two) terminals or computers. The distinction is with broadcast and multidrop lines.

pointer

In internal computer processing operations, a pointer is a number in a register which directs the processor to a particular address in memory — it points to that address.

In some operating systems you can have short pointers (usually 16-bits long) which provide an 'offset' (the distance from here to there), and long-pointers (usually 32-bits long) which provide an absolute memory address.

poison

To deliberately destroy data. This is not usually a malicious operation, but one done as a preventative measure with routers and routing table updates.

Usually when a network falls over or fails in some way, the routine routing-table information sent out to nodes will not include the address of the unreachable network. However in some cases it is necessary to transmit a poison (reverse) update to deliberately remove the address from all routing tables.

A network manager may need to do this to defeat large routing loops which are constantly ping-ponging packets.

poke

A programming command in Basic which allows the direct programming of the computer at the machine language level from within a higher-level program. By using the poke command, you can force a number into a certain memory address. You do this from within Basic and you can also peek at a location and read its value. See peek.

polar communications

This is more correctly called bipolar, to distinguish it from unipolar. See.

polar mount

A mobile satellite dish mounting system with the axis parallel to the axis of the earth. Astronomers would call this an equatorial mount. Note that polar mount is not necessarily the best with satellite tracking — it compensates in astronomical telescopes for the rotation of the earth — but that's not the problem with satellite tracking.

polar-orbiting satellites

Satellites which orbit around the earth, crossing over both north and south poles. The earth turns under these orbital planes, so usually six planes, each with at least eight satellites (at GEO heights), are needed for global coverage: Current equatorial orbits cannot provide coverage in the polar regions.

• Iridium and some of the other Big and Little LEO proposals use polar-orbiting satellites. These will not necessarily provide full coverage of all regions, all the time. Orbcomm's store-and-forward messaging system, for instance, only has two satellites in polar orbit.

polarization

• In satellite communications the beam to and from a satellite is often polarized to allow double use of the limited range of frequencies available.

By using polarization to distinguish one signal from another, the same frequencies can be reused in a nearby satellite without the risk of interference. The Intelsat transponders use circular polarization, while most of the other communications satellites use linear polarization.

There are two categories of polarization (linear and circular) and four main types of polarization — horizontal and vertical, right-hand circular and left-hand circular. These are known as orthogonal pairs.

• In terms of satellite reception, you should distinguish between a 'polarizer' which, at the feed horn, converts the dish to receive either linear or circular polarizations, and a 'polarizer rotor' which permits the selection of one of an orthogonal linear pair, horizontal or vertical.

• Any radiated electromagnetic wave has an electric field at right-angles to its magnetic field. Polarization is described by the direction of the electrical field. Put more technically, it is 'the property describing the time-varying direction and amplitude of the electric field vector'.

Polish Notation

A form of Boolean algebra where all operators precede all the variables. See Reverse Polish Notation.

poll

To check progressively — to make sequential enquiries — to interrogate systematically — to pass control to. The distinction is sometimes made between polling systems and token passing, and between both of these deterministic approaches and contention systems.

• In some electronic mail networks the central computer will systematically poll (call) each terminal to see if messages are waiting.

• In shared media, this is an access method where a primary network device inquires (by way of a message) whether secondary devices have data to transmit. Polling systems do this systematically, and the reply is (usually) passed back to the primary, which will then give permission. It is a host-centric way to run a network.

• In some LANs, a device can only transmit when given permission by a central computer, or when it holds a token. At one level, both these techniques can be classed as polling. The purists would argue that polling is always a query directed from a primary device to secondary devices — and the reply goes back to the primary device. Tokens give permission and can just be circulated among peers.

polymorphism

Many forms. This term is used in object-oriented programming to mean that the same message can be sent to many different objects, yet each will react in a different way.

polynomial

A polynomial is an algebraic expression of two or more terms. In algebra it is usually a multivariable expression such as:

$(x + 2y + 28)$ or $(1 + x + 2x^2 + 7x^3)$

• A 'generator polynomial' is a modulo 2 divisor.

PON

Passive Optical Network. A general term for a range of similar proposals for fiber to the home. These consisted of optical fiber as a main trunk feed in suburban streets, which were then split by a 'passive splitter' (half-silvered mirrors or gratings) into a number of streams distributed directly to the homes. Every home received everything on the cable. See T-PON and HFC.

pop/POP

- To take from the top of a stack. See push.

- **Population:** In cellular phone slang, this is the population within a cellular network coverage area. The result of the bids for the various PCS coverage areas (MTAs) put up for auction by the FCC in early 1995 showed price-per-capita ('pop') figures ranging from $0.61 (Guam) to $54.45 (Washington-Baltimore).

- **Point Of Presence:** In American teleco terminology, this is where a long-distance carrier (IXC) ends a circuit in a local carrier's (LEC's) exchange somewhere in a LATA. The IXC must negotiate an interconnection price if it is to use the local service.

However, it is at this point that a local company seeking to bypass the local LEC can make direct contact with the IXC via a microwave circuit or leased line, if it wishes.

Not all long-distance carriers have POPs in every exchange, by any means. The long-distance carrier selects these contact points, and they can have more than one in a LATA. The local phone company charges the IXCs for these POPs, and for interconnection costs on all calls diverted through their system, if the long-distance carrier bills the customer.

- **Post Office Protocol:** Internet e-mail protocol for linking mail server to mail client programs. You would use this on a Windows-based PC for contacting your e-mail host. There are three flavors: POP, POP2 and POP3, but the later versions are not compatible with early versions. POP has been superseded by IMAP. See.

pop up

- **Menus:** The opposite of pull-down menus, although the terms are often used interchangeably. Menus often appear only when you click on them with the mouse/cursor.

- **Programs:** TSR programs (also called 'hot key' programs)

POR

Pacific Operating Region — for satellites. See IOR and AOR.

porch

In television, the Horizontal Blanking Interval consists of the front porch (4.7μs) of the main sync pulse, plus the color burst lasting 2.3μs.

The front porch is the clean square-wave portion of the interval, and this is then followed by the back porch which is another 4.5 μs with the color burst. In most video scrambling systems, this is the portion most likely to be deliberately distorted so that the pirate viewer's TV will not lock on to the line.

See front and back porch, and HBI.

port

The connector to an external device. Port refers also to the associated circuits such as control, command and status registers, and also the drivers (amplifiers) which operate the port.

- The connection point for a cable in a computer, switch, bridge or router. The physical access for signals into and out of a computer or other equipment.

The term applies particularly to the physical plug interface, but in the case of a bridge or router it may also apply to the port-card itself. These are the plug-in cards which use the expansion bus (the 'backbone') of the unit for interconnection. Many cards will have two or more ports.

- In IP terminology, 'port' specifies a receiving upper-layer process.

- To port software from one system to another is to rewrite it (or modify it) to work on other hardware and/or other operating systems; 'on a different platform'.

- Port-switching suggests the type of patch-panel network reconfiguration process, rather than the fast LAN segment-switching of packets. See switching hubs and switch fabric.

portable computers

This is a general term which covers:

— Luggables: large, heavy and perhaps mains powered.
— Laptops: battery powered; larger than A4 in size.
— Notebooks and dockables: See notebooks.
— Palmtops: MS-DOS with QWERTY.
— Personal organizers (non QWERTY keyboards).
— PDAs.

portrait screens

Full-page display screens which can show the equivalent of a full A4 page vertically. Some IBM compatible portrait screens are now 768 pixels wide by 1024 pixels high, with four colors.

There are also 1280 x 1024 screens in a two-color mode. The Apple Portrait Display is 640 x 870, with 16 levels of gray.

The manufacturer of the special video adaptor cards needed, must also provide a suite of suitable software drivers. Not all applications will be able to exploit the new screen dimensions.

POS

Point-Of-Sale. Aka 'Electronic' (E)POS. This is a cash-register in a retail store, probably equipped with a bar-code reader and credit card swipe and keypad. It may be connected to a private or bank network to a) transfer product information, or b) conduct the transaction.

In many large stores the POS terminal will automatically update inventory files on the main computer at head office. POS technology is gradually being combined with EFT (Electronic Funds Transfer), but a distinction should be made between the two functions. POS equipment is not necessarily on-line to the financial institution, and therefore does not automatically debit the customer's account (or credit the retailers): that is the EFT function.

POS equipment assists the retailer through maintaining records of stock control, sales monitoring and the recording of individual transactions.

POSI

Promoting Open Systems Interconnect. A Japanese manufacturing group (includes NTT and suppliers) which was originally called SPAG Japan.

positional notation

The normal form of numbers. From the right, the first digit is the number of ones, the second is the number of tens, the third is the number of 100s, etc. Binary also has positional notation — but not Roman Numerals (i.e. XVIII).

Positive Intrinsic Negative

PIN photodiodes are optical-fiber receivers which are used to convert the optical pulses to electrical signals. The I stands for 'intrinsic silicon' which lies between the p- and the n-layers of the diode. PINs are not as sensitive as Avalanche Photodiode (APD) receivers, but they are cheaper.

positive polarity screens.

Paper-white screens, such as those used in the Macintosh and Windows PCs, where black letters are superimposed on a white background. The perception of flicker is increased in these screens, so refresh-rates need to be higher and often long-stick phosphors must be used. Most of the early character-user interface (CUI) screens used white, green, or amber characters on a black screen.

POSIX

Portable Operating System Interface eXtensions for computer environments. An ANSI-compliance standard now adopted as IEEE 1003.1 and ISO IS 9945-1; mainly associated with Unix. It has been adopted by the US Government for equipment tenders, and by most hardware vendors in both the OSF and Unix International camps.

Posix is based on Unix and incorporates elements of both System V and BSD. However none of the current Unix implementations will be totally compatible with Posix without modification. It has defined a new way of handling terminal I/O, and a new approach to record locking.

There are actually a number of Posix standards being formulated: networking, security, real-time operating systems, administration. IEEE 1003.1 is the system interface, 1003.2 is for shell and tools interface, and 1003.4 is real-time extension standards. XPG uses 1003.1 for its OS interface level.

post/POST

• To send a message to a Usenet newsgroup or a general forum elsewhere. You post signed electronic messages on these systems so that anyone can read them. The distinction is with one-to-one electronic mail.

• **Power On System Test:** The initial testing sequence which MS-DOS PCs runs automatically after power has been switched on. See diagnostic programs.

post-production

In the movie and TV production industry, the three main phases of making a program are:

— **Pre-production.** Scriptwriting, organizing finances, choosing cast and crew, finding locations.

— **Production.** The time in which shooting actually takes place 'on-locations' (away from the studio) or at the studio.

— **Post-production.** This is the selection of shots to be used, the editing process, sound re-recording, mixing, negative matching (film), credits and special effects, and the production of the (composite) Master copy, and the first Final Print.

In multimedia there will be additional stages here also, in the making of CD-ROMs, etc.

Post-production can also refer to the publicity, promotion, and distribution side of the movie business.

postalized rates

Postalized rates in telephony means that the rate is the same regardless of distance. It is a US telephony term, used in this way because their 29 cents cost for a one-ounce letter, is the same whether it only goes locally, or to Alaska or Hawaii.

postamble

See preamble.

posterization

The transformation of a full-tone image into stepped tones, to create a 'poster-like' image. The computer will deliberately constrain the color range or the tonal gradation, or both. You also get this effect when the pixel depth is limited. A 3-bit depth can only produce black, white and six tones of gray for instance.

In photography this is known as 'tonal separation'. Taken to the extreme, all whites and near whites will become white, all blacks and near-blacks will become black, and any mid-gray tones will become a perfectly flat and featureless mid-gray. Andy Warhol used this technique extensively and made a fortune out of it. In a computer, this is done by rounding-off, reducing and manipulating the pixel values stored in the bit-mapped computer image.

Postmaster

Every Internet-connected service is required to have a valid address for a 'Postmaster' who can handle e-mail queries and requests. You should always be able to contact someone in charge of e-mail, by using this address.

PostScript

Adobe Systems' page description language (PDL) which includes outline font handling, and later screen display techniques. This is still the *defacto* industry standard in the graphics art world, although simpler systems have outstripped it at the lower end of the market.

PostScript execution (the generation of the full-page bit-maps ready for printing) generally takes place in dedicated hardware within the laserprinter's own memory and CPU.

All the font-outline and graphic information is delivered to the printer in the most compact form, and the bit-map creation then occurs in the printer. For this reason, the early Apple Laserprinters had more powerful CPUs than their Macintoshes.

However, the PostScript execution process can also be emulated by special software in the PC, and the final full-page bit-map is then delivered to a non-PostScript laserprinter for the same end result. However, this approach is usually slower — especially if the cable link is serial and slow.

There are a number of variations in PostScript, but the translation between versions is usually handled automatically — although not always, when files are transmitted across a network. A few years ago there was a complete rewrite and revision of the system to produce PostScript Level 2. As a result:

- **PostScript Level 1** still provides the highest quality levels.
- **PostScript Level 2** supports both JPEG and LZW compression and decompression so that files can be sent faster to the printer. It is also totally backwards compatible (Level 2 files can't be run on Level 1 machines, but vice versa is OK).

Level 2 also includes support for later industry standards on color (CYMK color separation), faster font generation (by perhaps two or three times) and better file management. In Level 1, these were all add-on enhancements, but they have now been brought into the body of the program. Adobe says it has also sped up the program by optimizing the code, adding cache, improving memory management and improving the half-tone controls.

• Many of these new Level 2 changes have also been incorporated into Display PostScript which is a screen-based version used on the NeXt computer.

• The alternatives to PostScript are a number of proprietary and clone systems. TrueType/TrueImage was a joint effort by Apple and Microsoft to threaten PostScripts dominant position. See outline fonts, TrueType and PCL.

postulates

Unproven assumptions.

POTS

Plain Old Telephone Service — a half-humorous acronym used to distinguish analog telephone network services from ISDN (service, not system). Unadorned; plain vanilla.

• It generally means the old single line residential service with a rotary dial and a step-by-step exchange. No bells and whistles; a basic single analog telephone service for a subscriber. The term is very widely used in the industry.

• Sometimes the term is used to distinguish a normal line service from Centrex service, and sometimes it distinguishes the old dial-telephone set (a POTS-set) from a touch-tone set.

pound sign

See octothorpe.

power

• **Mathematics:** The number of times a number is to be multiplied by itself. So 'Two, to the power of four' is the equivalent of 2 x 2 x 2 x 2 = 16.

• **Mains power:** See power supply.

• **Computer sales:** The marketing man's rather confused idea of what lurks under a computer's bonnet. He imagines the CPU is akin to a twin V-12 with overhead cams and turbo-charge.

The use of the term 'power' has no practical meaning in computer comparisons other than to say, 'Mine's bigger and faster than yours.' And usually it carries the same level of penile credibility. See benchmark.

power supply

Mains power can be subject to numerous problems:

— **Sink.** Abnormally low voltage level. 240V systems will often sink to about 210V in local areas when load comes on the line. If the voltage measures more than 6% below the rated level, the power company is often legally obliged to do something about replacing or upgrading the local transformer.

— **Spikes or Transients.** Very short voltage surges of less than a 1/100th second. These are very common; they are caused by lightning strikes, nearby switching of highly inductive loads, automatic tap changing, and power factor corrections. They are characterized by high voltage peaks up to 2000V (and even up to 10kV) with a nanosecond leading edge and a duration of less than 100 microseconds.

— **Sag.** A drop of voltage of 20% or more, lasting from 1/50th to a few seconds. Usually this comes from a local machine startup, or from a car knocking over a power pole.

— **Surge.** A very brief overvoltage which is the opposite of sag. This usually happens when heavy machines are turned off. A surge can also come from load switching on the network and even from transient effects like lightning strikes at a great distance. Some surges can be hundreds of volts higher than normal, but most are in the +60% range and last for about 1/50th of a second.

— **Dropout and Dip.** When the voltage drops to zero for a period of 1/50th second or less.

— **Brownout:** Any unplanned voltage reduction which is noticeable. Usually overload of the local supply or storm damage are the main causes.

— **Hash or noise.** Higher frequency signals from 100Hz up, superimposed on the line. Some of this may be in bursts, but much of it is more regular. Electric motors are the main cause, especially motors with commutators.

power user

A self-styled, high-requirement, computer user who spends most of his time discussing problems at power lunches. Some power users actually do need their 1000 horsepower Super-Quad X5000 with the 10 gigaByte hard-disk storage, for CAD, image manipulation, etc.

PowerOpen

An Apple and IBM joint implementation of Unix. It will support both IBM's AIX Unix application software and Apple Mac applications software under Unix.

PowerPC

The name IBM, Apple and Motorola give to their joint RISC development. IBM's acronym, so they say, means 'Performance Optimization with Enhanced RISC.' There are numerous implementations:

— 601 is for mainstream desktop machines and servers.
– 603 is for low-end and portable machines.
— 604 is a high-performance chip.
— 610 family includes 613, 614 and 615 and are said to run Intel microcode.
— 620 is the highest performance.

PowerPC is a challenge to the dominance of the Intel Pentium.

- A pre-production version of the 610 was reported (Feb 1995) to be running at 400MHz. At an average of 2.5 instructions per cycle, this is five times the power of a Pentium.
- Apple's PowerPC computers are actually manufactured by Solectron in Milpitas, California. The new models are based on 604 with PCI bus architecture.

PPM/ppm

- **Pulse Position Modulation:** See Pulse Time Modulation.
- **Pages Per Minute.**

PPP

Point-to-Point Protocol. A single-route, data-link protocol (actually a suite of specifications) which provides a method of transmitting datagrams over serial links.

It is the key successor to SLIP as the serial-line protocol for wide-area synchronous and asynchronous communications between internetworking devices (router-to-router and host-to-network). It can do this because it has a 16-bit data field which identifies which protocol is encapsulated in the information field of the frame.

PPP is designed to overcome the problem of different encapsulation techniques being used for traffic over WANs. It provides a standardized mechanism for higher-level LAN protocols, and the encapsulation is done at the Data-link layer.

It is ideal where one side of the network connection is fast and the other is slow. It is byte-synchronous and uses HDLC framing and a 16-bit CRC. It can negotiate compression also.

PPP has two internal protocols — the LCP (Link Control Protocol) which is sent in the first packet and used to establish the link, and NCP (Network Control Protocol) used for transferring the data. Packets can be of three types: the LCP packets; higher-level protocol control frames; or data.

As defined by the Internet Engineering Task Force, PPP provides a standard way for routers to set up and terminate a session, and also allow them to monitor the link.

A PPP connection is often used to make a computer into a temporary Internet host. It will be carried between two modems over telephone lines, or used on high-speed links. It provides mouse and cursor control.

If you have PPP running, you will be able to use TCP/IP software such as Telnet, and WWW client-software. See SLIP.

pps

- **Pulses per second:** The telephone pulse dialing method uses both 10 pps and 20 pps in different parts of the world.
- **Packets per second:** The measure of the switching speed of a node on a packet-switching network. Ethernet rarely exceeds 1000 pps.

On X.25 networks, the default length of a packet is 128-Bytes, but an individual keyboard-generated packet could well be only a few bytes. The packet will then be padded. However, equally, many X.25 packets are now over 1000-Bytes.

FDDI over fiber has packet rates in the order of 175,000 pps.

PPV

Pay-Per-View. There is a distinction between the early PPV systems where you need to phone and pay (by credit card) well ahead of the event, and Impulse PPV which, today, probably uses a smartcard for stored value or on-line validation.

Smartcards may simply store the viewing details and transmit the data irregularly over the telephone network, or they may communicate interactively in real-time over a return channel on an interactive cable network.

Often with Impulse PPV systems, you will be allowed to watch the first few minutes of a program before the picture becomes scrambled. See Pay TV, VOD, NVOD, Premium TV.

PRA

Primary Rate Access. See.

PRAM

Parameter RAM. An area of memory which holds the information about the 'customization' parameters of a personal computer — the speed of movement of the mouse, pixel-depth for the screen, key repetition rates, etc. This will probably be held on disk or in an EEPROM.

PRC

Partial Response Coding. A coding scheme used in digital microwave transmission and recently applied to a new digital form of Data-Over Voice (DOV) transmission.

PRC involves the use of several pulses to indicate a coded bit. The combination of these pulses being received as certain patterns allows the system to disregard noise at the different frequencies.

preamble

A bit pattern used in communications systems to establish the beginning of a block of data. A postamble provides the end point.

• In synchronous TDM frames using a character-oriented protocol, usually two SYN-bytes are used as the preamble, and another identifiable byte is used as postamble. However, the frame may include frame-length information, as an alternative to using a postamble.

• In TDM frames using bit-oriented protocols, the preamble is always an 8-bit bit-pattern which is referred to as a flag.

• In Ethernet and 802.3-type LANs, an 8-byte (64 bit) preamble precedes the destination address in each frame (packet) of data. It is essential for synchronization.

• In paging systems, the preamble is the first transmitted portion of the signal formats when the transmitter changes to a new protocol. Because there is a preamble, the pager can operate in sleep mode while all other protocols are being transmitted. An identifiable preamble then wakes up the pager when it is the turn of this protocol to be broadcast. The preamble also re-establishes bit synchronization.

preamplifier

Many analog signals are too low in level to pass through cables, switches or pass directly to a power amplifier without first receiving a boost at the earliest possible moment. This is done via a preamplifier which is placed as close as possible to the transducer (originator). Most preamps provide both amplification and equalization (and sometimes pre-emphasis) and they are often built into devices like microphones.

precision

See accuracy.

predictive coding

A term used in some complex audio and video compression systems. It assumes that at a certain point, what lies ahead is very much the same as that in the immediate past. This may mean frames of a movie, or it may be the shape of a waveform.

With voice, you can predict that the next sample will be very like the last, since 180° phase changes do not happen with human vocal chords. This then makes compression (bit-reduction) easier since there are limited possibilities.

These systems 'predict' that the next stage will be the same as the last, and so they compare their prediction with the actual, and extract difference values. Provided the same practice is followed during the encoding and the decoding process, they can then simply transmit these 'differences'.

• In video coding, prediction is fundamental, and in more sophisticated systems prediction is also used to provide motion compensation vectors. Prediction occurs between frames in motion video (inter-frame coding).

• In some forms of video coding, prediction can also be used for intra-frame coding. The first block of coded image is used to 'predict' the second block, and differences are then extracted within the same frame.

• In audio, see linear predictive coding and DCC.

pre-emphasis

Many radio transmission signals are pre-processed prior to transmission as a form of 'pre-emptive' equalization.

For instance, standard FM broadcasts must have the high-frequency components of the signal artificially boosted prior to transmission because these high-frequencies are more attenuated than the low. There is a CCIR standard for FM radio pre-emphasis.

Satellite TV audio signals are often pre-emphasized at the high end to improve the signal-to-noise ratio, which is always worse in this part of the frequency spectrum. The receivers are then designed to de-emphasize these high-frequencies and restore the balance.

Pre-emphasis of this kind is used in most audio and video recording processes. Sometimes it is used as a form of equalization (correction of transmission defects), and sometimes with later de-emphasis to improve the signal-to-noise ratios.

Obviously the channel conditions need to be known and stable before pre-emphasis can be used successfully, or a training signal can be used to measure (reported back by the receiver) where, in the spectrum, the emphasis must be.

pre-emptive

Taking corrective action before the event. To occupy or establish a right to something.

pre-emptive multitasking

Pre-emptive multitasking is the mainframe and mini OS approach to handling multiple applications simultaneously. Pre-emptive multitasking is also (mis)named 'parallel' multitasking to distinguish it from 'serial' or 'task-switching' approaches.

Unix has long been a true multitasking operation system, and OS/2 brought pre-emptive multitasking to the desktop. With pre-emptive multitasking the processing time is fairly rigidly divided between the various applications and shared around by a resources manager — control of the CPU as a resource is external and handled by a schedular.

This is the main distinction between pre-emptive and co-operative multitasking which allowed the applications themselves to control the allocation of CPU time, and therefore required 'cooperation'.

• Unix, OS/2, Windows NT and 95 are pre-emptive multitasking operating systems. The Mac OS and Windows 3.x MS-DOS are task-switching. See task/tasking.

prefix notation

See Reverse Polish Notation.

preform

A large glass block which has been purified and is now ready to be drawn out into an optical fiber.

premastering

In CD-ROM manufacturing technology, this is the process of writing the information onto a 9-track tape and adding error correction and sector information. It is the exact data image of the final disk. This tape is then use to create a master glass disk, and from this, the stampers are made.

• Authoring is the name given to the process of logically constructing the premaster's format.

Premium TV

• In cable television terms, this is a cable channel which is scrambled and requires payment from a subscriber before the set-top box will be provided with the key to decode the signal. So this is a Pay TV subscription service.

The term does not generally include pay-per-view (PPV), but since PPV depends on a high-level, addressable, set-top box, both are often seen as Premium services. See Basic Cable and STV.

• In direct broadcasting satellite service, most or all the channels are 'premium' in the sense that you pay for the programming, not just for the delivery of a reasonable signal. Premium will then often refer to something above the normal. In BSkyB, Premium refers to movies and news, which are offered at a payment level above the normal diet of sitcoms, Star Trek and sports.

PReP

PowerPC Reference Platform. IBM's architecture for the PowerPC. It uses a little-endian approach. Unfortunately Apple has used a big-endian approach — which gives the two companies some compatibility problems on the same PowerPC platform.

preprocessor

A high-level language compiler which can create a source code in a specific format that can be easily read by another compiler.

Presentation layer

The sixth of the OSI's 7-layer model. This is a connection-oriented protocol whose main purpose is to provide a means by which the two computer applications can communicate across the channel, even if they are dissimilar.

It provides the syntax for the exchange between two Application-layer entities; the services that a particular application will need (such as the file command, 'Open File'), and it performs translation and byte-reordering so that, for instance, an IBM machine can talk to a DEC and a DEC to an Apple.

The ISO says that the Presentation layer 'provides for the representation of information that is communicated between, or referred to, by applications processes'. This is another way of saying that it handles the translation between the human interface and the machine's internal 'native' numerical format — the control codes, special graphics, and the character sets that define the features that appear on the user's screen.

This is the layer of Terminal Emulation (see).

So it includes functions which handle:
- the contents of the user messages (usually in ASCII No.5),
- features on the screen such as graphics, inverse video, etc.,
- terminal emulation procedures,
- peripherals (printers, plotters, disk drives),
- encryption processes.

Presentation Manager

Microsoft's graphics-user interface for OS/2. Windows 3.x is essentially a subset of Presentation Manager for DOS machines; and Presentation Manager is a superset of Windows for OS/2. Both provide a Mac-like interface for IBM-type PCs.

presentation services layer

This is Layer 6 in SNA.

Prestel

The British Post Office's interactive videotex (viewdata) system. Tests began in 1976 and the service was launched in Sep 1979. It was probably the first remotely-successful videotex system in the world — which doesn't mean much since only Minitel has proved to be really viable.

The original Prestel was an alpha-mosaic system, but the later Picture Prestel (II) has a much higher image resolution and used alpha-photographic techniques. Prestel and Prestel II are gradually being absorbed into the European CEPT standards. See Captain and alpha-mosaic.

PRF

Pulse Repetition Frequency.

PRI

Primary Rate Interface of ISDN. See Primary Rate Access.

primary

• **Primary devices** are those modules that are an integral part of the computer — as distinct from peripherals, which are usually on cables so that they can be easily removed.

• **Primary multiplexers** are those that handle the lowest group of channels in the hierarchy. They will usually take between 12 and 30 voice (or 64kb/s data) sub-channels and combine them into a single data stream. This is a term mainly used by carriers with frequency division multiplexers.
- In the USA, D-channel banks are the primary multiplexers for digital communications at the T1 rate of 1.54Mb/s. Then above this, the M1C multiplexer outputs at 3.152Mb/s and the M12 at 6.312Mb/s.
- In the ITU world standard the primary carrier system is E1, which runs at 2Mb/s.

See channel bank, high-order multiplexers.

• Primary memory is the main operating memory of the computer. The distinction is with secondary memory, which is disk or tape storage. However, see below.

• Primary storage is a term often applied to a hard disk, as against a back-up system such as sequential tape (DAT, etc) which is called secondary storage.

• **Primary Stations** on a network are those that send commands to the secondaries, and receive responses. HDLC and SDLC bit-synchronous systems need such a relationship.

Primary Rate Access

Primary Rate Access is the version of ISDN for larger organizations and between exchanges. PRA usually links a PABX to a local exchange. In America this is transported over T1 lines which permit 23 B-channels (64kb/s) and 1 D-channel (also 64kb/s). In Europe, Australia and most parts of the developed world where E1 lines are used, this is 30 B plus 1 D-channel (all at 64kb/s). Sometimes a cut-down PRA (with, say, 20 B-channels) is also offered. See BRA also.

primary storage

Main memory storage — RAM. However I have heard of it used for disk as against back-up tape storage.

prime numbers

Numbers that can't be factorized into smaller component parts — as in 3, 5, 7, 13, 17, 19, 23, 29, etc. Large prime numbers are used in almost all encryption techniques.

primitive

Small pre-defined routines or parts which can be put together to form larger structures. Some of these are like utilities or daemons.

• **Programming:** In FORTH and LISP the programmer can create new statements by defining them in terms of language primitives.

• **Graphics:** These are basic shapes used in object-graphics programs — lines, point-marks, arcs, circles, polygons, cones, cubes and squares. They are complex modules (objects) which can be further developed by juxtaposing other elements, scaling, and adding and subtracting.

In CAD, graphic primitives are the base elements from which an image is composed. Complex objects such as a computer-generated humanoid face will need many thousands of polygons, for instance, to be able to reproduce the shape in 3D — while a pyramid needs only four. Booleans can be used in some primitive graphics to, say, subtract a small square from the bottom-left corner of a large square. See also patch.

• **LANs:** In LANs standards, primitives are the basic functions necessary for tasks like network management. A networking primitive is something similar to a subroutine call or a subsection of a protocol specifying the function to be performed in an OSI layer.

Primitives are used to pass data and control information across the layer interfaces. At the user-IP interface, for instance, the Send primitive requests data from a data unit, and the Deliver primitive notifies the user of the data unit's arrival. These primitives will have parameters for source and destination address, protocol ID, service indicators and a range of other information including the data length.

• **Processor:** Operations which are fundamental to the function of the processor. Interrupts are the lowest-level primitives available in the operating system.

print server

A dedicated computer system which handles the print requests from the stations on the network. It will manage, buffer, prioritize, and send the data for printing. One server may manage a number of printers.

printer command languages

Languages like HP's PCL. These have evolved over the years to control a range of printer types from the early daisy-wheels, to dot-matrix, ink-jet and laserprinters. The basic commands are those concerned with printing ASCII characters line by line on the page. With laserprinters, more advanced forms handle bitmaps delivered, as a whole, by the computer.

Make a distinction between these earlier PCL types and page description languages (PDLs) like PostScript. Later additions to the languages allowed the printer to be instructed to create the page itself from a 'description' sent by the computer. True printer independence is possible only with PostScript-like PDLs. See TrueTypes also.

printer commands.

See escape sequence.

printers

There are many types:

• **Impact printers:** These are golf-ball, daisy-wheel, and 9-wire (pin) or 24-wire dot-matrix printers. They are either character or line printers.

• **Non-impact printers:** laserprinters, ink-jet and thermal. Ink-jet and thermal printers can also be character or line printers, although most ink-jets are page-printers as are laserprinters and LED-printers.

Printers are also sometimes categorized as having parallel or serial (RS-232 or SCSI) connections, but note that the term 'serial printer' can have two distinctly different meanings. The term 'line-printer' can also have three distinct meanings. See below.

• **Facsimile-type printers** use a 'raster-scan' and they print the page one image/pixel line at a time like a TV set, so it may take many lines of scanning to create a line of characters. These were once characterized as 'line printers'.

• **Serial printers** are those that print the data on the page as it is received, a character at a time (although the term is often also used for those which need an RS-232 serial connection). These are also called character printers.

However, if they take the full text of a line into memory, they can justify the text and print in a bi-directional (boustrophedon) way. They will begin printing from left-to-right, then returning on the next line from right-to-left. So these printers were then called 'line' printers, to distinguish them from the old left-to-right-only 'character' printers which didn't have much memory, and so couldn't justify the text.

• **True line printers** are high-speed impact printers which can print an entire line of text at one time. These printers stream out hundreds of fan-fold pages of low-quality print every minute.

• **Page printers** are those that create the whole page in internal memory, before the printing begins. Laserprinters are the best example. These are usually characterized by a long delay to get the first copy, then rapid printing of all subsequent copies of the same page.

priority

Various packet-based systems will provide priority queuing.

• In isochronous networks where voice and video need to travel across the network without delay, these packets will have one or more flag-bits set to show that they have greater priority than normal data packets which can afford to be delayed. The ATM standard cell has been provided with only one priority bit, so bits are either high-priority, or not.

• In LANs, the packets may be queued on the basis of packet size and interface type in the router, with some streams given priority.

• Note that priority can be interpreted in different ways. Voice packets will have a high-priority for crossing the network in terms of time, but if they are inadvertently delayed (congestion) then they have a low-priority, because it is better to discard them than deliver them late. A few packets won't be missed.

Data, however, has low priority in terms of time but high priority in not being dumped, because it must then be detected as missing and retransmitted across the network, thus increasing load.

Note also that interactive voice needs both regular delivery and low delays. One-way digital television needs regular delivery, but the delays can be substantial. It doesn't matter whether your TV news arrives a second late but, unless we have big memory buffers in each TV set, the data must be regularly delivered.

private line

The same as a tie-line or leased line. A rented unswitched channel between offices of a company to tie together PBXs or LANs.

• There's another sense where private line is applied to direct connection between an office and the fire-station or security firm, for security, telemetering or control systems. Some of these are shared with other customers in the same area in a party-line fashion.

PRMA

Packet Reservation Multiple Access. A form of TDMA used in wireless local loop and cellular phone systems, which relies heavily on the detection of voice (Voice Activated Detection).

It is also a LAN/WAN technique which uses a queued system for data, and yet provides priority isochronous traffic over real-time packet links. See E-TDMA also and VAD.

PRMD

Private Management Domain. This is a private X.400 e-mail network in the ITU-T terminology.

PRML

Partial Response, Maximum Likelihood. This is an error-control system developed for deep-space probes, and used for a couple of decades to improve modems.

It is now being used to increase the packing-density on hard-disks. IBM first used it on its disk drives in 1990 — it's a way of enhancing the reliability of the read operations.

PRML now replaces the earlier 'read channel' technique called peak detect, used since hard disks were first invented. By taking inter-symbol interference into account, the new scheme increases areal density by 30—40%. It is now being coupled with MR (Magneto-resistive) heads on new hard drives.

What is new is the high-speed digital signal processors which can now handle these tasks easily.

The algorithm allows the disk drive to filter out the noise associated with data which has been densely packed to the point where bits overlap. It analyzes the noise in the light of its knowledge as to how bits 'blur' when they are close together and overlapping.

As with RLL there are PRML constraints during the write operations. Constraints of, say, (**0 , 4 , 4**) mean:

— The **0**, says that logical 1s in the data stream can be concatenated. No separation is needed.

— The **4**, says there must be no more than four logical zeros in the data stream in any one place (and no more than six encoded bit-periods between flux changes).

— The last **4**, sets the maximum number of zeros which can occur in certain other data sequences.

PRMS

Private Mobile Radio Services. See PMR and trunked radio.

probabilistic

The shared-media LANs which use carrier-sense contention (such as Ethernet) rely on the stations waiting for a gap in the traffic before they transmit. When they detect a gap and transmit, they will often find another station has similarly detected and transmitted, so the two packets will collide.

This means that the transmission from any node on the network is erratic and unpredictable — rather than deterministic. These systems are known as carrier-sense contention, or probabilistic systems, and this makes them generally unsuited to real-time (isochronous) voice and video services, quite apart from bandwidth considerations. However, see Isochronous-Ethernet.

Token Ring and token bus systems can be provided with a strict order of transmission so that each node can be sure of access at some predetermined time in the cycle. Priority can also be given to special packets in some systems. This makes them deterministic (to varying degrees).

There are still problems of packet-length limits and how long each station can hold the token, which need to be taken into account.

probe

• In some networks problems are address resolution protocols. In the X.500 directory structure, probes are used to 'test the water' before sending a secure message to an unknown address. They will return with identifying information.

• In satellites, the probe is a short bar or wire inside the feed horn at the dish's focal point. It can be positioned to decide which of a pair of orthogonal polarized signals will enter the horn (H or V — RH or LH).

• In network management, probes are software 'agents' in a networked ('managed') device, monitoring device, or processes. These are now built-in into remote hubs, bridges and routers. They collect statistics on the traffic and communicate their findings to the management agent in the central management console (via the MIB).

The probe provides the data-gathering functions, and the management-client provides the data-display functions.

The name is also used for special portable test equipment which can be plugged into the cable and be left to monitor a network segment. The original problems were protocol-specific — Ethernet probes (from 1991) were stand-alone devices with a processor and software, which attached to the network and gathered statistics (defined by the agent) for the segment to which they were attached.

Some probes are still stand-alone, but many are now being integrated into network nodes. Some, like RMON probes, are also promiscuous, and they can be used for making configuration changes across the network. See RMON and SNMP.

proc amp

See processing amp.

procedural languages

These are well-known high-level computer languages like Basic, Fortran, Pascal and C which, when run, begin at the first line of instruction and move progressively through a long chain of coded instruction lines in sequence — except for loops, etc.

The contrast is with non-procedural, event-driven or functional programming languages. There are three types of programming components in procedural languages:

- Sequence.
- Selection/Decision.
- Iteration.

• In general use, the term 'procedural language' probably just differentiates it from object-oriented.

procedure

In programming, this generally means a subroutine or module which is relatively self-contained. It is probably an algorithm in its own right. Don't confuse procedure with process.

process (computer)

A process in a computer is a 'program' in execution — where 'program' means a task which must be performed in full. The management of these processes is the most important job of the operating system.

Even the older single-tasking operating systems had a number of different processes which needed to be managed in the course of running a program. A process is a defined task, such as saving files to disk, reading data from disk, updating the clock, and PCs are constantly doing these tasks in addition to running any application.

With multitasking operating systems, the number of processes running simultaneously and needing management, can be extensive. See task/tasking.

- A process can also have sub-processes called 'child processes' and these in turn can have child processes — so there can be a process tree structure. Pipes are used to connect two processes together.
- In a modern operating system, a process will generally have two parts: a user-part and a kernel-part. When a system-call is made, the process switches from the user-part to the kernel-part. The kernel-part then has access to many more objects (all pages in physical memory, the entire disk, etc.).

• Process tables are arrays which contain all necessary details to enable a suspended process to be restarted at exactly the same place as it left off. This is necessary in any multitasking operation — all current stages of progression are stopped while the CPU works on the other tasks.

process (general)

In printing, process colors are Cyan, Magenta, Yellow and Black. Process printing uses these three ink colors and black on four different printing plates. They are printed one at a time (in order Yellow, Magenta, Cyan, then Black) to produce the final pages. All photos are treated as half-tone separations. See CMYK. The contrast is with spot color, which is created by mixing inks.

process control/manufacturing.

The use of a computer to control some physical manufacturing process. These cover a wide range, from the data collected directly from instrumentation on the floor, to top-level planning and financial data.

processing amp

Aka proc amps. These are used when transferring video to clean up and amplify the signals. The sync is stripped, and replaced with a new square wave, and the pedestal and gain controls can usually be adjusted to give the best subjective results.

processor

The functional part of the computer that interprets instructions. Strictly the processor is the Arithmetic Logical Unit (ALU) within the Central Processing Unit (CPU), but in practice these are all lumped in together. In a PC, the processor is a single micro-processing chip. In mainframes it may be a complex series of chips and components.

processor speed.

Various measures are used for processor speed, most of them almost meaningless, and almost all of them are designed only for promotional applications.

- MIPS is a specific processor measure which is closely related to clock speed but it takes into account the 'wait states' caused (usually) by slow memory access. It is as good a benchmark as any, for general processor ratings.
- MFLOPS and GFLOPS are measures of algorithm processing times and include factors associated with the operating system, and so they depend on both the hardware and the software (although attempts are made to standardize the software). However different algorithms work differently on various processors, so these figures are also suspect.
- Other benchmarks measure more meaningful tasks — but these are also more limited in application. See Linpack, Dhrystones, Whetstone, iCOMP.

product

In maths, this means to multiply. The product of 2 and 3 is 6.

product numbers

There are a range of product codes: usually these are 13-digit codes associated with barcodes and retail scanning. They are either part of a universal product identification system or of a national system derived from the original UCC system.

As an example: the first two numbers identify the country issuing the number (Australia = 93); the next five, the manufacturer (allocated by the local association); the next five, the specific product of that manufacturer; and the last is a check digit. This can be in number form or as a barcode.

- The International Article Numbering Association (EAN — French acronym) administers the international system of product numbering in more than 50 countries, and the Uniform Code Council (UCC) administers a compatible numbering system in North America.

These standards allow for 37 different types of information, such as customer purchase order numbers, shipping container codes, product dimensions, internal applications, weight and size. These can all be carried in a barcode, and used in-house or for traded products. See barcode.

productivity software

The marketing term for software that does something (supposedly) useful. If you don't buy this software, you are told your business will surely fail. The only use I can see for the term is to distinguish business software from video games. But into which category do we then put spread-sheets?

profile

- In X.25 packet-switching, a profile is a full set of PAD parameters which can be chosen for a specific task. You use one profile when you are doing standard-bit ASCII information retrieval over the network, and another to send an 8-bit binary file.

A Profile change just resets a number of different parameters to new values. You will usually find a general-purpose default profile when you enter the PAD, and there may be a half-dozen variations from which to choose.

The command to see details of a profile is usually 'PROF?' and to change it, is PROFx, where x is the number you have chosen. Profiles default to a pre-set standard when you leave the system.

- In interconnectivity, profiles are suites of existing communications standards such as MAP, TOP and GOSIP.
- In MPEG specifications see MPEG levels and profiles.

PROFS

IBM's proprietary electronic mail system.

program

A general term that covers any sequence of instructions which can control a computer, both at the level of the operating systems and of the applications. A macro or a TSR is also a program. However it usually means substantial applications.

program trading

See programmed trading.

programmable function keys

These are special keys on the keyboard which can be made to issue complex, frequently used command sequences (macros) by a single keystroke. See function keys.

Programmable Logic Devices

PLD. A general term for three types of customizable VLSI chips which comprise thousands (between 2000 and 8000) of logical gates. These are usually a mix of AND and OR gates laid down in a grid pattern so that the connections can be changed in a large number of ways to create a wide variety of firmware.

In order to program the chip, the connections between the logical gates are selectively 'fused' (here meaning burned out), leaving the specified gate-array pattern required.

The difference between the three types of PLD is whether it is possible to fuse the AND gates, the OR gates, or both in the underlying stratum.

These chips are only really suitable for random logic circuits, not for implementing memories (apart from PROM), register banks, timing circuits, or useful in analog processing.

The three types are:

— **PROM:** Programmable Read-Only-Memory.

— **PAL:** Programmable Array Logic (only the AND gates can be programmed).

— **PGA or PLA:** Programmable Gate Array (aka Programmable Logic Array), which is the most versatile of the three.

- There's a variation on PGA called Field PGA (FPGA or FPLA) which is difficult to make and expensive. See FPGA.

See gate arrays, PROMs and PAL.

Programmable Read-Only Memory

A type of ROM chip which can be defined by the user, but which can't be erased and reprogrammed in the same way as an EPROM or an EEPROM. PROM contents can be read but not changed during a program's execution. PROMs are programmed in a special device, before they are used — and they can only be programmed once.

Like ROM, this is traditional firmware used for holding important code, but while ROM requires thousands of copies made to be economical, PROM can be programmed (but not made) on a one-off basis. This is done by 'burning' the information into a Programmable Logical Device (PLD) which consists of a matrix of AND and OR gates joined together by fusible links.

The 'burner' is usually a plug-in PC card which allows you to load the program and 'blow' the minute fuses to establish the non-volatile memory in the chip. See blowing, burner, EPROM and EEPROM also.

• Within the computer, the ROM or PROM can form part of the normal memory map and be addressed just as can any RAM location. However, PROM is often too slow to be used in this way, and so the contents are copied out of the PROM into normal RAM locations during the boot-up phase. The RAM then 'mirrors' the PROM.

• New CMOS PROMs have access times down to 20ns, and are produced with 0.8 micron element separations.

programmed trading

The use of computer systems to buy and sell shares automatically. There are two main types:

- The Crunchers download raw data from on-line commercial databases such as Dow Jones, and massage it themselves. They can then spit out ratios and recommendations on-demand.
- The Callers are telecommunications packages which connect to some large company mainframes and download previously massaged information.

• You can also categorize program trading as either 'stock-exchange arbitrage' or 'portfolio insurance' systems. The latter are believed to be the most destabilizing force in markets, because they attempt to minimize losses on portfolios by dumping shares as the market is falling. The more computers there are which react in the same way at the same time, the more the market becomes unstable and unpredictable.

programming environments

The term refers both to the operating system and to the collection of software (and CASE) tools that aid programmers in their job. The Unix environment, for instance, has special facilities for managing different program versions and for updating modified files, and it has sophisticated debugging techniques. See CASE and structured programming also.

progressive scanning

Progressive scanning is also known as 1:1, in contrast to the 2:1 interlace television system used at present. Progressive scanning systems only scan the image once per image frame — they begin at top-left, and scan every line until they reach bottom-right — then they begin again.

The main advantage in using this simple approach with computers is that the direct digital bit-mapping of images is then possible. Every location in video-RAM is polled sequentially, and the information fed to the screen driver. It is also easier for digital signal processors to deal with the signals if some manipulation is necessary. The contrast is with TV interlace scanning.

• New digital television systems will probably all be progressive scan in the future — once we have a full frame-store in the TV set, capable of holding one complete image. Once a frame is in memory, it can be scanned any number of times in any way the receiver wishes — interlaced, or progressive.

Project 21

This is Inmarsat's proposed satellite mobile communications system based on Inmarsat-P technology. The satellite segment will provide a short-hop link from the mobiles to the nearest PSTN gateway.

The current planning is that their constellation will have two intermediate circular orbital (ICO) planes, each with one spare and five operational satellites. The satellite-to-mobile links will be at 1.9/2.2GHz and the feeder links will be in the Ka-band. Each satellite will have 4500 communications circuits and about 500W of output power.

The choice of ICO constellation is said to offer high satellite elevation angles and therefore have multiple spacecraft simultaneously in view with most uses. At these heights the satellites are slow-moving which makes hand-off easier, while still providing acceptable transmission delays.

The space and ground segment will cost a total of about $2.6 billion in 1994 dollar terms. Don't expect this before 1999.

Project 802

A project of the IEEE which was established in the mid-1980s to standardize the various forms of local area networks (LANs) and metropolitan area networks (MANs). See numbers section.

Projectavision

A relatively new projection television system which may solve the problems of high definition TV screen size and flicker because it uses light-gate techniques. The light gate comprises an active matrix thin-film transistor (LCD) in the form of a 65 – 80mm glass plate, and there are special optics in the projector.

Prolog

• PROgramming in LOGic. An AI language which has been around for twenty years and which is still important (especially in Japan). It is very similar to LISP.

• In Windows, a prolog is a small piece of library code which ties a routine to a data segment.

ProLogic
Dolby's surround sound system. See Dolby.

PROM

- Programmable Read-Only Memory. See.
- Partial Read-Only Memory aka P-ROM. These are ISO-standard 3.5 inch magneto-optical (MO) disks which have been partitioned into two areas. One part is for pre-recorded read-only data, while the other is reversible-recordable. The pre-recorded data is stamped into the disk in the factory, and this can be done economically in reasonably short runs. Total disk space is about 128MBytes. See MO, MiniDisk and OROM.

promiscuous mode
Indiscriminate. Works with all types.

Bridges inserted between LAN segments immediately drop into promiscuous mode and begin snooping on all traffic. They start to keep track of all devices (MAC-layer addresses) so they can identify which Ethernet address is on which side of the bridge.

Once they learn all the addresses, they are more selective about which packets they handle. They will then use their internal tables to decide which packets should be bridged, and which should remain on one side only.

prompt
A special character or word string displayed on the screen to inform you that the computer is waiting for some input. The classic prompt is the famous > sign from MS-DOS and CP/M.

proof
A check copy of something. Usually the first print-out from a laserprinter, or a hand-rubbed printout from the printing plate before it is fully mounted on the machine. Color proofs are often called Cromalins. See.

propagation
Transmission from the source (the point of origination) to the sink (the final destination). This can be point-to-point, multipoint, or broadcast and through wireline media, optical fiber, or radio (or hybrids involving all three). See propagation delay below also.

- **Fiber cables:** In telecommunications systems even over 100km, the time taken for an optical signal to pass down the line can create problems — especially with today's ultra high-speed data. In glass led the light pulses travel at about 80% of the speed of light (300,000km/sec).
- **Copper cables:** The propagation rate in standard Ethernet coaxial cable is between 65% and 77% of the 'pure-space' speed of light. In twisted pair it is less than 60%.
- **Mode:** In radio transmission, there are five identifiable modes of propagation:
  - **Surface waves:** These follow around the curve of the Earth using a ground current effect. They are used by HF and longer wavelengths.
  - **Space waves:** These are the normal UHF, VHF waves we use for television. There are both direct space waves, and reflected waves. See multipath effect.
  - **Sky waves:** These bounce off the ionosphere. Short-wave (SSB HF) radio signals travel very long distances when conditions are right using these waves. See Sporadic E.
  - **Satellite propagation:** Using a satellite transponder for rebroadcast. This is actually a repeater function,
  - **Scatter:** This is used with high-powered, over-the-horizon systems. These use forward scattering (regeneration of signals) by the troposphere.

See all of the above, and see micro-meteor burst also.

propagation delay
Propagation delay is the time it takes for a bit to travel from source to sink. This includes the time taken to traverse the wires, optical fiber or satellite hop, plus the time taken to pass through switches, amplifiers and multiplexers, and the time taken in router-buffers in packet networks.

With the advent of frame relay and ATM cell-relay services, propagation delay becomes an important measure of network latency — of vital importance for networks carrying real-time isochronous services (voice and video). With ATM it is important that the delays be predictable, and fast-packet switching is designed to make this so.

- Propagation delay over a GEO satellite, is in the order of 240—300ms, which is nearly a third of a second. If two satellite hops are being used for an international call the delays can be considerable and quite disruptive. And the vocoder delay in an L-band satellite mobile communications link can add another 200ms to each end.

  If one hop is made over a GEO with the call originating from a digital cellular phone (where the voice-coding introduces its own delay) it can also be disruptive. If the call is from one digital mobile phone to another via two satellite hops, the delay will reach disruptive proportions.
- LEO satellites are favored for interactive communications because of the lower power requirement to reach the satellite, but also because the delay is very substantially less than GEO.
- Obviously X.25 networks, with store-and-forward computer-nodes at the switching points, will create highly variable and totally unpredictable propagation delays. Fast packet-switching makes ATM and frame relay networks more predictable.
- Propagation delay causes all sorts of problems in data error control and flow control. Satellite delay does nasty things to protocols like X.25 and TCP that are expecting an acknowledgment from the remote end. See sliding windows.

proportional

- **Spacing, Font and Pitch:** Typewriters usually print characters at regular intervals along a line (fixed pitch), no matter what the actual width of the character is. So an 'i' will be allocated the same space as a 'w'. This is why typewriter type is so readily identifiable.

Proportional spacing is when the width of the space between letters varies from character to character. The letter 'w' will receive much more space than the letter 'i'.

• **Font:** Proportionally-spaced fonts are much easier to read and occupy less space than the typewriter pica or elite fonts. The computer program or the page-description software within the printer itself, makes the decisions about the width of the space to allocate to each character based on tables of rules relating to letter combinations. The process of adjusting the letter space is known as kerning. See hint/Hint.

protected mode

Also called 'virtual address' or '386 extended' mode. In the later Intel microprocessors (80286 and on) the company maintained backward compatibility with earlier chips (and therefore PCs) by emulating the actions of an 8088 or 8086 chip. This is called the 'real mode'.

The protected mode was later introduced in the 80286 chip to provide memory protection (partitioning) so that multitasking operations were possible, and to allow the use of memory locations up to 16MBytes.

The changes also allowed applications that require more than the 640kB limit available to MS-DOS, to use DOS extenders (see EMS and EMM).

Unix, OS/2 and NetWare 286 (and later versions) also operated in this protected mode. See real mode and virtual 8086 mode also.

protocol

A set of conventions or rules that 'bind' two units. The word 'protocol' comes from a Greek root which translates literally as 'first glue'.

Before two computers can communicate they must agree on certain protocols — and creating these rules is the dominant work of standards committees, worldwide.

The major protocols in telecommunications are rules which apply to data formats, and those required in transmission timing, sequencing, and error handling. Protocols are formal agreements, and they often come in elaborate 'suites', or in even more complex protocol 'stacks'. Each layer in these stacks, or each individual protocol in a suite, will perform a very specific function.

All protocols comprise:

- Syntax: Commands and responses, including data format and signal levels.
- Semantics: Permissible structured sets of requests and actions, including control information for coordination and error handling.
- Timing: Types of events and their sequence, including speed matching and reordering.

Protocols are not required for communications if the lines are not shared and if no error detection is to be performed. A pair of wires with a telephone at each end does not require any protocols. So, in fact, all protocols constitute an overhead on communications systems – they are 'redundant' bits.

Protocols exist to resolve conflicts when a number of parties are using the same channels, and/or to specify the ultimate destination of the data (addressing), and also to detect and correct errors. There are three main classifications:

- **Data protocols.** ATM and frame relay are data protocols. They define what each bit means within the data stream, but they don't specific a single data speed or the media on which the bits will be sent.
- **Transmission protocols.** ISDN is a transmission protocol. It defines how fast each channel will operate, what type of connection can be made, over what distance the signal can be sent, what type of media can be used, and how calls will be established.
- **Low-level line protocols.** B8ZS is a low-level line-code protocol used for T1 and ISDN in the USA. It describes how a voltage on the line should be interpreted as binary.

• **Protocol Stacks** are multi-layered software architectures, where each layer needs to interwork with the one above and the one below. It is better to layer these architectures rather than design a complex large suite, because the modularity makes for easy modification and updating. See protocol stack and OSI reference model.

protocol address

A network address as distinct from a physical or MAC-layer address. This is the address used by routers.

protocol analyzers

These are more complex devices than LAN analyzers. Generally they are dedicated PCs (usually specially-built portables) with proprietary software and a hardware transceiver board allowing interconnection to the network cables for test and measurement.

They are designed to 'lurk' on the network, watching every packet traveling down the line, and to filter certain packets and analyze them. These are 'triggered' by a match to certain set parameters, and the 'events' are usually captured in a buffer for later analysis. They must be able to observe most of the packets being transmitted and do so at full LAN/WAN transmission rates.

Good protocol analyzers are intelligent enough to break out the layers of protocols in the ISO seven-layer mode, and examine each layer independently.

protocol boundary

This most likely refers to the API of LU6.2.

protocol converters

Boxes which allow interconnections between two dissimilar logical network standards. These may be stand-alone units; functions which are integrated into the terminal hardware; or software functions performed by the host's front-end (FEP).

Terminal emulators and packet-switching PADs all incorporate protocol conversion — although the term is now mostly reserved for dedicated equipment joining two dissimilar WAN networks. Gateways are protocol converters, but the term is mostly applied to smaller and cheaper boxes which allow alien equipment to be attached to a network. See spoofing.

protocol data

At one level this can just mean framing bits. At another, the distinction is probably being made between a frame (a string of bits identified usually by flags at each end) and a packet, which has addresses, sequence bits, length-bits, etc. The packet contains both 'protocol data' (header-overhead) and 'payload'.

Protocol Data Units

Packets. See PDU.

protocol overhead

Many networks are awash with a constant exchange of messages between various devices — especially with the management system. Much of this is to update routing tables, and some (up to 20% of traffic on many WANs) from SNMP agents reporting back to the MIB.

Different LAN types have different overhead problems. AppleTalk networks, for instance, distribute updated routing tables every 10 to 15 seconds. SNA networks engage in constant rounds of handshaking.

protocol stack

A general term for packages of protocols which comply with SNA, OSI layers, or some other similar layered structure. Network software, such as AppleTalk or DECnet will generally have a protocol stack. TCP/IP and OSI have even better known stacks.

The purpose of the protocol stack, as a whole, is to establish connections, route traffic and transfer data from one computer to another. Obviously, then, for computers to communicate, they must be able to run the same protocol stack.

A stack is not essential, however. All the features contained in a stack could be grouped into one very large, very complex protocol definition. However the definition of the stack as layers, allows easy modifications, and allows adoption to new media (say fiber instead of copper) or to a new application.

When defined as a stack, the emphasis becomes one of ensuring that each layer's boundary matches the next — the output of layer 2 must match the input of layer 3, and so on. This is why such emphasis is placed on 'interface' definitions. Provided the interfaces match, layers can be changed, yet the new layer will continue to work with the stack as a whole.

There are three types of protocol stack:

- Standard — the only real standard is the OSI's 7-layer model but TCP/IP is close.
- Quasi (or *de facto*) standard — primarily NetBEUI used by IBM and Microsoft, and perhaps NetWare.
- Proprietary.

protocol transfers

The transfer of data-files using an error-checking and correction technique, such as XModem. See file-transfer protocols, Kermit and Z-Modem.

protocol translator

A device which translates one protocol to another. This is often just software in an otherwise relatively standard network device such as a bridge or router. Usually the protocols are essentially similar. However, see translation bridging also.

proximity operators

In database searching, it is often possible to specify that two terms be separated by no more than x words. For instance you may instruct the system to search for 'JUST (1W) TIME' when attempting to find the phrase 'Just In Time' since the word 'In' is a stop-word that will not have been indexed.

Alternately you may look for the two words contained within the same line, sentence, phrase or record — depending on the search techniques/languages provided. These are all proximity operators.

proxy

Devices which pretend to be other devices. This is usually a technique used in spoofing. See.

proxy agent

This is software that handles the translation between an agent and a device, when they use different management-information formats or protocols. The proxy agent works on behalf of the network manager in communicating the information.

PRS

Personal Radio Service. A Japanese digital form of CB radio. See DSRR also.

PS

Program Service. This is the identifier used in RDS. It allows alpha message characters to be displayed on an LCD screen, and is used with the new European FM radio systems. See PI and RDS.

PS/2

Personal System 2. A range of IBM personal computers which emphasized the OS/2 operating system, 3.5-inch floppy disks, and the MCA bus architecture.

Generally, the PS/2 range was considered a bit of a disaster; it was IBM's attempt to block the clone makers (or make them pay very high licence fees) by standardizing the new MCA bus architecture. Instead, it drove them to create the rival EISA bus architecture. In the final wash-out, neither succeeded to any substantial degree.

PSC

Public Services Commission. See DPUC.

PSDN

Packet-Switched Data Network. See PDN, X.25 and frame-relay.

pseudo-code

In structured programming, this is a way of expressing the steps which need to be taken in the program, by using simple English language words. This is an alternative to flow-charts. It is a form of planning or outlining a program using a mix of programming language and human language.

- Also see p-System (UCSD).

pseudo-random

Most computer languages have a function which generates random numbers. But these numbers are never perfectly random; the sequence will repeat itself eventually. So strictly they should be known as pseudo-random.

pseudo-static RAM

A type of DRAM chip which has refresh circuits built-in to the chips themselves. To the computer, these chips appear to be static RAM; they require less power than SRAM, but only have the slower access time of DRAM.

pseudoternary

A multilevel binary line-code transmission system where zero, positive and negative voltage states are used (three conditions) but where only binary signals (0 and 1) are sent. There are a couple of variations, all with the three states which are used to create electrically balanced signals on the line (to prevent DC coupling). AMI is one form of pseudoternary.

However, the inverted form of AMI is now the most widely used — with binary 1 represented by an absence of a line voltage, and binary 0 represented alternatively by positive and negative pulses. In BRA ISDN you will find pseudoternary line coding used at the S and T interfaces as an alternative to NRZI.

PSI

Process to Support Interoperability. When an OSI compatible product conforms to certain defined standards, the SPAG (Standards Promotion and Application Group) issues a PSI mark. All products labeled with PSI must pass a battery of conformance and interoperability tests.

PSK

Phase Shift Keying. This has both a generic and a specific meaning.

- Specific: Simple Binary PSK (also called Phase Modulation, or PM) will use 180 degrees of change to represent 0s or 1s. This results in a very robust signal which can be handled by low-cost equipment. At this level one baud or symbol still transfers only one bit. Originally PSK just meant this type — now correctly called BPSK.
- Generic: A range of techniques used in the medium-to-high speed modems which transmit digital information by a change in phase of the carrier.
  - If 90° changes can be detected, then any one of four states can be simultaneously transmitted in each baud (usually called a symbol). So now each baud or symbol will transfer 2-bits (dibits). This is quadrature PSK or QPSK.
  - If 45° changes in phase can be detected then 3-bits (tri-bits) can be simultaneously transmitted with each symbol (baud).
  - If a change in the signal amplitude is added to this as well, then 4-bits can be simultaneously transmitted. PSK is often used in combination with ASK on modems above 4.8kb/s and the hybrid is then called QAM — and there are multiple levels of QAM (see M-QAM).
  - When the PSK and ASK components are used in combination according to strict rules, then a special form of parity-checking can also take place, using rules about phase and amplitude-change relationships. This is known as Trellis Code Modulation. See TCM.
- Differential Phase Shift Keying (DPSK) is when the measure of phase change is taken from the previous phase state, rather than from a fixed reference. Most PSK systems are of this kind. See BPSK, QPSK, QAM, and DAMQAM also.

PSN

- **Packet-Switched Node:** This is really a special router.
- **Packet-Switched Network:** This is usually an X.25 network. See PDN.

PSPDN

Packet-Switched (X.25) Public Data Network.

PSRAM

Pseudo SRAM — a type of DRAM that acts like SRAM. See pseudo-static RAM.

PSTN

Public Switched Telephone Network. The stock standard dial-up telephone network and exchange services — although the term probably includes ISDN in common use.

However, many people make a clear distinction between the analog and digital PSTNs, by calling the old analog system, POTS (Plain Old Telephone Service), and the newer digital service, ISDN (Integrated Services Digital Network).

- The original distinction was between PSTN and Telex networks. Later PSDNs and DDNs were added.

PTM

Pulse Time Modulation. See.

PTO

Public Telecommunications Operator. This is the same as a PTT, carrier, or telco.

PTT

- **Postal, Telephone and Telegraph** organization: The old term for a major telecommunications provider (carrier or telco) with a network.
- **Push To Talk:** A two-way radio which is half-duplex — both speakers use the same frequency channel, and press a button on the microphone to talk. See PMR and SMR.

PU

Physical Unit: The term is used in SNA both for the hardware itself and for the functions.

— Hardware. A node on the SNA network supporting one or more LUs.

— Logical functions. Physical Units are similar to the Logical Units (LUs) in that they are 'programs' which interact with the physical hardware; PUs manage the node, while the LUs manage the sessions.

Each terminal, controller and processor contains a PU which receives messages, orders startup, performs the testing, etc.

There are five types of PU defined for SNA nodes, and each must have at least one to manage and monitor the node's resources. They are numbered hierarchically.

- PU type 2.1 is the 'strategic' node type. All future IBM products will be classed as type 2.1, and all can be directly coupled. This is the type that is most important in networking because it can be directly connected to other PU 2.1 nodes.
- PU type 5 is for a mainframe.

public domain

A copyright term meaning that the property has been released for use without payment. It doesn't mean there's no copyright on the property — in most countries, anything created under the copyright laws remains under the control of the author unless formally assigned in writing. You can't resell a public domain program without permission. What you do have, is a license for free use under the conditions set by the author.

- Don't confuse this with shareware.

public key encryption

Public key encryption techniques allow messages to be sent to a person or device, without any previous arrangement being made for the exchange of encryption keys.

This is an asymmetrical (different length of codes) cryptosystem where the keys are in pairs and both are needed. One is used for encoding, and the other for decoding. The shorter key is kept secret, and the longer key (usually about 48-bits) is published.

The sender of a message looks up the encryption code under the proposed-receiver's entry in a public directory, and uses this to encode the message before it is sent. This is the public key.

The message can only be decoded by the matching private key (not by the key with which it was coded), and this is kept secret. This technique relies on complex mathematics, and is extraordinarily secure, if the public key is of sufficient length.

Public key encryption techniques also allow authentication of the sender, to the point of being able to guarantee legally the signing of contracts by electronic means. This is called 'digital signatures'.

The process is basically the same, but the keys are used in reverse order to establish authenticity. See RSA and PGP.

publish

In modern computer applications one file can 'publish' and another can 'subscribe'. This means that the published material can be inserted into the subscriber's file, and it will automatically be updated if at any time the published material is changed. This is hot-linking.

PUC

In the USA these are the Public Utilities Commissions which control telephone services, monitor service standards, and fix prices in each state. Members of the PUC are usually nominated by the State Governor, and so are political appointees. See DPUC.

puck

A form of mouse used in a digitizer. See graphics tablet.

pull-down menus

See pop up.

Pulse Amplitude Modulation

See PAM.

Pulse Code Modulation

This is the basic standard for analog-to-digital voice conversion. It was invented by Alec Reeve in 1938 but his patents ran out before the technology became feasible with the development of transistors and ICs.

Reeve's first PCM system sampled the audio frequencies at 8000 a second because this is twice the baseband frequency of a telephone system with a bandwidth 4kHz (in accordance with the Nyquist rule). However he only had 32 steps of quantization (2^5) rather than the standard 256 (2^8) now used. 8-bit PCM quantization has now fixed the modern 64kb/s channel bandwidth in digital telephone systems.

The term 'modulation' is a misnomer here, since nothing has been modulated — but rather 'measured', but this is the general way the term is now used. Almost all digitization techniques use PCM as the first stage of analog-to-digital conversion before other techniques are applied.

- Unless otherwise specified, PCM is a generic term which refers to the process of sampling the wave form at set intervals, and assigning to each sample a number related to its amplitude (quantization). The sample rate and the 'quantization' number can be any size/rate combination.

Some modern PC sound cards offer PCM with a sampling rate of 22kHz and a quantization level of 8-bits, or 16-bits for stereo.

- Log-PCM is the term applied to two closely related PCM techniques used in voice telephony — A-Law (International ITU standard) and μ-255 (USA). In both cases, the quantization generates digital values proportional to the logarithm of the waveform's amplitude, rather than a direct linear measure.

In practice, logarithmic quantization nearly doubles the subjective quality of the voice for the same bandwidth. See companding and A-Law.

• ADPCM is Adaptive Differential PCM which is becoming widely used in telephone systems around the world. It achieves a 2:1 or a 4:1 compression by coding only the difference between the sample and its predicted value. The prediction is done through a feedback loop, and the result is a very short binary number. See.

pulse dialing

The older technique of telephone dialing which uses the rotary dial mechanism (now available in a chip). The signals are transmitted by simply cutting the DC current on the line a set number of times, and for specified durations. You could pulse dial older networks by jiggling the on/off-hook key.

The correct name for this is decadic dialing, and the distinction is usually being made with DTMF or 'tone' dialing. See inband signals and out-of-band signaling also.

• **Make-break:** Different phone systems around the world use different pulse rates, different ratios between the make and the break, and different numbers of pulse for different numbers!

— In the USA and Canada the make/break ratio is 39% / 61% and pulses are dialed at 10 per second.

— In the UK and Hong Kong pulses are dialed at 10 per second, and the make/break ratio is 33% / 67%.

— In Japan, the make/break ratio is 33% / 67% and the pulses are dialed at 20 per second.

With modems, these systems are best handled by tone dialing.

• **Number sequence:** The sequence of numbers on a rotary dial is crucial since the shorter the rotation, the fewer the pulses. Some have the 0 before the 1, and others have it after the 9. However, while there is a general worldwide standard, there are still some strange variations:

`1234567890` — Most of the world.
`9876543210` — New Zealand and the city of Oslo.
`0123456789` — Standard Swedish dial.

pulse modulation

This is the general term for sending information along a line by trains of pulses.

The temptation is to assume that this is only a digital technique, but that assumption would be quite incorrect. There are numerous analog forms of pulse transmission.

A pulse can have amplitude and time (length) variations which are not necessarily integer values, and for a system to be digital, all of the pulse parameters need to be treated in an integer way.

The main pulse modulation techniques are:

— PAM: Pulse Amplitude Modulation.
— PDM: Pulse Duration Modulation.
— PPM: Pulse Position Modulation.
— PWM: Pulse Width Modulation.
— PTM: Pulse Time Modulation.
— PCM: Pulse Code Modulation.

See all of the above.

pulse spread

As data pulses pass down a copper cable or an optical fiber, variations in components of the signal cause minute delays which blur the edges of the square wave and make them more difficult to detect, to the point where intermodulation distortion can arise with one pulse overlapping the next.

Solid copper cables are particularly vulnerable to this effect because of induction. Co-axial wave-guides are much better because the signal actually runs on the surface in a wave-guide, rather than through the material.

• In fibers some of the photons travel a slightly longer path than the others, and some pass through different variations in the refractive index of the glass. So, over distance, the 'squareness' of the pulse is gradually lost. This 'pulse spread' is also called 'smear'.

Pulse Time Modulation

A variation on PWM invented by Alec Reeves who developed PCM. It uses a 'start-bit and stop-bit' (two short pulses) to represent the extremes of 'width' of the pulse in PWM. The signal is sampled at a higher frequency, and the pulses are transmitted as a varying voltage reference rather than a wave. PTM is also known as Pulse Position Modulation.

Pulse Width Modulation

This was an experimental technique for carrying analog voice across phone lines in a 'digital' (on-off) fashion. The width of the pulse correlated to the amplitude of the analog voice signal at the time of sampling. This basic technique was later used for analog LaserDisk using FM modulated analog video.

pulsed laser

As distinct from CW (continuous wave) lasers. The main purpose of pulsing a laser is to achieve higher peak powers; gigawatt levels of power can be achieved in some lasers. Continuous dye lasers are currently producing pulsed beams using a technique known as passive mode-locking.

pump

• To add energy to something.

• In laser terminology you pump a laser by cumulatively adding light to the device, resulting in a burst of high-energy at the output.

• In optical amplification, a special 'pump' laser is used to energize the erbium dopants, which then give up their energy to the signals passing through.

punch cards

Cardboard cards which had a series of punched holes. These were fed into the early computers in batches, both to program the machines, and to input data.

PUP

PARC Universal Protocol. Xerox's inter-networking protocol which is essentially similar to IP.

purge

The process of getting rid of unwanted stored material. However, sometimes you purge a device (buffer) because the data is wanted. In the case of RAM-disks or large memory stores which are temporarily holding working data, it is important to 'purge' them (force them to store the data in a more permanent form) before re-using the buffers or switching off the power.

push

The process of adding data to a stack. A stack is a LIFO structure (Last In, First Out), so the last byte to be 'pushed' onto a stack, will be the first to be 'popped' off.

PVC

- **PolyVinyl Chloride:** PVC is the plastic insulator used around most underground and undersea cables.
- **Permanent Virtual Circuit:** In cell and packet switching this is a permanently established circuit which is part of a private network and therefore doesn't require dial-up access. The distinction is with SVCs (Switched Virtual Circuits).

PVCs are easier to implement because the cell-nodes or routers can be given simple directions in look-up tables. With PVCs, there's not the added complexity (and bandwidth requirement) of establishing and tearing down circuits all the time where this may not be justified. See virtual circuit and ATM also.

PWM

Pulse Width Modulation. See this, and also bitstream.

pyramid coding

A form of layered coding used in modern video compression. The image is coded so that, at the base level, a full image of low-quality can be simply extracted, then each layer adds 'difference' information to provide higher grades of video image — even up to HDTV level. See DCT, UVC and layered (video) also.

Q931W

A wideband ISDN signaling system which allows an ISDN link to support multiple simultaneous calls at bandwidths between 64kb/s and 2Mb/s over a single network connection. This is the new channel-aggregation standard for video-conferencing over ISDN. See aggregation and BONDing also.

Q-band

Satellite systems. The term 'Q-band', which you'll occasionally find in communications literature, is the result of a verbal confusion. The writer has misunderstood the expression 'Ku-band'.

Q-parity symbols

A form of error checking used in CD systems.

Q-SIG

The European standard for digital (ISDN-based) PBX signaling which is replacing the UK standard called DPNSS. It is proposed by the European Computer Manufacturers' Association, and is based on the ITU Q9212 and Q931 standards for compatibility with Euro-ISDN signaling. There have been some complaints that it lacks all the features found on DPNSS.

QADM

A form of Analog (voice) and Data over Modems. See DSVD.

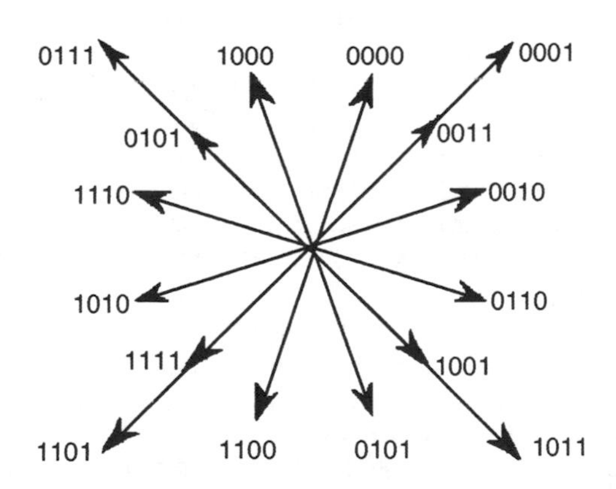

The Basic QAM Constellation
Only four of the phase changes are also associated with amplitude changes.

QAM

Quadrature Amplitude Modulation. QAM is a modulation scheme used in high-speed modems and many different transmission schemes (digital radio, television, etc.), including CAP (video over twisted-pair copper). Quadrature is a reference to the technique of modulating two sine waves independently, with a 90° phase shift (sine–cosine) between to create more subtle phase changes. Amplitude refers to additional changes in signal level.

The early QAM modem systems used eight phase changes (DPSK at 45°) and two amplitude levels (ASK) to transmit 4-bits during each baud (change of state). So a V.32 modem running at 1200 baud will transmit 9600 bits per second.

A better transmission system proved to be one using 12 phase changes with two amplitude levels, but amplitude variations were only provided on four of the differential phase changes. This meant that both a phase change and an amplitude change would be detected in some events (symbols or baud) but not in others, providing a self-checking self-correction mechanism when 'differential' changes were being used.

Multilevel QAM can be extended to 6-bits per baud by introducing finer divisions in the phase change and detection. Currently the limit appears to be1024 constellation points. There is a practical and theoretical limit to the fine divisions in phase which can be detected, especially in view of phase variations which normally occur on a telephone line.

Each extra bit carried doubles the required number of constellation points, and this requires a 6dB higher SNR in the received signal for the same BER. QAM is now usually coupled with Trellis Coding to provide error control. See TCM.

- With V.22bis modems, QAM is used to lift the 600 baud rate to 2400b/s. This standard has four phase angles and two levels of amplitude, creating tribits.
- On V.32 two-wire full-duplex modems, QAM handles four bits (quad-bits) at a time using eight phase angles and two amplitudes, giving 4800b/s.
- The 6-bit encoding scheme used in DAMQAM requires 64 Quadrature and Amplitude combinations.
- 1024 point (128–level) QAM modems can transmit 2Mb/s over a 240kHz (supergroup) analog phone channel. 64–level QAM can transmit 2Mb/s over 480kHz of bandwidth. See CAP.
- Differential QAM (DQAM) is a technique where it is the degree of change from the previous phase angle that creates the binary data — not the absolute angle from a fixed zero reference point. Most QAM systems use differential phase changes these days, so the differential term is usually discarded. See phase-change and PSK also.

See V.32, PSK, ASK and trellis code modulation also.

QC laser

See quantum cascade laser.

QCELP

Qualcomm's vocoder used in CDMA which has an adaptive threshold. This is Qualcomm's variation on CELP which introduces variable rate coding. It encodes voice at rates of 9.6, 4.8. 2.4 and 1.2kb/s for their CDMA cellular phone.

The average rate of the vocoder is about 4kb/s, and it steps between these various rates according to the requirements of the moment. When you are listening, it will wind down to the 1.2kb/s rate, just to provide comfort noise on the system.

Control can also be exerted by the basestation. In a civil emergency, the basestation can switch everyone down to the 4.8kb/s quality level, and effectively increase system capacity since the limit to the number of users on a spread spectrum system is the amount of noise they are willing to tolerate — and every user appears to the others as noise.

• CDMA is now being developed for 1.25MHz bandwidths and for 2.5MHz. There's a new 14.4kb/s wider-band Q-CELP vocoder for CDMA at 2.5MHz which adds a higher data rate and some in-between standards: 14.4, 7.2, 3.6 and 1.8kb/s.

QCIF

Quarter CIF. An intermediate ITU standard for H.261 videophone and videoconferencing images when conversion is needed between NTSC and PAL systems. It is based on half the CIF scan lines and half the horizontal pixel resolution — which is 144 scan lines each with 176 horizontal luminance samples. The chrominance resolution is 72 x 88 (half the lines and half the pixels).

Depending on the bandwidth available, QCIF will use variable frame-rates from 30 down to 7.5 frames per second.

CIF is considered to be optional in H.261, so QCIF is the only mandatory standard which ensures that all devices can communicate. See CIF and H.261.

QCP-1900

Qualcomm's CDMA cellular phones at 1.9MHz. This is the new PCS standard. See CDMA.

QDOS

• Quick and Dirty Operating System. The original name for the Seattle Computer Products' operating system, bought by Microsoft, which later became MS-DOS. It was basically CP/M.

• Quantum Duplex Optical System (Q-DOS) is a new single lens, 3D television idea that relies on colored glasses. It is being developed by Hanimex/Gestetner in the UK.

QDPSK

Quadrature Differential Phase Shift Keying. The 'differential' is now assumed. See QPSK and differential.

QDU

Quantization Distortion Unit. A subjective measure of voice quality on a network, developed by the ITU. It is applied mainly to regulate the quality of low-bit-rate voice on public telephone networks, and for the analog-to-digital conversion of speech.

The old 64kb/s A-Law PCM signal is taken as the base, and allocated a QDU of 1. Higher numbers are allocated to lesser quality signals by a panel of listeners who make a subjective judgement against standardized services.

ITU recommendation G.113 defines these QDU references.
- The standard 64kb/s PCM has a reference value of 1 QDU.
- The common 32kb/s form of ADPCM is assessed as being 3.5 QDUs.

There are QDU values assigned for any coding or transcoding equipment in the network.

QDUs are cumulative throughout a network, and the ITU has decreed that no more than 14 QDUs should be permitted end-to-end on a telephone call. Since noise and distortion (to varying degrees) is to be expected in all telephone network segments, a maximum of four QDUs is allocated to the international carriers and five QDUs are allowed for the local carriers at each end. See QOS also.

• At the other end of the scale are some of the new radio cellular phones which need special dispensation to be used on the networks.
- GSM's RPE-LPC vocoders add 7–8 QDUs.
- The new half-rate GSM vocoder is rated at 16–22 QDUs.

QFP

Quad Flat Packs. An integrated circuit package type. See SMT, dip/DIP, and SO also.

QIC

• **QI Consortium:** The consortium standardizing quarter-inch cartridge back up tape systems. This group of companies, each with their own proprietary standard, finally combined to develop a single quarter-inch tape back-up standard for computers in 1982.

• **Quarter-Inch Cartridge:** This now refers to the new standards for quarter-inch tape back-up. However, there are dozens of early standards also called 'Quarter-inch Cartridge'. The new standard has proved to be very popular for workstation backup. Don't confuse these QIC standards with the 8-mm tape, or the later Digital Linear Tape (DLT).

QIC uses multiple stationary heads and it streams the tape past these heads at a very fast rate (120-inches a second). Then it stops and reverses the tape, laying down a second linear track alongside, then it stops and reverses again — laying down another. So the tracks are recorded in a serpentine or boustrophedon way. The modulation method is GCR.

• This QIC approach offers slower file access than DAT (but faster than normal tape) and it has a faster rate of data transfer (over 1MByte/s) than DAT. The drives can also use hardware data compression. New 3.5-inch and 5.25-inch minicartridges will hold from 120MBytes to 5GBytes.

• There are two QIC tape cassette sizes:
- The 5.25-inch form called DC6000.
- The 3.5-inch form called DC2000 which is now the major PC backup standard in QIC (it usually holding 2GBytes).

• QIC has various sub-standards: 18 track (QIC-250), 26 track (QIC-525) and 30 track (QIC 1350) which are all proprietary standards.

The 44 track (QIC 950) set in 1982 is the only real standard. It can store 1GByte of data (transferred in 11 minutes) on a tape of 1200ft (360m) length. There are later developments also.

- See 8-mm and 4-mm DAT also.

QoS

Quality of Service. This is a term widely used in telecommunications and defined in a number of different ways. However, QoS standards are now being set globally by the ITU for most service types.

The term encompasses: the likelihood of a call being blocked on a call attempt, sound quality, response time, and image quality also. Generally QoS reflects both the transmission quality and the availability of the service. See QDU also.

- In analog telephony, QoS generally refers to the measurement of a user's ability to make contact through the network — the 'non-blocking' aspects, or successful call attempts. A system which has a 2% chance of blocking, will given a QoS figure of 0.02. See block.

- In telephony it can also refer to the quality of the carried audio signal in terms of bandwidth (audio frequencies carried) and the noise on the line. There are subjective measurements here. See QDU.

- In packet networks and LANs QoS can refer to many different things, including the access time, and the network congestion, latency, etc. measures.

- In ATM systems QoS also refers to the cell losses experienced by the network under various load conditions.

QPSK

Quadrature Phase Shift Keying (aka QuadraPhase Shift Keying). See PSK for full explanation.

Here the term quadrature (or 4-angle modulation) refers to the fact that the phase is shifted in 90° increments by manipulating both a sine and a cosine waveform. The theoretical limit of QPSK is to provide 2-bits for every Hertz of frequency.

This is a low-level form of Multilevel-PSK. Each symbol will carry one of 00, 01, 10, or 11. See the higher levels, 8-PSK and 16-QAM also in numbers section.

QPSK is now a very common form of modulation used in digital transmission (DAB and Digital TV), and it is often used in association with some spectrum-spreading technique such as COFDM.

- QPSK-C stands for QPSK-Compatible. It refers to a type of radio data receiver which can operate with a 4-value modulation today, yet which will have a migration path to even more spectrally-efficient modulation methods tomorrow. They say this next system will be 4-level FSK blended with pi/4 DQPSK. See CFDD.

QPSX

Queued Packet and Synchronous eXchange. See DQDB.

QT

- QuickTime: Apple's graphics engine in the Macintosh. See below.

- HDTV: Quincunx TV scanning systems. See.

QTC

QuickTime Conferencing. An application from Apple which permits many users to have video (and audio) conferencing. It is both a cross-platform protocol and a networking strategy, and it supports both TCP/IP and ISDN connectivity.

quad

- A form of plastic sheathed telephone cable often used in the last 100 meters from a street-pillar to a home or office. It has four wires, and can therefore support two normal telephone services. However, this is not true twisted-pair, and it is only suited to short runs where cross-talk isn't likely.

- Quads were the original 2-inch professional videotape machines. They recorded one quarter of a field in stripes across the tape. All later professional video recorders were helical scan.

quadbits

Four bits resulting from a single change of state ('event', 'symbol' or 'baud') in a modem or radio modulation system. The bit-rate will be four times the baud rate in these systems. See QPSK, PSK, dibits, tribits and baud.

Quadrature Amplitude Modulation

See QAM.

quadrature modulation

Modulation by phase-angle change. See QPSK and QAM.

quadtree coding

An image coding technique which divides the screen into square blocks (of, say, 16 x 16 pixels), then subdivides each of these into four (each 8 x 8), then subdivides each of these again into four (4 x 4) — and so on, down to single pixels.

This means that the areas of the screen needing fine resolution can be coded in more detail using a hierarchical tree structure, while those areas of image which are large, flat, untextured color (and which lack much fine detail) can be coded by longer codes since fewer are needed. This produces a Huffman-like variable length code, and bit-rate reduction.

quadword

Four-times the mainframe 32-bit words — or 128-bits.

quality of service

See QoS.

quantization

The process of converting a sample amplitude (usually of an audio or digital waveform) to a digital number.

- In the broadest sense this is the measurement of the amplitude in analog-to-digital conversion, where the results can be expressed in either decimal or binary representation.

• The quantization value is the number of bits required to store the value of each sample. The 16-bit quantization used in CD-Audio gives a range of 64k sound level values which can be used at each sampling point.

• Quantization does not often use a linear scale — all steps are not necessarily equal. Better results arise from the use of logarithmic scales. See log-PCM.

• Quantization distortion arises because of quantization errors, and these are inevitable in the quantization process because of the step-size — and this is always true, whether the steps number 126 (7-bits) or 32,000 (15-bits) — the first bit is a sign-bit, indicating + or –.

Obviously the smaller the step-size the smaller the errors, and the less the distortion. However decisions are also being made about whether to round-out errors beyond the limitation of the quantization (fractions of a step-size) or truncate them. Oversampling and bitstream techniques are also used to overcome some of these problems.

• In DCT video compression, the term 'quantization' is misused to mean the application of a 'resolution filter'. They set a minimum threshold to each coefficient, and eliminate those that represent spacial frequencies (image detail) that the eye can't see.

By abandoning high-frequency (fine detail), they can reduce the data required without any appreciable subjective loss of quality. This quantization stage occurs after the transform stage, and during compression. It is very similar to the Philips DCC masking approach with audio compression on tape. See.

quantization noise

When an analog signal is sampled and converted to digital, there will always be some 'rounding-off' discrepancies since the measurements can't be infinitely precise. This limitation creates some noise and distortion in the reproduced signal which is known as quantization error or noise.

The 16-bit quantization techniques used in CD-Audio give quantization errors which are less than one in 2^{16} (1:65,536).

The quantization error can be subjectively reduced by non-linear encoding (logarithmic) which is why companding is used with voice.

Oversampling and bitstream techniques also reduce the distortions due to quantization steps. See A-law, log-PCM and also bitstream and PDM.

quantum

A change by integer values — a jump from one to another without passing through intermediary stages.

Electrical and electronic circuits obey conventional rules (ohms law, etc.) until the separation between elements begins to approach the size of individual molecules. Below about 0.1 microns of separation, however, circuits begin to obey rules associated with quantum physics, rather than conventional physics — and many of these rules are still unknown.

However, there are now quite a few quantum devices which take advantage of quantum effects. The first was the tunnel diode invented at Sony in 1957 by Professor L. Esaki.

Later Fujitsu invented the HEMT (see) which appears to be very promising for a range of communications devices, such as low-noise satellite dish signal amplification. Quantum-well lasers are also now being used in communications, and Josephson use a variation on the tunnel diode to create LEDs. Hall's homojunction diode was also a quantum device which eventually became a very useful continuous-wave laser.

The combination of quantum devices and the Molecular Beam Epitaxy (MBE) manufacturing process, has accelerated the pace of development of these new electronic and optical communications devices.

• There are three main categories of quantum device:
- Enhanced semiconductors, such as HEMT and quantum-well lasers.
- Direct quantum mechanical devices, such as electron tunneling devices.
- Superconductor quantum devices.

quantum cascade laser

QC laser. A new development from Bell Labs for a tunable laser which can be tailored to emit light at any specific wavelength, over a wide spectral range. This has application in WDM.

quantum effect transistors

These are very small, very fast semiconductors which function according to theories associated with very small amounts of matter. Texas Instruments is leading the research into these devices.

quantum-well devices

See quantum and SEED.

quarter-inch

• **Reel-to-reel:** The original analog audio tape standard was quarter-inch, and it ran at rates of 7.5 and 15-inches a second for most professional recordings. Later, when tape and heads improved, the tape speed was reduced to 3.5 inches a second, and stereo was introduced. These were reel-to-reel recorders.

• **Digital video:** The new quarter-inch DVC (Digital Video Cassette) component video recording format is tipped to be the most important domestic and professional recording system for video and television in the near future. The DVC cassettes and tape will also be the basis for the professional D-5 recording standard. See DVC.

• **Computer back-up:** Quarter-inch is an old standard which was used for computer backup. The tapes (not cassettes) held about one gigaByte of data.

Later the QIC cassette standard became very popular for stand-alone systems and LAN backup. See QIC.

Even more recently, new quarter-inch cartridge digital linear tape (DLT) systems have begun to replace 8-mm helical scan systems. See DAT and DLT.

quarto

This was not originally a paper size; the term refers to a paper folding (quarto = 4 pages from one broadsheet). However in common use there is a paper size called UK quarto which is 210 x 265mm (8.25 x 10.5 inch) and an American quarto which is slightly smaller at 191 x 254mm (7.5 x 10 inch).

However, quarto is still used to mean a quarter of any broadsheet in some industries, and there are many of these quarter-fold sizes because 'broadsheet' is an equally vague term. The most common quarto comes from a Large Post broadsheet of 21 x 16.5-inch. See broadsheet and paper sizes.

quat

See quaternary below.

quaternary

Having four different states. (binary has two; ternary has three). This is a line-communications technique that takes pairs of binary bits and converts them to base four. So two binary digits become a single quaternary digit.

Then, by using two levels of positive voltage and two levels of negative voltage with no return to zero the data can be transmitted at a much higher rate. Each bit-pair (dibit) is coded into a four level symbol called a quat.

However, distinguish between the line-coding (or radio modulation) method used, and the quaternary approach — which is the base-2 to base-4 conversion.

— With NRZ line-coding this will probably use a line-code having something like +5 and +15 volts as two of the symbol condition, and –5 and –15 volts as the other two. See 2B1Q, ternary, bipolar and MDB also.

— Quaternary FM radio: This is a multilevel system with digital mobile radio where the quat signal alternates between four frequencies.

• Don't confuse quaternary with quadbits; these systems only produce dibits.

QUBE

An experimental interactive (two-way) video cable network installed by Warner-Amex at Columbus, Ohio in 1977 and later in Dallas, Houston, Pittsburgh, St Louis and Cincinnati. Homes with QUBE were able to interact with the cable head-office for selection of programs, voting, etc. It was a test-bed for a range of user controls and user interactive experiments, and was way ahead of its time. It closed in 1984 after the company lost $30 million trying to commercialize it.

query languages

These have simple standardized commands and syntax, and are used for interrogating computer databases in a request–response way.

SQL is the most popular of the query languages, and one of the simpler. There are a number of much more complex query languages which use English-like phrases, and probably come under the general category of 4GL.

And since query languages were, in effect, the interface of the PC application when interrogating a database, these 4GLs later became incorporated into application generators and developed even further. Eventually this extended the range well beyond simple querying, into report generation, and into applications development — to the point where they began to rival 3GL languages like COBOL. See CASE.

queue

A list of data units waiting to be processed or transmitted: a data structure — first in, first out (FIFO).

In routers, the queue represents the backlog of packets which need to be handled quickly — so queues result in network latency. When the network is entirely private and internal to a company, the problem of providing equal access to the network is relatively easy to handle. However when the network is owned by a carrier and provided as a service to a large number of different organizations, the problem of fair and equal access to network resources becomes critical; people always imagine that they are worse off than everyone else.

This problem of providing network users with demonstrable equal access, while still providing some priority for isochronous services on the large metropolitan area networks (MANs) stretching many dozens of miles, was finally solved by the Distributed Queue Dual Bus (802.6). See.

In effect, this implemented an automatic reservation system which prevented stations higher in the network (closer to the source of the cells) from hogging the network. The queues at all stations on the MAN were then accorded equal access to the network, no matter where they were physically located.

• In OS/2, queues are IPCs which provide messaging communications between applications. They are similar to named-pipes, but queues use shared memory within a single PC to communicate between applications. Queues only operate within one PC, and implement the hot-link relationship.

queuing theory

This is a part of the theory of probability which deals with delays and queuing — the 'waiting-line' phenomena. This has many applications in telephone systems, transaction processing, and multi-user systems.

There are three basic elements of a queuing system:

— the customer population (the source),

— the queue or waiting-line, and

— the service facility.

QuickDraw

QuickDraw is the Apple Mac's screen-drawing language, which works in cooperation with the Toolkit in handling Windows, etc. The latest version is 32-bit QuickDraw which allows the Macintosh to produce photo-quality images, and boosts the machine's value as a high-end workstation. Don't confuse this with QuickTime.

quicksort

This is the fastest sorting algorithm. It uses a recursive process. On each cycle it rearranges the list of items, so that one item (called the 'pivot') is in its final location. Those that should be before it, are in the correct part of memory, and those that should come after it in their correct part of memory.

Then both the front and rear sub-sections are treated independently in the same way, as sublists to be sorted. Quicksort may perform hundreds of these sublist sorts before the whole process is finished.

QuickTime

Apple's system enhancement for dynamic 'time-dependent' media (video, voice, animation, VR) which has now been taken up by IBM. It consists of system software, file formats, Apple compressors and human interface standards. Don't confuse this with QuickDraw.

• The system software component incorporates: The Movie Toolbox (for dynamic data such as sound, video and animation), the Image Compression Manager (which allows a number of compression standards to be used) and the Component Manager (which registers the capabilities of external system resources and software extensions).

The File Formats: are

— Movie for dynamic data, and

— PICT Extensions for compression and preview of cartoon-type material

Apple Compressors: include

— Photo Compressor based on DCT and the JPEG standard,

— Animation Compressor based on run-length encoding (RLE), and

— Video Compressor specifically programmed by Apple.

Human Interface Standards: are

— a standard movie controller,

— extended standard file dialog box with preview, and

— guidelines for compression, capture and more.

• Apple released QuickTime for Windows back in late 1993.

• QuickTime VR (virtual reality) lets users view a photographic or rendered representation of a scene in 360° pans and zooms, and to navigate from one scene to another. The panoramic files are about 800kBytes each. See panoramic video.

• QTC is QuickTime Conferencing. See QTC.

quincunx

Quincunx is a HDTV sub-sampling (not every pixel) technique used for bit-rate reduction in the MUSE Japanese (HiVision) television system.

It takes the odd pixels in one line, then the even pixels in the next, and so on. This is also called diagonal bit-reduction, and it naturally results in a one-half of the original bit rate. And since the original sampling was done at the Nyquist rate, this quincunx technique is known as 'sub-Nyquist' sampling.

This is best seen as a digital equivalent to interlace (2:1) scanning of an analog TV image, but with the pixels alternating on the line, rather than the lines alternating down the screen.

quotient

When one number is divided by another, a quotient is produced, together with (sometimes) a remainder.

QWERTY

The current type of English-language keyboard. There are both English-english and American-english variations. The typewriter was invented by Scholes, and he later redesigned the keyboard to the current QWERTY character positions in 1872. See Dvorak.

R-DAT

See page 549.

R-interface

Rate-interface. In ISDN this is the interface for those non-ISDN devices (called TE2 equipment) which were not designed to work with the ISDN network, and which therefore require a terminal adaptor (TA). It is the interface, for instance, between a standard G3 fax machine and the Terminal Adaptor which supplies the ISDN functions. See NT-#, T-, S-, and U-interface.

R2D

Aka CAS. An ITU-T inter-PBX signaling protocol. It is a way of exchanging voice features like call-waiting, camp-on and follow-me. It's used to transfer signaling information over ISDN channels between PBXs. However, it is normally translated into ISDN Q.931 signaling equivalents. See DPNSS.

R4000

A 64-bit RISC chip from MIPS which was to have formed the basis of the ACE workstation design. It has a 16kB cache with special controls for multiprocessing.

RACE

Research in Advanced Communications Technologies in Europe. The European Commission's main telecommunications research program, begun in 1985, which is directed towards a number of projects. Currently there are eight.

The 'mission' of RACE is to develop 'an integrated broadband communications network (IBCN) that has the capacity and universal compatibility to offer the telecommunications services the developed world will need as it approaches 2000'. Current mobile projects are:

MONET	The mobile network.
CODIT	Research into CDMA technology.
ATDMA	Advanced TDMA research.
PLATON	Research into cell-planning methods and tools.
MAVT	Work on a mobile multimedia terminal.
MBS	The future mobile broadband system.

Some other more general projects include:

OSA	Open service architecture for distributed services over telephone networks.

rack

In electronics and communications this is a metal frame or chassis into which electronic boards are plugged. It will normally have a 'backplane' bus connection of some kind and large connectors, to make contact with the cards.

Rack-mounted modems, etc. are usually cheaper than standalone products because they don't need the same marketing, packaging, or paper-work. They may also share power-supplies, fans, etc., and so be cheaper to manufacture.

radar bands

These bands are identified by alpha characters and, although they were originally defined for radar detection, some are now used for directional radio communications with satellites. There appears to be some flexibility in the spectrum limits claimed for these bands, but the approximate figures are:

P-band	200MHz to 400MHz (across VHF/UHF boundary)
L-band	400MHz to 1.7GHz (in UHF)
S-band	1.7GHz to 5GHz (across UHF/SHF boundary)
X-band	5GHz to 11 GHz (in SHF)
K-band	11GHz to 40GHz (across SHF/EHF boundary)
Q-band	40GHz to 55GHz (EHF)
V-band	55GHz to 70GHz (EHF)

There is also a C-band which is used in satellite communications, which is between 3.7GHz and 6.4GHz. Also the K-band has been subdivided into Ku (12.5—14 GHz) and Ka (20—30GHz) bands. See individual listing also.

radcom

Radio Communications. A term applied to any radio link in a telephone network, as distinct from a 'line link'.

- This can also mean mobile telephony: See RC2000.

radial

Along a line extending from the center of a circle outwards towards and at right-angles to the circumference.

radian

A radian is the angle subtended when the radius of a circle is marked out along the circumference of the circle. It is therefore just under $60^{\circ} - 57.295779^{\circ}$ to be exact.

radiation

A term used for both non-ionizing emissions (radio waves, light, UV) and ionizing emissions (X-rays, gamma-rays, etc.). See emission.

Radio Data Service

Don't confuse the 'service' with the broadcast 'system' (below). This is an FM paging technique which dates from the late 1970s and is still going strong. The idea is to transmit the paging information in the sub-carrier of existing FM radio broadcasting and TV stations.

The problem with these systems is that the pager can only use one local station without involving complex automatic tuning mechanisms to scan all the FM radio stations in order to find the signal. This would mean an expensive pager, so these multi-channel systems were not successful.

These systems therefore remain local and limited; but the single-station technique is used in some parts of Europe and the USA for local tone, numeric and alphanumeric paging.

- Recently, the term also seems to be used as a general term for radio paging in some publications.

Radio Data System

Don't confuse this with the paging 'service'. The 'system' is a broadcast service which uses basically the same transmission technology, but for different ends.

This is a BBC development finding wide acceptance in Europe as a radio-listener's aid. It is intended to make it easy to search the radio dial for a particular broadcasting station. The station transmits an inaudible search (beacon) signal at 57kHz above a channel's normal frequency, and this beacon carries digital identification data at a rate of about 1kb/s.

The code allows the FM radio receiver to automatically recognize the broadcast channel and lock onto it. It may also include station information, traffic information, and personal messages which will be displayed on a special LCD screen. See PS and PI and RBDS.

radio spectrum

At the lower end, the radio spectrum occupies frequencies generally treated as being audio frequencies. The bulk of the usable radio spectrum extends from 30 kiloHertz (generally taken to be above ultra-sonic) up to 3,000,000,000,000Hz (3 million megaHertz, or 3000 gigaHertz). At which point it butts up against the infrared part of the light spectrum. Visible light is in the region of 420 to 790 teraHertz.

The key radio bands are:

ELF	Extremely Low Frequency	300Hz — 3kHz
VLF	Very Low Frequency	3kHz — 30kHz
LF	Low Frequency	30kHz — 300kHz
MF	Medium Frequency	300kHz — 3MHz
HF	High Frequency (Short Wave)	3MHz — 30MHz
VHF	Very High Frequency	30MHz — 300MHz
UHF	Ultra High Frequency	300MHz — 3GHz
SHF	Super High Frequency	3GHz — 30GHz
EHF	Extra High Frequency	30GHz — 300GHz

Above this is Far Infrared (see). See all of these band categories, and radar bands also.

• Note that each band has exactly nine times the bandwidth of the total spectrum preceding it. This is why spectrum is now plentiful.

• Frequency can be translated to wavelength by simple calculation. Radio waves travel at 300,000,000 meters a second, which is the speed of light. So, for instance, the UHF band extends from wavelengths of 1 meter, to one-tenth of a meter.

radio text

This is text on a radio subcarrier in the television UHF and FM radio bands. Text can be carried at rates of 11.8kb/s according to CCIR Recommendation 643. Its application is for the transfer of financial and general information data to PCs.

Radiocom 2000

Aka RC2000. A French version of IMTS mobile phones. It had a cellular structure but no handoffs. It began in 1985 with a total of 256 channels in the 200, 450 and 900MHz bands. The payphones on French trains still use this system.

radix

A number's notational base. Binary is radix 2, our normal decimal counting system has a radix of 10. The term radix is out of favor because it is believed to be confusing. Use 'base' instead.

• The radix point in denary is the decimal point. To the left of this point are denary integers, to the right are decimal fractions.

• Radix sort is a complex sorting procedure involving various stacks and substacks of the records.

ragged right/left

In type-setting the copy in a column can be set to:

Justified: where the space between words (and sometimes letters) is expanded until the lines are of equal length and aligned at both the right-most and the left-most character.

Ragged Right: aka Flush Left, where the left-most characters are always aligned, but the right-most characters in the column finish wherever the natural break occurs between the words. As it does in this example.

Ragged Left: aka Flush Right, which is used for effect (often in poetry), and sometimes when captioning alongside a photograph. It has an untidy appearance, and is particularly hard to read in large blocks of text.

Centered: This is 'ragged-both' style which is usually used only for captioning, or sometimes for isolating specific text within the main body copy.

RAID

Redundant Array of Inexpensive Disks. The various RAID levels are ways of arranging data on disks and making the data more secure. There are six levels currently used to provide fault-tolerant storage systems, mainly for local area network servers. Higher data transfer rates are possible with RAID, because the data is split across several parallel channels.

The RAID specifications evolved at the University of California at Berkeley in 1987.

• **RAID 0** — for 3 to 7 parallel drives. This is the level used when simple disk-striping is the only requirement (see striping). It offers no data duplication/parity; it was designed for high speed access because it performs parallel reads and writes on the various platters, but it has low data reliability. Technically, this is not a true RAID system but is included as an option.

• **RAID 1** — for drives working in pairs. This is the level used for simple disk mirroring or shadowing; the same data is sent to two drives and two identical sets of data are kept. It is safe but slow (it has no striping), although it is sometimes combined with RAID 0 to move things along.

• **RAID 2** — for 3 to 7 drives. This technique writes across all drives by using bit-interleaving and Hamming error correction. It is slow but provides reliable data. It's not currently used.

• **RAID 3** — for 3 to 7 drives. One drive is for parity only, and the others are for data (always an even number). Data is transferred one byte at a time and read/writes are performed in parallel. It is slow, and it has major problems if the parity drive fails.

• **RAID 4** — for 3 to 7 drives. This technique also uses one drive only for parity. It differs from RAID 3 in using sector interleave; the data being written to the disks one sector at a time. Reads and writes occur independently. It has Hamming error correction and suffers the same problems as RAID 3. RAID 5 is better.

• **RAID 5** — for 3 to 7 drives. This does multiple simultaneous writes to all disks with parity. Data is interleaved and stored on each disk so that the error correction is distributed across the drives. This is an improvement on RAID 3 and 4. It is fast with high reliability (especially in reading), but requires substantial processing power.

• **RAID 6** and **RAID 7**. These are variations on RAID 4 with undocumented features; they are not accepted standards.

rain outage

Satellite signals in the Ku-band are often absorbed and lost by heavy rain. MDS microwave delivery systems also suffer from rain attenuation.

The amount of signal absorption by rain depends on the polarization, the frequency and the distance the signal travels through the rain. Up to 3GHz, the effect is negligible even with tropical downpours in the 100mm per hour range. For horizontal polarization the figures are:

2GHz — 0.05dB/km.
12GHz — 4.6dB/km.
29GHz — 20dB/km.

Vertical polarization reduces these figures by about 25%. See solar outage also.

Rake

A receiver used in CDMA cellular phone systems which allows the simultaneous reception of three or more signals at the same time. It should actually be called a 'correlator' since it is comparing incoming digital signals with a predefined random number code. This is digital modulation on a digital signal — an XOR process.

A rake has a number of 'fingers', each of which can lock onto a different multipath signal. These signals usually come from the same source via a direct path or reflection (multipath), but during hand-offs to a new base-station, one or more may come from a different transmitter entirely. It is this ability to disengage progressively with one base, and simultaneously make contact with the other, that allows CDMA to provide soft-handovers where the link is not broken between cells.

The rake correlation of multiple finger signals is a summation effect (it increases certainty), so CDMA gains a substantial advantage from multipath whereas other systems see it as a disadvantage.

In the Qualcomm CDMA design, the basestation has a four-finger rake receiver and the handsets have three. Each of the fingers is a full digital demodulator in its own right.

RAM

Random Access Memory: RAM now refers only to read/write volatile memory. The internal working memory of a computer is made up of 8 or 9 chips of RAM, which store:

— Most of the operating system (some small parts may be in ROM also).
— The video map being transferred to the screen (in video or VRAM).
— The BIOS (input and output controls — often shadowed from ROM).
— The device drivers.
— The application/s.
— The data being manipulated.

RAM (and sometimes ROM) will be in a linear 'memory stack' with addresses beginning at zero for the first location, and progressing upwards to the limit of the memory chips installed. The way this stack is divided is known as the memory map.

See random access memory, Dynamic and Static RAM, video RAM, and memory map also.

RAM-cache

A feature that sets aside and temporarily retains frequently used information in a section of memory. See disk cache and RAM-disk.

RAM cards

These are memory extension cards, and shouldn't be confused with RAM disk-cards which substitute (temporarily) for a disk drive. Some types of memory cards can be configured to act as either.

ramchips

The old 'cockroach' type of memory chip (correctly called DIP chips), which are now being superseded by SIMMs.

RAM-disk

The partitioning of a section of the computer's RAM memory (or a special RAM extension card) which is utilized as if it were a disk drive. The computer is fooled into using this section of memory for saving and off-loading programs (it can be used as a 'virtual disk').

Naturally it works extremely fast, much faster than a disk. Don't confuse this technique with RAM-cache where the cache area of memory is constantly reading from, and writing to, the disk.

• With RAM-disk, the computer thinks that this area of memory is the disk — and so the real disk is not updated until the process is forced deliberately (usually) by the user at the end of a session. But the danger is that you'll leave important data 'saved' to this drive when you turn your computer off. At that point you discover why it is called a 'virtual' disk! See flush.

RAM-resident

Held completely in volatile memory — as distinct from being 'disk-based' and bought only into memory as needed. The term is usually applied to programs rather than data. Some large application will leave certain functions on disk until they are needed; this is done simply to reduce the load on memory, and this term makes the distinction. See TSR also.

RAM shadowing

In a typical PC's memory map, various locations are used by RAM, and other sections are reserved for ROM. BIOS, for instance is almost always fully in ROM. However, since ROM is often a very slow form of memory when the CPU accesses it regularly, it is common for the contents of the ROM to be copied into RAM during the boot-up stages, and then the copy in RAM will be used for the session. This is RAM shadowing.

RAMDAC

See video-DAC.

RAMP

RAdio Mail Protocol. This is the two-way version of the APOC radio paging protocol, with which it is backwardly compatible.

random access

The term applies to the ability of the system to reach the required data on some storage medium in very quick time without needing to move progressively (linearly—sequentially) through other information (say, on tape).

For instance an LP audio disk has random access to a high degree, whereas an audio cassette does not. Random access makes possible branched programming, quick searching, and hypertext-type networking. This is why the video LaserDisk is often the preferred form of storage for educational video, rather than the sequential accessed videotape.

Note that, strictly, the term 'random access' can apply equally to RAM and to ROM (PROM, EPROM) with chip memory, since you can access either form instantly.

- The distinction is with sequential or linear access. See.

random access memory

I'm making a distinction here between the original use of the term generically (above), and RAM which is now specific — meaning a bank of eight or nine chips in a PC. See RAM.

In microprocessors and all modern computers the term RAM is now only applied to a type of memory chip which is totally volatile — it loses all memory when the power is switched off. But there's no logical relationship between the volatility of memory and the way in which a CPU 'accesses' data — it is just that this is the current state of chip technology, however FRAM could change this. See.

The misconception comes about because in the early days, both RAM chips and ROM chips were viewed (as they should be) as 'random access'. One was read-only random access memory and the other was read-write.

The major 'access' distinction which needed to be made, was therefore between magnetic tape and punched paper tape recording systems, which were classed as 'sequential' memory systems — and the core working memory of the mainframe that was 'random-access'.

We would now categorize this difference as one between primary working memory, and secondary 'storage' memory.

With the old tape-storage memory, you couldn't get at any module of data without beginning at the beginning of the tape, and working your way through in sequence.

So random access was associated with internal memory systems, and sequential access with external storage devices — and this terminology remained fixed by history even when disks (random access secondary storage) and the various primary devices, like ROMs, PROMs and EPROMs, appeared.

Remember that ROM, PROM and all the variations, are as much 'random access' as RAM — for read operations anyway.

random files

See sequential files.

random noise

The noise generated in an electrical circuit by the motion of electrons. See white noise and Gaussian noise.

random numbers

Random numbers are generated in a computer usually as decimal fractions between zero and just under one (0.99999999). They are never truly random (totally unpredictable) but they should show rectangular distribution, which means that every combination of the eight decimal/denary digits will have an equal chance of appearing. See pseudo-random.

random routing

See free routing.

range search

In a database, it is often necessary to search a limited number of records on the basis of a range. For instance: 'Find all records where the year is greater than 1984 but less than 1988'. Most Boolean search systems allow the specification of 'greater than' (>), 'less than' (<), 'between' (><).

- There are also string range functions. The command 'FIND JUST (1W) TIME' will search for the phrase 'Just In Time', even though the word 'IN' is a stop word (see) and is not normally recognized in a search. This search command will not find '... and just one last time...' in an abstract, because the range specified was only one word (1W) of separation.

RARP

Reverse ARP. It provides a method of finding an IP address based on knowing the media address. See ARP.

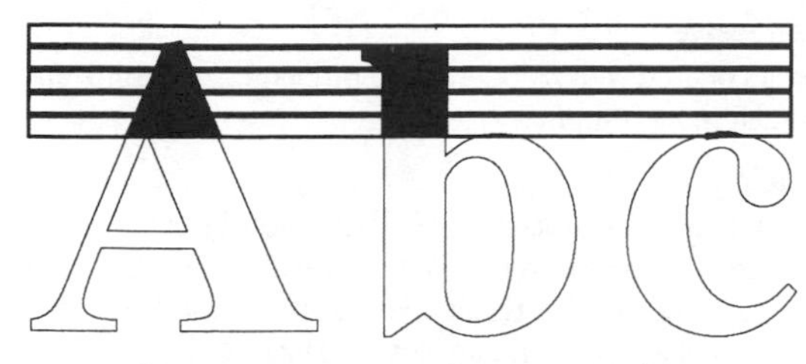

Raster scans — left, to right

raster

The raster progressively traces horizontal lines by varying the voltage of an electron beam as it scans the image into the picture tube of a TV set or monitor. With color monitors there are actually three (RGB) raster scans, but in general we treat them as one.

Raster is a general term which distinguishes the tracing of images by parallel horizontal scanning systems, from the vector systems which trace and fill objects, one at a time, using coordinates and plotting algorithms. Vector-scans use an object-oriented approach and have layers, while raster-scans are single, simple, flat, two-dimensional bit-mapped images.

Sometimes the term raster can also refer to the storage of the output bit-map in the printer or computer before it is sent to the scanning device; we talk about a 'raster image'. Here it distinguishes the bit-mapped image in Video RAM, from the source of that image — which might be ASCII code or vector-coded graphic information in the main area of memory.

Laserprinters also use a raster scan approach. Sometimes they take the ASCII and vector-coded information and rasterize the image themselves, and sometimes this job is done by the PC using special laser-driver software.

The term raster therefore has a flexible meaning:

- It can be the horizontal strip of image that the scanning beam of a video monitor draws as it travels across the screen. See Kell factor.
- It can be the full array of scan lines filling the whole screen (more correctly, a 'raster scan').
- It can also be the trace of the laser beam across the printing drum on a laserprinter.

raster image

Raster images are scanned or bit-mapped images — arrays of dots (pixels) arranged in sequential order, left to right, and from the top of the screen to the bottom. In memory, the bits are also stored in sequence (although see planar), and generally a single memory location will represent one screen pixel, and have one or more bits/bytes assigned to it, to control brightness and color.

The number of bits associated with each pixel is the pixel depth, and this represents color and tone with color systems, and tone-only with B&W.

The comparison is with vector graphics. See planar architecture and packed-pixel architectures.

raster scan

After scanning the screen, the raster-driver must switch off the beam while it repositions back to the top left corner, before it begins the next field or frame. The time taken for this repositioning is the 'vertical blanking interval' or flyback.

Similarly, at the end of each line, the raster must switch off, and reposition back to the left. This is the 'horizontal blanking interval' or horizontal flyback.

Raster scan systems can be either like TV with a 2:1 interlace, laying down only half the image each time (a field) — or they can be like computer monitors with progressive scan — creating the image from top to bottom, and painting all the lines as they go.

• With today's computers, raster scan systems are almost always controlled by the screen driver progressively scanning a bit-mapped area of memory stored in VRAM. See Kell factor, vector scan.

rasterization engines

In page description technology (for laserprinter), the rasterization engine is the last stage of the process which provides the sequence of bits sent to the printer for final output. It interprets the PDL, and 'compiles' it into a bit-stream for printing. This bit-stream will be stored in a bit-map, usually until the full page is compiled. The information is then written by the laser-driver and laser onto the printer drum.

• With many PCs, the rasterization engine may now be software running in the PC, rather than in the printer. The image is rasterized within the PC, then set to the printer as a graphics image, complete in every way and ready for printing. This transfer is a slow process over serial links.

rasterizer

An algorithm used to convert outline fonts (such as PostScript) to a bit output that the printer can use. It changes a vector-stored image into a raster-scan output. This is also known as 'ripping'.

At the present time different rasterization engines are needed to handle PostScript, TrueType and OS/2 fonts. The aim is eventually to create a single high-level imaging model for fonts using a single rasterization technique. See RIP, TrueType and ATM (Adobe).

rat's tail

The small external network adaptor box used on an Apple LocalTalk LAN. MACs have networking built-in, and only need these for direct connection.

rate adaptation

The change of data rate from one standard to another.

Most telecommunications channels come in fixed bandwidths — such as the standard 64kb/s B-channel of ISDN. But some terminals and services are unable to use the full 64kb/s; they may only be designed to product 9.6kb/s of output, so various forms of rate adaptation are required to marry the two rates. This is one function of the Network Termination (NT) boxes in ISDN.

• Packet switching systems (including frame and cell relay) are all rate adaptation systems, provided the cells/packets can be stored and forwarded. It is obviously easier to change from a slow rate to a fast rate than vice versa, because flow control will then be needed to prevent data from overflowing.

• Data overflow is a common problem in modern Ethernets with dedicated or shared 10Mb/s segments being switched to 100Mb/s server/backbone links. Overflow on the reverse path is a major problem. See back pressure.

• If an Ethernet 'switch' provides rate adaptation, then it isn't a true switch. It must be a store-forward device. It just happens to be acting very fast and switch-like, but it will probably be functioning as a bridge or router using RISC or ASICs for high-speed operations.

• A common use of the rate adaptation in the US is with ISDN where it is often necessary to link a switched or permanent 56kb/s data channel into a standard international 64kb/s channel. See V.110 and V.120.

rate one-half (two-thirds)

A general term applied to forward error-correction (FEC) systems to define the overhead in the system. A 'rate one-half' system will devote one in every two-bits to error-correction. When this happens, the true throughput-rate of the data will only be half the link-rate. However, it will be a very robust system — capable of correcting massive errors.

Some of the new satellite systems are able to vary the FEC overhead so that less important information has less correction. For audio or video (not data) the bits can be interleaved so that most-significant bits are grouped together, then second-most significant bits, etc. to least-significant bits.

They can then apply a sliding window of FEC which varies from 'rate one-half' back to 'rate one-third' or 'rate one-quarter', and so on. More protection can then be given to the most valuable bits.

Generally a combination of Viterbi, convolution coding, and Reed-Solomon error correction will be used. GSM digital phones do something similar to this.

rate rebalancing

The meaning differs depending on the country. This usually means that the monopoly or dominant carrier's rate will be changed to more closely reflect the cost of providing the service. They will probably move the cost of long-distance calls down, and local prices will go up. Alternately, they will increase domestic prices and drop business prices.

• Since optical fiber came into being in the early 1980s, long-distance links have been the milch-cows of telecommunications, but now, under competitive pressures, the carriers are sometimes being forced to keep tariffs more in line with real costs — but there's still a very long way to go.

Rauto-receive

A protocol used with ZModem (and others) for accessing unattended remote computer systems.

ray-tracing

A technique of constructing three-dimensional solid images on the screen by laboriously tracing the path of all light rays from a theoretical source. The result takes hours on even large computers, but it allows shiny surfaces to exhibit reflections, as in a real photograph. Parallel architectures are now widely used for this type of application. With color image rendering you can have three computers working in parallel on Red, Green and Blue.

Rayleigh channel

A radio channel which is subject to multipath problems. Cellular mobile phones receiving signals from terrestrial base-stations are using Rayleigh channels, and they all suffer from Rayleigh fading (interference). However when you use a direct-satellite mobile service, the channel is predominantly direct and therefore Ricean.

Both of these channel types are named after the people who first described the problems.

• There are also Rayleigh waves in ultrasonics. See.

Rayleigh fading

An interference problem found in all terrestrial radio systems, but more so with those using UHF and SHF frequencies.

A cellular mobile radio typically gets its signal indirectly because it will lack direct line-of-sight to the transmitter. The dominant signal is called the Rayleigh fading channel, and it is the result of many different signals arriving nearly simultaneously from many directions, with each component experiencing slightly different transmission delays. The phase difference between these signals can cause interference due to summation and subtraction, and this is heard as rapid changes in volume which reduces subjective quality.

• This type of interference is often very noticeable near hard reflective surfaces.

• Rayleigh fading in a moving vehicle can also produce a 'picket fence' variation due to the regular series of dips and highs in the received signal strength. It tends to adversely effect the control signals much more than voice.

Rayleigh scatter

This is an important attenuation factor in glass fiber. The light is dissipated after being scattered by variations in density of the glass. These variations are of very small magnitude (less than 0.1 wavelength).

RBDS

Radio Broadcast Data System. A digital radio technology now used by about 250 US FM radio stations. It adds digital data output to their conventional audio signals and is used to give the title of the song, a phone number to call, etc. to radios equipped with the decoder and LCD display. Any sort of supplementary information.

It is also part of the new Emergency Alert System; the radio can turn itself on and tune to an emergency channel in the event of a cyclone warning, etc.

The system is based on the European RDS, but it is not identical. However US systems are often labeled as being RDS when they strictly aren't.

RBOC

Regional Bell Operating Companies aka Regional Holding Companies (RHCs), local carriers, or telcos. RBOCs are the seven large telephone networking companies which now dominate the American telephone business and own the BOCs (the 22 local Bell Operating Companies, now called LECs). The RBOCs resulted from the forced break-up of AT&T (which remained only as a long-distance carrier) in 1984. See MFJ.

The name NYNEX gives away its BOC origins. It is made up of the New York local monopoly carrier and the major New England local carrier, which were merged to form the NY–NE–eXchange RBOC. The other six RBOCs are: Ameritech, Bell Atlantic, BellSouth, SouthWest Bell, Pacific Telesis and US West. GTE is not an RBOC but just the largest of the independent telephone carriers. Sprint, MCI and AT&T are long-distance carriers or IXCs.

Within the various regions serviced by each of the RBOCs there are still many independent phone companies, mainly in the smaller towns. These will have a relationship with an RBOC for non-local carriage.

See RHC, LATA, LEC, IXC, RHC and BOCs.

RC2000

An early analog cellular phone standard used in France. See Radiocom 2000.

RCA connector

This is a small round smooth plug about 6mm in diameter. The cable connector has a substantial central pin, and an outside sleeve which just slides on. This is the one which is very widely used in pairs for stereo audio connections.

RCP

Remote Control Program. A file transfer program used on the Internet. It is similar to FTP.

RD

Aka Rx. In modems this is Receive Data.

RD-LAP

Radio Data, Link Access Protocol. This LAP is used in packet radio networks. See DataTAC.

RDA

Remote Database Access. An ISO standards project for RDBMS computers to allow interoperability. The aim is universal database access and the ability to exchange data between different computer platforms. It is currently at the draft stage.

R-DAT

Rotary Digital Audio Tape. This is now better known simply as DAT. However there were once two forms of DAT: the current 'rotary' (helical scan) technology and Stationary DAT (S-DAT) which found some professional uses. See DAT and Data-DAT.

RDB

Relational Data Base. See RDBMS

RDBMS

Relational Database Management Systems. These are a later development than simple DBMS, and at a higher level of complexity in that they use a number of interlinked tables.

Relational DBMS avoid the problems of the repetitive storage of information by establishing relationships between different tables of data.

Records in one file (table) can be related to records in another — provided each file has a 'key field' (usually an account number) which ties the records together. Usually an RDBMS will also come with a range of associated tools.

Database management is now the core function of any large corporate computer system, rather than being seen just as an application.

The SQL Access Group is defining technical interoperability standards which will enable all RDBMS servers to service any client. The ISO is also developing RDA (Remote Database Access) standards.

Distributed database management systems (DDBMS) are the result of convergence between RDBMS and network technologies, and this appears to be the way we are heading, along with the developments in object-oriented databases. See Codd's rules, client/server, ODBMS and SQL also.

- The main players in this field are Oracle, Ingress and probably DB2.

rdev

Resource Devices. A term used with Macintoshes to indicate small driver programs that handle resources and are accessible through the Chooser.

RDI

Restricted Digital Information of US ISDN. RDI is a rate-adaptation process for a 64kb/s channel which can be used to transmit HDLC data at 56kb/s — but only by following some rules which avoid having six 1-bits in a row. The standards bodies have been trying to dump RDI, and prefer to use Unrestricted Digital Information. See UDI.

RDN

Relative Distinguished Names. The name by which an object (a person or a device) is known in the X.500 International Directory system. The distinction is with other entries which can be an alias, organizational person, or organizational role, etc. See X.500 and objects.

RDS

- Radio Data System. See.
- Radio Data Service. See.
- Remote Data Services in OS/2 EE. These are extensions to the Database Manager's ability to handle SQL requests.

RDSS

Radio Determination Satellite Services. These are generic position location services, mainly used for ships, aircraft and land mobile tracking. RDSSs compute the location of the ship or vehicle at the earth station and it will then transmit that location information to the mobile, if required. Some of these systems may also include data-mobile messaging and direct voice communications via satellites also. See GPS and GNSS.

read/READ

- **Memory:** To retrieve one or more bytes from memory. With RAM, this process does not normally destroy the information held in the memory location, but rather copies it and transports it to the CPU.

Disks are read to retrieve the data stored. This is always a non-destructive read. See RLL and MFM for magnetic disk technologies, and EFM for optical CD systems. See write.

- **Relative Element Address Designated:** The compression technique which checks the scan lines in a facsimile ahead of the one being sent, to see if duplication occurs. If the information is redundant it will send out a signal saying 'the next x dots are black (or white)' rather than sending the dot-code for black the require number of times.

Modified READ is used in Group 3 facsimile, and later versions will also have Modified Modified READ (MMR). In Modified READ the sender looks for correspondence between elements in one line with those in a reference line previously sent (i.e. two-dimensional).

The later MMR handles white or black tone in blocks; this requires greater memory buffering and more processing. The preliminary handshake establishes whether later systems can be used.

There's also a form of one-dimensional modified Huffman code in Group 3 facsimile which reduces the bit-stream even further after the READ coding.

- **CD-ROM:** (Read-Only) has a specific 'Read' family of commands.

read-ahead

In disk cache operations the disk controller brings into memory the required sectors of data. With 'read-ahead' it will then automatically bring the next sectors into cache on the assumption that these will soon be required. If a file is highly fragmented on the disk, the read-ahead operations may be of little value.

Read-Only Memory

See ROM.

real mode

Also called the real address mode. This is the mode whereby Intel 286 (and later) processors simulate the functions of the old 8086 or 8088. The 8088 or 8086 chips of the early PCs and XTs were only capable of real mode direct addressing of memory to a 1MByte limit, so the 'real mode' is a type of emulation where a more powerful chip mimics the functions of an older less powerful one. It is a non-protected mode where multitasking or multiuser operations aren't possible.

The distinction between real mode and protected mode (see) in the Intel 286 chip came about because the chip was originally designed to run Microsoft's Xenix (Unix) which didn't have the upper DOS limit.

- The 386 and 486 chips can run a number of MS-DOS applications simultaneously in virtual 8086 mode. This is little more than protected mode emulating a number of real-mode processes. See protected mode and virtual 8086 mode.

real numbers

The set of numbers that includes all integers and decimals (denary fractions) with a finite or infinite number of digits. The IEEE 754 standard for floating-point numbers allows for the following types:

— Single real (single precision 32-bit).
— Double real (double precision 64-bit).
— Extended real (double extended, 79-bits or more).

In Basic on the early PCs (i.e. the Apple II), integer numbers were represented by two bytes to a maximum of 65,535 values, while real numbers were represented by four bytes to a maximum of 4.3 billion.

real-time

- **Delay:** Almost instantaneous computing or communications — where the user doesn't notice any delay. The distinction is with batch processing and with store-and-forward systems. However, some store-and-forward telecommunications systems nowadays operate so fast they produce real-time results. See isochronous and ATM.
- **Time-sharing:** The real-time, when time-sharing on a large computer system, is the time actually involved in using the system in human terms, rather than the computer's time allocation. Time-sharing systems usually charge on computer processing time, measured in millionths of a second.
- **Voice/Video:** In ATM and fast-packet switching, real-time refers to isochronous services — those that can't afford to be delayed in delivery. We need to be clear on these differences:

— Some services, such as phone conversations, require almost-instantaneous delivery within, say, 1/10th second. Videoconferencing, which is interactive, requires the same.
— Some services, such as one-way broadcasting, can have cross-network delays of seconds, provided the data arrives at a reasonably regular rate in a constant stream.
— Some one-way services can handle some irregularity in the

rates, provided they have adequate buffers, and have guaranteed delivery within the time it takes to empty the buffer. These systems can tolerate long delays, but they must also be deterministic. Digital television could be of this kind if we had standardized buffer sizes.
— Some services are tolerant of both delays and irregularity. These are usually data file transfers and messaging services.

• **Clocks:** The real-time clock in a computer is the one that keeps track of the time and date like a watch, not the one that synchronizes the CPU and memory. See date.

reassembly
The reconstruction of packets or datagrams at the destination, after they have been fragmented (segmented), for transmission over packet systems. Sometimes this segmentation takes place at the source node, sometimes at intermediate nodes, and sometimes at both.

recalculation
Spreadsheets have three main forms of recalculation:
— **General recalculation,** where the whole spreadsheet is redone from beginning to end, working from the top-left to the bottom-right.
— **Minimal recalculation** where only those cells affected by a change are recalculated.
— **Background recalculation,** where you can continue to work on the spreadsheet while a full general recalculation occurs.

Minimal and/or background recalculation is important with large spreadsheets. The process can take many minutes and disrupt important work.

record
A standard data-structure — a set of fields in a database which relates to one item/entry. A collection of records makes up a file, and each record will consist of one or more fields.

For instance, in a company account database, the Name, Address, City, Credit, and Debit fields may all constitute one complete record, and each customer and supplier will have their own record.

If the Name, Address, and City are held in a 'Customer' database file, and the Name, Credit and Debit are held separately in an 'Accounts' database file, then the two records (one in each database) are 'related' by virtue of sharing the same Name field. This is then a relational database.
So a standard relational database hierarchy consists of:
— Databases which consist of a few related files.
— Files which consist of many associated records.
— Records which will each have numerous specific fields.
— Fields which will probably each have 'attributes', which restrict the type of data that can be entered.
— Data of the correct data-type, which is associated with each record, will be in each field.

This will all be managed by a database management system. See DBMS.

record locking
A method of protecting the integrity of shared data on a network. A multi-user database needs record-locking to prevent more than one user from concurrently accessing, and perhaps altering, a record that is in use elsewhere. This requires a relatively sophisticated server with good DBMS software.

• In simpler systems, file locking is used. This also prevents concurrent access — but to the whole file, not just the record. See file locking.

recovery
This is not the same as error correction. Recovery is the process of restoring the conditions that existed before a file-transfer or transaction process was interrupted. It may also involve picking up and continuing with the file-transfer task, but that is secondary. See ACID test and journal.

recto
Right-hand page. See verso.

recursion
To start again from the beginning (or from a defined point). A recursive program will have a sequence which 'calls' itself.

• **Recursive programs** are those in which the same operation is repeated one or more times. Usually each recursion (repetition) is based on the results of the previous one, so variables need to be maintained. Nested loops are a recursive.

• **Recursive functions** are those that call themselves. The function must always have a limiting condition so that it stops looping back after a finite number of calls, or it will lock into an eternal loop.

• **Feedback:** In electrical and general engineering, recursive means 'feedback' — some of the device's output is fed back to the input (usually negatively) as a means of control.

Red Book
• Compact Disk: See CD and CD-Audio.

• ITU specifications: Different colors are used on different occasions. There's often no other significance to the color other than the date of release.

red box
In phone phreaking, this is a small tone generator which creates tones which sound like money being inserted into a payphone. This once worked when we had manual long distance operators. See blue and black box.

redirect
In some ICMP and EI-IS systems a router can advise a host that the use of another router would be better. It can redirect the host to use a different path through the network.

redirector
• **Computer architecture:** A piece of software which forms the basis for virtually all network operating systems under DOS. It is a traffic policeman which checks any commands

issued by your keyboard (or peripheral) to system software. It decides which commands are appropriate for the local CPU, and which for the file server. It also captures requests/commands issued by applications and directs them to the service devices, printers, etc. It is also responsible for intercepting interrupts, and it checks to see whether the requested functions are DOS or file I/O operations. It also translates requests into lower formats.

• **Remote Access:** This is software which intercepts requests for resources and analyzes them for remote-access requirements. If this is a remote access request, the redirector forms a RPC and sends it down through the lower-layer protocols for transmission on to the network node which can best satisfy the request.

redlining

• The technique of marking text to show revisions (while not immediately removing it). This is often important in groupware, where the decision on the change needs to be a consensus, or referred to a higher authority.

In some word processors, the 'redlined' text is converted to a 'strike-through' font which is automatically removed during the printout.

• This also applies to telephone service providers who 'redline' various slum areas of cities, and avoid providing adequate services in these areas because of vandalism, etc. In the cause of Universal Sèrvice Obligations, this is frowned upon by the regulatory agencies. The redline comes from circling the area on a map.

Redmont

Smart journalese for Microsoft. This is where they have their headquarters.

redundancy

Not essential for operations (but could be necessary to salvage problems). In telecommunications, this is 'overhead' which is usually required for error checking or correction. It is part of the total data in a message packet which could be eliminated without loss of any essential information.

Checksums and parity bits are termed 'redundant' because they aren't essential to the data operations — they are there for error checking only.

• Duplication: Duplicated or back-up data is 'redundant', in the sense that it is not essential for normal operations.

• Parallel equipment on standby in case of failure, is redundant also. Disk mirroring (twin devices) is a redundancy approach.

• Redundancy checking, means systems like parity which increase the length of a byte, and exist for no other reason than to help find or recover errors.

• In information theory, redundancy is the difference between the actual rate of resources consumed to transmit something, and the best-possible rate that could be produced with a specially designed language:

The English language is said to be about 80% redundant, since the available symbols (letters) aren't used to maximum efficiency. A simple illustration of this redundancy is that we can read vowel-less sentences like:

```
MST PPL HV LTTL DFFCLTY N RDNG THS SNTNC
```

Obviously vowels are almost redundant on some occasions.

You find different statistical frequencies with different letters and letter groups (i.e. the QZ combination is probably unknown, while the TH combination is very common). So Huffman-like codes and LZW can reduce the overall rate by providing special codes to different combinations of letters.

It is said that if redundancy was removed, it would be possible to code English into binary digits at an average rate of only one bit for each letter of the original text. Currently English needs about 4.78 bits per character with a minimal ASCII-type code. With our 8-bit code, about half (4-bits per character) seems to be the limit for data-compression techniques unless the text is very repetitive.

• In text and graphics compression, the job of the compression algorithm is to remove redundancy. To reduce the datastream to the lowest possible level without destroying any of the original information/detail (lossless compression).

• Redundant array of inexpensive disks. See RAID.

redundancy checking

Error checking: see checksum, LRC and CRC.

Reed-Solomon code

A very effective system of error detection and correction used in CD-audio, CD-ROM, and by the Jet Propulsion Laboratories on the Giotto and Voyager space flights.

It relies on the application of checksums to bytes distributed in a matrix form which layers and interleaves the checking process. Usually they will also interleave the streams or blocks of data to reduce 'burst' errors to unbelievably small amounts. This is where CIRC (Cross-Interleaved Reed-Solomon Code) comes from. See CIRC and ECC.

re-engineering

• **General:** A concept introduced by Michael Hammer in the Harvard Business Review in mid-1990. He defined it as 'The fundamental rethinking and radical redesign of business processes to achieve dramatic improvements in critical, contemporary measures of performance, such as cost, quality, service and speed.' He later explained: 'At the heart of re-engineering is the notion of discontinuous thinking, of recognizing and breaking away from the outdated rules and fundamental assumptions that underlie operations.'

His philosophy involves phased planning and restructuring, with a greater emphasis on simulation, distributed computing, and electronic communications technologies.

Specific: The computer industry is now re-engineering legacy applications on mainframes and AS/400 computers to introduce modern screen-object manipulation, etc. This is often called downsizing or right sizing.

reference model
A graphical representation of a system showing the layers or modules, and describing their functions. The aim is usually to create a standard suite or 'model', where one layer (itself usually a suite of protocols) can be removed and replaced with another — provided the interfaces between layers remain the same. See OSI Reference Model.

reference sync
The output from a sync-pulse generator in a television studio. See sync-pulse generator.

referential transparency
This is a mathematical property which can be used to prove that a program is correct and bug-free — the El Dorado of programming. It is said to be only possible with functional programming languages. Variables don't change value in these languages, whereas in procedural languages they are constantly being updated — their original value is destroyed.

reflected binary
See gray binary

reflectometer
See TDR and OTDR.

refraction
Electro-magnetic waves can vary from the straight-and-narrow (never deviating) path either by reflection or by refraction. A mirror reflects, while a lens refracts.

Radio and light refracts whenever it propagates through a medium (glass, air or water) which is not completely homogeneous in its density. It will bend towards the more optically dense (or radio-dense) regions.

refractive index
RI. The measurement of optical density.

refresh
DRAM chips (standard RAM memory) need to be constantly refreshed (updated) to retain the charge in the capacitors which are the memory trace. This charge leaks away in 2ms.

The refresh cycle is performed by cycling through the address lines periodically, reading the locations one at a time, and writing back to them. The process reduces the speed of the computer. The loss in processing speed is called the 'refresh overhead' and is generally specified as a percentage. SRAM doesn't have this problem.

• The computer also refreshes the monitor screen by rewriting the image, usually 70 times a second.

• Flash memory, ROMs and FRAM don't need to be refreshed since they are non-volatile memory. SRAM is volatile, but it doesn't need constant refreshing (just power).

regenerator
It is possible to repeat analog signals by amplifying and retransmitting them, but each amplification stage cumulatively adds noise. However, digital signals can be either repeated (amplified and passed on) or regenerated (restored to their original square-wave shape) very easily.

The amplifiers (repeaters) used in the early analog undersea cables, gave way to digital regenerators, but now with optical fiber amplification we have gone back to repeaters once again. Generally, however, the signals in a 5Gb/s optical fiber will stay 'square enough' through these amplification stages over distances up to 2500km. See repeater.

Region #
The ITU and WARC have divided the world up into three regions for the purpose of radio spectrum control.

— **Region 1** covers Europe, Africa, Near East and USSR.
— **Region 2** is the Americas (North, Central and South).
— **Region 3** is Asia (excluding Russia), Australia and New Zealand, and the west Pacific Islands to the international dateline.

register
A series of volatile memory locations that store instructions, addresses and data, temporarily for subsequent use.

• **CPU:** Registers are used in the central processing chip for virtually all operations: there are memory address registers, shift-registers, program counters, status word registers, instruction registers and general purpose registers.

Don't confuse registers with cache. Cache is also built into CPUs, but a cache will store a much larger block of data from memory on the assumption that it might be required later. Registers store it while it is work-in-progress.

• **Modem:** In a modem, registers contain parameter values that will determine how your modem executes commands, and some that control the timing of various modem functions. In this case a register is nothing more than an extended array of variables — usually numbered; so you have the S0-register holding a parameter (from 0—255) which decides on which ring the phone will be answered. The S7-register holds a parameter (0—30) which is the 'Wait time for carrier' in secs.

You can change register values by using the Sx= y command. The difference here between a register and a normal variable is simply in the way they are named and the way they are changed.

• **Cellular:** In mobile cellular operations, registration is the process by which the mobile identifies itself to the base-station. It provides the network with its approximate location at the same time.

This reduces the number of basestations called upon to handle the paging messages when a call arrives intended for that mobile. If a phone hasn't been used for about a half-hour, the last basestation with which it is registered will poll the network to discover if it is still active.

• **Printing:** In printing, registration means to align one negative image with another. In color printing, four image 'separations' will be made (see CMYK), and these will need to be registered to ensure that they print exactly over the top of each other. The angle of each needs to be different. See moire.

• **DTP:** Desktop publishing programs can usually print registration marks outside the active area of the page which are used during the guillotine process, when the printed and bound booklets/reports are cut to size.

regulation

Telephone carrier regulation is generally unnecessary when telephone systems are run by government monopolies, as most have been until recently. However, as nations follow the US model and introduce competition into their long-line (and eventually local) networks, more and more regulation is generally necessary.

Rival companies must both compete and cooperate, which is a sure recipe for collusion. For some reason the press assumes that open competition inevitably means de-regulation — when all evidence points the other way.

The three main areas of regulation necessary are:

— **Standardization:** Free movement, total interconnection, and open competition are only possible when key technical standards are rigidly imposed. Left to self-regulate, the dominant carriers (cable or telephony) can set *de facto* standards (such as set-top decoders) which make it very difficult for customers to change to another supplier.

— **Antitrust:** By outlawing agreements which distort competition and create abuses of market power these are essential in such a singular culture as telecommunications. The complexity of the technology, plus some deliberate mystification, makes it difficult for customers to know when they are being bilked. Retail price maintenance doesn't need collusion in most duopolies.

— **Economic:** Government subsidies, tax advantages, commercial discounts, cross-funding, USOs, tariff imbalances, international relationships, etc. all work against a true free market operating, with open competition, in telecommunications.

relation

• A table in a relational database.

• In a hierarchical file system we talk of mother/daughter relations. This is a family tree-structure where each node will have one unique link to one node in the level above, but can have many links to levels below. Each person can only have one mother, but many daughters.

relational database

With relational databases you can have two or more files linked — which means that they must share at least one field in common. The term database then applies to all these files collectively.

The aim is to provide ways of linking information without placing all your data 'eggs' in one basket — but it creates new problems in record locking, since there is more than one related record that needs to be considered.

It is said that relational databases are 'inherently suited for parallel processing' because pieces of the database, and operations on them, can be readily partitioned and distributed among the multiple processors. This seems to be a bit of an overstatement.

• Very often, when people say 'relational' they simply mean 'multi-file'. This is especially common in PC environments.

• Codd's 12 rules of relational databases have long spelled out the required 'relational' features of a true RDMBS. See Codd's rules.

relational model

This is Ted Codd's 12 tests by which a database can be graded as conforming to his 'true' relational model. See Codd's rules.

relational operators

In computer programming, relational operators compare two sides of an equation. In Fortran there are six relational operators:

.EQ.	equals	**.NE.**	not-equal.
.GT.	greater-than	**.LT.**	less-than.
.GE.	greater-than-or-equals	**.LE.**	less-than-or-equals.

The Basic language uses the symbols =, <, >, for the same purposes. See arithmetical operators.

relative address

See absolute address and relocatable program.

relay

To pass along in sequence; as in a relay race. The implication is that packets/signals are passed along almost instantly, without a pause.

• An electro-mechanical device which switches a circuit in or out when a current is applied. Usually a small control current will be applied to a relay to switch a much larger current. It is a switch that can be worked by remote control.

• In networking, a bridge is classed as a Layer 2 'relay', and a router is classed as a Layer 3 'relay'. Here they just mean that packets get passed on at these different levels without modification.

• A remote relay is a remote bridge which links segments across a WAN link. In the OSI terminology, a relay is any device which connects two or more network segments or networks.

• Frame relay and cell relay are fast packet-switching systems where the frames and cells are passed along without the normal store-check-forward processes of the old packet networks.

reliable communications

The term usually means that, as the packets of data cross the network from one node to another they will be checked at every stage for errors and, if necessary, retransmitted. The alternative is to rely on the end terminal equipment to detect problems (end-to-end checking).

• Reliability in radio communications means 'availability'. At higher (GHz) frequencies, rain attenuation will affect performance. The reliability will be directly related to the fade margins, and it is usually measured in percent of time in which the system will be available (99% is usually needed).

relocatable

A relocatable program is one that will execute, no matter where it is located in memory. This assumes that the program was not written with fixed locations (absolute addresses) but rather with relative addresses. The base-address of the program can be checked, and offsets added to arrive at the location needed.

• MS-DOS .EXE files are relocatable, while .COM files aren't.

RELP

Residual Excited Linear Predictive coding. This is the basic voice technique used in GSM speech coding along with a long-term predictor (LTP). In combination, the vocoder is known as RPE-LPC (Regular Pulse Excited, Linear Predictive Codec). See.

The original sampling rate for RELP was 8k samples per second, with 13-bit 'resolution' (quantization) and the full-rate coder compresses this by a ratio of 8:1. The standard full-rate RELP coding is therefore 13kbit/s.

Later the half-rate coder is promised, and this will perform 16:1 compression and result in a 6.5kb/s rate. This compression is said to be possible with a new long term predictor which is better able to model the characteristics of the vowel sounds and remove the 'coarseness' produced by the coder. See CELP, LPC, MBE and vocoder as well.

Rembrandt

A name of a very popular video codec made by CLI which has become the American standard. The European standard which is still hotly in competition with this is the H.120. MPEG-based codecs will take over this field very soon.

remote bridge

A bridge that connects two segments via a WAN link. See half-repeaters also.

remote job entry

See RJE.

remote loopback

A modem testing procedure; the ITU-T has specified four loop-back tests that should be available in all modems. Two are local and two are remote. Remote analog loopback tests the carrier signal and sometimes the analog circuitry of the remote modem.

— Loop 2: The signals are reflected by the remote modem, but they do not including electronic regeneration of the signals (the remote modem acts passively).

— Loop 4: These actively include the remote modem.

See local loopback and loopback.

remote sensing

The term is applied mainly to satellite observation of the earth, although it could be applied to a wider range of technologies. The US Landsat program has probably provided the most information.

Landsats fly at 438 miles (700km) above the earth, and they can resolve blocks of the Earth's surface about 30 meters square. They record image strips of the Earth about 180km wide. Six sensors are used: Red, Green, Blue and three infrared frequencies. Each of the six provides a multibit sample for each pixel — and these numbers represent the 'spectral signature' of that block, which can be interpreted to show the crop or land type.

These images are usually shown in pseudo-color. The full image is composed of 42 million pixels and covers an area of about 8 million acres. These images are usually shown in pseudo-color.

Remote sensing is also used with aircraft-based systems. A few of the more interesting modern techniques are laser spectrometry and special cloud-piercing microwave radar systems. These systems are often coupled with GPS.

• The disciplines and organizations with an interest in remote sensing are geology and mineral exploration, oceanography and weather, botany and resource economics (particularly crop growth), international aid organizations and national disaster organizations (oil spills, fires, volcanoes).

• **Virtual Reality:** A different form of remote sensing entirely is used with virtual reality, where biological perceptions are fed back to the remote user or manipulator. This is the concept of medical operations by remote control using sophisticated communications and tactile feedback.

removable media

A term which can mean any form of cassette, magnetic disk or optical disk storage system. However it is usually applied to optical disks (and occasionally to Bernoulli and special types of hard-disk cards) where the disk can be removed from the read/write mechanism. Sometimes the whole miniature disk drive is removed also. See hard-card.

The problems are with mounting removable media (like CD-ROMS) on a network which has been designed to expect data to always be available. However most NOS now handle on-line, near-line and off-line media. There's a security benefit here with removable media, also.

REN

Ringer Equivalence Number. This provides a figure for the amount of electronic 'load' any telephone device puts on the ringing current of a telephone line — the power it draws. You can generally have a maximum of five REN-load equivalents on one circuit. Each device (modem or handset, fax, answering machine, etc.) should have a REN. If you add up the REN load on a single line the sum should not exceed REN5 or the exchange may not be able to handle the load.

• The AC impedance of a phone line should not be less than 1600 ohms, on-hook.

The traditional impedance of an on-hook phone is 8000 ohms, which is REN1. When the phone is off-hook, it will typically read about 600 ohms at the exchange.

• Unfortunately the REN is only a rough indication because it varies with the frequency of the AC current (usually 20Hz but up to 80Hz). Don't overload extensions with too many phones.

• Some modern phones also pass a certain amount of the 48V direct current during the ringing phase, and if too many are across a line, the flow of DC may be enough to make the exchange think the phone is off-hook.

rendering

• To draw an image as it would actually appear to the eye, rather than as a schematic, wire-frame or outline, etc.

• In text-processing and e-mail, rendering refers to formatting — indenting, type style/size/weight selection, etc.

• In 3-D CAD and graphics programs, the term is often treated as a generic to include the multi-stage conversion of wire-frame type graphic images into polygonal shaded images — then to smooth shading — and finally, to photo-realism.

• More specifically, rendering is the generation of surfaces, rather than the geometrical information used to wire-frame model.
— In animation packages, render only really refers to the last stage: the process is to model, animate, and then render. The rendered surface will probably be assigned some other properties (usually stored in libraries) such as texture, color, transparency, weight and substance.
— When realism is desired, after the rendering stage it is possible to add 'atmosphere' in the form of lights, reflections, textures and textured-movement. These are sometimes seen as rendering techniques, and sometimes as quite distinctive. See graphics modeling. See ray tracing.

• Pixar's extensions to Autodesk (called RenderMan) is the best-known rendering program.

rent

In the design of ICs, rent refers to cost in terms of chip-space, needed to provide some function. Since it costs about $20 million to fabricate even a small processing or DSP chip these days, the 'rent' for 1mm of surface space can be very high.

rent's rule

In chip design, this is an esoteric empirical rule and formula which tells us about the relationship between the number of connections and the number of gates on a chip. More logic needs more connections.

repeater (general telephony)

Repeaters just mimic the input and recreate it at the output at a higher level and usually with slightly more noise. If the signal is analog, repeaters are generally nothing more than amplifiers. If the signals are digital they may be either amplified or regenerated (the two are constantly being confused in the literature)

• In a communications satellite the transponder is just a large analog repeater in space. It accepts signals on one frequency, amplifies them, and retransmit them — sometimes on another frequency.

• Repeaters are used in optical fiber, twisted-pair and coaxial copper cables telephony also, and at a much closer spacing than is used in optical fiber. In copper, more stringent requirements are placed on gain, stability and linearity since there may be hundreds of them in tandem in a long link to minimize random noise, intermodulation and overload distortion.

Coaxial analog trunk cables with frequency division multiplexing up to 12MHz (2700 telephone channels), need transistorized repeaters spaced every 2—5km. With digital transmissions the spacing is much the same, but the carrying capacity is now much higher.

The term repeater is now being used casually to mean both signal amplifiers (analog and digital), and signal regenerators (digital only). The repeater is the whole unit, while the regenerator is the active digital electronics.

• Optical fiber used terrestrially usually has repeater spacings about every 40—50km but, under the sea, modern systems have regenerator spacings of 150km and data-rates of 24Gb/s. However, with the recent introduction of optical amplification (erbium) the spacing has come back down to 40—50km again.

• Regenerative repeaters perform two main functions: pulse reshaping and retiming. The copper-based repeaters used for PCM consist of impedance-matching networks, amplifiers, threshold bias circuits, timing circuits, pulse regenerators and a power supply. In terrestrial systems power is generally supplied down the same copper as the data traffic, at voltages of 48V or 130V DC.

• The closer regenerators are to each other, the less the error-rate experienced. In practices, adequate error performance is achieved with E1 and T1 (2Mb/s and 1.54Mb/s) channel rates in suburban systems over 22-gauge twisted-pair with repeater spacings of 2km.

repeater (LANs)

These are the simplest of LAN interconnection devices. They regenerate digital signals which may be required to travel long distances down a very long LAN cable. Repeaters are also used as the nodal points for connecting different segments in a tree-branch structure. They are central to modern hubs and concentrators. See stackable hubs and pileable hubs.

A repeater works only at the physical level (Layer 1): it forwards data in bits (as against packets or messages). So it has no comprehension of the protocol it is handling, and no knowledge of the address.

A repeater just reads signals on one side and re-broadcasts them on the other without applying filtering or any intelligent processes. This is, in fact, a true regeneration function.

A repeater has no inbuilt intelligence other than that needed by any LAN node to get access to the network. It has no filtering capability (apart from removing noise), no addressing functions, and it is transparent to upper layer protocols. However it does re-time and boost the signals.

• Stackable hubs will have their own repeaters to move the packets between the LAN segments. When many units are stacked and joined, the repeaters in all but one of the hubs will be disabled, so the stack acts as a single logical unit. This does not apply to pileable hubs where all continue to function.

• The Ethernet standard prohibits more than four repeaters in series between network nodes: this is the limiting hop-count. This also means that star topology 10Base-T network are limited to four hubs (hops) between nodes. See half-repeaters also.

• Non-reclocking repeaters allow you to add nodes to a 10Base-T network without exceeding the hop-count limit.

repeater (optical fiber)

Repeaters on an optical fiber transmission system usually provide four functions:
- Detection of the incoming signals.
- Amplification by electrical or optical means.
- Regeneration of the signals for onward transmission. This pulse re-shaping, may or may not happen.
- Service and maintenance support.

We should distinguish between repeaters and regenerators (aka regenerative repeaters). In all early analog cable systems, the repeater performed the functions of signal detection, amplification, and transmission. The term was applied both to the physical housing and the electronics: the general use of the term included all necessary power supplies and redundant electronics.

Later, with digital cable systems, the term 'repeater' strictly only applied to the housing unit. The active electronics inside the housing recreate the pulses — which is a regeneration function (restoration of shape of the original signal pulses, not just amplification of the old pulses).

This is an important distinction to make since a repeater amplifies both signal and noise (pulse distortions), while a regenerator amplifies the signal only.

With the advent of erbium-doped optical amplification (see) the term repeater is now correctly used once more for both the housing and the amplification stages. However, here there is now no conversion to electrical signals and very little added noise.

• Undersea optical fiber repeaters and regenerators require power which is supplied through the entire length of the cable — usually from each end as a multi-1000V DC current.

• Repeaters are costly (about $1 million each for any undersea unit handling, say, 4–6 fibers) and their electronics are the major cause of cable failure. So every effort is made to increase the distance between repeaters and thereby reduce the number required. A lot of redundancy is also built in; usually the repeater will automatically switch to back-up detectors and lasers if a component fails. See submarine cable.

repeater spacing (optical fiber)

In undersea optical fiber cables, the distance between the repeaters is a major factor in the cost and reliability of the cable system. It also sets the power (DC) which must be supplied at each end of the cable.

The most recent fibers and lasers allow a repeater/regenerator spacing of 150km which is close to the theoretical maximum for silica glass fibers. However, halide glasses can theoretically be made transparent enough for 1000km spacings and, at these distances, repeaters would never need to be submerged and inaccessible. Electrical power would then not need to be provided along the cable.

Coherent optical systems also promise a major improvement in repeater spacings although, initially, the new optical amplifiers in submarine cables are being spaced only about every 40km. However, there's no reason why optical amplification can't stretch the spacings back to 100km. See optical amplification and erbium amplifiers also.

replication

Replication engines and servers are now being widely introduced as backup and as load-lighteners for corporate database servers. Replication is the capacity to copy parts of a database (or the whole database) from one network server to another on a regular rotational basis. It is a way of relieving load, and of providing security while maintaining synchronization between the server files.

• Replication can also exist from one database platform to another quite different platform. Some of these are single direction, and some a bidirectional. Replication agents are included in with the proprietary standard to allow two database- types to communicate.

report generation

The ability to filter material from a database, consolidate it in some pre-determined way, format it, and present it — usually on paper. At the simplest level, this is the ability to print a selected group of records.

Repro

In DTP this is a high-quality printout of the page with all elements in place; it is always done on very good paper. It is to be used for photographic reproduction, and printing.

request/response

A technique used in some communications systems to carry warnings and other information across a network. (see SNA's RU). These messages may be sent periodically to check that the communications path still exists.

required space/hyphen

See hard space and hard hyphen.

reseek

When a disk controller is accessing a large file, the data is often transferred into a buffer at a rate faster than it can handle — and the buffer can overflow. It is then necessary for the disk head to be instructed to wait a while, then 'reseek' the next batch of data.

Reseeks are therefore a measure of how inefficient the controller is in handling the disk system. It is also associated with the sector interleaving and channel encoding on disks. See RLL

reseller

General telephony: A pseudo-carrier. These companies don't own their own lines and cables, but buy service in bulk from a true telephone carrier and resell it to other companies.

This is called 'value adding' for some unknown reason; the only value that is usually added is to the reseller's pocket.

• **Cellular:** Airtime reselling is a widely used practice in the mobile phone business. The reseller acts (or appears to act) as an independent retailer, giving independent advice to customers, selling them handsets, and signing them up as customers on commission for one of the local carriers. In fact, he is bulk-buying telephone call-time from the carrier, and reselling it to his customers. If he goes broke, they've got problems — they may not even own their handsets.

There are all sorts of nefarious practices conducted under this guise. Handsets are often subsidized and heavily discounted, then bundled together with air-time contracts which cost more in the long run. You'd need an electron microscope to read some of the fine print in these contracts.

reset

A reset reloads the operating characteristics of a piece of equipment and wipes out all information in cache, RAM cards, buffers and registers which have not been saved to disk (or stored in true non-volatile memory). This is a cold boot.

With most modern computers, you need to switch off for at least 10 seconds to make sure that memory has been wiped clean, with a deliberate reset.

resilience

The ability to recover from system failure. Resilience depends to a degree on redundancy, but other factors such as the overall network architecture and error-recovery systems, also play a part.

resistor color code

There are four-band and five-band resistor codes.

• For a normal five-band sequence, the colors are:
First digit – second digit – third digit – multiplier — (then a space) — followed sometimes by the tolerance allowance.

• With a four-band resistor (those made with high tolerance, and therefore highly-variable) the third digit is dropped.

There are always tolerance and multiplier bands, and these are usually separated from the digit-multiplier bands.

The color codes are:

Color	Digits	Multiplier	Tolerance
Black	0	1	
Brown	1	10	1%
Red	2	100	2%
Orange	3	1k	
Yellow	4	10k	
Green	5	100k	
Blue	6	1 million	
Purple	7		
Gray	8		
White	9		
Gold		0.1	5%
Silver		0.01	10%
None		N/A	20%

resolution

The ability to separate two events, pixels, shapes, frequencies.

• **Sound:** The ability to hear high-frequencies. A ten year old can hear frequencies between about 20Hz and 20,000Hz. However the ability to detect these high frequencies decreases progressively with age, so few people in their 60s (or 40s for the next generation) can detect frequencies above 15,000Hz and many not much above 10,000Hz. AM radio has a reasonable response to about 8kHz, and FM radio to 15kHz.

• **Magnetic domains:** The ability of a hard-disk to record the maximum amount of information on a hard disk is a product of its ability to resolve the small magnetic cells (domains) when reading the information. This depends on the head gap, the head width, the granular size and thickness of the magnetic particles, the height the head flies above the disk, the disk speed under the head, the modulation system used, and the intelligence which can be applied to the output waveform during interpretation.

• **Analog TV** resolution is always given in terms of lines. This is confusing because vertical resolution is set (to a degree) by the number of scan-lines the standard required (PAL had 625 and NTSC had 525). To roughly approximate the lines of image resolution, the number of active SCAN lines is multiplied by 0.6 to give a figure for vertical resolution. Horizontal resolution is measured by the ability of the scanning beam to switch very rapidly between full-black and full-white.

This was also quoted as 'lines', but here they meant the number of black-and-while lines (like alternating pickets in a fence) which could be shown across the width of the screen before the rapidity of the alternation overloaded the scanning system, and everything blurred into an intermediate uniform gray.

This is directly related to the bandwidth of the system electronics — to the highest frequency that can be handled through the electronic chain up to the drivers of the scanning beam.

Added to this was the problem of the stability of the image between one line and the next (interlace wander) and a horizontal shifting of the lines (time-base variation).

With B&W TV sets the viewed image resolution depended on these factors, however, with color TV the problems are compounded by having three independent beams (with Red creating the most problems) and a perforated plate behind the screen, which creates a screen resolution limit superimposed on the image resolution limit. See image resolution.

- **Digital video and television** resolution is also often quoted in terms of the maximum number (horizontal rows x vertical columns) of pixels that can be seen on the monitor as a whole. But this can be misleading. CRT devices are always fundamentally, analog devices even when they are being driven by digital controls and viewed through perforated plates having an integer number of 'physical pixels'. The bandwidth of the analog circuitry sets the limit of resolution — no matter how good the digital electronics are.

- With modern digital techniques, both screen and printer resolution should be measured in terms of pixels, but often they are given as dots per inch. For some reason, no one ever seems to use the metric equivalents. With monitor screens, the true dpi resolution will change with the size of the screen and with the window dimensions of the image projected.

- Screen resolution and image resolution are not the same. A computer video standard may switch screen pixels in groups of four, or eight, or sixteen (commonly done in the older computers), so the screen pixels then had a 4, 8, or 16 relationship with the 'image pixels'.

Similarly, some analog television proposals for so-called high-definition television sets used a line-doubling technique to get the magic 1000+ scan lines. But the image resolution was not improved one iota by this. The screen quality improved (you couldn't see the scan lines), but not the image resolution.

- Resolution in video is not the same as in photography. For instance, four black lines separated by three white lines of equal thickness would be regarded as four lines of resolution in photography, but seven lines in video. For this reason, video resolution is often quoted in cycles (i.e. a black line followed by a white) which brings it into line with photography.

- The normal eye is able to resolve about 150 lines per inch or 6 lines per millimeter at a normal reading distance (0.3 meters or 12-inches). Children can often resolve slightly closer lines (250 to the inch or 10 to the mm).

At ten times the distance (3 meters or 10 feet — across a room) a person with excellent vision will be able to resolve screen elements spaced 25 to the inch, or a millimeter in size.

resonant absorption

Within the band of radio spectrum known as the millimetric wavelengths (30—300GHz), and at certain characteristic frequencies, the molecules of oxygen and water vapor in the atmosphere absorb radiowaves very strongly. This is resonant- or molecular-absorption. The molecules then re-radiate the signals isotropically a fraction of a second later which creates further problems.

Don't confuse this with the signal scatter from hail and rain; this is non-resonant.

resonant tunneling transistor

The underlying principle of this device is the quantum well, an ultra-thin layer of semiconductor sandwiched between two cladding barriers, 1nm thick, and of a higher 'bandgap' material which will confine electrons in the well. A practical version of this device would have a huge impact on communications; it would permit switching speeds at least an order of magnitude faster than the 0.1 picosecond fastest speed today.

resource

Anything that is available to be used — hardware, software, data, cables, radio links — anything. In context, the term is obviously narrower:

- In operating systems, resources are usually system files and device drivers.

- In general networking, resources equipment that attaches the network — printers, file-servers, databases, etc.

- In programming, resources are library elements, modules or objects available for use.

- In education, resources are books, videotapes, overhead lecture transparencies and computer programs.

resource arbitrator

A database of information about the workings of a resource. See Plug and Play.

resource fork

Most Macintosh files have two distinct 'forks' or segments: the resource fork and the data fork. Each fork contains separate components of the file. Data forks contain the data (this is the part transferred as an ASCII file), while resource forks contain elements such as menus and dialog boxes. These are kept separately from the data to allow the easy creation of foreign language versions of the software. You use ResEdit to edit resources directly.

response frame

In videotex systems, this is a frame into which the user types details of a request for information. It is a form of 'direct mail' which transmits requests to information providers. The term is used generally also.

resting frequency

The central frequency of the carrier when using FM techniques. An FM radio bandwidth consists of the center-frequency with two primary sidebands, each equal to the modulation bandwidth. FM sidebands occur at integer multiples of the modulation frequency on both sides of the resting frequency.

retrieval set

The process of finding relevant data in a very large bibliographic database is that of progressively refining the search terms (using keywords and Boolean operators). The first attempt will probably select a large number of records in a database. This retrieval set will include the required data, but further refinement of the selection is necessary to reduce the number of citations. At each stage the retrieval set is culled until the data is found.

return channel

• **Voice and data:** Often the four-wire circuits needed for long-distance carriage of voice or data (or for radio telephony) are referred to as the 'go' and 'return' channels. This is always from the viewpoint of the user.

• **Data-transfer:** The back channel during full- and half-duplex operations where the bulk of the data is being sent one way. It is used for acknowledgments, requests, commands, etc.

• **Broadband cable:** The low-speed (low-bandwidth) channel provided on what is fundamentally a one-way communications system. This is also called a 'reverse channel' (see).

The return channel connects the receiver/consumer back to the sender/seller of information. In practice the term is usually applied to the home-owner's messaging channel back to the cable head-end which is used for ordering Pay-per-view service on modern systems.

The term is synonymous with reverse-channel and back-channel in most cases. In coaxial cable television networks, a return channel is usually provided by making available the spectrum below 54MHz (usually from 4 to 48MHz) which is most subject to impulse noise and interference.

• **Cellular:** In mobile cellular communications, the reverse or return channel is the one from the mobile to the base station, while the forward channel is from the base to the mobile. The engineers naturally see it from the viewpoint of the operator.

return path

This can mean a range of things. It normally applies to an acknowledgement channel on a file-transfer system, and more recently it applies to video broadband cable TV systems.

Here the return path can be from the home to the head-end via normal telephone circuits (dial-up), or via the 5—40MHz of unused bandwidth at the bottom end of the cable spectrum. From 40—54MHz in analog cable is usually left vacant (as a guard band) and the channel numbers begin at 54MHz. Digital cable systems will use other splits.

reverberation

A room without reverberation produces flat and dull sound recordings, because all reflected sound is absorbed by soft furnishings. Ideally a certain amount of reverb is needed to add 'life' to sound recordings, so the 'hardness' and 'softness' of the wall surfaces are strictly controlled in a recording or television studio. Artificial 'reverb' can be added later by running the sound tracks through an electronic low-delay echo process (or sometime through a 'live room' with loudspeaker and microphone), but it never sounds quite right.

Reverb and echo are the same, except that the reverb echos occur very closely behind the principle sound wave and provide a richness to the sound, rather than a destructive interference.

The reverberation time (RT) is the time it takes for the sound to fade away (decay) through 60dB — at which point it is generally inaudible. Typical reverberation times at mid-acoustic frequencies are:

— Open air	Nearly zero.
— Average sitting room	0.5 sec.
— Radio studio	0.4 sec.
— Live theatre	1.0 sec.
— Concert Hall	1.5 to 2.5 secs.
— Large cathedral	10 to 15 secs.

reverse

• In screen or print terms, this means to reverse the tonal qualities. Black becomes white, and white becomes black. Highlighting does this on a screen.

• In movie post-production and printing (or electronic processing of video), this means the physical reversing (Left becomes Right) of an image, also known as flopping or reflecting an image. It sometimes needs to be done to camouflage problems in the shooting stages, or for special effects.

• In film production there are two slightly different uses of the term. A normal 'reverse' is a shot taken from a mirror-image angle (without crossing the main line of action).

So a shot of the President being interviewed has him facing almost towards camera, but looking slightly to (say) the left of the camera's view-point. When the shot cuts to that of the Interviewer, the shot matches in size and angle, except that he/she is looking slightly to the right. This is a reverse, and it is interpreted by the audience as the Interviewer and President talking to each other — without this one-facing-left and one-facing-right mirror image, they would appear to be both looking off-camera at some unknown third person.

The reverse should be about the same size as the original, and same angle from the eyeline. The cameraman's rule is 'always to stay on the same side of the eyeline'.

• Reverse is also used in action-line sequences. If the cowboy rides across the screen left-to-right when riding into town, he should ride back to the ranch right-to-left. Again the rule is 'always stay on the same side of the action-line' (when the action is real direction). See eyeline.

• In a modem, the modem will dial using the 'originate' mode, then reverse the configuration setting to receive data in 'answer' mode after the connection is made. This is done by adding an 'R' at the end of the AT dialing sequence.

reverse-bias photodiodes

These are very small, very expensive devices which have high electrical resistance at an internal junction, and are used to detect light signals in optical fiber receivers.

reverse channel

A channel in a communications system used for supervisory or error control signals. This is a back-channel or return channel used for ARQ or for supervisor data.

reverse channel capability

The ability to interact with a system or device over a communications channel.

reverse directory

A phone directory listed by number. These are usually on CD-ROM, and they allow you to look up a number and find the name and address of the customer. Reverse directories are illegal in some countries. A very fast search engine could, in theory, produce much the same information from a standard directory.

• Reverse directories are highly prized by companies involved in telemarketing and phone sales because the incoming Caller Line ID (CLID) can be used to trigger a quick search, and throw up the name, address, credit details, and buying- and payment record of the customer.

• There are obviously very substantial privacy implications with reverse directories, but despite prohibition in some countries, it is always possible to modify a conventional Yellow Page CD-ROM to work in reverse. If the information exists, the computer can extract it.

reverse engineering

Poking and prying to discover how hardware or software does what it does. A company will take a competitor's product (software or hardware) apart to see how it works — with the obvious intention of copying it in the closest possible way without infringing patents or copyright.

Some recent copyright moves in Europe have attempted to make this practice illegal. It is hard to see how this would be policed, or why.

reverse interrupt

A transmission control character sent by a receiving station to request the termination of communications. It is used when a higher-priority message needs to take control of the system.

Reverse Polish Notation

A way of presenting mathematical equations where the equation is always executed from left to right. The arguments (values) to subroutines are written before the word (operator): for instance, to add 1 and 1 in the Forth language, you would write 1 1 +. The value of this approach lies in the way a CPU uses its stack. Each item must be placed onto a stack in strict order, and RPN represents the best order for performing calculations.

The only syntax rule in Forth is that 'words' must be separated by at least one space. RPN is also know as prefix notation, and the distinction is with normal algebraic syntax, called infix notation.

Infix notation in Basic

```
X = (3 * Y) + 5
```

Prefix notation in Forth

```
Y @ 3 * 5 + X !
```

reverse video

Inverse video. See.

Rexx

A high level IBM procedural language which may become part of SAA if SAA survives.

RF

Radio Frequency. The part of the electro-magnetic spectrum below the infrared (below about 400GHz) and above audio frequencies. Sometimes this is subdivided into radio and microwave sections, with the division being at about 1GHz.

RF signals (better called 'carriers') are always analog, and they need to be selectively tuned to the appropriate channel (passed through a bandpass filter) then demodulated (tuned) to extract any analog or digital information they may be carrying.

The history of radio has been that of progressively pushing the usable radio spectrum higher and higher. Currently we use the VHF bands as a matter of course, but equipment for the SHF band is still expensive and scarce. A decade from now the SHF band will be commonplace, and we may be pushing higher into the EHF band. (But equally we may not — the range may be too short to be useful.)

• Video recorders and laserdisk players often have RF outputs. This is a single connector port which carries broadcast frequencies. Behind it is, in effect, a very small transmitter which is modulating image and sound into R/F form. This output can then be fed to a normal TV set at the antenna point, and selected and tuned by a conventional TV tuner.

RF-ID cards

Radio Frequency-ID cards are contactless smartcards used for identification purposes.

RFC

Request for Comment. A common term in networking, and now in general business. A proposal.

• **Internet:** A document in the Internet which carries the specifications of a proposed protocol, technique, etc. RFCs are also used for proposals other than just technical specifications.

Since the Internet doesn't have any formal structure, these are the working notes of the Internet R&D community. Anyone can read them, and anyone can have an input; RFC proposals can be circulated around the Internet community by anyone who wishes to propose a change.

Although there are no formal standard-origination processes on the Internet, some RFCs are designated by the IAC as Internet standards. All of these are written up (and still called 'RFCs'), and the historic store of these documents (begun in 1969) now prescribe the suite of protocols which constitute the Internet. They also describe related matters needed to maintain the open Internet environment.

Copies of RFCs are kept and made available through Internet Network Information Centres, and new proposals are widely distributed to mirrored sites around the world.

RFI

Radio Frequency Interference. Aka EMI. Certain standards for RFI are laid down by various national and international regulatory agencies — both for RFI generated by electrical/electronic equipment and by transmitter. However, some European interference standards have only ever been set for 'non-transmitters' — and there's always an argument as to whether interference is an emission problem, or an immunity problem. With certain radio technologies (specifically TDMA) radio interference outside the transmission bands is inevitable.

See CENELEC, CISPR, EMC and EMI also.

RFP

Request for Proposal. A common term for a document put out by cable companies, telephone companies, etc. seeking a viable solution to a problem — such as the design of cheap set-top digital boxes. They are part of the tendering process.

RFS

Remote File Sharing. A network support product from AT&T. RFS is a filing system featured in Unix System 5 Release 3.0. It was introduced to makes it easier to support Unix workstations on a network, however, this area is dominated by Sun's NFS.

RFT

Revisable Form Text. See DCA/RFT. Don't confuse this with the very similar, RTF (Rich Text Format).

RG-#

- RG-58: The thin Ethernet coaxial cable used with Cheapnet (10Base-2). It has a 50-ohm impedance.
- RG-68: The 93-ohm impedance coaxial cable used by ARCnet.

RGB monitor

A color monitor (computer and television) which has separate inputs for Red, Green and Blue and for other controls as well. The analog version of these monitors will usually have very high resolution. In the final output to the electron-guns the signals must be separated into the RGB form. Usually three connectors are required to carry these signals, plus another for sound.

However, there are differences between RGB monitors. The early CGA digital RGB monitors needed to be supplied with separate signals which were labeled: Red, Green, Blue, intensity, horizontal-drive and vertical drive.

RGB is only one way of communicating color information to a monitor. The distinction is with more complex composite monitors which take the mixed sync and video signals in a blended form (just like a TV set), which then need separation into the RGB signals within the monitor. There are also other color component systems like Y/C and YUV also. See CMY, HSL, video-DAC and delta also.

- RGB analog monitors will normally receive three color signals, each in a voltage variation form. These signals can take on any value between upper and lower voltage limits.

TTL (digital) RGB monitors can only take signals which have only a few discrete values, depending on the pixel-depth supplied for each.

RHC

Regional Holding Companies. An older term for the seven Regional BOCs (RBOCs) which dominate the US local telephone service.

The seven RBOCs are: Ameritech, Bell Atlantic, Bell South, NYNEX, Pacific Telesis, SouthWest Bell and US West.

- GTE could be included here as a RHC also — although it is not an RBOC. It was never part of AT&T.

RI

Refractive Index.

rib/RIB

- RenderMan Interface Bytestream. A file format for the Pixar 3-D graphics program that has become widely used in architectural and animation circles.
- A rib is a link between a departmental network and a campus backbone.

ribbon cable

The flat multi-wired cable used for parallel local computer connections. See Centronics.

Ricean channels

See Rayleigh channel.

RIDES

The Raynet optical fiber-to-the-home system which is gaining popularity in the UK and some parts of the US. It has multiple fibers and copper in a common sheath.

RIF

Routing Information Field. A field in the Token Ring packet header which holds ring numbers, bridge numbers and other information. The RIF is used in source-route bridging to select the segments through which a packet must pass.

RIFF

- **Raster Image File Format:** A format for paint-style color images. It is an expanded form of TIFF and is used by many scanners.

• **Resource Interactive File Format:** A Microsoft multimedia file format which holds audio, image and animation in a common format for cross-platform use. This could well be the same as the one above (last page).

RIG

Re-use Interoperability Group. RIG has a specification before the IEEE for a standard library of reusable software. It has a single entry point from which users may select codes for multiple libraries within the system.

RIG standards use modules which have been tested and confirmed, so they eliminate a lot of rewriting and debugging. See CASE and module.

rightsizing

See downsizing.

RIM

Remote Integrated Multiplexers. These are multiplexer or concentrator units which exist in street cabinets and link back to an exchange over fiber or coaxial. See RSM and IDLC.

ring

• The alert signal to a telephone subscriber. It doesn't matter that this may now be a buzz or a chirp, it is still called a ring. Usually the ring tone is about 17Hz.

• This is also the name given to one of the twisted-pair in a telephone cable. It is often designated R. It is the least positive of the pair. (In the USA it is usually at –48V DC, while the tip is at ground potential.) See tip.

• Ring network topologies are closed loops. Data is put on to these networks by the active station (the one holding the token, if tokens are used) and then passed sequentially around the ring between all stations. Any station can examine and copy the data provided it is addressed to them. Finally the data returns around the ring to the originator which then removes it from the network.

There are single rings and dual rings. Some dual rings are used for full-time carriage of data in each direction, while in other systems, the second only provides backup.

• Dual ring networks are seen as 'self-healing'; the alternate path can be used to carry data if the major link fails. Most single rings are actually rings of active repeaters, and these repeaters need to be by-passed in the event of failure. This will happen automatically in most cases if a token is not received by a station within a defined time limit. Individual stations link to these repeaters to make connection to the LAN. See MAU.

• Interconnecting rings create a mesh architecture.

• Probably the first successful LAN was the Cambridge Ring. It has since disappeared, but see Token Ring, DQDB and FDDI.

ring indicator

In the definition of RS-232 this is an interface signal which indicates to the attached terminal that an incoming call is present. The Big-8 (most essential) connections don't include the ring indicator, but it is often listed as the 9th most common.

ring levels

Intel's 286 and 386 use a segmented memory architecture, and access to memory is controlled by four levels of privilege called 'rings'. The OS/2 kernel and device drivers run at the highest level (Ring 0) while OS/2 applications run at Ring 3.

ringover

See hunting.

Ringer Equivalence Number

See REN.

RIP

• **Raster Image Processor:** A program which translates the encoded instructions for a page (in a page-description language, or a graphics-output language) into the actual screen or printer bit-map. Hence 'ripping' is the same as rasterizing in the laserprinting process.

• **Rest in Peace!** An error code in Windows.

• **Routing Information Protocol** (aka Router Interchange Protocol). This was the first workable router protocol, and it is still the most popular variation of the Internet's Interior Gateway Protocol (IGP) suite — as used also by Novell and Berkeley Unix systems.

On the Internet, RIP passes routing information between gateways and it allows the dynamic interchange of data between routers. It has a distance-vector algorithm which calculates the shortest distance between the source and sink. This is the key routing protocol in most older routers, and if you are using native IPX then you must use RIP also.

The network administration requirements for table updating, however, generates high traffic levels. A RIP router maintains a table of hop distances to other routers as reported by all neighboring devices, and on a Novell network these updates are broadcast every 30 seconds — whether or not the topology has changed. See SAP, OSPF and IGP.

RIPE

French acronym for the European organization that coordinates Internet activity.

There are three main backbones in Europe:

- Ebone: Run by a consortium of research and commercial interests. It now has 23 member companies.
- Europane: No information.
- EUnet: The oldest commercial service provider with an installed base of 10,000 accounts in 40 countries.

ripping

Rasterizing. See RIP and rasterizer.

ripple

Residual wavelike variations in the DC voltage after AC-to-DC conversion in a power supply.

RISC

Reduced Instruction Set Computers. IBM invented RISC; it was founded on the principle that, in real life, developers can never totally come to grips with complex systems, and so never utilize these systems to full advantage. It is better to keep them simple.

RISC is said to be an expression of the Pareto principle in that 80% of all results come from 20% of resources. The philosophy behind RISC development is to identify that 20%, and optimize the chips to performs these functions as rapidly as possible. They are then content to allow the other 80% to be less efficient.

RISC is generally considered to offer better overall system performance, than complex instruction set computers (CIRC). This is certainly true now that much of the CPU load is handled by special co-processors and device controllers. RISC has proved to be ideal for performance-intensive applications because it offers more instructions per clock cycle.

With RISC, the distinction is between both the standard form of complex computer architecture (CIRC), and also some combined CIRC/RISC chips. See Pentium.

• The most obvious characteristics of RISC are: a relatively small number of instructions, fixed length instructions, and many general-purpose registers.

• The two classes of RISC CPUs which have evolved are superscalar and superpipeline. See Superscalar architecture, Superpipelining and SPARC also.

riser

• In corporate networking, this is the part of the backbone which carries the data between successive floors in a high-rise block.

• In television, it is the low flat box on which the 'talent' sits. It is an artificially raised floor to bring the seated interview or news-reader up to comfortable camera height.

river

When type is set in columns, sometimes the accidental juxtapositioning of the words vertically will result in a series of white word-spaces that may flow down the page. These 'rivers' of white space seem to often happen when justified type is used in columns that are too narrow — especially when hyphenation isn't being used.

RJ# connectors

• **RJ-11:** This is the very small (1cm^3) 4-wire modular connector used in modern telephones and modems. It is the most common now in use.

When the jack is held in the hand as it would be for insertion ((with the notch up), the six connectors number from right to left (1–6) with 3 and 4 (ring and tip respectively) being the main contacts in the center. The ring (red – 3) should have a negative voltage compared with the tip (green – 4). Note different companies often use different cable colors. The RJ-11W is the wall socket for an RJ-11C (connector).

• **RJ-12** and **RJ-13:** Four conductors are used in these for key-service connections.

• **RJ-14**: This is used for two single-line connections. The phone will have a switch mechanism so that it can take a call on either line.

• **RJ-45:** This is the 8-wire connector used by 1Base-5 Star-LAN and recently for 10Base-T Ethernet.

RJE

Remote Job Entry. There's both a general and a specific use of the term.

- General. Using a computer from a distance over communications links. This is on-line RJE.
- Specific. The protocol used for submitting batch jobs to a remote computer and for retrieving the results. This is old IBM terminology. In RJE environments you will submit your 'job' to the computing facility, and receive the results later. It is a batch process, and so the distinction is with real-time.

• This was an old term for a method of submitting work to a mainframe through card-readers and on-line printer terminals that were usually linked by telephone; it was a batch system. Later it was superseded by the 3270 terminal/cluster system. The term then migrated through this stage to today's PCs which can communicate with the host by emulating that old 3270 RJE equipment.

Today, VM host systems can only be accessed by a PC submitting RJE in a non-interactive way, or by the emulation of a terminal (typically a 3270).

RLAN

• **Remote LAN.** The process of transparently extending the full capabilities of a local area network, to remote sites and mobile users. You can identify three types of access:

- Remote control. This allows the remote PC to dial into another PC and virtually take control of that system on the network. Applications run on the host PC, with only screen and keyboard updates passing over the telephone lines.
- Remote node. The remote PC dials in, and is allocated a unique address which effectively extends the LAN to the remote site. This is effectively a bridging process. Special network software is needed to deal with a 'bridged LAN segment' which is sometimes there, and sometimes not.
- Total integration. The third stage, still to come, which will cope with ISDN, packet and frame relay services and even wireless technologies.

• **Radio LAN:** See wireless LAN and LAWN.

RLE

Run Length Encoding. A rudimentary, but very widely used data compression technique. It has applications in fax, graphics, video and probably any file where repetitive patterns are to be found.

With high-contrast black and white images (fax) it works by storing only the distance (in pixels) to a change in image (from black to white, or white to black). If the whole of a scan line is white, for instance, it will dispense with the whole line by transmitting only a single numeric. If a scan line matches that previously transmitted, it will just send a 'repeat' code. See READ and Huffman codes also.

RLL

• **Run Length Limited** channel encoding scheme for hard disks invented by IBM and introduced in 1986. This is a technique of reducing the clocking overhead inherent in MFM recording techniques, and thereby increasing the capacity of hard disks and reducing access times (both by up to 50%). This can be done without replacing the mechanical parts or increasing the actual disk packing density.

You can upgrade an MFM disk drive by changing the controller, but RLL will work on some drives only — and it works better on specially configured RLL drives. Like all disk systems, it requires the addition of redundant bits to guarantee that not too many logical 1s or 0s fall in a row.

RLL is only possible because disk drive mechanisms are now more precise than they used to be. It packs 26 sectors on to each disk (as against 17 on ST-506).

The current most popular technique is known as RLL 2,7 because of the two pause-intervals before every seven bits. Timing accuracy is essential on these drives because the RLL controller writes seven bits of data at a stretch, uninterrupted by clocking information. See interleave ratio/factor, ST506, ESDI and MFM also.

• RLL is also a technique used in CD-ROM as a way of providing self-clocking. The number of 1s or 0s possible in sequence is limited so that transitions regularly occur between pits and lands. The minimum length for a pit or land is 3 bits, and the maximum is 11 bits.

• A later version called Advanced RLL (ARLL) flourished for a while.

• **Radio in the Local Loop:** See WLL and all the cellular technologies.

rlogin

A Unix terminal emulation program similar to Telnet.

RM

Reference Model. This probably refers to the ISO's 7-layer reference model. However there are a few reference models. See OSI reference model.

RMA

Random Multiple Access. This is a contention (Carrier Sense) multiple access (CSMA) system for satellites, where there is no master station in control. After a station transmits, it listens on the downlink to see if a collision has occurred. See Aloha.

RMON

Remote Monitoring Of Networks. This is a relatively new specification to extend the SNMP as a management tool. It is a management language standard which controls the functions of agents and probes in remote hubs, bridges and routers.

There are two versions:

— The Ethernet version is RFC 1271 (Nov 1991). It has nine object groups.

— The Token Ring version is RFC 1513 (Sep 1993) and is an extension to the Ethernet version. It has 13 object groups.

RMON acts in a pro-active way — it gathers statistics and attempts to identify and isolate potential problems before they arise. It is fundamentally a probe (containing an agent) and a MIB (a subset of SNMP variables) which provides a common platform from which to monitor network traffic on local or remote sites. The RMON-client specifications define how SNMP-based management consoles can communicate with the remote monitors and receive alarms and statistics from them.

Other MIBs only keep transient information or cumulative totals, while RMON will keep past records of events. Without the RMON's MIB statistics, there may not be enough information for the manager to diagnose a problem. See MIB also.

• RMON provides:

— Observation of all nodes from a central console.

— The collection of statistics on incoming and outgoing packets from those nodes. These are filtered by the RMON agent rather than by the client.

— Segment-level statistics include packets, octets, multicasts, broadcasts, collisions, and errors.

— A history of past events, analysis, event reports, and alarms.

— The ability to make configuration changes by moving users between segments.

— The first step towards automatic fault-recovery.

• Currently, many devices (like bridges), can't support RMON. New RMON agents are being introduced by vendors.

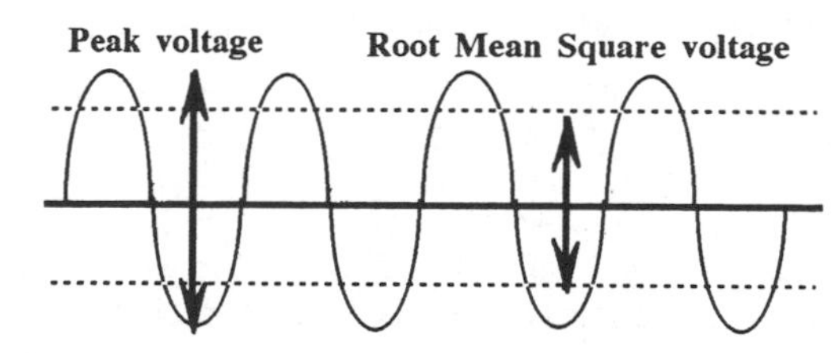

RMS

Root Mean Square. The nominal value of alternating (and varying) current in terms of DC resistance equivalents. It is a calculation which takes into account the fact that alternating current has constantly rising and falling voltages and, therefore, the current flowing through any resistive medium will be constantly varying between low and high EMF values.

RMS takes this into account by relating the total current flows to a DC equivalent. In effect it says that this varying waveform will cause the same power-dissipation in a resistor, as a non-varying DC voltage of X- volts.

The RMS for a sine wave is 0.707. To find the DC equivalent, you would multiply the peak voltage of the sine wave (AC) by 0.707. So a normal main's AC current with a peak voltage of 325 volts would have a RMS value of 230 volts.

• However the conversion value obviously depends on the wave-form. The RMS of a square wave is 1 (peak and RMS coincide), and the RMS of a saw-tooth wave is about 0.5.

See mains peak voltage.

roaming

In cellular telephony and paging this means to use a system outside your home area. The significance of this depends on the way the call is billed, and on the way your home carrier and the remote carrier exchange location and paging information — and conduct authentication and call diversion.

There are two factors in roaming:

— The companies involved in the authentication and billing processes will have some agreement on prices, and on the exchange of authentication information. In some cases the roamer will need to be included on a special register to be recognized by the alien system, while in most future systems this will be done automatically over messaging networks.

— The equipment must obviously be compatible. They may be compatible for voice, but not for data or for short messaging. Roaming is common with AMPS and NMT in analog, and in GSM and NADC in digital.

• In the USA with dual-mode AMPS/TDMA or AMPS/CDMA handsets, the AMPs network provides a common-denominator across most of the country. Therefore claims are now being made that you can 'roam' from a specific TDMA (NADC) territory to one where TDMA is not on offer by using the analog AMPS mode. This is a distortion of the term, but I guess it is as much 'roaming' as some of the other far-fetched claims; provided the billing agreements are in place for the analog system.

• Pseudo roaming is where the company agreements exist, but there's no equipment compatibility at any level. So, for instance, some American cellular companies have an arrangement which allows their customers to have a SIM card and use rented GSM equipment while in Europe. To claim that this is 'roaming' is stretching the term beyond reasonable limits.

• In analog cellular phone systems having both A and B bands (or more with GSM) licensed to different carriers, the term can also be used to mean inter-carrier roaming — being permitted to use the other band when your own channels aren't available. Some national regulators insist on this, and others don't. Some carriers allow it out of the home territory, but not within.

There are other issues when roaming in a country where both A&B are on offer, and both are provided by carriers having no connection with the home carrier.

robbed-bit signaling

In the transmission of digitized voice, reasonably high error-rates can often be tolerated, so bits are often stolen from the 8-bit quantization samples to be used for end-to-end signaling.

In the US T1 carrier, for instance, the least significant bit in each voice channel is removed from the 6th and 12th frames and 'in-band' signaling information is substituted. You can't do this when the channels are being used for digital data. See channel associated signaling, and A&B bit signaling.

robotics

A discipline that concentrates on mechanical and electronic engineering, but now includes high-levels of artificial intelligence and aspects of neural processing. Despite the common use of the term 'robot', the discipline of robotics suggests some sensory aspects (not just mindless action) to the machines, and some high-levels of machine intelligence. The ability to learn from mistakes would be nice!

See automation and numerical control.

Rock Ridge

CD-ROMs using the Unix file system are often created using system-specific extensions wrapped around an ISO-9660 core. This is known as the Rock Ridge protocol, and it is a superset of ISO-9660. This approach allows Unix systems to treat the CD-ROM as a Posix file system with its long-filenames, permissions, and symbolic links.

roll-back

In transaction processing, if something happens to damage the data in the process of changing it, the system needs to be able to roll-back the transaction and re-establish the data as it was before the transaction began. See ACID test and roll-forward. This is sometimes called back-out.

roll-forward

A database management system, in the process of recovering from a major problem, should be able to read the transaction log (aka the journal) and re-execute the transactions. See roll-back and ACID test.

ROM

Read-Only Memory. This is actually a generic term for:

— **MROM.** Masked ROM. This refers to the photograph negative which masks part of the material during manufacture, and allows the formation of the various interconnections. It is very expensive to set up and manufacture the first chip using this process — but very cheap when you make a few hundred-thousand. If the type of ROM isn't specified, they will mean MROM (however the M is rarely used).

— **PROM.** One-time ROMS which are programmed later by the user, rather than during fabrication. You fuse links on an off-the-shelf unprogrammed chip to add the program. Later you throw them away if they need changing.

— **EPROM.** Reusable PROMs, but they need to be taken out of the machine, exposed to UV light, then reprogrammed.

— **EEPROM.** Reusable, and reprogrammable within the device. They are reset electrically.

Flash memory should probably be included here also, and possibly FRAM — but these are more RAM-like.

• These chips in the ROM family all serve to hold the 'permanent' (non-volatile) memory within a computer. They hold their information without the need for constant power — the information is said to be 'burned in' at the time the chip was manufactured (or later in PROM, EPROM and EEPROM).

The generic ROM family is used to:
— store the basic parts of the operating system,
— store high level languages, and
— sometimes, store applications that are constantly in use,
— store customized parameters.

ROM tends to be slow to access, so often the information in ROM chips will be copied into the faster DRAM during boot up: this is called 'mirroring in RAM'. See memory map, RAM, . PROM, EPROM, EEPROM, non-volatile.

ROM cartridge

Plug-in ROM modules are often used as an alternative to disk-based application on a disk drive. This was a very common way to program home video-game computers and it increasingly has applications in portable computing now that the PCMCIA card bus can treat the card as part of the main memory map.

Roman

Typefaces: The term has lost most of its original meaning. It now means an upright typeface, as distinct from italics or oblique. In this usage, it can be serif or sans-serif.

• Sometimes can mean 'Times Roman' which is a specific typeface, very like the one used in this book.

root

The single highest point of access or control in any hierarchical system.

• The first or main directory. All directories other than the root should be called sub-directories.

• In a security system, the person in charge of maintaining the security procedures needs to have 'root access', so as to be able to fix problems when they happen: this provides them with the ultimate control denied to normal users. Hackers often seek out this root access, and if they break in via this 'back door', they can modify the security system at will.

ROSA

RACE Open Services Architecture. See RACE.

ROSE

Remote Operations Service Element. This is a CASE protocol in the OSI Application layer which helps establish peer-to-peer communications links and ensures the reliable transfer of the data from user-to-user in a distributed (open) system. The older (and better) term is 'Remote Procedure Calls' (RPC). See CASE, SASE, ACSE and CCR also.

rotary

• Rotary Dialing: See pulse dialing.

• Rotary Hunt or Call Diversion: See hunting.

rotational delay

One of the main reasons for slow access time (reading) on a disk. If two sequential blocks of information lie in adjacent sectors on a disk, the drive controller might not be able to react quickly enough after the first block is read, to read the second block. It takes a fraction of a second to process the first block, before it can begin reading the second.

If sequential blocks are in adjacent sectors on the disk, there will be a delay, equivalent to a full rotation, before the next segment can be read. For this reason sequential blocks of information are usually stored with some separation. See RLL.

round robin

A variation on network contention and polling systems, where each station in turn is given an opportunity to transmit in strict order. This is a very basic media-access control protocol.

route extension

A path from the destination node, through any peripheral equipment, back to the NAU.

routed protocols

Distinguish between routed protocols, and routing protocols. Routed protocols are those which can be routed by a router. To do this, the router must share the perception of the internetwork's logical connections with the protocol that is being routed. DECnet, AppleTalk and IP are all routed protocols. See routing protocols.

routers

Routers are special-purpose 'packet-switching' computers (usually specially built dedicated boxes) used in networks wherever there is more than one path for the data packets between network addresses. The router can choose the best path (by looking up a network address in a router table) and send its messages directly to that network destination (unlike bridges which function on individual physical MAC addresses). Routers send packets to a network, not to an individual station. See NSAP address.

Routers also talk to other routers and exchange information between themselves. This allows them to know which links are active and which nodes are available. See network address.

The router's natural habitat is complex peer-to-peer mesh-topology networks with multiple paths to any location. They can provide a built-in tolerance to WAN failures.

Routers are particularly useful for complex internetworks where bridged data can sometimes loop endlessly through different pathways: in large networks, the range of options available for routing can become extensive.

Routers act as bridges between LANs of the same type, and in many ways they are smarter than bridges (although they don't have the ability to self-learn). However routers provide much better traffic isolation than bridges.

Where segmentation and the division of the network into individual management domains is important, routing is the preferred technology. Where the segments are using different protocols, 'multi-protocol' routing is essential.

However router-based networks often have little in the way of bandwidth management capability.

Routers function at the Network layer (Layer 3) which is one level above bridges. Unlike bridges, they don't filter the information since they receive WAN traffic only, and they forward their data in message form (as against bits, or frames) to a global network address (not a MAC address). They are available in remote and local configurations, but they are slower than bridges (by about 10 times), and require substantial network management.

• Most early routers could only handle one type of network protocol such as DECnet, OSI or TCP/IP. But now we have multi-protocol routers which can deal with heterogenous environments. Some can link between Ethernet and Token-Ring, X.25 and SNA.

• Brouters (see) are a specific form of intelligent multi-protocol device which is both bridge and router. See transport relay, repeater, bridge, brouter, gateways also.

• The new high-speed Switched Ethernet and Switched Token Ring units function at the level of routers, so they can safely be called multiport routers. With these devices each workstation can have its own dedicated port of entry.

• Routers have traditionally been stand-alone devices, but increasingly they are being made as low-cost plug-in cards for servers. They then offer the bridge or routing function in the same device that is providing other services.

routine

Routine is usually synonymous with algorithm.

routing

The selection of a path through the inter-network for a packet of information. This is done in a way which consumes the least amount of the network resources possible, in the most reliable way, and with the least delays.

There are a number of different routing strategies:

— **Fixed routing.** The simplest method; usually the decision is made on the basis of a 'least-cost' algorithm. See.

— **Flooding.** Packets are sent to every neighboring node, and they then send to their neighboring nodes. This requires no network information to be kept, and all parts of the network will receive the data (important for routing table updates).

— **Random routing.** The outbound path is freely chosen from the available ports (not including the entry port), but sometimes a probability calculation is used to enhance the chance of this being the least-cost route.

— **Adaptive routing.** These react to network conditions. See adaptive routing.

routing bridge

This isn't a contradiction in terms. This is a MAC-layer bridge which uses Network layer (rather than Data-link) methods to determine a network's topology. See Transport relay also.

routing protocols

Distinguish between routing protocols, and routed protocols. Routing protocols are those network protocols responsible for distributing routing information throughout the network. They will often be tied to the use of other network protocols. The most common are:

— Novell's RIP (see) which is tied to IP.

— Open Shortest Path First (OSFP) used widely on the Internet (as is RIP also) with TCP/IP.

— Routing Table Maintenance Protocol (RTMP) which is used with AppleTalk.

Different approaches are used in these different protocols. For instance, RIP uses relatively simple distance-vectors to determine the best path. It does this by just counting the number of hops. OSPF is more hierarchical and complex, and it associates a cost with each segment; it then chooses the cheapest path. See IGP also.

routing table

Routers store tables which tell them which are the best routes to use when directing packets to a particular network destination. These routing tables often have other information (mainly load and latency metrics) about these routes.

Routing updates are distributed by routers to provide reachability and cost information. Depending on the protocol, some updates are sent every 10-15 seconds, and others less regularly. After a change in topology, tables are always updated. See poison also.

Royal fonts

The original name of Apple/Microsoft's TrueType outline fonts. See PostScript and PCL also.

RPC

• **Radio Paging Code:** See POCSAG which is RPC#1.

• **Remote Procedure Call:** RPCs are an essential part of client-server operations. They are needed to place 'calls' across a network (analogous to local processor-calls in PC programming), and also to retrieve the results.

They provide ways of storing and executing subroutines on remote computers, and are a means of invoking and executing requests between networked computers. RPCs are built or specified by clients, and executed on servers.

RPCs work at the Session level. You use them to build applications which can make use of individual procedures running on computers across a heterogenous network. They will let different kinds of computers process different parts of the same application, and they pass parameters and controls from one 'module' to another in a way that is completely transparent to the programmer.

RPCs are therefore very important in providing basic communications in distributed computing environments. An application will call a procedure across the network, provide information, pass a parameter list, and receive a parameter list in return. In OSI the same approach is called ROSE (see).

RPG

Report Program Generator. An early IBM business programming language.

• RPG II was a later more powerful version introduced in 1970.

• RPG III was a high-level non-procedural language

RPE-LPC

Regular Pulse Excited, Linear Predictive Codec. A family of voice coders which use long term prediction (LTP). This is the type of vocoder used in GSM mobile phones.

There are four distinct functions in the vocoder:

– Pre-processing.

– Short-term prediction analysis filtering.

– Long-term prediction analysis filtering.

– Regular pulse excited computation.

Broadly, they combine to code the conditions of the voice-tract as the speech occurs, by stripping out vocal information. To do this they use a process which is dependent on intelligent (computerized) filters. Since there is a limited range of change possibilities, future changes can be predicted from the past.

The vocoder then transmits the various filter settings and some indication of the voice frequencies. In the receiver (which is in the GSM switch) an inverted set of filters then reconstructs the speech.

Vocoders like this only work for speech, which is why modem tones can't be carried over most digital cellular networks. See RELP, linear predictive coding and source code.

• In GSM the coder input rate is 8000 samples of 13-bit PCM but this is then compressed 8:1. This makes the 13kb/s basic voice rate, which is segmented into 260 bit packages. When error correction is added, the data rate increases to 22.8kb/s and then with synchronization and guard bands, the voice is finally transmitted at a rate of 33.9kb/s. This is slightly over four times the rate you will often see quoted.

RPN

Reverse Polish Notation. See.

RPROM

Rewritable Partial ROM. See OROM.

RS

• ASCII No.5 — Record Separator (Decimal 30, Hex $1E, Control-^). An information separator also known as IS2 which usually delimits a record in a database file.

The four information separators (FS, GS, RS, US) are used in a variety of ways, but they should always be used in a hierarchical or nested manner with the FS being the most inclusive, and the US the least inclusive.

• **Reed-Solomon** codes. See CD-ROM also.

• **Recommended Standard:** The old EIA term. See RS-232, RS 449 etc.

• Record Separator.

RS-170

The current standard for NTSC monochrome and color (version A) video.

RS-232

The *de facto* standard for serial connection in PCs and peripherals now more correctly known as EIA-232 (but no one uses this).

The RS-232A standard was first finalized by the EIA in 1962, and later modified; the D-version is the current version, although C is by far the most common. It is the most widely used (but not well standardized) short-distance serial communications standard for computers, and the specification includes physical (plug shape), electrical (voltages, impedance), and functional (specific circuits) specifications. RS-232 only defines the local interface between DTEs (PCs mainly) and DCEs (modems mainly), however. Higher level computer protocols are needed for end-to-end signaling over phone lines.

There are hundreds of minor variations of RS-232C. The ITU also has equivalents, which are V.28 for the electrical standards and V.24 for the circuits, together with the DB-25 plug.

• **Data.** Data is transmitted in an 'unbalanced way' (all are referenced to a common signal ground) and 'upside-down' — a positive voltage (between +3 and +15V) represents 0, while a negative voltage (at –3 to –15V) represents 1. The idle state corresponds to a negative voltage on the line. The data is also transmitted backwards with the least-significant bit in each byte being transmitted first. This makes the parity bit the last to be transmitted (and therefore in the 'most-significant' place).

— Control. Apart from the two data-carrying lines (Tx and Rx) there are five other major control circuits (and many more minor ones) which are 'asserted' by a raised negative voltage to grant permission or to advise status.

• The standard defines 14 interface types requiring anywhere from 7 to 19 wires. However it is possible for a connection to operate with only 3 wires. Most have 8 wires.

• See the reference RS-232 Big-8 for pin assignments (below).

• The RS-232 standard is in four parts:

– Definition of the plug, sockets and allocation of pins.

– The electrical characteristics of the signals.

– The functional description of the signals.

– A list of subsets of the interface signals for specific applications (this is to take into account the variability).

• The main difference between RS-232C and D is that the D-version provides a test mode and loopback signals. There were also some semantic changes to the terms used.

• The EIA is planning a new standard to replace RS-232. See 562 in numbers section.

• For PC-to-PC (DTE-to-DTE) links, a sex conversion plug/cable and null modem must be used. See gender bender.

• The *de facto* standard for plug/sockets is the DB-25, but the smaller DB-9 is also widely used (see both for standard pin assignments). Recently the 8-wire RJ-45 modular phone connector has also been used, but there's no agreement among vendors as to how this should be wired up, so beware.

• Note that the Mac does not have an RS-232 port, it relies instead on a subset of the EIA/RS-422 standard which is almost identical. See RS-422.

• To conform to the EIA standard at maximum speed, no more than 20 meters of cable should be used with RS-232. In practice you can transmit at about 9.2kb/s over about 80 meters of cable, 2.4kb/s over 250m, and only at 300—600b/s over 1000m.

RS-232 Big 8

The Big 8 are the eight most important connections in RS-232 cables. These are from the viewpoint of the DTE (PC):

Pin 2	Tx	Transmit.
Pin 3	Rx	Receive.
Pin 4	RTS	Request To Send (DTE wishes to transmit or is ready to receive).
Pin 5	CTS	Clear To Send (response to RTS, or the DCE is ready to send).
Pin 6	DSR	Data-Set (modem) Ready (the modem has power and is ready to operate).
Pin 7		Common.
Pin 8	DCD	Data Carrier Detect (the modem tells the terminal it has received a valid carrier signal).
Pin 20	DTR	Data Terminal Ready (the terminal is ready).

Any reasonable cable or connection should have all of these links between the DTE connector at one end and the DCE connector at the other.

RS-422

The EIA's RS-422A specification is a balanced, high-speed multiple connector standard which provides a serial interface for computer and peripheral connections. It was designed to handle data at rates up to 10Mb/s over 12 meters, or 10kb/s over 1200 meters using NRZ-L line coding. It uses two wires for each signal (one wire of the pair is driven positive, the other negative), and unlike the RS-232C specification, RS-422 only deals with the electrical signal standards (not mechanical).

It is classed as a balanced implementation of RS-449. This is the standard serial port on the Macintosh and it is used in a way which makes it RS-232 compatible. See HSSI and V.35 also.

• One reference says RS-422 will handle rates to 1Mb/s only.

RS-423

The EIA's RS-423A specification is an unbalanced version of RS-422 which makes it more RS-232C-compatible. Only electrical standards are specified. It uses NRZ-L line coding with +2 to +4 volts being a binary 0, and –4 to –6 volts being a binary 1.

It only runs at one-tenth the speed as RS-422 (only at 100kb/s) over the same length cables and so this standard has generally lost popularity.

RS-423 uses a pair of wires for each signal. At the transmit end, only one wire is driven and the other is grounded ('single-ended' drive). At the receive end, a differential receiver is used for each wire pair. This reduces the noise immunity, since any noise which appears on both wires of the pair is cancelled out. This is the standard for DEC.

RS-449

This was originally intended to supplant the RS-232 standard and support data rates up to 10Mb/s. It was introduced as a 2Mb/s version of RS-232 which could handle longer cable runs.

RS-449 applied balanced RS-422 signaling for the 2—10Mb/s rates, and unbalanced RS-423 signaling for lower rates of 100kb/s (this lost popularity very quickly).

Two different connectors are given pin assignments as part of the standard — the DB-37 and the DB-9 (never used). So, in practice, this standard now means RS-422 signaling and the DB-37 connector. However to reduce costs even further, many vendors used the DB-25 connector and a subset of the RS-422 signals. This was later legitimized as RS-530 and RS-422A.

• RS-449 is now the physical and electrical interface underlying RS-422 and RS-423. The functionality of all three standards is very similar, although there are some subtle differences. From an everyday practical point of view, they are pretty much functionally equivalent.

Conversion between the three is reasonably straightforward in most cases. Sometimes the conversion will require a 'funny' cable however, while others will need line drivers.

• RS-449 specifications include mechanical, electrical, functional and procedural aspects. Differential (balanced) signaling makes RS-449 able to operate at higher data rates and over longer distances than RS-232. However, more pins are needed in the connectors because some signals are differentials and need pins in pairs, and there are more signals. The new signal names and mnemonics better describe their functions.

— **Mechanical.** A full implementation needs two cable connectors, one with 37 pins and one with 9 pins (see RS-449 connections). The small optional connector carries the secondary channel signals.

— **Electrical.** Two types of electrical signals are used. Some are single-ended, and follow RS-423A (similar to ITU V.10/X.27), and some are differential, following RS-422A (similar to ITU V.11/X.26).

— **Functional.** RS-449 defines extra signals that can be used for exerting better control of the DCE through the DTE.

— **Procedural.** These are reasonably apparent from their names.

RS-449 connections

RS-449 makes use of a 37 pin connector, but the size of this connector proved to be too great for the backplane of many computers and other devices, so EIA-530 was introduced in 1987. It was intended for lower rates (to 2Mb/s) using the old RS-232 standard DB-25 plug.

Currently a full implementation or RS-449 needs two connectors, one with 37-pins and an optional one with 9-pins. The 9-pin connector carries the secondary channel signals. This part of the specification is similar to ISO4902.

Below is a comparison between RS-449 and RS-232. Those signals with A/B indication are differential pairs:

- The 37-pin Primary Connector:

Pin	RS-449 signal	RS-232 equivalent
1	Shield	Protective Ground
2	Signal Rate Indicator	Data Signal Rate Selector in DCE
3	------nil-------	
4	Send Data A	Tx Data
5	Send Timing A	Tx Signal Element Timing in DCE
6	Receive Data A	Rx Data
7	Request To Send A	Request To Send
8	Receive Timing A	Rx Signal Element Timing in DCE
9	Clear To Send A	Clear To Send
10	Local Loopback	Local Loopback
11	Data Mode A	Data Set Ready
12	Terminal Ready A	Data Terminal Ready
13	Receiver Ready A	Rx Line Signal Detector
14	Remote Loopback	Remote Loopback
15	Incoming Call	Ring Indicator
16	Signal Rate Selector	Data Signal Rate Selector DTE
17	Terminal Timing A	Tx Signal Element Timing DTE
18	Test Mode	Test Indicator
19	Signal Ground	Signal Ground / Common Return
20	Receive Common	
21	---------nil-------------	
22	Send Data B	
23	Send Timing B	
24	Receive Data B	
25	Request To Send B	
26	Receive Timing B	
27	Clear To Send B	
28	Terminal In Service	
29	Data Mode B	
30	Terminal Ready B	
31	Receiver Ready B	
32	Select Standby	
33	Signal Quality	Signal Quality Detector
34	New Signal	
35	Terminal Timing B	
36	Standby Indicator	
37	Send Common	

- The 9-pin Optional Connector:

Pin	RS-449 signal	RS-232 equivalent
1	Shield	Shield
2	Secondary Rx Ready	Secondary Rx Line Sig. Detect
3	Secondary Tx Data	Secondary Tx Data
4	Secondary Rx Data	Secondary Receive Data
5	Signal Ground	Signal Ground / Common Return
6	Receive Common	
7	Secondary RTS	Secondary RTS
8	Secondary CTS	Secondary CTS
9	Send Common	

RS-530

This is probably better known as EIA-530. It is a cut-down version of RS-449. However, most EIA/RS-530 circuits use RS-422A. Those signals abandoned are all five secondary circuits, Ring Indicator, Signal Quality Detector and Data Signal Rate Detector. See both RS-449 and RS-422.

RS-RB

Remote Source-Route Bridging. This is source-route bridging over WANs.

RSA

- **Rivest, Shamir and Adleman:** The three people who devised the first popular public key encryption algorithm. The algorithm uses 'exponentiation modulo of large integer N' numbers. The main advantages of RSA (which is an asymmetrical/public key systems) over the older DES (which is a symmetrical private key system) are that with RSA:
  - You can deal securely with someone on first contact; without previously establishing a relationship and exchanging encryption keys.
  - There are no longer a problem of passing secret keys between the parties over unsecured systems.
  - The keys can be used to authenticate the service provider and the directory services also.
  - The system as a whole provides incontrovertible proof of the origin and authenticity of the message.

  The disadvantages are that the encryption and decryption process is very slow even with modern PCs, and the security of the message is only as good as the security of the third-party holding the public key. Eventually this will probably need to be a highly-regulated national directory service with very high security and privacy safeguards built-in. See DES, PGP, digital signature and public key encryption.
- **Rural Service Areas** in the USA cellular phone business.

RSI

Repetitive Strain Injury. A general name for a range of problems related to the fine control of muscle used rapidly on a continuous basis. Medically, RSI is difficult to establish and has often been treated as malingering or a sign of low-level hysteria or neurosis. But that's only because doctors don't often suffer from it themselves, and can't see external evidence of damage. Infrared imaging has recently been able to demonstrate abnormal functions in muscles with RSI. See Carpal Tunnel Syndrome.

RSM

Remote Subscriber Multiplexer. To reduce the capital and maintenance costs of small remote exchanges, it is common these days for a carrier to replace old analog exchanges with small stand-alone RSMs. These are cabinet-sized boxes which multiplex the local access network onto a few optical fibers (sometimes coaxial or even old twisted pair) and transfer the signals back to a larger digital exchange. See RIM and concentrators also.

RSN

'Real Soon Now'. See vaporware.

RSRB

Remote Source Route Bridging. Source-route bridging over WANS. See Source-route.

RSTV

Radiated Subscription Television. Subscription Pay TV, but not by cable or satellite. In the USA these were normally UHF channels which were able to provide a Pay service through renting decoder boxes. The term could also be applied to MDS and MMDS 'Wireless Cable', but generally it is reserved for UHF systems.

RSVP

ReSource reserVation Protocol. This is a draft Internet protocol to allow devices to reserve resources end-to-end, before they begin to communicate.

RT

- Radio-telephony.
- Remote Terminal: In telecommunications, an RT (remote terminal) is at the subscriber's end and an LT (local terminal) is at the switch end.

RTDB

Real-Time DataBase. The term is applied to databases used to control high-speed manufacturing and production equipment. Traditional disk-based database management systems aren't fast enough to capture this real-time data.

RTF

Rich Text Format. A widely used intermediate formatted file (a markup language) which allows you to transfer formatted documents from one platform to another, across software and hardware boundaries.

For instance, an IBM Wordperfect 4.2 file can be translated to RFT format, and then this file is sent by modem to a Macintosh, where it can be translated to Microsoft Word 4.0 format — with most of the major formatting commands still active. Most word processors can now create these files as an option.

These interchange files have an extensive preamble which provides details of the fonts used, colors, paper width, margins, etc. plus embedded (ASCII) commands within the document. This extends the file size by about 20% on average, although it can be much more.

RTF only uses the standard ASCII printable character set, and controls formats with identifiable strings such as <Bold> and \par. It can therefore be transferred over 7-bit communications networks in basic ASCII form.

Don't confuse it with IBM's RFT (see DCA/RFT). See Acrobat, PDF, SGML and HTML also.

rtfm

Read The Frigging Manual. Internet shorthand.

rtl

Return Loss. A code number, usually on an antenna or coaxial connector or cable splitter, which gives you and indication of the expected signal strength losses across the device. It is measured in dB, and larger numbers are better (20dB is good, 10dB is not so good).

These losses are due to impedance mismatch and are frequency sensitive, and they are across any connector.

RTMP

Routing Table Maintenance Protocol. The AppleTalk version of RIP. It updates tables every 10—15 seconds which imposes considerable load on the network.

RTS

Request To Send (Not 'Ready to Send') in RS-232. This is a raised voltage generated by the terminal (DTE) and it tells the modem (DCE) that it wishes to transmit and is able to receive. This is the Pin 4 connection on the DTE requesting clearance to transmit data. The signal is an output (a demand) from the DTE which asserts RTS as a voltage (–12V) on the connector. The response should be a CTS from the modem.

RTS is used primarily as flow-control, while DTR (which does much the same job) is the initial handshake control which establishes that the terminal is ready and functioning. But since both DTR and RTS are originated by the DTE, sometimes these two are bridged so that one can perform both functions. This is not a wise practice, but it is often done. The control of buffers is then performed end to end by XON–XOFF.

In common use with most modems, the PC assert RTS to the modem when it has free buffer space available and is ready to receive data. It will inhibit it when the buffer is almost full.

Programmers should be aware that a few characters may still be received after they drop the RTS voltage, so they must not wait until they are near the end of the internal buffers. Be aware also, that some modems react strangely to a low RTS; some will refuse to answer a ringing line even though DTR is asserted.

- In ITU terminology RTS is Circuit 105: in EIA terminology it is CA. See CTS.

RTT

Round Trip Time. This includes transmission of packets to and from an end point and any processing at the destination. It is used by some routers to calculate the optimal routes.

RTTY

RadioTeleTYpe. A once-popular form of radio transmission used by government agencies, companies, news services and radio amateurs. Originally Morse code was used, but later the 5-bit Baudot code of Telex became the standard. Now 7-bit ASCII codes are often used. The Baudot data-rate is usually either 45 or 50 baud.

RTV

Real Time Video. A data compression technique developed by Intel as part of Indeo.

RTVO

Receive TV Only. The domestic dish-reflector antenna system used for satellite direct broadcast television reception. This term is usually applied to the older C-band types, which were from 2 –3m across. See LNA and low-noise converters also.

RU

Request/Response Unit. In SNA different applications use different RU sizes. This is all part of the data flow control.

Small RUs are typically used in terminal emulation (3270) and larger ones are used for file transfers. RUs are exchanged between NAUs, and they come in many sizes: 256-, 512-, 1024- and 2048-Bytes mainly.

There are four modes of RU in SNA:
— Immediate-request. The LU can send only one request chain at a time. A response must be received before another is sent.
— Delayed-request. The LU can send many chains before receiving a response.
— Immediate-response. The request stipulates that responses must be returned in the order of the units received.
— Delayed-response. The responses can be in any order.

See chaining.

RUB

Router Hub. A device which combines both capabilities.

rubber bandwidth

A widely used slang term for bandwidth-on-demand. It can mean either, or both, of two concepts:
— Dial-up ISDN links which can be used to augment the private tie-lines, when required.
— Statistical multiplexing. See.

rubout

Slang for the backspace character, which is Control-H.

rule-based systems

• **AI:** An expert system consisting of antecedent-consequent rules. The executive consults the rules to find out how to modify the database whenever new patterns are recognized.

The three components of these systems are:
— The database which holds the information.
— The executive which performs pattern matching, and decides which rule to execute.
— The set of antecedent-consequence rules which guide the changes.

• **DBMS:** The type of software design approach used in the development of most conventional database applications which depend on tables — but not the new ODBMS which are seen as 'object-based'.

run

A Basic command that causes a program to execute.

run chart

These are flow-charts used in program planning. Often the term run-chart is reserved for a more complex diagrammatic representation of a major computer operation. It then combines a number of individual flow-charts, and indicates where files and data are to be input and output.

run-length

In computer programs, scanning systems, and compression systems, this is the number of bits that lie between two transitions or events.

run-length limited

RLL. A system of encoding data onto magnetic disk or tape. RLL means that the number of flux-reversals on the magnetic material are less than the number of bits stored. However the term is widely used for a technique of handling disk formatting which also incorporates this system. See RLL

run-on

The costs associated with increasing the number of items printed. A quote will be for, say, 20,000 booklets, with a run-on price per thousand above that number. Because of the cost of printing the first copy, the run-on costs are often surprisingly cheap by comparison.

run-time

• **Computer operations:** The time during which a program is being run, as distinct from the compilation time. This is the time it takes for the execution of a single, continuous object program.

• **Software licenses:** A cut-down version of an applications program that can only be used to run the application, not to program or alter it. This approach allows programmers to, say, customize a version of a database management program for a specific company, then provide this customized application on disk along with a run-time version of the database manager.

There is enough of the DBMS here for the program to run and be used normally, but not enough for it to be used to change the customization or re-used for further programming.

• Run-time licences are usually made available fairly cheaply to programmers, and are purchased by the dozen.

running head

A short phrase or key-word at the top of each page of a document which provides information about the document. This is usually the author's name, thesis title, chapter heading, etc.

In this book the running head (on this page 'running head') is either the first head-word on the left-hand page, or the last head-word on the right-hand page. This is a common dictionary standard to help you find words.

runts

LAN packets of less-than-legal length (Ethernet has a 64-Byte minimum). They are typically caused by malfunctioning adaptor cards.

RVI

Reverse Interrupt.

R/W

Read-write operations.

Rx

Receive. Aka RD. The receive pin and line in RS-232 and other cable systems. Tx (TD) is the transmission line.

RZ

Return to Zero. A baseband line-code where the logical 0 condition is represented by ground (zero) voltage, or in fiber optics, by no pulse. See NRZ.

S

S36

IBM's minicomputer, promoted as their 'strategic office mini'.

S100

An early micro architecture which was built around a relatively bare motherboard fitted with a number of slots, each with 100 connections. Each element of the system — the processor, memory, I/O interfaces — had its own independent card.

So the slots and motherboard bus did little more than tie the whole system together. However it was a very flexible design and had great expansion potential.

S-band

In satellite terminology, this is a radio frequency band between 1.7 and 5GHz (although only the lower end around 2.6GHz appears to be used) which is mainly reserved for signals from geostationary meteorological satellites. It is vulnerable to interference.

The S-band in the 2.5—2.7GHz range is also used for MMDS translators. See also radar bands, L-, X-, C-, Ku-band and Ka-band.

S-bus

The four-wire ISDN bus within a home or office to which, with Basic Rate Access, eight devices can be connected. The S-bus joins with standard twisted-pair cabling through Network Terminations (NTs) at the S interface. The bus is capable of handling 192kb/s (using 4B3T encoding) and it uses a CSMA/CD contention resolution mechanism to allow it to handle the two B-channels and the D signaling channel.

- There's also an SBus with SPARC. See page 578.

S-connector

See S-terminal.

S-DAT

See SDAT page 584.

S-interface

System-interface or reference point. In ISDN, the S-interface (or S-bus interface) allows a wide variety of TE1 equipment (those which were designed with ISDN interfaces) to connect directly with the ISDN network. This is the interface between the TA or TE1 and the NT2 (or NT1/2).

It guarantees a common connection for a wide range of office equipment. Internally, it uses S-bus with a four-wire cable. At the S-interface point a special ISDN tester can emulate an ISDN exchange, so terminal equipment can be plugged into the tester for checking prior to installation. See NT#, R-, T- and U-interface also.

S-MAC

See page 604.

S-register

In Hayes-compatible modems, a special array (single dimensional series of memory locations) which stores a single byte parameter under each variable, for the control of some customizable feature.

Each of these is associated with a variable name from S0, S1, S2,...to Sx and each of these parameters influences a minor action of the modem. These are usually timing values such as: How many rings before the line is taken off-hook: How long to wait with NO CARRIER before dropping the line. S-registers also control the loud-speaker ON or OFF, etc.

You change all of these with a Sx=Y command, not in the normal AT-command way.

S-SEED

See page 621

S-terminal

Aka S-video terminal. Usually a port (S-connector) behind a TV set or studio monitor into which the S-VHS Y/C (component) video signal can be input directly, so that you gain some advantage from having a high-quality camcorder/recording system. Without this, you will need to feed the signal to your equipment in composite form. See below, S-VHS and Y/C.

S-VHS

See page 636.

S-video

A form of semi-component video where the luminance (Y) signal is separated from the chrominance (C) signal for processing and storage. This separation and the extended bandwidth of the circuits, improves resolution and S/N ratio and dramatically reduces the cross-color problems.

However in true component video, the chrominance will be broken down further into Red-Green- Blue, or into the color 'difference' components Y, U and V (or YIQ).

S-VHS was the first S-video system on the market. Generally S-video cameras will generate both a higher quality Y/C output and a composite output for normal video recording.

SAA

Systems Application Architecture. This is IBM's successor to SNA and, at the time of its release, it was also an attempt to counter the market threat of the 'open' OSI reference model.

It was a philosophy of design to allow programs to work consistently across micros, minis or mainframes, but it was initially based on a mainframe host-centric view of the world, and its implementation was very expensive.

SAA was intended to supply many of the communications features which were needed to make IBM systems and networks universal and evolutionary, without making the existing SNA networks obsolete. The long-term objective of SAA was to create an architecture (a set of defined standards) which allows applications to be ported across the IBM product-line (between micros, minis and mainframes), while remaining masked from the individual operating systems.

It should therefore work with OS/2, OS/400, MVS and VM, and provide a consistent environment for programmers. SAA supposedly allows programmers to create a similar look and feel for applications running on any size machine.

An important part of SAA was the program-to-program interface which had, at its center, APPC and the protocol LU6.2. By adding this peer-to-peer capability to SNA, the networking efficiencies were improved, and the system migrated away from the earlier master-slave relationship.

• IBM's SAA specified a set of application interfaces, protocols and user interface conventions. There were four key elements:

- — Common Programming Interface (CPI) with languages, tools, and application interfaces for portability.
- — Common Communications Support (CCS) to provide easy communications for applications and devices between multiple systems.
- — Common User Access (CUA) which defines user interface conventions (Presentation Manager follows these).
- — Common applications.

• Multitasking is an important element of SAA philosophy.

saddle-stitch

The common and cheapest approach to magazine binding by the use of two or more staples forced through the 'spine' with the book open.

• Side-stitching is the binding process where staples are forced sideways through the backbone, with the book closed (often used with A4 reports). This makes magazines difficult to read and impossible to lay flat.

• The better magazines will have a glued square-back which uses a process called Perfect-bound.

sag

A long drop in the voltage of the mains power — in the order of 10 to 20 seconds. See surge and transients also.

salami method

A way of eroding some advantage in small regular incremental slices or changes. A way of attacking a problem by dealing with a small piece at a time.

• This is a brute-force method used in programming. You program in a recursive way, attacking the problem a little bit at a time, until you find a solution.

• A way of ripping-off customer's accounts in a way that will pass unnoticed, by charging a small fee; taking a few cents at a time. The banks do this with minor service charges.

• A form of technical embezzlement where an insignificant amount (often fractions of a cent) is taken from a large number of accounts. The famous computer scams involving 'rounding' have all used this approach. When any percentage is calculated with cash, it must always be round to the nearest cent.

So there will always be fractions of a cent remaining which effectively belong to no-one. The trick is to accumulate a few million fractional-cents a week and divert them into your own private account.

salutation

The Salutation Consortium was formed in Aug 1995 to produce an open architecture specification (the 'salutation standard') which will enable office machines, computers and communications equipment to talk to, and find out about each other.

This is not directly in competition with Microsoft's At Works, or Novell's NEST, but there are clear areas of overlap.

The aim is to design a salutation or handshaking framework which allows receiving and sending devices of all types to exchange information about themselves. It also includes a way to resolve conflicts so that the most suitable standards will be used in any exchange.

This is not just about fax machines, but about future digital telephones which may have variable bit-rate coders, and about intelligent photocopiers, and modems. You name it!

• IBM, Novell, Kodak, HP and a number of Japanese equipment manufacturers are backing the development.

• This could well be a very important development in telecommunications. It takes control of the use of communications channels out of the hands of the carriers. We no longer need pre-set (by the carriers) standards for voice, fax or data channels.

sampled-servo format

SS. One of the competing formats for 5.25-inch WORM disks; which should not be confused with the logical format of WOFS. The sampled-servo format is supported predominantly by US and European companies — HP, Philips, DuPont, Alcatel, etc. The alternate composite/continuous format is being promoted by Japanese companies.

sampling rate

The number of times every second that the amplitude of an analog signal's waveform is measured to produce digital values in ADC. Sampling rates are usually expressed in Hertz which is a bit confusing to the newcomer — it should be 'times per second' or something not suggestive of waves/cycles. See quantization.

SANE

Standard Apple Numeric Environment. SANE supplies the library of hardware independent maths routines for use by applications that perform complex calculations. With some Macs, (those without FPUs) this is done in software; in others it is performed by the FPU.

This is actually an IEEE floating-point operations standard adopted and renamed by Apple. The SANE resources reside in the Toolkit, and providing varying degrees of accuracy.

SANE calculates numbers to four different precision levels:
— Single precision (32-bits).
— Double precision (64-bits).
— Comp. (64-bits).
— Extended precision (80-bits).

sans serif

A style of typeface of a more modern appearance without the small 'serifs' (or protrusions) at the extremities of the letter strokes. Helvetica is a classic sans-serif typeface.

Abcdg Abcdg

Times Roman
Serif typeface

Helvetica
Sans-serif

Many people believe that sans serif typefaces are not as easy to read as serif styles.

However some of the san serif typefaces like Helvetica are probably easier to read at the same point-size than Times, mainly because of their x-height; the above two are actually the same size, but the Helvetica looks larger. See x-height.

SAP

• **Service Access Points:** The address codes of operating systems, which are necessary to know in the OSI reference model so that the OS can communicate with the various layers.

Each application on a computer must have an address that is unique within the computer, so that lower layers (mainly the Transport layer) can direct packets to the correct application.

• **Service Advertisement Protocol:** The way Novell network clients come to know what resources are available to them. But, as with RIP (see), SAP requires substantial WAN bandwidth. It announces the availability of servers, printers and all other network resources every minute. See OSPF.

SAR

• **Specific Absorption Rate:** The rate at which energy (radio energy) is absorbed by body tissues. It is measured in Watts per kilogram (W/kg). It is the standard way to record non-ionizing radiation dosage at frequencies above 100kHz.

SAR figures have long been associated with the now-discredited idea that the only possible biological effect of excessive exposure to radio waves, is in the heating of local tissue. We now know that there are many potentially dangerous biological effects, not associated with the thermal effect.

•**Segmentation and Reassembly:** The function of splitting long packets into shorter segments for onward transmission through a different packet protocol, and of reversing the process at the other end to restore the original packets. This is required whenever two dissimilar packet or cell systems interface. See AAL.

SAS

Single Attached Station. See DAS.

SASE

Specific Application Service Elements. This is a class of applications interfaces conforming to the ISO model which perform file transfer and distributed job processing. SASEs allow local-style operations to be performed on remote files. The SASEs under development or in use are:

DS Directory service
FTAM File transfer and management.
MHS Message handling and e-mail.
VT Virtual terminal — a standard terminal interface.
JTM Job transfer and manipulation for control of the distribution of services among servers.
MMS Manufacturer's message service — robotic control.
ODA Office document architecture.
IGES Initial graphics exchange system using GKS.
CMIS Common management information services. See CMIP also.

SASI

Shugart Associates Standard Interface. A *de facto* standard for micro-to-disk drive communications. See also SCSI.

satellite

This can mean either a secondary device or channel, which is dependent on a primary one, or a space-craft.

• **Secondary:** Subsidiary.
– **Satellite computers** are remote computers in communications with a host, under host control. They are a slave to the host, or they perform special off-line tasks for the host.
– **Satellite channels** are additional or secondary channels. They are usually channels for transmitting data in association with another used for voice.

• **Space-craft:** Until recently most of our key communications satellites, have been geostationary. Now different types of communications satellites occupy lower orbits. See GEO, LEO, MEO and TTM&C station. Remote sensing satellites are usually in low-earth orbits also.

satellite bands

A number of different frequency bands are commonly used in satellite communications. Some of these letter designations are for satellite systems and some come from radar and military sources, so sometimes the designated frequency range is not exclusive to the application.

• **L-band** is used for mobile communications. It is generally regarded as being from 400MHz to 1.7GHz (in UHF).

• **S-band** is not commonly used. It has low bandwidths and is vulnerable to interference. It is from 1.7GHz to 5GHz which is across the UHF/SHF boundary.

• **C-band** is the most common for TV, especially for receive-only (broadcast) use. It creates problems in cities and urban areas because of the large dish size (although there are now small C-band dishes) and these frequencies are affected by terrestrial microwave. It lies between 3.7GHz and 6.4GHz.

• **X-band** is at a higher frequency and is reserved for government use. It's in the range 5GHz to 11GHz (in SHF).

• **K-band** This is a very wide band that has subsections and is probably the key to future satellite services because of the high information channel rates it can provide, and the small direct-reception dishes it needs. It covers from 11GHz to 40GHz (across SHF/EHF boundary).

— **Ku-band** (12.5—14GHz) is widely used for DBS, and it is more flexible than C and uses smaller dishes. However it is vulnerable to outages in heavy rain.

— **Ka-band** (20—30GHz) has more capacity and is coming into use for direct video. It will eventually allow even smaller terminals, and has the potential to carry a lot more traffic.

• See radar bands also.

satellite dish

For direct home reception of TV you need a dish-reflector (or a phase array) which has a diameter partly defined by the frequencies being received — but mainly by the power of the received signal at the dish's location within the footprint of the satellite. see EIRP.

With C-band, a dish of about 3 meters diameter was long thought to be necessary for reasonable reception, but now with modern high power satellites and low-noise electronics, C-band can be received on a dish which is much the same size as Ku-band.

Ku-band signals generally need a dish in the 0.9 —1.2 meter range, and the Ka-band can get by with one about 30cm across if the transponder power is high enough.

• There are technologies other than the dish-reflector which can be used to receive satellite television signals — the 'squarial' phased array is one.

• Note that satellite dishes have a narrow angle of focus, and the standard type needs to be moveable if it is to receive signals from more than one satellite signal. However, off-set dual horn techniques can now handle two or more satellites in nearby orbital slots with a fixed antenna. See LNA.

satellite power

• **Low-power** refers to 10–20 watts per channel. This type of transponder level is used for telecommunications satellites and for TV relay to cable head-ends. If the downlink beam is narrowly focused, these signals can still be picked up by relatively small antenna.

• **Medium-power** refers to 50 watts per channel which are designed for direct feeds to the large cable systems, although some of the new TVRO dishes and the new small antennas can receive these signals. BSkyB in the UK uses a medium-power satellite transponder for direct-to-home PAL.

• **High-power** satellites are those transmitting 100–200 watts per channel. These are ideal for DBS.

saturation

• In colors, this is the intensity of the color. It is a measure of the amount of the hue (color) to the amount of white. Saturation distinguishes red from pink, so 100% red saturation means that the hue is pure in color (red) with no extra white component to make it more pastel (pink).

• In amplifiers, saturation is the operation point of a non-linear amplifier when the drive level has been adjusted to provide the maximum output power.

• In magnetic storage materials, the point of saturation is reached when further increases in signal strength don't produce corresponding increases in the recorded levels.

SAW filter

Surface Acoustic Wave.

• **A filter technology:** An intermediate frequency (IF) band-pass technology which provides good discrimination in modern TV sets and a range of other broadcast electronic devices.

The aim of the SAW filter is to provide very little signal loss ('insertion loss') on the required television channel, while providing very high suppression of both adjacent channels.

With SAW filters in 'Cable Ready' television sets, it is possible for the cable-provider to transmit signals in every available channel — unlike terrestrial VHF and UHF TV where adjacent channels are taboo.

The surface acoustic wave used here is a physical wave (at the molecular level like sound) which propagates along the surface of piezoelectric material. It causes the material to deform and therefore to oscillate in accordance with the input IF wave frequency, but only when the dimensions are right.

These devices are used in the 10MHz to 10GHz range.

• **A touch screen technology:** Surface acoustic waves (here, true ultrasonic waves) are transmitted across the horizontal and vertical face of a monitor screen. The signals received at the other side are modified by a finger touching the screen and from this, the x/y coordinates can be calculated.

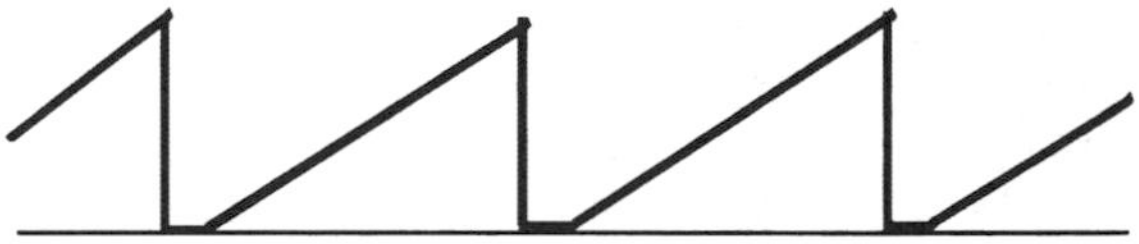

sawtooth

This is the 'shape' of the horizontal (and vertical) drive frequencies of a monitor or television screen. It is a form of square wave, with a slow and steady increase in amplitude. This means that there is no 'leading edge'. It rises to reach the peak level, and then the amplitude drops abruptly back to zero. It may stay at the lower level for a fraction of a second, before rising again.

In a TV set or monitor, this form of sawtooth current is applied to the drive-plates or coils which deflect the electron beam passing from the gun to the phosphors on the inside of the tube face — one rapidly repeated sawtooth change

(15.6kHz)creates the line-scan, and a much slower one (at 50Hz or 60Hz) provides the vertical scan.

The rising portion of the sawtooth represents the steady increase in the deflection force that causes the raster to scan from the left side of the screen to the right, and the sudden cut-off (which creates the tooth effect) is the voltage drop, which allows the raster to return to the left during the blanking interval.

This is the line driver-current (creating the scanning action), not the actually signal which is creating the image on the screen. However, both need to act in harmony.

With computer monitors, the horizontal sawtooth will have a frequency in the order of 50kHz to 100kHz (line frequency rate), while the vertical frequency will be from 70Hz to 100Hz.

SB-86

An old name for MS-DOS.

SBM

Super Bit Mapping. See.

SBS

• **Small Business Systems:** These are key systems; small telephone networks without a central PBX but which can exchange calls between themselves. The majority of SBS systems won't carry modemized data. They are also called NSUs.

• **Satellite Business System.** See Intelsat.

SBus

The SPARC bus used by Sun Microsystems. It runs at about 80 megaBytes a second — twice as fast as NuBus, MCA or EISA.

• There's also an S-bus for ISDN, see the entry at the beginning of the 'S' section.

SBX

Small Business Exchanges. The term should mean very small PBXs, but it is also used for SBS/key systems.

scalable

• **Typefaces:** Bit-mapped fonts aren't scalable, they can't be increased or decreased in size without appearing jagged. Fonts that are represented mathematically (such as PostScript fonts) can be scaled infinitely without these distortions. These fonts are usually known as outline fonts. See. See outline fonts and TrueType.

• **Computer Systems:** Larger systems that are modular in design and can be varied in power, etc. to fit requirements. The main technology which allows substantial scaleability at present, is parallel processing. Some of the new 'massively parallel' systems use conventional PC micro-processing chips. The NCR 3600, for instance, uses the Intel 486 CPUs in parallel configurations to deliver about 10,000 MIPS. Don't confuse this with scalar architecture.

• **MPEG compression:** There are both Signal-to-Noise Scalable profiles and Spacial Scalable profiles. In essence, the S/N Scalable profile is designed to wind back the data-rate transmitted automatically when the transmission conditions on a line or radio link are poor. The Spacial Scalable system is designed to allow a high-definition videoconference (or broadcast) sources to be received by a variety of different monitors — some of which may be of equal high-quality as the source, and some of which may be much lower in quality.

These scalable standards will be layered, so the receiver only receives the minimum number of bits needed by the display device. This is essential to cut communications costs.

Currently there are three layers — Base, Middle and High layers and these are defined for prioritizing the video data. Different forms of error correction will also be applied to these different layers. Two other modes of scaleability, Data Partitioning and Temporal Scaleability, have recently been added to the MPEG2 formats.

scalar architecture

Don't confuse this with scalable systems. Scalar processors perform single computations, one-at-a-time, operations. The distinction is with vector or array processors. This is a serial vs. parallel distinction.

Superscalar sends two or more instructions through the processor at once, but slightly staggered in time. See superpipelining (a similar technique), pipelining and microparallelism also.

scalar mode

The distinction is with the pipelined mode in parallel architecture processors. Scalar instructions, once issued, advance through the unit with each clock cycle until completed.

scaling

• Changing the size of an image in both (or either) the horizontal and the vertical dimensions.

• Specifically with chip construction, scaling involves reducing the dimensions of the active devices and connections on the chip's surface by a constant factor, and lowering input voltages proportionally so that potentials between the reduced-size elements remain the same as those in the original.

This concept is central to modern micro-electronics. CMOS has proved to be easier to scale than NMOS, which is why the emphasis is on CMOS chip fabrication below the 1 micron level. See Debye limit.

• In video compression. See scalable and MPEG2.

scan

The fine parallel lines of image laid down across a video screen by the sweep of the electron gun. This is a 'raster' process.

The beam of electrons from the CRT electron gun (three with color) travel across the screen (from L to R looking at the screen, front on) and lay down the picture information through variations in brightness during the scan.

These variations in brightness depend on variations in the voltage of the driver circuits — and the rate at which the voltage can change from maximum to minimum, is the bandwidth of the system — which is directly related to the horizontal image resolution.

The images show up on the screen because the varying energy of the electrons impact onto the phosphors inside the tube, causing them to glow for a very short period of time (50% loss of brightness in about 1/50th second). Color monitors have three different colored phosphors in a triangular dot pattern called a triad. A shadow mask behind the screen allows the appropriate electron gun to illuminate only the correct colored phosphor within each triad.

PAL systems scan 625 times from top to bottom of the frame (although about 40 of these can't be seen) and NTSC systems scan 525 times. Both systems paint two fields to every complete image. See fields, progressive scan, horizontal frequency, sawtooth.

scan code

See break.

scan rate

This is a confusing term when applied to monitors. There is a vertical scan rate (the number of times the screen is scanned from top to bottom each second — usually between 50 (PAL and SECAM), 60 (NTSC) and 70 or more (computer monitors) and a horizontal scan rate which is the number of lines (including those in the vertical blanking interval) being drawn every second. See horizontal frequency, dot-clock rate also.

scanner

• **Bar Code:** A bar-code reader is a scanner which just records one 'multi-bit word' as it is swept across the barcode. Modern barcode scanners can read from left-to-right, or right-to-left (they will know the difference) by detecting a start-code at one end of the sequence.

• **Publishing:** A device which raster-scans paper (and three-dimensional objects, sometimes) to produce a stream of bits which can be stored and processed by a computer system. The process involves raster scanning and quantization, and it produces image files (not ASCII, but usually TIFF or RIFF). To produce ASCII text output, a second stage is needed where the image file is passed to OCR software for interpretation.
Scanners come in a number of different physical forms:

— **Flat-bed scanners** which are much like the standard photocopier. These are probably the most common type because they can handle books and magazines as well as single pages.

— **Roll-through scanners** which need to have the paper fed a page at a time, like most old facsimile machines. These are cheap but limited in application.

— **Camera-type scanners** which 'photograph' a book or paper from above, like a small enlarger. These can even handle 3-D objects.

— **Hand-held scanners**. These are small T-shaped devices which are wiped across the area of the page required. They sometimes need multiple wipes down or across a page, and the software will then be required to seamlessly joint the various 'image strips' to create the full page image.

— **Fax.** I guess I should also mention using conventional fax machines as scanners, and sending the data to a PC fax card.

• **Electronic snooping:** Scanners are a popular radio hobbyist's eavesdropping device. You buy these for a few hundred or a few thousand dollars at your local radio store, and they will sweep large regions of the spectrum and lock onto signals.

Many people use them to listen in on aircraft radio frequencies, police emergency frequencies and all sorts of private conversations. In the USA it is illegal to use them to scan cellular phone frequencies, but everyone does. This is how the codes for pirate cellular phones are discovered.

• **Trunked radio** is also called scanning radio.

scanning radio

A trunked radio system is sometimes known as a 'scanning' system. More likely today, however, this will refer to the use of a scanner to listen in on telephone conversations. Scanners sweep a range of frequencies and lock on to those above a certain threshold.

SCART

A type of connector used on modern video equipment (and on some computer-type equipment) to provide video and audio links. It is named after a French set of standards, and is now very widely used in Europe.

Scart is also known as the Euroconnector or Peri-Television (Peritel) connector. It's common for European VCR-to-monitor connections and for links between video set-top decoders and TV sets. It can be used for interlinking and controlling several appliances.

The standard video Scart plug and socket carries 21 connections in two rows. Composite video (in and out) and the pure RGB components are carried separately along with two data lines, two control lines, the sync signals and stereo audio channels. A variation allows for separate chrominance and luminance signals (S-VHS or Y/C).

• There is now a new Scart upgraded version for high-definition television called Golden Scart.

• Scart pin numbers alternate left-to-right across the plug. The standard connections are:

1	Audio output 2 (R)	2	Audio input 2 (R)
3	Audio output 1 (L)	4	Audio earth
5	Blue earth	6	Audio input 1 (L)
7	Blue signal	8	Switching voltage
9	Green earth	10	Data 2
11	Green signal	12	Data 1
13	Red earth	14	Data earth
15	Red signal	16	Blanking signal
17	Video earth	18	Blanking signal earth
19	Video output	20	Video input
21	Plug screen earth.		

SCAT

Single Chip AT system controller. This was a famous single VLSI chip that replaces 94 discrete components in the original IBM AT personal computers.

scatter waves

This is a mode of propagation used by radio waves when the signal is scattered and effectively rebroadcast by the troposphere. It requires high power and VHF frequencies, but it does provide over-the-horizon transmission paths.

See propagation modes and micro-meteorite burst also.

SCB

Subsystem Control Block architecture/protocol.

schedular

The part of the operation system concerned with deciding which process is next in line for the CPU. With multitasking and multiuser time-sharing systems this job is complex and it must be balanced to provide: fairness to all processes, efficiency in use of the CPU, low response times, minimum turnaround time, and maximum throughput.

The early schedulars were 'run to completion'. Now many schedulars are 'pre-emptive' — meaning that any process can be temporarily suspended while another has a turn.

Processes can be suspended without warning, and virtually instantaneously. The major delay is in transferring every value, control, and status indicator to a process table. When the process is given permission to restart, these details will be extracted from the process table, and the process will continue as if it had never been interrupted.

- Pre-emptive scheduling can use different approaches:
- — Round robin gives every process a turn, and each has its own quantum (time interval).
- — Priority orders the processes according to some notion of importance. Some priorities are dynamically assigned by the system, others by the operators.

scheduled circuits

In telecommunications, these are private leased circuits for which a 'schedule' (performance guarantee) has been written.

Schottky

Bi-polar transistors which switch very quickly, These devices are now switching at 1 nanosecond rates.

scientific notation

Aka standard index form. Very large decimal numbers are best written in scientific notation. With only five bytes of memory (up to 2^{40} –1) very large numbers can be handled by the computer. Computers and calculators usually display scientific notation using the letter E.

- In decimal notation, a real number is expressed in the form 5.04E21 which translates to 5.04 x 10^{21}. Negative numbers are expressed slightly differently: 83E–21 is 83 x 10^{-21}.
- Floating-point numbers require more bytes than integers.

SCM

Security Control Modules. See.

SubCarrier Multiplexing. See.

SCMS

Serial Copy Management System. This is the famous Philips-designed, chip-based scheme which was to block all digital dubbing from a compact disk onto a digital tape machine (DAT or DCC) or recording optical disk, other than the first. This hardware copy-proof scheme was forced on the hardware manufacturers by the owners of music copyright to stop people using the new digital magnetic recording systems to copy CD-Audio disks in bulk. It didn't.

The term 'serial' refers to the creation of second and third generations from the master source. A first generation copy is permitted, but subsequent generations are barred. The DAT or DCC recorder must actively seek out coded permission (called the DAT-P function) to copy the incoming digital information, and then the recording machine modifies the permission code so that another copy cannot be made.

SCO

Santa Cruz Operations. A company which gave its name to a form of Unix V (Xenix) accepted as standard by the ACE group. SCO Unix or SCO Xenix is specifically for Intel-based systems.

scoreboarding

A technique used in parallel computing. The scoreboard records a history of the registers under load, and it guarantees proper operations of the code at the highest performance levels. It controls the ordering of instructions and use of registers.

SCPC

Single Channel Per Carrier. A satellite term for radio transponder sub-channels which have one source of signal each. This doesn't necessarily mean that only one voice channel or video channel is being carried over the circuit or via the transponder — just that each of the sub-channels retains a separate identity. However each shares the transponder with one or more other sub-channels from another source.

This is a variation on conventional FDM techniques. With a SCPC-divided satellite transponder, widely distributed transmitters can each share part of the transponder bandwidth by remaining within their own narrow frequency band.

This frequency-division technique is often used by small earth-stations (VSATs), working from remote areas. The contrast then is also with demand assigned (DAMA. See). However, the real distinction is with MCPC where numerous streams are interleaved into a single wide band of frequencies using all the transponder bandwidth. All are then obviously from the same earth-station.

- In satellite telephony this is called SPADE.

SCPI

Standard Commands for Programmable Instruments. A standard built on the early GPIB and later IEEE 488.2 standards. This specified a single comprehensive command-set. For instance, to measure voltage a primary node will issue the standard command :MEAS:VOLT? to its peripherals.

scramble

There are three quite different uses of the term.

• **Line coding:** A category of line-coding systems which overcome the synchronization problems associated with bipolar systems (such as AMI) but are not as wasteful in bandwidth as biphase systems (like Manchester).

The 'scrambling' techniques used are nothing more than special rules to ensure that there can't be long streams of zero-level line signals. See B8ZS and HDB3.

• **Security:** A security and payment guarantee technique of coding or jumbling signals to produce a message that is unreadable to anyone listening in without the 'key'. The term originated in the days of analog telephones scrambled for security — usually by taking segments in time and inverting them. These date from the years of the Second World War.

• **Conditional access:** Modern set-top boxes for Pay TV use a wide range of scrambling techniques, usually controlled by a key value (called an 'entitlement message' or EM) which can be transmitted to the set-top box on an individual basis. This is usually done over the air in the few seconds before the program begins — or, in some cases, while the program is being shown. Thousands of individual EMs can be transmitted over the air in a very short time, each individually addressed to a set-top box.

Some systems are suited to digital video (line shuffling) and some only to analog. There are also 'in-band' and 'out-of-band' systems; the out-of-band systems carry the sync on an entirely different (usually radio) channel. Sometimes the video sync is modulated onto the audio carrier. See sync suppression, and active video inversion.

• Usually the decoder will provide a number of different modes (video only — or both audio and video):

— **Clear Mode.** The signals are not scrambled and are available for everyone. A data channel is probably still active for message delivery alongside the video/audio channel. At the beginning of most programs, the picture will be broadcast for the first ten minutes or so in Clear mode.

— **Free Mode.** The image is scrambled and must be descrambled by the set-top box. Any decoder can unscramble this without needing special permission. This is used as a marketing tool for introduction.

— **Normal or Authorized Mode.** The decoder must be used for scrambled image, and it must have a valid EM for the channel it is viewing. The data channel carries the conditional access data at a relatively low rate (but in a robust channel).

• The usual scrambling method used in simple systems is 'active line, cut and rotate' where they break the active line (the one being scanned) and turn it over in a seemingly random fashion so that the picture doesn't make sense.

• The subscriber management system will authorize individual set-top boxes to decoding a program, over the air in most modern Pay TV systems. This is essential for Pay-per-view.

scratchpad memory

Temporary memory used to store partial results during processing. It is a form of register. Most CPU chips will need an area of memory designated as scratchpad RAM.

screen

• **Publishing:** In desktop publishing this refers to the coarseness of the grain, introduced into a continuous tone photograph by half-tone screening before printing on a conventional printing press.

There are two types of half-tone screen — glass and plastic-contact. The glass screens have 'lattice-like' crossed lines etched into a double-layer of glass and the lines block about half the light. The 'screen ruling' is the number of lines per linear inch.

Contact screens are made from a master glass screen, but they are transparent and plastic. They are made by photographing the lattice with a camera, which is set to deliberately produce a blurred image. Under high magnification the screen appears to be a series of minute dots with blurred edges.

A graphics camera is used to re-photograph the original material through one or other of these screens, and the size of the screen is chosen to match the quality of the printing press and the type of paper used.

Common screen sizes (in lines per inch) are:

65 to 85	For newspapers printed by letterpress.
100, 120, 133	For newspapers produced by offset.
120, 133, 150	For magazines and commercial letterpress.

At the top end of the scale it is possible for craftsman printers to use 200 for very detailed reproduction work. See halftone and gravure.

• **Monitors:** With computer monitors or TV sets the actual screen can be a phosphor-covered glass plate onto which the image is written by one (B&W) or three (color) electron guns at the end of a vacuum-filled cathode-ray tube.

Computer screens tend to have 'long-stick' phosphors because they aren't overly concerned with fast moving objects, but they do require a minimum of flicker. TV sets need shorter-stick phosphors to handle high screen activity, and they are also much lower in resolution. See monochrome, portrait screens and positive polarity screens also.

Color screens also differ in the type of shadow mask used just behind the glass to keep the various electron beams from hitting the wrong dot-colors within each triad. Monitors for computers and small TV sets, can also use LCD screen technologies, and more recently plasma technologies. See active-matrix screens and LCD.

screen dump

Many computer systems have a simple utility program which allows a graphic copy of the current screen to be taken and stored as a graphic file, or directed straight to a printer. This is a very quick and easy way to capture information for later use.

screen fonts

These are fonts designed to be displayed on the screen in a way that will closely match (as good as the system will allow) the final output of the file to the printer. Before outline fonts became available, WYSIWYG-type computers needed to have many different sizes and styles of screen fonts in the computer, and matching printer fonts (all bit-mapped) in the much-higher-resolution laserprinters.

The first screen fonts on the early CUI PCs, were even more primitive; they were character-based from ROM with no variation in size, style or format.

Later the full outline fonts like TrueType became available and all sizes and different resolutions could be created from the one mathematical character 'outline' (design). This is why we can now reproduce, almost exactly, the final screen appearance on a laserprinted page without needing special screen fonts or printer fonts. See outline fonts, PostScript and WYSIWYG.

screen popping

The use of an intelligent CLID device (usually a PC with software) to extract a caller's information from a company or general database, before the phone is answered. This is often tied to reverse directories (see), where the number is indexed sequentially, and provides access to the name and address of the caller.

screen priority recalc.

Spreadsheets. See sparse recalc.

screen refresh

Television images are 'refreshed' (redrawn) every 1/50th (PAL & SECAM) or 1/60th (NTSC) of a second. These rates are only marginal to prevent any flicker sensation: certainly 1/50th is near the lower limit. Computer monitors need a higher screen refresh rate if users are to work on them for many hours a day without suffering fatigue caused by slight flicker. Generally computer monitors refresh the screen at least 80 times a second, and they use progressive scanning, which improves resolution as well.

LCD screens don't need this constant refreshing — they stay in one state until a change is made.

screened

- **Cable:** Cables are screened (surrounded with a metallic material which is impenetrable to RF waves) to stop unwanted signals and noise getting in. See shielded cable.
- **Monitors:** Monitors are surrounded by a screen of metallic material to stop unwanted RF energy getting out.

script

A fancy name for programming code — popularized recently with object-oriented languages.

- Script/X is a multimedia object-oriented language which also has a run-time environment. It is platform-independent. See Bento also.
- A macro is often called a script. The name is often applied to short modules or command sequences written for communications programs, etc.
- In film and television, the script is the main blue-print for the drama production. It contains all the words that will be spoken, plus screen directions and explanations. There are standard formats used by different production houses for different types of production. See shot.
- Script typefaces are those that look like handwriting.

scrolling

The process of moving the screen image up or down so that it appears to pass the top and bottom borders. When a document image is larger than the screen on which it is displayed, the screen will either jump from 'page-to-page', or it will scroll progressively. Graphics and spreadsheet programs and some text based word processors, provide both vertical and horizontal scrolling. GUI windows then provide both horizontal and vertical scroll-bars for control.

SCSI

Small Computer Standard Interface. A development of SASI which allows a number of peripherals to be daisy-chained from a single computer port. This is an 8-bit parallel interface (a bus) which can connect eight systems (the computer itself + seven others) at rates of 5MB/s. It is now a widely adopted standard for many computers, although some minor differences exist in the implementations.

SCSI is a fully-functioning multi-peripheral bus which requires autonomous intelligence on the part of the peripherals — they are their own controllers. The operating system can handle requests from several different attached devices at the same time, and can process several commands to the devices feeding them into the host adaptor all at once. This solves the problem of sending a command, and then having to wait for a reply before sending the next, in a serial fashion, to different devices.

The standards also define a series of commands, and a technique of arbitration if many devices want access to the bus simultaneously.

SCSI differs from ST412/506 and EDSI in having all data, commands, and communications in parallel rather than serial form, so it is a true bus architecture. It is also extremely flexible in the type and timing of attached devices, so with SCSI you can mix hard-disks with CD-ROMs and slow old tape drives.

The advantages of SCSI are now fairly obvious, although it has remained fairly costly to implement. The Mac has standard software and hardware interfaces because it uses SCSI peripheral devices, while IBM PCs and clones are still cluttered by a confusion of options. However, SCSI adaptor cards are now available for ISA, EISA and MCA expansion slots. In the PC world, SCSI is not as easy to implement as it is in the Mac world. While SCSI has a single command set, at the chip level each product has a different register set, often requiring different hardware device drivers.

SCSI is a master-slave system; it has an 'initiator' which initiates requests, and 'targets', to which these requests are directed. The targets respond to the requests. There are now these SCSI variations:

- **SCSI-1** and **SCSI-2** differ in the type of connector used. SCSI-1 uses the standard DB-25 series connector, but SCSI-2 often needs the DB-50, a new 50 pin version of the connector. SCSI-1 can run over cable lengths of 25 meters, and SCSI-2 up to 6 meters. SCSI-2 has removed many of the inconsistencies found with SCSI-1, but retained the same data rate of 5MByte/s (40Mb/s).
- **SCSI-3 (Wide SCSI)** is already being designed with a 16-bit data path on a single 68-bit connector, or 32-bits on two connectors. It will have a fiber-option.
- **Fast SCSI-2** doubles the SCSI-2 data rate to 10MByte/s (80Mb/s). This is actually a set of restrictions applied to the standard SCSI specifications (termination mainly) for these faster rates.
- **Fast/Wide-SCSI-2** is listed variously as having a 20MB/s and a 40MB/s sustained transfer rate.
- **Doublespeed SCSI** is the same as Ultra SCSI.
- **Ultra SCSI** lifts transfer rates to 20 or 40MByte/s and has both 16-bit and 8-bit implementations. See also Ultra SCSI and FC-AL.

SCSI addressing

- With ISA and EISA SCSI, the boot device is the lowest number (Device 0), but with MCA SCSI, the highest number is used as the boot. In the Macintosh world the computer will be Device No.7, and the internal hard disk will be Device 0. The floppy disk will not normally be an SCSI drive, and a CD-ROM is generally set to Device 3 or 4.

SCT

SmartCard Terminal. See smartcard.

SD

- Single Density — floppy disks.
- Standard Definition — Television. See SDTV.
- Super Density CD. See.

SD-DVD

See Super Density CD and DVD.

SDAT

Stationary Digital Audio Tape. As distinct from R-DAT or Rotary DAT, which is now, generally called DAT.

S-DAT was a well developed technique and a hot favorite for high quality audio recorders. It was investigated by the Japanese DAT committee and rejected as a consumer standard, but it has since found wide use in professional applications and is closely related to the new Philips DCC digital recording system. There are data versions also.

The cassette was 86 x 54.5 x 9.5mm with a tape speed of 4.75cm/sec (the same as a compact cassette) giving 90 minutes of recording or 180 minutes at half-speed.

S-DAT had twenty-two parallel tracks across the width of this narrow tape. Twenty of these held data, together with some subcode, and there were two auxiliary tracks. The data transfer rate was 2.4Mb/s. See DAT and DCC.

SDC

Synchronous Digital Compression. This refers to modems and CSU/DSUs which operate over reasonably high-speed (56–64kb/s) circuits. Today with V.32, the link between the modems will run in a synchronous way, and data compression and decompression will be done in the modems.

SDDI

Serial Digital Data Interface. This is FDDI over shielded twisted- pair (STP — IBM Type 1 cable). It appears to be exactly the same as FDDI, except that it runs on copper with a different line-code. A logical 1 is a voltage pulse of about 0.5 volts, and a zero is the absence of a pulse. It will normally be used in a single-ring configuration. See FDDI and CDDI.

- SDDI is now being used as a backbone protocol in digital television studios. See SDI.

SDDVD

Super Density Digital Video Disk. See Super Density CD and DVD.

SDE

System Development Environments. These are special language environments used in the development of mini and mainframe applications.

SDH

Synchronous Digital Hierarchy. An international optical fiber telecommunications transmission standard adopted by the ITU in June 1988 (as a supplement to the USA's Sonet standard) and has since been enhanced in 1990. It is a digital transmission and multiplexing scheme which is used by the carriers to manage the basic raw transmission of high-speed bearer connections over long-haul links. PDH is carried over SDH links and when ATM arrives in the public network, it will initially be carried over SDH also.

SDH is based on a bit-rate of 155Mb/s which is three times the primary Sonet rate (OC-1). The relevant ITU standards are G.707, G.708 and G.709. The natural topology of SDH is a ring, but these rings can create a mesh.

In SDH, the 155Mb/s rate is denoted as STM-1 (Synchronous Transport Module level 1), with STM-2 = 311Mb/s, STM-4 at 622Mb/s, and STM-16 at 2.4Gb/s.

The aim of SDH is to provide a multiplex hierarchical system which can be a mesh network (rather than point-to-point backbone links) because synchronization and 'containerization' allows channels to be added or dropped using Add-Drop Multiplexers without needing to demultiplex the total bandwidth.

This is done through the use of virtual containers (aka 'tributaries'), and pointers which direct the system to the start point of a tributary. Eventually all public trunks will use SDH, probably carrying ATM cells.

• The SDH structure caters for the ITU's 2, 4, 8, 32 and 140Mb/s standard data rates, and the North American 1.5, 6 and 45Mb/s data rates.
See jitter, virtual containers, VCs and Sonet.

SDI

• **Strategic Defense Initiative:** This was President Reagan's pipe-dream — the futuristic space-defense program — a world without software bugs where high-tech defense equipment would always be reliable and safe.

• **Silence Deletion/Insertion:** A way of compressing voice files by removing all silent periods and replacing them with a numerical value for length.

• **Serial Digital Interface:** These are protocols standardized as SMPTE 259M for the carriage of baseband 4:2:2 digital television signals. SDI handles the digital data from a video source in a serial bit stream without compression at 270Mb/s. It can't tolerate delays needed for error checking, so it is not much used now. See SDDI also.

SDL

Specifications and Description Language of the ITU-T. An analytical tool used by telcos for system design. It resembles a very high-level language in computing, and it can be compiled and used for simulation.

Essentially, it is a technique for describing systems through abstraction in text and graphics, and the description process begins with a graphic outline of the system much like a flowchart. SDL provides a step-by-step iterative process (breaking down problems into smaller and smaller parts), and includes system-operating characteristics like process location and flows, interprocess messaging, and intercomponent channels.

SDLC

• **Synchronous Data Link Control:** A widely used code-independent IBM data-link control (DLC) protocol suite used in SNA synchronous networks (at the lowest level). It is an essential part of those networks.

SDLC is a complex transmission and error-control protocol suite which superseded the BiSync (BSC) protocols; it proved to be more efficient for framed data transmission between computers.

It's a bit-oriented full-duplex standard for remote hosts (point-to-point), and for multipoint communications. Various dialects exist. The ISO's modified version of SDLC is called HDLC, and LAB-P was also spawned from this protocol.

SDLC communicates in frames having a 1-byte flag, address and control bytes in the header, then an indefinite-length information/data field. This is followed by a 16-bit CRC and another 1-byte flag. Normally it requires specially-wired hardware or complex terminal emulation.

At the Physical layer, it uses the same electronic (pin and voltage) configuration as RS-232. It uses the NRZI methods of line coding and zero-bit insertion for synchronization.

• **System Development Life Cycle:** A concept in discussions of CASE and programmer productivity. The typical life-cycle phases of SDLC/CASE are: planning, specifications, system analysis, design, coding, testing, production, and maintenance.

SDLC pass-through and conversion

These are two ways of supporting SDLC/SNA networks on the same backbone that handles LAN traffic.

— **SDLC pass-through** is responsible for transparently passing the SDLC polling over the backbone. It has problems because of the master–slave relationship which generates excessive network traffic from unproductive polling, and it also has timing problems.

— **SDLC conversion** does a Data-link layer protocol conversion between the SDLC and LLC2.

SDMA

Space Division Multiple Access.

• A 'smart-cell' access technology which is used to increase the capacity of analog cellular mobile basestations. It relies on narrowing the angle of transmission of signals at the basestation (and therefore of the geographic reception area), to increase total capacity.

Sectorization of the antenna system into three or six sectors is a primitive and fixed form of space division.

• This is also a satellite term. See space division.

S/DMS

This is a Northern Telecom exchange switch product line. Don't confuse it with SMDS.

SDT

Synchronous Digital Transmission. The generic term means what it says: the transmission is both synchronous and digital.

• With SDH, this is a more specific technique which allows individual channels within a multiplexed signal to be extracted without demultiplexing the lot. It relies on the use of virtual containers. See.

SDTV

Standard Definition TV. This term is sometimes applied to both NTSC and/or PAL/SECAM. They are making the distinction between current standards and the new EDTV and HDTV systems. This term 'standard definition' stresses the 4:3 aspect ratio of the screen, the analog nature of the system, the 525 or 625 scan lines, and the single channel analog audio.

SDV

Switched Digital Video. It is both an Open System and media independent. It can work on tape, MO or CD optical disk, WORM, hard disk and floppy. The ECMA has accepted it as the standard.

This is related to Fiber to the Curb (FTTK or FTTC)

SEAL

Simple and Efficient Adaptation Layer. See AAL–5.

search

The ability to find a string of characters in a file.

SECAM

Systeme Electronique Couleur avec Memoire. The name applied to the French color television standard developed in 1957, which later spread to Russia, the USSR, the Communist satellites (the OIRT standard), some parts of the Middle East and North Africa.

American critics of SECAM translated the acronym as 'Supreme Effort Contre l'AMerique' which means 'Something Essentially Contrary to the American (NTSC) Method'.

The French took a more radical approach than PAL. They used two frequency-modulated subcarriers for the R-Y and B-Y respectively. Because they were using FM signals which do not mix, the SECAM designers were obliged to send the two chrominance signals on alternate lines.

SECAM sends the primary luminance and color signals sequentially, and holds the information of each scan line in a video delay — hence the 'Memoire'. See 4:2:0.

The original French E-standard television had 819 lines which put it almost in the high-definition class, but this was ridiculously expensive at the time and it was later replaced by the European standard of 625 lines to align SECAM with PAL. But the SECAM color approach is very clumsy in the post-production stages. Fortunately transcoding between SECAM and PAL is relatively easy.

For this reason most French TV programs are made in PAL and only translated to SECAM at the time of transmission (although the French often won't admit this). The two systems really only differ in the way they handle the color burst.

• There are now five technically different SECAM systems, SECAM I, II, III, IIIB and IV.

SECDED

See error.

second generation

• **Languages:** This is the assembly language level which is still directly coded by some programmers, using hex notation and mnemonics.. Certain commonly-used binary routines are given mnemonics (JMP for jump, ADD for add).

• **Computers:** The first to use transistors, rather than valves.

second-order multiplexers

See Plesiochronous Digital Hierarchy.

secondary channel

The backward or return channel in a communications interchange which is fundamentally one-way (half-duplex) or two-way but asymmetrical. It is used for supervision and acknowledgement (ACKs, etc.), not for data. Normally a secondary channel runs at a very low rate and handles device control, receipt acknowledgment of packets, diagnostics and management functions only.

secondary station

One that has a master–host relationship with a primary station. It responds to commands issued by that station. HDLC uses this technique.

secondary storage/memory

You need to make a distinction between secondary memory and secondary storage. The distinction isn't always clearly made.

• In memory terms 'secondary' is applied to disk, tapes, etc., as distinct from RAM, ROM and PROM memory devices. However EEPROM used in PCMCIA cards, flash-memory, etc. can be used for secondary 'memory'.

The distinction is based on whether these chips are part of the CPU's working area — whether they are part of the memory map — or whether they just supply the long-term storage which must then be actively bought into the working memory to be used.

This is an important distinction to make with flash and FRAM memory. Flash must often be changed only in blocks, and it has other problems which stop it being used for working memory. FRAM, however, can be based on working DRAM or SRAM, and it still holds its memory with the power off.

• In long-term archiving and storage terms, 'secondary' usually means back-up tape systems (and now optical disk) as distinct from the 'primary' hard disk. See DASS and primary also.

secret key

The key you keep private in a public key encryption system. Also called the private key.

sector

A block or recording area on storage media — traditionally a disk. This is the smallest unit of storage addressable.

In a standard magnetic disk the boundaries of a sector are represented by radial (logical) partitions, which break each track into (usually) 17 or 26 'segments' of a circle. Sectors are all of a standard fixed length in terms of the number of bits they carry.

Sectors have their own unique address. In many cases the term block and sector on a disk are used as if they are synonymous — but 'block' refers to the data unit, while 'sector' refers to the disk space.

• When multiple platters are used in a complex disk stack, the sectors on each recording surface sharing the same address, will be jointly known as a 'cluster'. A block of data will be simultaneously written to all sectors (by just invoking a single disk address), and here a single block will be recorded on many sectors. See cluster and RAID.

• The 'sector interleave' or 'sector map' is how sectors are numbered on a disk. Usually they will be numbered sequentially, 0 – 1 – 2 – 3, etc. but sometimes the numbering will be staggered, so physically (but not logically) the sequences are numbered 0 – 3 – 6 – 9 – 1 – 4 – 7 – 10, etc. See RLL.

sector interleave

With hard disks, the disk will read from a sector (or cluster), and then the controller will take a finite time to transfer this information to the processor. During this transfer, the disk will have moved past the following sector, and if the sectors of the file are contiguous (in strict order), it will need to wait a complete revolution before reading the next sector. For this reason blocks are interleaved (alternated in various ratios) to give the controller time between blocks but without wasting a whole revolution — this is the sector interleave factor.

An ideal interleave factor is 1:1, but in practice this allows no time for the disk controller to process the data before the next sector arrives (however, some drives can buffer data and handle this).

With a 4:1 interleave factor, the disk will read (on a single track) sectors 1, 5, 9, and 13, during the first revolution, then 2, 6, 10, 14, then 3, 7, 11,..., etc. on the next. It will take four revolutions of the disk to read all sectors on one track. See interleave factor, contiguous, and RLL.

security

The term encompasses a number of different factors in computers and communications. The OSI Draft Standard 7498 spells out the following mechanisms: encryption, digital signatures, access control, data integrity, authentication exchange, traffic padding, routing control, and notarization.

security control modules

SCMs are satellite (as in 'distributed') processors used in EFT-POS and other data security communications networks to take over the computational-intensive work of decrypting and authenticating messages, keys, etc.

security management

This is one of the five key factors identified by the OSI in network management, and part of the definition of CMIP. Security management involves the complex problem of creating alerts for unauthorized access attempts.

seed/SEED

• Self Electro-optical Effect Devices. Optical computing switches. Quantum-well devices are similar. See S-SEED.

• Hardware companies also talk about 'seed' units, by which they mean alpha-test equipment which is sent out to software developers. It allows the developers to get moving on creating software at a very early stage in the development of a new computer. Software companies also do this with operating systems.

seek time

The portion of total access time used by the disk-drive's access arm in finding the correct part of the disk before a read or write operation. If the data is not found, another seek may be necessary to refine the position. The quoted average seek time for a disk unit is usually taken as an average for a specified number (say 1000) of random 'read' seeks anywhere on the drive.

• In CLV videodisks and CD-ROM systems, the seek time also involves the time needed to alter the disk's rotational speed.

SEG

Special Effects Generator. (Usually called the 'Es-e-gee'.) An SEG is the main video-switching and image manipulation-control 'panel' in a television studio. All image sources from cameras, videotapes, film chains, etc. terminate here. Those selected by the director at the time are switched by the SEG to the 'line output' for transmission or recording.

The effects generator part is incorporated in the switch. Almost any SEG can generate, as a basic set:

— Fades (both to white, and to black).
— Dissolves (a fade out superimposed over a fade in).
— Wipes (progressive replacement of one image by another, across, up or down the screen or from a corner).
— Iris out or in. A circular wipe beginning or ending in the center.

Good SEGs have an enormous range of special effects of this kind, and they will also include key effects such as luminance and chroma-key, etc. All modern SEGs can be pre-programmed. See CCU and key.

segment

• In packet-switching, a segment is a part of a packet used to subdivide the data space. In most X.25 systems, segments are the data-volume units used for billing. Up to 64 characters (half a basic X.25 packet) is normally classed as a segment, and this occupies roughly one screen line on a display.

• In disk storage, a segment is the radial subdivision of a circular track. Tracks will normally have a fixed number of segments — often 13, 17 or 26 (but not in CLV or variable speed disk systems). Some hard disk units may have say 13 sectors for the inner tracks, 17 sectors for the middle tracks and 21 sectors for the outer tracks.

• In programming, segmentation is the process of dividing a large program into blocks of various sizes, and placing these segments in memory in non-contiguous locations.

• In coaxial LAN, segments are single continuous cable lengths.

• **In LANs**, a segment is a distinct self-contained logical networking unit which is joined to other segments via a repeater— it is a subdivision of a LAN. Repeaters just extend the reach of a logical LAN, but both sides of a repeater are segments of the same network. Ethernet can handle only up to four segments in a line because of delay limitations. See granularity.

Segment-switching (as distinct from port switching) is fast packet-based switching. This type of switching hub is changing the way we look at the segmentation of networks. In many cases each workstation will own its own segment.

• **In TCP**, a segment is a single Transport-layer unit of information.

• **In LCD or LED** character displays, a segment is the elements which create one character.

segmented memory

Memory segmentation to various degrees has been tried in numerous computer architectures over the years, mainly to solve problems with I/O. Usually each segment will automatically have spare capacity which allows for change without affecting the whole. So, if you add a word into the middle of a document file, the whole back end of the file doesn't need to shift location.

In segmented memory systems, the location within that segment is described in terms of its offset from the beginning of the segment.

• With Intel processors from the 8086 to the 80286 (and hence with most versions of MS-DOS) memory was segmented in 64kB blocks. This limited the size and complexity of the code that the machines could support in RAM. The basic problem was that the processor couldn't just specify a memory location; it needed to provide the segment number, and any given location within a segment was then described by its offset. Segmentation is unnecessary with 386-based machines or in the Mac architecture.

segmented virtual memory

The 80286 chip allows programs to use more memory than is actually present in the machine by swapping 64kByte segments of data to and from a disk. The distinction is with paged virtual memory.

segue

Pronounced 'Seg-way'. This is a dissolve between two sources of sound in video and film production. One source fades out, while the other fades in.

selection sort

A sort technique which selects the lowest-valued item and places it at the bottom of an array, then finds the next, and so on. This is a very slow way to sort.

selector

The type of switch used in a step-by-step exchange.

self-clocking

Coding schemes which automatically maintain synchronization. Phase shift keying and similar phase coding schemes on magnetic disk provide their own clock mechanism, so no external clock is required. See PSK, biphase and Manchester coding.

self-configuring

Plug and play. See NuBus also.

self-healing

Applied to networks which are point-to-point or bus topologies in a logical sense, but with a physical topology of a ring or mesh. If one section of the cable breaks, communications can still be maintained by sending the messages around the ring in the opposite direction.

self-view

On a videophone or videoconference unit this is a small 'vanity' screen which allows you to look at yourself (to check for spinach caught between your teeth) before transmitting your picture to the other party.

semaphore

Semaphores are data structures which signal a condition or status. They are often used to coordinate the shared access to data. They provide data integrity by operating as 'flags' governing each application's access to read shared data, and limiting the application's write-access to files. Each application or process must check a specific semaphore before proceeding. A 'lock' is the simplest form of semaphore.

Semaphores are also one form of Inter-Process Communications (see) and one will usually be attached to each I/O device. This is how permission to proceed is passed to devices.

• OS/2 uses semaphores as system flags to coordinate multiple-tasks within the operating system.

• This was also the name given to the first form of public-access telegraphy, invented by French Engineer, Claude Chappe. He built the first two-arm semaphore telegraph system linking Paris to Lille, in 1793–4.

semiconductor

A material that is neither a good conductor or a good insulator. The most common material is silicon, and the second is germanium. Semiconductors need to be 'doped' to be useful. See solid-state and transistor.

semiconductor lasers

The term covers a number of different types of solid-state lasers for various applications. The two most common designs are edge-emitting and surface-emitting lasers, which are also known as horizontal cavity and vertical cavity lasers (see).

All these laser types are built as flat layered chips on wafers, and manufactured in the same way as ICs. They have doped layers of n-type and p-type semiconductor material separated by a 'junction layer'. The n-layer has spare negatively-charged electrons, but the p-layer is deficient and so acts as if it has spare positively-charged holes.

When you power these devices, the electrons are excited to the point where they combine with the holes in the adjacent junction layer, and set off a cascade of electrons within this layer as each stimulates the others to emit light.

The more physically confined the active of these electrons are, the brighter and more efficient the light.

Vertical cavity lasers are stimulated only in a very tiny cylindrical cavity which exists between the two chip faces, and they have numerous layers of semiconductor, each with slightly different compositions (different amounts of aluminum) which act as mirrors to contain the light within the cavity.

• British Telecom has used basic semiconductor laser technology in a unique form as an amplifier in long-haul optical fibers where direct optical amplification is achieved by the full length of the fiber (not just a spliced in a length of erbium amplifier). In effect, the fiber itself is the cavity; there is no opto-electronic conversion. See erbium amplifiers also.

separations

• **Desktop publishing:** Before printing presses can handle color images, each image must be prepared as four different (CYMK) separations. These are full-size 'seps' (like X-ray negatives). Each is B&W only, but one of each will represent the Yellow, Magenta, Cyan colors and the Black.

Separations can now be generated by computer programs and output direct. Note, however, that the halftone screen orientation (the alignment of the dots) used for each image must not align. Generally Black is printed at 45°, Magenta at 75°, Yellow at 90° and Cyan at 105°.

• **Film archiving:** Color films of historical value can be printed through filters onto three B&W film reels (RGB) which, when processed for archival storage, provide probably the safest possible way to preserve important information for a Century or more. This process is called tri-separation, and each of the film reels is a separation.

Sequel

See SQL.

sequencing

• A computer music term. See MIDI.

• In programming languages. See assignment.

• A controller function which controls an industrial process from beginning to end, turning on and off services as required. Traditionally sequencers were electro-mechanical, but now they are more often computer-controllers.

sequential

Distinguish between serial and sequential. Sequence suggests that there is some rule controlling the relationship. Serial just suggests that the component parts are in a line. Usually the sequence is controlled by alphabetic ordering, or number sequence.

sequential access

Tapes have sequential access, while disks have random access. To find some data near the end of a tape, you've got to spool right through from the start. With a disk, the read-head can find the data almost instantly.

Computer memory chips are random access (both RAM and ROM), but not all forms of electronic-memory are of this kind. For instance, magnetic bubble memory devices take data in as very long sequences of bits. They can hold this data indefinitely and regurgitate it when required, but always as a serial output. The bits are actually stored linearly as small magnetic domains – like marbles in a very long tube.

• Registers and stacks, such as FIFO and shift registers, are also sequential access systems, not random access.

• Sequential access systems are ideally suited to linear media (novels, films, music) while random access is for reference (dictionaries, databases) and branched (depending on decisions) media.

sequential computers

The classic von Neumann architecture computers in any discussion about parallel computing.

sequential files

This was a term used in the early days of PC databases (aka serial files). In a sequential database file, each record was exactly the same size (all fields had predefined data lengths filled with nulls if no data was present). Computer memory was therefore strictly allocated in fixed amounts.

In performing a search on any field, therefore, the CPU could calculate how far to jump between records just by adding an offset equal to the fixed record length.

To jump to the second field in record 23, it would multiply the set record length by 23, and add the length of the first field. This made access very quick in days when CPUs were very slow and had no coprocessor support.

• With random file databases, the length of each record was variable and therefore unpredictable, so the computer needs to use either a look-up table of some sort, or cycle through every byte from the beginning of the file.

sequential logic

Sequential logic circuit have an output which depends on the previous states of the input. The distinction is with combinational logic.

serial

In line. Not using parallel techniques.

• A transmission system using a single channel where the data bits arrive in sequence. The distinction is with parallel connection which use multiple channels. With parallel cables the eight data bits which make up a byte are all sent at the same time — one along each of eight channels.

Serial bits are sent one at a time down a single common connection, and therefore require buffers and registers within a special processing chip (called UART, USRT, or USART) to organize the transmission and reception of the signal bits.

• Serial and sequential are not the same. See sequential.

• Bytes can be transmitted serially with the most-significant bit transmitted first (big-endian) or the least-significant bit first (little endian). RS-232 handles data back-to-front (little- endian)

• UARTs are parallel-to-serial devices (and vice versa). They also add and remove start- and stop-bits to/from each byte. See RS-232, handshaking and UART.

serial cache

See look-aside cache.

serial printers

This has two quite distinct meanings.

• Printers which require serial data from a serial port — as distinct from those needing parallel feeds from a Centronics port. Usually the transfer of data from a PC to a laserprinter is many times faster through a parallel port than through a serial port.

• Printers which print one character at a time as each is delivered from the PC. The distinction is with old line printers which could store more than one line and, therefore, justify the pages and print bi-directionally. Serial (character) printers couldn't do this.

serif

A typeface style used in desktop publishing. The serifs are the small protrusions at the extremities of a character in the older-style of typefaces (fonts). Times Roman is the classic serif typeface as is Garamond used in typesetting this book. See sans-serif ('without serif').

serpentine tape drives

A form of tape back-up system which records data in a back-and-forth, snake-like, 'S' pattern along the length of the tape. See QIC.

server

A computer (node) or software on a network that provides shared resources to everyone on the network. In practice it is either a storage device or an applications processor which is remote from the user — although the term can be used for a software resource on another nearby workstation, but only in low-level peer-to-peer file or applications sharing.

Specialized network servers range from standard PCs (even one occasionally used as a general workstation); to dedicated, purpose-designed boxes; to mainframes or minicomputers.

• In common LAN terminology 'server' is a general name for any network hardware/software combination with a dedicated function, which exists to supply some service to the connected workstation/PCs. Local area network often require PCs or minis dedicated to a wide range of network tasks. They provide file, print and electronic mail services, connection to mainframes, X.25 packet-switching services, to modems, and to ISDN, etc.

• In modern client/server systems, the server is a shared machine which is capable of recognizing and responding to a client's request for services. These can range from basic file and print services to support for complex, distributed applications. The server will generally also be responsible for security, record and file locking and atomizing transactions. See SQL and DDBMS.

Servers come with different functions built-in, and are given different (and often vague) titles:

— **Disk servers.** An old term for the supply of partitioned disk space for storage. Used with diskless workstations and the early PCs which didn't have their own hard disk.

— **File servers** provide shared files, and have either file or record-locking. On small LANs, the file server functions can also be distributed among the user terminals and no central facility might exist.

— **Network servers.** The generalized term for what is mainly a database server.

— **Database servers** are widely used in client/server applications. This is the most common form of server. It accepts requests for data, retrieves the data from its database (or sometimes from another node) and passes back the results to the client. Database servers almost always use record-locking rather than file locking.

— **Communications servers** provide gateways to other LANs or computers. Sometimes the term modem server will also be used, as will fax server.

• From queuing theory, a server is a generic term for any device that works on a communications task and passes it on.

service classes

You can categorize communications services in many different ways. Traditionally we emphasized duplex and cast.

• **Duplex:** Full-duplex (two-way simultaneously), half-duplex (two-way sequentially), and simplex (one-way only).

• **Cast:** How the signals are distributed.

— Point-to-point — one to one.
— Point-to-multipoint — narrowcast.
— Point-to-everyone — broadcast.

Traditionally these systems have been mutually exclusive. Point-to-point services were interactive, and provided over switched star-topology networks which were voiceband, and not suited for broadcast radio or television.

Broadcast and multicast services were not interactive, and were provided over shared-media networks with high bandwidth, which were not practical for isochronous services.

• **New categories:** Now these systems are merging, and the restrictions associated with each type are disappearing. We therefore have new types of networks which can be:

— Interactive: Capable of providing a two-way exchange of information — data, voice and video.
— Distributed: They can be used for broadcast, multipoint or multicast services, in addition to the one-to-one addressable services.
— Symmetrical: The interactivity on these networks can have equal bandwidth either way, if needed.
— Asymmetrical: When the information flow is predominantly one way (file transfers, or information retrieval), the bandwidth of the system isn't wasted providing a wide-band unused return channel.

— Conversational: The two-way links can be maintained for substantial periods of time without disrupting other users.
— Transactional: Suited to the delivery of secure short messages, which don't justify the establishment of a circuit before transmission.
— Constant bandwidth: Where a certain bandwidth can be reserved, and remain open for use for the session.
— Variable bandwidth: Where the bandwidth of the channel is dynamic, and provided on a 'needs' basis, a fraction of a second at a time.

These are all now capable of broadband (video) delivery, but are still likely to be economical for voice, data, and messaging. Interactivity and symmetry are important considerations, also. Telephone calls are generally symmetrical, fax is substantially asymmetrical, while free-to-cable TV signals are permanently asymmetrical (totally one-way)

The development of transaction and messaging services will also be important in the future — and these are generally very low bit-rate services. They will be needed for information retrieval operations, where information is being extracted from a database, and the flows are substantial and predominantly one-way. This is a query and answer operation.

We will also use networks for financial transactions, and for monitoring and metering. These are all short-messaging services which require a high degree of data security and are generally conducted in real-time. See ACID test.

Voice and video-conferencing services have real-time requirements to within 1/10th of a second if delays aren't to be obtrusive. These are therefore known as isochronous services. The data must be delivered with minimum delays across the network, and even more importantly, the delivery must be regular and predictable.

Video one-way services (video on demand, etc.) are also often classed as isochronous, but network latency isn't a problem. However the requirement of regular and predictable delivery is. Video packets can arrive late, but they must all arrive in sequence when required.

Data services are possibly the least demanding. The flows can be extensive, but the use of store and forward systems alleviates most problems, providing the delays aren't more than a few seconds.

Messaging services are also easy. The flows are short and time isn't a problem, however, the nodes need to be able to handle full message lengths of any kind and so they will often store on disk. These systems can take days to deliver a message around the world.

service integrator

The key company in the development of a large communications project, which brings together the various parts and does the overall planning.

There's a subtle difference between system integrators and service integrators. The first would be responsible for creating and integrating large corporate networks. They would probably write applications, and purchase hardware for the client. The service integrator, however, is more a consultant in the use of carrier-type services. Don't confuse this with service provider either (below).

service provider

In telecommunications this means an air-time reseller, or a reseller of capacity on a telecommunications network. There are three main categories:

• Facilities managers: These companies generally contract to large organizations or governments, and they will consolidate all network traffic into leased lines, and negotiate discounts with the carriers.

• Switched service providers: These are bandwidth resellers who will contract to purchase discounted capacity on a carrier network, and resell it to a wide range of customers. Very often they will class themselves as a telephone company, yet not own any telecommunications assets — at other times they may own their own switches and microwave links.

• Today the term is usually applied to the cellular phone market where the service providers package cellular phones and long-term service contracts, and perform the billing functions themselves. They may earn their money from (any or all of):

— The sale or lease of equipment.
— A commission from the carrier for signing each user.
— An on-going percentage from all calls made by that customer. Usually a service provider will bill you direct for the services, and you are his customer — not that of the carrier.

The theory is that competition between service providers keeps prices low — but since they already add a substantial markup to the discounted carrier cost, then it is hard to see how. Depending on the deal, when service providers go broke, the customer can lose both his phone number and his up-front payments.

The distinction here is with a dealer, who sells cellular phone handsets and perhaps receives a commission for signing you up with the carrier — but who does not perform the billing functions or receive further commissions.

servo

Servo mechanisms are synchronisation systems or tracking devices which use electronic feedback to match some physical condition or movement. A servo mechanism is one that is able to compare existing conditions (as measured) with a desired condition, and is able to adjust the first to match the second.

Thermostats are servo mechanisms because they use feedback to control and maintain the temperature in a room. The term is usually applied to a system such as the one applied to detectors and small tracking devices which keep a CD-ROM's laser beam following the line of pits and lands on the disk, or to another which handles vertical variations that could throw the laser beam out of focus.

session

A complex exchange of messages which involves the attention of two parties — a continuous dialog between intelligent devices over a single period of time.

• The term is used for a communications connection, say, between a PC which is emulating a terminal and the mainframe itself. The session begins at log-on and ends at log-off.

• In SNA two NAUs can communicate only after they establish a session — a logical connection.

• LANs gateways allow only a limited number of sessions to be conducted simultaneously. This number is set by the gateway software and hardware configuration.

• In EFTPOS, a session refers to the process of making numerous transactions while maintaining contact between the POS terminal and the bank's database. The distinction being made here is between EFTPOS devices which use key encryptions that are changed after every transaction, and those which are changed only after every session (usually at the start of the day).

The second approach is obviously less secure, since the session between a terminal and the bank may last all day, and this provides time for the key to be broken.

• In CD-ROM recording, single-session recording devices will only allow one table of contents on the disk, while multi-session recording devices provide a table of contents for each subsequent recording. See CD-R.

Session layer

The fifth layer in the OSI's seven-layer model. This layer sets up communications sessions between two computers and establishes the conditions needed for applications to communicate with each other across the network.

This is a connection-oriented protocol which exists between applications, rather than network components (like NetWare and NetBIOS). According to the ISO it 'provides the means to organize and synchronize the dialog between applications processes and manage their data'.

A session is a formal exchange of messages, or a dialog between two workstations or devices. If the communication session between two applications is disrupted, the Session layer will seek to restore it. It terminates the session on request, and informs the other workstation of the disconnect.

It also handles any accounting, and it administers the signal procedures needed for connection and disconnection. It supplies the security, name recognition, logging, administration and similar functions, and it coordinates 'who speaks when'.

This is the lowest of the 'applications' levels in the OSI model, and it is of great importance on PC-based networks. Session layer software needs to reside in every network station to pass calls between applications in the network and it is important in RPC. IBM's NetBIOS and APPC/PC are both important at this level.

set

• **Mathematics:** A collection of elements with a common property.

• **Logic:** As above. Mammals are a set, within the broader classification of animals.

• **Printing:** The measure of the width of a character of type in points. It is the amount of space required by each character. This varies slightly depending on the character itself and which character is adjacent to it. See kerning.

Note: 'set solid' means to set the copy with no leading space.

set-oriented languages

High-level, mathematically-oriented languages which provide for the definition and manipulation of sets at a level comparable to that used in mathematical text (SETL is an example).

set-top box

This is a wide term for the electronic unit which sits on top of a TV set and handles the incoming signals from cable, satellites, ADSL, or MMDS. These are also called IRDs for Integrated Receiver-Decoders if they perform both the tuning and the decoding function (they won't in ADSL).

Set-top boxes are in the process of evolving from dumb devices which just select and decode a single TV picture in hardware, to very smart communications controllers for the whole household.

In its eventual form, this device will handle the home-bus for delivery of video (multiple channels), telephone, data, security and monitoring services around the home, and be able to deal with the conditional-access systems from different service providers over satellite, cable and radio (including MDS and SHF) channels. It will be responsible for billing and authentication, messaging and security.

sex/Sex

• **Symbolic EXpression:** Also known as S-expression.

• **Male vs. female:** As applied to the pins and sockets in the two sides of a standard cable connector.

This is sometimes confusing because the 'sex' of a cable-connector should always be associated with the plugs (male) and sockets (female) of the connectors making the electrical contact. But it can also be misapplied to the outside casing of the connectors, and how they fit together.

• **sex converters/changes:** These are small dual-plug or dual-socket devices (a couple of inches long) which convert a 'female' cable connector into a 'male' or vice versa. With RS-232 you should always have one of each in your toolkit — a male-to-male and a female-to-female. Naturally they are also called 'gender benders'.

• Don't confuse gender benders with null modems. They may physically look the same but they perform totally different functions.

sexto

In printing, this is folding a nominated paper size into six. Octavo is the same in eight, and quarto was originally into four. This was irrespective of the former paper size (but always a broadsheet), which is why there is so much variation in the measured size of quarto paper. Strictly, there is none. However, sexto has retained the sense of just being folded.

SF

Single Frequency. A type of inter-exchange telephone signaling using a single signaling tone. Most commonly, 2600Hz is used for in-band signaling and 3700Hz for out-of-band signaling. The tone indicates on-hook when present, and off-hook when not. MF tones are usually used to carry numbers. See.

SFG

Second Harmonic Generation. This is a technique used to generate short-wavelength blue light from a red laser. It depends on a DuPont crystal substance known as Phosphoric Titanic Kalium (KTPm) which can double the frequency of the input 850nm red light, to produce a 425nm blue output.

SFT

System Fault Tolerance. The idea of having multiple storage (mirroring) devices (processors, memory and large disk drives) so that if one device fails, the data is available from another. There are several levels of tolerance in both hardware and software system fault-tolerance. Each level of redundancy (duplication) decreases the possibility of data loss.

• If a system is to automatically detect low-level problems and isolate them before they are disruptive, then three parallel systems are needed. All will then receive the same input, and all should generate the same output. But if the output of one doesn't match the other two, it is a pretty save bet that the odd-man-out is faulty, and an alarm can be sounded. With only two systems in parallel, it is impossible to tell which is faulty unless the system fails dramatically.

• See RAID also.

SGI

Silicon Graphics Inc.

SGML

Standard/Structured Generalized Markup Language. This is an extensible standard devised by book publishers to shortcut the typesetting process from word processing disks. Since 1986 it has been an ISO standard for describing richly-formatted documents. It was originally developed in association with IBM and later adopted and further developed by the US Department of Defense. It is now used worldwide in electronic documents.

SGML involves keying in codes which the typesetting system can later translate into the appropriate action: to create a header or footer, change font, use a symbol or create a new paragraph, etc. Basically the SGML code is embedded as character symbols like <p> for new paragraph, and <h1> for a headline — which then refers in a pre-set table of font size/-styles, to the one named Headline1.

Desktop publishing processes may make SGML largely unnecessary for in-house systems since the typesetting is done on the screen as part of the electronic desktop publishing program. However, many applications have incorporated SGML as a file-type because of the shift to electronic delivery of formatted documents.

The later versions of SGML provides us with an open standard which includes hypertext links, and which will work on most operating systems and many applications. It is an extensible system which will probably be improved in future years. It is therefore ideal for CD-ROM and electronic book publishing.

SMGL is now able to describes the document type, plus the markup language and the instances of each type (see HTML). The document has various parts which are not necessarily in the same file:

— The character set, and the identifying characters which distinguish text from tags.
— The document type, and the list of legal markup tags.
— The document 'instance' which has the text and tags.

HTML now has standardized the first two parts, so only the instance is variable.

• The DTD (Document Type Definition) is the equivalent of HTML.

• Most applications in the future will probably provide 'SGML-Save' as a menu option along with 'Text Only' (ASCII-Save) and the proprietary file-save formats. See HTML, Acrobat, RTF, and DCA/RFT also.

shading and shadows

In CAD, these are image modification processes which can make changes to duplicate the way light would be falling on the objects from one or more defined sources. See ray-tracing and rendering.

shadow mask

The very thin, highly perforated metal plate behind the screen in a color picture tube. See triad and vertical stripe.

shadow RAM

A speed-up technique being offered on most PCs. Typically, the BIOS routines found in ROM will be offloaded into a section of RAM within the memory map in the Upper Memory so as to provide faster access. ROM is quite slow, and accessing BIOS routines in ROM (which is done constantly) can slow down a PC noticeably.

• RAM shadowing is typically done in the area of RAM above the 640 PC limit where BIOS normally resides. It will be found in the identical location as normal BIOS.

• Cacheable BIOS is another similar technique that is even faster. Special SRAM chips are used for the BIOS operations.

Shannon–Fano

A compression coding system very similar to Huffman.

Shannon's equation

An equation which is used to calculate the theoretical data-rate limit of a communications channel. It was devised by Claude Shannon in the 1940s and it has stood the test of time.

The maximum theoretical speed is dependent on the end-to-end bandwidth of the system and on the signal-to-noise ratio of the channel being used. The variable C is in bits per second, the log is logarithm to the base 2, the bandwidth of the channel W (in Hz), while the S/N ratio is measured in dB (usually about 30dB for a telephone line). The equation in its most basic form is:

$$C = W \log2 (1 + S/N).$$

On this basis a typical analog phone line with a 3kHz bandwidth should be able to carry data at rates to about 32kb/s while a really top analog line might just go to, say, 38kb/s. This is more correctly called the Shannon–Hartly law.

This is the theoretical limit — but note that it is for a link right through an analog network. Not (like ISDN) just in the local loop.

Shannon's five criteria

In the 1940s during the Second World War, Shannon established five criteria for how difficult it is for code-breakers to break an encryption system. He said it depends on:

- The amount of secrecy offered.
- The size of the encryption key.
- The simplicity of the enciphering and deciphering operations.
- The propagation of errors.
- The extension (length) of the message.

It probably looked more profound at the time.

shaped beam

In satellite systems these are produced by shaped reflectors. They are beams of irregular shape which are designed to concentrate the power of the directed beam onto various parts of the country, at the expense of other parts. They use shaped beams with direct broadcasting satellites, so as to put the signal power into areas where the population is concentrated (and where, in cities, they can only use small antennas). These 'hot-spots' are sometimes called pencil beams.

shared

• Shared logic is a multiuser computer being shared between two or more terminals.

• Shared resources are peripherals which are used by two or more PCs.

• Shared DASD is a disk accessed by two or more computers within a data center. These are file or database servers attached to a mainframe or minis. See DASD.

shared-memory

An architecture used in parallel-based supercomputers. From the viewpoint of programming, this is a relatively easy form of parallel processing to handle. Cray Y-MP supercomputers are of this type as are some of the super micros. The main problem with shared-memory architectures is that the multiple processors spend far too much time contending for the bus. See macroparallelism. The distinction is also with message-passing parallel computers.

shareware

A semi-public domain approach to retailing software where the program is released openly, with permission for copies to be distributed freely. However, this is done on the condition that any user who decides to retain a copy for private use will pay a small fee by direct mail to the author of the software. The author retains full copyright.

SHDS

Switched High-speed Data Service. A UK term for SIDMS (see).

shell

The surface of a program, the look and feel — the I/O interface with which the user interacts, and which interprets commands before passing them onto the operating system. The part of a modular program that acts as the interface between the user and the program's kernel.

The more correct term is 'command interpreter'. A shell protects you from the complexity of normal operation or normal programming by providing a GUI, menu and/or toolbox.

The use of the term 'shell' suggests that this interface can be swapped or switched in some way.

• Windows 3.x is a shell providing a GUI interface for MS-DOS. It also provided such services as mouse, memory and printer management. This commonality provided by the shell also allows applications using Windows to exchange information through a clipboard and to share device drivers. Windows 3.1 is now the graphic shell for Windows NT 3.5 also. Windows 95 is not a shell, but a complete OS.

• In Unix, the two most popular shells are the Korn and the Berkeley C shells.

• With expert systems, the shell is the 'hollow' program structure before the knowledge base is entered. Sometimes it is little more than an expert system language.

• The most popular of the old DOS shells before Windows took over were Norton Commander and Xtree. They made it easier to copy, move, rename, delete or backup your file.

shell account

This is the most basic of Internet connections for a user. It allows a user to dial in and use the system for e-mail, Telnet, and Usenet material. See SLIP and PPP.

shell sort

This is a variation on insertions sort.

SHF

Super High Frequency. That part of the electromagnetic spectrum between 3GHz and 30GHz which is generally used (at present) for satellite and radar.

These are centimetric waves between 10 and 1 cm in wavelength, and are also known as microwaves. About 10GHz of this spectrum is used for Ka-band satellites, and this alone is equal to all the radio bands in use today.

SHF, with a total bandwidth of 29,000MHz, alone has nine-times the capacity of the total bandwidth in use to date, from VHF down.

More conventional communications services, such as short range radio and television links are now beginning to use these frequencies. However these bands suffer from rain-fade problems, so they must be used for less critical, short-distance devices.

Being line of sight only, this spectrum can be re-used in a cellular way with terrestrial transmissions over short distances up to 2–3km. It can have space-division diversity, in addition to the enormous increase in usable bandwidth — creating spectrum abundance.

SHF band promises to provide data and voice services to small hand-held mobile units, with the potential to make low-cost communications available to everyone. And, at these frequencies, it should also be possible to manufacture true pocket-size and wrist-watch sized communicators.

CellularVision has an analog MMDS trial service operating in this frequency band in New York.

The main users of this part of the spectrum at present are:
- Intelsat (4, 6, 11 and 14 GHz).
- Ku-band satellites (12, 14 GHz).
- Terrestrial microwave links (2, 4, 6, 13 GHz).
- Police radar (10.5, 24 GHz).
- Outside broadcast TV links (7 GHz).

SHG light

Second Harmonic Generation light. See laser.

shielded cable

Some forms of cable are wrapped with a metal wire mesh or aluminum foil to prevent electrical interference. Usually (but not always) this metal sheath will be grounded. A true shielded multi-conductor cable has the metal conductive shielding surround each pair (not just around a group of pairs) to prevent interference from cross-talk or electrical discharges.

Relative immunity from interference makes these cables ideal for higher data rates. Shielded twisted-pair can carry data at rates up almost to the Gigabit range for moderate distances. Don't confuse shielded with 'screened' cable, where the conductive sheath is only around the cable as a whole. See STP.

shift key

• **Generic:** Any key which changes the status (modifies the ASCII output) of other keys. The Shift and Control keys on the keyboard are both shift keys. The Alt key is a modifier, but it adds nulls to create a multi-byte special command.

• **Cap Shift:** The particular Shift key which produces capitals when coupled with an alpha key. The Shift key subtracts 0010 0000 from the output of the lower-case alpha keys, but this isn't consistent over the whole keyboard. Note that the Shift and Caps-lock keys don't necessarily have the same effect in modifying every character key — especially on the numerals.

• ShiftLock, CapsLock and NumLock are classed as shift keys.

shift(ing) register

• A special register or section of memory in a processor used for performing arithmetical operations. It shifts binary data sideways. Normally the process will be something like this:
- some binary mathematical operation will be performed on the data in the register, then
- each logical bit (0s and 1s) will make a one-step-to-the-left/right shift,
- the least- or most-significant bit will fall off the end,
- the mathematical operation will be performed once again, followed by
- another one-step-to-the-left/right shift, and so on....

Most binary mathematical operations need shift registers.

shock ratings

For a hard-disk unit, physical shock ratings are specified for operational, non-operational and vibrational shock. The figures provided are multiplications of gravity (G-forces).

Short-Messaging Services

SMS. A general term which includes one-way paging, but more often refers to the new two-way data transmission systems which are intended for messages under about 2000 characters. GSM cellular phones now have a SMS facility able to receive messages up to 140 bytes.

You can look on SMS as being a bidirectional e-mail service where the return signal confirms that the message was received. You could also see it as a two-way 'acknowledgment pager'.

shortest path routing

These are algorithms which calculate costs and distances between end-terminals or between routers, and minimize both. They will decide to use the cheapest available.

shortwave

See HF.

shot

A single 'take' with a still or movie camera. In the film and television production industry, a 'sequence' is broken down into 'scenes' (one continuous action within one location, at one time), then scenes are broken down into shots (a continuous run of the camera). Each shot may be repeated (each shot-attempt is then called a 'take') during the production stage until they get it right. Shots are spliced together in editing to create a sequence.

shot sizes

Camera operators and directors need a common language to indicate the relative size of the shots they need — especially in film production where the image is often not viewable on a monitor. This is more strict when related to humans (actors), rather than to scenes. Unfortunately, every national industry has evolved its own terminology, so no strict standard has resulted.

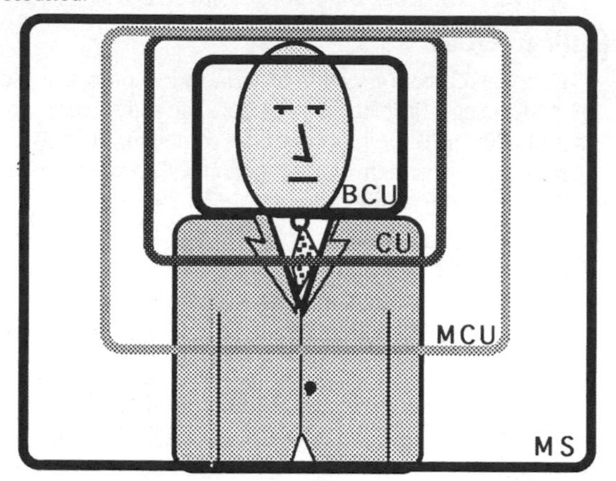

• **Humans:** The most widely used shot-size terms for interviews or for scripting drama involving people, are these:

— **BCU: Big Close Up.** The top and bottom of the frame cut across forehead and chin. This shot should be used very sparingly. It is interpreted as being inquisitory.

— **CU: Close Up.** The top of frame cuts at the top of the skull, and the bottom of frame cuts below the knot of the necktie.

— **MCU: Medium CU.** The frame cuts just above top of skull, and just below the nipple line.

— **MS: Mid-Shot.** Cuts just above top of head, and at the waist.

— **MLS: Medium Long Shot.** Above the head, and at knees.

— **LS: Long Shot.** Full height of figure. It cuts just above the head and slightly below the bottom of the shoes.

— **XLS: Extreme LS.** Of indefinite size. The figure is well in distance occupying probably less than half the frame height.

• **Groups:** When more than one person is involved, the shot may be described as:

— **Two-shot.** This is vague, except it specifies that both participants should be within the frame. Generally it will be a Side-on Two-shot, or an OS Two-shot, or a Wide-angle Two-shot. There is also a Three-shot, of course.

— **OS: Over the shoulder.** When two people are talking face to face, this means the frame should show the shoulder and side of face (from behind) of the closest participant, and a full face of the other.

— **POV: Point of View.** This is where the camera takes a position, as if it were one of the participants. Other people in the scene will then look (almost directly) into the camera lens, and talk to it. Usually the eyeline will be very slightly to one side, rather than directly into the camera. See eyeline.

This creates a form of subjective camera, rather than the normal objective camera, where the audience are disinterested viewers of the scene rather than participants in it.

• **Scenes:** General scenes can only every be described vaguely. Some of the terms used are:

— **ES: Establishing shot.** This is usually a wide-shot which takes in the overall scene, and gives viewers an idea of the setting.

— **WS: Wide Shot.** This just means a shot taken using a wide-angle lens. It is essentially the same as an establishing shot, but it may be used for other purposes.

— **CU: Close Up.** Other than with people, this can only be interpreted in context.

— **LS: Long Shot.** When used for scenes, this means the use of a long-lens (telephoto) to compress perspective, and isolate an object or subject against a more restricted background.

— **POV: Point of View.** Showing a scene from the position of the main character involved. If he/she is high up a radio mast, the camera will be placed high on that mast. If the character is lying on the ground, the camera will be placed on the ground.

shot transitions

In film and television production, two shots can be joined abruptly, or they can transition. This is very easy to do in television electronics, but more complex in film (it is always done in post-production, never in camera for professional productions).

— **Cut:** the last frame of one shot, is spliced to the first frame of the next.

— **Fade to black:** The last few frames (between 4 and about 72 frames) gradually decrease in brightness, and eventually become black. This is usually followed by a fade up from black of the next scene — but not always. It carries the idea that a sequence has ended, or that time has passed.

— **Fade to white:** This is rarely used, and is classed as a special effect.

— **Dissolve:** This is a Fade-to-black with a Fade-up-from black superimposed. It usually says 'meanwhile' or indicates that something related is happening elsewhere.

— **Superimposition:** One picture overlays the other — it may fade in and fade out, or cut in and cut out. Superimposition is used mainly for white titles. In film, you can only superimpose black in film negatives, so the process can be complex. Much superimposition requires a matte/key process.

— **Wipes:** This covers every imaginable shape and movement; horizontal, vertical, corner, iris in and out, etc. One picture progressively replaces the other on an area-basis.

— **Mattes and keys:** These are used in film for special effects where one picture is inserted into specially selected areas of the image. In television, the equivalent is internal and external keys, and chromakey. see.

— **Zip-pan, Zip zoom:** Very fast pans and zooms can be used to momentarily distract the eye while the image cuts from one scene to the other. In a high-action film, a zip-pan can mean 'Meanwhile, back at the ranch!'

— **Music and effects:** Don't under-estimate the importance of sound changes in signifying scene transitions.

Showscan
A film production system which uses 65mm film running at 60 frames per second. It is later printed onto 70mm film stock (with multiple sound tracks) and used for very large screen film projection in special theatres. See IMAX and Iwerks also.

SHSD
Switched High-Speed Data. The UK name for the SMDS-DQDB systems. See 802.6 and CBDS also.

si/SI

- ASCII No.5 — **Shift In** (Decimal 15, Hex $0F, Control-O). This control character is used in conjunction with SO and ESC to extend the graphic character set of the code. Characters after SI should be interpreted differently. See SO.
- **Stepped Index** in fiber optics. See.
- **Silicon:** In chip construction, Si is shorthand for Silicon which is used for IC substrata as distinct from GaAs, etc.
- **Systems Integrator:** The person or company (usually a contractor) who is responsible for the integration of business information and communications systems into one coherent whole. See system integrator.
- **System Index.** A computer benchmark. See Norton.
- **Systeme Internationale** units. This is the world standard for metric units in science and technology. These units are gradually replacing the British Imperial System in the UK, and the US Customary System in the USA. See SI Units.

SI-units

- For amounts above the fundamental unit:

kilo-	10^3	k (lower-case)
mega-	10^6	M
giga-	10^9	G
tera-	10^{12}	T
peta-	10^{15}	P
exa-	10^{18}	E
zetta-	10^{21}	Z
yotta-	10^{24}	Y
xenna-	10^{27}	X
vendeka-	10^{33}	V

- For amounts less than the fundamental unit are:

deci-	10^{-1}	d
centi-	10^{-2}	c
milli-	10^{-3}	m
micro-	10^{-6}	mu (μ)
nano-	10^{-9}	n
pico-	10^{-12}	p
femto-	10^{-15}	f
atto-	10^{-18}	a
zepto-	10^{-21}	z
yocto-	10^{-24}	y
xenno-	10^{-27}	x
deko-	10^{-33}	v

SIC
Standard Industrial Classification. A way of identifying a type of business using a numeric code. SICs are used with keyword searching during information retrieval. They will be in the Descriptor field. See Descriptors.

sideband
Whenever one analog frequency (A) is amplitude modulated by another (B), two sidebands are generated around the central carrier frequency. These correspond to harmonic frequencies A+B and A–B and are known as upper and lower sidebands.

The resultant wave can be thought of as consisting of three separate waves, each with a constant amplitude. This is why an AM signal (as with FM) is said to have a bandwidth, not just a frequency. These sidebands aren't just mathematical abstractions, they can be separated by filters and used individually.

With Frequency Modulation (FM) the sideband issue becomes even more complex. An FM wave can have more than one pair of sideband frequencies for each modulating frequency. You get sidebands at frequencies (Carrier plus and minus the audio baseband); Carrier +/– 2-times audio baseband; Carrier +/– 3-times audio, etc. However, the power of these harmonic sidebands drop off quickly and aren't often noticed. A phase relationship comes into this picture also.

- With amplitude modulation, only one of these sidebands is really necessary, so one can be discarded, which then produces single-sideband transceiver equipment. See SSB and VSB.

sidebar
A small (often lightly color-toned) box set into magazine or newspaper copy, which carries supplementary information about the article.

sidelobe

- **Radio transmitters/receivers:** This refers to the off-axis response of an antenna — the amount it emits as spurious signals not along the directional path intended. This applies both to the pattern of energy when transmitting, and to the sensitivity when receiving.
- **Microphones:** Sidelobes in microphone patterns are the same as above. They are unwanted areas to the side and behind a microphone where the sensitivity of the microphone is high, when ideally it shouldn't be. This problem is very frequency-dependent.

sidetone
Don't confuse this with sideband. A sidetone is part of the microphone signal in a telephone handset which is fed back into the earpiece to allow you to hear yourself speak. Without the correct amount of sidetone, the phone sounds dead.

SIDF
System InDependent File system. This is an open file system which is media independent. It can work on tape, MO or CD optical disk, WORM, hard disk and floppy. The ECMA has accepted it as the standard. It is based on Novell's SMS.

Sierra

- A fairly recent IBM mainframe series.
- CD-ROM. See High Sierra format.
- One of multiple layers in a stacked chip architecture — as distinct from planar (single-layer) chip architecture.

Sieve of Erathosthenes

An old, but once popular, general purpose benchmark test for computer number-crunching ability. It measures how a system will handle CPU-intensive calculations by deriving a series of prime numbers by brute force (so-called 'naively'!). It therefore isolates pure ALU/CPU performance from more sophisticated tasks.

The sieve works like this.

— Begin by writing an integer number sequence, starting at 1.
— Cross out all multiples of 2, but leave 2 itself in place.
— Cross out all multiples of 3, but leave 3 itself in place.
— The next integer is 5; cross out all multiples of 5, but leave 5 itself in place.
— Continue the above process to infinity.

Any integers remaining are prime numbers.

SIF

Source Input Format. This is a standard for video encoding based on 352 pixels and 240 lines of image at 30Hz. It is the basic image format for MPEG1.

SIG

Special Interest Group. This is usually a group within a computer club or association, or an on-line group/forum of people with a shared interest. The term is often used for computer conferences, but it is also applied outside the computer world.

signal

Strictly, this is the information-carrying physical waveform or pulse, as distinct from the carrier (although the term is also used widely for the carrier). Baseband analog and digital systems only have a signal. The term 'baseband' suggests that the signal has not been modulated onto a carrier.

- There's one complication here. A data signal may be modulated onto an audio tone carrier, by a modem. However, from the viewpoint of the telephone network, these audio tones are still 'baseband' (it doesn't know about the modulation). It depends on the viewpoint.
- Radio systems always have a signal modulated onto a carrier, and it is the carrier which is tuned (selected) to extract the one information stream from the mass.
- Signal-to-noise ratio. See SNR.

signal distance

See Hamming distance.

signal processing

The general term applied to both analog and digital signal manipulation techniques. Signal processing can extend from the pre-emphasis applied to analog FM radio during the broadcasting stages (to decrease the signal-to-noise figures), to the high-speed manipulation of digital motion video images by Fast Fourier Transform chips.

The pace of development of digital signal processing (DSP) chips has outstripped all other computer developments by a factor of at least two for the last half-decade. This has made MPEG, ATM and all the concepts that will drive the 'superhighway', now possible.

- Multimedia Processor is the term now being applied to the latest in a long line of very rapidly evolving DSPs that can encode and decode highly compressed wide-screen image and sound in real-time.

signaling

The transmission of control information through a network.

- In telecommunications there are four applications for the exchange of signals:

— **User-network signaling.** To establish and control calls.
— **Inter-office signaling.** By telecos between exchanges for the establishment and control of the connection, and for the exchange of billing and network traffic information.
— **User-user signaling.** End-to-end information. This is not currently used to any degree.
— **Network-user signaling.** Busy signals, etc. Also CLASS services like CLID, which provide users with information.

- Traditionally there are also four distinct categories of signal:

— **Supervisory.** On-hook, off-hook, disconnect, etc.
— **Addressing.** Carry telephone numbers.
— **Information.** Dial tones, busy signals, etc.
— **Maintenance.** Routing etc.

More recently however, with CCS#7 signaling and the development of CLASS we must include customer-information signaling such as CLID and ADSI (which can also include names), and customer-control information such as CLID call-blocking, call-trace, etc.

See CCS#7, D-channel, hook flash, out-of-band signaling, in-band signals, DASS, DPNSS and CLASS.

signature

- **Authentication:** With encryption systems, see digital signatures.
- **Printing:** Magazines and books are printed on large sheets, with many pages to the sheet. A signature is one of these sheets after it has been printed and folded (usually into 8, 16 or 32 pages).
- **Messaging etiquette:** A three or four line sign-off included by many people at the bottom of their e-mail or Usenet messages. They will often include an esoteric quote from some obscure Outer Siberian Buddhist monk — which the writer thinks is extraordinarily perceptive or humorous and wishes to share with us over-and-over-and-over again.

signature recognition

A technique used for virus detection. The program will check a disk, attempting to match sequences of code on the disk with a library of known viruses' object code. For signature recognition to be successful, the detection program must hold 'identikit pictures' of some of the key code strings of all major current viruses, so such a system can only be used for known viruses.

Generally these programs are out of date by the time they are purchased, but they can catch some of the more obvious and long-term viruses circulating.

• Other anti-viral techniques involve approaches such as the use of built-in checksums of an application's object code as a front-line of defense.

signed numbers

Numbers to which a plus or minus has been assigned. In many computer operations, the most-significant bit (the left-most bit) is designated as the sign. When the sign-bit is set to logical 1 the number is negative: if it is 0 then the number is positive.

significand

In the IEEE's standardization of floating-point arithmetic, the format for real numbers consists of a sign bit followed by exponent bits (8 or 11), and significand bits — up to 23 for single real numbers, and 53 for double-real and double-extended numbers. This contains either the mantissa or just the fractional part of the mantissa.

In scientific notation, numbers are represented as a significand and an exponent. In the number 2.1167×10^7, the 2.1167 is the significand and the 10^7 is the exponent.

significant digit

In decimal and binary notation, the digit to the left is the 'most significant' digit, and the one to the far right is the 'least significant' digit — for obvious reasons. See big-endian and little-endian also.

silent speed

The old silent movies were hand-cranked at a frame-rate of about 16 frames per second, but, as you would expect, this was rather variable. The early 8mm home silent movies were cranked at 18 fps.

• Modern movies run at 24 fps, as does US NTSC television (frames per second, not field rate), while PAL and SECAM countries run their movies at 25 fps.

silicon compiler

A VLSI chip design approach invented by Johannsen, and named in the tradition of a 'software compiler' — which allows a program to be written in a high-level language, then machine-translated down to low-level code.

The silicon compiler supposedly does the same thing for the design of chips. The designer specifies the requirements and outlines the functions to be performed, and the compiler program takes over the tedious steps of designing a chip's logic, circuit and chip-geometry layout. See PLD.

silicon disk/storage

This refers to the use of large banks of battery-backed RAM, EEPROM, flash EPROM or FRAM chips which can be used in place of a hard-disk unit. At present these are very expensive on a dollar per megaByte basis, but the costs are coming down. Flash cartridges are a form of silicon disk storage.

Silicon Valley

An area around the satellite city of Palo Alto and Stanford University, to the south of San Francisco (along the side and south-end of the bay). This is the Californian center of computer developments.

The area is a flat, continuously-urbanized sprawl of suburbs which have names that are readily recognisable to most computer buffs. San Jose is at one end of the area (and a city in its own right), while the other main 'suburbs' are Mountain View, Menlo Park, Los Altos, Cupertino, Santa Clara and Sunnyvale.

silo

Mainframe people often speak of tape silos. These are large banks of pigeon-hole storage for computer tapes which are now generally retrieved by automatic machinery. See jukebox and farm.

Silliwood

A pejorative term for the new 'virtual games' production industry growing up around San Francisco. It is SILIcon Valley plus HollyWOOD. This refers to virtual reality and video games companies.

SIM

Subscriber Identity Module. A smartcard standard incorporated into the GSM mobile phone. It holds key information, such as the international mobile subscriber identity (IMSI) which identifies the subscriber to the network, a PIN, and user-data such as abbreviated dialing numbers. The SIM provides the mechanism for authenticating itself to the network, and it has a number of security measures built-in.

In fact, the SIM is the actual customer-identification on a GSM network, so it can be removed from the handset and used with any other GSM unit. There are normal credit-card sized SIMs, and special 'broken-out' postage-stamp sized SIMs for small handsets. SIMs conform to the ISO 7816-1, -2, and -3 standards.

SIMD

Single-Instruction Multiple-Data. One of the two most popular types of parallel computer architecture. The other is MIMD. SIMD has processors which all carry out the same operation on many pieces of data at the same time. This is the approach used by Thinking Machines Corp.

This technique seems to be ideal for image processing. To get good performance, SIMD applications should use data in a way that allows it to be divided evenly between the processors so all are constantly active.

All the different forms of parallel processor are extremely difficult to program, but SIMD machines are said to be easier than MIMD. See MIMD also.

SIMM

Single In-line Memory Module. A convenient method of packaging memory chips to allow them to be plugged in and removed without the risk of damaging the delicate feet of the older chip designs. SIMMs are far more compact also and they have become the dominant type of memory chip.

A typical SIMM module has nine DRAM memory chips, eight of which are used for data and one for parity or error checking (8-only in Macs). The SIMM uses less power and generates less heat than DRAM chips also.

Currently you can purchase 256k, 1MB, 2MB, 4MB, and 8MB SIMM modules with 16MByte modules on the way. But:

- Check the SIMM access times against your computer's requirements.
- SIMMs must be plugged in the right way around. Check for the Pin 1 end-mark.
- 72-pin sockets are used with some devices and 30-pin sockets with others.
- Some SIMMS are 8-bits wide and some are 9-bits (written as 4Mx8 and 4Mx9).
- All the SIMM in a bank must be of the same type.
- There are both 'industry-standard' and 'non-industry standard' SIMM modules. Some of these modules are cards with SIMM contacts and standard low-rated DRAM memory chips on board. So be careful when buying.

simplex

• **USA:** In the US, this means the transmission of information in one direction only. There is no provision for back-channels carrying control information or acknowledgment, therefore there is no way to prevent buffer overflow or to signal when transmission should start and stop. Simplex was used in the early PC days for loading programs from audio tape recorders. Radio and TV are also a simplex, if you want to get technical.

• **Europe:** In Europe, simplex often means 'half-duplex' which isn't the same. Half-duplex does have a back-channel, although it is separated in time from the forward channel.

simulation

Don't confuse simulation and emulation. When you simulate something you are testing the design concepts, so all the essential parts are in place (if only temporarily). When you emulate something, the computer is a black-box. You don't care how the emulation is done provided the I/O interfaces match the original.

simulcast

To transmit simultaneously. This is usually a technique used for TV or radio channels.

• High-quality FM sound is often simulcast with TV to provide the best possible sound over the separate FM radio channels.

• In the earlier US HDTV context, simulcast means to transmit a normal NTSC 4:3 signal on one channel, and to reserve a second, quite distinct, channel for the HDTV (analog) version.

The distinction here was with augmentation systems which try to add the extra HD information to the basic NTSC signal (usually with an extra digital stream) so that the same channel remained in use by both old and new receivers.

SINAD

Signal to Noise and Distortion measurements.

sine wave

The wave created by a particle in a regular orbit when it is graphed against one dimension in time.

Pure audio tones of a single frequency (flute-like), or radio carrier waves before modulation, are both sine waves.

A sine wave has three characteristics: amplitude, frequency and phase — and frequency-change and phase-change are inter-related. With modems, the importance of this is that it provides the three main modulation methods — although only amplitude can be used with frequency or phase modulation systems. You can't use frequency and phase modulation together.

• Alternating current for mains power, in its purest form, is a sine wave. In audio, a near sine wave is produced by a flute. The sub-text here is that there is no distortion or harmonic modulation which contributes 'tone' to audio frequencies.

• Fourier tells us that all complex waves comprise integer harmonics of a fundamental sine wave, summed at different levels of amplitude.

single assignment languages

See functional programming languages.

single bit

In CD, see bitstream.

single current (telex)

A telex transmission method where the voltage is varied from –50 volts to earth potential. Early phones were of this kind also. See double current.

single density disks

See FM.

single-ended circuits

An old name for four-wire telephone circuits.

single-frequency laser diode

A laser diode used in coherent optical fiber transmission systems where it is important that only one laser 'mode' (frequency) is permitted. Theoretically a single-mode laser only has a single frequency output, but in practice it has a narrow line-width ('bandwidth') around this frequency. See Distributed Feedback (DFB) lasers.

single mode optical fiber

Also called monomode and used more for telephony than for LANs. This is optical fiber with a very fine core diameter. And, as a general principle, the thinner the fiber, the higher the data rate.

The 'mode' here should not be confused with the various laser modes (which refers to harmonic frequencies). This fiber mode is associated with the various physical distances that light particles can travel down a single length of fiber. This is largely dependent on the angle at which the photons enter the fiber and, therefore, on the amount of internal reflection (and therefore additional distance) involved as it travels.

If the acceptance angle of the fiber is wide (as it is with the thicker multi-mode fiber), the many photons involved in each pulse will be travelling over different distances to reach the end (i.e. different modes), and will therefore arrive at slightly different times, thus spreading the pulse. This can reach the point where one pulse can't be distinguished from another.

Single mode fiber restricts the range of distance-variations, and therefore the development of this pulse 'edge-smearing' — so it is possible to increase the data transmission rates down these fibers without the pulses merging over long distances.

However the narrowness of monomode fibers does not allow the cheaper LED sources to be used, so it is almost invariably associated with lasers. The distinction is with multi-mode and graded-index fiber.

single sideband

See SSB.

single tasking

Able to do only one thing at a time. See tasking.

sink

Telecommunications slang for the end point of a data transmission where it is consumed, or disappears from the network. Engineers often talk of sources and sinks, or ingress and egress points.

SIP

Single In-line Pin. The physical design of a type of electronic component for PCBs. Contrast with DIP/DIL. See also SIPP.

SIPP

Single In-line Pin Package. SIP packages. Usually memory chips which come plugged into an edge-connector card ready to be added to the motherboard. These are substitutes for SIMM memory modules. There are usually 8 or 9 chips to a SIPP or SIMM, with the 9th for parity. See SIP also.

SIR

Signal to Interference Ratio. All cellular phone systems are limited by the amount of radio interference as compared with the signal levels. It is quoted as a ratio in dB.

site licences

The sale of software usage rights to large organizations on a site-by-site basis, rather than as individual packages. Generally the deal will involve a considerable discount for between 50 to 100 copies — but all copies must only be used in the company's computers at the one location. See licences.

SIU

Subscriber Interface Unit. See OIU.

six exchange

A SPC crossbar exchange using CCS#6 protocols.

skeleton

In pattern recognition. See thinning.

skew

- In telecommunications, the skew is the time-delay between any two signals which are being compared.

- In VCRs, skew is a tape tracking problem which is usually adjustable. It causes the top few lines of the image to move substantially to the right.

skimmer

This is a phone thief who drives around late at night with a cordless phone listening until he detects another phone using the same frequencies. If this handset is off the base unit, the skimmer can often dial up and makes long distance calls.

Some of the older cordless phones allow you to do this even when the phone is on the base.

skin-effect

When any high frequency current flows along a conductor, the distribution is not uniform through the cross section of the wire. The current is concentrated on the surface or skin of the conductor.

This effectively reduces the area of the conductor carrying the current, and increases resistance to the alternating current. For this reason solid core wire will out-perform stranded cable.

- Higher-frequency signals ride on the surface of copper wire, and the losses are roughly proportional to the square root of the frequency. The thinner the wire, the smaller the circumference, and therefore the less the surface area — therefore, the greater the signal attenuation.

Skipjack

A family of encryption techniques (algorithms) proposed by the US government to permit a high level of personal security in voice and data messaging while allowing government agencies to decipher conversations under substantial legal constraint (escrow provisions).

The Clipper chip was to have been the first element of this family of hardware-based, government key-escrow encryption devices. Clipper itself was purely for voice and low-speed data, but other members of the Skipjack family, including Tessera and Capstone, were to be compatible with Clipper. They were intended to lead the way from a [CIA-breakable] encryption system for voice to a breakable encryption system for data.

sky waves

Radio waves which are reflected by the ionosphere. Both medium frequency (AM broadcast) and high-frequency (short-wave) signals use sky waves. Single-sideband HF travels extraordinary distances when reflected in the right conditions. See propagation, ionosphere, ground wave and surface waves.

slamming

The unauthorized switching of the default carrier. This is a technique of switching, say, a long distance default carrier on a phone line from one preferred supplier (say AT&T) to another (say Sprint) without the customer's permission or knowledge. Some of the local companies do this to get the benefit of better deals. Slamming also takes place in cellular phones.

slate interfaces

See pen interfaces.

SLC

• **Subscriber Loop Circuit:** The local loop. The connection between the user and the local exchange.

— SLC-96 and SLC-5 units are digital loop multiplexers used by carriers for pair-gain. This saves the company running new twisted pair down the streets, but it decreased customer service quality (sometimes disastrously with modems).

• **Subscriber Line Charge:** A US charge of about $4–6 per line which is imposed as an indirect tax by the FCC, and used to subsidize local call costs to keep basic residential telephone services affordable. This is part of the universal service obligation of carriers.

The SLC is federally mandated and is set annually for each channel but the amounts charged for SLC vary by jurisdiction.

SLED

Single Large Expensive Disk. See RAID.

sleep mode

Some computer equipment (those equipped with CPUs which are capable of retaining some functionality in a low-power mode) can self-activate to answer an incoming phone call.

This is the 'standby' or 'sleep mode'. The new AMD386 chip, for instance, has a sleep mode which consumes less than a milliamp of current (less than 1% of the Intel chip's normal rating). The computer needs to maintain the memory also, of course.

Sleep mode ('snooze') is common in most laptops these days. Remember that with laptops having hard-disk drive units (most of them these days), it may put less drain on battery power to leave a laptop in sleep mode for a few days, than to fire-up the hard-disk to re-boot.

• The first notebooks (designed by Microsoft) had no disk drives, and their applications were fixed in ROM. Everything was CMOS. Sleep mode was their standard mode, and they could remain asleep, powered by four AA-cells, for nearly a year.

SLIC

Subscriber Line Interface Card. The line card attached to the subscriber's line in the exchange. This can have analog or digital electronics, but nowadays digital will possibly be cheaper.

• In recent times the cost of analog line cards has begun to creep higher than digital (ISDN) cards. Since the inter-exchange network is now digital in most developed countries, it is more expensive to perform the ADC necessary for the analog cards and to supply the DTMF tone conversion. With ISDN this all happens in the customer's premises.

slicing

The opposite of concatenation. Strings can be joined together by concatenation, and split asunder by slicing — extracting the first x-characters in a string, or the last, or from the middle. In most languages there are special functions for string slicing.

• Time slicing is a technique used in multi-user computer systems and in multi-tasking PCs, to give each user/application some of the CPU time exclusively. See multi-user.

sliding dictionary

The LZW principle of data compression where the compressed data itself contains the dictionary of patterns. Pointers and tokens are used to later identify those patterns. The 'sliding' part here is that the dictionary only maintains information from the previous few megaBytes of data, so the dictionary actually changes as you progress through large files. Some of the early compression systems used fixed dictionaries. See LZW and DoubleSpace.

sliding windows

This is a way of relaxing flow control when ACK–NAC systems are being used to check blocks of data for errors.

Most of the early file transfer protocols would send a block of data, then wait to receive an ACK before sending another. This wasn't much of a problem at 300b/s over a local phone line, but at 30kb/s over a satellite link, the time wasted will be many times that used for sending, because of the satellite propagation delay.

Sliding windows allow a (variable) number of blocks to be sent before the ACK is received for the first. This has proved to be valuable in terrestrial systems as well as satellite — wherever propagation delays are experienced.

When an NAC is received by the sender, it will either:

- backtrack to find the block required, and begin transmission from this point on (most do this), or
- only retransmit the one faulty block, then pick up again at the point it left off. Z-modem seems to do this, but it requires a careful numbering of blocks in a sequence longer than the sliding window limits.

Obviously to handle sliding windows at all, the packet protocol must have a sequence identity number for each packet. This can't be too long because it adds to the overhead, and it then impacts on the transmission rate. This limits the sliding window size.

• A variation on the above is that some transmitters will keep track of unacknowledged packets and automatically retransmit them after a certain time. It doesn't need a NAC.

• TCP and many other Transport and Data-link layer protocols use sliding windows. Some modified X.25 systems use this also. Normally the window will only be 5—10 packets.

slip/SLIP

• Slip is a timing problem on a high-bandwidth synchronous data circuit — usually caused by long quiescent periods on the channel which give the receiver the opportunity to lock onto an incorrect bit in the frame. See zero-bit insertion and AMI.

• **Serial Line Internet Protocol:** An early *de facto* 'semi-standard' for point-to-point serial connections running TCP/IP over serial lines, such as telephone lines. It has now been upgraded to PPP in many sites.

A SLIP connection can be used to make a personal computer into a temporary Internet host by connecting it over the phone lines, using modems to extend the TCP/IP connection. For the duration of the connection, the user's local machine becomes a live 'host' and is allocated an IP address.

SLIP details the way in which data is to be encapsulated in IP packets. But no information on packet-type is held in the SLIP header, so a SLIP link is limited to one protocol at a time.

SLIP and PPP, are very similar (the terms are often joined) — PPP was derived from SLIP. But SLIP is a very simple, very basic packet-protocol, while PPP is based on HDLC and is designed to be much more flexible. See UUCP, PPP, mosaic and WWW.

• Despite the increasing popularity of dial-up IP connections to the Internet, the most common situation is still to have a dial-up shell account using SLIP/PPP.

SLM

Spacial Light Modulator. A general term for the key component of an optical computer. This is a discrete or continuous array of light modulators where the transmission/reflection of the surface is controlled optically/electronically using a computer-addressing technique. SLMs can be for input, interconnections, scratch-pad memory or processing devices. See POHM also.

slot

• **Computers:** Expansion slots are the multi-contact elongated sockets attached to the motherboard which are connected by the main computer bus to the CPU and memory. See NuBus, PDS, PCI, ISA, EISA, MCA and S100 also.

• **TDM:** In time-division multiplexing, a slot is a subsection of a frame into which data can be put. It is a fixed-size gap in the frame, directly identified with one input source and one output receiver. Sometimes it will be separated from the other slots by a few bits regarded as 'guard band'. Generally, however, it is only identified by counting from a frame flag-bit.

• **Satellites:** A satellite slot is an angular position in the geostationary (Clarke) orbit. Slots are ideally set 3° or 4° apart, but sometimes they are used with only 1° or 2° of separation, and sometimes satellites are co-located.

Ku-band satellites can be jammed closer together than C-band satellites. Co-located satellites use different signal polarizations or different frequency bands to maintain separation between their signals. See slot-sharing and DirecTV.

• **Film/TV:** Film and video cameras sometimes have slots at the 'gate' (the point of exposure) to allow for filter insertion.

slot-sharing

In geostationary satellites, it is sometimes possible for two (or possibly more) satellites to share the same slot position, provided they are using different parts of the radio spectrum or have different antenna polarizations. A C-band satellite may share an orbital slot with a Ku-band satellite, for instance, without too much of a problem.

• DirecTV in the USA is proposing to have its three direct broadcasting satellites in the same slot so that the receiving antenna can be fixed.

slot time

• **Ethernet:** This will refer to the random time an Ethernet NIC must wait after a collision before trying to transmit again. This is the gap in which it is possible to seize control of the shared network, and it can be up to 512 bit-periods.

This old standard was set to allow time for the transmitters to backoff after a collision. It doesn't mean that packets are limited to this slot-length.

• **Token Ring:** The time it takes for a station to receive a response from another station.

• **Synchronous:** In a time-division multiplexed bearer, this is the time available for an individual user's channel.

• **Asynchronous:** The above also applies to asynchronous systems also. Asynchronous access (as distinct from transport) systems still have synchronous time slots.

slotted ring

An IEEE 8802.7 LAN standard which circulates a series of fixed size slots around a ring architecture. Nodes may insert information packets into these slots as they pass.

• DQDB is actually a form of dual counter-rotating slotted ring system, but the slots are filled by asynchronous-access ATM-like cells.

slow-scan

This is video transmission over low bandwidth circuits. This may be for real-time videophones or for videoconferencing, or it may be for the transmission of high-quality images back to a TV studio for later replay at full speed.

• **Videoconferencing:** Instead of the image being scanned 25 or 30 times each second, as with normal television, to preserve the illusion of full-motion, the image is scanned at a much lower rate and appears jerky, smeared, or as a series of still pictures ('stuttervision').

Almost all videoconferencing is conducted with images repeated at less than 12 per second, and some systems get down to two a second with only parts of the image being updated. This is necessary when the bandwidth of the network is too low for even a highly compressed reduced frame image. See MPEG4, H.261 and H.320.

• **TV News:** Electronic News Gathering (ENG) for TV will sometimes involve the on-location compression of the digital video recordings using MPEG, then transmit two or three minutes back to the station over four to six phone links simultaneously. They often use AMPS analog cellular phones in the USA, but at a very slow rate. The original image will be regenerated in the studio at full rate. It may take a fifteen to send a few minutes of video — but this is quicker than battling traffic.

SLSI

Super Large-Scale Integration. These are chips with more components than ULSI chips. We'll see these in a few years.

S-MAC

Studio Multiplexed Analog Component television. The form of MAC video designed for studio production use. ACLE is a very similar technique. See MAC and N-MAC also.

small capacity systems

In telecommunications these are PBX networks that carry up to 24 telephone channels.

Smalltalk

The first real high-level object-oriented language, developed at PARC by guru Alan Kay's team. It introduced the point-and-click icon and other GUI concepts.

smart antenna

Antenna systems used with cellular mobile phones, where the antenna is intelligent enough to direct all the required energy of a channel only into the very narrow angular sector in which the vehicle lies. Some experimental smart antenna systems will pan the antenna 'beam' with the moving vehicle, while others will use special phased-array controls to handoff between narrow directional sub-sectors while retaining the same frequency.

These antenna systems are also used to produce vehicle-mounted antennas for mobile satellite communications.

smart card

See below

smart memory

See content-addressable memory.

smart power chips

Chips which combine logic and electrical power controls for the electric motors, refrigerators, air conditioners, etc. with the aim of increasing the efficiency of the devices. These are 'embedded' processors.

smart-quotes

Most computers use the same quotation marks at the head of a quotation as they do at the tail. However, in typesetting, one set is inverted. The provision of smart-quotes in some computer programs is a way of a creating these inverted styles. Smart-quotes (enlarged) are: ‘single’ and “double”.

If you intend to do a search and replace in a text program using smart quotes, be aware that both the single and the double forms have different ASCII values from normal quotes. You may need to cut and paste these into the find-replace box. Also note that smart-quotes come in pairs, whereas normal quotes are one each for the single and double form. Each end of the quote has a different ASCII value.

smart receivers

There are many types of 'smart-receiver'. The term applies to:

— Radio broadcast receivers which can find stations via a search function.

— Agile pagers which check for messages across a range of different paging frequencies.

— Intelligent HDTV and EDTV systems with ghost reduction.

Any receiver with an in-built processor will now be credited as being a 'smart' receiver.

• New TV systems: In order to reduce the bandwidth and retain the highest possible quality, our new TV receivers or set-top box will require intelligent digital signal processing. The smart receiver has become a key element in many advanced TV proposals; it must be intelligent, programmable, and have an open-architecture.

Ideally it will adapt to receive and decode a wide variety of signals, (see layered video) and it will obviously need a large frame-store.

The term seems to be applied to four main features:

— Anti-ghosting measures. Digital filtering to remove the late-arriving 'multipath' signals.

— Line doubling processes. The set receives 625 lines but shows 1250 lines by holding the image in a frame-store and double-scanning it.

— Field frequency doubling. Progressive scanning of the whole image field at a rate of 100Hz or 120Hz.

— Ability to handle motion compensation and conditional access techniques. This is now largely displaced by the world-standardization of MPEG2 which has motion compensation built-in, and possibly also of the DAVIC group.

smartcard

Aka Integrated Circuit Cards (ICC). A slightly thick credit-card which incorporates some form of memory and processing power. There are various levels, from just-smart to very-smart, and the amount of memory and the processing power is highly variable. Some manufacturers have recently considered changing from 8-bit processors (basically the old Apple 6502) to 16-bit architectures.

The technology has been developed mainly in France, although Japan and the US are now rapidly moving ahead. With large-scale manufacture, these cards can be produced very cheaply; ex-factory costs vary from a few dollars for basic memory cards to $100 for full-scale intelligent smartcards.

The main advantage of the smartcards for general retail use, is that the authentication of a person's PIN number will be performed within the card, so POS retail terminals do not need to be constantly on-line to a central computer for authentication purposes.

If the card is used as an electronic purse, deductions can also be made at the point-of-sale. The smartcard can carry both credit and debit figures, because the card has very high inherent security. See SVC and CAFE.

Smartcards are now also incorporated into the GSM digital mobile phone system (see SIM) and are widely used in some countries for banking, transport payments, and rechargeable phone cards.

The physical contacts of the cards have been standardized by the ISO and ETSI. The ISO standards include physical tests for bending and for both byte- and block-transmission protocols. They have eight electrical contacts which carry power and signals between the card and the SCT (SmartCard Terminal).

Contactless smartcards have also been developed, and seem to have a very rosy future. They can transfer information over distances of a half-meter using radio signals.

However, in smartcards generally, there is a noticeable lack of international standardization of file and directory structures to enable the one card to have multiple uses.

• In common use there are two sub-types of standard smartcard: CMM (micro and memory) and CLM (memory) however purists would see only the first as a true smartcard.

— Memory-only cards are widely used for prepaid telephone cards. Typically they contain a few hundred bytes of reusable memory coupled with some hardware security devices.

— True smartcards have an intelligent, single chip micro-controller embedded within the plastic. This gives them a high level of data security, and means that data can be securely updated — using EEPROM. RAM is also needed for processing memory, but this requires power which the card can't supply.

• AT&T Bell Laboratories has developed the contactless smartcard. It has no external contacts, and data is transmitted to a reader without metal-to-metal contact. The aim is both to improve reliability and to reduce risk of electrostatic discharge. See RF-ID cards.

See also SVC, STT, VME and CAFE.

SMATV

Satellite Master Antenna Television. See MATV.

SMB

Server Message Block. LAN Manager and some other network operating systems use this to package data. It is a distributed file-system protocol which lets a networked PC use remote devices, files and applications as if these were local.

This is a client-server protocol. The new LAN Manager uses SMB at a low level.

SMD

Surface Mounted Devices. See.

SMDF

Simplified Message Desk Format. See FCI.

SMDI

Standard (Simple) Message Desk Interface. A set of standards used between PBXs and voice mail systems. See below.

SMDI/VMI

Simple Message Desk Interface/Voice Messaging Interface. These are closely related Bellcore standards which are used for Centrex voice mail integration. The protocols use a standard 1.2kb/s–9.6kb/s RS-232 serial data link to provide Caller Line ID from the PBX or Centrex switch to the voice mail device. See FCI.

SMDR

Station Message Detail Reporting. These are electronic reports on the amount of traffic generated by PBX or key systems. They are used by management to check on private network usage.

Most small business telephone systems or PBXs will have their own proprietary SMDR format, and therefore each call-accounting system must be customized to match.

Typically an SMDR record includes call-start time, duration, calling extension, and number dialled. In some there will be fields for record type, route selected, and trunk selected — to check on the intelligent call-routing of the PBX. Each field occupies a specific range of columns in the record so a call-accounting system can easily process the data. See TIMS also.

SMDS

Switched Multimegabit Data Service. This is the name applied to the first US broadband MAN services. These began using the cell-relay DQDB (802.6) techniques for access, although some have since introduced FDDI.

SMDS is a 'service' name, not the name of a technology, and it is best thought of as a connectionless high-speed packet-switched service. Many SMDS services are already using HDLC and they will certainly migrate to ATM in the near future.

The fundamental data units defined for SMDS are self-contained (datagram) packets with a 40-Byte header and up to 9188-Bytes of payload (using E.164 addressing). These datagrams will be transported across the network over connection-oriented or connectionless HDLC, FDDI, ATM or DQDB services, since the core transmission system is undefined.

The key services that SMDS offers are: LAN interconnection, compressed video transfer, medical imaging, bulk data transfer, database transactions, and in some cases, voice. Note that these services all impose different technical requirements on the network:

— Interactive video (videoconferencing) and voice, are isochronous services which can't tolerate delays.

— LAN interconnection, imposes highly-variable loads (bursts) on the network, but can tolerate short delays.

— Database transaction packets are usually small in size, but need quick interaction, reliability and security.

• The Subscriber Network Interface is defined for IEEE 802.6 DQDB access. These systems will segment the SMDS datagrams into the free payload of DQDB cells (5 Byte header and a 48 Byte payload).

In this application the DQDB cell payload reserves an extra 2-Byte for header, and provides 44-Byte for data, with a 2-Byte trailer. This is nearly identical to an ATM AAL3/4 cell.

See DQDB, SHDS and CBDS also.

SME

Synchronous Modem Eliminator. The synchronous equivalent of a null modem. It is used when transferring files between two synchronous DTEs. It provides the normal null-modem crossovers of Tx/Rx, etc., but it must also be an active device and generate the clock signals. It is used for the direct transfer of SDLC and CSC data.

SMG

Special Mobile Group. This is the new name for the old GSM group within ETSI. They are now working on UMTS or FPLMTS.

SMI

Structure of Management Information. This is a standard set of rules (RFC 1155) which are intended to provide the guidelines in the definition of information models. SMIs are used to define managed objects in a MIB.

smileys

A light-hearted way of expressing feelings or conveying graphic information over a character-based data communications system. As an example (viewed sideways):

:-) an expression of pleasure.
:-(an expression of sadness.
8-) the sender wears spectacles.
:-] the sender is a machine (a robot).

There are thousands of these, and more than one book with useful and humorous lists.

smoothing

The process of eliminating the 'jaggies'. See anti-aliasing also.

SMP

• **Simple Management Protocol:** An enhancement to SNMP (actually, to a version called Secure SNMP) which incorporates security features and additional functionality. It is part-way towards SNMP-2. See SNMP.

• **Symmetrical Multi-Processing:** This is one of two key multi-processing architectures, used mainly for multimedia servers where multiple CPUs are working together within the same computer. The other is ASMP (Asymmetrical SMP).

The SMP system is tightly coupled; it combines the processing power of all the chips by distributing the process threads among the processors. However, to the operating system, an SMP architecture still looks like a single, powerful, CPU.

This approach is best for computer-intensive applications, but it has some drawbacks in task assignment. ASMP overcomes these by dedicating special processors to special tasks.

At the present time a reasonable SMP database-server will consist of two or four Pentium processors running (each) at say 100MHz. This will give the server a transaction rate of between 350 and 600 transactions per second.

In its first manifestation, Windows NT could handle two processors at once, and the advanced version can apparently handle four-way parallel processing. Version 2.2 of OS/2 can handle more simultaneous processes than Windows NT, which makes IBM's operating system extremely fast.

With the SMP approach, it is relatively easy to design a system to support up to eight processors, but SMP doesn't scale well when you attempt to add more processors. Above eight-way SMP, the task-assignment problem compounds to the point where the gains fall off quite rapidly. The limit for SMP architectures is believed to be about 32-way; from this point on, a true parallel architecture is needed.

• Note that SMP and ASMP aren't mutually exclusive; there are hybrids which combine both approaches.

• Loosely coupled architectures are either clustered, or massively parallel (MPP).

• IBM has released OS/2 with SMP 2.11 which boosts performance on 486 and Pentium PCs. Standard applications can be run on these systems.

SMPTE

Society of Motion Picture and Television Engineers. The main standards-setting body in the film and television industry. They will set the standards for the non-electronic side of film and TV.

SMR(S)

Specialized Mobile Radio (Service). A US term for most forms of analog trunked and private radio. The Enhanced version (ESMR) is digital and essentially cellular. See MIRS, PMR and Trunked radio also.

SMRP

Simple Multicast Routing Protocol. Apple's new routing protocol for multimedia over AppleTalk.

SMS

• **Short Messaging Service:** See.

• **Subscriber Management System** in Pay TV. See.

• **Systems Management Storage:** This is part of IBMs new ESA operating system extensions.

• **Satellite Multi Service:** These are business connections via a GEO using SCPC.

• **Storage Management Services:** Novell's system to aid backups — it is a standard architecture for supporting file systems operating under NetWare. This is actually a set of APIs which give the vendors of software a single interface into the operating system and file system. It 'abstracts the storage structures' and uses Target Service Agents (TSAs) to export file system data. Novell also offer HSM. See HSM and SIDF.

• **Systems Management Server:** Ex Hermes. This is part of Microsoft's Back Office integrated information system which includes Windows NT 3.5, SQL Server 4.21, SNA Server 2.1 and Mail Server 3.2. It is standard in Windows 95.

SMS is a front-end desktop tool (a 'collection of technologies') for hardware and software management at the enterprise level. It is scalable for networks, ranging from those with a half-dozen PCs, up to the largest. It needs a SQL Server with Windows NT, to provide:

— Centralized distribution and software support.
— Inventory of clients automatically collected and stored in an SQL database.
— Remote monitoring and control of nodes.
— Hardware configuration management (client/server) for a large enterprise (30+ nodes).

It uses Windows NT Server management features, such as performance monitoring, event logging, security, resource sharing, system backup, and networking.

Microsoft insists that SMS is not attempting to compete with high-end network management systems such as OpenView or NetView. They say it is for LANs and WANs running on DOS, Windows NT, Macintosh and OS/2 clients. On the server side, it can handle NetWare, NT, LAN Manager and DEC's Pathworks.

SMT

• **Surface Mounted Technology:** See surface mounted devices.

• **Station Management:** An acronym and term used in FDDI to mean the management of devices on an FDDI network. This is defined in the X3T9.5 specification.

SMTP

Simple Mail Transfer Protocol. The basic e-mail standard used with TCP/IP and the Internet. It outlines how two mail systems should interact, and specifies the format of the control messages to be exchanged when transferring mail. SMTP uses Telnet for the actual exchange.

This is a server-to-server Internet protocol which is not for general users. To access e-mail on their dial-up connections, users will need one of the POP versions or the IMAP protocols. See MOTIS, MHS and X.400.

S/N

Signal-to-Noise Ratio. See SNR.

SNA

Systems Network Architecture. IBM's layered approach to linking all micro, mini and mainframe products so that they can both communicate and share data with terminals. It has now become synonymous with IBM terminal-to-mainframe communications, and is very widely used.

SNA is old; it was introduced in 1973 — many years before the PC or Token Ring. But it is constantly being updated. It also predates OSI, and for a long time it was in competition with X.25 as the proposed standard for international public packet-switched data-networks. X.25 won.

However, SNA is still the *de facto* private data-network standard architecture for large corporations. There are said to be over 50,000 licensed networks, some serving up to 50,000 users. More than half all commercial data traffic still runs over SNA networks.

The term SNA actually refers to the proprietary specification of the protocols and data formats used on the network, and also to a more general set of guidelines.

The SNA standards suite specifies rules, procedures, and structures which dictate how networking units interconnect — both for host-centric (mainframe to dumb terminals) and, more recently, for peer-to-peer configurations. SNA is evolving into a more distributed system.

However, it retains its centralized, batch-processing lineage; it is fundamentally a master-slave type of communications architecture. SNA systems use host processors, cluster controllers and communications controllers in a hierarchical scheme. SAA, on the other hand, which came later, recognizes that today's networks are usually not hierarchical. See SAA also.

SNA is organized around domains, and an SNA domain is defined as a Systems Services Control Point (SSCP) plus all the resources that this SSCP controls.

The other two key components of SNA are Physical Units (PUs) which are essentially the nodes, and Logical Units (LUs). The SSCPs, the PUs, and the LUs, are all network addressable units (NAUs).

SNA establishes communications among Logical Units (originally mainframe applications and 3270 terminals), and it has protocols to handle the translation of network-names to addresses. Although LU6.2 is available for peer-to-peer links, LU2.0 is still the main logical unit in use on SNA.

• SNA also switches the circuits if one fails, and allows two mainframes to appear as if they were a single large system.

• SNA's logical architecture parallels the ISO's seven-layered outline to which it corresponds almost layer for layer. But SNA has eight layers and not all layer-divisions are identical to OSI:

— **Physical Controls.** The underlying physical/electrical connections.
— **Data-link Controls.** Reliable transmission between adjacent nodes along the route.
— **Path Control.** Routes data to the appropriate destination.
— **Transmission Control.** Security and data-pacing matched to end-terminals.
— **Data-flow Control.** End-to-end sync. of data-flow.
— **Presentation Services.** Format conversion.
— **Transaction/Application services.** E-mail and distributed database services. (This was added later.)

See FEP, NAU, Physical Unit and Logical Unit, SAA and APPC.

SNADS

SNA Distribution Services. An LU6.2-based architecture and network software that routes documents and mail through the network. It provides store-and-forward and synchronous transmission of messages and documents over SNA networks. Files can be transferred among dissimilar IBM devices.

It was once closely associated with PROFS, and more recently it was linked with SMTP and X.400; now it is associated with DCA and DIA.

This is one of three SNA Transaction Services Architectures (TSAs), along with Document Interchange Architecture (DIA) and Distributed Data Management (DDM).

snap/SNAP

Many drawing programs allow the automatic movement of a selected line, object or point, so that it touches the nearest grid intersection or connection point — or to some other object. As it gets close it will snap on, or lock onto that point.

This can be, for instance, a tangential line snapping onto a circle in a CAD program. This facility is often provided as a menu selection in drafting programs.

- **Systems Network Analysis Program.**
- **SubNetwork Access Protocol:** This specifies a way of encapsulating IP packets and ARP messages on IEEE networks (mainly Ethernet). The SNAP entity in the end system is responsible for the data transfer, connection management and the selection of QoS.

sneakernet

The cheapest and most foolproof local area network. You offload your files onto a disk, remove the disk from the computer, and walk through the office (stopping off for a coffee along the way) to the other person and hand over the disk. This is also known as a chat system.

SNG

Satellite News Gathering. Ku-band 'half-transponders' (27MHz wide) are now commonly used in the USA for satellite links from the news team back to the major station.

With the new digital video systems this bandwidth can easily accommodate a standard 36Mb/s wide video stream. Sony's new 'Son of Digital Betacam' ENG camera can output data at 18Mb/s at Level 3 operations, and the un-edited footage (material) can be replayed at double speed (called 'tape streaming') to pump it over the satellite quickly. It will still fit into this half-transponder channel. See Betacam and ENG also.

SNI

- Subscriber Network Interface. The demarc.
- SNA Network Interconnection. A gateway connecting many SNA networks.

sniffer

This is both a proprietary product and a concept. They are frame analysis tools.

Sniffers are hardware/software devices that are left connected to a LAN or WAN to monitor traffic. They read the frames or packets as they flow past, and retain those which appear anomalous — they need to be told what these frames should look like.

If a network suffers problems, the manager can refer to the sniffer and usually it will have a record of what went wrong. The main problem with sniffers is that you need to have a hypothesis — an idea of what fault to look for — so you can tell the sniffer which frames to trap.

Expert sniffers are now being used which are programmed with an 'expert system' (although they're not very expert!) to resolves some of these problems.

SNMP

Simple Network Management Protocol. The *de facto* standard network management protocol closely associated with TCP/IP. It was developed by the Internet Engineering Task Force (IETF) and it was once thought that it would eventually be replaced by the ISO's CMIP. That hasn't happened.

The original aim of SNMP was to identify fairly simple physical problems in the network, such as a faulty wiring hub, or a misbehaving router. It now can be used to both monitor and configure network devices.

The standard has been constantly developing over the years. It now defines how a network manager (a central console program which can manipulate and display information) can communicate with network agents (bridge/router software which can communicate with the manager) in the various devices scattered around a large network. It doesn't define 'what' to communicate, just how. It provides a language.

SNMP is a request-response protocol, and it is the one most commonly used to manage and control network devices today. It periodically polls the agents, and records the responses in a MIB (database).

The SNMP concept was first outlined in 1988 as a stop-gap measure while CMIP was being specified. However SNMP quickly became a major force in network management — due partly to the simplicity of the concept, and also because of its links to TCP/IP.

- The management software includes a management information base or MIB that resides in the management system. See MIB and RMON also.
- Reporting agent software resides in the 'managed devices' themselves. These are software entities which must implement SNMP, UDP and IP (although one version will run over IPX) in a stack above their network protocols. About 170 SNMP agents have been written for various devices, up to the present time. Not all of these agents support the full range of basic requirements, however, and most are for Ethernet rather than Token Ring and FDDI.
- What is lacking in SNMP, is standardized software agents for every kind of network device. Currently some standards exist

for bridges, routers and gateways only. Workstations and servers are still a problem.

• In the real world of the network manager, SNMP is simple to implement, but it lacks security features. Secure SNMP (1992) was an enhancement added to provide security.

• There was a later enhanced interim version called SMP (see) which attempted to correct some problems, and this later led to the development of SNMP-2 in 1992. This version has been slow to move.

SNMP 2 has manager-to-manager capabilities and increased security. It gained security but only at the cost of losing automatic discovery of devices on the network. Version 2 also has heavy requirements on the maintenance of management databases. Typically, SNMP-2 will be implemented on top of UDP which is part of the TCP/IP suite.

SNR

Signal-to-noise ratio. This is a ratio of the level of the signal, divided by the level of the noise on the system, usually expressed in dB (which is a logarithmic ratio).

In order for a channel to be usable, the perceived quality of the incoming signal must be an order of magnitude higher than the background noise of the system — and it must exceed it by a respectable margin. In cellular phones, an SNR of 12dB is only just enough. About 30dB is desirable.

• All electronic systems have a built-in noise 'floor' caused by molecular movement in wires, transistors, etc. And since satellite dish reception deals with extremely low-level radio signals, the amplifiers are constantly working near that floor. See LNB.

• A typical voice-grade telephone line will have a S/N ratio of about 1000 to 1.

snubberless triacs

This is something you probably don't want to know about — but I couldn't resist the term. They are high-speed switches which protect circuits against transient voltage spikes!

SO

• **Shift Out:** ASCII No.5 (Decimal 14, Hex $0E, Control-N). Originally this was the mechanical shift in impact characters of the old Baudot code. However, this control character is now often used in conjunction with SI and ESC to extend the graphic character set of the ASCII No.5 code.

In common usage, SO indicated that the code combinations that follow should be interpreted as being outside the normal ASCII characters set until an SI (Shift In) character is found.

— SO and SI are book-ends to special command code sequences.

• **Small Outline:** packaging. A popular type of IC package design. See DIP, SMT, MCM and QFP also.

socket

• In client-server architectures a socket is a dedicated and addressed communications channel.

• It is also a software structure which functions as an endpoint for communications within a device attached to a network.

SOCRATES

System Of Cellular RAdio for Traffic Efficiency and Safety. A 1991 European Union research program to provide telecommunications links between vehicles and information centers. The aim is to transfer dynamic routing guidance, traffic alerts, emergency information, etc.

soft

A general term meaning not fixed by hardware — easy to change and flexible.

• **Soft copy** just means the copy on the screen, rather than hard copy which is printed.

• **Soft Hyphen.** See below.

• **Soft key** is a function key or a key displayed on a touch-sensitive screen. Its functions can be readily changed.

• **Soft patch** is a quick fix made to a program running on a machine, but not made to the object or source code on the disk.

• **Soft return.** See below.

soft copy

When applied to facsimile it means a non-printed version of the file, probably stored in RAM or on a magnetic disk or tape. Hardcopy means a print-out onto paper.

soft fonts

In a laserprinter (and some other modern printers) fonts are often stored in non-volatile memory (ROM) or added to the printer by cartridge. However, other more temporary fonts can be used.

Soft fonts are those which can be sent ('downloaded') to the printer memory by the attached computer and then stored temporarily in RAM and then used by the printer. This process slows down the transfer substantially.

However it is now common for the PC to send outline fonts to the printer — so common, in fact, that the term soft font has dropped out of use. Soft fonts need to be installed in the computer, not in the printer.

soft/er handoff

In mobile telephony, this is the situation where the current basestation maintains its links with the mobile during the hand-off period (to another basestation) and only breaks the first link after the second link has been established. For a short period, the mobile will be receiving two signals.

This is currently only possible with CDMA systems using Rake receivers. Within a CDMA network, it is also possible to hand-off in an even 'softer' way, (without involving the central switching center) between sectors in the same transmission area. CDMA is able to do this because all cells and sectors use the same wide band of frequencies.

The distinction is with hard hand-off where the basestation discards the original link, and another basestation subsequently establishes the new link — there is always a fraction of time without a link, and for data transfer special protocols must be used to handle this (see ETC).

Hard hand-off is only necessary in CDMA systems when control is being passed from one switching center to another — or from one band of frequencies to another.

soft hyphen

Also called a discretionary hyphen. It is a special hyphen code inserted into a word to indicate the point at which to break the word, if a break is needed. It doesn't print unless the break takes place at this point.

It is added by the word-processor or DTP program. If you change the characters in a line, a soft hyphen may well shift or disappear. But hard hyphens are inserted into the text at the typing stage, for instance, in X-ray to avoid such a split as in X-ray between lines.

soft-keys

Control keys which are modified by the software. The early versions of soft-keys (in character interfaces) provided a line of key identifications across the bottom of the screen which corresponded to function keys on the keyboard. These could be changed on a minute-by-minute basis. Later it was felt that it was easier and better to handle these flexible commands by using mouse selections and pull-down menus.

• Small handheld computers of the PDA-type, don't have space to provide a set of alphanumeric keys below their large LCD screens, so they often use a text entry system which consists of a full keyboard image which is touch-sensitive. You key in the letters by touching the appropriate area of the screen with your finger or with a stylus. This is a soft-key system also.

soft-modem

Soft-modem technology uses general-purpose, reprogrammable Digital Signal Processing (DSP) chips rather than hard-wired ICs. Since these chips are reprogrammable, special software upgrades can take advantage of on-board RAM-memory and EEPROM. So when the standards change, new operational algorithms can be easily installed on-line, or systems can be quickly modified if bugs are found. The claim is made that, in future, you won't buy a new modem to upgrade to a new higher data-rate standard. You simply download new software.

• This is highly likely for two reasons:

— The tendency to have modem architectures designed around functions in software (using a standard CPU and RAM, rather than a ASIC) is now likely on both a price and performance basis.

— The limits of modems are now close to Shannon's theoretical limit with our noisy, bandwidth-limited, telephone lines. So there's probably not much point in devising even faster standards, since our present modems can't usually use even 28.8kb/s. The next stage is probably ISDN.

soft return

Soft returns are special carriage returns added by word processors to the end of each screen line, to force the insertion point to begin the next, at the left-hand margin. The Macintosh doesn't add these to document files, but MS-DOS machines do.

These will change if the document is reformatted. Hard returns signify the end of a paragraph and they don't change.

Soft-returns are applications specific. Some programs use ASCII 13 (the carriage return) and some use ASCII 10. Some always include a line-feed. When the ASCII 13 (CR) character is combined with a Line Feed (ASCII 10), this produces a hard return. There are probably other variations of soft return as well.

When saving the file as a text-only ASCII file, these soft returns should be removed automatically. See hard return.

soft-sector

Modern 5-inch floppy disks are usually 'soft sectored' which means that the management of the sectors on the tracks is under the control of the operating system and not fixed in place by mechanical means (usually holes punched through the disk). See hard sector.

software

Software refers to the program files (applications and system) — the coded instructions, as distinct from the hardware (physical) elements of a computer.

However there are some subtle variations in the industry's use of this term that beginners often find difficult to comprehend.

There are two main categories of software:

— System software, which manages the computer itself. This includes the operating system, but also system software which is not strictly part of the O/S.

— Applications, which solve problems for the user.

However there are two other less obvious categories:

— The stored data itself, as distinct from the application or operating system which allowed the data to be stored.

— Utilities and TRSs, which are very small applications.

• When people talk about movies being the 'software' for a cable network, they are really stretching the limits of the definition to ridiculous extremes. If we are going to use terms this widely, then they begin to lose useful meaning.

• I've also recently seen computer manuals being referred to as software. This is plain stupidity; the word 'manual' is already available and unambiguous.

• By extension, software has come to mean the media on which it is generally stored. So floppy disks, which are a physical (hardware) media, have come to be seen as the same as the applications, operating system and data. It is a silly extension, but a common one, so it has to be accepted.

• In general, you would expect the term 'software' to be applied mostly to the applications (which were probably purchased on floppy disk). If operating system was intended, the

term used is usually 'system software', and if data is intended then 'data disk' would be more common.
See operating system, system software and application.

software agent
See agent.

software driver
See device driver.

software engineering
The term covers the philosophies, methods, techniques and intellectual tools used in the design and development of computer software. See also structured programming and programming environments.

SOH
ASCII No.5 — Start of Header (Decimal 1, Hex $01, Control-A). This is often the first character in a header sequence. It is used to alert the receiver to the arrival of the header information (which usually contains addresses or other routing information).

SOHO
Simple Office: Home Office. A term used to suggest that certain computer and communications systems are designed specifically for small office use. It is trying to promote the idea of being something special for telecommuting.

SOJ
Small-Outline J-lead. A chip package (physical type) which was jointly developed by Hitachi and Texas Instruments for the 16Mbit DRAM chips. SOJ devices take up a large amount of space on a PCB when compared with some of the newer designs like ZIP and VSMP.

solar outage
Loss of satellite signals due to the sun passing through (or near) the receiving antenna's 'beam' (the narrow angle of its view). This happens with most direct-to-home satellite TV transmissions a couple of times a year. See rain outage also.

solid modeling
A CAD technique that produces a realistic appearance of an object on the screen as distinct from a wire-frame drawing. In these systems the object may then be treated as a solid, not just for drawing purposes, but also for calculating its mass. The first step past wire-frame modeling is hidden line removal, which is then followed by rendering, smoothing, etc.

solid-state
Semiconductor devices like transistors and photodiodes are solid-state, as distinct from the old vacuum tube devices. Thermionic valves still have many advantages in handling high power levels, however. The most powerful transistor currently made is about 150 watts, while electronic valve amplifiers routinely handle 40kW in television transmitters, and up to 1.5MW in long-wave radio stations. See CCD and thermionic tube also.

solid-state storage
Aka silicon disk. The use of memory chips as an alternative to hard-disk storage. This has only become economically feasible for larger files since memory chip prices began to drop — but hard-disk drive costs have dropped even faster still. The main advantage offered by solid-state storage is access speed, especially instant boot-up.

Note that solid-state storage devices can be used as primary memory (part of the working memory — within the memory map of the CPU), or as secondary memory (where it must be accessed and offloaded into primary memory).

- Some early notebook portables, used CMOS logic and special DRAM with battery backup. By dispensing with disk drives and large LCD screens (using eight lines of 40 characters only) The Tandy/Radio Shack 100 could keep 64kB of memory alive for months.
- SRAM was used later (when it became cheaper) because it didn't require constant refreshing. This made it possible to create small plug-in memory devices which could carry their own very small batteries.
- Later EEPROMs were used, but only for secondary storage because the electrical-erasable process is much too slow for working memory. Some PCMCIA cards still use EEPROM.
- Flash EPROM has now become popular for PCMCIA memory cards. The contents need to be offloaded to RAM, however. See flash.
- FRAM has the most potential because it can be used as either primary or secondary storage, and it can be superimposed over both DRAM and SRAM bases. See FRAM.

solitons
A specially shaped light wave which could carry digital signals through optical fibers for thousands of miles without repeaters. This is seen as the next stage in optical fiber technology after erbium amplification (EDFA), and wave-length division multiplexing (WDM).

The generation of solitons depends on the use of photonic technology such as pump lasers attached to special optic couplers. Soliton pulses hold their square 'shape' and therefore they have the capability of not spreading out and dissipating as they travel through the fiber. Tidal bores which sometimes travel extraordinary distances up rivers, are physical forms of a soliton.

Since the waves don't spread, they don't produce intermodulation distortion at high rates over long distances, so they may also be capable of handling data rates in the terabit/second range (1000Gb/s) over 1000km or more between regenerators.

• NTT and AT&T are both achieving fiber distances of 2000–4000km between modulators using solitons (but only in the laboratory).

solution
One of the most useless buzz words in computers — second only to 'supports'. It is an almost meaningless marketing term

which refers to some mythical productivity value (always unproven) which is to be gained by spending a fortune on a new applications.

• Sometimes the term means that software has been customized to serve a particular purpose.

SOM

System Object Model. See OpenDoc.

son

Backup files. See father

Sonet

Synchronous Optical NETwork (closely related to SDH with which it is merging). This is an attempt to standardize the transfer of data over high-speed telephone systems, and allow channels to be added to and dropped from multiplexed data streams without needing to demultiplex the whole stream.

It enables multi-vendor connectivity. It is said to be a cost-effective way of handling analog POTS (it includes a way of handling plesiochronous transmissions), is able to handle special services, and has enhanced OAM&P capabilities. Sonet/SDH paves the way for broadband services, and it will carry ATM during the introductory stages of fast packet-switching.

ANSI's original Sonet defined a hierarchy consisting of 1) standard optical signals, 2) an synchronous frame structure for multiplexed traffic, and 3) operational procedures.

It uses a byte-interleave multiplexing technique and it was based on OC-1 bit-rates (51.84Mb/s) which is known in Sonet as STS-1.

• The current defined bit rates for Sonet are: OC-1 – 52Mb/s; OC-3 – 155Mb/s; OC-9 – 466Mb/s; OC-12 – 622Mb/s; OC-18 – 933Mb/s; OC-24 – 1.2Gb/s; OC-36 – 1.9Gb/s; OC-48 – 2.5Gb/s.

• Sonet was originally proposed by Bellcore for American telephone networks entering the age of optical fiber. It thus took into consideration only the American transmission hierarchy at the time, which had established data rates only up to 2.5Gb/s. The international extension is known as SDH, and this differs in having two less bits of overhead and substantially higher data rates. SDH begins at 155Mb/s (STM-1) which is equivalent to Sonet's STS-3.

• The Sonet standard has STS-1 building blocks which are defined in the form of a 90 x 9-Byte matrix.
- The first three columns (27-Bytes) are used as transport overhead (section overhead, payload pointers and line overhead).
- The payload envelope is 87 columns each of 9-Bytes, and is subdivided into individual tributaries with the first column of each being path overhead. The first byte in each of these tributaries provides a virtual tributary payload pointer. This adds up to a lot of overhead!

The key advantage gained from using this approach is that the pointers allow an add-drop multiplexer to identify the start of data streams within the extraordinarily rapid and complex flow of composite data.

• Sonet's overheads come from data parity, frame and channel information, network maintenance information, and user-defined functions. The payload portion is carried in virtual containers, which in Sonet are called 'virtual tributaries'.

• The major differences between Sonet and SDH for 155Mbit/s rates and above, lies in the overhead and network management features. Terminology also differs slightly.

• There is considerable work being done in Sonet/SDH microwave radio systems (mainly using M-QAM), and in the use of these technologies in local access networks for Video On Demand.

• Sonet seems to be having significant problems from phase angle modulation and jitter. There are now complaints that the technology has been oversold, and that it doesn't offer the advantages once claimed.

• See STS, Sonet Signal Hierarchy, tributaries, SDH, and virtual containers.

Sonet Signal Hierarchy

The Sonet standard is limited to the US, and so Bellcore only took into consideration the North American transmission hierarchy for multiplexing. However the international SDH standard has some parallel STM modules. Sonet has the STS-n series of electrical bit-rates and the OC-n optical fiber rates:

Electrical	Optical	Bit-rate	SDH equivalent
STS-1	OC-1	51.84Mb/s	
STS-3	OC-3	155.52Mb/s	STM module 1
STS-9	OC-9	466.56Mb/s	
STS-12	OC-12	622.08Mb/s	STM module 4
STS-18	OC-18	933.12Mb/s	
STS-24	OC-24	1244.16Mb/s	STM module 8
	OC-36	1866.24Mb/s	
	OC-48	2488.32Mb/s	STM module 16

sonic delay

See acoustic delay.

sort

The sequential arrangement of records in a file according to some particular order requested by the user. The natural order of the computer is to sort on ASCII numbers, but this is often very confusing to the reader. Words beginning with a capital letter are sorted quite separately to those with lower case. Numbers precede alpha characters.

Date or time order, order of creation, or order or modification may also be used with files. See alphabetic order.

True ASCII sorts will handle lower-case letters as a distinct series from upper case.

There are many different techniques used in sorting algorithms (bubble sort, shell sort, quick sort). See bubble sort, selection sort, radix sort and insertion sort

Sound Blaster

This is a PC-compatible sound card that has now become the *de facto* standard. The Macintosh has sound built-in.

sound-in-sync

A technique of adding digital (and analog) sound to a PAL television signal by inserting data into the horizontal blanking interval. There's also a stereo version.

source

- The origination and the consumption of data are known as the 'source' and 'sink' in a communications system. These are also known as the ingress and egress points in a network.
- In a MOS transistor, the source is the source of the current. It is the equivalent of the emitter in a bipolar transistor.
- The 'source computer' is the one compiling the program — not necessarily the one that will eventually run it. See cross compiler.

source address bridges

Source address bridges are used with 802.5 Token Rings. They are much simpler devices than transparent bridges, and they can also operate at much higher data rates. These are probably the simplest interconnection devices to install on a WAN, and they provide for smaller network extensions than routers or gateways which are used for larger WANs. See source routing and transport relay also.

source code

- In computer programming, the original form in which the software was written — usually in a high-level language, and perhaps with many explanatory comments to aid debugging or alteration.

  Once a high-level program has been compiled, it is very difficult to unravel the machine code; debugging is often almost impossible without going back to the original source code. Naturally this new code will then need to be recompiled after the problem is found. The distinction is with object code and assembly code.
- In telecommunications, the source coding is the original processing of the material to be transmitted. For instance, a video signal will be digitized and compressed. This is the source coding process. However, before transmission, various error protection schemes may need to be added and interleaved/multiplexed with the audio and control signals — and this is called 'channel coding'.
- Voice source coding, refers to processing the PCM sound through a special predictive coder or vocoder, to extract the basic vocal tract parameters. The distinction here is with waveform coding, such as ADPCM. See vocoder and wave form.

source-explicit forwarding

A function of MAC-layer bridges. This will filter all packets not specifically listed as a valid source by the network manager.

source routing

IBM's Token Ring approach to bridging LANs. This is a dumb bridge approach, which is also used in APPN+.

With source-routing, the bridge is not required to supply route selection or network management information; each packet supplies its own in a routing information field. The bridges just do what they are told by the packets.

The routes will have been previously determined by the terminating devices working in cooperation. The sender broadcasts route-discovery datagrams which flood through all routes, and the first datagram to arrive at the destination device carries with it all its routing information (how it negotiated the network). This packet is then returned to the sender to be used as the guide for future packets.

Once the route has been selected, the sender then just lists the bridging sequence in the packet header, and all intervening nodes will check this list, rather than consult any independent address or topology tables.

This approach places the routing intelligence in the workstation, and the bridges or routers will just follows instructions.

- The contrast is with the transparent bridging and the spanning-tree approach. However source-routing and transparent bridging are often combined. See SRT.
- Source route translation bridging uses intermediate bridges to translate from one bridging protocol to the other. IBM is developing a dual device. This is SR/TLB, not SRT.
- Source routing is part of the IEEE standard 802.5, and it was initially rejected by the IEEE for anything other than Token Ring (where it is widely used). But IBM has since used source routing in the new APPN+ as a replacement for label swapping.

SP

- **Space:** ASCII No.5 — Space (Decimal 32, Hex $20, <space bar>). This is generated by the space bar. It is a character just like every other in the ASCII set. In fact, it is classed as a 'print-character' (as distinct from a non-print control character) since it affects the printing through the absence of a strike or mark.
- **Superior Performance:** A term Sony and other video equipment manufacturers have applied to up-graded Hi-band U-matic and Betacam equipment, modified to take advantage of the newer metal particle tapes. See BVU also.
- **Standard Play:** (As distinct from LP — Long Play.) This term is used with VHS machines when the machine runs at the standard tape speed. With the LP setting the tape generally runs at half-speed and picture quality is sacrificed.

space

- **Line-code:** An old telegraph term. Binary 0 is a space, and binary 1 is a mark. This is a very useful terminology when discussing the more fundamental technologies like line-codes.

  Be careful how you use the term in any system that used NRZ techniques. Very often a space (logical 0) is registered by a condition which is equally +ve or –ve.

• **Print:** The invisible character that exists between text characters on a screen, which is stored in computer memory as ASCII 32 (decimal) or hex $20. Space is a 'print' character which prints a gap — not a 'non-print' character. See SP above.

• **Space parity** is often provided as one of the parity selections in communications software, but it doesn't actually work in asynchronous communications. Space parity automatically adds a zero-bit to the end of each byte, but this is precisely the role of the stop-bit.

So space parity + 1 stop-bit is equivalent to 2 stop-bits. In fact, it is nothing more than an enforced delay of 2-bit periods. In most transfers, the use of space-parity would just slow down the file transfer program.

• **Radio**: See space waves.

space and mark distortion

A transmission problem caused by the receiver sampling the waveform at the wrong interval, and misreading. More likely, however, the sampling process is using an incorrect threshold (the point of decision; whether this is a mark or not).

A pulse will not remain as a perfectly square wave for any distance over a communications channel, so the receiver must deal with a shaped wave. It must decide that, at a certain stage and using a certain threshold-value, it is either reading a space or a mark. Problems arise when standards are not strictly followed (especially RS-232). See transmission impairments.

space-division/diversity

• **Multiple use:** The term makes a point that the same radio carrier frequencies can be reused in different physical domains, provided the radio signal doesn't spread everywhere and create interference.

If you don't transmit the signals strongly in all directions, you can reuse them again to service different locations by focusing (directing) the signal, or by reducing its strength.

This is more a marketing term than a technical one — and often falls into the category of the 'bleeding obvious'.

- If you put a dish behind a transmitting antenna and focus all the signals in one direction like a searchlight, you can then reuse the same frequencies in other directions nearby. This is how microwave links are used.
- If you don't broadcast them in a high-power way that causes them to deliberately spread everywhere, you can have many more low-power transmitters using the same frequencies in the same geographical region. This is how cellular radio frequencies are used.
- In the extreme case, if you push radio signals down cables, or light signals along fibers you can confine them to very narrow space-guides (wave-guides) and gain the ultimate in space diversity.

So if you pump a few hundred megaHertz of bandwidth down one coaxial line, you can reuse the same few hundred megaHertz in another coaxial lying immediately alongside, since the signals are contained within the cable sheaths.

• **Microwaves** will not have as much space-division as coaxial. You may need to keep high-powered microwave beams separated by a kilometer or so if they use the same frequencies.

• **Cellular phone services** use space division to divide their channels up into base-station clusters, and each cluster reuses the same channel frequencies. To increase the space diversity, cellular base-stations often use sectorized cells (directional antenna beams). See smart antenna also.

• **Geostationary satellites** exhibit a lot of space-diversity, provided they are physically separated in space by enough angle not to suffer mutual interference from uplinks. Downlinks depend on the type of antenna beam focus (spot, hemispheric, etc.). Polarity of the signals can be used to increase the separation.

• **Space-division switches** are matrix switches such as crossbars — they are many-input to many-output devices. What makes them space-division is that each connection between an input and an output is entirely dedicated to that signal carriage, and its use doesn't affect other connections within the same switch. Each signal path is entirely independent of the others.

The original manual patch-panels were true space-division switches because all the connections could be independently made. See fabric and crossbar. The distinction is with time-division or shared-backplane (bus) switches where only one packet can cross from one side to the other, at any one time. These rely on working at a very high data rate.

• Space-division is also applied to the current generation of decentralized digital switching units (such as the Alcatel System 12) which have the switching intelligence distributed over the various line-cards, rather than being under centralized control. This is really just distributed processing.

space segment

That part of a satellite network which includes the satellite itself and the means of controlling it. For some strange reason they include the TTC&M earth station as well as the satellites in this term — this distinguishes the satellite part of the network from the wireline feeder network.

space waves

Don't confuse these with sky-waves. Space waves (as distinct from ground waves) are radio waves that propagate terrestrially, roughly parallel with the earth's surface. They travel either directly from transmitter to receiver (direct space wave) or reflect off some nearby object (multipath space wave).

These signals consist of both pure space waves supplemented by some reflected from the ground (which are sometimes mistakenly called ground waves).

Normal terrestrial radio and TV transmission uses space waves. The distinction is with atmospheric bounce or sky waves as used by short-wave transmitters, and with ground/surface waves which travel around the curvature of the earth.

• See propagation.

spacial redundancy

A repetition of patterns in any two dimensional image. Every image has some lines or areas in which pixels are duplicated in brightness and color. These are 'redundant', in the sense that the information for each doesn't need to be separately stored or transmitted.

You can transmit the first, then effectively just say, 'the next 23 pixels (or lines) are the same'. In DCT compression systems the approach is more complex, but the result is similar.

The distinction is with temporal (time) redundancy where a pixel in one frame of image is compared with the matching pixel in the next image frame.

Spacial redundancy in video compression is used for intra-frame coding techniques, while temporal is used for inter-frame coding.

SPADE

The telephony version of SCPC. SPADE stands for: **S**ingle-channel-per-carrier (SCPC), **P**CM (digital telephony), Multiple-**A**ccess (distributed users seeking to share satellite facilities), **D**emand-assigned (no wasted bandwidth), **E**quipment.

There! It is quite self-explanatory!

SPAG

Standards Promotion Applications Group. A European computer manufacturer's group which issues the PSI mark for OSI conformity.

spaghetti code

Programs that lack a progressive and coherent structure. This is a program which contains so many GOTO statements that its trail would resemble a bowl of pasta. The criticism here is with the use of GOTOs, which structured programmers shun.

spamming

Posting large numbers of useless or promotional messages on the Internet (usually for commercial gain). Electronic junk mailing. There's now software called Cancelmoose which deals with spamming.

span

A full-duplex digital link.

spanning tree

A network topology which is conceptually tree-like (with a single trunk, many branches, and many more twigs) and which lacks any loops. This last point is important; it must not have loops.

The viewpoint is from that of a router. Each output port makes contact with many other network segments, and these in turn make contact with others. So to span from the router's location to any end point will involve negotiating through a tree-branch structure.

spanning tree algorithm

STA is now the IEEE 802.1d standard for detecting and disabling loops in such a bridged network.

The development of this algorithm was an important event in the history of LAN-bridging technology. Multiport bridges use routing tables, and they update those tables as the network topology changes: they can learn.

This approach was originally developed to make complex local Ethernets, often joined by many bridges, much more efficient. It can work out the best route between any two segments, and, in the process, remove any possibility that packets can drift endlessly around closed loops.

When there are often multiple paths through an internetwork (which means loop structures), and STA automatically specifies and uses the most efficient. If that path fails, it automatically reconfigures to bring another path to the fore. With spanning tree algorithms, there is only one best path 'spanning' the gap between LAN segments — and this has the structure of a constantly branching tree — whatever the physical topology.

IBM now has a similar system which is actually an improved version of the rival technique, source routing. Even more recently, the IEEE's Token Ring committee defined an approach to source routing which includes the spanning tree approach.

• The spanning tree algorithm consists of three mechanisms:

— **Frame forwarding.** The bridge maintains a forwarding table for each port. If the MAC address isn't on this list the packet is discarded, otherwise it is forwarded through the appropriate port.

— **Address learning.** Bridges often update their forwarding tables by just recording the source address of packets arriving at a port. A timer checks to see that any old address information is dropped if it is not regularly updated.

— **Loop resolution.** The spanning tree algorithm acts to (logically) cut the address-path for packets that could circulate around the loop.

spanning tree bridges

A spanning-tree bridge (also known as transparent or learning bridges) can automatically route packets without user intervention. It will note each packet's source-address and associate this with its arrival port, thus learning its location. These associations are then stored in a filtering database.

The bridge will check the destination address of a packet against these database tables, and if it doesn't find the information required (which port to transmit the packet on), it will forward copies of the packet to all ports, thereby flooding the 'Bridged LAN' (B-LAN).

When a response comes back from the correct recipient, the bridge learns the required location (associates a port with a destination address) and forwards subsequent packets only to that LAN.

• The distinction here is between source-routing bridges which depend on the packets carrying the route information.

SPARC

Scalable Processor Architecture. This term is applied both to the open architecture of the system (designed by Sun), and to the 32-bit RISC processors used.

SPARC was developed by Sun Microsystems in 1987, and it is now being promoted by AT&T, TI, Philips and Fujitsu. SPARC International was set up by these companies to push it as a *de facto* industry standard for Unix workstations.

SPARC is said to be 'scalable' — it can be reduced to laptop size or used in a super computer — and all SPARC systems should be able to run the same software. The key is to keep the chip small and simple; some SPARC chips have only 50,000 gates in an array, and this allows them to be constructed using GaAs and ECL materials. They are true von Neumann RISC processors, but cache, async floating-point processors, and pipelining all play an important part in the SPARC design.

• Sun has now released a CPU with Version 6 (full 64-bit instruction set) architecture for SPARC. This is known as UltraSPARC. See.

Sparc International

A consortium which has the rights (from Sun) to control the SPARC architecture. It sets conformance standards.

sparse

• **Files:** Any file which contains at least one empty block. These files can eat up megaBytes of a server's disk-space.

• **Arrays:** Any array where most of the entries are zero values. This is a waste of space, but some special data structures are available to deal with these.

• **Recalculation:** A process used with spreadsheets.

SPC

• **Stored Program Control:** See.

• **Semi Permanent Connection:** This is usually with ISDN, where a carrier provides a point-to-point connection on a long-term basis (a digital tie-line). However, since ISDN is a switched system, this can be 'spoofed', and actually be a dial-up feature. When you take the line off-hook, it automatically and instantly dials through to the other end.

This is then really a highly discounted normal connection, priced so that a company can afford to hold open the digital channel for many hours in the day.

SpE

Sporadic-E. See.

SPEC

The Systems Performance Evaluation Cooperative. This is an industry group that has developed a suite of ten useful benchmark tests for computers. The SPEC benchmarks for CPU performance have proved to be valuable in making more meaningful comparisons between CISC and RISC systems (MIPS is almost meaningless). The organization also has benchmarks for graphics and networking.

• Release 1 (1989) of the SPEC benchmark suite is specific to the CPU and memory subsystem test. There were 10 tests of which four are 'integer-based' and in the C language, and the remaining six are 'floating-point based' and are in Fortran. The SPECmark is a geometric average of these two. Since the suite was released in 1989, the results are known as SPECfp89, SPECint89, and SPECmark89.

• Release 1.1 (1992). In this more recent revision some tests were dropped and others added. There are now 20 tests in the suite. So we now have SPECfp92 and SPECint92. There is no SPECmark92; instead we have both the SPECrate_ftp92 and SPECrate_int92 which brings multitasking into contention.

• You can't compare SPECfp89 with SPECfp92; the suite has changed too much.

special characters

This generally means those characters above the numbers on a keyboard. The @ and # and % and &.

• In the IBM PC, see IBM also.

special effects generator

See SEG.

special services

In telecommunications, this term is often applied to the provision of fire-alarm, security, piped-music, telemetry, etc. over the telephone lines as distinct from the telephone, data and telex services.

SPECmarks

A benchmark test for computers which has been supported by a large number of vendors. See SPEC above.

Spectral Recording

SR is a Dolby development for improving sound quality on magnetic tape, especially for magnetic film-tracks used in cinemas. It intelligently monitors the signal to improve the dynamic range of analog sound. It is said to have sound quality equal to the best digital tracks.

spectral signature

In satellite (and aircraft) remote sensing, the image is split and filtered, then passed to a number of different sensors, each of which reacts to a different wavelength of light.

Currently three spectral colors are used from the visible band (R, G and B) and another three from the Far, Mid and Near Infrared band. These separations are then quantized for each pixel of image.

So the remote-sensed image consists of a unique pattern of digital quantization from six sensors (RGB + 3 IR), and these numbers are attached to a pixel corresponding (in satellites) to an area of earth about 30 meters square. These quantization groups can be displayed in false-colors (individually, or blended) and they allow the identification of crop, coverage and land-types — even down to the geology of the area. See remote sensing.

SPECTRE

Special Purpose Extra Channels for Terrestrial Radio communications Enhancements of the UK. This is a project to develop a complete digital television coding and delivery system that will operate around 13.5Mb/s. They've moved on now to MPEG compression, and joined the DVB group.

spectrum

If not specified further, this refers to the whole electro-magnetic spectrum, and it may include 'audio-frequency' radio waves in the 100—20kHz range, broadcast radio frequencies to 1GHz, then microwave radio frequencies into the teraHertz range, infrared, visible light, etc.

• The spectrum of a communications system is its bandwidth expressed in terms of the frequencies it can carry. The bandwidth of an analog telephone circuit is usually give as 4kHz, but the spectrum of voice frequencies it will carry is between 300Hz and 3600Hz in the best systems.

spectrum folding

This is analog component HDTV terminology. Spectrum folding and motion compensation were the two major digital techniques being added to analog television systems (D-MAC) in order to reduce bandwidth enough to make HDTV practical.

Spectrum-folding covers the use of space- and time-domain sampling and filtering techniques to reduce the chrominance bandwidth requirements. This information was then to be compressed and made available via an augmentation channel.

speculative execution

Undertaking an operation (the next in line) before being instructed to, on the assumption that it will be needed. This is a way to speed up CPUs. See micro-ops.

speech recognition

The very difficult task of having a computer understand the range of dialects, accents, slang-terms and sloppiness that constitutes human speech. Currently, speech recognition systems need to be 'trained' by one user to recognize a very limited range of commands, or they are restricted to only a few words.

The easiest form of speech recognition is a system based on dependent/discrete speech, but unfortunately, this type is also the least flexible and least useful. We still need to overcome the problems of speaker-dependence, and also find out how to handle continuous speech.

Constant research over the last decade is slowly making some progress, however. There are now a number of classes of speech recognizer:

— Speaker-dependent. These must be trained in one particular voice profile.

— Speaker-independent. These are trained in advance using very large numbers of speakers. The problem here, is that these appear likely to work well for the majority, but not for those with dialects, accents and impediments.

— Speaker adaptive. These attempt to combine both of the above by adapting to the speech variations as it progresses. These systems learn how each person's vowels are sounded, and adapts to suit.

— Isolated word recognizer. These operate to a limited but useful degree already. They recognize YES, NO, RIGHT, LEFT, UP, DOWN, etc. There are a few applications which find them useful — for instance, as a hands-free controller of a computer cursor. Remote bank information systems also use these.

— Continuous speech recognizer. This is the Utopian aim. The machine should reacts to us, rather than the operators being forced to modify speech patterns to suit the computer.

However, this doesn't initially mean people will be able to communicate with the machine in a natural language. They will have a limited range of available words which may, however, be uttered in a connected way.

— Keyword spotters. This approach is to look for a pattern-match in an otherwise uncomprehended blur of data. Only one word at a time is recognized, but it can be embedded in a natural sentence.

• Handwriting recognition is a trivial matter when compared with continuous speech recognition for dictation — yet an amazing number of people keep claiming that this form of computer control and entry is desirable, and available 'Real Soon Now' (RSN). Even after extensive training, I still can't get a Newton PDA to recognize half the words I print — and it can only handle about 10% of my cursive writing.

speech-synthesis

Computer generation of human-like sounds. See vocoder also.

speed

One of the top-ten most abused word in the electronic lexicon. Everyone claims to have a computer or software which is 'the fastest'. And yet in 95% of all applications the rate of processing doesn't matter a damn. Here are some atomic processing rates which may give you a point of comparison:

- National's 256kB SRAM chip (BiCMOS) has an access time of 10 nanoseconds (ns).
- AMD's 2 megabit EPROM chips operate at 100ns.
- The old 386-based computers run at about 8000 Dhrystones a second.
- The Sun SBus transfers data internally at 80MByte/s which is about twice the NuBus or EISA standards.
- The i860 chip running at 40MHz delivers 33 Vax MIPS and performs 85,000 Dhrystones. It also achieves 24 million Whetstones, with a peak of 80 MFLOPs in single-point precision, and 60 MFLOPs at double-point.

SPG

Sync-Pulse Generator. Deliberately generated sync-pulses (voltage inversions) are the primary means by which video signals are kept in phase. In a TV studio, with many image sources, if you cut between two images not using the same sync-pulse, the picture will roll until new sync timing can be established.

It is important, therefore, to have one master source of sync and to feed it to all devices (SEG, cameras, recorders, caption scanners, telecine units). They must then slave to this signal.
— In a large television studio, the SPG is a rack-mounted unit which generates a master sync pulse for the whole studio complex and this pulse is delivered by its own cable.
— In a small television studio, the sync pulse is generally taken from one of the video cameras. It is often output through a special socket on the side of the camera known as 'genlock'. This camera will then use its internal SPG and include it in the composite video output, while the other cameras will 'slave' to the first via a special cable.

• When a computer is inserted into these systems, the video card will have a genlock input to take the slave sync signal.

spike
A sudden very high voltage in the mains power. It can be many thousands of volts, although it may only last for a few thousandths of a second. With both 120 volt and 240 volt systems, spikes are normally identified as being changes of 6% in voltage for less than 1ms. Above this time, a spike becomes a 'surge'.

Spikes are 'transients' which are very disruptive to computer operations — and they can be high enough in voltage to jump across switch contacts. For this reason you should unplug (not just switch off) your computer when lightning is around. You should also unplug any modems or fax machines.

spike clipper
A device usually designed around a varistor which is said to prevent large amplitude voltage spikes on the power supply from damaging computer equipment. Varistors are very fast-acting and capable of dissipating energy quickly for short periods. See power supply.

spill
In magazine and newspaper publishing, this is to carry over the last section of a story or article to a later page ('Continued on page XX'). It is often frowned upon; readers find it annoying. But it also means that the design of the magazine or newspaper pages can be much more flexible. For instance, more headlines can be packed onto the front page.

spin stabilization
This is a way of keeping some satellites oriented in space. They spin the whole body of the satellite, and rely on gyroscopic action. Satellites of this type are drum-like in appearance, while the later three-axis stabilized types are usually box-shaped and use internal gyros, etc.

These should be called dual-spin satellites. While the main body spins to provide attitude stabilization, the antenna assembly (and often some of the electronics) must be de-spun by a motor and bearing system so that it continues to point accurately at the earth.

The second problem with spin stabilization, is that the body needs to be covered with expensive and weighty solar cells, but at any one time only one third of these are effectively facing the sun. See three-axis stabilized satellites.

SPIRIT
Service Providers Integrated Requirements for Information Technology, in Network Management Forum. See TINA.

spline
• See B-spline
• A flexible drafting ruler which is used to create complex curved shapes.

split-horizons
A router with two ports (A and B) is said to have one horizon on its A side, and another on its B side. Split-horizon techniques prevent routing information (table updating information) from being looped back to the source. The router is intelligent enough to know that it should only pass the updating information downstream to another horizon, and will not recycle it back upstream.

split-stack architectures
A term used to describe client-server philosophies. A split-stack architecture is where the client's PC 'splits' the SNA protocol stack with a gateway PC that is functioning as a Physical Unit (PU) controller. This is a 'LAN-centric' approach.

The distinction is being made with the 'downstream PU' or 'host-centric' method where each client PC on the LAN implements a full SNA stack and communicates to the host via a front-end processor or 3174 gateway. At present the 'downstream' method is much faster.

splitter
In optical fiber (and also in copper), a passive splitter allows two (or more) fiber communications channels to be generated from one source. The contrast is usually with switching technology, where the two outputs are mutually exclusive — when one is receiving a feed, the other isn't.

Splitters are generally passive devices (the light beam is subdivided), but the term can be applied to active components. The term passive splitters can also be used to mean a similar device used, in reverse, to combine two signals into one.

• Coaxial cable splitters. See directional couplers.

spluttered media
Actually, 'sputtered' media. See plated media.

spoofing
A way of handling dissimilarities between equipment on a network by an 'electronic white lie' — electronic false pretences.

• **Emulation:** Spoofing is making a network interface appear to the terminal as if it were the native (correct) host computer to which that terminal should be attached — and vice versa. It is a form of protocol emulation/conversion.

• **WAN spoofing:** Dial-up WAN connections aren't always on-line (obviously!). So they need a way to handle the constant delivery of routing tables and other management messages sent across the network. Network managers don't want a dial-up connection being made each time a routing table up-date comes along (sometimes every minute). So it is common to spoof the system, and have the dial-up unit reply as if it were the remote and store the tables for later delivery. ISDN D-channels are sometimes used now to carry these messages to avoid spoofing (they are always connected, via packets).

• **LAN spoofing:** Some devices on a network constantly send keep-alive messages, and some engage in constant handshakes (SNA does this). So it is common to spoof a remote device at a local concentrator, to save overloading the remote connection. Only the essentials are then transmitted.

• **Satellite:** On some extremely large WANs (especially with satellite links) the sender may need to spend a considerable time after transmitting each packet, waiting for an acknowledgment. In order to speed up the network processes, the network manager may implement 'acknowledgement spoofing'. The sender thinks the packets are ACKed, and that everything is OK, and quickly resumes transmission of the next packet. Obviously, there then needs to be a subsidiary mechanism to handle NACs.

• **Hacking:** IP source address spoofing has recently been used to take control of some of the network servers and modify their kernel. This has allowed the hackers to by-pass the firewalls in NFSnet.

spooler

Spool is supposed to mean 'Simultaneous Peripheral Operations On-line' but this sounds like a tall story to me. It is a device that allows a program using a slow output device to complete execution rapidly, and then return to other tasks.

The data is stored temporarily in queues, or on disk or tape, for later slow-speed transmission to the device.

In a PC, print spoolers are quite different from print buffers. A spooler is a memory-resident utility that intercepts the output from the application and stores it either on hard disk or in RAM. It then sends the data to the printer as a background task while the user returns to work on another program.

A print buffer is a hardware device (primarily RAM) which is inserted into the link between the PC and the printer. It doesn't require further operations of the CPU, so this approach frees the PC even more than spooling.

On a network, the spooling operation is now generally handled by the print-server. A good spooler will both be able to store large quantities of data, queue it in order (perhaps with priorities) and generally manage the operations of the printer.

• You could also have spoolers for modems and some other shared devices. But generally the term is only used for printers. The problem now with modems is that the PC isn't fast enough through the UART. Spooling wouldn't help.

Sporadic-E

The ionospheric Layer F is generally responsible for long-distance (short wave) reflection of radio signals. However occasionally (sporadically) the lower E Layer becomes highly reflective (mainly at night) to radio at the lower end of the H/F band. This produces unusual skip effects where radio signals turn up in areas where they are not supposed to be.

spot-beam

Communications satellites are usually able to select beam concentration. The controllers can decide whether to spread the beam across a whole hemisphere, spread it across a continental or regional area, or focus it on a much smaller area.

The smaller the footprint, the higher the signal level within the beam area. Spot beams are the most highly concentrated of the standard beam types and are used for point-to-point communications or where the receiving dishes are too small to take signals from the wider beams (although spot beams are often driven by the lower-power transponders).

Pencil beams are smaller in area, but the term is usually applied to a shaped beam which has multiple 'pencils'. A hemispherical beam is much wider in coverage (and lower in power) and a global beam is wider still. See shaped beam, hemi-beam and zone-beam.

spread ratio

In direct sequence spread spectrum (CDMA) the 'modulation' process is to replace all bits in the primary data-stream by a much faster pseudo-random bit-stream. The bits of this random stream are known as chips, and the spread ratio is that between the original data rate, and the chip rate. It is generally between 10 and 100. So what was one bit, may now be a recognisable pattern of 100 bits — which is the 'spread' concept in spread spectrum. This output then modulates the R/F carrier in a conventional way (usually BPSK).

At this higher rate, the channel obviously needs a much wider bandwidth (proportional to the increase in ratio), but this wider bandwidth is shared by other users. See orthogonal and Walsh codes also.

spread spectrum

SS. A technique of radio (particularly satellite and cellular phone) transmission that is reliable and automatically highly secure. It operates at very low power levels, so the satellite transmissions can be captured by small antennas and it is designed to operate in noisy environments.

There are a couple of forms, the most popular of which is the later invention called 'pseudo-noise' or 'direct sequence', which is often coupled with BPSK modulation. Interference is reduced and security is enhanced by the fact that all transmissions are encoded. That encoding process (digital-on-digital modulation) is what 'spreads the spectrum'. See above.

SS can mean CDMA in the direct sequence form or alternately the use of the older technique of frequency hopping. More correctly it should probably be referred to as SSMA

(Spread Spectrum Multiple Access) when used as a multi-user cellular radio access technique.

The three main types of spread spectrum are:

- **Time hopping:** Signals are transmitted in short pulses whose intervals are determined by a pseudo-random code sequence. This is also called 'chirp' and is not much used.
- **Direct sequence:** The digital message signal is modulated by a pseudo-random spreading code whose chip rate is much higher than the bit rate of the message. See CDMA.
- **Frequency hopping:** The message is modulated onto a carrier with frequency-hopping ability. The hopping pattern is determined by a pseudo-random code.

• See CDMA, Walsh codes, and frequency hopping.

sprite

A sprite is a graphics object of a specified pattern which appears in a position on the screen identified by a single pair of coordinates. If you change the coordinates (through program control) the sprite moves. It is treated as a whole.

The cursor shape, for instance, is a sprite. Sprites are commonly used in computer games and there are also standard patterns used in drafting. Think of sprites as small groups of pixels which always stay together.

SPS

Standby Power Supply.

sputtered media

See plated media.

SPX

Sequenced Packet Exchange. A Novell protocol by which two workstations or applications communicate across a network. SPX guarantees delivery of the message and maintains the order of the packet in the message-stream.

SQE

Signal Quality Error; aka a heartbeat. See.

SQL

Structured Query Language. SQL is now considered a 4GL, but it was invented by IBM in the early 1970s as a fairly basic, but common connectivity language for quizzing databases. It is widely known as 'Sequel' and it is the *lingua franca* of the relational database world. ANSI and ISO are involved in its standardization.

SQL is also the basic language/technique behind client-server systems. It puts the burden of file maintenance and query onto the file server, and so avoids the constant transfer of large files over the network.

The main problem with SQL is that each vendor's dialect is still slightly different from the next; communications protocols and message formats often vary significantly from one vendor to the next. Middleware such as ODBC may be needed to handle this. The SQL Access Group has now defined SQL Access Specifications to allow interoperability. See RDBMS, fourth-generation, client/server, DAL and RDA also.

square

• A slang name for the hash (#) key on some videotext terminals. The correct name for this key is octothorpe.

• Multiplied by itself.

squelch

In single-sideband radio systems and voice pagers, squelch is used to suppress the modulated signal while remaining aligned on the carrier.

SQUID

Superconductivity Quantum-Interference Device. A small, and extremely sensitive electro-magnetic field detector which is capable of picking up magnetic changes emanating from neurons in the brain, etc. SQUIDS have many medical applications; for instance, in a magnetic cardiograph they can detect changes in a fetus heartbeat. They are actually Josephson junctions which are being used in a detector mode. See.

SR

• **Sound:** Dolby's Spectral Recording.See Dolby.

• **Logic Gates:** With gates and flip-flops, SR refers to 'set/reset'.

• **Routers:** Source Routing. See.

SR/TLB

Source-Route/TransLational Bridging. A distinction is being made between three different techniques:

- SR/TLB is Source-route 'translation' bridging.
- SRT which is Source-route 'transparent' bridging.
- SRB which is the simplest form of Source-route bridging.

See source routing and SRT also.

SRAM

Static Random Access Memory. A newer and more expensive type of memory chip that works faster than DRAM. It is volatile (loses its memory when power is off) but it is often classed as semi-volatile since it doesn't need to be refreshed constantly.

SRAM is constructed around flip-flops using four to six transistors per bit. This is the reason for the increased complexity, greater cost and lower densities of SRAM memory chips over DRAM. The bit-density of current SRAM chips are only about a third of DRAM, and the chips cost about six times as much.

However SRAM chips have access times which are five or so times the speed of DRAM, and some are now being made with super-fast access times of 10 nanoseconds. These higher speeds allow SRAM to be very effective in RAM caching, etc.

• The fastest SRAM I have on file is IBM's experimental 128kb chip which operates at 6.5ns. Other companies have now announced 1Mb memory chips with 9ns access times, and Intel has said that it will make 10ns SRAM for use with the P6 processor. Intel's SRAM will provide 256kb of Level 2 cache which will be synchronous with the processor.

SRB

Source-Route Bridging. See source routing.

SRT

Source-Route/Transparent bridging. Don't confuse with SR translation bridging. SRT merges source-routing and transparent bridging. These are single bridges which pass source-routed data whenever appropriate. However, they will also transparently bridge the data when the packets don't carry the routing information.

SS

- Signalling System. SS7 is also known as CCS#7. See CCS#.
- Sampled Servo in optical disk techniques. See servo format.
- Spread Spectrum. See.
- The name given by hackers to the FBI's anti-hacker unit.

SS-TDMA

See below.

SSB

Single SideBand radio. This is a technique (widely used in point-to-point two-way radio) of suppressing the main carrier and transmitting only one of the sidebands.

The carrier frequency of an AM radio signal, can be considered to have an upper (USB) and a lower (LSB) sideband — both of which carry the modulated voice. Single sideband is one of these, and it therefore occupies only half the bandwidth of a true AM signal, yet offers nearly the same quality.

SSB has traditionally been avoided for many commercial communications purposes because it needs complex and expensive tuning systems and squelch techniques, and the technique cannot readily distinguish between signal fading and voice pauses because the full carrier is missing. 'Amplitude commanded SSB', linear modulation, and 'tone in-band' techniques are overcoming these problems.

- You can have both AM and FM single side-band, but usually only AM is ever used. See VSB and sidebands also.

SSCP

Systems Services Control Point. This is the domain-center and key control intelligence of an SNA networks. The SSCP is responsible for controlling the hardware in the Physical Units (PUs) and establishing communications between Logical Units (LUs) and controlling devices.

These three are all network addressable units (NAUs), and each part (domain) of a network is assigned an SSCP.

In SNA, the SSCP allows the operator to:

— Start-up and shut down the network.
— Establish communications sessions between NAUs.
— Control and manage all resources owned by the host.
— Manage network configuration (using PUs).
— Manage requests requiring problem determination.
— Provide directory and other session services for end-users.

- SSCP is a subset of VTAM in a host, and a SSCP-PU sessions is the process by which the SSCP uses PUs to send requests to, and receive replies from, individual nodes.

SSD

Shared Secret Data. A code transmitted by CDMA digital cellular phones along with the MIN and ESN to prevent cloning. The SSD is known by both the phone and the service provider, but it is never transmitted in the clear. It is a 128-bit number which is split into two 64-bit parts; SSD-A and SSD-B. SSD-A is used for authentication, and SSD-B is used for voice encryption.

SSEED

Symmetrical Self Electro-optical Effect Device. A device which provides photonic circuits with the equivalent of a transistor. It was invented at the Bell Laboratories. See BEAT also.

SSMA

Spread Spectrum Multiple Access. There are two types in common use:

— Frequency Hopping spread spectrum. See.
— Direct Sequence. See CDMA.

See spread spectrum also.

SSP

- A System/36 operating system.
- System Support Program. A utility on SNA networks which can load, dump and debug the Network Control Program.

SSPA

Solid State Power Amplifier. This is a type of amplifier used in satellite communications. See TWT.

SSTDMA

Satellite-Switched Time Division Multiple Access. A form of dynamic switching between transponders on a communications satellite to allow selection of either the hemi-or the zone-beams. It is controlled by on-board computer intelligence and memory, and they will often store 50 or more switch-states which can be triggered by commands.

Since the controls are in the satellite itself, they can change a switching and antenna pattern every few milliseconds if need be, and repeat it, if necessary, a few milliseconds later. This makes the satellite very flexible.

SS/TDM satellites can use a single transponder to send, say, six downlink beams by alternating the connections every two milliseconds, thus creating a complex slotted-TDM point-to-multipoint network.

The earth stations will compile and store the digitized signals before the uplink. It will sort the data into a large buffer, in the order in which they are to be relayed from the satellite. This is determined by the sequence in which the downlink connections are to be switched in space. The signal is then beamed up to the satellite in burst mode at a rate of, say, 120Mb/s, once every two milliseconds.

ST-412

A venerable standard interface for hard-disk drives which was invented by Seagate. It was later merged into the ST-506 standard to create a combined ST 412/506 standard. These protocols control the drives only at a very low-level by sensing the rotation of the platters through an 'index' signal generated by the drive. The controller then moves the head by transmitting a simple 'direction' signal (Swing In, or Swing Out) and it pulses a 'step' signal once for each track.

This is no longer the fastest way to access a drive, but it was simple and clever for the time. See ESDI.

ST-506

Seagate Technology: This was the standard disk drive and interface on the PC for so many years (from 1984) that it has become a benchmark standard (with a transfer rate of about 5Mb/s). It was originally designed for floppy disks, and later modified for iron-oxide on aluminum platters.

The standard was acceptable until 32-bit processing became common, but now it is clearly inadequate, and has largely been replaced by EDSI and SCSI.

Most ST-506 drives have a 4:1 interleave factor and they use MFM encoding with 17 sectors per track. However good ST-506 drives can be upgraded with RRL to improve performance (although with an increased likelihood of head crash).

See below, ESDI, IPI, IDE, SCSI, RLL and ATA also.

ST-506 RLL Interface

This is the RLL upgrade which allows the ST-506 drives to increase data density and have a higher data rate. The interface can handle two drives, to an addressable limit of 200 MBytes of data.

STA

Spanning Tree Algorithm. See.

stack

CPU: A section of memory which temporarily stores information in a last-in, first-out (LIFO) manner so the oldest data is always at the bottom of the stack and the newest at the top. Sometimes stacks are registers in the microprocessor, and sometimes they are reserved areas in RAM (software stacks).

Stacks are linear 'lists' for which all deletions and insertions are made at one end of the list (as distinct from a queue). You 'push' data onto a stack and 'pop' it off.

- **General:** A general term often used in application programs to mean a temporary store.
- **HyperCard:** Stacks are HyperCard files consisting of cards, backgrounds, fields and buttons. (Apple Macintosh)
- **Protocol:** A hierarchy or suite of protocols, such as those defined in the OSI reference model or SNA. See protocol stack.
- **Windows:** With Windows applications a stack is that part of the data segment used to hold the automatic variables and the arguments.

See also split-stack architectures.

stackable hubs

These are a later replacement for the conventional chassis-based star topology LAN hubs. They were cheaper mainly because they could be mass-produced.

3Com built the first stackable hub (LinkBuilder 10BT) in 1991 for Ethernet over UTP. Later they added versions supporting thick and thin coaxial, fiber optic cable, and Token Ring.

Rather than have everything joined by a single high-speed backplane bus in a large chassis (using plug-in cards), stackable hubs are separate modular units piled on top of each other, and joined by 'expansion cables' which link the backplanes.

Each hub will independently be a repeater, but when joined together at the backplanes, they only constitute one repeater from the viewpoint of the network. Generally users can combine different media types in a single stackable hub-stack and still link their backplane buses. In effect, all LAN ports are electrically connected to form a single logical LAN segment.

However they still lack some of the features of conventional hubs (such as redundant power supplies), but today, many stackable hubs will have built-in SNMP agents, and most will offer network management as an optional slide-in module — or as an extra box stacked on top.

A true stack of 'stackables' will always appear to the network as a single repeater, and all in a stack are managed as a unit.

- Pileable hubs look much the same, but they aren't linked via the backplanes. They are connected instead, by daisy-chaining the network ports. So a stack of pileable hubs appears as a number of different network segments — and with Ethernet the limit is set to four. See pileable hub.
- Before stackable hubs (pre 1991) the choice was between fixed-port or chassis-based hubs. See.
- Third generation hubs now feature port-switching, built-in fault tolerance and comprehensive network management. Network configuration is now stored and controlled in software.

STAIRS

STorage And Information Retrieval System. The IBM document (bibliographic) database management application for mainframes. STAIRS is used by some major information retrieval utility companies. It allows keyword searching using Booleans.

stamper

A special hardened metal plate which has been etched to hold a negative image of a disk (now CDs, but originally LPs). The stamper is pressed against the disk-blank's plastic surface (or the disk is injected moulded against it) to transfer the information to the disk.

The importance of the stamper may extend even further. For important information, it may be the item which is archived, and which lasts for many hundreds of years — outlasting disks, films and tapes.

- The stamper is also called the 'father' and is a negative copy of the glass master.

stand alone

Equipment that isn't connected to others through a network. It is independent. The term is usually used to distinguish a single computer from one that is connected to a network.

• A stand alone modem or disk drive is external to the computer, and connects to it via a lead. With modems this has important implications, since the stand alone fax-modem doesn't have all the bus connections of an internal expansion-card modem.

Standard A, B, C..., P

Mobile satellite systems. See Inmarsat.

standard-cell

A way of simplifying the design of VSLI chips by using pre-designed and standardized patterns for different functional 'cells' at the chip geometry level. These cells are combined like a jig-saw puzzle to create the final chip pattern.

standard index form

See scientific notation.

standards converters

These are digital devices used for transferring NTSC images to PAL or SECAM, or vice versa. The early analog machines used quartz delay-lines, while the present ones are basically frame-store digital computers.

The typical professional standards converter works like this:

- Analog input signals are encoded into digital YUV components. Between two and four fields are stored in computer memory.
- The images are analyzed in both spacial (intrafield) and temporal (between field and frame) ways. Every 10 lines of NTSC must become 12 lines of PAL, and so each new line will be derived by merging adjacent line values.
- New lines are constructed/deleted, and new field rates are calculated.
- The output is converted back into the original analog composite form in the new standard.

• Low-quality standard converter chips are now reasonably cheap (enough to fit into a TV set or VCR) — they just throw away lines, or duplicate them — and repeat every fifth field twice (or drop every sixth). However, high-quality standards conversion is still an expensive operation.

• PAL creates problems in the decoding process because, ideally, the comb filters must act over more than one field. This causes interline flicker problems.

standby

A receiving device in standby, is awaiting a call with only the minimum of electronics functioning. Pagers and some cellular phones actually wake-up from the standby mode every ten minutes or half-an-hour, and listen for a poll call.

Cellular phones will report their location. Pagers will listen for their address code.

The term is usually applied to measurements and marketing claims of battery life. The distinction then is for 'talktime' with a cellular phone.

• In most battery-run devices, the standard rule is to divide the claimed standby life by two — or four if you suspect the salesman is lying. For equipment being actively promoted and sold by non-technicians: divide by eight.

star

• **Key:** A telephone and videotext key type with the asterisk-like (∗) star symbol.

• **Service:** In the UK, 'star services' are CLASS services. See star code.

• **Asterisk:** Often a slang reference to the normal asterisk key on a computer.

• **Topology:** A network topology, with a controller at the center and radiating cables out to peripherals; this is a hub-and-spoke design. A variation on this, is the distributed-star network where each radiating cable then splits again at a 'node' or 'cluster controller', and radiates out once again in star fashion to another series of peripherals. Star clusters can also be an adjunct to a ring network, or reside as segments on a backbone bus network.

star code

A telephone command code used on DTMF phones for CLASS services. It is made usually by transmitting the 'star' (asterisk) key signal followed by two numbers. Here is a list of the recommended North American codes (there's no universal standard):

∗57 — call tracing request (or sometimes for call back).
∗60 — call blocking activated.
∗61 — priority ring activated.
∗63 — select call forwarding activated.
∗66 — repeat dialing activated.
∗67 — call number ID blocking (dialled before each call).
∗69 — call return activated.
∗70 — disable call waiting.
∗71 — three-way calling according to usage.
∗72 — enable call forwarding.
∗73 — disable call forwarding.
∗74 — modify speed-call directory entry (for 8# service).
∗75 — modify speed-call directory entry (for 30# service).
∗76 — call pickup.
∗79 — ring again.
∗ 80 — call blocking disabled.
∗ 81 — priority ring disabled.
∗ 83 — select call forwarding activated.
∗ 86 — repeat dialing disabled.
∗ 89 — call return disabled.

See CLASS.

StarLAN

The AT&T-designed 1Mb/s CSMA/CD star topology network, known as 1Base-5 and standardized by the IEEE in 802.3.

start-bit

In asynchronous data communications it is necessary to send a start-bit ahead of the 7- or 8-bit ASCII character (and usually one or more stop-bits after). So the byte appears to travel over the link between two control-bit 'bookends' — but this isn't the case. Only the start-bit actually exists. (See stop-bit.)

Start-bits provide the leading-edge marker — the reference for timing circuits sampling the changes in the line voltage to discover the value of each bit. They are added to each byte by the UART in the PC — and are stripped by the UART at the other end of the asynchronous link (today, at the local modem, before transmission over the phone lines). Start-bits both alert the receiver (today, the local modem) to the fact that the byte is about to arrive, and provide it with a reference for timing.

The main value of synchronous transmission is that it doesn't need start- and stop-bits. This is why, with modern higher-rate modems, the start-stop bit system is only used over the RS-232 link between the PC and the modem. The link between two modems over the telephone line will now almost inevitably be synchronous, transmitting the bare 7- or 8-bits.

Startex

A marketing name for PDC.

startup disk

The disk that contains the important parts of the operating system. During the boot process, the computer will load the information from the boot blocks into the lowest part of the memory map (RAM), then use this to control the loading of the whole operating system from the disk. It will then await instructions to load any application or data.

Stat Mux

Statistical multiplexer. See below.

static electricity

Insulated surfaces can build up very high voltages (tens of thousands) in static electricity, and an accidental discharge of these through a chip can permanently damage it. See ESD.

static pads

Static has always been a problem around computers because of the extremely high voltages generated — especially in air-conditioned offices with nylon carpets. Anti-static pads conduct this charge away before it can do damage. The antistatic pad should be connected to conductive wrist-straps, with a clip joined to the chassis (via the power supply) of the computer.

- It is very important (despite what many people will tell you) to always disconnect the power cable of the computer from the power point, so no earth lead exists to carry away static charges from the PC. This will increase potential difference.

Never work on a computer while the power lead is plugged in. With or without static pads and wrist straps, it increases the risk of component damage. See ESD.

Static RAM

See SRAM.

static raster

As TV sets get larger in size, you are more likely to see the individual scanning lines making up the image. These are called the 'static raster'. The less the eye can resolve these lines as lines, the better the picture quality appears. For this reason line-doubling techniques improve the subjective picture quality (when viewed close up) without actually improving the image resolution. See Kell factor.

static route

This is information entered into a routing table manually.

station

In communications, there doesn't seem to be any logic to the use of this term at all — other than in the sense of 'place'.

- In LANs terminology a station is any active network node which is a source or a sink — a server, a printer, a modem, a PC or workstation. It doesn't appear to apply to switches, bridges, repeaters or routers which pass data through.
- In telecommunications, a station is a local exchange (also called a 'central office' in the USA). Station equipment is local exchange equipment (switches and line-cards) which interfaces with the customers equipment over the local loop.
- In satellites, station-keeping is the process of making adjustments to the satellite's position. Each ground-based communications facility is also a 'station'. See below.

station-keeping

The job of maintaining a geostationary satellite in its fixed position in orbit. This is under the control of the TTC&M station — of which there is one for each GEO satellite.

This is done by firing thrusters for microsecond bursts every day or two, which obviously consumes thruster fuel. It is station-keeping which limits the life of a communications satellite.

Normally the ground control engineers will attempt to keep the satellite within +/– 0.1° (+/– 40km) in the Clarke Orbit. With recent satellites now being placed closer together the proposed requirement is to allow the satellite to drift no more than +/–0.05°.

Even when a satellite is nearing the end of its life and is permitted to drift North-South over 5° or so, the East-West station-keeping must be maintained. See inclined orbits.

stationery

A formatted document that can serve as a template. When you open a stationery file, a copy is immediately created and opened so that the original 'template' file remains unchanged.

You can usually save as stationery, any document created by a word processor or DTP program.

statistical multiplexing

There is both a general and a particular sense of the term.

- **General:** Any multiplexing system where the bandwidths of individual channels (or the availability of channels) vary according to the volume of the data content. In this sense, packet-switching 'channels' are statistically multiplexed, since packets are only transmitted only when filled with data.

• **Particular:** A statistical multiplexer is essentially a special-purpose computer with multiple I/O ports and a single trunk (bearer) I/O port to a similar device at the other end. Behind the multiple ports are a series of transmission queues (buffers) from which data is sent according to some pre-established priority.

These devices act like time-division multiplexers handling frames (not packets), but they have no fixed bit/byte assignments of slots in the composite data stream. This means a 'Stat Mux' can dynamically allocate increased trunk capacity to channels which are active, and reduce the capacity of those slots which are temporarily idle.

This means the interconnecting bearer is kept working at nearly full capacity, and that the system can handle many more connections than a standard TDM with fixed slot. For this reason, these devices are sometimes called 'concentrators'.

However, the control of the various channels and the dynamic allocation of bandwidth to the slots obviously increases the overhead significantly, and it requires substantial intelligence in the Stat Muxes.

They are very effective where data arrives on many channels, in an occasional fashion, or in bursts. But they aren't as efficient as TDM when data flows are continuous.

The difference is that:

- **TDMs** allocate time slots on the link irrespective of whether a channel has any data to transmit. They will transmit idle bits when a channel is idle.
- **Stat Muxes** identify which channels are active and then avoid transmitting these idle bits by reducing or removing the channel bandwidth. When an idle channel becomes active again, the information is transmitted across the link and the capacity is reallocated.

Obviously, a Stat Mux needs large buffers while channels are expanded and reduced, and it needs to signal constantly across the link. Usually one slot is set aside for control and slot 'mapping'. They superimpose a 'composite protocol' on the data being transmitted so that they can verify receipt at the end.

The individual frames traveling across a Stat Mux link will have headers, sequence numbers and error checking fields, so these are very like conventional packets — but the channels are reserved, so the system is also used to handle conventional circuit-switched (isochronous) data including PBX links — not only bursty LAN interconnection. They work with both asynchronous and synchronous channels.

• **Video:** The term is also applied to multiplexed digital video where a number of compressed TV channels are merged for transmission over a single satellite transponder or wideband terrestrial carrier. Satellite channels of a fixed bandwidth (say 20Mb/s within a standard transponder), can carry between four and six signals of various bandwidths. More bandwidth is needed for real-time action sports than for, say, B&W movies.

These interleaved stat-multiplexed channels can be either:

- Preset for the duration of the program, or
- Dynamically assigned progressively during the program — depending on the momentary needs (fast action requires greater bandwidth).

Dynamic assignment will increasingly be a requirement with the current moves to introduce the variable bit-rate MPEG2 video standards.

status

• One of three primary types of information handled by a CPU. These are:

- Status. Signals the current condition of the device.
- Control. Says what it should do with the data.
- Data. The raw working information.

Status information includes values stored in variables, the condition of all flags and semaphores, etc.

• There are also status flag-bits in packet headers. See flag.

status line

An additional line of text, at the bottom of the computer screen, which is not normally used as part of the active screen area. In the old character-based IBM PCs this was the 25th line. It held prompts, menus, and other information.

STC

Sinusoidal Transform Coding. A voice coding technique used in analog to digital voice conversion. See IMBE also.

STD

Subscriber Trunk Dialing, aka Direct Distance Dialing (DDD) in some countries.

STDM

This is a real trap for the unwary. These two are quite different technologies:

• **Statistical Time Division Multiplexing** — see Statistical multiplexing.

• **Synchronous Time Division Multiplexer** — see ATDM.

Steadicam

A camera balancing/stabilizing device used in feature film production. It is a body-worn harness which relies on springs, counter-balance weights and universal hinged movements to avoid transmitting the cameraman's body movements to the camera. The camera tends to 'float' because it averages out all movements.

It attaches to a belt around the operators waist, and he views through the eyepiece using a small video monitor.

steerable antenna

Large satellite reception dishes are always steerable. The larger the dish size, the more accurate the pointing angle, so if a satellite drifts in orbit, the dish must be able to follow. Satellites put out a beacon, and the dish can measure and track that signal.

• Most direct-broadcasting satellite receiving dishes are fixed because they are being sold as a package with a subscription to Pay TV. Now that more satellites are in orbit, it pays to have a dish which is steerable and can be directed at the satellite you want to watch. Many C-band TVROs are of this type.

• Some fixed dishes can point at two or more satellites simultaneously. The focal points are offset, and the feed horns are not aligned with the sighting axis.

steerable spot beams

Geostationary satellites often have a few steerable spot beams which can be turned to direct beam power to various areas of the hemisphere within their view. Some of the new LEOs also use steerable antenna, which can be programmed to direct their energies towards the settled land area, rather than wasted in the ocean.

Steinbrecher receiver

An approach to radio base-station receivers, where incoming multiplexed signals from the antenna are digitized immediately, and the tuning and bandpass filtering is then performed digitally as a computer process, rather than by normal means.

The advantage of this approach is that a Steinbrecher receiver does not really care what access techniques are being employed; apparently it can simultaneously receive and select signals from TDMA or CDMA or FDMA mobiles.

Step-by-Step

SxS: The common name for later modifications to the Strowger telephone exchange switch — the first fully automatic telephone switch invented in the late 1800s. SxS exchanges are still widely in use around the world, but usually in poor countries and the more remote outposts of civilization.

A SxS switch-unit handles two digits (pulse dialled) in sequence. The first digit moves the armature vertically through ten 'planes' of contacts, and the second rotates it through ten rotary contact positions on the selected plane.

This is sometimes called a 'rotary' contact switch — although there is often some confusion as to whether 'rotary' refers to the dialing method or to the way the switch operates. This technology was followed by the Crossbar, which was eventually upgraded and became the first Stored Program Control (computer controlled) switch.

step-index fiber

Aka 'stepped'-index fiber which is a form of multimode fiber. The term spells out that this multimode fiber has a definite boundary to the refractive index between its core and its cladding — so it is not graded-index fiber. The core RI will be high, and the cladding will be low to create the necessary condition of total internal reflection. Light will always tend to run down the glass at the slowest speed — that with the highest RI.

• In single- (mono-)mode fibers the core will be between 2 and 10 microns, while multimode fibers can be as thick as 200 microns. They will use LEDs, rather than lasers.

stepper

• A step-by-step exchange switch.

• A type of motor used in some of the cheaper hard disk drives to move the head for read-write operations. It is pulsed a number of times to 'step' to the required track number on the disk — a technique also called 'band-stepper'. These disk drives have an access time much slower (twice the time) than that of voice-coil systems.

However, if a stepper hard-disk drive has been constantly optimized and defragmented, it can be quicker than a voice-coil drive that hasn't. Voice-coil drives are also electronically more complicated, but mechanically they are simpler.

stet

Leave it as it is. Don't make the changes which were previously marked. This is a copy-proofing term.

sticky keys

Some keyboard commands need two keys to be pressed simultaneously (i.e. Alt-P). This may need two-handed operations, and not everyone can do this. Sticky key is an option on some computers that allows the effect of the first key-press to 'last' for just long enough for the second to be pressed.

STM

• **Synchronous Transfer Mode:** A type of packetized information frame-based communications which uses time-division multiplexing. It was a close race between ATM and STM as to which was to become the basis for Broadband ISDN, but fortunately 'asynchronous' ATM won because of the problems of providing global clock timing.

STM is an extension of traditional circuit-switching principles, providing fixed bandwidth channels and a fast packet-switched signaling mechanism. However, ATM's asynchronous approach offers the major advantage of flexibility.

• **Synchronous Transport Modules:** Level #. These provide the basic frame structure of SDH transmission techniques as used in Europe. Virtual Containers (VCs) are mapped into these modules. The current rates used are:

STM-1	155.52Mb/s
STM-2	311Mb/s
STM-4	622Mb/s
STM-16	2.4Gb/s

STM-1 is the European equivalent of STS-3c. After you extract all the overheads, the payload capacity is only 140Mb/s which provides 63 separate 2Mb/s signals which can be directly accessed and monitored.

Transport systems based on this standard can then use digital cross-connects and add/drop multiplexers. The STM signal is composed of frames which include standard monitoring and management signals, so a separate management network is not needed. Don't confuse these with the STS standards which are North American-only (STM-1 is equal to STS-3). See VC, Sonet, STS and SDH.

• **Scanning Tunneling Microscope:** The IBM invention which enables an operator to see and move molecules, at sizes down to 20 billionths of an inch. The technique can be used to pick up and change atoms on the surface, which means that it can theoretically provide an atomic-scale switch and storage system for information.

Matsushita is attempting to use this approach with phase-change optical disks to create terabit storage systems. Don't confuse the STM with the SEM (Scanning Electron Microscope).

STN

Single Twisted Nematic. A type of LCD screen. See.

stochastic

A statistical process which we would generally refer to as a branching-tree choice. File directory structures are of the stochastic kind. We choose between a finite number of choices at the first level (at the root level), then between another group for the second step, then another group for the third step, etc.

stop/start

• **Protocols:** See ARQ.

• **Flow control:** See XON/XOFF.

• **Transmission:** See asynchronous transmission.

stop and wait

See ARQ.

stop-bit

In asynchronous communications, 1, 1.5 or 2 stop-bits are added to the end of a byte as a termination mark. In fact, the stop-bit doesn't exist; this selection represents nothing more than a forced delay of one or more bit-periods before the beginning of the next byte. A space-parity selection is equivalent to one stop-bit — and both are the same as the idle state on the transmission link. See start-bit.

stop word

Words like 'to', 'and', 'the', 'by' etc. which are useless for text retrieval from a database because they are so common. They are normally rejected during the indexing process to reduce the index size. And since they are not included in any inverted file, they are therefore not searchable.

Sometimes, however, a stop word is an important part of a phrase: such as 'Just In Time', in which case the search parameter will require you to search for 'Just' and 'Time' separated by one word (unspecified). [FIND Just(1W)Time]

store and forward

A communication system where packets of information are passed between nodes on a network — each as a separate link. They will generally be held temporarily at each node before being passed along, and while being held they may also be checked for errors.

The distinction here is between circuit-switched communications systems like the telephone, and the older packet-switched networks like X.25 and Ethernet. Note that the new fast-packet networks of frame-relay and ATM are not store-forward systems, although they are packet-switched in nature.

It is just possible to use store-and-forward systems in real-time for voice, provided the end-to-end delay is guaranteed to be less than, say, 1/10th second. Unfortunately, with a number of switching nodes in a network, each subject to varying loads, it is impossible to guarantee delay limits of any kind — which is why most conventional LANs don't carry voice.

Fast-packet switching has, as a key aim, the ability to guarantee delivery of priority voice and video packets with minimum delays and maximum regularity. See ATM.

• The term store-forward is also applied to messaging systems like e-mail and EDI which transfer messages around the world through packet-switched services. E-mail may take a few seconds (up to a few hours) to get from the originator to the recipient's mailbox.

stored program control

A telecommunications term for the intelligent computer-control of an analog crossbar switch unit of the type used in telephone and telex exchanges. This is one stage removed from electro-mechanical (dumb) switching.

Computer controls were added to the old crossbars to extend their life and to integrate them into the modern digital network where signaling is carried as data packets on the CCS#7 signaling network (see).

Other analog switches were also built with integrated computer-controls, and these fall into the SPC category also. SPC improves the routing, load capacity, maintenance and administration of the switch and allows many of the modern CLASS services.

The distinction should be made between these SPC units and the newer fully-digital (solid-state) AXE switches — although these two generations can interwork reasonably well.

STP

• **Screened or Shielded Twisted Pair:** Standard telephone cable that has been screened to help reduce electro-magnetic interference — especially when the cable is used in a LAN. Most Ethernet cable is unscreened or unshielded (UTP) because it is cheaper. Screened or shielded cable (STP) is also widely used for in-company LAN wiring, usually as a four-wire system. See Category #, STP and shielded cable also

• **Spanning Tree Protocol.** See.

strained-layer laser

The problems of developing a laser which can operate at a specific wavelength demands careful selection of semiconductor materials and the accurate deposition of the crystalline layers to maximize performance. Each new layer must have precisely the same crystalline pattern as the one below. These are strained-layer lasers. See GRINSCH-SL-MQW.

streaming protocols

Communications protocols like TCP/IP's FTP, stream out the data without waiting for acknowledgments. Many of the older LAN protocols will send only one packet, and then wait for acknowledgment before sending another. If a satellite link is part of the network circuit, these systems bog down immediately. See sliding window and propagation delay.

streaming tape drives

Fast-recording tape recorders used as back-up devices on computers. They can often handle data rates of about 2 megaBytes per minute. One common standard uses the DC2000 data minicartridges which are ideal for desktop PCs; another uses DAT 4-mm audio tape cartridges (DDS).

'Streaming' means the information is transferred as a stream without any attempt to create identifiable blocks. As a consequence, you can always append new data to the end of a tape, but if you want to change a file anywhere other than the end, you must read and rewrite a completely new version.

The older tape system used blocks of information of a moderate size, and then left large interblock gaps so that only one block needed to be changed if data was upgraded or modified.

So the distinction here is between these so-called 'random access' tape systems where files can be updated in situ — and the 'streamers', which are designed specifically for fast backup. The main standards of streaming tape drives today are:

— QIC quarter-inch cassette.
— 4mm DAT (aka DDS).
— 8mm Exabyte.
— DLT linear from DEC.

street furniture

In telco talk, this refers to every box, pillar or pole that a carrier needs to park out in the street to provide a telephone service, other than cables (called 'plant') and underground installations. Active and passive pillars, RIMs and other multiplexing devices, power supplies, etc. are included.

StreetTalk

Banyan's global naming and directory service for VINES.

string

• In high-level programming, strings are series of characters, such as 'A164jklm' or 'Dog' or even 'True love'. (Note the space may still be a character in concatenation.)

Strings can be of any ASCII printable characters not being treated as values. The character sequence 12345 can either be a numeric value of twelve-thousand and something, or an numeric string (say, a zip code) which can't be added, divided or multiplied. 'Hello' can only be an alphabetic string.

• When discussing variables, the distinction is with values.

• A string at the CPU level is quite different. Here, the string is the flow of control of an application (a process) which is a sub-section of a task. A string is composed of many threads — which are discrete sub-sections. See process and thread.

strip

In video coding, the screen is often divided up into blocks of various sizes (often 16 x 16 pixels), and sometimes also into horizontal strips. Each strip may be treated independently for error-checking, etc.

striping

A disk storage technique used in RAID. It is a way of distributing the user-data across multiple hard disks. Disk sectors are recorded in vertical 'stripes' (as if all were stacked on top of each other).

Every time a data block is written to disk, the sectors within that block are simultaneously recorded elsewhere — but on identical locations on the other disk/s. To replay the block, only one data location needs to be provided, and all the sectors can then be accessed simultaneously.

stripping

This is the preparation and assembly of the film separations and half-tone images before the plates are made for professional printing.

Usually the camera-ready copy has all text and line drawings in place, but lacks the half-tone information needed for gray-tone images or colored pictures. These are stripped in (cut in physically) at the last stage before the plates are made.

stroke fonts

These are fonts which are half-way between bit-mapped and outline fonts. They are composed solely of line segments, and have no fill-in areas. They are infinitely scalable, but the quality is often very poor.

Strowger exchange

A step-by-step automatic telephone switch invented back in 1889 by the Kansas City Undertaker, Almon Strowger. This was the first type of automatic exchange switch. The Strowger originally was just a single rotary switch with ten positions, but later it had a vertical dimension added (giving 100 contacts) and this became the step-by-step switch.

The original phones had two buttons for dialing. To dial line 75, say, you pressed the first button seven-times for the first number dialled, then the second button five-times. This was considered to be enough numbers for any self-respecting American town at the time.

Later 10-number 'decadic' rotary dial systems were added to the phones.

structured

• **Graphics:** Geometric shapes which are stored as a set of descriptive codes rather than as bit-mapped images. If you defined a shape as 'a 40° isosceles triangle with a base of 10cm, with the bottom left corner at the coordinates x = 10, y = 80,' you would be describing it in structured form. See trace, vector graphics and object-oriented also.

• **Graphics metafiles:** A number of protocols which are used to store categories of images in a structured graphic form.

• **Data:** Database and spreadsheet files which have records and fields clearly delineated, as distinct from unstructured files such as those generated by word processing with variable-length paragraphs separated only by carriage-returns.

• **Cabling:** See structured wiring.

• **Networks:** Those using frames for the transmission of data, as against clear channels.

• **Language:** See structured programming below.

• **Wiring:** See below.

structured programming

A philosophy of programming which insists on a strictly organized pattern of sequential algorithms so that the solution to the problem is very visible and systematic. The contrast is with unstructured languages/programming (disparagingly known as 'spaghetti code').

This is a loosely defined IT discipline, but it is one that dominated programming for more than a decade (until OOPs).

Structured programming shuns the use of GOTO statements, and relies on independent program modules and a top-down design. It's often criticized as a concept for purists and pedants, and it dates from an article written in 1968 by an American guru John Dijkstra: ***GOTOs Considered Harmful***.

However, it is now recognized that there are degrees of structure, rather than absolutes. Commentator John Gantz (IDC) said it best: 'Structured programming works if it is done properly. But it is impossible to do it properly.'

However the disciplines of structured programming have been important in teaching programming techniques. Pascal, for instance, is much more structured than BASIC, which is why it is widely used as a teaching language. Students supposedly learn better programming habits with Pascal because this language forces them to develop and use a particular top-down design approach.

• Some of the more recent languages tend to impose structure and modularity on the program also. But, if anything, the trend is back towards more complex program design techniques — although with better controls. Object-oriented languages can be viewed as nothing more than numerous modules with a constant series of GOTOs. See OOP.

• As a rule, unstructured languages need an identifier for each command line. They need this so they can leave the strict procedural order, and return to the point where they left off. For this reason Basic generally had line numbers before each program statement.

Structured languages don't need line identifiers because the program proceeds in strict order. This is an oversimplification.

• The component parts of a structured program are functions—modules (groups of functions), and data structures.

Structured Query Language

See SQL.

structured wiring

The term refers to the systematic wiring of a building for voice and data, rather than ***ad-hoc*** wiring as was normal with LANs.

It means, top-down (designed and planned) wiring: it suggests that the system is planned and implies that a logical hierarchy will accommodate present and future requirements with the minimum of disruption.

The current best structured approach is for a 'vertical' backbone network to be dropped between floors (coaxial or fiber), then 'horizontal' twisted-pair or coax. cables will radiate out from hubs on each floor. The main servers for the building will be at the end of the verticals, to provide maximum bandwidth to the building.

The first structured wiring system was IBM's Cabling System for Token Ring. Later AT&T introduced Premises Distribution System. Now the term just seems to mean 'planned' and organized wiring — rather than *ad hoc*.

• Structured wiring generally involves the following sub-systems:

— **Horizontal subsystems.** Collapsed backbones or star networks are popular since UTP was introduced with 10Base-T. These provide better reliability and flexibility, although they can add cost and complexity. Star networks also make sense with the introduction of Switched Ethernet and/or the eventual change to ATM.

— **Vertical subsystem.** Fiber or coaxial between floors, dropping to a central equipment room in each building. These will generally run a backbone protocol such as FDDI, or one of the 100Mb/s Ethernets.

— **Campus subsystem.** Fiber or coaxial linking the various buildings on a site. This will connects to a carrier at a demarcation point (a demarc).

• The term structured wiring includes: these subsystems, the wiring closets, central equipment room connections and the demarc. See wiring hubs and concentrators.

STS

Synchronous Transfer Services or Synchronous Transport Signals.This is the US term for the electrical signal formats offered at various standard OC (optical fiber) and DC (copper) data rates as part of Sonet.

This is a family of synchronous electrical bit-rates for North America, which are the building blocks for Sonet — which is now part of the international SDH standard. STS-3 is the equivalent of OC-3 in the US, but in international terms (for SDH) it is also equal to STM-1.

— **STS-1** has a bit-rate of 51.84Mb/s and the frame structure includes section- and line-overhead bytes and pointers. The Synchronous Payload Envelope (SPE) which is used to carry the Sonet payload, is a 87 x 9-Byte envelope, but other pointers (overhead) exist within this envelope to identify tributaries. Its optical equivalent is known as OC-1, and this is the lowest-level optical signal handled by Sonet.

When overheads are removed, STS-1 has the data throughput capacity of 37.6Mb/s when framed, and when unframed, it provides more than 43Mb/s as a single pipe.

— **STS-3** has a bit-rate of 155.52Mb/s and is the equivalent of OC-3.

— **Higher** Sonet rates are obtained by byte-interleaving numbers of STS-1 frames to form the higher STS-'n' levels. Each will have an associated OC-'n'.

See Sonet Signal Hierarchy, virtual tributaries, DS-0 and SDH.

STT

Secure Transaction Technology. A technical specification being developed by Visa and Mastercard to allow low level transactions and payments to be conducted over networks. See SVC and CAFE also.

stuffed

Compressed — or perhaps, expired.

Stuffit

Stuffit is a LZ-based file compression program which has become the *de facto* standard for Macintosh. It is widely available as shareware, and has a commercial version which can handle MS DOS .ZIP files also.

STUN

Serial TUNneling. Tunneling involving encapsulation (see both) is a way of transferring packets of data across an intermediate network which uses different protocols. STUN is a router feature for SDLC or HDLC devices, which may need to be connected through a heterogenous network link.

stutter

• **Stutter dial-tone.** This is a modified 'dial-tone' you will get in some telephone services when you pick up the phone to dial. It usually indicates that a message awaits in your voicemail box. It is generated by the telephone company, and is seen to be giving them the advantage in the marketing of voicemail services. See MWI and SMDI.

• **Stuttervision.** A delightful term for low bit-rate videophones and video-conferencing systems which have once-a-second-if-you're-lucky screen updates.

STV

Subscription TV. Pay TV in the US is a Premium service, and this means subscription — you pay monthly. However, STV generally refers to a specific form of Pay TV in the States, which is Pay TV broadcast as conventional UHF television. It is scrambled before broadcasting, so only subscribers equipped with decoders can view.

STX

ASCII No.5 — Start of Text (Decimal 2, Hex $02, Control-B). This control character terminates the header and starts the text.

— STX and SOH are used as book-ends to the header.

— STX and ETX are used as book-ends to the actual text.

style-sheets

In word processing, these let you record standard ways of formatting and presenting your text. They are special documents or stationary files which contain sets of formatting instructions which can be applied to standardize documents or forms. The term can also apply to spreadsheets.

• In publishing businesses, there will be a company style sheet which sets out the typefaces to be used, the column widths, the use of capitals and acronyms, how names will be treated (always with a period after Aug. or Sep. etc.).

SUB

ASCII No.5 — Substitute (Decimal 26, Hex $1A, Control-Z), — Start Special Sequence. This is used to substitute for a character which has been identified as erroneous or invalid.

— SUB is used by MS-DOS as an End of File (EOF) marker.

sub-addressing

In telecommunications, this is 'suffix' addressing. After dialing a telephone number, you can dial extra digits which will then be transferred to the customer's premises, and used internally to switch the call to an extension. It requires that the extra digits be carried across the network, and delivered to the customer's PBX after the call has been connected.

• The term is also applied to Direct Inward Dialing, ISDN and distinct-ring services. With DID many hundreds of extension numbers can be attached to a few incoming lines, and each number will carry its own last few digits.

ISDN is often offered with eight distinct sequential numbers attached to a 2B+D connection. These can be split out and diverted to fax, modem, extensions, etc.

Distinct-ring systems can do much the same on analog connections; they use different numbers which trigger a different ring cadence. A box in the user's premises must identify the difference in the ring sequence and divert the call accordingly.

subarea node

The SNA communications network is formed by partitioned sections called subarea nodes (which can be communications controllers or hosts), each of which can be a host or a communications controller. Peripheral nodes (terminals and terminal concentrators) are attached to subarea nodes.

sub-band coding

A video coding technique that uses 'filters' to identify and remove unwanted information. There is both inter-frame and intra-frame forms of sub-band coding, and the filters are digital under intelligent control.

subcarrier

• In frequency division multiplexing, the carrier is divided into various subcarriers or subchannels, and each user on the system is allocated a single subcarrier.

• In time-division multiplexing a single slot is effectively a subcarrier or subchannel. Each user has one.

• A video or television signal consists of a number of different information-carrying subchannels:
— The luminance signal.
— The chrominance signals (usually two).
— The horizontal and vertical sync pulses.
— Teletex signals.
— One or more audio channels, and sometimes data.

subcarrier multiplexing

An optical fiber technique. The various information channels are each shifted to a different frequency then combined electrically onto an FDM modulated electrical subcarrier frequency. This is then used to amplitude-modulate the laser source.

This approach creates an analog fiber transmission system which can carry both modulated digital and analog in different parts of the spectrum. It is mainly used for one-way transmission of video (cable TV) and two-way digital signal (voice and data) services in the local loop.

subchannel

See subcarrier.

submarine cable

Early submarine cable systems were for telegraphy only. Trans-Atlantic telephone calls were carried by radio from 1927 until 1956 — although Florida to Cuba (90 miles) was spanned in 1930. It carried four two-way conversations on twisted pair.

Later twin-coaxial was used for long submarine cables, and now optical fibre pairs.

• TAT-1 across the Atlantic in 1956 carried 36 two-way conversations. In 1970, analog coaxial submarine cables were carrying 1000 two-way phone conversations.

• Analog coaxial submarine cable reached its limit with about 4000 duplex circuits in a cable 1.7 inches in diameter. These cables had repeaters every 5 miles, and equalizers every 100—150 miles. Power was 7000 volts DC at each end. They used frequency division multiplexing in the 1MHz—29MHz range.

• The first fiber-optic submarine cable systems were introduced in the late 1980s.

— **First** generation fiber cables provided 20,000 voice circuits per fiber pair, and used NRZ modulated multi-frequency 1.3 micron lasers in regenerative repeaters. Repeaters spacing was less than 100km, and the repeater units had multiple levels of redundancy.

— **Second** generation fibers (from 1991) doubled capacity and allowed routes to be split under the ocean to make two or more landfalls. This generation used 1.5 micron single-frequency lasers, with repeaters spaced at 150km and had rates of 560Mb/s. Typically they only had one level of redundancy and yet achieved the same reliability.

— **Third** generation fibers (from late 1994) boosted capacity to over 300,000 voice circuits per fiber pair (but with only 75,000 64kb/s PCM channels). These use erbium-doped optical amplifiers at distances of about 50km. The terminals operate currently (mid-1995) at 5Gb/s but it may be possible to upgrade these systems later (with WDM) without needing to modify electronics in the repeaters — since there isn't any, apart from the pump lasers.

— **Fourth** generation systems (1999) will use wavelength division multiplexing (WDM) and solitons for five or ten times the current fiber capacity. For shorter undersea distances, repeaterless systems with lengths up to 300km are possible with bit-rates of 622Mb/s or more. The reliability of optical amplification is several orders higher than any of the electronic repeater or regenerator systems.

See halide fiber also.

submarine patents

A common practice in the computer and communications industry. A company involved in the development of a complex new standard (such as MPEG) will hold back on its claims to patents over some aspects of the technology.

They wait until the standards are firmly in place, the chips have been made, and the industry is committed to taking this direction by implementing this standard — then they spring out with their claims. It is a common form of technical blackmail. It has been highly successful for the companies.

sub-micron chips

Chips in which the spacing between electrical elements has been compressed to the less-than-one micron level. These chips are very difficult to manufacture, given the fine tolerances required. See Debye limit.

subnet

This refers to a network segment.

• In IP networks it may refer to a subdivision with a particular 'subnet address'.

• In OSI networks, this is a single administrative domain with collections of end systems and intermediate systems using a single network access protocol.

sub-notebook

A very small portable computer, usually less than 9-in. (25-cm) wide and therefore with less-than-standard key separation.

sub-Nyquist sampling

The sampling method used in the MUSE HDTV system to reduce the number of information bytes being transmitted.

Nyquist sampling is when the sampling rate is twice the highest frequencies needed, and this is usually done during the A-to-D conversion stage. However to reduce the bit-rate further for transmission purposes, only every second byte (pixel) is selected and transmitted — this is sub-Nyquist sampling. It is the digital-pixel equivalent of line-interlace scanning. See MUSE and quincunx.

subrate

This usually refers to channels which are less than the standard PCM rate of 64kb/s. Some trunk traffic nowadays is carried at the subrate of 32kb/s using ADPCM for voice coding, but this low rate creates problems with modems.

• Subrate could also mean Fractional E1 or Fractional T1 rates. See fractional.

subrate multiplexing

Where two or more data channels share a single 64kb/s channel, such as using a single B-channel in ISDN for multiple PBX links.

subroutines

A section of a program which performs an action which is frequently used. So instead of repeating the code every time, the program will 'call' this subroutine from wherever it happens to be at the time. This is a GOTO or GOSUB call.

Some programs consist of little more than calls to sub-routines — which is anathema to structured programmers. When they approach this limit, they are acting very like object-oriented programs.

subsampling

The term is applied to the process of sampling a signal that has already been sampled (usually through A-to-D conversion).

After the initial sampling conversion, for instance, a HDTV image may consist of 1920 8-bit samples in each scan line. But this is too much for transmission, so each second bit is taken in the subsampling process. The full image is preserved for possible use in cinema projection. See quincunx and sub-Nyquist sampling also.

• A picture requiring 640 x 480 pixels is reduced by subsampling to an image of 230 x 240 pixels, or a video being shown at 30 frames a second, is reduced to 15 frames a second. Both of these are subsampling techniques. See CIF and QCIF.

subscriber

• A hot-linked file which draws information from a published file. See publish.

• The telephone carrier's customer. So the 'subscriber loop' is another name for the local loop, which is also known (collectively) as the Customer Access Network or CAN.

• The legitimate user of a public database.

• On a Cable TV network you first subscribe to the connection (which usually brings you Basic cable, free-to-cable programs); then you subscribe to Pay TV or Premium services (using an IRD). You don't subscribe to Pay-per-view or Video-on-demand services (not for the contents, anyway!).

subscriber access lines

A common term for the copper twisted-pair in the local loop.

subscriber management system

The key component in any form of Pay TV (cable, satellite or MMDS) service if the owners are to be paid. This involves the computerized systems at the head-end and the Integrated Receiver-Decoder (IRD) or set-top box in the customer's home, using some form of 'conditional-access' between.

Subscription allows the customer to order one or more services for a certain time. Control of the viewing is possible because the head-end computer system can individually address the set-top boxes and authorize them to decode certain programs. See scrambled video.

Some cable services are relatively basic, while some are quite sophisticated. Some of the more sophisticated now use smartcards to handle the descrambling code and authorization — others use the smartcard as a stored-value card (SVC).

Authorization to view may be granted following a phoned request over a phone link from the set-top box, or by the return channel in a cable network.

subscript

• **Printing:** A small letter or number printed below the baseline of the normal text, like $_{\text{this}}$ —usually to identify footnotes.

• **Programming:** In general, this is any symbol which identifies a set, or a particular element in a subset. See below.

• **Number:** The number used to identify an element in an array. A single dimensional array will probably use variable names such as A1, A2, A3, . . ., etc. A two-dimensional array will used variable names such as A1,1 and A1,2 and A1,3 . . . and later A2,1 and A2,2 and A2,3, . . ., Ax,y. In this case all the numbers are known as subscripts.

Subscripts in arrays are often surrounded by parentheses as in A(x,y).

subscription TV

• **Specifically:** A pay TV service in the USA which is broadcast in UHF channels in scrambled form. This is conventional TV with scrambling. See STV.

• **Generically:** A term for terrestrial, cable and satellite services where a monthly charge is made for viewing. As distinct from advertising supported networks and channels, and public broadcasting services.

It applies to both Basic Cable and Premium/Pay TV where payment is made on a monthly subscription basis for the service, so in this sense it is a vague term which covers both 'carriage' and 'content' subscription.

— **Basic Cable,** is a carriage/connection service where the monthly subscription is paid to the cable company for the delivery of a clean, high-quality electrical signal. The programming on Basic Cable is supported by advertising or financed by organizations (e.g. religious) with a vested interest in promoting a philosophy.

Usually most of these signals are stripped off air from nearby network or independent transmitting stations. Some could even be generated by the cable company itself, and supported by local advertising. These channels are rarely scrambled.

— **Premium or Pay TV** services are also sold by monthly subscription, and here the subscriber pays for content. A premium (depending on the number of channels selected) is added to the monthly Basic Cable subscription, and part of this goes back to the program maker. These premium TV signals are usually scrambled by the station before being transmitted so that they can only be viewed by using a decoder.

— **Pay-per-view** is not really a 'subscription service' (since you pay only when you view) but it is often included in the definition because you need to have the Premium set-top box before you can get PPV. See DBS, MMDS, SMATV, NVOD and VOD.

substrata

The physical material on which the circuits of a chip are fabricated. Most chips use a silicon substrata; others GaAs.

substring replacement

A data compression technique where particular words or groups of characters (substrings) are replaced by abbreviation codes (usually control characters, etc.). A dictionary must be held at the beginning of the file to define the meaning of the abbreviations. It is a very useful technique for repetitious material. See Memex.

Subsystem Control Block architecture

SCB is a relatively new IBM software protocol to allow add-in products to take advantage of MCA's peer-to-peer operations without involving the main processor.

subtractive color

See additive primaries.

subvector

A vector in SNA is a data segment of a message, 'comprising a length field, a vector-type key, and other vector-specific data'. A subvector is also a data-segment of a message, and it consists (would you believe!) of 'a subvector-type key, and other subvector-specific data'.

SunNet Manager

SunSoft (a division of Sun) introduced SunNet Manager as the first general network management platform in 1989. It is still the top selling Unix platform. See DME, NetView and OpenView also.

Super-#

• **Video:** Super-8 is a consumer VCR standard defined by the Japanese Electronics Industry, and supported mainly by Sony (rather than the VHS camp).

The standard supports two types of tape formulation:

- metal particle tape, which gives reasonable quality and is easy to produce, and
- metal evaporated tape which is difficult to make, expensive, erratic but of very high quality.

This basic system with enhanced electronics has become the Hi-band or Hi-8 video standard. See hi-band.

Super-8 uses PCM audio modulation, and it has been adapted to digital recording as a computer backup system. See 8mm tape in numbers section.

• **Film (amateur):** Super 8 was a very popular home-movie color film format for a while in the 1950s. Some of these films had a very narrow magnetic stripe which carried sound. Don't confuse Super 8 millimeter film with Standard 8 millimeter film which was made by splitting 16mm film lengthwise down the middle (after exposure and processing) and doubling the perforations.

• **Film (professional):** Super 16 is a development of normal 16mm professional movie cameras and film, where the camera gate is widened to incorporate the normal sound-track edge of the film. In professional production, the sound is always recorded separately onto quarter-inch tape or a DAT recorder, so this part of the film was just being wasted.

The lens is offset slightly in the lens holder to center it on the new frame area — but most of the lenses, motor-drives, tripods, etc. are stock-standard 16mm equipment. This is a very effective way of shooting low-cost, wide-screen documentaries and feature movies — but it is for acquisition only. It must be immediately dubbed onto a higher (35mm or video) standard for post-production.

Super 16 is likely to be an important format for production in the wide-screen (EDTV) and HDTV era.

super bit-mapping

A technique invented by Sony to improve the quality of CD audio by 'noise-shaping. This is the result of sound quality problems when a company records music as 20-bit samples on tape, then transfers it to 16-bit standards for CD disk. The abandonment of the last four-bits in each sample means that quiet music passages and background 'ambience' sounds are simply switched on or off — and this is disturbing to listeners.

SBM shapes the background noise by taking some of the sounds from the middle frequencies (where the ear is most sensitive) and reinserting it into lower or higher bands where human hearing is less acute. An alternate technology also used in CDs is to add 'dither' (see).

Super CD

Also known as DVD and CD-HD. It could also mean Super Density CD, below.

supercomputers

The term given to the fastest and most powerful general-purpose computers available at any given time. They are primarily number-crunchers, and are usually about 100 times faster than the biggest mainframes. Most are parallel computers nowadays. Normally the term supercomputer is only applied to machines capable of performing at 1 Giga FLOPS or more (1 billion calculations a second). Cray's Y-MP C90 runs at 16 GFLOPs and costs US$46 million.

Super Density CD

SD. This was Toshiba's high-density video CD system, which briefly rivalled Philips/Sony's Multimedia CD. The now have joined together to produce the DVD standard. See.

Toshiba, Time-Warner, Matsushita, Thomson, Hitachi, Pioneer, MCA, MGM/UA and Zenith were behind the SD standard and JVC, Mitsubishi, Nippon Columbia and Turner offered support.

Toshiba released details of their proposed new standard early in 1995. This was their second attempt; the first was based on two disks glued back-to-back (each read from behind), so this disk sandwich needed to be turned over to be read in full.

Later Matsushita greatly modified the laser unit, so it became possible to use a multi-layered system like Multimedia CD. These were half-thickness (0.6mm) disks glued face-to-face, and read through the transparent surface from only one side.

This later system used a dual-focus lens which could read the two surfaces simultaneously. It was possible to hold a full movie on this system (150 minutes of MPEG2). It was to be aimed at the home TV market rather than at the computer peripheral market.

The disks were to have been dual-sided with each holding 5GB (15x normal CDs). SD's MPEG-2 compression had variable bit rate coding, and the maximum data transfer rate (burst rate) was given as 10Mb/s. It could sustain an average of 5Mb/s, and the supplementary sound and data tracks could support 8 languages and 32 subtitles.

SD was to have provided Dolby AC-3 (5 channel) surround audio; and it had multi-aspect-ratio capabilities (4:3 and 16:9). Parental lock-out was to have been built in to the system.

• In September 1995 the Multimedia group and the Super Density group finally came together and formulated a new standard which incorporated the best of both. For details of the replacement standard, see DVD.

Super EGA

A minor monitor standard which had 640 by 350 pixels.

supergroup

A grouping of 60 analog telephone voice channels for the purpose of FDM. This is a cluster of five groups (each of 12 voice channels), and it requires 240kHz of bandwidth overall. The five carriers are spaced 48kHz apart in the spectrum between 312kHz and 552kHz. See group and hypergroup. Don't confuse this with the supermaster group.

superhigh-resolution (CD)

A technique used to increase the capacity of a CD or CD-ROM disk without changing the wavelength of the laser. It enables the reader to detect signals in a region much smaller than the diameter of the light beam, so the same basic standard can be effectively used with twice the recording density.

With Magneto-Optical systems the superhigh-resolution approach relies on a double layer of the magnetic-optical film. With conventional read-only disks they use a phase-change material with different reflectance characteristics.

This is the dual-layer approach now used on the new CD-Multimedia and Super Density disks. See.

superhighway

Data or Information Superhighway. The term was first used by US Vice President Al Gore in Dec 1993 and it became the buzzword of the 1990s. It applies to all forms of high-tech communications, from fiber-optics to the home, computer-to-computer communications, video-on-demand, and seems to be totally confused with the Internet. And since it encompasses everything, it defines nothing.

In reality 'Information Superhighway' is little more than a political catch phrase. However it carried the sub-text in the USA, of telephone and cable de-regulation. 'The High-Performance Computing Act of 1991' is the most significant of the new US government initiatives.

• Don't confuse the Superhighway and the Internet. The Internet is not a superhighway but a goat-track. It just happens to be the best damn goat track around — and it is probably more productive than all the other highways put together, except, perhaps for the telephone.

• And the Internet is probably not even vaguely a model for how superhighways will develop. It is totally anarchistic (where they will be regulated) and almost none of it has been purpose built (except for local services).

— The broadcaster's view of the superhighway is that of a one-way street where television consumers purchase entertainment services and products from head-end suppliers (the broadcasters) at the top of the hierarchy. Preferably they will purchase these on a subscription basis, as a package since this solves the problem of individual discrimination and selection.

—The Internet community's view of the superhighway, is as a meshed internetwork of independent networks, each acting non-commercially to exchange data in any bandwidth up to that occasionally needed for full motion video. They would also like to link their PCs into Cray Supercomputers when the urge overtakes them.

— The telcos view of the superhighway is a hybrid fiber coaxial cable network owned by them, running VOD servers, also owned by them. They will then control the middle ground between the wholesalers (video distributors) and the customers — and everyone knows which part of the chain makes the most money.

— The cable company's view is a cheaper and better version of what they already have, but able to carry local telephone traffic and interactive video games. This lets them also supplement their income with telephone call fees and interactive services.

— The public view of the superhighway, is that they don't care, provided it brings them the latest Arnie Swartzenegger films and the best football games for $5 a month.

— The truth is probably somewhere between.

• Disney President of TV, John Cooke recently listed the five major roadblocks he believes will need to be overcome to allow the development of the Information Superhighway:

— Security for intellectual property.

— Assurance for personal privacy.

— System integration obstacles that could cripple developers.
— The need for user-friendly designs.
— Government regulations.

Disney obviously sees problems arising from government involvement, but I would add at least another five:

— Corporate greed and desire for market dominance in the provision of information or entertainment.
— Lack of standardization enforced by regulation to ensure interconnectivity and accessibility.
— Need for the preservation of national cultural identities.
— Over-reaction against ultra-violence, pornography and materialistic philosophies propagated without any effective controls.
— Information and entertainment glut. Social changes leading to a simpler lifestyle.

supermaster group

An ITU term for 900 analog voice channels. It is a group of supergroups. See hypergroup.

super-micro

A powerful micro used mainly as a workstation or as a server.

SuperNOS

Novell's name for a new super network operating system which will replace NetWare and UnixWare. It will incorporate the NEST network embedded systems technology.

Super NTSC

See NTSC.

superpipelining

This is one approach to the design of new high-speed RISC chips. MIPS has focused on this architecture with their R3000 and R4000 CPUs. It provides a single pathway through the chip (rather than parallel), but splits the tasks into smaller units so that many instructions (strung out along the pipeline) can be handled in each clock cycle.

The R3000 is a five-stage pipeline and the R4000 is an eight-stage pipeline, handling eight instructions simultaneously. There is obviously a finite limit to superpipelining, since value cannot continue to be increased incrementally by smaller divisions of the tasks. See superscalar architecture also.

superscalar architecture

A form of pseudo-parallel processing architecture within a CPU, mainly achieved by instruction pipelining and vector processing. All of the major RISC chip designers (excepts MIPS) currently see this approach as the way of the future. Superscalar sends two or more instructions through the processor at once, but slightly staggered in time. See superpipelining (a similar technique), pipelining and microparallelism also.

superscript

The small digits or symbols that appear above the line. In contrast to subscripts which appear below uses superscripts.

superserver

A large high-capacity, fault-tolerant, computer installation used as a network server. It will usually have a number of CPUs, special parity-checks on memory, large cache, RAID disk arrays, backup power supplies, etc.

superstations

A broadcast term originally applied to Ted Turner's Atlanta TV station WTBS, which distributed its TV station signal free to cable operators via RCA's Satcom I satellite. He recovered the cost from increased advertising rates.

The term is now applied also to radio stations that distribute their signals via satellite to numerous 'satellite' re-transmission sites around the country.

Without these, the world wouldn't know Rush Limbaugh!

super Telex

See Teletex (no final 't').

Supertwist LCD

The more modern form of LCD screen found in many portables and notebook computers.

• This technology has a higher contrast ratio (4:1) than standard twisted-nematic types and a better viewing angle. The light is put through a 180° twist (as against 90° for single-twist LCDs), and triple supertwist puts the light through 270°, to improve contrast. Backlighting is often used with these screens, but they can function well without. However the higher costs, higher power consumption, and blue or yellow colorcast are problems still to be overcome.

• Double supertwist is an improved version (which can provide color also). It combines an active lower supertwist panel with a passive overlay panel of a different LCD material which compensates for the color-cast problem and improves contrast. These displays are often paper-white with contrast ratios as high as 12:1. They will often be hi-res VGA displays with 16 gray levels, but they draw more power and are more costly than normal LCD. They also require backlighting, which further reduces battery life. See double-twist also.

Super VGA

Super VGA was designed by the clone-makers to break IBM's dominance in standards-setting on the PC. However, it lacked strong efforts at standardization within the group, so there are variations. This is a 16-bit standard.

The Video Electronics Standards Association has since attempted further standardization and so the basic Super VGA is now 800 x 600, with an 'option' of 1024 x 768. See VESA.

However cards and monitors sold today as 'Super VGA' can provide a range of quite different image qualities: 640 x 480, 640 x 400, 800 x 600, 1024 x 768, 1076 x 768, and 1280 x 1024 pixels. At the middle level of resolution (1024 x 768), Super VGA generally uses interlace monitors and a graphics coprocessor (although non-interlace is also used).

This 'standard' had a text character-size of 8 x 16 pixels (for the old bit-mapped fonts) and the range of colors that could be supported depended on the image format. At the 640 x 480 size most would support 256 colors and, at the higher levels, 15-bit color is now being offered. A 16 million colors pallet is available for 800 x 600 pixel images.

Super VGA display cards were very often incompatible with each other — each manufacturer wrote custom display drivers, so a dozen DOS applications may require a dozen specialized drivers.

VESA attempted to clear up the mess by introducing the VESA Super VGA BIOS extension, and so a single VESA driver could then be expected to work with numerous DOS applications. Fortunately, a single Windows driver can also handle any Windows application – but often an accelerator card is needed.

- The 800 x 600 standard is often also called the 'Enhanced VGA' which adds to the confusion.
- To add even more confusion, the term Ultra VGA is sometimes used to mean 1024 x 768 with 16 colors but using non-interlace (progressive scan).

Super VHS

See S-VHS page 637.

supervision

Control and monitoring.

- In telephone terminology this refers to the mechanisms in a PBX or telephone exchange which tell when the phone is picked up, and what number was dialled, etc. It deals with the whole range of signaling information involved in call progression. See BORSCHT.
- In general communications, it can mean anything from a process which checks that a circuit is functioning, to one that handles quite sophisticated control messages and the error correction.
- In computing, the supervisor is a section of the CPU's control program which coordinates resources.
- A supervisory channel is usually a backward channel.

supervisory functions

In telephone networks these are all the call establishment and call clearing functions, together with some signaling and billing signals which are transparent to the user.

support

One of the most flexible and deliberately-vague terms in the engineer's, marketeer's and computer journalist's lexicon. It can be twisted and bent to mean almost anything it needs to mean.

It is intended to suggest that one product will work with another. But the conditions under which this may or may not happen are left unspoken — usually because the speaker doesn't know.

When everything is exactly to standard, the traditional phrase is 'compliant with the standard'. When things are only 99% to standard, the term is often 'compatible with the standard'. And when things are only vaguely around 95% or less, people say it 'supports the standard'.

- The fact that the technical journalist claims: 'Acme's Virtual Reality package supports our Super Hi-res Ajax monitors', may mean nothing more than he uses the box jammed under the monitor to tilt it to an ergonomically-correct angle. See compatibility, compliant and solution.

surface-emitting lasers

A new technique in small laser development made possible by MBE fabrication. The laser diodes can be fabricated, hundreds at a time, on a wafer of gallium arsenide. They are then tested, and then separated out for use. This makes mass production of lasers at very low cost, possible.

These lasers are particularly suited, both in their geometry and in their mode-matching, to optical fibers. And, if left as an array on a larger area of silicon wafer, they open up new possibilities for parallel optical switching, transmission and optical computing.

surface mounted devices

SMDs. These are very small devices carefully designed to conform to international standards. Automatic machinery can then carefully position these very small components on PCBs and solder them directly onto the conductors.

The distinction is with the older forms of discrete components (see DIP and SIP) which need to be inserted into holes in the PCB and soldered, or which fit into special multi-contact sockets mounted on the PCB.

SMD technology aims for size and thickness reduction, machine mounting of components on the board, and for increased reliability.

surface waves

- Some forms of radio wave propagate on the surface of conductors, using the skin-effect. Coaxial wave guides use this principle.
- Very long wavelengths used for low-rate signaling to submarines (VLF and LF) also travel around the earth using a surface-wave phenomenon. Don't confuse this with 'space waves' which is the way normal MF, VHF and UHF signals travel terrestrially, or with HF (short-wave) use of 'sky waves' which bounce off the ionosphere.
- See propagation modes and space waves.

surf

To cruise without purpose around the Internet (or zap around TV channels) looking casually at anything of interest.

surge

A rise in the power-supply voltage over a moderate term (say from 0.5 to 20 seconds). See spike and sag.

SVC

• **Switched Virtual Circuits:** These are dynamic network connections; they are set up for the duration of the call and then terminated. In cell-switching networks they are the ATM equivalent of the normal switched telephone call.

SVC will only be possible when the set-up and connection between any two nodes is made by a specific call-establishment process. This can be internal for a company, but in the long run it will be provided by the carriers.

The distinction is with the Permanent Virtual Circuits now being used (through tables provided to routers), and datagram (connectionless) operations. See ATM, PVC, virtual circuit and virtual path.

• **Stored Value Card:** These are 'debit' cards used as electronic purses. Cheap plastic phone cards are a form of SVC, but most disposable types will use a magnetic stripe. The smartcard form of electronic purse is more permanent; it can be reloaded from a bank account via an ATM.

However don't confuse this SVC utility with the smartcard's ability to function as an authentication tool, or as a standard credit card.

Smartcard SVCs are likely to be of a standard called VME (see) and use a secure transaction technology (STT). These cards can record transactions and adjust the internal cash balance each time the card is used in a card reader — say at railway stations to buy tickets, or at a vending machine.

Visa is trialing a worldwide scheme for SVCs which can be used in photocopy machines, payphones, vending machines, and public transport. See VME, STT and CAFE.

S/VGA

Super VGA. See.

S-VHS

Super VHS. A semi-professional video cassette standard which uses the same cassettes (the plastic, not the tape) as VHS.

This is a Y/C analog video semi-component recording system which is finding professional applications in small television stations for news and documentaries, and in non-broadcast video production areas. It was released by Panasonic in 1987 to challenge U-matic formats in these areas.

S-VHS uses higher and wider bands of video frequencies ('hi-band') to record the video signal on tape and keeps the Y (frequency modulated 'luminance') and C (QAM 'chrominance') signals separate. This improves the horizontal image resolution of the pictures to over 400 lines (in luminance only) — the domestic VHS is about 250 lines.

The system uses amorphous metal heads with 0.15 microns head gap, and a cobalt gamma ferric oxide tape (not metal particle or evaporated) with a greater coercivity than normal VHS tape. The hi-fi audio tracks are embedded in the video and can't be separately edited, although the linear stereo audio tracks can be. One of the stereo tracks is often used for time-code, but recently time-code has been added to the vertical blanking interval to free up the second linear track and prevent bleed-through.

Standard play, runs the tape at 23.4mm/s which is the same as normal VHS, while long-play is at 11.7mm/s. S-VHS machines can replay standard VHS pre-recorded cassettes and also record in the old VHS composite format.

More recently, some manufacturers have also proposed slipping 16-bit digital audio carrier signals into the space left by shifting the video frequencies to a higher frequency. The result is near-DAT quality, and a digital recording system that can record three hours of audio at CD quality. JVC have built-in a copy limitation scheme to keep the record companies happy.

S-VHS uses S-video techniques, but don't confuse S-VHS with VHS-C, which refers only to the compact cassette size — although there are now S-VHS-C camcorders.

SVID

System V Interface Definition. For AT&T's Unix, this is a device interface standard that defines how device drivers, system calls, terminals and printer drivers should be written, although it omits machine-specific and management utilities. XVS is the slightly modified alternative standard put up by the X/Open group. See ABI and Unix International.

SVVS

System V Verification Suite. See SVID.

SW56

Switched 56. See.

swapping

Disk-swapping: In Unix, this is the process of temporarily holding parts of a program on disk, then bringing them into memory when space becomes available. It is a multitasking feature that comes into play when there are more demands on memory than the available RAM can supply.

It is called virtual memory in PCs.

sweep frequency

In a TV set or monitor, the sweep frequency is the rate at which horizontal scan lines are drawn on the screen, expressed in Hz. In NTSC it is 15.75kHz, and in PAL it is 15.625kHz. In progressive scan monitors with 1200 lines on the screen and 70Hz refresh rates, it can be as high as 84MHz. See horizontal scan and line frequency.

sweetening

In film and video production, this is the technique of providing extra sound tracks, and of processing the available (recorded) sound to improve the audio tracks.

SWER

Single Wire, Earth Return. The early telephone systems were SWERs because this was the cheapest way to run copper on poles.

SWG

Standard Wire Gauge — the English standard. A wire of SWG gauge:
— 15 is 72 mils in diameter (1.83mm).
— 20 is 36 mils in diameter (0.91mm).
— 25 is 20 mils in diameter (0.51mm).
— 30 is 12.4 mils in diameter (0.31mm).
These days, wire is based on the IEC International annealed copper standard.

SWIFT

Society for World Interbank Financial Transactions. This is an international organization and an EFT system used since 1979 by the international banking community to transfer money. It now provides bank-to-bank transaction links to virtually every developed and newly-industrialized country.

Currently there are about 1.8 million transactions a day conducted over the SWIFT network between 3000 institutions in 72 different countries. Language and interpretation problems are eliminated by the use of standard message types.

Originally SWIFT used a telex EFT interchange system, but now with SWIFT II over a X.25 packet network, it is a full EDI system. Naturally it has a high degree of security.

• A parallel system for foreign exchange is called ACCORD.

SWIM

Sanders-Wozniac Integrated Machine. The famous disk-handling chip from Apple which pays homage to the legendary Woz, designer of the Apple II and inventor of the GCR disk format. Apple Macs have been forced into line with the rest of the world, and SWIM chips now use the MFM format.

SWIMs control disk access on the Superdrive which is why they can read such a wide range of Apple and IBM formats.

• The earlier WIM chip was designed by Steve Wozniac for the early Apple II disk drives, and it was a major breakthrough in component reduction (and cost saving) at the time.

switch (telephone)

Over the years there have been a number of different automatic telephone switching technologies used in the circuit-switched public network. In order of implementation, they are:
— Strowger.
— Step-by-Step (a Strowger modification).
— Cross-bar (electro-mechanical).
— Cross-bar with stored program (computer) control.
— Full digital (ISDN) Integrated Circuit switches with centralized processor control.
— Digital switches with distributed processor control.
We are now moving into an era of:
— Broadband fast-packet switching with SDH and ATM. See Banyan and delta switching.
Parallel to the circuit-switched network, has been the development of packet-switched networks using switching nodes, which are really routers (computers with multiple I/O ports).

• British Telecom has suggested that the use of WDM over optical fiber may eventually lead to them abandoning exchange switching altogether. A recent report suggests their current 6000 switching exchanges around the UK, could be reduced to 40 broadband switches in a few years. And then later, switching may disappear forever.

George Gilder makes the same point when he says 'bandwidth is a substitute for switching'.

Of course switching is also a substitute for bandwidth.

switch (LAN or WAN)

Network switching of relatively standard LAN protocols appears to be offering us a stepping stone in the transition from legacy LANs to the future ATM networks.

LANs can be switched at the backbone 'enterprise level', the 'departmental level', or at the 'workgroup level', and LAN switching is fast replacing the traditional 10Base-T hub.

True LAN/WAN switching is 'cut-through' or fast-packet switching, and it requires banyan-type matrix switches. However true switching which does not retard the passage of the packet or cell, cannot perform rate adaptation, either.

What is often called LAN switching is really nothing more than RISC- or ASIC-based bridges and routers which have been souped up to transfer data faster. These 'switches' are either shared-bus or shared-memory devices, and they always hold the data temporarily in store while they read the address.

They are able to provide rate adaptation, and are usually designed to handle 10Mb/s segments, with 100Mb/s backbones. However, if they don't have adequate flow control, they can overflow and lose data packets.

Problems have been experienced with these pseudo switches, mainly on backbones because of inadequate buffering and flow control, but this appears to be the way of the future, when segment capacity is a problem.

• Switches (both true and pseudo) tend to reduce network latency for isochronous traffic, which makes them more suitable for multimedia applications than networks using standard-speed bridges and routers (which both use store and forward techniques).

Multiport switches are also an ideal way to handle dedicated segments to power workstation users. These switches can be half-duplex or full-duplex — providing 10Mb/s per segment, or 20Mb/s.

• See switching techniques, Switched Ethernet, FDDI and Token Ring.

Switched 56

Switched 56 (kb/s) links are popular in North America as an alternative to full ISDN. Essentially it is switched DDS using T1 links where bits are being robbed for supervision.

Switched 56 appears to have higher installation costs than ISDN in many regions, and calls are charged at usage rates about the same as normal analog business phone lines.

The popularity of Switched 56 depends mainly on the variability of implementation of ISDN, and also on its availability and costs. Recently, however, ISDN rates have dropped.

With Switched 56 you pay only for connection time, and each connection has its own telephone number. With ISDN you get a higher data rate, two lines (each of which can have its own number if required) plus the packet-switched D-channel.

Both are carried over the same twisted pair in the local loop, but with ISDN you get a clean 64kb/s per channel, while the Switched 56 line robs bits, leaving only 56kb/s clean.

Both synchronous and asynchronous connections are possible through the CSU/DSU. Twin Switched 56 is popular for videoconferencing, and it provides a wide-band connection for dial-up LAN links.

- Dedicated 56 services are also offered.

switched circuits

Telecommunications engineers have a different concept of switched circuits than lay people. The general public imagines that for telephone, there is an electrical connection from one end to the other. However circuits can have segments at baseband frequencies over copper wires, others at low radio frequencies multiplexed over coaxial or microwave, and other segments as one of a half-million multiplexed calls travelling on light-beams down fiber. All of these combine to create a circuit.

A switched circuit requires a mechanism to set up the call (trigger the sequence of switches and reserve the necessary bandwidth), provide supervisory information while the call is in progress, and a mechanism to take down the circuit when the call is completed (end-to-end also). It requires a billing mechanism also.

- Switched 'virtual' circuits can exist in packet-switching networks. The circuit then established is nothing more than a chain of special routing instructions which are communicated to the intermediate switching nodes. These instructions advise them of the action they should take when transferring packets with a special Virtual Circuit Identifier (VCI).

Packet networks use either virtual circuits or datagrams, and the distinction is in whether a circuit is established end-to-end before the data begins to be passed.

There are three main categories of switching:

— Circuit or line switching. This is the standard telephone approach until recently. This is not generally a physical (electrical circuit) but rather a reserved ('provisioned') channel.

— Packet-switching, where small packets (usually 128-bytes) of binary digits are transmitted over a shared network system. Each packet has a header which will carry either a full address or a VCI — and these control the switching of the nodes. ATM is a fast-packet version of this approach using 53-byte cells.

— Message-switching. This is a 'large' packet system where the full text of each message is contained in a single packet. This is forwarded progressively from sender to recipient's mailbox. See X.400.

Switched Ethernet

Switched Ethernet uses a radiating star cable topology from a switching hub where each workstation (if required) has exclusive use of a segment, and therefore of the full 5—10Mb/s of dedicated bandwidth. This is a much more scalable technology than Fast Ethernet and can provide adequate bandwidth to the desktop for most power users.

The NIC and cables are also reasonably compatible with existing Ethernet, and so this change has low equipment and conversion costs.

- See Ethernet switching, and intelligent switching hubs. See multilayer switches also.
- Full Duplex Switched Ethernet. This is a sub-set of Switched Ethernet which allows bi-directional (10Mb/s each way) transmission/reception of data.
- Kalpana now has a switch which has both cut-through and store-forward functions.

Switched FDDI

Following the success of Switched Ethernet and Switched Token Ring, Switched FDDI were introduced. There are both half- and full-duplex versions. Some people believe that these switched systems, together with 100Base-X and 100BaseVG-AnyLAN will eliminate the need for ATM on most LANs.

I think they misunderstand the value of a single cell-based system like ATM.

switching fabric

The term used for many-to-many interconnects (multiple cross-connects) where more than one signal may be passing across the switch 'fabric' at any one time. The metaphor is of the warp and weft of woven fabric with many lines going in many directions.

This is an extremely general term, but there are a few major architectural categories:

— The n-log-n multi-stage fabric uses a chip with few ports, but which can be cascaded to any desired size.

— Crosspoints (aka complete mesh structures) are similar to the old crossbar matrix.

— Bus-based, shared media switches which used time-division to shunt packets across the common backbone.

— Shared memory switches where RAM is driven by a complex addressing scheme to route packets between ports.

Switching fabrics can be multistage (as are banyan switches) with small switching elements assembled into large arrays of input and output ports. See banyan and delta.

switching hub

Many switching hubs are conceptually bridges or routers with high port densities, and low latency. They are usually a plug-and-play device which replace the traditional hub — but there are both modular hubs and stand-alone devices.

Most switching hubs will provide high performance with very little set-up or maintenance overhead, and today, they are increasingly being used to provide workgroup and departmental bandwidth and high-speed connections to file servers.

• The term 'switching-hub' was first coined by Chipcom back in 1994. It was then a hub that allowed network managers, using a remote network management console, to reassign ports from one segment of a hub to another. This is now called 'port switching' to distinguish it from the forms that followed.

• Later Kalpana started developing its high-performance 'on-the-fly' switching hubs (cut-through switches) which went by the name of EtherSwitch. This was a totally different approach to the above, where each port on the switch represented a separate Ethernet segment.

switching techniques (LANs)

Three key techniques of LAN switching have evolved:

— **Minimum-latency** or **cut-through** switches. These switch cells or packets on the fly, and provide minimum delays on the packets passing through. The header has been read, and the switch thrown, before the full length of the packet has entered the switch.

This type of switch can provide minimum latency, because the data effectively travels through the switch without stopping. The disadvantage of this technique is that it doesn't provide for rate adaptation — the rate of the data coming in, must be the same as the rate going out. Buffers generally aren't provided. These switches provide fast throughput to light loads where isochronous services are essential.

— **Collision-free switches.** These check the first 64-Bytes of message for evidence of fragmentation or collision. These are not really store and forward devices because the hold time is kept to an absolute minimum by only inspecting the first 64-Bytes. They are particularly suited to backbone operations on heavily-used networks.

— **Fully-buffered switches.** These are full store-forward devices where the packets can be checked to prevent errors from being propagated throughout the network and where rate adaptation is possible. These switches will perform a full CRC on the message, and discard the data if it doesn't compute. The output of these switches can be ATM, Fast Ethernet or FDDI at ten-times (or more) the rate of the segment ports.

• 3Com has defined the following switch-types:

Class 0: These switches support a single network protocol and a single port speed. They are designed for workgroups and usually have a shared backplane architecture.

Class 1: These have high speed ports and provide either backbone or server connections.

Class 2: High powered switches that support multiple protocols and have redundant features for spanning tree bridging. They will also have multiple data paths.

Class 3: High-speed switches configured as routers. These allow broadcast domains to be created.

SWR

Standing Wave Ratio. This is the ratio of the maximum to the minimum current/voltage along any line. It is a measure of the mismatch between the load and the line. Loading coils are used to correct problems.

At the antenna output of transmitters, standing waves are often created. How this effects performance, depends on the length of the cable and the antenna. The SW ratio will equal 1 when the line is perfectly matched.

SX

See 386SX.

SxS

Step-by-step exchange. Originally called a 'Strowger' exchange.

syllable

In computing this is a string of characters which are part of a word.

SYLK

Symbolic Link. A type of interchange file for spreadsheets. It can translate formulas, values, formatting, field names and display settings, as well as print areas, borders and charts. Macros are also translated in some exchanges.

symbol

In the most general sense it means any identifiable character, shape or graphic on the screen. In a more limited sense it means a mark which describes a mathematical operation or statement. e.g. +, –, / * are symbols of arithmetical operators, as are =, <, >, etc. The distinction here is with alphanumerics.

A symbol set is also a set of specific national characters (with diacritics, etc.) or special ASCII characters (usually the 8-bit upper 126 characters) for specific applications.

• In modems and communications systems in general, it means the 'event' which conveys information. In modems the symbol-rate is the baud-rate — but since with QAM it is possible to transfer 4- or 6-bits of information with each phase-change (event or symbol) then there is a 4:1 or a 6:1 ratio between baud and bits per second.

• In CAD, an object may be created once, saved as a symbol and stored in the symbol library. It can then be inserted into a drawing as many times as necessary.

• In CD-audio terminology, a symbol is half a 16-bit sample. One symbol contains the eight most significant bits, and the next, the eight least significant bits.

symbolic address

This is a variable name, or an item label in a computer program. For instance, you may use NAM1 for First Name, and ADDR for address.

symbolic programming

This just means a computer system which can use high-level programming languages.

Symmetric Self Electro-optical Effect Devices

S-SEEDS. Optical switching devices used in optical computing and communications. They vary between an opaque and a transparent state under the influence of electrical voltage. Rates of one billion changes a second are possible. See Josephson junction.

symmetrical

• In data communications, symmetrical suggests roughly equal traffic loads traveling in both directions across a link, or through a network. Telephone conversations and interactive data have symmetrical requirements — broadcast TV doesn't.

• In some packet-switching terminology, symmetrical means that the devices are communicating DTE-to-DTE, while asymmetrical means DTE-to-DCE.

• In the older modems, it meant that half the bandwidth of the link was reserved for the bi-directional channels. Two different frequency pairs were needed. See FSK.

• In encryption systems, symmetrical key is the use of only one key for both encoding and decoding. The sender and the receiver must first exchange the key. See DES, public key encryption, asymmetrical and RSA.

symmetrical binary

A form of binary notation where the hex numbers from 0 to 15 are changed so that the bit sequence for zero is the mirror image of 8, and that of 1 is the mirror image of 9, etc. To do this, the bit values are changed: zero = 0111, 1 = 0110, 2 = 0101, 3 = 0100 ... 8 = 1000, 9 = 1001, etc. See gray binary and Hamming distance also.

symmetrical multiprocessing

See SMP.

SYN

Synchronous-Idle in ASCII No.5 (Decimal 22, Hex $16, Control-V — pattern 0001 0110). A control character transmitted by a synchronous system used to achieve synchronization, and often sent when data signals are absent (idle) to maintain synchronization.

sync-pulse generator

See SPG.

sync suppression

This is a family of fairly basic, but popular, video scrambling techniques. Usually the horizontal sync pulse of each line will be replaced by a 3-Byte digital word which will contain video and audio information, plus encryption data. The key to the encryption may be transmitted in a separate carrier. See active video inversion and line shuffling.

synchronicity

Synchronicity is the solution to the problem of using two different computers, each holding copies of the same file. The problem is that one may be modified, but the other may not — typically address files stored both on a desktop and on a portable notebook.

Which is up-to-date?

What happens if new records have been added to both?

How will each find the exclusive information that only the other has?

Special programs have been developed to maintain synchronization between files by checking for changes, and updating the earlier files. This is becoming increasingly important with executives using notebooks and PDAs away from the office, which is why the IR links are being added to these devices.

synchronous

Sync. To marching in step — sharing and using coordinated time events to match actions. However the 'time' sense of the term is often distorted in some terminology just to mean matching images or files.

Some synchronous devices will use internal clocking (where they provide the clock pulse themselves) while the partner will use external clocking (derived as a 'slave' from the first). The clock signal may be generated locally or remotely, and some new devices take synchronization from GPS satellites.

Most modems now use phase-shift techniques which are called 'self-clocking'. The occurrence of the shift in phase provides all the clocking and synchronization they need. The measurement of the angle of phase shift and perhaps the amplitude of the signal gives them the bit information.

• **Synchronous links:** A normal serial connection consists of a DTE at one end, and a DCE at the other. With external clocking, by convention, the DCE is the clock source and the DTE takes its clock synchronization from the modem.

In the older full-duplex systems there would generally be two separate clocks, one for the data going in each direction, and the DCEs would provide both.

• Synchronous circuits are sometimes incorrectly called 'isochronous' circuits. Usually these circuits are used in a way that allows real-time carriage with a low latency, so it is easy to see how these ideas became confused.

• Synchronous circuits can carry asynchronous data (see ATM) and they are not always suitable for isochronous (real-time) carriage. In this case we must distinguish between the synchronous line-code functions, and the asynchronous access functions. See ATM.

• Plesiochronous means 'almost in sync'. At one level it will be synchronous, but not at another. See PDH.

• In the CPU operations. The system must operate in sync with signals produced by the CPU during memory or I/O (read/write) access.

• In Unix disk-caching, sync refers to the image of the data in RAM always matching the image held on disk. See coherence.

• Computer companies are using the term 'synchronization' to refer to the parallel updating of files. See synchronicity.

• The problems with synchronous systems in telephony are mainly to do with network size. A single clock signal needs to be propagated world-wide to synchronize a global network — or there must be some timing buffers at international gateways. SDH has been designed to overcome many of these problems.

• In B-ISDN and ATM terminology: 'synchronous' suggests the regular transmission of data packets so that the user receives a regular and reliable supply of data for real-time (voice and video) use. This is a misapplication of the term — they really mean 'isochronous'.

The original intention of the Broadband ISDN designers was to use STM (Synchronous Transfer Mode) because it was easier to guarantee isochronous services with STM. But they finally settled on ATM (Asynchronous Transfer Mode) because of the problems of providing world-wide synchronous clock pulses over the global telephone network. With the higher data rates now possible, any slight delays in ATM aren't really significant.

synchronous demodulation

For coherent optical fiber, where the heterodyne detection system uses an electrical carrier at an intermediate frequency as a way to demodulate the data signal. The two frequencies are beat against each other and the difference is extracted.

Synchronous Digital Hierarchy

See SDH.

synchronous file

Some very large files — motion video and audio, specifically — need to be fed to presentation devices at a designated transfer rate. Special device-drivers and circuits make sure that these synchronous files maintain their data flow.

The distinction is with asynchronous files (normal data files) where the rate of delivery isn't critical.

'Synchronous' is quite incorrect here. It has been confused with isochronous — but the term is widely used this way,

synchronous modems

The distinction between asynchronous and synchronous modems has now partly been lost. In the early days the cheap modems were asynchronous across all three links:

— From the terminal-to-modem (RS-232).

— From modem-to-modem across the telephone lines.

— From the remote-modem to the end-terminal (RS-232).

With V.21 and V.22 modems, stop- and start-bits would be carried as 'book-ends' to each byte (making these 10-bit bytes), and each byte would be passed through the modem almost instantly and transparently.

Later, with the higher rates offered by PSK and QAM techniques, the stop- and start-bits were not needed across the telephone line — but they were still needed between the PC and the modem. On the telephone link, the phase change provided a source of self-clocking, so these extra book-end bits were just an increase in overhead.

And since modems are now independent computers with applications performing data compression and error correction, they can also be used to strip (and add at the other end) the start- and stop-bits. So now the asynchronous links between the PC and the modem (still at 10-bits) both need to operate much faster than the 8-bit link over the telephone.

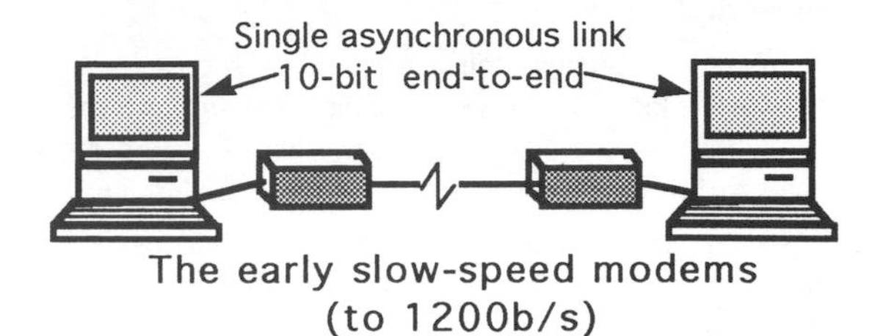

The early slow-speed modems
(to 1200b/s)

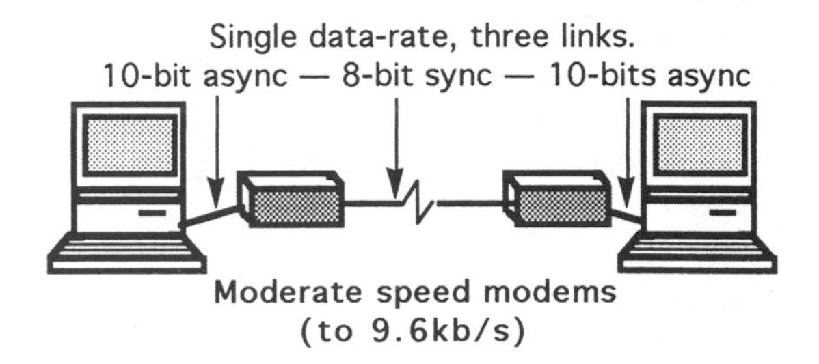

Moderate speed modems
(to 9.6kb/s)

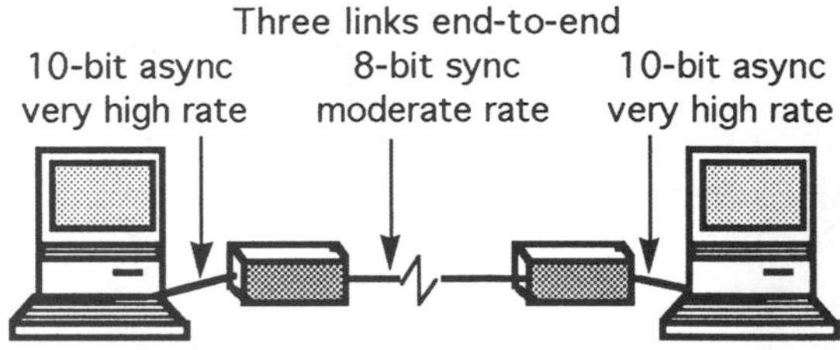

High-speed compression modems
V.42/bis 14.4kb/s — 34kb/s

For this reason, high-speed modems are useless without high-speed links between the PC and the modem, and the limitation is generally in the UART.

If you have a V.42bis modem providing 2:1 compression of text, then you will need a UART capable of handling the asynchronous operations at twice the rate of bytes on the telephone link. This may be 2.5 times the bit rate (20-bits vs. 8-bits).

Note that the link across the telephone line operates quite independently of the two RS-232 connections, and it is self-clocking in modern modems.

This means that the distinction between an 'asynchronous' and a 'synchronous' modem now lies entirely with the two modem-to-terminal links. The so-called 'asynchronous' modems all use RS-232 cables between UARTs in both the modem and PC.

synchronous satellite

The same as geosynchronous or geostationary satellite. A satellite that sits in the Clarke orbit around the Equator and takes 24 hours to make each orbit — the same time it take the earth to spin. These are almost always communications satellites.

synchronous transfer mode

STM is the current way digital data is transferred in systems like ISDN. A fixed number of bits is periodically available to each connection existing in the network — they have slots which are exclusively theirs to fill. If the channel is idle, these slots will be wasted.

When traffic is bursty, the fixed capacity of an STM connection leads to waste, and it is difficult to handle (especially switch) connections which require substantially different bit-rates (video and voice, for instance).

The distinction is with the new telephone standard; Asynchronous Transfer Mode. See ATM.

Synchronous Transfer Modules

The hierarchy of data rates in the new SDH transmission standard begins with STM-1 at 155.520Mb/s, and higher-order bit-rates are integer multiples of this first level.

Virtual Containers (VCs) are mapped into the STM-1 frame, and these can be concatenated to form higher-bandwidth pipes. So, for instance, four VC-4s in an STM-16 (a 2.4Gb/s channel) can together constitute a 622Mb/s channel to connect future ATM switches. See SDH, SONET and VC.

Synchronous Transport Signals

See STS.

syndetic

The interconnections within a system, or the cross-referencing within an index.

synopsis

A concise report (a subsidiary) which contains the main ideas of another article or report. It is an abstract, but it may also include diagrams and references.

syntax

The rules governing the use and structure of a computer language. It applies to a set of programming rules for creating a statement that the computer will understand. See grammar and keywords.

• Syntax errors: Errors caused by the computer not being able to understand an instruction in a program. For instance the word 'PRINT' may have been mistyped as 'PRIMT'. The distinction is with logical errors.

synthesizer

A computer-created musical instrument where the sound is generated by electronic means. This is now almost entirely a computer operation although some systems use libraries of pre-recorded sounds. See MIDI and LPC.

sysop

Systems Operator. The person responsible for a bulletin board in a computer club or a keyboard conferencing forum. He's the one the authorities will jail if you exchange pornography across his board.

system

This generally means the operating system. See.

System #

• **System 34:** IBM's mid-range mini computers very widely used for database management.

• **System X:** This is a stored program control exchange technology widely used in the UK.

• **System 7.x:** The current Macintosh operating system. More correctly called MacOS 7.x.

system board:

The planar or motherboard.

system extension

These are device drivers, and programs that add features such as the ability to run CD-ROMs.

system file

The main operating system; it provides all the resource controls necessary to get the computer up and running, and puts it in control of the keyboard, screen and disk drives.

system folder

In the Macintosh system, a high-level directory containing the System file, fonts, DAs, etc.

system integrator

There are three views:

• A group/company of specialists hired to establish a complex system for a large organization. Their job is to link together disparate items of information technology into one coherent working network.

The SI puts the package together, probably writes many of the special applications, and makes sure that it all works before handing the completed job over as a turnkey system.

The 'integration' role is to oversee the purchase of all equipment (although it may supply some itself), negotiate all the contracts, select the sub contractors, and manage the project.

• A contractor who claims no allegiance to hardware or software vendors, but contracts to advise/consult on the best technology available for a client's requirements. The term has now been debased enough to include product sales staff who give advice. See VAR and service integrator.

• A person or company which purchases unbadged equipment or software, (occasionally modifies it) and packages it together with some other items, as a 'solution' for a niche market. Often they will sell it under their own badge. See OEM.

system library

The files and data-sets in which sections of the operating system are sometimes held.

system program

Usually a compiler or linkage editor.

system software

System software includes utilities such as text editors, copy programs, interpreters, compilers, and linking editors. The distinction is with application programs.

This is a general term for operating systems, device drivers and special utilities which enable a computer to function and control its own operations. It also includes GUI shells like Windows 3.x, which are 'layered' over the operating system.

The main categories of system software (not including the OS):

- Shells or command interpreters which handle the user-interface.
- Compilers translate high-level programs to machine code.
- Editors, which enable changes to be made to code.
- General utilities. Copy programs, etc.

systems analysis

A task that was widely separated from programming at one time, but the jobs now seem to have merged. The systems analyst would do the preliminary research into the requirements, and prepare flow-charts, and a detailed breakdown of the programming requirements.

The programmers would then be given their tasks, and would work independently. They were often expected to program without knowing what the overall application was for.

T

T-1

Trunk 1: An American multiplexed bearer standard with 24 multiplexed voice channels (DS-0) carried on two twisted-pairs. It was designed by Bell Telephone 22 years ago.

Currently, T-1 is more a service standard than a technical standard. The service (with signaling, etc) is called T-1, while the combined data rate is known as DS-1.

However, you'll find that the term T-1 is also applied to:

- a specific type of equipment,
- a general carrier system,
- a standards committee,
- the 1.54Mb/s data rate, and
- various line-codes and framing conventions.

This is all very confused.

Originally T-1 was an analog standard for voice, based on 24 voice calls each of 4kHz, multiplexed by frequency to 96kHz, and carried on copper twisted pair (this was about the maximum for twisted pair).

But from 1962 on, as digital voice was being introduced, T-1 became a convenient measure for a standard bandwidth with 24 time-divided PCM channels requiring a bearer of 1.54Mb/s. At the time this rate was the best that two twisted pair could carry over 1 mile — the rough distance between manholes, and therefore amplifiers.

These digital channels are time-division multiplexed and they use either bipolar (B8ZS) or ternary line-codes.

— Older systems: The older T-1 services were also closely related to D1 and D2 channel-banks (see) used as street concentrators. Twenty-four channels may be used through a channel-bank to service four times as many homes when access lines are scarce.

The D1 and D2 channel banks provide 8000 frames a second with 24 slots of 8-bits, but since they steal the least-significant bit in each byte for signaling and supervision, only 'clear' 7-bit bytes are delivered. This is why many older American networks only guaranteed 56kb/s clear channels when the actual bandwidth is still 64kb/s. Most of these older systems used AMI line coding.

— Newer systems: The newer T-1 services will have D3 or D4 channel banks, with D4 or ESF framing which gives clear 64kb/s channels. These can be channelized (multiple channels) or non-channelized (using full bandwidth) services. The latest T-1 systems use B8ZS line coding.

See A&B bit signaling and channel banks.

• When 24 T-1 channels are framed at 8kHz, there's a total of 193 bits to each frame (a single 'framing bit' is added to each frame for control). The signal is amplitude modulated (pulse amplitude: 2.4—3.0V; pulse-width 324 +/–45ns) onto a single 772kHz carrier, with no separate tones required for receive or transmit. T-1 signals need to be regenerated and retimed at intervals of about 1.2 miles (in some cases 2 miles).

• **ITU:** T-1 is sometimes used as a general term for high-speed digital channels in the 2Mb/s range. In Europe, Asia and Australia these are 64kb/s channels multiplexed to 2.048Mb/s which is the ITU E-1 standard. T-1 should refer only to the American standard; the E-1 European equivalent is 32 x DS-0.

• **Committee:** T-1 is also the American committee for defining these digital telecommunications standards, and so it is the equivalent of the European CEPT. T-1 is a committee of the ECSA and is accredited to ANSI. It has many subcommittees including T1C1, T1D1, and T1X1. T1D1 is developing the US ISDN standards with the support of T1X1.1 which addresses Signaling System No 7.

T-1 carrier system

This covers the range of digital transmission services introduced by AT&T in the US in 1963. They were primarily for voice, and so used bit-robbing techniques for signaling and supervision. See T-1, A&B bit signaling

These are primary extensions of the T-carrier.

- **T-1** (1963) transmits data at 1.544Mb/s (DS-1) with 24 PCM TDM channels over two twisted pair. The maximum distance between repeaters is about 1 mile.
- **T-1A** (1963) transmits data at 3.152Mb/s (DS-1C) with 48 PCM TDM channels over two standard twisted pair.
- **T-1C** with 48 PCM circuits.
- **T-2** (1972) transmits data at 6.132Mb/s (DS-2) with 96 PCM TDM channels over special high-quality wire pairs. The distance between repeaters can extend up to 25 miles.
- **T-3** transmits data at 44.736Mb/s (DS-3) and can handle 28 DS-1 lines, or 672 voice/data channels at 64kb/s. This is an important network rate which is further expanded below.
- **T-4M** (mid-70s) is a coaxial system with data rates of 274.176Mb/s (DS-4) with 4032 PCM TDM channels. This handles 168 DS-1 signals.

• T-1 and T-3 can also refer to fiber-optic carriage these days. They apply the term to the data rate and signaling, rather than to the physical medium.

- **FT3** is the fiber system equal to T-3. It can handle 672 PCM channels at a data rate of 44.736Mb/s over distances of four miles between repeaters.
- **FT3C** is the fiber system which can handle 1344 PCM channels, at a data rate of 90.52Mb/s over distances between repeaters of 4 miles.
- **FT4E**-144 is the fiber system able to handle 2016 PCM channels at a data rate of 140.0Mb/s over distances between repeaters of 8—12 miles.
- **FT4E-432** is capable of 6048 PCM channels at a data rate of 432.0Mb/s over distances between repeaters of 8—12 miles.

T-1 specifications

This is the most common bandwidth requirement for US companies and universities, so it deserves further examination.

• **Framing:** A T-1 circuit is a bi-directional serial connection, sending and receiving 8000 data frames per second. Each frame consists of 193 bits: there are 24 8-bit channels + a framing bit.

Framing allows the T-1 carrier to be sub-divided into the 24 PCM (64kb/s) DS-0 channels, but it has no bearing on the data carried over those channels. Framing can be either D4 (aka SF) or ESF. ESF is by far the better, and it should be selected if your equipment can support it. Unframed T-1 may also be specified.

• **Line Code:** The line-coding can be either AMI or B8ZS.

— **AMI:** If the line-code is AMI, then there is a requirement that the line must carry a mark (logical 1), on average, once in every eight bits, and there is an upper limit of 15 consecutive zeros in a row (a 'ones-density' requirement).

On AMI T-1, individual 64kb/s channels can be run if some of the adjacent channels are unused, but normally this system is only provided for voice and n x 56kb/s data channels.

— **B8ZS:** If the line-code is B8ZS, there is no 'ones-density' requirement, and 64kb/s wide clear-channels can be provided.

T1S1

The title of the American committee of ANSI charged with overseeing the definition of ISDN and B-ISDN standards. The European equivalent is ETSI.

T-3

Trunk 3. T-3 is a US digital line that operates at 44.736Mb/s. When used for general communications, it carries 28 multiplexed T-1 channels, or one unframed high-speed data link.

Generally a multiplexer is used to divide the DS-3 frames into 28 DS-1 signals and these are then connected to a PBX. DS-3 frames repeat at 8kHz, so typically this provides 672 voice or analog modem connections (PCM channels). However, in many applications, the T-3 line is connected directly to a channel bank or PBX that can group the bits into DS-3 frames.

Over copper, T-3 uses a bipolar (B3ZS) or ternary line code, however many current installations are carrying DS-3 signals over optical fiber, which is operating as an OC-1 interface.

T.30

This is a fax handshaking protocol. T.30 is now being augmented to include fax routing information but naturally this facility won't be available with older fax machines. See fax.

T.130

A data conferencing standard which gives Windows a multipoint collaboration capability. It defines the way multipoint data transmissions can take place over a variety of media types. It also regulates collaborative tasks and procedures for document-conferencing and file transfer.

T-connector

The standard coax connector used for LAN node connections. It has two female and one male BNC connectors.

T-interface

Terminal-interface or 'T' reference point. In many implementations of ISDN, this is the point of connection between the carrier's NT1 and the customer's NT2. The main job of NT1 is to convert the four-wire T-interface into a two-wire cable which carries the link back to the exchange.

At this reference point the customer will get the minimal ISDN service (the carrier is providing only the least equipment possible), but the carrier's equipment interfaces here with the customer's equipment.

• ISDN PBXs interface to the ISDN network at the T-interface. See NT#, S-, R- and U-interface also.

T&M

Test and Measurement. Equipment used for installation and maintenance of computer sites and networks. This covers a wide range. See protocol analyzer, probe and reflectometer.

t-stop

This is the 'transmission' stop of a camera — some would say, the 'true' stop for exposure. It is a variation on the f-stop calculation which compensates for the absorption of some of the light by the lens elements.

In most professional lenses, the f- and t-stops are almost the same, and the t-stop will exist simply as a colored reference point (a dot or a line-marker) slightly to one side of the f-stop reference (usually in white).

However in some lenses, particularly those with a half-silvered mirror and a reflex viewfinder, the difference between the t-stop and the f-stop can be almost a full stop. You should always set exposure using t-stops if there is a difference.

TA

Terminal Adaptor of ISDN. See ISDN terminal adaptor, and also TE # and NT.

tab

In ASCII No.5 (Decimal 9, Control I, Hex $09). Unless otherwise specified this is Horizontal Tab. The tabulation key is used in word processors to shift the insertion point to a new 'tabulated' (previously calculated) position along a line. In databases and spreadsheets it will usually jump to the next field, cell, column or row.

• Formatted tabs can be set: flush-right, flush-left, centered, or decimal.

tab-defined

It is common in ASCII output from database files to have the fields separated ('delineated') by tabs, and the records, by carriage returns. Another common approach is to use commas.

Tabs are usually a better form of delineation than commas because commas are often required to be used in text, while tabs are usually limited to charts or formal tables.

• You often need to enter these in decimal or hex form: the comma is decimal 44 or hex $2C. The carriage return is decimal 13 or hex $0D. The tab is decimal 9 or hex $09.

table

A matrix or array with only two dimensions — rows and columns. A spreadsheet is a table, and databases are seen as tables by programmers — having rows (records) and columns (fields).

Computer fanatics delight in inventing new and exotic names for the components of tables, so a 'turple' is also a row or record, and an 'attribute' is also a column or field. The whole table ('file') may be called a 'relation' when it is part of a more complex relational database.

Spreadsheets are more obviously tables to the human eye because they are laid out in a form which makes the two dimensions obvious. However, a table can also be a special partitioning of a spreadsheet — a spreadsheet domain within a spreadsheet. Tables can often be manipulated with special spreadsheet commands.

- There are hundreds of other 'tables' used in computing and communications. Routers have routing tables, color screen drivers have color look-up tables (CLUTS).

tablet

An electronic graphics board which allows you to enter free-hand designs into the computer. The computer tracks the movement of a special pen across the tablet and provides a digitized version of the drawing in memory. Rand invented the idea, way back in 1956.

Digitizing tablets are now being incorporated into the LCD screen as the main I/O device of the new pen-based computers and PDAs. Four technologies are currently in use (there are many more): electromagnetic grids, electrostatic grids, resistive films and capacitive/electrostatic films.

taboo channels

Broadcast television channels which aren't available for use in a certain location — although they may be in use nearby. In analog TV, the concept of taboo channels is quite confused. Prohibited channels include:

- Those used in one location which can't be reused without substantial geographical separation. They are more correctly called 'co-channels' and the potential problem is same-channel interference. They are 'taboo' within a certain distance, depending on the transmitted power and antenna pattern.

- The most common use of the term, however, is to mean those channels which are neighbors of the ones in use ('adjacent channels'). They are left vacant to prevent cross-interference problems (mainly the sound of one affecting the video of the next). Adjacent channels aren't half the problem today that they were in the past; old transmitters used to wander in frequency a bit, and old receivers were not able to filter-out adjacent frequencies. Don't assume that adjacent channel 'numbers' are necessarily taboo, because there may be a deliberate gap in the spectrum (in Australia, between Channels 9 and 10, for instance) with some other use being made of the space between (say, for radio location), which then makes both channels available for use.

- Those which can't be used for transmission because they are reserved for other applications. Specifically, one or two channels will be reserved for linking VCRs to TV sets, and these may be treated as taboo in certain areas, or in certain countries. Some are also reserved for FM radio.

- When all transmitters are beaming from the one mast-head, and at the same power (as they are in MDS) then the problem of adjacent channels doesn't arise these days. Nor does it arise in cable networks (which are just gigantic antenna systems) when 'channel-ready' TV sets are used. These have greatly improved sideband filtering available today.

The key is to maintain uniform signal strength so that one signal doesn't intrude into the bandwidth of another.

- The problem of taboo channels decreases substantially with digital transmissions. Both adjacent channels and co-located channels can be used for digital terrestrial broadcasting at much closer physical spacings. So the current 60-odd usable analog VHF and UHF TV channels can probably provide 120 or so digital conventional channels or about 60 digital wide-screen channels, into a single location, in the near future.

tabulate

This means to arrange in rows and columns. It has also come to mean to 'add-up' — to sum the elements and create totals.

TAC

Terminal Access Controller on the Internet. These are the permanently connected host computers which can accept dial-up connections. See SLIP and PPP.

- **TACACS:** This is the TAC Access System.

TACS

Total Access Communications System. The European (UK, Italy and Spain), and Japanese 900MHz FM analog cellular telephone system which was a Europeanized version of the American AMPS. It used different frequencies and handled data differently in the control channels.

The UK devised the first TACS networks which went into operation in 1984. This is said to be only an 'air-interface' specification; it doesn't include specifications for the complete cellular network system (it doesn't include how and where to place location registers, switches, etc.).

TACS originally occupied the spectrum at 890—905MHz and 935—950MHz with individual 25kHz channel spacing and a 45MHz duplex offset, for 600 channels. When the UK systems came under load, the government withdrew some military frequencies and allowed bandwidth extensions to create E-TACS, which can use channels in the 872—888MHz and 917—933MHz regions. The band 950—960MHz was also reserved for E-TACS.

Excluding some GSM reserved channels, there are now up to 1320 E-TACS channels, and recently Motorola has developed a narrow-band TACS which could triple the current analog cellular capacity. See AMPS (Nth. America), NMT (Scand.), NTT (Japan) J-TACS, DAMPS, CDMA, GSM, CT# and DECT.

tag

A label or name. Usually this applies to the name applied to a database field, but it can be a name assigned to any data structure.

- A tag field is the key field in a relational database.
- A tag specifies a format style that can be applied by a DTP program to characters or blocks of text.
- TIFF is Tag Image File Format (see).

tail

- The other end of 'head'. Any data block, packet, or frame will have a head and a tail. The head bytes in a packet are called a 'header' while the tail bytes are called a 'trailer' (note 'r').
- A tail circuit is a drop or feeder circuit, or an access line to a network node.
- In satellite terminology it is the terrestrial link used to connect the earth station to the customer.

Taligent

A new operating system being developed by Apple, IBM and HP. It is an object-oriented system, which will probably be componentware (see). This once had the code name Pink. Don't confuse it with OpenDoc which is similar, but aimed at a lower level — however both are to be componentware systems.

Taligent is being built around a microkernel (originally Opus, but now Mach 3) It will be a true multitasking system, and it will run the Mac OS, Windows and OS/2.

talk

Both an Internet protocol and a program used for keyboard chatting.

talktime

In cellular telephony this refers to the battery power available to run the transceiver in full operational mode. The time the battery will last while the unit is being used actively. See standby.

tandem

Dual mode — or in pairs — or in parallel.

But also 'in series' (Don't ask me why?), as in 'There may be hundred of repeaters in tandem in a long cable system!' Now obviously these aren't in parallel, so I guess they mean tandem in the sense of a pair of 'go' and 'return' repeaters lying alongside each other.

- **Computers:** Tandem computers are two machines hooked together for multi-processing. See.
- **Exchange:** This is an automatic transit exchange for toll switching. It switches calls from one exchange to another and doesn't directly service subscribers. Tandems are Class 4, which is one exchange-class higher than local exchanges (5).

They act as concentrators, but don't provide direct customer services. Originally, tandems were always grouped into pairs so they could provide backup for each other.

- **Dialing:** Most intelligent modems can be set to switch from pulse dialing to tone dialing (or vice versa) in the middle of a dialing sequence to handle the special requirements of some PABXs. This is tandem dialing (again, a serial function!).
- **Mobile links:** Tandem calls are from one cellular mobile phone to another mobile phone. In digital systems, this may result in the voice being coded by a vocoder in the handset, decoded in the base-station, transported by analog wireline, converted to PCM digital for the long-haul, re-coded in another vocoder standard entirely, and transmitted for decoding in the second mobile. This chain of incompatible coders is likely to be a major source of high echo and poor voice quality in future telecommunications.

tap/TAP

- **Connection:** To make a physical connection with a trunk cable, using a drop cable.
  — A baseband LAN tap is the connector that attaches the transceiver to the main cable. Sometimes 'tap' refers to an active device which contains electronics — but it should mean just the wire connector.
  — A broadband LAN tap removes part of the signal and feeds it to the drop line. It is usually a 'passive' device that consists only of a connection block with no active electronic components. Broadband taps are also known as splitters, directional couplers (see) or multitaps.
- **Network problems:** Half taps and bridge taps are a constant problem on old telephone networks. These are remnants of past connections which create echo problems because they are un-terminated. See bridge tap.
- **Telocator Alpha Paging (TAP)** protocol. This is the industry-standard version of IXO combined with Motorola's PET alphanumeric paging protocols. It transfers all the printable characters (7-bit only), and is unlimited in message length, although the provider usually limits messages to 2000 Bytes. There are two versions— a manual flavor and an automatic one. TAP commands and messaging can be emulated with scripts in PC communications applications like MicroPhone and ProComm. Also see POCSAG, ERMES, TDP and FLEX.
- **TAP Magazine:** A well-known radical hacker magazine.

tape back-up

Magnetic tape is the oldest data storage technology in use today — and it is still viable for back-up or mass storage, provided access speed isn't expected to be critical. The most common tape standards are the various DAT systems, data tape cartridges, 8-mm helical-scan tape (ex-video), and the 9-track reel-to-reel computer tape used on mainframes and minicomputers. Data Linear Tape (DLT) is growing strongly.

- Reel-to-reel tape was standardized in the '60s using 2400 ft reels, and 6250 bpi GCR techniques to hold about 170MBytes of data. Extended length tape reels of 3600 ft (1200 meter) hold over 200MBytes.

• Cassettes and cartridges are considerably easier to handle, and the newest 8mm systems with helical scan can store more than 2GBytes of information. See DataDAT, DLT, Digital Data Storage, and streaming tape drives also.

TAPI

Telephone Applications Programming Interface. Intel and Microsoft's new (1993) telephone API for Windows on desktop machines. It requires a special telephony card in each PC.

TAPI will ship with Windows 95, and all the major switch vendors (Northern, AT&T, Siemens, Rolm, etc.) have written TAPI service-providers for their hardware. Microsoft has also written a service-provider for standard modems.

Initially TAPI will only allow users to dial out automatically from their Windows PCs or screen incoming calls, but later more sophisticated controls will be added. This is seen as a first-party call-control system only, it enables the terminal to do little more than can be done from a standard in-house desktop handset (redirect calls, etc.).

TAPI is a front-end interface for a range of applications on the PC which we will use for telephone communications (i.e. dialing utilities, or desktop conferencing programs). It bridges the gap between computer and telephone worlds for such applications as voice conferencing, electronic whiteboards, faxing, voice processing and videoconferencing. A TAPI developer's kit is available. TAPI is a WOSA . See CSTA and TSAPI also.

• Microsoft is releasing a NT server version of TAPI in 1995.

Targa

See TGA.

target

Any remote point of interest.

• When sending data across a network, the target is the machine or node to which it is being sent.

• When seeking data from some remote database, the target is the data keyword or term; it is the one being used to define the search strategy.

• In cross compilation they often talked about 'host' computers (on which the program is currently being executed) and the 'target' computer on which the program will eventually execute.

tariff

In telecommunications this usually means the published rate-list of prices available for a metered service; the combination of time and distance costs.

• Tariffed services are those which are set and published by the carrier in association with the local regulator. Usually these are basic telephone services (perhaps reserved services which the carrier supplied on a monopoly basis).

• Non-tariffed services, in this context, are those which the carrier can set at will, and which are deemed to be outside the control of the regulator.

TASI

Time Assigned Speech Interpolation. This is an effective approach to reducing the bandwidth demands on multiplexed analog telephone channels — especially international circuits. TASI is almost universal on analog international (submarine-cable) calls.

With TASI, channel capacity is only allocated when the hardware detects that the speaker is speaking — in effect, it produces a 55% (roughly) reduction in the bandwidth by reusing the pauses in a large number of speech channels.

Unfortunately this bandwidth stealing often resulted in clipped speech (the result of 'freeze out' when channels temporarily aren't available), and so during high-load conditions, whole words were lost. However, experience shows it is possible to lose about 1 – 2% of speech without real annoyance.

TASI was fitted to submarine cables worldwide in the period around 1965 to overcome load problems as phones became popular. The digital version is called DSI. See DSI, DCME, Adaptive PCM and VRAM also.

task/tasking

A task is a single, self-contained, process which is executing at a certain time. It can be a basic function being run by the operating system, or it can be an application in total. The term is used quite ambiguously in many cases.

In a true multitasking environment, any task may be regularly suspended while other tasks are performed, but each will continue to take its share of CPU time until the process is finished. In this case, 'task' refers to the application in total.

However, they also say that a word-processor running on a computer is a single 'task' or 'process', while shifting data in and out to a disk may also involve another task. In this case task refers to a subdivision of an application — a process.
See multitasking, cooperative processing and pre-emptive multitasking.

• Single-tasking operations were those provided by the first personal computers with the early operating systems. Only a single application could be run at any one time. While the disk drive was running, the CPU could only sit and wait — it couldn't even use the idle time.

• True multitasking is when more than one application is running concurrently on the computer (note 'running' — meaning it is active). Of course a CPU can really only handle one job at a time, so in practice this means that the processes are interleaved and concurrently executed. This is sometimes called 'parallel multitasking' or 'pre-emptive multitasking'.

• Pseudo-multitasking operating systems provide 'serial' multitasking. They allow desk-accessories and TSRs, and multiple programs to co-exist in memory in what is essentially still a single-tasking machine. See multitasking and co-operative processing.

• When the claim is made that a Unix system 'runs DOS as a task', they simply mean that this multitasking environment treats the DOS operating system as if it were an independent application. DOS can ride on top of Unix and use this operating system to handle the various machine resources.

TBC

Time-Base Correctors. (See time base also.) There are both analog and digital TBCs, but now only digital devices are used.

TBC are needed whenever timing degradation begins to affect signal quality. This is in measurable whenever:

- The signal is replayed from mechanical devices like video-records.
- When multiple copies have been made (symptoms include a 'tearing' at the top of the picture).
- When broadcasting (high timing stability is a regulatory requirement).
- When editing and introducing other image sources to produce dissolves, fades, wipes, and special effects.

If two images are to be on the same screen, both require a high time-base stability.

Low-cost digital TBCs can hold only a line or two of image, while other more sophisticated models hold 16—32 lines (very common). At the top of the range are the 'infinity window' types which hold a full image frame. Infinity window is actually a frame-grabbing function in addition to the TBC function, and it allows the unit to perform double duty by providing freeze frames for live TV broadcasts.

TBCs are now so cheap at the bottom end, that some are being built-in to camcorders and VCRs, while others are incorporated into switchers and special effects generators (SEGs). Top of the range professional units are usually sold as stand-alone units.

• To handle gross errors, the cheaper TBCs must provide feed-back to the VCR, while the infinity window type can operate by simply dropping (or doubling) an occasional full frame of image, and so they don't need to control the VCR source.

• Component TBCs are needed for video in component form. There are Y/C and YUV/YIQ types. See 2-wire and 3-wire.

• Most professional TBCs use 8-bit and 10-bit quantization and sample the signal at around 14MHz. Older TBCs used 7-bit resolution and lower sampling rates. TBCs are now being incorporated into mechanical recording equipment rather than being purchased as a stand-alone. TBC boards are available as an option for BVU series of U-matic recorders, for instance.

TC

• **Transmission Control:** A group of control characters which control telecommunications networks. In the ASCII No.5 character set these are ACK, DLE, ENQ, ETB, ETX, NAK, SOH, STX and SYN.

• **Tele-Cine:** See telecine and flying-spot scanners also.

TCAM

An SNA access method. See Data Transfer layer.

TCM

• **Time Compression Multiplexing:** One of two techniques use in ISDN to support basic rate access on twisted-pair wiring. In TCM, the transmission for each direction is 'bursted' in alternate intervals to simulate full-duplex communications. The data rate on the loop is slightly more than twice 144kb/s, but the TCM end-circuits restore the original rate.

North American ISDN uses 2B1Q line coding at 160kb/s, over a single pair with alternating data bursts (+ guard times to allow the line to settle). There are six TCM channels (both-ways for 2B+D) on this wire pair. See ECH.

• **Trellis Code Modulation:** This is an FEC standard which is closely coupled with QAM modulation techniques. See trellis code modulation, QAM and CAP.

TCP

• **Transmission Control Protocol** is a suite of software standards which bundle and unbundle the data into packets, send and receive packets, and manage the transmission of packets on a network. They also check for errors and look for packets out of sequence.

TCP was designed for ARPAnet in the early 1970s, It appeared in the public domain in 1983 and was later taken up by Sun for its workstations. It is still commonly used on workstations along with Sun's NFS. See IP.

The TCP part of the TCP/IP combination is very similar to the ISO Transport Layer 4 (ISO 8073). It provides a relatively error-free, connection-oriented transmission service for applications. It can collate all the connectionless datagrams of the IP layer (Network Layer 3), sequence them, discover errors and request retransmission. See TCP/IP and IP also.

TCP/IP

Transmission Control Protocol/Internet Protocol. An inter-networking pair of transport protocol suites championed by the US Department of Defense and adopted by the US Government for connectivity between dissimilar micros, minis and mainframes.

It originated in ARPAnet and took root in both the Unix and the Ethernet worlds. It was designed for ARPA by Berkeley University, using Unix to modify X.25.

It is now the basic network engine for Unix networks and it is widely used to connect multiple LANs into a WAN — partly because there are hundreds of free implementations of TCP/IP. It is now the core of the Internet, and is clearly the most widely used transport protocol except for Ethernet.

If you want to define 'the Internet', then the only realistic definition is that of a loose confederacy of networks all using TCP/IP to interconnect.

TCP/IP does not conform entirely to the OSI model. However it is typical of Network layer and Transport layer protocols. There are probably over 100 data communications protocols now included in the double suite.

• The TCP part, like NetBIOS, provides peer-to-peer communications at the Transport layer 4. It performs end-to-end error control, flow control and packet-sequence control.

• The IP part handles messages routing over (and between) networks at roughly the Network layer 3 level. It also handles problems of fragmentation and reassembly of packets, and provides multilevel security for packets traversing a network.

The other major higher-level application protocols associated with TCP/IP are SNMP, SMTP, FTP and Telnet. These are all defined in RFCs. See all these.

• There are four layers in the TCP/IP protocol stack:

- The lowest, the **Network layer**, provides network access and contains specific protocols for LANs and other types of subnets. There is no strict requirements at this level.
- The second is the **Internet Protocol**, providing routing and relaying between subnets. Only one protocol, the IP exists at this level; it is a datagram protocol.
- The third is the **Transport Layer** which controls end to end communications links between systems. Both the virtual connection-oriented protocol TCP exists here, and also a connectionless UDP. See.
- The fourth and last is **Application,** which provides such things as remote log-in and file transfers. The most commonly-used applications are Telnet, FTP and SMTP.

• In Internet terminology, a TCP/IP connection means that the host is permanently connected to the Internet, and that real-time (fast) responses are possible. 'Restricted TCP/IP' means that there's a firewall between the Internet and the host (to prevent hacker attack) and that the IRC and other real time services are not generally available. SLIP/PPP gives a dial up contact the temporary status of being a TCP/IP connection.

TCU

• **Trunk Connection (Coupling) Unit:** In Token Ring LANs, these are electronically controlled switches (relays) which take their power from the workstations and allow the station to connect to the trunk cable. If the workstation fails, the relay closes and isolates it from the network. TCUs lie at the concentrator end of the connection.

• **Transmission Control Unit:** A communications control unit under the control of a computer. It doesn't execute its own instructions, and has no internal programs. These are large units like FEPs, which are totally slave to the host.

TD

Aka Tx. For modems, 'Transmit Data'. In RS-232, this a is an input for DCE and an output for DTE on Pin 2, and an output for DCE and input for DTE on Pin 3.

TDCC

• Time Division Cross Connect. A data switch unit for digital data networks.

• TDCC was also an old US standard for EDI. See ANSI X.12.

TDD

• **Time Division Duplex:** A method of time-dividing a single channel into 'go' path and 'return' paths to emulate a full-duplex circuit.

This technique is used in FDMA systems such as CT2, and with TDMA systems such as DECT and PHP — they only ever have one radio channel. The main cost and complexity value is in simplifying the R/F stage by using one set of components alternately for both sending and receiving.

Another major advantage of TDD is that there is no 'duplex separation', so it is easy to extend the operating spectrum without running into the duplex allocation. See FDD also.

• **Telephone Devices for the Deaf:** This includes a number of messaging devices, some of which have their own special codes. Many still use Baudot (see V.18) but most of these devices will also handle ASCII No.5 code at 300b/s FSK (most likely the Bell standard).

TDM

• **Time-Division Multiplexing.** See. (Aka Time-Domain Multiplexing).

• **Tuner DeModulator** in a set-top box or even a TV set.

TDMA

Time Division Multiple Access. A protocol used for cellular telephones and VSATs. Don't confuse it with TDM (although it is obviously closely related). TDM is use with one-to-one or one-to-many links, while TDMA is many-to-one.

TDMA allows data and voice communications on a multipoint to point basis (mobiles to base-stations) by allocating a single slot (fractions of a second long), within discrete time-segments (called frames), to each user. The 'multiple access' function is the control of transmission timing by each remote mobile, so each burst of signal fits into its allotted slot and doesn't ride into the next user's slot. This is the key problem to be overcome with TDMA systems because radio signals take a finite time to cover distances of even a few kilometers.

Stations wishing to transmit are often at varying distances from the base, and these distances have varying propagation-times. More distant transmitters therefore need to be advanced in their synchronization over closer transmitters.

With GSM, special training sequences are transmitted to the mobiles, and the mobile immediately turns them around and retransmits them back to the basestation. From this, the base can calculate how much the remote mobile should advance its synchronization.

Each transmitting station has the entire channel bandwidth available to it for the slot-duration and so the various signals are interleaved in time. See FD, FDMA and CDMA also.

• In the American cellular vernacular, TDMA is a specific term applied to the cellular IS-54 standard which most of the world refers to as Digital AMPS or D-AMPS. This is more correctly known as North American Digital Cellular (NADC — see).

• TDMA cellular systems transmit in bursts. They therefore are similar to strobe-lights, in that they put out an intense signal for a short period of time — and it is this burstiness of energy which seems to produce most of the problems associated with this form of radio communications.

The power envelope of the bursts produces interference in many normal electronic devices including telephones, hearing aids, TV sets, cassette recorders — and it may have some significant, but low-level (perhaps long term), biological effects.

TDP

Telocator Data Protocol. A later variation on the TAP paging protocol which can handle 8-bit files. TDP is actually a suite of five protocols which look after various aspects of the paging operations:

— **TME:** Telocator Messaging Entry. This protocol is used to get 8-bit data and long messages into a paging system.

— **TFC:** Telocator Format Conversion. Provides the rules needed to pass TME data through an existing paging system which has not yet been upgraded for TDP.

— **TRT:** Telocator Radio Transport. This is the protocol establishing data packetization and housekeeping methods over a channel from the paging system to a data transceiver.

— **TMC:** Telocator Mobile Computer. This provides the rules needed to extract and interpret information passed from the wireless data receiving device to the host mobile computer.

— **WMF:** Wireless Messaging Format. This is a method for formatting and specifying how to use data sent end-to-end, and it is very similar to the X.400 approach of using both envelop and contents.

See TAP, POCSAG and FLEX also.

TDR

Time Domain Reflectometer. These are cable testing devices. They send a signal through the cable and read any reflected signals from breaks or variations in impedance. Anything a TDR can detect and measure is likely to impact on performance since it represents a departure from the ideal state of having no impedance changes, and therefore no reflected signals. Optical TDRs (OTDRs) perform the same functions in fiber.

TE#

In ISDN terminology, Terminal Equipment (TE) must be physically connected to Network Terminating (NT) equipment by a physical line using standard 8-pin connectors (ISO 8877). This corresponds roughly to the DTE–DCE relationship between a PC and a modem with RS-232.

• Terminal Equipment standards of ISDN.

— **TE1.** This type of equipment has a true ISDN interface and directly handles ISDN calls, setup, etc. Digital phones, integrated voice/data terminals and digital fax machines will be TE Type 1 devices.

— **TE2.** This is non-ISDN equipment. It has an X- or V-series interface. Most of these devices will use RS-232 for connection, or have an X.25 interface with a computer. TE2 equipment needs to have a terminal adaptor to perform the ISDN-specific functions. The TA (Terminal Adaptor) allows a TE2 device to be connected to ISDN because it acts as the signal translator. See terminal adaptor.

• The 8-pin connector standards (many duplicated) are:

Pin	TE side	NT side
1	Power Source 3	Power Sink 3
2	Power Source 3	Power Sink 3
3	Transmit	Receive
4	Receive	Transmit
5	Receive	Transmit
6	Transmit	Receive
7	Power Sink 2	Power Source 2
8	Power Sink 2	Power Source 2

tear-down

The phase, at the completion of a call (aka 'call clearing'), where the circuit is disconnected and the billing mechanism is stopped. This is an active process and it requires an exchange of signal packets.

teaser

In most television programs (but not in the movies) an action scene often precedes the opening title. This is the teaser, and it is important in locking in the audience interest, directly after the commercials — and before the title sequence begins.

telco

TELephone COmpany. A widely used abbreviation for those organizations which supply telephone carriage services in a country. Sometimes the term is applied to those companies which manufacture telecommunications equipment.

It also seems to be spelled occasionally as 'teleco', and with the variation 'telecom' (as a generic term), and it sometimes seems to be applied to resellers as well as the primary carriers (who own the plant and switches).

tele-writing

The transmission of free-hand manuscripts or graphics transmitted directly from a tablet, not facsimile. See electronic ink and Jot.

telecine

TC. Studio equipment used for translating film to video. Telecine is a video-camera plus film-projection system locked in synchronization. TCs use claw mechanisms to hold each frame temporarily in the 'gate' while the camera scans the picture, so they have problems with synchronization. The distinction is those systems using optical compensation, and particularly with flying-spot scanners.

Telecine is more important in NTSC than PAL countries because of the need to scan the frame twice, then three times, then twice, etc. to convert 24 frames to a 60Hz field rate.

• See teleconverters, flying spot scanners and optical compensators.

telecommuting

Working from home or perhaps at a local office in the country, by using the phone lines, a computer and a range of other communications technologies. Telecommuters are employees of a larger company, and their business is conducted through that company. This is said to be the distinction between telecommuting and teleworking. See telecottage also.

teleconferencing

A generic term which encompasses videoconferencing, voice/telephone conferencing, and computer conferencing. The term is often misapplied to mean videoconferencing only.

Teleconferencing is any system (including, presumably telepsychic phenomenon) which allows two people or groups to confer while not in the same room. It includes voice-to-voice systems, PC-to-PC — on a one-to-one, one-to-many, and many-to-many basis. Electronic whiteboards, for instance, are real-time teleconferencing equipment.

The term doesn't appear to be constrained to 'real-time' conferencing either. One of the substantial advantages of keyboard conferences is that they can take place over time, and therefore give those involved time to think before they speak.

teleconverters

American TV series, which are sometimes shot on film at 24 frames per second and then converted to NTSC video for the editing, are particularly difficult to convert to PAL. The problem doesn't exist if the series is made completely on film; it is the combination of film, plus the 3:2 field sequence of NTSC telecine conversion, then the secondary conversion to PAL that creates the problems. Field dominance is an important issue here. See.

• Standards converters are designed to expect video images where there is a change between every field. Film only produces a change between every frame (two or three fields).

telecosm

George Gilder's term for the future world where computers, broadcasting and telecommunications are integrated — the 'information superhighway', if you insist! He talks about the bandwidth tidal-wave.

Gilder's Law of the Telecosm states that the society will find an exponential gain when linking computers over the new high-bandwidth networks. Interconnect any number ('n') of computers and their total value rises in proportion 'to the square of n', he says.

The larger the network grows, the more efficient and powerful are all its parts. See microcosm and Metcalfe's Law also.

telecottage

This is the result of a movement in semi-rural areas to provide work for local citizens, and also to reduce the amount of commuting necessary for people living on the outskirts of a city.

The local council, shire, or community organization (sometimes religious or a self-help group) will rent a few rooms in the center of town and fit it out with the essentials needed for people to telecommute — numerous phone lines, a fax machine, photocopier, and especially computers with modems. This allows people to try teleworking or telecommuting without a heavy financial outlay; they can test the waters.

Many people prefer to work in such a cottage environment anyway; some can't handle being alone in a house, rather than in an office or shared environment.

teledensity

The number of phone lines per 100 inhabitants of a country. Between 1983 and 1992 teledensity increased by 10% in developed countries, and only by 1.4% in the rest of the world.

Typical teledensities are:

Argentina	11.1%
Australia	48.0%
Brazil	6.3%
Canada	57.8%
China	0.9%
Finland	53.5%
Hong Kong	42.8%
Indonesia	0.5%
Japan	44.8%
Mexico	7.0%
South Africa	9.4%
Sweden	68.3%
United Kingdom	45.5%
United States	54.5%

Teledesic

A concept to fly 840 LEO satellites after the year 2000. This will be a broadband for fixed installation — mainly data and videoconferencing — and the promoters are Bill Gates and Craig McCaw.

telefax

An old name for a facsimile device. The name remains in the publishing business to mean a high-quality fax for page images.

telegraphy

The old Morse code way of transmitting information. Telegraphs were sent between 'telegraph offices' and they were then printed out, and hand-delivered by 'telegraph boys' riding bicycles. Later the Baudot code was used for telegraphy and Teletypewriters (TTYs) superseded telegraphy. An international telegram was known also as 'a cable' — and it was called a cable even when transmitted by radio to a ship at sea.

• Telex is the switched form of telegraphy which was connected, like the public phone network, to your own office. See Telex.

• In some countries the term telegraphy was used for any messaging system which operated at a low rate (below 200b/s), including ASCII-based systems. In other countries the term Telex gradually took over.

telemarketing

The selling of goods and services by phone. Automatic diallers are now being used: these run through the list of numbers, dial the next automatically, pre-announce using voice synthesis 'Please hold the line', then leave you waiting until the human salesman/woman finishes the last call and comes on the line to sell you something.

Some telemarketing systems are totally automated, even to the message. If you put the phone down on them in the middle of their spiel, they are programmed to redial and continue the message until you've heard it all.

In the USA, these companies have successfully resisted all attempts to limit their activities, quoting the Constitutional right of free-speech in support.

telematics

The integration of computer technology and telecommunications. It is a general term for a range of text services, but it's not now much used. It once described Teletext, Videotex and Teletex, in particular, and sometimes interactive information services in general.

telemetry

Sensing and measuring at a distance. This is a vague term which is applied to two quite different concepts:

- The measurement of distances and position using radio measuring systems.
- The transmission of monitoring and control information over distance.

In satellite systems, this means the status data of the satellite itself. So you find a TTC&M (Tracking, Telemetry, Command and Monitoring) satellite earth station which is using telemetry to maintain the satellites in position.

You'll also find the term used for electricity- and water-meter readings where they are automatically transmitting data back to the utility supplier over phone lines. It is now being used also for the remote monitoring of mechanical equipment, water levels in dams, etc.

Telenet

A value-added network services company in the USA which operates a widely used packet-switching network. It is now owned by US Sprint and known as SprintNet.

telephone industry

As a rough guide to the worldwide telephone scene:

- At the end of 1993 there were 641 million telephone 'access points' in use around the world. This was made up of 607 million fixed lines, and 34 million wireless (cellular) subscribers.
- Telecommunications equipment and services sales in 1993 was estimated to be US$575 billion. Of this US$455 was for services and US$120 billion was for equipment. Revenues were rising at about 6%, twice the general industry average.
- Mobile phones now represent 6% of total service revenues in developed markets.
- Japan's NTT is the biggest telco in the world with 58.46 million lines, a staff of 248,000 and US$60.135 billion in revenues.
- The RBOCs in the USA had total revenues of $84 billion in 1993-94. A quarter of their sales and a third of their profits came from access charges paid by the IXCs. Another $10 billion in sales comes from intra-LATA toll calls.

The RBOCs get about 10% of their revenues from cellular, which is about $8 billion a year (and increasing rapidly).

telephone service

The telephone companies apply this term specifically to the provision of a line connecting you to the exchange. They mean the CAN or customer access network which may include the twisted-pair, the line-card in the switch, and possible customer premises equipment (a telephone).

Telepoint

The generic name given to the commercial public pedestrian-cellular phone service which was offered a few years ago in London by four companies using the CT2 technology. It failed because too many companies were servicing an unresponsive market at a time when cellular phones were being given away.

The term has now come to mean low-cost pocket--sized cellular phone services with very limited functions — essentially the same as wireless PBX equipment, but used outside the office with direct links to the public switched network.

teleport

A technology park set up to provide a whole range of special communications services as a way of attracting high-tech. business. Many of these business start-ups are small, and cannot get access to these services in any other way.

The emphasis is probably on providing high bandwidth satellite links in many teleports.

telepresence

This sometimes just means a 'virtual attendance' at a video-conference — perhaps for consultation — but it can also mean those future virtual reality techniques which aren't just for games, and this can include a range of high-bandwidth imaging services also. It is anticipated that virtual reality techniques will provide the control of medical operations by remote specialists. Ultrasonic scans and X-rays will be read and interpreted by remote specialists.

teleprocessing

An old term for data communications.

telepublishing

This is a very confusing term. It is sometimes being used to mean Internet publishing.

At other times it deals with electronic means (computer and communications) of handling the production of printed material through all stages: preliminary layout, typesetting, imaging, make-up, plate reproduction and printing, without using physical layout, CRC and the physical transport of all this to the printer. It probably also includes problems of distribution control and payment/billing systems.

teleputer

The combination of television and computer. It's not quite clear what they mean. A teleputer could range from computer functions integrated into your home TV set, to a redesigned intelligent set-top box which acts as the home security and communications controller.

It is hard to see why you'd integrate a computer (a single-user device) with a television set (multiple viewers) — or why you'd want to integrate your hi-fi with either computers or TV sets. When one breaks or is in use, you also lose the other.

This urge to integrate everything digital has generally failed in the past. Remember when all PCs were going to have a phone built in? And when the fax, photocopier and laserprinter were about to merge? (HP has one now!) Sometimes it makes sense to integrate, and sometimes it doesn't.

Telescript

A programming language for communications developed by General Magic. They say it supports various kinds of messages, including sound, coded voice, data, fax/image and text. It will help developers create software applications that can send and receive messages via LANs or the public-switched network.

This is also both an agent standard and a special agent language. It is a way of creating small software modules which can perform agent functions over networks. See UAA also.

teleservices

The ITU has defined the following standardized teleservices (applications) for ISDN and digital phones. It is now a year or two out of date, but for what it is worth, here's how they see it:

- **Telephony.**
- **Teletex** (without the last 't'). This is fast Telex which is disappearing from favor. E-mail has taken its place.
- **Telefax 4.** This is Group 4 digital fax.
- **Mixed mode**. This is a combination of Teletex and Telefax 4. It is the exchange of Teletex messages (ASCII symbols) which can contain Telefax images.
- **Videotex.** Text and graphic based information-retrieval and messaging.
- **Telex.** The old steam driven protocol which, in this case, can share the network with the latest in ISDN services.
- Others include file transfer, desktop conferencing, video-conferencing, etc.

Teletel

The famous French viewdata system based on Antiope technology and Minitel terminals. This is easily the most successful videotex system in the world. It used the Antiope alpha-mosaic standard, and it is very like Prestel.

Teletel is gradually being absorbed into the CEPT European videotex standards, and it has also evolved into the more general Minitel services, which were originally based on videotext technology for phone directories. See Minitel.

Teletex

(Capital T, no final t.) The Siemans invented and promoted upgrade on the old steam-driven Telex system. It was taken up briefly in an attempt to revitalize the old Baudot-based system.

Teletex was designed to interconnect word processors, and it did have a brief run of popularity as an ITU standard before someone remembered e-mail. The Teletex transmission rate is 2400b/s (250 characters a second) using an 8-bit code which provides both upper and lower case letters. It also allows binary file transfer and Telex (with answerback) and can interconnect to the Telex network.

Teletex is now incorporated into the ITU's X.400 Message Handling System, so there's not much reason for it surviving independently for much longer. At present it provides nothing more than Telex with some EDI applications.

- Don't confuse Teletex with teletex**t**, below.

teletext

(Lower-case t and final t.) A generic name for broadcast one-way videotex systems where the TV station sends text and block-like graphics to your screen in the same signal it uses to broadcast conventional TV programs.

Teletext uses up to 12 lines per channel in the 'vertical blanking interval' of a television signal (seen as a black bar at the top/bottom of your television picture) to carry digital data. These signals can be captured, held and decoded in special circuits added to an otherwise conventional television set. You have a remote control which allows you to request the appropriated pages of data (there may be hundreds from which to choose) and display them progressively on the screen.

The digital text information is constantly being transmitted in a rotation of frames, so the user keys in the required page-code and waits. When the desired page comes around in the cycle, it will be grabbed (latched) from the blanking interval by the teletext adaptor in the TV set and displayed on the screen. Each page of text needs about 1kByte of RAM, and the limit of teletext is set by the time users are willing to wait for a page.

- The British Ceefax and Oracle systems are the most successful teletext services, but nowhere in the world has the system proved to be highly successful. The main interest appears to be among punters.

Teletype

A trade-name and contraction of Teletypewriter that became a generic term for old printer terminals (teleprinters). These were originally used for printer terminals on time-share computer installations before video screens became popular. Most of the later machines used the ASCII No.5 code although you can apparently still buy them with EBCDIC and Baudot code. The term is also used for associated equipment.

- Teletype Corp. also made many Telex machines (ASCII No.2 code) and, in some places, the names are synonymous.

television

There are three major world standards in analog television at present. They are all 'composite' standards, although there was an attempt in recent years to improve the quality of satellite systems by introducing analog component TV (see).

The current composite standards are:

— **PAL** with 625 lines and a 50Hz field rate (25 frames per second), in the European countries and ex-colonies — except for France and old Eastern Bloc countries.

There are many sub-standards and variations which are identified by an alpha code. Australia, for instance, uses PAL B/G. This refers to the B specification for audio separation in the VHF band, and G in the UHF. Other countries use different variations, including one with 525 lines and a 60Hz scan rate.

— **SECAM** with 625 lines and a 50Hz field rate (25 frames per second) in France, the French colonies, the Middle East and the Eastern Bloc. This is effectively PAL with a different color burst. (There was a previous 915 line SECAM system.) There are many variations on SECAM.

— **NTSC** with 525 lines and a 60Hz field rate (using a 24 frame per second film rate in a 3:2 field pattern) in the USA, Canada, Japan, ex-colonies of the US (Philippines), Middle and South America.

The aspect ratio of all of the above is 4:3 (called 'Academy') and all are designed around mono-audio. PAL also can has a supplementary NICAM digital stereo audio system.

• See all of these independently. See active lines, horizontal retrace-time, vertical blanking interval, teletex, PAL, NTSC, SECAM, HDTV also. Also see MAC (component systems).

• For digital TV see MPEG, COFDM, DVB, Grand Alliance, and DAVIC.

television spectrum

Television around the world typically occupies three areas of the radio spectrum, and the cable systems usually mirror VHF and UHF so that the same sets can be used.

However since TV signals tend to be confined to the national air-space of each country there is substantial variation in the subdivision of the frequency bands. These are usually not contiguous bands, and certain channels will be taken out of TV and applied to other services. This means the channel numbering schemes are fairly chaotic.

NTSC signals require 6MHz of bandwidth, PAL and SECAM require 7MHz, although 8MHz is allocated in many countries, and this allows the introduction of NICAM digital stereo sound.

VHF was the original television spectrum, and it had good long-distance carrying capabilities. Later, UHF was added, but this was seen as an inferior spectrum for many years. Now it is equally as good as VHF.

MDS and the new SHF are much shorter in range, and tend to be used for cellular-like transmissions in some areas, where a city will have a dozen or so transmitters all using the same frequencies. When multiple channels are transmitted from a single tower at exactly the same power, adjacent channels can be used with MDS and SHF, which creates 'Wireless Cable' systems known as MMDS and LMDS.

The general divisions of TV spectrum are:

VHF	45—54MHz (once used, but not now).
	54—88MHz
	88—108MHz for FM radio
	174—216MHz
UHF	470—890MHz
MDS	2.15— 2.686GHz. Various allocations within this area, usually in the 2.4—2.5GHz range.
SHF	26.5—28.5GHz. This is experimental at present.

telewide

The rocking see-saw type button controls for the zoom on some automatic (motor driven) zoom lenses.

teleworking

Teleworkers are not telecommuters. They work from home using the phone lines, a computer, and communications technologies, as do telecommuters, but teleworkers work for themselves. This is an important distinction to make since the problems are different. The teleworker's home is also his/her office, and it is unlikely they will require high-speed remote access to a LAN. Their needs are more likely to be in e-mail, file transfer, information retrieval, and EFTPOS. It's sometimes a fine distinction, but worth making.

Telex

TEleprinter (or TELetypewriter) EXchange (aka Teletype). A switched text service which uses entirely different lines and exchanges from the normal telephone service. It was first introduced in the early 1930s and was as revolutionary as the telephone in its day. The USA finished up with two competing, but very similar standards, Telex and TWX, while the rest of the world had only one.

Telex uses the Baudot code (ASCII No.2), which runs at 50 baud/bps, and transmits messages at a rate of 6.7 characters a second or about 66 words a minute. It uses 5-bit bytes, 1 start-bit, and 1.5 stop-bits.

Internationally, Telex is still a very important means of communication; in a large part of the world, it is still the only reliable form of text communications generally available. If you want to book a week's holiday in a Central African lodge, or a Samoan hut, it will probably be done by Telex.

Telex can be used conversationally in real time, or messages can be composed off-line and sent during the night. Telex is designed to receive messages even when unattended.

Early machines all produced hard-copy print out, and most had punched-paper-tape storage for automatic operation. Special input machines were available to generate the punched paper tape, or the Telex machine would generate the tape itself; tapes were very often generated for 'broadcast' transmissions (head office to all branches).

Telex runs over a worldwide dedicated network of automatic exchanges (many of which are still electro-mechanical) which are extraordinarily reliable.

• There is a more modern version with which the Telex network interconnects, called Teletex (no final t). See. Telex also now interconnects to many electronic mail systems.

• When sending a Telex through an electronic mail service, remember that the maximum length of a Telex character line is 69 characters. It is best if you truncate the line length yourself, unless you know the e-mail service and the recipient. Send CAPS only to be sure.

Most e-mail gateway services will make these translations for you, but some don't.

Telidon

A Canadian videotex system which uses an advanced form of graphics (alpha-geometric). It provides person-to-person communications in addition to the broadcast and interactive information facilities. See NAPLPS also.

TeLinc

An attempt to create a world standard for PABX-to-PABX communications. See DPNSS, DASS and Q-SIG.

Telnet

A terminal emulation protocol which is associated closely with TCP/IP and used over the Internet to contact many remote host computers. It is one of the basic set of protocols which make up the TCP/IP suite. Telnet will always provide DEC VT-100 emulation as a basic, however it is more than just a simple terminal emulation protocol since it can package commands using the Internet Protocol (IP).

Telnet allows you to log onto remote computers, access public files and databases, and sometimes run remote applications. It allows one TCP/IP system to emulate a character-based terminal on the other (a virtual terminal interface) and so it allows PCs to act as if they were directly connected through the serial port of a time-sharing system.

• Programs using the Telnet terminal emulation protocol are also often called Telnet as well. There are many free implementations of this protocol.

• Basic Telnet uses 7-bit bytes, so you are restricted in the use of Z-modem. Z-modem is an 8-bit protocol so it won't work unless you establish an 8-bit Telnet connection, but not all Telnet clients support an 8-bit mode. See VTP also.

tempest

These are standards for electro-magnetic radiation from computer equipment and for buildings as a whole. They are designed to reduce the chance of electronic eavesdropping rather than for any health reasons.

To comply with the top Tempest specifications, a building will need to have fine metalized mesh over every glass window, and metal mesh in the walls.

You won't use a cellular phone in these buildings.

template

In general applications this is a 'form' or 'structure' used in an application as the basis for further work. Templates are stored in libraries, and used many times as the foundation for a new design.

For instance, a pre-designed convention-registration form can be saved as a template, then reused hundreds of times, each time with new data being added. Similarly quite complex spreadsheets can be established as templates, with headings and formulas all in place for later use.

The term here is almost synonymous with stationery (page design) and glossary (standardized text) in word processing, but templates are more elaborate and complex.

• In voice recognition systems, a template is one of a series of stored digital matrices which make up the machine's vocabulary. A sound is matched against these templates until it is identified. In many systems, 'training' is the process of building these templates. See parsing.

temporal

Pertaining to time, as distinct from spacial (space).

• **Temporal aliasing:** Effects that occur through time-mismatch (such as wagon-wheels appearing to rotate backwards on TV). See alias.

• **Temporal redundancy:** The repetition of the overall image patterns between one frame of a motion picture (or video) and the next. Obviously, most of the frame area is repeated in every successive frame (unless this is very fast-action) and so, by matching successive images and only sending the 'difference' between them, compression schemes are able to reduce the bandwidth enormously. MPEG1 and 2 differ from JPEG in exploiting temporal redundancy. See spacial redundancy, interframe coding and intraframe coding.

• **Temporal prediction:** The prediction comes from the fact that following compression, decompression often doesn't fully restore the original. Therefore, to provide the best possible results, the system will compress and transmit the first frame. It will then decompress a copy of the first frame itself, and use this as a 'predictor' of the second frame.

If the image is changing, that prediction won't be 100% correct, so it compares the two and extracts the difference. But note: the difference is between a decompressed first image and the second, not just between clean first and second images.

In motion compensation, prediction is just a means of comparing successive frames on a block-by-block basis. The equipment is trying to guess where a block has shifted to, from one frame to the next.

These systems have a search pattern to follow, with some block-shifts more likely than others.

temporary files

Computer programs often build temporary files for their own use and later convert these to 'permanent files,' which are saved under another name (or another extension). The old temp file is then usually destroyed — but not always.

A disk file which has been changed and is being rewritten, is also designated as a temporary file. The process follows a pattern which will be similar to this:

— The old file extension name in the directory is converted from .TXT to .$$$.
— The updated file is written to a new part of the disk and verified.
— The file name with the .TXT extension is entered into the directory.
— The old file name with the .$$$ extension is erased, or in some systems, it is renamed .BAK.

• The Macintosh creates temporary files at the drop of a hat. If the operating system suspects that something has gone wrong, it will offload the clip-board, or perform a secondary store operation which will result in a temp file. Often these are found in the Systems folder.

TENs

Trans European Networks. This is the EU version of the NII (data superhighway) vision of the US Clinton/Gore team.

TEP

Terminal Emulation Program.

ter

In CCITT modem standards this is French for three or 'third attempt'. The addition of 'ter' and 'bis' to numbers in the V.## modem standard simply means that these are later variations on the basic standard. Often they will be another layer of technology added on top of previous layers — so they are not so much versions as upgrades. See bis also.

tera-

The SI prefix for 10^{12}. It should be abbreviated as capital T.

A teraByte is a thousand gigaBytes or a million megaBytes. See peta also.

teraFLOP

Floating point operations of 10^{12} per second. This is the current aim of supercomputer manufacturers.

termid

An SNA cluster controller ID for switched lines. Also called Xid.

terminal

A device on the end of a communications chain — it is either a sink or a source. A terminal is generally a DTE device and it terminates the link (it is the sink). The distinction is with DCE devices which are in the middle of the link.

But note, the distinction may be subtle. A multiplexer is a terminal for two links (it stores the data temporarily), not a DCE in the middle of one.

For computers, the first terminals were printing machines with keyboards; the most notable of which was the Teletype 33. Video terminals then followed, but these were 'dumb'. Even though they looked like personal computers, they were often only smart enough to print one line of text at a time on the screen, and they couldn't even scroll when they reached the bottom. The screens would change as a series of pages.

Smart terminals came later; they could scroll, clear the screen, insert or delete text, position text, and protect areas of the screen from accidental modification. You could move the cursor around these screens under your own control.

Personal computers (with emulation) have tended to make 'true' terminals obsolete.

— A 'dumb' terminal, these days is one that is running a program which does nothing except communicate with the software. It has a very limited ability to process the information. In this context, a modern high-powered Pentium PC with the latest in high-speed modems, will still be a dumb terminal when it is running a terminal emulation program.

— A 'smart' terminal is now one capable of more than just communicating; terminals used in client-server applications are smart terminals. The processing task in client-server, is shared between the host and the terminal.

• **Terminal emulation,** unspecified, usually means TTY or V100 (ANSI) emulation. See VT-## terminals.

• **Video terminal** could suggest the VT-## terminal emulation, or it could mean equipment capable of running full-motion video.

• **Terminal access** often suggests that an on-line service can be accessed via a standard character-based keyboard, as distinct from some super-new GUI applications on a workstation.

• **Peripherals:** Terminal could refer to printers, plotters, etc. but usually not these days in the PC world. If you are lurking in academia or mixing with mainframe programmers, however, this is highly likely.

• **Terminal Adaptors:** See.

• **Terminal telephone exchanges** are local exchanges as distinct from transit or tandem/trunk exchanges.

terminal adaptor

TA. See ISDN terminal adaptor.

terminal emulation

A range of widely used applications which are used on PCs when you wish to replace relatively dumb terminals with the PC.

When PCs are linked to a mainframe they almost always need to mimic a dumb terminal of some sort in order to be able to communicate. Terminal emulation is used in a master-slave relationship; the mainframe has total charge of the conversation.

The most important terminal emulation types are: TTY, RIP, VT-52, VT-100, VT-102, VT-220, VT-230, AT&T 4410, IBM 3101, Minitel, Prestel, Avatar, Televideo 925 series, Wyse 50, Heath 19 and the so-called ANSI standards.

The problem with this terminal emulation approach in the modern day, is that the combined power of all the PCs attached to the host is not being used. However, each additional PC/terminal imposes a load on the mainframe's resources. The mainframe therefore needs to be much more powerful to handle these loads than it otherwise should be.

- TTY provides the universal ASCII character standard.
- If you don't know which to use, select TTY for most e-mail or information retrieval systems, and VT-100 (or VT-52 or VT-102) for Telnet.
- If you have a choice, choose RIP over ANSI and ANSI over TTY.
- When using DEC's VT standards, choose the highest compatible number.

Terminal Management

TM is an ISO standard governing the user interface on workstations, and the way data is represented on the network. See X Windows also.

terminal mode

In the early days, a terminal simply communicated keystrokes to a remote computer using the special protocol required by that mainframe. Later PCs began to be used as the 'terminal' or 'front-end' to the remote mainframe, and they therefore needed to appear to the mainframe as if they were dumb terminals — hence the need for terminal emulation.

This was the original 'terminal mode' in communications software (also known as 'conversational mode').

But now, since many complex communications programs can emulate many different types of terminals, and often deal relatively transparently with intelligent modems, the term 'terminal mode' has come to mean 'non-automatic' functions.

In terminal mode you can send commands directly to the modem (probably using the AT command set) from the keyboard, by using a typed sequence such as:

```
ATDT123456<CR>
```

(to tone-dial the number 123456).

In automatic mode, this will usually be done for you by the software. You would only need to provide the number and pre-select tone dialing.

You will also use terminal mode to change the parameters in your modem. The AT-command set has specific command sequences for this purpose.

- Terminal mode provides a simple way to check that the PC and software are both linking to an active modem if you're having communications problems. Just type 'AT' and hit the carriage return while in terminal mode, and the modem should respond with 'OK'. If it doesn't, then start checking for set-up or wiring or power-supply problems.

The modem should respond even if the local echo is set off; it usually over-rides the setting (but check your modem to be sure).

terminal server

This is a device (a communications processor) that sits on a LAN and supports asynchronous terminal access from a number of (usually between four and eight) terminals being used for basic data-entry and/or retrieval. The terminal server acts like a post-office between the terminals and the mainframe, LAN or WAN.

This technique is often used to reduce the number of terminal connections on a mainframe. Terminal emulation software will be used with a host, and network software will be used for LAN/WAN access.

Terminate and Stay Resident

See TSR.

termination

- To end a session or sequence.
- The terminal at the end of a communications link. See Token Bus.
- A small resistance/capacitor device (terminator) is often added to the end of a communications link (i.e. SCSI) to provide termination and prevent unwanted echo.

ternary

A digital counting system based on three. In electronic line-coding terms these are usually positive, negative and zero — but three voltage states may also be used in binary line coding. See pseudo-ternary and AMI.

Ternary is a technique used in digital communications to reduce the baud rate while retaining a high data rate (ISDN). Three binary digits (eight variations) can be translated into two ternary states (nine variations) which represent a 33% saving in 'events' — and it leaves one combination for supervision or signaling. See modified duo-binary and quaternary also.

terrestrial link

- **Broadcasting:** Transmission across the ground. Public radio is generally in the MF band for AM transmissions, and in the UHF band for FM.

Most of today's free-to-air broadcast TV stations use terrestrial transmissions in the VHF and UHF range, although the high-end of the UHF (called MDS) is coming into use, as is the new SHF band. The distinction is with satellite line or cable.

The lower frequencies tend to curve around objects such as buildings and hills, while the higher exhibit line-of-sight limitations — although they can reflect into 'shadow' areas.

- **Satellites:** This probably means the links between the ground station and the local public network, as distinct from the space segment.

tertiary storage

Archival storage systems.

tessellation

Tile-like or mosaic. Patterns made of juxtaposed interlocking pieces, usually forming a coherent whole. Cellular phones are theoretically planned on the basis of interfitting hexagonal cells. The term is also used for computer screens to mean a form of window control where windows don't lie one on top of the other, but strictly side by side.

Tessera card

A special PCMCIA card holding the Clipper chip which could plug into a portable PC and perform the encryption and decryption.

TETRA

Trans-European Trunked RAdio. The ETSI digital trunked radio (PMR) standard due to come online by 1995–6. It will come in two versions: a voice and data version, and a data-only version, This will be packet-switched data.

A few years ago CEPT recommended that European countries leave spectrum in the band from 870–888MHz paired with 915–933MHz and/or 410–430MHz paired with 450–470MHz.

TETRA is estimated to provide a gross data rate of 36kb/s corresponding to four channels per carrier. They may shift to 16-level QAM modulation later to get higher data rates. They have currently adopted $\pi/4$ DQPSK modulation techniques with a carrier separation of 25kHz.

• Interest in Tetra has been dropping over the last few years, and there is still no common frequency band in Europe. It appears to show very little productivity value over Mobitex and DataTAC.

TeX

A typesetting language which differs in concept from WYSIWYG word processors. There's an expanded version called LaTeX. These 'languages' are more in the nature of document styles.

text

The distinction is usually with data and graphics. Text is words, characters, punctuation and conventional symbols. Text can be data — but data also includes a variety of other forms, such as numerical and statistical data, numerical machine controls, graphics, music, etc.

• Text mode, means text can be displayed, but not graphics.

text display

In computer monitors this was originally quite distinct from a graphics display. All the early PCs had text-display systems where the text characters were pre-defined in firmware (ROM). It was possible to generate only one size and style using the text system, and each character was confined to a standard pixel-map of, say, 8 x 14 (EGA) or 8 x 8 (CGA).

The advantage of these systems was that they were very economical on CPU processes and video memory — they didn't need to store a full bit-mapped image of the screen in the Upper Memory blocks. This approach is still used at the MS-DOS level, but it was abandoned in Windows.

Usually these early PCs also had a 'graphics mode' but the bit-maps then occupied most of the main working memory.

The Macintosh was the first to use a totally graphic display where a full screen bit-map was held quite separate from the working ASCII text and applications.

Graphics displays now show images of the text characters which are generated by software integrated into the System as outline fonts — not the fixed output of a ROM chip.

textural mapping

The process of applying (wrapping) a two-dimensional image onto a three-dimensional object. A CAD and VR term for algorithms which can create the appearance of many varieties of textural surfaces — from shiny metallic surfaces to woven cloth.

TFLOP

See FLOP.

TFT

• **Thin Film Transistor:** Thin film technology is relatively new, and it involves the deposition of molecular layers which are microscopically thin.

— **Chips:** A new technology used in chip construction.

— **LCDs:** A general term applied to active-matrix LCD flat-panel screen displays. See active-matrix.

• **Trivial File Transfer** (Protocol). See TFTP.

TFTP

Trivial File Transfer Protocol. This is a reduced version of FTP. It can transfer files but it has no password protection or user-directory facilities.

TFTS

Terrestrial Flight Telephone System. This is being designed by ETSI to allow telephone services for airline passengers.

TGA

Targa File Format. An image file format used by many image-processing packages, originally developed for the Targa/Vista graphics boards. It has a relatively simple file structure, which is extremely reliable when transferring images between different computer platforms. There are several versions: some are for simple B&W and color-mapped images, while others are in the common 16-, 24-, or 32-bit formats. Images may be uncompressed or coded by RLE.

THC

TCP/IP Header Compression. This is sometimes done over X.25 links for increased link efficiency.

THD

Total Harmonic Distortion. See distortion.

thermal noise

Noise on any electronic network or device, caused by non-random movements of electrons. It produces white noise. See transmission impairments.

thermal printers

A common form of cheap printer using special paper which darkens under localized heating. Stand-alone fax machines are usually thermal printers unless otherwise specified. Some new high-quality color printers also use thermal systems.

• Thermal fax paper retains its image for about 18 months under good conditions.

• For color, see thermal wax below.

thermal wax

This was the first major breakthrough in producing computer printers able to reproduce high-quality color. Heat is applied to special mylar transfer ribbons which are coated in a thin layer of colored wax in many varieties of colors, tones and saturations. The wax softens and is transferred to the surface of the paper or transparent film. See dye sublimation also.

thermionic tube

The original type of valve invented by Lee de Forest (the Audion) in 1907. They are still used in radio and television transmitters for power amplification. The distinction is with solid-state or CCD devices.

Thermionic means that ionized particles are produced around heated elements within the rarefied tube.

thermo-magnetic disks

The magneto-optical disk recording techniques. See MO.

thesaurus

• A reference for writers where words of a similar meaning are grouped together. The original was Roget's Thesaurus (1852). In some thesauruses, words of the opposite meaning are also grouped in close proximity.

• In information retrieval, this implies a controlled vocabulary used for searching. See identifier and descriptor.

• An electronic list which groups terms by meaning.

THF

Tremendously High Frequencies. The WARC must have run out of prefix terms when they decided on this one. These are radio waves with frequencies between 300 and 3000GHz. These waves are between 1 and 0.1 millimeters in wavelength and they overlap into the extreme infrared part of the light spectrum.

thimble printers

A variation on the daisy-wheel type printers and similar to the famous IBM golf-ball typewriter. Daisy-wheel and thimble printers are 'formed-character' printers, where the shape of each character exists in sculptured form, and it is hammered against the paper in some way.

thin film

• **General:** These are recently developed manufacturing processes which can generate layers of exotic materials only a few atoms thick for chip and LCD screen production.

Layers of these dimensions create numerous mass production problems, including the need to be protected from atmospheric moisture, etc.

Often the distribution of molecules is not even across the surface, and in many small areas, spaces between the new molecules can be readily contaminated. In this context the term 'packing density' relates to the percentage of the surface covered by the thin-film layer.

• **LCD screens:** Thin film screens are active-matrix screens. See.

• **Hard drives:** Thin film hard disk drives make it possible to exceed the gigaByte range in single platter data storage for PCs.

To achieve the magnetic domain-densities needed for gigaByte recording, the heads must fly at 25 millionths of a millimeter above the disk surface — about two air molecules — so the disk surface cannot afford to have mountainous surface features or 'arid' areas without magnetic material. Prototype thin-film hard disk units seem to promise 10-fold or more improvements over current storage methods.

thin-film transistor LCDs

See active-matrix screens.

Thinnet

See 10Base-2 and Ethernet.

thinning

Thinning is an important pre-processing stage in pattern recognition systems. It is the technique which extracts the 'skeletons' (basic shapes) from images. Thinning techniques attempt to remove all redundant points while maintaining the basic structure and connectivity of the original patterns. See eigenface.

third generation

• **Languages:** These are the common programming languages like Cobol, Fortran, Basic, C, C++, and Pascal. All have complex commands and structures, and all use a rough approximation of English.

• **Computers:** These are computers made with discrete integrated circuits, rather than the current generation which use Very Large Scale Integration (VLSI).

third-order multiplexers

See PDH.

thrashing

Constantly changing — shifting in and out.

• **LAN:** Thrashing is a LAN condition when a data buffer overload causes retransmit-timers to fail. Datagrams caught up in the congestions are retransmitted, causing even more network congestion. This, in turn, creates more problems.

• **Virtual memory:** Thrashing is the constant shuffling of application segments or pages in and out of memory in badly-designed large applications.

• **Mobiles:** In cellular mobiles thrashing can occur at the junction between two cells. The mobile may be handed off to a new basestation then immediately handed back again — backwards and forwards between the two cells, as it moves down a street and finds different levels of signal.

thread

The flow of control within a process which allows 'pseudo' concurrency. The CPU can seemingly handle more than one task at a time, but it swaps between in a controlled manner. This is a term used in multitasking, multiuser and parallel processing systems.

A thread is a discrete section of a string. The string is the flow of control of an application process, and is composed of many threads. In multitasking, the computer must complete a thread in one string, before picking up a thread in another, since the thread is defined as being the smallest subdivision of the task possible.

Threads therefore control the flow of programs: they are 'virtual instruction pointers' within the operating system which keep track of where the machine is within a program. They exist at the atomic level — you can't split a thread. But you can split a higher level process or task.

Processes, by comparison, are much more complex actions. Threads can share a processor between applications because they are able to keep track of where several programs are at the same time.

• This approach is also used in client-server control, where the server may be called upon to initiate and support many threads for clients.

• In distributed databases, it is more efficient to manage the transaction at the level of the threads, than to use an entire process. Generally a thread shares with other threads the instructions and data in memory, file descriptions, disks and directories, etc. — but in other ways they are quite separate control functions.

• Multi-threading means that the system is capable of pre-emptive multitasking.

threaded interpretive languages.

These are programming languages like Forth which consist of assemblies of subroutine modules. The program itself is a list of the addresses of these subroutines, and each of these is composed of addresses of other subroutines.

This gets down to the level of 'primitive' language operations. Forth can be interpreted or compiled, but there's very little difference in running speed.

three-layer model

Modern computer communications can be viewed as involving three agents: applications, computers and networks. The applications execute on the computers, and the computers are connected to the network.

So the problem entails getting the required data across the network to the computer, then through the computer to the application. This task can be best organized as a three-layer model. In practice a 7-layer model is used, but for a basic grasp of the fundamentals, the 3-layers outline provides the major framework, with the additional layers being sub-divisions:

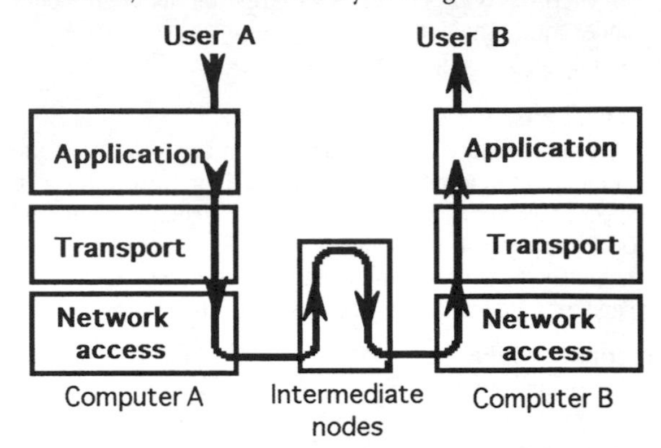

— **Network access layer** (the Communications Subnetwork layer), is concerned with the exchange of data between the computer and the network. The computer must provide an address (and the network must know how to route it), and the computer may wish to exert priority for its packets (if that service is provided) and use some of the other facilities the network may have to offer. So this layer depends on the services of the network; it is network-dependent.

At intermediate nodes through a network, this is the only layer required, although sometimes packets will be referred to the transport layer for checking.

— **Transport layer**, is concerned with getting the data from one end to the other reliably. This is a common requirement for all network types, and for all application types, so it makes sense to keep this as a separate layer and treat the problem independently. This is the layer responsible for error-detection and retransmission, etc.

— **Application layer** (the User-Service layer) is concerned with many different applications which may be running on the computer (many simultaneously), so this layer is highly specific to the application. To transfer formatted files between two similar applications requires a file transfer layer (format-control protocol) suited to that application.

• See Transport layer and then OSI reference model for a more detailed explanation of the seven-layer model.

• When an application communicates over the network with another application, it does so at a peer-to-peer level. But it passes its packets down through the Transport layer (which adds its own protocols), then to the Network layer (which formulates the packets to suit the network). Then, at the other end, the packets flow up through the layers, where the reverse procedures are performed.

• The unique physical address of each computer (node) is known generically as a MAC address — and the network itself will have an address if it connects over a wider area (WAN). This is often called the protocol address.

• In the OSI seven-layer model, the so-called Network layer and Applications layer in this model, are each sub-divided into three layers (using the names Network and Application to apply to more restricted functions).

three-axis stabilized

See 3-axis in numbers section

threshold

In satellite signal reception there is a Gain/Temperature ratio below which the signal is lost in noise and the video image is covered by 'sparklies'. This is the noise threshold. The antenna and block converter system should always work at a G/T margin a few decibels above this threshold.

Thresholds can always be lowered by the use of super low-noise amplifiers; some major installations will have cryogenic cooling (liquid hydrogen) to inhibit the molecular noise. A larger dish will also help by increasing the amount of signal received. Alignment of polarization is also important here.

throttle-back

See back pressure.

Thumb

A cut-down, 16-bit RISC version of the ARM chip.

thunking

See Win32#.

THX

A surround sound system invented by Lucas Films for 'Return of the Jedi' and now converted by Technics for home use.

— **Professional THX** is a complex film post-production facility which needs to control everything, from the original recording to the softness of the cinema furnishings.

— **Home THX** is a superset of Dolby Pro-Logic. It is not an encoding process so much as controlled enhancement of the reproduced sound with a special electronics box. It provides left, center and right front channels, left and right rear channels, and special sub-woofers for low bass. See Frox also.

TI

• Texas Instruments Co.

• Terrestrial Interference.

— **Satellites:** In satellites this is interference caused by signals originating from the ground. Usually it will be a microwave signal sharing a common down-link frequency.

— **Set-top box:** A TI filter is needed on many North American set-top boxes because terrestrial microwaves in the 3700–4200MHz spectrum are used also for point-to-point TV and telephone relay. These signals cause significant interference in the reception of programs from C-band satellites.

TIC

Token Ring Interface Coupler. A n IBM 3725 or 3745 mainframe device which allows the use of Token Ring for communication with local (terminal emulation) workstations.

tie line

Aka private lines. In telecommunications, this is a direct line joining two separate switchboards or LANs so that the traffic between them doesn't pass through the normal switched telephone network. You lease tie lines by the year, and they are said to be 'nailed up' (not variable). Semi-permanent connections are now a valid option in some cases. See.

• Since tie lines are physically tied to cables and interexchange channels, it is possible to condition these lines. See conditioning.

tied

A link is 'tied' when it is dedicated to one channel for the duration of a connection.

TIES

• **Time Independent Escape Sequence:** Hayes, the modem makers, have a patent over the use of the standard escape sequence in the AT command set. This is the:

```
<1 sec delay> +++<1 sec delay>
```

sequence, which is one of the few which do not need the preceding AT command.

To bypass this patent problem, some modem manufacturers use a sequence of three + signs followed immediately by AT:

```
+++AT (with no pauses).
```

This is TIES, and it switches the modem from Data mode, to Command mode. You've just got to be careful you don't use this +++AT character sequence in text. See escape sequence.

• **Telecom Information Exchange Services** of the ITU. This is an information service available through the ITU in Geneva. Gopher info.itu.ch or Fax +41 22 730 5337.

tie trunk

A long distance tie line.

TIFF

Tagged Image Format File. A bit-mapped graphic file format created by Aldus, Adobe, Apple and Microsoft especially to transfer files between software packages in both Mac and IBM environments. The original TIFF was for B&W uncompressed images, but newer versions support color and compression.

There are now about six different TIFF formats, and most software will only support a limited number of these. They are known as Class A to Class F TIFF files. Most image scanners will produce a TIFF file, and the format is very widely used in other areas because of its transferability. However the drawbacks with the TIFF format are:

— Its complexity.

— Not all graphics programs are 100% compatible (considerable variation).

— Lack of compatibility across PC–Mac platforms.

The third problem is due to byte-order differences in files for the PC, from those for the Mac. However, most of the better TIFF graphics packages save TIFF files either way.

TIFF provides two-tones (B&W) gray-scale, palette color and RGB color images to be described. One, 8- and 24-bit pixel-depth is available to be stored for each pixel. It handles monochrome, gray-scale, pseudo-color, color-mapping, true color RGB, CMY or CMYK.

Class E (24-bit RGB TIFF) provides the 16.7 million color possibilities. For palette colors, they use a look-up table.

TIGA

Texas Instruments Graphics Architecture. A Texas Instrument screen standard using a family of TI programmable graphics processors for higher resolution graphics (above 1024 x 768) on an IBM PC. TIGA has become a standard applications interface between computers using Intel microprocessors and special hi-res graphics boards. These TI-340 boards use the TI 34010 and 34020 processors.

The system is extremely flexible since it is resolution- and color-independent, but it is able to support a wide variety of products with a single software driver.

TIGA defines a core set of primitives which can be replaced with routines customized for windowing, etc. It is now widely used in CAD areas. New functions can be downloaded and added at any time.

Tiger

- Tiger is a video-server system designed by Microsoft to allow Pentium-based microcomputers running Windows NT 3.5 Server (the network operating system) to replace mainframes and supercomputers for video-on-demand (VOD) applications. The Tiger has Cubs also; this is a hierarchical architecture of cascading very-fast PCs. The Tiger is the controller, while the cheaper and more modest Cubs are the continuous media-servers. This approach depends on gigaByte RAID arrays for storage, and numerous PCs which handle the data using PCI slots and the special Tiger software.

Tiger software is designed to transfer continuous media (video, audio). Data files from VOD servers need to have a guaranteed regular flow, even though each may originate from a storage device servicing dozens of other users simultaneously.

Delivery is network neutral, but it will probably be over ATM networks for the switching ability. For compression it will use MPEG2.

- Tiger teams: These are hackers who are hired to try to break into a system as a way of checking its vulnerability.
- Microsoft is designing some variations on Tiger: City Tiger will be used by public service providers (cable companies); Corporate Tigers will be used by large corporations and universities; Personal Tigers will be used by video-professionals for non-linear editing.

tight coupling

This means that the two devices (computers, processors, or whatever) are mutually dependent on each other.

- **Computers:** One may control the other, or one may depend on the other for monitoring or pre-processing purposes. A PC and a LAN server would be loosely coupled.
- **Electronics:** Those requiring direct electrical connection. You de-couple two devices by putting condensers, opto-isolators or transformers between, to break the direct connection.

tilde

A diacritical used above the letter n in Spanish. In ASCII we give the name to the ~ key which prints it in isolation.

tile

An area which is broken up into smaller sub-areas is said to be tiled. Sometimes this is used to handle small areas which will eventually be combined back into a larger whole. Sometimes the term is just applied to the break up of the screen into smaller non-overlapping windows.

The distinction is then made between 'tiled' windows and 'overlapping' windows. See tessellated.

time-base

Time-base refers to the regularity of the lines scanning a video image. As you can imagine, with 15,625 scan lines being drawn every second on a TV screen, the exact moment at which each scan-line begins its path across the screen must be extremely precise. If a line holds, say, 1000 image pixels, a delay of 1/15 millionths of a second would shift each pixel sideways by one step, and it would then be out of alignment with the pixels in the line above and below it — lowering resolution.

So if the time-base of a scan line varies, the line will appear to shift horizontally (imperceptibly) against the others, and the horizontal resolution of the image will suffer. Yet any mechanical replay device can have rotational (wow and flutter) variations of this degree.

Time-base errors are caused by mechanical reproduction devices like video disk and tape machines. The minute variations in rotation speed of the heads are enough to reduce the image quality substantially. So time-base correctors (TBCs) are employed whenever a signal is replayed from a mechanical source — especially where a mechanically-sourced image is being combined with an electronically-sourced one. See TBC.

Electronically generated images are generally very stable and do not have appreciable time-base errors. So a TBC is either an analog delay device, or, these days, a digital device which takes the image as input, digitizes and buffers it momentarily, then replays it with strict electronic timing. Various models provide different levels of correction, and so prices vary enormously.

time-code

In professional film and video editing, computer controllers are used to synchronize the image tracks, and to log and match audio cues with visual actions. Film and video professionals now use a standardized 'absolute' time reference called the SMPTE timecode to identify image frames. This is adaptable to color and B&W, NTSC and PAL, film and video standards.

The time-code is an electronic number (in hundredths of a second) recorded on a tape's linear track (usually one of the spare audio tracks), or it can be stored in the vertical blanking interval (VITC) of the helical scan. The 'sub-frame accuracy' of 1/100th of a second, is often essential for audio.

The time-code is brought onto the screen only as a superimposition during the editing process. If 'off-line' editing decisions need to be made, the timecode will be incorporated into a low-quality image on VHS tape.

Time-codes allow the director and editor to make most major editing decisions off-line. They will produce a list of edit points, then this decision-list will later control a larger CMX online editing suite. Most decisions are therefore taken off-line.

• The American NTSC television standard has a problem of 'rounding' because NTSC does not run at exactly 60Hz — it is slightly below (59.97Hz), and this leads to frame errors. See offline, and CMX.

• The SMPTE time-code is the world standard; previously it was known as ECCO numbering.

MIDI also has a time-code system which divides time into beats per quarter note. It can be locked to SMPTE by use of a converter.

Time Division Duplex

See TDD and FDD also.

Time-Division Multiplexing

The very common technique of allowing multiple users (digital only) to share a bearer by time-dividing it into slots and assigning a fixed slot to each channel.

In effect, time-division multiplexers are pairs of synchronized devices which quickly connect the terminals end-to-end (but through buffers), one at a time at regular intervals — providing each connection with a burst of data traffic.

TDM systems are of three types:

— **Bit** TDMs which interleave the data one bit at a time.

— **Byte** (also called character) TDMs which interleave bytes.

— **Block** TDMs which interleave larger blocks.

Bit and byte are the most common, with byte being best suited for asynchronous data (the start and stop bits can be removed before transmission and restored at the other end). Byte is also better for voice, since each sample is 8-bits with PCM.

For synchronous links, however, Bit TDM is better, and it is also better with delta modulation systems (see Delta).

The main distinction is between TDM and FDM (Frequency Division Multiplexing), and also with a special type of variable TDM called Stat-Mux (Statistical Multiplexing). See all.

• Statistical TDM is also sometimes called 'asynchronous TDM' while the normal fixed-slot TDM is known as 'synchronous TDM'. This is very confusing.

• In general, a TDM frame is a consecutive set of cyclic time-divisions. Frames are subdivided into 'slots, and are the primary division of any time-division multiplexed system.

A slot is that section of the frame assigned for use by one station (node). The digital position of each time slot within the frame is related to the frame-alignment signal. If a device has no data for its channel, in TDM its slot remains empty.

• A typical TDM frame structure will consist of a link header, TDM control, pre-defined fixed slots for each channel, and a link trailer.

time line

A graphic device (pre-computers) which was used to represent the passage of time. In its most basic form it is simply a horizontal (or vertical) line which is marked off in days, months, or (usually) years, with the key events noted alongside.

More elaborate computerized time lines include images, sound etc. These display formats are finding applications in hypertext multimedia presentations. They will use picons along the time line to serve as linkage points to key events.

time-multiplexing

Time division multiplexing. See TDM.

time-out

• A technique where a computer or terminal device will drop out of connection after a preset time of inactivity, or if the link is disrupted.

Time-outs may be used for the non-arrival of ACK/NAKs in a file-exchange, in which case the sender will step back and retransmit the blocks. Alternately, the time-out may result in the session being abandoned and the link dropped.

• In telephony, time-outs serve a number of different purposes. You will get a time-out which drops dial-tone, for instance, if you don't dial within a preset time after lifting the handset. This is to ensure that an accidentally dislodged handset doesn't hog exchange equipment.

Time-outs are also needed in some dialing systems. Some telephone networks don't use international or long-distance access codes, or special code structures which enable the exchange to distinguish dialed local numbers from long-distance or international numbers.

For instance a local number could be 345 6789, while a long-distance number may be (349) 678 9111. The only way the exchange can identify when the dialing sequence has been completed, is to wait for a few seconds of time (using a time-out) and assume that no more numbers will be dialed if they aren't dialed in that period.

This is not ideal, but it was a technique widely used in the early days. See NXX and access code.

time-sharing

An old mainframe computer service where people had dumb terminals linked by dedicated lines to central mainframes, and paid the computer bureau for the time they used the mainframe system. The CPU actually clocked time on a microsecond basis — although you were also charged for connect-time. The term was applied mainly to data crunching, although much the same system was used for on-line data retrieval.

Time T

This is the date (Dec 31 1996) set by the ITU, by which all telephone switches should be able to handle international numbers of 15 digits in length. The current international limit is 12 digits, and this has already been exceeded in some East European countries.

timing jitter

The clocks used in synchronizing communications lines are often subject to instabilities, and some timing problems are caused by noise. If, instead of a sharp edge to the clock pulse, the edge is spread because of delay distortions, the synchronization will be vague. A number of similar factors all contribute to timing jitter. See transmission impairments.

TIMS

Telephone Information Management System. A management system which can provide an analysis of telephone traffic, costs, etc. These are used in company PBXs to analyze and diagnose your bills. The communications manager can call up and quantify under various cost-centers, the calling destinations and the traffic patterns. See SMDR also.

TINA

Telecommunications Information Networking Architecture. A very large (40+) and influential consortium of world telecommunications and computer vendors, operators, suppliers, and software developers (TINA-C – for Consortium), is designing an open, next-generation architecture for telecommunications (by 1997). They are putting together TINA as a software architecture for the delivery of multimedia services, with the aim of allowing the efficient introduction and management of telecommunications, worldwide.

The consortium members include Bellcore (US RBOCs), AT&T, European telecommunications carriers and companies, NTT, DEC, IBM, HP, Fujitsu, NEC, OKI, Siemens, and others.

The TINA ideas draw on computing and broadcasting, and attempt to migrate many of the old core functions of the network out to the periphery. Eventually, this will make the networks less dependent on the underlying hardware. They plan to use object-oriented software, and the ODP design method.

TINA will provide ways to bridge the gap between telecoms and computing on a global scale. The processing of a task will be able to be distributed between any computers attached to any TINA-compliant teleco, so computers in widely distributed locations can work together.

TINA-C says it is 'developing an open-network architecture for communications superhighways of the future'. They will address the deficiencies of the current architectures, intelligent networks (INs), and management networks (TMN).

TINA is also said to end the divisions between communications, computing and broadcasting, 'by combining the intelligent network and network-management concepts' of the telcos with a distributed computing model. They seem to be adding a layer of middleware, some of which is based on CORBA, ODP, ROSA and SPIRIT.

TINA has four main component parts, each with its own distributed environment:

– Service Architecture (with Session Service, and the Subscription Model);
– Network Architecture (with Net Resource Model);
– Management Architecture (with Fault and Configuration Management);
– Computing Architecture (with data processing for the distributed execution of the components).

See DAVIC, ODP and CORBA also.

tip

In a telephone twisted-pair cable, one wire is known as the ring, and the other as the tip. These were named after the old phone jacks used in manual exchanges.

The ring connected to the shielding (generally earthed), while the tip was the single connector down the middle. In voltage terms, the tip is the more positive of the two. It is usually shown in diagrams as 'T'.

TL-1

Transaction Language 1. Bellcore's standard language for writing exchange switch-control applications. This has become a *de facto* standard for telephone companies wishing to integrate and interact. It is the language of the Operational Support System (see OSS).

TLA

Three Letter Acronyms. The plague of this industry. Actually most of these are abbreviations, not acronyms.

TM

- Terminal Management. See.
- Telemetry. See.
- Time Management.

TMAP

See TSAPI.

TMN

Telecommunications Management Network. That part of the signaling and bearer network responsible for billing, fault reporting, etc. This usually applies to common carriers, but some large corporations also have billing systems for charging departments for network use., consultancy time, etc.

These systems will use a substantial suite of management tools such as HP's Overview. They will have SNMP agents in their bridges and routers. See TINA. and GDMO.

TMR

Triple Module Redundancy. See.

TMTV

Time Multiplexed Television. This was developed by Comsat under a contract from Intelsat. It allows the simultaneous transmission of three high-quality TV signals in 36MHz of bandwidth (half a transponder), which is the capacity usually reserved for a single analog video carrier. The signals are time-compressed and multiplexed. See NMAC also.

toggle

To change between two states. Meaning if you activate it once, it will change from state A to state B, then if you activate it again it will change back again to A.
— Toggle switches are everyday two-state light switches.
— Toggle fields are fields that can only hold a Yes or a No, a True or False, a tick or a cross. These are also called logical fields, or Boolean fields.

token

• **Networking:** An electronic identifier (the 'token') which is passed between users of a network to ensure that only one station ever transmits at the one time. It's actually a short packet of data, passed around the network sequentially, and only the station with the token can transmit. There are busy tokens and free tokens. Busy tokens are circulated around the network while data is being transmitted, and when this transmission finishes the token is changed to 'free' and passed on. The next device can then capture it, and begin to transmit — or pass it on.

• **Programming:** A token is a code which takes the place of ASCII command codes for compressed storage in some languages.

It is also just a distinguishable unit in a set of characters.

• **Compression systems:** A token is a pointer which points back to some previous occurrence of the word string. See LZW.

token bus

The IEEE's 802.4 transmission system which used a linear topology like Ethernet with a token passing system. It differs in function from the IBM Token Ring mainly because it explicitly sends a permission slip from one station to the other — each station knows the address of neighbors on both sides.

While token bus systems are physically buses, they are logically rings. They use broadband transmission over coaxial to provide two-way forward and reverse channels within a single length of cable (FDM), and the head-end terminator completes the loop, returning the packets of data down the other side to create the logical ring. MAP is the most common implementation of token bus.

Token Ring

Designed originally by Olof Soderblom in 1969 (then called the Newall ring) and later taken up by IBM. This is now the IEEE 802.5 standard for a 'star-wired ring' LAN at both 4 and 16Mb/s.

There are many types of token ring LANs, but IBM's is the best-known and most widely used. The ring circulates an electronic 'token' past the front door of all stations in the ring and it returns to the originator.

This token is effectively a 'permission-to-communicate' slip which is generated by one workstation on the network called the monitor. It is actually a frame, three bytes in length: the first byte is a start delimiter, followed by an access control byte, and an end delimiter. Data and addressing can be added to this token to create a packet or frame.

The station which is currently transmitting controls the token (by setting it to 'token busy'). It then adds its destination and source addresses, plus the message, and despatches the packet on the media. It then waits for the token/packet to return before the token is reset and handed on.

A new form of 'early token release' system is also available which doesn't require all stations to wait until the token has circulated. They can put packets onto the ring, but only when the first message has passed.

A Token Ring LAN has a limit of 255 stations per ring and the maximum distance between stations is 100 meters. It uses the conventional twisted-pair wiring and it provides either of 4Mbit/s or 16Mbit/s data rates.

The physical ring is actually a ring of repeaters called MSAUs (see) in a closed unidirectional loop. The MSAUs repeat the signal and provide a recovery function. Stations on the network connect to a MSAU. Stand-alone repeaters can be used to extend the LAN's distance, but they can't increase the number of stations.

The frame structure for IBM's Token Ring packets consist of a 1-Byte start delimiter, 1-Byte access control, 1-Byte frame control, destination and source addresses of between 2- and 6-Bytes, variable data bytes, followed by a 4-Byte frame-check sequence, then 1-Byte each for end delimiter and frame status.

The maximum payload of a Token Ring frame is 16,384 (for 16Mb/s mode) and 4096 (for 4Mb/s mode). See MAC, MSAU and MAU also.

Token Ring NetBIOS

This is the Network Basic Input/Output System which allows IBM's Token Ring LAN to run most programs written on NetBIOS, the interface offered with the IBM PC Network.

TokenTalk

A Macintosh low-level protocol (replacing LocalTalk in the AppleTalk stack and the lower two layers) that allows most of the modern Macs to connect to Token Ring networks. It requires an additional card in the Mac, and operates at 4Mb/s.

toll

In most of the world this means long-distance: the UK would call it a trunk.

However, in the USA it is a term for calls which are not charged at local rates — which includes relatively short-distance calls between say, a city and its suburbs (called an intra-LATA toll). Long-distance tolls are those outside the LATA area (inter-LATA tolls).

The first form of toll has been a very lucrative monopoly of the Baby Bells (LECs), while the second has been a lower-profit competitive area between IXCs. This is all now changing.

• Toll-quality is internationally accepted as requiring the delivery of voice frequencies between 300 and 3400Hz, all at roughly the same amplitude (flat response).

toll office

A US telephone exchange which handles toll calls (probably long distance, not intra-LATA).

tonal separation

A technique from photography which is now used with computerized images, to create what appears to be high-quality ink-line art work from photograph images. A point between off-white and dark-gray is chosen as the point of separation, and any tone lighter than this becomes white, while any tone darker becomes black. You end up with only two (or sometimes three) tones.

tone

• **Images:** The relative shades of light and darkness. Tone illustrations are distinguished from line-drawings, which have black and white only (shading may be by hatching). Tone illustrations have continuous tone variations (as with a photograph).

Half-tones are actually pseudo-tone illustrations. The appearance of gray shading is achieved by changing the dot size — but the dots themselves are black (or with colour, CMY).

• **Sound:** The term is vaguely used in music to mean 'character', and this includes notions of pitch, timbre and reverberation. In voice, it can also include the concepts of volume, modulation and word stress.

• However a 'pure' or 'fundamental tone' means a note without overtones or harmonics. This usage is therefore emphasizing the pitch or frequency aspects.

• A tone control on an amplifier is a broad-spectrum variable filter which boosts base, against treble notes, or vice versa.

tone dialing

More correctly called Single or Multi-Frequency dialing. See DTMF, MF and SF.

toolbox/kit

A set of routines (program modules) and utilities which help a programmer develop software. The Macintosh toolbox lets the programmer create the Mac-style icons and menus.

These routines are basically managers of menus, windows, and other interface elements, while some are used as part of the operating system. High-level toolbox routines may use low-level routines. See workbench also (client-server).

TOP

• **Technical Office Protocol:** Not to be confused with TOPS which is a popular proprietary type of multi-vendor baseband LAN.

TOP and MAP are closely related OSI network protocols aimed, respectively, at the office and the manufacturing environments. TOP was primarily the result of work by Boeing Computer Services in 1985 to provide a universal standard for office machines, in the same way that GM's MAP did for manufacturing robots.

Boeing's aim was to use existing communications standards and to be compatible with existing equipment. At the lowest level, therefore, TOP uses the Ethernet CSMA/CD protocols (802.3). At higher levels it is using MAP protocols.

• **Time-division Over Packet:** A technique used for sending isochronous (voice/video) real-time data over private packet networks.

top-down

• **Design:** Starting with the idea, outlining it graphically, then working systematically down to the lowest (detail) level.

• **Programming:** This is a structured programming approach which requires the systems analyst to first broadly outline the problem, identify the basic program functions, create tasks each with a clearly defined purpose, then subdivide these into smaller sub-functions, called modules. Modules are then written, independently tested, and coupled.

topology/topography

The physical layout of something.

• In LANs and other network types, these terms refer to the layout of the wiring — the overall architectural design. The basic LAN topologies are ring, star, bus and mesh, but there are combinations of these, such as IBMs Token Ring, which is a star-wired ring. The word topology actually refers to the geometric designs, while topography is a geographic term for the study of particular localities.

• The term topology is sometimes also applied to the logical structure of a network. This deals with the possible connections between network nodes; whether they are able to communicate with their neighbors — even if they don't have direct physical connection. A token bus network is a physical bus (laid out in a line with a beginning and end), but a logical ring (the tokens circulate automatically from the end, to the beginning).

• In chip design, the topology of the chip is also closely associated with 'real estate'. The topology refers to the layout, while real-estate refers to available area.

TOPS

• DEC's multiuser, multitasking operating system that ran on the PDP minicomputers.

• An open network designed by Centram Systems West. It uses a totally distributed approach (no dedicated server) and allows IBM PCs, Macs and Sun workstations to share the same network and use each other's resources. It was based on the AppleTalk protocols.

TOR
Telex Over Radio.

torus
A hybrid dish shape for satellite antenna. It has a parabolic geometry along one dimension, and a spherical geometry along the other (at right-angles). Torus shaped dishes are used in multiple beam antenna systems — where it is possible to receive signals from more than one satellite.

ToS
Type of Service. This sometimes refers to established Quality of Service (QoS). See QoS.

touch screen
Touch-sensitive screen. A special type of screen which records the position of a touch either directly against the surface, or very close to it. There are five different technologies used today for touch screens:
— **Matrix sheet.** Two transparent plastic sheets overlay the screen. One has very fine horizontal, and the other vertical, conductors etched into the contact surfaces. When the finger touches the screen, good contact is made at the junction. This is the cheapest touch-screen technology.
— **Analog capacitance.** These require a sophisticated controller, usually linked to an RS-232 serial line. A capacitance field is generated by an oscillator, and the system measures the change in capacitance from the corner to the point of finger contact. This system is widely used for kiosks.
— **Analog resistive.** These are similar to the capacitance system, and to the matrix sheet approach, except that the controller simply measures a single dimension of resistance, rather than detecting an x/y coordinate. It has a coarse resolution.
— **Infrared.** These use arrays of IR LEDs and photo-transistors. The finger cuts the multiple light beams.
— **Surface Acoustic Wave**. See SAW filter.

• The problem with touch screens is that they all have fairly coarse discrimination, although the new capacitance and SAW screens are better than the old. In kiosk situations they are a simple means of providing coarse control for menu-type selection, but they get sticky, dirty and damaged. They aren't generally seen as a substitute for a mouse and cursor in an office.

touch tone
See DTMF.

touchpad
An attempt to emulate a mouse for laptops. Above or alongside the keyboard is a relatively large area (say, 12 x 5cm) covered with a touch-sensitive electrically-resistive membrane. You were expected to use a finger as a pointer/cursor controller, but the control was far too coarse.

Touchstone
The joint Intel and US Defense Agency DARPA's $27.5 million project to develop a parallel computer using more than 2000 of Intel's 860 chips (they may now use the Pentium Pro). The aim is to achieve a processing rate of 150 billion floating-point operations a second which is about 100 times the current Cray Y-MP speed.

tower machine
A PC or mini which is designed (physically) so that the main box stands vertically under the desk, rather than horizontally on top of it. Apart from the obvious value in removing the large box from your desk, tower machines generally have better air circulation and so better cooling.

TP

- Telephony
- Transaction Processing. See and OLTP also.
- Teleprocessing
- Transport-layer Protocol.

TP#

• **Transport Protocol:** The OSI's transport protocols, of which only two 'classes' really matter.
— **TP0** (Class 0) is the connectionless protocol for use over reliable subnetworks defined in ISO 8073.
— **TP4** (Class 4) is the connection-based protocol defined in ISO 8073.

• **Transaction Processing:** TP1 was a benchmark used to compare the ability of on-line transaction systems to handle large numbers of users. It is mainly used in banking and share-trading operations. The first in the TP# series was set in 1985; now there are many, and they are called TPCs. See TPC.

TP-DDI
Twisted-Pair; Distributed Data Interface. This is a FDDI-over-copper cabling standards. It allows the use of either 100 ohm UTP Category 5 cable (TP-PMD standard) or 150 ohm STP for distances of about 100 meters. See TP-PMD.

TP-PMD
Twisted-Pair, Physical Media Dependent. The ANSI X3T5.9 term for 100Mb/s FDDI over copper. It uses stream-cipher scrambling and a three voltage level bit-encoding scheme called MLT-3 which transmits 2-bits per Hertz. To achieve the line rate of the FDDI (125Mb/s including the 1-in-5 control bits), the cable must carry 62.5MHz.

TPAU
Twisted-Pair Access Units. This is a fancy name for a wiring block where the building wires are connected to the external cables — the demarc(ation) point in many cases.

TPC

• Transaction Performance Council. The standard benchmarks produced by this group of 45 hardware and software vendors and analysts are mainly for on-line transaction processing, primarily with large computer systems. IBM, HP, Compaq, Oracle, DEC and another 40+ companies are involved.

The benchmark results are recorded in transactions per second, and the tests are assumed a loading of ten users for each system. Price-performance figures are also calculated by taking the estimated overall five year costs as a ratio of transactions processed.

— **TPC-A (1989).** This is a simple OLTP benchmark for bank ATMS, etc. It models overall performance, I/O, database activity, etc. It was designed to test how well a client-server application would compare to a stand alone.

— **TPC-B (1990).** Industry standard database stress test. This is a test which stresses general databases. It is similar to A but without I/O terminal activity.

— **TPC-C (1992.** A complex OLTP tests for general purpose workload on unit-processors only in departmental situations. This will be a strategic benchmark in the future; it uses a more complex transaction model than A.

— **TPC-D (1994).** This is the current release for large computing environments. It covers uni- and multi-processor systems for databases of a size between 1 gigaByte and 1 teraByte. The benchmark measures 'queries per hour'. It is being designed for decision support.

— **TPC-E (late 1995).** This is for enterprise computing, including client-server complex OLTP.

TPI benchmark

Transaction Processing Interactive. This is listed as benchmark for LANs. It tests how fast debit/credit transactions can be added to a database. I suspect that this is either the old TP1 (with the 1 misprinted as I), or it could be any of the TPC's benchmarks under a different name.

TPON

Telephony on Passive Optical Network. A fiber-to-the-home distribution system devised by British Telecom which uses a single fiber cable to the street pillar, and two progressive passive splitter levels in a hierarchical fashion (4 splits, then 8) to feed 32 households.

The basic TPON system only carried one telephone channel to each home (time-division multiplexed in the fiber cable), but it could be upgraded by frequency division multiplexing (WDM) to provide 18 non-interactive channels of TV.

This has now evolved into Lamba PONs, but digital compression and HFC seems to have overtaken all of these PON techniques. See also Lamba PON, B-TPON and BIDS.

tps

Transactions per second.

TPM

Total Process Management. This combines the principles of re-engineering and TQM (see both). It requires radical change to be followed by continuous improvement and benchmarking.

TQM

Total Quality Management. The business practice of managing production systems in a way that keeps defects to a minimum at all times through the process, as against simply relying on detecting faulty end products. The philosophy includes the need for continuously managed improvements.

trace

• **Graphics:** Tracing is a preliminary process for converting a bit-mapped screen image into a vector-based form. Both automatic and manual tracing can be used around any visual entity on a bit-mapped screen as a way of identifying that object to the computer.

Automatic tracing systems look for obvious discontinuities in the bit-map (obvious edges). There are stand-alone programs which can perform this function.

Trace algorithms can take a scanner image (bit-mapped) and transform it into a scalable vector format using Bezier or polygonal curves, and create distinct objects on the screen which can then be moved or resized. This creates the difference between draw-style images and paint-style images; in draw-style files each object is distinct and separate.

• **Electronic scan:** The term trace is also applied to the electron scan on a VDU or TV set. The trace is the line it 'writes' on the screen. See Kell factor also.

• **EDI:** In EDI, each electronic form traveling through the system must leave a 'trace' (called the audit trail) if the transaction is to have legal standing. EDI is little more than formalized electronic mail, plus the trace and authentication functions.

• **Diagnostics:** In de-bugging new applications, a trace program is one that monitors the execution of the algorithms (a stage at a time) and stores the variables, etc. for examination.

track

• **Storage:** Any linear, helical, spiral or circular path onto which information is coded.

— Audio cassettes use linear tracks (except for R-DAT).

— Videotapes use linear tracks for audio and control, but have helical-scan (diagonal) tracks for the image.

— LP disks and CD systems use spiral tracks.

— Magnetic disks and some of the new optical record systems use circular tracks.

• **Film/video:** In film and video production it is common to have many audio tracks in the pre-production stages, which are later 'mixed' (merged at various amplitudes) into one (or a stereo pair) final audio track for release.

Even a small documentary would normally need three or four audio tracks, and feature films may have twenty or more. Special multi-track recorders are used in the editing studios to hold this material while mixing. See Foley also.

• **Entertainment:** In music and video, the word 'track' can refer to a tune or sequence from the producer's viewpoint.

track servo

The feedback and control mechanism that keeps the read and write head of a disk system (usually an optical disk) centered on the track. Old video recorders had a manual tracking control, but this is now automated. VCR have special head-servo systems which feed signals to a piezo-electric element which can react in millionths of a second to move the head back onto the helical track if it wanders off (in stop-motion, for instance).

trackball

A large round ball set into the computer keyboard (or attached alongside) and used as a substitute for a mouse. The main advantage is in reducing the amount of desk space required to operate the machine.

Trackballs are a good mouse substitute for laptops when used on the lap (Leg-hairs tend to get caught in the mouse-ball!), but on a desk, trackballs don't give you the feedback of a mouse and so they are often more fatiguing. See joystick and isopoint also.

tracking

• **Text:** Tracking is the process of adding and/or subtracting equal amounts of space between every letter in a word to stretch it (usually for justification reasons). Distinguish between tracking and kerning; kerning is the process of varying spaces between different letter combinations by different amounts to produce a more balanced appearance to the eye. See ligatures also.

• **Video:** Tracking is the technique of keeping the read-head aligned with previously written tracks. When mistracking occurs, the read-head will detect a drop-off in signal and a fair amount of hash will appear on the screen. A good tracking servo system will move the head back over the center of the track.

When the video-tape is stationary (stop-motion or freeze frame) and the read-head is still spinning, the head is no longer able to read the length of a single track without substantial head realignment. This is because the track was recorded along a line which was the resultant of both the helical scan movement, and with the tape's forward motion.

This is why VCRs often show three or four bars across the screen when they try to show still frames; the head is skipping diagonally across three or four tracks and showing bits of each.

The new piezo-electric tracking mechanisms can handle this amount of head realignment, and produce 'clean' still images.

• **Audits:** See trace and audit trails.

tractor fed

The sprocket mechanism for continuous fan-fold paper in some printers. The distinction is with sheet feeders. See continuous stationary and perfory.

traffic

A general term for the total data movement of all kinds over a communications link.

trailer

• In packet protocols, this is the other end of the packet to the header — aka the 'tail'. It usually follows directly after the 'payload' of useful data. The trailer mostly contains error checking and an end-flag only.

• In film and television, this is a publicity 'short'. It will use the key action sequences, or humorous extracts from the production. Trailer production is a specialized art.

training

• **Modems:** A training tone sequence is used in some high speed modems. After the original handshaking has taken place the modems will exchange a series of tones containing standard data to test for line quality, to quantify any equalization necessary, and to measure near- and far-end echo timings. The modems may continue to monitor and retrain whenever necessary through the session.

Automatic equalizer modems adjust to line parameters at the start of each communications session through this training process.

• **Cellular:** In digital mobile telephony (GSM), this is a short sequence of bytes included in every transmitted frame of data. Since this sequence is fixed and known to the receiver, it can apply its knowledge of the timing and the condition of the data to construct an 'inverse filter' to be applied to the data frame as a whole. It will do this to adjust synchronization, and to remove any disruptive components resulting from multipath propagation. In cellular telephony, this process is known as equalization or negative filtering.

TRAM

Transputer and RAM. This refers to a single plug-in card carrying both modules — usually with some special ASICs as well.

transaction

Any communications of information which requires that a file be updated, or a record created. Payment and receipt; reception and acknowledgement.

This can be the two-way exchange of either value or information. The significance is that a transaction must be run to completion or it must fail — one or the other — with no in between possibilities.

It must have auto-locking and it must have atomic updating. In transaction processing (finance and banking), a transaction is a unit of work characterized by the ACID properties. See ACID test.

• In general company databases, the change made to a record (or the creation of a new record) is often called a transaction, but the emphasis is also on the exchange of information across the network. The changes are initiated from a remote terminal, and the process is 'result-based'. See OLTP.

• If transactions aren't being conducted with the major database records, there will be a transaction file which contains the modified (updated) information. In batch mainframe operations, this file will later be merged with the master file which contains the permanent data. See journal also.

• The term 'transaction' also sometimes misused to mean that changes to the records are made totally in the terminal, and are then transmitted to the mainframe as a single burst, when the transmit button is pressed.

The distinction being made here is with character-by-character, or packet-by-packet terminals. However this is a misuse of the term; transactions are often transmitted progressively in packets or bytes.

• There is no common measure of transaction-handling capabilities in computers because vendors use the terminology differently. One vendor's transaction is another vendor's I/O; one vendor measures internal response times and the other measures external. See TCP.

transaction file

This is a file which records the activity. The data in the transaction file is used to update the master files of the organization. Transaction files are always retained, since they provide an audit-trail which allows checks to be made later, if the data is important. See journal.

transaction-oriented entry

This simply means the data is keyed directly into the computer system from the place where the transaction occurs — in contrast with batch operations where the data will be entered at some later time.

transaction processing

A general term applied to the communications of financial (specifically banking) data — although it includes booking systems, and similar two-way exchanges. Transaction processing is typified by the exchange of short messages, which need to be handled in a highly secure manner.

The use of permanently on-line computers to buy and sell shares, book airline tickets, make theatre reservations, transfer funds between bank accounts, and pay bills and accounts, is now usually called On-Line Transaction Processing (OLTP).

There are three main components of a good transaction processing system: throughput, responsiveness and cost — and the ACID test should be applied to all TP systems. A special terminology is used in TP discussions: 'transactions', 'associations', 'ACID', 'dialog'. See all of these.

transaction services layer

This is the SNA's Layer 7. It is really an Applications layer.

transaction sets

In EDI — these are groups, all of one document type — such as statements, invoices, etc. within one functional group.

transborder data flows

The flow of information across national boundaries. With the Internet, it is very difficult for national governments to control these flows today but the US, for instance, implements a control over encryption which would come under this category.

transceiver

A transmitter and receiver in one unit.

• In baseband networks the transceiver monitors the network for in-bound packets and transmits packets from the device on the network. See MAU and AUI.

• In Ethernet, this is the MAU. In Token Ring it is the MSAU.

• Transceivers are now large ASIC chips. It is this hardware that provides the physical connection between a device and the network cable and supplies the facilities to handle the packets and deal with line-coding. You can buy single-port and multiport transceivers, and special types for thick and thin Ethernet coaxial, and for UTP, fiber, etc. See NIC.

• In broadband network terminology, the transceiver must always be a RF-type modulator/demodulator and provide band-pass filtering (tuning) because of the radio frequency signaling methods used in these systems.

• Transceiver is also the general name given to two-way taxi-phones, CB-radio handsets, cellular phones, etc.

transcendental functions

Sines, cosines, tangent, arc-tangent, exponentiation and logs.

transcoder

A device used to change one coding system to another. This can be a line-code translation, or a higher-level type translation. Some transcoders can also be called an encoder, or even a gateway — but the point is being made that the signal is already in one digital format, and is being translated to another.

• **Computers:** This is also known as a data transfer coder, and it is used to transfer text between computers and/or terminals. Generally 7-bit ASCII or 8-bit EBCDIC is the character code used today, but in some older systems 6-bit transcoders were used to reduce the data rate requirement.

• **Communications:** In digital communications systems it is often necessary to transfer the output of, say, an ADPCM voice-coder operating at 16kb/s, to a 64kb/s PCM ISDN channel. A transcoder performs these rate-adaptation and standards-conversion changes.

• **Video:** In video, transcoders are devices that translate composite video signals to component video — or vice versa. Transcoders also translate one form of component signal to another, although then they are called standards convertors. The term 'encoder' is used for a transcoder which converts RGB or YUV component video to composite NTSC or PAL.

transcribe

To convert from one form to another — usually the transfer from one form of storage medium to another (which may, or may not, require translation). When speech is recorded as short-hand, and the short-hand notes are then typed into a computer, we have two stages of transcription.

Between electronic devices, this is transcoding, but more likely the word transcribe will be used at the logical or intellectual level, rather than for line-codes.

When, in the near future, a digital TV signal is received, yet the household is still using an analog TV set, a transcriber will be incorporated into the electronics of the set-top box to make the translation. It can equally be called a transcoder.

transcript

The recording in text form of words spoken in an interview, lecture, television program, etc.

transcript file

In some CAD programs, each addition to the drawing is recorded automatically in a transcript file. If a major mistake is made, these module additions can be replayed from the beginning and stopped at a certain point. In effect, it is a form of Undo for complex processes.

transducer

The general name for any device that converts one form of energy to another, for instance, from physical pressure to an electrical signal, or from electrical signals to physical motion. So a loudspeaker is a transducer, as is a microphone, and an electronic pressure sensor. Transducers are both input and output devices.

The term is more commonly used in measuring systems to mean any device which generates an electrical signal from real-world physical measurements. When it is interrogated, the device will relay back to an instrument the changes it has experienced. Pressure gauges, light-meters, etc. all use transducers.

A transducer is generally an analog device, but it can be coupled with ADC to produce a digital output.

transfer modes

There are three main transfer modes in use on today's public and private telephone networks:

- **STM** (Synchronous Transfer Mode) is a slot-synchronous, TDM system with both scalable bandwidth and dedicated bandwidth. It uses dedicated circuits and is circuit-oriented. See STM.
- **Packet TM** is widely used in private networks. It has packet addressing, and has various sizes and formats in the packets. The protocols are Ethernet, Token Ring, X.25, and FDDI. See all.
- **ATM** (Asynchronous Transfer Mode). This is a cell synchronized, packet addressed system providing scalable and flexible bandwidth. ATM is the telephone technology of the future — and also the WAN technology. See ATM.

transfer orbit

The intermediate stage between the low-earth orbit of a GEO communications satellite just after launch from the Shuttle, and its final resting place in the Clarke orbit. The satellite is often held in the first orbit for equipment tests, then a kick motor is fired to 'transfer' it to its final orbit (35 000km out).

Many rocket launches now put the satellites directly into an elliptical transfer orbit. The transfer orbit will reach into the Clarke orbit at the apogee and drop to the low-earth orbit at the perigee. At precisely the right time an apogee motor will fire in reverse-thrust to slow the satellite at the top of the loop, and 'round' the orbit so that it circles the Equator at a constant distance. See Clarke's orbit.

transfer rate

The speed at which data can be transmitted to and from a storage or transmission device. Communications systems tend to quote the rate in kilo- or megabits-per-second or octets-per-second (where an octet is a byte), while computer systems tend to quote data rates in kilo- or megaBytes per second.

transform coding

This refers to a complex change in the way complex cyclical functions are represented in a digitally-encoded form.

Usually it means the transformation from a sampled waveform or audio or video, to a frequency-component form (Fourier transform). The complex waves are represented by weighted values attached to each component (harmonic) which made up the complex form.

These ideas were originally worked out on analog audio waves, but they apply equally to any complex waveform, including digital video images treated in blocks.

In transform coding of video images:

- The original picture is divided into blocks — typically 8 x 8, or 16 x 16 pixels.
- The scans of each line of pixels are treated as if they were a single line of 8 or 16 'waves' (cyclical changes).
- A two-dimensional transform of these lines results in an 'uncorrelated coefficient block'. The transform process removes first the overall 'fundamental' in the waveform, then progressively examines the line for harmonics.
- The most significant information is thereby concentrated in only a few of the leading coefficients, with finer image detail being added later. The first few coefficients are sufficient to describe the original block in low-resolution (reveal its most prominent features).
- The least significant information (the highest resolution elements of the image) will be extracted and added to the linear hierarchy, only if the processor has time. The more processing that can be applied in the time available, the finer the detail recorded.
- When decisions need to be made to discard data because of bandwidth constraints, it is now relatively easy to discard those of the least-significance (finer detail).

When objects in the scene are moving rapidly, the eye cannot see fine detail, so the later transforms can be abandoned — which is important because when the images change, there is more difference between two frames to be transmitted.

This is only one form of data reduction used in video and videoconferencing. Motion compensation is also important.

• In MPEG, the difference between two frames is calculated before the results are transformed.

• In audio, transform coding allows the system to discard high-frequencies (high resolution elements) which would be blanketed (masked) by louder low-frequency sounds, and so reduce bandwidth. See DCC.

• There are three main forms of transform in fashion at present:

— **DCT** (Discrete Cosine Transform) is a frequency transform process which is mathematically a continuous function.
— **Wavelet bitstream transforms** are spacial transforms, and are discrete mathematical functions.
— **Fractal transforms** code an image as a hierarchy of elements, each of which is a fractal and can be recreated by a simple algebraic equation.

See image processing, MPEG, DCT, DWT, intra- and inter-frame coding, cosine transform coding, Fourier transform and fractal.

transient

A short-term change in mains voltage (milliseconds). The term 'fluctuation' is applied to longer period changes. Transients are caused by dirty electrical contacts, change in voltage, dialing noise — and ten thousand other things. See surge, spikes, line hit, line noise and transmission impairments.

transient area

The area of RAM available for user programs and system utilities. This is the working memory area of a computer, not including the space occupied by the operating system.

transistor

These are semiconductor devices made by sandwiching one type of semiconductor between two layers of the opposite kind of semiconductor. The most common, the p-n-p transistor, has a thin layer of n-type material between two p-type layers.

An n-p-n transistor has the opposite construction, and all voltages are reversed.

In both cases, the middle layer is known as the base, and the two outer layers are the emitter and the collector. The voltage across the base-to-emitter contacts controls the flow of current across the collector-to-emitter contacts. This is why amplification is possible; small voltage changes in base-emitter are translated into substantial current flows in the collector-emitter.

transistor–transistor logic

This is an important family of logic ICs used in computers. They have multiple emitter transistors on the input gates.

TTL is characterized by high speed, moderate power consumption, low heat and low cost. The technology was invented by Texas Instruments back in 1962 and it has set the power-supply voltages of most computer devices at +5 volts over the thirty years since.

The first series of TTL ICs was called the 54 series, and the later series the 74 series. You can usually recognize TTL-type ICs in a computer because they will have a '74' or '74LS' in the middle of the type-number.

• TTL signal levels have become standard for virtually all internal signals in a mains-powered computer. Each signal line can assume either one or two values represented by two different voltages on the line.

However, the current 5V standard will soon be replaced by the 3 or 3.3V standard ICs because of the need for smaller component sizes with lower heat output and battery drain — especially for portables.

• TTL memory chips using bipolar technologies are potentially faster than current SRAM or DRAM chips (which now use MOS techniques), but the high power requirement usually limits the number of bipolar memory cells on a chip's real-estate to about 1000 bits.

translation bridging

This is bridging between network segments which use different MAC-layer protocols. See source-route translation bridging.

translator

• See transcode and transcribe also.

• In computers, a compiler or assembler translates.

• In telephony, a device which converts DTMF dial tones into signalling packets, is a translator.

• In TV and MDS, a translator is a signal booster which receives on one frequency, and re-transmits on another. It is used for local in-fill. Don't confuse it with a signal bender or booster which retransmits on the same frequency.

• In LANs, a translator can be a dedicated device built to allow one protocol device (or network) to interface with another of a different type, or it may be a bridge or router which performs translation as a secondary function. See source-routing, translator bridging.

transmission control layer

The fourth layer in the SNA architecture, responsible for the establishment, maintenance and termination of an SNA session. It also has responsibilities for sequencing messages, and for session-level flow control.

transmission group

In SNA, this is a group of channels which are routed together.

transmission impairments

The major transmission impairments in a communications circuit are (see all):

— Electrical noise
— Echoes
— Crosstalk
— Transients
— Delay distortion
— Thermal noise
— Phase and timing jitter
— Intermodulation noise
— Space and mark distortion
— Harmonic distortion

transmission loss

The loss of signal strength during transmission — the attenuation of the signal. This does not mean the loss of the transmission itself — that would be transmission failure.

transparent

Automatic and invisible. Something is transparent if it has no visible effect. The term has slightly different applications:

- **Communications:** A link is transparent when everything required to pass over it, does so without change or limitation.

  The distinction is with systems which have in-built filters (or restrictions) of some kind. Transparent links will ignore control characters (like flow-control). They will also handle the 8-bit patterns of extended ASCII (binary files), without attempting to interpret the bytes. See data transparency and HDLC.

- In most cases the implication is simply that the message has passed through a technically difficult environment — but that the complexity of the system is not apparent to the user.
- Transparent bridging is spanning tree bridging.

transparent bridging

Aka spanning tree bridging. This is the preferred technique used in Ethernet and 802.3 networks. The bridges pass the packets along, one hop at a time, using tables within each bridge. These simply associate an end-node address with a specific bridge port — the bridge has no overall view of the network or the topology. See spanning tree algorithm.

transpiler

A new term used in parallel architecture for a special compiler. The term is a combination of transputer and compiler. The transpiler will take a program written in, say, Occam, and configure it to run on a network of transputers.

transponder

A self-contained receiver/transmitter unit — this is a single repeater unit in a satellite but you also find transponders in aircraft for identification, and in trucks for vehicle tracking.

The term is most closely associated with satellite communications. Transponders don't modify the signals in any substantial way, the most they do is shift the carrier frequency, and rebroadcast the signals — they are 'dumb' repeaters. (Some aircraft and truck identification transponders, however, do rebroadcast a code when polled.)

Within a geostationary communications satellite, the transponder will receive a wide band of signal carriers from one or more earth stations in an uplink frequency band, amplify them, perform a frequency-shift operation, and switch the signal to one of the focusing dish reflectors at a downlink frequency. Most satellites will have about 20 transponders.

Transponders can transmit video, audio, data, etc. — anything transmitted on the up-link will be retransmitted on the down-link — and they make no distinction between analog and digital signals (they handle everything at radio frequencies).

- Bent-pipe transponders rebroadcast on the down-link using the same frequency as the up-link.
- Unless 'bent-pipe' is specified, assume that the transponder is performing a frequency shift. Generally this is necessary because the strong downlink transmissions will otherwise interfere with the weak uplink signals.

  Usually the two frequency bands will then be written with the uplink frequency band first — as in 14/12GHz.

- There is no standard bandwidth for a single transponder; you get them with bandwidths of 27, 36, 40, 54 and 72MHz, so they are often capable of handling multiple TV signals. The limitation is generally in the power the satellite can provide.
- If a transponder has a bandwidth of, say, 72MHz it can often be used as two half-circuits of 36MHz each. The common 54MHz Ku-band transponders have a half-transponder bandwidth of 27MHz, and these are now widely used for satellite news gathering (SNG) and for transporting TV images around the country. When a second TV signal is added to a transponder, the power available for the first TV signal will generally drop by 3—5dB.
- Transponders differ also in polarity and power output. Many older communications satellites will have both 12 watt and 40 watt transponders and new ones will have 50 to 150 watt transponders. Polarization can be right-hand or left-hand circular, or vertical and horizontal.
- The reception power on earth (measured in EIRP) will depend on the power of the transponder and how concentrated it can focus the signal. See hemi-beam, zone-beam, spot-beam, shaped beam and footprint.
- The capacity of a satellite is expressed in the number of transponders it can provide. Typically, a video channel (a half-circuit of 27MHz or 36MHz) can carry one analog TV signal or about 1000—1200 telephone links (half conversation) in one direction only. The television signals are not normally transmitted in the narrow AM band of 6—8MHz, but are frequency modulated and require a much wider band.
- Digital compression techniques can multiply the television channel capacity for both voice and video circuits by between four and eight. See N-MAC and TMTV for analog, and SCPC and MCPC for digital.
- See traveling wave tube.

Transport layer

The fourth layer in the OSI's 7-layer model, and the highest of the hardware layers. Its primary function is to guarantee the reliable delivery of the packets of data by performing error-checking functions and providing other reliability services.

When a number of packets are 'in process' at any one time, the Transport layer controls the sequencing and provides flow-control. Any duplicate packets will be recognized and discarded by this layer.

One way to look at the Transport layer is as a 'local' version of the Network layer. It is the control center of the network which maps the general user-services (in the top three layers) into the communications subnetwork (the bottom three layers). In doing so it provides a guaranteed delivery service — as distinct from the lower layers which can only provide a 'best effort' service. It also:

— ensures the data packets are in order when received,

— saves data if the network is broken in any way,

— redirects data around a break, if it can,

— allows different computer systems to talk to each other,

— establishes and releases the communications channel.

Transport is usually implemented using a connection-oriented protocol (although a connectionless version is defined). With most simple LANs, the data can't get out of order (even if the Network layer is connectionless, using CLNS) because packets are transmitted one at a time over a single shared channel and are rarely lost or misdirected. So this layer has little to do in LANs unless the network fails.

However, for the larger national or corporate networks, there are a variety of key protocols at the Transport layer level: TCP/IP is probably the most important. These are involved in collating the data, re-sequencing datagrams, checking for errors and requesting retransmission.

transport relay

A form of LAN bridging which passes data between two segments of a network. Unlike a bridge, it operates at the Transport layer. Before transport relay can be used, the ISO layers above Data-link (Network and Transport) must be identical.

Transport services

Transport layer protocols need to cope with many different networks, each of which may be offering a different quality of service. Five service classes are available:

• **Class 0** is the basic class with no additional enhancements to the network service. It was designed for teletex. Flow control is by the network layer.

• **Class 1** has error recovery. This should be used with any network subject to frequent interruptions. It was designed for basic X.25 services which may need minimal error recovery. Flow control is by the network layer.

• **Class 2** is Class 0 with multiplexing. Flow control is improved.

• **Class 3** is Class 1 with multiplexing. It is therefore an amalgam of Class 1 and 2.

• **Class 4** has enhanced error detection and sequence confirmation. This is for unreliable networks.

Note: the transport service specification is the same for all classes. The service exists to provide end-to-end data transfer, quite independent of the underlying network.

transputer

TRANSistor + comPUTER. A supposed 'computer-on-a-chip' development which had promises of becoming the mainstay of parallel processing systems until challenged by some of the modern conventional RISC chips.

Modern transputers have multiple RISC-based processors all working in parallel. Each CPU has its own ROM and RAM memory and ports. The basic idea is that the transputer would include a run-time kernel which allows concurrent tasks to be carried out without the need for external software.

The earliest Inmos transputer was the T414, and the current top-of-the-range is the T800 which combines 32-bit processing with on-chip memory and a 64-bit floating point unit, all on a VLSI with 84 contact pins. Many thousands of T800s can be linked together for super-computer performance.

trap

Special actions only initiated when something happens that would normally interrupt a program.

• Many programming languages provide error trapping. If the program attempts to do something impossible (divide by zero) the attempt will be trapped. Control will then pass to a special routine created by the programmer, and the normal functions are restored. Basic has a trap statement ON ERROR ....

• When an agent in an SNMP-managed device detects the occurrence of a 'significant event', it will trap the event information, and send a 'trap' (unsolicited) message to the network management system.

• Trapping is an important part of desktop publishing when you are printing high-quality color. You need to do trapping for type and objects. The main concept in trapping is to provide an overlap between adjoining colors by adjusting the 'stroke' (the external outline of an object) so slight mis-registration of the separations won't leave white lines. It also adds a slight 'edge' to color junctions, and gives the image a cleaner appearance. When the stroke is told to overprint adjoining color, it is partially allowed to enter the stroke (half the specified stroke thickness). This process includes chokes and spreads.

trapdoor

When complex code is being written for large systems, it is usual to provide ways for diagnostic tools to be later used on different modules of the programs. So 'trapdoors' are left which make it easy for the programmer to enter the system at a later date and check for problems. These also become the entry-point into an operating system, which can allow a hacker to enter.

traveling wave tube

TWT. A type of amplifier/transponder used in satellites. TWTs depend on a helix pipe which is capable of wide-band amplification — typically 40 to 80MHz. The key point about TWTs is that they can retransmit any type of signal directed at them — whether digital or analog in form.

These devices are particularly useful in microwaves and satellites because they can handle the required wide-band of frequencies with a minimum of circuitry. They also have very high reliability (although a limited life) and high power efficiency (typically 30—50%), which is essential in satellites.

• TWTs cannot be run at full power for any length of time, so typically the contract for service will specify the power which can be used. This is reflected in the cost.

• The other major type of transponder used is the klystron narrowband amplifier. See bent-pipe transponders also.

tree

A basic structure of data widely used in computing. The analogy is that of a single 'root' or 'trunk', which branches into two ('binary-tree') or more branches, then each of these branches splits into two or more twigs, leaves, etc. This is the standard hierarchical architecture of an MS-DOS directory which results in a path-name consisting of a sequence of directory/sub-directory/filenames representing the different layers of the hierarchy.

• Tree structures are used as a way to manage data in databases. These are usually binary trees, making two splits at each level in the hierarchy.

• An LZW approach used in V.42bis data compression techniques (and others) is called the 'trie' (*sic*). It is a tree-like sliding window 'dictionary/library'.

• At the logical level, tree and bus network architectures are the same, except that the tree has more complex branching. In both cases, the transmissions from one station will propagate to the extent of the network, and be received by all stations.

Trellis-Code modulation

TCM is both a modulation technique and an error detection and forward error-correction (FEC) scheme. It is used in many new communications systems, especially V.32 modems, to detect and correct data errors. Trellis Coding is the 'parity-checked' version of multi-level QAM, and it is generally held to reduce the error-rate by three orders of magnitude.

TCM refers to the use of amplitude and phase modulation in a way that links certain phase changes and amplitude changes, in a way that should always follow the rules. Not all amplitude changes are available with every phase-change, so it is possible for the TCM device to detect 'violations' of the rules, and take a best-guess at ambiguous changes.

Trellis Code modem scheme currently works on quadbits of data which have parity attached to create quinbits. Both phase and amplitude are QAM modulated to create a 32-point constellation, and TCM then adds a parity bit to help the modem sort and decode the information more effectively.

With V.32, the modem examines the signals of previous data packets and predicts the likely form of current packets. This then provides an immediate check if phase-hits occur or if the receiver slips out of line.

Trellis Coding's real value is in interpreting data transmitted over systems subject to echo problems, where these phase/amplitude mistakes are most likely to be made. It is now tightly coupled with automatic equalization schemes. Also see QAM.

• Trellis Coding is a form of convolution code.

• TCM uses a 'state diagram' which rules on what are permissible transmissions during each signaling interval, and it also checks on the legality of the states by reference to prior states. There are both proximal and temporal checks on each symbol.

Both sender and receiver are pre-programmed to know which bit-patterns are legal, following past permissible patterns. If any illegal pattern is detected, the receiver assumes an error. But since it knows both the transmitted states and the permissible states, it can usually analyze the pattern and deduce ('best guess') what the original must have been (using a Viterbi algorithm and decoder). If the next symbol proves to be legal, based on this deduction, it can correct rather than just detect the error.

Note that this adds a substantial processing delay to the output — often in the order of 15 to 20 symbol intervals (baud).

triad

The triangular groups of three phosphor color-dots (Red Green and Blue) which create a single pixel on a color television or monitor screen. A combination of these three at varying brightness produces the effect of the full-color range up to white (all three at maximum brightness).

The average PAL screen will have about half a million triads or 'pixels' (picture elements) and a color monitor may have between 1 and 4 million.

In a color computer monitor, one triad creates one pixel of image at the highest possible resolution. There is often some confusion about the triad-pixel relationship because computer software can sometimes group triads into clusters and switch them on and off together — and this cluster is then 'logically' a pixel. Pixel is a measurement of visual resolution of images, while triad is a technical term for physical dots on a screen.

• A triad is formed by three electron-guns in the picture tube which all beam electrons simultaneously through the same hole in the shadow-mask. The mask is a thin mild steel plate mounted about 10mm behind the screen, and it is perforated by a pattern of tiny holes (one for each pixel). The three (RGB) beams converge on a hole and pass through momentarily during a scan, and strike the colored triad phosphors dots.

The beam intensity of each gun is modulated in amplitude by the three color electronic drive circuits according to the waveform. If all are at full intensity we see a white triad; at other intensity combinations we see colours. With zero intensity, the triad remains black. See vertical stripe also.

• A drop of water on the screen will often let you see the triads. The water acts as a powerful magnifying glass.

triax

A new type of professional cable developed for TV studios in 1992. It handles fully multiplexed component video signals, to (sync and controls) and from (RGB) the cameras. It replaces the old 1-inch thick, 80 copper-conductor cable.

tribits

In PSK and QAM modems more than one bit is sent with each 'angular' phase change (known as an event or symbol). If three bits are transmitted with each change then these are known as tribits. This is why 'baud' (the number of changes per second) and 'bits-per-second' aren't necessarily the same; here the ratio would be 3-bits per second to each baud. See dibits and quadbits.

triboelectric

Static electricity. See.

tributaries

Secondary in a master-slave relationship, or sub-bearers or subchannels in a multiplexing operation.

When a signal is extracted from a multiplexed bearer so that it can be treated alone, it becomes a tributary. A single PDH tributary can be demultiplexed out of a SDH bearer.

- When digital voice circuits are multiplexed in stages, the stages generally go from 64kb/s (the basic PCM level), to 2Mb/s, 8Mb/s, 45Mb/s then 140Mb/s (there are different hierarchies in different countries). Each stage represents a tributary to the next. So there are thirty tributaries of 64kb/s PCM voice in a multiplexed 2Mb/s stream.
- The term 'virtual tributary' is also popular in Sonet optical fiber networks, while in SDH they are called 'virtual containers'. With SDH/Sonet you can extract one tributary, without needing to demultiplex the whole signal. There are four standard sizes in Sonet: VT1.5, VT2, VT3 and VT6 which have respectively 1.5Mb/s, 2Mb/s, 3Mb/s, and 6Mb/s payload capacities. See add-drop multiplexers.

trie

A data structure very similar to the standard hierarchical 'tree' structure, but with some important differences which make it easier to implement. In V.42bis data compression a trie consists of three pointers (only two are used for encoding, while the 'parent pointer' is also required for decoding). The pointers identify the pieces of information by referring to their location.

The literature makes a distinction between 'tree' and 'trie'. They say that in any tree structure, the information is specified by its proximity — but not in trie — a rather esoteric distinction which boggles the mind. In effect, both 'trie' and 'tree' structures perform the same functions, but 'trie' is apparently easier to program.

Trinitron

This was a Sony picture tube invented way back in 1968. It is characterized by being perfectly flat in the vertical face dimension of the tube, and less curved than normal in the horizontal. It has superb blacks, and in the early days it was clearly superior in vertical resolution and color clarity.

The difference between Trinitron and normal picture tubes, is that Sony discarded the standard perforated shadow mask behind the phosphor surface, and replaced it instead with vertical wires under tension. Instead of the triads of phosphor dots, Sony uses triads which are defined by vertical lines on the screen and the horizontal scan of the raster. The wires act as the shadow mask, blocking each of the vertical colored phosphor strips from being activated by two of the three electron guns. The guns are on the same horizontal, separated by angle.

Vertical resolution is improved and moire decreased because the match of the scan lines with the perforated triads (in the alternative system) is no longer an issue.

triple module redundancy

A fail-safe method of configuring important equipment using three units — each of which is redundant.

When important equipment is duplicated with one in use and the other redundant, the backup device can be triggered to cut in if the first fails. However, detecting anything less than a major failure is extraordinarily difficult in complex devices. If one device is just providing the occasional miscalculation, this won't be detected just by checking its output against the redundant partner; you won't know which is wrong.

However if three duplicate sets of equipment are automatically monitored, one failing unit can always be identified since it will be the 'odd-man' when it begins to play up. It will be two-against-one. The faulty gear can therefore be shut down automatically and alarms sounded while one of the others takes over. The chance of any two devices suffering the same fault at the same time is so remote as to not be worth considering.

Some applications such as fly-by-wire aircraft controls cannot afford to fail, so two full parallel backup units will be provided at a minimum. See graceful degradation also.

triple-X

In X.25 packet-switching networks, these are the three essential ITU-T X standards — X.3, X.28 and X.29 used for dial-up asynchronous access. These are all access standards.

- X.32 is a later alternative to using the Triple-X protocols for dial-up two-way-calling synchronous X.25.
- Hayes also uses the term 'Triple-X PAD' to mean a PAD which implements X.32 (not X.3), X.28 and X.29.

Trojan horse

A Trojan horse is a virus-like program that may or may not replicate itself or cause damage. The term appears to be applied to two slightly different concepts:

- A program that purports to be an application (usually a game or utility), but which is basically designed to damage or disrupt the computer system. A Trojan horse can install a virus, but it is not a virus itself.

• A hidden algorithm which executes a function secretly and surreptitiously. A programmer with criminal intent may add a Trojan Horse to a mainframe operating system to store all entered identification codes and passwords at a specific memory location. Later he can retrieve and use these passwords.

tromboning

A call is made in London to reach an Australian visitor staying at a London hotel. The call is placed to his GSM mobile phone.

The London exchange switches it internationally to his home town, Sydney, Australia, where it gets diverted back through another long distance circuit to London.

It then reaches the mobile via the London cellular network. This is tromboning. It needs network intelligence, international agreements and standards, and flexible electronic directory structures to overcome these problems.

TRON

A Japanese designed computer architecture and operating system which was an attempt to establish a non US-dominated PC industry. Under US Government pressure, the Japanese have removed the preferences (in schools, mainly) given to the purchase of TRON equipment. The G-Micro group have developed three kinds of 32-bit TRON chips, however the architecture seems to be dead. It probably lives on in some of the video games.

tropo scatter

Radio waves scatter in the troposphere (above 50km), and this phenomena can therefore provide over-the-horizon radio links up to about 1000km. Tropo scatter techniques are widely used for telephone links to remote areas — to islands, for instance.

These systems use large dish antennas (typically 18 to 36 meters) and high power microwave circuits. Often 72 voice channels can be provided, but the system is subject to fading and atmospheric variations. See propagation mode and micrometeor burst also.

troposphere

The atmosphere above the Earth to a height of about 10km. It is below the ionosphere, and much more stable. See tropo scatter and ionosphere.

TRS

Trunked Radio System. The terms PAMR, PMR, SMR are also used here.

TrueImage

The page description language developed by Microsoft. It is associated with the outline font called TrueType, and is in competition with ATM/PostScript. It currently supports both Adobe PS Type 2 and TrueType fonts.

TrueType

Apple/Microsoft's joint effort at developing common outline fonts for IBM and Mac computers, and a page-description language for printers. The aim was to produce an alternative to PostScript and break Adobe's control of the *de facto* standard.

TrueType was once a direct competitor to Adobe's Type Manager (which is the core of PostScript), but the differences have since been settled and the two have merged. Apple and Microsoft have made some agreement with Adobe to ensure joint compatibility with PostScript, and Apple has announced that it will continue to support both font technologies.

• There is a parallel page description language for screens, called TrueImage which is in competition with Display PostScript. See multiple master, ATM and outline fonts.

• The original name for the fonts was Royal.

truncate

To shorten, by dropping off the end.

• Unless otherwise stated, to truncate a decimal number, you drop off the decimal fractions and only an integer remains.

• When you truncate binary numbers, you will most likely drop off the least-significant bits — but not necessarily. In error checking systems you often drop off the most significant bits.

trunk

• In telecommunications, this is a multichannel transmission bearer which interconnects switches (telephone exchanges) in the telephone network over long distances. These exchanges are variously called trunk exchanges, transit exchanges, or tandem exchanges, and the calls may be known as inter-LATA toll calls in the USA.

The term trunk is generally only applied to the connection when switching units are in different physical exchanges — hence it carries the implication of distance and of 'tolls' (premium charges).

• In LANs terminology, it means any high-capacity link between two switches or multiplexers.

• In Cable TV networks, the trunk is the main feeder cable, usually using a thick coaxial cable (or optical fiber in the new HCF systems). Coaxial trunks or street feeders will have trunk amplifiers every kilometer or so, depending on the load. Feeder cables will split from these trunks, and supply the drop feeds to individual houses.

• Trunking is the process of call-routing through a network — and this will involve intelligence and the need for route diversity. The routing of calls from a wireline exchange to a cellular base station, via the cellular switch and location register, is also trunking. This is the way the term is used in 'trunked radio'.

Trunk Connection Unit

See TCU.

trunked radio

Aka ESMR, PAMR or generically PMR or SMR. A form of mobile radio communications which is the direct successor of 'two-way' dispatch mobiles. It is now very closely related to cellular telephones, but it is designed for use by companies with large vehicle fleets — those with dispatch and control problems.

Trunked radio systems are used for both private and public-access communications. Emergency services also use special systems (see APOC). The largest of the shared public trunked networks, such as one in the UK, cover the whole of the British Isles with hundreds of base-stations.

In a trunked radio system many users share a common pool of channels which are allocated on demand for the duration of the call — but each company can make general 'fleet calls' and have direct contact with their drivers in the simplest possible way. Some operate only on a single channel (press-to-talk) while others provide a full-duplex phone-like channel (but perhaps of more limited bandwidth).

Modern automatic trunking use computerized switching to reduce waiting time and boost the efficient use of the bandwidth. They will allow call queuing, automatic user location and call routing as well as data transmission and direct dialing into the telephone network. Pre-formatted messaging is often included, and in some systems, status reports also.

The Philips-designed MPT1327 protocols are being widely accepted as the international standard, and the access method is a variation on Slotted Aloha.

Automatic trunking was developed as far back as 1964, so it predates cellular systems by seven years. Europe still uses the 410—430MHz PMR band for trunking although elsewhere 800MHz is popular. The key frequencies in use in much of the world seem to be 865—880MHz, or 815—825MHz.There is now both digital and analog trunked mobile radio. See TETRA, DSRR, APCO-25 and MIRS.

- Cellular phone systems use the sub-audible tones to signal when they should shift channels to prevent crosstalk, but most trunked radio systems don't have this sophistication, so they tend to be less effective at frequency-reuse. However, the boundary between a cellular system and a trunked radio system is now blurred.

- The value of the trunked approach is that many commercial users can share the scarce radio channels; normally (except for taxis) many mobile two-way radio channels are only used for minutes in every day.

trunk exchange

The non-US terminology for a 'toll exchange'. Trunk carries the implication of being able to handle signal bearers (with multiplexed channels) more than the idea of distance. So trunk exchanges don't necessarily only handle long-distance calls, but they do handle inter-changes between exchange switches, rather than having contact with local customers.

trunk junction

This is a multiplexed bearer between a trunk exchange and a local exchange.

trunkside

Trunkside is that side of the exchange connected to other exchanges. The distinction is with lineside.

truth tables

Small tables used in Boolean algebra to demonstrate the functions of logic gates. Two variables A and B are given the values true (T) or false (F), and all the combinations are calculated. There is a truth table associated with each of the primary binary Boolean operations (AND/NAND, OR/NOR, XOR/NXOR) and one with the unary operations (NOT).

TS

- Time Sharing.
- Test Structure.

TSAPI

A Telephone Switch API (similar to Microsoft's TAPI) written by Novell and AT&T for network servers. It is designed for multi-user network-oriented applications — those geared towards corporate LANs and large-scale applications.

The distinction is with TAPI which is designed for single users on the desktop. However both will provide APIs for computer-based answering of calls, transfer of calls, and for dialing preloaded numbers.

This is a third-party call-control system which allows any user to control any telephone call from any terminal (initiate conference calls, for instance). TAPI doesn't have third party call control.

TSAPI aims to link networked LANs with the features and functions of telephone PBXs. It includes a client/server API, a telephone-server NLM, and an open PBX-driver interface. Virtually all major telephone switch vendors have now undertaken to write TSAPI drivers.

- The key to future TAPI/TSAPI integration is a translation layer called TMAP which is to be provided free of charge by Northern Telecom and Intel. TMAP will allow a TAPI application to run on a TSAPI host, so that telephony software vendors don't then need to write two versions of an application. See CSTA and TAPI also.

TSOP

Thin small-outline package. A standard for surface-mounted components.

TSR

Terminate and Stay Resident. The MS-DOS users' term for desktop accessory programs. The name comes from the fact that they can run, stop and stay in RAM until they are called upon again. If you are pushing the limits on memory, keep the number of TSRs in your computer to a minimum since each seizes and occupies memory even when it is idle.

- These programs are also called memory-resident or RAM-resident, although both of these terms are wider than TSR.

TSS

Telecommunications Standardization Sector of the ITU. The name of the CCITT was changed initially to ITU-TSS, but now it has become ITU-T. See CCITT and ITU.

TST

Transparent Spanning Tree. See spanning tree algorithms and bridge.

TSTN

Triple-SuperTwist-Nematic LCD screens. These use three LCD panel/layers (Cyan, Magenta and Yellow) to reproduce good quality color by way of a polaroid subtractive process. Each layer subtracts color from the white (backlit) background. Essentially this is an electronic version of the color printing process — it is a light-gate. The distinction is with some other color LCD systems (usually active matrix) where the colors are additive. See passive matrix.

TT&C

Tracking, Telemetry and Control. Earth stations used to control communications satellites. Also called TTC&M (for 'monitoring') or just plain 'controlling earth stations'.

TTC&M station

In satellite terminology — a Tracking, Telemetry, Command and Monitoring station which controls the position of the satellite/s in orbit. Each satellite or satellite constellation needs to have at least one of these. The other earth stations can be for data only.

The aim these days, is to control a satellite within a box which is within +/– 0.1° (viewing angle from the ground) or within +/–40km of actual space in the Clarke orbit.

The latest Ku-band satellites are subject to even more strict controls. They must be maintained within +/– 0.05°, which is an almost imperceptible change of angle to all but the largest ground station.

The earth station will exert this control by firing thrusters about once a day over a lifetime of 10 years. See NCC and gateway also.

TTD

Temporary Text Delay. This is a two-byte control sequence of STX + ENQ which is sent when the sending terminal wants to retain the connection, but is not yet ready to transmit. It blocks the automatic time-out.

TTL

Transistor-Transistor Logic. See.

TTL monitors

A monitor which operates on digital signals using the standard internal computer operation's TTL logic format. The signals are carried to the monitor on lines which switch between 0 and +5 volts. You can buy monochrome TTL, or TTL RGB color monitors. If only two digital states (a pixel depth of six) are used to control each of the R, G or B elements, the monitor is often known as an RrGgBb monitor.

The distinction is with analog monitors where the drive states are infinitely variable — but only because the DAC is done in the driver, before the signal is sent to the monitor.

TTS

• **Teletype Text Service:** This was an old 6-bit character code used by wire-services for the transmission of text to newspapers. It was a variation of Baudot.

• **TeleTypeSetting.** See above.

• **Transaction Tracking Systems.** If a transaction is incomplete through failure in a network component, then the TTS forces the system to back out, thus ensuring the integrity of the database. See ACID test.

TTY

Teletypewriter. Usually applied to a terminal type and its emulation by a PC and communications software. This is the common form of dumb terminal, with only the basic ASCII Alphabet No. 5 set of characters. It is the terminal you emulate, when you're not emulating a specific terminal type.

• TTY terminal emulation does not have cursor control keys or any of the later sophistications developed with the ANSI VT-standards. If you can, always use VT-100 (or a higher number) emulation in preference to TTY.

tube

In the USA this tends to mean the cathode ray tube or screen of a TV set. In other parts of the world it carried the meaning of thermonic valve.

Tundra orbit

An inclined highly elliptical orbit used by Russian communications satellites. It has a 24 hour cycle and is inclined (crosses the equator) at 63.4°. When the satellite is at its nearest point (perigee) it is 25,000km above the Earth, and at its highest (apogee) it is 53,000km. This means that, unlike Molniya orbits, these satellites do not need to pass through the high radiation environment of the Van Allen belt.

The rapid movement of the satellite during the perigee phase (and a little in apogee) does cause Doppler shift of the frequencies, but much less than Molniya and LEOs. Tundra satellites have one apogee a day and if two satellites are provided, separated in space by 180°, a full-time coverage of the region high into the poles can be provided. See Molniya orbit also.

tunneling

• Passing something through another system. When you pass a packet from an Ethernet LAN through a frame relay interconnection to another Ethernet LAN, you are said to have tunneled through the frame relay cloud. See encapsulation.

• Tunneling electrons operate at the quantum level in chips — which means that they don't function according to the normal everyday rules of physics and electronics. They travel in an almost 'ballistic' way through the silicon, without bouncing off the internal atomic structures.

They therefore generate almost no heat and are probably capable of creating very fast-acting ICs. See STM, Josephson junction, SQUID and quantum effect transistors.

tuple

A row in the table of a relational database. It is a mathematical term from set theory. For instance, ('company', 16, ?x) is a tuple with three fields — two of these fields are values (one a variable), and one is a string.

turbo

— Mode: A processing chip can often run at a faster clock-speed than many applications can handle. So rather than advertise a computer with a fast clock-speed that 'can be slowed down' to be usable, the marketeers advertise one that runs at a usable speed and has a 'turbo mode'. The term originally referred to PCs and XTs that used the 8088-2 or the 8086-2 Intel chips which run at 8MHz rather than the normal 4.77MHz. It now also means systems upgraded by an accelerator (or turbo) card.

— Ball: A large trackball.

— Everything else: Marketing hype!

Turing machine

This was the imaginary machine that Turing devised to work out some of the ideas we now use in computers. His machine consisted of a reel of paper of infinite length, which rolled through the machine. The machine could perform only a limited number of operations on the paper. It could:

— Move the paper one space either way.

— Place a mark in one space.

— Erase a mark from one space.

— Halt.

With these limited functions, Turning showed how to build a 'thinking machine.'

In effect, he showed with this thought experiment that very complex problems could be solved by a machine using algorithms and data which were recorded as sequential bits of information in a linear store.

Turing test

This is the famous 1950 thought experiment which Alan Turing called the 'imitation game'. It seeks to define when a machine can be considered 'intelligent'. The scenario is this:

An interrogator sits in a room alone. He seeks to distinguish between the teletype feeds coming from a man, and another from the machine. He asks questions over his teletype, and both man and machine can reply to those questions in any way.

The question is: Can the interrogator recognize which is which? If he can't consistently distinguish the human from the machine, then Turing believed the machine to be intelligent.

• Note the machine only has to act intelligently, it doesn't have to be intelligent. Turing thought computer makers would win this game by the year 2000. In fact, ELISA, showed that it can be convincingly won by quite primitive machines, given public gullibility. See ELISA and Chinese room also.

turn-around time

• In half-duplex communications, this is the time taken to change over from the process of transmitting the file in one direction, to the transmission and reception of ACK or NAK reply packets in the other.

• In general business, it is the time from the placement of the order, to the delivery of the goods or services.

turnkey systems

This means that the contract calls for the supply, installation, provisioning and testing of the equipment, so it is ready to go without any further technical involvement from the customer.

Usually in computing, it is applied to dedicated mini computer systems set up and configured for only one job. It suggests that the user will only have to learn the minimum about the system — 'turn the key, and begin working'.

turtle

• In an unsorted list of records, some which should be at the top of the list will inevitably be found at the bottom. These require many passes with a bubble-sort routine before they reach their correct position, so they are known as turtles.

• Turtle graphics were used in Logo programming to teach kids to be aware of how computers are controlled.

TVRO

Television — Receive Only. Strictly it applies to any satellite ground station of any size, used only for receiving data, voice or television pictures from satellites. In practice it means a 'backyard' antenna system.

It is common for US amateur enthusiasts and people in rural areas to use small direct satellite receivers for DirecTV/DSS, and larger dishes for C-band systems.

The DSS dish is 18-inches in diameter and the C-band dishes are normally 10-ft (3 meter) in diameter and on steerable mounts. They can therefore look at a large number of satellites which are transmitting a wide variety of channels. Many are free, but it is possible to subscribe to some scrambled services. DirecTV has sold over 1 million dishes in the year 1994-95.

• The Cable companies will generally use 25-ft diameter C-band TVRO dishes to receive their signals, although many are now swapping over to digital MPEG, on Ku-band satellites.

twin cable

The basic form of telephone cable, called twisted-pair. See.

twinaxial cable

A type of coaxial cable with two center conductors and the outer sheath.

twisted-pair

The term applied to the standard telephone wiring cable which uses two wires with an impedance of about 600 ohms. The term actually applies to both shielded and unshielded varieties used in LAN connection, and to the bundles of telephone wires used in street ducting.

Don't confuse twisted-pair with quad, which is a cheap form of domestic-feeder cable with two-pair of wires enclosed within a thin plastic sheath. The telephone cables down the street are most likely twisted-pair, but the link from the last pillar to your home is probably quad. LAN links use true twisted-pair.

With twisted-pair, the wires rotate around each other in a controlled way to provide some protection from interference and crosstalk, but the cable is still susceptible to some crosstalk, noise pick-up and capacitance problems with other wires in the bundle.

Twisted-pair was commonly thought to have a bandwidth limit of about 100kHz over distances up to 4km. But views have changed:

- For telephone use, twisted-pair now carry ISDN bit-rates of 198kb/s (with all overheads) for up to 2km — and more with repeaters.
- Some twisted-pair cables are used to carry multiplexed analog signals at bandwidths of over 1MHz between trunk repeaters spaced at about 2km.
- With ADSL, twisted-pair capabilities have been pushed into the 6Mb/s range for distances over 2km for video-over-copper. Reliability is doubtful, however.

• For data network use there are some required standards for twisted-pair. See PDS, IBM Type 3 and 10Base-T also.

• Old phone links on poles were single-wire (see SWER).

twisted-nematic

Unless otherwise specified, this means the single twisted-nematic screens which were very popular for LCD displays in portables until 1986. These screens put the light through a 90-degree twist (a single twist) which produces a relatively low-contrast display.

Single twisted-nematic is a relatively cheap technology to make, and the screens are light on power requirements (usually under 1 watt). However they have a very narrow viewing angle and low contrast ratio, which makes them difficult to read. See LCD and Supertwist LCD.

twitter

See interframe flicker.

two-dimensional parity

This is the use of vertical redundancy checking with longitudinal redundancy checking. See VRC and LRC.

two-phase commit

A technique used in distributed database management operations, to ensure that a transaction which updates more than a single database does so in full, or not at all.

The component files of a relational database need to be updated in an all-or-none way; either every related record in every associated file is updated, or the transaction must fail completely and obviously. See ACID test.

two's complement

A way of changing a binary number to a negative form (usually for subtraction). Each bit in the number is negated (1s become 0s, and vice versa) then you add 1 to the result.

two-way

In communications the term is totally confusing:

— 'Two-way simultaneous' refers to full-duplex operations.

— 'Two-way alternate' refers to half-duplex operations.

But 'two-way' can mean anything.

• Two-way radio is a half-duplex operation, where a single channel is shared for both directions (press-to-talk).

two-wire

See wire.

TWT

Traveling Wave Tube. See.

TWTA

Traveling Wave Tube Amplifier. See traveling wave tube.

TWX

A public messaging service once used in the USA in competition to Telex. See Telex.

Tx

Transmission or transmitter. In RS-232, this is the pin and line used for transmission. Rx is the receive line.

TXL

A high-speed cellular mobile data protocol very like ETC.

Tymnet

An old value-added network service company in the USA which operates a widely used packet-switching network. It was later owned by British Telecom. Tymnet used proprietary protocols for their packet-switching network which were very similar to X.25, but not exactly the same. It was connectionless at one time.

typamatic

A type of keyboard (used by the IBM PC) which sends one code to the computer when pressed and a second when released.

type

Originally a small piece of special cast metal with raised characters. Nowadays, the term usually refers to a typeface or what is often called in computing, a 'font'.

Type #

• Typefaces. See ATM and PostScript.

• IBM Cabling. See IBM cabling system.

• IEEE 802.2 LLC protocols. See.

type codes

• In the Macintosh, a type code is a four character mnemonic which is used to identify the file-type. For instance 'INIT' is a startup document, 'APPL' is an application, 'FONT' is a typeface. These codes are normally hidden from the user, but they are the Apple equivalent of MS-DOS and CP/M filename-extensions. See also creator codes.

• In network connections. See LLC.

typeface

There are different ways to classify types (fonts). One is to divide them into:

— Oldstyle Garamond, Caslon,
— Modern Times Roman, Caldonia
— Square Serif Clarendon, Cairo,
— Sans Serif Helvetica, Futura,
— Script, Text Letters and Decorative Types.

Another way is to treat them as three basic typeface groups:

— Body copy type, which is generally either serif or sans serif.
— Diplay typefaces, which are used for titles, headlines and for novelty or emphasis.
— Ornamental — used for decoration.

Don't confuse typeface with typestyle. Typeface refers to the whole group of characters (sizes and styles) which are variations on the same basic design.

These are often incorrectly called 'fonts' in computer graphics and word processing, but the name has stuck so we need to live with it. Typestyle refers to those variations within a typeface — italics, roman, bold, expanded, condensed, etc.

Here are some standard type examples, all at the same point-size (10 point). Note the length of the lines:

- This is **Garamond Narrow** which is used throughout this book. It is not as easy to read as some of the Bookman styles, but it can compress more words into a page.
- This is **Times Roman**, which is a classic typeface, but only designed in 1932. It is probably the most widely used type in the world — especially for newspapers.
- This is **Helvetica.** This sans-serif style was known as grotesque. It has been popular since 1957, and it is widely used for books.
- This is **Bookman,** which is a very readable type, and popular for children's books.
- This is **New Century Bookman** which is very common in book publishing.
- This is **Courier** . This type is not proportionally spaced, and therefore deliberately looks like a typewriter output.
- This is **Old English Text** which is now only used as a display or feature font.
- *This is Swing, which is a cursive style font, which looks very much like handwriting.*
- These are Zapf Dingbats — a collection of symbols which are treated as fonts. ✳✼❄▲ ❁❒❉ ✺✈✉❃✕✄ ✼❆ ✤✻■✳ ❂❁❞ ❝✼ ✍☞ ✿✚☛ ✖✆▼▲

typestyle

The variations within a typeface — italics, roman, bold, etc. See typeface.

U-interface

The U-interface in ISDN provides a test point for the two-wire circuit of the local loop. It is the interface between the ISDN exchange and the network terminating unit, NT1. The main job of NT1 is to convert the four-wire S- or T-interface into a two-wire U-interface. See NT#, R-, T-, and S-interface also.

u-Law

Actually the Greek μ-Law — see mu-Law and A-Law also.

U-matic

A videocassette standard devised by Sony. In designing this standard Sony broke the original EIAJ standards agreement, but U-matic proved to be so popular that most Japanese videocassette recorder manufacturers took it up as a *de facto* standard. U-matic uses 3/4 inch tape in a side-by-side cassette, and it became widely used in universities, colleges, etc., with the early 'low band' recording systems. Low-band U-matic recorders are no longer manufactured —VHS has replaced them.

The later development of the (largely incompatible) hi-band form, made U-matic the mainstay of TV news and small country stations until cheaper Betacam and S-VHS systems came along.

UA

User Agent in the X.400 Message Handling System.

UAA

Universal Agent Architecture. A standard for 'wandering' communications and personal agents on the Internet.These agents are sent out into the network to do your bidding, but many hosts see agents as potential viruses and so reject them.

Marc Belgrave of Bell Northern Labs has developed UAA as an open standard. It identifies the agents to hosts and allows them to cooperate without the agent being treated as a virus. See Telescript also.

UART

Universal Asynchronous Receiver/Transmitter. The UART is an integrated circuit which is essential to computer communications: in fact, it was probably the first LSI device to be widely manufactured. It takes the parallel signals from the computer's internal bus and creates the serial output at the serial port through use of a 'shift register', and can simultaneously perform the reverse action.

It is the key asynchronous device for communications, and it performs several important functions:

— It takes a parallel input from the data bus, buffers the bytes, and then outputs the data in serial form.
— It adds start- and stop-bits to each byte of data.
— It checks and sets parity.
— It handles the assertation and inhibition of control lines.
— It sends data in serial form.
— It receives data and handles CPU interrupts.
— It strips start- and stop-bits from each incoming byte.
— It convert serials bytes into parallel form for the data bus.

You will find a UART behind any RS-232 or MIDI interface — where parallel data is being converted to serial, or vice versa.

The problems with UARTs arise from the fact that the processor must be interrupted 240 times a second to handle data through the UART at 2.4kb/s (async bytes are 10-bits). This was relatively easy in the early PCs, but few UARTs could cope with 960 interrupts a second for 9.6kb/s serial rates. Also characters began to get lost because the small buffers of these early UARTs would overflow.

So the original 8250 UART chips were later changed to 16450 UARTs, but these new chip families still only had single-byte buffers.

In recent years, the new 16550A UARTs (with two 16-Byte buffers) have solved some of the problems. However, with the new modem data rates and compression schemes, even these are now well behind the modem requirements.

A modem running on top-quality lines at 28.8kb/s with 4:1 compression (only possible with, say, an assembly language file) will need a UART feed of 144kb/s (in 10-bit bytes) which is well beyond current PCs. See Hayes ESP, USRT and USART.

UCSD

University of California at San Diego, which gave its name to both the widely used form of Pascal, and an operating system (P-system) based on Pascal principle. See P-system.

UCT

Universal Coordinated Time. A widely used, but wrong abbreviation for the international time-scale which replaced Greenwich Mean Time (GMT) in 1928. See UT and UTC.

UDI

Unrestricted Digital Information. In the ISDN world this is also called 64U, 64-Clear or 64C. There is also RDI (Restricted Digital Information, which is also called 64-Restricted or 64R or 64I). Together, these two request 64kb/s ISDN data calls.

This terminology seems to be tied to the North American practice of providing 56kb/s data channels over 64kb/s ISDN B-channels. Both restricted and unrestricted services use V.110 rate adaptation (bit stuffing into the 8th bit).

UDP

User Datagram Protocol. This is normally part of TCP/IP which is handled at the IP layer. It's a Transport layer protocol on the Internet. The protocol describes how the messages will reach an application within a destination computer and it adds a level of reliability and multiplexing to IP datagrams. UDP is the connectionless version of TCP.

- UDP/IP is a faster but less reliable set of network and internetwork protocols than TCP/IP.

UDTV
Ultra-Definition TV. A 1993 Japanese Ministry of Post and Telecommunications project for a super-fidelity, super-resolution digital television transmission system which can provide an image quality comparable to 70mm motion picture film.

This is to be used for the transmission of library images and high-resolution graphics including CAD/CAM and videotex, and for electronic cinemas.

The funding of the UDTV project is estimated at $US100m, and the design work is anticipated to take seven years. The quoted bandwidth requirement is 2.5—3GBytes/sec, which seems high by current standards.

The key domestic application is 'personalized television' which allows viewers to select motion pictures, museum displays, or other exhibitions-on-demand.

UHF
Ultra High Frequency. The part of the electromagnetic spectrum between 300MHz and 3GHz used for television transmission (among other things). These are decimetric waves from one meter to 10-cm in wavelength.
The main uses for this part of the spectrum is:

— Military satellites 300MHz.
— Land mobile 403—520MHz.
— Cellular telephones 450 and 800MHz.
— Television 470—890MHz.
— Microwave distribution 2GHz.
— Microwave ovens 2.45GHz.
— Electronic news gathering 2.5GHz.

Part of the normal TV spectrum between 620GHz and 790GHz has been made available for dual use. It can be used both for terrestrial television, and for direct broadcasting satellites.

Despite popular opinion, UHF television is not just line-of-sight, but, at the higher MDS frequencies, it is getting reasonably close.

UI

- User Interface — see CUI and GUI.
- Unix International.

UL
Underwriter's Laboratory (US) cable standards. See Category # cable.

ULA
Uncommitted Logic Array. This is a gate array or PLA, before burning.

uLaw
Actually the Greek μLaw. See mu-Law and A-Law.

ULP
Upper Layer Protocol. This refers to any layers which are higher than the OSI layer being discussed at the time. ULP is often used as a general term to signify nothing more than the next layers higher in the stack.

ULSI
Ultra Large Scale Integration. This is the next stage past VLSI. ULSI chips have more than 100,000 components, and feature separation as small as 0.2 microns. We are just entering this level of miniaturization.

USLI calls for processing-precision down to several tens of nanometers, and thus it represents the first step into the possibilities of manufacturing chips at nanometer scales. This order of device miniaturization will allow the development of one gigabit (one billion bits) RAM chips, and immense-scale ATM switches for telecommunications.

Intel and some of the Japanese manufacturers are already showing experimental processors and memory at this level.

Ultra SCSI
This is a new higher speed version of SCSI which became available in mid-1995.

It was devised by the ANSI XT310 committee to double a disk drive's burst-data transfer rate while requiring only minor changes in firmware and physical connections. It has a buffer-to-host data rate of 40MByte/s in its 16-bit implementation, and half of this in an 8-bit implementation.

Ultra SCSI uses the same connectors as Fast SCSI, and it is a reasonably low-cost upgrade to the older system.

Ultra VGA
A term sometimes applied to Super VGA with non-interlace monitors. See Super VGA.

ultrafiche
A special microfiche which can hold about 1000 documents on a single film negative (about 100 x 150mm).

ultrasonics
Audio frequencies above 20,000Hz — the generally-accepted limit of human hearing. These can be longitudinal (in air) or transverse waves (in solids). Ultrasonic waves propagate through the medium or along the surface of solids (Rayleigh waves), or as Lamb waves through thin rods.

UltraSPARC
Sun has been falling behind in RISC for a number of years after having pioneered the technology with SPARC. UltraSPARC is a high-end 64-bit chip that could put them back in the race.

It runs about five times as fast as the 32-bit SuperSPARC chip (which, at 60MHz, was slower than a Pentium), and it is being manufactured by Texas Instruments. The new SPARC chips all compensate for their inherent slowness by using symmetrical multi-processing (SMP).

UltraSPARC can support video data-streams, and it has a SPECint92 rating of 200—400. See SPARC also.

ultraviolet light
UV. The boundaries of UV light are generally taken to be 390nm (the end of the visible spectrum) and about 10nm, (where it abuts X-rays). However some would limit the

definition of UV to wavelengths of about 100nm while others would include those frequencies up to the obvious X-ray limit. The extent of the main UV subdivisions are:

End of Violet — 390nm.
Near UV — 390 – 300nm.
Far UV — 300 – 200nm.
Extreme UV — 200 – 100nm.
X-rays from 10nm.

Ultrix

DECs version of AT&T's Unix operating system for the VAX and PDP-11 minicomputers. It hasn't been highly successful.

UMA

Upper Memory Area in MS-DOS. This is the memory space between 640kB and 1MB. Most of this space is used by system hardware such as the display adaptor and system BIOS. Usually this part of memory will be allocated in the following way:

640—768kBytes — Video RAM.
768—800kBytes — Video ROM.
(UMB 896—960kBytes, used by EMS and XMS).
960—1024kBytes — System BIOS.

With 386 and 486 computers the UMBs can also be used for device drivers and TSR utilities. See conventional memory.

UMB

Upper Memory Blocks. These are unused blocks of memory space in the Upper Memory Area above 640kB and below 1MB — often 200kB of memory space can be found here. Your hard disk and each plug-in peripheral card must also be allocated part of this UMB space.

It is the job of the memory manager (within DOS or a utility) to ferret out unused UMBs and make them available for utilities, TSRs, etc. This leaves more conventional RAM free.

Information is then written into AUTOEXEC.BAT, CONFIG.-SYS and Windows' SYSTEM.INI so that these blocks are re-established and made available after a re-boot.

umbrella cells

These are cells in a cellular phone system which cover a very wide area, and may have a number of much smaller cells within their coverage range. In some of the new satellite systems, the idea is to provide umbrella coverage via the satellite, to fill any holes left by the smaller terrestrial cells. The NMT cellular standard and some implementations of GSM also use umbrella cells.

UMTS

Universal Mobile Telecommunications System. The original aim of UMTS was to provide a standardized environment for mobile cellular, cordless, paging and private mobile services to a large number of subscribers. It was a European RACE project directed towards producing a single communicator which could be used in all environments. It was to support GSM, DECT and ISDN services as well, and have a network component based on CCS#7.

Later these ideas were extended to include video-telephony, wide-area paging, and possibly videoconferencing, so it became a dream-list rather than a practical plan. The availability date was pushed back until 'well after 2000'. It is now impossible to distinguish UMTS from FPLMTS, and it shares many characteristics with UPT. See FPLMTS and UPT.

unary

A single entity. A single operation. A single change (in volts).

unbalanced

A term that is used in a couple of different ways:

• Two-wire **audio** and **data** circuits are said to be 'balanced' when each of the wires carries a signal of opposite polarity, and the voltage of each differs equally from earth potential. These are generally twin cables, with both conductors surrounded by a metal earth-potential sheath.

As one conductor moves into its positive half-cycle, the other will be mirroring these changes in a negative half-cycle — both relative to earth. Any EMI that affects one, equally affects the other, and the two should balance out.

Unbalanced circuits usually have only one inner cable, with the return path through the outer metal sheath. This is satisfactory for short distances, but it is subject to interference and capacitance problems if the cable length is greater than a few meters. The unbalanced signal always consists of voltage changes on the wire relative to some 'signal ground' — which may be at 'earth ground'. See ground loop.

• Four-wire **telephone** circuits are also called 'single-ended' or 'unbalanced', because voice/data are carried over one pair in only one direction. This is an old 'wire' term left over from the days when channels were electrical end-to-end connections. The more correct term is half-duplex.

• A **line-code** transmission system which sends more one-bits than zero-bits (or vice versa, depending on the system). This can create capacitance and coupling problems because the sum of the extra bits is represented as a DC component on the line. These lines are said to be unbalanced.

• The **RS-232** standard used for most computer and communications equipment (for short distance connections only) is said to be unbalanced. Although there are multiple wires within the shielding, and a single (Pin 7) common return path, there is only one wire for each transmission or receiver channel. See RS-422 and RS-449.

• In **HDLC**, this is a network configuration with one primary node and multiple secondary stations.

unbundling

Breaking the component parts out of a previously 'bundled' package of products or services and charging separately for each. This much is reasonably obvious; but less obvious is that the software companies often mean they are no longer willing to provide support in the form of a free help desk with the product they are selling.

You'll purchase support services separately. This would be conscionable if the software came bug-free and then was only released after careful testing with a wide range of operating systems, utilities, other applications and hardware. But most software companies use customers to test their products for them.

The other way to look at this, is that perhaps we should each receive a credit voucher against our next purchase every time we report a legitimate problem to the help desk.

UNC

Universal Naming Convention.

undelete

Files that have been 'deleted' or 'trashed' still remain intact on the disk for some considerable time, so they can be undeleted with the correct software. The quicker, the better, because later file changes may overwrite these disk sectors.

The 'delete file' function simply changes a few bytes of directory in the system area to tell the drive controller that the space is available again — it does not actually erase a disk as does a format command.

undocumented features

Writers of operating systems and applications often leave some features undocumented — either accidentally, or deliberately.

There is often a suspicion that sometimes these features are deliberately undocumented because the originating company wants to gain benefit from easier applications programming.

But there's also the argument that OS programmers may not want the feature used for legitimate reasons: future versions of the same program/system may discard them. And, sometimes they have been included but not fully tested.

However there's little doubt that features are sometimes undocumented because the company wishes to put obstacles in the way of other software companies — and to give itself a competitive advantage. See API.

unerase

See undelete.

UNGTDI

United Nations Guidelines for Trade Data Interchange. The basis for EDIFACT.

UNI

User-to-Network Interface. This is also used as a general term. More particularly, this is an implementation agreement which specifies an interface between the end-user device and a public or private ATM network. The key aspect of UNI is in providing a signaling protocol for call set-up and management with switched networks. See NNI also.

- In ATM this refers to standard ways of designing access devices such as routers and adaptor cards so they can connect to the switch. The 1992 UNI specification for ATM only defined permanent circuit connections (PVCs), not switched (SVCs). The 1993 UNI version 3.0 included SVCs.

unicast

The obverse of broadcast. It is a message which specifies only one network address.

Unicode

An alternative to both the ASCII No.5 code alphabet and the new ISO 10646 proposal for a standardized code capable of handling most of the world's languages. The alternate ISO 10646 proposal was for a 4-Byte code capable of storing all Japanese, Chinese, Korean, Burmese, characters/ideographs, etc. However most people felt that 4-Bytes for each character was excessive.

Apple and Xerox PARC developed the Unicode 2-Byte per character version and now most US computer manufacturers have fallen in behind this standard. Unicode can handle Chinese Han characters and most Asian languages, but it doesn't extend to the more exotic scripts like Burmese.

unidirectional microphone

This is a microphone which has its greatest sensitivity in a given direction. If the direction is 'uni', but the angle of acceptance is fairly wide, it will probably be known as a cardioid mike. If the angle of acceptance is extremely narrow, it will probably be called a shot-gun or rifle mike. The distinction is with omnidirectional (all) and bi-directional mikes.

UniForum

A worldwide association of Unix users who started the move to get Unix standardized. They handed the job over to the IEEE.

uninstalling

While installing a program in Windows is usually a breeze, uninstalling one can be a nightmare. Deleting the icon from the Program Manager does almost nothing — the main program remains behind, perhaps wreaking havoc on other applications. And the program will have files hidden away in all sorts of places. There's a essential program-suite called Uninstaller.

uninterruptible power supplies

See UPS.

unipolar

- **Communications:** Signals having only one voltage — the other is zero. A signal which pulses at +3V for a logical 1 (mark), and at no voltage for a logical 0 (space), is a unipolar signal. This technique has a significant failing in that it produces a DC voltage on the line (See DC coupling).

Bipolar or so-called 'polar' signals will use +3V for the logical 1, and –3V for the logical 0 — or some similar two-sided approach which removes the DC component and effectively replaces it with a square-wave AC form. This can pass through transformers, capacitors and the like. See AMI.

- **Magnetic material:** Magnetic domains on tape or disk usually have both north and south-poles (bi-polar), but with vertical recording techniques, the surface may only detect one.
- **Transistors:** These are semiconductors which are formed

from n-type material or p-type material, not both.

unique identifier

An identification code burned into ROM in a plug-and-play device. This is often called a fingerprint. See.

uniselector

A line finder in an older telephone exchange. When you lift your handset before dialing, a line-finder must first find a vacant switch before you receive dial-tone. You can often hear them clicking away in the background trying to locate one.

unit buffered

This effectively means 'non-buffered'. It can only handle one bit at a time.

unit cost

This is a charging mechanism for telephone calls, which often uses metered pulses (but need not). Rather than charging on a per-second or per-call basis, the carrier will establish a fixed charge per unit, and then vary the length of the unit depending on the distance of the link and the time of day.

The link between metered pulses and unit costs is not rigid. For instance, the UK system used by British Telecom sends 10 pulses down the line for each increment on the meter, and it adds an extra pulse at the time of connection.

So while the call is being metered in 30 second units, the call-charging increments are actually only 1/10th of this. However, BT's competitors make much of charging by the second.

unit system

In printing, this is the way of measuring line-length, etc. by typesize rather than absolute measure. You use this when you ask for a one-Em indent. The Em, has the dimensions of a theoretical box, the square of the point-size.

unity gain

In broadband networks, this is the balance through amplifiers, between signal loss and signal gain.

universal service

The philosophy which says the telephone is now so fundamental to our society that it should be available equally to everyone.

In other words, it is believed that phones are important enough to the maintenance of a functioning society (and for reasons of social equity) that pure market forces are not adequate. The disadvantaged should be subsidized by the advantaged in the society.

This has led to the concept of Universal Service Obligations or Community Service Obligations. See CSO.

• Universal service concepts first arose in 1913 in a US Federal antitrust suit against AT&T. As part of that settlement AT&T guaranteed that anyone in any community who wanted a telephone would be able to have one.

• Part of the national information infrastructure (NII) debate in the USA and elsewhere, is whether this concept of USO should extend to providing higher-grade information and electronic education services, in addition to basic telephony.

Unix

A family of true multi-user, multi-tasking operating systems, originally developed at the Bell (AT&T) Laboratories in the 1969 for the PDP-7 minicomputer. Unix was written by Ken Thompson and Dennis Riche, and first released commercially on 3 Nov 1971. It came with a Basic interpreter.

It proved to be very popular with computer scientists because it provided a range of programming tools, and this created a productive environment for the development of software by teams of programmers.

The current AT&T version is called Unix V, Release 4 (V.4), but 80% of the world Unix market is shared between Unix V, Release 3 (V.3), Berkeley University's BSD 4.3, Sun OS and Xenix. See BSD and Xenix.

The major value of Unix is its portability and flexibility, which is due largely to it being written in C language, however there is little true portability across the many versions. As an installed operating system, it is third behind MS-DOS and Apple's MacOS, and ahead of OS/2.

The V.4 version has added POSIX 1003.1 and XPG3, but it does not support mirrored disks or OSF/Motif. See POSIX, Motif, OSF/1, SCO and SVID.

• The key features behind the success of Unix are:
- A modular, structured design. It is designed around a small number of subroutines.
- The operations are based on a small number of consistent principles; there is little arbitrary detail.
- The system is extensible. Unix provides a shell and programming language.
- I/O simplicity and redirection. Filters are powerful tools.
- It used a hierarchical directory structure at a time when this was unusual.
- It was small enough and good enough for the minicomputers of the day, and these were mainly in universities. The message spread with their graduates.

• AT&T's major 1984 revision of Unix established System V, which incorporated all the important features added in previous versions. The SVID (System V Interface Definition) contains the specifications needed for compatibility.

• XVS is the slightly modified alternative standard put up by the X/Open group. See ABI and Unix International.

Unix International

The Unix Wars are now over, but UI was the AT&T sponsored non-profit group (including Sun, NCR, Unisys, Motorola and about 130 other vendors) standardizing and promoting Unix System V and OpenLook in opposition to the Open Systems Foundation's OSF/1 and Motif.

The UI laboratory was called USL (see). USL Unix had a kernel called SVID and they were attempting to unite Unix at the Applications Binary Interface (ABI) level.

The old rivalry between UI and the OSF group has now died, and UI with it. In April 1993 the USL and OSF publicly agreed on a range of common standards, and they also agreed to conform to each other's published interface specifications. See Posix.

unjustified

Without line justification. The length of the lines of text have not been stretched to provide even line length. The copy in this book is justified. See ragged right.

unmeasured lines

See measured lines.

unmovable blocks

These are blocks of data which the operating system expects to find at a certain location on a disk. For instance, the DOS system files (directories, FAT, etc.) must always be in a certain track and sector. Defragmentation programs sometimes identify these blocks as problem areas, but they aren't.

unnumbered frames

HDLC uses this type of frame for starting up a link, closing it down, specifying modes, and for general housekeeping.

unrouteable protocols

SNA and NetBIOS traffic. These protocols need special source-route bridging, DLSw and spoofing techniques. See.

unstructured languages

Basic, Fortran and Cobol languages and programming techniques are considered to be unstructured, while C, Pascal and dBase are generally regarded as structured languages. See structured programming.

UNTDI

See EDIFACT.

UPC

Universal Product Code. This is a standard barcode which is printed on retail merchandise. By scanning this code you can read the vendor's identification and product number which jointly identify the product. UPC was first adopted as European Article Number (EAN) in 1974, then as Japan Article Number (JAN) a few years later. The most common form has 13 digits:

— The first two are the national code.

— The next five are the manufacturer's code.

— The next five are the product code.

— The last is a parity check.

upbanding

In VCRs and radio cellular systems, this just means shifting to a higher frequency.

• VCRs are upbanded by increasing the bandwidth which can be recorded and thereby extending the resolution and the color separation. This produces hi-band recording systems.

The luminance is still at baseband, but the chroma has been modulated to a higher frequency which keeps it well out of the way of the luminance signal.

Both signals also have a wider bandwidth, and are therefore capable of producing higher resolution. See S-VHS and Hi-8.

• CDMA and TDMA cellular phones are up-banded from their original 900MHz frequency band to the new PCS/PCN bands of 1.8 and 1.9GHz.

uplink

In satellite communications, this is the signal from the earth station to the satellite transponder. The distinction is with downlink. Don't confuse up- and downlinks with forward and backward channels; over a satellite, these will always have one uplink segment and one downlink segment.

• With C-band, the frequency difference between the uplink signal and its corresponding downlink signal (duplex partner) is always 2.2GHz, with the uplink at the higher frequency.

uploading

This is the process of transferring data from your computer to another. Downloading is when the data file comes to you.

You usually upload from a PC to a mainframe host. If it is from one PC to another, you would probably call this a transfer.

upper case

Capital letters. This also would include the special form which is often listed in DTP as Small-Caps.

upper memory

This is the now common name for the area of DOS-controlled memory between 640kBytes and 1MByte which isn't available (normally) for applications or data. At one time this was called high-memory, but the term 'Hi-mem' was later appropriated for that area above 1MByte.

Upper Memory Area (UMA) is reserved for video RAM, BIOS, and for the drivers which run the system hardware. However there has always been a couple of segments which were not used by the system, and which became known as the Upper Memory Blocks (UMB). These are used for bank-switching.

Later generations of MS-DOS and chips running in protected modes have use UMB to provided work-arounds to bypass the old limitations. Now with DOS 5.0, it has become possible to also load small MS-DOS programs into this area, thus freeing up more general RAM. See UMA and UMB.

upright fonts

Roman letters as distinct from sloping italics and oblique fonts.

UPS

Uninterruptible Power Supply. A battery-backed power supply that will sustain the computer for a short time if the power fails. A good UPS will allow an orderly shut-down of the system after a few seconds without mains power.

The two main types of UPS are:

— **Reverse transfer (on-line) systems.** These are on-line systems which provide power through a DC/AC inverter, which is itself powered from a mains-driven rectifier/charger and battery supply. The battery continues to supply power through the inverter when the mains supply fails.

This approach means that the system experiences no break during change-over, since the load is only ever powered from the inverter. The supply is also totally isolated from mains fluctuations and mains-borne noise.

— **Forward transfer (off-line) systems.** These are 'standby' systems which normally draw load directly from the mains, but which will switch in and provide the output from a battery-driven inverter in the event of mains failure. This approach is cheaper. There are 'active' systems where the inverter is synchronized with the mains and runs continuously to give fast, smooth change-overs, and 'cold' systems which take one or two seconds to switch on.

Cold systems are generally only used for emergency lighting, etc. When power drops below a preset level the UPS needs to cut in quickly if the computer memory is to be saved, A good UPS will take-over in 1ms; more than 6ms can cause problems.

• Note that Bell Laboratory says the main problems with computers and power is under-voltage, not loss of voltage.

• Reverse transfer systems are the most common in large installations because they completely separate the operating load from utility power. See inverters also.

upstream

This means 'in the direction opposed to the main data flow'. It usually refers to the return or back-channel used for flow control, ACKs, and supervision.

UPT

Universal Personal Telecommunications. A Utopian 'third generation' in the international telecommunications system where everyone will be issued with a personal telephone number (which may last them for life). This is currently being discussed in the ITU and ETSI, and there are some recommendations in the pipeline.

The standards-definition process is directed mainly towards extending the functions of the present PSTN, ISDN and GSM networks. These recommendations were promised by mid-1993, with early services to begin in 1995, and advanced services in 1997. But the ITU is dragging its feet.

There are four major features of a UPT network:

— **Personal access numbers** (not terminal numbers) which are location-independent.

— **Personal service** profiles which are user-definable and determine how incoming and outgoing calls are to be handled (at different times of the day, different days, call-diversions and call forwarding, etc.).

— **Ubiquitous services.** PSTN, ISDN, B-ISDN, mobile (GSM) or packet-switching, must all be treated as one.

— **Personal charging** and billing systems which are independent of the terminal used.

Identification and authentication of users will be of prime importance in UPT, and these functions must carry across different networks and national boundaries. Both PINs (over DTMF phones) and smartcards are being considered, and there was some thought of designing a special small UPT 'key', which was a smartcard with keypad, display and DTMF tone generator.

Mobile links are an intrinsic part of UPT, and ETSI sought frequency allocations for UMTS/FPLMTS from WARC in 1992. The ETSI has a special section looking at UPT specifications, and the US is studying the ideas in ANSI's T1P1 subcommittee.

The proposed services extend from narrow-band paging up to broadband services.

• One recent definition of UPT is 'the ability to allow users to direct their calls through any terminal they choose, to make outgoing calls from any terminal, and have it charged to their own account'. UPT has also been defined as 'a service where the user can be reached as well as make calls through any telecommunications network'.

See PCN, UMTS, FPLMTS, B-ISDN and IBCN also.

upwardly compatible

This term is applied mainly to software and operating systems. If a program will work with later versions of an operating system than the one for which it was written, then it is 'upwardly' compatible. If it works with older versions, then it is 'backwardly' compatible.

URL

Uniform Resource Locator. A standard used on the Internet 'for specifying an object which can be accessed by any of the standard protocols (WWW, Gopher, Telnet, etc.)'. Translated, this means nothing more than the full WWW address.

This is the standard way to refer to resources. It specifies both the type of service and the location of the file or directory.

A URL is a long address sequence such as:

```
http://info.cern.ch/hypertext/DataSources
        /bySubject/Overview.html
```

US

ASCII No.5 — Unit Separator (Decimal 31, Hex $1F, Control-_). An information separator also known as IS1.

• The four information separators (FS, GS, RS, US) are used in a variety of ways, but they should always be used in a hierarchical or nested manner, with the FS being the most inclusive, and the US the least inclusive.

USART

Universal Synchronous/Asynchronous Receiver/Transmitter. This is a semi-conductor chip designed to provide both synchronous and asynchronous communications support at any serial port.

Previously, these chips would only support asynchronous communications, and so were known as UARTs (a name that has stuck, despite the change).

In fact most modern UARTs chips behind the serial ports in most personal computers, modems and printers, are actually USARTs these days, and are therefore capable of handling either asynchronous or synchronous data.

- The Intel 8251 handles async data at rates between 110b/s and 19.2kb/s, and synchronous data up to 56kb/s. It would need to handle async data at 144kb/s to service a V.34 modem operating at 28.8kb/s with a file which is compressed at a 4:1 ratio (which is rare, but possible). See UART and USRT also.

USASCII

This usually means the US Extended ASCII — the 8-bit version of ASCII used by MS-DOS machines. See EBCDIC also.

USAT

Ultra-Small Aperture Terminals. See VVSAT.

USC

A US standard for EDI (document exchange). See ANSI X12.

USDC

Yet another acronym for the North American TDMA digital cellular technology. It is also called Digital AMPS, and IS-54. In this book you'll find details under NADC and IS-54.

usec

Actually the Greek 'mu' (micro); or millionth of a second.

Usenet

A feature of the Internet devoted to worldwide discussion groups (keyboard conferences) and the exchange of 'news' (in the sense of up-to-date information).

Usenet is a cooperative network which began in 1979 and now has over 11,000 hosts and millions of users. Usenet files flow around the world in a store-and-forward way, creating a global community. These are all keyboard forums, and there is no central source, repository, or control of distribution. If you contribute to a Usenet forum, your message will be read by thousands of people.

Not all Internet hosts will subscribe to all 11,000 Usenet groups ('subscribe' means to have the updated data automatically sent) and not all Usenet groups are on the Internet list. However most are.

Most shell connections to the Internet will offer e-mail and Usenet groups (you specify which ones you want). E-mail is sent to your address and stored in the host for you, but you need to actively go into a Usenet forum and read the updated information. Updates are only sent as far as your news server.

See listserv, UUCP, and kill also.

- In 1980 there were 15 Usenet sites passing 10 articles a day; in 1987 there were 5000 sites passing 2.5MBytes a day.
- By 1992 the number of sites was unknown. However, it was estimated that about 2.5 million people were getting a selection of 35MBytes a day in 1992, so it is undoubtedly much larger now.

User Agent

UA. In an X.400 Message Handling System (MHS) the User Agent is the means by which users access the system and retrieve messages from the Message Store. In almost all cases this 'agent' will be a PC or with communications software.

user data

- **Packets:** The user data in a packet is the payload, as distinct from the header.
- **General:** Often the term refers to data being sent to the user's application, rather than to the user him/herself. Sometimes it carries the implication of being personal messaging, and sometimes just that of being data.
- **ISDN:** In ISDN, the B-channels carry switched voice and data, but the D-channel is specifically designated for signaling. This is a packet-switched channel and only lightly used for the primary switching function. In the exchange the D-channel connects to the CCS#7 packet-switched network which carries the signaling information between exchanges. This only carries a light load of switching-packets, and it is often based on X.25 and it offers User Data services for messaging.

  ISDN carriers are therefore able to offer a packet-switched, data-transfer facility over the D-channel which can be used by the public for facsimile, telex, and computer-to-computer data exchange. The carriers generally charge by the byte or by the packet, but you are given a preset number of free packets for signaling. See X.31.

  Many ISDN implementations also make direct connection through B- and/or D-channels to X.25 networks — quite distinct from the CCS#7 network.
- **Optical Disk:** User data is often provided on a data-base type CD-ROM or similar disk by the information provider. Here the term means special retrieval software — as distinct from the data itself.

user-data field

In the CD-ROM specifications, this is a 2048-Byte section of the data field (in an addressable sector) that is dedicated to providing access to user data (see above).

user-friendly

If you believe advertisements, 'user-friendly computer' is a tautology — obviously every computer and every piece of software is so easy to use that you don't really need a manual!

It is a claim made so often for all computer systems, software and peripherals that it has become part of the background noise of the industry.

In practice, the most it tells you is that, perhaps, you won't actually need a college degree in higher mathematics to be able to use it — and anyway, if you run into problems, there's always the 'easy-to-read' 400 page manual to fall back on.

• One reference defined 'user-friendly' as: 'Intended for users who know nothing about computers, and have no intention of learning'. Now that's optimism for you!

user interface

The keyboard, mouse control, screen, and perhaps a touchpad or tablet — anything that allows a user to see, hear, and enter data or queries into a computer system. Aka the man–machine interface.

It may also be more narrowly defined as the software functions (the I/O part of the operating system and design of the application) that govern the ease with which a user can interact with the computer.

In the near future it may include voice-recognition (continuous and discrete), voice-synthesis, pen-based (cursive-writing and print-character) recognition, etc.

• The two main categories of PC user interfaces today are:

– **Character User Interfaces** (CUI) which all of the early computers used, and which require typed commands.

– **Graphic User Interfaces** (GUI) which use icons, mouse and pull-down menus. See WIMP.

USL

Unix System Laboratories. See Unix International. Novell has now bought USL with the aim of enhancing interoperability between NetWare and Unix. They see Windows NT as a threat.

Recently (late 1995) they've gone in with Hewlett Packard and SCO to develop a high-volume Unix, with NetWare and Unix enterprise services.

USO

Universal Service Obligations. See universal service and CSO.

USRT

Universal Synchronous Receiver/Transmitter. See USART.

USSB

United States Satellite Broadcasting. See DirecTV.

UT and UTC

Universal Time and UT Coordinated. Universal Time replaced GMT in 1928, and in 1955 the International Astronomical Union created a number of kinds of UT. There are now UT0, UT1 and UT2, which progressively take into account irregularities in the rotation and wobble of the earth.

Coordinated (synchronized) time is between the UK and the USA, and it dates from 1960. It involves the insertion of leap-seconds and other calendar calculations. This is the replacement for GMT 'Greenwich Mean Time' but it relies on the use of atomic clocks rather than solar calculations.

The UTC seems to be written and expressed in many different forms around the world. It is sometimes like this:

`49273 286 143645 UTC`

The '49273' (the Modified Julian Day Number) is the number of days since the start of the Julian calendar on January 1, 4713 BC (with two leading digits left off); the '286' is the Xth day of year; and the '143645' is the hour (14), minute (36), second (45) in groups of two digits. If the last group only has four numbers then seconds have been abandoned.

• A 'Z' is sometimes substituted for UTC or GMT, so 1344Z, is the same as 1:44 pm.

utilities

Small ancillary programs to the main operating system that perform 'house-keeping' tasks — usually common routine tasks. Some utilities are kept with the system, and some remain separate, to be used as small applications, when required. They are used for copying and initializing disks, finding text words in files, listing directories, etc.

UTP

Unshielded Twisted-Pair.

• **Public Telephony:** The standard type of telephone cable in large bundles in the streets; they are twisted in a very precise fashion in order to limit cross-talk. Note that the cable from the pillar into your house is not 'twisted-pair' but 'quad' (if it has four wires). The twisting is not as tightly controlled as it is in the large street bundles.

• **Corporate wiring:** There are two basic types of UTP, one for voice and one for data; the main difference is in the number of twists per meter. Data is generally sent down a two-pair (four-wire — send and receive) system while telephony only uses one-pair.

• **LAN and data:** There are now a number of special categories defined. UTP cabling commonly consists of four pairs of 24 AWG (American Wire Gauge) solid or stranded copper wire surrounded by a thin insulating jacket.

Category 4 and 5 are probably the most popular (because they are cost-effective) for LAN cabling at present; Category 5 in particular. This is now the preferred choice for LANs, especially for horizontal wiring.

Category 5 is now rated for signals up to 100Mb/s, but special balancing techniques are needed for these high data rates. See Category # cable and STP also.

UUCP

Unix-to-Unix Copy Program, a simple file transfer protocol like Z-modem which has long been used on the Internet. It was designed by Bell Laboratories in 1977 and is included in every version of Unix.

This copy program has since been adapted for use on almost any platform, including PCs and Macs. It is mostly modem-based and it allows any computer to automatically dial and connect via a modem to any other computer using UUCP for the exchange of files.

• The term is also used for a Unix-based network closely related to UseNet. UUCP sites are relatively basic Internet connection points.

• With UUCP, news is stored on site and transferred to the Internet in batches at predetermined times. This means that a UUCP connection can't give you access to real-time links (as can the IRC) — and it is slow, but also relatively cheap.

See NNTP and MHSnet also.

UUEncode

See MIME.

UV

Ultraviolet. See.

UVC

Universal Video Coding. A form of video compression and coding which is based on a multi-layered approach. At the lowest layer, the image satisfies the requirements of QCIF for video phones. At the next layer, supplementary information provides a higher video quality, and so on — up to HDTV standards at the fourth layer. The proposed layers are:

— Layer 0 Videophones with resolution of 176 x 144 pixels.
— Layer 1 Videoconferencing at 352 x 288 pixels.
— Layer 2 Conventional TV at 720 x 576 pixels.
— Layer 3 HDTV at 1920 x 1152 (or 1250) pixels.

Only Layer 0 can provide a picture alone. The others supplement the basic image.

This layered approach integrates the coding of all services and hierarchies, and provides an economical way of interworking between systems having different requirements. Layering can also provide a degree of resilience to packet loss on broadband networks.

• UVC has been partly superseded by the development of Levels and Profiles in the MPEG standards, but layering may be added later. See Pyramid coding.

• See video transfer levels also.

V.# modem

A series of ITU-T standards for analog communications over phone lines, and therefore closely (although not exclusively) associated with modem data rates, error-correction, and data compression. For more details see each V.standard.

In general, these are the main standards used for normal two-wire, full-duplex modems:

V.21	300b/s
V.22	1.2kb/s
V.22bis	2.4kb/s
V.23	1200/75 (for videotex use)
V.32	9.6kb/s
V.32bis	4.4kb/s
V.38	28.8kp/s

The addition of 'bis' (second or encore) indicates a variation on the basic standard, and 'ter' (third) is the next variation.

- V.42 is an error correction standard which is only a minor variation on MNP 2-4.
- V.42bis is a data compression standard which is substantially better than MNP 5 (by about 35%).

• The USA used the Bell standards (see) for a number of years, but now they have aligned with international standards.

• Note: The V.# order often indicates very little in the ITU series, since old numbers are often reused. V.8 is a relatively new standard, while V.29 is quite old, for instance. However all those listed above are in order of their implementation.

V.# terminals

DEC's video terminals. See VT-## terminals.

V.3

This is the ITU-T standard for the standard International Alphabet code ASCII No.5 (known as 'Askey')

V.8

A handshaking extension for new high-speed modems which can be added to any of the new V.standard modems. It is included in the startup sequence of all new V.34 modems, and it outlines a pre-handshake series of tones and data messages which exchange information on each modem's capabilities. It uses the V.21 (300b/s) FSK protocols just for the initial contact.

V.17

This is a half-duplex version of the V.33 used for leased-line modems. V.17 is now used in Class 1, Group 3 fax machines and fax modems-for data rates up to 14.4kb/s. These modems will also provide the V.29 protocol which is the standard 9.6kb/s version used by most current Group 3 fax modems. The standard uses QAM and Trellis Code. See V.8 also.

V.18

A modem standard adopted in mid-1994, for text-phones for the deaf. This is a 5-bit Baudot code at 45.5 b/s. See TDD also.

V.21

This was the original standard for a 200b/s two-wire modem for switched-telephone use. Later it was boosted to 300b/s which is now the standard rate. It provides full-duplex operations using Frequency Shift Keying over two carrier tone pairs.

The originate tone has a center frequency of 1750Hz and the answer tone has a center frequency of 1080Hz. The signal is modulated onto these by frequency offsets of +/–100Hz to produce marks (+ logical 1) and spaces (– logical 0) — so the originate pair will be 1850Hz/1650Hz and the answer pair will be 1180Hz/980Hz.

V.21 is designed to be used without any form of line-equalization, but the standard includes a handshaking tone of 2100Hz which will disable the echo suppressors found in long-distance telephone circuits. This standard isn't used much for modems any more, but it is for fax handshaking (see V.8).

V.22

This is the standard for full-duplex 600—1200b/s dial-up modems which introduced Phase Shift Keying (PSK). Two sub-standards (600b/s is fallback) and two different frequencies are used. In both sub-standards, the frequencies are 1200Hz (Originate) and 2400Hz (Answer).

- At 600b/s, the modem shifts the phase by 180° and so the symbol rate (baud) has a one-to-one relationship with the bit rate (b/s). This is Binary PSK (BPSK).
- At 1200b/s, the shift is in 90° increments and it is 'differential' with two-bits being transmitted each baud (dibits). This is strictly known as Quadrature Differential PSK (QDPSK).

A phase-shift occurs every second cycle on the 1200Hz frequency, and every fourth cycle at 2400Hz. Line equalization is provided, but it is non-adaptive (fixed). V.22 is similar to the Bell 212A standard but the two are incompatible because of the audio frequencies used.

V.22bis

This is a standard for 1200—2400b/s full-duplex dial-up modems. They operate at 600 baud ('events' or 'symbols' per second) using Quadrature Amplitude Modulation (QAM). As with V.22, there are two carrier frequencies of 1200Hz (Originate) and 2400Hz (Answer).

- At 1.2kb/s this standard uses 90° Differential PSK shifts, with only a single amplitude level. So each 'symbol' transmits dibits, (which allows V.22bis modems to interwork with V.22 modems).
- At 2.4kb/s a V.22bis modem uses 45° Differential PSK and two amplitude levels. So each 'symbol' transmits quadbits (four bits per baud). It is the change in phase, not any absolute angle which defines the symbol.

The standard included adaptive line equalization, but this was not always provided by the modem makers.

V.23

This is a rather odd frequency-shift modem standard which was adapted for videotex. There are actually three sub-standards.

V.23 was originally designed as a half-duplex 600b/s system using FSK, but then later it was boosted to 1.2kb/s on the forward channel. Even later, it was given an optional back-channel of 75b/s. It was originally intended for teletypes, and later it was used for videotex because the channel from PC-to-host only needs to carry slowly typed characters, while the host-to-PC needs plenty of bandwidth for file downloads.

The sub-standards are:

— At 600b/s it is half-duplex only, using a center frequency of 1500Hz and a +/– offset of 200Hz for both sending and receiving. (After signaling a turn-around, the other modem uses the same frequencies.)

— At 1.2kb/s half-duplex, it still uses the center frequency of 1700Hz, but now it has a +/– offset of 400Hz.

— At 1200/75 it is a hybrid. It is still 1.2kb/s, but with a slow-speed back-channel of 75b/s. For the backchannel it now adds another tone with a 420Hz center frequency and +/– 30Hz offsets. There is no provision for line equalization.

V.24

This is not a modem standard, but rather the ITU's attempt to define the 'functionality' of the RS-232C interface. The functions are those necessary between the modem and the terminal (more correctly, between a DCE and a DTE).

It must be read in association with V.28 . It doesn't define electrical characteristics, plug type, or pin configuration.

V.25

Among other things, this standard defines the 2100Hz tone which is sent down the phone lines to disable the echo suppressors/cancellers on long-distance telephone lines. These echo-suppressors work well for voice, but they make it impossible for full-duplex data operations when 'Echoplex' (deliberate echoing of characters) is required.

Echo-suppressors must be temporarily disabled if Echoplex is to be used. Note that echo suppressors will switch back to on, if they detect a loss of carrier on the channel for more than 100 milliseconds. So the carrier tone must be maintained constantly for full-duplex data (but you want the reverse for half-duplex fax).

V.25bis

This is an ITU standardized command set used to control a synchronous or asynchronous modem (but it is used more for synchronous operations). It is the international equivalent of the Hayes AT command set (but nowhere near as popular), and it provides dialing functions also.

There are three main V.25bis 'modes' which can be applied to commonly-used modems:

— Asynchronous.

— Synchronous with bit-oriented framing (HDLC).

— Synchronous with byte-oriented framing (BSC).

V.26

This is a 2.4kb/s full-duplex modem for four-wire leased (conditioned) lines.

V.26bis

This is a standard for a half-duplex PSK 1200—2400b/s modem on the switched telephone network. It is a synchronous standard using a single audio frequency of 1800Hz.

— The fall-back standard is 1.2kb/s PSK with 180° BPSK (single bits transmitted each baud).

— The primary 2.4kb/s data rate uses a four 90° phase changes to send dibits. An optional 75b/s back-channel is provided. The standard has a fall-back to the slower rate.

- V.26bis has fixed line equalization.
- Note that the V.26bis 1.2kb/s PSK standard is not compatible with the V.23 1.2kb/s FSK modems in any way.

V.26ter

This is a modem standard identical to V.26bis except that it provides both full- or half-duplex, and adds end-hybrid echo cancellation to improve performance.

V.27

This is a 4.8kb/s full-duplex modem standard for leased (conditioned) lines.

V.27bis is the above with the addition of 2.4kb/s fallback.

V.27ter

This is a 2.4—4.8kb/s standard for switched telephone networks. It is half-duplex , but with an optional 75b/s back-channel. It transmits on a single audio frequency of 1800Hz with a symbol rate of 600 baud.

— At 2.4kb/s it uses 90° Differential PSK modulation to transmit dibits.

— At 4.8kb/s it uses basic QAM modulation (45° differential phase-changes and two-levels of amplitude) to transmit quadbits

These modulation techniques are also used by Group 3 facsimile when they fall back to 4.8 or 2.4kb/s. The important additional technology introduced by the ITU-T here was the inclusion of automatic adaptive equalization to improve the effective line quality.

V.28

This is the ITU standard for the electrical connections of RS-232C — but not the pin configurations. It is an extension of V.24 which defines the electrical characteristics (what the controls mean) but not pin configurations either. See RS-232.

V.29

This is a half-duplex synchronous standard basically used for Group 3 facsimile at speeds up to 9.6k/s in half-duplex mode, and for leased-line modems over the same range of speeds. It uses QAM (phase and amplitude) modulation and has fallback rates of 7.2kb/s and 4.8kb/s.

V.29 became popular for leased line modems before V.32 appeared, and it's sometimes seen as a cheap version of V.32 because the chips are made in enormous quantities for fax machines.

There are three sub-standards all using a single audio frequency of 1700Hz (both ways after turn-around), at a symbol rate of 2400 baud:

- At 4.8kb/s this is a PSK modem transmitting dibits (with 90° shifts only) at each event. At this rate the modem is virtually the same as V.26.
- At 7.2kb/s this is a PSK modem transmitting tribits (with 45° differential shifts and a constant amplitude).
- At 9.6kb/s this is a QAM modem transmitting quadbits (45° shifts and one amplitude change).

It has automatic adaptive line equalization so it works well on bad local or international circuits. See V.27ter also

V.32

This is a 9.6kb/s full-duplex dial-up modem which is the first of the modern breed. The standard was adopted way back in October 1984, but it took years to iron out the bugs.

It uses QAM with 45° phase changes and two levels of amplitude. There are three sub-standards all using a single 1800Hz carrier both ways (and note this is full-duplex, not half):

- At 4.8kb/s the modem uses PSK as a fall-back.
- At 9.6kb/s the modem uses QAM with quadbits. It has a symbol rate of 2.4kbaud.
- At this same rate, the modem can also use an error-checking variation on QAM called Trellis Coding (TCM) which provides an extra bit (quinbits) in each baud as a form of parity checking.

The breakthrough with these modems is that they have highly efficient adaptive equalization circuitry. A V.32 full-duplex modem must have true digital echo-cancellation (a computerized function capable of constructing a delayed subtractive version of everything transmitted) since each modem simultaneously transmits and receives data on the same frequencies over the same dial-up link — and it must be able to not listen to its own transmissions, either at the time of transmission, or from a far-end echo.

- V.33 is a version of V.32, but without echo cancellation.

V.32bis

This standard was adopted by the ITU-T in February 1991 for high-speed, full-duplex modem communications at 14.4kbit/s, and with fall-backs to 12, 9.6, 7.2 and 4.8kbit/s.

The V.32bis handshaking procedures negotiate these data rates during the initial contact phase, but if errors mount during the data exchange, they can also re-negotiate further fall-back or step-up during transmission.

V.32bis uses two-dimensional 8-state TCM (QAM plus trellis coding), with a symbol rate of 2400 baud. The modem transmits six bits per symbol.

Almost all V.32bis modems will also include MNP2-4 (error correction) and MNP5 (data compression), and most of the better ones will also provide the ITU's V.42 (error correction) and V.42bis (data compression) standards also.

The V.32bis handshaking procedures negotiate data rates, compression and error correction during the initial contact phase. By performing the error checking in the modem, the two ends can also re-negotiate fall-back or step-up rates during the data-transmission phase by measuring BER.

- Some of the claims of higher data rates with these modems, come about because people misunderstand (or prey on misunderstanding) about the effect of data compression taking place in the modem. The useful throughput limitation is likely to be set by the rate at which the UART can deliver uncompressed data to the modem (in the local or the remote RS-232 link) if the line quality is good.

V.32ter

V.32ter was a proposed ITU standard which was to have pushed V.32bis modems further into 19.2kb/s range (with fall-back to 16.8 and 14.4kb/s) — but the standard was never approved. Instead they decided to develop V.Fast which then became V.34.

V.32terbo

A non-adaptive interim 'rogue' standard before V.fast becomes available: AT&T Paradyne and National Semiconductor were behind this scam. It was called V.32terbo as a play on words — on the term 'V.32ter' which means third and it therefore sounds as if it might be an official ITU-T standard. It is not.

V.32terbo offers two-dimensional 8-state trellis coding (the same as V.32bis), and it has a single symbol rate of 2400Hz. However the standard pushes older technology up to 16.8 and 19.2kbit/s range (with a fall-back to 14.4k) which the ITU had already decided was beyond the acceptable level of reliability. So this standard was not approved by the ITU — they considered and rejected it.

However, the chips being used for V.32terbo modems would generally provide a combination of:

- The old V.32bis, but stretched to the higher 19.9kb/s rate. It only works if the phone line is near perfect (it never is).
- V.42bis data compression.
- V.32 modulation (9.6kb/s/4.8kb/s).
- V.22bis and Bell 212A (2.4kb/s/1.2kb/s) modulation.
- V.42 (LAPM) and MNP 4 error control.

V.33

An ITU standard for 14.4kb/s leased-line modems. It has a fall-back to 12kb/s. It is a version of V.32 without echo cancellation. It is synchronous, and uses QAM and TCM. See V.32 and V.17 .

V.34

This is the current record-holding ITU-T modem standard (previously called V.Fast) for 28.8kbit/s full-duplex, dial-up telephone lines.

It was standardized, finally, in mid-1994 after many delays. These modems are highly adaptive and able to take into account poor-quality connections (adaptive equalization) because they are effectively intelligent communications processors in their own right.

V.34 modems are the first to 'probe' the line conditions to determine data rate before sending data, and they perform an extensive series of protocol negotiations, yet they managed to cut the handshaking period from 7 sec. (V.32bis) to about 5 sec. See V.8.

V.34 uses multi-dimensional Trellis Coding (QAM with error correction), having between 16 and 24 states. It has a variable symbol rate of 3200 or 3429 baud.

It can send up to nine bits per symbol, and 3200 symbols per second, which is the way it achieves 28.8kb/s. This is twice as fast as V.32 bis and getting very close to the theoretical limit over today's noisy lines. Even with good lines, the 4kHz bandwidth of the telephone network will not permit much above 38—40kb/s (discounting compression gains).

• Not every phase-change is associated with an integer number of amplitude changes — the V.34 constellation is quite complex. The preliminary negotiation period is characterized by 'line probing' which calculates which of these symbol rates is to be selected. V.32 can use any one of six symbol rates: 2400, 2743, 2800, 3000, 3200 or 3249 baud.

• New techniques introduced into the V.34 standard include:
— Multi-dimensional constellation coding.
— Precoding; a subset of adaptive equalization processes.
— Non-linear coding (they separate out those elements which are liable to interfere).

• About 2000 computations must be devoted by the CPU in the modem to the definition of each symbol. Compare this with the 200—300 needed for Group 3 fax.

V.35

This is a high speed interface specification for exchange-to-exchange modems using '60—108kHz group band circuits' (bearer band frequencies, not normal domestic telephone frequencies). It was originally intended only for the 48kb/s data rate, but now it is often used for interfaces operating at multiples of 56 and 64kb/s.

The data and clock signals are balanced, while the control signals are unbalanced. The *de facto* connector standard for V.35 is the Winchester-type 'M-series' which has 34 pins, however DB-25 connectors are also used.

• This was also once a 19.2kb/s leased line standard — but this is obsolete, so the ITU have recycled the file numbers.

V.41

The ITU-T's recommended CRC (error checking) procedure.

V.42

The ITU's V.42 protocol is a highly standardized point-to-point error-checking protocol. It defines error-detection methods using LAP-M as the basic function (which is, itself, based on the Link Access Procedure/D-channel used in ISDN, which was also based on HDLC).

To be fully compliant with this standard, the modem must implement LAP-M and also some alternative protocols based on MNP Classes 2-4. The key-word here is 'fully' compliant — modems can be 'compliant' without implementing everything.

During the handshake negotiations, two V.42 modems will attempt to apply LAP-M, and if this fails they will fall back to the MNP-4 protocol.

V.42bis

A standard for data compression selected by the CCITT for modems — especially those using the V.42 LAP-M error-control standard. It is based on the Lemple-Ziv-Welsh algorithms, and is stable enough to work with ISDN direct links.

In V.42bis modems, there are always two accepted data compression techniques; V.42bis itself and MNP Class 5. This came about because of a patent dispute. The V.42bis standard is better, with a theoretical capability of 4:1 compression, while MNP 5 is probably only half as good with the average text file.

However line impairment levels, modem processor power and data-file construction all affect compression efficiency. A V.42bis modem probably averages considerably less than 3:1 on normal text, although spreadsheets and assembly language programs will be at about this level or higher.

V.52

This is an ITU standard for measuring distortion and error rates in data transmissions.

• V.52 is also an ANSI terminal. See V.### terminals.

V.54

This is a relatively new loopback procedure now incorporated into many modems to allow self-testing. In the simplest form of local loopback testing, transmitted test signals are stripped from the phone connection and immediately reinserted into the receive circuits of the modem, where they can be checked for accuracy. This is a local loopback procedure.

V.56

An even more comprehensive series of loopback tests for modems on dial-up lines.

V.58

A modem management standard of the ITU adopted in mid-1994. This is to allow SNMP and CMIP control and reporting.

V86 mode

See Virtual 8086 mode. This has nothing to do with the CCITT's V.## series. V86 just means Virtual 8086 mode, which is the way the IBM PC XT machines operated a few years ago.

V.110

An ITU standard technique for rate-adaptation (matching one data rate to another) between POTS and ISDN adaptors. In V,110, a TDM frame is used together with end-to-end control signals, and both synchronous and asynchronous data can be supported.

For low speed channels it is possible to map more than one stream into a single B-channel, but all the contents of the B-channel must be heading to the same end-location since ISDN switches the B-channel as a whole. There are no error detection or correction functions.

V.120

An ITU standard for rate-adaptation which uses HDLC frames with 56kb/s links. Interframe flags are used to pad these out to the full 64kb/s. Only seven of the eight bits are used, with the last bit being always set to 1.

This technique supports asynchronous, HDLC-framed synchronous, and bit-transparent synchronous. Error detection (and some correction) is provided. Some statistical multiplexing is also possible.

V-chip

A US Senate proposal for a chip which, when fitted to television sets, would allow parents to block the viewing of violent and sexually explicit programs. The broadcasters would be required to code-identify the programs, and the V-chip would then accept or block the viewing. It needs to be coupled with compulsory identification codes.

V.Fast (Class)

You must differentiate between V.Fast, and the variant called V.Fast Class or sometimes V.Class.

— V.Fast was the ITU's 28.8kb/s preliminary standard which has now become V.34.
— V.FC (Fast Class) was a Rockwell proprietary product/standard which used a different technology to V.34 but which was to have been made compatible when the final version of V.34 was released.

V.FC was more a design than a standard, although many OEMers of Rockwell chips announced their support. It uses V.34's line-probing techniques and multi-dimensional trellis coding.

V.FC

See V.Fast (Class).

V&V

Verification and Validation.

vac/VAC

• As in '60vac'. This is a confusing use of two abbreviations: 'v' for voltage, and 'ac' for alternating current.

• Virtual Access Circuit switch. This is used in telephone systems; I don't know why or how.

vaccine

A computer program which attempts to protect you from virus attack. It checks the integrity of the operating system, but it can only really detect well-known viruses. Unfortunately, the most troublesome are viruses which are relatively new.

However, it still pays to have a good vaccine program in your PC, and to upgrade it regularly. Many install programs insist that the vaccine be removed before installation, so make sure you keep the up-to-date copy of a good commercial vaccine product close at hand.

vacuum tube

Amplification and detection devices used before the invention of the transistor. The first really practical amplifier was probably the Audion. Key stages in this important technology were:

— 1904: Sir John Fleming invented the vacuum-tube diode detector for Marconi.
— 1906: Lee DeForest invented the non-vacuum Audion diode tube. It became a vacuum triode system in1907.
— 1908: Prof. Boris Rosing of the Petrograd Institute began promoting the idea that the cathode-ray tube was the only solution to practical television. Vladimir Zworykin who later headed Westinghouse/RCA research in 1923, was his lab. assistant.
— 1909: For the first time, long distance calls of almost a thousand miles could be made between key cities in the US, using Audion triode amplifiers virtually every mile along the route. These came into common use from 1913 on.

We still use vacuum tubes in the 1990s, but generally we don't call them that. We have:

• Cathode ray tubes for television sets and computer monitors. These are vacuum tubes.

• Television and radio transmitters still use special vacuum tubes for amplification. Transistors can't handle the power loads. See klystron and thermionic tubes.

• Satellites and microwave generators use vacuum tubes.

VAD

• Voice Activity Detection. A technique used in GSM mobiles to reduce self-generated radio interference in the cells by turning off the transmission (discontinuous transmission) whenever the voice is not active. See DTx also.

• Value Added Dealer. A value-added reseller. See VAR and O&M.

VADIS

Video-Audio Digital Interactive System. An earlier collaborative European project on Digital TV and HDTV, known also as Eureka 625.

validation

The checking of data and programs to ensure they are correct — or rather, that they lie within certain boundaries, or conform to certain rules — that they are valid within the rules.

Don't confuse validation with authentication. Validation just checks for internal inconsistencies; authentication checks with external authoritative sources.

• **Data entry:** The term is usually applied to the checking of data at the time of entry. This can be automated by assigning attributes to record fields, and by trapping figures, dates, etc. which fall outside pre-set boundaries. Many copy programs also include a validation (cross checking) process.

• **Networking:** Validation services are electronic security systems that police the network. They check to ensure users have permission to access the resources. The basic validation service is still password checking — and this is an internal check. See call-back and biometrics also.

value-added

This is more a marketing term than anything else, although it sometimes relates to something real. 'Value-added' is often a claim made to justify premium-charged services in telephone and data networks. They are supposedly more than the basic POTS (so-called Basic Services) or DDS services.

For instance, a packet-switched network is value-added when compared to POTS, and a data network may provide automatic error detection and correction for users, which is also an added value.

The other side of this coin is that they also mean 'extra charges apply' — often out of all proportion to the value added. This is often because the value-added services are outside the strict price-controls and universal service obligations (USOs) imposed by regulators.

• The added value may be provided within the network itself, or by it providing connection to peripheral resources (such as mainframe computers). Internal value-adding would be best illustrated by CLASS services (call forwarding, for instance), while external value would be electronic white pages. The first could only be offered (at this level) by the network provider, but the second could be offered by any computer-owner (provided they had access to the information).

• More specifically, packet-switching networks are known generically as value-added services, although it is hard to see why. I guess the added value is in handling packets on a pay-as-you-use basis.

• ISDN probably isn't classed as a value-added service in most countries.

• EDI, EFTPOS, and e-mail services are certainly value-added.

• In paging systems, the term means additional broadcast services such as regular stock-prices or news updates.

• Electronic directories are often value-added, while operator-assisted directory information services are usually classed as 'basic' services.

VAN

Value-Added Network. In the wider sense these are enhanced telecommunications networks, and the term usually includes the services they provide — hence VAS (Value-Added Services) and VANS (Value-Added Network Services).

In practice it means anything above the level of 'POTS' (Plain Old Telephone Service) which may be provided by a carrier or a reseller of rented capacity. It includes such services as the forwarding and storage of messages, and the provision of news, sport, weather and other forms of information over paging systems, and access to mainframe applications.

The term was first applied to Telenet and Tymnet in the USA which provided packet-switching services by renting lines from AT&T. They adding the value of the packet-switching nodes and the error-checking of the packet-services.

In this more restricted sense, VANs are private resellers of capacity rented from the dominant carriers. They provide some of the following:

— Special protocols not normally available.
— Bandwidth subdivisions (fractional T1, etc.) not provided by the carriers, plus aggregation and rate-adaptation.
— Complex billing, not provided by the carriers (detailed breakdown, and international aggregation, payment in one country, etc.).
— Company-wide numbering plans, over shared networks.
— Protocol conversion, encryption, usage statistics, security features and network management facilities.
— Conditioned lines, error checking, and fault bypass (redundant) networks.

The aim of private VAN suppliers is to give large corporations a 'one-stop-shop' for national and international telecommunications. This is especially important for international companies who may find it impossible to deal with a range of intransigent monopolistic national carriers, and to support installation and supervision staff in small countries. See VANS also.

vanilla

Anything basic, or cut down to the bare essentials: without enhancements. 'Plain vanilla telephony' means POTS services.

vanity numbers

Phone numbers with an easy-to-remember or easy-to-associate patterns. Those ending with 1111 or 1234, or, for a florist, those alpha-keys that spell out FLOWERS. These are numbers for which the carrier will probably demand a premium.

VANS

Value-Added Network Services (or Networks and Services). VANS is a generic term which covers both VANS and VASs. These terms are all used vaguely and very widely to cover a range of different service provisions. The VANS market can be categorized as VANs, Messaging Services, EDI/EFTPOS, Information and Processing Services.

• The term VAN (no 'S') surfaced in monopoly countries when carriers began experimenting with packet-switching and videotex services, over and above POTS and telex. It was the carrier's way of saying; 'These will cost you extra.'

In the USA, Tymnet, Telenet and ARPAnet were established to provide packet-switching outside the normal carrier control. In much of the world the term VAN now suggests a communications facility which is not necessarily a 'reserved' service (not within the carrier's exclusive control) but perhaps offered by a reseller of bandwidth with some value-added (perhaps just much more detailed billing).

• The term VANS and VAS later also included large computer bureaus which offered centralized time-sharing facilities, accessible through leased lines. Swift and SITA were cooperative forms of these International VANS (IVANS).

There's always some aspect of either system integration or managed networks in a VANS. These may also be integrated with messaging services and application services also.

vapor phase

A technique for making liquid glass materials (rather than solids) used as the starting point for early optical fiber production. This technique was a breakthrough in silica glass fiber clarity because the material could be easily purified.

vaporware

The world's most dominant product in the area of computer hardware and software marketing. Vaporware is a claimed 'leading-edge', super-new product that never eventuates.

As one critic said about a loudly-promoted new database manager: It is 'the ultimate vaporware, since it's unannounced, undesigned, undeveloped, unknown, has no marketing plan, sales plan, packaging plan, nor any release date or pricing.' Vaporware is often also tagged with the acronym RSN, for 'Real Soon Now'!

Don't confuse vaporware with the delayed introduction of software which happens because the software is going through a long debugging process. Instead, fall on your knees every day that it is delayed, and thank the Almighty that another 50 bugs have probably been removed today, before the commercial release.

VAR

Value Added Reseller (aka VAD). It's a term used in two slightly different ways:

• These are corporations which purchase equipment from original manufacturers, re-package it with possibly customized software, add a new badge, and sell it to a customer. This is an OEM operation. See OEM also.

• These are consultants who coordinate and centralize equipment purchases for large corporate networks or large systems. The 'value-added' part of the term is used very generously in many cases.

Sometimes the term VAR is used to suggest a higher ethical level than so-called system- or service-integrators, who may have a proprietary interest in promoting one vendor's line. (But, so do many people and companies claiming to be VARs.)

variable

Any identified container in algebra or programming which can assume any set of values — one at a time. Variables can hold any value, factor, quantity, or condition that can be measured, and each is identified by one or more alphabetic characters or by a word. Variable can be nominated to hold one particular data-type only:

– There are string variables (usually identified in Basic by the $ sign) which will take both alphabetic and numerical characters, but will treat them both the same as alpha.
– Number variables will only take numbers, and these can be used for calculations.
– Boolean variables will only take Yes or No values.

• Within the depths of the computer architecture, a variable is simply a named memory location where data will be stored and referenced via a table.

The contents of this location are changeable and will probably be modified by the program during run-time. There are two types of variables in most languages: numeric (for numbers only) and string (for any and all alphanumeric characters).

• The distinction is between variables and constants. In the equation $y = 2\pi r$, the y and the r are variables, while both the 2 and pi are constants.

What's more, to confuse matters further, the data that will be added into a variable will be a constant — either a numeric-value constant, or a string-word constant.

variable length

• **Fields and Records:** As distinct from fixed or constant-length fields and/or records. Variable length fields and records in a database save storage space but make it more difficult for the computer to skip to a calculated record number. This was important in the early PC days, but it isn't significant now.

The computer handles the variable length record by adding a length-code at the beginning of each field. See fixed length field/record.

The use of text compression largely replaces the advantages gained through variable length fields. The long runs of 'nulls' added to fixed length fields to pad them out, will be reduced probably to two bytes with RLE compression.

• **Packets:** In packet-switching and frame-relay networks the packets can be fixed or variable in length. In some systems the variations are only permitted in fixed amounts (say, 50-byte increments). A variable-length packet will need either:

– a field within the header which identifies the length, or
– have highly identifiable end flags, so that the terminal can isolate the packet modules.

Varitype

A medium-grade typesetting device used by printing companies for offset printing. Varitype produces image quality of 600 dots per inch as against the 300 to 600 dpi for laserprinters, and 1200 dpi for the best of the Compugraphics typesetters.

So Varitype is probably four times better than the best laser-printer, and only a quarter as good as Compugraphics in terms of dot densities. Generally, this is only important for graphic reproduction; you won't tell the difference with text.

VAS

Value-Added Services. There is no consensus as to what sets the boundaries between value-added networks and value-added services. You would assume VASs would be peripheral services offered without special network connections, but the term doesn't seem to be used in this way.

VAS includes airline reservations systems, time-share computer bureau operations, on-line database utilities; and it can also include EDI and e-mail, but these are often placed in a separate category under VANS.

All of these terms are very vague and are used very generally, so you'll often find that VAS is also a generic term to cover VAN and VANS.

In the narrow sense, you can usually break down VASs into two categories:

- **Information Services:** On-line databases, videotex, directory services.
- **Processing services:** EFT, EDI, SWIFT banking, SITA telecommunications, airline reservations, etc.

Some people would include e-mail messaging and public-access packet-switching here, while others would claim these as VAN functions — so if you need to play safe, include them all under the generic VANS.

VAX

A Digital (DEC) mini or mainframe computer with a distinctive architecture; they actually range down to desktop versions now. Originally the term meant 'Virtual Address Extension' — but it is hard now to see why!

Digital has consistently used one operating system (VMS) and one processing architecture (VAX) across a wide range of computer hardware over the years, and this has proved to be a very viable strategy. It preserves corporate investments, and allows features such as common file naming, etc. (which probably accounts for the name) and easy interconnection.

The current strategy is to make VMS compliant with Posix and XPG3. The latest version is known as 'VMS Open'.

- This OS is often known as VAX/VMS. See VMS.
- VAX Rdb/VMS is DECs relational DBMS.
- VAX clusters are groups of VAX computers which are coupled into a multiprocessor environment. This is a very effective way to provide fault-tolerance and also mainframe power from mini components.
- DEC's VAXmate was an AT-clone introduced in 1986.

VAX MIPS

An attempt to provide a meaningful benchmark for PCs by comparing them with the VAX-equivalent in millions of instructions per second. SPEC has taken over here. See MIPS.

VBI

Vertical Blanking Interval. See.

VBR

Variable Bit Rate. This is usually applied to codecs, both voice coders (vocoders) and video (mainly MPEG2). The distinction is with Constant Bit Rate (CBR) and Available Bit Rate (ABR).

Wherever compression is being performed on a data stream, the result will be variability in bandwidth requirements — unless the source of the data is entirely bland and featureless.

However, most codecs buffer their output (to even it out), and some will pad-out the data stream with null characters to fill-out the stream to a constant, fixed traffic channel.

A much more sensible approach in these days of fast-packet switching is to only supply the essential data and not to delay it in buffers. As a result, the output will be highly variable.

However, since these new fast-packet public networks will be shared by thousands of different data streams, the Law of Large Numbers will act to absorb any impact of variability — your burst of bandwidth will be matched, on average, by my quiescent period.

- MPEG2 was designed to handle both fixed rate and variable rate services. Obviously, you must pay for these services in volume terms (bits delivered), not connected time.
- In ATM, the VBR method of traffic control involves providing a fixed bandwidth for a call (as with CBR – see), but it is variable because it allows this rate to be exceeded in short bursts if the bandwidth is available. Available Bit Rate, is now a distinct possibility in ATM networks. See ABR and leaky bucket also.

VBX

Visual Basic eXtensions. These are the custom controls, modules and components of the Visual Basic system. This is a component architecture and it is tied to both Windows and Visual Basic. See OLE, DLL and componentware.

VC

- **Virtual Containers:** See.
- **Virtual Channel:** This is defined in ATM as a fast-packet switched channel between two end-points.
- **Virtual Circuit:** The X.25 term for a channel. This is also defined in ATM as being between two nodes (a single virtual connection) along which a Virtual Channel passes.
- **Virtual Connection:** Frame relay version of virtual channel.

VCC

Virtual Channel Connection. This is an end-to-end chain of Virtual Channels (links between two switch points) which exist between end-nodes in an ATM network. The VCC is defined by the routing information — by the combination of VPIs and VCIs across the network. See VPC.

VCELP

Vector CELP. A variation on the CELP vocoder used in the North American TDMA (Digital AMPS) system. See CELP.

VCI

Virtual Channel Identifier in ATM cells and in packet-switching in general. The term Virtual Circuit Identifier also seems to be used. This is the primary identification applied to one route for a cell traveling through the network from one switching node to the next. The VCI specifies the route, but it is not an address, in the sense of identifying an end-terminal.

In the call establishment phase, the full address will be carried across the network in a connectionless way, in the payload of the first cells. But once the call is connected, the cells all flow in a virtual circuit using VPI and VCI only.

In ATM, the VCI occupies 2-Bytes in the cell header, and is used in conjunction with a 1-Byte VPI. Both VPIs and VCIs are assigned by the network for each hop between nodes when the channels are being established and provisioned. They last for the duration of a call only. See virtual circuit and VPI also.

VCII

VideoCipher II. An analog video set-top box and decoder standard. See VideoCipher.

VCPI

Virtual Control Program Interface. This is a specification for how programs should behave with extended memory in an 80386 machine. It only applies to those programs that access extended memory direct, and it became the *de facto* standard.

VCPI is designed to solve problems of conflict with extended memory, and thus allow multitasking. Programs that conform to this standard can run in protected mode simultaneously.

- Lotus uses a VCPI-compatible DOS extender. See EMS, XMS and VDM also.

VCR

- **Video Cache RAM.**
- **Video Cassette Recorder:** Before VCRs, we had VTRs (video tape recorders or 'open-reel'). The first true VCR was the 3/4-inch U-matic from Sony.

Later Philips introduced a range of color recorders (well ahead of their time) for schools and home use, and later still, a VCR that they called the '2000'; it has an under-and-over cassette which gave a lot of trouble.

Cartridge (supply side only) machines were also popular in the transition from reel-to-reel to modern VCRs. At one time VCR stood for 'Video Cartridge Recorder'.

The most successful of the half-inch domestic cassette recorder standards were Sony's Beta and JVC's VHS (with some slight competition from Philips 2000). Sony and JVC fought a war for domestic dominance for many years, and VHS won.

VCR-Plus

See Video Plus.

VDD

Virtual Device Driver. A device driver written in ANSI's VDI format. This is incorporated into the CGI standard. See VDI.

VDI

Virtual Device Interface. (Aka CG-VDI.) This is an ANSI standard for graphic device drivers (called VDDs). It was designed mainly for interactive applications to 'normalize the program interface to various input and output devices'.

The idea was to have a standard format and interface which could accommodate screen displays and adaptor cards with proprietary resolutions, just by including a virtual device driver (VDD). Software packages that support the VDD (VDI format) would then be able to work with the proprietary screen system. The ideas have been incorporated into the CGI standard. See.

VDM

- **Virtual Device Metafile.** See CGM.
- **Virtual Direct Memory** access. This is a service specification supported by some memory management systems. It helps software obtain the required information before performing DMA operations while running in a virtual or protected environment.

VDT

- **Visual Display Terminal:** A CRT or video monitor with keyboard. See VDU and CRT.
- **Video Dial-Tone.** See.

VDU

Vision Display Unit. This is probably just the monitor, but the term is often used for the whole terminal, keyboard and all. See VDT, CRT or monitor.

vector

- In matrix algebra, a vector is a single row matrix.
- Vectors are geometric quantities that contain both magnitude and direction. Vector notation generally consists of a starting point, a length and direction — calculated sometimes from a fixed point, and sometimes from a previous point — there are absolute and relative vectors.

In computing applications, especially drawing and drafting programs, vectors are used in this way constantly. However, for computer I/O devices the term also carries other implications. The distinction is usually being made between vector approaches to input and output, and the alternative raster-scan or bit-mapped approach.

The vector's advantage is mainly in providing flexibility. Here vector implies that the intelligent instructions (how to do something) are being used for I/O, rather than just a 'facsimile' or scaled copy of an image which has been stored in memory.

- The vector process requires intelligence and instructions to create a hard-copy or a screen display, while the raster-scan just requires a relatively dumb repetitive action — like a TV set.

vector architecture

Vector processing performs simultaneous multiple operations on numerical elements. This approach to designing parallel computing systems creates a type of array processor that performs simultaneous multiple operations on data in matrix/array form. Complex problems (weather prediction, simulation of air-flow over aircraft wings, etc) can often be solved by these matrix calculations.

With both vector processing and an object-oriented approach to the software, very complex displays of vector-graphical objects (virtual reality) can be moved around on the screen in a very fluid way, by simultaneous calculations of coordinates using a powerful computer based on vector architecture. See vector processing.

vector compression

In compression techniques, vectors are sets of samples which are stored in a look-up table (see vector quantization). This is a bit of a cheat on the term.

Suppose I send you by code: 'John 3.16'. If you have a New Testament bible at hand, you can 'decode' (not decipher) this to reveal a large number of characters beginning with 'For God so loved the world....' The vector here is the code which 'points' to a library of information.

The question is this: Is 'John 3.16' a vector compression of the paragraphs it references?

Do we regard this as a valid compression technique just because it greatly reduces the amount we need to transmit?

If you use a V.42bis data-compression modem, GIF files, or any of the data compression systems based on Lemple-Ziv, you've probably got to say; Yes! They all use a modification of this approach.

vector fonts

These are typefaces which are stored as structured graphics rather than in bit-maps. Vector fonts can be scaled up or down to any degree without creating jagged edges (jaggies). See outline fonts.

vector graphics

- Vector graphics code images as coordinates and straight lines, rather than as a series of dots (raster or bit-mapped graphics).

- Vector graphics create objects which exist only as related collections of 'references' — coordinates and primitives. For a line you need three bits of information: an x/y start point, a direction, and a length.

More complex primitives, such as circles, squares, triangles — and three dimensional spheres, pyramids, and cones, can be coded in a slightly more complex way using mathematical formulae and reference libraries (collections of 'objects').

At a higher level still, these graphic representations can include fill patterns, colors, textures, and have scaling and rotational information. These are then the fundamentals of vector graphics, and they form the basics of CAD (2D and 3D) and virtual-reality applications.

- If you store a graphic image in vector form, you store the instructions which will allow the computer to redraw it — not the bit-map of the final drawing. This means that easy scaling up and down is possible, and with the right instructions and adequate information, you can probably view it from another angle, or from a different perspective.

If you store the external surface appearance of objects in vector form, you can simulate the way in which light and shade will fall on the surfaces, and change the 'virtual lighting' at will. You can do this up to photo-realistic levels of subjective quality, provided you've got enormous processing power, and/or a lot of time.

- For a draw-program to store information in vector form, it must have algorithms and formula which describe the objects and their position on the final page (including which object is in front of the other). This means that the draw-program can be modified, or even 'wound-back' to an earlier form, but a paint program can't be.

- Vector graphics is a form of object-oriented computer graphics. Objects can be relocated by recalculation of the coordinates.

- The term 'vector graphics' is often applied to the creation of the 'wire-frame' model. See rendering.

- Tracing programs are used to convert bit-mapped images to basic one-dimensional vector graphics. See trace.

- Vector Graphics Inc. produced microcomputers in the early 1980s.

vector printers/plotters

A vector printer (a plotter) sees the objects on the page in much the same way as described above — as mathematically described lines and shapes. And it draws these in order. While a raster-scan printer (most of the popular types) sees the page as a single scannable bit-map, and so it will draw the whole image at one time.

vector processors

A processor which performs simultaneous calculations on numeric elements. See array processor and vector architecture.

vector quantization

A non-uniform technique used for compression of text. Both sender and receiver either have, or progressively construct, a library of text strings. The compression then consists of transmitting a pointer (called a token) which references the library reference. The token 'represents' a set of samples (called a vector) which have been stored both at the source and the sink. See LZW.

vector recognition

See trace and also OCR and ICR.

vector scan

An alternative to the standard TV image 'raster scan' technique. Vector scan is sometimes used in video games and special CAD equipment. It is a coordinates-based system which will draw each object on the screen by scanning the electron beam using x/y coordinates, rather than in a fixed way from side to side.

- An oscilloscope uses vector scan.
- Plotters draw images using a vector scan technique.

vectorize

To convert bit-mapped graphics to structured graphics.

velveeta

This is an undesirable commercial posting on the Usenets, similar to a spam.

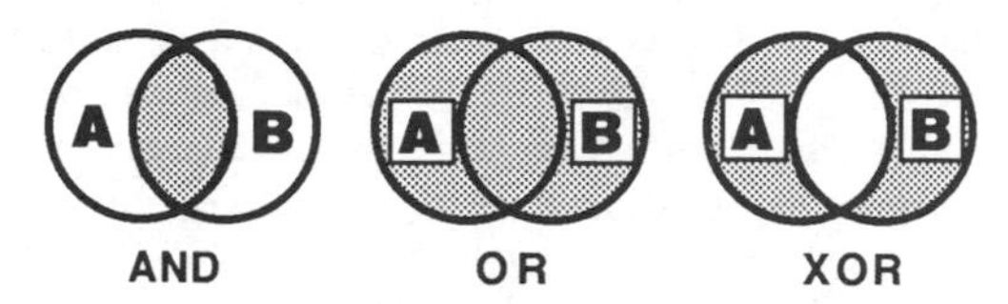

Venn diagram

A graphic representation of Boolean algebra and truth tables. Venn diagrams use overlapping circles to illustrate the AND, OR, XOR, NOT, NAND and NOR functions.

verify

To check for accuracy. In disk copying, this is an instruction to the computer to check the copy of a file against the original, to ensure the transfer has been accurate.

- Verifying a disk before recording, consists of recording a known pattern of signals across the whole disk, then reading them to check for disk faults. The IBM approach is to mark the faults, and avoid them in future recordings. The Macintosh approach is to reject the disk if faults are found.

Veronica

Very Easy Rodent-Oriented Net-wide Index to Computerized Archives. This is an Internet menu-based facility which helps you find the Gopher servers which may contain the information you need. It is an aid to finding the titles of files or documents.

It's often available from Gopher sites. You browse Veronica in the same way (using menus and lists) as Gopher. See Archie, WAIS and WWW also.

version number rule

The First Law of Common Sense says: 'Never buy any hardware or software with the version number 1'.

Unfortunately, ever since dBase II appeared (there was no dBase I), the marketeers have made it a rule to always begin numbering at version 2, 3 or even higher.

- The original rule for version numbering is not now always followed. If you had a product: Acme SuperWrite 2.3.4

– The first number in the sequence (the 2) referred to the last major rewrite of the software.

– The second number (the 3) was a version number where there had been a substantial change.

– The third number (the 4) was for bug-fixes.

So Acme SuperWrite Version 3.2.3 was the third complete rewrite, and the second upgrade of that rewrite, with a couple of later minor bug-fixes.

This numbering scheme had economic implications also. Originally it was intended that you should be required to buy a new copy whenever the version number changed (after a complete rewrite). However with lesser changes, you should, at most, be required to pay a minor handling fee.

Minor bug-fixes are probably specific to applications, or to CD-ROMs, or something too minor to worry about except in special circumstances.

Versit

This is the infrared communications standard being developed by a US consortium called IrDA which involves Apple, IBM and most of the PDA makers. Currently the work is directed at the exchange of simple file information between PDAs and desktop computers, but later it is to be extended to include PDA-to-PDA exchanges of business-card information, and also to exchanges with telephone handsets.

Recently a lot of work has been done to standardize Versit's API around the TSAPI programming interface.

verso

This means the left-hand page. Recto means the right. Left-hand pages are numbered with even folios.

vertical applications

This refers to niche applications used by a particular industry sector, and of little use outside that industry. Horizontal applications are widely used in all businesses — such programs as spreadsheets and word processors.

Note that desktop publishing was originally seen as a vertical application area (for use by graphic artists, typesetters and printers) and later became a horizontal application because it is now used by almost everybody.

vertical blanking interval

When a video screen is raster-scanned from top-left, to bottom-right (as in TV and most monitors), there is a reversal in voltage of the drive electronics which sends the electron beam back to the top-left of the screen, ready for the next scan.

Electrons won't flow from the electron gun when the voltage is reversed, so no lines are drawn during the repositioning. This period of reversal appears as a black bar at the top and bottom of the screen, which is only seen if the picture rolls.

Of the total 625 lines available in the PAL system, for instance, only about 575 are used for image, a few are hidden behind the 'bezel' (the screen surround) and the other 46 are 'wasted' in allowing time for the retrace and repositioning of the beam at the top of the screen again. However these wasted lines can be used for information.

The VRI is said to exist at the beginning of the field (rather than at the end). So in each field transmitted (at 50 fields per second), lines 1 to 21 are available for data. Line 22 is a mandatory black line, and line 23 marks the first line of vision. With two fields, this accounts for lines numbered from 0 to 48.

VBI can be used for captioning and teletext by using special digital techniques. You require a teletext adaptor for the home TV set. See horizontal retrace time also.

vertical cavity

Aka 'surface-emitting lasers'. These are the earlier type of semiconducting lasers which beam at right-angles to the flat plane-surface of the chips. The other type has a horizontal cavity geometry (also called 'edge-emitting') where the units beam out from the edge of the chip.

Vertical cavity lasers have the advantage that their beams are more symmetrical and more focused, so they are usually preferred for optical fiber communications systems. They are also cheaper to produce.

IBM has recently invented a self-focusing form of vertical cavity laser called a zone laser, or z-laser. See z-laser and semiconductor lasers.

vertical frequency

With monitors, this is the number of times each second the image is scanned from top to bottom — the 'vertical scan rate'.

American NTSC TV has a 60Hz rate and PAL/SECAM has a 50Hz rate. With MDA and Hercules screen displays this was 50Hz, with CGA and EGA it was 60Hz, and with VGA it is 70Hz. American TV has a 60Hz rate and PAL/SECAM has a 50Hz rate. For computer monitors, you generally need a higher frequency to prevent flicker. This is easier to tolerate on a TV set.

Note that to reduce flicker, some of the new European television sets hold the full field of image in memory and scan it twice, so these have a 100MHz refresh rate. See scan rate and dot-clock rate also.

vertical market

A marketing term which means that equipment or software is created specifically for a narrow market niche. Dentistry and legal services are vertical markets, while business administration and education are horizontal markets. Word processors, spreadsheets and databases are all horizontal market applications. Dental-record applications are a vertical market.

vertical recording

A magnetic technique where the magnetic domains are oriented vertically through the recording medium, rather than horizontally. Heads are needed on both sides of the disk or tape, but the technique was said to allow for a much greater packing density (ten times or more) than normal. It is an idea from the 1980s which never really caught on.

The problem was that vertical recording required a special double-layer magnetic material, while higher density standard recording was achieved with thinner magnetic layers. The vertical technique produced 4MB on a 3.5-inch floppy (back in 1985) and a few hours of recording on a standard VCR cassette.

This idea has been under development for a decade, but so far, the promises have not been converted into products.

vertical redundancy check

This is basic parity checking, using simple checksums. See VRC.

vertical resolution

The resolution of a monitor or TV screen vertically. Paradoxically, this is measured and expressed as the number of horizontal lines of an image that can be distinguished on the system.

You must distinguish between the number of scan-lines (a digital function in the vertical dimension) and the resolution of the image (which in its original form is infinitely variable and therefore analog).

In a 625 line PAL TV system, there will only be about 585 active lines available for image. Some of these will be lost behind the bezel, so the maximum that can be seen will be about 560 — but these are 'scan lines'. Most engineers would then apply the Kell factor to this, to arrive at a vertical image resolution of about 340 lines (here they mean the ability to distinguish black lines drawn closely together on a white background — one of each creates a 'cycle').

The use of Kell in this calculation doesn't make sense with replay devices like TV sets. But for other reasons (coincidence), the figures they calculate turn out to be reasonably close. When TV sets are tested subjectively, the image resolution seems to be equal to about 60% of the available scan lines, provided the image is moving a little.

Translated into 'cycles' this means that in the vertical dimension, we should be able to resolve about 180 'cycles' (black-white line pairs), and since the screen has a 4:3 ratio, this would translate, ideally, to 235 cycles (470 lines) horizontally. This is not far off that found in a TV studio, but the bandwidth after transmission limits even the best home reception to about 400 lines.

- Progressive scan screens (rather than 2:1 interlace) give an appreciable improvement in vertical resolution for the same number of active lines mainly because of reduced time-base errors and interline artefacts. See also Trinitron.
- After recording on a VCR, the video playback will be reduced to about 250 horizontal lines.

vertical software

Software designed (or modified) specifically for one industry — as distinct from general purpose software (like word processors) which are used across the board (in 'horizontal markets').

vertical stripe

Aka Trinitron. A form of color picture screen which has largely superseded the triad type delta-gun picture tube. The vertical aperture or vertical stripe tube still uses a shadow mask but the color phosphors are arranged in thin vertical stripes instead of triangular dot patterns. The shadow mask is a series of vertical slits and the electron guns are arranged in-line in the horizontal plane. This means simplified deflection circuitry and increased picture brightness. See Trinitron.

VESA

Video Electronics Standards Association. This association has the support of a number of major graphic hardware makers, so its recommendations have tended to become standards.

The association defined graphic modes for Super VGA monitors. These now range from 640 x 400 pixels (256 colors) to 1280 x 1024 pixels (256 colors) and they added support for five extended-text modes — especially for 132-column text modes.

The VESA Super VGA BIOS Extension gave DOS applications a way of determining the type of display. Their drivers tend to be standard for many DOS applications, and a single Windows 3.x driver runs every Windows application. See VM channel also

vesicular film

A UV-sensitive type of self-developing negative that is used to make copies of microfiche.

vestigial sideband

VSB. In terrestrial television transmission the picture is carried by an AM signal which originally had two sidebands (totaling 11MHz in bandwidth). In order to conserve the limited bandwidth available in the spectrum, one of these is suppressed as much as possible. Thus the picture is carried by the vestigial sideband — the main one that remains.

For this reason, in a typical PAL system, the 7MHz channel is not evenly divided. The main picture carrier is located only 1.25MHz above the lower limit of the channel frequencies (only 1.25MHz of this lower sideband is used vs. 5.25MHz of the upper sideband). The vestigial sideband occupies the 5.25MHz above the carrier, and it carries the main picture information. A 0.5MHz FM audio carrier fills the space above this. See Sideband, SSB and VSB also.

VF

- **Voice Frequency:**
  - **Telephony.** Voice frequencies over analog telephone systems are between 300Hz and 3400Hz, although 4kHz of bandwidth is provided.
  - **General:** Voice frequencies are in the range 150Hz to 15kHz in good quality audio recordings.
- **Very Fast:** In modems. See V.Fast and V.34.

VFW

Video For Windows.

VGA

Video Graphics Array standard. The 8-bit IBM color monitor standard which was released in 1987 with the PS/2 range. It was a superset of EGA with a color resolution up to 640 x 480 pixels, and a range of color modes (monochrome 720 x 400 pixels).

In the standard high resolution color mode, it could display 16 out of a 256k palette of colors, and with 320 x 400 pixels (low-resolution) it can display 256 colors simultaneously.

VGA was backward compatibility with EGA and CGA, but the standard is difficult to clone with complete compatibility.

There are both analog and digital VGA monitors. The horizontal frequency of VGA monitors is 31.49kHz and the vertical frequency is 70Hz.

Super-VGA took the resolution up to 800 x 600, then Extended VGA (1989) increased it to 1024 x 768 with the ability to display 256 colors from a palette of 262,000. This extension was later known as 8514/A and it was incompatible with CGA, EGA and VGA. See video-DAC, Super VGA, XGA and 8514/A.

- The pin connections for a VGA monitor are: Pin 1 Red; Pin 2 Green or mono; Pin 3 Blue; Pins 4 & 5 not used; Pin 6 Red-ground; Pin 7 Green-ground; Pin 8 Blue-ground; Pin 9 not used; Pin 10 Ground; Pin 11 Color ground; Pin 12 not used; Pin 13 Horizontal sync; Pin 14 Vertical sync; Pin 15 not used.

VHD

- Video High Density. A form of grooveless capacitance video-disk invented by the Victor company and promoted by a number of Japanese companies. It is still used in Japan.
- Very High Density: Floppy 3.5-inch disks which can store 5MBytes or more. These drives require special servo mechanisms using magnetic or optical patterns on the disk to help locate the tracks (see cylinder and floptical), and they generally use barium-ferrite media.

VHDL

Very High Density Language: (aka Visual HDL and previously Verilog HDL). This is an advanced ASIC modeling technology developed by the US Department of Defense, and now being standardized by the IEEE.

VHDL is basically a hardware description language which is analogous to software programming languages. It allows logic circuit designers to work at a high level of abstraction using automated techniques, including CAE tools, for hardware synthesis and circuit optimization. It therefore aids in the rapid development of complex electronic systems, by using circuit libraries and a structured approach.

This has become the *de facto* standard for chip design; there are now over 90 companies which offer VHDL products and services. See HDL also.

- **Very High Density Logic:** The term also seems to be used to mean Very High Density Logic to describe chips made by the VHDL design process.

VHF

Very High Frequency. The part of the electromagnetic spectrum between 30MHz and 300MHz used for television transmission among other things.

These are metric waves between 10 and 1 meter in wavelength. These waves tend to pass through the ionosphere and are therefore relatively immune to ionospheric interference.

VHF has traditionally been associated with frequency modulation in radio, but note that the video signal in television transmission is amplitude modulated, only the sound is FM.

The main uses of this part of the spectrum are:

Cordless phones	30, 39 and 40MHz.
Television	45 to 222MHz.
FM radio	88 to 108MHz.
Land mobiles	70 to 85MHz.
(and also)	156 to 174MHz.
Aircraft communications	118 to 136MHz.
Paging	148 to 150MHz.
Military	230 to 300MHz.

VHS

The half-inch video-cassette recorder designed by JVC and used widely around the world. At the time, there were two viable half-inch alternatives to VHS — Sony's Beta and Philips' System 2000. While both of these were technically better, the marketing and promotional side of VHS proved to be more effective. Philips and Sony tried to keep their designs under their company control, while JVC produced an open standard.

VHS-C is the compact cassette version used in many VHS camcorders. More recently, Panasonic has produced Super VHS (S-VHS) which is an upbanded or hi-band version (see) and now all of the half-inch systems are under challenge from very high quality Super 8 and Hi-8 video systems.

There has been a number of improvements to the basic VHS standard over the years; long-play (doubled the recording time), hi-fi sound, and later the 'HQ' picture enhancements.

VHSIC

Very High Speed Integrated Circuits. These chips will have a million logic gates using 0.5 micron technology. The electronics work at the molecular level and permit the movement of electrons at close to the speed of light. See quantum effect.

video (high quality)

This 1980's term has been applied by the CCITT/ITU to video systems that provide picture quality at least equal to current TV standards. The CCITT suggested that to be classed as 'high-quality' the picture should be supported by 15kHz stereo sound and user-to-user data. They thought, at the time, that a digital data rate of about 30Mb/s would be needed.

• With MPEG2 digital compression, the above can be achieved in the mid 1990s at a data rate of 2Mb/s if sufficient time is taken over the compression process (for movies), or about 6Mb/s for real-time sports.

• The studio output of uncompressed video (4:3 PAL) using the 4:2:2 coding scheme is about 250Mb/s, but with virtually loss-less compression, this can be brought down to about 34Mb/s for studio-to-studio transmission.

Video-8

A Japanese standard for good quality analog video recording equipment using an 8mm tape in a cassette not much bigger than a Philips standard audio cassette. Video-8 tape can also be used for the storage of computer data, and, in fact, the system makes an ideal cheap back-up medium.

See Hi-8 for video, and 8mm tape in numbers section for computer backup.

video adaptors

Add-in cards which are used to provide the computer monitor feed. This term was widely used before computers were able to handle motion video, so the term just refers to the 'screen', rather than to any ability to run motion images.

The video adaptor card must match the monitor type and must have the correct software drivers. All early IBM PC video adaptors (MDA, CGA and EGA) were designed for TTL digital monitors. In the IBM world, PGA was the first with analog signals and this was later followed by VGA. The Macintosh has always used analog monitors.

The categories of basic standards are:

— Monochrome display systems.
  - Analog baseband video.
  - Digital TTL video.

— Color display systems.
  - Analog composite baseband video.
  - Analog RGB (direct color component feeds).
  - Digital RGB (direct color component feeds).
  - Digital HSV monitors (rare).

• Some will also provide PAL or NTSC video input and output. These tend to be known as 'video graphics adaptors'.

• With the advent of multimedia and video compression systems, video adaptors are taking on a new role in compression and decompression of images in real time. Many of these cards will have a graphics coprocessor on-board.

Generally the standard bus of the computer is much too slow to handle data at these rates, so local buses are used — the PCI bus, VL-bus, or VMC. PCI is the clear winner.

video buffering

The term 'video' here is used in two ways; one to mean a single electronic image (as in Video RAM image — a bit-map) and the other to emphasize moving (multiple successive) images.

• This was a slang term at one time for Video RAM (VRAM). The area in the upper memory (in IBM systems) which holds the current bit-map being displayed on the monitor.

• With double-buffering, two mirrored images are held in video memory, and the switch is made between the banks of RAM. This technique is used to speed up screen refresh rates.

• In computer image technology, a frame buffer is still synonymous with a display buffer. It is a block of memory that holds the bit-image being used to create the video signal.

However, some systems will have many frame buffers that can hold intermediate images at various stages of processing, so that real-time moving displays are possible.

• Z-buffering is an advance on frame buffering, but it needs more memory to hold 3-D images. It is able to distinguish foreground objects from objects further back in the layers.

Z-buffering and processing techniques can be used to vary the intensity of pixels which represent areas which appear to be further away, or closer to the viewer. For instance, they can add an illusion of depth to the scene by superimposing a light blue cast to the farther objects,. and by reducing their image resolution. This adds to the illusion when they are also enhancing the brightness and sharpness of near objects.

video CD

This is a general term covering one old specific disk format, and a few new, disk standards.

- The old CD-Video standard didn't last long. It was used only for music clips and such-like and it only held a few minutes of video along with the audio.
- Two new temporary standards (to conform to the DVD requirements) were Philips–Sony's High Density CD system which was later called Multimedia CD (MMCD). The other was Toshiba's development of a similar but incompatible system called Super Density CD (SD). These new standards were both two-layer bonded disks, and they both offered replay-only and recordable disks units.
- The latest development is that these two rival groups have combined, and are now producing a 4.75GByte twin-layered disk with the ability to store 142 minutes of MPEG2 compressed video. When blue lasers come available, the capacity can be increased, and so the disks will be able to handle wide-screen, and eventually even High Definition TV (HDTV).

See DVD, CD-Multimedia, and Super Density CD.

VideoCipher

The analog version of DigiCipher (see). VideoCipher I from GI, uses DES encryption of the audio and video, but it is used for transmission to mastheads only. There is no home descrambler version at the time of writing.

• VideoCipher II was to have been a home system, but it is now obsolete.

• VideoCipher II+ is the current version with basic video encryption and DES audio. It will shortly be replaced by VideoCipher IIRS, because the plus version has been widely pirated.

video codecs

See codec.

video compression

The digital transmission of studio PAL television (in 4:2:2 form) requires a bandwidth of 216Mb/s. A High Definition TV system with a 16:9 ratio image would, in raw digital form, require a data-stream approaching 1Gb/s.

Compression schemes have been devised to reduce this to a manageable level — and so video compression is one of the great success stories of the digital revolution.

The new MPEG2 compression standards can provide a completely lossless compression to one-sixth the original bandwidth. If relatively minor image degradation is accepted, then the compression ratio can be 100:1. This means the communications channel requires only one-hundredth of the full bandwidth.

Motion video compression is more efficient than still picture compression, in terms of the percentage reduction in the data transmitted. This is because motion video is able to take advantage of 'temporal' (time) factors — in motion images, one image is pretty much like the next.

The main compression technique applied to video, is therefore to only transmit blocks of image pixels which have changed between one frame of picture and the next. This is called inter-frame coding (between frames) — and since only differences are transmitted, these are often called 'delta' techniques.

However, the worse-case test of any video compression technique is when it is required to transmit images taken while the camera is moving (i.e. the total area of background is changing between each frame), or when the camera is zooming rapidly (all elements are moving in different directions at different rates), and when a scene changes completely after a cut. Here all elements of the image need to be updated between frames.

Fortunately, the eye ignores detail when images move, and it takes a few frames to comprehend a new scene. So with DCT coding techniques, redundant image detail can be thrown away. It is possible to rebuild the new image after a cut by providing large-scale image detail only, for the first few frames, and then progressively adding detail in subsequent frames.

Similarly, when images are moving, the eye is less critical since it can't see fine detail, and the systems can dump these codes and only transmit the larger-scale image detail.

Buffering of images also helps flatten out the load, and in future variable-bit-rate (VBR) transmissions, will dynamically provide wider data-streams to a video when it is needed.

Most video compression is asymmetrical in processing effort. While it takes a lot of crunch power to compress a movie, decompression is comparatively easy. This will make MPEG decoding chips in a home TV relatively cheap, even while the encoders cost half-a-million dollars.

For this reason, digitized images of old movies can be compressed down to 1.5Mb/s streams with very little loss in image-detail by using large computers, which are able to take their time. However real-time sports may need 6Mb/s of bandwidth because there is less processing time and more action. See MPEG, DCT, motion compensation.

• The first widely used video compression scheme was 2:1 interlace as used by analog PAL, SECAM and NTSC. It reduced the bandwidth needed for analog television by 50%. The use of VSB modulation reduced it by nearly another 50%.

• The term video compression is still applied to the compression of analog TV signals by firstly digitizing them, then performing various bit-reduction processes, and later converting them back to analog for retransmission. The various MAC systems took this approach. See.

- The main intra-frame digital compression techniques are:
  - DPCM which was one of the earliest (used for H.120 videoconferencing).
  - Subband decoding which uses bandpass filtering.
  - DCT; now the *de facto* standard.
  - Wavelet compression has also been suggested.
  - Fractal compression.
  - Pyramid coding. This is a layered approach for the future.
- With modern MPEG2 compression, the major techniques used are:
  - Pre-processing to remove unnecessary image detail.
  - Quantization of the image blocks.
  - DCT transform to distinguish large-scale image detail from the fine resolution factors in each block.
  - Prediction–comparison between frames, resulting in the difference signal.
  - Motion compensation. Block movement calculations.

videoconferencing

Videoconferencing and video phones have been the next 'killer applications' for computers and communications since they were first shown at the 1938 New York World Fair. They have been offered as a carrier service since that date, and every year we are promised the industry has made a breakthrough, and that every home will have one 'Real Soon Now'.

It is the most oversold and under subscribed technology ever developed. The only thing you can be absolutely sure about in videoconferencing, is that it is not a replacement for business travel. It is a poor substitute.

Other factors are more psychological. Any TV producer knows that the on-camera personalities projected by documentary or interview participants are manufactured and artificial. This is not a 'natural' medium.

Participants also lack direct eyeline clues, sound is often controlled, and interruptions are not permitted (in switched conferences), and the slow-scan visual images do not convey (unconsciously perceived) body language. So these meeting are not a substitute for reality.

Videoconferences are often more about posturing and power-plays, than about corporate performance.

However, for one-to-one conferencing and consultation, or for work-groups who meet regularly, these problems can be overcome. There's no doubt that there is a valid place for video-conferencing in business and commercial communications — provided its benefits aren't oversold, and the problems inherent in these techniques are understood.

Both analog and digital videoconferencing systems have been devised over the years — in fact, probably more research dollars have been poured into this area, with less productive effects, than any other area of computers and communications. Yet this is seen as the vanguard of 'convergence'!

Video phones and videoconferencing fundamentally use the same techniques. These are interactive systems which impose greater limitations on delay than one-way (broadcast TV). Delay is inherent in any codec, since it is required to process each frame of image (and also the sound) to compress it to a manageable bandwidth. It must also be decompressed at the other end.

The voice communication is hopefully half-duplex (one person talking at a time), while the video is full-duplex and continuous. However video delay is less stringent than voice, and in many cases people don't appear to be too critical of lip-synchronization. Delay differences of 120ms between image and sound are common. The limit is probably about 150ms.

Videoconferencing can be two-way/two-participant, or it can have multiple participants — in which case, someone must be nominated to be in control of the image switching.

The term videoconferencing is also used for essentially one-way broadcast systems where the CEO of a multinational company may be speaking to his employees via satellite, with questions returned via telephone to the studio location.

- Many proprietary standards have been popular over the years, including Rembrandt and other CLI products. The international standards for digital videoconferencing have moved through a number of stages: the first was H.120 and H.130; then came H.261 and a number of related standards; and more recently H.320 for videophones. The next stage is probably MPEG4. See all of these.
- Intel's ProShare proprietary desktop videoconferencing products are not being taken up with much enthusiasm. ProShare is not just the chip version of what was once the DVI standard; it is actually three products:
  - A very low rate software-only conferencing application running over a modem (using standard telephone lines) or a LAN, but without video capability.
  - An 'electronic blackboard' type application for phone lines, which lets you to see details of the other party's computer screen. You can run macros and edit data across the link.
  - True slow-scan videoconferencing on the desktop, which needs an ISDN link. This is Indeo, which was once DVI.

ProShare is a proprietary standard, and the lack of its success probably shows that today's customers are too sophisticated to fall into this trap with two-way devices.

- Sound is usually more of a problem than video in videoconferencing. It is important to have a good sound system. If it is switched, then it won't permit normal conversational interruptions; if it is un-switched, then it will often have feedback and excessive background noise problems.
- TI now has a new H.320 chip which is about four-times as fast as anything before. ISA cards will soon be available in this standard.
- See also video transport levels, video phones, CIF and QCIF, MPEG4, H.120 & H.130, H.261 and H.320

video-DAC

Computer monitors can be digital or analog, and the difference is primarily where the D-to-A conversion takes place. This decision influences what type of signal is being carried in the cable connecting the two devices.

With digital monitors, the baseband digital output of the computer travels across the cable directly to the drive circuits of the monitor's electron guns. Color monitors will need at least three signals, one each for R, G and B (some had six).

With analog monitors, it is necessary to buffer the digital output in the computer, and convert the digital signals to analog form before passing them onto the monitor. This approach has now proved to be much more flexible, and it provides better image quality.

Video-DAC chips (also called RAMDACS) have a large video buffer into which will be mapped the three DAC circuits (RGB). These chips can be in the monitor if it is a digital monitor, or in the machine if it is an analog monitor. These converters establish the quality of color on the screen; the more bits they can handle simultaneously, the more colors that are possible.

VGA systems initially used 6-bit DACs and provided a palette of 256K colors. Recent developments have substantially increased that number and many systems now routinely handle 32-bit (4 x 8-bit) color. (One byte is called the alpha channel.)

video dial-tone

VDT. The term is applied in the US to video-on-demand feeds supplied over telephone networks or by the telephone carriers. Its a vague term which has recently been promoted by the FCC, but its use is sometimes confused.

The concern of the regulators when allowing the telephone carriers to build cable networks, appears to be the suspicion that they will cross-subsidize the construction and operations of entertainment networks from their telephone revenues.

US cable-management regulations, imposed by the FCC, say that if the carriers are to provide video services, they must make channels available to all providers on equal and reasonable terms. The carrier companies can provide some programming themselves, but before doing this they must own a local cable franchise.

- The FCC defines VDT as a 'common carrier' service which allows multiple video programmers to supply video on a non-discriminatory basis. So obviously the FCC is not seeing video dial-tone specifically as a VOD service, but just as carrier-supplied cable. It says VDT is 'a common carrier access and transport network service for video information' — and leaves it at that. Under FCC regulations, a US carrier may not be a content provider, but it may have a subsidiary that is.

- The main technology proposed here is HFC cable or Fiber-to-the-Curb (FTTC). Some companies are still looking at the possibility of ADSL. Almost certainly, these systems will all eventually use ATM switching and multiplexing protocols.

videodisk

There have been a number of different forms of videodisk — some analog and some digital. The first was the original Selfridge/Baird system using black-shellac 78 rpm records in the late 1920s. The main modern systems are:

- **Laser-optical videodisks** — specifically the analog Philips/MCA/Pioneer LaserVision (see) which is still widely used, and which created the technologies of our modern digital CD systems. A very-similar Thomson's see-through optical laser system has now been discontinued.

- **Capacitance electronic** — RCA produced a grooved analog capacitance system called SelectaVision which has now been discontinued.

- **Very High Density capacitance** — JVC developed the VHD analog grooveless capacitance system still used in Japan.

- Videodisk can also mean the CD-V and CD-i standards, and now to the Multimedia CD, Super Density and DVD standards.

video graphics adaptors

This suggests a level above the normal screen adaptor (see video adaptors). Usually it suggests that the input or output will be motion video in the NTSC or PAL format, or perhaps more recently in one of the digital compression formats.

These are often expensive video adaptor cards which can handle a range of screen formats, together with the video information from a camera or VCR. They may also have sufficient memory to act as a 'frame-grabber' — to hold and digitize an analog TV frame.

videography

- A seldom-used generic term for systems that display text and graphics — videotex, teletex, teletext, etc.

- The term is used as a generic for amateur or home-video.

video memory

This is where the computer stores the graphic (bit-mapped) information which will be seen on the screen. The information here may be directly manipulated by the application, or it may be a secondary stage which is copied to the video memory, only after changes have been made in another part of memory.

It is the screen driver's job to create the video memory bitmap from the vector and ASCII screen elements. See VRAM.

The video monitor drivers then run progressively through video memory, reading the information a location at a time, and transferring the information in a raster-scan form (one line at a time) to the monitor in the form of variable voltages on the RGB lines. See packed-pixel architecture and planar also.

video modes

Along with the development of the PC and the introduction of different monitor types and different adaptors, we have also had the progressive introduction of video modes. These were a vital part of PCs in the days of character user interfaces (CUIs) — and the modes specified the standard character set, which at this time was always held in a bit-mapped matrix of a standard size. Generally each new development would maintain backward compatibility with at least the major modes of the past.

The first was the MDA adaptor with two text modes and no graphics modes. This was followed by CGA adaptors which introduced three color graphics modes, etc.

- The IBM modes were numbered.
- — Mode 0, was monochrome text in 40 columns.
- — Mode 1 was the same in 16 colors.
- — Mode 2 was 80 column text in monochrome.
- — Mode 3 was 80 column text in 16 colors.
- — Mode 4 was graphics in 4 colors with 320 x 200 pixels (CGA).
- — Mode 5 was Mode 4 in B&W.
- — Mode 6 was graphics with 640 x 200 pixels ... and so on.

Different adaptors would provide different modes, although the modes gradually accumulated in later chips.

The mode numbers seem to have fallen into disuse recently, mainly, I suspect, because the VGA adaptors provided all of the first nineteen.

Windows may also have had something to do with it, since it makes modes superfluous.

See EGA, CGA, VGA, PGA and Hercules.

video monitor (color)

- **Analog:** Theoretically there is an infinite number of levels an analog monitor can produce for each of the Red, Green and Blue signals. However the practical limit is set by the number of bits of pixel depth the video adaptor card (actually by the video-DAC chip) can handle.
- **Digital:** These have changed greatly over the years. In the early color PCs, the digital signals simply turn the RGB guns on or off. However later digital adaptors provide both 'RGB' and secondary 'rgb' signals. When Red was on and the secondary Red was also on (high intensity), the electron gun fired twice the number of electrons.These were called RrGgBb monitors.

This gave greater color flexibility, but the result was nowhere close to analog quality. Now digital monitors are made with an inbuilt video-DACs capable of handling 16 million colors. See video-DAC.

video monitor resolution

An NTSC image has 525 scan lines, but only 483 of these are active. Under the best possible circumstances we could therefore expect to resolve (in the vertical plane) 241 black–white line pairs (called 'cycles') or 482 lines. But in fact this is not achieved because artefacts and the interlace factor reduce the subjective resolution to about 330 lines at most (165 cycles) — about 60%, depending on the image and movement.

Allowing for the wider horizontal dimension (4:3), this is roughly equivalent to a horizontal resolution of, say, 440 lines (220 black–white cycles). This second type of 'line' is produced by a camera pointing at a test card. You can see them in test cards transmitted by the TV stations.

- With PAL systems the reductions are proportionally much the same and the overall resolution is slightly higher (but with more flicker). And in both cases the overscan of the image causes a further reduction of about 5% in the horizontal and vertical image size. This is image lost behind the bezel.
- Note also that horizontal resolution is directly related to the bandwidth of the baseband video (the maximum rate at which the scanning beam can change from a white to black). It is also reduced significantly by transmission modal problems and signal reflection. Interlace also creates a loss of vertical detail, and time-base errors reduce horizontal resolution. See TBC.

Interlaced systems have about 0.6 the vertical resolution of progressive scan systems (and slightly less horizontal resolution) with the same number of scan lines. This is partly because the interlace scans wander slightly in the vertical direction, and also because the lines jitter slightly in a sideways direction (time-base problems), thus blurring edges.

video-on-demand

VOD. This refers to the system of cable television where a user can choose what to watch and when — and control the replay (pause, restart, etc.). There must, therefore, be a one-to-one relationship between the video source (the server) and the customer.

There are two basic approaches being developed:

- — The server will replay the program through buffers feeding individual channels and homes in real-time. This will require a 2Mb/s dedicated channel to each home using the system. This is possible with HFC networks linking, say, 500 homes per fiber.
- — The server will replay the program at high speed to a memory storage device in the home taking, say, ten minutes over a channel of about 50Mb/s of bandwidth. You will then replay it at will. Some satellite delivery systems may use this approach.

Video-on-demand also has some relationship with both the new highly interactive HFC cable TV systems, and with video dial-tone (VDT) services using ADSL.

The term is often misused to include 'near' video on-demand services (NVOD), where a major movie will be run on a half-dozen different cable channels, with a start-time staggered by, say, 15 minutes per channel. However this is not a one-server-to-one-home interactive system, but a multi-channel broadcast service.

True VOD needs a single output port from a video server, delivering the information, under the exclusive control of one home viewer — and it assumes that an exclusive channel will be available between the server and the viewer. This will probably be a cell-relay ATM channel.

VOD has only become remotely feasible because of the development of MPEG video and audio compression, hybrid fiber-coaxial (HFC) cable networks, and high-speed, gigaByte video servers. ADSL also offers an alternate route to VOD through switched networks.

The real significance of VOD is two-fold:

- It breaks the delivery of video away from the traditional subscription approach (Buy a month ahead!), and allows for purchase only on demand. This opens the market to a wide variety of suppliers if open standards are used in the set-top box and the cables are owned by common carriers.
- A video server is just a data-server carrying video files. If one is economically feasible, then so is the other. These systems will eventually carry a wide range of information and a multiplicity of services other than video.

See video server and NVOD.

video overlay

Video overlay cards allow graphics to be mixed with PAL- or NTSC-encoded video signals, taken off-air or from a VCR, etc.

The video signal is routed directly to the monitor, not through the computer memory. As a consequence, you can't modify the signal — hence the use of the term 'overlay'. However, the combined output can be captured by a second VCR and so the system can be used for captioning and titling video.

video phones

Small-screen video conferencing facilities built into a relatively normal handset terminal. Video phones use slow-scan techniques, and require a video bandwidth in the order of 64kb/s, which can be provided over a single ISDN B-channel. The other B-channel is used for the sound.

There have been many announcements of breakthroughs in analog video phones over the last few years, but despite the brand names (like AT&T) attached to these 'breakthroughs' they have been little more than scams. The public has had the good sense to ignore them. See MPEG4 and H.320.

Video Plus

Aka VCR-Plus, Gemstar, and G-code. A European proprietary identification system invented by Gemstar to help viewers set the recorder-timer of a VCR from newspaper codes.

The codes vary between 3 and 8 digits, and are not directly related to dates or times — you need to know the special calculating algorithm. Gemstar both sells the devices and licenses the printing of the codes in TV guides.

Three-quarters of the new Japanese VCRs now have the Video Plus system. The American ABC is multiplexing the program data on to their TV signals, and European and Japanese broadcasters may soon follow.

Gemstar has also developed an automatic indexing system for VCR tapes called 'Index Plus'. It holds the identification of each tape in SRAM in the VCR together with the titles of all programs on the tape. This means you can index your tapes.

• VPS is a quite different system which relies on codes broadcast in the vertical blanking interval. See VPS and PDC.

video presentation

In videoconferencing terminology, 'video presentation' means one-way video with two-way audio (usually using a combination of television broadcasting facilities and telephone lines). It is usually analog at high (broadcast) quality to a large audience in an auditorium.

This is not so much a conference as a pep-talk or a lecture. It will usually a keynote speaker at a conference, or the President of some wealthy corporation trying to inspire the troops. The problems differ from the normal point-to-point or multipoint two-way vision and sound conferences.

video processing

Video processing amplifiers are often added to editing and post-production circuits to sharpen the edge of pre-recorded video images. They are sometimes known as 'proc-amps'. Be careful when buying these; most of them do very little (some actually degrade the picture) although sometimes a boost of video level can help a bit.

• Time-base correction (TBC) is also a form of video processing, but it is generally known by its own name. A TBC is by far the most useful device for maintaining video image quality through the post-production stages. See TBC.

video RAM

There is a generic and a specific use of the term:

- The generic use is for that part of the computer's memory reserved for the storage of pixel information in a raster-scan form (bit-mapped), which will control the image on the monitor screen.
- The specific use is for a special type of Dual-ported DRAM. See VRAM.

video server

Video-on-demand (VOD) requires the development of large video servers capable of handling a full motion picture (100 minutes at least). The movie will be stored in MPEG2 compressed form, on a single reliable replay device. Some of these will be optical disk units replaying CDs in the new DVD disk format, but generally they will need to be gigaByte-sized hard disks.

This is because the disk's read-access will need to be rapid enough to handle multiple users simultaneously through large buffers. Developing such disks, and the required multi-user controllers is a key task at present.

This is seen as important for a number of reasons. Firstly because the VOD servers can provide interactive access to audio, graphics and text-information as well as video data.

And secondly because, along with VOD, there must be an economical way of paying for usage on a one-shot, time or volume basis. This is the key to the information superhighway — the way global access to information is distinguished from Subscription Pay services. See Tiger.

videotex

(Note: no final 't') A generic for many different home-oriented, electronic information systems (usually with color graphics) which have changed constantly over the years.

The term is correctly applied to both broadcast and wired systems. However, it is often confused with the wireline form only (with an extra 't').

So, to be pedantic, the term 'teletext' is best used for the broadcast (one-way) form, and videotext (with a 't') is best applied only to the wireline (and therefore interactive) form.

However both one- and two-way systems use essentially the same technologies. There are three main types:
— Alpha-mosaic — Prestel, Teletel, Viatel,
— Alpha-geometric — NAPLPS, Telidon,
— Alpha-photographic — Captain (Japan).
See all of these 'alpha' terms.

• The ITU-T established a recommendation (T.101) which identified these three data formats and named them Data Syntax I, II, and III. These correlate with the UK-Europe (DS I) alpha-mosaic; North American (DS II) alpha-geometric, and Japanese (DS III) alpha-pictorial developments.

• The European CEPT organization has established its own standard 'Levels' which represent progression towards their ideal system. Levels 1, 2 and 3 roughly coincide with DS I, but the emphasis is on migration from one level to the other, while Levels 4 and 5 relate to the true DS II and DS III levels.

The original Prestel is classed as CEPT Level 1, while German videotex has now move to Level 3 (it is alpha-mosaic but with dynamic character sets (see DRCS). Prestel II is now also at Level 3 (called Picture Prestel). Level 4 is an alpha-geometric system, and Level 5 is alpha-photographic.

videotext

(Note: with a final 't' — see above.) This is the wireline, two-way (interactive) form of videotex.

The value of using on-line videotext access, against the broadcast teletex is:

— Payment can be made for usage, so access can be provided to an enormous range of computer-stored information, not just to that paid for by commercial companies as part of their advertising or promotion.

— Much more information can be stored (through gateways, the amount is infinite). Broadcast systems are limited to a few hundred pages because of the cycle time — how long are you prepared to wait before the information you requested comes around in the broadcast cycle.

— Interactive systems don't just provide information, they can also be used for e-mail, asking and answering questions, creating orders, and making payments.

— Quality of image can be much higher without overloading the system. Videotext could take advantage of new technologies much more readily than broadcast systems. For instance, you can have different phone numbers for different data rates, and for different image file-types, if necessary.

• Videotext was the data 'super-highway' of the 1980s, except that no one thought to give it such a catchy name.

• See videotext data rates and videotex.

videotext data rate

The ITU-T V.23 standard provides for 75b/s output from your keyboard (a fast typing speed) while reserving 1200b/s of the channel bandwidth for the higher-speed transmission from the host computer to you.

This became the standard modem speed for the accessing of videotext and, for a while, 1200/75b/s proved popular for accessing e-mail and packet-switching networks (which have no relationship with videotext).

However, V.23 created problems because many modems couldn't distinguish between the full-duplex 1200/1200 V.22 standard and the hybrid 1200/75 V.23 so it quickly dropped out of favour. However V.23 is still used as a standard for the French Minitel.

video transport levels

International agreements have set a few standard levels of broadband video digital information carriage:

- **Level 4** carries 274.176Mb/s (4032 PCM channels) in North America; 97.728Mb/s (1440 PCM) in Japan, and 139.265Mb/s (1920 PCM) in Europe. This is broadcast TV.
- **Level 3** is 44.736Mb/s (DS-3) in North America, 32.064Mb/s in Japan, and 34.368 in Europe. Good videoconferencing codecs are available at these rates, and television is often transported in a compressed form also.
- **Level 2** is 6.312Mb/s in both North America and Japan, and 8.448Mb/s in Europe.
- **Level 1** is 1.544Mb/s (DS-1 or T-1) in North America and Japan, and 2.048Mb/s in Europe (E-1).
- **Level 0** is a single PCM channel which is nominally 64kb/s, but in North America sometimes only 56kb/s is available.

In-between rates of 128kb/s and 384kb/s are also now popular with the H.261 and H.320 standards.

• H0 is the name sometimes given to the 384kb/s standard codecs, and H4 to the 140Mb/s higher broadcast rates.

vidiplex

A technique used by TV stations to cut the cost of satellite signal transfers by halving picture quality — and to hell with the viewers. They will transmit only odd fields of one image and the even fields of another (simultaneously), then split out these fields and field-double each in the receiver. This is common in the international distribution of news.

When vidiplex is coupled with a minor misalignment in the standards converter, these images have a glorious cross-hatched herringbone pattern.

view

A formatted screen display of some underlying data in a database or text file. For instance, a mailing list may have a Form view (with only one name to a screen) or a Tables view (with a scrollable list of names in column/row format). The term view can be taken to mean a bar-chart graphic presentation of spreadsheet data, or a page-setup of text and graphics in a desktop publishing program.

Viewdata

Interactive videotex in the UK. This refers to videotext — the system using wired (telephone-based) communications and a normal TV-type display terminal or a computer terminal. The distinction is with 'broadcast videotex' ('teletex') systems. See videotex and videotext (with a final 't').

VIM

Vendor Independent Messaging interface. This is an upgrade to what was previously known as OMI. It is a common set of cross-platform messaging specifications agreed between Lotus, IBM, Novell and Apple for electronic mail services. Microsoft has something similar (MAPI), but it is not part of this standard, and Apple already has OCE which is very similar to VIM.

In effect, VIM will integrate e-mail into the Mac Finder or other OS GUI and tell applications how to access mail directories, and how to send and store messages. However, VIM does not define the mail-server interface itself since the aim is to allow applications to communicate with disparate mail-servers.

So VIM specifications are multiplatform sets of APIs for Windows, DOS MacOS and OS/2. They will all sit above the level of the basic X.400 'plumbing' (between Applications and Transport levels), and they will help software developers write mail-enable applications which can operate across multiple computing platforms. VIM will work with Novell's MHS also.

VINES

VIrtual NEtwork System. Banyan Systems' popular NOS.

violation

Some line-coding schemes for digital data have strict rules which must always be followed. For instance, the AMI technique calls for a binary 1 to be signaled by a positive voltage in the first instance, and by a negative voltage in the second instance, then by a positive again. This is the 'Alternate Mark Inversion'. So if the receiver finds a sequence of a positive voltage (binary 1), followed by a zero (binary 0) then a positive again (binary 1), it knows that this is against the rules — the second 1 should have been a negative voltage.

These violations can be caused by errors, and therefore violations signal the detection of an error, or they can be introduced deliberately as a way of carrying signaling and control information over the network.

In many modern telecommunications optical fiber networks where errors are now rarely a problem, rules are often violated as a way of signaling, so violations are no longer necessarily faults. See data transparency also.

VIP

• **Visual, Intelligent, Personal:** This is a buzz word for a Japanese view of future telephone networks. It promotes the idea that future-networks must have the characteristics of being Visual (carry video and graphics), Intelligent (find you wherever you are, and perform other intelligent functions) and Personal (numbers are tied to the person, not the terminal).

• **Video Information Provider:** The source of Video-on-demand programs These are content providers on a common-carrier (video dial-tone) network. See MSO also.

virtual

In the context of telecommunications, it is something that appears to exist, but which is actually being emulated by the system. It probably has a logical appearance but not a physical existence. You'll find the term used mainly with packet-switching systems (including ATM).

A circuit-switched network establishes an 'end-to-end' circuit which is reserved with guaranteed bandwidth for the duration of a call — and it handles all this at the Physical layer through switches slow acting switches.

A packet-switched network can provide the same reserved and guaranteed bandwidth, but over a shared medium. So it controls the network via fast-acting switches, which require a layer of computer-intelligence. The software handles the reservation and guarantee of bandwidth. This is therefore a 'virtual' channel.

Note that packet switching systems often don't reserve bandwidth, but just pass along any packets that arrive at the router port. This is a datagram service, and it doesn't require the establishment of virtual channels.

• In fact, the use of 'virtual' in front of anything involving packet-switching doesn't have a lot of significance, other than that above. Treat virtual channels as if they were actual channels — the problems are those of the carrier, not the user.

virtual 8086 mode

Intel's 80386 and 80486 chips are capable of running in a mode not available on the 286 chip. This is called the virtual 8086 mode (a subset of the protected mode) which allow it simultaneously to run a number of 8086-type programs using a special multitasking operating environment. Each program is given access to its own limited address space, and this appears to the application to be the same as the normal 1MB address space needed by MS-DOS. Each application has its own 1MB of address space.

The point is that this virtual-mode allows a 386 machine to be used as multitasking environment for a number of the old 8086-based applications. See VPAM, VM/386 and real mode.

virtual addressing

• This is relative addressing, as distinct from real addressing. The early PCs generally used a 'real' addressing system where the actual address location was specified in the software. Virtual addressing is used by computers which can support multitasking (real or pseudo) where programs may be loaded into a different part of memory each time. So data and instructions are referenced by offsets from a memory-location which may change from program to program. The memory management unit will convert the relative addresses to real addresses — the programmer and the software itself need never know the real addresses. See absolute address.

• On heterogenous LAN/WANs, problems with the Network layer often hamper routing. Gateways get around these problems by using the Transport layer to supply 'virtual addressing'.

• This is the use of virtual address space (virtual memory) with hard disks. See virtual memory.

virtual address mode
This is the protected mode which first became available with the 286 chip. Don't confuse this with the virtual 8086 mode which came later with the 386.

virtual-call
See virtual circuit.

virtual channel
See virtual circuit.

virtual circuit
A telecommunications channel or circuit that appears to the transmitter and receiver of the information to exist as a dedicated line end-to-end, but which actually uses shared facilities in the process.

It is a logical construct — a reservation of some system resources at all intermediate nodes — and it is established before communications begin to ensure reliable transport between the two end devices.

The terms virtual circuit, virtual route, virtual call and virtual channel often seem to be used indiscriminately.

• A virtual circuit in a packet-switching network actually consists only of an erratic serial-sequence of packets identified to the switching node by a common virtual-channel identifier (VCI). This identifies a route, not an address.

A preliminary circuit-establishment packet is first transmitted through the network to establish a fixed logical path to the destination, and a call set-up packet is returned from the end node. In between nodes are provided with the routing information by the return-advice packet.

All subsequent data packets then follow along this path, and are identified to the switching nodes by a Virtual Channel Identifier (VCI) rather than by an absolute address. When virtual channels are being routed to the same premises, they may be grouped together and switched as a virtual path using a Virtual Path Identifier (VPI).

So, in essence, the VPI and the VCI are two parts of a route identifier — like a zip-code which gets the packets to a geographic region, and the street-code which takes it to a house.

• In ATM systems the VCI may only be used between two nodes, so the cell will transit a number of independent links end-to-end and use a number of VCIs along the way. Each node will, however, know the correct VCI to use for each stage.

• The contrast here is with 'connectionless' datagrams where each packet travels over the most convenient route available at the time, to its destination. This requires a full address, and that address is often carried as a payload in the data section of cells. See ATM, virtual path, VCI, VCC and VPI also.

• X.25 defines two types of virtual circuit:
— Virtual calls, which are dynamically established using call set- up procedures.
— Permanent virtual circuits (PVC), which are fixed and network-assigned.

See logical channel.

virtual community
See cyberspace, WWW, WELL, MUD and CMC.

virtual containers
In SDH transmission, it is the network rather than the traffic which is synchronized. All transmissions at the current transmission rates are packaged (multiplexed and framed) into standard-speed electronic 'virtual containers' with the multiplexing hierarchy.

Each of these virtual containers is located at a readily identifiable point in the transmission (a control called a 'pointer' is available to identify them) and therefore each traffic channel is electronically accessible within the overall multiplexed structure. A computer can use the pointer to calculate which bits in the millions passing through every second, belong to a certain data-stream.

This is only possible because the transmission is synchronous, so PDH channels must be buffered and padded to fit into the stream.

However, a single optical fiber carrying data at 2.4Gb/s, can have 35,000 voice streams within it, so a hierarchy of access is needed to identify just one.

The principle is much the same as having standard-sized boxes which fit into standard-sized shipping containers for freight. These virtual containers are carried in the Synchronous Transport Module and handled independently of their contents. The payload of high-level containers can be a group of lower data rate channels or some other type of signal, including plesiochronous (not strictly synchronized) channels.

Non-SDH signals are mapped into small-VCs, which in turn are mapped into larger-VCs which can carry different signals simultaneously. Finally the high-order VCs are mapped into the STM-1 frame — and each level carries its own pointers. See SONET, B-ISDN and SDH.

virtual disk
See RAM-disk.

virtual DOS
This is a mode of operations, available with the new RISC-based PowerPCs and some other chips, where the system emulates the old MS-DOS and is therefore able to run DOS applications.

virtual LAN
This is the technique of using switches and routers on a network to create a 'logical topology' (of node addresses) which is quite independent of the physical network topology. It requires intelligent switches and routers.

This approach allows them to combine distributed LAN segments across a network into one autonomous user or workgroup. This will look to the users as if it is a single LAN and allow them to share information and resources without using WAN (routers) equipment and processes.

The idea is to segment the network into different broadcast domains, and the packets are then switched between ports which are part of the same virtual LAN. Not every part of the network sees all the packets, so this approach reduces traffic on the network as a whole, while allowing distributed segments to be joined seamlessly.

Naturally, the backbones will need to carry more load if these segments are widely distributed. Some of the new smart hubs, especially those with switching capability, can assign any port to any LAN segment. So workstations on different physical segments can be joined in a way that allows them to act as a virtual LAN. This is often the ultimate in micro-segmentation — one workstation to each LAN segment. See virtual networking.

- The IEEE is specifying these protocols in both 802.10 and 802.15. It has applications in both Ethernet and Token Ring.

virtual locks

Aka logical record locks. This is a new development in database management. It satisfies the need to prevent local access to records which are currently in use across a distributed and perhaps mirrored network — where the records at the remote site may be in the process of being changed.

This is a powerful feature in some APIs because it permits synchronized access to records on a file-server across a distributed system without needing physical locks. Virtual locks are faster than physical locks; they don't require the program to open files. They permit mnemonic record identifiers.

virtual machines

There are two totally different uses of this term.

- **Interface:** Often the term emphasizes that the 'look and feel' of our PCs is entirely artificial. It points out that we are protected by a shell (command interpreter) with perhaps a graphic user interface (GUI) from the 'real machine' which functions at the binary code level. So we deal with virtual machines every day. We are protected from the complexities of the CPU and operating system.

- **Emulation:** In emulation, a virtual machine is any combination of hardware and software that creates an instruction environment – which mimics another platform. This is environment in which applications can be run which would not normally run without specific hardware and/or a different operating system.

In mainframes, IBM is the main promoter of this idea through Virtual Machine (VM), which is a method of distributing machine resources between applications. The idea is to create an environment which presents a complete hardware configuration to the application. The mainframe is then capable of emulating an environment quite different from the underlying 'real' machine. The value of this technique is in maintaining support for legacy software; hardware manufacturers can build new machine architectures without compromising existing software or operating systems.

- When the OS/2 operating system is running on 386-based machines in 'virtual 8086' or 'real' mode, a number of DOS applications can run simultaneously — each supposedly in its own 'virtual machine'. Each partition acts like a separate DOS PC with its own memory-resident programs and up to 640kBytes of its own memory (and its AUTOEXEC.BAT file).

virtual memory

There are two applications of the term.

- Bank-switching memory beyond the theoretical limit of the hardware and the operating system. See expanded memory.

- A system that allows you to extend the available working RAM space by treating the hard-disk as a 'virtual' extension. The disk can be used temporarily to store program segments or data not currently being used. In the jargon, parts of RAM are 'mapped' onto the disk.

This is an operating system technique which allows a computer to run applications much larger than the available RAM — but naturally these programs will tend to run very much slower, since data will be shifting to-and-from disk, constantly.

Virtual memory was originally a mainframe technique only. It is now available on PCs and Macs. They will need to be equipped with Memory Management Units (MMU) hardware or special software.

- With virtual memory, instead of accessing physical RAM locations directly, all applications page the memory through virtual addresses. This is a bank-switching operation. The virtual address space is divided into pages which get swapped between disk and memory in fairly large chunks. The MMU keeps a record (a set of tables) which maps the virtual page addresses into physical locations.

- There are also segmented virtual memory systems and paged virtual memory systems. The difference is in the size of blocks being moved in an out of RAM.

virtual network

Network is vague enough to mean LAN, WAN, or even public switched network, so it is often not clear what is being discussed with this term.

The problem is not so much that the terminology is incorrect — rather that it is unnecessary.

In current usage, virtual networking suggests something more than virtual LAN — mainly in the use of higher level addressing. But the distinction is a bit vague since all forms seem to require the same intelligent hub-switching (high-speed routing or bridging) functions.

Virtual LANs can be created on virtual networks, and they are then based on MAC addresses. The user will have access to his/her workgroup no matter where he/she is located.

- **ATM:** The term was almost becoming synonymous with ATM. The point being made was that each ATM workstation has its own switched link through an enterprise-wide network (as with a PBX voice-call) and therefore the old concept of one-packet-at-a-time LANs disappears.

• **LAN:** When the term is used for Switched Ethernet, it emphasizes that users communicate without contending for bandwidth.

Virtual LANs have segment switching with only one workstation on each segment. They are connected to the rest of the network by a router, and so they require a different network address. Don't confuse this level with virtual workgroups which are generally connected to the network via a bridge and have the same network address as the network.
Note: 'switches' here, are often just fast bridges or routers based on RISC or ASIC chips.

• **WAN:** Virtual enterprise networks address the need for extensive high-speed access, and they facilitate the dynamic allocation of available circuits. Today, key corporate functions are being distributed or outsourced, and corporate users are being relocated to remote sites or to working from home. So building a virtual enterprise-wide network becomes critical to the success of the enterprise.

The network is 'virtual' because it can automatically establish 'work-group' connections no matter how distributed the participants, and it dynamically allocates resources to meet the enterprise's fluctuating requirements.

virtual path
In ATM connection-oriented switching, the 5-Byte header of each cell contains both a virtual path and a virtual circuit identifier. Both are needed. A virtual path consists of an 'ensemble' of virtual channels transported through the network to the same destination. See VPI.

virtual private networks
See VPN.

virtual reality
VR. The concept of providing all the supplementary visual and physical orientation clues that distinguish 'watching an image' from 'being immersed in an environment'. Virtual reality provides people with real-time sensations (vision, touch and sound) which are not part of their immediate physical environment.

This concept includes the futuristic idea of operating within computer games or learning environments using full-face helmet and datagloves — where gestures, head and eye movements recreate a 'virtual' world in which you exist.

While VR simulation was an extension of the work done with air force fighter pilots on gun-aiming devices, when coupled with hyperspace concepts, it can also be seen as a possible telepresence work tool (long-distance surgery), or as a learning and experiential environment in the future. Total immersion may prove to be an important aid to concentration.

Virtual reality is available at a very low level with 3-D CAD type applications where you can 'walk through' an architect's drawing of a building and get some ideal of its final appearance from a range of different perspectives — but this is more correctly called 'artificial' reality or 3D-modeling.

Real VR needs powerful parallel processors with 'rendering' engines to be capable of computing two whole scenes (one for each eye) at least at a rate of 25 frames per second, and of controling a wide range of 'objects' (including VR 'humanoids') in the scene.

• There are four different factors which characterize true virtual reality:
- **Immersion.** You aren't watching an image, you feel immersed in that image environment.
- **Navigation.** You can move around within the environment and change your point-of-view.
- **Manipulation.** By using gloves and other devices, you can make changes to the environment.
- **Feedback.** The environment effects you.

• True VR always exists as a model within the computer, not a series of stored files of images taken from various viewpoints (as was Aspen). Don't confuse VR with cyberspace. See artificial reality, cyberspace, hyperspace and rendering also.

• VRML is a platform-independent open file structure for virtual reality. See.

virtual route
This means a virtual channel in SNA.

virtual SRAM
Virtual Static RAM. This is actually a form of Dynamic RAM which has such a low consumption of power that it retains its data for a week, or perhaps even a month. It therefore appears to work much like Static RAM.

virtual storage
This means virtual memory.

virtual terminal protocols
See VTP.

virtual tributaries
The Sonet term for what is, in the ITU's SDH specification, a virtual container. See virtual containers and tributaries.

virtual workgroup
These are established often to provide a security wall between a department and the rest of the network. If the department grows and needs to occupy space, say, on a new floor, this new section may need to remain logically a part of the old workgroup but not compromise security by being permitted to breach the security wall.

Switching is generally then used to create a virtual workgroup with both groups logically on the same network, behind the same security wall.

virus
A malicious program designed to infect and modify data in your computer, often with disastrous results. A virus attaches

itself to the operating system or an application, and it spreads from one machine to another when programs are exchanged.

Note: it is currently considered impossible for viruses to be transferred by data.

They can probably only infect a machine if they are in executable code (applications, operating systems, utilities, games, etc.). They can't be carried by normal e-mail messages (only by executable file transfers).

However, there is the possibility that they can be carried within macros and agents (knowbots, etc.) and there's some suspicion about active files like Acrobat. So be careful.

Viruses cause random, selective, or systematic destruction of computer functions, extending from relatively harmless practical jokes up to the complete destruction of the system. Some just completely swallow up the memory of a machine or hog network resources.

• The best way to ensure your computer doesn't ever get a virus, is to never allow a program to run on the machine that doesn't come from a legitimate publisher. If you must occasionally run suspicious programs, open the computer box and disconnect the hard-disk drive; then boot using your back-up system disks and run everything on floppies. This is 100% safe.

• There are about 6500 documented virus types, and they will only ever be wiped out when totally secure operating systems are written. See logic bomb and Trojan horse.

visible light

This is light between the wavelengths of 770nm and 390nm. The boundaries between the more obvious colors in the spectrum can be taken as:

Near IR-to-Red	770nm
Red-to-Orange	622nm
Orange-to-Yellow	597nm
Yellow-to-Green	577nm
Green-to-Blue	492nm
Blue-to-Violet	455nm
Violet-to-Near UV	390nm

• It is important in all forms of media to understand something about the nature of light. Light is subject to:

— **Reflection.** Angle of incidence = angle of reflection. Reflected light is always polarized in the transverse plane.

— **Refraction.** On passing from a less-dense to a more-dense medium, the light will diverge towards the normal (vertical to the surface). The degree of divergence will depend on the refractive index ratio between the two media.

— **Diffraction.** The spreading (divergence) of light into shadows and the wider distribution of light passing an edge, or through a hole. Diffraction is an interference phenomenon.

— **Dispersion.** Splitting light into its component colors.

— **Polarization.** The distinction between planes at right-angles in a light beam. Reflected light is polarized.

• The speed of light in open space is 300,000km per second (186,000 miles per second). In glass it is about 80% of this speed. For many years light and radio waves were thought to travel through a mysterious substance called the Ether. Hence 'Ether'net, which originated as a radio networking protocol.

• Light from normal sources is emitted by atoms which are not phase-correlated, so there are random irregularities or 'incoherences' between the waves. See laser.

• Light waves are transverse waves (at right-angles to the propagation) and the two components of these waves (electric and magnetic vectors) are at right-angles to each other. Sound is a longitudinal wave.

• Wavelength and frequency are inversely related.

See group velocity.

Visual Basic

Microsoft's Visual Basic is available for Windows 3.x as a substitute for using the C-language. There is both a standard and a professional version. With Visual Basic you can use Windows to program for Windows. See VBX.

Visual Basic allows programmers to define screen objects — forms, buttons and list boxes — and to attach code to them to define their behavior. This is event-driven programming.

visual programming

Modern 4GL programming tools allow a lot of the development to be done by point-and-click, or drag-and-drop techniques. They will also include libraries of code snippets that can be cut and pasted into programs, together with pre-formatted 'forms' and 'reports' which have the basic layout elements already in place. See fourth-generation.

VITA(sat)

Volunteers In Technical Assistance. This group has taken some practical steps to providing basic communications services to underdeveloped countries. VITA first began experimenting with LEO satellites in 1984 when its Digital Communications Experiment was launched in UoSat-2; later a prototype payload was launched in UoSat-3 in 1990. The FCC has now granted VITA pioneer's preference status for such a LEO.

A new purpose-built VITAsat-A is due to be launched at the time of writing (Apr 1995). This is a joint venture with the satellite builder CTA Inc. of Rockville, Md. VITA will use the satellite for development and humanitarian communications, while CTA recovers its costs by using it for commercial purposes.

VITC

Vertical Interval Time Code. See timecode.

viz

In printing, this means 'namely' and is used in footnotes.

VL

Video Lens. A lens mount standard for small video camera systems (Video-8, Hi-8, VHS and S-VHS). It is an interchangeable bayonet lens mount which includes electrical connections for serial communications. This allows the camera-body and the micro-chip in the lens to exchange information about iris, zoom and focus (and therefore have standard auto-exposure, motor zoom and auto-focus controls).

VL-bus

Video Local. A local bus architecture being promoted by the VESA. It uses a different connector from that of the ISA, EISA or MCA expansion connector, but these are all kept in line so dual-purpose cards (longer than normal) can be inserted, and make contact with both slot types.

Not every VL-bus based card is compatible with every VL-bus system, and cards are specific to both the microprocessor speed and the particular expansion-bus architecture.

The aim of the VL-bus was to transfer data in a 32-bit wide path at rates up to 200MB/sec (at 50MHz), but in practice only about half the theoretical rate is achieved. The specification allows for three peripheral cards running at 33MHz, but fewer at a higher speed. Contention for the address and data lines is the limiting factor since the CPU, cache and cards all need to use these same connections, often at the same time.

• There is now a version 2.0 VL-bus specification with a 64-bit wide data-path.

• Pentium-based machines seem to have preferred the PCI approach to the VL-bus. Apple has decided to use PCI also.

See expansion bus, local bus, PCI and ATA also.

VLAN

See virtual LAN and virtual networking.

VLF

Very Low Frequencies. The radio frequencies between 3kHz (actually about 10kHz in practice) and 30kHz. These are 'myriametric' waves from 100km to 10km in wavelength. The ionosphere tends to trap radio waves at these frequencies, however, at these wavelengths, surface waves carry around the curvature of the Earth for long distances.

In this band you will find submarine communications and the omega navigation system. See LF also.

VLSI

Very Large Scale Integration. VLSI chips have between 10,000 and 100,000 transistors. They are only possible because of metal-oxide semiconductor (MOS) technology. VLSI chips usually need a rectangular multi-pin package.

VM

• Virtual Machine. An IBM mainframe operating system. See IBM and also VM/SP.

• VM-channel. This is the VESA Media channel for video adaptors, decoders and video capture boards. The VM-channel connector replaces the VGA features connector. The VM-channel is processor- and bus-independent, and it allows 16- and 32-bit devices to be mixed. It can be implemented on VGL-bus, PCI, MCA, EISA and ISA systems.

• VM/386: This was a multitasking 'environment' that used the 386's virtual mode to run numerous MD-DOS applications.

VM/SP

Virtual Machine/System Product. An operating system control for larger IBM mainframes. The idea here was that a machine could simulate multiple copies of itself. This is another way of making the computer capable of interleaving several tasks — each 'virtual machine' will run an application designed only for single tasking. The VM/SP part allows users to log on and off the machine, and be allocated their 'virtual machine'.

Most VM/SP virtual machines run CMS, and serve one user at a time. But they can also run MVS and service a few dozen.

VMC

VESA Media Channel. This architecture provides a fast path for passing data to video peripherals. There are various implementations of VMC however. It is seen as a guideline, rather than a standard.

VMD

Virtual Manufacturing Devices. This is part of MMS for interfacing different types of devices and robotics equipment.

VME

Virtual Money Exchange. A new standard being developed to allow global interoperability between smartcards and stored-value cards (SVCs). Visa, MasterCard and Europay are behind this group. See STT and CAFE.

VMI

Voice Messaging Interface. See SMDI/VMI.

VMS

Virtual Memory System, aka VAX/VMS. A very widely used operating system for DEC VAX mini computers. It uses a command language very similar to MS-DOS. The current version (called OpenVMS) being promoted by DEC is being integrated, through APIs, with Windows NT.

• VMS keeps previous versions of the files. When you don't specify a version number it will give you the latest. To remove all but the current version, you use the Purge command. To Delete all versions, you can use the asterisk wildcard.

• DEC systems also use Digital Unix, which is OSF/1.

VMWI

Visual Message Waiting Indicator. A new system (related to CLID) where the telephone company is offering voice mail services. It is a substitute for stutter-tone for people with VMWI devices. It transmits a FSK analog modem sequence down the line without going off hook. The device reads this, and switches on a light.

vocoder

VOice COder/DecodER. This is a type of voice coder which depends on speech analysis, and a speech synthesizer for reproduction. A vocoder models the voice using an analog waveform, and it derives 'parameters' which allow low bit-rate transmission.

The term vocoder is applied both to source coders and to hybrid coders, but not to waveform coders (PCM and ADPCM).

Vocoders take an analog voice waveform, break it up into frames. They then extract from each frame, parameters that describe the sound during that period. They can work at rates of 50 or 100 frames per second — and can produce comprehensible output at rates as low as 2.4kb/s, but 4.8kb/s is currently the limit for 'telephone-quality' speech.

Vocoders depend on various models which have been constructed for the action of the human vocal tract — and so different types work in different ways. In general, they extract parameters for pitch (created by the vocal cords), filtering (created by the shape of the mouth), and energy (how loud).

Predictive coders then calculate the most likely change in the waveform (based on past experience) and reconstruct the sound by decoding these parameters. This predicted waveform is then compared with the incoming signal, and the difference is coded and transmitted. These digital vectors are then added to the basic synthesizing information.

• Early vocoders tended to have a machine-like quality but recently they have been dramatically improved. They are now able to operate effectively at data rates as low as 2.4kb/s, and they can provide very acceptable telephone quality at 9.6kb/s — about one-seventh that needed for PCM.

• Note that voice codec is a general term, and that there is a difference between voice coders and vocoders. The term vocoder applies to a chip or device which uses source coding techniques (or hybrid coding techniques), whereas voice coders are compression (bit-rate reduction) systems which require prior digitization (usually into PCM).

See source code, waveform digitization and phoneme encoding also. Also check VSELP, CELP, LPC and MBE.

VOD

• **Video-On-Demand.** See.

• **Voice Output Device.** Mainly toys. These are usually programmed with phonemes. See.

voice

• Voice reproduction in analog telephone systems requires about 3kHz (300—3400Hz) of bandwidth to carry the words, the variety of tones, and the particular sense of the speaker.

• The original standard set for voice reproduction in digital telephony was 64kb/s PCM (see A-law). However, voice quality which is almost as good can be achieved at half that bandwidth, with 32kb/s ADPCM. Even at 16kb/s, ADPCM quality is very good — it is used at this rate on transatlantic cables.

• At rates below 16kb/s, vocoders are very much favored for voice reproduction, but some of these have troubles with the higher pitch of women's voices. They will not handle musical notes or modem tones.

However good vocoder speech can now be coded at 4.2kb/s using the MBE vocoder, and at 4.8kb/s using QCELP. At the rate of 2.4kb/s, voice is still recognizable and understandable, but this is approaching the limit of tolerance for most people, even on a private network.

• A voice in music synthesis is a distinct sound. Many synthesizers will simultaneously generate many (say, 20 or more) voices.

• In a normal telephone conversation, the voice-duty cycle is taken as about 35%. This is the amount of time a channel is being used. For half the time each side is listening, and for some of the time neither is speaking.

voice activity ratio

A figure used in telephony and cellular radio calculations to represent the 'bursty' nature of voice conversations. Typically a speech channel (in interactive two-channel conversation) is used for only one third (0.33) of the available time.

This unused space can be used to transfer other voice calls in international cable systems (see DCME), and the suppression of the carrier when it isn't needed is an important part of maintaining the capacity of some cellular phone systems such as GSM. See DSI.

voice bandwidth

In telephony, this is usually taken to refer to the telephone frequencies from 300 to 3400Hz. The full hearing range for young people is from 100Hz to 18kHz, and frequencies up to 20kHz are needed for the 'sparkle' on music recordings. The detection of the high frequencies drops off regularly throughout life — and also from heavy-metal abuse.

voice codec

This is a hardware device which translates an analog voice waveform into a highly-compressed digital output. 'Codec' here just means 'coder-decoder' which is the opposite of 'modulator-demodulator' or modem.

There are two types of voice codecs: waveform codecs and source codecs. ADPCM is currently the favoured waveform coding technique for reducing telephone bandwidth (to 32- and 16kb/s) while Linear Predictive Coding (LPC) and Multi-Band Excitation (MBE) are the two current source-coding techniques that provide the best results for the least bandwidth. MBE voice codecs at 4.2kb/s provide good quality telephone speech. Source codec are also called vocoders. See voice coding, waveform digitization and vocoder.

voice coding

The digitization of voice in telephony. There are five main techniques in use today:

— **Pulse Code Modulation (PCM)** which is a relatively standard process of sampling the waveform (8kHz) and measuring the deviation from zero (quantized to 8-bits). Variations called 'log PCM' quantize the signal using non-linear measures and this improves the quality-to-data rate ratio. See A-law and mu-Law. This is the base waveform-coding technique used everywhere — not just in telephones. It is now treated as the first stage, from which most of the other forms of waveform (difference and delta techniques, and source coding) are derived.

— **Adaptive Differential PCM (ADPCM)** is a variation on the above which transmits the difference between two samples, rather than the absolute value. It is also adaptive in the sense that it can adjust the sampling technique in the light of experience. ADPCM codes voice waveforms at excellent quality at half and a quarter the PCM standard 64kb/s rate.

— **Delta Modulation** is a variation on the above which codes the direction of the waveform in terms of bits (not bytes). The sample rate must be much faster than PCM, but a logical 1 signals that the waveform is rising, and a logical 0 signals that it is descending. At a high-enough rate, this 'delta' system can carry good quality voice at 16kb/s, but it is useless for modems.

— **Source code systems (vocoders)** use a variety of techniques. Basically, they break the voice down into constituent sounds (vowels, consonants) and code these. The receiver then synthesizes the sounds using these transmitted parameters. Vocoders now produce very good quality at 8kb/s and acceptable quality at 4kb/s. See vocoder and phoneme encoder.

— **Hybrid systems** use both source coding and waveform coding techniques. This appears to be the way for private line telephony.

voice coil

Voice-coil actuators on hard disks are used to control the arm with the read-write head; they are actually linear induction motors. They are quite complex electronically because they rely on logic control, however they are very quick in action and have few moving parts. Voice coil hard disk drives typically have access times of 40ms or less. Average access times of 28ms are common and 12ms is not unknown. See stepper also.

voice-grade

A circuit with a bandwidth of 4kHz. See grade of service also.

voice mail

The term covers everything from simple answering machines, to carrier offered services, and sophisticated voice command/-voice recognition call diversion and recording systems.

Discounting simple tape answering machines, voice mail now exists on everything from a PC up. The main limiting factor is the mass-memory available to store the messages in digital form. Some are provided through the company PBX, some come as a stand-alone unit, and some are offered as a bureau service.

Telephone carriers are now offering voice mail as a standard service on cellular phone systems (so that calls aren't missed when the phones are not in contact) and as a premium service for wirelines. When a message has been received and is being held at the local exchange, the phone dial-tone will change to a 'stutter-tone' as an indicator that a message is waiting. With future ADSI systems, a special text-command message will be delivered to your phone, and trigger an indicator light.

Also coming soon are:

- Message forwarding, where a recorded message can be referred to another mail box (as distinct from call forwarding).
- Append-and-forward systems, where a voice-mail message can be added to a text e-mail message, either as annotation, or for referral.
- International voice mail. Your recorded message is stored locally, then batch transferred (usually at night) to the computer storage facilities in the other country. This is now being widely used in academic circles; it only requires the use of a touch phone to enter the address.
- Features such as personalized greetings, distribution lists and automatic diversion of calls may also be included.

voice processing

• **Systems:** A generic term applied to a range of telephone system extensions to the normal PABX which include:

- Voice mail/messaging.
- Customer information systems.
- Voice bulletin boards.

• **Technologies:** See vocoder, waveform digitization, source coders, voice recognition.

voice-quality

A high-quality voice telephony system must have the following characteristics:

Clarity and Loudness.
Freedom from background or extraneous noise.
Lack of distortion which affects intelligibility.
Timbre which affects recognition.
No perceivable delay (latency).
No annoying echo or reverberation.
No breakup.
No disconnection of link.
No crosstalk with other conversations.
No variation in pitch (frequency).

See QDU, which is a subjective measurement of digital quality, and also intermodulation distortion.

voice recognition

See speech recognition.

voiceband

The standard bandwidth of a telephone channel. This is between 300 and 3400Hz in analog, although with guardbands, a full 4kHz is usually left. For digital transmission, this probably means the PCM standard rate of 64kb/s, although good voice can be carried much below this. See broadband.

voiced sounds

In speech synthesis, these are vocal sounds such as vowels, semi-vowels, voiced-stops and nasals which are created by vocal-cord vibrations. The distinction is with fricatives and other unvoiced sounds.

See also formant synthesis and phoneme encoding.

volatile memory

Memory that loses its information when the power is switched off. See Dynamic RAM and non-volatile memory.

voltage signaling

This refers to the technique used in RS-232 and other cable connections, where the control signals are exerted by raising or dropping the voltage on a line. When the voltage is raised (in RS-232 it is 'raised' to a negative voltage of about –12V) it is said to be 'asserted', and when it is not it is 'inhibited'. Usually it is asserting something like 'I am ready to transmit', or 'My buffers are full', or it may be asking a question 'Are you ready to receive?'

volume

An old computer term that meant one coherent spread of data — usually a physical area on a disk space which was under the control of a single disk directory. In practical terms, a floppy-disk usually holds only one volume, but large hard disks may be partitioned to contain many (usually two or three) volumes. Each partition will have a different volume name, and each can effectively be treated as a totally different system.

• Each volume can also have its own operating system, and operate quite independently of the others. This is a smart move to make when you are upgrading your operating system; create a partition with a copy of the old system inside. If the new system begins to crash your applications, you'll be able to swap back again at a moment's notice.

volumetric storage

This means future storage devices which don't just store bits on the surface of some disk or tape, but within it — in planes within the medium at x/y/z coordinates.

These are three-dimensional devices — memory cubes. Tamarack's holographic system has this potential.

von Neumann architecture

First developed theoretically by John von Neumann. The expression refers to the basic control-flow used by most traditional computer designs. These comprise:

— A CPU (program controller) with microcode and an ALU.
— A single large bank of memory.
— Input-output devices.

The key to the success of von Neumann architectures is that the processing chip performs one function at a time and takes both data and instructions from a single bank of memory.

It does this very fast, but always in a strict one-step-at-a-time sequence. See Turing machine.

With this technique here are three conceptual 'black-boxes':

— **The program counter** tells the processor which activity is next. It does this by pointing to an instruction location.
— **The processor** decodes and executes each instruction. The instruction will provide the coded address of the data, and so the processor first fetches data from:
— **Memory** which stores both instructions and the data. Another part of the instruction tells the processor where in memory to store the results when it has finished executing the instruction.

• Adding a coprocessor doesn't really change the von Neumann computer's linear design; rather it extends it. Coprocessors are not parallel processors.

• The distinction is now with full-scale parallel processing (non-von), some of the new superscalar chips, and with neural networks.

von Neumann bottleneck

In a standard PC design, the computer's top processing rate is set by the amount of time it takes the chip to access data and instructions from the working memory.

The clock speed of the CPU often outpaces the ability of the memory to deliver data over the data bus. The bottleneck is overcome to varying degrees by the use of Static RAM (or TTL/ECL memory), cache systems, wider data buses, reduced wait states, and pipelining. See all.

von Neumann machines

This refers to some futuristic thinking of von Neumann. He was discussing the theoretical possibilities of constructing a machine capable of remaking itself in its own likeness. This is Arthur C. Clarke's Universal Replicator.

VP/ix

A way of running DOS applications under Unix.

VPAM

Virtual Protected Address Mode. This is the mode introduced with the 80286 which provided an address space of 16MBytes. Any DOS programs written for the 8088/8086 could only run on the 80286 in the 'real mode' which was limited to 1MB of direct addressing. Xenix 286 and some other operating systems were able to use the full 16MBytes of VPAM, but DOS needed work-arounds. See expanded memory, extended memory, EMS and XMS.

VPC

• **Virtual Path Connection:** In ATM this is the chain of Virtual Paths along which the cells travel, identified by the VPI (identifier) in the header. A VPC is a collection of VCCs routed together as one unit. See VCC also. Don't confuse this term with PVC which is a permanent virtual circuit.

• **Virtual PC mode.** See VPAM.

VPI

Virtual Path Identifier. In ATM cell headers, the virtual path field is 1-Byte only. So the CPI must be used in conjunction with a 3-Byte VCI to establish the full 'address'.

In effect the VPI is a group route/address for a bunch of VCIs traveling to the same location. The virtual path actually only exists from one node to the next, where it is then exchanged for another VPI before the cell is passed on.

Each switch/node maintains a VPI translation table that is updated every time a new connection is established. The switch table also maintains a record of the priority for each VCI.

VPN

Virtual Private Network. The use of normal public-access switching and network facilities in such a way that the customer thinks the network is private — even though channels have not been reserved for exclusive use.

This is possible with modern CCS#7 switching systems, ISDN, and intelligent networking (IN). The customer is permitted to assign his own numbering scheme through the company's virtual network, and these calls are processed through the intelligent network, using look-up tables.

The local exchange facilities will have been programmed to perform any necessary number translations, and they will allow various PBX-like features such as camp-on, call conferencing, and class-of-service level (including subscriber service management). See Centrex also.

- VPNs account for about 8% of the US long haul market, but they are growing at 13% compound. Essentially, VPNs are proliferating because large corporations are now able to negotiate special discounts (35% or more) and service deals with the carriers, and so they get priority rights to what pretends to be public-access facilities.
- In some countries VPNs are offered by the carriers in order to stem the tide of private networking which is gradually stealing public-switched services. They are selling services rather than capacity. This is why they often won't sell dark fiber.
- VPN also includes the use of shared-media services, such as X.25 packet-switching and frame-relay, as if it were a private network link. Some VSAT services are also classed as VPN.
- VPNs are usually tariffed on a pay-as-you-use basis (distance x time) plus some recurring costs such as access line charges. Some carriers, however, charge flat-rate. Often a different charge will apply to on-net (within the company), as distinct from off-net calls (over the VPN then out into the wider world).

VPS

Video Programming System. A teletext-based timing system used in Germany to enable teletext-equipped VCRs to identify programs and their start and end times. VPS uses a unique identifier for each program which is transmitted as a data signal in a spare teletext line.

Your VHS timer is set by using a bar-code called a Program Delivery Code. See PDC.

VR

Virtual Reality. See.

VRAM

- **Variable Rate Adaptive Multiplexing:** A technique used in multiplexed telephone networks to handle potential overload conditions by reducing the quality of the audio. See TASI, DCME and Adaptive PCM.
- **Video RAM:** VRAM refers to special fast memory chips used in that section of computer memory which holds the video information in graphics-oriented computers.

The correct name for the specific chips called VRAM, is Dual-Ported Dynamic RAM. The dual ports allow the video information to flow to-and-from different sources, simultaneously, over two buses. The chip can be both outputting information to the video display, and still allow the CPU to have access through the other port.

VRAM can also write a bit to a single pixel location (masked write) while normal DRAM must handle the data a byte at a time. In the new HDTV systems, frame-store RAM requires even faster techniques.

— In computers, frame buffering and double-buffering techniques are now used to speed up screen refreshment rates. See video-buffering.

VRC

Vertical Redundancy Check. This is a fancy name for what is basically parity checking. See parity.

VRML

Virtual Reality Modeling (Mark-up) Language. This is an open platform-independent file format which can generate 3-D images. It is an extension of OpenGL, which was itself a development of Iris GL.

VRML provides for 3-D graphics on Windows and it has a command language for geometry, motion, and WWW links. It is available on the Internet and WWW, and is now supported by a couple of dozen of the major companies — DEC, Netscape, NEC, etc.

VSAT

Very Small Aperture Terminals. A small-dish (1.2 to 2.4 meters diameter) system used on C-band and Ku-band geostationary satellites for voice, video and low-rate data transmission. This is a packet-oriented, communications system for banking ATMs, personal computers, LAN links, and some low-quality voice.

VSATs use a star network with small earth stations predominantly communicating through a large central hub, however mesh (hubless) VSAT networks can communicate with each other directly. There are both one-way (receive only) and two-way (interactive) VSAT terminals. Most VSAT terminals have a dish in the order of 1.2 to 2.4 meters in diameter.

Interactive VSATs usually handle data at 128kb/s on the uplink and 512kb/s on the downlink. They will generally use SDLC or X.25 connections, but they can be configured for other packet or circuit protocols. Various systems use FDMA, TDMA, or CDMA access techniques.

In the USA, there has been a rapid growth in the use of VSATs for satellite news gathering, business videoconferencing, and for data-distribution. However, the techniques are still used mainly for remote telemetry, data-gathering, and for links to ATMs, POS equipment, VDUs and for RJE.

The main problem is that VSAT dishes tend to be subject to adverse weather conditions, and they are expensive. New technology has led to the construction of smaller and cheaper antennas (down to 0.75 meter) which transfer data at 64kb/s on the uplink and 128kb/s on the downlink.

There are international problems with VSAT when they are used to by-pass local telephone carriers and local regulators. They are likely to find the new LEO systems highly competitive.

VSB

Vestigial sideband. When any amplitudinal signal is modulated onto an analog carrier wave, the signal which results requires twice the bandwidth of the baseband signal. We speak of the modulated signal as having two 'sidebands', each a mirror image of the other — and therefore one of these sidebands is redundant and can (theoretically) be removed without loss of information, but with a 50% saving in bandwidth.

This is known as SSB (Single Sideband). In practice the SSB signal is of lower quality than the full AM original, so, for this and other reasons, the amplitude modulation of a TV picture signal is subject to a partial SSB operation. This 'economizes' on bandwidth without removing the sideband completely.

The major band remaining, is the vestigial sideband (see). See sideband and SSB also.

- A PAL signal has a bandwidth of about 11MHz if both sidebands are available. VSB reduces this to about 6MHz, and another 1MHz is made available for the sound carrier, resulting in the standard PAL channel-width of 7MHz. Because the two sidebands of the carrier are asymmetrical, the center frequency of the PAL video signal is only 1.25MHz from the edge.

- VSB has recently come back into prominence as one of the modulation methods being considered for future American digital HDTV.

In the Grand Alliance HDTV trials, Zenith's 16-level VSB system transmitted 43Mb/s in a single 6MHz cable channel — one-third more data, and at lower cost, than competing 64-QAM proposals.

At the time of writing, the Grand Alliance is still committed to VSB, although Canada and Mexico have aligned with the European DVB group in promoting either the 4000-FFT or the 8000-FFT (disputes are still raging) form of COFDM. These will have thousands of carriers.

VSD

Video Signal Decoder. Generally this chip decodes a video data stream (say, from a CD-ROM disk) and sorts out the interleaving video data and/or stereo channels. It will generally provide an RGB output.

VSE

Virtual Storage Environment. An IBM mainframe operating system, which is the 'little brother' of MVS. The new version is called VSE Version 5.1. It is designed for the lower end of the System/370 family. It does not conform to the SAA standards.

VSELP

Vector Sum-code Excited Linear Predictive. A vocoder technique used in very low bit-rate voice digitization. The 8kb/s VSELP codec is used in American NADC mobile phones where it doesn't have a good reputation.

Current VSELP algorithms can reconstruct (with reasonably accuracy) a 'radio' voice quality with only 4.8kb/s of digital information and give telephone quality with 8kb/s — but the mechanical quality of the sound has led to NADC being called 'underwater' phones in Hong Kong. The vocoder technology is a variation on LPC which was developed by Motorola.

V.series

See V.## at the front of this section.

VSMP

Vertical Surface Mounted Package. An LSI package design invented by TI for surface mounting where space is important on the PCB. See ZIP and SOJ also.

VT

ASCII No.5 — Vertical Tab (Decimal 11, Hex $0B, Control-K). A format effector which advances the cursor to the next set line in a series of pre-assigned printing lines.

VT-# terminals

Video Terminal standard No.# (aka ANSI terminal standards.). These are popular terminal standards from Digital Equipment Corp. (DEC) designed originally for the VAX computer.

VT-52 and VT-100 selections appear commonly in communications software as emulation standards. They are the preferred alternative to dumb TTY terminal emulation.

The VT-standard provides cursor controls and other special functions, so if you have a choice when dealing with a remote host, you should always try to use VT-102 or VT-100. VT terminals have different ways to perform screen-clear, split screen scrolling, wrap-around and answerback, and they can send strings in response to an inquiry.

- The main models are:
  - **VT-52.** The base-level standard. It is basically TTY emulation with formatting controls so that you can control the movement of the cursor around the screen.
  - **VT-100.** This terminal had a major impact on the computer industry because it provided control codes for positioning the cursor, clearing the screen, and selecting normal, bold or underline type. VT-100 also has pass-through printing, visual attributes, and other features. The basic VT-100 terminal will have four LEDs, so these need to be simulated using four PF# keys in terminal emulation.
  - **VT-101.** This was only a minor upgrade on V.100.
  - **VT-102.** This version added special function keys known as PF1 to PF4, and arrow and pad keys. It also had automatic printer controls, but otherwise it is identical to V.100.
  - **VT-125.** The full ANSI standard is mostly a subset of VT-125, and this mode is backwardly compatible with VT-100.

— **VT-220 and VT-320.** These have additional character sets and keys which include Find, Select, Insert, Remove, Previous Screen, Next Screen, an arrow cluster and F1 to F20. VT-320 includes the 25th display line as a status line. These are also backwardly compatible with VT-100.

• Although these terminals were originally defined by DEC, they were later taken up by ANSI. This created an ANSI series of control standards for terminal screens based on the ESC[control code sequence (see ESCape). However, the original DEC terminology has stuck.

• If in doubt, emulate VT-100.

VT/VTP

Virtual Terminal/Virtual Terminal Protocol (ISO 9040 and 9041). This is the application standard for EDI.

VTAM

Virtual Telecommunications Access Method. That part of the NetView IBM software package that mainframe applications use to send and receive messages over an SNA network; it is the mainframe's communications subsystem. It runs on the host mainframe and cooperates with the Network Control Program (on the FEP) to establish links between the host and the cluster controllers. It also sets the pacing and LU characteristics, and implements layers 2, 3 and 4 of SNA.

VTP

Virtual Terminal Protocols. These are OSI application protocols (DIS 9041) at the Application layer. These are only partly defined.

They enable different computers to communicate with different types of terminals by establishing virtual terminal connections across the network.

VTP presents the user with a local terminal interface which the OSI network interpreters can translate into a different terminal type, and so emulate another type for the host application. On TCP/IP networks, Telnet performs this function.

The four main types of terminal emulation are:

— **Scroll-mode terminals.** These have no local intelligence and the characters are transmitted immediately on entry. Incoming characters are printed (or sent to the screen) in order; the top line constantly scrolls off the screen and is lost.

— **Page-mode terminals.** These are keyboard and display terminals which are cursor-addressable. Both the user and the host can randomly access and modify characters on the display. The display can be one page at a time.

— **Form or data-entry terminal.** These are the same as page-mode terminals but with predefined forms for data entry. They have fixed or variable field length and field attributes are used for validity checking.

— **Graphics terminal.** These are fully addressable and able to create two-dimensional patterns on the screen.

VTR

Video Tape Recorder. These use reel-to-reel tapes rather than cassettes. Most 1-inch production videotape machines are of this kind. See B- and C-format video.

Vtx

Videotex. See.

VU

Volume Unit (meter) which measures the relative loudness of a sound. It is a recording level indicator used on professional audio recorders and sound-mixing equipment which indicates the average signal levels in dB, relative to a fixed (0dB) point — the limit of distortion-free recording. This is a standardized system used in professional equipment.

VVSAT

Very, Very Small Aperture Terminal. These are very small sized VSAT terminals used on Ka-band satellites. MSAT and USAT terminals are also terms applied to terminals smaller in size than conventional VSATs.

VxD

Virtual Device Drivers. These handle caching, installable file systems, networking communications, etc. Windows runs its applications as a virtual machine. In this environment, the virtual drivers, along with true device drivers, are installed in conventional memory. VxDs in a multimedia device don't use memory above 1MB, except for code. The data buffers and hardware buffers also remain in low memory.

W-band

A vague term for the part of the radio spectrum above 90GHz.

W3

World Wide Web. See WWW.

W4W

Windows for Workgroups.

wafer

- The large (3- to 8-inch) circular slice of silicon on which chip manufacturing depends. Hundreds of chips are etched onto each wafer, then the wafer is cut up for use. 4Mbit DRAM, which is the most common memory chip made today, is manufactured on 8-inch wafers using 0.8 micron line-widths.
- Wafer stacks were once proposed as 'silicon disks'. By using the whole surface of the wafer (which holds dozens of memory chips) and layering these together, it was possible to get many gigaBytes of data into a device, together with a battery power supply.
- Wafer cartridges are small continuous-loop cartridges of tape.

wafer-scale

An idea, widely promoted by a number of companies, to construct very-very large-scale integrated circuits using the whole surface of a silicon wafer (2 to 3-inches in diameter) as the base. This giant chip could, in theory, perform faster and better than a lot of discrete components which need to be soldered and wired together.

The problem is that defects and design difficulties rise as the square of the size. At present it is cheaper and better to make VLSI components independently, test them, then solder them together on a PC board.

WAIS

Wide Area Information Server (pronounced 'ways'). This is a distributed information service available to search Internet database indexes, using simple natural language input. It allows you to perform keyword searches of the full-text using electronic forms. The search strategy is much more sophisticated than Archie and Veronica, which only allow you to search for keywords in titles.

WAIS also has a feedback mechanism which enables you to refine your search strategy progressively. There are many free implementations available.

WAIS requires special servers, but these are also made available from WWW; they can also be accessed through Gopher menus. See Archie, Gopher and WWW.

wait-states

RAM memory chips can't be accessed instantly. Therefore the speed of the CPU will not necessarily define the overall speed of the computer — it may be limited by slow memory (slower memory chips are cheaper). If the processor needs to wait a cycle or two for the memory chips to supply information along the data bus, then the overall processing speed of the machine will decrease.

A computer with one wait-state will normally be about 10–20% slower than one with zero wait-states.

Zero wait-state is therefore the desirable end, but it is not achievable in many computers without upgrading the RAM. In practice, the wait-states of most computers are set in BIOS to an optimum number and if you reset it to zero the computer may simply stop working. See pipelining also.

- The 80386 chip requires two clock cycles for a standard no-wait-state memory access operation.
- The 80486 reduced this to one cycle, and it added a block-mode transfer as well.
- You might also need to change the wait-states of a computer to allow older, slower adaptors to work with a new machine.

Walsh codes

These are the primary transmission patterns used in CDMA mobile phone systems. A few dozen CDMA handsets can be operating within (sharing) the same 1.25MHz of radio bandwidth, and it is important that the interference effects are minimized. Walsh codes serve to identify each transmitter to the base (or vice versa) and they spread the code (the chip rate) — which is the spread spectrum effect.

However to reduce interference, each, on average, must counteract the effect of the others — if 50% are transmitting a positively phased pulse at any moment, the other 50% should be transmitting a negatively phased pulse. This is possible only because all transmissions are synchronized, and all use orthogonal Walsh codes.

If you toss a coin 50 times you will most likely have an orthogonal Walsh code which is 50 bits in length. Do this a dozen times, and you'll have a dozen different orthogonal 50-bit codes, each quite distinctive in its pattern. This is the basis of direct-sequence code-division multiplexing systems.

WAN

Wide Area Network. The correct use of the term is for a single private digital interconnection network for common use in a company; generally it will be nationwide or international. Its main functions will be to join LANs to LANs, and to join them to minis and mainframes. The distinction here would be with Metropolitan Area Networks (MANs) which are carrier-supplied services of a fixed protocol (see DQDB, SMDS and FDDI-II), and LANs. LANs are defined as that part of the network which can be locally addressed (by a MAC address) while a WAN is that which can be addressed using a network address.

WANs will usually be linked by standard leased telephone circuits using routers and gateways. They will interconnect the LANs and form a single homogeneous or heterogenous network which can exchange data. The most commonly-used protocol today would be IP, although SNA still has a very strong following in the IBM mainframe environment.

• In telecoms, the term WAN is sometimes applied to very large public packet-switched networks. See X.25.

wand

• A hand-held barcode reader.

• A special hand-held OCR reader for special OCR fonts.

WARC

World Administrative Radio Conference. This organization is now a subsection of the ITU (ITU-R) and it oversees radio communications, allocates radio spectrum in a general way, and controls direct-broadcasting satellite frequencies — among other things.

warm-boot

To restore the operating system and re-establish all the variables, without totally switching the computer off first. The only difference between a cold start and a warm-boot is that less physical strain is placed on the hard-disk, since it doesn't need to come to a halt and park the head. In IBMs and compatibles a warm-boot is performed when you simultaneously press Control, Alt and Del.

warm-link

See hot-link.

Warp

IBM's 'leaner and meaner OS/2'. It has been produced to challenge Windows 95 and provide an alternative for potential low-end users of Windows NT. Its main advantage seems to be that it can fit into 4MB of RAM. It will run 16-bit Windows applications, but it won't run the new 32-bit applications being designed for Windows 95 and NT. It is seen as a competitor to Windows 95.

• The LAN Client Connectivity version will have peer-to-peer capabilities and a TCP/IP toolkit.

Watermark cards

These are magnetic cards in which a number has been embedded into the structure of the magnetic tape during manufacture. This number can't be erased or changed, so it acts like a watermark in a banknote. Additional magnetic data can be added or erased in the normal way, over the top of the Watermark, but any attempt to erase the Watermark itself will destroy the card.

WATS

• Wide Area Telephone Service: An older US term for free phone 800 service. See InWatts.

• WATS extender. This is a call diverter (see).

WAV

Windows Audio/Visual file-type. Microsoft says this type of file can store audio messages, and be embedded in applications such as e-mail and spreadsheets.

waveform digitization

Also called waveform coding. This is the range of techniques such as PCM, ADPCM and Delta PCM, which treat audio as audio — not audio as speech. The distinction is being made between voice codecs which can handle any form of audio equally (waveform coders) and those which are designed specifically to mimic speech (source-coders or vocoders).

Waveform digitization is independent of signal type, and so it can be used for speech, music, DTMF tones, modem carriers, etc. But reasonable speech quality can generally only be handled down to about 12kbit/s. ADPCM at 16-bits is widely used in submarine cables and the quality is very acceptable.

Vocoders synthesize speech at even lower rates by using parametric models of the vocal tract, where the values of certain parameters are transmitted, rather than the waveform variations. However modems and DTMF signaling won't work through these systems, which is why you can't send data through a digital phone microphone circuit. See vocoder, source code, hybrid coding, LPC, MBE and formant synthesis.

waveguide

A general term for electrical and optical 'pipes' carrying radio-frequency and light signals.

• The early waveguides were copper tubes which carried broadband radio channels. The signals actually flow across the inner surface of the conductor (the skin effect) and the tube confines and protects them from external electrical disturbances. Before the days of optical fiber, large numbers of phone calls qould be frequency-division multiplexed and carried in this way.

There are two main types of radio waveguide still in use: rectangular waveguides are used in microwave antenna installations where distances from the transmitter to antenna are only a few hundred meters. They are often about 30cm in width and height. Circular waveguides are generally about 5cm in diameter and typically carry radio frequencies up to 100GHz.

• Sometimes cheap coaxial cable is classified as a 'waveguide' but the term should only be applied to hollow-tube constructions (however, some coax. is of this type).

• Optical fibers are light-waveguides. See.

wavelength

The length of the wave, which is an inverse of frequency (given that the propagation of the signal is at a constant speed). Light and radio waves travel at 300,000km/sec, which is 300,000,000 m/sec. Divide 300 million by the frequency in Hertz to calculate the wavelength in meters. The sign for wavelength is lamba λ.

• The UHF band (300MHz to 3GHz) extends from wavelengths of one meter, to one-tenth of a meter.

- The HF band (3MHz to 30MHz) extends from wavelengths of 100 meters, to 10 meters.
- Visible light has wavelengths between (Red) 900 and (Violet) 400 nanometers. It is often expressed in angstroms.
- Infrared is in the region between 1000 (upper end of microwave band) and 0.9 microns (Visible Red is 900nm).
- Wavelength determines the optimal size of transmitting and receiving antenna. Normally a half-wave, quarter-wave or full-wave antenna will be used; it is exact integer divisions which provide the best results. See loading coils.

Wavelength Division Multiplexing
See WDM.

wavelet compression
See DWT and transform coding.

wavetable synthesis
A synthesis technique which has recently migrated to some of the better new sound-cards for PCs. This approach is widely used with MIDI electronic keyboards in the music industry.

Wavetables are stored, digitized samples of the actual waveforms of musical instruments. These are used as a source when reproducing the sound-quality — rather than synthesized approximations of the instrument's notes. See FM synthesis.

To play a particular wavetable note corresponding to an instrument, the computer will look-up and extract its digitized sample and make frequency and MIDI modifications. Currently only one sample is held for some instruments, but more are stored for those instruments with complex harmonics.

The changes which are still under MIDI control are the pitch, volume, attack, sustain, and decay — so the sound is part 'recreated' but also part synthesized (or at least modified).

When more than one recorded note is available for pitch-shifting, the two closest will be averaged for merge harmonics.

WD
In OSI and ITU standards definition, this is applied to the working draft. Later it becomes the DP (draft proposal), then the DIS (draft international standard).

WDM
Wavelength Division Multiplexing. An optical fiber transmission technique which provides multiple channels by using lasers at slightly different wavelengths. It is a form of frequency division multiplexing which needs very stable lasers at very specific and narrow frequencies.

For a time WDM was seen as a cheap way to get fiber-to-the-home distribution for both telecommunications (interactive) and distributed (broadcast) radio and TV services. The different wavelengths would serve to separate the interactive from the one-way services. However, this idea has now generally been abandoned (it was overkill).

Now it is seen as a cheap way to get multiple use out of major fiber trunks (especially submarine systems). Why pay the cost of two or ten fibers, if you can project two or ten separate lasers down the same filament?

The main problem here, is that the glass used to make the fibers, has a very narrow window of maximum transparency, so these lasers need to be highly coherent, and relatively close together in wavelength.

Two main techniques are used in WDM detection; optical gratings or 'beat' (heterodyne and homodyne) techniques.

- A recent US experimenter claims to have pushed 700 distinct laser beams over a fiber in a laboratory.
- AT&T has recently (Sept 1994) demonstrated four 2.4Gbit/s laser beams down a 2000km fiber under the Caribbean Sea.

Web
- The World Wide Web. See WWW.
- WebSpace is a 3-D viewer which supports VRML.

WELL
Whole Earth 'Lectronic Link. A Californian-based social network which is part of the Internet. It is free and has dozens of public and private conferences. Howard Rheingold enthuses about it in his book, *The Virtual Community*.

Whetstone
A measurement of processor performance used for floating-point (developed in Whetstone UK in 1976). Whetstone I series tested 32-bit FP operations, and Whetstone II tested 64-bits.

The tests are a synthetic mix of floating-point and integer arithmetic, transcendental functions, floating-point array computations and subroutine calls. The idea was to mimic the behavior of the common applications of the time.

Whetstone benchmarks were based on Algol and then Fortran and were suitable for small machines and workstations.

Whetstones are probably a reasonable indicator of a computer's ability to perform the maths-intensive calculations needed for CAD or graphics. The score is in thousands or millions of Whets, and the higher the numbers, the better the performance. See benchmark, Linpack, SPEC also.

white balance
Our eyes accept a great deal of variation in the balance of spectrum coloration, in what we interpret as 'white'. When compared to the direct light at noon (the standard), early-morning or late-evening sun will have a distinct red cast, as will tungsten light-bulbs indoors. A white page under tree-shade will be lit predominantly by scattered blue sky light and reflections from green grass and trees — yet we still see it as white.

Our eyes adapt to these color casts, but the camera needs to make an adjustment either by adding filters, changing the film type (indoor vs. outdoor), or rebalancing electronically if it is a video camera. Normally video white standards are set by switching on the white-balance circuits while the camera is zoomed in on a 'standard' white object (a piece of paper), or sometimes directly from the light source through a translucent white lens cap. See color temperature.

white list

Many mobile phone carriers around the world will not permit a phone to be used on their network unless bought through their reseller chains or authorized dealers. This is generally because handsets are highly subsidized in some areas, and not in others. White lists are kept of the authorized phone codes, and the phone's ESN (or the equivalent) will be checked against the list before entry is permitted.

white noise

This is noise generated virtually across the spectrum. Like white light, it contains a balance of all the spectral frequencies in equal amounts. The term applies to both audio and electrical noise

In many systems white noise is Gaussian or random noise generated by the movement of electrons. White audio noise sounds like the hiss of a steam jet. See Gaussian noise.

Whois

An Internet directory service (and command) which provides information about people, addresses, domains, networks, hosts, etc. It is used to find e-mail addresses of people involved in the workings of the Internet.

Wide Area Network

See WAN and also MAN.

wideband

This is generally now applied to those telecommunications services with rates above 64kb/s but below 2Mb/s.

However it is also often used as a level below broadband. An AM or FM radio signal would probably be regarded as wideband since these are in the 8-to-15kHz (not TV's MHz) range.

widescreen

• In film projection, widescreen formats are a compromise between the old Academy (4:3) and the extreme Cinemascope (2.35:1) aspect ratios. There are two main widescreen ratios, 1.85:1 and 1.65:1.

• Super-16 film (where the sound edge is also used for image) is considered a widescreen production format.

• Widescreen television uses 16:9 ratios, as does High Definition TV. However, don't confuse the two; HDTV can be in any aspect ratio since the term only refers to the use of more than one-thousand true scan lines on the screen.

The 4:2:2 standard (Rec. 601) was not designed with widescreen in mind, but these aspect ratios have now been included. See also PALplus for analog widescreen.

• A Dutch station, TC Plus, has been on air with widescreen (D2-MAC) images since the 1992 Winter Games.

widget

A mythical product which is closely related to thingamajigs, whatchima-callits and boondoggles. Actually, if you read Roald Dahl's airforce mythology called *The Gremlins* (published during WWII) you will find that widgets are offspring of (male) gremlins breeding with (female) fifinellas.

widow

In desktop publishing, a widow is a part-line of type (the end of a paragraph) which has been left dangling and appears at the top of the next column (as at the top of this column). The remainder of the paragraph remains in the last column or last page.

It looks particularly ugly when the columns are justified, and it is frowned upon in good typography. Most publishing programs will automatically block one line 'widows and orphans'.

• The term is sometimes also used for very short words (such as 'on') or hyphenated suffixes (such as '-tion'), remaining alone on a line at the end of a normal paragraph within a column.

(As with the above.) These are generally caused by problems of hyphenation, and, if they look particularly ugly, they can be adjusted by adding or deleting words in the paragraph, changing word order, or by setting the particular paragraph to 'hyphen off'.

wildcard

A symbol which is used to mean 'any normal alphanumeric symbol'. Both the asterisk sign and the question mark are widely used as wild cards. MS-DOS took the asterisk over from CP/M, and this has made it the *de facto* standard for file-names. So TRIAL.* will retrieve TRIAL.TXT or TRIAL.DOC

In information retrieval systems, it is often necessary to be more specific with the number of characters the search should cover. Sometimes both * and ? will work, with the asterisk meaning 'any number of characters' and the question-mark meaning 'one character only'. In other retrieval systems you are expected to use ? ? (with a space between) to indicate any number of characters, or ??? to indicate three characters only.

wildcard transfers

Transferring a whole range of similar files as a batch, by selecting the file names (or extensions) using a wildcard — which is an * (asterisk) in MS-DOS. Any command referring to A:*.* will select all files on volume (disk) A:

WIM

Wozniac Integrated Machine. See SWIM.

WIMP

Windows, Icons, Mouse, Pointer. Sometimes the P is said to mean 'Pull-down menus' and sometimes 'WIMPS' (with the 'PS') is translated as 'Pointing Software'. This is the type of interface developed originally by Xerox PARC, but licensed, popularized ,and brought to current standards, by Apple with the Macintosh.

WIMPS was a derogatory term commonly used by MS-DOS fanatics ('Real Men Don't Use WIMPish Interfaces') until Windows 3.x managed to clone the look and feel of the Mac — to the best of Microsoft's abilities at the time. Only with Windows 95 have they nearly caught up.

WIN

Wireless In-Building Network. The name Motorola has given to a radio-based LAN system. The term appears also to have a generic use, but it is being replaced by LAWN (Local Area Wireless Network) which is also a proprietary term.

Win32#

Win32 is a fully-featured API for Windows 95. There's a version called 'Win32c' which provides most of the calls needed with Windows NT. This has some multithreading capabilities.

Another migratory aid — a non-thread version called 'Win32s' is available for Win95 applications, since thesealso work with 16-bit Windows 3.x on the desktop. When you install Windows 95, the same applications should work.

WA Win32s application in a 16-bit system provides the DLLs which handle the API, plus all translation between 32-bit calls and 16-bit calls (called 'thunking'!). This takes up CPU time and increases the overhead of the system.

Winchester

A slang name originally for a very large multi-platter hard-disk unit on the early IBM mainframes which became a generic term for all hard-disk drives. The story behind the name is that the model number for the disk drive was IBM 30/30 — which is also the famous Winchester rifle. So the name stuck.

window

One of those all-purpose words — it has time, size, place, overlap, and transparency applications.

- **Screen:** This is the area of the monitor screen which is devoted to one particular file. Multiple windows can be either tiled (side-by-side), or stacked (layered).
- **Buffer:** This is a reserved area in main memory.
- **Operating systems:** Different operating systems handle windows in different ways. On the Macintosh, the window is a view into a document, while in OS/2 it is said to be 'a task which requires task-termination handling'. Tasks can be in the form of windows or icons. See windowing also.
- **Communications:** The time in which data can be sent, without an ACK being received. See sliding windows.
- **Optical fiber:** The glass used today has two light-frequency 'windows' where the fiber exhibits the least attenuation of the signal. The easiest to use is one at 1300nm, while the best (most transparent) is at 1550nm. There was an older one, used in legacy systems, of 850nm.
- **Clocking:** In terms of computer clock speed, a window is the time between two successive clock pulses. A 10MHz clock rate provides windows of 100 nanoseconds: a 386 computer at 25MHz has windows only 40ns wide, which is faster than the access time for even the best DRAM.
- **Satellites:** A very common use of the term is for 'window of opportunity'. In satellite launching, this is the relatively short period in a day when the launch is possible.
- **Paging:** In paging, it is a cyclic period of time in a data sequence, when something can occur. Different paging protocols are transmitted over the same channels but in bursts at different times, so a pager only wakes up to look for its messages within a certain 'window' of time.

window dub

This is an off-line video 'workprint' — a quick copy made to allow editing decisions. The window is a small area of the screen into which the editing time-code is inserted.

The director can then view this copy and make editing decisions, using only a cheap VHS recorder without the need for special time-code viewing electronics. The timecodes allow him to be 'frame-accurate' in his decisions — and these are later keyed into the CMX computer which controls the high-quality editing process.

windowing

- In data communications this is the transmission of a number of blocks of information (usually about 6 or 8) before acknowledgment is received. In many simple ARQ systems, the transmitter will wait for an acknowledgment after each block, before sending the next. Naturally, it is essential for a reasonable window size to be set if reasonable data rates are to be maintained over satellite hops.
- Windowing systems in PCs are Microsoft's Windows, GEM, X-Windows and the OS/2 Presentation Manager.

Windows (Microsoft)

Windows versions, until Windows 95, were all shells. They were graphic user interfaces and programming environments with multitasking capabilities which sat on top of MS-DOS. DOS provided the basic operating system functions.

The main purpose of Windows 3.x was to make interaction with application programs more user-friendly, by way by icons and symbols rather than by DOS's cryptic commands and eight-character filenames.

Within Windows, the application can share and swap data, but you need Windows versions of all software. Windows popularity can also be traced to the fact that it allowed all applications to share a single device driver — instead of each needing its own (which was the curse of MS-DOS).

There are a number of versions of this shell:

- **Windows 1.0**, released in 1984, was not very impressive (to say the least). No one used a mouse in those days except 'wimpish' Mac users.

The original Windows was hamstrung by slow processors, low-resolution screens, and the 640kB limit set by MS-DOS, and it received very little support from software developers.

There were a couple of versions of the original Windows standard: Windows/286 gave limited pre-emptive multitasking, and Windows/386 was designed for 80386 systems with memory between 2 and 4MB.

- **Windows 2.0** isn't talked about in polite company.

• **Windows 3.0**, shipped in May 1990, became an instant success. It introduced the GUI to MS-DOS-users by providing a shell overlay of the old operating system, and it also solved many of the driver problems with add-on hardware — thus making the machines device independent.

Windows 3.0 had perhaps 80–85% of the functions of the Macintosh interface but it still lacked true object-orientation, links, and agents. For this reason, New Wave 3.0 was often added as another layer to provide these functions, and to support extended memory and protected mode applications.

MS-DOS then required about 60kB of the working memory and the original Windows 3.0 kernel required another 45kB.

• **Windows 3.1** is the most successful GUI to date. This is still not an operating system, but a shell.

• **Windows for Workgroups 3.11** is the small multi-user, peer-to-peer networking version. It makes an excellent front-end to Novell Network, and it can run on Novell ODI.

• **Windows 4.0:** This was code-named Chicago, then it later became Windows 4.0. In 1994, after constant delays, Gates changed the name to Windows 95.

• **Windows 95** is Microsoft's first 32-bit desktop OS. It has pre-emptive multitasking and multithreaded code operations. It also has some basic network support and PnP.

There is still some argument about how much Win95 has abandoned the old MS-DOS features, and how much it is a transitional system on the way to a fully independent operating system.

For instance Windows 95 still allocates the lower 1MB of conventional memory in the same way as the old Windows 3.x, and this was set in the 1980s by MS-DOS design limitations.

• **Windows Daytona** later became Windows NT 3.5 Advanced Server (the word 'Advanced' has since been dropped). This is the company's new 32-bit high-end server system. They call this an 'applications server' but it has been positioned to challenge NetWare 4.0

• **Windows NT 3.5** is a true operating system, and it therefore doesn't need MS-DOS 5.0. It will run in both 16-bit and 32-bit modes on Intel-based PCs and on ACE-compliant machines.

It has file-handling and disk management routines, and like Windows 95, it has pre-emptive multitasking and multi-threaded code operations. It includes an MS-DOS emulator.

• New features are Plug-and-Play, OLE 2.0, TAPI and MAPI.

Windows Sockets

See Winsock.

wink

A wink is a quick reversal of voltage on a line, or a quick on-off flash of the line-hook — typically for less than 500 milliseconds. It is generated in some telephone systems when the attached equipment is ready to accept dialed digits. If a wink is not sent, there are no registers available to accept the numbers.

This is an old analog technique of signaling between telephone exchanges, or between telco exchanges and PBXs. It is used for trunk, rather than line signaling.

• **Wink start** techniques are used in association with E&M and loop/ground starts. See ground start and loop start.

WinNV

A common news reader for Usenet.

WINS

A old US standard for EDI (document interchange). See ANSI X12.

Winsock

Windows Sockets (Windows Network Transit Protocols). These are part of WOSA (see) and they allow transparent operations of Windows applications over networks which are using a different type of data transport protocol. Winsock is loaded as a Windows application and makes the connection to an Internet host by emulating TCP/IP.

• The Winsock 2 Forum is adding a wireless software extension to Windows through a standardized API.

Wintel

A relatively new buzzword for the 'Windows + Intel' standard which the vast majority of PC users now have. You'll find the term still being used on clones not running an Intel chip.

wire

The use of the term 'wire' in two-wire and four-wire circuits, is a hang-over from the days when electrical connections existed end-to-end across the telephone network. These days the term probably means 'channels' or 'circuits'.

In general, two-wire circuits are used in local loops (between the home telephone and the exchange) and carry 'full-duplex' or bidirectional traffic. While four-wire circuits (connected through hybrids) are used for exchange to exchange links — with each pair carrying traffic in only one direction.

• In practice the 'four-wire' circuit is probably pulses of highly multiplexed light down pairs of interconnecting fibers.

• In the US, four-wire voice-grade leased connections are described as 3002 channels.

wire-frame

In three-dimensional CAD and graphics, the objects are first designed and displayed as outlines constructed of wire. The second stage is 'hidden-wire' removal or erasure (taking out those 'wires' which would normally be cut off in vision by closer planes or objects); and the third stage is rendering. The distinction is with solid modeling.

The ACIS (American Committee for Interoperable Systems) has a 3D wireframe kernel which has recently become the de facto standard for CAD/CAM. It is an open standard, and it provides an object-oriented geometric modeling toolkit for building 3D applications. It has wire-frame, solid, and surface functions as hierarchical C++ class libaries.

ACIS runs under DOS, Windows 3.x and Windows NT, and it offers a new object oriented API for AutoCAD — called AutoCAD RX (Run-time eXtension). AutoCAD (now release 13) is the standard in this area and it is also now based on ACIS. ACIS has been embedded in the products of 28 other companies.

wireless

This encompasses radio and microwave, infrared and the use of the visible spectrum — anything that isn't wire linked. So the term is probably synonymous with 'cordless', rather than with 'radio' these days.

- **Spectrum:** See radio spectrum.
- **Local loop:** Wireless local loop services use microwaves or VHF radio signals as a replacement for copper twisted-pair. See WLL, CT2, DCS-1800, PACS, PHP and CDMA.
- **Wireless LAN:** See LAWN.
- **PDA links:** See IR, Jot and Versit.
- **Telephony:** See AMPS, TACS, NMT, GSM, NADC, CDMA.
- **Cable**: See MMDS, LMDS, SHF, MDS and MVDS.

wireline

This term distinguishes traditional twisted-pair telephone services from the new radio-based services such as cellular and PCS. Wireline includes optical fiber, satellite and microwave links as part of the normal POTs and digital phone network.

- **Cellular:** In the USA, one of the two AMPS cellular phone frequencies in each area was given automatically to the local telephone carrier. So, paradoxically, this cellular phone service is often referred to as the 'Wireline' service.

wiring center

With star networks (phones and hubbed-LANs) each node has its own discrete cable to a hub (switched hub, concentrator or wiring panel) which is usually located in a wiring cabinet.

With today's 10Base-T Ethernet LANs this star-approach has become essential, since these are often one-device—one-cable networks. So switched Ethernet has accelerated the trend towards centralized star networks. However, active electronics have now replaced the traditional dumb wiring hub.

wiring closet

The physical room in which the wiring hubs, concentrators, and repeaters are housed. Usually both voice and data networks will terminate here.

wiring hubs and concentrators

Star networks have proved to be more reliable and flexible than the old bus wiring scheme. They allow easy upgrading and help with individual device management to port level.

Hubs/concentrators can have various degrees of sophistication in network management. In 10Base-T and Token Ring they will act to isolate a node creating problems.

Active hubs at the wiring centers will also regenerate and retime packets, and more recently they will bridge and route them. Most importantly, they provide a central point for network management, troubleshooting and security. Monitoring can be conducted in these centers by firmware which is accessed across the network from a central console.

Token Ring systems have long used wiring closets with MSAUs. However the original Ethernet connections were spread out over a long 'bus-backbone' of coaxial.

Later, for management reasons, the long-run of Ethernet coaxial backbone, shrank to only a few meters connecting transceivers in a closet. This was known as the 'collapsed backbone', and it paved the way for switching..

The current interest in hubs and concentrators follows the introduction of 10Base-T with UTP wiring. This is essentially multiple twisted-pair phone cable, so it made sense to parallel the LAN links to the radiating phone links and share the cable.

- So the terms hub and/or concentrator can mean:

— **Wiring hubs** are the cheapest and simplest. Originally they were just frames with connection points, to which the cables would be soldered.

— **Port-switching hubs** are generally closed boxes (repeaters) with fixed internal wiring , linking say, 12 ports (RJ-45). They don't have slots for internal expansion.

— **Concentrators** are typically modular cabinets with a common backplane and slots into which cards can be added. These will provide different wiring connections, management modules and devices such as bridges. This seems to be a vague term for large rack-mounted equipment.

— **Pileable hubs** are basic repeaters, and each is logically a segment. See pileable hubs.

— **Stackable hubs** are more sophisticated, and they can handle rate adaptation, and link to a shared backbone. They have bridging and routing functions. See stackable hubs.

— **Switching hubs** can be true switches (no rate change) or high-speed bridges or routers (rate-adaptable 'switches').

WKS

A type of file developed by Lotus for its 1-2-3 spreadsheet, and now widely used because this product has a dominant position in the marketplace. There are .WKS and .WKS1 file types.

WLL

Wireless in the Local Loop. This is another family of technologies which promise much and deliver little. The problem is one of cost, and the break-even point where cellular phone technology can be applied economically must be close. However, there are other factors to be considered also.

- Wireline links have moderate broadband capacity, now supposedly 6Mb/s for distances of a few kilometers with ADSL. WLL links are limited to the bandwidth supplied.
- WLL can only really be cheaper when no wireline alternative exists. If the twisted pair cable is already down the street then there is no real cost saving. (It's over-pricing!)
- WWL is said to be ideal for developing countries, but this is only true while it is servicing a very few wealthy inhabitants and corporations. WWL involves expensive imported

technologies, while wireline costs are mainly local labor.

— WWL is far more susceptible to interference and outage problems, theft and eavesdropping, and the basestations need constant maintenance.

• The capacity of a radio technology used for WLL may be well above that experienced when the same technology is used for cellular. This is because the basestation antenna system can be highly sectorized, and the base doesn't need hand-off and location facilities if the users are always stationary.

• The main technologies being considered in the local loop are CT2, DCS-1800, PACS, DECT, CDMA and Ionica.

WMF

Windows Metafile Format. This is a vector graphics metafile format developed by Microsoft for Windows. See metafile.

WOFS

Write Once File System. This is similar to the High Sierra format for CD-ROM. It establishes a path-table ('directory list') which functions as a database for every directory on the list, and supplies a technique for quickly locating the most recent versions of a file on the disk. See WORM, sampled-servo format, and composite/continuous format also.

word

• Binary: A computer word is one unit of information which is treated as a whole by the hardware; if the CPU is 16-bits wide, then a word is 2-Bytes; if it is 64-bits wide, then a word is 8-Bytes. This often means a chain of related bits in memory which are addressed under a single location address.

• In general use, the meaning of 'word' is often rather vague. These are the variations:

— since this term is hardware-oriented, generally the word 'word' refers to 8-bits in the older PCs, 16-bits in the modern micros and minis, 32-bits in mainframes, and 64-bits is a word in supercomputers. This simplistic categorization falls down because many modern micros, like the Mac, have 32-bit words.

— Some people apply the term 'byte' to 8-bits, 'word' to 16-bits and 'long-word' to 32-bits.

— In VAX terminology, a word is 16-bits, despite the fact that this is a 32-bit architecture.

• In general asynchronous communications, you can directly translate modem data-rates (measured in bits per second) into average 'English-language' words per minute. A word is roughly 6-Bytes long, including the space.

• With printers, wpm means average English-words per min.

word addressable

See byte addressable.

word form

It is generally agreed that any useful speech recognition systems (apart from basic command systems) will need to identify about 20,000 words or word forms. 'Watch' is a word, while 'watches' and 'watched' are word forms — variations on the root. Often discussions on speech recognition don't make this distinction, or make clear what was meant by 'word'.

word width

The number of bits upon which the processor can act at any one time.

word wrap

The early word processors held either 40 or 80 characters to each line of text on the screen. At the end of the line (the 41st or 81st character) they would automatically have a line-feed and carriage return inserted to force the insert point down to the beginning of the next line. This happened anywhere — between words and in the middle of words.

Word-wrap was a major breakthrough in word processor sophistication because it could search back from the end of the line (by then, more than 80 characters) and find the first 'space' character, and insert the break. It inserts a command token which represents Line Feed + Carriage Return. The space may be removed at the end of the line, and it will then be reinserted automatically if the word wrap position changes.

• When transmitting ASCII text files, it is important that they first be saved using the 'Text-only (no line formatting)' option if one is available. This removes the hard CR+LF at line-end.

workbench

A type of toolkit for client/server application developments. It handles the design, coding, screen building, testing and version controls.

work-group

A network term applied to a group of people in a company or organization who, although perhaps geographically dispersed, work together and need to share similar computer resources.

Typical work-group divisions in a large company will be accounting, personnel, inventory control, marketing, etc.

Work-groups are sometimes served by small LANs, such as the original AppleTalk, and this may then be linked through a gateway into a large company-wide network. Most work-groups, however, will use the standard Ethernet or Token Ring, and make connection to the main network by bridge or router.

Some LANs work-groups are involved in specialist tasks which may need high security, so they may be separated from the main corporate network via high-speed switching and routers configured as a firewall.

workstations

A very vague term for a range of powerful dedicated PCs and terminals used for specific tasks.

In the early time-sharing days, the term workstation was applied to dumb terminals — usually printer terminals.

The term was also applied originally to the 'carrel' (or low-partitioned work area) surrounding the terminal. It is still used in this way in schools.

Then, for a time, the term was applied only to diskless workstations on a LAN which had no local disk storage and required a dedicated file server on the network to boot. These units were usually built around an 8088 processor and Hercules-compatible monochrome video with special BIOS ROMs.

Today the term is mainly applied only to very powerful microcomputers used for CAD-type operations, research and development, and image manipulation. These workstations are highly disk-intensive and usually come with a 1GB hard disk, floating-point processor, a big color screen, Unix, and with software giving high-resolution graphics — and quite probably with built-in file-sharing and networking multitasking access.

• Purists would insist that Mac, Windows and OS/2 machines are not workstations; that the term should only be applied to Sun, Apollo (H/P) and Intergraphic type machines which are optimized for intensive single-user processing.

• Servers are not classed as workstations even if they are virtually identical in processing power, memory, etc

WorldScript

Apple released this non-ASCII No.5 character coding system with System 7.1, to provide languages (other than English) with a 2-Byte character set. It is now known as Unicode. See.

World Wide Web

See WWW.

worm/WORM

• **Virus:** A worm can be an algorithm or program embedded inside another program (application or system). It burrows into a section of the computer memory and often reproduces itself until the memory overflows and the system crashes.

It may be designed deliberately to have a destructive intent or it may just be a prank message. If it self-replicates and can pass to another machine via a disk exchange, then it would be more correct to call it a virus.

• **Parallel programming:** A worm is used to explore a transputer network. It is loaded into the first transputer and inspects all links to connecting transputer/s. The worm then transfers to this second level and repeats the process until the whole network has been defined. Finally it reports its findings back, usually through a graphic display.

This would more correctly now be called an agent.

• **Optical disks:** Here WORM means Write Once, Read Many-times. You may write to this disk and read this data back any number of times, but you cannot erase and reuse the sectors. (You can, however, erase by overwriting or destroying directory pointers.) See write-once (generic) also.

There are three techniques used for WORM: ablative pit, bubble formation, and dye-polymer. See all. Usually all types of WORM disks are read-back by the same record drives.

The value of WORM is that all disk activity leaves an audit-trail (unless deliberately destroyed). No one can actually destroy evidence without this destruction being obvious.

The disadvantage is that you soon use up valuable disk space if you write and rewrite changing files, so WORM is mainly used for archiving. Most WORM drives use CAV techniques for maximum access speed, but at the cost of storage capacity. See write-once medium also.

• **ROM:** With chips, the term WORM is sometimes applied to ROM, to distinguish it from EPROM or other erasable/rewritable forms of non-volatile memory.

WOSA

Windows Open Services Architecture. A set of APIs which are Microsoft's key to workgroup computing; a model for providing applications with access to multiple back-end services, which may be provided by different vendors. WOSAs allow any desktop application to work seamlessly with a variety of these services. The five main components of WOSA are:

— **MAPI** (Messaging API), which gives Windows applications transparent access to a variety of messaging services on several network types.

— **ODBC** (Open Data Base Connectivity), which allows Windows applications to access multiple types of database systems anywhere across an enterprise network.

— **RPC** (Remote Procedure Calls). This is OSF/DCE-compliant. RPC will allow Windows applications to call up a wide range of applications on other platforms across an enterprise network.

— **Windows Sockets** (Windows Network Transit Protocols) which allow transparent operations of Windows applications over networks using a different type of data transport protocol.

— **License Services API.** This is a software-licensing interface to help companies monitor and control the use of their networked applications.

— **TAPI** (Telephone API), which gives windows applications transparent access to telephone equipment and networks.

— An SNA API has also been defined.

wow

A problem of older turntables and mechanical reproductive devices which produce noticeable frequency variations. It is due to slight variations in the rotational speed. See flutter.

wpm

Words per minute. A measure often used in radio teletype and telegraphy, and at one time with computer printers.

write

To place an item of information in some form of temporary or permanent storage. A CPU is said to 'write' to a memory location when it sends data, instructions or an address to be stored in that RAM location. It can also write the bytes to more permanent memory on a floppy disk. The contrast is with read.

• Write-protect, is to use hardware or software to prevent some part of a storage device from being overwritten.

write-inhibit

• All 3.5-inch floppies now come with one or two sliders which are checked for position by the drive, and which can be set to the read-only position. In addition, most computer systems have a means of logical control; the disk or files can be locked from write-access, while permitting read.

• Magento-optical disk drives can now handle WORM. To stop people writing to the disks, a small write-inhibit hole is added to the case, which inhibits both write and erase.

write-once medium

A form of disk storage (inevitably optical) which can record data, but which cannot then be erased and re-recorded.

There are two forms of write-once disks:

— those which can only be recorded on one single session,

— those to which data can be appended in many sessions.

Some disk recording systems have a table of contents (TOC) file which can only be written once, and so any further recordings wouldn't be listed or found.

However, the Kodak Photo CD disk format allows the player to search beyond the last track to see whether space is available, and record in this. You can take a Kodak disk back to the photo-finisher and have extra scenes added to it. See WORM.

write-through/back

In RAM-cache systems, important information which needs to be retained on the hard disk is often being held (for fast access) in RAM. If you turn the computer off without saving the cache information, this will be lost. And since, in some cases, the disk's own directory may be in cache, this creates problems — more than somewhat!

'Write-through' cache overcomes this problem by sending write instructions to the disk-controller immediately, while 'write-back' cache stores the write requests temporarily, and accumulates a few before it writes to disk. This is slightly more dangerous, but also less time-wasting.

writing boards

The general term used for teleconferencing technologies where an image of material written on a large whiteboard (and sometimes just on a computer screen) is transmitted over the lines to the other party for conferencing purposes.

The aim of true electronic white-board technology is to create a system which doesn't require cameras, and where the image is reproduced in real-size at the other end on a duplicate white-board. Most electronic white-board systems, however, only reproduce a fax output on paper.

A Japanese development called Clearboard comes closest to the ideal. It has a large area of screen (desktop size) on which free-hand drawings can be integrated with camera and computer output, and the whole is electronically transferred to the remote unit. Each participant should equally have the opportunity to create or modify the image displayed, and transmit it to the others. See electronic whiteboards.

writing tablet

An emerging technology, which has more to do with OCR software than with the hardware. It allows the entry of data through handwriting using a special stylus on a tablet. The Apple Newton PDA uses this as the primary input.

The current writing tablets need RISC processors to make sense of cursive styles, and even then they require 'training'. Printed characters are easier to interpret, but not as useful.

WWW

World Wide Web. This is an Internet file indexing and information distribution facility which uses the HyperText Mark-up Language (HTML). There are Web magazines and newsletters, but these aren't downloaded (because of the hypertext links) but are rather bowsed on-line.

This is a client-server system which requires an Internet browser which uses a client application such as Mosaic, Netscape, WinWeb, Cello (all graphical user interfaces) or Lynx (text tool). Users can create, edit or browse these hypertext documents.

The front end to the Web is a familiar point-and-click interface, however the back-end is a highly complex globally linked system. The WWW hypertext browser gives you a summary of part of a document with certain words highlighted or numbered. You select one of these to jump to another area of interest (which may be a picture file, rather than only text).

WWW is also compatible with Gopher, e-mail and ftp, and so information stored at any Internet site can be made available through the Web. However it takes a different approach to Gopher (which is menu-based) and it uses a hypertext-like linking process to browse Gopher servers.

This is the preferred interface for accessing WAIS information. A new development called Hyper-G (see) takes the hypertext links further.

• You request a destination from WWW in the form:

```
<tool>://<computer.name>/<directory/
        name>/<file.name>
```

In the directory-name section, the subdivisions are separated by slashes, rather than periods (dots) and there can be more than one of them.

• The Web is organized by CERN in Switzerland and they provide an index to WWW at:

```
http://info.cern.ch/hypertext/DataSources
      /bySubject/Overview.html
```

• The tools can be those like http, Gopher or ftp; the last group of characters will give you an indication of the file-type. HTML is Hypertext Markup Language, txt is ASCII text, and gif is a graphics file.

WXmodem

A faster version of Xmodem. It doesn't wait for ACKs.

WYSIWYG

What You See Is What You Get. Pronounced 'Wiz-ee-wig'. Don't take this term too literally — it is a marketing term, not a technical specification! Screens can never display exactly the same image as a laserprinter output, but they'll be very close.

X

- Often used as short-hand for X Windows and the range of systems around this windowing standard. This system is now a standard on Unix workstations. See X Windows and X Terminal.
- X file format: This is a bit-mapped format for X Windows.
- The x-axis is the horizontal axis.
- X-height is the height of the main body of a character (without ascenders or descenders). See page 739.

X.# standards

These are the ITU-T's main standards for digital communications systems and some related services. X.25 for instance is the most popular access standard for digital packet-switching networks, while X.400 is the new Message Handling Service, and X.500 the International Directory Service. There is little logic to the numbering.

- The ITU-T's V-standards are for analog services — mainly for modems and interfaces.

X.3

The CCITT standard which controls the interface of asynchronous terminals on an X.25 packet-switching network. This is the protocol of the PAD facility parameters. The ITU has defined specific 'Profiles' (standardized groups of 18 PAD parameters) which makes it easy for users to change their parameter list, when necessary.

You will generally use one Profile as the default, and another when transferring binary files.

X.12

EDI: This has no relationship to the ITU's X.## series of digital communications standards. X.12 was defined by ANSI as a North American standard for Electronic Document Interchange. It defines a large number of electronic forms which can be used for formal business messages. See EDIFACT.

X.16

A pin standard for ROM, PROM, and EPROM devices.

X.21

A protocol used when sending serial data along a link. This is an ITU-T digital data protocol which provides a standard synchronous interface for circuit-switched public networks — for data network equipment (DCEs) and data terminals (DTEs).

It covers the control signals which allow the computer to dial its own calls when connected to a digital line.

X.21 provides the Network layer procedures needed, to set up circuits in a switched network. The standard supports low-speed transmissions up to about 9.6kb/s, and the DB-15 connector is specified in the standard.

X.21bis

This is very close to RS-232.

X.25

The ITU standard which defines interfaces to packet-switched networks. It is an international set of rules and procedures (protocols) for the interface of the host computer with the network — generally WANs and public-access data networks

Note that X.25 does not define the way in which the network will transmit information between the main switching nodes as the packets pass across the network — it is a user-access protocol (like ISDN) only. Different X.25 networks use quite different transit technologies. X.25 networks are connection-oriented systems, but there have been some 'almost X.25s' which were connectionless also. Connection-oriented networks will set up a virtual circuit before transmission begins.

Historically, X.25 has long been the major packet-switching standard used around the world, and, in fact, our most widespread data networking technology. It was devised by ARPA and later ratified by the CCITT (now ITU) in 1976. It was later refined in 1980, 1984 and 1988, so there are many variants on X.25.

Some systems still only conform to the 1984 specification, while others conform to 1988. Any reference to X.25 standards should be followed by the date at which those standards were formulated, such as X.25 (1980) or X.25 (1984).

However most X.25 systems will interwork through X.75 (for interconnecting X.25 networks), and they all use the triple-X protocols of X.3, X.28 and X.29. These cover access, and do not dictate how data will be organized and routed. See PAD also.

Technically, X.25 describes the subscriber interface (DTE–DCE) to a connection-oriented packet-switched network. It is a three layered (transmission, organization and routing) protocol, which, at the lowest level, is almost identical to SDLC:

— **Layer 1.** The Physical level of the interface's mechanical and electrical characteristics is a refined version of RS-232C (X.26 and X.27) using noise-reduction techniques. X.21 and X.21bis are also important at this level.

— **Layer 2.** This is the Data-link procedure which controls the data on the line between the user and the network. This is the level which adds error detection and control/recovery information to the frames, and HDLC is important here.

— **Layer 3.** The Network (packet) level establishes and releases virtual circuits, creates the packets, and controls flow across the circuits. This is the Packet Level Protocol (PLP) which includes logical channel numbers, clear, reset and restart procedures, flow-control and windowing.

See PAD and X.75, and X.32. For X.25 numbering, see X.121.

- X.25 has also recently introduced the concept of 'fast select' (see) which is a replacement for connectionless datagrams.
- Generally public access to X.25 has often been restricted to

about 9.6kb/s, but access is now being provided to some X.25 networks at 256kb/s and soon at 2Mb/s. Packets of a size up to 2kBytes are now possible, so Ethernet frames can be encapsulated without segmentation.

- The X.25 standard has provision for both switched virtual circuits (SVCs) which are initiated by the DTE sending a Call Request packet to the network, and permanent virtual circuits (PVCs) which are, effectively, private lines.

- Strictly, the X.25 protocol is only used between full-time, fully connected hosts on a network. When communicating from PCs through a public-access PAD, the dial-up version is called X.28.

- The original standards have been progressively modified to ensure compatibility with the OSI model and allow for more extensive addressing.

- The ITU standards allow X.25 data calls to be placed over ISDN (D- and B-channels) and to use ISDN phone numbers (E.164) instead of the X.25 addresses (X.121), even when the traffic transverses non-ISDN X.25 networks.

X.28

These are the dial-up protocols for linking an asynchronous non-packetizing terminal (usually a PC with modem) to a public-access PAD (Packetizer And Depacketizer) on an X.25 packet-node (host). Note that the X.28 interface completely disregards the parity bit, so it is possible to transparently transfer any 8-bit binary file. (X.3 Profiles don't have a facility to modify parity, or the number of data or stop bits).See C-DTE.

X.29

In X.25 packet networks, this protocol defines the PAD-to-PAD and PAD-to-packet-mode terminal interfaces. Broadly speaking, X.25 packet networks provide for two classes of user terminal: packet-mode terminals and asynchronous character-mode terminals. Packet-mode terminals are expensive but they connect directly to the host. But most dial-up users will use character-mode terminals which need to connect to a PAD (which then handles the packetizing and depacketizing for them). See X.28 also.

X.31

These are the current packet-switching recommendations for ISDN. These protocols carry the initial signaling to set up the bearer channel call to the X.25 host. X.31 protocols are carried 'out-of-slot' (in the D-channel), and the subsequent virtual-circuit signaling is carried 'in-slot' by either a B-channel or a D-channel — depending on which is being used for X.25 access.

This new method of packet handling has advantages because the procedures required to set up and release the virtual circuits are essentially the same as those used for circuit switching, and therefore the packet handling is much simpler. X.31 also allows the aggregation of B-channels

X.32

Also known as 'Dial-up X.25'. This is a companion protocol for X.25 packet-switching which has been adapted for indirect synchronous access to X.25 hosts. It bypasses the network entry PAD at the first node — so the packetizing takes place in the X.32 device. See X.28.

X.32 can be provided over leased lines or over circuit-switched data networks. Unlike X.28, this is fundamentally a two-way calling system (you can call out, and others can call in) which makes you appear to be a full-time node, even when you aren't.

The idea was to allow a normal PC to become a 'virtual' working network node on a packet-network through two-way dial-up access; you'd use this if you only required X.25 access erratically and couldn't justify the cost of a full X.25 host.

The PC becomes a 'satellite' host with X.32, so this is an ideal protocol for file transfers and for database interrogation.

At the time of writing, the standards haven't been finalized to allow the X.25 node to dial-out to X.32 users through the public-switched network (but this will come). Obviously you can dial in, but currently, if two-way calling is required, the X.32 PC must be on a leased line.

The recommendation includes a highly secure procedure for identifying calling terminals (for charging and security). Most implementations are at rates up to 64kb/s, although some have restricted access data-rates of 9.6kb/s.

- X.32 is also likely to provide a cheap way for PC users to access ISDN networks without needing a permanent ISDN connection. See X.3 also.

X.75

These ITU-T data protocol specify the way in which packet-switching nodes act as part of an international chain of connections. X.75 defines how the packet-switched signaling system is to be used between public networks to allow international links to be established even where some incompatibilities exist. It is a gateway interface that exists at network borders.

X.75 uses X.25 as the foundation, but adds new internetworking facilities which are needed to pass alien traffic through to a final destination point — which could be three or four networks along in an international chain. It is only really needed because X.25 has undergone many local and general modifications, and no two networks are now exactly the same.

For instance, X.25 packets are reasonably flexible in size, so X.75 allows the packets to diverge from the original standard of 128-Bytes. If a longer packet needs to pass through a network which only permits smaller packets, X.75 will provide segmentation and reassembly.

It can handle packets from the shortest control-message standard of 16-Bytes, to the new long packet format of 4096-Bytes which is used to encapsulate Ethernet.

X.75 is expensive to implement and, unfortunately, the main value is not directly to the carrier itself, but in maintaining cross-border data exchanges when the boundary networks are implemented in slightly different ways. It hasn't been as widely implemented as one would hope.

See X.25 and X.121 also.

X.121

The international standard for data destination codes or 'addresses' used by public data networks (such as X.25 packet-switching). There are two forms:

- A four digit Data Network Identification Code (DNIC) followed by a Network Terminal Number (NTN). The DNIC itself consists of a Data Country Code (DCC) and a Network Digit.
- Within a country, subscribers will specify just the network digit followed by an NTN. See E.163, E.164 and F.69 also.

• DNICs are specific to a network within a country. Some of the key ones around the world are:

Australia	Austpac	5052
Canada	Teleglobe	3025
	Datapac	3020
	Infoswitch	3029
France	Transpac	2080
	NTI	2081
Germany	Datex-P	2624
Hong Kong	IDAS	4542
	DAS	4544
Japan	DDX-P	4401
	Venus-P	4408
Singapore	Telepac	5252
United Kingdom	IPSS	2341
	PSS	2342
USA	Tymnet	3106
	Telenet	3110

X.400

The ITU's standard suite relating to electronic mail. More correctly it is the standard of Message Handling Service (MHS) which carries all sorts of information. X.400 handles fax, telex, EDI and EFT as well as standard e-mail.

The standards involve the definition of User Agents (US) and a Message Transfer System using Message Transfer Agents (MTAs) as intermediaries in the network. Message Stores (MS) hold the information until it is collected by the end-user, or until the MS acts in accordance with pre-stored instructions.

• The X/Open APIs for X.400 and the X.500 Directories (ASN.1) have now become IEEE 1224. See SMTP and X.500.

X.500

The ITU-T's international directory and distributed database standards. These directories will eventually be electronically accessible from any part of the world, and they will contain a very extensive range of information — including postal addresses, personal and job-related telephone numbers, telex, fax, e-mail addresses, and the public-keys needed for secure encryption of messages and for secure transfer of money.

You will be able to change your personal details yourself in an X.500 directory, and so have call- and message-forwarding based on priorities and message types.

See X.400, ASN.1 syntax, RSA and public key encryption.

X.800

The security architecture standards for X.400/X.500 open system messaging. This suite of protocols is still being finalized, but it will probably be based on RSA algorithms and the use of these for public-key encryption and digital signatures.

X-band

A satellite frequency band between about 5 and 11GHz (mainly in the 8/7GHz range). X-band is reserved solely for government use in many countries. See radar bands, L-, S-, C-, Ku- and Ka-bands.

x-copy

This utility comes with MS-DOS and OS/2, and it allows the copying of a subdirectory and all its files. It will create a new subdirectory at the target, if one doesn't already exist.

X-DOS

A utility which takes IBM PC/Intel assembly level code and converts it to run under Unix.

x-height

The height of a small 'x' character in a typeface at a certain size. This is not the same as point-size, which is measured from the top of the highest 'ascender' to the bottom of the lowest 'descender'; x-height is only of the main body itself.

X-height gives you an indication of the readability of different font styles and sizes. A 12-point font in one style may have a smaller x-height than a 10-point font in another style — if the 12-point font's style has long ascenders and descenders.

For instance, these are identical lines of 10 point type:

- **Times:** Now is the time for all good men
- **Garamond:** Now is the time for all good men
- **Helvetica:** Now is the time for all good men
- **Bookman:** Now is the time for all good me

The x-height of the Helvetica and Bookman is much larger than Times or Garamond, so even though these are the same point size, the text looks larger.

Also the Helvetica and the Garamond are slightly compressed versions, — they are reduced lengthwise, which alters their appearance even more. Overall, the Bookman is about twice as readable as the Garamond for the same point size, which is why it is used in school text books.

However, if we had used Bookman in this book, the extra length of each line would have added nearly 200 pages, and about 30% to the costs and retail price.

X-On/X-Off

See page 741.

X/Open

X/Open is a Unix 'vendor-neutral' software standards organization (originally of European vendors) which doesn't develop standards, but adopts them, publishes guidelines, and sets specifications and conformance-testing procedures. It aims to ensure that applications are portable across various systems.

Both OSF and Unix International companies are members of X/Open, and membership has now extended from Europe to include most important US firms. X/Open has now adopted Posix, and has defined the Common Applications Environment (CAE). It is one of the main unifying forces in Unix.

The current standard is called XPG4 which is probably the closest to a current world standard. You can only brand your products with the XPG4 Base Profile brand when it has undergone an extensive set of conformance tests. These components then form the foundation of the CAE (Common Applications Environment). See Unix International also.

• The X/Open APIs for ASN.I, and their X.400 MHS and X.500 Directory Services, have now become IEEE 1224.

• X/Open's XA transaction processing system standard is currently before ISO. The transaction processing monitor standard is completely dominant in on-line transaction processing. See OLTP.

X.PC

An error checking protocol used for X.25 packet-switching.

X-Remote

See page 742.

X Terminal

Intelligent workstations specifically designed to run the X Windows system. These terminals offer higher resolution than dumb terminals or PCs. Recently, the more powerful Wintel and Mac PCs have eroded this market.

X Windows

A set of open networking protocols developed (by the MIT, IBM and DEC) originally for Unix but now extended to MS-DOS and other OS environments. It is based on graphic workstations and client-server network architecture.

X Windows is vendor-independent and network-transparent, and since it is designed to run on many operating systems, it became a *de facto* industry standard for graphics-orientated windowing workstations.

The protocols govern the user-interface of workstations, and the way in which data is represented and organized on networks. So it allows graphics which have been generated on one computer system to be transferred and displayed on a remote workstation.

X Windows allows PCs and workstations to access various applications on multiple hosts across a network, and display them in separate windows on the one screen simultaneously. However, the host must be running a multitasking operating system such as Unix, MVS or VMS.

Unlike most windowing systems where the user's program makes calls directly to the local display, an X Windows application communicates its interface over the network to a remote display terminal. This makes it network-transparent, but it consumes substantial amounts of RAM and disk space.

Colors and gray-scales are handled through a color mapping facility, and graphics and text operations are specified at high-levels in an attempt to minimize the bandwidth requirements.

However, X Windows is still a low-level graphics standard which doesn't dictate the form of the graphics interface seen by the user. It is not tied to any particular network protocols or hardware, and it is often used with a GUI like HP's New Wave and Nu-Vue. Both DEC and IBM supported the development, and the code is freely available.

• This is not an ISO standard. The ISO has its own version called Terminal Management (TM).

• The success of X Windows can be summed up in the less-than-enthusiastic statement: 'It's free, it's portable, and it works — but it is also awfully slow'. See X Terminal and Motif.

x/y/z

This is any three-dimensional matrix or data-structure. The z adds the third dimension of depth to the x/y matrix.

XAPIA

X.400 Applications Programming Interface Association. See CMC also.

Xbar

A cross-bar exchange switch. See crossbar switch.

XBase

The ANSI standard built on the dBase procedural language.

XCMD

External commands. A term used in HyperCard which appears to be coming into general use. It refers to object-oriented scripts used for accessing operations external to the computer — such as videodisks, CD-ROM, etc. See XFCN.

XDMCP

X Display Manager Control Protocol. This is used for communications under Unix with X Terminals and workstations.

XDR

External Data Representation. See NFS.

XDS

X/Open Directory Services. See X/Open and X.500.

Xenix

Microsoft's 'enhanced' (read 'stripped down') version of Unix which has become a *de facto* standard for PCs. It needs a 286-based AT at least.

Xenix has many special features for PC-type hardware and for integration with MS-DOS (copy utilities, etc.), however it does not conform to the Common Object File Format (COFF) of Unix System V. Unix System V/386 is a merged product which is able to execute files in either format.

• AT&T, SCO, Microsoft and others now contribute to the development of Xenix.

XFCN

External functions. See XCMD.

XGA

eXtended Graphics Array. An IBM graphics standards released in 1991. It was bundled with the first IBM 486 machine for the PS/2 line, and at that time it ran on the MCA bus with 486, 386 or 386SX chips. Unlike 8514/A it was VGA-register compatible. It was an interlace system which needed 512kBytes of VRAM.

XGA ASICs now allow tasks to be processed independently of the CPU for faster applications.

Unfortunately, XGA appeared at about the time the clone makers were breaking away from IBM's dominance (the MCA bus fiasco). They created Super-VGA instead of using XGA which caused the standard to be still-born.

XGA supports three distinct modes: true VGA compatibility, 132 column VGA-compatible text, and extended graphics. It has a resolution of 1024 x 768 pixels and supports 256 colors. The resolution is 'extendable' and, to compete with Super-VGA, it can be pushed to 1024 x 1024 pixels.

Xid

See termid.

XIP

eXecute In Place. A name applied to a technology used in memory cards which allows the computer to treat the card RAM as if it were normal memory space, and execute programs in this space. The popular PCMCIA 2.0-standard cards can act as extended/expanded memory in multi-megaByte chunks, and therefore store applications and entire computing environments as plug-in units. See PCMCIA.

XModem

Probably the world's most widely used form of error checking, and the default standard.

The protocol was devised by Ward Christensen and so is often called 'Christensen's protocol'. However some of the later protocols, such as ZModem, are now much faster and can handle batches of files, and some like Kermit can be used more effectively with mainframe hosts.

There are a couple of variations:

— The original **XModem** ('Standard') transmits a 128-Byte block with a simple checksum (not a CRC).

This is the base standard which is supported by most bulletin boards. However many BBSs now use XModem/CRC even though they may still call it 'XModem'.

— **XModem/CRC** adds CRC error checking to the XModem frames for increased reliability. It still uses 128-Byte blocks but the CRC is much more reliable.

— **XModem-1K** is another CRC version with the size of the data blocks increased to 1024-Bytes to speed up file transfers. Sometimes this version is called YModem. (It is virtually identical.)

— **ACK-Ahead** is another faster form of XModem.

See Kermit, YModem and ZModem also.

XMS

The eXtended Memory Specification. This has changed over the years; XMS was devised by Lotus, Intel, Microsoft (LIM) with some later input from AST (the LIM group plus AST). The initial aim was to allow MS-DOS to use an extra 64kB of extended (not expanded) memory for a total of 740kB of user-usable RAM space.

This extra 64kBs of XMS memory sits just above the 1MB limit of the old 8086 bus in what is now known as the High Memory Area. It is, in fact, the first 64kB of extended memory. Part of MS-DOS can be loaded into this area to free up conventional memory, but this only became possible with the 80286 chip in the IBM AT.

Later the LIM standard was modified by AST to extend the extension beyond this 64kB limit, and this became XMS. Windows/286 was the first program to recognize it, but now most programs that use extended memory will follow the XMS specification.

If you want to use extended memory for some DOS applications, you'll need a memory manager like HIMEM.SYS (there are others also). See extended, expanded memory, UMB, and memory managers also.

xmt

Ancient shorthand for 'transmit'.

XNS

Xerox Network System. This is a widely used (almost ***de facto***) standard for networking, designed for Ethernet. It has been adopted by Novell and other LAN vendors as a distributed file protocol which allows any network station to use other computer's files and peripherals as if they were local. Recently it has fallen out of favor. See TCP/IP also.

XOM

X/Open APIs for ASN.1 Messaging. See X/Open.

XON/XOFF

This is the most basic of all software flow controls for file transfers, and it is so commonly used that it is often just called 'flow-control' (or something similar) in communications programs.

XON/XOFF is used to prevent buffer overflow. The receiving device (PC or peripheral) issues a request on the back-channel to begin transmission by sending the XON signal (Control-Q — DC1) which is part of the normal 7-bit ASCII set. When its buffer is nearing saturation, it will signal a request to stop transmission by sending XOFF (Control-S — DC3).

Often a long multi-section packet-switching link will use XON/XOFF controls independently over a number of stages.

However note that the XON/XOFF control signals are part of the seven-bit ASCII character set, but in point-to-point transfers they are only sent from receiver back to file sender.

It gets a bit more complex in some packet-switched networks where XON/XOFF may also be used between intermediate nodes, which are handling bi-directional packet transfers.

See ETX/ACK also.

XOR

Abbreviation for Exclusive-OR in Boolean algebra. See Venn.

XPG

X-Open's Portability Guide. The current standard is XPG4 and this is close to being the current world standard for Unix.

XRemote

A protocol used with X-Windows over serial link connections. It reduces some of the overheads.

XT

The IBM-XT personal computer was the souped-up version of the IBM-PC. In the XT, the original 8-bit 8088 chip was replaced by the 8/16-bit 8086 running at 4.77MHz (initially). In its day, this was seen as a major step forward. The next major change was with the AT using the 286 chip.

XTALK

Crosstalk. There's both near-end crosstalk (NEXT) and far-end crosstalk (FEXT). See crosstalk.

- Don't confuse a telephone systems crosstalk problems with its echo problems. Many people do, because this exists in near- and far-end forms also.

XTP

eXpress Transfer Protocol. A possible new international standard for high-speed Internet-type operations. This combines the functions of transport and Internet protocols. It requires a minimum amount of processing and allows some parallel processing. There are two categories of packets and a number of 'types':

- **Information packets:** First, Path, Data, Diag(nostic) and Route-type.
- **Control packets:** Cntl (returns status to sender) and RCntl (router generated), Maint (collects route and hop information).

A common fixed-size header (10 x 4-Bytes) and trailer (4-Bytes) format is used in both. Response messages (ARQ, flow control, etc.) are under the control of the sender.

XVS

X/Open System V Specification. The alternative specification to Unix International's SVID, promoted by the X/Open group. It describes system calls, library routines, command and utilities.

XVS is very close to SVID but with special emphasis on portability — mainly through the inclusion of database handlers, SQL, and an indexed sequential file access.

x/y

- **Coordinates:** Graphics and programming term for determining a point on a screen by providing the distance along the horizontal plane (the x coordinate) and the vertical height (y). Note that some computer systems measure the y from the top down for programming purposes.

x-y-z coordinates are used with 3-D modeling in CAD programs.

- **Matrix:** Any two dimensional matrix. The x is always the horizontal dimension, and the y the vertical.
- **Monitors:** A vector graphics screen display is called an x/y display.
- **Satellite dish:** A type of satellite dish which is not polar mounted, but simply moves in two directions — horizontally and vertically.
- **Polarization:** The two planes of polarization in satellite transmission — horizontal and vertical (not circular).

XY resolution

The number of locations (x/y coordinates) or 'pixels' that a screen can handle, horizontally and vertically.

y-axis
In any x/y matrix, the y-axis is the vertical.

Y signal
The luminance signal in a TV system, giving black, a dozen identifiable tones of gray, and white only.

YAG
Neodymium-Yttrium Aluminum Garnet. A type of laser used in optical fiber systems where transmission distance is important. Most YAG lasers operate in the 1550nm glass-transparency window. They need external modulation. See DFB.

yaw
To rotate around the primary axis.

• **Satellites:** This is the unwanted rotation around the imaginary axis between the satellite and Earth. The distinction is with pitch and roll.

• **Disk drives:** The head in all disk drives can be misaligned. If they are bent or moved in such a way that the head is rotated around the primary imaginary axis between the head and the tape or disk, the head is said to have a yaw error.

This means the gap in the head is misaligned with respect to the tracks of previous recordings.

Y/C
Color video signals, as recorded by all high-band 'semi-component' video cassette recorders, have the 'Y' (luminance) carrier which is handled at a higher frequency than the combined colors of Red, Green and Blue known as the 'C' (chrominance) carrier.

Both composite and semi-component video recorders handle the color images in a Y/C format, with the luminance signals at the low (baseband) frequencies and modulated combined-color information at the high end of the same spectrum.

In normal composite video systems these are jointly handled as a single signal, and the Y and C tend to overlap — the high-end of the luminance bandwidth (fine resolution) becomes confused with the low end of the color. Unfortunately, this results in cross-luminance and cross-color (moire) problems.

The term Y/C is now applied to upbanded versions of this equipment which has enough bandwidth to retain the separation. This makes them 'semi-component' systems because they are able to maintain complete separation between the color and the luminance — to the point where they can transmit the two signals over separate cables and keep them isolated throughout the editing and post-production processes (provided the correct Y/C editing equipment is available).

This is not broadcast-quality production equipment, but it can produce excellent results in semi-professional area (news and corporate video production) with medium-cost gear.

The best known of these standards is Super VHS, which is sometimes known as the 'poor-man's Betacam'. Sony also has a Hi-8 Y/C high-band eight millimeter camcorder.

To maintain the advantage of Y and C separation, these two signal components must be processed through the camera, recorder, editing stages, TBC, etc. with separation being maintained for as long as possible. Eventually the twin outputs will need to be merged into composite form for television transmission or for normal distribution.

This is half-way between 'composite' (PAL/NTSC) video, and true 'component' (RGB, YUV or YIQ) systems. See composite and component video also.

YCC
In video, this means luminance and two color difference signals. The term only appears to be used with Kodak's Photo CD so it must signify something. Everyone else labels the two color-difference signals as R-Y & B-Y, or U & V, or as I & Q.

See Photo YCC and Photo CD.

Yellow Book
This book sets the CD-ROM standards. It also includes the extensions for CD-ROM/XA. See CD and CD-ROM.

YIQ
A component video recording technique ('color space') for NTSC which separates the Y (luminance) signal from R-Y and B-Y (the two color-difference signals).

This is very similar to the European YUV process (see YUV for explanation) except that it uses a different vector axis for encoding and decoding.

The R-Y is recorded In-phase (the 'I'), and the B-Y is recorded 90° out-of-phase (Quadratic, or 'Q'). This system is popular in analog U-matic recorders and is also used in the D-2 digital composite standard.

YModem
A popular file-transfer protocol which extends the XModem concept. Unless otherwise specified, YModem-1K is the type generally used. It includes CRC error-checking and a batch file mode, and it also has a fall-back on noisy lines to 128-Byte blocks.

The variations are:

— **YModem:** The original YModem used a block size of 128-Bytes. This was virtually a straight copy of XModem, but neither of these protocols is used much these days.

Note: The term 'YModem' is often mis-applied to both the YModem-1K and to XModem-1K protocols.

— **YModem-1K** has a block size of 1kBytes. This is an extension of XModem-1K which allows for larger files and multiple files transferred as a batch. It automatically transmits filename, size, file-date and time-stamp information.

— **YModem-G** has no error checking, and is used only with MNP or V.4 2 modems. It can rely on them for error checking and correction. It has discarded the 1024-Byte block acknowledgment, and it speeds up the file transfer process by eliminating line turnaround after each block. You can transfer multiple files.

• Because it sends, then waits for acknowledgment before sending another block, YModem is not good for satellite transmissions; this is why YModem-G was introduced. See Kermit, XModem, ZModem also.

yoke

A group of read or write heads along a radial arm, in a magnetic disk or tape system. These act together.

YUV

A widely used component color system in PAL, where the luminance and the chrominance are handled separately as 'components'.

In YUV systems the luminance (Y) signal usually occupies the maximum bandwidth, while the chrominance signals (U and V) will each occupy half that bandwidth, since the eye is less sensitive to color detail.

These U and V signals are 'difference' signals; they are constructed by extracting the luminance signal alternatively from the color signals (Red and Blue). The U color-difference signal is Blue–Y and the V is Red–Y. The luminance image (Y) will usually be transmitted 25 or 30 times per second, while the two difference signals will alternate at half that rate.

This system is also known as 'equiband' and also as Y, R-Y, B-Y. Note that to transport a YUV signal between machines, you will need three cables (plus at least one more for the sound). See also YIQ, color-difference and component video.

Z

Z (time)
See UTC.

Z-buffering
See video buffering.

z-laser
A zone laser. See.

Z-Modem
ZModem was written by Charles Forsberg for TeleNet. It is an excellent communications protocol for modems, and it was designed to overcome the problems of the X-Modem and Y-Modem file transfer protocols — specifically with satellite communications, where delays in the acknowledgment of blocks was unacceptably high.

Z-Modem is able to adjust its block size dynamically (depending on the condition of the link) from 1024-Bytes per block back to 64-Bytes. Each block includes a 16-bit or 32-bit CRC. The sender doesn't wait for the acknowledge signal at the end of each block, but continues to send. If a defect is signalled by a NAK, it steps back to the faulty block and retransmits from this point again.

This is a batch or multi-file protocol, and it is very efficient and able to adapt to high levels of line noise. It can resume an interrupted file transfer at any later time, beginning again from the point of interruption. So if the line is too noisy you can afford to drop the connection, redial, and establish the link once again on a better line — and you won't have lost previous data.

Some Z-Modem systems have Auto Receive which allows file transfers to take place with an unattended remote system.

Don't use Z-Modem with 7-bit transport systems. It needs 8-bit bytes. It also seems to have problems in some countries when used on networks with XON/XOFF.

Zapf
A company that produces typefaces. Zapf Dingbats are a range of exotic symbols treated as if they were characters.

zero
The zero in computers can be positively or negatively signed.

zero address
An instruction which has no address part. In some circumstances, it can be an instruction generated from the stack rather than from memory.

zero-bit insertion
In SDLC and HDLC and many other data transmission systems, extra zero bits need to be added to prevent extended periods of unchanged voltage on the line. There are dozens of rules about when, after a run of logical 1s in an idle state, a zero must be inserted to maintain clocking. See Non-Return to Zero.

zero fill
This is the process of filling vacant spaces in a file or a data base with null characters ($00) In effect, this destroys any previous characters.

Be careful here: this is actually the Null character being used, not the ASCII No.5 character for the number 0 or spaces.

See zeroization.

zero page
A section of the main RAM space which, in some of the early computers, acts faster than other memory locations (pages) because it can be accessed with a truncated address. The page-size depends on the computer data bus. In the 8-bit days it was usually the first 256 locations above $0000.

zero suppression
Numbers: The removal of unwanted extra zeros from the end of a numeral. With decimal dollars, for instance, you will probably only want two zeros, at most, to the right of the decimal point, in the result of any calculation.

Line code: Actions taken to ensure long strings of zeros on a synchronous transmission link won't cause the timing to slip. See 0TLP. This is not a good use of this term.

Zero TLP
See 0TLP in numbers section.

zero wait-state
The most desirable condition with a CPU, where it is working at its maximum pace, and every clock-cycle is being used for processing. Scc wait-state.

zeroization
Overwriting material with nulls to ensure no one can ever read the material again. This is a security measure which should be taken on all hard-disks before an old computer is sold. See zero fill.

ZIF
Zero Insertion Force sockets. These are soldered onto the system board so that you can upgrade your CPU (or other device) without breaking the legs of the chips. The ZIF sockets have a small locking handle which locks and releases the chip legs.

zig-zag
The term often applied to progressive scanning of monitors and television screens, as distinct from the more usual television method of interlace scanning. The distinction is also with boustrophedon (see) in some printers.

zine
See e-zine.

zip/ZIP

• **Compressed MS-DOS files:** The .ZIP file has replaced the older .ARC compression system. There are many decompression utilities around including PKUNZIP.EXE which is very popular (shareware). PKZIP is the file compression version. WINZIP is a windows tool that will work with .ZIP files. Other incompatible forms of compression result in extensions: HQX, ARJ, LHA, ZOO, ARC and Z.

• **Zigzag Inline Packet:** A form of chip design which can be mounted vertically on PCBs with surface mounted technology.

zone

• Duplicate component areas in a fault-tolerant computer system.

• Logical subsets of nodes on an AppleTalk network.

zone beam

A transmission beam from a communications satellite which is intermediate in width and signal concentration. It is wider than a spot beam, and narrower than a continental or hemispherical beam. Normally a zone beam is said to cover less than 10% of the earth's surface. Sometimes this is a shaped beam (see).

See spot beam and hemi-beam also.

zone codes

In network numbering schemes, a zone code is the first number in what most people call the country code. North America doesn't have country codes, only the zone code 1.

The zones are:

Zone 1 — USA, Canada and some Caribbean Islands.
Zone 2 — Africa.
Zone 3 — Some parts of Europe (very mixed).
Zone 4 — The other parts of Europe.
Zone 5 — Central and South America.
Zone 6 — Oceania, Australia, Indonesia, Philippines.
Zone 7 — The old Soviet Union and Eastern Bloc.
Zone 8 — China, Japan, South East Asia, Mongolia.
Zone 9 — Middle East, Pakistan, India.

See country code.

• Some of the Zone 8 codes have been given to Inmarsat for specific codes covering the Atlantic, Indian, and Pacific Oceans.

zone laser

A zone or z-laser is a vertical cavity laser (see) which is self-focusing. It needs no lenses to focus light on a very tiny point.

This IBM invention promises to be the way to produce highly efficient lasers for optical fibers, optical disks, and for use in interconnecting (coupling) chips on an optical logical board. Z-lasers are made of layers of indium-gallium-arsenide and aluminum-gallium-arsenide.

zoom

To change the focal length of the camera lens progressively. It is an optical effect that has only been possible since the 1950s. The early professional cameras only had zoom ratios of 4:1, but by the early 1960s, under the influence of sophisticated computer simulations and calculations, lens makers have been able to produce professional zoom lenses with zoom ratios of 10:1 and f-stops of about f-2.

The difference between a still camera zoom and a movie camera zoom is that the focus must remain perfect throughout the zoom range of a movie camera.

• Zoom-out means to change from a long-focus lens (a high focal-length — or 'telescopic' lens) progressively to a wide-angle lens. This is from a binocular-type view of the scene, to a panoramic view. Zoom-in is the opposite: it finishes on a close-up.

ZVS

Zero-Voltage Switched power supplies.

ZWS

Zero Wait State. See.